DISCRETE ALGORITHMIC MATHEMATICS

DISCRETE ALGORITHMIC MATHEMATICS

Stephen B. Maurer
SWARTHMORE COLLEGE

Anthony Ralston
STATE UNIVERSITY OF NEW YORK AT BUFFALO

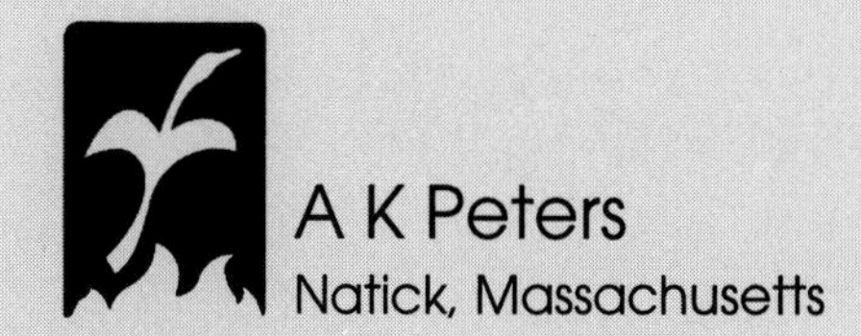
A K Peters
Natick, Massachusetts

Editorial, Sales, and Customer Service Office

A K Peters, Ltd.
63 South Avenue
Natick, MA 01760

Cover illustration: The Baltimore Hilton Problem. For more information, please see page 197.

Library of Congress Cataloging-in-Publication Data

Maurer, Stephen B.
Discrete algorithmic mathematics / Stephen B. Maurer, Anthony Ralston. — 2nd ed.
p. cm.
Includes bibliographical references and index.
ISBN 1-56881-091-1
1. Mathematics. 2. Computer science–Mathematics. I. Ralston, Anthony. II. Title.
QA39.2.M394 1998
510–dc21 98-23066
CIP

Printed in the United States of America
02 01 00 99 98 10 9 8 7 6 5 4 3 2 1

Preface

In the years since its original publication, *Discrete Algorithmic Mathematics* has developed a devoted following. We are pleased by the steady sales, a good part of it outside the United States, and by the comments we get from users. Some use it as a text, some as a reference.

Therefore, we are particularly pleased that A K Peters has assumed publication. We thank Alice and Klaus Peters for their belief in our book, and thank them and their staff for their prompt and careful work.

We have used the occasion of republication to correct errata and make a few other small changes. Plans for more substantial changes later are discussed below.

The Preface to the First Edition, which follows, explains in detail our goals in writing the book and pathways through the book for one- and two-semester courses.

Our book remains special among discrete mathematics books in a number of ways. First is its attention to *algorithmics* (algorithms as objects of mathematical study) and the associated inductive and recursive paradigms. While the coverage of algorithmics in other discrete books has increased since our book originally appeared, our book still addresses this theme more centrally than other books.

Second, DAM (as we like to call it) works especially well with better prepared students, and is more sophisticated than most discrete texts. For instance, while there are many routine problems, there are plenty of nonroutine ones too, often with lengthy discussion within the problems. If you teach stronger students, both you and your students may be happier with this book.

Third, DAM has topics of interest to a wide variety of students, not just computer science majors and math majors.

Fourth, if you do teach a full-year discrete course, ours is one of the few books with enough material.

The following people have been especially helpful in providing corrections for DAM and ideas for future changes: David Flesner, Joseph Halpern, David Kincaid, Paul Klingsberg, David Levine, Udi Manber, Malcolm Mason, and Paul Stockmeyer.

The future. We are planning a substantially revised text for a one-semester course, to appear in 2000–01. We plan to pare down the preliminaries in Chapter 0 and the discussion of algorithmic language in Chapter 1. We also plan to delete Chapters 8, 9 and the Epilogue, while bringing some material forward from them. A few other topics may be added, and a few deleted. Any suggestions *you* have for improvements are most welcome.

Stephen B Maurer
Swarthmore PA
smaurer1@swarthmore.edu

Anthony Ralston
London
ar9@doc.ic.ac.uk

May, 1998

Preface to the 1st Edition

We have written this book because we believe there should be a discrete mathematics course which is central and satisfying in the same way calculus is intended to be. Specifically, such a course, like a good calculus course, should have the following attributes.

1. It should be a *core* course for all mathematically capable students regardless of their intended majors.
2. Consequently, it should be a *broad* course, touching on all the discrete topics of notable use in a variety of disciplines, not just in mathematics and computer science. (See our chapter summaries below for a clearer idea of what we mean by this.) Similarly, there should be illustrative examples from many applications areas.
3. It should be an introductory college course, to be taken by *well-prepared* freshmen and sophomores, the same students who are well prepared for the calculus. Which year they take it would depend on whether they take calculus before or after, and both orders should be possible.
4. The course should have a *central theme* rather than be a hodgepodge of topics.
5. The emphasis should not be on the theoretical structure of the subject; as with continuous mathematics, this should be treated in upper level courses. Rather the course should be fairly informal and should concentrate on problem solving and modeling using the standard techniques and concepts of the subject.
6. Consequently, there should be a great variety of problems to solve, ranging from easy to very hard.
7. Nonetheless, students should develop some mathematical maturity. They should gain some idea of what a good mathematical argument is and they should take first steps toward learning how to devise and present such arguments themselves.

In various speeches and published articles, we have argued for such a course. This book is our way of making our position concrete by showing in detail how we would do it. In recent years discrete math books have been moving in the direction of the seven points above, and our book continues that trend.

Our vision is of a freshman/sophomore mathematics program which contains a year's worth of calculus and a year's worth of discrete mathematics. We think an integrated two-year discrete-continuous course might be best, but for now this is not practicable. Thus, our book is intended for a year discrete course. Indeed, it has more material than can be covered in a year. This allows instructors considerable choice. It can also be used for a semester course; indeed, one of us does just that. Later we discuss various paths through the book for semester and year courses.

Most colleges have at least two versions of calculus, often three—slow, regular and honors. If discrete math truly catches on, there will come to be slow, regular and honors discrete math, too. We intend our book to be usable in at least the regular and honors versions. Clearly, honors sections will cover more material and tackle the harder problems.

Algorithms, Paradigms, and Discrete Mathematics

We have said that discrete mathematics needs a central theme. In fact, our book has both a central *object*—algorithms—and two central *methods*—the inductive and recursive paradigms. We now say a bit about these.

There have always been algorithms in mathematics. Furthermore, with the importance of algorithms increasing, it is now recognized that humans need a precise way to communicate to each other (not just to machines) about algorithms. Hence the increased use of "algorithmic language" in discrete (and other) mathematics books, including this book. (We don't call it "pseudocode" because from the point of view of communication between humans there is nothing "pseudo" about it.)

Our algorithmic notation is more formal than in most discrete mathematics books, but less formal than that in books which use an actual computer language. We think it is important that students learn early to treat algorithms as mathematical objects, with appropriate precision. Nevertheless, our language is sufficiently simple, and the meaning of its structures sufficiently transparent, that even students with little or no computer experience should be able to understand it, read it, and write simple algorithms with it. In particular, a programming course is *not* a prerequisite for this book.

Many mathematicians still regard algorithms as afterthoughts—first you prove that something exists and later, as a lesser activity, you find some way to compute the thing. In our philosophy, algorithms are more central than that. Algorithms can be used to prove that objects exist and simultaneously provide a way to construct them—thus providing the elegance of a stone which kills two birds. This idea of "proof by algorithm" shows up repeatedly in our book. Algorithms are also an object of mathematical study as well as an engine for proofs. We devote much attention to the analysis of algorithms and some to the verification of correctness of algorithms.

Our attitude towards algorithms is summarized by our title. It is not "Discrete Mathematics"; it is not even "Discrete Mathematics with Algorithms"; it is "Discrete Algorithmic Mathematics". "Algorithmic" is an essential modifier of "Mathematics".

But problems come before algorithms. How do you devise an algorithm which solves the problem at hand, and how do you convince yourself you are right? This is where our central methods come in. The inductive paradigm is the method of building up cases, looking for patterns, and thus obtaining a (usually) iterative algorithm. The recursive paradigm is the method of supposing you already know how to handle previous cases, finding a connection to the current case, and thus obtaining a (usually) recursive algorithm. We emphasize repeatedly how much mileage one can get with these elaborations of the simple ideas "build from the previous case" and "reduce to the previous case".

Problem Sets. Ours are large and varied, both in gradations of difficulty and in topics covered. We even think some of our problems are interesting! We also tie a number of the problems closely to the exposition by referring to them from the text (using numbers in square brackets, e.g. [14]). We provide hints and answers in an appendix to about half the problems (those whose problem number or part is in color). There will also be two solutions manuals—one for instructors giving solutions for all problems, and another for sale to students giving detailed solutions to those problems for which there are hints or answers in the text.

Chapter by Chapter Summary

Prologue. This gives an application which foreshadows both the subject matter and the point of view of the rest of the book.

Chapter 0, Mathematical Preliminaries. The goal is to make sure students are comfortable with standard concepts, notations and operations used throughout the book (e.g., the function concept, set notation, matrix multiplication, logical implication). To fend off tedium, we have avoided long abstract discussions of these matters and instructors should not dwell on this chapter longer than necessary. This material needs to be included because few students arrive familiar and comfortable with all these topics. For very well prepared students, much of it may be skipped, or merely referred back to when appropriate.

Chapter 1, Algorithms. This chapter introduces our language through a number of examples, several of which we return to repeatedly in later chapters. Also introduced through examples are key issues in the analysis of algorithms and an introduction to complexity ideas. Our algorithmic language includes recursion. The use of explicit recursion in algorithms and its close connection to mathematical induction are essential parts of our recursive paradigm theme. The instructor must decide how much algorithm *writing* to expect from students. If only modest writing fluency is expected, portions of this chapter can be covered fairly quickly.

Chapter 2, Induction. This is a very thorough presentation of mathematical induction. Induction is presented in all its aspects (not just a method of proof, but also a method for definitions, discovery, and algorithm construction), and students are asked to devise inductive definitions and proofs themselves. This may seem out of keeping with our emphasis on problem solving over theory, but we think not. First, we think induction proofs are in many ways the easiest type of proof to ask students to do. Induction has a pattern which varies little from one proof to another (although sometimes it requires thought and ingenuity to apply the pattern). Second, we go to great lengths to explain to students how to do it—both how to discover the induction and how to write it up. Third, a reading knowledge of induction is absolutely necessary for the text, as induction is the foremost method of solution and proof in discrete mathematics.

Chapter 3, Graphs and Trees. Graphs are a primary construct for modeling discrete phenomena. After a discussion of standard terminology and matrix methods, we give a selection of topics which allow us to emphasize algorithms and the inductive/recursive paradigms. We use path growing to provide algorithmic proofs of the standard Eulerian graph theorems. We introduce various graph and tree searching paradigms and use these to determine shortest paths, connectivity, two-colorability, and minimum spanning trees. We also discuss graph coloring more generally as well as various aspects of binary trees.

Chapter 4, Fundamental Counting Methods. This chapter provides a fairly traditional development up through Inclusion-Exclusion. However, there is significant emphasis on combinatorial arguments (conceptual approaches instead of computation) and the two final sections are very algorithmic (construction of combinatorial objects and algorithmic pigeonholes!).

Chapter 5, Difference Equations. These are a fundamental modeling tool, on par with differential equations. Difference equations (a term we use synonymously with "recurrence relations") are also the part of counting which is closest to the inductive and recursive paradigms. For these reasons we devote a whole chapter to them. We devote considerable space to the standard solution methods for linear difference equations, but first we work on how to model problems with difference equations and how to conjecture and verify solutions. In the long run, with standard problems solved by symbolic algebra routines on calculators and computers, the modeling and conjecture skills will likely prove to be the most important. Two important approaches to difference equations, finite differences and generating functions, are left to Chapter 9.

Chapter 6, Probability. Informed citizens need to know more about probability than they are currently taught. Consequently we include this chapter, which focuses on discrete probability and includes applications as far flung as screening tests for disease and guessing on the College Boards. Probability also has specific connections to other parts of the book. We use it to justify some average case analyses of algorithms done informally earlier in the book. We also show how the recursive paradigm allows easy solutions to certain types of probability questions.

Chapter 7, Logic. Usually a logic chapter comes much earlier, with the goal of helping to develop logical thinking. We have addressed this general goal through various means in our early chapters, particularly through the section on logic concepts in Chapter 0. Chapter 7 is of a more technical nature. Most of the chapter focuses on the propositional calculus, including discussions of natural deduction, Boolean algebra, and the application of the propositional calculus to the verification of algorithm correctness. There is also an introduction to the predicate calculus and its application to algorithm verification.

Chapter 8, Algorithmic Linear Algebra. Chapters 8 and 9 are the ones most unlike what appears in other discrete math books. Why include a very substantial portion of linear algebra when linear algebra is already a standard (sophomore) course? Remember, we advocate a new freshman/sophomore curriculum with a year of continuous and a year of discrete. Then where does the current semester of linear algebra go? Our answer: The essential material for all students should go in the discrete course; the rest should go in a new, upper level linear algebra course of a more abstract sort, which only some students would take. (Actually, this would be a return to the sort of linear algebra course given 25 years ago.) Our idea of the essential linear algebra material is Chapter 8. Of course, the chapter is highly algorithmic; the theory is there but it is developed out of the algorithms. Gaussian elimination pervades the chapter, but we also go on to eigenvalues (the most important use of elementary linear algebra in science and engineering) and use eigenvalues to analyze Markov chains. We hope that faculty will look carefully at what we accomplish in this chapter. We think it's a viable approach.

Chapter 9, Infinite Processes in Discrete Mathematics. This is a bridge chapter between discrete and continuous mathematics. It contains all the standard material on limits in the sequence context. There are also sections on series, order notation, generating functions and finite differences, and a final section on approximation algorithms. Why include so much material that spills over into continuous mathematics? First, we wish to make the point that there isn't a wall between the continuous and the discrete. (This point is made earlier in the book, too. In fact, this chapter firms up several arguments done informally earlier.) Second, for students who study this book before calculus, this material will make limits of functions more approachable, and will allow more material to be covered in the year available for calculus. The possibility of deleting sequences and series from introductory calculus may help in the reorganization of calculus now under way on the national level.

Epilogue. Through a fairly detailed study of one application, sorting, this chapter recapitulates many of the themes introduced heretofore and provides a bridge to later mathematics courses. A challenging final problem set attempts to do the same by leading the reader into several other applications.

Regrettably, in order to keep this book to a size students and instructors can lift, we had to omit many things from earlier drafts, in particular, sections or subsections on double induction, front-end and back-end recursion, planar graphs and the five-color theorem, nonlinear difference equations (chaos), some continuous probability and standard statistics related to the normal curve, and linear programming. All of these items have been preserved on disk, and we would like to hear if readers think they should be included (but instead of what?).

Dependencies and Pathways Through the Book

The chapters of this book are fairly independent after the first few. Specifically:

1. Almost everything in later chapters depends on elementary knowledge of Chapter 0. Therefore, as noted earlier, cover as much or as little of it as your students' preparation requires, and cover it at the beginning of the course or just when it seems needed.
2. All subsequent chapters depend on Chapters 1 (algorithms) and 2 (induction), particularly on Sections 1.1–1.4, 2.1–2.3, and 2.7. These two chapters can be taught in series, as one of us does, or in parallel, as the other does.
3. Chapters 3 (graphs) and 4 (counting) are so much the guts of discrete mathematics that significant portions (3.1–3.3, 3.6, 3.8, 4.1–4.6) should be taught

before later chapters. Although the dependence of Chapters 5–9 on 3–4 is not always strong, graphs are used as examples in many places and simple counting arguments show up repeatedly.

4. Difference equations, being recursive entities, appear throughout the book. Solution methods from Chapter 5 reappear in Chapters 6, 8 and 9, so at least Sections 5.5 and 5.7 should be covered before those chapters.
5. The Epilogue is expressly intended to depend on almost all that preceded it. Therefore, before trying to teach any section of it, you should assure yourself that you have covered the necessary prior material.

Dependencies *within* chapters. Generally, each section depends on the preceding one. We now list the exceptions, which allow sections to be skipped. Let $x.z$–$x.w \leftarrow x.y$ mean that Section $x.y$ is the last one that must be covered before any of Sections $x.z$ through $x.w$. Then the exceptions are

$1.6 \leftarrow 1.4$
2.5–$2.7 \leftarrow 2.3$
3.6–$3.7 \leftarrow 3.2$, $3.8 \leftarrow 3.6$
4.7–$4.10 \leftarrow 4.4$
$5.4 \leftarrow 5.1$, $5.8 \leftarrow 5.4$
$6.7 \leftarrow 6.5$
7.4–$7.5 \leftarrow 7.2$, $7.6 \leftarrow 7.3$, $7.8 \leftarrow 7.2$
$8.5 \leftarrow 8.2$, $8.6 \leftarrow 8.4$
9.4 and $9.7 \leftarrow 9.2$

Now here are two suggestions for possible one-year courses.

For very well-prepared students: First semester: Prologue, Chapter 0 (quickly and not in complete detail), Chapters 1–4. Second semester: Chapter 5 or 6, Chapter 7, Chapter 8 or 9, Epilogue. Of course, other choices are possible, too, including some material from all chapters but omitting some sections from most chapters.

For less well-prepared students: First semester: Prologue, Chapters 0–3. Second semester: Chapters 4–7, Epilogue. In this case too it is not expected that every section in each chapter would be covered.

Now here are two suggestions for one-semester courses.

For very well-prepared students: Parts of Chapter 0 (perhaps only Sections 5 and 6), Chapters 1 and 2, then Chapters 3–5 or parts of 3–6.

For less well-prepared students: Chapters 0–4 or parts of 0–5.

Acknowledgments

This book has taken a long time—we started in 1980! It has been a happy collaboration, despite many delays on both sides. Each of us appreciates how much he has learned from the viewpoint, experience, and detailed criticism of the other.

Many people have helped us over this period and we want to thank them. First, we thank the Addison-Wesley staff for the strong interest they have taken in this book from the beginning. David Chelton, Developmental Editor, was especially helpful; his detailed reviews included many valuable remarks about our mathematical presentation as well as about more typical in-house concerns (use of color, format, etc.). We also thank Addison-Wesley for letting us have much more say about the design of this book than authors usually get; in particular, we thank them for accepting several nonstandard conventions about formatting and grammar.

Next we thank the many other people who have helped us with comments, suggestions and corrections at various stages of writing: Kevin Cherkauer, Dov Gabbay, Charles Grinstead, Jeff Horvath, Ellie Johnson, Paul Klingsberg, Russ Miller, Bill Rapaport, Gerald Rising, Eileen Schoaff, Jeff Zucker, the many reviewers for Addison-Wesley (we don't know most of their names) and our many students who confronted numerous preliminary versions and lived to tell us their opinions. Surely this book is much better than it would have been without the help of all these people.

Many thanks to Donald Knuth for inventing TeX, which allowed us to produce the format we wanted. Thanks to Fred Bartlett and his staff for providing some of the TeX macros and doing some of the keyboarding.

Finally, we each have some personal acknowledgments.

One of us (SM) says: I thank my wife Fran for her patience. This book predates our marriage, and for her it's been as if I have a former lover who won't go away. As she has put it, "I think that being a TeX widow must be worse than being a football widow." My kids Leon and Aaron, ages 4 and 1, have also on occasions played second fiddle to this book, although more often the book has played second fiddle to them. While they are too young to understand the conflict, in time they may appreciate what I hope is the resolution: a good book *and* a good daddy. Most gratefully, I dedicate my work on this book to my mentor and thesis advisor, Albert W. Tucker (of the Kuhn-Tucker Theorem in mathematical programming). I was fortunate to learn the algorithmic point of view at his knee, but he had to come to it on his own—through years of trying to really understand what the simplex algorithm was trying to tell us about linear optimization problems. Still keen in his 80s, Al will be disappointed that linear programming had to be omitted from our book, but I hope he will see his influence on almost every page and be pleased.

The other of us (AR) says: My wife, Jayne, has become used—too used, she would probably say—to being a book widow. But once again she has uncomplainingly played that role, this time, however, since they have flown the coop, without the solace—or is it the additional burden?—of having children abandoned by their father with whom to share the pain.

Swarthmore PA
Buffalo NY
May, 1990

Contents

Symbols, Notation, and Conventions

The list which follows contains the symbols and notation used in this book. Definitions or descriptions, if necessary, are given at the first use of the symbol or notation. The reader is cautioned that some symbols may serve double duty but on each occasion the way such symbols are used should be unambiguous.

Meaning	Symbol or Example	Page or place first defined or used
A. Algorithms		
Main features of notation		Table 1.1 (p. 73)
Description of notation:		
Except for procedures and recursion		Section 1.3 (p. 88)
Procedures and recursion		Section 1.5 (p. 110)
References to algorithms		
"Algorithm" sometimes followed by a number and then by the algorithm name	Algorithm 1.1, MULT	72
Just the name of the algorithm in large and small capitals	MULT	73
Notation specific to algorithms:		
Assignment (of expression value to variable)	$\leftarrow$	73
Exchange of two values	$\leftrightarrow$	89
(See also Appendix 1)		
B. Notation related to problems		
References to problems		
At the end of a section: numbers in square brackets	[19]	15
In the supplementary problems of a chapter	[7, Supplementary]	88
In another section or chapter	[27, Section 2.2]	11
Indication of when a hint or answer will be found in the Hints and Answers—number of problem or part of problem in color	1.	7
C. Numbering conventions		
Equations: Numbered consecutively from (1) in each section		
Definitions: Numbered consecutively in each section		

Meaning	Symbol or Example	Page or place first defined or used
Theorems, lemmas, corollaries: Handled as a single group and numbered consecutively in each section		
Figures: Numbered consecutively in each chapter		
Tables: Numbered consecutively in each chapter		
D. General mathematical notation (alphabetized on *Meaning* column within each category)		
1. Miscellaneous notation		
Big Oh	O	739
Concatenation	$\|$	816
Ellipsis	$\ldots$	9
Difference operator	Δ	746
Divisibility (of q by p)	$p\mid q$	18
End of example or theorem	▮	11
Equal mod m	$=_m$	19 [3]
Indivisibility	$\nmid$	41
Infinity	∞	10
Little Oh	o	741
Order	Ord or Θ	739
Product	$\prod$	47
Summation	$\sum_{i=1}^{n}$	39
	$\sum_{i=1}^{8}$	39
2. Sets		
Cardinality	$\mid\ \mid$	10
Cartesian product	$\times$	15
Complement (see also Event complement under 6. below)	$\overline{S}$	12
Containment	$\subset$	10
Difference	$-$	12
Enumeration	$\{1, 2, 3, \ldots\}$	9
Integers from 1 to n (set of)	$[n]$	48 [5]
Intersection	$\cap$	12

Meaning	Symbol or Example	Page or place first defined or used
Membership	$\in$	8
Sequence	$(a_1, a_2, \ldots)$	16
	$\{a_n\}$	725
Set builder	$\{\ldots \mid \ldots\}$	9
Set of n-tuples of real numbers	R^n	667
Tuple	$(a_1, \ldots, a_n)$	16
Union	$\cup$	11
3. Functions		
Absolute value	$\mid \ \mid$	27
Ceiling	$\lceil \ \rceil$	27
Congruence	$\equiv$	37 [28]
Factorial	!	28
Falling factorial	$x_{(k)}$	749
Floor	$\lfloor \ \rfloor$	27
Greatest common divisor	gcd	81
Inverse	f^{-1}	23
Least common multiple	lcm	109 [3]
Limit	$n \rightarrow \infty$	31
Logarithm base	$\log_2 n$	30
Mod (or remainder)	mod	28
Rising factorial	$x^{(k)}$	756 [15]
4. Graphs		
Adjacency matrix	A	216
Chromatic number	$\chi(G)$	257
Clique number	$c(G)$	260
Complete bipartite graph	K_{mn}	214 [24]
Directed edge	(v_i, v_j)	202
Edge (undirected)	$\{v_i, v_j\}$	201
Edge set	E	201
Empty graph	$\emptyset$	202
General graph	G	201
	$G(V, E)$	202
Path matrix	P	219
Vertex set	V	201

(continued on inside back cover)

Prologue

What Is Discrete Algorithmic Mathematics?

What is discrete algorithmic mathematics? If we could answer that in a section, it wouldn't be a subject worth a whole book.

Nonetheless, you deserve an inkling of what's in store. So this section is devoted to a problem illustrating several of our themes. As you read, concentrate on these themes. The problem itself, and the fine points of its solution, are not themselves so important right now.

The problem is: Find the minimum amount of time needed for an intermediate stop on an airline flight given that

a) we have identified the various tasks which must be done during the stop;

b) we know how long each task will take;

c) we have people available to carry on as many tasks simultaneously as we want; but

d) some tasks must be completed before others can begin.

To simplify things in this example, let's assume the only tasks are the following, where the numbers in the second column are the times needed in minutes. (We'll explain the third column in a moment.)

1. Unload luggage	20	
2. Unload deplaning passengers	10	
3. Load new luggage	20	after 1
4. Clean the cabin	15	after 2
5. Load on more food	10	after 2
6. Load new passengers	25	after 4

This example is typical of problems in which we need to determine a minimum amount of time to complete a complicated project.

What makes such problems particularly interesting is d) above: Some of the tasks can be done simultaneously—unloading passengers and unloading luggage—but some must be done after others—new passengers cannot be loaded until the cabin is cleaned, which in turn can't be done until the deplaning passengers are off. This *precedence relation* is specified by the third column. For instance, "after 2" in line 4 means that cleaning the cabin can be started as soon as the deplaning passengers have left.

The key to solving many mathematical problems is to find an appropriate way to *represent* or *model* them. The best representation here turns out to be a very common one in discrete mathematics, a *directed graph*. *Caution*: This is not the graph of a function but an entirely different concept. It is a collection of vertices (dots) with edges connecting some of them. Directed means that the edges have arrows on them. We will discuss such graphs at length in Chapter 3.

Figure P.1 shows the graph representation. There is a vertex for each task. Vertex v_1 stands for task 1, and so on. We have put the time needed to complete each task above its vertex. We have also added two more vertices, S for start and F for finish. These represent fictitious marker tasks, so we assign them both time 0. Finally, for every i and j, we put an edge from vertex i to vertex j if task i must directly precede task j. Naturally, we declare that task S must precede what would otherwise be the earliest tasks, and task F must follow what would otherwise be the last tasks.

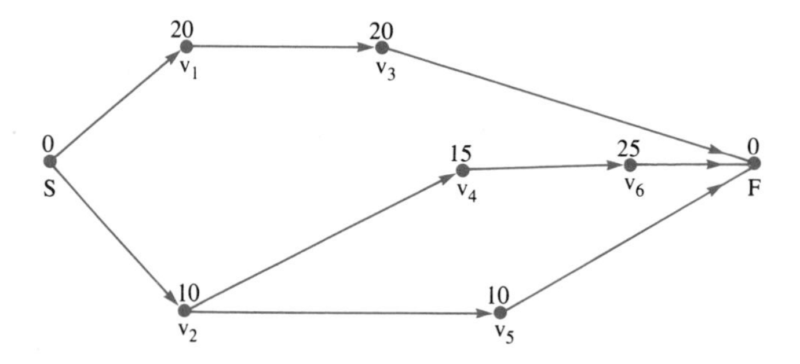

FIGURE P.1

Graph Representation of the Airline Problem

For a small example like this, you can probably compute the minimum time to complete all the tasks from looking at the graph. But we want a general method, because a task analysis for an actual project would be much more complicated. Imagine, for instance, the graph for the project declared by the United States in 1961, called "Put a man on the moon". (That project took eight years and used the mathematical method we are developing here.)

To help us discover a general method, we need a graph just complicated enough that we can't see the answer and wouldn't want to look for it by brute force. Figure P.2 provides such an example. Never mind what project it represents. We need only schedule the tasks so that the final task finishes as soon as possible.

To begin, we expand the problem. Instead of asking how soon the final task can be finished, we ask how soon *each* task can be finished. This may seem to be

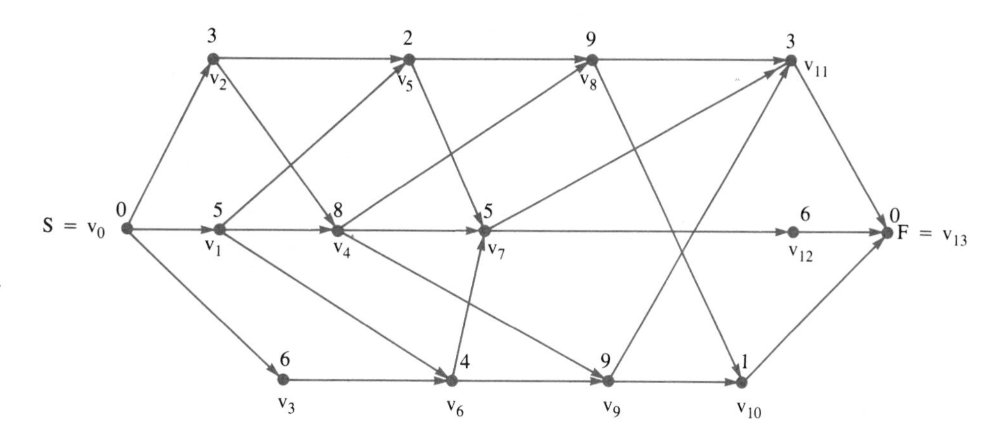

FIGURE P.2

A Graph for a More Complicated Scheduling Problem

multiplying our woes but, in fact, it is not. Indeed, such expansion or *generalization* of a problem often makes the solution easier to find.

To solve this expanded problem, we will use the **Recursive Paradigm.** This is one of the most powerful tools of discrete mathematics and of problem solving more generally. It is a major theme of our book.

The first step is: Jump into the middle! Suppose you already knew the answer for all the vertices up to a certain point, say, up to but not including v_7 in Fig. P.2. Ask: How could I use this information to figure out the answer for v_7 itself? Recursion means essentially reducing to previous cases, and that's just what we're doing here. (*Note*: The "middle" above doesn't mean the exact middle, but rather a typical, generic point.)

In Fig. P.3, we repeat enough of Fig. P.2 to show v_7 and all the vertices immediately preceding it: v_4, v_5, and v_6. We imagine we have found the minimum completion times of these tasks. Since we haven't actually calculated them, let's just call them A, B, and C. Now the *earliest* that task 7 can be completed depends upon which of tasks 4, 5, and 6 is completed *latest.* Thus the minimum time to finish task 7 is

$$5 + \max(A, B, C),$$

where $\max(A, B, C)$ represents the largest (*max*imum) of A, B, and C.

A better and more general notation is

$$T(k) \;=\; \text{minimum time to complete task } k,$$

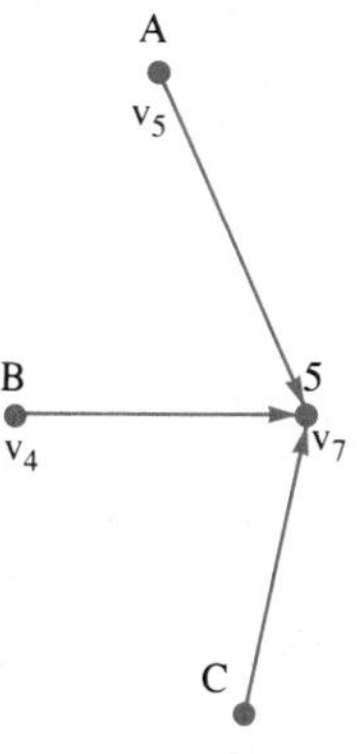

FIGURE P.3

A Portion of the Graph of Figure P.2

and

$$d(k) \;=\; \text{time } \underline{\text{d}}\text{uration of task } k \text{ itself.}$$

Then

$$T(7) \;=\; d(7) + \max\{T(4), T(5), T(6)\}.$$

We did this analysis for v_7 to be concrete. But by now it should be clear that a similar analysis is correct for any vertex. Indeed, we can state a general relationship. If we let

$$\max_{i \to j}\{T(i)\}$$

mean the maximum of the $T(i)$'s for all vertices v_i with edges directed to v_j, then

$$T(j) \;=\; d(j) + \max_{i \to j}\{T(i)\}. \tag{1}$$

Equation (1) is a kind of recursive equation called a *recurrence relation* because it expresses the value we wish to calculate, $T(j)$, in terms of other values $T(i)$.

All right, we've jumped into the middle of our problem, but how does this help us figure out what happens at the end? After all, we want the minimum time to the Finish. That is, we want $T(13)$. Suppose we set $j = 13$ in Eq. (1). We obtain

$$T(13) \;=\; 0 + \max\{T(10), T(11), T(12)\},$$

since $d_{13} = 0$. We don't know $T(10)$, $T(11)$, and $T(12)$, but we can analyze them in the same way. For instance, setting $j = 12$, we have

$$T(12) \;=\; 6 + \max\{T(7)\} \;=\; 6 + T(7),$$

since v_{12} has only one predecessor, v_7.

In short, using Eq. (1) we can keep working backwards. This would be completely futile—just digging a deeper and deeper hole—except for one thing: *The process eventually stops.* We already know that $T(0) = 0$, and eventually we work back to that. Look at Table P.1. The first column gives us a complete list of the backward steps using Eq. (1). Once we get down to $T(0) = 0$, we can start plugging in numerical values and working back up.

Indeed, that's what we have started to do in the second column. Starting at the bottom, we fill in the second column by simplifying the equation in the same row of the first column. We can easily obtain the last three rows of the second column, since the corresponding expressions in the first column depend only on $T(0)$. The next row up, $T(4)$, requires knowledge of the lower rows in the second column, since $T(4)$ depends on $T(1)$ and $T(2)$. That's why we work *up* in the second column. It's easy. Finish it! [2]

We started in the middle to obtain the key recursive relationship Eq. (1). Then we applied this relationship to the top, i.e., the final value we wanted, $T(13)$. This led us, in order to actually compute anything, to work up from the bottom, i.e., from the initial values. So why didn't we just start computing with the initial values in the first place?

We could have. To do so would have been an example of the second major problem solving technique in our book, the **Inductive Paradigm,** in which we start computing at the beginning and continue until we either get to the end or see

TABLE P.1
Times needed to complete tasks of Figure P.2

$T(13) = 0 + \max\{T(10), T(11), T(12)\}$	
$T(12) = 6 + T(7)$	
$T(11) = 3 + \max\{T(7), T(8), T(9)\}$	
$T(10) = 1 + \max\{T(8), T(9)\}$	
$T(9) = 9 + \max\{T(4), T(6)\}$	
$T(8) = 9 + \max\{T(4), T(5)\}$	
$T(7) = 5 + \max\{T(4), T(5), T(6)\}$	
$T(6) = 4 + \max\{T(1), T(3)\}$	$T(6) = 4 + 6 = 10$
$T(5) = 2 + \max\{T(1), T(2)\}$	$T(5) = 2 + 5 = 7$
$T(4) = 8 + \max\{T(1), T(2)\}$	$T(4) = 8 + 5 = 13$
$T(3) = 6 + T(0)$	$T(3) = 6$
$T(2) = 3 + T(0)$	$T(2) = 3$
$T(1) = 5 + T(0)$	$T(1) = 5$

a pattern. Then we make a conjecture about what the end result will be and prove this result without computing any more. We will consider this technique in depth in Chapter 2. In this example we could have proceeded inductively from the bottom of Table P.1 and worked our way to the top, as shown in the second column. Here no conjecture about the final result is possible, so we would just have to compute until the end. In using both the recursive and inductive paradigms we'll always be looking for ways to get the final result *analytically* rather than *computing* all the way to the end.

In this problem the two paradigms are not much different. Both quickly lead to computing from the bottom, and both get nowhere unless you hit on Eq. (1) pretty fast. But, in later examples, you'll see that the two paradigms often diverge. Applying the recursive paradigm does not always lead to marching up in order from the bottom. Applying the inductive paradigm does not always require understanding a recursive relationship at the beginning; sometimes you compute the early values to *discover* the relationships you need. Also, the inductive paradigm seems to come naturally to most people—at least the try-some-small-cases aspect of it. The recursive paradigm, however, seems very unnatural at first to many people. Jumping into the middle seems like a crazy idea. Yet, in our experience, the recursive paradigm is as powerful as the inductive paradigm.

In any event, the two paradigms are related, and as a pair they are *very* powerful. In some sense, this whole book is about what you can do with them.

We set out to illustrate discrete algorithmic mathematics with a problem. Now it's time to stand back from our problem and see what it does illustrate:

- It involves distinct ("discrete") objects, i.e., specific tasks and the precedence among them, as opposed to continuous objects, like the unbroken path of a satellite moving through space.

- Although it would have been nice to find a *formula* for the answer, we were satisfied to find a method for efficiently analyzing and computing the answer.
- We started with a real-world problem (admittedly simplified) and concerned ourselves with a mathematical *model* or *abstraction* (i.e., our representation as a directed graph) in which the essential details of the actual real world problem were removed. This representation helped us analyze the problem.
- The problem concerned *optimization*, i.e., how to achieve the best possible result, which in this case is the *minimum* time needed to complete the task.

The last two points are not unique to discrete mathematics. But discrete mathematics is especially rich in the variety of applications it can treat, especially optimization applications. It is particularly powerful for treating applications in some areas where continuous mathematics has been less helpful (e.g., computer science, social science, and management science), but it is also useful in the biological and physical sciences.

The second point in the preceding list—about formulas—deserves more explanation. We don't deny that we used formulas in our analysis, notably Eq. (1). But this is not a formula from which you can immediately get an answer. It's not, say, like $A = \pi r^2$, into which you can immediately plug numbers and get the area of a circle. That's why we called Eq. (1) a recurrence relation—we could also have called it a *recursive equation*—in contrast to more familiar equations, which are said to be in *closed form*, or *explicit*.

Furthermore, you wouldn't expect a closed-form formula to exist for the sort of problem we have discussed: There's too much variability in the shape of the graph and the numbers at the vertices. True, if we made the problem much simpler and symmetrical, we might be able to derive a formula. For instance, suppose there were n vertices, $d(k) = k$ for each vertex k, and each vertex k preceded vertex $k+1$ and no others. Then there is indeed a formula for $T(n)$ [3]. But this is too restrictive to be of general interest. Sometimes formulas in discrete mathematics are of general interest and we'll be interested in some of them in this book. But more often than not there are no such formulas. That is why discrete mathematics might be called the study of *organized complexity.*

The fact is, we typically use formulas like Eq. (1) only as stepping-stones to organized procedures for computing answers. Such procedures are called **algorithms**. Whereas classical mathematics is about formulas, discrete mathematics is as much about algorithms as about formulas.

One reason mathematics has not, until recently, been much concerned about algorithms is this: You can't analyze something mathematically unless you have a precise language for talking about it, and there hasn't been such a precise but easily readable language for algorithms. But now, with the advent of computers and computer languages, such *algorithmic language* has been developed. (Consequently, algorithmic language is often called "pseudocode" by computer scientists. But from our point of view, calling algorithmic language pseudocode is precisely backwards. For us, programming languages are the pseudo-objects, because they are not the real thing for precise communication among people.) In Chapter 1, we'll define carefully what we mean by an algorithm and introduce the algorithmic language we'll use throughout the book. The method we have developed for solving our minimum-time problem, the method illustrated by Table P.1, really is an algorithm. But until Chapter 1 we won't have the language to describe it as such.

Note: Algorithmic language is not a synonym for programming language. The former is simpler to learn and use, cannot be processed directly by a computer, and is intended for communication between *people*.

Here we are in the Prologue, and we have already referred you to Chapters 1, 2, and 3. Obviously, we have gotten far ahead of ourselves! As we said at the start, this section is simply an illustration, which you need not grasp in detail—yet. We just want to give you a feel for what discrete algorithmic mathematics is all about.

Problems: Prologue

1. What is the minimum time for our airplane stopover?

2. Finish filling in Table P.1. What is the minimum time to finish the project?

3. Find a formula for $T(n)$ when, for each vertex k, $d(k) = k$ and the only edge from vertex k is to vertex $k+1$.

4. Our solution procedure outputs, for each task, the optimal time at which that task *ends*. For example, if the procedure assigns 37 to task 4, that means task 4 can be completed at best 37 time units after the project begins. However, what your subcontractors probably want to know is when they can *start* their tasks.

   a) It's easy to figure this out since you have already computed the finish times of all the tasks. How?

   b) But suppose you want to figure out the optimal starting times directly from the graph. You do this by applying the recursive or inductive paradigm. What key relationship replaces Eq. (1)?

5. In a slightly different project scheduling model, the vertices merely represent checkpoints and they take essentially no time. The time-consuming tasks are represented by the edges between checkpoints. Thus we get a directed graph model with the weights on the edges. We can still ask: What's the minimum time to complete the final check? Solve this problem. What is the analog to Eq. (1)? How should the path lengths be defined?

6. In our minimum-time algorithm, it was essential that the tasks be numbered so that preceding tasks always have lower numbers. This was essential because, without it, there would be no "up" in the table; there would be no organized way to know, for any task j, when you had already computed all the $T(i)$'s needed before computing $T(j)$.

   Yet, we never said anything about where the numbering came from. Indeed, in any real problem the tasks come with names, not numbers. It won't be at all obvious how to number them in order to be "coherent" with the precedence relations. It won't even be obvious that any such numbering exists!

   There is a theorem that says such numberings always exist, so long as there is no "vicious circle" of tasks such that each task on the circle must precede the next one. Naturally, we prefer a proof of this theorem which actually shows how to construct a numbering. Such an algorithm could serve as a preprocessor for our minimum-time algorithm. Any ideas on how to construct the numbering? The recursive and inductive paradigms might help. (See [16, Supplementary, Chapter 3].)

7. In solving the airline problem, where did we use assumption c)—that there are enough workers to carry out simultaneously as many tasks as we want?

CHAPTER 0

Mathematical Preliminaries

0.1 Sets

A **set** is what mathematicians call a collection of objects—*any* objects—when the order of the objects is irrelevant. Few ideas are more important in mathematics than that of a set. In this book you will need to know the *language* of sets and how to *manipulate* sets.

The objects in a set may be physical (a set of students or a set of states of the United States) or abstract (a set of numbers or a set of philosophical ideas). The *members* or *elements* of a set, that is, the objects in it, may be homogeneous—all of the same kind—as in a set of students, but they need not be, as in a set consisting of an apple, the number 3, and Einstein's general theory of relativity. In general, we will use uppercase letters to denote sets and lowercase letters to denote the elements of a set. The symbol $\in$ will designate set membership. Thus

$$b \in S$$

is a statement that object b is a member of the set S.

One way to list the members of a set is by putting them in braces, separated by commas. Thus we can denote the set of all positive even integers less than 20 by P and enumerate it as

$$P = \{2, 4, 6, 8, 10, 12, 14, 16, 18\}.$$

Since the *order* of the elements between the braces is irrelevant, the preceding set is *precisely* the same set as

$$P = \{6, 12, 2, 18, 10, 4, 14, 16, 8\}.$$

If we had to list all the elements of a set whenever we wished to enumerate them, things would get tedious pretty fast. Luckily, there are other notations which enable us to avoid this problem. One is **set builder notation**, an example of which is

$$S = \{m \mid 2 \le m < 100 \textit{ and } m \text{ an integer}\}. \tag{1}$$

The letter before the vertical bar is a variable representing the name of a typical element. The conditions after the bar give the properties which each element must satisfy. Thus in Eq. (1), S is the set of all integers from 2 to 99. We may replace the "and" in Eq. (1) with a comma:

$$S = \{m \mid 2 \le m < 100,\ m \text{ an integer}\}.$$

Sometimes we wish to give a choice of conditions, in which case we use "or". Thus the set of positive integers less than 100 that are even or divisible by 3 could be denoted by

$$T = \{k \mid 0 < k < 100,\ k \text{ even } \textit{or } k \text{ divisible by } 3\}.$$

So far, our examples of sets have had a *finite* number of members, but this need not be the case. Discrete mathematics also deals with infinite sets such as the integers. To denote the set of nonnegative integers—remember that zero is not positive but is nonnegative—we may write (with N standing for *nonnegative numbers* or *natural numbers*)

$$N = \{0, 1, 2, 3, \ldots\}.$$

You're probably used to seeing three dots employed as in the definition of N. They are called an *ellipsis* and indicate that you may add elements to the group to the *left* of the dots by simply extending the rule used to form those elements. Thus, the numbers after 0, 1, 2, 3 are 4, 5, 6 etc.

To distinguish N from the set of positive integers (alas, sometimes also called the natural numbers), we'll use N^+ to denote the latter. Thus

$$N^+ = \{1, 2, 3, \ldots\}.$$

Other important sets that have a standard notation are:

i) The set of all integers, or

$$Z = \{\ldots -2, -1, 0, 1, 2, \ldots\}.$$

Note the use of the ellipsis on the left, which represents here an extension of the rule used to form the elements to its *right*. An ellipsis with nothing preceding it is the only case in which the ellipsis refers to the elements following it.

ii) The set of rational numbers, that is, numbers which are the ratios of integers, or

$$Q = \{a/b \mid a, b \in Z,\ b \neq 0\}.$$

When the ratio is required to be positive, we get

$$Q^+ = \{a/b \mid a, b \in N^+\}.$$

iii) The set of all real numbers (i.e., all numbers which can be expressed in decimal form), which is denoted by R and the set of positive real numbers R^+.

One other very important set for which we specify a standard notation is the **empty set**, which contains no members at all and is denoted by $\emptyset$. Thus

$$\emptyset = \{\}$$

On the other hand, $\{\emptyset\}$ represents the set whose only member is the empty set. The idea of an empty set may seem artificial, but it is very useful. It plays a role in relation to sets very much like the role that 0 plays in relation to numbers (hence the similarity of notation).

We often wish to talk about the number of elements in a set, called its **cardinality**. We do this by placing vertical bars around the set. (Compare this notation with that of absolute value in Section 0.4.) Thus

$$|\{1,3,9,15\}| = 4 \qquad \text{or} \qquad |N| = \infty.$$

As another example of cardinality consider the set

$$S = \{p/q \mid p,q \in N^+,\ p,q \leq 10\},$$

which is the set of all fractions with numerators and denominators which are integers from 1 to 10. What is $|S|$? [18] Since there are 10 choices for p and 10 for q, you might think that $|S| = 100$. But, for example, 2/4 and 4/8 represent the same number and so do not both contribute separately to the cardinality of S.

May a set have repeated elements as in

$$\{2,4,4,4,7,9,9\} \qquad ?$$

We shall usually reserve *set* to mean collections of *distinct* objects and shall use **multiset** when we want to emphasize that we are referring to collections in which one or more elements may be repeated.

Let S be a set. Then T is said to be a **subset** of S if every element of T is also in S. Thus if $S = \{1,4,8,9,12\}$, then $T = \{4,8,12\}$ is a subset of S but $V = \{1,8,10,12\}$ isn't, because 10 is not a member of S. An immediate consequence of this definition is that S is a subset of itself. Another consequence of this definition is that the empty set $\emptyset$ is a subset of every set. This is reasonable because, if $\emptyset$ were not a subset of some set S, it would be because it contained an element not in S, which is impossible because $\emptyset$ contains no elements.

If S is a subset of T and $S \neq T$, then S is called a *proper* subset of T. If S is a subset of T (proper or otherwise), we denote this by

$$S \subset T.$$

Caution: In some other books this notation means that S is a proper subset of T. In the case when S might be a proper subset or could be equal to T, the symbol used in such books is

$$\subseteq .$$

Don't confuse the symbols $\in$ and $\subset$. The symbol $\in$ is used to denote a relationship between an element of a set and a set, whereas $\subset$ denotes a relationship between two sets.

How many subsets does a set have? Consider, for example,

$$S = \{a, b, c\}.$$

We can find the subsets of S by brute force enumeration:

$$\emptyset \ \{a\} \ \{b\} \ \{c\} \ \{a,b\} \ \{a,c\} \ \{b,c\} \ \{a,b,c\}. \tag{2}$$

(Do you understand why we wrote $\emptyset$ and not $\{\emptyset\}$ here?) Similarly, the subsets of $S = \{1,3,7,8\}$ are

$$\emptyset \ \{1\} \ \{3\} \ \{7\} \ \{8\} \ \{1,3\} \ \{1,7\} \ \{1,8\} \ \{3,7\} \ \{3,8\}$$
$$\{7,8\} \ \{1,3,7\} \ \{1,3,8\} \ \{1,7,8\} \ \{3,7,8\} \ \{1,3,7,8\}.$$

From studying these two examples you should be able to conjecture how many subsets there are for a set with n elements. (Recall the inductive paradigm discussed in the Prologue.) The first set had 3 elements and 8 subsets; the second had 4 elements and 16 subsets. A natural conjecture is that the number of subsets of a set with n elements is 2^n.

We can also apply the recursive paradigm to this problem. Suppose we "jump into the middle" and assume we know the number of subsets when $|S| = n$. What if $|T| = n+1$? Suppose without loss of generality that $S = \{1,2,3,\ldots,n\}$ and $T = \{1,2,3,\ldots,n,n+1\}$. Every subset of T is also a subset of S, or it has $n+1$ as a member. But if $n+1$ is in a subset, the remaining members of the subset are a subset of S. Thus T has twice as many subsets as S since, for each subset of S, there is an identical one in T and one in T with $n+1$ also. Since a one-element set has $2=2^1$ subsets (why?), we conclude again that a set with n elements has 2^n subsets. We'll ask you to give a rather more careful proof in Chapter 2 [27, Section 2.2].

We conclude this discussion with the remark that the elements of a set can themselves be sets. Thus, for example, we may consider the set P of all subsets of a given set S. We call P the **powerset** of S. Thus the powerset of $\{a,b,c\}$ is given by putting braces around the sets listed in Eq. (2).

Set Operations

Let S and T be two sets. What kinds of operations might we wish to perform on S or T or both, which might show the useful properties of these sets? The four most common and useful operations are:

1. Set **union**, which we denote by $\cup$. We define $S \cup T$ to be the set of all elements which are in either S *or* T (but without repeating elements which are in both). Thus

$$S \cup T = \{x \mid x \in S \text{ or } x \in T\}.$$

EXAMPLE 1 Let $S = \{3,5,6,9,47\}$ and $T = \{4,6,7,9,82\}$. Then

$$S \cup T = \{3,4,5,6,7,9,47,82\}. \ \blacksquare$$

EXAMPLE 2 Let S be all the odd positive integers—$\{1, 3, 5, \ldots\}$—and let T be all the even nonnegative integers—$\{0, 2, 4, 6, \ldots\}$. Then $S \cup T = N$, the set of all nonnegative integers. ∎

2. Set **intersection**, which we denote by $\cap$. We define $S \cap T$ to be the set of all elements which are in *both* S and T. Thus

$$S \cap T = \{x \mid x \in S,\ x \in T\}.$$

EXAMPLE 3 With S and T as in Example 1, $S \cap T = \{6, 9\}$. ∎

EXAMPLE 4 With S and T as in Example 2, $S \cap T = \emptyset$, since the two sets have no elements in common. ∎

EXAMPLE 5 With S as in Example 2 and $T = N$, $S \cap T = S$, since all the elements in S are in N. ∎

3. Set **complement**, which we denote by an overbar, as in $\overline{S}$, is defined to be the set of all elements *not* in S. But what can that mean? For example, if S is the set of all odd integers, then not only are 4 and 114 not in S, but neither is an apple nor the set of all subsets of S nor, for that matter, the kitchen sink. In order to make the idea of set complement meaningful, we must introduce the notion of a **universe**. This is the set U of all elements which we wish to consider. Then the complement of a set S is all members of the universe not contained in S. Thus

$$\overline{S} = \{x \mid x \in U,\ x \notin S\},$$

and we can define U in any way appropriate (as long as the reader is informed!).

EXAMPLE 6 If $S = \{1, 4, 7, 9, 10\}$ and the universe is the set of all positive integers no greater than 10, then $\overline{S} = \{2, 3, 5, 6, 8\}$. ∎

EXAMPLE 7 If the universe is N, the complement of S in Example 2 is T and, vice versa, the complement of T is S. ∎

4. Set **difference**, which we denote by $-$. If S and T are two sets, then we define $S - T$ to be the set of all elements in S that are not in T. Thus

$$S - T = \{x \mid x \in S,\ x \notin T\}.$$

EXAMPLE 8 If $S = \{3, 6, 7, 10\}$ and $T = \{7, 10\}$, then $S - T = \{3, 6\}$. ∎

EXAMPLE 9 If $S = \{3, 6, 7, 10\}$ and $T = \{5, 10\}$, then $S - T = \{3, 6, 7\}$. Note that we just ignore elements of T which are not in S (5, in this example). ∎

EXAMPLE 10 If S and T are as in Example 2, then $S - T = S$. Note that, using set difference, we can define the complement of a set S as $U - S$ (why?). ∎

If you're familiar with the programming language Pascal, you'll recognize that the four operations just described are precisely those incorporated in Pascal. Note that three of the four, namely, set union, intersection, and difference are *binary* operations in that the operators $\cup$, $\cap$, or $-$ require *two* sets as arguments. By contrast, set complement is a *unary* operation because it requires only one argument (although a second set, the universe, is implicitly involved). Also, union and intersection are *commutative* in that the order of the sets is immaterial; for example,

$$S \cup T = T \cup S.$$

However, set difference is not commutative.

In mathematics, as elsewhere, a picture *is* often worth a thousand words. The most useful visual way to display set operations is the **Venn diagram** (after John Venn, 1834–1923, a British logician), as shown in Fig. 0.1. The outer shape in a Venn diagram, usually a rectangle, represents the universe and the inner shapes, usually circles, represent sets. The shaded areas (which are not necessary in Venn diagrams) represent the result of the set operation being displayed. The four set operations we discussed are displayed in Fig. 0.1.

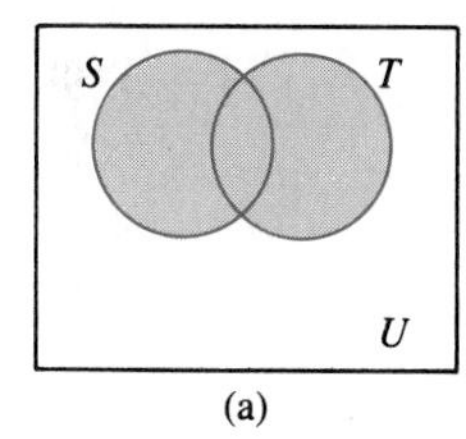

(a)

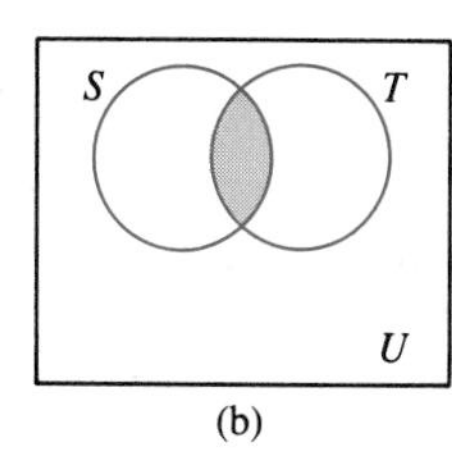

(b)

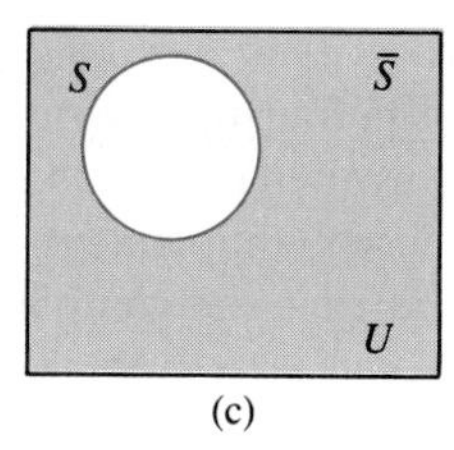

(c)

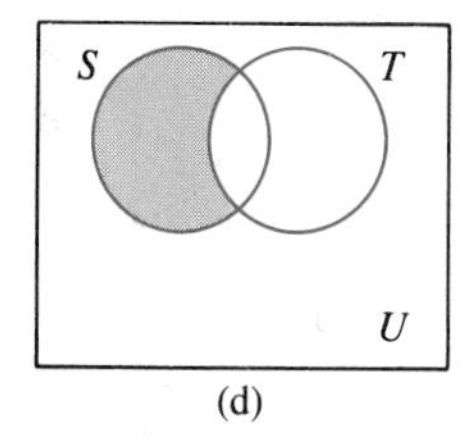

(d)

FIGURE 0.1

Venn Diagrams: (a) Union $S \cup T$; (b) Intersection $S \cap T$; (c) Complement $\overline{S}$; (d) Difference $S - T$.

As with arithmetic operations, set operations can be used to form more complex expressions than just $S \cup T$ or $S \cap T$, etc. Examples are

$$R \cup S \cap T, \qquad A \cup B \cup C \cup D \cup E \cup F \cup G, \qquad I \cap J \cup K \cap L.$$

Only the second example has an obvious meaning: The union of a number of sets contains the elements which are in any of the sets. But is the first example the union of R and the set $S \cap T$ (see Fig. 0.2a), or is it the intersection of $R \cup S$ and T (see Fig. 0.2b)? This is analogous to the question in ordinary arithmetic of whether

$$2 + 3 \times 4 \tag{3}$$

equals $2+(3\times4)=14$ or $(2+3)\times4=20$. There is no "correct" answer to this question. It is a matter of convention. That is, we establish the order in which various operators are to be applied. The usual (but by no means universal) convention in arithmetic is to apply multiplication before addition, so that the expression in (3) would evaluate to 14. With sets, the standard convention is to apply intersection before union. Thus Fig. 0.2a expresses the meaning of the first of the three examples. Similarly, Fig. 0.2c gives the standard interpretation of the third example. Note, by

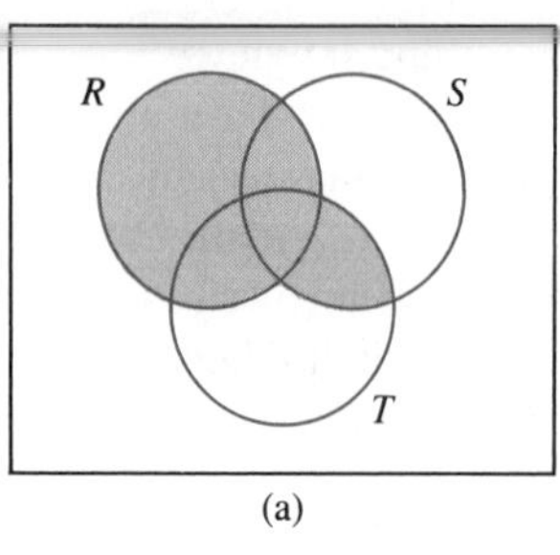

(a)

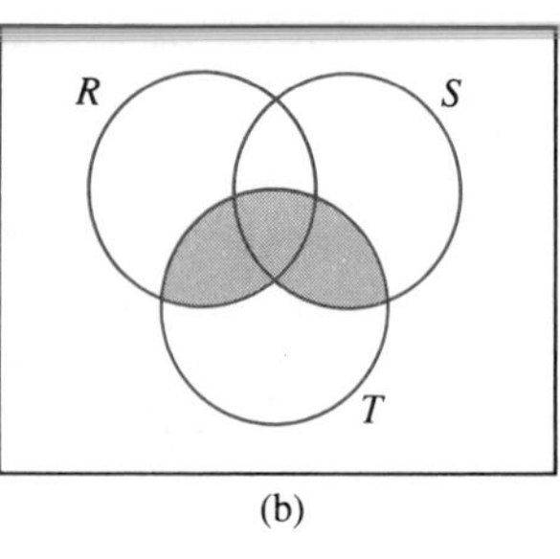

(b)

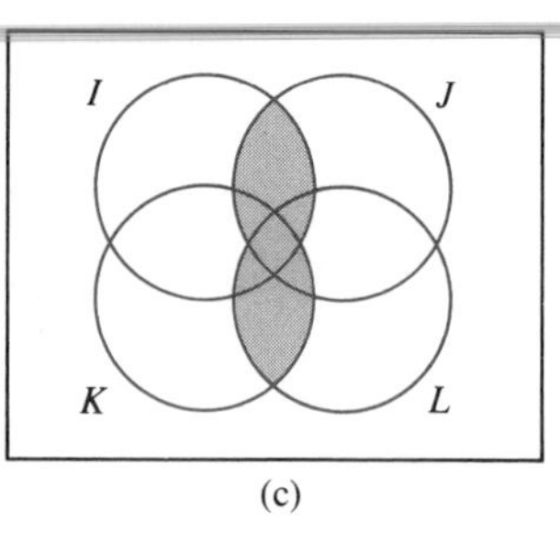

(c)

FIGURE 0.2

(a) $R \cup (S \cap T)$;
(b) $(R \cup S) \cap T$;
(c) $I \cap J \cup K \cap L$.

the way, that once the number of sets becomes greater than three, Venn diagrams are no longer very useful ways to portray set expressions.

To avoid ambiguity we normally write (3) as

$$2 + (3 \times 4),$$

even with the convention that multiplication is performed first. We *must* use parentheses if we wish (3) to be interpreted as $(2+3) \times 4$. Similarly, with sets we use parentheses to override the standard interpretation. Thus to represent the intersection of T and the union of R and S, we would write $(R \cup S) \cap T$.

This discussion implies that we need some kind of ordering or *hierarchy* of the four operators, $\cup$, $\cap$, $\overline{}$, and $-$, so that we can correctly interpret expressions involving more than one of them. The usual ordering—and the one we'll use here—is set complement, then set intersection, and then either set union or set difference, so that the complete hierarchy is

$$\begin{array}{ll} \overline{} & \text{(overbar)} \\ \cap & \\ \cup, \ -, & \end{array}$$

with the proviso that we can always use parentheses to override the hierarchy, as illustrated in the following example. Such an ordering of operators is also often called a *precedence relation.*

EXAMPLE 11 If $S = \{4, 7, 8, 10, 23\}$, $T = \{5, 7, 10, 14, 20, 25\}$ and $V = \{2, 5, 10, 20, 30, 36\}$, evaluate the following expressions using the operator hierarchy.

a) $S \cup T \cap V$ **b)** $S - T \cap V$ **c)** $S \cap T - V$

Solution

a) Since $\cap$ precedes $\cup$ in the hierarchy, we first compute $T \cap V = \{5, 10, 20\}$, then $S \cup \{5, 10, 20\} = \{4, 5, 7, 8, 10, 20, 23\}$. If instead we had written

$$(S \cup T) \cap V,$$

then the parentheses would have required that we first evaluate

$$S \cup T = \{4, 5, 7, 8, 10, 14, 20, 23, 25\}$$

followed by

$$\{4,5,7,8,10,14,20,23,25\} \cap V \;=\; \{5,10,20\}\,.$$

b) We again have $T \cap V = \{5,10,20\}$ and $S - \{5,10,20\} = \{4,7,8,23\}$.
c) $S \cap T = \{7,10\}$ and $\{7,10\} - V = \{7\}$. ∎

But do the hierarchy and parentheses rules solve all our problems? How about expressions such as

$$S \cup T \cup V, \qquad S \cap T \cap V \cap W, \qquad S \cup T - V, \qquad S - T - V?$$

For the one on the left, does it matter whether you first compute $S \cup T$ and then take the union with V or whether we first take $T \cup V$ and then compute the union of this with S? The answer is No. You can compute any sequence of unions in any order; you always obtain the set of all elements that appear in any one of the sets whatsoever. Similarly, you can compute any sequence of intersections in any order [19]. But for set differences and for combinations of set differences and unions, this is not true.

EXAMPLE 12 If $S = \{5,6,7,8,9,10\}$, $T = \{6,7,8,9\}$, and $V = \{8,9,10\}$, then

$$S \cup T \;=\; \{5,6,7,8,9,10\} \qquad \text{and} \qquad \{5,6,7,8,9,10\} - V \;=\; \{5,6,7\}\,,$$

but

$$T - V \;=\; \{6,7\} \qquad \text{and} \qquad S \cup \{6,7\} \;=\; \{5,6,7,8,9,10\}\,.$$

Similarly,

$$S - T \;=\; \{5,10\} \qquad \text{and} \qquad \{5,10\} - V \;=\; \{5\}\,,$$

but

$$T - V \;=\; \{6,7\} \qquad \text{and} \qquad S - \{6,7\} \;=\; \{5,8,9,10\}\,.$$ ∎

Thus for set difference and union operators we need one further rule to avoid ambiguity: When there are two or more consecutive set difference operators or consecutive set difference and union operators, they should be evaluated from left to right. Therefore the correct interpretation of $S \cup T - V$ is $(S \cup T) - V$ and that of $S - T - V$ is $(S - T) - V$.

One additional set operator will play a role in Section 0.2: the **Cartesian product** operator denoted by $\times$. If S and T are sets, then $S \times T$ is the set consisting of all *ordered* pairs where the first element of the pair is a member of S and the second element is a member of T.

EXAMPLE 13 If $S = \{2,5,7,9\}$ and $T = \{4,5,8\}$, then the Cartesian product of S and T is

$$S \times T \;=\; \{(2,4),\ (2,5),\ (2,8),\ (5,4),\ (5,5),\ (5,8),\ (7,4),\ (7,5),\ (7,8),\ (9,4),\ (9,5),\ (9,8)\}\,.$$

Note that, although the order of the elements in the set (i.e., the pairs in parentheses) does not matter, the order of the elements in each pair does matter (i.e., $(2,4)$ is not the same as $(4,2)$). ∎

The Cartesian product of S and T has a simple interpretation when both S and T are the set of real numbers R. If we think of S as containing all the points on the x-axis (see Fig. 0.3) and T all the points on the y-axis, then $S \times T$ is the set of all points in the x-y plane.

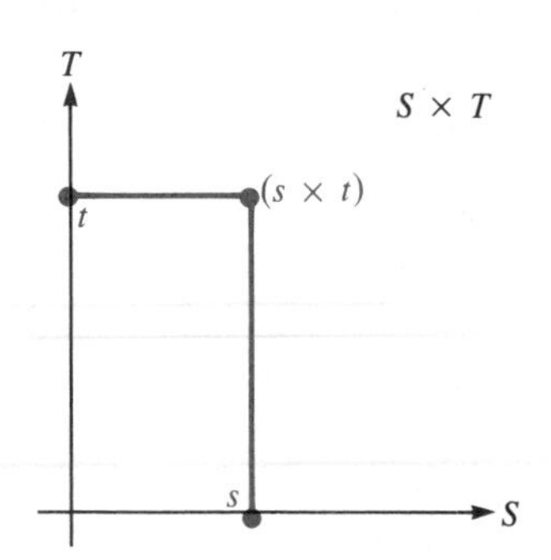

FIGURE 0.3

The Plane as the Cartesian Product of the Set of Real Numbers with Itself

The idea of a Cartesian product suggests the idea of an **ordered set** in which the order of the elements *is* relevant. From the ordered pair of a Cartesian product we can naturally generalize to an ordered *tuple* which, if it contains n elements, is called an n-tuple. To distinguish between ordered sets and sets, we will use parentheses rather than braces. Thus

$$\{1, 3, 5, 9\}$$

is the set of four elements 1, 3, 5, and 9, but

$$(3, 9, 5, 1)$$

is a 4-tuple. Two n-tuples are equal if and only if they are equal entry by entry. Thus $(3, 9, 5, 1) \neq (3, 1, 9, 5)$, even though $\{3, 9, 5, 1\} = \{3, 1, 9, 5\}$. We call the infinite tuple

$$(a_1, a_2, a_3, \ldots)$$

a **sequence**. Sequences will be an important topic of discussion in Chapters 5 and 9. We can easily extend the definition of a Cartesian product of two sets to n sets so that

$$S_1 \times S_2 \times \cdots \times S_n$$

is the n-tuple $(s_1, s_2, \ldots, s_n)$, where s_i is a member of the set S_i.

Problems: Section 0.1

1. Display all subsets of

a) the set $S=\{x,y,z\}$;

b) the set $\{2,4,6,8\}$.

2. Display the powerset of the powerset of $\{a\}$.

3. Describe in English each of the following sets.

a) $\{m \mid m=2n \text{ for some } n\in Z\}$.

b) $\{s \mid s=n^2 \text{ for some } n\in N^+\}$.

c) $\{q \mid q=mn \text{ for some } m,n\in N,\ m,n>1\}$.

4. Use set builder notation to describe the set of

a) all odd numbers between 100 and 200;

b) all points on the graph of the function $y=x^2$.

5. Let N, N^+, Z, and Q be as defined in the text. Let E be the set of even integers and F the set of integers divisible by 5. (Note that $0,-5\in F$.) Use set operators to name the set of

a) positive integers divisible by 5;

b) even integers divisible by 5;

c) nonintegral rational numbers;

d) integers divisible by 10;

e) positive integers divisible by 10;

f) odd positive integers;

g) odd negative integers;

h) nonpositive integers;

i) negative integers;

j) ordered pairs, where the first entry is even and the second odd.

6. Let $R=\{1,2,3,4\}$, $S=\{1,3,5,7,9\}$, and $T=\{3,4,5,6,7,8\}$. Compute:

a) $R\cup S$, $R\cup T$, and $S\cup T$.

b) $R\cap S$, $R\cap T$, and $S\cap T$.

c) $R-S$, $R-T$, and $S-T$.

d) the complements of R, S, and T assuming the universe is the set of all integers from 1 to 10.

e) $R\cap S\cup T$.

f) $R-S\cap T$.

g) $R-S\cap T-S$.

h) $R\cap S\cap T$.

7. (a)–(h) Draw Venn diagrams to illustrate each of the expressions in [6].

8. List all the elements of $\{b,c,d\}\times\{e,o\}$.

9. Is the Cartesian product commutative? That is, is $S\times T$ the same as $T\times S$? Why?

10. Let R, S, and T be sets and let U denote the universe. Prove the following equalities using for each either a logical argument in English or a Venn diagram.

a) $U-R=$ the complement of R.

b) $R-(S\cup T)=(R-S)\cap(R-T)$.

c) $R-(S\cap T)=(R-S)\cup(R-T)$.

d) $R\cap(S-R)=\emptyset$.

11. If $A\cap B=\emptyset$, what does that tell you about $|A\cup B|$?

12. If $A\subset B$, what does that tell you about $A\cap B$ and $A\cup B$?

13. If $A\cap B=\emptyset$ and $B\cap C=\emptyset$, what does that tell you about $A\cap C$?

14. If $A\cup B=D$, must $D-B=A$? If not, what can you conclude if $A\cup B=D$ and $D-B=A$?

15. If $A\subset S$ and $B\subset S$, what can you say about $A\cup B$?

16. If $|A|=m$, $|B|=n$, and $m\geq n$, what is the least $|A\cup B|$ can be? The most?

17. Consider $|A\cap B|$ and answer the questions posed in [16].

18. Find the cardinality of the set

$$S = \{p/q \mid p,q\in N^+,\ p,q\leq 10\}.$$

19. Starting from the definition that $A\cup B$ is the set of all things which are in at least one of A and B, give an argument in English why $(X\cup Y)\cup Z$ means the set of all things which are in at least one of X, Y, and Z. (*Hint*: Begin by using the definition to get a complicated English sentence describing $(X\cup Y)\cup Z$ and then use your understanding of English to simplify the sentence.)

20. a) Using the same method as in [19], show that $X\cup(Y\cup Z)$ is also the set of all things in at least one of X, Y, and Z. Thus set union is associative. (By induction—see Chapter 2—we could go on and prove that parentheses make no difference when unioning n sets: Any order gives the set of things which are in at least one of the n sets.)

b) Similarly, show that intersection is associative.

0.2 Relations

Relations are as ubiquitous in mathematics as they sometimes seem to be in life. For instance, in the previous section we discussed $\subset$, which is a relation between sets, and $\in$, which is a relation between an element of a set and a set. Intuitively, any property which can hold between objects is a relation. In this section we give a formal definition of a relation which may seem odd at first, but it has the advantage of generality and precision. In particular, it gives us a language in which to discuss certain properties which relations may have: **symmetry**, **reflexivity**, and **transitivity**.

Definition 1. A **relation** on sets S and T is any subset of the Cartesian product $S \times T$.

A relation as defined here is called a *binary* relation because it involves the Cartesian product of two sets.

EXAMPLE 1 The "larger than" relation of real numbers is the set

$$L = \{(x, y) \mid x, y \in R,\ x > y\}.$$

Here $S = T = R$. To say that $(x, y) \in L$ is just a formal way of saying that x is larger than y. Similarly, $(x, y) \notin L$ is just a formal way of saying that x is not larger than y. ∎

EXAMPLE 2 Let S and T both be N^+. Then the relation of **divisibility** is defined as the set

$$D = \{(s, t) \mid s, t \in N^+,\ s \mid t\}, \tag{1}$$

where the vertical bar in $s \mid t$ denotes that s divides t with a remainder of zero (i.e., t/s is an integer). So $(s, t) \in D$ is just a formal way to say that s divides t. Thus (2, 4) and (47, 235) are in D but (17, 52) is not. (Note the unfortunate double use of vertical bars in Eq. (1), but both uses are standard in mathematics. Since the number of symbols available for mathematical notation is limited, some multiple use of such symbols is inevitable. The trick is to avoid two uses of the same symbol which might lead to ambiguity.) ∎

EXAMPLE 3 The relation

$$E = \{(s, t) \mid s, t \in N,\ s = t\}$$

is the definition of the equality relation between nonnegative integers. We may define inequality relations similarly [1]. ∎

The most important properties of relations are these:

Reflexivity. A relation R on $S \times S$ is *reflexive* if $(x, x) \in R$ for every $x \in S$. Thus equality is reflexive (since $x = x$ for every x), as is $\leq$ (since every $x \leq x$).

However, $<$ is not reflexive (since $x < x$ is not always true; in fact, it's never true!). Set inclusion is also reflexive (why?).

Symmetry. A relation R on $S \times S$ is *symmetric* if $(x, y) \in R$ implies that $(y, x) \in R$. Of all the relations we have mentioned so far, only equality is symmetric. Here are three other examples of symmetric relations:

- Set A is related to set B if $A \cap B \neq \emptyset$.
- Person p is related to person q (in the mathematical sense) if they are related in the social sense. (If A is B's parent, then A is related to B and B is related to A.)
- p is related to q if and only if $2 \mid (p-q)$ (i.e., 2 divides $p-q$). This is called the **parity relation**, because we say that two integers have the same parity if they differ by an even number.

Transitivity. A relation R is *transitive* if, whenever $(x, y) \in R$ and $(y, z) \in R$, then $(x, z) \in R$. That is, whenever x is related to y and y is related to z, then x is related to z. The relations

$$= \qquad < \qquad > \qquad \leq \qquad \geq \qquad \subset$$

are all transitive. The relation p is the parent of q is not transitive but p is an ancestor of q is. What about the parity relationship [4b]? What about the other two symmetric relations above?

When a relation is reflexive, symmetric, and transitive, the relation is called an **equivalence relation**. *Equality* is an equivalence relation.

Naturally, if there are binary relations, there are also ternary, quaternary, etc., relations. For instance, the set of all real (x, y, z), such that $z = x + y$, is a ternary relation, which we call "sum". The 3-tuples which satisfy this relation are a subset of $R \times R \times R$. However, while n-ary relations, such as in this example, are obviously important in mathematics, the three properties of relations we have introduced are really useful only for binary relations.

Problems: Section 0.2

1. Express the following relations in set builder notation.

a) One number is less than or equal to another.

b) One integer is a factor of another.

c) Two integers are unequal.

d) One set is a subset of another.

2. For each of the relations in [1], determine whether it is symmetric, reflexive, or transitive.

3. We define the relation $=_m$ (read "equal mod m") to be the set

$$\{(p, q) \mid m \mid (p-q)\}.$$

a) Give two pairs which are in the relation $=_4$ and two pairs which are not.

b) Determine whether $=_m$ is symmetric, reflexive, or transitive.

4. a) Is the relation $\in$ transitive? Why?

b) How about the parity relationship?

5. We define the **successor relation** on $N \times N$ to be the set

$$\{(m, n) \mid m = n+1\}.$$

Give two pairs which are in the successor relation and two pairs which are not.

6. The **reverse** of a relation R on $S \times T$ is a relation R' on $T \times S$ defined by

$$(t, s) \in R' \iff (s, t) \in R,$$

where the double arrow should be read "if and only if". What are the reverses of

a) $>$? b) $\subset$? c) $\geq$?
d) $=_m$ (see [3])?
e) the successor relation?

7. If R is symmetric, is R-reverse symmetric? Answer the same question for reflexive and transitive relations.

8. The **transitive closure** of a relation R on $S \times S$ is another relation on $S \times S$ called $\mathrm{Tr}(R)$ such that $(s, t) \in \mathrm{Tr}(R)$ if and only if there exists a sequence $s = s_1, s_2, s_3, \ldots, s_n = t$ such that $(s_i, s_{i+1}) \in R$ for each i.

a) What is the transitive closure of the successor relation?
b) What is the transitive closure of the $>$ relation?

9. Let S be the set of all people. The "parent" relation is just what you think: People a and b are related by it if and only if a is the parent of b.

a) What is the reverse of the parent relation?
b) What is the transitive closure (see [8]) of the parent relation?

10. Given relations $R_1 \subset S \times T$ and $R_2 \subset T \times U$, the **composition** $R_1 \circ R_2$ consists of all pairs $(s, u) \in S \times U$ for which there is a $t \in T$ with $(s, t) \in R_1$ and $(t, u) \in R_2$. What is the composition of

a) the parent relation (see [9]) with itself?
b) the parent relation with its reverse?
c) the reverse of the parent relation with the parent relation?
d) the relation $|$ and $\leq$?
e) the relations $=_3$ and $=_5$ (see [3])?

11. Given a relation R on $S \times T$, we define $R(s)$, the *image* of s in T to be

$$\{t \in T \mid (s, t) \in R\}.$$

Similarly, we define $R^{-1}(t)$, the *preimage* of t in S, to be

$$\{s \in S \mid (s, t) \in R\}.$$

For the relation $p|q$, what is the image of 2? The preimage of 36?

12. When S and T are finite, there is a useful visual representation of a relation on $S \times T$. Draw a dot for each element of $S \cup T$, and for each $(s, t) \in R$ draw an arrow from s to t. This representation as a *directed graph* (see the Prologue) is especially useful if $S = T$. (We say much more about directed graphs in Chapter 3.)

a) Let $S = T = \{2, 3, \ldots, 14, 15\}$. Draw the directed graph of the relation $p|q$ on $S \times T$. (*Suggestion*: Draw all arrows so they angle down the page. You should be able to find a placement of the dots and arrows so that the graph looks "clean" and so that it visually highlights certain facts about factoring.)
b) Let $S = T = \{0, 1, 2, \ldots, 12\}$. Draw a graph on S of the relation $=_4$ (see [3]). The striking nature of this graph is due to the fact that $=_4$ is an equivalence relation.
c) Let $S = T = \{$your maternal grandfather and all his descendants$\}$. Draw the graph of the parent relation on S. (What you get is called a "family tree" by sociologists, a "rooted, directed tree" by mathematicians, and just a "tree" by computer scientists.)
d) Let $S = T =$ the powerset of $\{a, b, c\}$. Draw the graph of the relation $\subset$ on $S \times T$. If you position things cleanly, your graph will include a well-known geometric figure.

13. Going back to the airplane loading problem of the Prologue, the precedence discussed there is a relation. Name the set S (which equals T) and list all the ordered pairs. Also draw the graph of the relation. How does it relate to the diagram of the problem shown in the Prologue?

14. Let S be the set of concepts

$$\{\text{rectangle, rhombus, parallelogram, quadrilateral, square}\}$$

Define a relation R on $S \times S$ by $(a, b) \in R$ if and only if concept a is a special case of concept b. List all pairs which are in R. Draw the graph.

0.3 General Properties of Functions

You are probably familiar from high school mathematics with the concept of a function. Functions play an important role in both algebra and trigonometry. But perhaps you haven't realized that a function is nothing more than a special form of a binary relation, as defined in Section 0.2. If a relation R on $S \times T$ has the property that for each $x \in S$ there is exactly one $y \in T$ such that $(x, y) \in R$, then R is called a **function** and y is referred to as a function of x. Moreover, there is a common notation $y = f(x)$.

For example, let S and T be N, the set of nonnegative integers. If $x \in S$, let the corresponding element of T be the square of x. Note that while this function is defined for every element in S, it is not necessary that every element in T be a member of some ordered pair in R. In this example only the subset

$$\{0, 1, 4, 9, 16, 25, \ldots\}, \tag{1}$$

of T contains elements of N which form the second part of (x, y) in R. Sometimes you will want something more specific than the generic name $f(x)$ to use when referring to a function such as this. You are probably used to writing this function as x^2, but another notation is $\text{sqr}(x)$. The latter corresponds to the notation commonly used in programming languages.

The formality of defining a function as a relation can serve to obscure how we normally use functions. We're concerned mainly with the fact that y is *produced* from x by f rather than that (x, y) is in the relation. Thus it is more useful to consider an equivalent definition of a function as a **mapping**, which takes each element in one set, the **domain**, and associates with it one—and only one—corresponding element (the one it *maps* into) in a second set, the **range**. The element corresponding to x, namely, $f(x)$, is called the **image** of x. *Image* is also used to describe the set of all elements in the range which are mapped into by elements in the domain. This is illustrated in Fig. 0.4. The image may be a proper subset of the range, as in the squaring function example above. For the squaring function the set shown in (1) is the image set.

The requirement that each element of the domain map into a unique element of the image means that certain common "functions" are not really functions at all. For example, suppose that the domain is the set of all nonnegative real numbers R^+ and that the range is the set of all real numbers R. Then the square root "function" is not a function in our terms because, corresponding to each element in the domain (say, 1.44), there are two elements in the range $(+1.2$ and $-1.2)$ which are the images of 1.44. To make the square root a function as defined here, we must define it to be either the positive square root of an element in the domain or the negative square root, but not both.

You may be used to thinking of functions as formulas, for example, $y = 3x + 2$. But you should be aware that not every formula gives a function and not every function has a formula. Thus

$$x^2 + y^2 = 1 \tag{2}$$

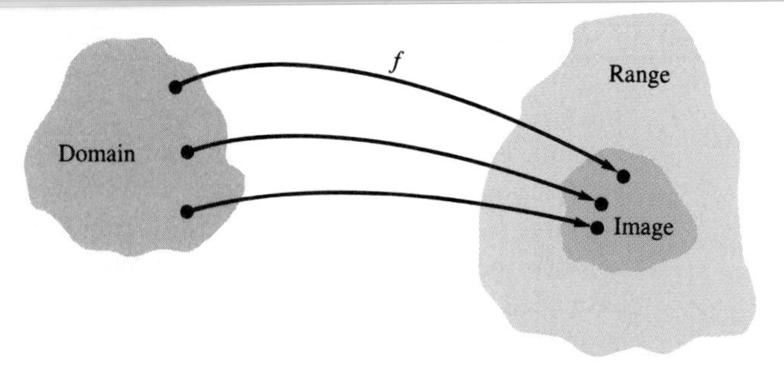

FIGURE 0.4

Function Terminology. The image is the set of points in the range mapped into by the function f operating on each point in the domain.

with domain and range R is a formula, but it does not define a function because two y's correspond to each x. (However, Eq. (2) does define a relation.) In discrete mathematics, as opposed to algebra and calculus, functions sometimes do not have algebraic formulas because the subject matter of discrete mathematics often deals with aggregate objects—sets, statements, strings of symbols, etc.—instead of numbers. An example of a function which does not have an algebraic formula is the function whose domain is all finite subsets of Z and whose value is 1, if the size of the set is even and 0 if it is odd. (This function should remind you of the parity relation in Section 0.2.)

Nevertheless, most of the functions you will see in this book have formulas and we'll generally use the $f(x)$ notation. A common need will be to **evaluate** $f(x)$ (i.e., find the point in the range into which x is mapped by f) when the argument is a general symbolic expression rather than just a point in the domain. For example, suppose

$$f(x) = 3x^2 + 4x + 1.$$

Then

$$\begin{aligned} f(3t+1) &= 3(3t+1)^2 + 4(3t+1) + 1 \\ &= 27t^2 + 30t + 8. \end{aligned}$$

Also,

$$\begin{aligned} f(3x+1) &= 3(3x+1)^2 + 4(3x+1) + 1 \\ &= 27x^2 + 30x + 8. \end{aligned}$$

Does it matter whether we use t or x? The answer is No because the specific name we choose for a variable point in the domain is of no significance and, even if we use x to define the function, we can use it again as we've just shown. The convention is to use x, but we can use any symbol whatsoever. Our only caveat is: Don't reuse a symbol if you thereby create some confusion for yourself or, perhaps, for the instructor who will grade your work.

A function defines a mapping from a domain to an image. But how about the mapping from the image back to the domain? Is this a function, too? Yes, if corresponding to each point in the image, there is one and only one point in the domain which maps into it. But even if the mapping from the image to the domain is not a function, it is still well-defined:

> Associate with each point in the image set that point or points in the domain which map into it.

If the mapping from the domain to the image is denoted by f, then, when it exists, the **inverse mapping** from the image to the domain is denoted by f^{-1}. (Why "when it exists"? Because, if more than one element in the domain corresponds to a single element in the image, then f^{-1} is not a function.) Don't be misled by the -1 notation. It doesn't mean the reciprocal of the function f:

$$f^{-1}(x) \neq \frac{1}{f(x)}$$

This is another of those cases where we make mathematical notation do double duty. You should recognize that this notation is a reasonable one and not just because we call f^{-1} the inverse function. What the inverse mapping really does is *undo* the mapping f. That is,

$$f^{-1}(f(x)) = x,$$

which we normally write as

$$f^{-1}f(x) = x.$$

This just says that if you map a point x in the domain to a point $y = f(x)$ in the image and then apply the inverse mapping to y, you get back the original x. You would be quite wrong to interpret the preceding notation as *multiplying* f^{-1} times f, but it is *analogous* to multiplying p^{-1} times p to get 1. Hence the justification for the notation.

Function Terminology

A considerable amount of terminology is associated with functions, which we might as well get out of the way now. To make things a bit more annoying, there is an old set of terminology and a new set, either of which you may run across in other books. We'll illustrate the meaning of each term with some examples.

1. **Onto functions**. A function is said to be onto or *surjective* if its image set is the entire range (see Fig. 0.5). Since more than one point in the domain can map into the same point in the range, the inverse mapping of an onto function may not be a function.

EXAMPLE 1 You've seen that the squaring function is not onto if the domain and range are both N. But suppose the domain and range both consist of all nonnegative real numbers.

FIGURE 0.5

Onto Function. A function is onto (surjective) if its image contains all points in the range.

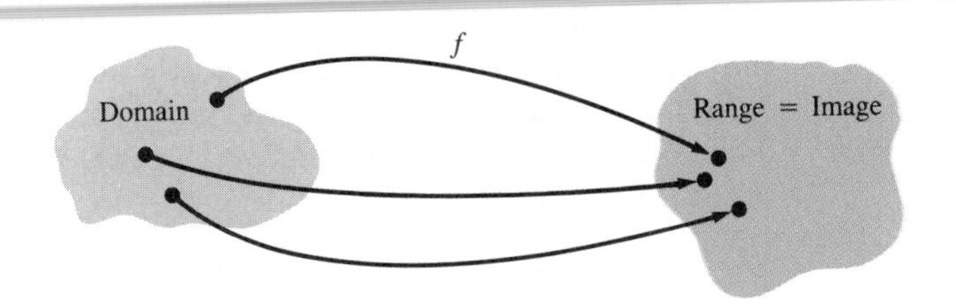

Then the squaring function is onto because, corresponding to every nonnegative real number y, there is another such number x (its nonnegative square root) whose image is y. Thus whether a function is onto depends on the definition of its domain and range. ∎

EXAMPLE 2 Suppose $S = \{1, 4, 7, 9\}$ and $T = \{1, 5, 8\}$. Then the set of ordered pairs

$$\begin{array}{ll} (1,5) & [f(1) = 5] \\ (4,1) & [f(4) = 1] \\ (7,8) & [f(7) = 8] \\ (9,1) & [f(9) = 1] \end{array}$$

defines a function which maps $1 \in S$ into $5 \in T$, $4 \in S$ into $1 \in T$, etc. This function is onto because all elements in T are the images of some element in S. Note that whereas each element in the domain maps into one and only one element of the image, an element of the image (1 in this example) can be mapped into by more than one element of the domain (see Fig. 0.6). Since 1 in T is mapped into by both 4 and 9, the inverse mapping is not a function. ∎

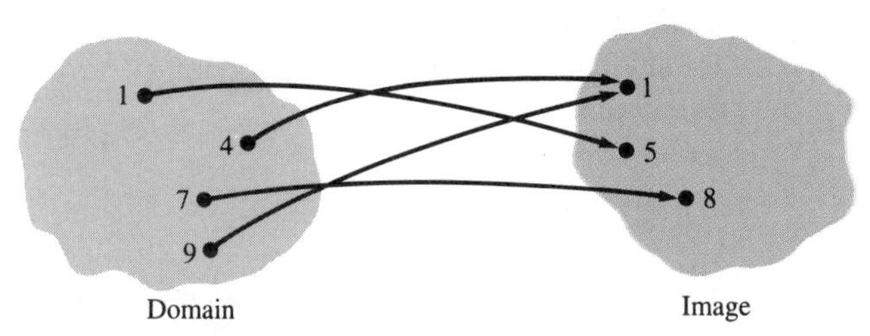

FIGURE 0.6

A Range Point as the Image of More Than One Domain Point. In Example 2, the range point 1 is the image of two domain points (4 and 9). Thus while the mapping is an onto function, the inverse mapping is not a function.

2. **One-to-one functions**. A function is said to be one-to-one or *injective* if no element of the image is the image of more than one element of the domain. Thus the inverse mapping of a one-to-one function is a function.

EXAMPLE 3 The function defined in Example 2 is not injective because 1 is the image of two elements in the domain. But if the domain $S = \{1, 6\}$ and the range $T = \{3, 5, 9\}$, the function defined by $(1,3)$ and $(6,9)$ is one-to-one because no point in the image set—$\{3,9\}$—is the image of more than one point in S. (It's all right for an element of the range to be the image of no element in the domain.) This function is, however, not onto. ∎

EXAMPLE 4

So far our examples have had domains which are sets of integers or real numbers. But the domain can be any set. For example, consider the keyboard of a terminal used as input to a computer. For each key (except for a few like the control key) or allowable combination of keys (e.g., the shift key and a letter) pressed, a corresponding *code*, usually of 7 or 8 bits, is stored in the computer. The most common such codes are ASCII (American Standard Code for Information Interchange), which has 7-bit and 8-bit forms, and EBCDIC (Extended Binary-Coded Decimal Interchange Code), which is an 8-bit code. The mapping from the keyboard combinations (the domain) to the computer codes (the range) is injective. ∎

EXAMPLE 5

If the domain and range are both N, then the function

$$y = f(x) = 2x + 3$$

is injective because each x in the domain maps into one and only one y in the range. The image is the set $\{3, 5, 7, \ldots\}$. The inverse function is

$$x = f^{-1}(y) = \frac{(y-3)}{2},$$

since

$$f^{-1}f(x) = f^{-1}(2x+3) = \frac{2x+3-3}{2} = x.$$

[Recall that we defined the inverse mapping to be from the *image* to the domain so that $(y-3)/2$ is always an integer. Some authors define the inverse mapping to be from the *range* to the domain, in which case $f(x)$ in this example would not have an inverse (why?).] ∎

3. **One-to-one and onto functions**. A function is called a *one-to-one correspondence* or *bijective* if it is both one-to-one and onto. None of the functions in Examples 1–5 is bijective. (The function in Example 4 isn't bijective because some possible computer codes will not correspond to any keyboard combination.)

EXAMPLE 6

Whenever the domain and the range are the same set, there is a trivial example of a bijective function called the **identity function**, the function which maps each element of the domain into itself. ∎

EXAMPLE 7

Let $S = N$ and let $T = \{0, 2, 4, 6, \ldots\}$, the set of even nonnegative integers. Then the function $y = 2x$ from domain S to range T is bijective. The inverse function is $x = (1/2)y$. ∎

Problems: Section 0.3

1. In the following functions let the domain be N, the set of nonnegative integers. What is the image of each of these functions?
   a) $n+2$ b) n^2 c) $2n+5$

2. At a university, each student has a user-id (identification) on the computer. This amounts to a function with domain the set of students and range the set of user-ids. Which of the special properties of functions (injective, surjective, or bijective) is appropriate for this function? Why?

3. At a residential college, each student is assigned to an available dorm room. Which of the special properties of functions is appropriate here?

4. The United States assigns a Social Security Number to every person in the country who works or has a financial account (e.g., a bank account). Social Security Numbers are 9 digits. What is the range and domain of the associated function and what special property of functions does it satisfy?

5. Are the following statements true or false? Give reasons.
   a) If a function is injective, its domain and range are the same set.
   b) If a function isn't injective, its domain must contain more points than its image. (Assume the domain is finite.)

6. Repeat [5], replacing "injective" with "surjective".

7. Repeat [5], replacing "injective" with "bijective".

8. Which of the following functions (each followed by its domain and range) are injective, surjective, and/or bijective?
   a) $2n$, N, N b) $n+1$, N, N c) $n+1$, Z, Z.
   d) 2^m5^n, $N \times N$, N e) 2^m6^n, $N \times N$, N

0.4 Some Important Functions

In this section we discuss a variety of functions which we'll use—some of them over and over again—later in this book. Some are much more important in discrete than in continuous mathematics; others are important in both realms.

Basic Functions

In defining various functions, we will use x to refer to an element of the domain of a function and y to an element of the image or range. (Remember that if we specify the image of a function, we need not specify its range, which can be any set of which the image is a subset.) The element x is called the **independent variable** because its choice is unconstrained, and y is called the **dependent variable** because its value "depends" on the value of x. We will sometimes call x the *argument* of the function being defined over the domain.

1. **Absolute value function.**

   Domain: R, the set of real numbers.

   Range: The nonnegative elements in R.

   Definition:

$$y = \begin{cases} x & \text{if } x \geq 0. \\ -x & \text{if } x < 0. \end{cases}$$

Notation: $|x|$.

In effect, the absolute value function takes a real number x and leaves it unchanged if it is nonnegative but strips off the sign if it is negative. This very simple function is valuable throughout mathematics whenever you wish to ensure that a quantity cannot be negative. Unfortunately, the notation for this function is the same as that for the cardinality function defined in Section 0.1. But there should be no confusion because the argument of the cardinality function must be a set and that of the absolute value function must be a real number.

An important relationship involving the absolute value function is the **triangle inequality**:†

$$|a+b| \;\leq\; |a|+|b|,$$

where a and b are any elements in R. You can prove this by considering all possible combinations of values of a and b (i.e., both positive, both negative, etc.) [12]. A generalization of this inequality, whose correctness can be proved using the tools of Chapter 2 [2, Section 2.3], is

$$|a_1+a_2+\cdots+a_n| \;\leq\; |a_1|+|a_2|+\cdots+|a_n|.$$

2. **Floor function.**

Domain: R.

Range: The set of all integers Z.

Definition:

$$y \;=\; \text{the largest integer not greater than } x.$$

Notation: $\lfloor x \rfloor$.

In discrete mathematics generally, the most common function domains are N and Z. However, when performing operations on integers, we often end up with nonintegral quantities. For example, $\sqrt{3}$ is not an integer. The floor function and its companion, the ceiling function, considered next, provide us with convenient ways to convert real numbers to integers when the fractional (nonintegral) part of the number isn't of interest to us. Thus $\lfloor\sqrt{3}\rfloor = 1$ because 1 is the largest integer less than $\sqrt{3} = 1.732\ldots$. Other examples of the floor function are $\lfloor -4.67 \rfloor = -5$, $\lfloor 86.739 \rfloor = 86$, $\lfloor .602 \rfloor = 0$, and $\lfloor 77 \rfloor = 77$.

3. **Ceiling function.**

Domain: R.

Range: Z.

Definition:

$$y \;=\; \text{the smallest integer not less than } x.$$

Notation: $\lceil x \rceil$.

---

† This name derives from an analogous formulation involving vectors.

This function is the obvious analog of the floor function. Whereas the floor function finds the integral bottom or "floor" of its argument, the ceiling function finds the integral top or "ceiling" of its argument. Some examples of the ceiling function are $\lceil -4.67 \rceil = -4$, $\lceil 86.239 \rceil = 87$, $\lceil .302 \rceil = 1$, and $\lceil 77 \rceil = 77$.

From the definitions of the floor and ceiling functions and from our examples you should be able to verify the following [8]:

$$\lceil x \rceil = \begin{cases} \lfloor x \rfloor + 1 & \text{when } x \text{ is not an integer.} \\ \lfloor x \rfloor & \text{when } x \text{ is an integer.} \end{cases}$$

You are probably familiar with the **rounding function**, which replaces a decimal number by the next greater integer if the fractional part is 1/2 or greater and by the largest integer not greater than the number otherwise. Can you express this function in terms of the floor and ceiling functions [14]? Some other properties of the floor and ceiling functions are considered in [9–10].

4. **Factorial function.**

Domain: N.

Range: N.

Definition: For a positive integer n, we define the factorial of n, usually just called n factorial, as

$$n(n-1)(n-2)(n-3)\cdots 3 \times 2 \times 1.$$

In other words, it is the product of all the integers from 1 to n. If $n = 0$, then by convention, its factorial is defined as 1.

Notation: $n!$

Factorials appear many times in this book. One of the most important properties of a factorial is how fast it grows. Whereas $0! = 1$, $1! = 1$, $2! = 2$, and $3! = 6$, by the time we get to 10 we have $10! = 3{,}628{,}800$. And 1000! is a number with over 2500 digits in it.

5. **Mod** (or *remainder*) **function.**

Domain: $N \times N^+$.

Range: N.

Definition:

$$m - \lfloor m/n \rfloor n,$$

where m is a nonnegative integer and n is a positive integer.

Notation: $m \bmod n$.

The effect of this function is to compute the integer remainder when m is divided by n. Thus

$$\begin{aligned} 8 \bmod 3 &= 8 - \lfloor 8/3 \rfloor 3 = 8 - 2 \times 3 = 2; \\ 57 \bmod 12 &= 9; \\ 7 \bmod 9 &= 7. \end{aligned}$$

The name of this function is an abbreviation of

$$m \text{ modulo } n,$$

where "modulo" means "with respect to a modulus", that is, the value of m with respect to the size (or modulus) of n, which is just the remainder when m is divided by n.

A motivation for presenting it here, in addition to its later usefulness, is to make sure that our examples of functions thus far haven't misled you. We noted in Example 4 of Section 0.3 that the domain of a function need not be a set of numbers like integers or real numbers. Remember that the domain of a function need only be a set, *any* set. In case you feel that the domain of keyboard combinations in Example 4 is a little special (it isn't really), the mod function provides another example of a function whose domain is "numerical" but is neither just the set of integers nor real numbers.

Exponentials and Logarithms

Any function of x of the form $f(x) = a^x$, with a a positive constant, is called an **exponential function**. The most important property of exponential functions is

$$a^x a^y = a^{x+y}.$$

We say therefore that an exponential function converts multiplication into addition (of exponents) and vice versa.

An important property of exponential functions is that, if $a > 1$, then a^x grows faster than any power of x, x^n, no matter how large the constant n. This means that for $a > 1$ and for any $n > 0$,

$$\frac{x^n}{a^x} \tag{1}$$

will be as close to zero as you wish if you make x large enough. We'll return to this result in Chapter 9 (see Theorem 1 of Section 9.3), but you might find it useful now to use your hand calculator or computer to compute some values of (1) for a variety of values of a, n, and x. A useful notation to express the behavior of (1) as x becomes large is

$$\lim_{x \to \infty} \frac{x^n}{a^x} = 0$$

with *lim* shorthand for "limit". More generally, we write

$$\lim_{x \to \infty} f(x) = L$$

to indicate that, as x becomes arbitrarily large, $f(x)$ gets arbitrarily close to L. We will use this notation informally at various places in this book and then, in Chapter 9, we'll discuss more formally the notion of a limit.

Make sure you understand the difference between an exponential function where the variable x is the exponent and a **power function**, where the quantity raised to a power is the variable and the exponent n is a constant. If we were to write y^x, it would be ambiguous, but either

$$f(x) = y^x$$

or

$$f(y) = y^x$$

is unambiguous, with x being the independent variable in the first equation and y a constant whereas their roles are reversed in the second equation.

Exponential functions are closely related to logarithms. For over three centuries after Napier invented logarithms in about 1600, logarithms were an important computational tool. Generations of high school students were taught how to use tables of logarithms; unfortunately, some still are. Long ago the slide rule eliminated the need to use logarithms for computation except where accuracy to more than 2 or 3 decimal places was needed. Now the hand calculator has eliminated entirely the computational use of logarithms.

Therefore it's understandable, but unfortunate, if you think logarithms aren't very important and have forgotten what you learned about them. However, the **logarithm function** is very important in mathematics, particularly in the study of algorithms (an important theme throughout this book and the subject of Chapter 1). Let's review the basic properties of logarithms.

The logarithm of a positive real number x to the base $b > 1$ is that power y to which b must be raised to get x. In other words,

$$y = \log_b x \iff b^y = x. \tag{2}$$

The traditional base for computations is 10. Hence when we write $\log x$ without a subscript, you may assume for the time being that 10 is understood. However, as we'll soon show, the interesting properties of logarithms are truly independent of the base. Thus when we leave the base off, that really means that the base doesn't matter.

EXAMPLE 1 Evaluate $\log_3 9$ and $\log_2 64$.

Solution Since $3^2 = 9$, we have $\log_3 9 = 2$. Since $2^6 = 64$, it follows that $\log_2 64 = 6$. ∎

In Fig. 0.7 we show the graph of $f(x) = b^x$ for a typical $b > 1$. We also display the graph of $f(x) = \log_b x$ for the same value of b. Note that $\log_b x$ is positive if and only if $x > 1$.

From Eq. (2) you can see that a point (r, s) is on the graph of $y = \log_b x$ if and only if the point (s, r) is on the graph of $y = b^x$ since $r = \log_b s$ if and only if $s = b^r$. Geometrically, the two graphs in Fig. 0.7 are mirror images of each about the 45° line $y = x$; rotating about this line flips the horizontal axis into the vertical axis and vice versa. Using the terminology of Section 0.3, whenever the graphs of two functions have this property, the functions are inverses of each other (why?).

Thus it follows that the exponential and logarithmic functions are inverses of each other. That is, either one undoes the other in the sense that

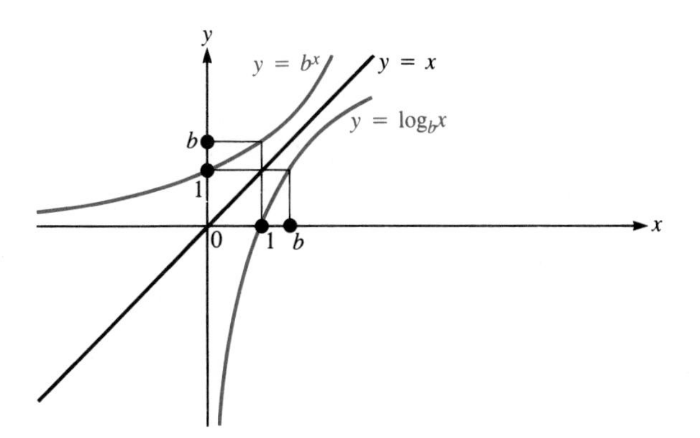

FIGURE 0.7

Graphs of the Functions $y = b^x$ and $y = \log_b x$.

$$b^{\log_b x} = x \qquad \text{and} \qquad \log_b b^y = y.$$

These identities follow by substituting the left side of Eq. (2) into the right and vice versa.

Here are the properties of logarithms which we'll need:

1. $\log xy = \log x + \log y$ (i.e., logarithms "turn multiplication into addition").
2. $\log x^r = r \log x$ (i.e., logarithms "turn powers into multiples").
3. $\log_c x = \log_b x / \log_b c$ (i.e., changing the base from b to c merely changes the logarithm by a constant multiplicative factor).
4. $\lim_{x\to\infty} \log x = \infty$, but $\lim_{x\to\infty} [(\log x)/x] = 0$.

The first three properties are explained in traditional treatments of logarithms in high school texts; property 4 isn't. But it is property 4 which accounts for the special importance of logarithms in this book.

Property 4 is a statement about the *rate of growth* of the logarithm. Yes, the logarithm does grow; in fact, as indicated by both Fig. 0.8 and property 4, it goes

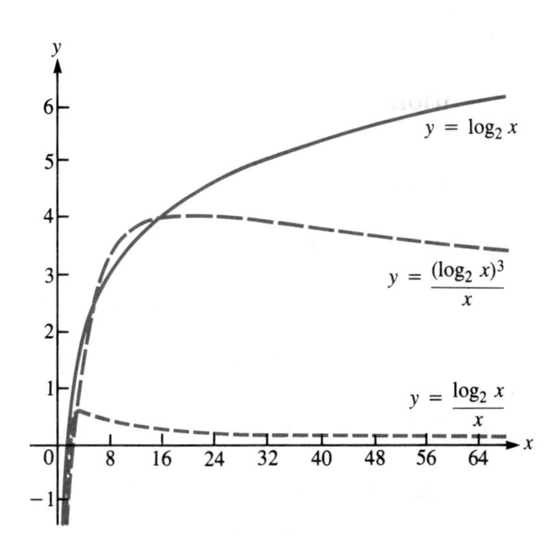

FIGURE 0.8

The relative growth rates of three functions.

to infinity as x gets arbitrarily large. But—and this is the key point—*the logarithm gets large very slowly.* For instance, it grows so slowly in comparison to the growth of x that (see property 4 again) the ratio of $\log x$ to x decreases to zero. Even if we beef up the logarithm by raising it to the thousandth power, in the long run the result is negligible compared to x. More generally,

$$\lim_{x\to\infty}\left[\frac{(\log x)^m}{x}\right] = 0$$

for any positive m whatsoever.

Property 4 is also independent of the base. If $\log_b x$ goes to infinity, then any constant multiple of it also goes to infinity. Also if $(\log x)/x$ goes to 0, then for any constant k, $(k\log x)/x$ also goes to zero. But replacing $\log x$ by $k\log x$ has the same effect as changing the base (why?).

Figure 0.8 shows graphs of $\log x$, $(\log x)/x$, and $(\log x)^3/x$. Do you see that property 4 is illustrated by these graphs? We won't prove property 4, but we'll discuss it further in Chapter 9 (see Theorem 1 of Section 9.3).

Why are these rates of growth interesting? Because a major concern in this book is to analyze how many steps it takes to compute various results as a function of the initial data. Very often, the answers involve logarithms. For example, methods for solving many types of problems depend on a parameter n. Often the number of steps required by the solution method is a function n like:

$$n, \qquad n\log n, \qquad n^2, \qquad n(\log n)^2. \tag{3}$$

Suppose you have four possible methods for solving the same problem, each of which takes a number of steps equal to a different one of the functions in (3). A natural question is: What is the relative size of these functions as n gets large? The answer will determine which of the methods is the most efficient when it really counts, namely, when n is large and there is a lot of computation to do. Using property 4 we can answer the question as follows: If we reverse the order of the last two, then the four will be in descending order of efficiency. To see this, note that

$$\frac{(n\log n)}{n} = \log n \to \infty,$$

so n is more efficient than $n\log n$ by an arbitrarily large factor as n gets large. Next,

$$\frac{n(\log n)^2}{n\log n} = \log n,$$

so $n\log n$ is more efficient than $n(\log n)^2$. Finally,

$$\frac{n(\log n)^2}{n^2} = \frac{(\log n)^2}{n} \to 0$$

Figure 0.9 shows the relative growth rates of the functions in (3).

EXAMPLE 2 Evaluate the following:

a) $\log_{125} 125$ **b)** $\log_{10} 0$ **c)** $\log_2 -12$
d) $\log_7 1$ **e)** $\log_9 3$ **f)** $\log_2(1/4)$.

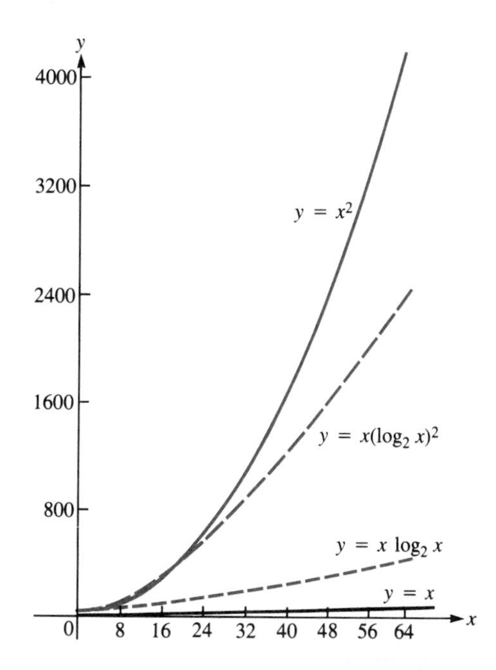

FIGURE 0.9

The Relative Growth Rates of the Four Functions in (3)

Solution

a) $\log_b b = 1$ for any b because $b^1 = b$, so $\log_{125} 125 = 1$.
b) $\log_{10} 0$ doesn't exist because no finite power of any base is 0.
c) $\log_2 -12$ doesn't exist either because b^y is never negative for any $b > 0$ and any real y. No negative number is in the domain of any log function.
d) $\log_b 1 = 0$ for all b since $b^0 = 1$, so $\log_7 1 = 0$.
e) $\log_9 3 = 1/2$ because $9^{1/2} = \sqrt{9} = 3$.
f) $\log_2(1/4) = -2$ because $2^{-2} = 1/(2^2) = 1/4$. (If this isn't clear, review negative exponents.) ∎

EXAMPLE 3 Simplify $\log_2(2^{10}3^5)$.

Solution From properties 1 and 2:

$$\begin{aligned} \log_2(2^{10}3^5) &= \log_2 2^{10} + \log_2 3^5 && \text{[Property 1]} \\ &= 10\log_2 2 + 5\log_2 3 && \text{[Property 2]} \\ &= 10 + 5\log_2 3 \end{aligned}$$

This is as far as we can go without looking up $\log_2 3$. But how could we look it up? Tables of logarithms to the base 2 are not generally available, and few calculators have a button for it. The answer is to use property 3, which tells us that

$$\log_2 3 = \frac{\log_{10} 3}{\log_{10} 2}$$

which, approximately, is $.477/.301 = 1.585$. ∎

Actually, logarithms to the base 2 have become quite useful in recent years because computers work in the binary system. Also, many algorithms involve repeated division of problems into two smaller problems. Indeed, on the few occasions when we will wish to use a specific base for a logarithm, it will always be 2. Some books use $\lg x$ to denote $\log_2 x$.

Polynomials

Without exaggerating, we can assert that polynomials are the most important functions in mathematics. A **polynomial** is a function, such as

$$y = P_n(x) = a_n x^n + a_{n-1} x^{n-1} + \cdots + a_1 x + a_0, \qquad [a_n \neq 0] \tag{4}$$

whose domain is normally R, the set of real numbers, and whose range is also R. We restrict a_n to be nonzero to ensure that the term in x^n actually appears. In this book we will sometimes restrict the domain to be N or N^+. The nonnegative integer n is called the **degree** of the polynomial and the members of the set

$$A = \{a_n, a_{n-1}, a_{n-2}, \ldots, a_1, a_0\}$$

are the **coefficients** of the polynomial. Normally the coefficients are also elements of the domain of the function. Sometimes, however, the coefficients may be restricted to be integers even when x can be any real number. Thus a polynomial is a function of x with **parameters** consisting of the degree n and the coefficients A.

When x is a variable which can take on values from some domain, we can evaluate the polynomial just like any other function. Sometimes in this book, however, we'll use x as a purely symbolic quantity associated with no domain whatsoever. That is, it will never be given or presumed to be given a value, numerical or otherwise. The polynomial then becomes a **place holder** for its coefficients. As such it gives us a symbolic way of combining the coefficients into a single entity in which they can be manipulated without losing their individual identities. For example, it may be convenient to manipulate the tuple $(1, 3, 3, 1)$ as the coefficients of $(1+x)^3$.

From Eq. (4) the simplest case of a polynomial is when $n = 0$, in which case the polynomial is just the constant value a_0. This means that, for all values x from the domain, the image is the single value a_0 (see Fig. 0.10). The simplest nontrivial case is $n = 1$, when the polynomial has the familiar form

$$y = P_1(x) = a_1 x + a_0 \tag{5}$$

and is called a **linear polynomial**. The familiar straight-line graph for a particular case of the polynomial in Eq. (5) is shown in Fig. 0.11. (Note that Fig. 0.9 also contains the graph of two polynomials, n and n^2.)

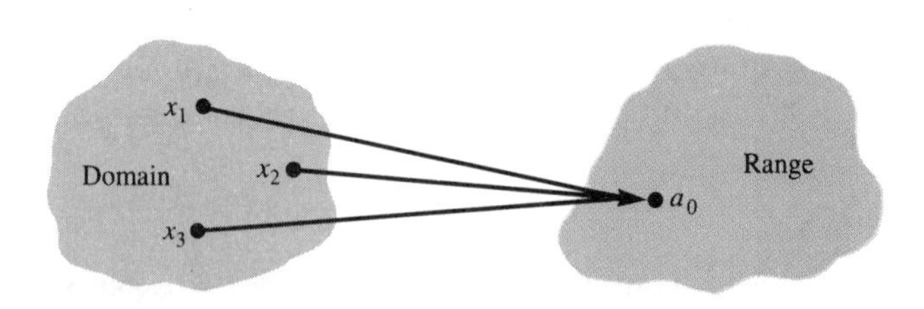

FIGURE 0.10

A constant polynomial, $P_0(x) = a_0$, has a single image value for all domain values.

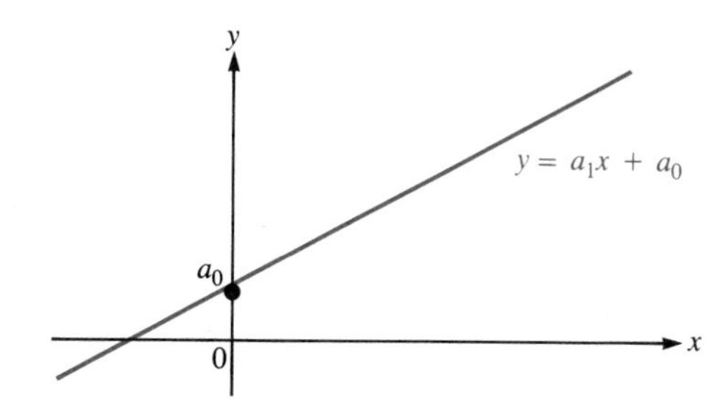

FIGURE 0.11

The Graph of a Linear Polynomial, $P_1(x) = a_1x + a_0$

When the degree of the polynomial is $n = 2$, we have the also familiar quadratic polynomial. You probably remember how to "solve" linear polynomials by setting y in Eq. (5) equal to 0 and finding the corresponding value of x. Similarly, you will recall how to "solve" quadratic polynomials using the famous quadratic formula. Although we'll solve (i.e., find values of x for which y is zero) a considerable number of polynomials in this book, our stronger interest is in manipulating polynomials as algebraic quantities. We'll assume you remember how to add or subtract two polynomials. Since we won't have any significant need to divide polynomials in the remainder of this book, we'll conclude this section with a discussion of the multiplication of polynomials.

Suppose we have two polynomials $P_n(x)$ and $Q_m(x)$ whose degrees, n and m, could be different. Using the notation in Eq. (4), we write these two polynomials as

$$P_n(x) = a_nx^n + a_{n-1}x^{n-1} + \cdots + a_1x + a_0$$

and

$$Q_m(x) = b_mx^m + b_{m-1}x^{m-1} + \cdots + b_1x + b_0.$$

We assume that the domain is the same for both polynomials. To find the product of $P_n(x)$ and $Q_m(x)$ we must compute

$$(a_nx^n + \cdots + a_1x + a_0)(b_mx^m + \cdots + b_1x + b_0) \qquad (6)$$

term by term and combine terms which have like powers of x (as we'll show in Example 4). The notation we've introduced thus far is insufficient for writing (6) in a more convenient form. In Section 0.5 we'll introduce some notation to remedy this and then return to (6) to show specifically how to write it more conveniently.

EXAMPLE 4 Given the two polynomials

$$P_3(x) = x^3 + 4x^2 - 3x - 2 \qquad \text{and} \qquad Q_2(x) = x^2 + 2x + 1,$$

find their product.

Solution We use (6) and multiply term by term to get

$$\begin{aligned}(x^3+4x^2-3x-2)(x^2+2x+1) &= x^3(x^2+2x+1) + 4x^2(x^2+2x+1) \\ &\quad - 3x(x^2+2x+1) - 2(x^2+2x+1) \\ &\qquad \text{[By the distributive law]} \\ &= x^5 + 2x^4 + x^3 + 4x^4 + 8x^3 + 4x^2 \\ &\quad - 3x^3 - 6x^2 - 3x - 2x^2 - 4x - 2 \\ &= x^5 + 6x^4 + 6x^3 - 4x^2 - 7x - 2. \quad \blacksquare\end{aligned}$$

You need not become very proficient at polynomial arithmetic. In fact, you can now do such arithmetic symbolically on hand-held calculators. But it is important that you *understand* what it means to perform arithmetic on two polynomials.

Problems: Section 0.4

1. Give the values of

a) $|4.73|$ **b)** $|-2.531|$ **c)** $|-174|$.

2. Graph

a) $y = |x| + 4$ **b)** $y = ||x| - 4|$.

3. Give the values of

a) $\lfloor 3.99 \rfloor$ **b)** $\lceil -6.21 \rceil$
c) $\lceil \lfloor -7.31 \rfloor - 6.43 \rceil$.

4. Graph

a) $f(x) = \lfloor x \rfloor$ **b)** $f(x) = \lceil x \rceil$
c) $f(x) = |x|$.

5. Graph

a) $f(x) = |\lfloor x \rfloor|$ **b)** $g(x) = \lfloor |x| \rfloor$.

6. Give the solution set for

a) $\lfloor x \rfloor = 0$ **b)** $\lceil 2x \rceil = 3$.

7. Explain in words why $|x||y| = |xy|$.

8. Verify the display on page 28, which describes ceiling in terms of floor. (You just have to talk through both cases. For instance, when x is an integer, what are the values of the two expressions claimed to be equal? When x is nonintegral, we may write $x = n + a$ where n is an integer and $0 < a < 1$. Now what are the values of the two expressions claimed to be equal?)

9. For each of the following equalities, state for which real numbers it is true (no proof needed).

a) $\lfloor x \rfloor = \lceil x \rceil$ **b)** $\lceil x \rceil = \lfloor x \rfloor + 1$
c) $\lfloor x+1 \rfloor = \lfloor x \rfloor + 1$ **d)** $\lfloor -x \rfloor = -\lfloor x \rfloor$
e) $\lfloor -x \rfloor = -\lceil x \rceil$ **f)** $\lfloor 2x \rfloor = 2x$
g) $\lfloor 2x \rfloor = 2\lfloor x \rfloor$ **h)** $\lfloor \lfloor x \rfloor \rfloor = \lfloor x \rfloor$
i) $\lfloor x/n \rfloor = x/n$, n an integer in parts i,j,k
j) $\lfloor x/n \rfloor = \lfloor \lfloor x \rfloor / n \rfloor$ **k)** $\lfloor x/n \rfloor = \lfloor x \rfloor / n$.

10. In each of the following problems, two expressions are given. Name an inequality between the two which is always true, if there is one. If the inequality is not strict (e.g., $\leq$ not $<$), give an example where the two expressions are equal and an example where they aren't equal. If no inequality is always true between the expressions, give an example where the first is greater, a second example where they are equal, and a third example where the second is greater.

a) $\lfloor x+y \rfloor$, $\lfloor x \rfloor + \lfloor y \rfloor$ **b)** $\lfloor x+y \rfloor$, $\lfloor x \rfloor + \lfloor y \rfloor + 1$
c) $\lfloor x-y \rfloor$, $\lfloor x \rfloor - \lceil y \rceil$ **d)** $\lfloor x-y \rfloor$, $\lfloor x \rfloor - \lfloor y \rfloor$
e) $n\lfloor x/n \rfloor$, $\lfloor x \rfloor$ **f)** $\lfloor xy \rfloor$, $\lfloor x \rfloor \lfloor y \rfloor$
g) $\lceil x/y \rceil$, $\lceil x \rceil / \lceil y \rceil$.

11. If x, y, and z are real numbers, show that

$$\lfloor x+y+z \rfloor \;\leq\; \lfloor x \rfloor + \lfloor y \rfloor + \lfloor z \rfloor + 2.$$

What generalization of this result would you propose?

12. Prove the triangle inequality for real numbers: $|a+b| \leq |a| + |b|$. There are four cases; the first is that both a and b are nonnegative.

13. Prove the three-term triangle inequality: $|a+b+c| \leq |a| + |b| + |c|$. (*Hint*: Temporarily substitute d for $a+b$.)

14. The rounding function $R(x)$ rounds any nonnegative number x with decimal part .5 or greater to the next larger integer. Show that $R(x) = \lfloor x + .5 \rfloor$.

15. Rounding .5 to the next larger integer is not the only rounding convention that might be used. Some people believe in rounding .5 down and only rounding up decimal parts greater than .5. Call this rounding function $R'(x)$. For nonnegative x,

a) express $R'(x)$ in terms of ceiling or floor;

b) express $R'(x)$ in terms of $R(x)$. (*Hint*: Consider $-x$.)

16. Show that $f(x) = \lfloor 10x \rfloor / 10$ rounds to the nearest tenth less than or equal to x.

17. Use floor or ceiling to define a function which rounds off to the nearest tenth. State what your function does in the case that x is exactly halfway between tenths (e.g., $x = 23.45$).

18. Generalize the previous problem to rounding to the nearest n decimal places.

19. Simplify $(n+1)!/n!$.

20. a) Compute the values of $n!$ for $n = 1, 2, 3, \ldots, 10$.

b) Can you infer from these results how many zeros there are at the end of 20!? How did you get your answer?

c) Can you generalize the result of b) for $n!$? (This is not easy and is not just a simple extension of the result for 20!)

21. Explain why $\dfrac{(2n)!}{2^n n!} = 1 \times 3 \times 5 \times \cdots \times (2n-1)$.

22. Which grows faster, 100^n or $n!$? Answer by explaining intuitively what you think $\lim_{n \to \infty} 100^n/n!$ is.

23. Define f on the domain N^+ by $f(n)=$ the remainder when $n!$ is divided by 1000.

a) What is the smallest n such that $f(n) = 0$?

b) What is $f^{-1}(6)$? The hard part is to show that you have all members of the set.

24. A function f with domain N^+ is defined implicitly by

$$f(1) = 4$$

and $f(n) = f(n-1) + 3$ for $n > 1$.

a) Determine $f(2)$, $f(3)$, and $f(4)$.

b) Determine a formula for $f(n)$ as a function of n.

25. Compute

a) $10 \bmod 3$ b) $127 \bmod 10$ c) $25 \bmod 5$.

26. Determine

a) $(127 \bmod 10) \bmod 4$ b) $127 \bmod (10 \bmod 4)$.

27. Graph the function $f(n) = n \bmod 4$ for $0 \le n \le 10$. Unlike the graphs in the earlier problems, this one is defined just on integers n.

28. If $m \bmod k$ and $n \bmod k$ are the same, then we say m is *congruent* to n mod k and we write this as a **congruence**:

$$m \equiv n \pmod{k}.$$

Express in set builder notation all integers x which solve the following congruences:

a) $4x \equiv 3 \pmod{5}$,

b) $5x - 2 \equiv 7 \pmod{6}$;

c) $6x - 3 \equiv 1 \pmod{10}$;

d) $187x + 22 \equiv 115 \pmod{87}$.

29. If m is fixed, then $a \equiv b \pmod{m}$ with $a, b \in Z$ defines a binary relation on $Z \times Z$. Prove that this relation is reflexive and transitive.

30. a) If k is an integer, show that $k^2 \equiv 0$ or 1, modulo 4.

b) Show that $x^2 \equiv 45 \pmod{100}$ has no integer solution. (*Hint*: Consider the form that the square of a number ending in 5 must have.)

31. Give the values of the following logarithms:

a) $\log_5 125$ b) $\log_{12} 1$

c) $\log_3(1/9)$ d) $\log_{17} 17$

e) $\log_2 1024$ f) $\log_4 1024$

g) $\log_8 1024$ h) $\log_{16} 1024$

i) $\log_{64} 1024$ j) $\log_2 65536$.

32. If $\log x$ to base a is denoted by L, express the following logarithms in terms of L. Assume that a and b are positive and that k is any integer.

a) $\log x^k$ to the base a

b) $\log x^k$ to the base b

c) $\log(1/x^k)$ to the base a

d) $\log(1/x^k)$ to the base b.

33. How large does x have to be before $\log_{10} x$ is

a) at least 2?

b) at least 3?

c) at least 1000?

34. Without using a table or a calculator,

a) show that $1 < \log_2 3 < 2$;

b) find $\lfloor \log_3 24 \rfloor$;

c) find $\lceil \log_4 100 \rceil$

35. Show that $\log(x/y) = \log x - \log y$. Does the base matter?

36. The function $\log\log x$ is defined by $\log\log x = \log(\log x)$.

a) What is the domain of loglog? The range?

b) Show that $\log\log(c^k) = \log k + \log\log c$.

37. This problem is for people who can find an old slide rule in the attic and want to figure out how it worked.

a) Numbers were multiplied on slide rules by moving two "log scales" relative to each other. On a log scale, each number x is marked at the distance $k \log_{10} x$ from the left of the scale, where k is the length of the entire scale (see the scale in the figure). Explain how property 1 allows us to multiply numbers by using such scales.

1 2 3 4 5 6 7 8 9 10

Since $\log_{10} 3 = .477$, the point labeled 3 is .477 of the way from 1 to 10.

b) Explain how you could find c^k by moving a "loglog scale" relative to a log scale. (*Hint*: Use the result from [36].)

38. Find the relative order of growth of n^3, $n^2(\log n)^5$, $n^4 \log n$.

39. Find the relative order of growth of $\log^2 n$, $n^2/\log n$, n, $\log n$, $n \log^3 n$.

40. Have you wondered why we limited the base for logarithms to $b > 1$? The key property we need in order to define $\log_b x$ is that the exponential function b^x is defined for all real x and provides a one-to-one correspondence between R and R^+. Do other b's have this property? Clearly, $b = 1$ and $b = 0$ do not. Neither does negative b (unless you're willing to allow complex numbers to be used). This leaves only those b's satisfying $0 < b < 1$ as additional candidates. Show that logarithms *can* be defined by Eq. (2) for b's between 0 and 1. What parts of properties 1–4 remain true for such logarithms?

41. Let

$$P_5(x) = x^5 \quad \text{and} \quad Q_1(x) = x - 4.$$

Compute the sum, difference, and product of these polynomials.

42. Find the sum, difference, and product of $x^2 + x + 1$ and $x - 1$.

43. Multiply $(x^3+x^2+x+1)(x-1)$.

44. Evaluate $7x^3+4x^2+9x+3$ at $x = 10$ in your head.

45. If $f(x)$ is quadratic, show that $f(x+1) - f(x)$ is linear.

46. Let $f(x) = x^2$ and $g(x) = 2^x$.

a) Evaluate $f(x+1)/f(x)$ and $f(2x)/f(x)$.

b) Evaluate $g(x+1)/g(x)$ and $g(2x)/g(x)$.

47. A value c is called a **fixed point** of the function f if $f(c) = c$.

a) Find all the fixed points of the function $f(x) = 3x + 8$.

b) Find all the fixed points of the function $f(x) = |x|$.

c) Let $U = \{1, 2, 3, 4\}$ and let f be defined on subsets of U by $f(A) = U - A$. Find all the fixed points of f. (*Note*: In this case "points" means "sets".)

d) Let f have as its domain the set of strings of lowercase roman letters, and let f map each string into its reverse. For instance, $f(xyz) = zyx$ and $f(\text{love}) = \text{evol}$. Name a fixed "point" of length 3. Name a fixed point of length 4.

48. Let f be a function from $(n+1)$-tuples to numbers defined by

$$f(a_0, a_1, \ldots, a_n) = a_0 + a_1 x + \cdots + a_n x^n.$$

The function f is not a polynomial, because now x is a constant and the a's are the variables, but clearly f is related to polynomials.

a) With $n = 3$ and $x = 8$, find $f(2, 5, 1, 6)$.

b) With $n = 3$ and $x = 8$, find $f(4, 1, 0, 3)$.

c) With $n = 3$ and $x = 10$, find $f(4, 1, 0, 3)$.

d) In words, what is the function f all about?

0.5 Summation and Product Notation

In Section 0.4 we wrote a polynomial $P_n(x)$ as

$$P_n(x) = a_n x^n + a_{n-1}x^{n-1} + \cdots + a_1 x + a_0. \tag{1}$$

Perhaps you find reading this as tedious as we found typing it on our word processor. In any case, some shorthand notation other than the use of ellipsis for writing long sums like this would clearly be useful and would save a lot of space, if nothing else. **Summation notation** is such a shorthand. It is one of the most powerful bits of mathematical notation there is, and we'll use it often. Although summation notation is simple in concept, students often have trouble using and understanding it. Since it is so important and since its use involves considerable subtleties, we devote an entire section to it and its close relative, product notation.

Summation notation allows us, for example, to write the polynomial in Eq. (1) as

$$P_n(x) = \sum_{i=0}^{n} a_i x^i. \tag{2}$$

Instead of writing a sequence of terms, we have written a single *generic term* $(a_i x^i)$ using a new subscript, the letter i. (We could just as easily have used j or k or l or even n, if we weren't already using it for another purpose in Eq. (2).) The letter i is called a *dummy variable* in that, if we were to write out the entire expression, this variable wouldn't appear at all. (Section 1.5 discusses a related use of this terminology.) Using the dummy variable we have, in effect, captured the explicit essence of the *pattern* in Eq. (1), which is only implicit using an ellipsis. If you substitute $i = 0$ into the pattern term, you get $a_0 x^0 = a_0$, since a nonzero number raised to the zero power is 1. If you substitute 1 for i, you get $a_1 x^1 = a_1 x$, the linear term. And so on, until substituting n for i you get $a_n x^n$.

It is probably already clear that the $\sum$ in front of the pattern term indicates that you are to add all the terms you have gotten by substituting into the pattern. ($\sum$ is the Greek capital letter Sigma, which corresponds to a Roman "S"; hence Sigma for Sum.) The line below the $\sum$ gives the lowest value of the dummy variable, whereas the line above indicates its highest value. (Sometimes, if we want to write a summation in a paragraph rather than setting it off on a line by itself, we write the upper and lower limits to the right of the $\sum$, one below the other, as in $\sum_{i=1}^{8}$.) What we have called so far the "dummy variable" is more commonly called the **index of summation** (or just the *index*). We will use this terminology from now on. Implicit in this example is that the index takes on all integral values between the lowest and highest values.

If you understand our explanation of Eq. (2), you'll immediately see that it had two advantages over use of the ellipsis:

- It is more compact.
- Whereas the pattern in the use of an ellipsis may be obvious to the writer, it will sometimes be less than obvious to the reader; summation notation gets rid of any possible ambiguity.

Note also the power of using subscripts in Eq. (2) where a_i stands for a generic value among $a_1, a_2, \ldots, a_n$ and x^i stands for a generic power of x. Without such use of subscripts, summation notation would not be possible for how could you possibly represent, say, $a, b, \ldots, z$ by a single symbol?

In general, suppose we wish to express

$$c_j + c_{j+1} + \cdots + c_{k-1} + c_k$$

with summation notation. We do this as

$$\sum_{i=j}^{k} c_i. \tag{3}$$

This general example includes our polynomial example: just set $j = 0$, $k = n$, and $c_i = a_i x^i$. This example indicates that the index of summation can appear in the quantity being summed as a subscript, an exponent, or indeed, in any way you wish. Or it need not appear at all, as in

$$\sum_{i=1}^{N} 1,$$

which represents the sum of N 1's since, for each value of i, the quantity being summed is the constant 1.

EXAMPLE 1 Evaluate:

a) $\sum_{i=3}^{6} i^2$ b) $\sum_{p=0}^{5}(2p+3)$ c) $\sum_{i=1}^{4} i^{2i}$.

Solution For a) we get

$$3^2 + 4^2 + 5^2 + 6^2 = 9 + 16 + 25 + 36 = 86.$$

For b) we have

$$(0+3) + (2+3) + (4+3) + (6+3) + (8+3) + (10+3) = 3+5+7+9+11+13 = 48.$$

As noted above, the index of summation need not be i; it can be anything, but i, j, or k is commonly used. Note also that, if we had left the index of summation as p but had written $2i + 3$ instead of $2p + 3$, i.e.,

$$\sum_{p=0}^{5}(2i+3),$$

then this sum would have called for the addition six times of $2i + 3$ (with the result $12i + 18$). Why? Because anything after (i.e., to the right of) the summation sign which does not include the index is a constant with respect to the summation.

For c), where the index of summation appears more than once under the summation sign, we get

$$\begin{aligned} 1^2 + 2^{2\cdot 2} + 3^{2\cdot 3} + 4^{2\cdot 4} &= 1^2 + 2^4 + 3^6 + 4^8 \\ &= 1 + 16 + 729 + 65{,}536 \\ &= 66{,}282. \ \blacksquare \end{aligned}$$

The set of values taken on by the index of summation is called the **index set**. There is no requirement that this set consist of a sequence of consecutive integral values (although this is by far the most common case). For example, we may write

$$\sum_{\substack{i=2 \\ i \text{ even}}}^{8} a_i, \tag{4}$$

which represents $a_2+a_4+a_6+a_8$. Or we may write

$$\sum_{\substack{i=1 \\ i \text{ not divisible} \\ \text{by 2 or 3}}}^{11} 5^i, \tag{5}$$

which represents

$$5 + 5^5 + 5^7 + 5^{11} = 5 + 3125 + 78{,}125 + 48{,}828{,}125 = 48{,}909{,}380.$$

Note the rather awkward way we had to describe the condition on i in (5). If, as earlier, $a \mid b$ means that the integer a divides the integer b with a remainder of zero, then we may write that condition as

$$2 \nmid i \quad \text{and} \quad 3 \nmid i,$$

where the slash through the vertical bar should be read "not."

In full generality we may write

$$\sum_{R(i)} a_i, \tag{6}$$

where $R(i)$ is a function, often called a **predicate**, whose domain is usually N and which has the value *true* for each i included in the index set and the value *false* for all other i in the domain. (Are you disturbed at all by a function whose values are not numbers but rather *true* and *false*? No need to be. Remember that, like the domain, the range of a function may be any set whatsoever. In Chapter 7, when we discuss logic, we'll be dealing almost entirely with functions whose values are *true* or *false*.) For example, in part a) of Example 1, $R(i)$ would be true when $i = 3$, 4, 5, and 6 and false otherwise. In (4), $R(i)$ would be true only when $i = 2$, 4, 6, and 8. In (5), $R(i)$ would be true only when $i = 1$, 5, 7, and 11. Indeed, we could have written all these summations using predicate notation. For example, we could have written part a) of Example 1 as

$$\sum_{3 \le i \le 6} i^2,$$

although we wouldn't usually write it that way when the lower limit–upper limit notation is natural. By the way, we write the predicates *below* the summation sign merely because of convention.

In the predicate $R(i)$, i need not even be an integer index. For example, we may write

$$\sum_{x \in T} f(x),$$

where T is a set and f is a function defined on elements of T. Or we may write

$$\sum_{S \subset T} f(S)$$

where T is a set and f is a function defined on subsets of T.

One implication of (6) is that, if $R(i)$ is true for all elements in N, then that summation contains an *infinite* number of terms—one for each nonnegative integer. This is indeed possible. We'll meet such summations in later chapters, and we'll discuss them at length in Chapter 9.

Suppose $R(i)$ is not true for *any* value of i, as in

$$\sum_{8 \le i \le 6} i^3.$$

Then the sum is *empty* and, by convention, its value is zero. Or suppose we write

$$\sum_{i=8}^{6} a_i.$$

Since, again by convention, we *always* step up from the lower limit to the upper one (as opposed, for example, to what is allowed in some programming languages), this sum is also empty and has the value zero.

The compactness of summation notation allows us to manipulate it easily in a variety of ways. For example, if c is a constant, then

$$\sum_{R(i)} ca_i = c \sum_{R(i)} a_i, \tag{7}$$

because the constant c multiplies every term and may therefore be factored out independently of $R(i)$ [16a]. More generally, if c and d are constants, then [16b]

$$\sum_{R(i)} (ca_i + db_i) = c \sum_{R(i)} a_i + d \sum_{R(i)} b_i. \tag{8}$$

Double Summation

Suppose we have a set of *doubly subscripted* quantities a_{ij}, where the range of values of i is $m, m+1, \ldots, n-1, n$ and the range of j is $p, p+1, \ldots, q-1, q$. Now suppose we wish to add the a_{ij}'s for all possible pairs of values of i and j. We can do this with a double summation, which we write as

$$\sum_{i=m}^{n} \sum_{j=p}^{q} a_{ij} \tag{9}$$

and interpret as follows:

> For each value of the index in the outer (i.e., leftmost) summation let the index in the inner summation range over all its values and sum the terms generated; then increase the outer index by 1 and repeat, always adding the result to the previous sum until i reaches n.

In effect, this means that (9) contains implied parentheses, as in

$$\sum_{i=m}^{n}\left(\sum_{j=p}^{q} a_{ij}\right).$$

Thus (9) represents the sum

$$\begin{aligned}\sum_{j=p}^{q} a_{mj} + \sum_{j=p}^{q} a_{m+1,j} + \cdots + \sum_{j=p}^{q} a_{nj} &= a_{mp} + a_{m,p+1} + a_{m,p+2} + \cdots \\ &+ a_{mq} + a_{m+1,p} + \cdots + a_{m+1,q} + \cdots + a_{np} + a_{n,p+1} + \cdots + a_{nq},\end{aligned} \tag{10}$$

which is what we wanted.

EXAMPLE 2 Evaluate:

$$\sum_{i=1}^{3}\sum_{j=2}^{4} i^j.$$

Solution We obtain

$$\begin{aligned}&(1^2+1^3+1^4) + (2^2+2^3+2^4) + (3^2+3^3+3^4) \\ &\qquad = 1+1+1+4+8+16+9+27+81 = 148.\end{aligned}$$

When the index values for the two sums are defined by two general predicates $R(i)$ and $S(j)$, the principle is the same: For each value of i for which $R(i)$ is true, evaluate all the terms for which $S(j)$ is true and add them.

Suppose that the term under the summation sign in (9) is $a_i b_j$, that is, the product of two terms, one with subscript i and one with subscript j. Since Eq. (10) indicates that the double summation includes all possible combinations of the subscripts i and j, it follows that in this case we may write (9) as

$$\sum_{i=m}^{n}\sum_{j=p}^{q} a_i b_j = \left(\sum_{i=m}^{n} a_i\right)\left(\sum_{j=p}^{q} b_j\right), \tag{11}$$

since the product of the two sums includes all possible combinations of i and j. Another way to see this is to note that in Eq. (11) a_i is a constant *with respect to the summation with index j* and therefore can be brought outside that summation as a constant, as in Eq. (7). For example, with $n = 2$, $q = 3$, and $m = p = 1$,

$$\left(\sum_{i=1}^{2} a_i\right)\left(\sum_{j=1}^{3} b_j\right) = (a_1+a_2)(b_1+b_2+b_3)$$

$$
\begin{aligned}
&= (a_1b_1+a_1b_2+a_1b_3) + (a_2b_1+a_2b_2+a_2b_3) \\
&= \sum_{i=1}^{2}\left(\sum_{j=1}^{3} a_ib_j\right) \\
&= \sum_{i=1}^{2}\sum_{j=1}^{3} a_ib_j.
\end{aligned}
$$

If you understand this example, you should be able to see that the right-hand side of Eq. (11) follows in general [17].

So far, then, double summation hasn't been much more difficult than single summation. And you should be able to extend the preceding discussion without any trouble to triple summation or to any number of sums in succession. However, suppose we write

$$\sum_{i=1}^{4}\sum_{j=1}^{5-i} \frac{j^2}{(2i-1)}. \tag{12}$$

Here one of the limits for the inner sum is not a constant but contains a variable, namely, the index of the outer sum. Still, it should be fairly clear that we should evaluate (12) as follows:

i) Set $i=1$.
ii) Set $j=1,2,3,4$ (since $5-i=4$), evaluate $j^2/(2i-1)=j^2$, and sum.
iii) Set $i=2$ and $j=1,2,3$ (since $5-i=3$), evaluate, and add to the result of (ii).
iv) Repeat for $i=3$ and $j=1,2$.
v) Finally, repeat for $i=4$; since here the upper and lower limits are the same, i.e., 1, we have only $j=1$.

The result is

$$1+4+9+16+\frac{1}{3}+\frac{4}{3}+\frac{9}{3}+\frac{1}{5}+\frac{4}{5}+\frac{1}{7} = 35\frac{17}{21}.$$

We could give other, even trickier examples, but all we want to do here is introduce you to summations with variable limits. From Eq. (10) it should be clear that, when all the limits are *constants*, (9) represents the same sum regardless of whether the i summation or the j summation comes first. Thus

$$\sum_{i=m}^{n}\sum_{j=p}^{q} a_{ij} = \sum_{j=p}^{q}\sum_{i=m}^{n} a_{ij}. \tag{13}$$

But what about summations like (12)? Clearly, we can't just interchange the order of summation, because then the outer summation would have a limit depending on a summation to its right, which doesn't make any sense (why?). We won't discuss just how you go about interchanging the order of summation when one or more of the limits is variable. But the need to do so does occur occasionally. Be on your guard if it ever does occur, because great care is required to handle such cases.

We conclude our discussion of summation notation with three important examples of manipulating summations. First, we consider how to change the limits of the index of summation or, to put it another way, how to change the index of summation from one variable to another. A common case occurs when you have a sum of the form

$$\sum_{i=1}^{n} a_i \tag{14}$$

and you would like to change it so that the lower limit is 0. You do this by defining

$$j = i - 1 \qquad \text{or} \qquad i = j + 1. \tag{15}$$

When $i = 1$, $j = 0$ and when $i = n$, $j = n - 1$. Using these limits and replacing the subscript i by the equivalent $j + 1$, you may rewrite (14) as

$$\sum_{j=0}^{n-1} a_{j+1}. \tag{16}$$

If you write out (14) and (16), you will see that both have precisely the same terms, namely,

$$a_1 + a_2 + \cdots + a_n,$$

and thus both represent exactly the same thing. In actual practice, after we've made this change of variable, we often then go back to the original index and write

$$\sum_{i=0}^{n-1} a_{i+1}.$$

Remember that the index is a dummy variable and its "name," therefore, is irrelevant.

With a change of variable like that in Eq. (15) you'll be able to change the limits of summation in almost any way you wish. You must remember, though, to change all places where the index of summation appears in the term under the summation sign.

Our second example concerns the case where you wish to reverse the order of the terms in the summation. That is, instead of having the summation (14) represent

$$a_1 + a_2 + \cdots + a_n,$$

we wish to have it represent

$$a_n + a_{n-1} + \cdots + a_2 + a_1.$$

We do this by making the change of variable

$$j = n - i,$$

which changes (14) to

$$\sum_{i=1}^{n} a_i = \sum_{j=0}^{n-1} a_{n-j}. \tag{17}$$

Our third example concerns a common occurrence where we have two separate summations which we would like to combine into one, e.g.,

$$\sum_{i=0}^{n} a_i + \sum_{i=1}^{n} b_i.$$

How can we combine these expressions, which have different limits, into a single summation? By removing the first term of the first summation to obtain

$$a_0 + \sum_{i=1}^{n} a_i + \sum_{i=1}^{n} b_i.$$

Now, since the limits on both summations are the same, we can write this as

$$a_0 + \sum_{i=1}^{n} (a_i + b_i).$$

We mentioned in Section 0.4 that we didn't have a very good notation for expressing the product of two polynomials. Now we do. As in Section 0.4, let the two polynomials be

$$P_n(x) = \sum_{i=0}^{n} a_i x^i \quad \text{and} \quad Q_m(x) = \sum_{i=0}^{m} b_i x^i. \tag{18}$$

Our aim is to express the product of these polynomials in a simple, compact form. Here it is:

$$P_n(x)Q_m(x) = \sum_{i=0}^{m+n} \left(\sum_{j=0}^{i} a_j b_{i-j} \right) x^i, \qquad \begin{cases} a_j = 0, & j > n \\ b_j = 0, & j > m. \end{cases} \tag{19}$$

For example,

$$(a_1x+a_0)(b_2x^2+b_1x+b_0) = a_1b_2x^3 + (a_1b_1+a_0b_2)x^2 + (a_1b_0+a_0b_1)x + a_0b_0.$$

Note that each term in parentheses on the right-hand side corresponds to one term of the summation in parentheses in Eq. (19). For each i, the inner sum in Eq. (19) consists of a sum of products of two coefficients of powers, one from $P_n(x)$ and one from $Q_m(x)$, whose sum is i. This sum of products is then the coefficient of x^i.

Product Notation

We have much less to say about product notation than summation notation for two reasons:

- Everything we've said about summation notation carries over pretty directly to product notation.
- Product notation isn't nearly as common or useful as summation notation.

Suppose we wish to compute the product of many terms, as in

$$a_1 a_2 \cdots a_n.$$

The direct analogy with summation notation is to write this as

$$\prod_{i=1}^{n} a_i,$$

where we use $\prod$ because Pi in Greek corresponds to "P" (for product) in the Roman alphabet. Thus, for example,

$$\prod_{k=1}^{n} k \;=\; 1\times 2\times 3\cdots(n-1)n \;=\; n!,$$

$$\prod_{i=2}^{7} \frac{(i-1)}{i} \;=\; \left(\frac{1}{2}\right)\left(\frac{2}{3}\right)\cdots\left(\frac{6}{7}\right) \;=\; \frac{1}{7},$$

$$\begin{aligned}\prod_{n=1}^{4} 2^n &= 2^1 2^2 2^3 2^4\\ &= 2^{1+2+3+4}\\ &= 2^{10}.\end{aligned}$$

If the product is *empty*, that is, if the predicate which defines the values of the index is true for no value of the index, then by convention, the product is 1. Why do you suppose we use this convention for products when the corresponding convention for sums is to replace an empty sum by 0?

In [18–19] we consider how to extend the idea of summation and product notation to other operators, such as set union and intersection.

Problems: Section 0.5

1. Write the following using $\sum$ and $\prod$ notation.

a) $1+2+3+\cdots+100$

b) $1\cdot2\cdot3\cdots100$

c) $2+4+6+8+\cdots+100$

d) $1\cdot3\cdot5\cdot7\cdots99.$

2. Write the following polynomials using $\sum$ notation.

a) $x+2x^2+3x^3+\cdots+10x^{10}$

b) $1-x+x^2-x^3+\cdots+x^{10}$

c) $x+x^2+x^3+\cdots+x^{14}$

d) $x+2x^2+3x^4+4x^6+\cdots+8x^{14}.$

3. Evaluate the following sums.

a) $\sum_{i=2}^{7}(2i^3+3i)$ b) $\sum_{i=2}^{4}(2i)^{-i}$

c) $\sum_{i=3}^{5}(3i+2p).$

4. Evaluate the following sums.

a) $\displaystyle\sum_{\substack{i=3\\ i\text{ even}}}^{8} 1/i^2$ b) $\displaystyle\sum_{\substack{i=3\\ 3\,|\,i\text{ or }5|i}}^{20} (2i^2+6).$

5. Evaluate each of the following sums for the values of n indicated. If the answer is particularly simple, see if you can explain why.

a) $\sum_{j=1}^{n}\left(\frac{1}{j}-\frac{1}{j+1}\right)$ $\quad n=4,5.$

b) $\sum_{S\subset[n]}(-1)^{|S|}$ $\quad n=2,3$ and $[n]=\{1,2,\ldots,n\}.$

c) $\prod_{j=1}^{n}\frac{j}{j+1}$ $\quad n=5,6.$

d) $\prod_{j=2}^{n}\frac{j^2-1}{j^2}$ $\quad n=5,6.$

e) $\sum_{d\,|\,n}\left(\frac{d-n}{d}\right)$ $\quad n=6,30.$

f) $\sum_{S\subset[n]}2^{|S|}$ $\quad n=2,3$ and $[n]=\{1,2,\ldots,n\}.$

6. a) Express the following inequality in summation notation.

$$(a_1^2+a_2^2+\cdots+a_n^2)(b_1^2+\cdots+b_n^2) \geq (a_1b_1+a_2b_2+\cdots+a_nb_n)^2.$$

This is called the Cauchy–Schwarz inequality, and it is always true.

b) Express the following inequality using sum and product notation.

$$\frac{a_1+a_2+\cdots+a_n}{n} \geq (a_1a_2\cdots a_n)^{1/n}.$$

This is called the arithmetic–geometric mean inequality and is true for all nonnegative a_i's.

7. Explain why $\sum_{j=1}^{n} j2^j = \sum_{k=0}^{n} k2^k$.

8. Rewrite $\sum_{k=1}^{10} 2^k$ so that k goes from 0 to 9 instead.

9. Rewrite $\sum_{i=0}^{n}(2i+1)$ in the form $\sum_{i=?}^{?}(2i-1)$. You have to figure out what the question marks should be.

10. Rewrite $\sum_{k=0}^{n}3k+\sum_{j=1}^{n+1}4j$ so that it involves just one $\sum$ sign. There may be some extra terms outside the $\sum$ sign.

11. Evaluate the sums:

a) $\sum_{i=2}^{4}\sum_{j=3}^{5}(2i^2-3ij+j^3)$

b) $\sum_{i=1}^{3}\sum_{\substack{j=1\\ ij\text{ even}}}^{3}\lfloor(i/j)+(j/i)\rfloor.$

12. Evaluate the sums:

a) $\sum_{i=2}^{4}\sum_{j=i}^{2i}(i^2-2ij)$

b) $\sum_{i=1}^{9}\sum_{j=1}^{\lfloor 9/i\rfloor} ij^2.$

13. Evaluate:

a) $\sum_{j=1}^{n}\sum_{i=j}^{n}\frac{1}{i}$ $\quad n=3,4.$

b) $\sum_{k=1}^{n}\sum_{j=1}^{n}\frac{(-1)^k}{j}$ $\quad n=3,4.$

c) $\sum_{j=1}^{n}\sum_{i=1}^{n}\cos\left(\frac{\pi i}{2j}\right)$ $\quad n=3.$

14. If $\sum_{i=0}^{100}x^i$ is squared and rewritten as $\sum_{i=0}^{n}a_ix^i$, what is a_{50}? What is n?

15. If $\left(\sum_{i=0}^{100}x^i\right)\left(\sum_{i=0}^{25}x^i\right)$ is multiplied out, what is a_{50}?

16. Use the distributive law to prove the correctness of

a) Eq. (7) **b)** Eq. (8).

17. Use the idea suggested by the example which follows Eq. (11) to prove the correctness of Eq. (11) in general.

18. For this problem we use the following **interval notation**:

$$(a,b) = \{x \mid a<x<b\}$$

and

$$[a,b] = \{x \mid a\leq x\leq b\}.$$

Also, $\bigcup_{i=1}^{n}S_i$ means $S_1\cup S_2\cup\cdots S_n$ and $\bigcap_{i=1}^{n}S_i$ means $S_1\cap S_2\cap\cdots\cap S_n$. Evaluate the following expressions.

a) $\bigcup_{i=1}^{10}(0,i)$ **b)** $\bigcup_{j=1}^{5}[2j-2,2j]$

c) $\bigcup_{i=1}^{\infty}(-i,i)$ d) $\bigcup_{k=1}^{10}[0,1/k]$

e) $\bigcup_{k=1}^{10}(0,1/k)$ **f)** $\bigcap_{k=1}^{10}[0,1/k]$

g) $\bigcap_{k=1}^{10}(0,1/k)$ **h)** $\bigcap_{i=1}^{\infty}[0,1/i]$

i) $\bigcap_{i=1}^{\infty}(0,1/i)$.

19. Let P_k be the statement that the integer k is a prime. Let $\vee$ stand for "or" and let

$$\bigvee_{i=1}^{n}P_i$$

mean $P_1\vee P_2\vee\cdots\vee P_n$. Similarly let "$\wedge$" stand for "and" and let

$$\bigwedge_{i=1}^{n}P_i$$

mean $P_1\wedge P_2\wedge\cdots\wedge P_n$. What do the following expressions assert and are they true?

a) $\bigvee_{k=102}^{106}P_k$ b) $\bigwedge_{k=1}^{3}P_{2k+1}$

c) $\bigwedge_{k=1}^{4}P_{2k+1}$.

0.6 Matrix Algebra

In this section we introduce you briefly to a very important branch of mathematics: **linear algebra**. We'll devote Chapter 8 to algorithmic linear algebra.

A **matrix** is a two-dimensional, rectangular array of numbers which we write as follows:

$$\begin{bmatrix} a_{11} & a_{12} & \cdots & a_{1n} \\ a_{21} & a_{22} & \cdots & a_{2n} \\ \vdots & \vdots & & \vdots \\ a_{m1} & a_{m2} & \cdots & a_{mn} \end{bmatrix}. \tag{1}$$

We write the elements of the matrix, a_{ij}, in doubly subscripted form, with the first subscript representing the *row* (horizontal line) in which the element occurs and the second subscript representing the *column* (vertical line). Thus the matrix in (1) contains m rows and n columns. As with sets, we use uppercase letters to denote matrices and often write a matrix as $A = [a_{ij}]$. For example, if

$$A = \begin{bmatrix} 2 & 1 & 3 \\ 4 & -2 & 6 \end{bmatrix},$$

then $a_{13} = 3$, $a_{22} = -2$, and so on.

Matrices play many roles in mathematics, and you will see a variety of them later in this book. When m in (1) is 1, the matrix consists of a single row. Therefore the row subscript conveys no useful information, and we may write a single row with a single subscript:

$$\begin{bmatrix} a_1 & a_2 & \cdots & a_n \end{bmatrix},$$

which is called a **row vector**, or just a **vector**.

Our object in this section is simply to define how two matrices of form (1) may be added, subtracted, and multiplied, as well as to define some other simple operations on and facts about matrices. (We discuss division, which is usually referred to as multiplication by the "inverse" matrix, at the end of this section and in Chapter 8.)

In order to perform any operation on two matrices, A and B, they must be **conformable**. In the case of addition and subtraction, this means that the number of rows and columns in one matrix must be the same as in the other matrix. If this is the case, then we can find their sum by adding corresponding elements. That is, if $A = [a_{ij}]$ and $B = [b_{ij}]$, then $C = A+B$ has elements c_{ij} such that

$$c_{ij} = a_{ij} + b_{ij}. \tag{2}$$

EXAMPLE 1 Let the rows of the following two matrices refer to the freshman, sophomore, junior, and senior classes at Podunk College, and let the columns refer to its humanities, social science, and science divisions. Each entry is the number of students in a particular class (freshman, etc.) studying in a particular division. The matrix on the left represents the fall semester and the matrix on the right the spring semester.

$$\begin{bmatrix} 212 & 475 & 624 \\ 273 & 419 & 602 \\ 263 & 410 & 560 \\ 290 & 408 & 553 \end{bmatrix} \qquad \begin{bmatrix} 253 & 423 & 587 \\ 295 & 386 & 549 \\ 280 & 400 & 535 \\ 301 & 376 & 540 \end{bmatrix}$$

Since both matrices have four rows and three columns, they are conformable. Using Eq. (2), we find that their sum is

$$\begin{bmatrix} 465 & 898 & 1211 \\ 568 & 805 & 1151 \\ 543 & 810 & 1095 \\ 591 & 784 & 1093 \end{bmatrix}.$$

What does this matrix represent [5]? This example illustrates the usefulness of matrices for tabulating and manipulating data. ▮

To subtract two conformable matrices we just use Eq. (2) with the plus sign replaced by a minus sign. Another simple operation which can be performed on any matrix is to multiply the matrix by a constant. If we are given $A = [a_{ij}]$, then we define rA, where r is a constant, to be the matrix whose elements are

$$ra_{ij}.$$

That is, we multiply each element of A by r. Thus if $r = 2$, multiplying the result of Example 1 by r would give

$$\begin{bmatrix} 930 & 1796 & 2422 \\ 1136 & 1610 & 2302 \\ 1086 & 1620 & 2190 \\ 1182 & 1568 & 2186 \end{bmatrix}.$$

Another simple matrix operation is to take the **transpose** of a matrix, which means only that we interchange the rows and columns. When transposed, therefore, a matrix $A = [a_{ij}]$ of m rows and n columns becomes a matrix $B = [b_{ij}]$ of n rows and m columns such that

$$b_{ij} = a_{ji}. \tag{3}$$

At times we'll denote the transpose of a matrix A by A^{T}.

EXAMPLE 2 Find the transpose of the matrix

$$A = \begin{bmatrix} 3 & 18 & -9 & 0 \\ 11 & 14 & 7 & 8 \end{bmatrix}.$$

Solution Using Eq. (3), we obtain

$$A^{\mathrm{T}} = \begin{bmatrix} 3 & 11 \\ 18 & 14 \\ -9 & 7 \\ 0 & 8 \end{bmatrix}. ▮$$

Note in Example 2 that a row in the original matrix becomes a column in the transposed matrix and vice versa. Thus, in particular, the transpose of a *row vector*,

$$\mathbf{x} = \begin{bmatrix} x_1 & x_2 & \cdots & x_n \end{bmatrix},$$

is a *column vector*:

$$\mathbf{x}^{\mathrm{T}} = \begin{bmatrix} x_1 \\ x_2 \\ \vdots \\ x_n \end{bmatrix}.$$

Next we define and give some examples of matrix multiplication. In so doing we'll show an instance where the use of summation notation is vital. Let $A = [a_{ij}]$ have m rows and k columns and let $B = [b_{ij}]$ have k rows and n columns. For two matrices, A and B, to be conformable for multiplication, the number of columns in the first matrix (k) must be the same as the number of rows (k) in the second. When two matrices are conformable for multiplication, we define their product

$$C = AB,$$

so that the elements c_{ij} of C are

$$c_{ij} = \sum_{r=1}^{k} a_{ir} b_{rj}. \tag{4}$$

In Fig. 0.12 we illustrate how to interpret Eq. (4). In matrix multiplication the (i, j) element in the product matrix C is formed by taking the row vector formed by the ith row A and multiplying it element by element times the column vector formed by the jth column of B and then adding all the products.

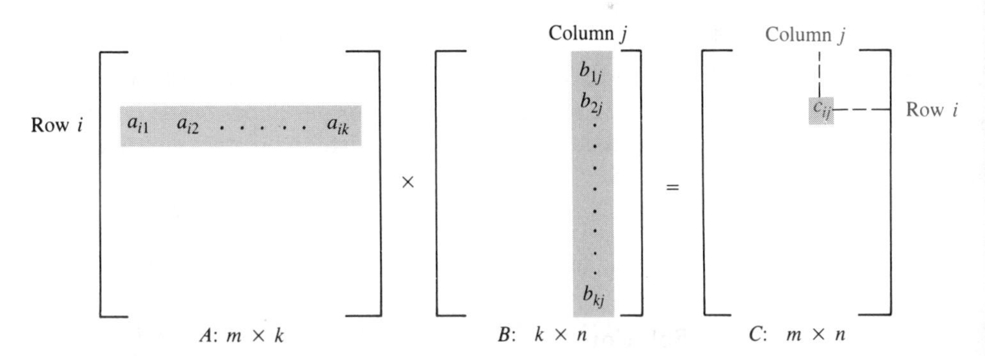

FIGURE 0.12 Matrix Multiplication. The multiplication of $A = [a_{mk}]$ by $B = [b_{kn}]$ gives matrix C, where

$$c_{ij} = \sum_{r=1}^{k} a_{ir} b_{rj}$$

EXAMPLE 3 Calculate the product of

$$A = \begin{bmatrix} 6 & 7 & 8 \\ -5 & -3 & 17 \\ 2 & 0 & 4 \end{bmatrix} \quad \text{and} \quad B = \begin{bmatrix} -5 & 10 \\ 12 & 6 \\ 2 & -1 \end{bmatrix}.$$

Solution To calculate the first row of the product matrix C, you take the first row of A and multiply it term by term by the first column of B, obtaining

$$6 \times (-5) + 7 \times 12 + 8 \times 2 = -30 + 84 + 16 = 70,$$

and then multiply the first row of A times the second column of B, obtaining 94. Proceeding similarly with the second and third rows of A—make sure you understand which calculations need to be done—you obtain finally

$$C = \begin{bmatrix} 70 & 94 \\ 23 & -85 \\ -2 & 16 \end{bmatrix}. \blacksquare$$

The vector times vector product expressed by the sum in Eq. (4) is called an **inner product**. Typically, we write vectors as boldface letters, such as **a** and **b**, and often denote inner products by

$$\mathbf{a} \cdot \mathbf{b} \qquad \text{or} \qquad \langle \mathbf{a}, \mathbf{b} \rangle .$$

Another common notation is

$$\mathbf{a}\mathbf{b}^{\mathrm{T}},$$

which emphasizes that the first vector is a row vector and the second is a column vector (the transpose of a row vector). (Another common notation is $\mathbf{a}^{\mathrm{T}} \cdot \mathbf{b}$, when **a** and **b** are both column vectors.) Therefore the inner product is just a special case of matrix multiplication when the first matrix is $1 \times k$ and the second is $k \times 1$. Using the notation of inner products, we may express matrix multiplication as

$$AB = [\mathbf{a}_i \cdot \mathbf{b}_j]$$

where $\mathbf{a}_i$ is the ith row of A and $\mathbf{b}_j$ is the jth column of B.

You may be somewhat curious about our explanation of matrix multiplication. Why use this strange way of defining matrix multiplication when we could just have defined

$$c_{ij} = a_{ij} b_{ij},$$

that is, analogous to matrix addition and subtraction, just multiplied them term by term? The answer, as we hope you've guessed, is that our definition is more useful than this alternative. Here's an example illustrating why. Let

$$\begin{aligned} z_1 &= 2y_1 + 3y_2 + 4y_3 & \qquad y_1 &= x_1 + 2x_2 \\ z_2 &= 3y_1 + 2y_2 - y_3 & y_2 &= 3x_1 + 4x_2 \\ & & y_3 &= 5x_1 + 6x_2. \end{aligned}$$

The matrices of the coefficients of the z's in terms of the y's and the y's in terms of the x's are, respectively,

$$A = \begin{bmatrix} 2 & 3 & 4 \\ 3 & 2 & -1 \end{bmatrix} \quad \text{and} \quad B = \begin{bmatrix} 1 & 2 \\ 3 & 4 \\ 5 & 6 \end{bmatrix}. \tag{5}$$

To express the z's in terms of the x's algebraically, we substitute:

$$\begin{aligned} z_1 &= 2(x_1+2x_2) + 3(3x_1+4x_2) + 4(5x_1+6x_2) \\ &= (2 \times 1 + 3 \times 3 + 4 \times 5)x_1 + (2 \times 2 + 3 \times 4 + 4 \times 6)x_2 \\ &= 31x_1 + 40x_2; \\ z_2 &= 3(x_1+2x_2) + 2(3x_1+4x_2) - 1(5x_1+6x_2) \\ &= (3 \times 1 + 2 \times 3 + (-1) \times 5)x_1 + (3 \times 2 + 2 \times 4 + (-1) \times 6)x_2 \\ &= 4x_1 + 8x_2. \end{aligned}$$

Thus the coefficient matrix of the z's in terms of the x's is

$$C = \begin{bmatrix} 31 & 40 \\ 4 & 8 \end{bmatrix}.$$

But you can see by referring to Eq. (5) that the upper left entry in C is just the first row of A multiplied term by term by the first column of B (i.e., it's the inner product of this row and column), and similarly for the other entries in C. Therefore, using our definition of matrix multiplication,

$$C = AB.$$

Indeed, matrix multiplication was originally defined by the mathematician Arthur Cayley [(1821–1895), a leading British mathematician] just for the purpose of succinctly expressing substitutions such as the preceding.

Ordinary multiplication is commutative. That is, $ab = ba$. Is matrix multiplication commutative? In general, it can't be because, if A and B are conformable for multiplication of AB, then B and A won't be conformable unless the number of rows in A is the same as the number of columns in B (why?). So suppose A and B are conformable for both multiplications AB and BA. That is, assume A is $m \times k$ and B is $k \times m$. Will $AB = BA$? Again the general answer must be, No, because AB is $m \times m$ and BA is $k \times k$. So the question about commutativity makes no sense unless A and B are both *square* matrices, that is, with the same number of rows as columns. Let's assume, therefore, that A and B are both $m \times m$. Is $AB = BA$? The following example answers this question.

EXAMPLE 4 Let

$$A = \begin{bmatrix} 2 & 3 \\ 4 & -1 \end{bmatrix} \quad \text{and} \quad B = \begin{bmatrix} 5 & -2 \\ 1 & 0 \end{bmatrix}.$$

Then

$$AB = \begin{bmatrix} 13 & -4 \\ 19 & -8 \end{bmatrix} \quad \text{and} \quad BA = \begin{bmatrix} 2 & 17 \\ 2 & 3 \end{bmatrix},$$

so $AB \neq BA$. ∎

Therefore matrix multiplication is not commutative for conformable square matrices. Note that this doesn't mean that AB can never equal BA but just that it need not. Indeed, it is possible to find pairs of matrices for which matrix multiplication is commutative.

We conclude this section with some definitions which will be useful in Chapter 8:

1. An identity matrix I is a square matrix with 1's on the main diagonal and 0's elsewhere. Thus it has the form

$$I = \begin{bmatrix} 1 & 0 & 0 & \dots & & 0 \\ 0 & 1 & 0 & \dots & & 0 \\ 0 & 0 & 1 & 0 & \dots & 0 \\ & & & \ddots & & \\ 0 & 0 & 0 & \dots & 1 & 0 \\ 0 & 0 & 0 & \dots & 0 & 1 \end{bmatrix}.$$

   For each m there is an identity matrix with m rows and columns; sometimes we'll use I_m instead of I. If you've understood our definition of matrix multiplication, you'll see that if M is $m \times k$, then $I_m M = M I_k = M$.
2. A zero matrix is a matrix all of whose entries are 0.
3. Let M be a square matrix. If there exists a matrix L (there doesn't always) such that $ML = I$, then L is called the *inverse* of M, and is denoted by M^{-1}. This terminology and notation should seem natural to you from our discussion of the inverse of a function. Now suppose A is a matrix and B is its inverse so that $AB = I$. Then it is true, although we won't prove it here (but see Section 8.6), that $BA = I$. In the language of matrix algebra we say that any *right inverse* of a square matrix is also a *left inverse*.

Problems: Section 0.6

1. Evaluate

$$\begin{bmatrix} 1 & 2 \\ 3 & 4 \end{bmatrix} + \begin{bmatrix} 5 & 6 \\ 7 & 8 \end{bmatrix}.$$

2. Evaluate

$$\begin{bmatrix} 1 & 2 & 3 \\ 4 & 5 & 6 \end{bmatrix} - \begin{bmatrix} 6 & 4 & 2 \\ 5 & 3 & 1 \end{bmatrix}.$$

3. Write down

$$4 \begin{bmatrix} 1 & 2 \\ 3 & 4 \end{bmatrix}.$$

4. Write down

$$\begin{bmatrix} 1 & 2 & 3 \\ 4 & 5 & 6 \end{bmatrix}^{\mathrm{T}}.$$

5. In Example 1 what is the interpretation of the sum matrix?

6. Let $\mathbf{a} = \begin{bmatrix} 1 & 2 & 3 \end{bmatrix}$ and $\mathbf{b} = \begin{bmatrix} 4 & -1 & 6 \end{bmatrix}$. Evaluate

a) $\mathbf{a} + \mathbf{b}$ b) $4\mathbf{a}$ c) $\mathbf{a} \cdot \mathbf{b}$

d) $\mathbf{a}\mathbf{b}^{\mathrm{T}}$ e) $\mathbf{a}^{\mathrm{T}}\mathbf{b}$

7. Let the matrix A have elements $a_{ij} = i + j$, and let B have elements $b_{ij} = i \times j$.

a) Write 3×3 matrices for both A and B.

b) Compute $A + B$, if both A and B have 4 rows and columns.

c) Compute AB, if A has 2 rows and 3 columns and B has 3 rows and 2 columns.

d) Compute AB if both A and B have 3 rows and 3 columns.

e) Compute the transposes of A and B in c).

8. Let

$$A = \begin{bmatrix} 1 & 2 & 3 & 4 & 5 \end{bmatrix},$$
$$B = \begin{bmatrix} 1 & -4 & 9 & -16 & 25 \end{bmatrix}.$$

Compute

a) AB^{T} b) $A^{\mathrm{T}}B$.

9. Evaluate the following products:

a) $\begin{bmatrix} 1 & 0 \\ 0 & 1 \end{bmatrix} \begin{bmatrix} a & b \\ c & d \end{bmatrix}$

b) $\begin{bmatrix} a & b \\ c & d \end{bmatrix} \begin{bmatrix} d & -b \\ -c & a \end{bmatrix}$.

10. On the side of a box of Yummy-o's, there is a chart that says what percentage of minimum daily food requirements are provided by 1 ounce of the cereal:

protein	5%
fat	10%
carbohydrates	20%

If you eat 2 ounces, what percents do you get? What matrix and what matrix operation were involved in your calculation?

11. Assume the following matrix shows the number of grams of nutrients per ounce of food indicated. (Actually, the numbers are made up.)

	meat	potato	cabbage
protein	20	5	1
fat	30	3	1
carbohydrates	15	20	5

If you eat a meal consisting of 9 ounces of meat, 20 ounces of potatoes, and 5 ounces of cabbage, how many grams of each nutrient do you get? Use matrix multiplication to obtain your answer.

12. If A is a column vector of n elements and B is a row vector of n elements, describe the form of AB.

13. Let

$$A = \begin{bmatrix} 1 & 1 \\ 0 & 1 \end{bmatrix}, \quad B = \begin{bmatrix} 1 & 2 \\ 0 & 1 \end{bmatrix},$$
$$C = \begin{bmatrix} 1 & 0 \\ 1 & 1 \end{bmatrix}.$$

Determine which pairs of these three matrices commute under multiplication.

14. Consider the four matrices

$$I = \begin{bmatrix} 1 & 0 \\ 0 & 1 \end{bmatrix}, \quad \begin{bmatrix} 1 & 0 \\ 0 & -1 \end{bmatrix},$$
$$\begin{bmatrix} -1 & 0 \\ 0 & 1 \end{bmatrix}, \quad \begin{bmatrix} -1 & 0 \\ 0 & -1 \end{bmatrix}.$$

Show that the square of each one of them is I. This shows that the identity matrix, unlike the identity number, can have many square roots.

15. Verify that

$$\begin{bmatrix} 2 & 5 \\ 1 & 3 \end{bmatrix}$$

is the inverse of

$$\begin{bmatrix} 3 & -5 \\ -1 & 2 \end{bmatrix}.$$

There are two products you have to check.

16. Let

$$M = \begin{bmatrix} 1 & 1 & 1 \\ 0 & 1 & 1 \\ 0 & 0 & 1 \end{bmatrix}.$$

Compute M^2 and M^3. What do you conjecture about the form of M^n?

17. Let

$$\begin{aligned} c &= b_1 + 2b_2 + 3b_3; \\ b_1 &= a_1 + 2a_2; \\ b_2 &= 3a_1 + 4a_2; \\ b_3 &= 5a_1 + 6a_2. \end{aligned}$$

Express c in terms of a_1 and a_2 two ways:

a) By direct algebraic substitution.

b) By matrix multiplication.

Show how various calculations in the two methods correspond (say, by drawing arrows between corresponding parts).

18. What sizes must A and B be for AB^{T} to exist? Assuming the sizes are right, write a formula using summation notation for the typical entry of AB^{T}. Start by letting $A = [a_{ij}]$. Do you see that B should be denoted $[b_{kj}]$, not $[b_{jk}]$?

19. What sizes must A, B, and C be for ABC to exist? Assuming it exists, write a formula using summation notation for the typical entry of ABC. Start by letting $A = [a_{ij}]$.

20. If $A_1, A_2, \ldots, A_n$ are matrices, state a rule that guarantees that the product

$$A_1 A_2 \cdots A_n$$

can be computed.

21. Let A be defined as in [7]. Compute A^2 and A^3. When is it possible to compute powers of a matrix? When is it not possible?

0.7 Proof and Logic Concepts

Our purposes in this section are (1) to make sure you understand certain key words used in mathematical reasoning, like "necessary" and "converse"; and (2) to introduce a little notation, like $\Longrightarrow$. Our approach is informal. In Chapter 7 we'll introduce a formal algebra of logic and use it to (among other things) review and further clarify all the points about logic and proof we will have made by then.

The most common type of sentence in a mathematical argument is an **implication**. This is anything of the form A implies B, that is, *if* A *then* B. For instance,

$$\textit{if } x = 2, \textit{ then } x^2 = 4.$$

The *if* part of the sentence ($x = 2$) is called the **hypothesis**. Synonyms are **assumption** and **premise**. The *then* part ($x^2 = 4$) is called the **conclusion**.

Sometimes an implication is broken into two or more sentences. Indeed, this is quite common when the implication is asserted as a theorem. For instance, you might find

> Theorem. Let n be an odd integer. Then n^2 is also odd.

This theorem is really the assertion of the implication

> If n is an odd integer, then n^2 is odd.

Also, neither the hypothesis nor the conclusion of an implication needs to be a simple statement. For instance, consider the theorem

> Let ABC be a triangle. Suppose angle A = angle B. Suppose further that angle $C = 60°$. Then ABC is equilateral. Furthermore, all three angles are $60°$.

In this theorem the first three sentences together form the assumption(s) and the last two are the conclusion.

There is a nice notation for A implies B:

$$A \Longrightarrow B.$$

Some authors write $A \rightarrow B$ instead, but single-shaft arrows are also used for limits of sequences and functions, e.g., $x_n \rightarrow 0$ as $n \rightarrow \infty$ (see Chapter 9, but also see our use of this notation in Section 0.4). We prefer to keep the different uses distinct.

In this book, $\Longrightarrow$ is a **logical operator**†, making a single statement from two separate statements, whereas $\rightarrow$ is an abbreviation for the verb "approaches."

Why are implications so useful in mathematics? Because they chain together so nicely. That is, if $A \Longrightarrow B$ and $B \Longrightarrow C$ are both true, then clearly so is $A \Longrightarrow C$, and so on with longer chains. This transitivity property (see Section 0.2) means that, using implications, a mathematician can try to show that A implies Z by reducing it to many smaller steps.

Many different English phrases boil down to $\Longrightarrow$. We have already mentioned "implies" and "if–then." Here are some others. Each sentence below means $A \Longrightarrow B$.

Whenever A is true, so is B.

B follows from A.

A is sufficient for B.

B is necessary for A.

A only if B.

It's not supposed to be obvious that all of these mean $A \Longrightarrow B$, especially when stated so abstractly. So, first, let's restate them with $A =$ "It is raining" and $B =$ "It is cloudy."

Whenever it's raining, it's cloudy.

It follows that it's cloudy, if it's raining.

It's sufficient that it rain in order that it be cloudy.

It is necessary that it be cloudy for it to rain.

It's raining only if it's cloudy.

Some of these constructions make awkward ordinary English, but we hope you see that they mean the same thing as the original.

Did it surprise you that, in some of these equivalent statements, the order of appearance of A and B in the sentence changed? It may seem particularly odd that "A only if B" should mean the same as "if A, then B" — the "if" has moved from A to B! But that's the way it is—natural languages are full of oddities. As we'll see in a minute, if we say "B only if A," it does not mean the same thing at all.

The use of "necessary" and "sufficient" is particularly important. We'll return to these words later in this section.

Closely related to the implication $A \Longrightarrow B$ is its **converse**, $B \Longrightarrow A$. (This is sometimes written $A \Longleftarrow B$.) The converse has a separate name because it does *not* mean the same thing as the original. For instance,

---

† There is substantial subtlety in the notion of implication, most of whose elucidation we'll leave to Chapter 7. As an example here of this, $x = 2 \Longrightarrow x^2 = 4$ is a stronger implication than "if it is the Sears tower, then it is the tallest building in the world." Although the former is always true, the latter, presumably, will at some time in the future not be true.

If it's raining, then it's cloudy

is true, but

If it's cloudy, then it's raining

is false because you can't conclude that it's raining just because you know it's cloudy. If we'd used "when" instead of "if," you might have found this easier to understand. The first statement would still be true, and you might find it easier to agree that the converse is false. When you give examples in ordinary English to illustrate mathematical concepts, the nuance of meaning in a natural language like English can make things difficult.

We hope your high school geometry teacher drilled into you the difference between a statement and its converse. If you are told to assume A and prove B, you must demonstrate $A \Longrightarrow B$; if you demonstrate $B \Longrightarrow A$ instead, you have wasted your time, because $A \Longrightarrow B$ could still be false. If you start with A and end up with B, you have done the job. If instead you start with B and end up with A, you have shown $B \Longrightarrow A$ and goofed. This is called "assuming what you are supposed to show."

There is another implication closely related to $A \Longrightarrow B$ which *does* mean the same thing: its **contrapositive**, i.e.,

$$(\textbf{not } B) \Longrightarrow (\textbf{not } A),$$

or using the notation for "not" that we'll use in Chapter 7,

$$\neg B \Longrightarrow \neg A$$

In the contrapositive, the order of statements is reversed and they are both negated. Thus the contrapositive of

If it's raining, then it's cloudy

is

If it's not cloudy, then it's not raining.

When we say an implication and its contrapositive mean the same thing, we mean that either they are both true or they are both false. So whenever you wish to show one you can instead show the other. For instance, if you want to prove $A \Longrightarrow B$, one method is to assume that the desired conclusion is false ($\neg B$) and from there somehow show that the hypothesis is false ($\neg A$).

We hope your high school geometry teacher also drilled contrapositives into you. If not, here is a brief explanation as to why a statement and its contrapositive mean the same. Suppose $A \Longrightarrow B$ is true. Then if $\neg B$ were true, we couldn't have A true because then $A \Longrightarrow B$ would make B true at the same time as $\neg B$. This would be a **contradiction**. Therefore $\neg B$ implies $\neg A$. Next we must suppose $\neg B \Longrightarrow \neg A$ is true and argue that $A \Longrightarrow B$ follows. The argument is similar to what we just did [8].

Biconditionals and Equivalence

If both $A \Longrightarrow B$ and $B \Longrightarrow A$ are true, we abbreviate by writing

$$A \iff B.$$

Such a statement is called a **biconditional** (or *bi-implication*) and it says that statements A and B are **equivalent**.

Another way to write $A \iff B$, without symbols, is "A if and only if B." Indeed, that is the usual way to say $A \iff B$ out loud. For those who don't like arrows, there is yet another accepted shorthand for "if and only if": iff. When you must pronounce iff, really hang on to the "ff" so that people hear the difference from "if."

Equivalence is a very useful concept, even when logical symbols like $\iff$ aren't used. Two statements are equivalent if they are always both true or always both false. In other words, wherever we have said that two statements mean the same, we could instead have said they are equivalent. For instance, an implication and its contrapositive are equivalent. Another way to say that is to say

$$(A \Longrightarrow B) \iff (\neg B \Longrightarrow \neg A).$$

A **definition** is a declaration that henceforth a certain term will mean the same as the phrase that follows the term. For instance, a square is defined to be a four-sided figure with all sides equal and all angles 90°. Thus we could also write definitions in the form of an equivalence: "A figure is a square iff it is four-sided, with all sides equal and all angles 90°." Logically speaking, it would be incomplete to write "if" instead of "iff", but we regret to inform you that, by tradition, this is exactly what is done. This substitution of "if" when "iff" is meant is allowed *only* in definitions. You can always tell in this book when a sentence is a definition: either the word "define" is present, or else the term being defined is printed in boldface.

"Equal" and "Equivalent" are *not* the same. Equivalence is a relation which holds between *statements*. Equality can hold between any sorts of objects, but only if they are identical. Thus the statements $A \Longrightarrow B$ and $\neg B \Longrightarrow \neg A$ are equivalent but not equal. A good rule of thumb is: When you are talking about statements use "equivalent"; when you are talking about other sorts of objects (numbers, functions, graphs, etc.) use "equal".

(*Warning*: "Equivalent" can and often is defined to have meanings other than the one just given. For instance, integers might be defined to be equivalent if they have the same remainder when divided by 3. Then 2 and 11 are equivalent, although they are not equal. Indeed, any relation which is reflexive, symmetric, and transitive (Section 0.2) is called an equivalence relation, and it is therefore natural, when such a relation is being considered, to call x and y equivalent if (x, y) is in the relation. Still, the moral of the preceding paragraph still holds: Equivalence is not the same as equality, and you should be careful not to mix up the two words. Also the usage of equivalent in this book will almost always be in reference to statements.)

Let's now return to the words "necessary" and "sufficient". These are very useful words. Remember,

> $A \Longrightarrow B$ means the same as A is *sufficient* for B and also the same as B is *necessary* for A.

For instance, since being a square implies being a rectangle, we can say that being a square is sufficient for being a rectangle and being a rectangle is necessary for being a square. "Sufficient" is a useful word because it makes clear that A need not be the only thing which guarantees B—squares are not the only rectangles—but it suffices. "Necessary" is a useful word because it makes clear that B may not be enough to get A—being rectangular is only part of being square—but you can't do without it.

Since the usual goal in mathematics is to prove one thing from another, the value of knowing that A is sufficient for B is clear: Once you get A you are home free with B. But suppose instead you know that A is necessary for B. The implication is going the wrong way as far as deducing B is concerned, so what good is this knowledge of necessity? A related question is: Since every statement containing the word "necessary" can be rewritten using "sufficient", what's the point of using both words? The answer to both questions is: The concept of necessity is natural and useful when you are after negative information. That is, suppose you know that A is necessary for B and you also know that A is *false.* Then you can conclude that B is false, too. That is, instead of using $B \Longrightarrow A$ directly, you use the equivalent contrapositive $\neg A \Longrightarrow \neg B$.

To illustrate further the proper usage of "necessary" and "sufficient" (and a few other terms), we prove a few tiny theorems.

EXAMPLE 1 Prove the theorem that the number 2 has a real-valued square root.

Solution For a number to have a real square root, it is sufficient that the number be positive. The number 2 is positive. ∎

Note: It would be wrong to use "necessary" in place of "sufficient" in Example 1 for two reasons. First, the claim would be false (see Example 2 for a hint, if you don't see why). Second, even if it were true, it would be irrelevant. We are starting with a positive number and trying to get to the existence of a root. For positivity to imply something means it must be a sufficient condition.

EXAMPLE 2 Prove the theorem that the number -2 does not have a real-valued square root.

Solution For a number to have a real square root, it is necessary that the number be nonnegative. But -2 is not nonnegative. ∎

EXAMPLE 3 Prove the theorem that a quadrilateral is a square if and only if it is both a rhombus and a rectangle.

Solution If: Let Q be a quadrilateral which is both a rhombus and a rectangle. From the definition of a rhombus, all four sides are equal in length. From the definition of a rectangle, all four angles are right. A square is a four-sided figure in which all angles are right and all sides are equal. Therefore, Q is a square.

Only if: Let Q be a square. By definition, all four sides are equal, so Q is also a rhombus. Again by definition, all four angles are right, so Q is a rectangle. ∎

Note: We could replace the words "if and only if" in the theorem statement by "iff" or by "$\Longleftrightarrow$". Also, we could replace the "If" at the start of the proof by "$\Longleftarrow$" and "Only If" later by "$\Longrightarrow$" (not the other way around!). In Example 4, we show how to restate Example 3 using "necessary" and "sufficient."

EXAMPLE 4 Prove the theorem that for a quadrilateral to be a square, it is necessary and sufficient that it be both a rhombus and a rectangle.

Solution Necessity: same as "if" part in Example 3. Sufficiency: same as the "only if" part above. ∎

Since we have been discussing the logic of proofs, this is a good place to discuss some of the terminology of proofs. There are three names that are commonly used for mathematical facts; which name is used for a given fact depends on its logical relation to other facts. A **theorem** is an important fact which has a proof. A **lemma** is a proven fact which, while not so important in itself, makes the proof of a later theorem much easier. A **corollary** is a result which is quite easy to prove given that some theorem has already been proved. In other words, lemmas imply theorems, and theorems imply corollaries.

Strictly speaking, lemmas are unnecessary. The statement of a lemma, and its proof, can be incorporated into the proof of the theorem it is intended to help. However, doing this may make the proof of the theorem so long that the key idea is obscured. Also, if a lemma is helpful for several different theorems, it clearly is efficient to separate it out.

Quantifiers

Mathematicians are rarely interested in making statements about individual objects (2 has a square root), but prefer to make statements about large sets of objects (all nonnegative numbers have square roots). Words which indicate the extent of sets are **quantifiers**. Here are some true statements illustrating the use of the key quantifiers in mathematics. The quantifiers are in boldface.

Every integer is a real number.

All integers are real numbers.

Any integer is a real number.

Some real numbers are not integers.

There exists a real number which is not an integer.

No irrational number is an integer.

In Section 7.6, we'll discuss the subject of quantifiers in a more formal mathematical setting.

Sometimes a quantifier is not stated explicitly but is understood. For instance, the statement

Primes are integers

means, "All primes are integers." Or consider the earlier example, "If it's raining, then it's cloudy." Although someone saying this could mean

If it's raining now, it's cloudy now.

it's much more likely that the person means

At all times, if it's raining, then it's cloudy.

In other words, there is an implicit quantification (*all* times).

"Some" has a somewhat special meaning in mathematics. Whereas in ordinary discourse it usually means "at least a few but not too many", in mathematics it means "at least one, possibly many, or possibly all." "There exists" has exactly this same meaning. Thus the following peculiar mathematical statements are technically correct:

Some even integers are primes.

Some positive numbers have square roots.

There exists a positive number with a square root.

Why "some" is used this way we will explain shortly—when we get to negation. For now let's admit that mathematicians would rarely write the three preceding statements. These statements don't say everything we know, so why not tell the whole story? That is,

Exactly one even integer is prime,

or even better,

There is exactly one even prime: 2.

Similarly we would write

All positive numbers have square roots,

or even better,

A real number has a real square root $\Longleftrightarrow$ it is nonnegative.

However, if all we know is that at least one "smidgit" exists, and we have no idea how many there are, then we do not hesitate to write "There exists a smidgit" (even though there might be many) or "There are some smidgits" (even though there might be only one).

A word of caution. We have not drawn any distinctions between "every", and "any" and "all". Generally, in mathematics, they do have the same meaning. But in ordinary English there are sometimes differences, especially between "every" and "any". For instance, you will surely agree that the following sentences mean the same:

Everybody knows that fact.

Anybody knows that fact.

But just as surely the following sentences don't mean the same thing at all:

She'll be happy if she wins every election.

She'll be happy if she wins any election.

In general, "any" is a very tricky word, meaning "all" in most instances but "one" in some.

Why then don't mathematicians simply avoid "any" and always use "every" or "all"? Perhaps the reason is that "any" highlights a key aspect of how to prove statements about an infinite number of objects in a finite number of symbols. Suppose we wish to prove that "Any square is a rectangle." The word "any" suggests, correctly, that it suffices to pick *any one* square and show that *it* is a rectangle—so long as we don't use any special knowledge about that square but only its general squareness. And indeed, that's exactly what we did in Example 3. It's also how most proofs proceed. The words "every" and "all" don't make this point well.

This sort of proof is sometimes called arguing from the "generic particular." In such a proof, it is important to check that the example really is arbitrary. It cannot be obtained by a special construction. As you read through our proofs in this text, check to see that we have paid attention to this. The discussion of buildup errors in Section 2.5 is particularly relevant to this concern.

Negation

The **negation** of statement A is a statement which is true precisely when A is false. For simple statements, it is very easy to create the negation: just put in "not." For instance,

The negation of "x is 2" is "x is not 2."

Thus we have been using negations all along whenever we wrote (not A) or $\neg A$.

You should be warned though, that negating can get much trickier, if the original statement contains quantifiers or certain other words. For instance, the negation of

Some of John's answers are correct

is not

Some of John's answers are not correct,

because it is not the case that the latter statement is true if and only if the former is false. They are both true if John gets half the questions right and half wrong. The correct negation of the first sentence is

None of John's answers is correct.

We said earlier we would explain why "some" is used in mathematics to mean "one up to all". The reason is so that it will mean precisely the negation of "none". If "some" meant "a few", it would only cover part of the situations included in the negation of "none". Similarly, if "there exists" was limited to meaning "exactly one exists", then it, too, would not negate "none".

You'll be happy to hear that there are mechanical rules for correctly negating a sentence in symbolic form, including sentences with quantifiers. We'll discuss them in Section 7.6.

Finally, be careful about the word "or". In English it is often used to mean that exactly one of two possibilities is true:

Either the Democrats or the Republicans will win the election. (1)

On other occasions it allows both possibilities to hold:

You can get there via the superhighway or by local roads.

The first usage is called the **exclusive or** the second the **inclusive or**. In mathematics, "or" always means inclusive or unless there is an explicit statement to the contrary. Thus if for some strange reason statement (1) was the object of mathematical analysis, it would have to be prefaced by "exactly one of the following holds" or followed by "but not both".

Finally, here are two more words and one phrase whose special mathematical uses you should know about: trivial, obvious and vacuously true. Many people think "trivial" and "obvious" mean the same thing. However, writers on mathematics use them differently. An argument is **trivial** if every step is a completely routine calculation (but there may be many such steps). A claim is **obvious** if its truth is intuitively clear (even though providing a proof may not be easy). For instance, it's trivial that

$$(x+2)(3x+4)(5x^2+6x+7) = 15x^4 + 68x^3 + 121x^2 + 118x + 56$$

but it's hardly obvious. On the other hand, it's obvious that space is three-dimensional but the proof is hardly trivial (see Section 8.8).

Students sometimes think that writers or lecturers who use these words are trying to put them down. And perhaps sometimes they are. But, used appropriately, these words are useful in mathematical discourse and should not seem threatening.

A claim is **vacuously true** if there is nothing that need be done to prove it. For instance, it is a true statement that:

> Any set of n squares can be cut along straight lines so that the pieces may be reassembled into a single square.

The case $n = 1$ is vacuously true: make no cuts! (On the other hand for $n > 1$ the claim is neither vacuous nor obvious nor trivial—see [28, Section 2.3].) Is time spent on vacuously true facts a waste? No, indeed. At various places in this book we'll show that such facts have value, particularly in Section 7.7.

Problems: Section 0.7

1. Try to think of some more ways in English to say $A \Longrightarrow B$.

2. Explain the meaning of the words

a) implication **b)** hypothesis
c) equivalence **d)** negation.

3. Find an implication which is

a) true but its converse is false;

b) false but its converse is true;

c) true and its converse is true;

d) false and its converse is false.

4. Which of the following statements are true? Why?

a) To show that $ABCD$ is a square, it is sufficient to show that it is a rhombus.

b) To show that $ABCD$ is a rhombus, it is sufficient to show that it is a square.

c) To be a square it is necessary to be a rhombus.

d) A figure is a square only if it is a rhombus.

e) A figure is a square if and only if it is a rhombus.

f) For a figure to be a square it is necessary and sufficient that it be an equilateral parallelogram.

5. State the converse for each of the following.

a) If $AB \parallel CD$, then $ABCD$ is a parallelogram.

b) If a computer program is correct, it will terminate.

c) If it's Tuesday, this must be Belgium.

d) When it's hot, flowers wilt.

e) To snow it must be cold.

6. State the contrapositive for each sentence in [5].

7. The *inverse* of $A \Longrightarrow B$ is the statement $\neg A \Longrightarrow \neg B$.

a) Give an example to show that a statement can be true while its inverse is false.

b) Give an example to show that a statement can be false while its inverse is true.

c) What statement related to $A \Longrightarrow B$ is its inverse equivalent to? Explain why.

8. Finish the proof in the text that $A \Longrightarrow B$ and $\neg B \Longrightarrow \neg A$ are equivalent.

9. Which of the following statements are equivalent? Why?

a) $x = 3$. **b)** $x^2 = 9$. **c)** $x^3 = 27$.
d) x is the smallest odd prime.
e) $|x| = 3$.

10. Define the relation R by $(m, n) \in R$, if m and n have the same remainder when divided by 3. Show that R is an equivalence relation.

11. Which of the following are true? Interpret the statements using the conventions of mathematical writing.

a) Some rational numbers are integers.

b) Some equilateral triangles are isosceles.

c) Some nonnegative numbers are nonpositive.

d) Some negative numbers have real square roots.

e) There exists an odd prime.

f) There exists a number which equals its square.

g) Either $\pi < 2$ or $\pi > 3$.

h) Not all primes are odd.

12. Is the following a valid proof that "an integer is divisible by 9 if the sum of its digits is divisible by nine"?

> Proof: Consider such a number n in the form $10A + B$. Since $10A+B = 9A+(A+B)$ and since $9\,|\,(A+B)$ by hypothesis, then $9\,|\,n$.

13. Which pairs from the following statements are negations of each other?

a) All mammals are animals.

b) No mammals are animals.

c) Some mammals are animals.

d) Some mammals are nonanimals.

e) All mammals are nonanimals.

f) No mammals are nonanimals.

14. In some of the following sentences, "every" can be replaced by "any" without changing the meaning. In others, the meaning changes. In still others the replacement can't be made at all—the resulting sentence isn't good English. For each sentence, decide which case applies. (When it comes to what is good English, there may be disagreements!)

a) He knows everything.

b) Everybody who knows everything is to be admired.

c) Everybody agreed with each other.

d) Not everybody was there.

e) Everybody talked at the same time.

f) Every square is a parallelogram.

15. There is always a way to negate correctly a sentence using *not*. What is it? (*Hint*: Use "not" as part of a phrase at the beginning of the sentence.)

16. Which of the following are vacuously true; which not? Explain. When the answer is no, is the statement trivial or obvious or both or neither? Explain.

a) For all k such that

$$1 \le k \le 0, \quad 2k^3 + 2k^2 - 3k = 1.$$

b) For all k such that

$$1 < k \le 2, \quad k^9 - 3k^7 + 4k^5 - 30k^3 = 16.$$

c) Every prime which is a perfect square is greater than 100.

d) The case $n = 2$ of "The sum of the degrees of the interior angles of every planar n-gon is $180(n-2)$."

e) The same proposition as part d) when $n = 3$.

f) The same proposition as part d) when $n = 10$.

g) The case $n = 1$ of: If all $a_i > 0$, then

$$\left(\sum_{i=1}^{n} a_i\right)^2 \ge \sum_{i=1}^{n} a_i^2$$

h) The same proposition as g) when $n = 2$.

i) The same proposition as g) when $n = 10$.

Supplementary Problems: Chapter 0

1. Are the following statements true? If not, why not?

a) $[x = 2] \Longrightarrow [x^2 = 4]$.

b) $[x^2 = 4] \Longrightarrow [x = 2]$.

c) $[x = 2] \Longleftrightarrow [x^2 = 4]$.

d) $[2x+3 = 5] \Longleftrightarrow [x = 1]$.

e) $[6+\sqrt{x} = x] \Longrightarrow [x = 4 \text{ or } 9]$.

2. The **greatest common divisor**, $\gcd(m, n)$, of two nonzero integers is the largest positive integer which can be divided into both without leaving a remainder. In Chapter 1 we'll derive a method for computing the greatest common divisor of two integers. For this problem, however, just use any (brute force) method you wish. Find the gcd of the following pairs of integers.

a) 315 and 91 b) 440 and 924
c) 322 and 3795.

3. The **least common multiple**, $\operatorname{lcm}(m, n)$, of two nonzero integers is the smallest positive integer into which both m and n can be divided without leaving a remainder. Find the lcm of the following pairs of integers.

a) 6 and 70 b) 33 and 195
c) 35 and 143.

4. a) Find the greatest common divisor of each pair of integers in [3].

b) Using the results of part a) and those of [3], conjecture what the product of the gcd and lcm of two integers always is.

c) Try to prove your part b) conjecture.

5. Let $f(x) = x^2 - 3x + 2$.

a) Show that $f(-x) = x^2 + 3x + 2$.

b) Show that if c is a root of

$$x^2 - 3x + 2 = 0,$$

then $-c$ is a root of

$$x^2 + 3x + 2 = 0.$$

c) Generalize part a). If

$$f(x) = \sum_{k=0}^{n} a_k x^k,$$

what is $f(-x)$? (If you can't figure out how to write this with summation notation, write it with dots.)

d) Generalize part b). If $f(c) = 0$, what equation does $-c$ solve?

6. Let

$$f(x) = \sum_{k=0}^{n} a_k x^k,$$

where all the a_k are integers.

a) Show: If 0 is a root of $f(x)=0$, then $a_0=0$.

b) Show: If $f(c)=0$, where c is an integer, then $c \mid a_0$. (*Hint*: Subtract a_0 from both sides of the identity $f(c)=0$.)

c) Show: If $f(b/d)=0$, where b/d is in lowest terms, then $d \mid a_n$ and $b \mid a_0$.

d) What are the only *possible* rational roots of $2x^2-x+6=0$? That is, answer the question *without* solving the equation.

7. a) Verify that the zeros of $2x^2-x-6$ are the reciprocals of the zeros of $-6x^2-x+2$.

b) Show: If $x \neq 0$ is a root of

$$\sum_{k=0}^{n} a_k x^k = 0,$$

then $1/x$ is a root of

$$\sum_{k=0}^{n} a_{n-k} x^k = 0.$$

8. Descartes's **rule of signs** says that the number of positive zeros of the polynomial $P_n(x)$, as given by Eq. (4) of Section 0.4, is the number of variations in sign (from plus to minus and vice versa) in the sequence

$$a_0, a_1, \ldots, a_n$$

(with zeros ignored) or less than this by an even number.

a) State an analogous rule for the number of negative zeros.

b) Deduce a rule for the number of real zeros.

9. Apply Descartes's rule and your results in [8] to find an upper bound on the number of positive, negative, and real zeros of

a) $P_5(x)=x^5+3x^4-x^3-7x^2-16x-12$;

b) $P_{10}(x)=x^{10}+x^9+x^8+x^7+x^6+x^5+x^4+x^3+x^2+x+1$;

c) $P_6(x)=x^6-3x^4+2x^3-x+1$.

10. Find a general formula for

$$\sum_{S\subset[n]} |S|,$$

with $[n]=\{1,2,\ldots,n\}$. (*Hint*: You can write this as the double sum $\sum_{S\subset[n]}\sum_{i\in S} 1$.)

CHAPTER 1

Algorithms

■

■

1.1 Introduction

Mathematics requires both precision of thought and precision of language. To do mathematics by yourself, precise thought may, perhaps, be sufficient. But to communicate mathematics to others, precise language is essential. Our emphasis in Chapter 0 on the importance of good mathematical notation reflects the need for precise language. Notation, however, is to mathematics what an alphabet is to a natural language like English: It only provides the building blocks for constructing meaningful entities. In this chapter we'll focus on one of the most important entities used to communicate mathematical thoughts: the *algorithm.*

After an introductory example in this section, we'll develop the idea of an algorithm through a series of examples in Section 1.2. Then in Section 1.3 we'll describe much of our notation for expressing algorithms. In Section 1.4 we'll introduce the subject of recursive algorithms, and in Section 1.5 we'll describe how our notation can be used to express recursive algorithms as well as nonrecursive procedures and functions. Finally, in Section 1.6 we'll introduce the analysis of algorithms in which we consider algorithms as mathematical objects whose properties can be studied.

Let's begin with an example which at first glance will probably seem trivial:

Given a positive integer I, how do we find its successor (i.e., $I+1$)?

The obvious answer is: "Add 1 to I," which is certainly satisfactory for communication between human adults speaking English. But suppose you need to describe the process of finding the successor of an integer to a *robot* whose understanding of mathematics is limited to a knowledge of

- the digits from 0 to 9;
- how to add 1 to the digits 0, 1, ..., 8 (note the implication that our robot cannot handle two-digit numbers); and

- basic mathematical notation including the use of subscripts.

Now the answer to our question is considerably more difficult. Think about how you might answer it, given the constraints on the robot's intelligence, before you look at Figs. 1.1 and 1.2 on the next page. In these figures we present two prescriptions which could be given to our robot. Several things are worth noting about these two figures:

1. Both figures implement the idea that, starting from the least significant digit of I (i.e., the right-hand end), we work to the left replacing 9's by 0's until the first digit not equal to 9 appears; we then increase it by 1 and leave all subsequent digits (continuing on to the left) unchanged. If all the digits in I are 9, we change them all to 0 and insert a 1 at the left-hand end. Note the importance of handling this special case if we are to have a truly general solution. This feature is a necessary part of all algorithmic thinking: *All possibilities must be considered.*

2. Each of the two solutions assumes a certain linguistic ability on the part of our robot, which enables it to understand the notation in the flowchart and the "language"—i.e., **repeat**, **if**, **then**, **else**, etc.—in Fig. 1.2. Indeed, we hope that the notation and the terminology in both figures are fairly clear. In particular, in Fig. 1.1, $a \rightarrow b$ means that the value of expression a replaces the value of variable b. In Fig. 1.2, $b \leftarrow a$ also means that the value of expression a replaces the value of the variable b. We use arrows pointing in different directions to mean the same thing because right-pointing arrows are commonly used in flowcharts but left-pointing arrows are commonly used in notation such as that in Fig. 1.2. We devote Section 1.3 to a detailed explanation of the notation used in Fig. 1.2.

3. Although the procedures in both Figs. 1.1 and 1.2 can be applied to *any* positive integer I, our robot can't follow the instructions if I is too large. That is, if $I \geq 10^{10}-1$, the instructions themselves require adding 1 to 9, the very thing our robot can't do (and the thing that necessitated these instructions in the first place). This incomprehensible instruction occurs when the robot is told to move left from the 9th to the 10th digit ($k+1 \rightarrow k$ in Fig. 1.1 or $k \leftarrow k+1$ in Fig. 1.2).

4. The algorithms in Figs. 1.1 and 1.2 are not quite literal copies of each other, although both do the same task. Do you see that the questions in Fig. 1.1 are asked slightly differently than the tests after **until** in Fig. 1.2?

5. We don't know whether you find the approach in Fig. 1.1 more pleasing than that in Fig. 1.2 or vice versa. We hope it's the latter because, essentially, we'll be using the notation and language of Fig. 1.2 throughout the rest of this book. If you like Fig. 1.1 better, perhaps you can see that its notation could be quite unwieldy for large algorithms (i.e., algorithms much more complicated than this one). The advantage of the type of notation in Fig. 1.2 for large algorithms is not the only reason we prefer it. We believe that, once you get used to it, you'll find it easier to read and understand than

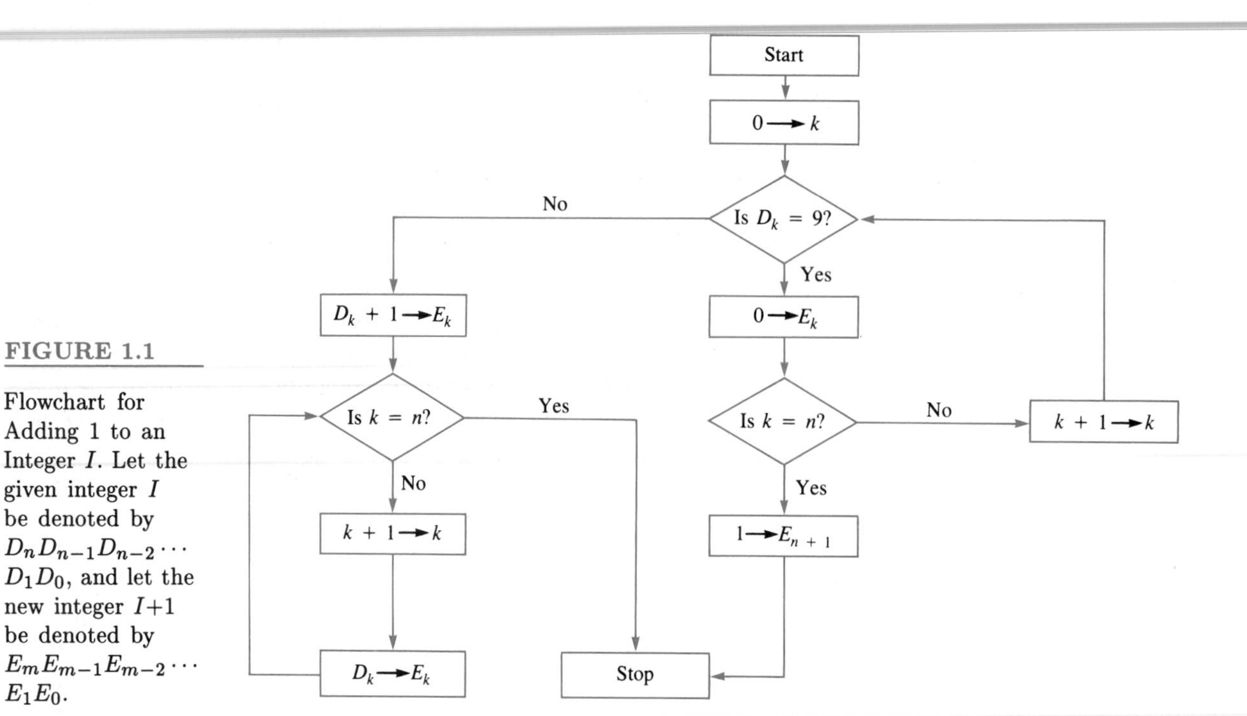

FIGURE 1.1

Flowchart for Adding 1 to an Integer I. Let the given integer I be denoted by $D_n D_{n-1} D_{n-2} \cdots D_1 D_0$, and let the new integer $I+1$ be denoted by $E_m E_{m-1} E_{m-2} \cdots E_1 E_0$.

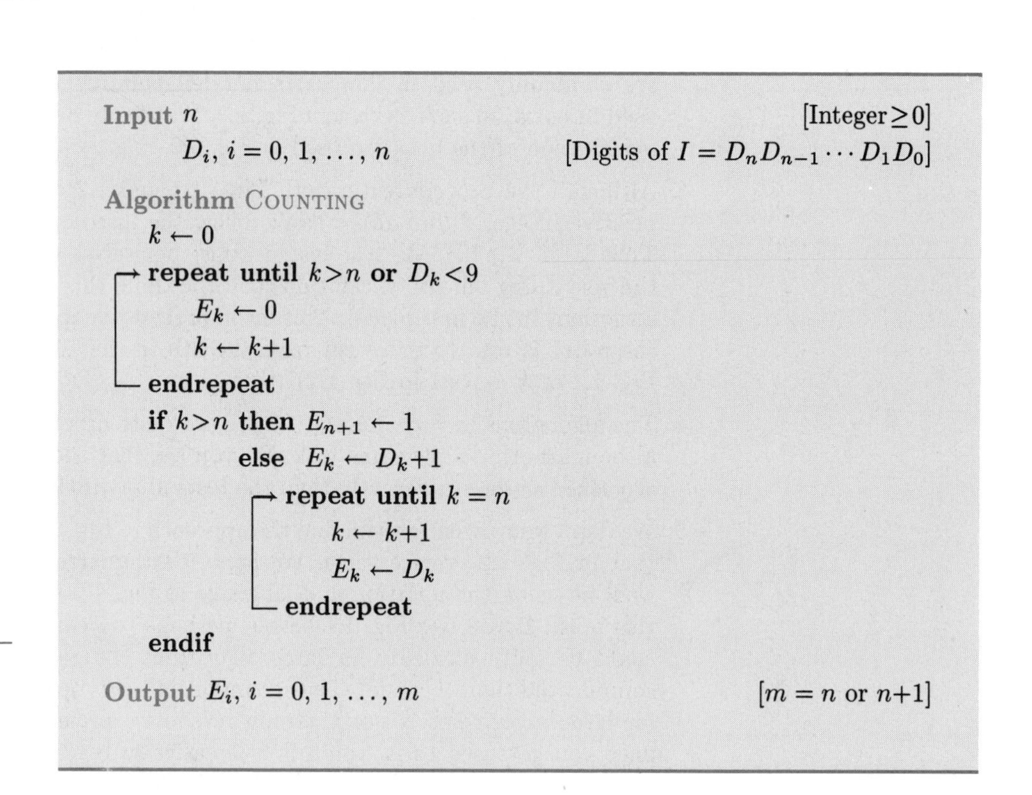

Input n [Integer ≥ 0]

$D_i,\ i = 0, 1, \ldots, n$ [Digits of $I = D_n D_{n-1} \cdots D_1 D_0$]

Algorithm COUNTING

$k \leftarrow 0$

repeat until $k > n$ **or** $D_k < 9$

  $E_k \leftarrow 0$

  $k \leftarrow k+1$

endrepeat

if $k > n$ **then** $E_{n+1} \leftarrow 1$

  else $E_k \leftarrow D_k + 1$

   repeat until $k = n$

    $k \leftarrow k+1$

    $E_k \leftarrow D_k$

   endrepeat

endif

Output $E_i,\ i = 0, 1, \ldots, m$ [$m = n$ or $n+1$]

FIGURE 1.2

Algorithmic Notation for Adding 1 to an Integer I

flowcharts. From our perspective, this type of notation enables us to reason about algorithms much more effectively than we could with flowcharts.

You probably found that it took a few minutes to understand the two figures because you were unfamiliar with the notations used. You also may have had some difficulty because we introduced a robot with limited capabilities. This made a problem, whose solution you knew immediately, more difficult by requiring that you break the solution down into a number of very simple parts. Why did we do this? Why make a simple problem more difficult on purpose? That certainly is not the aim of mathematics; quite the contrary. There are two answers:

1. Whereas ordinary English suffices to communicate simple mathematical thoughts ("add 1 to I"), more complex mathematical ideas are more easily understood by other humans when stated in a more formal language (like that of Fig. 1.2), which consists of a relatively small number of precise terms.
2. Our "robot," as you may have guessed, bears some resemblance to a computer. Since mathematical ideas must often be communicated to computers rather than to people, it is clearly desirable to use a language for mathematics similar to that required by computers.

With this introduction we may now proceed to the main topic of this chapter: algorithms and a notation in which to express them.

1.2 The Notion of an Algorithm

According to *Webster's Third International Dictionary*, the word *algorithm* refers to "the art of calculating with any species of notation." Not very useful. More helpful is the definition in the *Encyclopedia of Computer Science and Engineering*, which defines an algorithm as "the precise characterization of a method of solving a problem" and further notes that any algorithm must have the following properties:

1. Finiteness. Application of the algorithm to a particular set of data must result in a *finite* sequence of actions.
2. Unique initialization. The action which starts off the algorithm must be unique.
3. Unique succession. Each action in the algorithm must be followed by a unique successor action.
4. Solution. The algorithm must terminate with a solution to the problem, or it must indicate that for the given data, the problem is insoluble by the algorithm.

Usually, it is the last of these properties which is the most difficult to verify. Note that the algorithm in Fig. 1.2 has all four properties.

As its definition implies, an algorithm is nothing more—or less—than a recipe for carrying out a sequence of operations to solve a problem. The phrase in the definition which needs emphasis is "precise characterization." Our aim in this chapter is not only to give you a feel for what an algorithm is, but also to develop a language which will enable you to achieve precise characterizations.

The *word* algorithm itself derives from the ninth century mathematician al-Khorezmi (i.e., native of Khorezm).† However, the *notion* of an algorithm goes back much further than Al-Khorezmi, at least to Euclid who lived about 300 B.C. and whose algorithm for finding the greatest common divisor of two integers is presented in Example 5 later in this section. Still, the use of algorithms and precise methods for expressing and studying them did not become commonplace in mathematics until the development of computers and computer science during the past three decades. Some computer scientists, indeed, define computer science to be the "study of algorithms." But the algorithmic idea is too important to be of interest only to computer scientists. In recent years mathematicians have come increasingly to see that algorithms provide an excellent vehicle for presenting many mathematical ideas. Hence our emphasis in this book on algorithms.

We begin by presenting a number of different algorithms to give you a feel for what they are. In these examples we use a notation like that in Fig. 1.2, again relying mainly on its transparent meaning. To provide some additional help, however, we summarize the main features of the notation in Table 1.1. In Section 1.3 we'll discuss this notation in detail. If you have any difficulty understanding the examples in this section, you may find them easier to understand if you come back to them again after reading Section 1.3.

EXAMPLE 1 **Multiplication**

The actions of this algorithm are as trivial as those in the example in Section 1.1, but this algorithm does provide another useful illustration of the ideas and the notation. Suppose you have to perform multiplication as a sequence of additions. How can you express these operations as an algorithm? One answer is Algorithm 1.1 MULT.

To show how Algorithm MULT works, suppose $x = 4$ and $y = 11$. Then the values of *prod* and u each time we enter the **repeat** ... **endrepeat** construct are:

prod	u
0	0
11	1
22	2
33	3
44	4

† His full name was Mukhammad ibn Musa, and he lived from about 780 to 850. He was born near the present site of Tashkent in the USSR, but he spent most of his life in Baghdad. Al-Khorezmi described some of the manipulations of arithmetic in what we would now call algorithmic terms.

TABLE 1.1
The Main Features of the Algorithmic Notation Used in This Book

Assignment

```
variable ← expression                  [Value of expression becomes
Example: D ← b²−4ac                          new value of variable]
```

Selection

```
if condition then statement1            [If the condition is true then
             else statement2   statement1 is executed and statement2
endif                                  is not; otherwise statement2 is
                                         executed and statement1 is not]
Example: if n is even then n ← n/2
                      else n ← (n+1)/2
         endif
```

Iteration

```
repeat until condition               [Until the condition becomes true,
    statements to                   the statements between repeat and
    be executed                     endrepeat are executed repeatedly]
endrepeat
Example: repeat until n = 0
             a ← a + n²
             n ← n − 1
         endrepeat

for C = Cstart to Cstop             [With C equal successively to Cstart,
    statements to                        Cstart + 1, Cstart + 2, ..., Cstop
    be executed                    the statements between for and endfor
endfor                                           are executed repeatedly]
Example: for i = 1 to 15
             a_i ← b_i + b_{i+1}
             c_i ← a_i²
         endfor
```

Do you see why we required x to be nonnegative but allowed y to be any integer? We leave the answer to [2b]. ∎

MULT illustrates one other property of any good algorithm, namely, that it solves not just a single problem (say, 7×17) but a whole class of problems (the product of x and y for any integer y and any nonnegative integer x). Although we certainly could write an algorithm just to multiply 7×17 as a sequence of additions, you'll surely agree that it wouldn't be of any interest at all.

The **repeat** ... **endrepeat** *loop* in MULT is a construct you'll meet often in algorithms, since repetitive calculations of this type are a feature of most algorithms.

Algorithm 1.1

Mult

```
Input  x                                      [Integer ≥ 0]
       y                                      [Integer]
Algorithm MULT
    prod ← 0; u ← 0;
  → repeat until u = x                        [When u = x, skip rest of loop]
        prod ← prod + y
        u ← u + 1
    endrepeat
Output prod                                   [= xy]
```

An aspect of all such loops is that there is some relation which is true when the loop is first entered, as well as on each subsequent entry into the loop and at the final exit from the loop. This relation, which is really just a *predicate*, as defined in Section 0.5, is called the **loop invariant**, and plays an important role in analyzing the behavior of many algorithms. In this algorithm the loop invariant is

$$prod \ = \ uy.$$

This means that, while the values of *prod* and u change during the execution of MULT, each time **repeat** is encountered the value of u multiplied by y gives the value of *prod* (as shown in the list). And, when the end is finally reached after the final execution of the loop, *prod* still equals uy. If you think about it, you'll see that this is just the equation you want satisfied each time the loop is entered. Since $u = x$ when you finally exit from the loop, the loop invariant is such that, at the end of the loop, $prod = xy$, which is just what you want. Often the most effective approach to designing an algorithm is to construct a loop which has the proper loop invariant. Thus the loop invariant concept is very important in algorithmics (i.e., the study of algorithms).

Algorithms need not be used just to solve mathematical problems. Example 2 illustrates the application of algorithms to everyday problems.

EXAMPLE 2 **The Washington Metro Algorithm**

When you ride the Metro (the subway) in Washington, D.C. (see Fig. 1.3), you don't pay by putting money or a token in a turnstile or by buying a ticket. Instead you purchase a *farecard* from a machine for an amount which may be greater than the fare you need to pay. The farecard is debited when you finish your trip and, if its value hasn't been reduced to zero by the trip just completed, you retain it and use it again the next time you travel. That's really all you need to know about the Washington Metro in order to understand Algorithm METRO:

Now don't laugh. We're not suggesting that anyone who rides the Washington Metro does—or should—explicitly follow this algorithm. Nevertheless, algorithms like this one do have an important role, namely, in the *design* of large systems like the Washington Metro. At some point, before any construction was begun on the

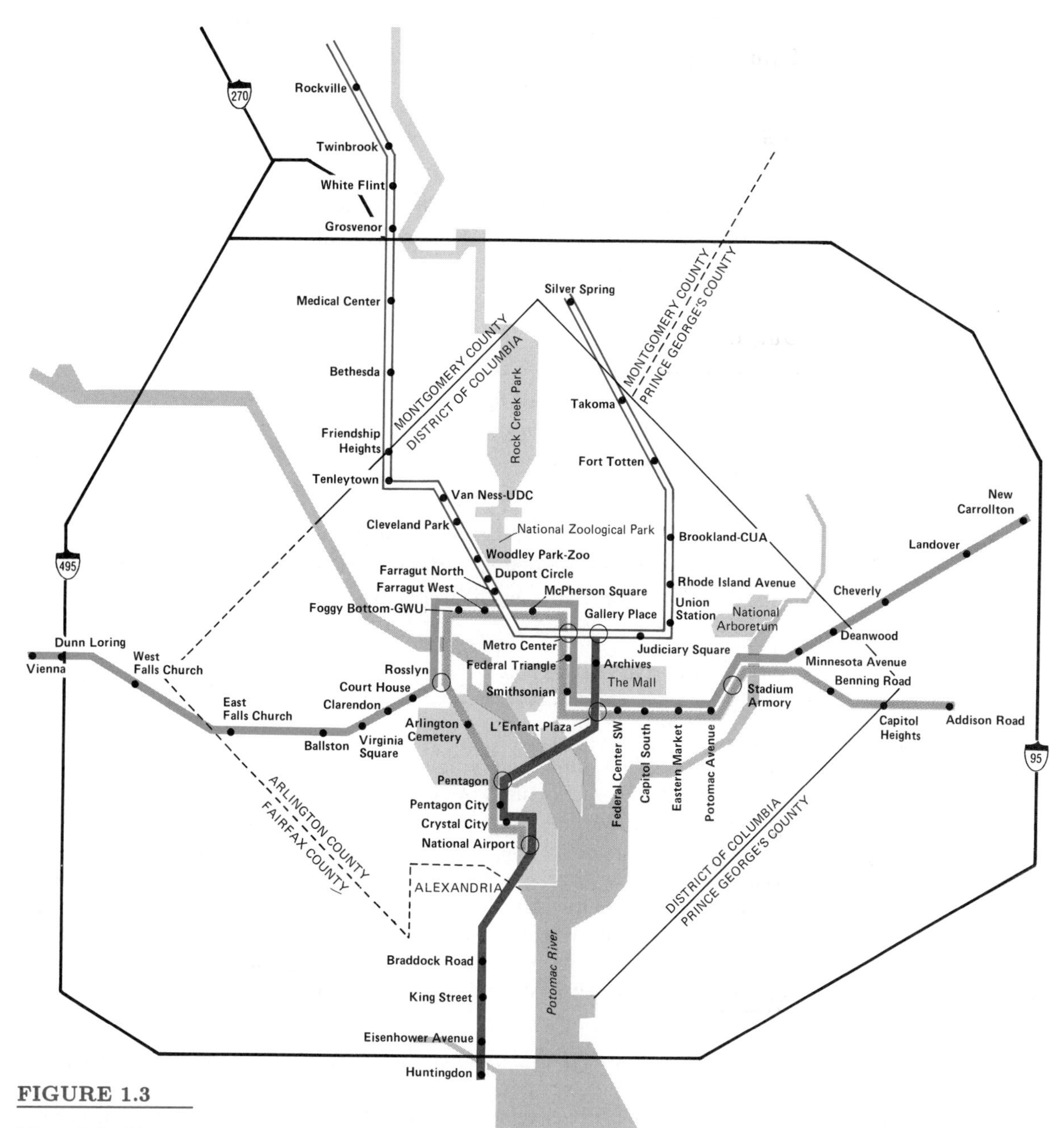

FIGURE 1.3

Map of the Washington Metro system.

Algorithm 1.2
Metro

```
Input Rider at start of trip
Algorithm METRO
    if rush hour then read rush-hour fare to destination from table of fares
                 else  read regular fare to destination from table of fares
    if have no farecard then go to farecard machine and buy farecard
    pass through entrance gate (by inserting farecard in appropriate slot
      and then retrieving it)
  → repeat until reach destination
        if there is a choice of trains (i.e., different lines)
                 then check map and board first correct train
                 else  board first train
        if need to change trains then exit at appropriate station
    endrepeat
    exit train at destination station
    if farecard has insufficient value for trip
        then go to addfare machine and increase value of farecard enough
          (or more than enough) to pay for trip
    pass through exit gate
Output Rider at destination
```

Metro, someone had to think in very much the same way as in Algorithm METRO about how the Metro would be used by riders in order to convince the planners that all possible eventualities had been considered. Indeed, we simplified the problem by ignoring aspects of riding the Metro that the planners had to consider. (For example: How does the rider get information on the stop nearest to where he or she wants to go? If the rider catches the last train at night, how does he or she make sure there will always be a connecting train available?) The point is that algorithmic thinking can be very valuable in planning large projects, because it helps planners organize their thinking and avoid errors caused by not considering all possibilities. ∎

One noteworthy aspect of METRO compared to MULT is the preponderance of **if** statements but only one **repeat** ... **endrepeat** structure. Although there are exceptions, algorithms which refer to actual life situations—filling out an income tax form is another example—tend to involve lots of decisions (i.e., **if**s) and few loops, while (significant) problems to be solved on computers tend to involve a much higher ratio of loops to decisions.

EXAMPLE 3 **Prime Numbers**

A prime number is a positive integer greater than 1 which is divisible only by 1 and itself. Thus 2, 3, 5, 7, 11, ... are prime, but 4, 6, 8, 9, 10, 12, ... are not. Our objective in this example is to generate the first K prime numbers, where K is some given positive integer. We could do so by taking each integer i and trying to divide it by

$$2, 3, 4, \ldots, i-1.$$

But this approach is clearly wasteful; since all primes after 2 are odd, it's enough to test only odd integers and to try to divide them by

$$3, 5, 7, 9, 11, \ldots, i-2. \tag{1}$$

For all integers > 2, if i is not divisible by any integer in (1), then i is prime.

A key expression in the following algorithm is

$$\lfloor i/j \rfloor * j, \tag{2}$$

where j is an integer, $*$ represents multiplication and the floor function is as defined in Section 0.4. This expression is equal to i if and only if i is divisible by j (i.e., i leaves no remainder when divided by j). If you don't see this, go back and review the meaning of the floor function.

Algorithm 1.3
PrimeNum

Input K [Number of primes]

Algorithm PRIMENUM

```
p_1 ← 2                                  [First prime]
k ← 2                                    [Counter for primes]
i ← 3                                    [First integer to test]
repeat until k > K
    j ← 3                                [First test divisor]
    repeat until j = i
        if ⌊i/j⌋ * j = i then exit       [Exit if i has divisor]
        j ← j + 2                        [Next test divisor]
    endrepeat
    if j = i then p_k ← i; k ← k + 1     [Prime found]
    i ← i + 2                            [Next test integer]
endrepeat
```

Output p_k, $k = 1, \ldots, K$ [First K primes]

Table 1.2 contains the results of using Algorithm 1.3 PRIMENUM when $K = 7$. A few remarks are in order:

1. When, for example, $k = 5$, i is initially 9 but since $j = 3$ divides 9 evenly, the **exit** is taken and i is increased to 11 (k isn't increased because j isn't

TABLE 1.2
Results of Algorithm PrimeNum when $K = 7$.

k	i	j	p_k	k	i	j	p_k
2	3	3	3			7	
3	5	3				9	
		5	5			11	
4	7	3				13	13
		5		7	15	3	
		7	7		17	3	
5	9	3				5	
	11	3				7	
		5				9	
		7				11	
		9				13	
		11	11			15	
6	13	3				17	17
		5					

Note: The values of i and j are those taken on for each value of k inside the outer loop.

equal to i). Then successive values of j up to $j = 9$ do not divide i. Finally, j is set equal to 11 and the **exit** is taken, this time with $i = j$.

2. The statement **exit** has the effect of passing control of the process to a point directly *after* the innermost **repeat** ... **endrepeat** loop in which the **exit** is contained (i.e., to the line **if** $j = i \ldots$).
3. We could have written the first line of the inner loop as

   repeat until $j = i$ **or** $\lfloor i/j \rfloor * j = i$

   and thereby avoided the next statement with the **exit** entirely. We didn't do it this way because the essentially different nature of the two criteria (i.e., whether j equals i and whether j divides i) argues, for reasons of readability, against including them in the same test.
4. This algorithm is correct; that is, it does the job of finding the first K primes. However, it does the task very inefficiently because it performs many more computations than are necessary. Can you see ways to improve it? ([31, Section 1.3]) ∎

EXAMPLE 4

Building a Telephone Network

Let Thirdworld be an undeveloped country with no telephone system. Suppose it wants to develop one and decides to start by connecting all its major cities at the minimum possible cost. To build such a network, you don't have to have direct lines between every pair of cities. All you need is a **path** between any two cities (which may pass through some intermediate city or cities). We'll assume that the only cost we need to consider is the construction cost of the telephone poles and wires between cities. (Each city will need a switching network independently of how we connect the cities with wires.) Suppose also that the construction cost for each possible link between two cities has been estimated. Then, given the costs shown in Fig. 1.4(a), what is the most economical way to build the network? That is, what pairs of cities should be linked directly?

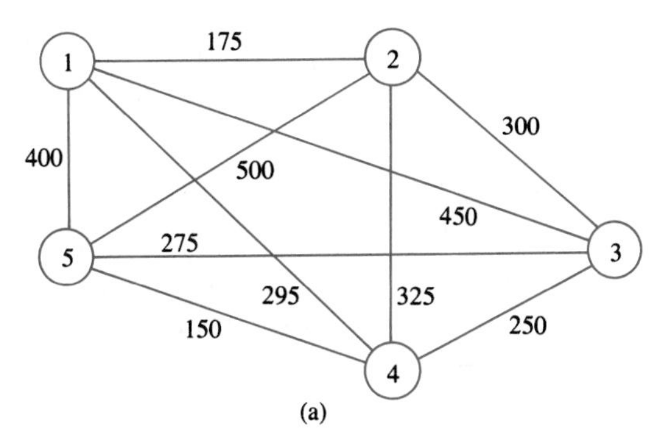

Edge	Cost (= Distance)	Tree
(4, 5)	150	5 – 4
(3, 4)	250	5 – 4 – 3
(1, 4)	295	1 – 4; 5 – 4 – 3
(1, 2)	175	1 – 2; 1 – 4; 5 – 4 – 3

(b)

FIGURE 1.4

A network algorithm, showing a) a map of the cities in Thirdworld and the distances between them and b) how the algorithm works for this map.

Our algorithm is based on the almost simple-minded idea that we should start by joining those two cities whose link is the least costly of all. Then at each subsequent stage we should build the link which is least costly among all links from previously joined cities to cities not yet part of the network. Algorithms of this type, which are based on immediate gratification, so-to-speak, are called **greedy algorithms**. In this case we may state our idea in algorithmic form as in Algorithm 1.4 TELENET.

Algorithm 1.4

TeleNet

```
Input n                                          [Number of cities]
      c_ij, i,j = 1,2,...,n  [c_ij is the cost of the link from city i to city j]
Algorithm TELENET
    Find c_lm, the minimum cost for a link between any two cities; if there
      is a tie, choose arbitrarily any link of minimum cost
    S ← {l,m}                 [Initialize S, the set of cities in the network]
    T ← {{l,m}}                    [Initialize T, the set of pairs of cities
                                           joined together in the network]
    repeat until |S| = n        [That is, until all cities are in the network]
        Find k ∉ S and q ∈ S such that link {k,q} has minimum cost
          among all links joining a city in S to one not in S
        S ← S ∪ {k}                                           [Add k to S]
        T ← T ∪ {{k,q}}                                    [Add {k,q} to T]
    endrepeat
Output T          [The set of pairs of cities to be joined in the network]
```

Note our use of the set notation introduced in Section 0.1. The set T contains elements from the Cartesian product of the set $\{1,2,\ldots,n\}$ with itself, except that the order of the elements in each pair is irrelevant.

Let's see how Algorithm TELENET works for the "map" of Fig. 1.4(a). The linkages are illustrated in Fig. 1.4(b), which shows how each city is successively added to the network. It is worth emphasizing that, at each step, we do not add the cheapest link not already in the network but rather the cheapest link from a city not in the network to a city which is in the network. For instance, after we picked link (4,5), we did not pick (1,2), even though it is the cheapest link not yet used, because it does not include a city already in the network. Also, after we picked (3,4), we picked (1,4) instead of the cheaper (3,5) because both 3 and 5 are already in the network. The final network shown at the bottom of Fig. 1.4(b) is called, technically, *a minimum spanning tree* for the *graph* in Fig. 1.4(a). Don't worry about this terminology now. We'll come back to it in Section 3.8, where we'll discuss this algorithm again. ∎

We don't apologize for our use of English in Algorithm TELENET. When we haven't developed more formal tools or when it's just plain convenient to do so, we will not hesitate to use English in our algorithms. Our main objective is to make algorithms readable and understandable.

How convincing do you find TELENET? We hope you're convinced that it does result in *a* network which enables all cities to communicate with all others. But can such a simple-minded idea result in the *best* (i.e., minimum cost) network? It doesn't seem likely, does it? We'll return to this question in Section 3.8. We hope that you wouldn't expect greedy choices to be best all the time. Indeed, we all know cases where taking the path of least resistance gets us into trouble later on (e.g., see Example 6). A vital concern in the study of algorithms is analyzing them to see how close they come to being the best possible. More on this in Section 1.6.

EXAMPLE 5 **The Euclidean Algorithm**

The greatest common divisor (gcd) of two positive integers is the largest positive integer that divides both of them. For example, $\gcd(225, 945) = 45$. If either integer is zero, the gcd is defined to be the other one. Thus $\gcd(14, 0) = 14$ and $\gcd(0, 0) = 0$.

The Euclidean algorithm is a method for calculating the gcd. It is an approach based on the recursive paradigm introduced in the Prologue, because we will calculate $\gcd(m, n)$ by reducing it to a "previous" case, that is, a calculation of $\gcd(p, q)$ where (p, q) is "smaller" than (m, n) in the sense that

$$p \leq m \qquad \text{and} \qquad q \leq n,$$

with the inequality being strict ($<$ instead of $\leq$) in at least one case.

To discover how to reduce the calculation of $\gcd(m, n)$ to the calculation of the gcd of a smaller pair, we only need to note that any common factor of m and n is also a factor of $m - n$ (assume for now that $m \geq n$). For example, $\gcd(945, 225) = \gcd(945-225, 225) = \gcd(720, 225)$. Also, any common factor of n and $m - n$ is a factor of m, since $m = n+(m-n)$. Thus (m, n) and $(m-n, n)$ have the *same* common factors. In particular, they have the same largest factors, that is, the same gcd. Therefore $\gcd(m, n) = \gcd(m-n, n)$.

But, as long as we are going to subtract one factor of n from m, why not subtract as many factors of n as we can, so long as the difference remains nonnegative? To find out how many factors of n we can subtract from m, we divide m by n, to get a quotient q_1 and a remainder r_1. Throughout this book we will observe the convention that *the quotient of two integers is always the integer part of what results from the division.* That is,

$$q = \lfloor m/n \rfloor.$$

Thus, for example, the quotient of $5/2$ is 2 and the remainder is 1. With this convention we may then write

$$m = nq_1 + r_1, \qquad q_1 \geq 0; \quad 0 \leq r_1 < n. \tag{3}$$

It should be clear why the quotient is nonnegative, but are the bounds on the remainder equally clear? As an example of Eq. (3), if $m = 36$ and $n = 14$, then $q_1 = 2$ and $r_1 = 8$. The quotient q_1 in Eq. (3) is just the number of times we can subtract n from m with a nonnegative result, which is the remainder r_1. Thus by the argument above, we obtain the equation which is the essence of the Euclidean algorithm:

$$\gcd(m, n) = \gcd(n, r_1). \tag{4}$$

You should be able to see directly from Eq. (3) that any factor of m and n is also a factor of n and r_1 and vice versa.

Now we come to the key step of the algorithm. Don't be confused by the notation. Since m, n, and r_1 are just a way of writing three distinct integers, suppose we write Eq. (3) again with n and r_1 playing the roles of m and n, respectively, and with a new integer r_2 playing the role of r_1. We get

$$n = r_1 q_2 + r_2, \qquad q_2 \geq 0; \quad 0 \leq r_2 < r_1. \tag{5}$$

Then, just as we argued about m, n, and r_1 to get Eq. (4), we can argue about n, r_1, and r_2 to write

$$\gcd(n, r_1) = \gcd(r_1, r_2). \tag{6}$$

What we have done twice we can continue to do. At each stage we let the divisor at one stage (e.g., n) become the dividend at the next stage, while the remainder in the present stage (e.g., r_1) becomes the divisor at the subsequent stage. Doing this we obtain

$$\begin{aligned}
m &= nq_1 + r_1, & 0 &\leq r_1 < n;\\
n &= r_1 q_2 + r_2, & 0 &\leq r_2 < r_1;\\
r_1 &= r_2 q_3 + r_3, & 0 &\leq r_3 < r_2;\\
r_2 &= r_3 q_4 + r_4, & 0 &\leq r_4 < r_3;\\
&\vdots\\
r_{k-1} &= r_k q_{k+1} + r_{k+1}, & 0 &\leq r_{k+1} < r_k;\\
r_k &= r_{k+1} q_{k+2},
\end{aligned} \tag{7}$$

with the final remainder $r_{k+2} = 0$. Can we be certain that sooner or later some remainder will be zero? Yes, because as Eq. (7) indicates,

i) the remainders form a *strictly decreasing* sequence (i.e., each remainder is strictly less than its predecessor), but
ii) all are nonnegative.

These two conditions are compatible only if, at some point, one of the r_j's becomes 0 (why?). The reasoning that led to Eqs. (4) and (6), when applied to Eq. (7) gives us

$$\begin{aligned}
\gcd(m, n) &= \gcd(n, r_1) = \gcd(r_1, r_2) = \gcd(r_2, r_3) = \cdots\\
&= \gcd(r_{k-1}, r_k) = \gcd(r_k, r_{k+1}).
\end{aligned} \tag{8}$$

Now observe that the last line in Eq. (7) implies that r_k is a multiple of r_{k+1}, so that

$$\gcd(r_k, r_{k+1}) = r_{k+1}.$$

(Why?) Thus from Eq. (8),

$$\gcd(m, n) = r_{k+1}.$$

To illustrate our algorithm, we let $m = 315$ and $n = 91$. Then, as in Eq. (7), we compute

Algorithm 1.5

Euclid

Input m, n	[Integers ≥ 0]
Algorithm EUCLID	
$num \leftarrow m$; $denom \leftarrow n$	[Initialize *num* and *denom*]
repeat until $denom = 0$	
$quot \leftarrow \lfloor num/denom \rfloor$	[Quotient]
$rem \leftarrow num - quot * denom$	[Remainder]
$num \leftarrow denom$; $denom \leftarrow rem$	[New *num* and *denom*]
endrepeat	
Output *num*	[$= \gcd(m, n)$]

$$\begin{aligned} 315 &= 91 \times 3 + 42 \\ 91 &= 42 \times 2 + 7 \\ 42 &= 7 \times 6 \end{aligned}$$

and therefore $\gcd(315, 91) = 7$.

Using Eq. (7) we can now state the Euclidean algorithm as Algorithm 1.5 EUCLID. This algorithm has some noteworthy features:

1. Instead of using subscripts as in Eq. (7), we have used *num* and *denom* to represent the dividend and divisor (i.e., the *num*erator and *denom*inator) at every step. Subscript notation is often useful in the statement of algorithms. However, when quantities play the same role each time through a loop and, most importantly, when the values at each stage do not have to be saved for the next stage, it's convenient and efficient to call the quantities by the same names each time through the loop.
2. Nevertheless, we do have to do some saving from one pass through the loop to the next. This is accomplished using the *temporary* variable *rem*. If, instead of $rem \leftarrow num - quot * denom$, we had written

$$denom \leftarrow num - quot * denom,$$

then the old *denom* (the one on the right-hand side, which becomes *num* in the next step) would have been *replaced* by the new *denom* (the one on the left-hand side), since the arrow, $\leftarrow$, indicates that the quantity on the right is to replace the quantity on the left. The result would be that *num* would be set equal to the new *denom* rather than the old *denom*. Note that we couldn't first do $num \leftarrow denom$ because then the old *num* would not be available for $num - quot * denom$.

Can you determine the loop invariant for Algorithm EUCLID? It is

$$\gcd(m, n) = \gcd(num, denom),$$

which is just the property of the gcd we used to motivate this algorithm in the first place. That is, the greatest common divisor of m and n is equal to the gcd of *num* and *denom* at each entry into the loop and when we finally exit from the loop. Since $denom = 0$ at the end and since $\gcd(x, 0) = x$, we have $num = \gcd(m, n)$ at the end just as we wish. When we use the recursive paradigm, the loop invariant is often just the relation used to motivate the algorithm in the first place.

You should make sure that not only do you understand how Algorithm EUCLID works, but also that it works when $m < n$ and when either m or n or both are zero. Although Algorithm EUCLID isn't complicated, its *analysis* (i.e., determining how efficiently it finds the gcd) isn't simple at all; we'll come back to it in Section 1.6. ∎

EXAMPLE 6 **A Scheduling Algorithm**

Scheduling is not only one of the most common of all human activities, but also one which normally requires step-by-step planning. It is therefore a natural candidate for an algorithmic approach. Our very first example in the Prologue was a scheduling algorithm: Minimize the total time spent on a multitask activity if you can work on an arbitrary number of tasks at the same time, but with the constraint that some tasks must precede others.

In this example we consider a conceptually simpler scheduling problem in which there are no precedence relations and in which the capacity for simultaneous work is strictly limited. This is the problem of scheduling tasks (i.e., programs to be executed) on computers in a multicomputer system. Assume that we know in advance the length of time required by each task. Each computer can process only one task at a time and we assume that, once started, a task is carried to completion by a computer before it starts another task. Our problem is to schedule the tasks on the computers so as to complete all the tasks in the shortest possible elapsed time from the start of the first task until the completion of the last one. We assume that each computer is identical, so that a task takes the same amount of time no matter which computer executes it. Although our assumptions idealize the problem considerably, what we wish to do is still quite similar to what actually needs to be done on *multiprocessor* computer systems.

How should we proceed? First, some notation. Let

$$T = \{t_1, t_2, \ldots, t_m\}$$

be the set of times required by the m tasks to be performed and let the number of computers be n. Only when m is substantially greater than n is this problem really interesting. (If $m \le n$ the solution is just to assign each of the m tasks to a different computer.) A plausible plan ("scheduling algorithm") is the following:

1. Begin by assigning the n longest tasks to the n computers (for convenience, COMPSCHED assumes $t_i \ge t_{i+1}$.).
2. Whenever any task is completed on any computer, assign to that computer the task with the longest required time among those not yet assigned to any computer.

This is another example, like that in Example 4, of a *greedy algorithm.*

Let's see how this algorithm works on an example. Suppose that $n = 3$, $m = 13$, and

$$T = \{25, 23, 22, 22, 20, 20, 18, 18, 14, 14, 13, 10, 10\}.$$

(The times could be seconds or minutes or even hours; it doesn't matter so long as the units are all the same.) Figure 1.5(a) illustrates how the algorithm would work on this data. If we define b_i as the time that processor i is busy, then $b_3 = 82$ is the largest and indicates how long it takes to process all the tasks.

Algorithm 1.6 CompSched describes the process in our algorithmic notation. This algorithm has a structure common to many algorithms. It consists of an **initialization phase** in which variables and sets are given initial values. Following this is a **processing phase** in which the desired results are computed. In Algorithm

a) How the method works for $T = \{25, 23, 22, 22, 20, 20, 18, 18, 14, 14, 13, 10, 10\}$.

Processor 1: 25 20 14 14	Total time: 73
Processor 2: 23 20 18 13	Total time: 74
Processor 3: 22 22 18 10 10	Total time: 82

b) How the algorithm calculates the solution.

Processor 1		Processor 2		Processor 3	
Task	Elapsed time	Task	Elapsed time	Task	Elapsed time
1	25				
		2	23		
				3	22
				4	44
		5	43		
6	45				
		7	61		
				8	62
9	59				
10	73				
		11	74		
				12	72
				13	82

c) A better solution.

Processor 1: 25 22 18 10	Total time: 75
Processor 2: 23 22 18 13	Total time: 76
Processor 3: 20 20 14 14 10	Total time: 78

FIGURE 1.5

A Scheduling Algorithm

Algorithm 1.6
CompSched

Input n [Number of computers]
m [Number of tasks; $m > n$]
$t_i, i = 1, \ldots, m$ [Times of the tasks; assume that $t_i \geq t_{i+1}$]

Algorithm COMPSCHED

```
for i = 1 to n
    b_i ← t_i          [Assign task i to computer i and set busy time b_i to t_i]
    S_i ← {i}          [S_i is set of tasks done on computer i]
endfor
for i = n+1 to m
    Find the minimum busy time b_j        [If tie, choose arbitrarily]
    b_j ← b_j + t_i    [Assign task i to computer j and
                        compute new busy time]
    S_j ← S_j ∪ {i}    [Update set of tasks for computer j]
endfor
t ← Maximum busy time
```

Output t [Total elapsed time]
$S_i, i = 1, \ldots, n$ [Set of tasks done on computer i]

COMPSCHED we have again interspersed formal notation with English to make the algorithm readable—but not sloppy. Figure 1.5(b) illustrates the operation of this algorithm for the given data.

How good is our solution? Is it optimal? The answer to the second question is No, as illustrated by Fig. 1.5(c). But Algorithm COMPSCHED (sometimes called the LPT algorithm for "least processing time"), even though it isn't optimal, does tend to give quite good results. To get a better solution we could use brute force (i.e., by trying all possible schedules to find the best one). For our simple example here, brute force might seem a reasonable and not too time-consuming approach. But as the number of tasks gets large, a brute force approach would soon result in a prohibitively large amount of calculation—sooner or later, larger than the calculation required by the tasks being scheduled! Finding the best solution (i.e., the one in which all tasks are completed in the minimum possible elapsed time) is a *very* hard problem. It is so hard, in fact, that there is no known algorithm which does this efficiently when the number of tasks is large. Indeed, probably no efficient algorithm exists, although this has not yet been proven. ∎

As you read our descriptions of these six algorithms, did you ask yourself: "Is each really an algorithm? Does each satisfy the four properties listed at the beginning of this section?" The answers to those questions are as follows:

- All satisfy the second and third properties—unique initialization and unique succession—because our notation enforces both.

- All six algorithms are certainly finite because the sequence of steps inevitably leads to the end of the algorithm.
- The most difficult property of an algorithm to verify, as we noted earlier, is whether it really leads to a solution of the problem. Later in this book we will return to some of these algorithms and verify in a fairly formal sense that they do in fact lead to solutions. For the moment, we hope that our informal discussions of these algorithms have convinced you that they do indeed do the jobs for which they are intended.

Algorithms play a very important role in problem solving. But you shouldn't get the impression that using an algorithm is the right approach to solving *all* problems. Let's consider briefly two questions:

1. Is there an algorithm that will solve any problem if only we can state the problem in some unambiguous fashion?
2. If we know that an algorithm exists to solve a particular problem, can we necessarily find it?

Perhaps it will surprise you to learn that the answer to the first question is No. There are, in fact, problems for which it can be proved mathematically that no possible algorithm could solve them. The answer to the second question is also No. For example, we know that an optimal strategy (i.e., an algorithm) exists in chess for both players. That is, no other strategy followed by one player could lead to a better result if the other player follows an optimal strategy. (Interestingly, however, we don't know whether the result of using these optimal strategies would be a win for white, a win for black, or a draw.) But to find this algorithm, even using a computer which could evaluate 10^6 (= one million) board positions per second, would take on the order of 10^{120} years—far longer than the life of the universe.

Instead of an algorithmic approach, good chess players use what is usually called a *heuristic* approach. Webster's describes a **heuristic**, usefully this time, as something which serves "to guide, discover or reveal". That is, a heuristic approach points in what seems to be a useful direction even if you can't *prove* that it leads to a solution. In effect, we used a heuristic approach in algorithms in Examples 4 and 6 since in both cases we argued, but didn't prove, that our approach was plausible. An obvious heuristic in chess is to have more pieces than your opponent does. A somewhat more sophisticated heuristic is to strive to control the center of the board with your pieces.

Finally, we mention briefly the relation between algorithms and computer programs. Without stretching the truth, it is fair to say that almost everything that computers do consists of implementing algorithms in computer programs. We sometimes say that the computer program is a *realization* of an algorithm. Thus, whereas ideas about algorithms—and many algorithms themselves—are useful in noncomputer contexts, they are crucial to understanding how to solve problems on computers.

Problems: Section 1.2

Note: There are also problems at the end of Section 1.3 which refer to the examples presented in this section, but which require a precise knowledge of the details of the algorithmic language discussed in Section 1.3.

1. Apply Algorithm MULT to the following data:
a) $x = 12$, $y = 4$ **b)** $x = 9$, $y = -5$
c) $x = -9$, $y = 3$

2. **a)** Describe in your own words why MULT is correct. That is, give an argument using ordinary English, but with mathematical notation as appropriate, which proves the output value *prod* is, indeed, xy.
b) How does your argument show that x must be nonnegative for the algorithm to work, but that there need be no such restriction on y?

3. Suppose that PRIMENUM has been executed up to the point where $p_7 = 17$ and i has been set to 19. Show how p_8 is calculated by indicating each value of i and j and each calculation in the outer loop.

4. Apply TELENET to the following data:
a) $n = 4$:

i \ j	1	2	3	4
1		210	590	620
2			330	450
3				520

b) $n = 6$:

i \ j	1	2	3	4	5	6
1		150	175	125	200	150
2			225	250	150	125
3				200	125	200
4					175	200
5						175

5. Apply the Euclidean algorithm to find the greatest common divisors of
a) 315 and 91 **b)** 440 and 924
c) 322 and 3795
(Compare with [2, Supplementary, Chapter 0].)

6. Prove that Algorithm EUCLID works when $m < n$, or when either m or n is zero.

7. Assume that $m = 15$ and the times for doing a set of tasks are

i:	1	2	3	4	5	6	7	8	9	10	11	12	13	14	15
t_i:	10	30	55	15	40	20	60	50	25	25	15	60	40	40	45

a) Apply COMPSCHED to this data when $n = 3$.
b) Apply COMPSCHED when $n = 5$.
c) For both parts a) and b), is your schedule optimum? If not, see if you can find the optimal schedule.

1.3 Algorithmic Language

In this section we discuss in detail the notation, or *language*, used to express algorithms. This discussion should clear up any confusion or uncertainty you felt in reading the algorithms in Section 1.2, as well as being useful in understanding the algorithms in subsequent chapters.

Before beginning our discussion of the algorithmic language, we stress again the importance of good notation for achieving precision in mathematical communication, in general, and in the statement of algorithms, in particular. At the same time we must note, alas, that, although good notation is necessary for precision, it is not sufficient. As in all mathematical reasoning, clear and accurate thinking must underlie the statement of any algorithm.

Our notation resembles—and has been strongly influenced by—programming languages like Pascal. However, no prior knowledge of any programming language is necessary to understand our algorithmic notation. Moreover, while our notation does have a relatively formal structure, our approach—and, we hope, yours, too—to using this notation is quite informal. We are most interested in precise thought in the expression of algorithms. We are less interested in such things as whether semicolons appear in the correct places because such details, while vital in writing computer programs, do not contribute much to understanding an algorithm.

All algorithms presented in this book have the following three parts.

Input: A listing of all quantities, values of which are assumed to be "available" (i.e., able to be used) at the start of the algorithm itself. Therefore the input is the *data* provided to the algorithm.

Algorithm: The sequence of steps which defines the *transformations* of (i.e., changes to) the input data which will provide the desired results.

Output: A listing of the results of the algorithm.

Each algorithm in this book has a name, which appears after the word *Algorithm*, to enable us to refer to it conveniently.

We first state two general precepts about our algorithmic notation:

1. Any mathematical notation in normal usage may be used in an algorithm. Obvious examples are subscripts, exponents, and the summation notation introduced in Section 0.5. In addition, we will use other notation as convenient. One example is

$$m \leftrightarrow n,$$

which denotes the *exchange* of two values (i.e., the new value of m is the old value of n, and the new value of n is the old value of m).

2. As in Section 1.2 when we want to emphasize conceptual understanding rather than mathematical precision, we will not hesitate to use English phrases and sentences in our algorithms.

Syntax and Semantics

The notation itself may be thought of as a *language* (albeit a generally more formal one than, say, English). In what follows we list the **syntactic structures** of our language—that is, the entities of the language to which we ascribe some meaning—and we discuss the **semantics** (i.e., what that meaning is) of each structure.

Assignment. An assignment has the form

$$\text{variable} \leftarrow \text{expression},$$

where the **expression** can be any mathematical expression (but see the second precept above) and the arrow implies that the value of the expression (evaluated using the current values of all the variables in it) becomes the new value of the variable (i.e., it *replaces* the old value). Examples:

$i \leftarrow 3;$ [Replace whatever value i has by 3]

$i \leftarrow i+1;$ [Increase i by 1]

$diff \leftarrow d_j - d_{j-1};$

$D \leftarrow b^2 - 4ac;$

$P(x) \leftarrow \sum_{i=0}^{n} a_i x^i.$ [Compute the value of a polynomial]

Selection. The selection, or *decision*, structure has the form

if B_1 **then** S_1
B_2 **then** S_2
⋮
B_n **then** S_n
[**else** S_{n+1}]
endif

where $n > 0$ and

- the B_i are mathematical entities called **logical expressions** whose values, as with the *predicates* introduced in Section 0.5, are *true* or *false*. Examples are:

$$\lfloor i/j \rfloor * j = i;$$
$$D \geq 0;$$
$$i = n.$$

- the S_i are any of the structures we are now defining or any sequence of these structures (see "compound structures" on p. 96); often the S_i will be one or more assignment structures.
- the [...] notation implies that what is contained within it is an *optional* part of the structure which need not be present.

The meaning of the selection structure is as follows:

- The B_i are evaluated one after another, using the current values of all variables which appear in them; as soon as a B_i is found whose value is true, the corresponding S_i is executed and calculation with the selection structure is terminated.
- If all B_i are false and the optional **else** S_{n+1} is present, S_{n+1} is executed and calculation with the selection structure is terminated.
- If all B_i are false and **else** S_{n+1} is not present, no action is performed.

The flowchart in Fig. 1.6 explains the selection structure.

Usually at most one of the B_i's is true, and there is no **else** clause, in which case the selection structure just explained has the same effect as

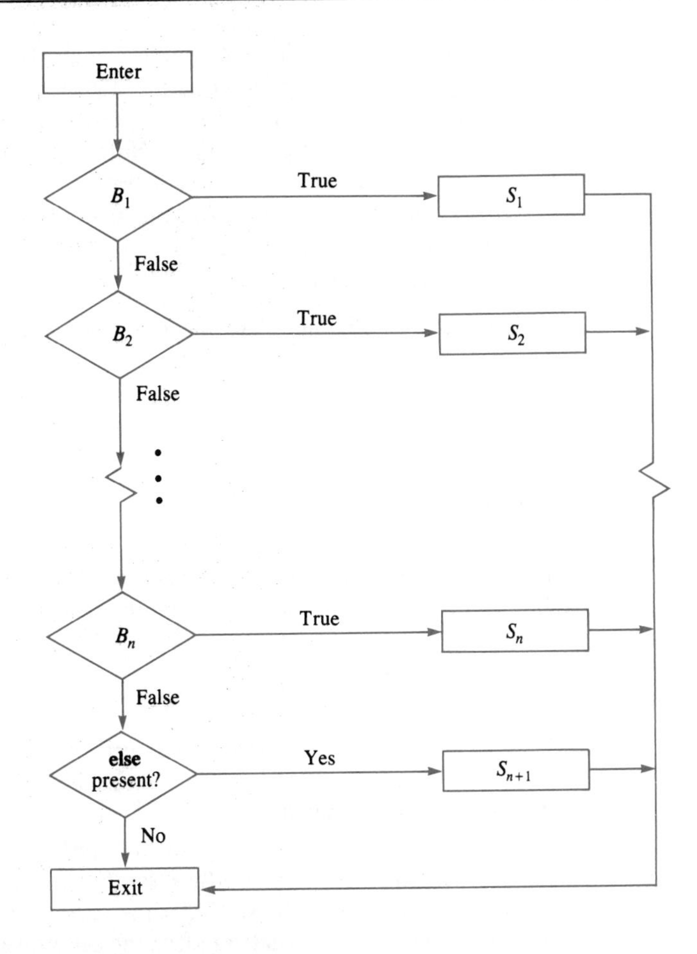

FIGURE 1.6

The Semantics of the Selection Structure

if B_1 **then** S_1 **endif**
if B_2 **then** S_2 **endif**

$\vdots$

if B_n **then** S_n
endif

But if more than one of the B_i's is true, or there is an **else** clause and B_n is false, then the two structures imply different computations. Do you see why?

We used selection structures in most of the algorithms in Section 1.2. Another example is

if $D \geq 0$ **then** $x_1 \leftarrow (-b+\sqrt{D})/2a$; $x_2 \leftarrow (-b-\sqrt{D})/2a$
$D < 0$ **then** $x_1 \leftarrow -b/2a + i\sqrt{-D}/2a$; $x_2 \leftarrow -b/2a - i\sqrt{-D}/2a$
endif

This structure means that if D (which would be the value of the discriminant if this were part of a quadratic equation algorithm) is nonnegative, then x_1 and x_2 are calculated using the formulas for two real zeros. However, if D is negative, the formulas for complex zeros are used. Note that no **else** structure is needed in this example because either $D \geq 0$ or $D < 0$ must be true.

We will occasionally omit the **endif** when no visual ambiguity results. Note that the structures S_i may themselves be **if** ... **endif** structures, which results in a **nesting** of selections.

Iteration. The simplest form of the iteration structure (also called a *loop* structure) is

repeat
    S
endrepeat

where S is any structure or sequence of structures as defined above or in what follows. S may also contain one or more structures of the form

if B **then exit** (1)

where B is a logical expression and **exit** means that the next statement to be executed if B is true is the one after the *innermost* iteration structure in which statement (1) is contained.

The semantics of **repeat** ... **endrepeat**, when S contains a structure of the form (1), are that S is repeatedly (*iteratively*) executed until at some point B is true when (1) is executed. At that time execution of the loop is terminated (i.e., execution jumps to the structure after the **endrepeat**). See, for instance, Examples 1, 3, 4, and 5 in Section 1.2. As another example, let I be a positive integer and consider

$j \leftarrow 1$
repeat
    if $j > I$ **then exit**
    $j \leftarrow 2j$
endrepeat

What is the value of j at the time of exit from this loop [21]?

Without an **if** ... **exit** structure or a **stop** (see p. 96), the **repeat** ... **endrepeat** structure would execute forever (a so-called *infinite loop*).

Since S may contain **repeat** ... **endrepeat** structures itself, they, like selection structures, may be nested. (Also, S in a loop structure may contain selection structures just as S in a selection structure may contain loop structures.) As with **endif**, we'll occasionally omit **endrepeat** if there's no ambiguity.

The combination

repeat
    if B **then exit** (2)

that is, where the **if** ... **exit** is the first statement in S, as in the preceding example, is very common. Thus we allow the following alternative to statement (2):

repeat until B

which means that S is repeatedly executed **until** B is true, at which point execution of the loop is terminated. The test of the truth of B occurs each time you come to the beginning of the loop, including when the loop is first entered. Thus the *body* of the loop—that is, the portion between **repeat** and **endrepeat**—will not be executed at all if B is initially true. We used this iteration structure in Examples 1, 3, 4, and 5 in Section 1.2. Using **repeat until**, we could write the preceding example as

$j \leftarrow 1$
repeat until $j > I$
    $j \leftarrow 2j$
endrepeat

Actually, we do more than "allow" the **until** alternative to statement (2); we positively encourage it and the other alternatives to **exit** (see below). Indeed, you should view skeptically the use of **exit** where none of our possible alternatives applies. More often than not it means that something has been written which will be hard for someone else to read, since it may require the reader to look (rather) far ahead to the end of the loop to see what is happening.

Sometimes, instead of continuing to execute a loop **until** some condition becomes true, it is convenient to continue to execute **while** a condition B is true. We could do this by writing something like

repeat
    if not B **then exit**
    ...
endrepeat

since **not** B will be true only when B is not true. Instead of this form, we provide the shorthand

repeat while B

...

endrepeat

One last variant of the **repeat** structure takes care of the case when it is convenient to end execution of the loop by a test at the end rather than at the beginning of the loop. This situation is addressed by the construct

repeat

...

endrepeat when B

which terminates execution of the loop **when** B is true at the end of the loop. This could also be implemented using the **if** ... **exit** structure (how?) but the preceding is a little more convenient.

Another common iterative situation occurs when we want the execution of a loop controlled by a **counter**. The syntax we use for this is

for $C = Cstart$ **to** $Cstop$

S

endfor

where the variable C is a counter which is assumed to take on only integer values, *Cstart* is the initial value of the counter and *Cstop* is its final value. The semantics are that S is executed repeatedly with C having the values

$$Cstart,\ Cstart + 1,\ Cstart + 2,\ \ldots,\ Cstop$$

The counter C may—and usually does—appear in the structure S. Also an **if** ... **exit** structure may be part of S, in which case the execution of the **for** ... **endfor** loop may terminate before C has taken on all possible values.

We could use the **for** ... **endfor** structure to express the summations and products discussed in Section 0.5 in algorithmic form. But since we allow free use of mathematical notation in our algorithms, we will seldom do this.

We used **for** ... **endfor** in Example 6 in Section 1.2. Another example is

Input $a_0, a_1, a_2, \ldots, a_n$ and x

Algorithm FRAGMENT

$P \leftarrow a_n$

for $k = 1$ **to** n

$P \leftarrow Px + a_{n-k}$

endfor

Can you determine what is computed by this loop [22a]? (See also Section 5.9.) And could you rewrite this loop to accomplish the same task using **repeat** ... **endrepeat** instead of **for** ... **endfor** [22b]? In fact, you can always use **repeat** ... **endrepeat** to accomplish the same thing as **for** ... **endfor** [22c].

In the preceding presentation of the **for** ... **endfor** loop, we assumed implicitly that

$$Cstop \geq Cstart$$

But sometimes it's convenient to count down (e.g., from n to 1) instead of counting up. We allow this with the syntax

for $C = Cstart$ **downto** $Cstop$
  S
endfor

We allow an additional form of **for** ... **endfor**, which we'll make use of in Chapter 3:

for $x \in T$
  S
endfor

where x is a variable and T is a set. The statement S is then executed once for each member x of the set T with no specification of the order in which the members are to be considered. We defer examples of this construct until Chapter 3.

In Section 1.2 we used vertical arrows on the left to illustrate the flow of control using **repeat** ... **endrepeat** and **for** ... **endfor**. Having explained the semantics, we'll usually omit these arrows from now on.

Output. Although the output quantities are always listed in the **Output** portion of the algorithm, in order to increase readability we allow output to be specified in the algorithm itself by

print list of expressions

where each expression itself is either a variable, a mathematical expression, or a string of characters (called a **literal**) inside quotes. In the first two cases, the value of the variable or the value of the expression (using current values of the variables in it) is output. In the last case, the output is the string of characters itself. Examples are:

print m, $n+1$
print 'No solution'

Termination. The physical end of an algorithm, as well as the normal place for completion of its execution is indicated by **Output** at the beginning of the output section. Sometimes it will be desirable to indicate termination of execution somewhere in the middle of an algorithm in case some unusual condition is satisfied. This is indicated by inserting

stop

Compound Structures. A compound structure is *any* sequence of the preceding structures. Each new structure may be started on a new line or a new structure may begin on the same line as its predecessor, but separated from it by a semicolon. We will use semicolons only with very brief assignment structures which conveniently fit on one line. The following is an example of a compound structure (assume that the variables which appear have been appropriately input or calculated previously):

$n \leftarrow 1$; $m \leftarrow 0$
repeat
    $a_n \leftarrow b_m^2 - c_m$
    if n even **then** $m \leftarrow 2n$
                else $m \leftarrow 2n+1$
    endif
    $n \leftarrow n+1$
endrepeat when $n > N$
print a_N

All the algorithms in Section 1.2 contain compound structures.

This completes for now our discussion of algorithmic language. Because our discussion has been informal and because we allow English where convenient, you may, when working on problems later in this book, be in some doubt as to just how formal or informal to be. Don't worry about this too much. The important thing is the accuracy of your thinking and not how you use our language. Our general advice is: Use the language without English where you can do so without too much tedium and where it doesn't obscure the main ideas of the algorithm.

It is also important that you write the notation in as readable a way as possible. You have probably noticed that we have done this by

- indenting under each main construct; thus inside a **repeat** ... **endrepeat** loop we indent all the statements in the loop; if there is another **repeat** ... **endrepeat** inside the outer one or a **for** ... **endfor** or an **if** ... **endif** construct, we indent the statements in this construct farther and so on;
- printing each word which closes a construct directly under the word which opens it and similarly making sure that corresponding **then**'s and **else**'s are printed under each other.

When you write algorithms, you should generally follow these conventions. But the rules for an algorithmic language need not be as rigidly obeyed as those for a programming language for computers. So long as clarity is maintained, the rules may be bent.

The one remaining thing we must discuss in this section is, strictly speaking, not part of notation itself, but is nevertheless a vital part of good algorithm presentation. Much as we might like to believe that algorithms stated using our language are transparently clear to all readers, we know better. (Sometimes they're not even clear to us!) Some algorithms presented in this book are inherently quite complex; for others, we undoubtedly have failed to find the most felicitous presentation. In any case, annotations or *comments* on the algorithms in ordinary English should enhance a reader's understanding of what is intended. Our mechanism for providing such comments, as in our previous examples, is to put them in square brackets and, to the extent that typography allows, flush right on the same line as the algorithm step being annotated.

The only way to become comfortable with the notation for algorithms introduced here is to become familiar with it and to use it yourself. You will be forced to do both as you proceed through this book. Since our language allows the occasional use of English phrases, our algorithms are not always as precise and unambiguous as those expressed in a computer language. But readers are not computers, and so we believe that the risk of ambiguity is small. The trade-off for this risk is a notation that is easier to learn, as well as being more readable and flexible, than a computer language. And our notation avoids unnecessary peripheral issues such as the details of input and output or the details of declaring variable types.

A final word of guidance about our language: It is worth stressing that an algorithm, like other forms of written communication, typically has one writer but many readers. The construction of an algorithm using this language should always be approached with the goal of creating something which will be meaningful to its readers.

Problems: Section 1.3

The following problems are divided into two sets. The first set should be considered warm-up drills. Like finger exercises on the piano, they are meant to get you in shape for more interesting things to come. The second set relates to the specific algorithms of Sections 1.1 and 1.2, or asks you to devise new algorithms for other applications.

Set 1

Write algorithms in the format described in this section, except that you may skip the input and output sections, since the problems in this set tell you what the input (if any) and output are to be. For ease in grading, name each algorithm after its problem number. For instance, your algorithm for [10] should be called ALG-10 (unless you are asked to use a different, more informative name).

1. Given an integer N as input, write an algorithm which prints N (just once) if it is even or positive. (*Suggestion*: In this and several later problems, use the mathematical symbol "$|$", which means "is a factor of" or just "divides" (see Section 0.5). For instance, "**if** $3 \mid n$ **then print** n" outputs the current value of n if it is a multiple of 3.)

2. Change the algorithm of [1] so that it prints N if N is even *and* positive.

3. Change the algorithm of [1] so that it prints 0 if N is not printed.

4. Change the algorithm of [2] so that it prints 0 if N is not printed.

5. Change the algorithm of [1] so that it prints N if N is even and also prints N (thus sometimes for the second time) if N is positive.

6. Write **if** ... **endif** structures to do the following tasks:

   a) Set a variable L equal to the larger of two numbers a and b.

   b) Set a variable L equal to the largest of three numbers a, b, and c.

   c) Set a variable P equal to the smallest of 2, 3, or 5 which divides (with no remainder) a variable a, or set P equal to 0 if none divides a.

7. Write a **repeat** ... **endrepeat** structure to find and output those integers from 1 to N which are divisible by both of two primes p_1 and p_2.

8. Write an algorithm which, given an integer $N > 1$, prints N itself a number of times equal to the largest power of N which divides one million. For example, 2^6 divides one million and 2^7 does not, so if $N = 2$, the algorithm should print 2 six times in a row.

In [9–15], let $L(i)$ represent the ith letter of the alphabet. Thus the line "**print** $L(3)$" causes the letter C to be printed.

9. Write an algorithm which prints ABCDEF ... XYZ.

   a) Use **for** ... **endfor**.

   b) Use **while** (and not **for** ... **endfor**).

   c) Use **until**.

10. Write an algorithm which prints ZYX ... CBA.

11. Write an algorithm which prints ABBCCCDDDD ... Z ... Z (26 Z's).

12. Write an algorithm which prints

    A/AB/ABC/ABCD/.../ABCDEFG ...XYZ/

    (*Note*: To print a specific symbol or string of symbols, use quotes. For instance, to print a slash, write **print** "/". To print "$L(1)$", write **print** "$L(1)$". This is different from **print** $L(1)$, which prints the value of $L(1)$, which is A.)

13. Write an algorithm which prints

$$\underbrace{AA\ldots AA}_{26\text{ times}}\underbrace{BB\ldots BB}_{25\text{ times}}\ldots XXXYYZ$$

14. Write an algorithm which prints the alphabet through the letter P, but without you determining in advance which letter of the alphabet P is. (*Hint*: You can use conditions like **if** $L(i) = P$.)

15. Write an algorithm which prints the alphabet in order, except that it skips the letter L.

16. Write an algorithm to sing "The Twelve Days of Christmas". In addition to the usual constructs of our language, you may add **sing** "X", where X is any string of English words. If you don't know this song or have forgotten it, you should be able to find it in any book of Christmas songs. (Actually, to keep from being tedious, it will be enough if your algorithm sings the first four verses.)

17. Write an algorithm to output all the divisors of a positive integer N. For instance, if $N = 12$, the algorithm should output 1, 2, 3, 4, 6, and 12.

18. Write an algorithm to output the odd positive integers less than 30.

    a) Use the **for** ... **endfor** construct.

    b) Do not use the **for** ... **endfor** construct.

19. Write an algorithm to print $\sum_{i=1}^{k} i$, for each k from 1 to 10, with and then without using the **for** ... **endfor** construct.

20. Given a list of numbers $x_1, x_2, \ldots, x_n$, write an algorithm which, for each number in the list in turn, prints that number if it is even and also prints it (thus sometimes for the second time) if it is positive.

21. What is the value of j when the loop in the algorithm fragment on page 92 is exited?

22. a) What is computed by the loop in Algorithm FRAGMENT on page 94?

    b) Rewrite that loop using **repeat** ... **endrepeat**.

    c) Show that a **for** ... **endfor** loop can always be replaced by a **repeat** ... **endrepeat** loop.

23. Suppose that the selection structure in algorithmic notation was *nondeterministic* in the sense that

the order in which the B_i are evaluated is chosen randomly (and assume that the **else** option, if it is present, is one of the random choices). Which of your solutions to [6] would still work? Which could be easily modified so that it would work? (Although you may find it surprising, such nondeterminism leads to some interesting and important results in the study of algorithms.)

Set 2

24. Revise the robot adder algorithm (Algorithm COUNTING, Section 1.1) so that it works in base 5 instead of base 10.

25. Assume that the robot adder (Section 1.1) also knows how to subtract 1 from any single digit from 1 to 9. Teach the robot how to subtract 1 from any positive integer.

26. Just as it is possible to do multiplication by repeated addition (Algorithm MULT, Section 1.2), it is possible to do division by repeated subtraction. Write such an algorithm. It should take as input positive integers m and n and output the integer quotient and remainder of m/n.

27. Going a step further (or rather, further back into the development of arithmetic), it is possible to do addition by repeated addition of 1. That is, imagine that you have finally taught the robot of Section 1.1 how to add 1 (to any positive integer whatsoever, not just those of 9 or fewer digits!). Now teach it how to add arbitrary positive integers. That is, write an algorithm which it can follow, which does this momentous task. Assume that the robot understands mathematical notation, as it did in Section 1.1.

28. The obvious way to compute x^m when m is a positive integer is to multiply x by itself $m-1$ times. However, if m is a power of 2, there is a more efficient way. First multiply x by itself to get x^2. Then multiply this by itself to get x^4, and so on.

a) Write up this idea in algorithmic language.

b) Compare the number of multiplications to compute the 2^n power of x with the obvious way.

c) Even if m is not a power of 2, the method of a) can be used to compute x to the highest power of 2 less than m; then the remaining powers can be computed the obvious way. Put all this into an algorithm.

d) The powers of 2 idea can be exploited further in computing x^m. How? Do it. (*Hint*: Think recursively.)

29. One part of Algorithm METRO (see Example 2, Section 1.2) was

"go to farecard machine and buy farecard"

An algorithm actually to do this, which would be a procedure (see Section 1.5) in a fully elaborated Algorithm METRO, is Algorithm 1.7 FARECARD.

Algorithm 1.7 FARECARD

```
Input w                          [Ticket value you
                             want in multiples of 5 cents]

Algorithm FARECARD
  d ← 0                          [money deposited so far]
  repeat while d < w
    deposit more money
  endrepeat
  if d > w then
    repeat until d = w
      hit decrease button        [Decreases
                                  d by 5 cents]
    endrepeat
  endif
  push finish button
  if change then take it
  take farecard
Output farecard and change, if any
```

Rewrite Algorithm FARECARD so that it does exactly the same thing, but use only one **repeat** ... **endrepeat** loop.

30. Buying a farecard for the Washington Metro is not the only occasion in modern life that a person must perform a clearly algorithmic procedure. Write an algorithm for the following situation at about the same level of precision as in Algorithm FARECARD in [29]. You must balance your checkbook. You start with the previous balance and work through a list of entries, each one of which is either a debit (a check) or a credit (a deposit or interest).

31. Algorithm PRIMENUM (see Example 3, Section 1.2) is not very efficient. Can you think of

any ways to improve it so that the primes would be calculated with less computation? Display your answer as an algorithm.

32. Here is another greedy approach to the telephone network problem of Example 4, Section 1.2. Instead of picking the cheapest edge between so-far visited and so-far unvisited vertices, as in TELENET, pick the cheapest edge *anywhere*, so long as it doesn't connect two vertices between which there is already a path of chosen edges. Call it Algorithm TELENET′.

a) Apply TELENET′ to the graph in Example 4.

b) Apply TELENET′ to the data in [4b], Section 1.2. What can you conjecture about the differences between TELENET and TELENET′?

33. Sometimes a person wants to find the network having the *maximum* cost. For instance, suppose that you own a construction company in Thirdworld (see Example 4, Section 1.2). Suppose further that you have computed for each possible telephone link between the two cities the difference between the price the government will pay and the price the construction will cost you. Then you would propose building those links for which the sum of the profits would be greatest. This naturally suggests two algorithms, TELENET-MAX and TELENET′-MAX, which are just like TELENET and TELENET′ (see [32]), except you go for the biggest available edge each time.

a) Apply TELENET-MAX to the data of [4b], Section 1.2.

b) Apply TELENET′-MAX to this data.

c) We will show in Chapter 3 that TELENET always finds the minimum weight spanning tree. Assuming this, can you argue that TELENET-MAX always finds the maximum weight spanning tree?

34. In the scheduling problem of Example 6, Section 1.2, perhaps it is better to finish the small jobs first. Run COMPSCHED again, with the same data as in Example 6, but this time input the times in increasing order. What do you conclude?

1.4 Recursive Algorithms

The recursive paradigm which we introduced in the Prologue is going to play a major role throughout this book. Many of our algorithms will be direct translations into our algorithmic language of recursive paradigm approaches to problems. In this section we'll discuss two such algorithms written in our algorithmic language. The algorithmic expression of the essence of the recursive paradigm—reducing to a previous case—requires a feature of the language not needed for the algorithms of Section 1.2. We will discuss this feature—**procedures**—informally here and in more detail in Section 1.5.

Warning: The material in this section and the next will seem quite difficult compared to that in Sections 1.2 and 1.3, particularly if you are not familiar with a modern programming language which allows recursion. What is important, however, is not that you grasp all the details now but rather that you get a feel for the idea of recursion and how it is expressed in our algorithmic language. That will be enough to enable you to understand later chapters.

Our first example is one that we discussed in Section 1.2: the Euclidean algorithm. Now, however, we discuss it in recursive terms in contrast to the iterative approach previously taken.

EXAMPLE 1 Derive a recursive formulation of the Euclidean algorithm and express it in algorithmic language.

Solution The essential formula in our previous derivation of the Euclidean algorithm was Eq. (4), Section 1.2:

$$\gcd(m,n) \;=\; \gcd(n,r_1), \tag{1}$$

where r_1 is the remainder when m is divided by n to get an integer quotient. That is,

$$m \;=\; nq_1 + r_1, \tag{2}$$

with $q_1 = \lfloor m/n \rfloor$. We may rewrite Eq. (2) as

$$r_1 \;=\; m - n\lfloor m/n \rfloor.$$

Substituting into Eq. (1), we get

$$\gcd(m,n) \;=\; \gcd(n, m{-}n\lfloor m/n \rfloor) \tag{3}$$

Equation (3) expresses the gcd of two integers m and n in terms of two others n and $m{-}n\lfloor m/n \rfloor$. In particular, comparing the two second arguments:

$$n \;>\; m - n\lfloor m/n \rfloor$$

because the right-hand side is the remainder when m is divided by n and is therefore less than n. So the second argument on the right-hand side of Eq. (3) is less than the corresponding argument on the left-hand side. (How about the two first arguments? [1]) Thus Eq. (3) embodies the essence of the recursive paradigm: reducing to a previous (i.e., smaller) case. Using the recursive paradigm we continue to reduce to the previous case, thereby getting successively smaller—but always nonnegative—second arguments, until finally we arrive at a case we know by definition. For the Euclidean algorithm the case we know by definition is that $\gcd(m,0) = m$ for any m.

Here's an example of how Eq. (3) works, for $m = 63$, $n = 27$:

$$\begin{aligned}
\gcd(63,27) &= \gcd(27, 63{-}27\lfloor 63/27 \rfloor) && \text{[Using Eq. (3)]}\\
&= \gcd(27,9) && [\text{Evaluating } 63 - 27\lfloor 63/27 \rfloor]\\
&= \gcd(9, 27{-}9\lfloor 27/9 \rfloor) && \text{[Using Eq. (3) again]}\\
&= \gcd(9,0) && [\text{Evaluating } 27 - 9\lfloor 27/9 \rfloor]\\
&= 9 && [\text{Since } \gcd(m,0) = m]
\end{aligned}$$

Now, how do we implement the calculation of the gcd via Eq. (3) in our algorithmic language? Algorithm EUCLID-REC supplies the answer but needs some explanation:

1. Focus first on the lines from **procedure** to **return**. These embody nothing more than an algorithmic representation of Eq. (3) with (m,n) replaced by (i,j). To compute the $\gcd(i,j)$ you first check to see if $j = 0$, in which case the answer is i. Otherwise, you *call* the procedure with two new arguments given by the right-hand side of Eq. (3). When somehow you've unraveled all these calls to the same procedure, you *return* the final value assigned to *answer*.

Algorithm 1.8
Euclid-Rec

```
Input m, n                                        [Integers ≥ 0]

Algorithm EUCLID-REC
    procedure gcd(i, j)
        if j = 0 then answer ← i
                 else gcd(j, i−j⌊i/j⌋)            [Recursive call]
        endif
        return answer
    endpro
    gcd(m, n)                                     [Main algorithm]

Output answer                                     [= gcd(m, n)]
```

2. The portion of the algorithm beginning with the line commented "Main algorithm" initiates the whole process by calling the procedure with the given values m and n.

This discussion hides lots of details, which we'll discuss in Section 1.5.

Figure 1.7 illustrates how Algorithm EUCLID-REC works for the same data as on p. 101. Make sure you understand that the procedure is called three times (once from the main algorithm and twice more by itself) and that each of these calls is completed (i.e., we get to **return**) in the reverse order in which the calls were made. Think of it this way: Each time Procedure gcd is invoked, it is as if a new copy of its definition is made, which uses the requested values for i and j. The old copy is interrupted but is still there to be returned to later at the place it was interrupted. Each such copy is enclosed in a box in Fig. 1.7.

As this example shows, a procedure is just a portion of an algorithm which performs some specific task when called upon by another portion of the algorithm. When a procedure calls itself, it is called **recursive** and the entire process is called **recursion**. But procedures do not have to be recursive (see Section 1.5). ∎

Now we're ready for our second example of a recursive algorithm. It is an example which, in a variety of forms, will recur many times in later chapters, starting in Chapter 2.

EXAMPLE 2 **The Towers of Hanoi**

The Towers of Hanoi puzzle was first posed by a French professor, E. Lucas, in 1883. Although commonly sold today as a children's toy, it is often discussed in discrete mathematics or computer science books because it provides a good example of recursion. In addition, its analysis is straightforward and it has many possible variations.

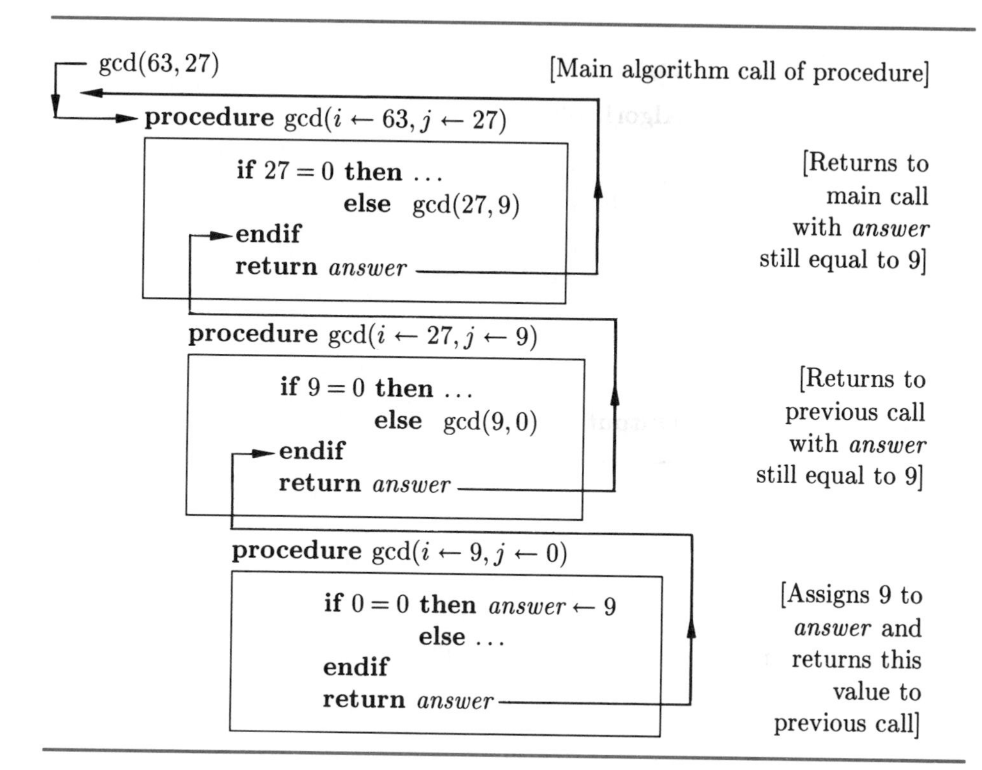

FIGURE 1.7

Operation of Algorithm EUCLID-REC for $m = 63$, $n = 27$

The object of the Towers of Hanoi (see Fig. 1.8) problem is to specify the steps required to move the disks (or, as we will sometimes call them, rings) from pole r ($r = 1$, 2, or 3) to pole s ($s = 1$, 2, or 3; $s \neq r$), observing the following rules:

i) Only one disk at a time may be moved.

ii) At no time may a larger disk be on top of a smaller one.

The most common form of the problem has $r = 1$ and $s = 3$ (Fig. 1.8).

Solution Our algorithm to solve this problem exemplifies the recursive paradigm. We imagine that we know a solution for $n-1$ disks ("reduce to a previous case"), and then we use this solution to solve the problem for n disks. Thus to move n disks from pole 1 to pole 3, we would:

1. Move $n-1$ disks (the imagined known solution) from pole 1 to pole 2. However we do this, the nth disk on pole 1 will never be in our way because any valid sequence of moves with only $n-1$ disks will still be valid if there is an nth (larger) disk always sitting at the bottom of pole 1 (why?).
2. Move disk n from pole 1 to pole 3.
3. Use the same method as in step 1 to move the $n-1$ disks now on pole 2 to pole 3.

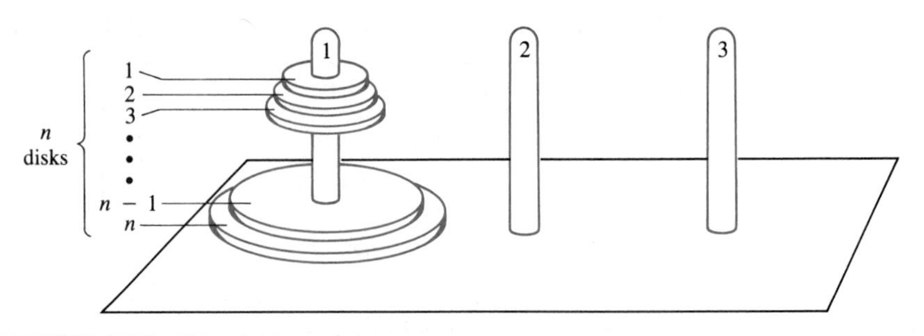

FIGURE 1.8

The Towers of Hanoi Problem. The problem is shown with the initial pole $r = 1$. In the most common form of the problem, the final pole would be $s = 3$.

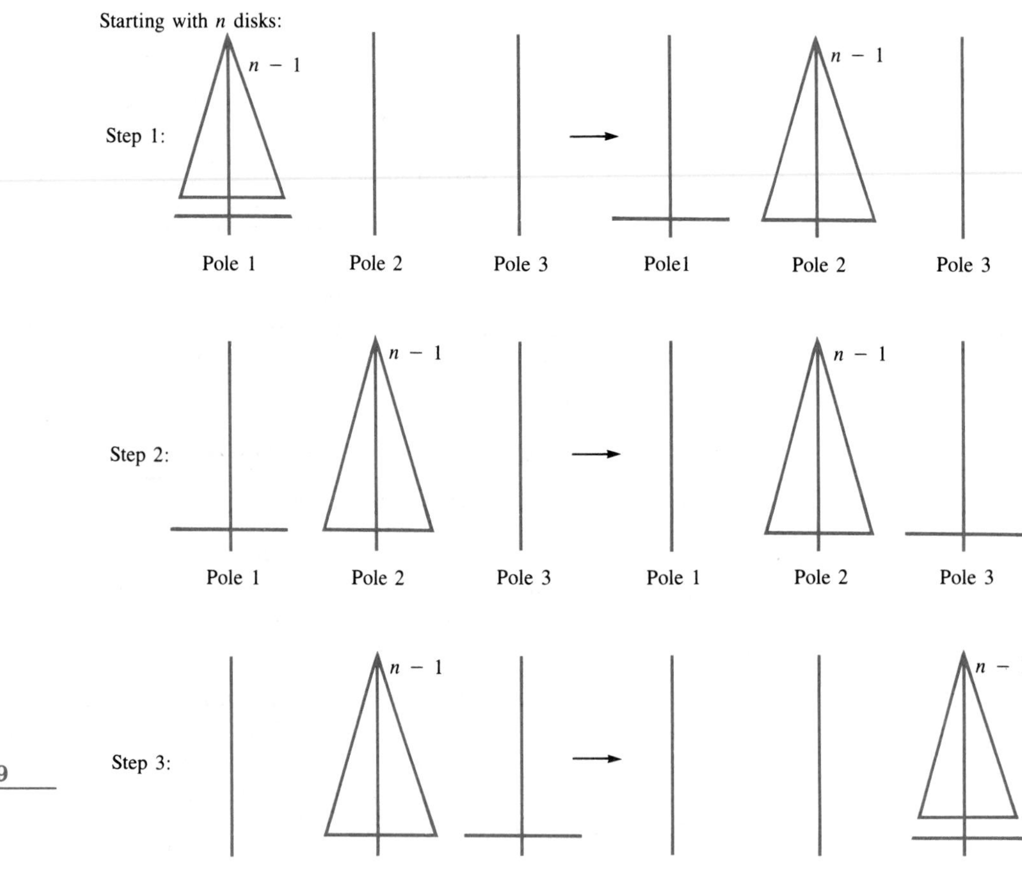

FIGURE 1.9

The Towers of Hanoi—A Recursive Solution

Figure 1.9 illustrates these three steps. Even so, this solution may seem a little like sleight of hand. But now we'll show that it works.

Let's suppose that, being lazy, we don't want to move the disks ourselves but have contrived a robot arm to do this for us. All we have to do is tell the robot what to do. We assume that our robot understands instructions of the form

$$\text{robot}(r \to s), \tag{4}$$

which should be interpreted as a request to move the top disk on pole r to pole s. Using this instruction we can express the solution embodied in steps 1–3 as Algorithm 1.9 HANOI.

Algorithm 1.9
Hanoi

```
Input num                                    [Number of disks]
      Pinit                   [Initial pole; 1 ≤ Pinit ≤ 3]
      Pfin        [Final pole; 1 ≤ Pfin ≤ 3, Pinit ≠ Pfin]
Algorithm HANOI
  procedure H(n, r, s)       [Move n disks from pole r to pole s]
     if n = 1 then robot(r → s)
             else H(n−1, r, 6−r−s)
                  robot(r → s)
                  H(n−1, 6−r−s, s)
     endif
     return
  endpro
  H(num, Pinit, Pfin)                       [Main algorithm]
Output The sequence of commands to the robot to move the disks from
       pole Pinit to pole Pfin
```

Algorithm HANOI looks short and sweet, but you may find it difficult to understand. As in EUCLID-REC there is a procedure at the beginning of HANOI. Its name is H and its arguments are n, r, and s. Also like procedure *gcd* in EUCLID-REC, procedure H is called from within itself. Once again, the main algorithm starts the execution. It consists of only one line—$H(num, Pinit, Pfin)$—which invokes the procedure H. In effect, $H(num, Pinit, Pfin)$ tells the procedure to solve the Towers of Hanoi problem with $n = num$ disks, which start on pole $r = Pinit$ and are to be moved to pole $s = Pfin$.

One other remark before we give an example of HANOI: The reason for the use of $6-r-s$ is that, if r and s are any two of the three poles 1, 2, and 3, then $6-r-s$ is the third pole (e.g., if $r = 1$ and $s = 3$, then $6-r-s = 2$).

To show how HANOI works, let's consider what happens when $num = 3$, $Pinit = 1$, and $Pfin = 3$ (i.e., we wish to move three disks on pole 1 to pole 3). Figure 1.10

a) Algorithm HANOI for *num* = 3, *Pinit* = 1, and *Pfin* = 3.

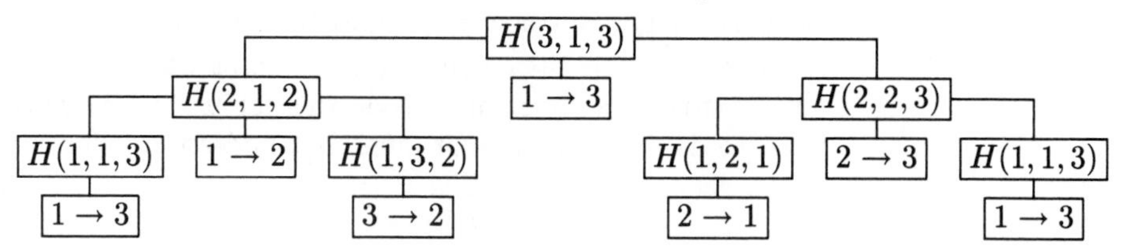

b) Trace of $H(3,1,3)$. Each call of $H(num, Pinit, Pfin)$ takes the trace one column to the right and each return takes it one column to the left.

[Initial call]

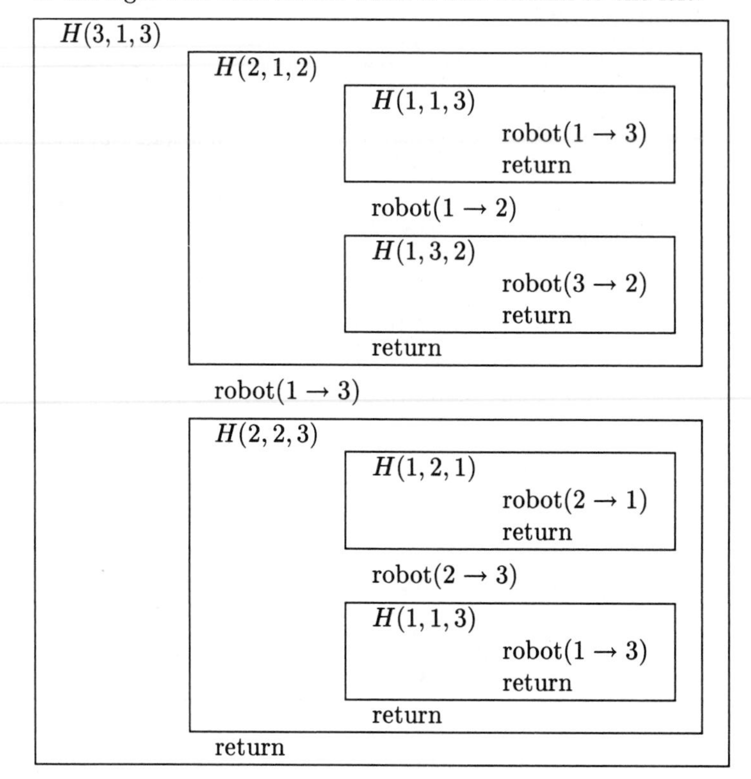

FIGURE 1.10

contains two ways of viewing the process that we're about to describe. Execution of the algorithm begins in the main algorithm with the call $H(3,1,3)$ to the procedure in which the three *calling arguments*, 3, 1, and 3, are assigned, respectively, to the variables n, r, and s. Since $n = 3 \neq 1$, we go initially to the **else** portion of the algorithm, which becomes

$$\begin{array}{l} H(2,1,2) \\ \text{robot}(1 \rightarrow 3) \\ H(2,2,3) \end{array} \tag{5}$$

The first invocation of H within $H(3,1,3)$, namely, $H(2,1,2)$, causes us to *interrupt the normal sequential execution* of steps of the algorithm and *go back to the beginning* of a new copy of the procedure (indicated by indenting and by a new level of boxing in Fig. 1.10b), now with $n=2$, $r=1$, and $s=2$. Again, since $n \neq 1$, we go to the **else** portion, which now becomes

$$\begin{aligned}&H(1,1,3)\\&\text{robot}(1 \to 2)\\&H(1,3,2)\end{aligned} \tag{6}$$

Once more the invocation of the procedure—$H(1,1,3)$—causes us to interrupt the sequence and return to the beginning of a new copy, with $n=1$, $r=1$, and $s=3$. Since now $n=1$, the **then** portion applies:

$$H(1,1,3): \qquad \text{robot}(1 \to 3)$$

We then reach **return**, which means that we have completed $H(1,1,3)$, and so we return to the interrupted execution of (6). First, we add robot$(1 \to 2)$ to our list of instructions to the robot. Then we execute $H(1,3,2)$, whose total activity, since $n=1$ again, consists of a single instruction, robot$(3 \to 2)$. This then completes the execution of $H(2,1,2)$, the result of which is

$$\begin{aligned}&\text{robot}(1 \to 3)\\&\text{robot}(1 \to 2)\\&\text{robot}(3 \to 2)\end{aligned}$$

You should verify that this sequence of moves solves the problem when $n=2$, $r=1$, and $s=2$. But, for our problem, we must now return to the interrupted execution of (5). Having completed $H(2,1,2)$ we add robot$(1 \to 3)$ to a list of commands to the robot and then execute the call $H(2,2,3)$. Since $n \neq 1$, we proceed as above and compute

$$\begin{aligned}&H(1,2,1)\\&\text{robot}(2 \to 3)\\&H(1,1,3)\end{aligned} \tag{7}$$

Since $n=1$ in both invocations of H in (7), we get for $H(2,2,3)$:

$$\begin{aligned}&\text{robot}(2 \to 1)\\&\text{robot}(2 \to 3)\\&\text{robot}(1 \to 3)\end{aligned}$$

Finally, putting all these steps together, we get for $H(3,1,3)$:

$$\begin{aligned}H(2,1,2)\left\{\begin{array}{l}\text{robot}(1 \to 3)\\\text{robot}(1 \to 2)\\\text{robot}(3 \to 2)\end{array}\right. & \qquad \begin{array}{r}\text{[Small disk to 3]}\\\text{[Middle disk to 2]}\\\text{[Small disk to 2]}\end{array}\\ \text{robot}(1 \to 3) & \qquad \text{[Large disk to 3]}\end{aligned} \tag{8}$$

$$H(2,2,3)\begin{cases} \text{robot}(2 \to 1) & \text{[Small disk to 1]} \\ \text{robot}(2 \to 3) & \text{[Middle disk to 3]} \\ \text{robot}(1 \to 3) & \text{[Small disk to 3]} \end{cases}$$

You should verify that the list of instructions in (8) really is a solution. You might take a quarter, a nickel, and a dime and use them as the disks on imaginary poles. Figure 1.10(a) graphically illustrates the steps in solving this problem. Figure 1.10(b) is a *trace* of this algorithm, in which the "depth" of successive calls of $H(n,r,s)$ is illustrated by the indentations from left to right across the page and by the boxes surrounding the successive calls of H. ∎

Self-invocation of a procedure is what makes an algorithm recursive. It embodies the essence of the recursive paradigm: Reduce to a previous case. But perhaps you can see that such self-invocation might get out of hand. You might dig a hole so deep—that is, invoke the computation so many times without ever completing an execution of it—that you could never climb out with a solution to the original problem. A necessary condition for avoiding this situation is that there be at least one value of an input parameter ($m = 0$ in EUCLID-REC and $n = 1$ in HANOI) which does not involve a call to the algorithm itself. But this alone is not enough to ensure that you will be able to "climb out of the hole". For EUCLID-REC it's rather easy to see that you do, indeed, climb out, but it's not nearly so easy to see for HANOI. You should convince yourself for this algorithm—and for all other recursive algorithms in this book—that no matter what the input parameters, the process eventually terminates. In Section 2.2 we'll prove the correctness of HANOI and, thus, that it terminates.

The use of the recursive paradigm need not lead to a recursive algorithm. Sometimes the implementation of a recursive idea as an iterative algorithm allows a more felicitous presentation of the algorithm than would a recursive algorithm. Additionally, an iterative implementation will often lead to a more efficient algorithm (see Section 1.6 for what we mean by "efficient") than the recursive implementation. Algorithm EUCLID (as given in Section 1.2) is an example of this, and we'll exhibit another in Section 1.6. When a straightforward iterative approach is available, you will usually understand it more easily. Often, however, no straightforward iterative approach is possible. This is the case with the Towers of Hanoi problem, which we expressed naturally and implemented recursively. Although it can be implemented iteratively (see [20, Section 2.6]), it's trickier and not so easy to understand.

Understanding Examples 1 and 2 is worth some effort. Try working through some cases of both algorithms. For the Towers of Hanoi, you'll find that solving the problem when $n>6$ is very tedious. But solving the problem for $n \leq 6$ should be fun. Success will surely give you some insight into the Towers of Hanoi problem in particular and recursion more generally.

Problems: Section 1.4

Note: There are more problems about recursive algorithms at the end of Section 1.5.

1. We argued that Eq. (3) fits the recursive paradigm because the *second* argument is reduced at each stage. But how about the *first* argument? Isn't it also reduced at each stage? Explain.

2. In Example 5, Section 1.2, we derived another relation besides Eq. (3) which equates the gcd of two different pairs of arguments:

$$\gcd(m, n) \;=\; \gcd(m-n, n)$$

when $m \geq n$. Write a recursive algorithm analogous to EUCLID-REC which uses this equation. Show how your algorithm works when $m = 63$ and $n = 27$ by displaying the calculation and by drawing a figure analogous to Fig. 1.7.

3. The least common multiple (lcm) of two nonzero integers m and n is the smallest positive integer which is a multiple of both m and n. For example, $\mathrm{lcm}(15, 24) = 120$. Although there is no simple relationship like Eq. (3) or that in [2] for the lcm, we can derive efficient methods for computing the lcm using its relationship to the gcd.

   a) Show that $\gcd(m, n)\,\mathrm{lcm}(m, n) = mn$ if $m, n > 0$. (This is the conjecture you should have made if you attempted [4b, Supplementary, Chapter 0].) *Hint*: Express both m and n as the products of their prime factors. Thus, $15 = 3 \times 5$ and $24 = 2^3 \times 3$. Then you can find both the gcd and the lcm by comparing the exponents of the same prime in the two factorizations and choosing the maximum for the lcm and the minimum for the gcd. Thus $\mathrm{lcm}(15, 24) = 2^3 \times 3 \times 5 = 120$ and $\gcd(15, 24) = 3$.

   b) Use the relationship in a) and the recursive procedure in Algorithm EUCLID-REC to write a recursive algorithm to compute the lcm of two nonnegative integers m and n.

4. Use the relationship in [3a] to write an algorithm to compute the lcm, using Algorithm EUCLID in Section 1.2 for the computation of the gcd.

5. List the set of moves in the Towers of Hanoi problem for

   a) $n = 2$ b) $n = 4$

6. In the Towers of Hanoi problem suppose that there are four pegs instead of three. Then the problem of finding a solution with *the minimum number of moves* is quite difficult. But try doing it for

   a) $n = 3$ b) $n = 4$

7. You are given n different rings as in TOH, but now there are only two poles. Your goal is to move them from the smallest-down-to-largest configuration on the start pole to a largest-down-to-smallest on the other pole. (Clearly, larger rings may now go on smaller ones.) There is an obvious way to do this: Just transfer the rings one at a time.

   a) Describe this solution recursively in ordinary English. That is, use a phrase like "suppose that we already knew how to do this for $n-1$ rings".

   b) Write your solution as an algorithm using a recursive procedure. You may use the command: robot($r \to s$).

8. There are actually two different recursive representations of the solution of [7]. In one version, you move a single ring and then do the recursive step; in the other, you move a single ring after the recursive step. Whichever you did in [7], now do the other. (The former solution is an instance of *tail-end* recursion; the latter, an instance of *front-end* recursion.)

1.5 Algorithmic Language—Procedures and Functions

In Section 1.4 we introduced procedures informally and used them in algorithms. In this section we'll give you the details of the syntax and semantics of procedures and their close relatives, functions.

Procedures

Both recursive algorithms in Section 1.4 had the form

Procedure

Main algorithm with
call to procedure

The procedure defined the recursive task to be performed. Then the call in the main algorithm activated the procedure for particular initial values of the arguments. This model is also characteristic of nonrecursive algorithms in which procedures are useful. We'll give several examples of these in this section, but first we need to specify carefully the syntax of a procedure. After we've done that you may wish to go back to Section 1.4 to verify that the procedures used there conform to this syntax.

The structure of a procedure is

procedure Name(variable list)
  procedure body
endpro

where the Name is, as with gcd and H in Section 1.4, just something used to refer to the procedure. The *variable list* is a list of variables separated by commas.

The *procedure body* is any sequence of statements in the language as described in Section 1.3 but also may include, as in Section 1.4, calls to the procedure itself or to other procedures. An additional rule is that the procedure body must contain at least one—and, normally, there is only one—**return** statement whose syntax is just the word **return** followed optionally by a list of variables separated by commas.

The **return** statement often is—but need not be—the statement before **endpro**. The **return** statement has two purposes:

- It returns control to the portion of the algorithm from which the procedure was called.
- It provides a mechanism, via the list of variables, for communicating results from the procedure back to the main algorithm (or other algorithm) which called it.

We'll return(!) to the second of these purposes shortly.

The *call* of a procedure, as in the case of EUCLID-REC and HANOI, consists of the name of the procedure followed by a list of arguments in parentheses. Each argument may be the name of a variable, a value (i.e., a constant), or any mathematical expression. These **calling arguments** are then associated injectively (i.e., by a one-to-one correspondence) with the *defining* (or *dummy*) *variables* in the definition of the procedure. Thus when we called Procedure gcd(i, j) in EUCLID-REC with

$$\text{gcd}(m, n)$$

i was given the value of m and j was given the value of n. Note that we will consistently use the word *arguments* when referring to the call and the word *variables* when referring to the procedure definition.

The variables in the **procedure** statement following the procedure name are called **defining variables** because they serve only to *define* the procedure and are always replaced by the calling arguments when the procedure is actually executed. The term **dummy variable** denotes that such a variable only reserves a place for the arguments actually used when the procedure is called. In our algorithmic language, communicating values from the place of call to the procedure called via the calling arguments is, with one exception, the *only* way in which values can be transmitted to a procedure. This exception is that all variables named in the **Input** section of an algorithm are assumed to be accessible from any place in the algorithm. (Such variables are sometimes said to be *global.*) Note that the names of the calling arguments and defining variables have always been different in our examples thus far. But these names may—and sometimes will be—the same in our algorithms when this usage causes no loss of readability (see, for example, Algorithm STRINGPROC in Example 1).

Results are communicated from a procedure back to the portion of an algorithm from which it was called through the values of the variables in the **return** statement; for example, in EUCLID-REC the value assigned to *answer* is communicated back to the main algorithm; in general, if a variable appears in a **return** statement, the name of the variable and its value are known and therefore can be used at the place from which it was called. This means that, if in the main algorithm there is no variable of the same name as that appearing after **return**, a new variable will be *created* in the main algorithm as with *answer* in EUCLID-REC. On the other hand, if there already is a variable of that name in the main algorithm, the value returned will merely replace the previous value of that variable.

A variable in the list after **return** can be one of the defining variables. In that case, our convention is just that the value returned replaces the value of the calling argument corresponding to the defining variable. For example, consider the procedure:

procedure anything(x)
    procedure body
    return x
endpro

The value transmitted to the procedure, which becomes the value of x, would normally be changed by the execution of anything so that a different value of x would be returned. If anything is called by

anything(y)

the current value of y would be assigned to x and then the value of x returned would become the new value of y.† One necessary constraint when defining variables are returned to the caller is that the calling argument *must* be a variable name and not a value or an expression. For example, the call

anything(z+4)

would not be allowed because there is no variable named "$z + 4$" to return the value of x to.

The only exception to returning values via the **return** statement is that the values of variables named in the **Input** section may be changed by a procedure without being explicitly returned. The reason is that these variables are assumed to be accessible from anywhere in the algorithm. However, it is bad practice to change the values of the input because in real life, the same input values may be needed for more than one algorithm, so you shouldn't tamper with them. When you do calculations which, effectively, change input values, do what we did in EUCLID in Section 1.2: Immediately assign the inputs (m and n in that case) to other variables (*num* and *denom*) and then change the latter in subsequent computation.

Allowing a list of variables rather than a single item after **return** makes it possible for a single procedure to create many results rather than just one (see, for example, [14]). On the other hand, the list after **return** can be empty. An empty list could occur, for example, when the procedure itself computes the output of the algorithm and has no other results to communicate back to the calling point. This is the situation in HANOI because the output, namely, the list of commands to the robot, is computed in Procedure H.

Perhaps the hardest thing about procedures is learning how values get transmitted back and forth between a procedure and the place it was called from. Summarizing the preceding discussion, except for the **Input** variables which are accessible and, therefore, changeable from anywhere in an algorithm:

- Values can be transferred to a procedure *only* by the calling arguments.
- Values can be communicated to the place of call from a procedure *only* via the list of variables after **return**.

Another important reason for our conventions is that, for algorithms which include recursive procedures, they enable us to protect one level of the recursion from another without the paraphernalia needed in programming languages. For example,

---

† Those of you who are knowledgeable about programming languages will recognize this as, effectively, a *call by reference* mechanism, whereas our normal communication between calling algorithm and procedure is a *call by value* mechanism.

consider Algorithm HANOI. The algorithm consists of one call of $H(n,r,s)$ with initial assignments:

$$n \leftarrow num, \qquad r \leftarrow Pinit, \qquad \text{and} \qquad s \leftarrow Pfin. \tag{1}$$

Inside the call we have

$$\begin{aligned} &H(n-1, r, 6-r-s) \\ &\text{robot}(r \rightarrow s) \\ &H(n-1, 6-r-s, s) \end{aligned} \tag{2}$$

The procedure call itself, $H(n-1, r, 6-r-s)$, then initiates a new assignment of values to the dummy variables n, r, and s:

$$n \leftarrow n-1, \qquad r \leftarrow r, \qquad \text{and} \qquad s \leftarrow 6-r-s.$$

The new calls of H then spawned will produce still other assignments to these variables. So, when the first line of (2) has been executed, what reason do you have to trust that n, r, and s have the original values which are needed for execution of the second and third lines of (2)? The answer is: Each call of H creates entirely new *local* variables which are oblivious to the values in all other calls, as no values of n, r, or s are **return**(ed) from one call to another. In other words, when you have finished with the first line of (2), you are back at the main call $H(num, Pinit, Pfin)$ and n is once more num, etc.

More generally, any recursive procedure creates many variables with the same name (one for each call) but our strict "disclosure laws" for passing variables down and back ensure that these variables do not modify each other in unintended ways.

You should be able to see how the procedure syntax and semantics we have described apply to the recursive algorithms in Section 1.4. Now we'll give some examples of nonrecursive procedures.

EXAMPLE 1 Suppose that you have a string S, consisting of characters $s_1, s_2, \ldots, s_n$, each of which is either a letter, a digit, or a blank, except that the final character, indicating the end of the string, is some special character, say, ǂ. Suppose also that, while processing (i.e., performing some operations on) S, you wish to ignore all the blanks, which you know may come in groups of more than one. This task is common to many computer applications and is related to a problem we will discuss in Chapter 7. Algorithm 1.10 STRINGPROC is the outline of such an algorithm.

Here is the procedure skipblanks for Algorithm STRINGPROC:

```
procedure skipblanks(i)                [s_i is the next symbol to look at]
    repeat while s_i = b̸                     [b̸ represents a blank]
        i ← i+1
    endrepeat
    return i
endpro
```

Note that S need not be a defining argument because, as an input, it is automatically accessible in skipblanks. Procedure skipblanks does nothing if the character s_i is not a blank; otherwise, it moves the *pointer* i along the string S until a nonblank character is found. The value returned to the main algorithm is i, which will be the index of the next nonblank character. This value replaces the previous value of i in the main algorithm.

We won't display a body for the procedure processcharacter because its form depends on what you wish to do. In [23] we'll ask you to write some procedure bodies for processcharacter to accomplish various tasks.

The complete algorithm (except for the **Input** and **Output** sections) looks like this:

Algorithm 1.10
StringProc

```
Algorithm StringProc
    procedure skipblanks(i)
        repeat while s_i = b̸
            i ← i+1
        endrepeat
        return i
    endpro
    procedure processcharacter(i)
        body of processcharacter
    endpro
    i ← 1                                   [Start of main algorithm]
    repeat while s_i ≠ ‡
        skipblanks(i)
        processcharacter(i)
        i ← i+1
    endrepeat
```

Note that in all our examples thus far we have first *defined* procedures at the beginning of the algorithm and then later *used* them in the main algorithm. When writing an algorithm, it is natural to write the main algorithm first and then write the necessary procedures. But, in presenting an algorithm, it's often better to do it the other way around because, for the *reader* of an algorithm, it's usually nice to know what the procedures are supposed to accomplish before he or she reaches the calls to them in the main algorithm. Still, we set no firm rule about this. When it seems more natural to present the main algorithm before the procedures, we'll do so.

Even when a procedure is as short as skipblanks, it may make an algorithm quite a bit more readable to structure it as in Example 1 instead of—as we could have done—replacing the *call* to skipblanks in the main algorithm with the three statements:

```
repeat while s_i = b̸
    i ← i + 1
endrepeat
```

EXAMPLE 2 Suppose that you wanted to evaluate

$$\sum_{i=1}^{n} a_i. \tag{3}$$

We denote by A the vector of constants $(a_1, a_2, \ldots, a_n)$. Here is a procedure to evaluate the sum in (3):

```
procedure Summation(A, n)
    sum ← 0                                  [Initialize sum]
    for i = 1 to n
        sum ← sum + a_i
    endfor
    return sum
endpro
```

This procedure could then be called any time the sum Eq. (3) needs to be calculated. If Summation were called by

$$\text{Summation}(B, m),$$

the sum of the elements of $(b_1, \ldots, b_m)$ would be calculated. Here, arguments provide the flexibility which makes it possible to use the same procedure to sum vectors of any length. ∎

Functions

Now suppose that in your main algorithm you wanted to calculate

$$\sum_{i=1}^{n} a_i + \sum_{i=1}^{m} b_i. \tag{4}$$

You could do so as follows:

```
Summation(A, n)
result ← sum
Summation(B, m)
result ← result + sum
```

but that's not very esthetic. (*Note*: The assignment between the two function calls is needed so that *sum* returned by the first call is not lost when a new value of *sum* is returned by the second call.) Therefore we allow a special kind of a procedure called a **function**, whose syntax differs from that of procedures only in that:

- **procedure** and **endpro** are replaced by **function** and **endfunc**; and
- there must be one and only one variable in the list following **return**.

As we'll illustrate in the following example, we allow functions to be used a bit differently from procedures.

EXAMPLE 3 Implement the calculation of Example 2 as a function and use this function to calculate the value of the sum (4).

Solution Following the rules for functions and changing the name to make the function easily distinguishable from the procedure, we have

```
function Sumfct(A, n)
    sum ← 0
    for i = 1 to n
        sum ← sum + a_i
    endfor
    return sum
endfunc
```

Functions are called (as are procedures) by giving a name and a list of arguments, except that the call *must be part of an expression on the right-hand side of an assignment statement.* The value of the variable after **return** is then the value given to the function in the expression. A call to a function may be placed anywhere in an expression where a variable or constant could legally be. Thus we could calculate the sum (4) as:

$$result \leftarrow \text{Sumfct}(A, n) + \text{Sumfct}(B, m). \tag{5}$$

Each time Sumfct is called, the value of *sum* returned is the value used in calculating the expression on the right-hand side of (5). We think you'll agree that using (5) is easier than doing the same thing with Procedure Summation—and prettier. ∎

EXAMPLE 4 We could write Algorithm EUCLID (see Example 5, Section 1.2) as a function as follows:

function Euclid (m, n)
    $num \leftarrow m$; $denom \leftarrow n$
    repeat while $denom \neq 0$
        $quot \leftarrow \lfloor num/denom \rfloor$
        $rem \leftarrow num - quot * denom$
        $num \leftarrow denom$; $denom \leftarrow rem$
    endrepeat
    return num
endfunc

Alternatively, we could write this algorithm recursively using the idea of Example 1 of the previous section:

function Euclid-Rec(m, n)
    if $n = 0$ **then** $answer \leftarrow m$
            else $answer \leftarrow$ Euclid-Rec$(n, m - n\lfloor m/n \rfloor)$
    endif
    return $answer$
endfunc

Either of the two preceding functions could then be used in an algorithm to find the gcd of k nonnegative integers, r_1, r_2, ..., r_k, which is defined to be the greatest integer that divides all the integers. Our next algorithm is based on the fact that [19]

$$\gcd(r_1, r_2, \ldots, r_k) = \gcd(\gcd(r_1, r_2, \ldots, r_{k-1}), r_k).$$

We can use this equation to build up the gcd of k numbers by calculating repeatedly the gcds of two numbers as follows:

$$\begin{aligned} \gcd(r_1, r_2, r_3) &= \gcd(\gcd(r_1, r_2), r_3) \\ \gcd(r_1, r_2, r_3, r_4) &= \gcd(\gcd(\gcd(r_1, r_2), r_3), r_4) \\ &\vdots \end{aligned}$$

Thus we can find the gcd of k integers by successively calculating the gcd of the two quantities r_i and $\gcd(r_1, \ldots, r_{i-1})$. Using the function EUCLID, our algorithm is Algorithm 1.11 KGCD. We could just as easily have used the function EUCLID-REC in this algorithm. ∎

We'll use procedures and functions at various places throughout this book. You cannot learn to use any programming language well until you have learned how to use its facility for procedures. Similarly, with algorithmic language, you must have procedures in your arsenal if you are going to write more than quite simple algorithms (and if you wish to write *any* recursive algorithms).

Algorithm 1.11
kGCD

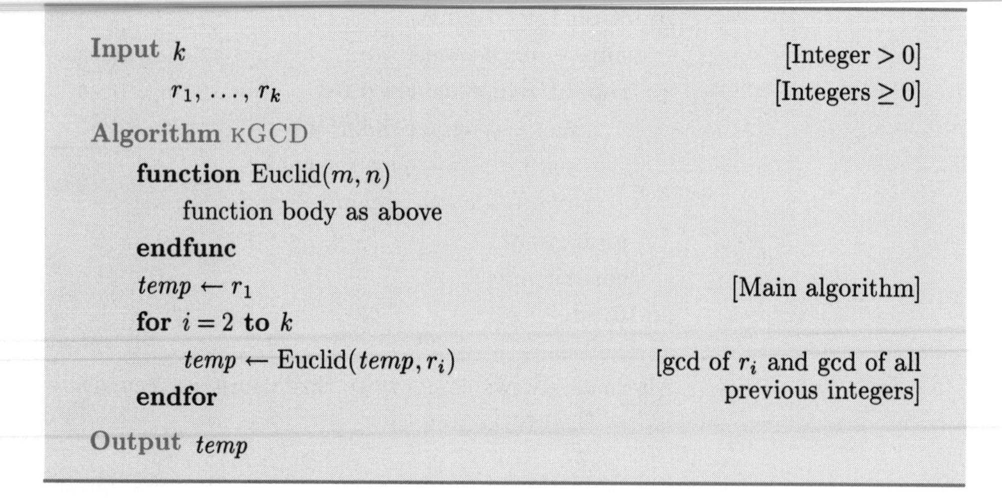

Input k [Integer > 0]
$r_1, \ldots, r_k$ [Integers ≥ 0]

Algorithm KGCD
function Euclid(m, n)
function body as above
endfunc
$temp \leftarrow r_1$ [Main algorithm]
for $i = 2$ **to** k
$temp \leftarrow$ Euclid$(temp, r_i)$ [gcd of r_i and gcd of all previous integers]
endfor

Output $temp$

Problems: Section 1.5

Some of these problems require you to use procedures or functions. Others are just intended to test whether you have generally learned how to use our algorithmic language.

1. A professor has n grades for a student: $g_1, g_2, \ldots, g_n$. (Assume that they are all numerical.) Write an algorithm to average the grades if:

a) The professor simply takes the "straight average" (all grades count equally).

b) The professor wishes to throw out the lowest grade and average the rest.

c) The professor uses the following scheme: A grade is included in the average only if it is higher than the average of the previous grades.

2. Suppose that our algorithmic language did not include the absolute value notation (but did include all the other mathematical notation we have used). Write a function ABS(x) which returns $|x|$.

3. In some computer languages, INT(x) is the integer part of x closer to 0. That is, INT$(-3.4) = -3$ and INT(4.9)=4 . (In other languages, INT$(x) = \lfloor x \rfloor$, so that INT$(-3.4) = -4$.) In a) and b), we want you to use the "closer to 0" meaning.

a) Define INT(x) as a function in our algorithmic language. You may use $\lfloor\ \rfloor$ and $\lceil\ \rceil$, since they are part of our language.

b) Now suppose we assume that INT(x) is part of our language [i.e., could be used without writing the function in a)] and, further, suppose that $\lfloor\ \rfloor$ and $\lceil\ \rceil$ are not allowable notation. Write a definition of Floor(x) in our language.

4. According to one convention, rounding off should be done as follows. Any number not exactly halfway between two integers should be rounded to the closest integer. Those numbers which end in .5 should be rounded to the *even* integer which is $\frac{1}{2}$ away. Call this rounding function $R(x)$. Use our algorithmic language to define $R(x)$ in terms of floor and/or ceiling (cf. [14–15, Section 0.4]).

5. Use algorithmic notation to write an algorithm which calculates the roots of the quadratic equation

$$ax^2 + bx + c = 0,$$

where a, b, and c can be any real numbers. When you have done this, compare your solution with that in the Hints and Answers to Problems section. Explain any differences between your solution and the recommended solution.

6. a) Write a function min(x, y) whose value is the smaller of x and y.

b) Use this function in an algorithm which prints the minimum of three inputs, a, b, and c.

7. Suppose that you weren't allowed to use the notation $x \leftrightarrow y$ to denote the interchange of the values of x and y. Write a procedure, Switch(x, y), which does this. Use this procedure within an algorithm, REVERSE, which takes a sequence and reverses it. (Let the input be n, a_1, a_2, ..., a_n.)

8. Return to the minimum completion time problem of the Prologue. We didn't write up the solution in algorithmic language because we hadn't introduced the language yet.

a) Write the solution method iteratively. (This should produce just the "work out of the hole" part of the solution.)

b) Write the solution method recursively. (This should produce both the "dig a hole" and the "work out of the hole" aspects of the solution.)

9. Here is an algorithm which multiplies by repeated addition, and does so by using a recursive procedure:

Algorithm

```
Input x, y                                  [y ∈ N]
Algorithm MULT2
  procedure MR(p, q)
    if q = 0 then prod ← 0
             else MR(p, q−1)
                  prod ← p + prod
    endif
    return prod
  endpro
  MR(x, y)
Output prod
```

a) Draw a tree diagram (see Fig. 1.10a) for the calls of MR with inputs (i) $x = 2$, $y = 3$, and (ii) $x = 3$, $y = 5$.

b) This problem is slightly easier if you use a recursive function instead of a procedure, so redo it that way. (A recursive function is often better if the current value is obtained by an algebraic manipulation of the previous value. Actually, for this problem an iterative algorithm, namely, Algorithm MULT (Example 1, Section 1.2) is better than either recursive type.)

10. Algorithm MULT3 (at the right) is a variant of the algorithm presented in [9], which also correctly multiplies. Why isn't an **else** clause needed for the **if** structure?

11. Look at Algorithms MIN1 and MIN2 (p. 120), each of which employs a recursive function to find the smallest of n inputted numbers.

Algorithm

```
Input n, a1, a2, ..., an
Algorithm MIN1
  function Min(k)
    if k = 1 then min ← a1
             else if ak < Min(k−1)
                     then min ← ak
                     else min ← Min(k−1)
                  endif
    endif
    return min
  endfunc
  small1 ← Min(n)
Output small1
```

a) Min′ takes an additional line and introduces an additional variable, *temp*. Yet it is more efficient than Min. Why?

b) Write an Algorithm MIN3 which does the same thing as MIN1 and MIN2 but uses a recursive procedure instead of a recursive function; [10] may help. (Using a recursive procedure is often better if you have to make a lot of conditional decisions before you can use the results of the previous case to decide the current case.)

Algorithm

```
Input x, y                                  [y ∈ N]
Algorithm MULT3
  procedure MR′(p, q, prod)
    if q > 0
      then MR′(p, q−1, prod)
           prod ← p + prod
    endif
    return prod
  endpro
  prod ← 0
  MR′(x, y, prod)
Output prod
```

Algorithm

```
Input Same as MIN1
Algorithm MIN2
  function Min'(k)
    if k = 1 then min ← a_1
          else temp ← Min'(k−1)
               if a_k < temp then min ← a_k
                             else min ← temp
               endif
    endif
    return min
  endfunc
  small2 ← Min'(n)
Output small2
```

Algorithm

```
Input n, a_1, a_2, ..., a_n
Algorithm MAX2
  procedure Max(m)
    if m = 1 then print a_1
            else
              if a_m > Max(m−1)
                 then print a_m
                 else Max(m−1)
              endif
    endif
    return
  endpro
  Max(n)
Output largest number
```

12. For each of the following quantities, write two algorithms to compute it, one iterative and one recursive.

a) The sum of n inputted numbers, $a_1, a_2, \ldots, a_n$.

b) The product of the same n inputted numbers.

c) The union of n inputted sets, $S_1, S_2, \ldots, S_n$.

13. Write two functions in our algorithmic language for each of the following expressions. One function should be iterative and one recursive.

a) $n!$ b) $H(n) = \sum_{k=1}^{n} 1/k$.

c) $T(n) = \sum_{k=1}^{n} k$. [$T(n)$ is called the nth *triangular number*. Why?]

d) s_n, where $s_0 = 0$ and $s_n = 3s_{n-1}+2$.

14. Write a recursive procedure BN(m) which looks at the first m available inputs and assigns to the variable B the value of the *Biggest* of those inputs and assigns to N the value of the *Next biggest*. Put this procedure in an algorithm which, given specific inputs $a_1, a_2, \ldots, a_n$ and n, prints the biggest and next biggest of the a's.

15. Algorithm MAX2 is supposed to find and print the largest of n numbers. What's wrong with it?

Fix Max(m) without turning it into a function.

16. What's wrong with MAX3, an attempt to fix the algorithm in [15]? Does it not work at all, or does it do something other than what is intended?

17. Write two versions of an algorithm which outputs all the prime factors of a positive integer N. In the first version, do not account for the number of times a prime is a factor. Thus for $N = 12$, the output should be 2, 3. In the second version, each prime should be printed once for each time it is a factor. Thus for $N = 12$, the output should be 2, 2, 3.

Algorithm

```
Input n, a_1, a_2, ..., a_n
Algorithm MAX3
  function Maxnew(m)
    if m = 1
      then print a_1
             max ← a_1
      else
             if a_m > Maxnew(m−1)
               then print a_m
                      max ← a_m
               else max ← Maxnew(m−1)
             endif
    endif
    return max
  endfunc
  large ← Maxnew(n)
Output large
```

a) Write the first version iteratively.

b) Write the first version recursively.

c) Write the second version iteratively.

d) Write the second version recursively.

18. Write an algorithm which, given integers b and c, determines whether x^2+bx+c can be factored as $(x+d)(x+e)$, where d and e are integers. The algorithm should output d and e, if they exist, and output "no factors" otherwise.

19. **a)** In Example 4 we claimed that the greatest integer which divides k integers, $r_1, r_2, \ldots, r_k$, is such that

$$\gcd(r_1, r_2, \ldots, r_k) = \gcd(\gcd(r_1, r_2, \ldots, r_{k-1}), r_k).$$

Give an argument to justify this equation.

b) Write a recursive algorithm k′GCD which does the same calculation as Algorithm kGCD in Example 4.

20. Here is a procedure to do something very simple with the TOH rings.

```
procedure TOH?(n, r, s)
    robot(r → s)
    if n > 1 then TOH?(n−1, r, s)
    return
endpro
```

What does this procedure accomplish? Write an iterative procedure to do the same thing.

21. Here is another TOH procedure.

```
procedure TOH??(n, r, s)
    robot(r → 6−r−s)
    if n > 1 then TOH??(n−1, r, s)
    robot(6−r−s → s)
    return
endpro
```

What does this in fact accomplish? Write an iterative procedure to do the same thing.

22. Draw a "tree diagram" of procedure calls, similar to Figure 1.10(a) for TOH, for the following algorithms.

a) Algorithm FIB-RECUR to compute the Fibonacci numbers when $n = 5$. [The *Fibonacci numbers* are defined in Section 2.7 (see also Section 5.2) using ordinary mathematical equations.]

Algorithm

```
Input n
Algorithm FIB-RECUR
    function Fib(n)
        if n = 0 or 1
            then fibnum ← 1
            else fibnum ← Fib(n−1) + Fib(n−2)
        endif
        return fibnum
    endfunc
    answer ← Fib(n)
Output answer
```

Note: This is not a very efficient algorithm for generating the Fibonacci numbers; your tree diagram should show you why. For this problem, and indeed for most recursively defined sequences of numbers, an iterative algorithm is better.

b) Algorithm SUM-RECUR for $n = 4$.

Algorithm

```
Input n, a1, a2, ..., an
Algorithm SUM-RECUR
    function Sum1(k)
        if k = 1 then sum ← a1
                 else sum ← ak + Sum1(k−1)
        endif
        return sum
    endfunc
    sigma ← Sum1(n)
Output sigma
```

c) Algorithm SUM-RECUR2 (p. 122) for $n = 4$. (Your diagram should help to clarify the difference between this algorithm and the one in b).)

23. Write procedures for processcharacter in Algorithm STRINGPROC which do each of the following:

Algorithm

```
Input n, a1, a2, ..., an
Algorithm SUM-RECUR2
   function Sum2(k)
      if k = n then sum ← an
               else sum ← ak + Sum2(k+1)
      endif
      return sum
   endfunc
   sigma ← Sum2(1)
Output sigma
```

a) Count the number of each of the digits 0,1, ..., 9 while ignoring all other characters.

b) Count the number of words, assuming that a word contains no blanks and is followed by a space.

c) Replaces each letter by the sixth letter alphabetically following it (c by i, w by b, etc.) as you might in simple encoding; leave blanks alone but also replace each digit by the sixth following digit (2 by 8, 7 by 3, etc.).

24. You are hired to raise a million dollars for a political candidate. This candidate runs a grassroots campaign, appealing to people who can't be expected to contribute more than $10 each. You develop the following recursive strategy. You corral 10 friends and delegate to each of them the job of raising $100,000. They in turn each corral 10 of their friends and give them the task of collecting $10,000. And so on, until 100,000 people are contacted, each of whom can give $10. Write this up as a recursive algorithm.

1.6 The Analysis of Algorithms

Designing algorithms for solving a variety of problems is one of the central themes of this book. It isn't enough, however, merely to develop an algorithm which satisfies the four properties listed in Section 1.2. We must also reckon the *cost* associated with actually using the algorithm. Even though computers are becoming faster and cheaper, if an algorithm is to be used again and again on a computer, this cost may still be considerable. Therefore we must be able to *analyze* the properties of an algorithm.

We need to ask two general classes of questions about algorithms:

i) *Analysis Questions.*

We are always interested in how well algorithms perform. Usually our measure of performance will be how fast the algorithm works. This means that we are interested in the number of steps required to execute the algorithm to completion. And our interest is usually not for a particular set of input data, but for *any* data which might be presented to the algorithm. Since the number of steps almost always depends on the *amount* or *size* of the data, we will be trying to find the speed of an algorithm as a function of some parameter which represents the amount or size of the data. Also we are interested in how fast this function increases as the amount or size of the data increases. The ideas introduced in Section 0.4 on the rates of growth of functions will be important here. Occasionally, instead of how fast an algorithm executes, we'll be interested in how much (computer) storage it requires.

The branch of mathematics and computer science concerned with the performance characteristics of algorithms is called the **analysis of algorithms**.

ii) *Complexity Questions.*

Instead of asking how well a particular algorithm does its job, suppose we ask instead how good the *best* algorithm for a particular problem is, given a suitable definition of "best". Your intuition probably will tell you that this is usually a much more difficult question than asking about the performance characteristics of a particular algorithm. After all, if you don't even know what all the possible algorithms for a problem are (and you almost never will), how could you determine how good the best one would be? And even if you could determine that, would you necessarily be able to find an algorithm which realizes the best performance? As with the analysis of specific algorithms, the most common definition of "best" is fastest, but sometimes the least-storage criterion is used here, too.

Discovering the properties of best algorithms and finding algorithms which have these properties is the concern of a branch of mathematics and computer science called **computational complexity**. When the concern is with fast algorithms, we speak of **time complexity**. If minimum storage is the object, we call this **space complexity**. The study of computational complexity generally requires mathematics beyond the scope of this book, although occasionally (as just below) we will consider it.

Unless we explicitly say otherwise, we will always be interested in speed, not storage, considerations when analyzing algorithms. The performance of an algorithm is almost always a function of the properties of the data presented to it. In this context we use three different kinds of analysis:

- *Best case analysis*—how fast is the algorithm under the most favorable circumstances (i.e., for data which allow the algorithm to run to completion with the least total computation)?
- *Worst case analysis*—how slow is the algorithm under the least favorable circumstances?
- *Average case analysis*—what is the algorithm's average performance over all possible sets of data, usually under the assumption that all data are equally probable?

Your intuition will probably tell you—correctly—that average case analysis is usually much more difficult than best case or worst case analysis. The reason, obviously, is the necessity of somehow considering exhaustively all possible sets of data.

We will use Algorithm EUCLID (which we have reproduced in Fig. 1.11) to illustrate the analysis of algorithms. The efficiency of this algorithm depends on how many times the statements between **repeat** and **endrepeat** are executed or, equivalently, how many times the division must be performed in the step

$$quot \leftarrow \lfloor num/denom \rfloor$$

```
Input m, n                                        [Integers ≥ 0]
Algorithm EUCLID
  num ← m; denom ← n                  [Initialize num and denom]
  repeat until denom = 0
    quot ← ⌊num/denom⌋                                [Quotient]
    rem ← num − quot * denom                         [Remainder]
    num ← denom; denom ← rem              [New num and denom]
  endrepeat
Output num                                         [= gcd(m/n)]
```

FIGURE 1.11

before *denom* becomes 0. A worst case analysis might involve answering the question:

> For all nonnegative m and n which are both no greater than K, what is the maximum number of divisions, as a function of K, required to find $\gcd(m,n)$?

It turns out that this question is too difficult for us here. A worst case question which we *can* answer, however, is:

> For nonnegative m and n such that $n < m$, can we find a useful upper bound on the number of divisions required to find $\gcd(m,n)$?

We reason as follows:

1. At the jth stage the divisor is r_{j-1}—see Eq. (7), Section 1.2—and the remainder r_j is either $\le r_{j-1}/2$ or $> r_{j-1}/2$.
2. If $r_j > r_{j-1}/2$, at the next stage the quotient will be 1 (why?), and the remainder r_{j+1} will be the difference between r_{j-1} and r_j, which is $\le r_{j-1}/2$.
3. Thus after each pair of two steps, the remainder will be reduced by at least a factor of $\frac{1}{2}$. That is, either $r_j \le r_{j-1}/2$ or $r_{j+1} \le r_{j-1}/2$. For example, after the first two steps (with n playing the role of r_0) $r_2 \le n/2$. What, then, is an upper bound on the number of divisions to reduce the remainder to 0? If n is a power of 2 and is halved repeatedly, the number of divisions to make the result 1 is $\log_2 n$ (why?). Therefore, if n is a power of 2, an upper bound on the number of steps of the Euclidean algorithm to reduce the remainder to 1 is $2\log_2 n$ (if it always takes two steps to halve the remainder—actually, this can't *always* happen) and one more to reduce the remainder to zero. Thus, the number of divisions is not more than

$$2\log_2 n + 1.$$

This is also an upper bound, even when n is not a power of 2 [8]. Note that this upper bound is quite conservative. That is, it is usually far larger than the actual number of divisions, since the remainder will often be halved in one step rather than two.

For a given K, many different pairs of m and n values need to be considered in analyzing the Euclidean algorithm. By contrast, Algorithm HANOI (Example 2, Section 1.4) is much easier to analyze. There is no input as such which affects the execution of the algorithm other than n itself. Thus for a fixed n, there is only one case. The worst case, the best case, and the average case are the same. So let's now try to answer the question:

> For a given n, how many moves are required to move the disks from pole r to pole s?

Let's denote the number of moves we are trying to calculate by h_n. When $n = 1$, it is clear that $h_1 = 1$, since the solution consists of the single move

$$\text{robot}(r \rightarrow s)$$

When $n > 1$,

$$h_n = 2h_{n-1} + 1, \tag{1}$$

because the calculation of $H(num, Pinit, Pfin)$ consists of two calls to the algorithm with first argument $n - 1$ (this gives the $2h_{n-1}$ term) plus one move between these two calls. In Chapter 5 we will discuss general methods for solving *difference equations*, of which Eq. (1) is an example. But here we can resort to the discovery (i.e., the "compute") portion of our inductive paradigm.

Suppose we use Eq. (1) to compute some values of h_n, using the known value $h_1 = 1$ to get started. Then using $n = 2$ in Eq. (1), we calculate h_2. We use this result with $n = 3$ to calculate h_3 and so on. We obtain

n	h_n
1	1
2	3
3	7
4	15
5	31
6	63
7	127

What formula can you conjecture for h_n as a function of n from this brief set of values? How about

$$h_n = 2^n - 1? \tag{2}$$

Is this correct? For $n = 1$ it gives 1, which we know is the correct answer. For $n > 1$ we check it by substituting into Eq. (1). With $h_{n-1} = 2^{n-1} - 1$, the right-hand side of Eq. (1) becomes

$$2(2^{n-1}-1)+1 \;=\; 2^n-2+1 \;=\; 2^n-1,$$

which is precisely what we have conjectured the left-hand side, namely h_n, should be. Thus, since h_n and 2^n-1 have the same value when $n=1$, they always have the same value. This demonstration that our conjecture is correct is an example of a proof by mathematical induction (the subject of Chapter 2) that, for all positive n, Eq. (2) is, indeed, the formula for the number of moves in Algorithm HANOI. Recall, by the way, that in Example 2, Section 1.4, $n=3$ and required seven moves as predicted by Eq. (2).

For the TOH problem we can even answer the complexity question by proving that no possible algorithm can require fewer moves than Eq. (2). Let h_n^* be the minimum number of moves for the best possible algorithm. Consider now the relation of h_{n+1}^* to h_n^*. For the $n+1$ disk case, the bottom disk must move at some time; before it does, the other n disks must all move to another pole. This must take at least h_n^* moves and, similarly, after the bottom disk moves, there must be at least h_n^* more moves to put the other n disks on top of it. Thus h_{n+1}^* must be at least $2h_n^*+1$. Since we know that $h_1^*=1$ (why?), it follows from Eq. (1) that $h_n^* \geq h_n$ for all n. But since h_n^* is the number of moves for the best algorithm, $h_n^* \leq h_n$. Thus it follows that $h_n^* = h_n$, so that our algorithm is optimal. Moreover, our algorithm is *uniquely* the best one, because any algorithm which requires a minimum number of moves *must* do precisely the sequence of moves of our algorithm.

An important function of the analysis of algorithms is to allow us to compare the efficiency of two different algorithms intended for the same task. We will illustrate this with two algorithms, both of which have the purpose of finding an entry in an ordered list of, say, words.

EXAMPLE 1 **Sequential Search**

Suppose that we have the following alphabetized list of words, where each word is indexed to indicate its position in the list, and w_i denotes the word with index i.

i	w_i
1	Are
2	Body
3	Career
4	Computer
5	Dam
6	Mathematics
7	Science
8	Ugh
9	Why

Suppose further that we are given a new word, w, and that our problem is to determine whether the new word is on the list and, if it is, where. We denote by $w=w_i$, $w>w_i$, and $w<w_i$, respectively, the situations where w is identical to w_i;

w follows w_i alphabetically; and w precedes w_i alphabetically. (If w is not on the list, we might wish to insert it; see [9].)

Algorithm 1.12 SEQSEARCH provides a solution for this problem.

Algorithm 1.12

SeqSearch

```
Input n                                          [n > 0; number of items in list]
      w_i, i = 1, 2, ..., n                      [Alphabetized list of items]
      w                                          [Search word]

Algorithm SEQSEARCH
   i ← 1                                         [Initialize index of position in list]
   repeat
      if w = w_i then print i; stop              [Success]
         w < w_i then print 'failure'; stop
         w > w_i then i ← i+1                    [Try next word]
      endif
      if i > n then print 'failure'; stop
   endrepeat

Output i                                         [Index of word if in list]
       or
       'failure'                                 [If word not in list]
```

An alternative to Algorithm SEQSEARCH might seem to be to delete the "**if** $i > n$ ..." statement and replace the **if** ... **endif** by

```
if w < w_i or i > n then print 'failure'; stop
   w = w_i          then print i; stop           [Success]
   w > w_i          then i ← i+1                 [Try next word]
endif
```

But there is a subtle error here. If w is not on the list but is greater than (i.e., alphabetically after) w_n, then at some point i will be set to $n+1$ and, in the first condition above, w will be compared with w_{n+1}, which doesn't exist!

To analyze the speed of this algorithm, we focus on the step in it which most nearly determines how much computation must be performed. In this algorithm the number of times the loop is executed is the same as the number of times we must *compare* w and w_i. We will count this 3-way comparison, which must be accomplished on a computer by two 2-way comparisons, as one comparison. Thus we reason as follows:

Best case: 1 comparison; $w_1 \geq w$.

Worst case: n comparisons; $w_{n-1} < w$.

Average case: For simplicity we focus on the case where w is on the list. Suppose that w is equally likely to be equal to any w_i (a crucial assumption). (Why?—see [11].) Then it is equally likely that there will be 1, 2, 3, ..., n comparisons. Thus the *average* number of comparisons is

$$\left(\frac{1}{n}\right)(1+2+3+\cdots+n). \tag{3}$$

To find a closed form expression for this, let's consider just the sum of the first n integers in (3). Writing this sum twice as

$$1+\ 2\ +\ 3\ +\cdots+n-1+n$$

and

$$n+n-1+n-2+\cdots+\ 2\ +1,$$

we see that each of the n columns sums to $n+1$. Therefore

$$2(1+2+\cdots+n) = n(n+1),$$

so that

$$1+2+\cdots+n = \frac{n(n+1)}{2}.$$

Substituting this expression into (3), we obtain for the average number of comparisons

$$\left(\frac{1}{n}\right)\left[\frac{n(n+1)}{2}\right] = \frac{(n+1)}{2}.$$

Why might you have guessed this answer without any analysis at all? (Note, by the way, that the method we used to derive the sum of the first n integers is a trick of limited applicability. You should be skeptical of such tricks, which have little use for solving other problems. We'll derive this formula several times later in this book by methods which have much wider applicability.)

This completes our analysis of sequential search. The algorithm is very simple and the analysis was fairly easy, too. As predicted previously, the average case analysis was the most difficult. ▮

The sequential search algorithm does its job; willy nilly it finds w if w is on the list. But we hope you aren't very happy with this algorithm. If the list had 10,000 words in it, finding w would be rather tedious (even for a computer) if w were near the end of the list. And, for example, if the "list" were a dictionary, you wouldn't look a word up by starting at the beginning and examining one word at a time until you found the word or determined that it wasn't in the dictionary. In Example 2 we consider a better procedure, one that is somewhat closer to what we actually do when using a dictionary.

EXAMPLE 2 **Binary Search**

Remember the game where someone picks a number in the range 1 to 100 and you have to guess it in the fewest number of tries? Every time you guess, you are told

Algorithm 1.13

BinSearch

Input n [$n > 0$; number of items on list]
w_i, $i = 1, \ldots, n$ [Alphabetized list of items]
w [Search word]

Algorithm BINSEARCH
```
F ← 1; L ← n                          [Initialize pointers to first (F)
                                        and last (L) items on list]
repeat
    i ← ⌊(F+L)/2⌋                     [Index of approximate midpoint]
    if w = w_i then print i; stop     [Success]
       w < w_i then L ← i−1           [Search first half]
       w > w_i then F ← i+1           [Search second half]
    endif
    if F > L then print 'failure'; stop
endrepeat
```

Output i [Index of word if on list]
or
'failure' [If word not on list]

whether you are correct, high, or low. A good strategy is to guess about in the middle of the range (say, 50) and, if this isn't correct, guess about 25 if you were high and about 75 if you were low. You then continue to narrow the range in which the number must lie by always guessing in the middle of the range where you know the number is. **Binary search** implements this idea for searching an ordered list.

As in Example 1, suppose we are given an alphabetized list and a search word w. We will apply the recursive approach outlined in the preceding paragraph by first looking in (approximately) the middle of the list. If we don't find the right word (as we normally won't for a list of any significant length), we then look in that half (approximately) of the list where w must be, if it is there at all, based on whether w is alphabetically before or after the word with which it was compared. Thus by jumping into the middle of the list, we either succeed at once or we immediately reduce our problem to half its original size. Then if we don't succeed, we repeat this process by comparing w with a word in about the middle of the half of the original list where it must be if it is there at all. And so on until we find w or determine that it isn't in the list. The name of this algorithm comes from the fact that the list is divided into *two* parts at each step. Although we could present this naturally recursive idea as a recursive algorithm, its iterative form, as given by Algorithm 1.13 BINSEARCH, is so simple that it is probably the easier one to follow. An example of how this algorithm works is shown in Fig. 1.12.

Before going on to the analysis of the algorithm, you should understand that binary search works by 1) each time through the loop setting i to the approximate midpoint of the sublist defined by the two (moving) pointers F and L; and 2) if

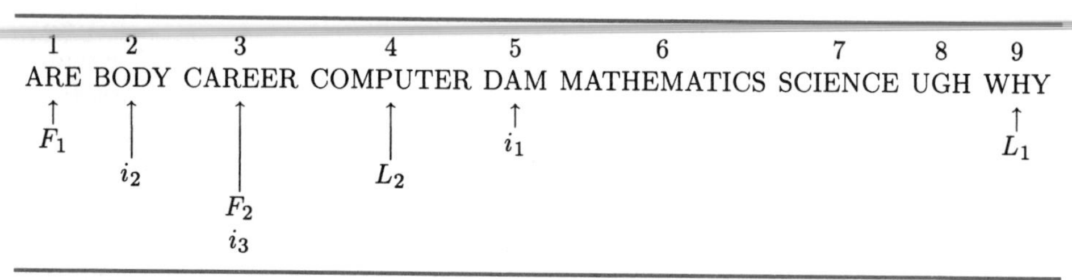

FIGURE 1.12

An Example of Binary Search. Subscripts distinguish the successive values of F, L, and i. Let the search word be CAREER.

$w_i \neq w$, then changing either F or L to define a new sublist. Also you should satisfy yourself that you understand why $F > L$ is the correct test for terminating execution of the algorithm with an indication of failure. Indeed, looking at the algorithm you may be surprised that F could ever become greater than L since, when either is changed, it is moved to about the middle of the current sublist. But $F > L$ will eventually occur if w if not on the list. Our analysis follows.

Best case: 1 comparison; $w = w_j$, where $j = \lfloor (n+1)/2 \rfloor$.

Worst case: This is not so easy. Since each pass through the loop reduces the size of the list to be searched by approximately half and since (make sure you understand why) the last list to be searched in case of failure has length 1, we expect that the worst case should require a number of comparisons related to how many halvings of n are required to get to 1. For $n = 16$ this is four $[16 \to 8 \to 4 \to 2 \to 1]$. (We go from 16 to 8 because, if a list has 16 items, after the first test the *longer* sublist has 8 items.) For $n = 21$ it is also four $[21 \to 10 \to 5 \to 2 \to 1]$. (We go from 21 to 10 because, if a list has 21 items, after the first test each sublist will have 10 items.) If you understood the discussion of logarithms in Section 0.4, you'll realize that this number of halvings is, in general, related to the logarithm of n and, specifically, should be something quite close to $\log_2 n$. This is so because, as we argued when discussing Algorithm EUCLID, if $n = 2^m$, $\log_2 n = m$ is the number of times 2^m must be successively halved until the value reaches 1. If $2^m < n < 2^{m+1}$, then $\log_2 n$ *approximates* the number of times n must be halved to get a value nearly equal to 1. In [13] we ask you to conjecture an exact formula for the worst case for all n, and in Section 2.6 we prove this formula.

Average case: This is quite difficult, and we will not treat it. ∎

Finally, just to emphasize that the analysis of algorithms is a difficult subject, particularly average case analysis, we present one more, very simple algorithm.

EXAMPLE 3 Maximum of a Set of Numbers

Our problem is to find the maximum of a set of numbers which we shall denote by

$$x_1, x_2, \ldots, x_n.$$

Algorithm 1.14
MaxNumber

```
Input n                                    [n > 1; number of items]
      x_i, i = 1, ..., n                   [Items in set]
Algorithm MaxNumber
  m ← x_1                                  [Initialize maximum value m]
  for k = 2 to n
      if x_k > m then m ← x_k
  endfor
Output m                                   [Maximum value]
```

Algorithm 1.14 MAXNUMBER solves this problem. Can you state the loop invariant for the **for** ... **endfor** loop in this algorithm? It's simply

$$m = \max(x_1, x_2, \ldots, x_{k-1}).$$

You should be able to prove that this is correct [16]. When you use this loop invariant, it is straightforward to prove that Algorithm MAXNUMBER is correct because, when the loop is exited for the last time, k is implicitly $n+1$ and m is then just the maximum of the n numbers, as desired.

Our analysis focuses on the number of times the replacement $m \leftarrow x_k$ is performed as a function of n, since this is the only operation which depends upon the actual values of the numbers on the list.

Best case: 0 replacements; $x_1 \geq x_i$ for all i.

Worst case: $n-1$ replacements; $x_1 < x_2 < x_3 < \cdots < x_n$.

Average case: We assume that all the x_i's are distinct, because equal values considerably complicate the analysis. Then without loss of generality, we may assume that the n values of the x_i's are 1, 2, 3, ..., n. And we naturally assume that all possible permutations (i.e., orderings) of the integers 1, 2, 3, ..., n are equally probable sequences x_1, x_2, ..., x_n. What now? Somehow we must consider all $n!$ permutations of 1, 2, ..., n and count the number of replacements for each. Brute force—that is, considering each permutation separately—is out of the question, except for very small values of n.

But we can make some progress by reasoning not about each permutation but rather about the *position* of the replacements in the permutations. The algorithm starts by making the first item the current largest one. What, then, is the probability that the second item will replace the first one as largest when it is compared to it? The answer is $\frac{1}{2}$, because it is 50–50 whether the first item or the second is larger. Thus on the average, there is $\frac{1}{2}$ of a replacement when $k=2$. Now how about the third item? The probability that the third item is larger than either of the two that precede it is $\frac{1}{3}$, because 2 of the 6 possible orderings of three numbers have the largest number last. Therefore on average there is $\frac{1}{3}$ of a replacement when $k=3$. Working our way toward the last element, we find that—all told—the average number of replacements is

$$\sum_{k=2}^{n} \left(\frac{1}{k}\right) \tag{4}$$

Caution: This analysis was very informal. Why, for example, is it OK to sum these average numbers of replacements in (4)? You'll have to wait for Chapter 6 for a good answer.

Thus once again, even for a very simple algorithm, average case analysis is not easy. It led to (4), which is not in closed form and whose derivation was not very satisfactory. ∎

As a final note in this chapter, we refer back to all the algorithms we have presented thus far, with the exception of the recursive algorithms. Each of these other *iterative* algorithms has a similar structure, consisting of

1. an *initialization* in which some variables are set to their initial values (e.g., *num* and *denom* in the Algorithm EUCLID and m in Example 3 of this section); and
2. a *loop*, in which a sequence of steps is applied repetitively, each time to *different data*, until some condition is satisfied.

This paradigm pervades all *iterative algorithms*. It is one you will see again and again in this book, and one you will construct in almost all the iterative algorithms you develop yourself.

Problems: Section 1.6

1. Here is an algorithm for evaluating a polynomial:

a) How many additions are needed to run this algorithm? How many multiplications?

b) This algorithm is very inefficient because it computes each power of x from scratch. Write an improved algorithm, POLY-EVAL2, which saves the previous power and uses it to compute the next higher power. How many multiplications does POLY-EVAL2 take?

Algorithm

```
Input n, a_0, a_1, ..., a_n, x     [Degree, coefficients,
                                     and variable value]
Algorithm POLY-EVAL
  V ← 0                             [Value of polynomial]
  for k = 0 to n
    T ← a_k                         [Current term]
    for j = 1 to k
      T ← T * x
    endfor
    V ← V + T
  endfor
Output V
```

2. You are given 10 gold coins and are told that all but one weigh the same. You are also given a known good coin, and a pan balance. Your problem is to identify the bad coin. There is a simple-minded algorithm for this: Weigh in turn each of the 10 against the test coin until the pans don't balance.

a) Write this in algorithmic language.

b) What is the best case outcome? Assume that the only step that takes time is doing a balance.

c) What is the worst case outcome?

d) What is the average case?

Note: There is a much smarter algorithm for this problem. Its best case is a bit worse, but its worst and average cases are much better (see Section 4.3).

3. Again you have 10 gold coins and a pan balance and are told that exactly one coin has a different weight from the others. But this time you don't have a test coin. Again, there is a naive algorithm for solving the problem: Weigh two coins; if they balance they are good and you are back to the previous case; if they don't balance, weigh either one of them against another coin. Repeat a)–d) in [2] for this situation.

4. Algorithm FINDPOWER2 finds the highest power of 2 which divides a positive integer N:

Algorithm

```
Input N
Algorithm FINDPOWER2
   k ← 0
   repeat
      q ← N/2
      if q ≠ ⌊q⌋ then exit
      k ← k+1
      N ← q
   endrepeat
Output k
```

Assume that the inputs are taken at random from 1 to $1024 = 2^{10}$. Also, assume that the only time-consuming step is division.

a) What is the best case?

b) What is the worst case?

c) What is the average case? (*Hint*: Determine the fraction of the number of inputs which lead to each possible number of divisions.)

5. Algorithm FINDPOWER2′ is a modification of the algorithm in [4]. In some ways, FINDPOWER2′ is preferable to FINDPOWER2. It has fewer lines, and it puts the exit condition at the start of the loop, where you can find and appreciate it

Algorithm

```
Input N
Algorithm FINDPOWER2′
   k ← 0
   repeat until 2∤N          [Where ∤ means
      k ← k+1                "does not divide"]
      N ← N/2
   endrepeat
Output k
```

quickly. But how does it compare in efficiency to FINDPOWER2? Answer this question by considering which operations are the most important in the two algorithms. Explain. (*Hint*: What has to happen in order to check whether $2 \nmid N$?)

6. Write algorithms to do the following:

a) Compute the dot product of two vectors of length n.

b) Compute the product of an $m \times n$ matrix and an $n \times 1$ column vector.

c) Compute the product of an $m \times n$ matrix and an $n \times p$ matrix.

d) Assuming that each real-number multiplication takes 1 step, what are the best, worst, and average case complexities of your algorithms in a)–c)?

e) Suppose A and B are $n \times n$ matrices and X is an $n \times 1$ column vector. The product ABX can be computed either in the order $(AB)X$ or $A(BX)$. Compare the complexities of these two alternatives.

7. For the Euclidean algorithm, tabulate the number of divisions required for all combinations of m and n such that both are less than 10. What pattern do you see, if any, in the results?

8. Show that an upper bound on the number of divisions in the Euclidean algorithm when $n < m$ is $2\log_2 n{+}1$. Compare this bound to the results of [7]. Can you improve on this bound by using the floor function?

9. Modify Algorithm SEQSEARCH so that if w is not in the list, it is inserted into the list in correct alphabetical order. Why is this process quite inefficient? Can you think of any way to organize the list of words so that when a new word is inserted, you don't have to "move" all the words which follow it?

10. a) Modify Algorithm SEQSEARCH again, so that it works even if the list is *not* in alphabetical order.
    b) Give the best and worst case analyses of the revised algorithm.
    c) Give the average case analysis, assuming that the search word w is *not* in the list.
    d) Give the average case analysis, assuming that the search word has probability p of being in the list and is equally likely to be in any one of the words in the list.

11. Suppose the probability that $w = w_i$ in sequential search is p_i, and the probability that w is not equal to any p_i is p (with $\sum p_i + p = 1$) . How must the analysis of this algorithm be modified? (Assume the list is alphabetized.)

12. Suppose the first step after **repeat** in Algorithm BINSEARCH is changed to

$$i \leftarrow \lceil (F+L)/2 \rceil$$

    a) Is the algorithm still correct? If not, what must be done to correct it?
    b) How must the analysis of the algorithm be changed?

13. For Algorithm BINSEARCH and $n = 2, 3, 4$, and 5, compute:
    a) The number of comparisons in the worst case. Can you make a conjecture from this data of the answer for a general n?
    b) The number of comparisons on the average if the search is successful. Assume that all search arguments are equally likely.

14. Rewrite Algorithm BINSEARCH as a recursive algorithm.

15. a) Modify Algorithm MAXNUMBER to start the search from the end of the list.
    b) Suppose the numbers need not be distinct. Modify Algorithm MAXNUMBER with the search starting at the beginning of the list, so that the output is the *smallest* subscript j such that x_j equals the maximum value.
    c) Modify the algorithm again to output the *largest* subscript j such that x_j equals the maximum value.

16. Prove that the loop invariant given for MAXNUMBER is correct. That is, show that it really is a loop invariant.

17. Compute the average number of replacements for Algorithm MAXNUMBER for $n = 2, 3$, and 4.

Supplementary Problems: Chapter 1

1. Conversion of integers in base b to base 10. Let's analyze this problem using the recursive paradigm. Suppose you already know how to convert an integer with n digits in its base b representation and you now wish to convert an integer with $n+1$ digits. As usual, index the digits from left to right, as $a_n, a_{n-1}, \ldots, a_1, a_0$.
   a) The easiest approach is to think of $a_{n-1}, \ldots, a_0$ as the sequence you already know how to convert. Write an iterative algorithm to implement this approach.
   b) Write a recursive algorithm to handle the approach in a).
   c) You could also think of $a_n, a_{n-1}, \ldots, a_1$ as the sequence we already know how to convert. To convert the entire sequence, you must "shift over" the known subsequence and then add in a new 1's digit, a_0. Write an iterative algorithm to do this.
   d) Write a recursive algorithm to handle the approach in c).

2. What is the relation of Algorithm FRAGMENT on p. 94 to the base conversion problem discussed in [1]?

3. Conversion of integers in base 10 to base b. Let's analyze this problem using the recursive paradigm. Suppose you are given the integer N in base 10. Suppose you already know how to convert all integers $M < N$ to base b.
   a) You could try to determine the leftmost digit of N in base b first. This depends on finding the

highest power of b which is $\leq N$. Once you find it, you can subtract it from N the maximum number of times and then proceed to convert the difference to base b. Write an iterative algorithm based on this idea.

b) Write a recursive algorithm to handle the approach in a).

c) Or you could try to determine the rightmost digit of N in b first. This is the remainder when N is divided by b. Why? Also, the quotient, if converted to base b, tells you the other digits of N in base b. Why? Turn these observations into an iterative algorithm to convert to base b.

d) Write a recursive algorithm to handle the approach in c).

4. The Knapsack Problem. You have n knapsacks with capacities $c_1, c_2, \ldots, c_n$. You have m items of sizes $s_1, s_2, \ldots, s_m$. The goal is to get all the items into the knapsacks—in fact, into as few of the knapsacks as possible. For instance, let's have $n=3$ knapsacks with $c_1=25$, $c_2=30$, and $c_3=27$; let's also have 9 items with sizes

$$20, 15, 14, 9, 8, 7, 4, 3, 2.$$

a) One approach is "largest first fit"; i.e., stuff the largest item into the first knapsack in which it fits. "First" means lowest index, so you put the item of size 20 in knapsack 1. Then you put the next largest item (size 15) into the first (lowest index) knapsack in which it now fits, namely, knapsack 2. And so on, until all items are in the knapsacks or some item won't fit in any knapsack. Carry this algorithm out for the rest of the given data.

b) Another approach is "smallest first fit". Now you start by putting the smallest item into the lowest index knapsack in which it fits, then the next smallest item, and so on. Carry out this algorithm for the given data.

c) A variation on largest first fit is to first renumber the knapsacks in order of decreasing size. Thus $c_1 = 30$, $c_2 = 27$, and $c_3 = 25$. Carry out largest first fit again with this change.

d) Another approach might be called "largest roomiest fit". Items are stuffed in order of decreasing size, but you use the knapsack with the most remaining room. (In case of a tie for remaining room, use the knapsack with the lower index.) This is clearly not a good algorithm if you are trying to use the minimum number of knapsacks (why?), but if the goal is simply to fit the items into the knapsacks available, it is at least as plausible as the other approaches described. Carry it out for the data given at the beginning of this problem.

e) Is it possible to get all 9 of the given items into the 3 knapsacks? Why?

f) The knapsack problem is very similar to the computer scheduling problem in Example 6, Section 1.2. What does the size of an item correspond to? What are the mathematical differences between the two problems? Can you give a slightly different computer scheduling problem which is identical, mathematically, to the knapsack problem?

5. The assignment problem. You are the manager of a large consulting firm. You have n jobs to be performed and n employees to do them. You have rated each employee for each job, with 10 being outstanding and 1 being miserable. You wish to assign the people to the jobs so that the work gets done as well as possible. The usual mathematical formulation of this is: Find the assignment which maximizes the sum of the ratings. There is an obvious, greedy approach to doing this. State it. Apply it to the following situation, where the number in the ith row and jth column is the rating of person i for job j.

$$\begin{bmatrix} 10 & 9 \\ 8 & 1 \end{bmatrix}$$

What do you conclude about the success of the greedy algorithm for the assignment problem? (See also [14, Section 3.8].)

6. Maximum-weight independent set of vertices. Imagine that you have a graph (i.e., a set of vertices and edges between some pairs of them, as in Fig. P.1 in the Prologue), with each vertex having a number.

You wish to pick a set of vertices, no two of which have an edge between them, so that the sum of the numbers on the selected vertices is as large as possible. (The vertices might represent people, an edge might mean that the people hate each other, and the goal might be to select a team for a project where close cooperation is essential. The numbers might represent the quality of the work a person does when other people aren't bothering him or her.) There is an obvious greedy algorithm for this problem. State it and then apply it to the following graph. What do you conclude?

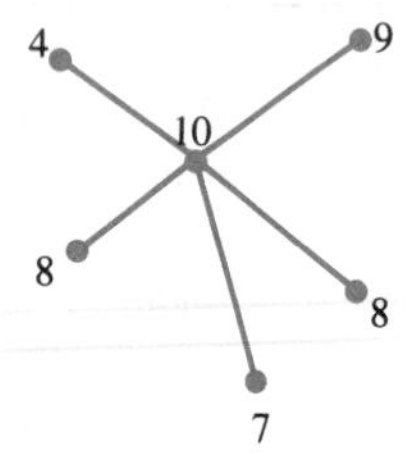

(*Note*: There is no known efficient algorithm which solves this problem in general.)

7. The worst case analysis of Algorithm EUCLID on page 124 did not provide the strongest upper bound possible. This is not surprising since it broke the problem into 2 uneven cases: In one case the size of the remainder goes down by $\frac{1}{2}$ in one step, in the other case by $\frac{1}{2}$ in two steps. Thus all we can be certain about is that the remainder halves in at most two steps. Let's consider an analysis which reduces the problem by the same factor per step. Suppose we pick some fraction x between $\frac{1}{2}$ and 1 and break into two cases as follows: Case 1 is if $rem < xn$, and case 2 is otherwise. (As in the text, we assume that $n < m$.) Then in case 2 the new remainder one step later is at most $(1-x)n$. Thus we either decrease the size of the smaller number by a factor of x in one step or by a factor of $1-x$ in two. So suppose we choose x so that $x^2 = 1-x$. This makes the reduction factor *per step* the same either way.

a) Find the x which meets this condition.

b) The argument in the text gave a reduction by $\frac{1}{2}$ in two steps. What reduction does the method of this problem give in two steps?

c) Collect some data (more than you were asked to collect in [7], Section 1.6) and see how close our new bound is to the actual worst case.

8. Consider the problem of determining whether an input integer N is a prime. You can find out by simplifying Algorithm PRIMENUM in Example 3, Section 1.2. Namely, try dividing N by 2, then by 3, 5, 7, 11, etc., until you find a divisor or until N itself is reached. Assume that the only step which takes time is division.

a) What is the best case?

b) What is the worst case, if N must be < 100?

c) The average case is hard. But what fraction of positive integers will require just one division? What fraction, 2 divisions? What fraction, 3 divisions? Can you use this information to determine the average case for $N < 100$? (*Hint*: At worst, every $N < 100$ will either be shown to be composite by one of the first 4 attempted divisions, or it will end up being a prime and require every prime division from 2 to N.)

9. For the robot adding algorithm in Section 1.1 (Fig. 1.2), assume that the only hard work is changing each digit. That is, the steps

$$E_k \leftarrow D_k + 1,\ E_k \leftarrow 0, \text{ and } E_{n+1} \leftarrow 1$$

take work, but other steps don't. Assume that the robot is given an integer I with at most 9 digits, and I is picked at random. Determine the best, worst, and average cases for this algorithm.

10. Merge Sort. Suppose you have two lists of words, $w_1, \ldots, w_m$ and $x_1, \ldots, x_n$, and each list is in alphabetical order. You wish to merge them into a single alphabetical list. There is a simple way to do this. Start by comparing the first words in both lists and picking the one which comes first. Then what?

a) Write up this approach in our algorithmic language. The only tricky part is what to do when one of the lists is exhausted.

b) Assuming that the relevant thing to count is the number of comparisons, determine the best and worst cases for this algorithm. (The average case analysis is quite tricky, for it depends on determining the probability that one list will be exhausted when there are k words left in the other; see Section 5.9.)

CHAPTER 2

Mathematical Induction

■

2.1 Introduction

A row of dominoes, all standing on their ends, extends beyond the horizon. The dominoes are spaced close enough so that, if one falls over, the next falls down too. (See Fig. 2.1.) Someone knocks the first domino over. The result: They all fall down.

FIGURE 2.1

The parable of the dominoes: The essence of mathematical induction

This parable of the dominoes describes the essence of **mathematical induction**, the most important method of proof in discrete mathematics. Mathematical induction applies when you have an infinite sequence of statements,

$$P(1),\ P(2),\ \ldots,\ P(n),\ P(n+1),\ \ldots,$$

and you want to prove they're all true. The statements are the dominoes. To be proved true is to be knocked down. If you can show that the truth of any one statement $P(n)$ would imply the truth of the next statement $P(n+1)$, then each domino is close enough to knock down the next. If you can also show that the initial statement $P(1)$ is true (falls down), then all the statements are true (fall down).

Mathematical induction is intimately related to the inductive and recursive paradigms. The recursive paradigm emphasizes finding a relation between one case and previous cases. This approach corresponds to showing that the previous domino is close enough to knock down the next domino. The inductive paradigm emphasizes building up a pattern iteratively from the beginning. This approach corresponds to showing that the initial domino falls down and that the others fall down in turn.

So far we have thought of our two paradigms as being methods of discovering relationships, devising algorithms, and solving problems. However, we also want to be certain (prove) that our algorithms and solutions are correct. This is where mathematical induction comes in. It is the standard proof technique for both paradigms. Specifically, anyone interested in verifying that algorithms are correct must necessarily be interested in mathematical induction.

In this book, you will see many proof techniques. We hope you will develop a general understanding of them all. For mathematical induction we have a higher goal. We want you to understand it so well that you can *do* it yourself, even in this first course. Not only is mathematical induction the most useful proof technique but, fortunately, it is also the easiest to learn. The reason is that induction has a specific form. You always have to show that the initial domino falls; then you have to show that each domino knocks over the next.

This is not to say that inductions are all the same, or that all are easy. There are variations, and we will deal with several of them as we go along. For instance, some inductions require more than one starting case, and other inductions require more than one previous case to get the next case. Also, at the end of this chapter we will discuss yet another associate of our two paradigms: inductive definitions. It turns out that every inductive proof relies, usually implicitly, on an inductive definition. In the final section we will make this reliance explicit.

One last comment: The fact that this proof technique is called induction is in some ways unfortunate. In general discussions of reasoning (by lay people, psychologists, etc.) the word "induction" is contrasted with "deduction". Deduction describes reasoning that draws airtight conclusions by logical analysis. Induction refers to the discovery of patterns from examples; these patterns may be believed, but they have not been proved.

It is in this general sense that we have used the word "induction" in the phrase "inductive paradigm"—at least with respect to the first, discovery stage of that paradigm. It is also the general sense that is meant when scientists refer to "scientific induction". But this is not at all the sense we mean in the phrase "*mathematical* induction". Mathematical induction is a form of rigorous proof and is thus actually a special case of deduction. Thus the word "mathematical" in front of "induction" makes a crucial difference. Nonetheless, in mathematics books, when it is clear that a method of proof is being discussed, it is customary to drop the word mathematical and simply say induction. The lay person will be terribly confused, but you will understand!

2.2 Examples of Induction

Before you can *do* inductions, you need to see a few. We start by proving

$$\sum_{k=1}^{n} k = \frac{n(n+1)}{2} \tag{1}$$

for all positive integers n. (We already showed this in Section 1.6, Example 1, while analyzing sequential search, but the proof was by a trick.)

If you happened to study mathematical induction in high school, proving Eq. (1) is probably the sort of example you did. Indeed, some people think that's *all* mathematical induction is about—proving formulas for sums. Nothing could be further from the truth! The rest of this chapter shows why. However, it doesn't hurt to start on familiar ground.

Before we start the proof, though, you have a right to object: "How are we supposed to come up with the statement of theorems like this? What good does it do to teach us how to devise a proof if we don't know how to conjecture what to prove?"

Good questions—and in this book we promise not to ignore them. However, in this section we are only going to say a little about how to conjecture what to prove; learning how to do inductions is already a mouthful, and we don't want you to gag. We save most of our advice about how to make conjectures for Section 2.4, which is devoted entirely to that subject.

For now, let us simply say that the most straightforward way to conjecture Eq. (1) is to use the inductive paradigm. Write down the first few cases:

$$1 = 1, \qquad 1+2 = 3, \qquad 1+2+3 = 6, \qquad 1+2+3+4 = 10, \ldots,$$

and see if you can find a pattern. We have already told you that the right-hand sides are of the form $n(n+1)/2$. If we hadn't said this, maybe you would see the pattern, maybe you wouldn't. As we said, Section 2.4 will help; also see [5].

Now, turning to the proof, there are three steps. The first step is to name explicitly the statement form $P(n)$ on which to do the induction. In this case it's easy. Eq. (1) is really an infinite collection of statements, one for each positive integer n:

$$\begin{aligned} P(1): &\quad 1 = 1(2)/2, \\ P(2): &\quad 1+2 = 2(3)/2, \\ P(3): &\quad 1+2+3 = 3(4)/2, \end{aligned}$$

and so on. Thus $P(n)$ is just the statement $\sum_{k=1}^{n} k = n(n+1)/2$ *for a particular* n. We wish to prove $P(n)$ for *all* n.

The second step is to prove the initial case $P(1)$. This is easy: We simply check that 1 does indeed equal $1(2)/2$, as stated.

The third step is to prove that *any one* statement $P(n)$, *should it be true*, implies the truth of the next statement $P(n+1)$. In other words, if $P(n)$, then $P(n+1)$. This implication is usually abbreviated as $P(n) \Longrightarrow P(n+1)$. To show that $A \Longrightarrow B$, the most straightforward way is to start with statement A and manipulate it until we end up with B. Thus we need some manipulations that look like:

$$\sum_{k=1}^{n} k = \frac{n(n+1)}{2}$$

$$?$$

$$\sum_{k=1}^{n+1} k = \frac{(n+1)[(n+1)+1]}{2},$$

because the first line is $P(n)$ and the last line is $P(n+1)$.

To help us fill in the middle, we note that the only change on the left-hand side (LHS) from the top line to the bottom line is the addition of one more term, $n+1$. Following the Golden Rule—do unto one side as you would do unto the other—it's clear that we must also add $n+1$ to the RHS. Then, it's simply a matter of massaging this sum until it looks like the RHS of the bottom line. The result is:

$$\begin{aligned}
\sum_{k=1}^{n} k &= \frac{n(n+1)}{2}, \\
\left(\sum_{k=1}^{n} k\right) + (n+1) &= \frac{n(n+1)}{2} + n+1, \\
\sum_{k=1}^{n+1} k &= \frac{n(n+1)+2(n+1)}{2} \\
&= \frac{(n+1)(n+2)}{2} \\
&= \frac{(n+1)[(n+1)+1]}{2}.
\end{aligned}$$

We have indeed shown that $P(n) \Longrightarrow P(n+1)$ for a generic n. In other words, we have shown that $P(1) \Longrightarrow P(2)$, $P(2) \Longrightarrow P(3)$, $P(3) \Longrightarrow P(4)$, and so on. Since earlier we showed $P(1)$, we now know $P(2)$, from it $P(3)$, then $P(4)$, and so on forever. In other words, $P(n)$ is true for all positive integers n. The proof by induction is complete. (Thinking in terms of dominoes, we have shown that all the dominoes fall down because the first fell and each one knocks over the next.)

Remark 1. We have described mathematical induction as having three steps. Most books give only two; they leave out our first step. The reason they leave it out is that often the choice of $P(n)$ is obvious—it's just the nth case of the statement the problem started with, as in this example. However, in other cases the choice of $P(n)$ is far from obvious. Indeed, figuring out $P(n)$ is often the hardest part of the proof, especially in proofs of the correctness of algorithms. Please, *always state your $P(n)$ explicitly.*

Remark 2. Our step 2 is called the **basis step**. Step 3 is called the **inductive step**. The $P(n)$ of the inductive step is called the **inductive hypothesis**. During the inductive step we do not know that the inductive hypothesis is true. Only when the whole proof is finished do we know whether it is true. During the inductive step we merely assume temporarily that it's true and see what follows.

Remark 3. To prove $A \Longrightarrow B$, it is not necessary to use A at the very beginning of your proof. You must end up with B, but you can use A anywhere along the way that is convenient. For reasons that will only become clear later, in many induction proofs the best place to use $P(n)$ is not at the start of the inductive step, but in the middle. So you might as well practice putting it in the middle even when it's easy enough to put it at the start.

We are now going to prove Eq. (1) again. Doing so lets us illustrate how to put $P(n)$ in the middle. Also, now that we have explained the format of induction at length, there is no need to explain it within the proof. Indeed, typically such explanations of induction are not given within proofs, so doing this proof again allows us to present a more standard write-up.

Theorem 1. For all positive integers n,

$$\sum_{k=1}^{n} k = \frac{n(n+1)}{2}.$$

PROOF By induction. Let $P(n)$ be the statement $\sum_{k=1}^{n} k = n(n+1)/2$.

Basis. $P(1)$ asserts that $1 = 1(2)/2$, which is true.

Inductive step.

$$\sum_{k=1}^{n+1} k = \Big(\sum_{k=1}^{n} k\Big) + (n+1) \overset{P(n)}{=} \frac{n(n+1)}{2} + n+1 = \frac{n(n+1) + 2(n+1)}{2}$$
$$= \frac{(n+1)(n+2)}{2} = \frac{(n+1)[(n+1)+1]}{2}.$$

Thus the first expression equals the last, which proves $P(n+1)$. ∎

Note the $P(n)$ sitting over the second equals sign. Putting it there is a brief way of saying that $P(n)$ is the reason why the equality is true. Indeed, in this case the RHS of that equality is obtained by substituting $P(n)$ into the LHS. There is another way to indicate that $P(n)$ is the reason: Write "$P(n)$" in brackets at the right-hand end of the line where it is used, in the same way we have put reasons in brackets in algorithms. We'd like you to get comfortable with both formats. When using the bracket method, it is a good idea to have just one equality (or inequality, or whatever) on that line; otherwise, which equality is being referred to will be unclear. In other words, in bracket format the first line of the preceding display would best be broken up as

$$\begin{aligned}\sum_{k=1}^{n+1} k &= \left(\sum_{k=1}^{n} k\right) + (n+1)\\ &= \frac{n(n+1)}{2} + n+1. \qquad [P(n)]\end{aligned}$$

Note also that in everything we have done so far in this chapter, $P(n)$ is the name of a *statement*. (P is the traditional choice here. It stands for "proposition", another word for statement. See Chapter 7.) $P(n)$ is *not* a number, nor a polynomial function with real values, nor an algebraic expression. Thus consider the last sentence of the proof. It is *not* an acceptable alternative to write "Thus $P(n) = n(n+1)/2$." This alternative is not correct because it's complete gobbledygook—statements can't ever equal numbers.

EXAMPLE 1 Prove that

$$(1+x)^n \geq 1+nx$$

whenever n is a nonnegative integer and x is a nonnegative real number.

Solution First, why should an inequality of this form be interesting? Because the LHS might be tedious or ugly to compute exactly (say, if $x = .3972$); so it is nice if we can approximate or bound the value by an expression that *is* easy to compute (the RHS). Since this particular LHS shows up in lots of problems, both inequalities and equalities for it are worth obtaining. (We prove an equality, the "Binomial Theorem", in Section 4.6.)

Why should we conjecture this particular inequality? You probably already know that

$$(1+x)^2 = 1+2x+x^2,$$
$$(1+x)^3 = 1+3x+3x^2+x^3,$$

so in the cases $n = 2$ and $n = 3$ the first two terms of the exact expansion are $1+nx$. Since every term in the expansion is nonnegative when x is, the inequality $(1+x)^n \geq 1+nx$ seems reasonable.

In any event, let $P(n)$ be the statement $(1+x)^n \geq 1+nx$. Note that the initial statement this time is $P(0)$, nct $P(1)$, because we were asked for a proof for all nonnegative n.

Basis. $P(0)$ states that

$$(1+x)^0 \geq 1+0x, \qquad \text{or} \qquad 1 \geq 1. \checkmark$$

(Henceforth, we use the check $\checkmark$ as shorthand for "which is true".)

Inductive step. We start with the LHS of $P(n+1)$, use $P(n)$ in the middle, and obtain the RHS of $P(n+1)$.

$$\begin{aligned}(1+x)^{n+1} &= (1+x)^n(1+x)\\ &\geq (1+nx)(1+x)\end{aligned}$$

[By $P(n)$ and positive multiplication]

$$= 1 + nx + x + nx^2 = 1 + (n+1)x + nx^2$$
$$\geq 1 + (n+1)x. \qquad \text{[Since } x^2 \geq 0\text{]}$$

Thus

$$(1 + x)^{n+1} \geq 1 + (n+1)x,$$

which is $P(n+1)$. (By "positive multiplication" we mean the rule that if $a \geq b$ and $c > 0$, then $ac \geq bc$.) ▮

Note. The only hard part in the inductive step for Example 1 is the first equality. Of all the ways that we might manipulate $(1+x)^{n+1}$, why do we choose $(1+x)^n(1+x)$? Because we know that we have to use $P(n)$ somewhere, so we must "break down" what we start with in order that some key aspect of $P(n)$ shows. Since the key aspect of this $P(n)$ is the expression $(1+x)^n$, the factorization $(1+x)^n(1+x)$ is the thing to try first.

EXAMPLE 2 Find and prove a formula for the sum of the angles of a convex polygon.

Solution A **convex set** is a set such that every line segment connecting points of the set lies entirely in the set. A **convex polygon** is a polygon whose perimeter and interior together form a convex set. See Fig. 2.2. A polygon has at least $n = 3$ sides (a triangle) and a triangle has 180°. The triangle case is all you need to know; the break-down approach allows you to both discover the more general formula and prove it. As the convex polygon in Fig. 2.3 shows, we can always break down an $(n+1)$-gon into a triangle and an n-gon. Therefore an $(n+1)$-gon has 180° more than an n-gon. That is, for an n-gon we get 180° for each side beyond the second, obtaining

$$180 + 180(n-3) = 180(n-2).$$

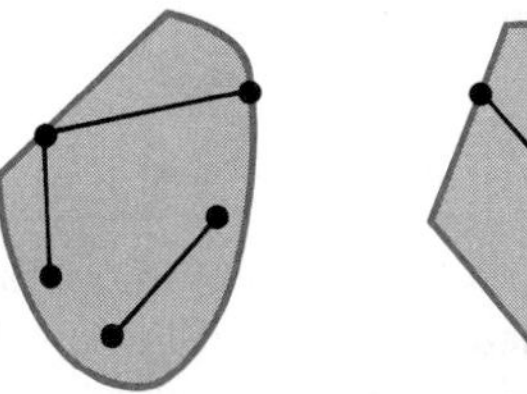

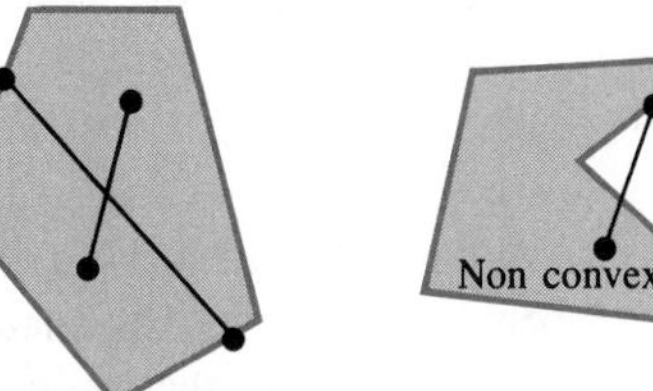

FIGURE 2.2

A convex set, a convex polygon, and a nonconvex polygon

We may summarize this analysis in the form of a tight and careful proof.

PROOF Let $P(n)$ be the statement that any convex n-gon has $180(n-2)$ degrees as the sum of its interior vertex angles. Clearly $n \geq 3$, and the basis case $P(3)$ is well known. For the inductive step, given any arbitrary $(n+1)$-gon, slice it into an n-gon and a triangle with segment V_1V_n as in Fig. 2.3. By $P(n)$, the sum of the marked angles in the n-gon is $180(n-2)$. The sum of the marked angles in the triangle is 180°. The sum of both is the sum of the angles for the $(n+1)$-gon. This sum is

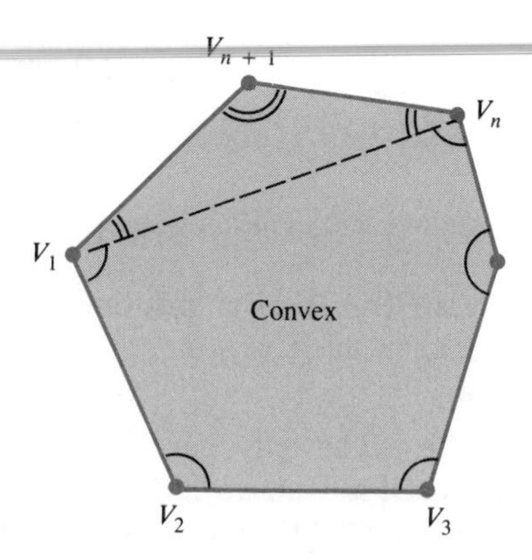

FIGURE 2.3

Break down of an $(n+1)$-gon into an n-gon and a triangle

$$180(n-2) + 180 = 180[(n+1) - 2], \tag{2}$$

which gives us $P(n+1)$. ∎

Did you notice that we actually used more than one previous case to prove $P(n+1)$? In addition to $P(n)$ we used $P(3)$. Such multiple use should not surprise you; in the recursive paradigm one often has to relate the next case to several previous cases. However, we have to admit that this use is not in keeping with the specific formulation we gave to mathematical induction in Section 2.1—there each domino was strong enough *by itself* to knock over the next. So is this proof valid?

Yes, it is. We have shown $P(3)$ and $[P(3) \text{ and } P(n)] \Longrightarrow P(n+1)$. Substituting $n = 3$ in the implication, we conclude that $P(4)$ is true. Now substitute $n = 4$. Since we know both $P(3)$ and $P(4)$ at this stage, we get $P(5)$. Next, $P(3)$ and $P(5)$ imply $P(6)$, and so on. We have another valid induction principle.

There are many other variants of induction—three previous cases are needed, a different two previous cases are needed, and so on. Fortunately, they can be lumped together by allowing all previous cases to be used. This results in the principle of **strong induction**, which we discuss in the Section 2.3.

Induction and Tournaments

We devote a subsection to tournaments because they provide a fine opportunity to break away from the idea that induction is solely a method for proving formulas. In fact, this subsection contains no algebraic computation at all. Tournaments also provide another preview of the broad usefulness of graphs, the subject of Chapter 3.

We will discuss ranking in round-robin tournaments. A **round-robin tournament** is one in which each team plays every other team exactly once. There is a real-world difficulty with ranking teams in such tournaments: How many games a team wins might not be the best indication of relative strength. (Why?) Consequently, there has been considerable research into other possible ways to rank teams. One approach that seems quite reasonable (but see [29]) is to look for a "chain of command"—an ordering of the teams so that each team beats the next team in the order. If such an order exists, one might declare this order to be the ranking. You might suspect, however, that in general no such ordering exists, especially if bizarre things happen in the tournament, like team A beating team B but then losing to every team that B beats. Surprisingly, such an order *must* exist:

Theorem 2. Suppose that n teams play a round-robin tournament in which there are no tie games. No matter what the individual game outcomes are, one can always number the teams $t_1, t_2, \ldots, t_n$, so that t_1 beat t_2, t_2 beat t_3, and so on, through t_{n-1} beat t_n.

Figure 2.4 illustrates this phenomenon for a particular tournament with five teams. The tournament is represented by a directed graph. The teams are the vertices, and the arrows on the edges point from winners to losers. For instance, 1 beat 2. Sure enough, there is a ranking of the sort that Theorem 2 claims: 5,3,4,1,2. (That is, $t_1 = 5$, $t_2 = 3$, and so on; when the teams are numbered to begin with, the numbering in the theorem amounts to a renumbering.) The theorem claims that this is not luck—any round-robin tournament with any number of teams will have such a ranking!

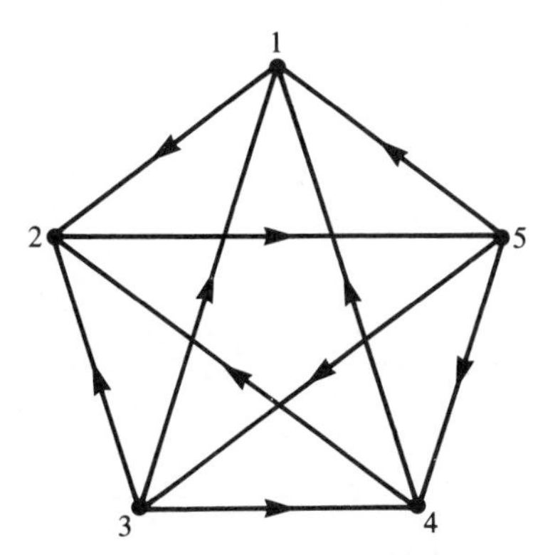

FIGURE 2.4

A particular tournament with five teams.

PROOF OF THEOREM 2 Let $P(n)$ be the statement of the theorem for n teams. The smallest meaningful value of n is 2, for a tournament consisting of one game. Thus $P(2)$ is obviously true: Call the winner of that game t_1 and the loser t_2.

The heart of the proof is the inductive step. Assume the inductive hypothesis: Any tournament with n teams has the desired ranking. We must show that any tournament with $n+1$ teams does also. So, imagine that somebody presents us with the results of an arbitrary $(n+1)$-tournament (tournament with $n+1$ teams). We must break down the new case to make the old case show. How? Temporarily, ignore any one team, call it t^*. The games played between the remaining teams form an n-tournament, so by $P(n)$ there is a ranking $t_1, \ldots, t_n$ such that t_i beat t_{i+1} for $i = 1, 2, \ldots, n-1$. This is illustrated by the directed path of arrows along the bottom of Fig. 2.5.

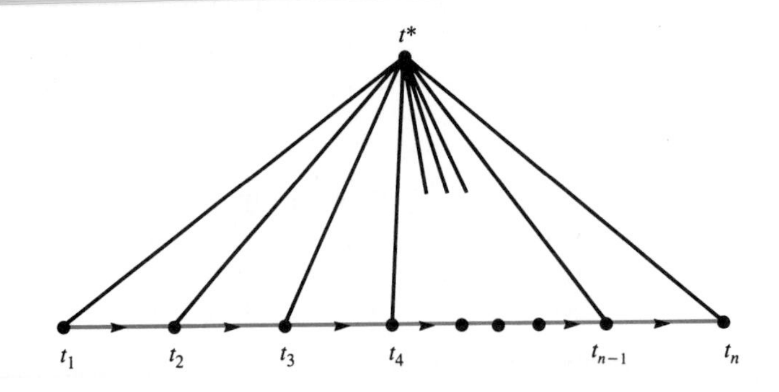

FIGURE 2.5

The graph of an arbitrary $(n+1)$-tournament, assuming that all n-tournaments have rankings

Now remember t^*. It plays against each of the other teams; we just don't know which of those games it won. Thus we have indicated those games in Fig. 2.5 as lines without arrows. In terms of the figure, the question is: Is there some directed path which includes all the vertices? For instance, is there somewhere we can insert t^* into the directed path along the bottom and keep all the arrows going the way they should? Fig. 2.6 illustrates a particular way the games involving t^* might have gone, and an insertion which works in that case. But we must give an argument which demonstrates the existence of an insertion no matter how the arrows in Fig. 2.5 are filled in.

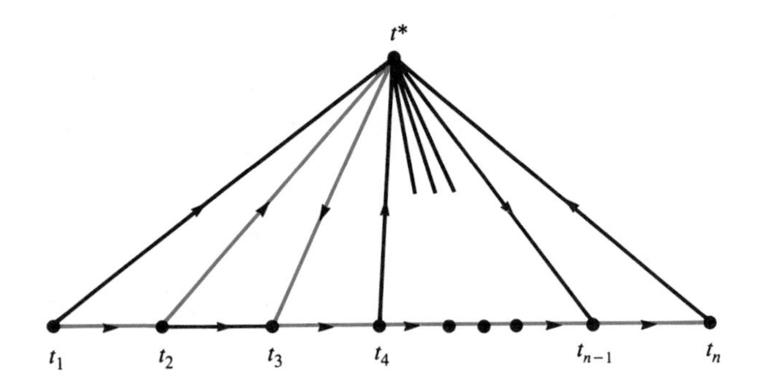

FIGURE 2.6

One possibility for inserting t^*

Suppose that the arrow between t^* and t_1 points down. Then we are done: Insert t^* at the start of the ranking. So consider the remaining case where the arrow points up. Look from left to right at the other edges from t^*. If we ever find an arrow pointing down from t^*, then we are in business: Let t_i be the first vertex for which the edge from t^* is directed down and insert t^* between t_{i-1} and t_i in the ranking. If there aren't any arrows pointing down from t^*, we are also in business: Use the last arrow to insert t^* at the end of the ranking. Since we have found a ranking in every case, we have shown that $P(n) \Longrightarrow P(n+1)$, completing the induction. ∎

Remark 1. Note the intimate relation between this proof and the recursive paradigm: We suppose we know the truth of the previous case (when, in fact, we don't yet understand at all why it should be true) and then we show that this assumption implies the truth of the current case. In short, after the initial case of an inductive proof we jump right in the middle—the recursive paradigm. Heretofore, we have used that paradigm to compute the solution to problems, usually by an algorithm. Now we find we can use it to do proofs. Indeed, we can use the recursive paradigm to do both at once. For instance, now that we have proved Theorem 2, wouldn't it be nice to have an algorithm that actually finds the ranking? The recursive relationship between cases n and $n+1$ in the proof should suggest an algorithm to you [35, 36].

Remark 2. Note that at no point in the proof did we have to think about the whole n-tournament or the whole $(n+1)$-tournament. The inductive step (recursive paradigm) forced us to look only at what the nth conclusion provides, namely, the bottom row of Fig. 2.5. Given this limited information about the n-tournament, we are led to look for a directed $(n+1)$-path that differs from the n-path only by an insertion. If we had been left to imagine the whole tournament and look for any sort of directed path, we would have been incapacitated by the lack of focus.

Finally, the proof above is quite long. This time it's because we tried to motivate our approach. *You* don't have to provide the motivation with your proofs (although, if your approach is particularly clever, motivation will help even wise old professors to read it). To help you see what good, succinct mathematical writing style is, we again provide an appropriately short rewrite.

TIGHT PROOF OF THEOREM 2 By induction. Let $P(n)$ be the statement of the theorem for n. Thus $P(2)$ is clear: Rank the winner of the one game before the loser. For the inductive step, consider an arbitrary $(n+1)$-tournament. Temporarily ignore some one team, t^*, leaving an n-tournament. By $P(n)$, there is a ranking $t_1, t_2, \ldots, t_n$ of the remaining teams so that t_i beat t_{i+1} for all $i = 1, 2, \ldots, n-1$. Now consider the games between t^* and the others. Either (1) t^* beat t_1; (2) some $i > 1$ is the first i for which t^* beat t_i; or (3) every other team beat t^*. In the first case, insert t^* at the start of the ranking. In the second, insert t^* between t_{i-1} and t_i. In the third, insert t^* at the end. ∎

In this tight version, we have left out certain important things. For instance, the proof could well continue:

> Thus we have shown $P(n) \Longrightarrow P(n+1)$. Therefore the inductive step is complete, so, by mathematical induction, $P(n)$ is true for all $n \geq 2$.

Working mathematicians tend to leave such things out because—once the writer has announced that the proof is by induction—to them such remarks are obvious. For now, though, it may help you get a grasp on what is going on to leave such sentences in.

In any event, please note that, even in this tight version, most of it is words. *Mathematical arguments do not consist of unconnected calculations.*

Induction and Algorithms

To show the close relation between induction and algorithms, we consider at length the Towers of Hanoi (TOH) puzzle. When we discussed this puzzle before (Example 2, Section 1.4), we gave a recursive algorithm, HANOI, for solving it. Though we motivated the algorithm at length, we never really *proved* that it works.

Temporarily ignore the algorithm. Let's just prove that the puzzle has a solution. Then we turn back to Algorithm HANOI (restated in this section as Algorithm 2.1) and prove that it constructs a solution. Comparing the proofs will illustrate an important point about recursion.

Theorem 3. Towers of Hanoi can always be solved.

PROOF For the first time, the definition of $P(n)$ does not stare us in the face from the statement of the theorem. The theorem appears to be a single assertion rather than a sequence of related assertions. But it's not hard to split up. Let $P(n)$ be the proposition that TOH can be solved when there are exactly n rings—whichever pole they start on and wherever they are supposed to go. If we prove $P(n)$ for all $n \geq 1$, we have proved the theorem.

$P(1)$. It's obvious: Move the ring directly to the finish pole.

$P(n) \Longrightarrow P(n+1)$. We are given $n+1$ rings on some first pole and told to move them to a second pole. See Fig. 2.7, in which we illustrate the case with the first pole on the left and the second pole in the middle. Temporarily ignore the bottom ring and think of the others as a unit (shown as a triangle in the figure). By $P(n)$, somehow there exists a way to move the remaining n rings to *either* other pole. Imagine they are moved to the *third* pole. Now remember ring $n+1$. At this point it may be moved to the second pole. Finally, by $P(n)$ again, the top n rings may be moved from the third pole to the second, on top of ring $n+1$. Thus, assuming $P(n)$, we have shown $P(n+1)$. ∎

Admission: There is a small gap in this proof. We didn't fill it in so as not to obscure the main ideas of the proof. Curious readers should see [44].

At this point you should look at Algorithm 2.1. In order to make the proof of Theorem 4 below read more naturally, we have made some cosmetic changes from the version in Section 1.4. We have given the parameters in the recursive procedure H the same names as the corresponding input variables. (As we explained in Section 1.5, the parameters of a procedure are dummy variables, so we can call them what we please.)

Theorem 4. Algorithm HANOI works.

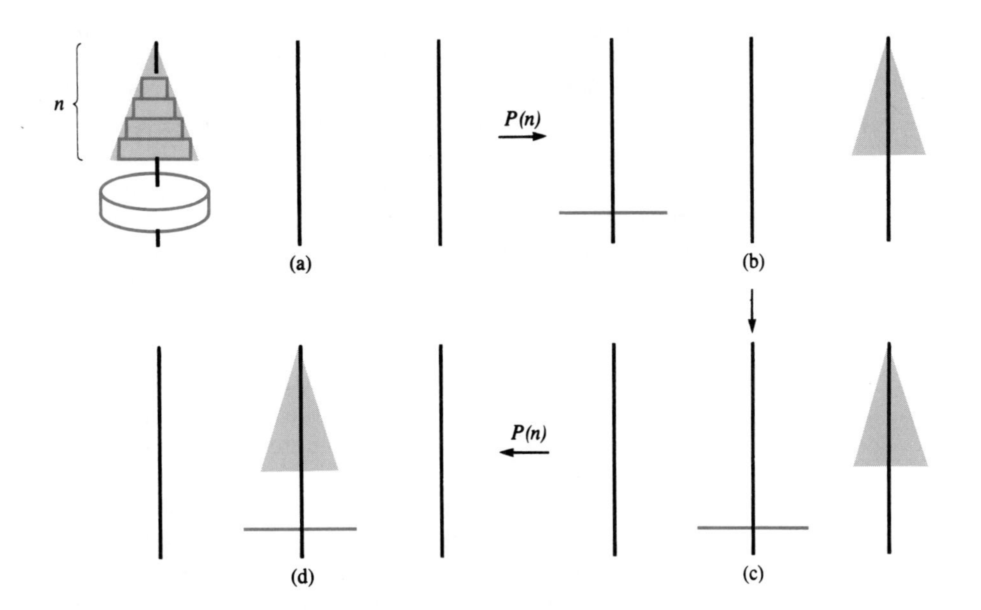

FIGURE 2.7

The inductive step in the proof that Towers of Hanoi is solvable

Algorithm 2.1

Hanoi

Input n [Number of rings]
r [Initial pole; $1 \le r \le 3$]
s [Final pole; $1 \le s \le 3$]

Algorithm HANOI

```
procedure H(n, r, s)
    if n = 1 then robot(r→s)
             else H(n−1, r, 6−r−s)
                  robot(r→s)
                  H(n−1, 6−r−s, s)
    endif
    return
endpro
H(n, r, s)                                  [Main algorithm]
```

Output The series of commands to the robot to move the rings from pole r to pole s.

The approach we will use is the standard approach for validating algorithms that rely on a recursive procedure. We do induction on the statement that the recursive procedure does what it is supposed to. From there it is usually easy to show directly that the main algorithm is also correct.

PROOF OF THEOREM 4 Let $Q(n)$ be the statement that procedure $H(n, r, s)$ gives a correct sequence of moves whenever it is invoked for this n, arbitrary start pole r, and arbitrary finish pole s. Since the main algorithm of HANOI is just a single call of H for the actual input values, proving $Q(n)$ correct for all n will prove that HANOI is correct.

Basis. To prove $Q(1)$, simply walk through the procedure when $n = 1$. Since the **then** clause takes effect instead of the **else** section, the robot moves the ring from pole r to pole s, and then the procedure stops. The correct "sequence" of one move has been made.

Inductive step. This time we assume $Q(n{-}1)$ and prove $Q(n)$, because that's the way the indices in this typical recursive algorithm are set up. Clearly, proving $Q(n{-}1) \Longrightarrow Q(n)$ for $n > 1$ is the same as proving $Q(n) \Longrightarrow Q(n{+}1)$ for $n \geq 1$, because both claim that, starting with $Q(1)$, each proposition Q implies the next proposition Q. This is so small a change in the induction format that we won't even refer to it as a variant form.

So suppose we already knew that procedure H works whenever it is called for $n{-}1$ rings, where $n{-}1 \geq 1$. Let us invoke the procedure with n rings, and any r and s, and walk through it to see what happens. Since $n > 1$, the **then** clause is skipped. The first thing that happens in the **else** section is that $H(n{-}1, r, 6{-}r{-}s)$ is called. By $Q(n{-}1)$, the result of this call is a sequence of moves that somehow correctly gets the top $n{-}1$ rings from pole r to pole $6{-}r{-}s$. See Fig. 2.7(b) again! The next line of the algorithm says to move the top ring currently on pole r to pole s. This is ring n, since it is the only ring on pole r. So now we have Fig. 2.7(c). The next line calls $H(n{-}1, 6{-}r{-}s, s)$. By $Q(n{-}1)$ again, this call correctly moves the $n{-}1$ rings onto pole s—Fig. 2.7(d). This completes the call of $H(n, r, s)$, and the rings are just where we want them to be. ∎

Figure 2.7 illustrates both proofs at once. Indeed, although we marked the key transitions in the figure with $P(n)$, we could just as well have marked them by $Q(n{-}1)$. This is no accident. The essence of the recursive paradigm—jump in the middle—is the key to all of the following: Discovering how something can be done; proving that it can be done; writing an algorithm that does it; and proving that the algorithm does it. Put another way, if you discover how to do something recursively, then you already know how to prove your method and write an algorithm to do it—they are all related facets of recursion.

EXAMPLE 3 Show that the following is *not* an algorithm.

Algorithm WHAT'S-WRONG

$a \leftarrow 4$
repeat while $a \neq 2$
$a \leftarrow (a+2)/2$
endrepeat
Output a

What's wrong with WHAT'S-WRONG? It looks like an algorithm. And even if it isn't, how can we break the assertion that it's not into an infinite set of statements $P(n)$ to be proved by induction?

Recall that part of our definition of an algorithm is that it must terminate in a finite number of steps. We will show that WHAT'S-WRONG does not. This time let's use the inductive paradigm to figure out why. Walk through the "algorithm" for a while. You will find that a gets a new value each time through the loop, and that the first few values are

$$4, \quad 3, \quad 2.5, \quad 2.25, \quad 2.125.$$

The pattern should be clear [45], but the only important thing about the pattern is that the values keep *not* being exactly 2 (although they get closer), so the exit condition for the loop is never met. Hence the process never terminates.

We have discovered a **loop invariant**, namely, $a \neq 2$. (See Example 1, Section 1.2 for an earlier discussion of loop invariants. We discuss them further in Section 2.6 and Chapter 7.) Loop invariants are intimately tied to inductive proofs about loops in algorithms, because the inductive step of such a proof amounts to walking through a typical iteration of the loop.

In this example, let $P(n)$ be the statement "when the loop of WHAT'S-WRONG is entered for the nth time, $a \neq 2$". If we can prove $P(n)$ for all positive integers, then we have proved that the exit condition for the loop is never met, so the process never terminates. We prove $P(n)$ by induction.

$P(1)$. The first time the loop is entered, $a = 4$ from the previous assignment statement.

$P(n) \Longrightarrow P(n+1)$. If A is the value of a when the loop is entered for the nth time, then the value of a when the loop is entered for the $(n+1)$st time is $(A+2)/2$. Thus

$$P(n) \implies A \neq 2 \implies A + 2 \neq 4 \implies \frac{(A+2)}{2} \neq 2 \implies P(n+1). \blacksquare$$

Remark. In computer science, you usually want to prove that things *do* terminate (and compute what you claim they compute!). In some sense, then, proofs of algorithms are "finite inductions"—you are not trying to prove something for an infinite number of iterations. But the fundamental ideas of basis step and inductive step are the same. What you do is prove by induction that *if* a loop is entered for the nth time, then the loop invariant is maintained. Furthermore, a good loop invariant has the property that for some n the exit condition will be met.

Summary. The Principle of Mathematical Induction (in the "Weak" or "First" form) is the following.

The Principle of Mathematical Induction, Weak Form

To prove the proposition $P(n)$ for all integers $n \geq n_0$, it suffices to prove both

i) the Basis Step: $P(n_0)$, and

ii) the Inductive Step: for all $n \geq n_0$, $P(n) \Longrightarrow P(n+1)$.

Caution. This standard way of stating the principle hides the fact that formulating the proposition $P(n)$ is often a major part of the work.

Problems: Section 2.2

1. Discover and prove a formula for the sum of the first n odd positive integers.

2. By doubling the equation in Theorem 1, we get a formula for the sum of the first n even positive integers. For practice, prove this formula directly by induction.

3. A sequence $u_1, u_2, u_3, \ldots$ is defined by $u_1 = 1$ and, for $n \geq 1$, $u_{n+1} = u_n + 8n$. Compute the first few terms in the sequence. Guess a closed formula for u_n. Prove it by induction.

4. An arithmetic sequence with initial term a and difference d is a sequence $a, a+d, a+2d, \ldots$, where each term is d more than the preceding term. Show by induction that the nth term is $a+(n-1)d$. (This fact is pretty obvious; the point is to practice your induction skills.)

5. In high school geometry, you learned: If two triangles are similar, with the ratio of their respective sides being k, then the ratio of their areas is k^2.

a) Represent $\sum_{k=1}^{n} k$ as several rows of dots (k dots in row k) so that the fact just stated suggests that the closed formula for this sum is a quadratic in n. What do you think the coefficient of n^2 will be?

b) You have now conjectured that

$$\sum_{k=1}^{n} k = an^2 + bn + c,$$

for some yet unknown constants a, b, and c. However, you also know that when $n = 1, 2, 3$, the sums are 1, 3, and 6, respectively. Thus a, b, and c must satisfy

$$1 = a1^2 + b1 + c = a + b + c,$$
$$3 = a2^2 + b2 + c = 4a + 2b + c,$$
$$6 = a3^2 + b3 + c = 9a + 3b + c.$$

Thus you have three linear equations in three unknowns. Solve them.

In other words, if your conjecture that there is a quadratic formula is correct, a, b, and c as solved from these three equations must be the constants. This approach is called the **method of undetermined coefficients**. It gives you exactly the formula $n(n+1)/2$ of the text (expanded out). It does not prove that the quadratic assumption is correct, but it does tell you exactly what to conjecture. Now you could give an inductive proof.

6. Compare $\sum_{k=1}^{n} k^3$ with $\left(\sum_{k=1}^{n} k\right)^2$. Conjecture a relationship and prove it by induction.

7. Prove by induction the standard formula for the sum of the first n terms of an arithmetic series:

$$\sum_{k=1}^{n}[a+(k-1)d] = an+\tfrac{1}{2}n(n-1)d.$$

Then prove it again using Theorem 1 and facts about sums from Section 0.5.

8. Prove by induction the standard formula for the sum of the first $n+1$ terms of a geometric series with ratio $r \neq 1$:

$$\sum_{k=0}^{n} ar^k = \frac{a(r^{n+1}-1)}{r-1}.$$

9. If you write out $\sum_{k=1}^{n}(a_{k+1}-a_k)$ for $n = 1,2,3,4$, it becomes pretty obvious that, in general, the sum simplifies to $a_{n+1}-a_1$. This simplification is called the **telescoping series** property. Prove it rigorously using induction.

10. Use the telescoping series property to find a simple expression for $\sum_{n=1}^{m} 1/(n+1)n$. You will need to reexpress $1/(n+1)n$ in the form $a_{n+1}-a_n$.

11. Find and prove a formula for

$$\prod_{n=2}^{m}\left(1-\frac{1}{n^2}\right).$$

12. Formulate and prove a **telescoping product** property; see [9].

13. Let a and b be integers. Prove that $a^n - b^n$ is a multiple of $a-b$ for all positive integers.

14. State the formula for the sum of angles in a convex polygon (Example 2) using radians. Redo the inductive proof.

15. Given a line segment of unit length, show that you can construct with straightedge and compass a segment of length $\sqrt{n}$ for each natural number n. To start, recall how you would construct $\sqrt{2}$.

16. Discover and prove a formula for

$$\begin{bmatrix} 1 & 1 \\ 0 & 1 \end{bmatrix}^n.$$

17. Prove that $(1+x)^n > 1+nx$ for all $x > 0$ and $n > 1$.

18. Prove by induction that $2^n > n+1$ for all integers $n \geq 2$. Then prove it again as a special case of [17].

19. The following questions pertain to the discussion of "positive multiplication" in Example 1.

a) For which real numbers c does $a > b \Longrightarrow ac > bc$?

b) For which real numbers c does $a \geq b \Longrightarrow ac \geq bc$?

c) What can you conclude about ac and bc if $a > b$ and $c < 0$?

d) What conditions must be put on x and y so that $x > y \Longrightarrow 1/x < 1/y$?

20. For $n > 1$, consider

$$1+\frac{1}{\sqrt{2}}+\frac{1}{\sqrt{3}}+\cdots+\frac{1}{\sqrt{n}}.$$

Compare this sum to $\sqrt{n}$; then conjecture a relationship and prove it.

21. Let $x_1 = 1$, and for $n \geq 1$ let

$$x_{n+1} = x_n + \frac{1}{x_n}.$$

Prove: For $n > 1$, $x_n > \sqrt{n}$.

22. Consider the sequence

$$x,\quad x^x,\quad x^{(x^x)},\quad x^{\left(x^{(x^x)}\right)},\ldots$$

Prove: If $x > 1$, then this sequence is increasing. *Hint*: Let $P(n)$ be "$a_{n+1} > a_n$", where a_n is the nth term.

23. Show that, for each integer $n > 1$,

$$\underbrace{\sqrt{1+\sqrt{1+\sqrt{1+\sqrt{1+\cdots}}}}}_{n\ 1\text{'s}}$$

is irrational. The innermost radical sign contains only the integer 1. You may assume that $\sqrt{2}$ is irrational; if you have never seen a proof of this, get someone to show you.

24. Fact: $n^3 - n$ is divisible by 3 for all integers n.

a) Prove this fact for nonnegative integers by induction.

b) Prove it for nonpositive integers by "downward induction"; that is, show $P(0)$ and then prove $P(n) \Longrightarrow P(n-1)$.

c) Show that the nonpositive case follows from the nonnegative case by algebra alone—a second induction is unnecessary. *Hint*: Let n be any positive integer. You know that $n^3 - n$ is divisible by 3 and want to show that $(-n)^3 - (-n)$ is divisible by 3.

d) Prove this fact all at once for all integers n by factoring.

25. Prove that the sum of the cubes of any three consecutive integers is divisible by 9. See [24] for several methods.

26. Prove that if n is even, then $13^n + 6$ is divisible by 7.

27. Conjecture a formula for the number of subsets of an n-set and prove it by induction.

28. Prove by induction that the number of even cardinality subsets of an n-set is equal to the number of odd-cardinality subsets.

29. Theorem 2 says less than you might think.

a) For instance, it does not claim that the ranking is unique. Find another such ranking for the tournament of Fig. 2.4.

b) It also does not say that each team beats *all* teams farther down in the ranking, or even most teams farther down. Find a ranking in Fig. 2.4 in accord with Theorem 2 and in which the first team in the ranking beats *only* the second team in the ranking.

30. Suppose that tie games are allowed in a round-robin tournament. Prove that there is a ranking so that each team either beat or tied the next team.

31. Again suppose that ties are allowed, but that we still insist on a ranking in which each team beat the next team. Give an example where such a ranking does not exist.

32. In an alternative form of round robin, each team plays every other team k times, where k is some fixed positive integer. (In many professional sports leagues, each season consists of such a round robin.) We say that team i "won its series" with team j if i won more than half the games against j. Prove that, if k is odd, there is always a ranking such that each team won its series against the next team.

33. A tournament is not a round robin if some teams never play each other. Show by example that a ranking as in Theorem 2 need not exist if a tournament is not a round robin.

34. Show that, if a round robin tournament has an odd number of teams, it is possible for every team to win exactly half its games. *Hint*: Do induction by 2; i.e., assuming there is a tournament verifying $P(2n-1)$, show how to add two players and fix their games so that you have a tournament verifying $P(2n+1)$.

35. Write an iterative algorithm that accepts as input the outcomes of the games of an n-tournament, and outputs a ranking as in Theorem 2. *Suggestion*: Input the data as an $n \times n$ 0–1 matrix, with $a_{ij} = 1$ iff team i beat team j. The hard part to do efficiently is the updating of the ranking. If you know about pointers, use them.

36. Same as [35], but make the algorithm recursive.

37. Prove that Algorithm HANOI moves individual rings $2^n - 1$ times. (This was already argued by induction in Section 1.6, but it was an informal presentation because we hadn't explained induction yet. Now do it carefully.)

38. Prove that Algorithm HANOI solves TOH in the minimum possible number of steps. (See Section 1.6.)

39. Prove that in TOH the minimum sequence of steps to move n rings from pole r to pole s is unique.

40. Let a TOH clutter be any arrangement of different sized rings on the three poles so that a larger ring is not above a smaller ring. Show that, using the rules of the game, it is always possible to reassemble a clutter into a single pile.

41. Straightline TOH. This version has an additional rule: In a single step, a ring can be moved only to an adjacent pole. Since the order of the poles is now important, it makes a difference where we start and finish. Let's say we want to move the rings from one end pole to the other.

a) Prove that Straightline TOH can be solved.

b) Write an algorithm that solves it.

c) Prove the algorithm correct.

d) Prove your algorithm takes the minimum number of moves.

42. Repeat [41], but now move the rings from an end pole to the middle.

43. Answer [40] again for Straightline TOH.

44. There is actually a subtle problem with our proofs of Theorems 3 and 4. We invoke the previous case in the inductive step (for instance, that the $n-1$ game can be solved in Theorem 3) but that case only asserts a solution when there are exactly $n-1$ rings on the board. Now there is an additional ring. Maybe that ring messes things up—by making some move illegal or by getting moved by some call of robot(r, s) that was meant to move another ring between those poles.

The usual way around this problem is to argue that the additional ring, being largest and being initially at the bottom, can never cause the moves from the previous case to be illegal and will never be moved during a previous call. Although this is pretty obvious, to assert it is actually an implied second induction. For instance, to assert that the new ring doesn't move is to assert a loop invariant through the set of moves—that, having stayed at the bottom during the last move, it stays at the bottom again.

To have an airtight proof, it's best to treat this subtlety head on by putting everything you need in the original induction.

a) Prove Theorem 3 again with $P(n)$ redefined to be: Whenever there are $N \geq n$ rings on the board in a legitimate configuration, and the smallest n are together on one pole, those n may be moved to either other pole without moving any other rings.

b) Prove Theorem 4 again with $Q(n)$ redefined to be: Whenever there are $N \geq n$ rings on the board in a legitimate configuration, and the smallest n are together on pole r, $H(n,r,s)$ correctly moves those n rings to pole s without moving any other rings.

These two adjustments are examples of "strengthening the inductive hypothesis". Sometimes, when your original proof isn't really legitimate, this is what you have to do to make it work.

45. State a formula for the value of a at the end of the nth time through the loop of Algorithm WHAT'S-WRONG. Prove it by induction.

46. We have proved that, as a theoretical procedure, WHAT'S-WRONG doesn't terminate. What actually happens if you implement it on a real computer? Why?

47. Let $Q(n)$ be the claim "When WHAT'S-WRONG is run, the loop is entered at least n times." Prove that WHAT'S-WRONG is not an algorithm by proving $Q(n)$ for $n > 0$ by induction.

48. Suppose we modified WHAT'S-WRONG so that you get to input whatever initial value of a you please. It would still not be an algorithm, because for at least one input value, 4, it would not terminate. For exactly what input values does it terminate? Prove what you claim by induction.

49. Suppose we changed WHAT'S-WRONG further, replacing

repeat while $a \neq 2$

with

repeat while $a \neq b$

and replacing

$$a \leftarrow (a+2)/2$$

with

$$a \leftarrow (a+b)/2,$$

where the user gets to input both the initial value of a and the value of b. Now, when does the algorithm terminate?

50. Prove that the following is not an algorithm:

Algorithm ERROR
  $a \leftarrow 4$
  repeat while $a \neq 2$
    $a \leftarrow \sqrt{a+2}$
  endrepeat

2.3 Strong Induction and Other Variants

Imagine again our infinite row of upright dominoes. This time, however, imagine that each domino after the second is huskier than the preceding one—if the preceding domino alone falls on it, that force alone will not be enough to knock it down. Imagine it takes the combined force of all the preceding dominoes falling over to topple the next one. Again, someone knocks over the first one. Will they still all fall down?

We reason as follows. Call the dominoes $P(1)$, $P(2)$, and so on. When $P(1)$ falls on $P(2)$, that *is* the combined force of all preceding dominoes. (That's why we said all dominoes *after* $P(2)$ were huskier.) So $P(2)$ falls over. Now $P(3)$ has the force of $P(1)$ and $P(2)$ on it, so it falls over. Now $P(4)$ has the force of all preceding dominoes on it. And so on. The answer is Yes.

Our physics here is a bit shaky. When $P(100)$ is hit, $P(1)$ has long since come to rest far down the line and isn't contributing any force. Fortunately, it's the mathematical principle that we're interested in; the parable is simply a convenient analogy. We hope we have convinced you of the

---

Principle of Strong Mathematical Induction

To prove $P(n)$ for all integers $n \geq n_0$, it suffices to prove both

i) the Basis Step: $P(n_0)$, and
ii) the Inductive Step: for all $n > n_0$, the truth of all $P(j)$ from $P(n_0)$ to $P(n-1)$ implies the truth of $P(n)$.

We need strong induction because, as noted in Example 2, Section 2.2, ordinary (weak) induction doesn't always do the trick. It's nice if we can get the conclusion of the inductive step by assuming just the immediately preceding case—we have fewer balls to juggle—but sometimes that case doesn't give us enough punch. Strong induction is the opposite extreme: We assume every preceding case.

Often, we don't need that much punch; just a few preceding cases will do. Fortunately, further "intermediate" induction principles are not needed. The inductive step of strong induction *allows* us to use all preceding cases, but we are not *required* to use them. Any situation where we use one or more preceding cases is therefore justified by strong induction. For instance, ordinary weak induction is actually a special case of strong induction. In other typical cases, we need

- the immediately preceding case and one early case; or
- the k immediately preceding cases (k is usually 2); or
- one or more preceding cases, but you don't know which ones.

We now give three examples, one for each case.

EXAMPLE 1 **The Generalized Distributive Law**
Assuming the distributive law

$$a(b+c) = ab + ac, \tag{1}$$

prove the generalized distributive law: For $n \geq 2$,

$$a \sum_{i=1}^{n} b_i = \sum_{i=1}^{n} ab_i. \tag{2}$$

Solution Let $P(n)$ be the nth case of Eq. (2). The basis, $P(2)$, is Eq. (1), and we were told to assume it. Inductive step: We break down $\sum_{i=1}^{n} b_i$ into $\left(\sum_{i=1}^{n-1} b_i\right) + b_n$.

Thinking of the sum to $n-1$ as a single term B, we have the start of a proof beginning with the LHS of $P(n)$:

$$a\Big(\sum_{i=1}^{n} b_i\Big) = a(B+b_n).$$

Our immediate impulse is to distribute on the right, after which we can expand aB using $P(n-1)$. However, in distributing, we are using $P(2)$. That is, the entire argument is

$$\begin{aligned} a\Big(\sum_{i=1}^{n} b_i\Big) &= a\Big(\sum_{i=1}^{n-1} b_i + b_n\Big) \\ &\overset{P(2)}{=} a\sum_{i=1}^{n-1} b_i + ab_n \\ &\overset{P(n-1)}{=} \sum_{i=1}^{n-1} ab_i + ab_n = \sum_{i=1}^{n} ab_i, \end{aligned}$$

where again we have inserted P's over the equal signs to indicate where we have used what. In other words, this is a strong induction using one early case and the immediately preceding case. ∎

Remark. Both the distributive law and the generalized distributive law are probably obvious to you, in which case you probably feel that we should treat them equally—either assume both without proof, or if we insist on proofs, then prove $P(2)$ also. We agree with you! As far as having confidence in the truth of the distributive law is concerned, we don't claim that Example 1 has accomplished anything. Instead, it is intended as a first example of strong induction, one in which the statement being proved is so familiar that you can concentrate on the method.

How does one prove $P(2)$? By induction—aided by many other things. We don't do it here because it is very subtle. Indeed, the more elementary the arithmetic, the harder it is to prove things rigorously—we have less to work with. For a proof, take a course with a title like "Foundations of Analysis" or "Philosophy of Mathematics".

EXAMPLE 2 Suppose we are given a sequence $a_1, a_2, \ldots$, where for $n > 1$,

$$a_{n+1} = 2a_n + a_{n-1}. \tag{3}$$

Show that if a_1 and a_2 are integers, then all terms are integers.

Solution Again, this is pretty obvious and we do it simply as a first example of its type. It's obvious because, when you add integers to get new terms, you get integers. But this is just the gist of the inductive step. Realizing that, let's go right to a proof.

Definition. Let $P(n)$ be the statement that a_n is an integer.

Basis. $P(1)$ and $P(2)$ are given.

Inductive step. Assume that all terms through a_n are integers. To show $P(n{+}1)$, we simply note that a_{n+1} is an integer by Eq. (3). That is, we actually use the preceding cases $P(n)$ and $P(n{-}1)$. ∎

To be very precise, this valid induction is not an example of strong induction as we defined it at the start of this section. As defined, for each n the inductive step may use all preceding cases, but when $n = n_0 + 1$, there is only one preceding case. However, the inductive argument using Eq. (3) always requires two distinct preceding cases. That's why we gave two basis cases (a_1 and a_2 are integers), whereas strong induction starts with one. Nonetheless, in practice any induction using more than one preceding case at least some of the time is referred to as strong.

EXAMPLE 3 Prime factorization. Every kid in junior high knows that every positive integer greater than 1 can be factored into primes; e.g.,

$$18 = 2 \cdot 3 \cdot 3, \qquad \text{and} \qquad 1001 = 7 \cdot 11 \cdot 13.$$

Prove this fact by strong induction.

PROOF Define $P(n)$ to be the statement that n can be factored into primes. Then the basis is $P(2)$, which is true because 2 is a prime. As for the inductive step, suppose that all numbers before n have prime factorizations. Consider n. If it is a prime, we are done—n is a factorization of itself into a single prime factor. If n is not prime, then by definition it has some factorization $n = rs$. We do not claim that r and s need be primes, but clearly they are less than n. Therefore, by $P(r)$ and $P(s)$, they themselves have prime factorizations $r = \prod p_i$ and $s = \prod q_j$. By strong induction

$$n = \left(\prod p_i\right)\left(\prod q_j\right)$$

is a prime factorization of n. ∎

Remark 1. You may know an even stronger theorem, the *unique* factorization theorem for integers. Its proof also involves induction, but is much harder. We discuss it further in Section 2.5.

Remark 2. The difference between Example 1 and Examples 2 and 3 is that here we don't know the relation of r and s to n. One of them may be small and the other large, or they may both be middling—we don't know and don't care. This sort of situation arises quite frequently in number theory.

Finally, we introduce one more induction variant. This one is really several weak inductions in parallel, rather than a strong version.

EXAMPLE 4 Prove that any item costing $n > 7$ kopecks can be bought using only 3-kopeck and 5-kopeck coins.

Solution The key idea is: If you can buy an item costing $n - 3$ kopecks with just such coins, you can also buy an item costing n kopecks—use one more 3-kopeck

FIGURE 2.8

The domino parable for split induction

coin. So as long as we can get back to the basis by multiples of 3, we can prove the nth case. But that means we need three basis cases and three parallel "inductive rows", as illustrated in Fig. 2.8.

Here's a good write-up:

Definition. Let $P(n)$ be the statement that n kopecks can be paid using 3-kopeck and 5-kopeck pieces only.

Basis. $P(8)$ is true since $8 = 3+5$. $P(9)$ is true since $9 = 3+3+3$. $P(10)$ is true since $10 = 5+5$.

Inductive step. For $n > 10$, assume $P(n-3)$. However $n - 3$ kopecks are paid using 3- and 5-kopeck pieces, use one more 3-kopeck piece and n kopecks have been paid. ∎

Let's call this variant a **split** (weak) **induction**. Here we split three ways.

There is one more variant that we should mention. When you have a problem involving several variables, and you can't figure out how to do an induction on just one of them, you have to do a **multiple induction**. Most commonly, there are two variables, in which case this variant is called **double induction**. We postpone a discussion of double induction until we first need it, in Section 5.3, Example 7 Continued. For those who want to figure out double induction for themselves, see [13, Supplementary Problems].

Problems: Section 2.3

1. Almost every rule of algebra has a generalized form. Assuming the standard form, use strong induction to prove the following generalized forms. The sums and products below are assumed to run from 1 to n. The standard form is the case $n = 2$.
   a) $\left(\sum x_i\right)/m = \sum(x_i/m)$
   b) $\left(\prod x_i\right)^\alpha = \prod\left(x_i^\alpha\right)$
   c) $b^{\sum a_i} = \prod b^{a_i}$
   d) $\log\left(\prod x_i\right) = \sum(\log x_i)$

2. For $n > 2$, prove $\sum_{i=1}^n |x_i| \geq \left|\sum_{i=1}^n x_i\right|$, assuming the case $n = 2$. Then, prove the case $n = 2$ also.

3. It is well known that in a triangle with sides a, b, c, we have $c < a+b$. State and prove a generalization of this inequality to arbitrary convex polygons.

4. It is a fact that if a prime p divides the product ab, where a and b are integers, then either p divides a or p divides b. Assuming this fact, prove that if p divides $\prod a_i$, where each a_i is an integer, then p divides at least one of the a_i's.

5. Let A and B be sets. Assuming
$$A \cap (B_1 \cup B_2) = (A \cap B_1) \cup (A \cap B_2),$$
prove
$$A \cap \left(\bigcup_{i=1}^n B_i\right) = \bigcup_{i=1}^n (A \cap B_i).$$

6. Prove the dual of [5]; that is, prove the statement with unions and intersections interchanged.

7. Assuming DeMorgan's first law of sets,
$$\overline{A \cup B} = \overline{A} \cap \overline{B},$$
state and prove its generalized form. Recall from Section 0.1 that $\overline{A}$ is the complement of A relative to some universal set U.

8. State and prove the generalization of DeMorgan's second law,
$$\overline{A \cap B} = \overline{A} \cup \overline{B}.$$

9. Use [5,7,8] to give another proof of [6] that does not directly use an induction.

10. Explain why
$$\lfloor x \rfloor + \lfloor y \rfloor \leq \lfloor x+y \rfloor \leq \lfloor x \rfloor + \lfloor y \rfloor + 1.$$
State and prove a generalization that bounds $\left\lfloor \sum_{i=1}^n x_i \right\rfloor$.

11. For the ceiling function, find and prove similar results to those in [10].

12. Every odd integer ≥ 3 has an odd prime divisor. Given the factorization theorem, this statement is obvious and easy to prove. (What integers have even factors only?) For practice, prove it from scratch by strong induction.

13. Suppose you can prove that the truth of $P(n)$ implies the truth of $P(n+2)$ for all positive n. What else must you be able to prove before you can assert that $P(n)$ is true for all positive n?

14. Let $P(n)$ be a proposition defined for each positive integer n.
    a) Suppose you can prove that $P(n+1)$ is true for $n > 3$ if *all* of $P(n)$, $P(n-1)$, $P(n-2)$ and $P(n-3)$ are true. What else must you prove to assert that $P(n)$ is true for all positive n?
    b) Suppose you can prove that $P(n+1)$ is true if *any* of $P(n)$, $P(n-1)$, $P(n-2)$ or $P(n-3)$ are true. What else must you prove to assert that $P(n)$ is true for all positive n?

15. Prove the result of Example 4 again, using a 5-kopeck piece in the inductive step.

16. Prove that
$$(-1)^n = \begin{cases} 1 & \text{if } n \text{ is even} \\ -1 & \text{if } n \text{ is odd} \end{cases}$$
by a split induction.

17. Suppose that a sequence of numbers $a_1, a_2, a_3, \ldots$ has the property that for each $n > 1$,
$$a_{n+1} = 2a_n + a_{n-1}. \tag{4}$$
    a) Prove: If *any* two consecutive terms in the sequence are integers, then every term is an integer.
    b) Prove: If a_1 and a_2 are increased, and all other terms are changed to maintain Eq. (4), then all terms are increased.

18. Suppose that a sequence $a_1, a_2, \ldots$ satisfies

$$a_{n+1} = 2a_n - a_{n-1}. \tag{5}$$

Prove: If a_1 is increased, a_2 is increased more than a_1 is increased, and all other terms are changed so that Eq. (5) remains true, then every term is increased.

Caution: This proof is much trickier than that in [17b]. If you make the obvious choice for $P(n)$ — "a_n is increased"—you won't be able to do the inductive step, even assuming the two preceding cases. Try it! Then find a "good" $P(n)$.

19. Suppose that a function T is defined on the nonnegative integers by

$$\begin{aligned} T(0) &= 0, \\ T(n) &= T(\lfloor n/3 \rfloor) + T(\lfloor n/5 \rfloor) + \\ &\qquad T(\lfloor n/7 \rfloor) + n, \qquad n > 0. \end{aligned}$$

Prove that $T(n) < 4n$ for $n > 0$.

20. Suppose a sequence $a_0, a_1, \ldots$ is defined by $a_0 = a$, $a_1 = b$ and

$$a_n = 1 + \max\{a_{\lfloor n/2 \rfloor}, a_{\lceil n/2 \rceil}\}, \quad n > 0.$$

Prove: If $a \leq b$, then $\{a_n\}$ is an increasing sequence. (This problem is relevant to all sorts of binary division algorithms, e.g., BINSEARCH in Example 2 of Section 1.6. Do you see why?)

21. How many strictly increasing sequences of integers are there that begin with 1 and end with n? Prove it. (A strictly increasing sequence $a_1, a_2, \ldots$ is one in which, for all i, $a_i < a_{i+1}$. For instance, when $n = 6$, two of the strictly increasing sequences from 1 to n are 1,2,6 and 1,3,5,6. Notice that different lengths are allowed.)

22. Consider the directed graph with vertex set $V = \{1, 2, \ldots, n\}$ such that there is an edge from vertex i to vertex j iff $i < j$. (Our diagrams of tournaments in Section 2.2, for instance, Fig. 2.4, are directed graphs.) How many directed paths are there from 1 to n? Discover a pattern and prove it. Then look at [21].

23. Let $Q(n)$ be the statement "Every positive integer from 2 to n has a prime factorization." Prove that every positive integer has a prime factorization by *weak* induction on Q.

24. Round-robin tournaments revisited. Here is another attempt to come up with a way to determine the strength of a team in a round robin. Let a team be called a "Real Winner" if, for every other team, it either beat that team or beat some third team which beat that team. Prove that every tournament has a Real Winner. Try both weak and strong inductions.

25. In [24], it happens to be easier to prove the theorem without induction. Given a round robin, show that any team which wins the most games is necessarily a Real Winner.

26. Alas, the real winner concept of [24] is not very satisfactory, for there can be too many winners. Indeed, it's a theorem that, for every $n > 4$, there exists an n-tournament in which *every* team is a Real Winner. Prove this theorem. *Suggestion*: It is possible, but tricky, to prove this by ordinary induction. It turns out to be much simpler to prove it by the induction principle of [13], modified to start at $n = 5$ instead of $n = 1$.

27. Let θ be an angle for which $\sin\theta$ and $\cos\theta$ are rational. Prove that $\sin n\theta$ and $\cos n\theta$ are rational for all integers n.

28. Fact: Given two squares, it is possible to dissect them into a finite number of polygonal pieces (all sides straight) that can be reassembled into a single square. Fig. 2.9 shows one way to do it: If the squares have sides a and b, with $a \leq b$, then don't cut the smaller square at all, cut the larger square into four pieces as shown, and reassemble.

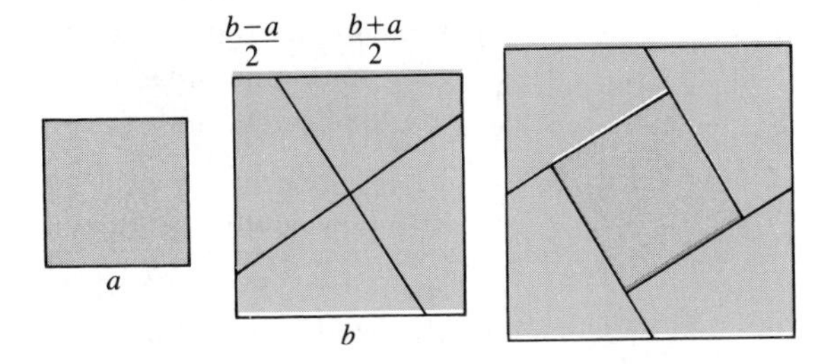

FIGURE 2.9

a) Verify that this method works, i.e., the reassembly really does fit together just right (no overlaps or gaps) and forms a square.

b) Prove that *any* number of squares may similarly be dissected and reassembled into one square.

c) Make three squares of sides 4, 5 and 6, respectively, from different colors of construction paper. Actually cut them up into the polygonal pieces indicated by your proof, and reassemble. (No, this isn't art class, but there is something to learn from this hands-on approach about the relation of inductive proofs to explicit constructions.)

29. The **Cauchy-Schwarz Inequality** says: For any n-tuples $(a_1, a_2, \ldots, a_n)$ and $(b_1, b_2, \ldots, b_n)$ of real numbers,

$$\left(\sum a_i^2\right)\left(\sum b_i^2\right) \geq \left(\sum a_i b_i\right)^2.$$

In words, the product of the sums of the squares is greater than or equal to the square of the sum of the products. Try to prove this inequality by induction. It can be done, but the algebra is very involved. There are much simpler proofs without induction, but you need to know some vector algebra to do them.

30. Another form of the Triangle Inequality (see Section 0.4) is:

$$\sqrt{\sum a_i^2} + \sqrt{\sum b_i^2} \geq \sqrt{\sum (a_i+b_i)^2}\,.$$

Why Triangle Inequality? Because (taking $n = 2$ for convenience) if we consider the points $D = (0,0)$, $A = (a_1, a_2)$ and $C = (a_1+b_1, a_2+b_2)$, then this inequality says that in triangle ACD, $DA+AC \geq DC$. Show, using just algebra, that the Triangle Inequality follows from the Cauchy-Schwarz Inequality.

31. (Uses calculus) Let $[n] = \{1, 2, \ldots, n\}$. Assuming the derivative product formula $(fg)' = f'g + fg'$, prove the generalization

$$\left(\prod_{i=1}^{n} f_i\right)' = \sum_{i\in[n]} \left(f_i' \prod_{j\neq i} f_j\right).$$

2.4 How to Guess What to Prove

So far, we've emphasized *how* to prove things by induction. But how do you conjecture *what* to prove?

Unfortunately, there is no sure-fire answer. In many cases, though, the first part of the inductive paradigm can help—collect data and look for a pattern. There are other techniques, too. Sometimes it pays to jump right in and try to prove *something*, even a wild guess; you'll probably get stuck, but how you get stuck will give you some ideas about what is more reasonable to prove. We begin with an example of the inductive paradigm approach.

EXAMPLE 1 Into how many regions do n straight lines divide the plane?

Solution Figure 2.10 shows four straight lines dividing the plane into 11 regions. You should imagine all the lines going off to infinity in both directions. Let us

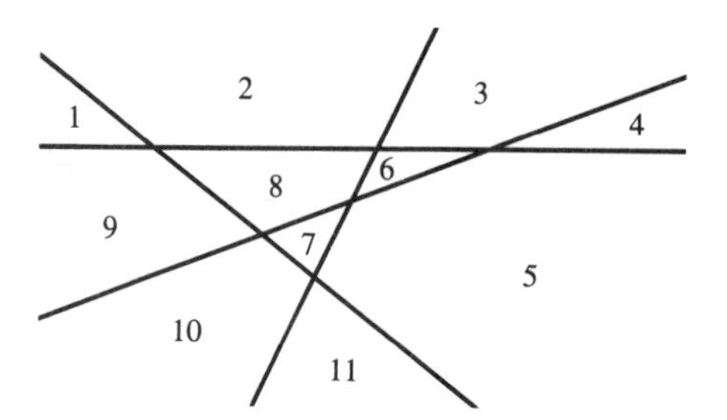

FIGURE 2.10

A plane divided by four straight lines

assume, as in this figure, that no pair of the lines are parallel and no three pass through the same point. Such lines are said to be in **general position**. We assume the lines are in general position because we are hoping for a unique answer in terms of n; even with two or three lines, without this assumption the same n can lead to different numbers of regions.

Collect some data. Draw figures with $n = 1, 2, 3$ and 5 lines. In fact, draw several different cases for each n, and some more cases with $n = 4$, to convince yourself that the number of regions does seem to be determined by n alone. Let r_n be this number that you believe exists. You should get the first two columns of Table 2.1. (Ignore the other columns for now.)

TABLE 2.1
Number of Regions formed by n Lines

Number (n) of Lines	Number (r_n) of Regions	n^2	r_n/n^2	$r_n - (n^2/2)$
1	2	1	2.00	1.5
2	4	4	1.00	2.0
3	7	9	0.78	2.5
4	11	16	0.69	3.0
5	16	25	0.64	3.5
6	22	36	0.61	4.0

Now you need to guess a formula for r_n. This is hard to do directly. So first make some rough observations, and then try to refine them.

To start with, r_n is clearly growing faster than n. How much faster? The simplest thing to try is n^2. We've put that in column 3. Well, that's too big. But perhaps only by a multiplicative factor. Indeed, in column 4 we have computed r_n/n^2. This looks promising. It's decreasing, but seemingly more and more slowly. Maybe it doesn't reach 0. Indeed, believing that a simple problem like this ought to have a simple solution, we guess the nicest thing that looks consistent—that the limit of the ratio is 1/2.

Next, since in each particular case r_n is more than $\frac{1}{2}n^2$, it makes sense to look at the difference. Look at the last column, where we've written these differences down. The pattern there is obvious—we're home! Clearly,

$$r_n - \tfrac{1}{2}n^2 = \tfrac{1}{2}(n+2),$$

in which case

$$r_n = \tfrac{1}{2}(n^2 + n + 2). \tag{1}$$

Remember, we haven't *proved* anything yet. But now we have our conjecture. We set $P(n)$ to be: Whenever n lines are placed in general position on the plane, they form $\frac{1}{2}(n^2+n+2)$ regions. We can now proceed to a proof by mathematical induction.

Basis. Clearly, any one line forms two regions, and

$$\tfrac{1}{2}(1^2+1+2) = \tfrac{1}{2}(4) = 2. \checkmark$$

Inductive step. The difficulty with having obtained our formula by guessing is that guessing doesn't give us any clue as to how to break down $P(n+1)$ to involve $P(n)$. But perhaps, when you drew your more complicated examples, you did so by adding another line to your simpler examples, in which case you may already have observed the key idea we exploit next.

Assuming $P(n)$, we must show that any general configuration of $n+1$ lines has a certain number of regions. So imagine someone gives us an arbitrary $(n+1)$-line configuration. We need to see an n-line configuration in it. Temporarily color one of the $n+1$ lines invisible. The remaining lines are still in general position, so by $P(n)$ there are $\frac{1}{2}(n^2+n+2)$ regions. Now make line $n+1$ visible again. Since it crosses each of the n previous lines once, it necessarily passes through exactly $n+1$ previous regions, dividing each of them in two. (See Fig. 2.11, where $n = 3$ and the fourth line is bold.) Hence r_{n+1} is also uniquely determined:

$$\begin{aligned} r_{n+1} &= r_n + (n+1) \qquad (2)\\ &\overset{P(n)}{=} \frac{n^2+n+2}{2} + n+1 = \frac{n^2+3n+4}{2}\\ &= \frac{(n^2+2n+1)+(n+1)+2}{2} = \tfrac{1}{2}[(n+1)^2+(n+1)+2], \end{aligned}$$

which gives $P(n+1)$. ∎

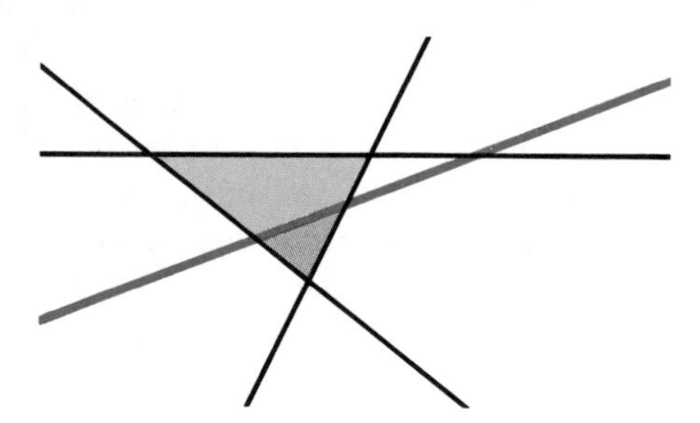

FIGURE 2.11

Putting back line $n+1$.

Remark 1. In the solution to Example 1 we have illustrated all three steps of the inductive paradigm:

1. Compute the results in some specific cases;
2. Conjecture a generalization on the basis of these results; and
3. Prove the correctness of the conjecture.

Remark 2. Guessing is not the only way to approach this problem. For instance, had we started with the recursive analysis in the proof, we would have gotten Eq. (2) before we wrote up Table 2.1. In that case, it would have been useful to write the first two columns as in Table 2.2. The point is, by *not* combining the sums, we get to observe a simple pattern. We conjecture immediately that $r_n = 1 + \sum_{k=1}^{n} k$. Since we already know the formula for this sum, our conjecture becomes $r_n = 1 + \frac{1}{2}n(n+1)$, which is indeed equivalent to Eq. (1).

TABLE 2.2
Another Way to Count the Regions from n Lines

n	r_n
1	2
2	$2+2$
3	$2+2+3$
4	$2+2+3+4$
5	$2+2+3+4+5$

The proof by this second method would still be by induction, but we would have arrived at our conjecture more quickly—assuming that we already knew the formula for $\sum k$. If we didn't know that, we would have to apply the same sort of table analysis to it that we did to r_n above [1]. ∎

EXAMPLE 2 Find the number of moves in Straightline Towers of Hanoi [41, Section 2.2].

Solution In this version of TOH, the rings must get from one end pole to the other, and each move must be between *adjacent* poles. In the problem in Section 2.2, you were asked to prove that this version could be solved and to write an algorithm to do it. Now we analyze the number of moves involved.

Let s_n be the (minimum) number of moves for n rings. Let's build a table. Just as you drew pictures to help you fill Table 2.1, you could fill this table by actually playing the game for $n = 1, 2, 3, \ldots$. Indeed, by this method we easily find that $s_1 = 2$ and $s_2 = 8$, but then the method gets tedious—even if you don't get confused and make a mistake. Straightline TOH takes many more steps than ordinary TOH.

Instead, it's better to use recursion from the start. Without loss of generality, let's say that the rings start on the left. The way to handle $n{+}1$ rings is:

1. By the previous case, transfer the top n rings from left to right (this takes s_n moves).

2. Then move the $n+1$st ring to the middle (1 move).
3. Then transfer the top n back to the left (s_n moves).
4. Move the $n+1$st to the right (1 move).
5. Finally, transfer the top n to the right again (s_n moves).

We conclude that

$$s_{n+1} = 3s_n + 2. \tag{3}$$

From this equation it is an easy matter to fill out the top two rows of Table 2.3.

TABLE 2.3
Number of moves in Straightline TOH

n	1	2	3	4	5	6
s_n	2	8	26	80	242	728
s_{n+1}/s_n	4	3.25	3.08	3.03	3.01	
3^n	3	9	27	81	243	729

What shall we compare s_n to this time? It seems to grow very fast, so a low-power polynomial seems unlikely. We could try higher powers, but it seems unreasonable to expect, say, n^{17}, to appear in a simple problem like this. So we try exponentials instead—they grow faster than powers. That is, we conjecture that the most significant term of s_n is Ab^n, where the coefficient A and the base b are some constants. To find A and b, we could start by comparing $\{s_n\}$ to specific sequences such as $\{2^n\}$ and $\{3^n\}$, but there is a nice trick which allows us to determine b right away—and from that, A. Namely, if the most significant term really is of the form Ab^n, then

$$\text{as } n \text{ increases,} \quad \frac{s_{n+1}}{s_n} \quad \text{approaches} \quad \frac{Ab^{n+1}}{Ab^n} = b.$$

Furthermore

$$\text{as } n \text{ increases,} \quad \frac{s_n}{b^n} \quad \text{approaches} \quad \frac{Ab^n}{b^n} = A.$$

Therefore, in the third row of Table 2.3 we compute s_{n+1}/s_n. The results suggest very strongly that $b = 3$, so we put the values of 3^n in the fourth row. At this point it's obvious that $A = 1$. More than that, it's obvious that $s_n = 3^n - 1$. (Had A been different from 1, the exact form of s_n wouldn't be obvious yet, but we could then subtract $A3^n$ from s_n in order to see the pattern, just as we subtracted $n^2/2$ from r_n in Example 1.) In any event, a proof by induction [3] confirms that $s_n = 3^n - 1$. ∎

In a certain sense, exponential formulas as in Example 2 are more common than polynomial formulas as in Example 1. Why? Because the recursive formulas associated with most counting problems have exponential solutions, as you'll see in Chapter 5 (at which point you won't need to guess as much). Therefore it is usually best to look first at successive values of s_{n+1}/s_n, as in Example 2. Even if the answer is polynomial, looking at s_{n+1}/s_n will do no harm. The ratios will simply go to 1, indicating that the "exponential part" of the formula is $1^n = 1$; that is, the "real action" is due to a nonexponential part. For instance, the successive ratios r_{n+1}/r_n in Example 1 are

$$2, \quad 1.75, \quad 1.57, \quad 1.45, \quad \text{and} \quad 1.38,$$

which do seem to head towards 1 already. We summarize the method of these two examples as follows: Assuming the sequence is called $\{s_n\}$,

1. Find the *type* of the highest order term (exponential, power function, etc.). If the type is exponential (b^n), then you can find b by guessing a limit for successive ratios s_{n+1}/s_n.
2. Find the *coefficient* of the highest order term. For instance, if the type is 3^n, then the actual term is $A3^n$ for some coefficient A. If the type is n^2, then the actual term is An^2. To find A, guess a limit for the successive ratios of s_n to 3^n (or to n^2, or to whatever type you have determined). In Example 1, we did this in the fourth column; in Example 2, in the third row.
3. Subtract your highest order approximation from the actual value s_n. For instance, in Example 1, the difference is $r^2-(n^2/2)$. This will be a simpler sequence. If you cannot guess what it is, find the type of its highest order term and iterate steps 1–3 until you can.

Warning. We don't claim that this method is foolproof, though you'll be surprised how far you can get with it. There are lots of useful sequences which are neither exponential nor polynomial. For instance, sequences describing the complexity of algorithms often involve base 2 logarithms—recall the analysis of BINSEARCH in Chapter 1. In the problems you will have to test your ingenuity against sequences of a number of different types.

Both preceding examples concern sequences for which we are trying to find formulas. This is the sort of problem for which guess and check works best. It does not work so well for Theorem 2, Section 2.2 (rankings in round-robin tournaments). In the examples so far in this section, we had the limited task of turning numbers into formulas. For the tournament problem we have the much vaguer task of turning diagrams into an assertion. Furthermore, in the examples so far each value of n corresponded to just one formula value. In the tournament problem, for each value of $n > 2$ there are several different tournaments, and the number grows rapidly.

So how does one discover something like the tournament theorem? One way is to make a conjecture, even with the flimsiest evidence, and try to prove it. For instance, if we are interested in rankings, it is natural to *want* an ordering in which t_i beats t_{i+1} for each i. So we say: "I'll try to prove this. If I succeed, I have a surprising result; if I don't, the place where my attempt breaks down will suggest a more restricted situation where the ordering I want exists, or a weaker ordering result that is always true." In other words, when it is hard to make conjectures from data, mathematical induction (or other forms of deduction) may be used as a *discovery technique*.

Specifically, when we try to prove the ranking theorem, we very quickly get to "Suppose that I already knew it was true for n-tournaments". As soon as we say those magic words, we get Fig. 2.5 and the proof follows. But if we hadn't dared to try a proof, we would never have said the magic words.

But is this daredevil approach useful when it fails? Watch and see. Look at Fig. 2.12(a), which repeats Fig. 2.4 from Section 2.2. We pointed out there that in the order 5,3,4,1,2, each team beat the team to its right. But more is true: Team 2 also beat team 5. In other words, if we wrap the ordering 5,3,4,1,2 clockwise around a circle, as in Fig. 2.12(b), each team beats the team directly clockwise from it. So maybe we should have conjectured more: that every tournament has such a "circular order".

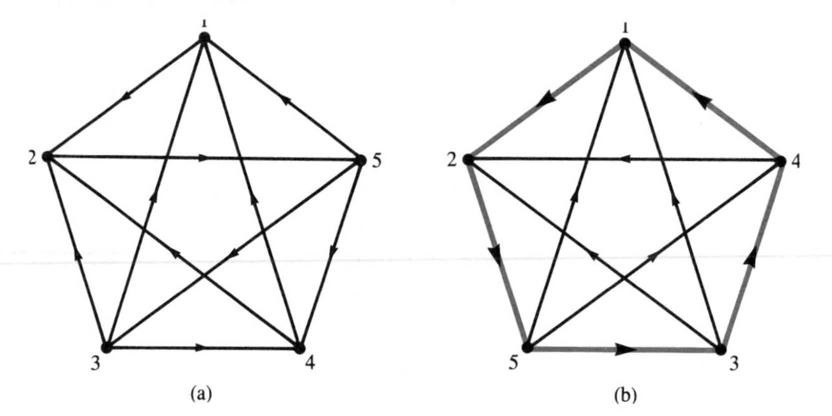

FIGURE 2.12

A tournament and a circular ranking for it

Rather than draw more tournaments, we jump right in and try to prove this conjecture. Let $P(n)$ be: Every n-tournament has a circular order. Right away the basis case, $P(2)$, shows that this conjecture is false—the unique 2-tournament has no cycles because there is only one game.

All right, our conjecture is wrong. But maybe it's wrong only for small tournaments. Maybe it's true for all tournaments with at least some number n_0 of teams. Let's turn to the inductive step, where the value of n_0 is irrelevant.

So suppose that $P(n)$ is true for some n; will $P(n+1)$ be true? Let's see. Consider any $(n+1)$-tournament. Deleting any one team t^*, we are left with an n-tournament, so by hypothesis there is a circular order $t_1, t_2, \ldots, t_n$. Can t^* be inserted into this circle? Yes, if and only if there is some i such that t_i beat t^* and t^* beat t_{i+1}. (When $i = n$, $i+1$ means 1.) Well, so long as t^* beat some teams and lost to others, there must be such a t_i and we would be done. But what if t^* lost to *all* other teams, or beat all others? We could certainly make up such a tournament, so this attempted inductive step is invalid.

Moreover, this reasoning shows that the conjecture is false for every n. Whenever some team wins all its games or loses all its games, there cannot be a circular order involving that team, for such an order requires it to both win and lose.

But we have hardly reached a dead end. The inductive step broke down because of a special case. If n is large, you wouldn't expect to find a team which loses all its games or wins them all. In other words, maybe almost all n-tournaments have a circular order, or maybe all those without "total winners" or "total losers" do. In short, there is much more to look into. We won't, because tournaments per se are not our purpose here. (Consult one of the several mathematics books on tournaments to find the answers, or work on it yourself.) Our purpose was merely to convince you that trying to prove things is a good discovery method.

Let us summarize the message of this section. Mathematics books typically devote little space to the issue of how results are discovered, and they rarely even hint that experimenting and data collection have any role in the process. Yet all mathematicians use these methods. You should do the same. In fact, generating such information has gotten to be quite a lot of fun now that computers can help. However, mathematicians also try to employ more powerful techniques—techniques that allow them to discover what is true and prove it almost simultaneously, and techniques that allow them to attempt proofs early on and develop a sense of what is reasonable to conjecture as their deductions fail. You should try these approaches, too.

Problems: Section 2.4

1. Pretend that you've never seen the closed-form formula for $\sum_{k=1}^{n} k$. Use the method of Example 1 to find the formula.

2. Discover and prove a formula for $\sum_{k=1}^{n} k^2$.

3. Prove the formula $s_n = 3^n - 1$ for Straightline TOH.

4. "Rediscover" the formula for the number of moves in ordinary TOH using the method of Example 2.

5. Come up with formulas for the following sequences. *Note*: In one sense there is no right answer for these sequences, because we give you only a finite number of terms. It is a theorem that, for any finite sequence, there are infinitely many formulas that correctly describe it. (The formulas would start giving different values if you added more terms.) But we have in mind a simple formula for each sequence. With the methods of this section, you should be able to find it. Assume that the first term corresponds to $n = 1$.

a) 3, 6, 11, 20, 37, 70, 135, 264, 521, 1034

b) 1, 0, −1, 0, 7, 28, 79, 192, 431, 924

c) 1, 5, 13, 41, 121, 365, 1093, 3281, 9841, 29525

d) 0, 0, 6, 24, 60, 120, 210, 336, 504, 720

e) 2, 8, 24, 64, 160, 384, 896, 2048, 4608, 10240

6. Find and prove a formula for

$$1 \cdot 1! + 2 \cdot 2! + 3 \cdot 3! + \cdots + n \cdot n!$$

7. Find and prove a formula for

$$\frac{1}{1 \cdot 4} + \frac{1}{4 \cdot 7} + \frac{1}{7 \cdot 10} + \cdots + \frac{1}{(3n-2)(3n+1)}.$$

8. For the expression

$$\left(1 - \frac{4}{1}\right)\left(1 - \frac{4}{9}\right)\left(1 - \frac{4}{25}\right)\cdots\left(1 - \frac{4}{(2n-1)^2}\right),$$

conjecture a closed form and prove it correct.

9. Find and prove a closed formula for

$$\sum_{k=1}^{n} \frac{1}{k(k+1)}.$$

Compare with [10, Section 2.2].

10. Discover and prove a formula for $\sum_{S \subset [n]} |S|$, where $[n] = \{1, 2, \ldots, n\}$.

11. Find and prove a formula for a_n if $a_1 = 1$ and

$$a_n = a_{\lceil n/2 \rceil} + a_{\lfloor n/2 \rfloor} \quad \text{for } n > 1.$$

12. Find the solution to

$$L(n) = 1 + L(\lceil n/2 \rceil), \qquad L(1) = 0.$$

13. Find the solution to

$$L(n) = 1 + L(\lfloor n/2 \rfloor), \qquad L(1) = 0.$$

14. Consider again the regions formed by n lines in general position in the plane. Now, instead of counting the regions, we merely want to color them—so that no two regions sharing a common boundary are the same color. (Regions that share just a point in common—for instance, "opposite" regions at a point where two lines cross—need not be different colors.) What is the minimum number of colors necessary when there are n lines? Prove it.

15. Same as [14], but now the lines need not be in general position; several may share a point in common or be parallel.

16. Where in the proof for Example 1 did we use the hypothesis that the lines are in general position?

17. State a conjecture for the value of $\sum_{k=1}^{t} \sum_{j=k}^{s} 1$ and prove it correct.

18. What does the following algorithm do? Prove it.

```
Input natural numbers n, m
Algorithm MYSTERY
   function F(n, m)
      if n = 0 then F ← 0
               else F ← (m+n−1) +
                             F(n−1, m−1)
      endif
      return F
   endfunc
   print F(m, n)
```

19. Given just the midpoints of the sides of a triangle, the triangle can be determined. That is, if all that is given are three dots on the paper, there is a unique triangle for which they are the midpoints of the sides, and this triangle can be constructed with straightedge and compass. Given just the midpoints of the sides of an arbitrary n-gon, is it similarly possible to determine the original n-gon? Determine, with proof, the values of n for which this can be done. (Specifically, assume that you are given points $M_1, M_2, \ldots, M_n$. You are asked to find the vertices $V_1, V_2, \ldots, V_n$ of a polygon so that M_1 is the midpoint of segment V_1V_2, M_2 is the midpoint of $V_2V_3, \ldots,$ and M_n is the midpoint of segment V_nV_1.)

2.5 Faulty Inductions

One theory of pedagogy holds that an instructor should never deliberately put something false on the board, even if it's announced. The students will mindlessly copy it in their notebooks and memorize it!

Maybe—but it's hard to grasp precisely what something is without contrasting it with similar things that it is not. This is particularly true when the thing in question is subtle. In this vein, we provide several arguments that look like successful induction proofs but are not. The first is a classic—every mathematician knows some version of it. The others are our own contributions.

EXAMPLE 1 Prove that all women are blondes.

"Proof": Let $P(n)$ be the statement: For any set of n women, if any one of them is blonde, then all n of them are blonde. First, we will prove this for all n by induction. Then, taking n to be the number of women on earth, we get the special case: For the set of all women on earth, if any one of them is blonde, then all of them are. Since it is well known that at least one woman is blonde (look around, or maybe you are), we conclude that all women are blonde.

Now for the induction. $P(1)$ asserts: For any set of 1 woman, if any one of them is blonde, then all 1 of them is blonde. This is just a contorted way to say "a blonde is a blonde", and is clearly true.

Next, assume $P(n)$. We must show that, given an arbitrary set S of $n+1$ women, including a blonde, all the women in the set are blonde. To apply $P(n)$ we must look for sets of n women, including a blonde, within the $n+1$ set. Figure 2.13 shows how to do it. We let A be some subset of S that contains the known blonde and that has n members. We let B be any other n-subset of S that contains the known blonde and that includes the one member left out of A. By $P(n)$, set A is all blonde, and by $P(n)$ again, set B is all blonde. Therefore $A \cup B$ is all blonde. But $A \cup B = S$, so we are done. ∎

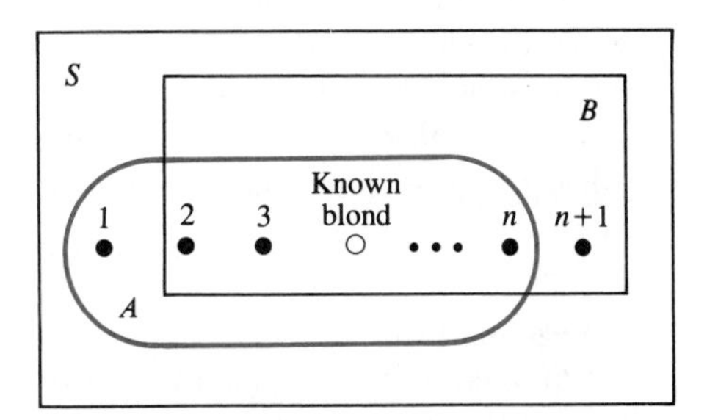

FIGURE 2.13

All women are blondes, by induction?

Where's the flaw? We've hidden the answer later in this book, so you might as well argue about it in class! The explanations of the next three faulty inductions are given after Example 4.

EXAMPLE 2 Prove that every integer > 1 has a unique prime factorization. (We really should say unique *up to order*: The factorizations $5 \cdot 2 \cdot 3 \cdot 2$ and $2 \cdot 2 \cdot 3 \cdot 5$ are not identical factorizations of 60, but they contain the same primes each used the same number of times, so they are considered the same "up to order".)

This time the claim is true—but the following proof is wrong. Why?

"Proof": Let $P(n)$ be the statement that n has a unique factorization. For the basis, $P(2)$ is clearly true since 2 is prime. By strong induction, assume $P(k)$ for $k < n$ and consider n. If n is prime, by definition there is no other way to factor it and we are done. Otherwise, it factors somehow as $n = rs$, where $r < n$ and $s < n$. Therefore, by assumption, r has a unique factorization $\prod p_i$ and s has a unique factorization $\prod q_j$. Thus $\prod p_i \prod q_j$ is one prime factorization of n, and since none of the factors in either piece can be changed, none of the factors of the whole thing can be changed. That is, the prime factorization of n is unique. ∎

EXAMPLE 3 Prove that in a regular n-gon, the sum of the internal angles is $180(n-2)$ degrees.

"Proof": Let $P(n)$ be the statement for a particular n. The basis case is $P(3)$, which is true by the famous theorem that a triangle has 180°. For the inductive step, given a regular n-gon, lop off vertex 2 by cutting along the diagonal from v_1 to v_3.

By assumption, the remaining $(n-1)$-gon has an interior angle sum of $180(n-3)$ degrees. Now put back the triangle, which has 180°. Adding, we get $180(n-2)$ degrees. ∎

Didn't we do the preceding proof in Section 2.2? So where's the flaw?

EXAMPLE 4 Consider 0–1 sequences in which 1's may not appear consecutively, except in the rightmost two positions. For instance, 0010100 and 1000011 are sequences of this sort, and 0011000 is not. Prove that there are 2^n "allowed" sequences of length n.

"Proof": Let s_n be the number of such sequences of length n and let $P(n)$ be the statement that $s_n = 2^n$. The basis case, $P(1)$, asserts that $s_1 = 2$. This assertion is clearly correct: Both 0 and 1 are allowed binary sequences of length 1. Now, to show that $P(n) \Longrightarrow P(n+1)$, take any allowed sequence of length n. We may append either 0 or 1 at the right end—in the latter case, we may create 11 in the last two positions, but that's okay. Therefore,

$$s_{n+1} = 2s_n = 2 \cdot 2^n = 2^{n+1},$$

finishing the proof.

Unfortunately, 2^n is the number of *all* binary sequences of length n. (Such facts will be justified at length in Chapter 4.) Yet some binary sequences (e.g., 01101) are not allowed because of the consecutive 1's. Where did we go wrong? ∎

Here are our explanations for the flaws in Examples 2–4.

Example 2 explained. The flaw is a **breakdown error**, one that is easy to fall into if you are asked to prove that something is unique or optimal. We broke n down into rs. There was nothing wrong with that. There was also nothing wrong with asserting by induction that none of the factors in either r or s can be changed. But who says that the only factorization of n is one which in the process factors r and s? Maybe n also equals tu, and maybe the prime factorizations of t and u are also unique, but the product of those factorizations looks nothing like the product of prime factorizations of r and s. To be sure, such a situation can't happen for the integers, but the argument in Example 2 doesn't prove it—and indeed it *does* happen in some other number systems (beyond the scope of this book). See also [14].

Moral: If the pieces are uniquely constructed, that doesn't mean the whole thing is uniquely constructed. The decomposition into pieces may not be unique.

Example 3 explained. The kicker here is the word "regular". It has resulted is another sort of breakdown error, one that is sometimes very subtle, although here it was pretty blatant. When we take a regular n-gon and lop off a corner, the resulting $(n-1)$-gon isn't regular and so $P(n-1)$ doesn't apply. That's the error: not actually obtaining the previous case when you break down the current case. Indeed, in this example it's hard to imagine what you could do to a regular n-gon that would make a regular $(n-1)$-gon appear naturally.

But, you object, the angle formula *is true* for the $(n-1)$-gon we obtain by deleting a corner, so what's wrong with using it? What's wrong is that in the method of induction you are allowed to use only the previous case(s) of the P you are trying to prove, and $P(n-1)$ is the statement that a *regular* $(n-1)$-gon has an angle sum of $180(n-3)$ degrees. True, we are allowed to use general knowledge and previous theorems as part of the argument in an induction, as we may in any proof. However, if we are going to use the general result about polygons from Example 2, Section 2.2, then we wouldn't now be trying to prove the special case of regular polygons.

In other words, we don't know any way to fix this faulty induction, despite the fact that the result is true. The only way to use induction is to prove the stronger result in Example 2, Section 2.2. This is an instance of **strengthening the inductive hypothesis**. It points up a curious subtlety about induction: To prove $P(n)$ for all n by induction, $P(n)$ can't be too strong, but it also can't be too weak. With other sorts of proofs, if you can't succeed in proving that Q implies R, you either strengthen Q (assume more) or weaken R (aim for less) and try again. But in induction, P is both the conclusion (in the case n) and the hypothesis (in the case $n-1$), and so you have a delicate balancing act.

How do you decide the right strength for P? As always, experience helps. Noting carefully how you get stuck also helps. For instance, suppose you really didn't know that the interior angle formula holds for all convex polygons and you were trying to prove it for regular polygons only. Presumably, you would start writing down a proof as in Example 3. However, seasoned by the experience gained from this section, you would catch the flaw in the inductive step. Since you would see that lopping off a corner does not preserve regularity or other special properties, you would suspect that the only way to make the induction work is to strengthen P by not assuming any special properties. Once you got the idea to strengthen P, you might think to strengthen it all the way to arbitrary polygons.

This example illustrates the important point that generalizations are sometimes easier to prove than special cases.

Example 4 explained. The flaw this time is a **buildup error**. We asserted that, by appending either 0 or 1 to all the allowed sequences of length n, (1) all the allowed binary sequences of length $n+1$ are obtained and (2) only the allowed ones are obtained. Claim (2) is false. For example, take $n = 3$ and the allowed sequence 011. Appending either 0 or 1 gives a disallowed sequence. In the problems there are some examples where building up case $n+1$ from case n leads to too few things instead of too many.

It is now time to explain the comment made earlier (in Remark 3 just before Theorem 1, Section 2.2) that the best place to use the previous case in the inductive step is generally in the middle, not at the beginning.

In the simplest inductions, $P(n)$ is a statement about a single object. For instance, in the proof that $\sum_{k=1}^{n} k = n(n+1)/2$, $P(n)$ is one specific equation, the equation just given for that particular n. Therefore, in proving $P(n) \Longrightarrow P(n+1)$, there is no problem about starting with $P(n)$ and working up to $P(n+1)$. There is one specific equation you are heading for.

However, in many inductions, $P(n)$ is a statement about a large set of things, even when n is fixed. For instance, $P(5)$ might be a statement about the set of all round-robin tournaments with 5 teams or the set of all allowed binary sequences of length 5. To prove something about all items in a set, it is necessary to argue about an *arbitrary* item in the set. That is why we have often started our inductive steps with lines such as "so imagine that somebody presents us with the results of an arbitrary $(n+1)$-tournament" (from the proof of Theorem 2 of Section 2.2), or "so imagine someone gives us an arbitrary $(n+1)$-line configuration" (from the proof for Example 1, Section 2.4). If instead we start the inductive step by looking at the previous case $P(n)$ and building up to $P(n+1)$, we are taking control of how example $n+1$ is constructed, and there is a danger that we will not construct an arbitrary example.

Similarly, if we want to count the elements of a set, it is necessary that we consider all the elements and nothing else. If we start the inductive step by looking at the previous n-set and building up to the $(n+1)$-set, there is a danger that we will not construct all the elements or will construct other objects as well.

To avoid falling into such traps, we strongly recommend avoiding the buildup approach entirely. If $P(n+1)$ is about all objects of "type" $n+1$, begin the inductive step by considering an arbitrary such object. Break it down to obtain an object of type n, use $P(n)$, and work back up. For instance, had we started the inductive step in Example 4 by considering an arbitrary allowed sequence of length $n+1$ and had we worked *down* by lopping off the final bit, then we would have realized that the remaining sequence can't have consecutive 1's *anywhere* and that therefore the simple relation between s_n and s_{n+1} required to obtain $s_n = 2^n$ is false. (There *is* a formula for s_n, but deriving it depends on finding a relation between s_{n+1} and two previous terms. Stay tuned for Chapter 5.)

Even when $P(n)$ is a single formula, you can use this "down and back" approach. We illustrated this with the second proof in Section 2.2 of $\sum_{k=1}^{n} k = n(n+1)/2$ (the proof after this fact was stated as Theorem 1). Namely, start with the LHS of $P(n+1)$, break down to the LHS of $P(n)$, assert the RHS of $P(n)$ by induction, and then build back to the RHS of $P(n+1)$.

To illustrate the importance of avoiding buildup errors, we give one more example.

EXAMPLE 5 A faulty proof of Theorem 2, Section 2.2 (ordering teams in a round robin).

$P(n)$ is defined as in the earlier proof, and the basis case, $P(2)$, is the same. To prove that $P(n) \Longrightarrow P(n+1)$, consider an arbitrary n-tournament. By assumption, there is an ordering of the teams so that t_1 beat t_2 and so on. Now add a new team t^*. Since the results of its games with the previous teams are arbitrary, we may assume that it loses to t_n, in which case we may simply put t^* at the end of the list. ∎

This "proof" is a bad joke on the word "arbitrary". We must prove the existence of an ordering for an arbitrary $(n+1)$-tournament, but "arbitrary" here does not mean *we* can choose the outcome. To the contrary, it means we must cope with any outcome that someone *else* chooses. Suppose the someone had chosen to make t^* beat t_n. Then we have not named an ordering that works.

Problems: Section 2.5

Each of problems 1–14 consists of an assertion and a "proof". Most of the proofs are invalid, or at least badly muddled. Nonetheless, some of the assertions are true. For each problem, identify any errors in the proof. If there are errors but the assertion is true, provide a good proof.

1. Assertion: Every positive integer is either a prime or a perfect square.

   "Proof": The number 1 is a square. By strong induction, assume the theorem for $m < n$ and consider n. If it is prime, we are done. If not, it has some factorization $n = rs$. By assumption, r and s are squares, say, $r = u^2$ and $s = v^2$. Thus

   $$n = (u^2)(v^2) = (uv)^2,$$

   which shows that n is a perfect square.

2. Assertion: Every integer > 6 can be represented as a sum of 7's and 11's.

   "Proof": Let $P(n)$ be the statement that n can be represented as a sum of 7's and 11's. Basis: P(7) is clearly true. Inductive step: By strong induction, we may assume that $P(n-7)$ is true. Add one more 7 and n is represented as a sum of 7's and 11's as well.

3. Assertion: $\sum_{k=1}^{n} 2^k = 2^{n+1}$.

   "Proof": Let $P(n)$ be the case n of the assertion. Assume $P(n-1)$ and add 2^n to both sides:

   $$\sum_{k=1}^{n-1} 2^k = 2^n$$
   $$\sum_{k=1}^{n-1} 2^k + 2^n = 2^n + 2^n$$
   $$\sum_{k=1}^{n} 2^k = 2 \cdot 2^n = 2^{n+1},$$

   which is $P(n)$.

4. Assertion: If the sequence $\{a_n\}$ satisfies

   $$a_0 = 2, \quad a_1 = 3, \quad \text{and}$$
   $$a_n = 2a_{n-1} - a_{n-2} \quad \text{for } n > 1,$$

   then no term of the sequence equals 1.

   "Proof": Clearly $a_0 \neq 1$. Now, for the inductive step, it suffices to show that $\{a_n\}$ is increasing, since it starts at 2. Assume that $a_{n-1} > a_{n-2}$. Then

   $$a_n = 2a_{n-1} - a_{n-2} > a_{n-1} > 1,$$

   so $a_n \neq 1$.

5. Assertion: Define $\{x_n\}$ by:

   $$x_1 = 1, \quad x_2 = 3, \quad \text{and}$$
   $$x_n = \tfrac{1}{2}(x_{n-1}+x_{n-2}) \quad \text{for } n > 2.$$

   Starting with $n = 2$, $\{x_n\}$ is strictly monotonic decreasing (that is, $x_{n+1} < x_n$).

   "Proof": We prove by induction that each term is smaller than the preceding term. Since $x_2 = 3$ and $x_3 = \frac{1}{2}(x_1+x_2) = 2$, the basis is true. Now we want to show that $x_{n+1} < x_n$. Substituting the definition twice, we get

   $$\begin{aligned} x_{n+1} &= \tfrac{1}{2}(x_n + x_{n-1}) \\ &= \tfrac{1}{2}(\tfrac{1}{2}[x_{n-1}+x_{n-2}] + x_{n-1}) \\ &< \tfrac{1}{2}(\tfrac{1}{2}[x_{n-2}+x_{n-2}] + x_{n-1}) \\ &\qquad\qquad [\text{Since } x_{n-1} < x_{n-2}] \\ &= \tfrac{1}{2}(x_{n-1} + x_{n-2}) \\ &= x_n. \end{aligned}$$

6. Assertion: Every positive integer $n > 1$ has a unique prime factorization.

   "Proof": Let $P(n)$ be the statement that n has a unique factorization. $P(2)$ is true since 2 is prime. As for the inductive step, assume that every integer $< n$ has a unique factorization and consider n. By Example 3, Section 2.3, we know that n has at least one prime factorization. Thus it suffices to show that any two prime factorizations are the same.

Let q be a prime that appears in the first factorization. Thus q divides n. By [4, Section 2.3], q divides one of the factors in the second factorization. However, all the factors in the second factorization are primes, and the only way for q to divide a prime is for q to be that prime. Thus both factorizations have at least one q. Temporarily delete one q from both factorizations; that makes them factorizations of n/q. By assumption, the prime factorization of n/q is unique, so these two factorizations of n/q are the same. Putting back the q, we find that both factorizations of n are the same.

7. In a round-robin tournament, call a team a "Total Winner" if it wins every game. Call it a "Total Loser" if it loses every game.

   Assertion: If a (round robin) tournament with $n \geq 3$ teams has no Total Winners or Total Losers, then there is an ordering of the teams so that Team 1 beat Team 2, 2 beat 3, ..., $n-1$ beat n, *and* n beat 1.

   "Proof": Let $P(n)$ be the assertion for n teams. The basis case is easy to check: Each team must win one game and lose one, which forces a cycle. As for the inductive step, assume $P(n)$ and consider an arbitrary $(n+1)$-tournament without Total Winners or Losers. Temporarily ignore one Team, t^*. By induction, there is an ordering $t_1, t_2, \ldots, t_n$ of the remaining teams so that each beats the next cyclically. Consider t^* again. Since t^* is neither a Total Winner nor a Total Loser, t^* beats some t_i and loses to some t_j. Go around the cycle of n teams starting with t_j. At some point we reach for the first time a team t_k that t^* beats. Then insert t^* in the cycle after t_{k-1} and before t_k. (If $k = 1$, then $k-1$ means n.)

8. Assertion: Any adult can lift at one time all the sand in the world.

   "Proof": Let $P(n)$ be: Any adult can lift n grains of sand at one time. We prove this by induction. The assertion is simply the case where n is the total number of grains of sand in the world. Basis: Even the weakest adult can lift 1 grain of sand. Inductive step: Consider any adult. By assumption s/he can lift $n-1$ grains of sand. One more grain is an imperceptible difference, so s/he can lift n grains as well.

9. Assertion: In any convex polygon, the longest side is shorter than the sum of the lengths of the other sides.

   "Proof": Let $P(n)$ be: In a convex n-gon, the longest side is shorter than the sum of the others. The basis case, $P(3)$, is the well-known fact that every side of a triangle is shorter than the sum of the other two. To show that $P(n) \Longrightarrow P(n+1)$, take any convex $(n+1)$-gon P. Label the sides $1, 2, \ldots, n+1$ in clockwise order, with side 1 the longest. Slice off sides 2 and 3, replacing them by a single side $2'$, and thus obtain a convex n-gon P'. By $P(n)$, side 1 is shorter than the sum of the other sides of P'. If we now replace the length of side $2'$ by the sum of the lengths of sides 2 and 3, the sum of the lengths of sides other than 1 gets even greater.

10. Let a_n be the number of ways to place n distinct balls into three different boxes, if each box must have at least one ball.

    Assertion: For $n \geq 3$, $a_n = 6 \cdot 3^{n-3}$.

    "Proof": For the base case $n = 3$, the claim is that $a_n = 6$. This is right because exactly one ball must go in each box. There are 3 choices for which ball to put in box 1, times 2 choices for which of the remaining balls to put in box 2. As for $P(n) \Longrightarrow P(n+1)$, since $n \geq 3$, each box already has at least one ball when there are n balls, so ball $n+1$ may go in any of the three boxes. Therefore $a_{n+1} = 3a_n = 3(6 \cdot 3^{n-3}) = 6 \cdot 3^{(n+1)-3}$.

11. Let b_n be the number of sequences of n digits in which no two consecutive digits are the same.

    Assertion: $b_n = 10 \cdot 9^{n-1}$.

    "Proof": Clearly $a_1 = 10$, because any one of the 10 digits may appear by itself. That's the basis. As for the inductive step, for any n-long sequence without equal consecutive digits, we may add on as the $(n+1)$st digit any digit except the one which is currently last. That is, for each n-long sequence there are nine $(n+1)$-long sequences. Therefore

    $$b_{n+1} = 9b_n = 9(10 \cdot 9^{n-1}) = 10 \cdot 9^{(n+1)-1}.$$

12. A **derangement** of order n is a listing of the integers 1 to n so that no integer is in its "right" place. For instance, 312 is a derangement of order 3 because 1 is *not* in the first position (from the left), 2 is not in the second, and 3 is not in the

third. Let d_n be the number of derangements of order n.

Assertion: For $n > 1$, $d_n = (n-1)!$

"Proof": The basis, $P(2)$, says that $d_2 = 1$, and this is correct since 21 is the only derangement of order 2. To show $P(n-1) \Longrightarrow P(n)$, take any derangement of order $n-1$ and place n at the right end. This is not a derangement, but we may now switch n with *any* of the $n-1$ previous numbers. Afterwards, the integer n will surely be in the wrong place, and so will the number it switched with. All other numbers remain unchanged and thus are in the wrong place. Therefore, there are $n-1$ derangements of order n for each derangement of order $n-1$. Thus

$$d_n = (n-1)d_{n-1} = (n-1)[(n-2)!] = (n-1)!$$

13. Assertion: For every positive integer n, $n = n+1$.

"Proof": Let $n+1$ be called the successor of n, or $s(n)$ for short. Let M be the set consisting of 0 and all positive integers that equal their successors. Let $P(n)$ be the statement that $n \in M$. We prove the assertion by induction on $P(n)$. Basis: $P(0)$ is true by definition of M. Inductive step: Suppose $P(n)$. This means n is in M. That is, $n = n+1$. But that means $n+1$ is in M, so $P(n+1)$ is true.

14. A polygon is **triangulated** if it is divided by chords into nonoverlapping triangles. For instance, a rectangle with one diagonal drawn in is triangulated, and a triangle is triangulated using zero chords.

Assertion: Every convex polygon has a unique triangulation.

"Proof": Let $P(n)$ assert that each convex n-gon has just one triangulation. The basis, $P(3)$, is true because a triangle is already triangulated, and there are no chords you could draw to make other triangulations. For $n > 3$ use strong induction. Given an n-gon N, draw in one chord, forming an r-gon R and an s-gon S, with $r, s < n$. By induction, R and S have unique triangulations, thus so does N.

Remark: Note the similarity in the *form* of this argument to that in Example 2.

15. Critique the following claim about the basic format of mathematical induction: In the inductive step it is all right to prove $P(n-1) \Longrightarrow P(n)$, but it is invalid to prove $P(n) \Longrightarrow P(n+1)$. The latter is invalid because it assumes the very thing the whole mathematical induction is trying to show, namely, $P(n)$. The former is okay because it concludes $P(n)$ rather than assuming it, thus completing the whole proof.

16. Suppose that $P(n)$ is of the form $Q(n) \Longrightarrow R(n)$. For instance, $P(n)$ might be: "If S is a set of n integers, then the difference between some two integers in S is even." If you want to attempt a proof of $P(n)$ by induction, with inductive step $P(n) \Longrightarrow P(n+1)$, explain how to use the "down and back" approach when $P(n)$ is of this if-then form. Specifically, in the inductive step, what is the best thing to assume in your first sentence? Should it be $Q(n)$, or $Q(n) \Longrightarrow R(n)$, or $Q(n+1)$, or $R(n+1)$, or what?

17. In [34 Section 2.2], we asked you to prove that for any odd n there exists an n-tournament in which every team wins the same number of games. We told you to show this by assuming such a tournament for $n = 2k-1$ and constructing from it such a tournament for $n = 2k+1$. Isn't this precisely the sort of buildup argument that we now say is all wrong? Why not?

18. Consider the sequence defined recursively by

$$a_n = a_{\lfloor n/2 \rfloor} + a_{\lceil n/2 \rceil}, \quad a_0 = 0.$$

Which of the following proofs is correct, if either?

a) Claim: $a_n = n$ for all $n \geq 0$. Proof by strong induction. The basis $P(0)$ is given. For the inductive step,

$$\begin{aligned} a_n &= a_{\lceil n/2 \rceil} + a_{\lfloor n/2 \rfloor} && \text{[Definition]} \\ &= \lceil n/2 \rceil + \lfloor n/2 \rfloor \\ & && [P(\lceil n/2 \rceil),\ P(\lfloor n/2 \rfloor)] \\ &= n. \end{aligned}$$

b) Claim: $a_n = 2n$ for all $n \geq 0$. Proof by strong induction. The basis $P(0)$ is given, since $2 \cdot 0 = 0$. For the inductive step,

$$\begin{aligned} a_n &= a_{\lceil n/2 \rceil} + a_{\lfloor n/2 \rfloor} && \text{[Definition]} \\ &= 2\lceil n/2 \rceil + 2\lfloor n/2 \rfloor \\ & && [P(\lceil n/2 \rceil),\ P(\lfloor n/2 \rfloor)] \\ &= 2(\lceil n/2 \rceil + \lfloor n/2 \rfloor) \\ &= 2n. \end{aligned}$$

19. Okay, so we can't prove that all women are blondes. Convince a friend of the weaker result that all women have the same color hair.

20. Find an unsuspecting friend and convince him or her that all men are bald; that all pigs are yellow.

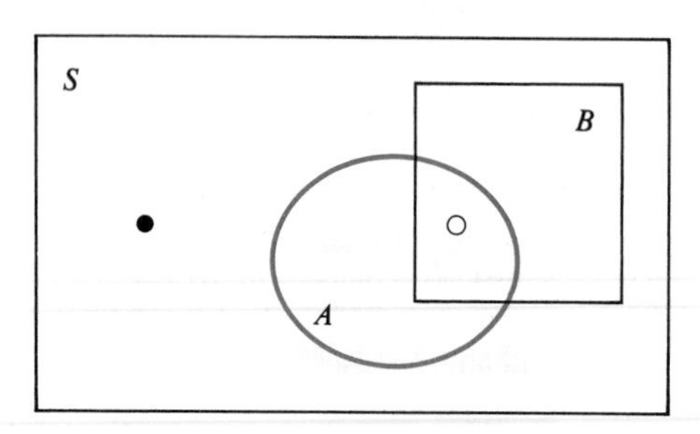

FIGURE 2.14
Why not all women are blonds

Example 1 explained. (You didn't really think we would make this too hard to find, did you?) Since $P(1)$ is true, but in general $P(n)$ is false, the error has to be in the inductive step. Figure 2.13 shows that the inductive step is all right generally, so it must break down in some particular early case. In fact, it must already break down for $P(1) \Longrightarrow P(2)$, because $P(2)$ is false—take any set consisting of a blonde and a brunette. So let's try to draw Fig. 2.13 in the particular case $n = 1$. We can't! If the known blonde is in both sets A and B, making them both 1-sets as they are supposed to be, then the other woman can't be in either set and so we cannot conclude that she is blonde. See Fig. 2.14.

Moral: Make sure your inductive argument really works for all n beyond the basis case(s).

2.6 Induction and Algorithms

We have already discussed induction and algorithms in the subsection of that name in Section 2.2. We proved a recursive algorithm correct by induction and an iterative algorithm incorrect (it didn't terminate) by induction on a loop invariant. In this section we

1. illustrate the use of induction in analyzing the complexity of algorithms, and
2. discuss induction and iterative algorithms further; specifically, we show how to prove iterative algorithms correct.

We don't discuss correctness of recursive algorithms further because there's basically one good approach and we've already explained it: Do induction on the claim that the recursive procedure is correct. For iterative algorithms, things are not quite so straightforward.

EXAMPLE 1 **Binary Search revisited**

In Example 2, Section 1.6, we began the worst case complexity analysis of Algorithm BinSearch (repeated here as Algorithm 2.2). Use mathematical induction to finish the job.

Solution Our strategy is standard enough: Conjecture a formula and prove it by induction. However, the proof will be algebraically more complicated than others we have done so far. As messy algebra is quite typical for proofs about the run time of algorithms that divide the data in half each time, and as such algorithms are quite common, we might as well face the music now.

Algorithm 2.2
BinSearch

```
Input      n                                      [Number of items on list]
           w_i                                    [Ordered list of items]
           w                                      [Search word]
Algorithm BINSEARCH
      F ← 1; L ← n                    [Initialize pointers to First and Last]
      repeat
           if F > L then output 'failure'; exit
           i ← ⌊(F+L)/2⌋
           if                                     [Selection structure]
                w = w_i then output i; exit       [Success]
                w < w_i then L ← i−1              [Search first half]
                w > w_i then F ← i+1              [Search last half]
           endif
      endrepeat
Output     i                                      [Index of word if on list]
           'failure'                              [If word not on list]
```

Algorithm BINSEARCH searches through an *ordered* list of inputs to determine whether a specified item, w, is in the list. It begins by looking at the middle item, w_i. If this is not w, the algorithm thereafter looks at only the earlier items or the later items, depending on whether $w < w_i$ or $w > w_i$. So, let C_n be the worst case number of comparisons for binary search of an ordered list of length n. We noted in Chapter 1 that C_n is approximately $\log_2 n$. If you combined this knowledge with the specific worst-case data from [13] Section 1.6, you may well have conjectured the exact formula:

$$C_n = \lfloor \log_2 n \rfloor + 1. \tag{1}$$

Why might you conjecture Eq. (1)? We surely need floors or ceilings, since logarithms are generally not integers. (It's the floors and ceilings that make proofs for this sort of algorithm messy.) Moreover, C_n jumps from one integer to the next when n moves to a power of 2, that is, a number at which $\log_2 n$ is an integer. Thus logarithms that are just below an integer must be distinguished from those that equal an integer, which is what floor does. (There is another formula, involving ceilings, that is also correct; see [4, 5].)

Let $P(n)$ be Eq. (1). We now prove $P(n)$ by induction. The basis case is $P(1)$, and sure enough,

$$C_1 = 1, \qquad \text{and} \qquad \lfloor \log_2 1 \rfloor + 1 = 0 + 1 = 1.$$

For the inductive step, we need a formula relating C_n to earlier C's. When there are n items, the middle one is item $\lfloor (n+1)/2 \rfloor$. If this isn't the searched-for item, BINSEARCH tries the first half or the second half of the list. The length of the first half is

$$\left\lfloor \frac{n+1}{2} \right\rfloor - 1 \;=\; \left\lfloor \frac{n+1}{2} - 1 \right\rfloor \;=\; \left\lfloor \frac{n-1}{2} \right\rfloor, \tag{2}$$

and the length of the second half is

$$n - \left\lfloor \frac{n+1}{2} \right\rfloor \;=\; n + \left\lceil -\frac{n+1}{2} \right\rceil \;=\; \left\lceil n - \frac{n+1}{2} \right\rceil \;=\; \left\lceil \frac{n-1}{2} \right\rceil. \tag{3}$$

(We have just used several properties of floor and ceiling from the problems in Section 0.4, and we will shortly use several others. See [1, 2, 3] at the end of this section.) Thus, the worst case with n inputs consists of making one unsuccessful comparison in the middle and then going to whichever side has the bigger worst case. In other words,

$$C_n \;=\; 1 + \max\{C_{\lfloor \frac{n-1}{2} \rfloor}, C_{\lceil \frac{n-1}{2} \rceil}\}. \tag{4}$$

We now have the recursive relation we need.

(Since $\lfloor \frac{n-1}{2} \rfloor \le \lceil \frac{n-1}{2} \rceil$, why can't we immediately replace Eq. (4) with

$$C_n = 1 + C_{\lceil \frac{n-1}{2} \rceil}\,?$$

Because we don't really know yet that C_n, the worst-case complexity, is an increasing function of the number of inputs. Certainly it should be, and it will be if our conjectured explicit formula, Eq. (1), is right. But we don't know that yet.)

In any event, to use Eq. (4) requires a strong induction. Assume that $P(k)$ is true for all $k < n$. Then we may use the earlier cases of Eq. (1) where we replace n by $\lfloor \frac{n-1}{2} \rfloor$ and $\lceil \frac{n-1}{2} \rceil$. Let us substitute these earlier cases in Eq. (4) and try to make the case n of Eq. (1) come out. The first good news is that, since $1 + \lfloor \log_2 n \rfloor$ *is* an increasing function, we now are justified in writing Eq. (4) as

$$C_n \;=\; 1 + C_{\lceil \frac{n-1}{2} \rceil} \;=\; 1 + \left\lfloor \log_2 \lceil \tfrac{n-1}{2} \rceil \right\rfloor + 1.$$

Somehow the +1 at the end has to be used to simplify the ugly expression inside the log. The trick is to realize that $1 = \log_2 2$. We can then use the fact that $\log x + \log y = \log xy$. Specifically, we have

$$\begin{aligned} C_n \;&=\; 1 + \left\lfloor \log_2 \lceil \tfrac{n-1}{2} \rceil \right\rfloor + \log_2 2 \\ &=\; 1 + \left\lfloor \log_2 \lceil \tfrac{n-1}{2} \rceil + \log_2 2 \right\rfloor \\ &=\; 1 + \left\lfloor \log_2 (2 \lceil \tfrac{n-1}{2} \rceil) \right\rfloor \\ &=\; 1 + \begin{cases} \lfloor \log_2 (n-1) \rfloor & \text{if } n-1 \text{ is even} \\ \lfloor \log_2 n \rfloor & \text{if } n-1 \text{ is odd.} \end{cases} \end{aligned} \tag{5}$$

We are done if $n-1$ is odd—the last line of Eq. (5) is $P(n)$. To finish when $n-1$ is even, observe that

$$\lfloor \log_2(n-1) \rfloor = \lfloor \log_2 n \rfloor, \quad \text{when } n-1 \text{ is even and } n > 1,$$

because $\lfloor \log_2 n \rfloor > \lfloor \log_2(n-1) \rfloor$ only when n is an integer power of 2, and no odd number $n > 1$ is an integer power of 2. ∎

Now let us turn to inductive proofs of correctness of iterative algorithms. We continue with binary search. We will spell things out in much more detail than is typical, so that you really see what is going on.

As we have said before, the most natural way to prove correctness for algorithms with loops is by using loop invariants. So let's look for a loop invariant for Algorithm BINSEARCH. We want something that remains true from iteration to iteration and has something to do with finding the search word. If you had to explain intuitively to someone why BINSEARCH works, you might say that it keeps narrowing its search, always narrowing it to an appropriate sublist. The right loop invariant amounts to stating this idea as an invariant:

> Statement S: Each time the top of the **repeat** loop is reached, either the search word w is in the current sublist (from w_F to w_L inclusive, for the current values of F and L), or else w is not on the entire list at all.

Statement S is true because it's true at the start, and each time through the loop BINSEARCH either finds the search word w or else reduces the sublist to the half of the previous sublist in which w must be if it is on the list.

Given the truth of this invariant, correctness of the algorithm follows: Look at the termination conditions. There are only two ways that BINSEARCH can terminate. Either (1) it announces it has found w because it just has, or (2) it announces it has not found w because it was just at the top of the **repeat** loop and discovered that $F > L$. We need to show that, in case (2), w is not on the original list. In case (2), Statement S is true at termination with an empty search list. Statement S is an either-or statement, and the first clause can't be true when the current list is empty. Therefore the second clause is true: w is not on the entire list.

This completes the proof—almost. First, where was the induction? The answer is: In the argument that Statement S is invariant. If you look carefully at the paragraph after Statement S, you'll see that it was a tersely stated induction on the statement:

> $P(n)$: The nth time the start of the **repeat** loop is reached, Statement S is true.

Sure enough, that brief paragraph has a basis, $P(1)$, and an inductive step: If Statement S is true at the start of one pass, it is true at the start of the next. The inductive step is valid because of the **if** clause in BINSEARCH. With a good loop invariant seeing the invariance is so simple that writing it up as a formal induction seems heavyhanded. Nonetheless, you need to understand that it *is* an induction.

Second, the proof is not really complete until we show that BINSEARCH *must* terminate. This fact is pretty obvious: Each time through the loop the number of items on $w_F, \ldots, w_L$ is smaller than the previous time, so eventually the number is

0 (if w has not been found in the meantime) and thus the algorithm will terminate because $F > L$. Many proofs of termination depend on such a **finite descent argument**, which is often implicit. If you think about it, a finite descent argument is just another induction [10].

To summarize, proof of algorithm correctness has two parts: a proof that a certain loop invariant holds and a proof that the loop will terminate. Both parts are inductions, perhaps disguised.

If you don't like the fact that the loop invariant method splits the proof in two, you can usually do a single induction on something else. For instance, BINSEARCH can also be proved correct [13] by strong induction on

> $Q(k)$: If the top of the **repeat** loop of BINSEARCH is ever reached at a point where $k = L-(F-1)$ (for the current values of L and F), then the algorithm eventually terminates, outputting the index of w if w is on the sublist $w_F, \ldots, w_L$ and outputting 'failure' if w is not on this sublist.

Note that the inductive variable here is the number of items still under consideration, whereas in the loop invariant method it is the number of times the loop has been entered.

Proofs of iterative algorithms are easy, *if* you find the right loop invariant. But how do you find it? There is no simple answer. For instance, in the problems we challenge you to prove correct an iterative TOH algorithm for which we don't know a simple loop invariant [20]. (You do an induction on something else, but it is very tricky to make your argument complete.) We have but two pieces of advice about loop invariants. The first we've already said: Familiarize yourself with the loop until you see intuitively why it works, and then try to formalize your intuition as a statement about invariance.

Second and more important: In real life the algorithms whose correctness you are most responsible for are the ones you create yourself. You should think about invariants *before* you write the algorithm. In other words, try to think of a solution method that works because it maintains some property from beginning to end. Ask yourself, "Suppose I already reached a stage where such and such a property holds; could I maintain it to the next step?" (Recursive paradigm!) Then make the algorithm fit the method.

Problems: Section 2.6

1. In the three-term sequence w_1, w_2, w_3 there is an exact middle element; in the sequence w_1, w_2, w_3, w_4 there is not. When there are n items on the current list, Algorithm BINSEARCH picks item $\lfloor (n+1)/2 \rfloor$. When is this the exact middle item? If it isn't the exact middle, what is it?

2. Explain why $2\lceil \frac{n-1}{2} \rceil$ equals $n-1$ when $n-1$ is even and equals n otherwise. This fact was used in Eq. (5).

3. What general facts about floors and ceilings from the problems in Section 0.4 were used in Eqs. (2) and (3)?

4. An equivalent formula to the one in Example 1 for the worst case of Algorithm BINSEARCH is

$$C_n = \lceil \log_2 (n{+}1) \rceil.$$

Prove this formula correct by induction.

5. Prove the formula in [4] correct by using properties of floors, ceilings, and logs to show that

$$\lceil \log_2 (n{+}1) \rceil \;=\; \lfloor \log_2 n \rfloor + 1$$

for all integers $n \geq 1$. Can the word "integers" be replaced by "real numbers"?

6. Consider the following divide-and-conquer algorithm for finding the maximum of n numbers. Divide the set of numbers as close to in half as possible, find the maximum of each subset by recursion, and then compare the two maximum values.

a) What is the basis case(s) for this algorithm?

b) Find a recursion for C_n, that is, the number of comparisons of inputs this algorithm requires when given n numbers.

c) Discover an explicit formula for C_n and prove it by induction.

7. Find and prove the explicit solution to

$$L(n) \;=\; 1 + \min_{n/2 \leq k < n} \{L(k)\}, \qquad L(1) = 0.$$

This recursion arises in finding a lower bound for the number of multiplications needed to compute x^n.

8. Find the solution to

$$L(n) \;=\; 1 + \max_{n/2 \geq k \geq 1} \{L(k)\}, \qquad L(1) = 0.$$

9. Prove that if we start with $u = 0$ and repeatedly add 1, eventually we can reach any positive integer. *Hint*: Let $P(n)$ be the statement that we can reach n. (What you are asked to prove is obvious, but since incrementing a counter is done repeatedly in algorithms, the fact that any n can be reached is implicit in the proof that many loops terminate. Thus it is worthwhile to appreciate that this fact boils down to an induction.)

10. Prove by induction that if we start with any nonnegative integer u, no matter how large, and repeatedly replace it with a smaller nonnegative integer, eventually we reach 0.

11. In our proof of correctness of Algorithm BinSearch, we put no restrictions on n, the number of data. Suppose we start with $n = 0$ that is, no data, or even with $n < 0$. Does the algorithm correctly state that w is not found, or does it crash? If the latter, then there is a gap in our proof, because as stated it seems to prove that BinSearch works for any n. In general, algorithms do crash unless the input is restricted in some way. Thus, in a more detailed treatment of proof by loop invariant, one always treats input restrictions on data explicitly. See Chapter 7.

12. We said that BinSearch, as displayed in Algorithm 2.2, is a repeat of BinSearch as presented in Section 1.6. We told a little lie! We've moved one line. Had we not moved it, the proof of the loop invariant would have been more complicated in annoying ways. Find out what we moved and give a loop invariant proof for the original algorithm (thus discovering the complications).

13. Prove Algorithm BinSearch correct by induction on the statement $Q(k)$ displayed in this section.

14. The iterative Algorithm SeqSearch from Example 1, Section 1.6 finds whether w is on a list by simply marching straight through it. Prove the algorithm correct by a loop invariant.

15. Consider the problem of summing n numbers, $a_1, \ldots, a_n$. There is a natural iterative algorithm for this summation. Write it. Prove it correct by a loop invariant.

16. Consider the multiplication Algorithm Mult (Example 1 of Section 1.2).

a) Rewrite it recursively, and prove this version correct by induction on x.

b) Prove the original algorithm correct using the loop invariant mentioned in the example.

c) Prove the original algorithm correct by an induction not directly involving a loop invariant.

17. Consider the Euclidean algorithm for finding the gcd of two positive integers m and n. An iterative version, Algorithm Euclid, was discussed in Example 5, Section 1.2, and a recursive version, Algorithm Euclid-Rec, was given in Example 1, Section 1.4.

a) Prove the recursive version correct.

b) Prove the iterative version correct with the loop invariant discussed in Section 1.2. (The proof was done informally there.)

c) Prove the iterative version correct using the following proposition $P(j)$: If the loop of Euclid is ever entered with num $= i$ and denom $= j$, then Euclid outputs $\gcd(i, j)$.

18. In our loop invariant for BINSEARCH (Statement S), we were particular about where in the loop we claimed the invariant is true. It would be simpler to say that the invariant is always true (that is, true everywhere in the loop for every pass), and informally we often speak this way. However, invariants are rarely true throughout loops. The invariant for the gcd Algorithm EUCLID was (speaking informally) that the value of gcd(num,denom) remains the same throughout the algorithm.

a) Find a place in the **repeat** loop of EUCLID where gcd(num,denom) changes. Fortunately, the gcd is not recomputed there, and later it changes back.

b) Name a specific place in the loop that this invariant should refer to.

19. Consider the problem of dividing a polynomial

$$c_n x^n + c_{n-1} x^{n-1} + \cdots + c_1 x + c_0$$

by $ax + b$, where $a \neq 0$.

a) Write up the standard long-division approach as an iterative algorithm. The inputs should be a, b, and the c's, and the output should be the coefficients $d_{n-1}, d_{n-2}, \ldots, d_0$ of the quotient polynomial and the remainder r.

b) Argue the correctness of long division by a loop invariant.

20. If we put the three poles of Towers of Hanoi on the circumference of a circle, we can describe which pole to move to next by saying "move clockwise" or "move counterclockwise". Algorithm 2.3, ITERATIVE-TOH, makes use of this circular setup. Not only is it iterative, but it is easy for a person to act out, say, at parties. The reason is that it requires almost no memory, whereas the recursive algorithm HANOI requires keeping track of where you are in a lengthy series of procedure calls.

While performing ITERATIVE-TOH is simple, proving it correct is not. Try to give a complete proof.

Algorithm 2.3. ITERATIVE TOWERS OF HANOI

```
Input all rings on one pole
Algorithm ITERATIVE-TOH
  Move smallest ring clockwise
  repeat until all rings are on one pole
    Move 2nd smallest exposed ring to the one
        place it can go
    Move smallest ring clockwise
  endrepeat
Output all rings on another pole
```

2.7 Inductive Definitions

Let's look again at the very first induction in this chapter, the proof that $\sum_{k=1}^{n} k = (n+1)n/2$. The key step, breaking down case $n+1$ to get n, was accomplished by using the equality

$$\sum_{k=1}^{n+1} k \;=\; \left(\sum_{k=1}^{n} k\right) + (n+1). \tag{1}$$

Why is this equation correct?

We hope it's obvious that Eq. (1) is correct, but when someone asks a question challenging something obvious, he or she is usually looking for an explanation based on a theorem, or in the case of notation like $\sum$, on its definition. Here is where things get interesting: We've never given a good definition of $\sum$. The reason is that $\sum_{k=1}^{n}$ represents infinitely many things, one for each positive integer n, so it's hard in a finite amount of space to define them all. To be sure, we have explained the sigma notation with dots,

$$\sum_{k=1}^{n} a_k = a_1 + a_2 + \cdots + a_n,$$

and this is quite satisfactory. However, we can imagine that for other more complicated notation with infinitely many cases such "dot dot dot definitions" might be confusing.

Have we ever had to treat a problem with an infinite number of cases before? Of course: In proving inductions. This suggests that the same "basis and inductive step" approach can be applied to definitions as well. Indeed, from our point of view, sigma notation is *defined* by the following display:

$$\begin{aligned}\sum_{k=1}^{1} a_k &= a_1,\\ \sum_{k=1}^{n+1} a_k &= \Big(\sum_{k=1}^{n} a_k\Big) + a_{n+1}\end{aligned}$$

Not surprisingly, this sort of thing is called either an **inductive definition** or a **recursive definition.** It should also not surprise you that it can be written equally well going from $n-1$ to n:

$$\sum_{k=1}^{n} a_k = \begin{cases} a_1 & \text{if } n = 1 \\ \Big(\sum_{k=1}^{n-1} a_k\Big) + a_n & \text{if } n > 1. \end{cases}$$

In any event, the answer to why Eq. (1) is correct is now, simply, "by definition".

Many other notations can be defined recursively.

EXAMPLE 1 **Factorials**

One often sees $n!$ defined by

$$n! = n(n-1)\cdots 3\cdot 2\cdot 1,$$

but one can avoid the dots by writing

$$n! = \begin{cases} 1 & \text{when } n = 1 \\ n[(n-1)!] & \text{when } n > 1. \end{cases}$$

In fact, it is common to define $0! = 1$ (see [9]), in which case we may restate the inductive definition as

$$n! = \begin{cases} 1 & \text{when } n = 0 \\ n[(n-1)!] & \text{when } n > 0. \end{cases} \blacksquare$$

Just as proofs by induction come in a variety of forms, so do inductive definitions. The next example shows a case where there are two terms in the basis. The thing being defined is a sequence. We've given recursive definitions of sequences earlier in this book, but we do it again to indicate that they are similar in spirit to recursive definitions of symbols and of operations, as introduced in this section.

EXAMPLE 2 **Fibonacci numbers**

Define the sequence $\{f_n\}$ by

$$f_0 = f_1 = 1,$$
$$f_{n+1} = f_n + f_{n-1}, \quad n \geq 1.$$

Thus

$$f_2 = f_1 + f_0 = 1 + 1 = 2,$$
$$f_3 = f_2 + f_1 = 2 + 1 = 3,$$

and so on. This is a famous sequence, which we will return to in Section 5.2. ∎

In this book, we first introduced recursion in the context of algorithms. Thus you should not be surprised to learn that recursive definitions can easily be put into algorithmic form. For instance, here is $\sum$ written as a recursive function. It is assumed that the main algorithm in which it sits has $a_1, a_2, \ldots, a_n$ among its inputs:

```
function Sum(n)
    if n = 1 then sum ← a_1
              else sum ← Sum(n−1) + a_n
    endif
    return sum
endfunc
```

In the case of $\sum$ it is easy to write an equivalent iterative fragment as well:

```
S ← a_1
for k = 2 to n
    S ← S + a_k
endfor
```

The appearance of $k = 2$ in the iterative version may seem a little odd; the only two numbers one would expect are 1 and n. This suggests a slight revision:

```
S ← 0
for k = 1 to n
    S ← S + a_k
endfor
```

The recursive version of Sum can also be rewritten to correspond to this second iterative version [3].

This second iterative version suggests that there is a reasonable way to define the empty sum $\sum_{k=1}^{0} a_k$, or more generally, $\sum_{k\in\emptyset} a_k$. It should be defined as 0, because if you want to get the right answer in the end, that's what you must start with before you add anything. Perhaps this point seems obvious—of course you have 0 before you do anything—but actually it's more subtle than that. Consider the empty product [7, 8] and the empty factorial [9] and you will see that you *don't* always have 0 before you do anything.

Inductively Defined Sets

So far, we have used the basis and inductive-step approach to define *values* for expressions and functions. The same approach can be used to define sets. Our goal here is to show you through examples how to interpret and create inductive set definitions. Just as ordinary inductive definitions are a key ingredient in many proofs, so are inductive set definitions, but developing facility at such proofs is not our goal.

EXAMPLE 3 Devise an inductive definition for the natural numbers.

Solution The essence of the natural numbers is that they are all the numbers you get (and only the numbers you get) by starting with 0 and forever adding 1. In other words, we have:

Definition 1. The set N of **natural numbers** is the smallest set such that

i) $0 \in N$, and
ii) if $n \in N$, then $(n+1) \in N$.

The word "smallest" is crucial here. For instance, the set of all *real* numbers also satisfies (i) and (ii). ∎

This definition is typical of the form of inductively defined sets. There is a basis—one or more specific objects are declared to be in the set—and there is an inductive step—one or more **production rules** are given, which say that if such-and-such is in the set, then so-and-so must also be in the set. (In the preceding definition of N, there is just one specific object, 0, and just one production rule: If n is in the set, then $n+1$ is produced for the set, too.) Finally—and this is where inductive definitions of sets differ from other inductive definitions—there is also the assertion that the set in question is the smallest set satisfying the basis and the inductive step.

Recursive definitions of sets are especially valuable whenever you have an infinite set that is hard to describe in a direct way. Consider, for instance, any computer language you know. There are quite exacting rules about what is a legitimate "sentence" in that language, yet there are infinitely many legitimate sentences, and you would be hard put to define them directly. If you study language design at some point, you will see that their definitions are recursive—with a big basis and lots of production rules. To give the idea, we introduce some simple "toy languages" in Examples 4 and 5 below. In any event, the subject of **formal languages** is an area of interest to both computer scientists and linguists, for it has a bearing on natural language, too. We might add that constructing the languages is the easy part; figuring out how to parse expressions in them is the bigger challenge. (To parse means to figure out how the expression was put together, so that you then can figure out what it means.)

EXAMPLE 4 Let S be the set of items defined recursively by

a) every lowercase roman letter is in S, and
b) if A and B are expressions in S, then so is $(A+B)$.

Describe the set S directly.

Note 1. The phrase "the smallest set" does not appear above, but in a recursive definition of a set it is always understood.

Note 2. In (b), A and B are generic names for any two beasts in S; A might be (x+(x+y)). There is no claim that the capital italic letters A and B are symbols in S. On the other hand, the parentheses and plus sign in "$(A+B)$" mean exactly themselves. That is, the production rule requires that they be included, even in cases where parentheses are optional in ordinary algebra.

Solution to Example 4 Let's look at some typical elements of S. Applying (b) once to the elements from (a), we get expressions like

$$\mathrm{(x+z)}, \quad \mathrm{(a+b)}, \quad \mathrm{(b+c)}, \quad \mathrm{(n+n)}.$$

Applying (b) again to everything we now know to be in S, we get things like

$$\mathrm{(a+(b+c))}, \qquad \mathrm{((a+b)+c)}, \qquad \mathrm{((x+z)+(n+n))}.$$

In the next round we get things like

$$\mathrm{((a+(b+c))+d)}, \quad \mathrm{((a+b)+((x+z)+(n+n)))},$$
$$\mathrm{(((x+z)+(n+n))+((x+z)+(n+n)))}.$$

Therefore we can describe S by saying it consists of all "fully parenthesized" expressions involving addition and using lowercase roman letters. (See Section 7.8 for further discussion of fully parenthesized expressions.) If + represented an operation that is not associative, this is just the formal language you would want to study.

EXAMPLE 5 **Palindromes**

A palindrome is an expression that reads the same backwards and forwards, e.g., "abcba" or "able was I ere I saw elba". Let P be the set of all palindromes using only lowercase roman letters and no blanks. Devise an inductive definition for P.

Solution An inductive step will require that we express longer palindromes in terms of shorter ones. This is easy: The first and last symbol of a palindrome must be the same, and if we strip them off we still have a palindrome (if anything is left). Since this stripping operation relates a palindrome of length n to a palindrome of length $n-2$, we need two basis cases: palindromes of lengths 1 and 2. Therefore we have the following definition: P is the set of symbol strings defined recursively by

a) single and double roman lowercase letters are in S (e.g., "q" and "hh"), and

b) If A is in S, and $*$ represents a single lowercase roman letter, then $*A*$ is in S. ∎

Problems: Section 2.7

1. Write the recursive definition of $n!$ using n and $n+1$ instead of $n-1$ and n. Let $n = 0$ be the basis case.

2. Write $n!$ as a recursive function $f(n)$ in our algorithmic language. Let $n = 0$ be the basis case.

3. Using our algorithmic language, write a recursive function for $\sum_{i=0}^{k} a_i$ that has $k = 0$ as its base case.

4. We don't have to invoke algorithms to explain why mathematicians have defined $\sum_{k=1}^{0} a_k$ to be 0. Let A and B be disjoint, nonempty sets. Then clearly

$$\sum_{i\in A\cup B} a_i = \sum_{i\in A} a_i + \sum_{i\in B} a_i.$$

How should $\sum_{i\in\emptyset} a_i$ be defined so that the identity is true even if either A or B is empty?

5. Write a recursive function for $\prod_{k=1}^{n} a_k$ in our algorithmic language. Use $n = 0$ as the base case.

6. Write an iterative algorithm fragment to compute $\prod_{k=1}^{n} a_k$ so that the loop goes from $k = 1$ to n.

7. Arguing on the basis of [5, 6], figure out how mathematicians define $\prod_{k=1}^{0} a_k$.

8. Using an argument similar to that in [4], figure out how mathematicians define $\prod_{k=1}^{0} a_k$.

9. Why should $0!$ be defined as 1?

10. Consider the definition of Fibonacci numbers in Example 2.

a) Write it as a recursive function in our algorithmic language.

b) Write it as an iterative algorithm fragment.

11. Give a recursive definition for the set of even natural numbers.

12. The **empty string** (no symbols) is often considered to be a palindrome, but it is not included in the set defined in Example 5. Revise the definition to include it. (The empty string is usually denoted by Λ, the Greek capital letter Lambda. One solution is simply to add Λ to the list in condition (a) of the definition, but you can do better.)

13. Let m and n be specific integers. Define the set U recursively by

a) $m, n \in U$, and

b) $p, q \in U \Longrightarrow p+q \in U$.

Give a direct (i.e., nonrecursive) description of U in terms of m and n. Give a direct, numerical description of U when $m = 7$ and $n = 11$.

14. Give a direct description of the set of numbers V defined recursively by

i) $2, 10 \in V$, and

ii) if p and q are in V, then so are pq and p^q.

15. Assume that we are building the foundations of mathematics and have succeeded in defining and proving all the standard properties of addition for nonnegative integers. The next step is to define multiplication inductively as follows:

i) $0 \cdot m = 0$,

ii) $(n+1) \cdot m = n \cdot m + m$.

Prove from this definition that $3m = m+m+m$ for all nonnegative integers m.

16. Assuming the generalized distributive law

$$a\sum_{i=1}^{n} b_i = \sum_{i=1}^{n} ab_i$$

(Example 1, Section 2.3), prove the further generalization:

$$\sum_{i,j} a_i b_j = \Big(\sum_i a_i\Big)\Big(\sum_j b_j\Big).$$

What definition of the symbol $\sum_{i,j}$ is implicit in your proof?

17. Use the inductive definition of the natural numbers N (Example 3) to prove the Principle of Weak Induction (as stated at the end of Section 2.2, but you may assume that $n_0 = 0$). *Hint*: Whatever proposition P is, let S be the set of *real* numbers x for which $P(x)$ is true. Show that S satisfies conditions (i) and (ii) from the inductive definition of N. Since N is the smallest set satisfying these conditions, what can we conclude? (This is a typical example of a proof using an inductive set definition.)

Supplementary Problems: Chapter 2

1. Let the three poles for TOH be on a circle, and suppose we add a new rule that a piece can only move to the pole immediately clockwise from its current position. Can this game be won for every n? Give a proof or counterexample.

2. Here is another iterative algorithm for ordinary TOH. Color the rings alternately red and white from top to bottom of the original tower. Also color the platform under the largest ring differently from that ring, color the second platform differently from the first platform, and color the third platform the same as the first. In addition to the usual rules, never move the same piece twice in a row, and never let a ring move directly on top of another ring or platform of the same color. Prove that these rules force a unique sequence of moves that result in a winning game.

3. Let f be a function that, for every pair of real numbers x and y, satisfies

$$f(x+y) = f(x) + f(y). \qquad (1)$$

   a) Name such a function; check to see whether you are right by substituting into Eq. (1).

   For any such function

   b) Prove by induction that for integers $n > 0$, $f(nx) = n\,f(x)$.

   c) Prove that $f(x/n) = f(x)/n$ for integers $n > 0$. Hint: Induction isn't helpful directly (try it and see why). Instead, substitute y for x/n; now induction is helpful—in fact, you've already done it.

   d) Show that $f(\Sigma x_i) = \sum f(x_i)$.

   e) See what else you can discover about such functions. For instance, what is $f(0)$? How does $f(-x)$ relate to $f(x)$?

4. Consider the curious function defined as follows:

$$f(n) = \begin{cases} n-3 & \text{if } n \geq 1000 \\ f\big(f(n+6)\big) & \text{if } n < 1000. \end{cases}$$

   What is $f(1)$?

5. You are given n bricks, all the same rectangular shape, and all with the same uniform distribution of weight. If you pile them up, one on each level, but use no mortar, what is the maximum overhang you can achieve?

   (The relevant principle from physics is: The configuration is stable if, for each brick B, the center of gravity of the set of bricks on top of B is directly over some point of B. For instance, in the left-hand configuration in Fig. 2.15, the center of gravity of all 1 bricks above the bottom brick is directly over the point marked 1, which is a point on the bottom brick (just barely!). In the right-hand configuration, the center of gravity of the bricks above the bottom brick is also over point 1. Also, the center of gravity of the top brick is directly over point 7/6, and therefore over the middle brick, which extends from 1/3 to 4/3. Thus, if we declare that the length of each brick is one unit, we find from the figure that for $n = 2$ the maximum overhang is at least 1/2, and for $n = 3$ it is at least 2/3. Is it possible to do better? How much?)

6. N identical cars are placed randomly along a long circular track. Their gas tanks are empty. Then

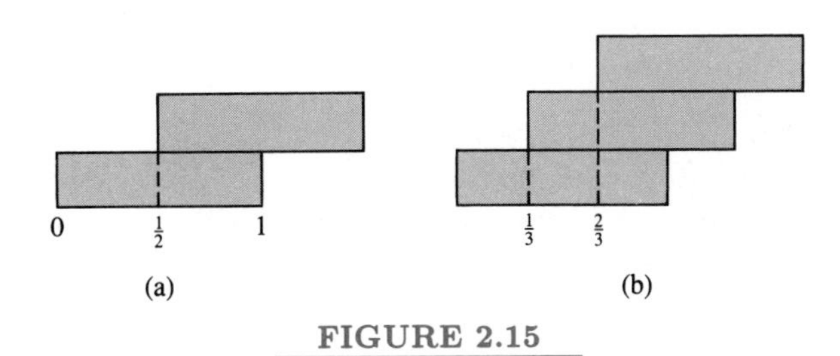

FIGURE 2.15

just enough gas for any one of these cars to make exactly one circuit of the track is divided up in a different random way and distributed to the cars. You are given a siphon and challenged to get all the way around the track clockwise in the cars. You get to pick which car you want to start in, and if you reach the next car, you may siphon gas between them before starting out again. You are told the positions of the cars and the amount of gas each has received.

a) Prove that challenge can always be met.

b) Devise a recursive algorithm for determining which car to start in. Prove your algorithm correct.

c) Devise and verify an iterative algorithm to determine which car to start in.

7. Jump Solitaire (Frogs). You have a board consisting of $2n+1$ squares in a row. The leftmost n squares have one sort of piece on them, call it an X, and the rightmost n squares have another sort of piece on them, call it O. The middle square (marked below by a dot) is empty. For instance, with $n = 3$ you have

X	X	X	·	O	O	O

(Or, think of black frogs and white frogs facing each other on a log.) The pieces can move like checker pieces—by sliding into an empty square or jumping over a single adjacent piece of the opposite type into an empty square. However, X's can only move right and O's can only move left. The object is to reverse the arrangement, i.e., to obtain

O	O	O	·	X	X	X

a) Prove that the game can be won for every n. (But try the next part first; if you succeed you will have a proof by algorithm that subsumes this part.)

b) Devise a recursive algorithm that wins Jump Solitaire and prove it correct.

c) Devise and verify an iterative algorithm for Jump Solitaire.

8. The **Arithmetic-Mean/Geometric-Mean Inequality** (**AGI** for short) says that, for any positive real numbers a and b,

$$\sqrt{ab} \leq \frac{a+b}{2}.$$

a) Prove this inequality by algebra. Start by considering the square of the inequality.

b) The **generalized AGI** says: For positive numbers a_i,

$$\sqrt[n]{\prod a_i} \leq \frac{1}{n}\sum a_i.$$

Try to prove this inequality directly from (a) by a standard induction—$P(2)$ and $P(n)$ together imply $P(n{+}1)$. If you succeed, let us know—we don't see how. But see [9].

9. The generalized AGI can be proved by induction by using the following ingenious approach devised by the famous French mathematician Cauchy. Let $P(n)$ be the nth case. We wish to prove $P(n)$ for all $n \geq 2$. Note that the basis, $P(2)$, has already been proved in [8].

a) Show that $[P(2) \text{ and } P(n)] \Longrightarrow P(2n)$.

b) Show that $P(n) \Longrightarrow P(n{-}1)$ as follows:

i) Given $a_1, \ldots, a_{n-1}$, define

$$a_n = \Big(\sum_{i=1}^{n-1} a_i\Big)/(n{-}1).$$

Show that, therefore,

$$\frac{1}{n}\Big(\sum_{i=1}^{n} a_i\Big) = a_n.$$

Note: This sum goes to n, not $n{-}1$.

ii) Write down $P(n)$, raise both sides to the nth power, and simplify.

Why does all this prove $P(n)$ for all $n \geq 2$?

10. (Requires calculus) Rolle's Theorem says:

> If g is differentiable on $[a,b]$ and $g(a)=g(b)=0$, then there exists a point c between a and b where $g'(c) = 0$.

The Generalized Rolle's Theorem says:

If g is n times differentiable on $[a,b]$, $g^{(k)}(a) = 0$ for $k = 0,1,\ldots,n-1$, and $g(b) = 0$, then there is a point c between a and b where $g^{(n)}(c) = 0$.

Prove the Generalized Rolle's Theorem by induction from Rolle's Theorem.

11. Define the **harmonic numbers** recursively by

$$H(1) = 1$$
$$H(n) = H(n-1) + (1/n).$$

Show that

a) $\sum_{i=0}^{m} 1/(2i+1) = H(2m+1) - \frac{1}{2}H(m)$.

b) $(m+1)H(m) - m = \sum_{i=1}^{m} H(i)$.

c) $H(2^m) \geq 1 + (m/2)$.

12. This problem should suggest yet another form of induction, **simultaneous induction**. Let sequences $\{s_1, s_2, \ldots\}$ and $\{t_1, t_2, \ldots\}$ be defined by

$$s_n = t_{n-1} - t_{n-2} + 4,$$
$$t_n = t_{n-1} + s_{n-1}.$$
$$s_1 = 3, \quad s_2 = 5, \quad \text{and} \quad t_1 = 1,$$

Discover formulas for s_n and t_n and prove them by induction.

13. You are given an increasing sequence of numbers $u_1, u_2, \ldots, u_m$ (u for "up") and a decreasing sequence of numbers $d_1, d_2, \ldots, d_n$. You are given one more number, C, and asked to determine if C can be written as the sum of one u_i and one d_j.

The brute force approach of comparing all mn sums $u_i + d_j$ to C works, but there is a much more clever way. See Algorithm SUM-SEARCH.

a) Figure out what's going on in SUM-SEARCH and explain it.

b) Use induction to prove that SUM-SEARCH is correct. (This is an opportunity to invent double induction.)

14. The **associative law** of addition says that, for all real numbers x, y and z,

$$x + (y + z) = (x + y) + z.$$

The **generalized associative law** says that, no matter how you parenthesize the addition of real numbers $x_1, x_2, \ldots, x_n$, the sum is the same—so long as the numbers appear in the same order. Thus, for instance, the generalized law says that

Algorithm 2.4. SUM-SEARCH

```
Input u_1,...,u_m increasing,
      d_1,...,d_n decreasing, C
Algorithm SUM-SEARCH
  i ← m; j ← n
  pair ← (0,0)                    [C not found yet]
  repeat until i = 0 or j = 0
    if                                  [3-case if]
       u_i + d_j = C then pair ← (i,j)
                          exit
       u_i + d_j < C then j ← j-1
                                   [Peel off a d_j]
       u_i + d_j > C then i ← i-1
                                   [Peel off a u_i]
    endif
  endrepeat
Output pair          [Indices of summands of C,
                      or (0,0) if C not obtainable]
```

$$(x+y) + (z+w) = [(x+y) + z] + w$$

and

$$(x_1 + (x_2 + (x_3 + (x_4 + x_5)))) = (x_1 + x_2) + ((x_3 + x_4) + x_5).$$

Of course, equality would still be true if the order of numbers was changed, but that involves the generalized commutative law [15].

Assuming only the associative law, prove the generalized associative law. (Despite how familiar these facts are to us, the proof is subtle.)

15. The **commutative law** of addition says that, for any real numbers x and y, $x+y = y+x$. The **generalized commutative law** says that, no matter how you order any real numbers $x_1, x_2, \ldots, x_n$, the sum $x_1+x_2+\cdots+x_n$ is the same. (Note that we must assume the generalized associative law from [14] just to state the generalized commutative law, for we have left out all parentheses.) Assuming the commutative law, prove the generalized commutative law.

16. Is this following principle of "strong induction for the *real* numbers" valid?

In order to prove $P(x)$ for all real numbers $x \geq 0$, it suffices to prove

i) P(0), and

ii) for all $x > 0$,
$[P(y)$ is true for all y in $0 \leq y < x] \Longrightarrow P(x)$.

17. The **Well Ordering Principle** says that every nonempty set of natural numbers has a least element. In an advanced course, all of induction is typically developed from this Principle. Prove that the Well Ordering Principle is equivalent to the weak induction Principle of Section 2.2 as follows:

a) Prove that Well Ordering $\Longrightarrow$ Weak Induction by supposing that Weak Induction is false and getting a contradiction. So suppose that there is some sequence $P(0)$, $P(1), \ldots,$ of propositions such that conditions (i) and (ii) from the end of Section 2.2 hold, yet the set

$$S = \{\, n \mid P(n) \text{ is false}\}$$

is nonempty. By the Well Ordering Principle, S contains some smallest number n'. Then $P(n'-1)$ is true. Apply condition (ii).

b) Prove that Weak Induction $\Longrightarrow$ Well Ordering by contradiction. Suppose that the natural numbers contained a nonempty subset S without a least element. Let $P(n)$ be the proposition that none of the integers from 0 to n is in S. Show that $P(0)$ is true and that $P(n) \Longrightarrow P(n+1)$ is true. By Weak Induction, what contradiction can you obtain about S?

CHAPTER 3

Graphs and Trees

3.1 Introduction and Examples

One way to judge the importance of a branch of mathematics is by the variety of unrelated problems to which it can be applied. By this measure, the subject of this chapter is surely one of the most important in discrete mathematics. To illustrate, we begin by presenting three quite different problems to which graph theory can be applied. One of these is a children's puzzle; one is a problem on the properties of sequences; and the last is a common problem in computer science. (By the way, the first example in this book (in the Prologue) also was an example of graph theory.)

First we'll state the three problems and then consider their solutions.

EXAMPLE 1 **The Wolf, Cabbage, Goat, and Farmer Problem**
A farmer is bringing a wolf, a cabbage, and a goat to market. The farmer arrives with all three at one side of a river that they need to cross. The farmer has a boat which can accommodate only one of the three. (It's a big cabbage.) But if the wolf is left alone with the goat, the wolf will eat the goat. And if the goat is left alone with the cabbage, the goat will eat the cabbage. (The wolf can be left alone with the cabbage because wolves don't like cabbage.) How can all three get across the river intact?

EXAMPLE 2 We understand that when checking into the Baltimore Hilton Inn, a guest receives a four-digit "combination", not a key. On the door of each room is a keypad. Any time those four digits are entered in the proper sequence (regardless of what has been entered previously), the door opens. The problem is this: A burglar wishes to break into a certain room. What is the minimum number of digits the burglar must use to be *certain* of entering that four-digit key in sequence?

EXAMPLE 3

You are presented with a sequence of distinct words or names, one at a time, and wish to alphabetize them. Design an algorithm to do so.

This is surely a disparate set of problems. They are susceptible to a variety of approaches. However, we can approach all three using a similar mechanism—*graphs*—as we will now demonstrate.

EXAMPLE 1 Continued

This problem can be solved without too much trouble by informed trial and error. But we are interested here, as throughout this book, in systematic (i.e., algorithmic) approaches. Denote the four principals by W, C, G, and F. According to the rules of the problem, the following combinations are legal on either side of the river:

$$\begin{array}{lll} \text{WCGF} & \text{WC} & \text{W} \\ \text{WCF} & \text{GF} & \text{C} \\ \text{WGF} & & \text{G} \\ \text{CGF} & & \emptyset \end{array} \tag{1}$$

where $\emptyset$, the empty set, represents the situation where none of the four is on one side of the river (and therefore all four are on the other side). Note that WF and CF are not included because, while both are allowable, WF on one side implies CG—which is illegal—on the other side, and CF implies WG—which is also illegal.

Figure 3.1 displays our solution. In part (a) we show the 10 legal combinations on each side of the river with lines showing how a trip in the boat with the farmer and, perhaps, a passenger leads from one configuration to another. In each case the trip could proceed in either direction. For example, the line from WCF on the near side to CGF on the far side could represent the cabbage being transported in either direction.

In part (b) we display the same information without showing the river explicitly. Each point or *vertex* in (b) represents one of the 10 cases in (1) on the near bank of the river (i.e., the bank where all four start). Also shown for each vertex is the state on the far bank of the river at the same time, assuming that no one is in transit across the river. Each line or *edge* represents a trip by the farmer from one side to the other with or without a passenger (e.g., the edge between WGF and G represents the farmer taking the wolf from one bank to the other). The label on the edge designates who besides the farmer is in the boat. Since we start with WCGF on the near bank and wish to end with $\emptyset$ on the near bank, our object must be to find a *path* in (b) from WCGF to $\emptyset$ because this will be a solution to the problem. Or will it? Is it possible that during transit of a passenger from one bank to the other, a disallowed state will occur on either bank? No, because while the trip is in progress, the state on the bank from which it began must be allowed since, when the trip is over, the states on both banks will be allowable. And, while the trip is in progress, the state on the destination bank is unchanged from its previous, allowable state.

From Fig. 3.1(b), you can see that there are two solutions:

$$\text{WCGF} \rightarrow \text{WC} \rightarrow \text{WCF} \rightarrow \text{W} \rightarrow \text{WGF} \rightarrow \text{G} \rightarrow \text{GF} \rightarrow \emptyset$$

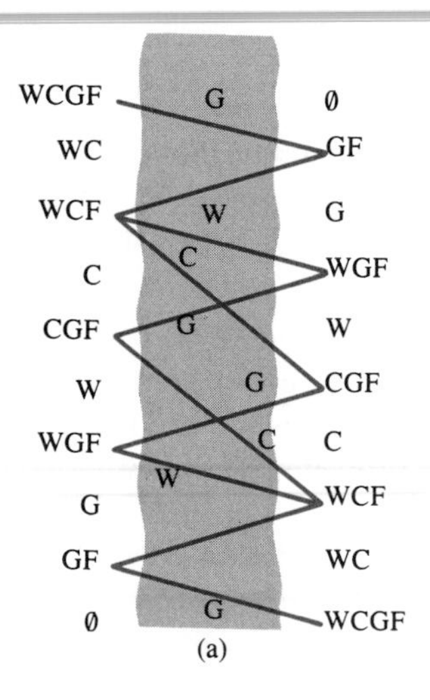

(a)

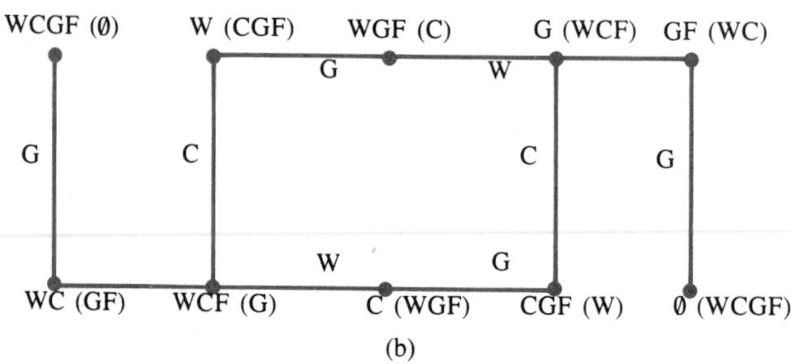

(b)

FIGURE 3.1 The Wolf, Cabbage, Goat, and Farmer Problem. (a) The allowable states are shown on both sides of the river. The labels in the river are the passengers in the boat besides the farmer. (b) Representation of allowable states as a graph. The label at each vertex is a state on the near bank. The label in parentheses corresponds to the state on the far bank. Each edge label represents a passenger in the boat besides the farmer.

and

$$\mathrm{WCGF} \to \mathrm{WC} \to \mathrm{WCF} \to \mathrm{C} \to \mathrm{CGF} \to \mathrm{G} \to \mathrm{GF} \to \emptyset.$$

The structure shown in Fig. 3.1(b) is a *graph* (an *undirected* graph because the edges have no arrows on them). Structures analogous to this are the subject of this chapter. The meaning of "graph" which you are used to from high school mathematics—namely, a picture you draw to display the shape of a function like a straight line—bears little relation to our meaning here. The graph in Fig. 3.1(b) is usually called a **state graph** because the vertices represent the various configurations (states) possible, and two vertices are joined by an edge if a single move gets you from one state to the other. In this example, each state describes the whereabouts of the four participants, and each edge represents a single boat trip. ∎

EXAMPLE 2 Continued

Since there are 10,000 four-digit combinations (0000 to 9999), the total number of digits in all the combinations is 40,000. But surely the burglar doesn't have to dial all 40,000 digits to get into the room. For example, after dialing, say, 6083, then by dialing a 7, the burglar can check whether the combination is 0837. This idea of overlapping combinations should make it possible to test all the combinations with a lot fewer than 40,000 digits. But what is the minimum?

To make the discussion easier, let's simplify the problem by restricting the digits in the combination to 0's and 1's (the *binary* case). (This simplification means that the hotel could have at most 16 rooms because there are only 16 four-digit binary numbers!) We denote the binary digits in the combination by the usual contraction *bits*.

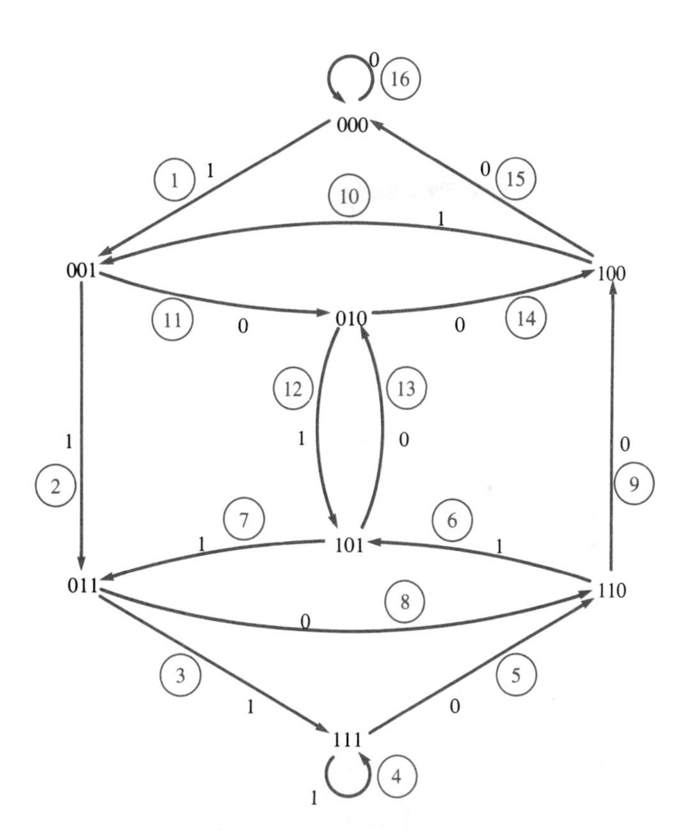

FIGURE 3.2

The Baltimore Hilton Problem (Binary Case).

Figure 3.2 displays a graph in which each vertex has a *label*, which is one of the eight three-bit sequences of 0's and 1's. Each edge has a direction, so we call this graph a directed graph, or *digraph*. Each edge also has a label which is 0 or 1. The digraph is constructed so that there is an edge between two vertices if and only if the rightmost two bits of the label of the vertex at the *tail* of the arrow are the same as the leftmost two bits at the vertex at the *head* of the arrow. The label on each edge is the last bit of the vertex label at the head of the arrow. This digraph contains two *loops*, which are edges that begin and end at the same vertex.

Now consider the edges in the graph in the order given by the sixteen circled numbers. This gives us a *path* which traverses each edge of the digraph once and only once and ends back at the same vertex (000) from which it started. (A path which ends where it begins is called a *cycle*.) The bits of the labels of the consecutive edges from 1 to 16 are

$$1111\ 0110\ 0101\ 0000. \qquad (2)$$

The interesting thing about sequence (2) is that it contains all 16 four-bit binary sequences (i.e., room numbers in the hotel) if we allow *wraparound* from the end of the sequence to the beginning (e.g., 1001 is in the middle; 0001 starts at the end and ends at the beginning). It follows therefore that to open a door whose "combination" is a four-bit sequence, we need only dial the 19 bit sequence:

$$1111\ 0110\ 0101\ 0000\ 111$$

(19 because, of course, wraparound isn't possible on an actual lock). Do you see that 19 bits is the shortest possible sequence which could test all 16 combinations?

Can this solution be generalized to the Baltimore-Hilton case where the digits 0, 1, 2, ..., 9 are allowed? Can the idea of using cycles on graphs be generalized to any set of digits and any length sequence of the digits? The answer to both questions is "yes". We don't discuss the general case in this book, but we will return to the binary case in Section 3.4. The sequences that we've been discussing in this example are usually called de Bruijn cycles, or **full cycles**.

Since we're interested in sequences of four bits, could we have used a graph whose vertices have four-bit labels instead of the graph shown in Fig. 3.2, which has three-bit labels? Sure, but it would have had 16 vertices instead of 8 (because there are 16, four-bit binary numbers) and 32 edges instead of 16. The graph would therefore have been quite a bit more complicated than Fig. 3.2. ∎

EXAMPLE 3 Continued

Consider the list of words in Fig. 3.3(a). To sort these words we have constructed (Fig. 3.3(b)) a special kind of digraph called a *tree*, a structure we introduced in Example 4, Section 1.2. A tree is a digraph because—although no arrows are shown on the edges—all edges implicitly point *down*. Algorithm BUILDTREE shows how to construct a tree like the one in Fig. 3.3(b). Figure 3.3(c) illustrates the "growth" of the tree at various stages.

How can we use the tree in Fig. 3.3(b) to alphabetize the list? All we have to do is list the words in the tree from left to right regardless of their height (or *level*) on the tree. But the success of this idea depends on a carefully drawn tree in which, for example, the "computer" node is to the left of the "mathematics" node. In Section 3.8 we'll present an algorithm to alphabetize the words which doesn't depend on how carefully the tree is drawn.

The alphabetizing method illustrated in this example is only one of many possible ways of *sorting* a list, a problem which is ubiquitous in computer science and

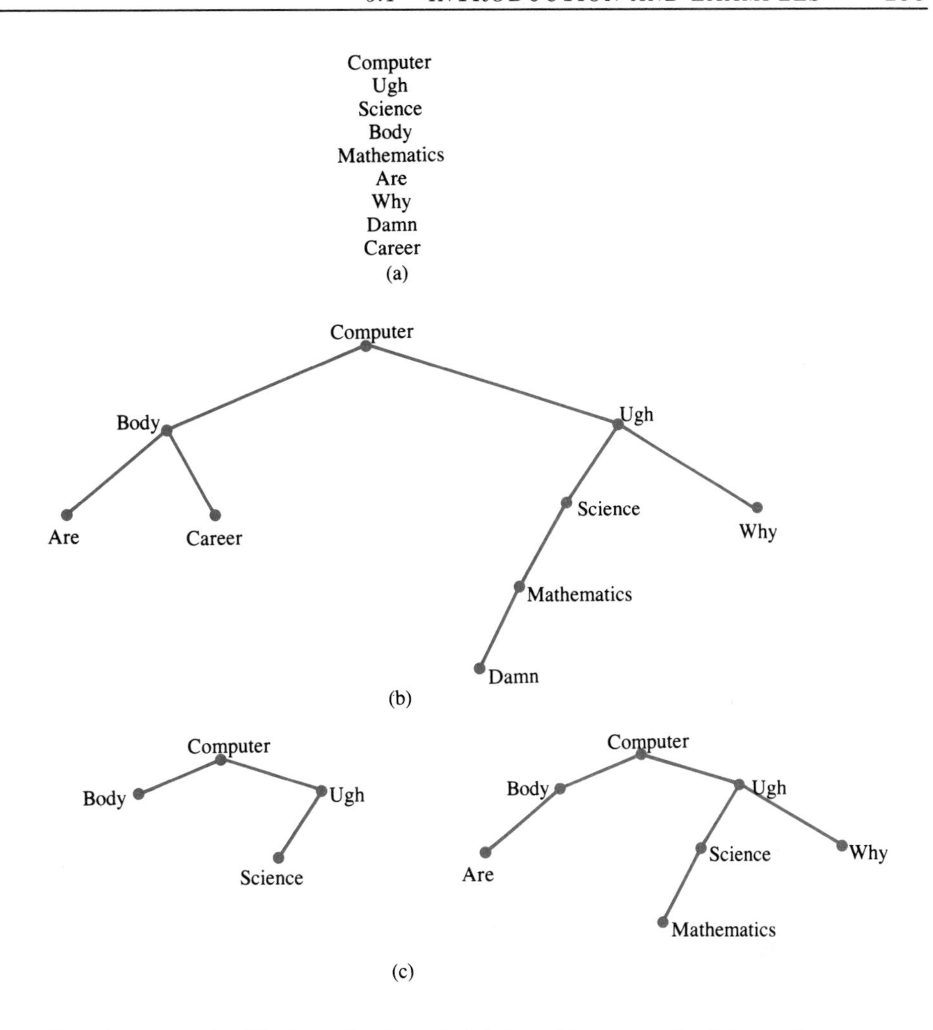

FIGURE 3.3 Tree Structure for Words to be Sorted: (a) list of input words; (b) complete tree; (c) tree at intermediate stages.

which we'll return to in Section 7.7 and the Epilogue. The method shown here is the basis of important and efficient techniques of sorting. ∎

These three very different problems are only representative of those that can be solved—or, at least, approached—using the structures we call graphs. In the remainder of this chapter we will discuss the main properties of graphs and consider various problems and algorithms related to the use of graphs.

Algorithm 3.1
BuildTree

Input n [$n > 0$; number of words]
w_i, $i = 1, \ldots, n$ [Words to be alphabetized]

Algorithm BUILDTREE
for $i = 1$ **to** n
  if $i = 1$ **then** construct a tree root and place w_1 at root [Root is top-most vertex]
    else starting at the root move left if $w_i <$ word at vertex and right otherwise until a place with no vertex is reached; construct a new vertex and place w_i at it [Assume that no two words are the same; $w_i <$ word means prior alphabetically]
  endif
endfor

Output Tree of words [As in Fig. 3.3(b)]

Problems: Section 3.1

1. Suppose that we add a lion to the wolf, cabbage, goat, and farmer problem. The lion eats wolves and goats but not cabbages (and, of course, not farmers either). Can you find a way to get all four to the far side of the river or prove that it can't be done?

2. The following diagram is a much simplified version of the "15 puzzle", shown in two positions. Each square with a number in it is a little chip of plastic and the blank square is an empty space. A move consists of shifting one of the adjacent plastic chips into the empty space. You are given the game in the configuration on the left. The goal is to get it into the configuration on the right.

1	
2	3

3	
1	2

You can probably solve this game in your head, but the purpose here is to develop a general approach using graph theory. Define a "state" of the game and what it means for two states to be adjacent. Draw the associated state graph. Use it to show that winning the game is possible. What is the minimum number of moves needed to win?

3. The Jealous Husbands Problem. Two married couples come to a river, which they must cross in a boat big enough to carry one or two people. (It can't be empty because there's no ferryman.) The men are very jealous. Neither will allow his wife to be with the other man unless he himself is there (either on shore or in the boat and whether or not the other man's wife is also present). Using graph theory, determine whether all four can get across the river. If so, what is the minimum number of crossings?

4. a) Suppose that there are three couples, with everything else the same as in [3]. Again, no husband allows his wife to be with one or more other men if he is not there, too.
   b) Repeat a) for four couples.

5. The game of tic-tac-toe has plenty of different states but the number becomes manageable if the symmetry of the three-by-three array is taken into account. Draw all states which are not related to each other by symmetry after the first move and after the second move and draw the digraph in which these states are vertices and an edge going from one state to another indicates that a move can take you from the first state to the second. (See also [30, Section 3.8].)

6. Define and draw a state graph for ordinary TOH with two rings. Use it to show that, under TOH rules, you can get from any configuration to any other configuration.

7. Find another sequence besides sequence (2) which includes all 16 four-bit sequences (allowing wraparound) by traversing a path in Fig. 3.2. This sequence should *not* be just a circularly permuted version of (2) (i.e., one in which digits are moved from the end to the beginning as in 0110 0101 0000 1111).

8. Suppose that the Baltimore-Hilton problem specifies three-digit instead of four-digit "combinations", and we restrict ourselves to the binary case. For this case draw a graph corresponding to Fig. 3.2 and, by finding a cycle, display a full cycle.

9. Display the tree structures corresponding to Fig. 3.3(b) if the input words are:

   a) Discrete, Mathematics, Is, Easier, Than, Continuous, Mathematics. (Make sure that "Mathematics" appears twice in your tree.)

   b) You, Have, To, Be, Crazy, To, Study, Mathematics, Or, Computer, Science. In this part don't put repeated words on the tree twice. Show how Algorithm BUILDTREE must be modified to do so.

10. For a given set of words, Algorithm BUILDTREE builds different trees depending on the order in which the words arrive. For simplicity, assume that there are just three distinct words in the set and that they arrive in any order with equal likelihood.

   a) Draw all possible trees which can be built.

   b) Define the *height* of a tree to be the number of edges in the longest path from the root. Determine the average height of the tree built by BUILDTREE for three words by considering the trees for all possible orderings of the words.

3.2 Terminology and Notation

To a degree, the terminology and notation of graph theory depend on which book you read. That which we'll use here is, if not universal, at least quite common and accepted. We begin by formalizing the informal notion we presented in Section 3.1 of what a graph is.

Definition 1. A **graph** G is a finite set V of points called **vertices** or **nodes** and a finite set E of **edges** or **branches** such that each member of the set E is a pair of members of the set V.

If a graph has n vertices and m edges, we can write the sets V and E as

$$V = \{v_1, v_2, \ldots, v_n\} \qquad \text{and} \qquad E = \{e_1, e_2, \ldots, e_m\} \tag{1}$$

where each e_k has the form

$$e_k = \{v_i, v_j\}, \qquad v_i, v_j \in V. \tag{2}$$

Note the use of the set braces $\{\ldots\}$ in Eq. (2). Since an edge does *not* have direction in a graph, we denote an edge by a set of two vertices without any ordering of the two. Thus $\{v_i, v_j\}$ and $\{v_j, v_i\}$ represent the same edge.

As an example of this formalism, the graph in Fig. 3.1(b) has a vertex set

$$V = \{\mathrm{WCGF}, \mathrm{W}, \mathrm{WGF}, \mathrm{G}, \mathrm{GF}, \mathrm{WC}, \mathrm{WCF}, \mathrm{C}, \mathrm{CGF}, \emptyset\}$$

and an edge set

$$E = \{\{\mathrm{WCGF}, \mathrm{WC}\}, \{\mathrm{WC}, \mathrm{WCF}\}, \{\mathrm{WCF}, \mathrm{W}\}, \{\mathrm{WCF}, \mathrm{C}\}, \{\mathrm{W}, \mathrm{WGF}\}, \{\mathrm{C}, \mathrm{CGF}\}, \{\mathrm{WGF}, \mathrm{G}\}, \{\mathrm{CGF}, \mathrm{G}\}, \{\mathrm{G}, \mathrm{GF}\}, \{\mathrm{GF}, \emptyset\}\}.$$

Sometimes, we will denote a graph not just by G but by $G(V, E)$. When V is empty and therefore E is also empty, we have a **null graph**. As with the empty set, we denote it by $\emptyset$. In this chapter we will call a member of V a *vertex* when discussing graphs but a *node* when discussing trees. Similarly, a member of E will be called an *edge* in a graph and a *branch* in a tree. Note that our definition allows G to contain an edge $\{v_i, v_i\}$ which, as in Section 3.1, we call a **loop**.

Definition 1 and the discussion following it present a graph as a purely abstract concept. But they suggest the kind of pictorial realization we used in Section 3.1 and which we'll continue to use throughout this chapter. As alluded to in Section 3.1, Definition 1 defines what is often called an *undirected* graph. We simply call it a graph. When each edge has a direction, we speak of a *directed* graph or **digraph**. For a digraph we write an edge from v_i to v_j as

$$(v_i, v_j)$$

instead of as in Eq. (2) because the parenthesis notation implies an order. Thus the edge just above has a direction from v_i to v_j. Often our definitions and theorems will obviously apply equally well to digraphs and graphs. When they do, we won't bother to mention it. Figure 3.4(a) and (b), respectively, illustrate a graph and a digraph.

When are two graphs, which look different when they are drawn, really the same? We say that two graphs are *isomorphic* (*iso* meaning equal; *morph*, structure) if they have the same number of vertices and if the vertices of the two graphs may be assigned labels so that the edge sets are identical. Figure 3.4(c) shows a pair of isomorphic graphs.

Back in Section 0.2 we introduced the notion of a binary relation. A digraph may be used to represent any binary relation on a finite set. Conversely, any digraph may be interpreted as a binary relation [6]. The vertices of the digraph represent the elements in the set and the edges represent the relation itself. Thus if a and b are two elements of a set, then there is an edge from vertex a to vertex b if and only if $(a, b) \in R$ for the relation R. A reflexive relation is one for which there is a loop at every vertex. A symmetric relation is one for which whenever there is an edge from a to b, there is a corresponding edge from b to a. Thus a symmetric relation may be represented by a graph instead of a digraph, as is illustrated in Fig. 3.5. How would the transitivity of a relation be expressed graphically [5]? Because of the advantages of pictorial representations generally, depicting a relation as a graph is often a way to glean insights about the relation not discernible from its definition as a set of ordered pairs.

We say that two vertices, v_i and v_j are *adjacent* if $\{v_i, v_j\} \in E$. For a digraph two vertices are adjacent if either (v_i, v_j) or (v_j, v_i) is in E. For example, in Fig.

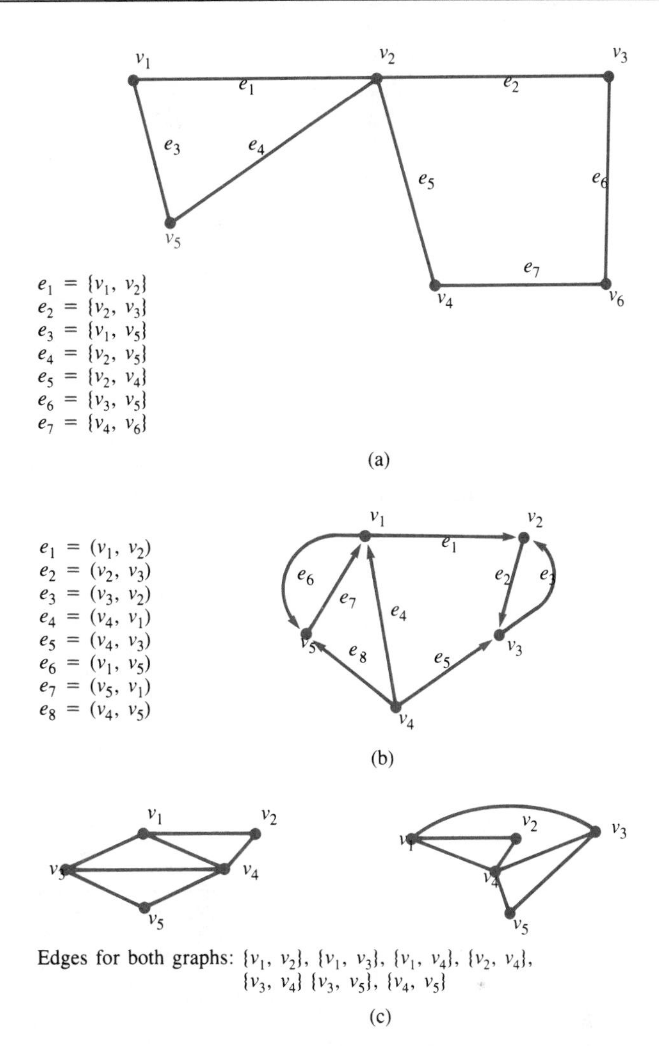

FIGURE 3.4

(a) An undirected graph with six vertices and seven edges; (b) a directed graph with five vertices and eight edges; (c) two isomorphic graphs.

3.4(a), vertices v_1, v_3, v_4, and v_5 are adjacent to vertex v_2 and, in Fig. 3.4(b), vertices v_1 and v_3 are adjacent to vertex v_2. If v is any vertex of a graph, we define $A(v)$ to be the set of vertices adjacent to v. Thus in Fig. 3.4(a) $A(v_2) = \{v_1, v_3, v_4, v_5\}$.

Because our definition of a set in Section 0.1 does not allow for repeated elements, our definition of a graph really allows only a single edge between a pair of vertices. Graphs with this restriction and no loops are called **simple graphs**. Sometimes **multigraphs**, in which there can be more than one edge between a pair of vertices, are useful. Our definition will also provide for multigraphs if we allow E to be a *multiset*, that is a set in which elements are allowed to appear more than once. In fact, we used the idea of a multiset when we defined a loop to be $\{v_i, v_i\}$.

A simple digraph is one in which there is at most one edge *in each direction* between each pair of vertices. Thus a simple digraph may have two edges, one in each direction, between two vertices. Figure 3.6 shows some examples of graphs and multigraphs. In this chapter, you may assume that *all* graphs and digraphs are simple unless we explicitly indicate otherwise.

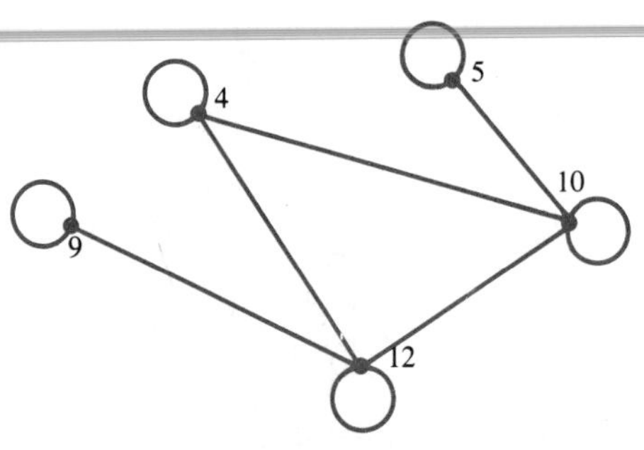

FIGURE 3.5

Relations as Graphs Let $S = \{4, 5, 9, 10, 12\}$ and the relation R be such that $(a, b) \in R$ holds only when a and b have a prime factor greater than 1 in common. This graph represents the relation R.

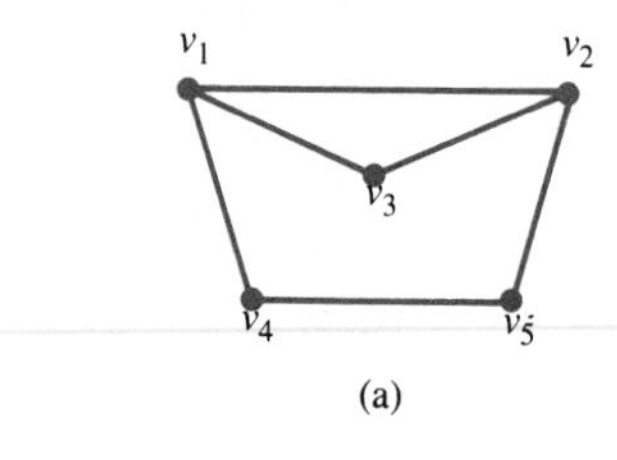

(a)

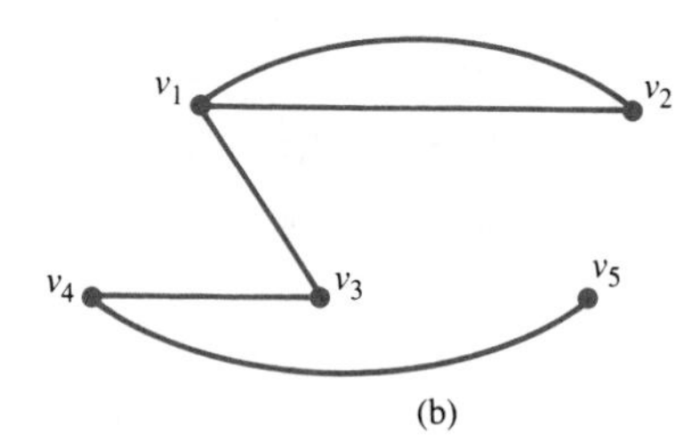

(b)

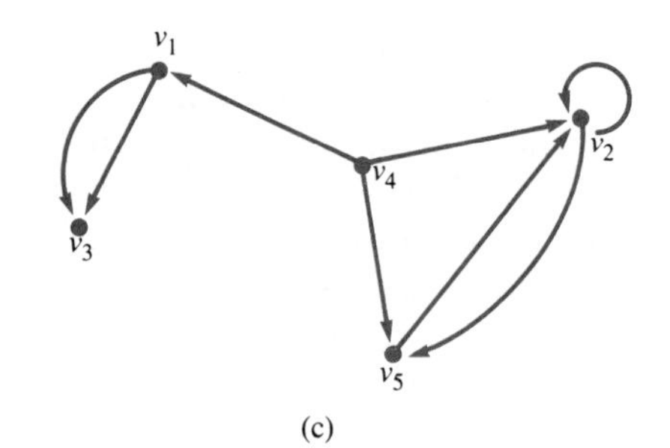

(c)

FIGURE 3.6

(a) A simple undirected graph; (b) an undirected multigraph; (c) a directed multigraph with a loop.

Let $G(V, E)$ be a graph. Then $G'(V', E')$ is a *subgraph* of G if V' is a subset of V *and* E' is a subset of E whose edges join only vertices in V'. If either $V' \neq V$ or $E' \neq E$, then the subgraph is said to be *proper*. If $V' = V$, the subgraph is said to *span* G. (Recall the definition of a *spanning tree* in Example 4, Section 1.2, which was about building a telephone network. We will discuss spanning trees in Section 3.8.) Figure 3.7 illustrates the idea of a subgraph.

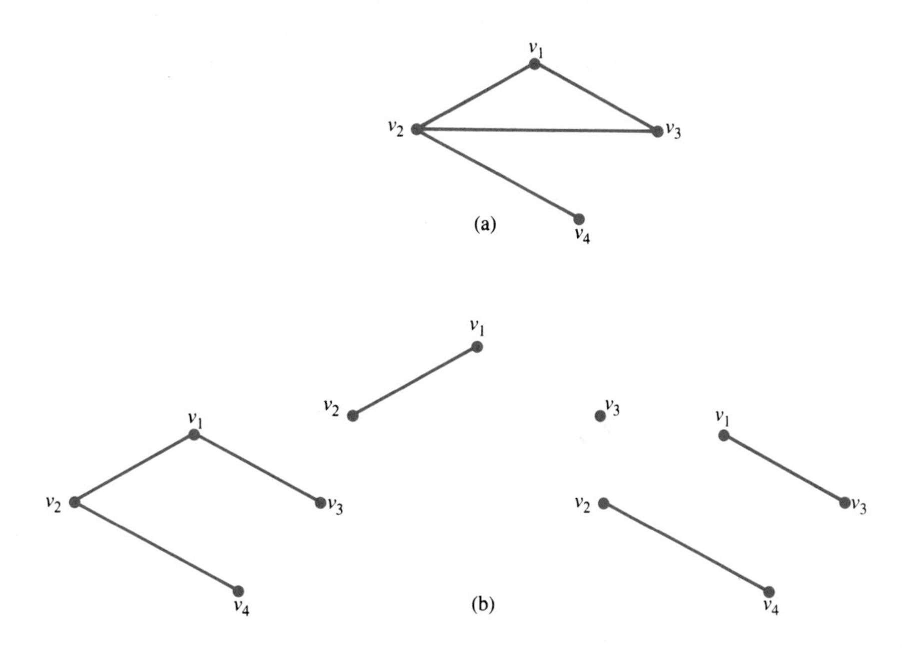

FIGURE 3.7

(a) A graph and (b) three of its subgraphs.

The *degree* of a vertex v is the number of edges *incident* on it (i.e., the number of edges for which v is one of the pair of vertices which define the edge). For a digraph we have also the **indegree** and the **outdegree** of a vertex which are, respectively, the number of edges which terminate on or originate at the vertex. In Fig. 3.6(c), the vertex v_5 has degree 3, indegree 2, and outdegree 1. We will sometimes denote the degree of a vertex v by $\deg(v)$. When $\deg(v) = 0$, the vertex is said to be **isolated**.

An important result which follows from our definition of a graph is:

Theorem 1. Let $G(V, E)$ be a graph, where $V = \{v_1, v_2, \ldots, v_n\}$. Then

$$2|E| = \sum_{i=1}^{n} \deg(v_i) \qquad (3)$$

PROOF Since each edge of a graph joins two vertices, if we add up the degrees of all the vertices [the right-hand side of Eq. (3)], we will have counted each edge twice. Hence this sum equals the left-hand side of Eq. (3). Simple, isn't it? But wait a minute. How about the graph that consists of a single vertex and a loop at that vertex? According to our definition, such a vertex has degree 1 because there is only one edge "incident" on it. For this graph, Eq. (3) is certainly not true. What's the problem? It is that we should have said that the degree of a vertex is the number of edge *ends* at that vertex. That is, the degree of a vertex is the number of edges which are not loops incident on that vertex, plus 2 for each loop at the vertex. For instance, vertex v_2 in Fig. 3.6(c) has degree 5, indegree 3, and outdegree 2. Then Eq. (3) and thus Theorem 1 is true for any graph. The lesson here is that you must always take care when defining any term to cover all possible cases. ∎

To see whether you really understand what Eq. (3) says, show that the number of people at a party who shake hands with an odd number of people must be even. Let each vertex of a graph represent a person. Then, to answer this question, what should an edge of the graph represent? This problem is explored further in [7].

We can now use the basic vocabulary of graphs to define some important concepts in graph theory, some of which we introduced informally in the previous section.

Definition 2. A **path** P in a simple graph is a sequence of vertices

$$(v_0, v_1, \ldots, v_n)$$

such that

$$\{v_i, v_{i+1}\} \in E \quad \text{(for a graph)} \qquad \text{or} \qquad (v_i, v_{i+1}) \in E \quad \text{(for a digraph)},$$

for $i = 0, 1, \ldots, n-1$. The edge $\{v_i, v_{i+1}\}$ or the directed edge (v_i, v_{i+1}) is said to be *on* P and P is said to *traverse* this edge. (Why doesn't this definition work for multigraphs [9]?)

If $v_0 = v_n$, then the path is a **cycle**. A path or a cycle with no repeated vertices (except the first and last for a cycle) is said to be **elementary** and a path with no repeated edges or loops is said to be **simple**. An **Eulerian path** is one which traverses every edge in E once and only once. An Eulerian path which is also a cycle is called an **Eulerian cycle**. A **Hamiltonian path** is one which passes through every vertex of V once and only once. A Hamiltonian path which is also a cycle is called a **Hamiltonian cycle**.

In a graph a path $v_i, \ldots, v_j$ can be traversed from v_i to v_j or vice versa. But in a digraph, such a path can be traversed only in the direction of the arrows and not in the reverse direction. A graph with no simple cycles is said to be **acyclic**. Figure 3.8 illustrates the terms in Definition 2.

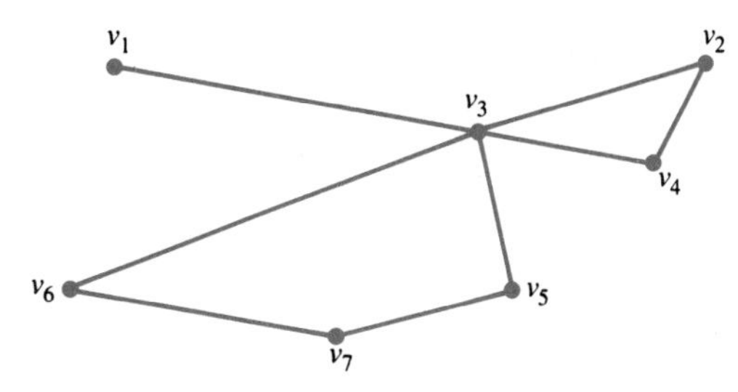

Some paths between v_1 and v_5: (v_1, v_3, v_5)
$(v_1, v_3, v_6, v_7, v_5)$
$(v_1, v_3, v_2, v_4, v_3, v_6, v_7, v_5)$

Some cycles: (v_3, v_2, v_4, v_3)
$(v_3, v_6, v_7, v_5, v_3)$

An Eulerian path: $(v_1, v_3, v_6, v_7, v_5, v_3, v_4, v_2, v_3)$

(a)

Some paths between v_1 and v_4: (v_1, v_4)
(v_1, v_3, v_6, v_4)

Some cycles: (v_1, v_4, v_1)
(v_4, v_6, v_4)
$(v_1, v_3, v_6, v_4, v_1)$
(v_4, v_5, v_4)

A Hamiltonian path: $(v_3, v_6, v_4, v_1, v_2, v_5)$

(b)

FIGURE 3.8

Paths and Cycles: (a) in an undirected graph; (b) in a directed graph.

Let's consider how the idea of a cycle can be useful. Suppose that you had some data from biological experiments that indicates when two segments of the same strand of DNA which control different traits of some species, overlap (i.e., intersect). Suppose further that you want to test the hypothesis that the DNA is, in fact, a *linear* structure (i.e., that it has no branchings or crossing portions). You are interested, therefore, in the *intersection structure* of the DNA, that is, in determining which segments overlap which others. As "intersection" is just a relation between sets, you may represent the known segments as vertices of a graph with an edge between two vertices meaning that the segments do overlap (i.e., intersect).

The intersection structure of linear sets (that is sets, which are segments of a line) is limited. That is, if segment A intersects segment B which intersects segment C but A does not intersect C, then no segment D can intersect A and C without also intersecting B (Fig. 3.9a). This is the same as saying that the graph of the intersection relation among linear sets cannot contain a cycle with four edges (a

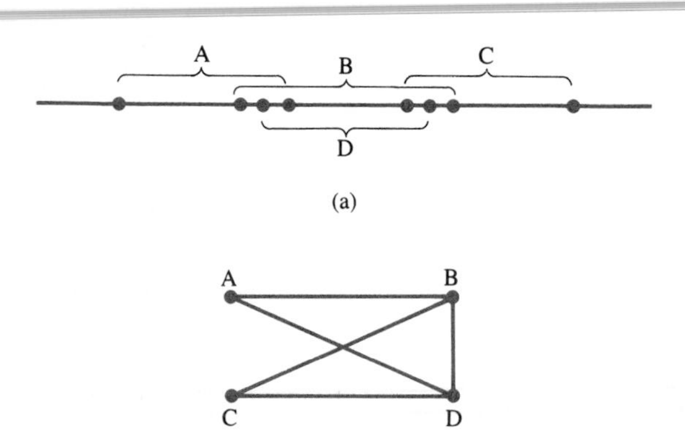
A
B
C
D
(a)
A
B
C
D
(A, B, C, D, A) is a four cycle.
(A, B, D, A) and (B, C, D, B) are three cycles.
(b)

FIGURE 3.9

Linear Sets. (a) Four overlapping segments on a straight line. (b) A graph in which vertices represent segments and an edge between two vertices means that the segments overlap.

four-cycle) unless there is also a cycle with three edges involving the same sets. To see whether your biological data is consistent with the hypothesis of linearity, you could draw the intersection graph to find out whether there are any four-cycles with no related three-cycles. If you found a four-cycle without a related three-cycle (Fig. 3.9b), that would mean the structure of the DNA could not be linear. What if you didn't find such a four-cycle? Actually, its absence would *not* prove that the structure *is* linear. A few more conditions (which we won't state here) must be added before you have a necessary and sufficient condition for linearity.

Definition 3. A graph is **connected** if, for every distinct u and v in V, there is at least one path (which could be a single edge) from u to v. A digraph is **strongly connected** if, for any pair of vertices u and v, $u \neq v$, there is a directed path from u to v and another from v to u; if there is a path from u to v or from v to u for all distinct pairs of vertices but not in both directions for at least one pair, the digraph is **weakly connected**, or just *connected.* A **connected component** (or just a *component*) of a graph G is a connected subgraph G', which is not itself a subgraph of any other connected subgraph of G.

Figure 3.10 illustrates the idea of connected graphs and components. Note that in Figure 3.10(d), the triangle formed by v_4, v_5, and v_6 is not a component because it is a subgraph of the portion of the graph labeled Component 2. A term sometimes used as a synonym for component is **maximal connected subgraph**.

You probably already have an intuitive notion of the difference between a general graph and a tree from the examples in Section 3.1. A formal definition of this distinction is:

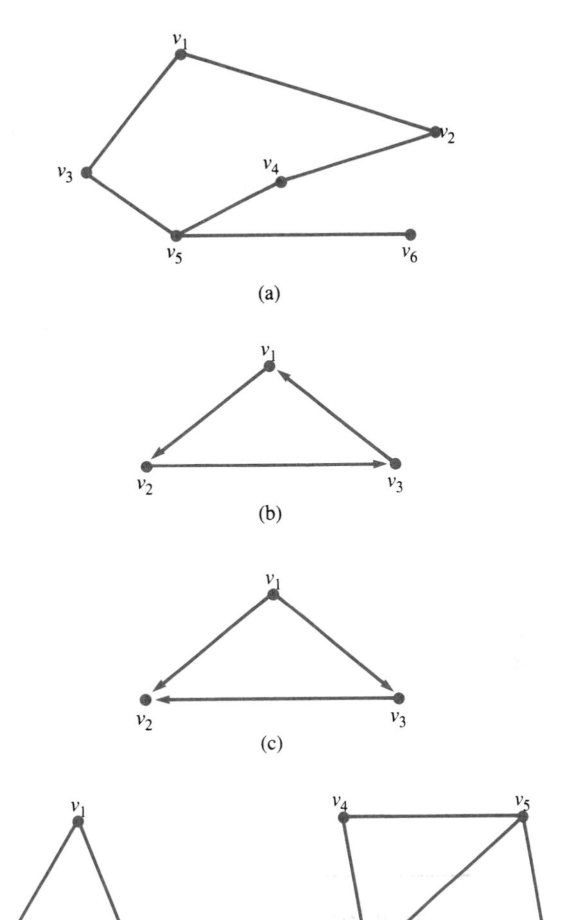

FIGURE 3.10

Connected Graphs and Components. (a) A connected simple graph. If the edge (v_5, v_6) were deleted, the vertex v_6 would be isolated and the graph would no longer be connected. (b) A strongly connected digraph. (c) A weakly connected digraph. (d) A graph with two components.

Definition 4. A **tree** is a digraph

a) which, with edge directions removed, is connected and acyclic;

b) in which every vertex v except one, called the *root*, has indegree 1; and

c) in which the root has indegree 0.

Actually, in a) requiring both connectedness and no cycles is redundant; we leave the explanation to [18].

As in Fig. 3.3, we will almost always display trees with the root at the top and with all edges proceeding from top to bottom. We don't do this just to be perverse, but because it is convenient to "grow" trees from top to bottom. We will therefore always omit the arrows on a tree.

Trees can be used to represent a wide variety of situations. One example is the structure of a game in which the root represents a position in the game and each vertex represents a later position after a move by one player. (See [27–32, Section 3.8] and [42–46, Supplementary].) Another familiar use of trees is for family trees. In this case, a tree is used to represent (quite literally!) relations among people.

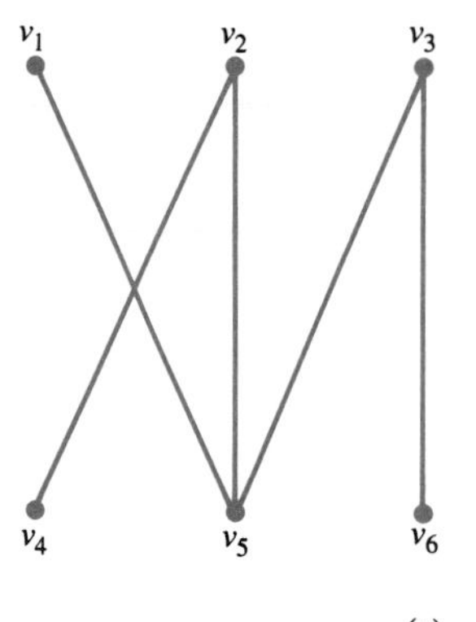

$V_1 = (v_1, v_2, v_3)$

$V_2 = (v_4, v_5, v_6)$

(a)

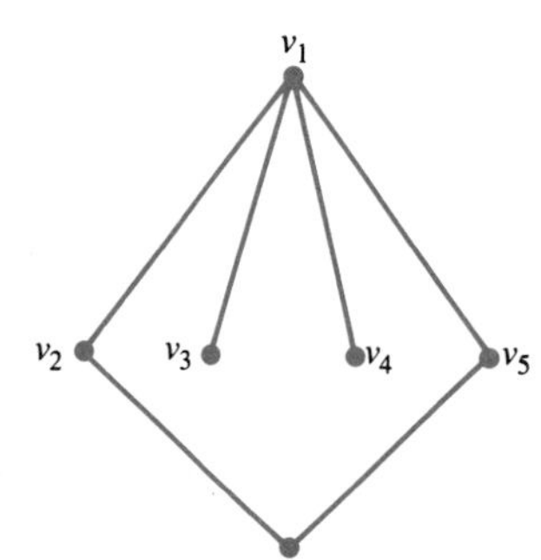

$V_1 = (v_1, v_6)$

$V_2 = (v_2, v_3, v_4, v_5)$

(b)

FIGURE 3.11

Two Bipartite Graphs.

Some authors define a tree to be just a connected, acyclic graph (not a digraph) in which there is no distinguished node called the root. Because trees defined this way are much less important in computer science than those defined in Definition 4, computer scientists usually call acyclic, connected graphs **free trees**. Conversely, connected, acyclic graphs are often called trees by discrete mathematicians, who refer to those in Definition 4 as **rooted directed trees**. Because the latter phrase is somewhat tedious and because most of our tree examples will satisfy Definition 4, we'll use "tree" as in that definition.

Our next definition concerns an important type of graph, which we will discuss in Section 3.7.

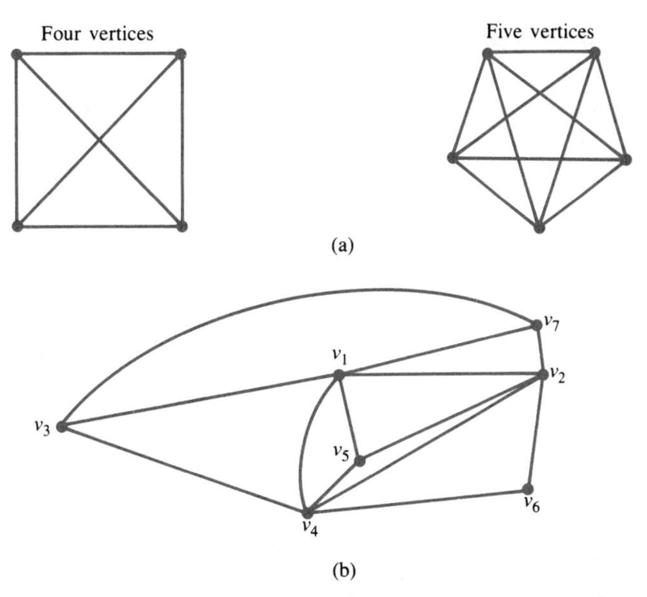

FIGURE 3.12

Complete Graphs and Cliques (a) Two complete graphs; (b) The clique number for the graph is 4. The following sets of vertices form cliques:

$\{v_1, v_3, v_7\}$ $\{v_1, v_2, v_5\}$ $\{v_1, v_4, v_5\}$
$\{v_1, v_3, v_4\}$ $\{v_2, v_4, v_5\}$ $\{v_2, v_4, v_6\}$
$\{v_1, v_2, v_4, v_5\}$ $\{v_1, v_2, v_7\}$

Definition 5. Suppose that the vertex set V can be partitioned into two disjoint sets V_1 and V_2 such that every edge in E joins a vertex in V_1 to a vertex in V_2. Such a graph is called **bipartite**.

Figure 3.11 illustrates some bipartite graphs.

Definition 6. A simple, undirected graph is **complete** if it has no loops and if, for distinct u and v in V, $\{u, v\}$ is in E. That is, a complete graph contains every possible edge between two vertices. A complete subgraph of a graph is called a **clique**. The **clique number** of a graph G is the maximum number of vertices in a clique of G.

Figure 3.12 illustrates complete graphs and cliques.

Problems: Section 3.2

1. a) Draw a graph $G(V, E)$, with $V = \{u, v, x, y\}$ and $E = \{\{u, v\}, \{v, x\}, \{x, u\}, \{v, y\}\}$.

   b) For the graph below, what are the sets V and E?

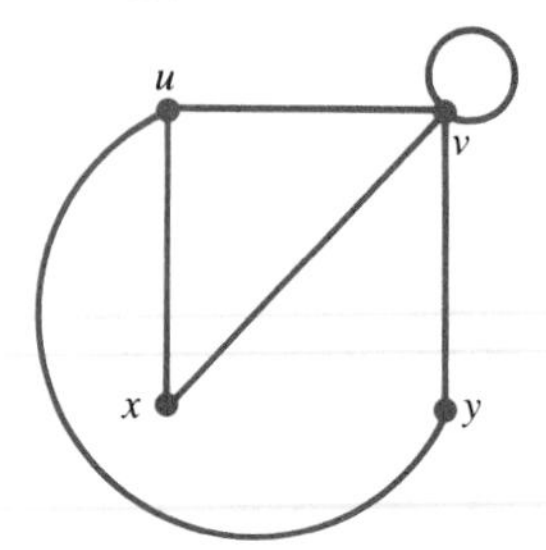

2. a) Draw a digraph $G(V, E)$, with $V = \{u, v, x, y\}$ and $E = \{(u, v), (v, x), (x, v), (y, y)\}$

   b) For the following digraph, what are the sets V and E?

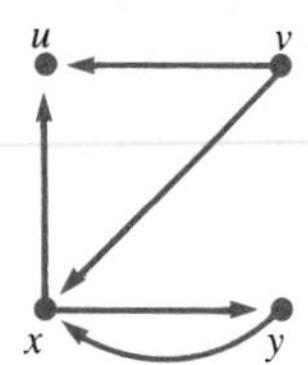

3. a) Draw all the simple, undirected, nonisomorphic graphs having four vertices.

   b) Draw all the simple, nonisomorphic directed graphs having four vertices, at most three edges and no edges in both directions between any pair of vertices.

   All these graphs can be drawn without having edges intersect except at vertices; to save the grader from eye strain, avoid edge crossings.

4. a) Draw all the simple, undirected, nonisomorphic graphs having six vertices for which the degrees of the vertices are 1, 1, 1, 2, 2, and 3.

   b) Similarly, draw all the simple directed graphs having five vertices for which the outdegrees and indegrees of the vertices are (2, 1), (0, 2), (3, 0), (1, 3), and (1, 1) (i.e., the first vertex has two edges leading from it and one going toward it).

5. Suppose that G is a digraph representing a relation R. How would you determine from the graph whether the relation is transitive?

6. Consider the digraph below.

   a) For the binary relation R defined by this graph, list all pairs $(a, b) \in R$. Is the relation reflexive, symmetric, or transitive?

   b) What familiar relation between integers is represented by this graph for the integers 1 to 6?

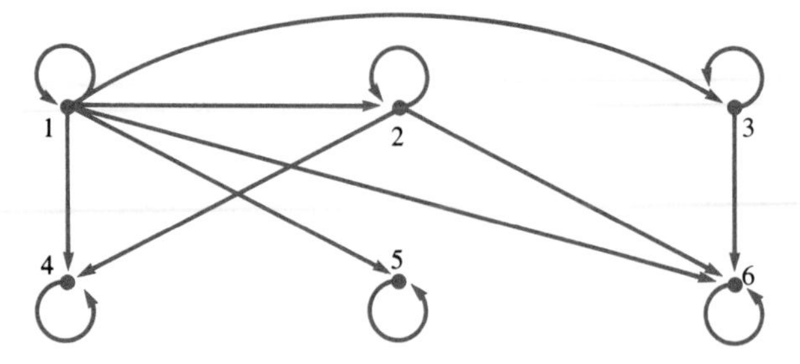

7. a) Use Theorem 1 to show that any graph has an even number of vertices of odd degree.

   b) Use this result to show that, in any group of people shaking hands in pairs, an even number of people each shake hands with an odd number of people.

8. Prove that a simple graph without loops has at least two vertices of the same degree. *Hint*: Let G have n vertices. Then what is the largest degree that any vertex can have? If this maximum is attained, what is the smallest degree that some other vertex can have? What happens when there are isolated vertices? Your argument will boil down to the **Pigeonhole Principle**: If p pigeons are put in $q < p$ holes, then some hole contains at least two pigeons (see also Section 4.10). (*Note*: Induction might seem like a natural for this problem, but we don't see how to make it work. If you do, let us know!)

9. The definition of a path in Definition 2 doesn't work for multigraphs. Why not? Come up with a definition of a path in a multigraph which does work.

10. Derive a formula for the number of edges of a complete graph with n vertices.

11. If G is a simple graph with n vertices and no two edges share a vertex, what is the maximum number of edges that G can have?

12. In the complete graph with five vertices, how many

distinct (i.e., not containing the same vertices in the same order) Hamiltonian cycles are there?

13. a) Find all the simple cycles in the following graph.

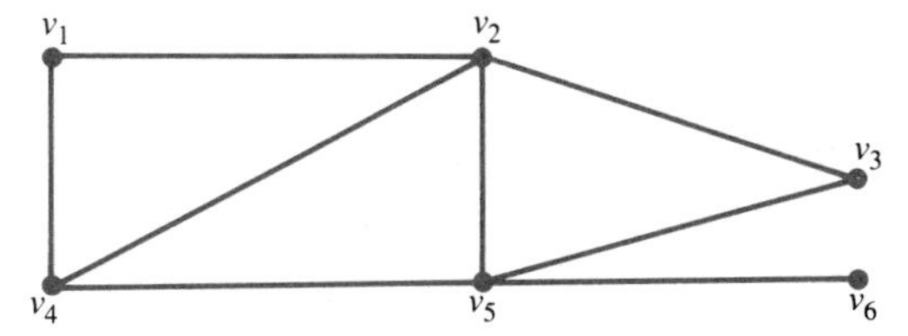

b) In the following graph, find three different simple cycles, each of which has at least five edges.

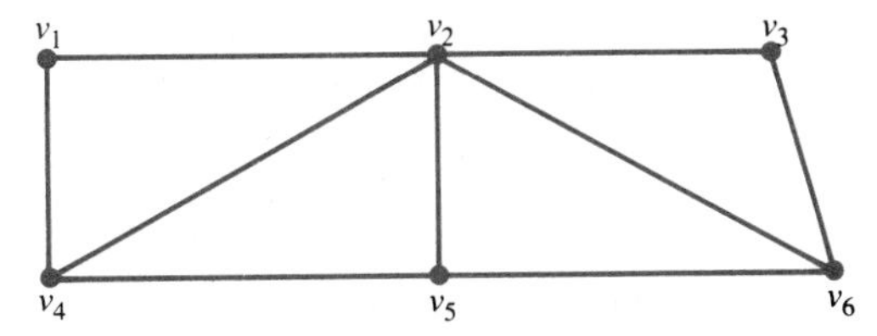

14. Cut-Sets. If the vertices of a graph are divided into two sets V_1 and V_2, the set of edges which join a vertex in V_1 to a vertex in V_2 is called the **cut-set** of the graph relative to (V_1, V_2).

a) For the graph in [13a], with $V_1 = (v_1, v_2)$ and $V_2 = (v_3, v_4, v_5, v_6)$, what is the cut-set?

b) For the graph in [13b], with $V_1 = (v_1, v_3, v_5)$ and $V_2 = (v_2, v_4, v_6)$, what is the cut-set?

c) Explain why a graph is bipartite if and only if there is some partition of the vertex set into V_1 and V_2 so that all the edges of G are in the cut-set relative to (V_1, V_2).

15. Draw all the complete graphs you can without any edges which intersect except at vertices.

16. A **tournament** is a digraph in which, for each pair of vertices, there is exactly one directed edge between them. (See Induction and Tournaments in Section 2.2.)

a) Draw all the nonisomorphic tournaments having three vertices.

b) Is there a tournament on five vertices in which the outdegrees of the vertices are 3, 2, 2, 2, and 2? Why?

c) Is there a tournament on five vertices in which the outdegrees of the vertices are 3, 3, 3, 1, and 0? Why?

17. Trees, as we have defined them, are directed graphs. Therefore we can ask whether they are weakly or strongly connected.

a) Except for the tree consisting of only the root vertex, no tree is strongly connected. Why not?

b) Very few trees are weakly connected. Can you describe all those that are?

c) For each of the two trees shown, what is the minimum number of directed edges which must be added between existing vertices to make the tree weakly connected? Strongly connected? (Of course, they will no longer be trees.)

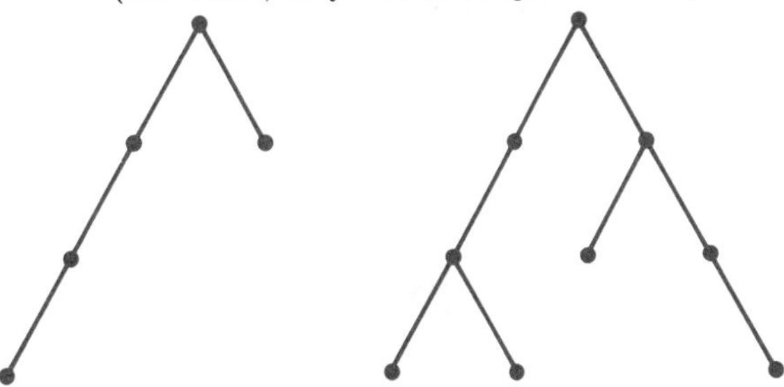

18. a) Show that a consequence of Definition 4 is that there is a path from the root to every vertex in a tree.

b) Why is it redundant in part a) of Definition 4 to require that a tree be both connected an acyclic? That is, if either "connected" or "acyclic" is deleted from the definition but parts b) and c) are left the same, the same graphs are defined.

19. The following sentences are *mathematically* ungrammatical because they misuse graph terminology. However, by changing just a word or two, they can be made mathematically grammatical. Do it.

a) The number of degrees of vertex v (in some graph) is six.

b) Edge e is adjacent to vertex v.

c) To say that vertices u and v are connected means that $\{u, v\}$ is an edge.

d) The complete graph with $V = \{u, v, x, y\}$ has six paths: $\{u, v\}$, $\{u, x\}$, ..., $\{x, y\}$.

e) In the complete graph with five vertexes, each vertice is adjacent to four others.

20. Let G be the complete graph on four vertices. Give

a) a connected subgraph with three edges.

b) a disconnected subgraph with three edges.

c) a spanning subgraph with two edges and no isolated vertices.

21. Labeled graphs. A graph is **labeled** if each vertex has a name. (We have often given vertices in our graphs names to make it easy to refer to them but the names have been arbitrary. Here, however, we mean that the labels are fixed symbols which are parts of the graphs.) Two labeled graphs G and H, with the same set of vertex names, are isomorphic if and only if, for any two vertices u and v, whenever $\{u,v\}$ is an edge in G then $\{u,v\}$ is an edge in H and vice versa. Thus in the following figure, G_1 and G_2 are isomorphic, but G_3 isn't isomorphic to either of them.

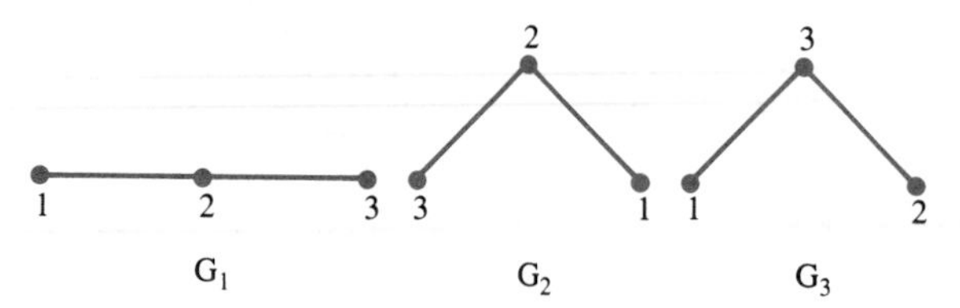

a) Draw all the nonisomorphic labeled simple graphs on the vertex set $\{u, v, w\}$.

b) If the names of the vertices are now erased (so that the graphs are *unlabeled*, that is, ordinary graphs), how many nonisomorphic graphs are left?

22. The **divisibility graph** $D(n)$ is defined by

$$V = \{1, 2, \ldots, n\}, \quad E = \{\{i, j\} \mid i \text{ divides } j\},$$

where "i divides j" means that there must be no remainder. Draw $D(12)$. How many components does it have? What is its clique number?

23. What is the clique number of the graph defined by

$$V = \{1, 2, \ldots, 25\}, \quad E = \{\{i, j\} \mid |i - j| > 4\}?$$

24. A **complete bipartite graph** is a bipartite graph in which there is an edge between *every* pair of vertices in the sets V_1 and V_2. The complete bipartite graph K_{mn} has m vertices in V_1 and n vertices in V_2.

a) Draw K_{14} and K_{23}. The former is an example of a *star*. Why do we use this name?

b) In general, how many edges does K_{mn} have?

c) What is the size of the largest clique in any bipartite graph?

3.3 Paths and Cycles—The Adjacency Matrix

Two of the three examples in Section 3.1 involved finding paths or cycles in a graph. Practical applications of graphs are often related to paths or cycles because, by their very nature, graphs describe in some way how data associated with one vertex is related to data associated with other vertices. How else could such relationships be explored or exploited except by studying the connections—that is, the paths—between vertices? In this section and the next two we consider various aspects of paths and cycles.

Reachability and Representations

An obvious question to ask about a graph $G(V, E)$ is whether you can find a path from one vertex to another. For any of the graphs illustrated thus far in this chapter, you can easily determine by inspection whether there is a path from one vertex to another. But suppose that you were confronted with a graph having hundreds of vertices. It would probably not be drawn at all, because the page would be too cluttered, but rather might be presented by enumerating the sets V and E. Answering the question about paths for such a graph would be decidedly nontrivial.

A vertex of a graph (or digraph) is said to be **reachable** from another vertex if there is a path from the second vertex to the first vertex. The reachability of one vertex from another is what the wolf, cabbage, goat, and farmer problem (Example 1, Section 3.1) is all about.

Reachability is also an important concept in all sorts of *network* problems in which entities are *linked.* For example, in a communications network you are interested in whether a message can get from one vertex to another and, perhaps, in how many different (communication) paths there are from one vertex to another. A second example is a road system linking cities from which you may want to find the shortest route from one city to another. A third and quite different example concerns a food web, which is a directed graph in which the vertices represent species and there is an edge from vertex u to vertex v if species u eats species v. Then the question of whether vertex v is reachable from vertex u is the same as asking whether the two species represented by u and v are on the same food chain.

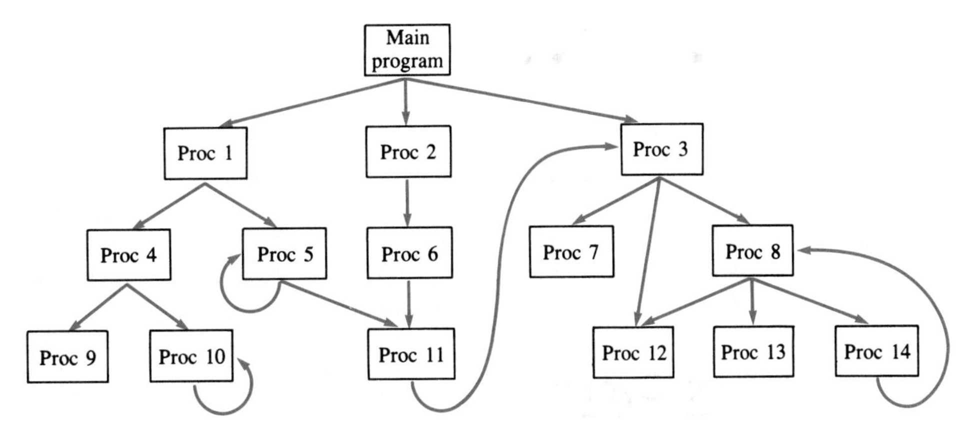

FIGURE 3.13

A Typical Program Call Graph.

As a final example, consider a large and complex computer program organized as a *main program* and *procedures* which call one another (sometimes recursively, sometimes not—see Fig. 3.13). This graph is known as the **call graph** of the program. Often you may need to know whether there is a path in the call graph from one procedure to another and, if so, how many separate paths there are. The result may determine, for example, how the data in the program should be organized. Or you may need to know whether there is a path from a procedure back to itself, that is, whether there is a cycle (e.g., Proc 8 to Proc 14 and back to Proc 8 in Fig. 3.13, which results in **indirect recursion**). Typically, all you know to begin with is, for each procedure, the procedures it may call. That is, for each vertex you know all the edges leading from it. From these data you then must construct the overall call graph and determine which vertices are reachable from which others. When the program is very large and complicated, this can be difficult.

Finding the best *representation* of a problem is often crucial to finding an effective way to solve the problem. Nowhere is there a more graphic (pun intended) example of this than in problems in graph theory. It may have occurred to you that if a graph has hundreds of vertices and edges, long lists of them may not be the most useful way to deal with the graph. Nor is a drawing of the graph very useful. A representation of graphs which is often very effective when you're *thinking* about graph problems and which is also often the best one to use when you're *computing* with graphs is called the **adjacency matrix**.

For a simple graph $G(V, E)$, its adjacency matrix $A = [a_{ij}]$ is a square matrix with the number of rows and columns n, equal to the number of vertices in the graph. The element in the ith row and jth column is the *number* of edges from v_i to v_j. This number will be 1 or 0 for simple graphs but could be greater than 1 for multigraphs. Using our algorithmic notation, we can easily define the adjacency matrix for a simple graph with vertex set $V = \{v_1, v_2, \ldots, v_n\}$ as follows:

for $i = 1$ **to** n
  for $j = 1$ **to** n
    if $\{v_i, v_j\} \in E$ **then** $a_{ij} \leftarrow 1$
                else $a_{ij} \leftarrow 0$ (1)
    endif
  endfor
endfor

This definition works for digraphs, too, if we replace the set braces by parentheses. For graphs, A is **symmetric** (i.e., $a_{ij} = a_{ji}$ for all i and j), because there is an edge from u to v if and only if there is one from v to u. Thus for graphs we need only to know the **main diagonal** $(a_{ii}, i = 1, \ldots, n)$, which gives the loops in the graph, and either the triangle below it (the **lower** triangle) or the one above it (the **upper** triangle) to know the entire matrix. For digraphs, the adjacency matrix is not, in general, symmetric (when will it be?) and must usually be given in its entirety.

Figure 3.14 displays three graphs and their corresponding adjacency matrices. As these matrices indicate, for a digraph the sum of the entries in the adjacency matrix is the size of the set E (why?). But for a graph, the size of the set E is the sum of the entries on the main diagonal and one of the triangles (why?).

How does the definition of the adjacency matrix get us any closer to determining the reachability of one vertex from another? To answer this question, let's define **path length** to be the number of edges on a path. If the same edge is traversed more than once on a path, the edge is counted in the path length the number of times it is traversed. Thus the path length is one less than the number of vertices in Definition 2, Section 3.2. When the path is a cycle, the number of distinct vertices on the path equals the path length.

Sometimes, considering a vertex itself to be a path of length 0 from itself to itself is convenient, whereas a loop is a path of length 1 from a vertex to itself. The usefulness of such "null paths" is explored in [17–18]. In the text, however, we won't use this convention; for us, the shortest path will be a single edge.

One way of interpreting the element in the ith row and jth column of an adjacency matrix is that it represents the number of paths of length 1 from v_i to v_j. The reason the adjacency matrix is so useful in path problems is implied by the following theorem.

Theorem 1. Let A be the adjacency matrix of a graph G. Then the

	v_1	v_2	v_3	v_4	v_5
v_1	0	1	1	1	0
v_2		0	0	1	0
v_3			0	1	1
v_4				0	0
v_5					0

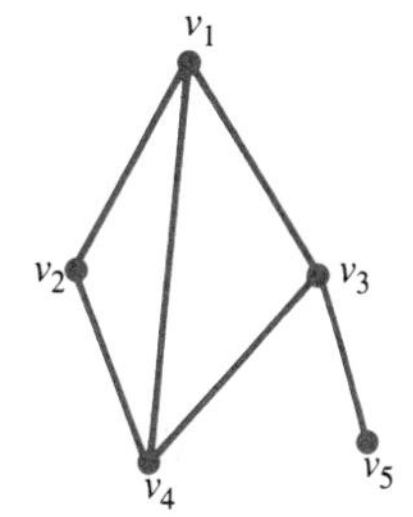

(a)

	v_1	v_2	v_3	v_4	v_5
v_1	0	0	2	2	1
v_2		1	0	1	0
v_3			0	0	1
v_4				0	0
v_5					1

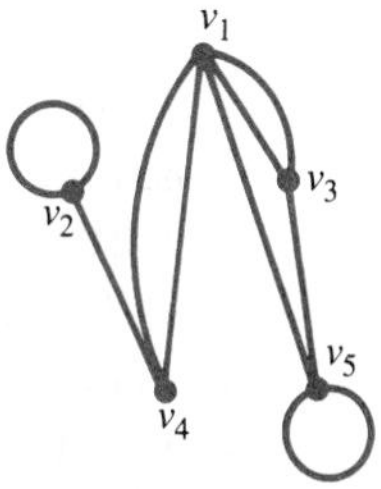

(b)

	v_1	v_2	v_3	v_4	v_5
v_1	0	1	1	1	0
v_2	0	1	0	0	0
v_3	0	0	0	0	0
v_4	0	1	0	1	1
v_5	0	0	2	1	0

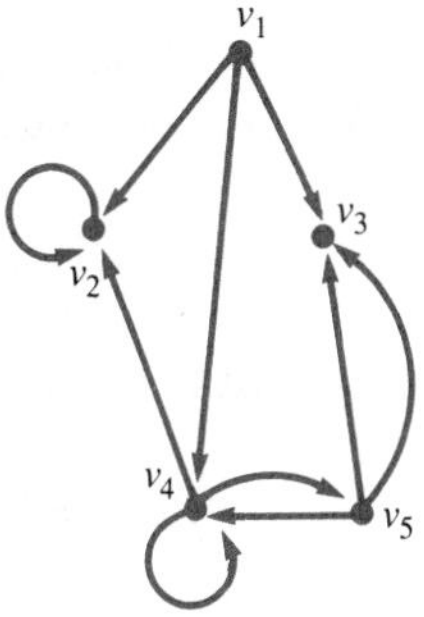

(c)

FIGURE 3.14

Adjacency Matrices for Three Graphs.

element in the ith row and jth column of A^m is equal to the number of paths of length m from vertex i to vertex j.

The notation A^m means the mth power of the matrix A. In Section 0.6 we defined the product of two matrices A and B. It is therefore natural to define A^2 as the matrix product of A with itself. (Note that a square matrix is always conformable with itself.) Similarly, A^3 is the product of A^2 and A (or of A times A^2). In general, $A^m = A^{m-1}A = AA^{m-1}$. In fact, the laws of exponents work with powers of a matrix just as they do with powers of a number.

PROOF By induction. We have already noted that a_{ij} is the number of paths of length 1 from v_i to v_j. Therefore the theorem is true for $m = 1$, which provides the basis step for our induction.

For the induction step we write the adjacency matrix as

$$A = [a_{ij}].$$

Similarly, we write for the rth power of A ($r > 1$):

$$A^r = [a_{ij}^{(r)}].$$

Then, using the definition of matrix multiplication in Eq. (4), Section 0.6, the element $a_{ij}^{(2)}$ in the ith row and the jth column of A^2 is

$$a_{ij}^{(2)} = \sum_{k=1}^{n} a_{ik}a_{kj}, \tag{2}$$

and in general, since $A^m = AA^{m-1}$,

$$a_{ij}^{(m)} = \sum_{k=1}^{n} a_{ik}a_{kj}^{(m-1)}. \tag{3}$$

Now let the induction hypothesis be that the theorem is true for $m-1$ and consider each term in the summation in Eq. (3). The product of a_{ik} and $a_{kj}^{(m-1)}$ gives the number of paths of length m from v_i to v_j *which go through v_k as the second vertex on the path.* Why? Because $a_{kj}^{(m-1)}$ is, by the induction hypothesis, the number of paths of length $m-1$ from v_k to v_j. If $a_{ik}=1$ (i.e., there is an edge from v_i to v_k), then appending this edge at the beginning of each path of length $m-1$ gives a path of length m with v_k the second vertex. Are there any paths of length m not counted by the sum in Eq. (3)? No, because *any* path of length m may be viewed as a path of length 1 to *some* vertex v_k adjoined to a path of length $m-1$, and the sum counts all of them. Are any paths counted more than once by Eq. (3)? No, again, because each term in Eq. (3) counts only those paths starting at v_i and going to a particular v_k as the second vertex. Therefore $a_{ij}^{(m)}$ gives all the paths of length m from v_i to v_j. ∎

We can easily verify that the theorem also is true for digraphs and multigraphs [11].

EXAMPLE 1 Find the number of paths of length 3 in the directed multigraph of Fig. 3.6(c), which we reproduce here.

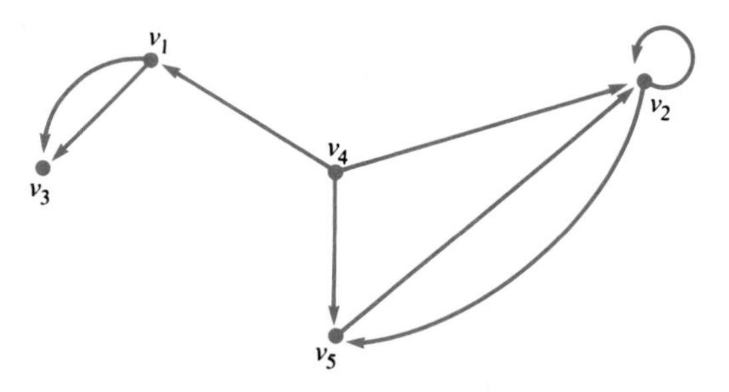

Solution Directly from the figure we have

$$A = \begin{bmatrix} 0 & 0 & 2 & 0 & 0 \\ 0 & 1 & 0 & 0 & 1 \\ 0 & 0 & 0 & 0 & 0 \\ 1 & 1 & 0 & 0 & 1 \\ 0 & 1 & 0 & 0 & 0 \end{bmatrix}.$$

Multiplying A by itself we obtain

$$A^2 = \begin{bmatrix} 0 & 0 & 0 & 0 & 0 \\ 0 & 2 & 0 & 0 & 1 \\ 0 & 0 & 0 & 0 & 0 \\ 0 & 2 & 2 & 0 & 1 \\ 0 & 1 & 0 & 0 & 1 \end{bmatrix}.$$

Multiplying again by A we have

$$A^3 = \begin{bmatrix} 0 & 0 & 0 & 0 & 0 \\ 0 & 3 & 0 & 0 & 2 \\ 0 & 0 & 0 & 0 & 0 \\ 0 & 3 & 0 & 0 & 2 \\ 0 & 2 & 0 & 0 & 1 \end{bmatrix}.$$

Thus there are, for example, three cycles of length 3 from v_2 to itself ($a_{22}^{(3)} = 3$). Referring to the illustration above, we determine that these cycles are

$$(v_2, v_2, v_2, v_2),$$
$$(v_2, v_5, v_2, v_2),$$
$$(v_2, v_2, v_5, v_2).$$

Also, there are two paths of length 3 from v_4 to v_5 ($a_{45}^{(3)} = 2$), which are

$$(v_4, v_5, v_2, v_5),$$
$$(v_4, v_2, v_2, v_5). \blacksquare$$

We can use Theorem 1 to determine whether vertex v is reachable from vertex u, that is, whether there is *any* path from u to v. Let the vertex set $V = \{v_1, v_2, \ldots, v_n\}$. Then we first define the *path matrix* $P = [p_{ij}]$ algorithmically as follows:

for $i = 1$ **to** n
  for $j = 1$ **to** n
    if there is any path from v_i to v_j
      then $p_{ij} \leftarrow 1$
      else $p_{ij} \leftarrow 0$
    endif
  endfor
endfor
(4)

Note that the entries on the main diagonal, p_{ii}, are 1 if and only if there is a path from v_i to itself (which could be a loop). Like the adjacency matrix for simple graphs, P is a **Boolean matrix**, that is, one whose entries are all 0's and 1's.

It isn't hard to show [15] that, if a graph with n vertices has no path of length n or less between u and v, then there is no path at all. Thus to find the path matrix, we could just compute A^2, A^3, ..., A^n. The path matrix would have a 1 in any position where any of these matrices or A itself has a nonzero quantity.

Are you satisfied? We discussed why knowing the path matrix may be important. Then we defined a representation—the adjacency matrix—which is a convenient starting point for computing the path matrix. Next we proved a theorem to show how the adjacency matrix may be used to find the number of paths of any length between two vertices. Finally, we showed how this theorem may be used to design a simple algorithm to compute the path matrix. Isn't that enough? No, it's not.

It's one thing to have *an* algorithm to accomplish a task; it's something else to have a *good* (i.e., efficient) algorithm. The one we presented just above is not a good algorithm if all we want to know is what is connected to what. As Example 1 implies, for all but the smallest values of n, the computation of powers of A is quite tedious. In fact, if there are n vertices, each matrix multiplication takes n^3 multiplications and additions. Since there may be $n-1$ matrix multiplications, we may have to do about n^4 operations, a very large number for all but quite small n. Is there a better way? Yes. It is embodied in an algorithm designed by S. Warshall.

Warshall's algorithm also starts with the adjacency matrix. Then, one at a time, we add vertices in V to a set of vertices V' *through which paths are allowed to pass* when going from one vertex in V to another. Putting it another way, suppose that we already knew, for any pair of vertices, when one was reachable from the other by a path which was allowed to pass only through vertices $v_1, v_2, \ldots, v_k$. If there is no such path, then we try allowing a path which could also pass through v_{k+1}.

Thus at the zeroth stage (when V' is still empty and so *no* intermediate vertices are allowed), for each ordered pair of vertices (u, v), we consider only edges from u to v, that is, paths with no intermediate vertices. Then, at the first stage, for each ordered pair of vertices (u, v). we consider only edges from u to v and paths from u to v which have only v_1 as an intermediate vertex. At the next stage we allow paths which pass through v_1 or v_2 or both. In this way, we can avoid much unnecessary computation.

To implement this idea we first define the **modified adjacency matrix** $A^{(1)}$ to be the adjacency matrix with all terms greater than 1 replaced by 1. For a simple graph, the adjacency matrix and the modified adjacency matrix are the same. Then we define a sequence of matrices

$$P_0(= A^{(1)}), P_1, P_2, \ldots, P_n$$

such that each $P_k = [p_{ij}^{(k)}]$, $k = 0, 1, \ldots, n$ has the property that

if there is an edge from v_i to v_j or a path from v_i to v_j
  including only vertices from the set $\{v_1, v_2, \ldots, v_k\}$
    then $p_{ij}^{(k)} \leftarrow 1$ (5)
    else $p_{ij}^{(k)} \leftarrow 0$
endif

Algorithm 3.2
Warshall

```
Input n                                          [Size of V]
      A^(1)                      [Modified adjacency matrix of G]

Algorithm WARSHALL
   P ← A^(1)
   for k = 1 to n                                [Index for P^(k)]
      for i = 1 to n                             [Row index]
         for j = 1 to n                          [Column index]
            if p_ij = 0 then p_ij ← p_ik p_kj    [As in (7)]
         endfor
      endfor
   endfor
Output P = [p_ij]                                [Path matrix]
```

Algorithm fragment (5) considers at each stage all pairs of the n vertices in V, but at the kth stage, allows paths to go through *only* the vertices in the set $\{v_1, v_2, \ldots, v_k\}$.

It should be clear from (5) that

$$P_n = P, \tag{6}$$

the path matrix. As k gets close to n most values of p_{ij} which will eventually be 1 are 1 already. This enables us to reduce the total amount of computation considerably. Let's see how we can embody this idea in an algorithm.

If P_k is to have the property defined by (5), in going from P_{k-1} to P_k, the only time $p_{ij}^{(k)}$ will differ from $p_{ij}^{(k-1)}$ is when there is no path from v_i to v_j including at most vertices from

$$\{v_1, \ldots, v_{k-1}\}$$

as intermediates (i.e., $p_{ij}^{(k-1)} = 0$) but, there is a path from v_i to v_j including at most vertices in

$$\{v_1, v_2, \ldots, v_{k-1}, v_k\}.$$

This latter case can occur only when there is a path from v_i to v_k including only vertices in $\{v_1, \ldots, v_{k-1}\}$ (i.e., $p_{ik}^{(k-1)} = 1$) *and* a path from v_k to v_j including only vertices in this same set (i.e., $p_{kj}^{(k-1)} = 1$). Therefore $p_{ij}^{(k)}$ is defined by:

$$\begin{aligned} &\textbf{if } p_{ij}^{(k-1)} = 1 \textbf{ then } p_{ij}^{(k)} \leftarrow 1 \\ &\qquad\qquad\qquad\quad \textbf{else } p_{ij}^{(k)} \leftarrow p_{ik}^{(k-1)} p_{kj}^{(k-1)} \end{aligned} \tag{7}$$

with the product after **else** equal to 1 only when the addition of v_k to the set of vertices allows a path from v_i to v_j, which was not possible previously.

Putting all these ideas into algorithmic form gives Algorithm WARSHALL. Remarks:

- We dropped the superscripts of algorithm fragment (7) to indicate that you can store successive P_k on top of each other. But wait a minute. Suppose that p_{ik} and p_{kj} in the algorithm, which should be values in P_{k-1}, have already been changed to their values in P_k earlier in the kth pass through the outer loop. Could this cause a problem? The answer is No. If you don't understand why, think about this again after you read our proof that this algorithm is correct.
- When $i = k$, then (7) becomes

$$\begin{aligned} \textbf{if } p_{kj}^{(k-1)} = 1 \textbf{ then } & p_{kj}^{(k)} \leftarrow 1 \\ \textbf{else } & p_{kj}^{(k)} \leftarrow p_{kk}^{(k-1)} p_{kj}^{(k-1)} \end{aligned}$$

  so that $p_{kj}^{(k)}$ always has the same value as $p_{kj}^{(k-1)}$ (why?). A similar remark holds when $j = k$. Therefore we could have tested whether i or j equals k at the same time we tested whether $p_{ij} = 0$. We didn't to keep the statement of the algorithm as simple as possible.
- Once $p_{ij} = 1$ for any i and j, it is not necessary to consider this i or j for any further values of k. Thus the test "$p_{ij} = 0$" may be made many times for a given i and j. But it is easier to perform this test than it is to keep track of (and test) those values of i and j for which p_{ij} has already been set to 1.
- For graphs rather than digraphs the algorithm can be simplified because $p_{ij} = p_{ji}$ [24].

Have we convinced you that WARSHALL really calculates the path matrix? Perhaps, but our discussion has been so informal that we really should give a proof. As with so many proofs of the correctness of algorithms, the crucial step is to find the loop invariant. Here, finding the loop invariant is pretty easy because our idea was that P in the algorithm should be P_k as defined in (5) after the kth pass through the loop in WARSHALL. Then when we exit from this loop the last time, P will be P_n, the path matrix, which is what we want.

Therefore we hypothesize that the loop invariant is

$$P = P_k.$$

We prove this by induction (of course!). The basis step is that, before the first entry into the loop (really the $k=0$ case), $P = P_0$. But this is true because we set $P = A^{(1)}$ before entry into the loop and $A^{(1)}$ is just the modified adjacency matrix, which by definition is P_0.

Now for the inductive step, we assume that, after exit from the $k-1$ trip around the loop, $P = P_{k-1}$. That is, we assume that p_{ij} tells us whether there is a path from v_i to v_j using at most the set

$$S = \{v_1, v_2, \ldots, v_{k-1}\}$$

as intermediate vertices. If the value of p_{ij} is 1 (yes, there is a path), then it remains 1, as it should, during the next pass through the loop. If $p_{ij} = 0$ at the beginning of the pass, then it should only change to a 1 if there is now a path from v_i to v_j with v_k on it. This can only happen if there is a path from v_i to v_k and a path from v_k to v_j with both paths using at most vertices from S. This is the same as saying that p_{ij} will change from 0 to 1 if and only if

$$p_{ik}p_{kj} = 1,$$

which is just what the algorithm says is computed during the kth pass. Thus after the kth pass, P does equal P_k, which is what we wanted. This completes the proof.

What remains to be done is to verify that WARSHALL really is more efficient than the brute force approach we outlined earlier. A worst case analysis is both easy and instructive. Clearly, the total computation for this algorithm depends entirely on how many times the statement

$$p_{ij} \leftarrow p_{ik}p_{kj} \tag{8}$$

is executed. The worst case of WARSHALL is a graph with no edges! (Why?) In this case, (8) is executed n times for each j. As there are n values of i and k, the total is

$$nn^2 = n^3$$

Now suppose that G is a digraph with m edges. Then there are m 1's in the adjacency matrix A and (8) is never executed for these m pairs of i and j. Therefore

$$n(n^2-m) \tag{9}$$

is at least as great as the number of executions of (8) The bound in expression (9) compares quite favorably with the n^4 estimate of the number of operations in our earlier, brute force algorithm. Actually, we would expect (8) to be a quite *conservative* (i.e., substantially greater than the actual) upper bound because paths of length greater than 1 will gradually create more nonzero p_{ij} terms. As you might expect, the average case analysis of WARSHALL is very difficult, indeed.

EXAMPLE 2 Apply Algorithm WARSHALL to the following digraph.

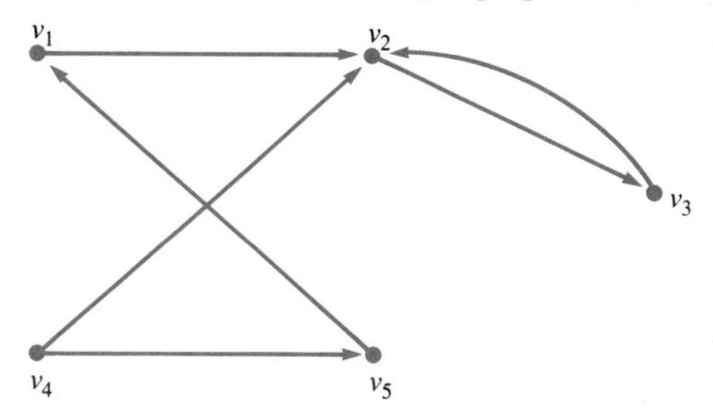

Solution The digraph is simple, so the modified adjacency matrix is the same as the adjacency matrix, or:

$$A = \begin{bmatrix} 0 & 1 & 0 & 0 & 0 \\ 0 & 0 & 1 & 0 & 0 \\ 0 & 1 & 0 & 0 & 0 \\ 0 & 1 & 0 & 0 & 1 \\ 1 & 0 & 0 & 0 & 0 \end{bmatrix}.$$

For successive values of k, we obtain for the P matrices:

$$k = 1: \begin{bmatrix} 0 & 1 & 0 & 0 & 0 \\ 0 & 0 & 1 & 0 & 0 \\ 0 & 1 & 0 & 0 & 0 \\ 0 & 1 & 0 & 0 & 1 \\ 1 & 1 & 0 & 0 & 0 \end{bmatrix}; \qquad k = 2: \begin{bmatrix} 0 & 1 & 1 & 0 & 0 \\ 0 & 0 & 1 & 0 & 0 \\ 0 & 1 & 1 & 0 & 0 \\ 0 & 1 & 1 & 0 & 1 \\ 1 & 1 & 1 & 0 & 0 \end{bmatrix};$$

$$k = 3, 4: \begin{bmatrix} 0 & 1 & 1 & 0 & 0 \\ 0 & 1 & 1 & 0 & 0 \\ 0 & 1 & 1 & 0 & 0 \\ 0 & 1 & 1 & 0 & 1 \\ 1 & 1 & 1 & 0 & 0 \end{bmatrix}; \qquad k = 5: \begin{bmatrix} 0 & 1 & 1 & 0 & 0 \\ 0 & 1 & 1 & 0 & 0 \\ 0 & 1 & 1 & 0 & 0 \\ 1 & 1 & 1 & 0 & 1 \\ 1 & 1 & 1 & 0 & 0 \end{bmatrix} = P.$$

You should be able to verify the correctness of this result by inspecting the preceding figure. For example, there is a 1 in the (5, 2) position when $k=1$ because there is a path from v_5 to v_2 which passes through v_1. The reason that the matrices for $k=3$ and $k=4$ are the same is that the addition of v_4 to the set of vertices does not allow any new paths since there are no edges into v_4. If you count how many times (8) is actually executed in this example, you will see that expression (9) is, indeed, a very conservative bound [32]. ∎

Problems: Section 3.3

1. Give the adjacency matrix for the following graphs and digraphs.

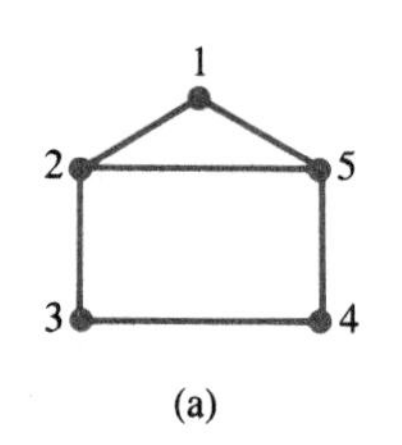

(a)

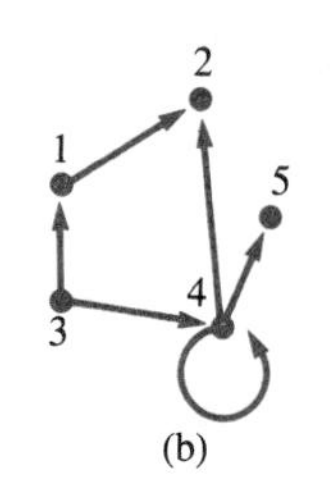

(b)

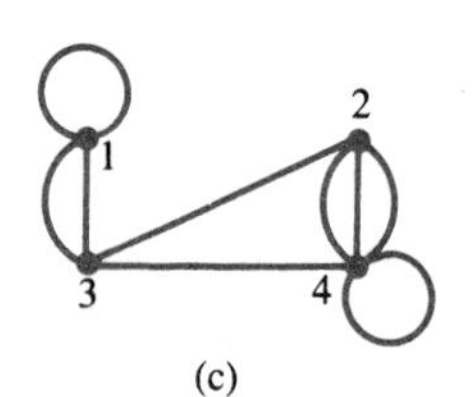

(c)

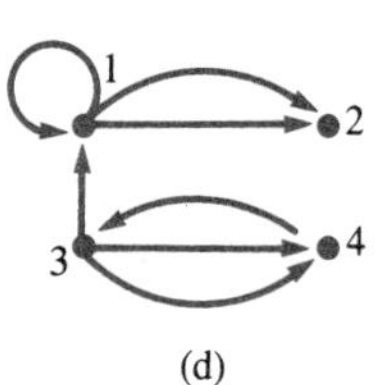

(d)

2. a) Draw the graphs or multigraphs with adjacency matrices (i) and (ii).

$$\begin{bmatrix} 0 & 1 & 0 & 1 & 0 \\ & 1 & 1 & 0 & 1 \\ & & 1 & 0 & 1 \\ & & & 1 & 0 \\ & & & & 0 \end{bmatrix} \qquad \begin{bmatrix} 1 & 1 & 0 & 1 & 1 \\ & 0 & 0 & 2 & 1 \\ & & 1 & 1 & 0 \\ & & & 1 & 0 \\ & & & & 0 \end{bmatrix}$$

(i) (ii)

b) Draw the digraphs with adjacency matrices (i) and (ii).

$$\begin{bmatrix} 1 & 1 & 0 & 1 & 1 \\ 1 & 0 & 1 & 1 & 1 \\ 0 & 1 & 0 & 0 & 1 \\ 1 & 1 & 0 & 1 & 1 \\ 0 & 0 & 0 & 0 & 0 \end{bmatrix} \qquad \begin{bmatrix} 0 & 2 & 0 & 1 \\ 1 & 1 & 0 & 0 \\ 0 & 1 & 1 & 2 \\ 2 & 0 & 1 & 0 \end{bmatrix}$$

(i) (ii)

3. The **incidence matrix** of a graph $G(V, E)$ without loops is a Boolean matrix (all entries 0 or 1) with $|V|$ rows and $|E|$ columns such that the (i, j) entry is 1 if vertex i is at one end of edge j and 0 otherwise.

a) Write the incidence matrix for the following graph. Call this matrix M.

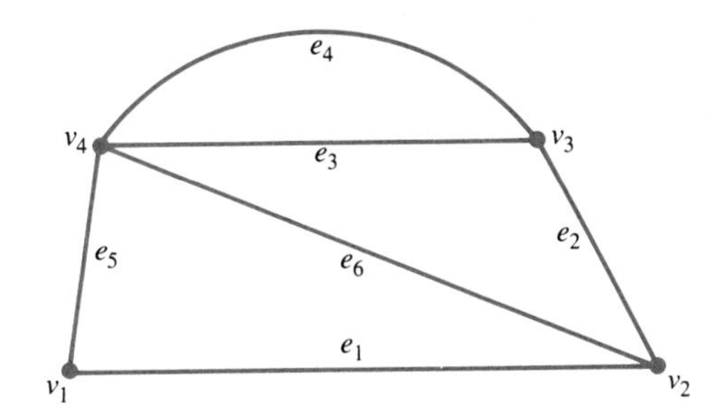

b) Write the adjacency matrix A for the same graph.

c) What can you say about the row sums of an incidence matrix? What about the column sums?

d) Compute MM^{T}. How do you interpret this result?

e) Prove that this interpretation holds for arbitrary graphs without loops.

f) What goes wrong (or needs to be redefined) if there are loops?

4. The incidence matrix is also defined for directed graphs. The entries are now $+1$, 0, or -1.

a) Come up with a definition yourself. Apply it to the following digraph.

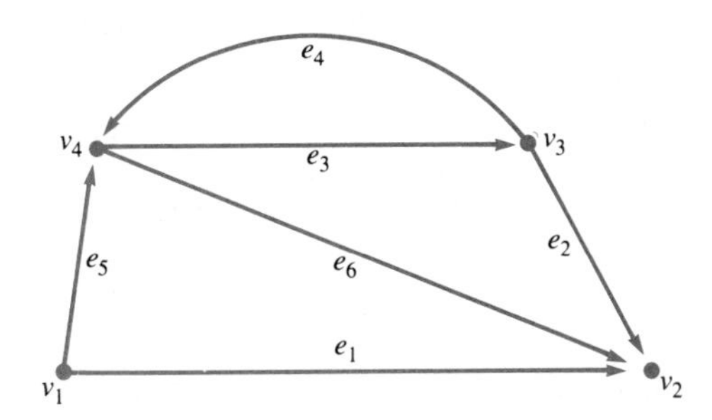

b) Answer [3c] for digraphs.

c) Answer [3d] for the digraph above.

5. Write an algorithm whose input is an adjacency matrix and whose output is a list of edges.

6. Write an algorithm whose input is a list of edges and whose output is an adjacency matrix.

7. For the accompanying graph, let A be the adjacency matrix. Compute A^2 two ways:

a) Directly.

b) By counting visually the number of paths of length 2 between each pair of vertices.

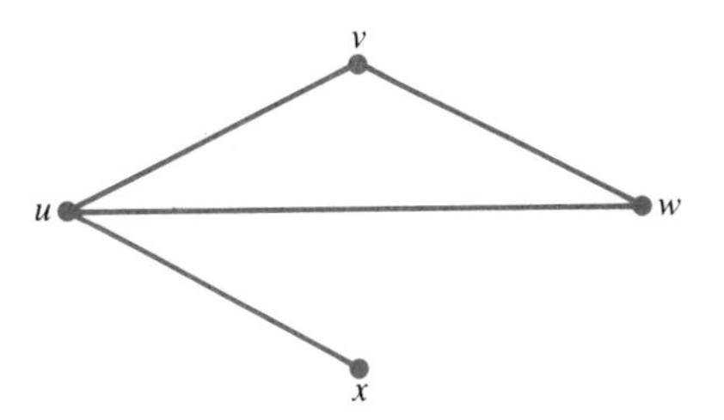

8. For the graph in [7]:

a) List all paths of length 4 between u and w. (The simplest way to record your answer is to denote each path by its vertex sequence. For instance one such path is $uvwuw$.)

b) Check your answer using the adjacency matrix.

9. Consider the graph with five vertices 1, 2, 3, 4, and 5 and edges $\{1,2\}$, $\{2,3\}$, $\{2,4\}$, $\{3,4\}$, $\{3,5\}$, and $\{4,5\}$.

a) Write the adjacency matrix.

b) Compute the matrix which, for all pairs (i, j) of vertices, lists the number of paths of length 2 between them.

c) Compute the number of paths of length 3 between vertices 2 and 3. (This is a single entry from another matrix. Show how you can avoid computing the entire matrix.)

d) Check your answer to c) by listing all the paths of length 3 from vertex 2 to vertex 3. Name each path as a sequence of vertices.

10. a) Explain how to use the adjacency matrix of a simple graph and matrix multiplication to determine the number of triangles incident on any given vertex (i.e., the number of triangles which have the given vertex as one point).

b) Explain how to use the adjacency matrix to determine the number of triangles in a graph.

11. State and prove a version of Theorem 1 for
a) digraphs. b) multigraphs.

12. Repeat [7] for the following digraph.

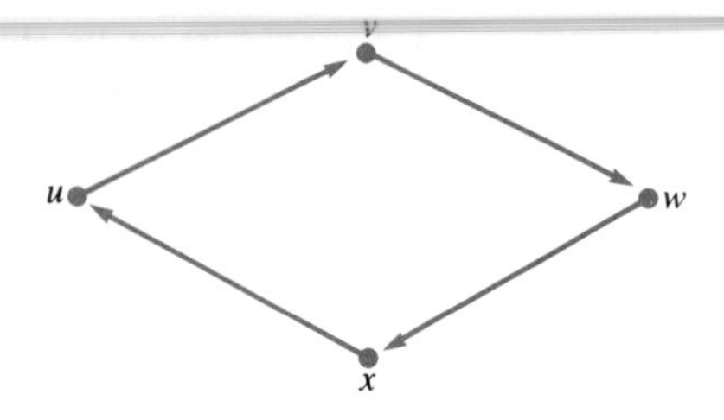

13. Repeat [8] for the preceding digraph.

14. Repeat [9] in the case where each edge is directed (i.e., (1, 2) instead of $\{1,2\}$).

15. Let G have n vertices. Explain why, in order to determine whether a pair of distinct vertices is connected, it is sufficient to compute powers of A only through A^{n-1}.

16. If a graph has at least one edge, it has infinitely many paths. Explain.

17. a) If we allow null paths, then it makes sense to ask for a matrix which counts the number of paths of length 0. Ideally, Theorem 1 would hold for $m = 0$ as well as for $m = 1, 2, \ldots$. Show that it does.

    b) In the text we argued that the path matrix is just the matrix
    $$\sum_{k=1}^{n-1} A^k,$$
    with all nonzero entries replaced by 1. If we allow null paths this statement is still correct if revised slightly. How?

18. Let A' be the same as the modified adjacency matrix $A^{(1)}$ of Algorithm WARSHALL, except that all diagonal entries of A' are 1. (Thus if G is a simple graph, $A' = A+I$.) Suppose that we replace $A^{(1)}$ by A' but otherwise leave WARSHALL unchanged. What happens? Are there any graphs for which the final output is different from that of WARSHALL as shown in the text? If so, is one of the versions wrong, or is it a matter of a slightly different definition of when vertices are joined by a path?

19. Consider the directed graph with five vertices 1, 2, 3, 4, and 5 and edges
    $$(1,4), (2,1), (2,4), (3,2), (4,5), \text{ and } (5,3).$$
    Determine the path matrix using WARSHALL. Show each of your intermediate matrices. In each matrix circle each 1 which appears where there was a 0 in previous matrices.

20. Consider the undirected graph obtained from the digraph in [19] by removing all the arrows. Use WARSHALL to determine the path matrix for this graph. You should find that you will be able to quit before computing all the matrices. What lets you know that you can quit? What one-word property that we like graphs to have can you attribute to this graph when you are done?

21. For the accompanying graph, carry out the following part of WARSHALL. Start with the adjacency matrix and do three iterations of the matrix update. That is, your final matrix should have a 1 in the (i, j) entry only when there is a path from vertex i to vertex j, using as intermediate vertices at most vertices v_1, v_2, and v_3. Show that you are really computing all the entries in the path matrix by explaining how you get the (1, 4) and (5, 6) entries of your final matrix. (To do the entire algorithm on this graph is tedious.)

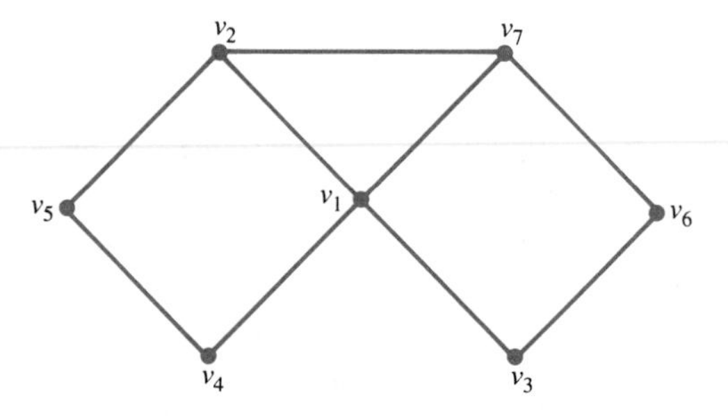

22. Repeat [21], but first make every edge directed from left to right. Now 1's in the matrix should indicate the existence of a directed path.

23. Suppose that the input to Algorithm WARSHALL contains a list of those rows and columns of the adjacency matrix which are all 0. How could you use this information to modify the algorithm to increase its efficiency?

24. Show how the computation of Algorithm WARSHALL can be halved for undirected graphs. Rewrite the algorithm to show what you mean.

25. Suppose that the edges of a simple graph have nonnegative weights associated with them. For such a graph define the length of a path to be the sum of the weights of the edges on it. Use the idea behind Algorithm WARSHALL to devise an algorithm which produces a matrix C in which the element c_{ij} is the length of the shortest path between vertices i and j. (*Hint*: Start with the matrix which

is the adjacency matrix, except that 0 elements are replaced by ∞ and nonzero elements by edge weights. Think recursively.)

26. Find the shortest path length from v_1 to v_6 in the following undirected graph using the shortest-length variant of Algorithm WARSHALL from [25].

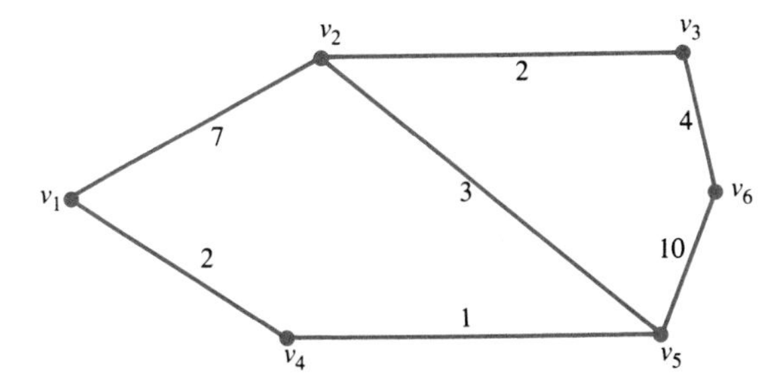

27. Extend Algorithm WARSHALL so that, for each pair of connected vertices, it outputs a path between them. Think of it this way: The output now becomes two matrices. The entries of the first are 0's and 1's as before. The (i, j) entry of the second is empty if there is no (i, j) path and otherwise is the sequence of vertices in such a path.

28. a) [27] involves an excessive amount of work if all you want to do is look at a few paths. For that purpose, you need the algorithm to output only the first internal (or the last internal) vertex of the connecting path. Explain.

b) Write a postprocessor algorithm which takes this more limited output and, given i and j, outputs the entire list of vertices forming a path between them.

29. Carry out the modifications of [27] and [28] on the shortest-length version of Algorithm WARSHALL in [25].

30. Show how to halve the computations in [25] for an undirected graph.

31. The shortest-length version of Algorithm WARSHALL still runs if the graph has negative weights. Does it still succeed in finding the shortest simple paths? If it does, explain why. If it doesn't, give an example of why you might have expected it not to work even before you found a counterexample. (One way to do this is to show that the recursive thinking which led to Algorithm WARSHALL breaks down; that is, knowing the previous case does not make the current case easy to compute.)

32. Count how many times (8) is executed in the computation of Example 2. Compare this result with expression (9). What do you conclude?

3.4 Eulerian and Hamiltonian Paths and Cycles

One of the most famous problems in graph theory is the Königsberg Bridge Problem. Solved by Leonhard Euler in 1736, this problem is usually credited with initiating graph theory as a mathematical subject.

There were seven bridges across the river Pregel at Königsberg, which joined the two banks and two islands in the river as illustrated in Fig. 3.15(a). (Königsberg was in East Prussia in Euler's time but is now in the Soviet Union and is called Kaliningrad.) The story (legend?) goes that the townspeople of Königsberg asked whether it was possible to take a walk in which each of the seven bridges is crossed once and only once.

If we replace land masses by vertices and the bridges joining them by edges, we get the graph shown in Fig. 3.15(b). The problem therefore is equivalent to asking whether this graph has an Eulerian path. Alternatively, since we don't have to end where we begin, an equivalent question is whether the graph has an Eulerian cycle or whether the addition of *one more edge* between any two vertices (which could take us from where we ended our walk back to where we started) would result in a graph with an Eulerian cycle.

Some practical problems also involve Eulerian paths and cycles. One class of

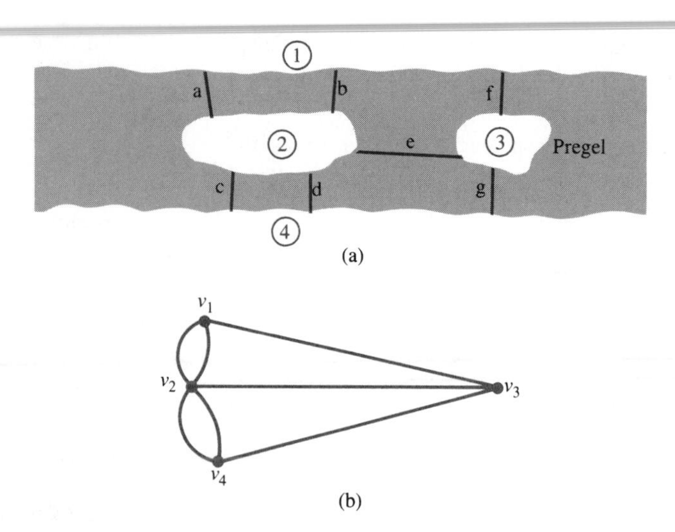

FIGURE 3.15 The Königsberg Bridge Problem. (a) The river Pregel at Königsberg, with seven bridges connecting two islands and the two river banks. (b) Graph of the Königsberg bridge problem, with land areas as vertices and bridges as edges.

problems is based on the graph formed by the streets of a city or town, where the edges are the individual streets and the vertices are the intersections. A postal worker delivering letters would like to travel an Eulerian cycle from the post office and back along the sides of streets on which mail has to be delivered. Such a route would be a minimum-length path since no side of a street would be traversed more than once. Similarly, a street cleaning crew would like to travel an Eulerian cycle from the garage and back along the streets which have to be cleaned. In such problems we want to know whether there is an Eulerian cycle and, if there is, how to find it. (If there is no Eulerian cycle, then we would like to find the shortest cycle which traverses each edge at least once; this is quite a hard problem, which we won't discuss in this book.)

How might we go about determining whether a graph has an Eulerian cycle? Might we try the path matrix? Well, if there is a 1 in position i on the diagonal of the path matrix, we know that there is a path from the ith vertex back to itself. But this may or may not be an Eulerian cycle. Alternatively, if there are m edges, we might compute A^m, which tells us whether there are any paths of length m. But a path of length m need not be an Eulerian path because it could traverse one edge more than once. In fact, the path matrix just isn't much use in trying to find Eulerian paths and cycles. A more effective approach is to use the following theorem.

Theorem 1. A connected, undirected multigraph with no loops has an Eulerian cycle if and only if it contains no vertices of odd degree.

Why the restriction to connected graphs? Well, there are disconnected graphs with Eulerian cycles, but they can occur only when there are one or more isolated vertices (why?). Let's call a component that is not an isolated vertex a **nontrivial component**. Then if a graph has two or more nontrivial components, it cannot have an Eulerian path. This follows because an Eulerian path traverses all the edges of a graph, but no path can traverse an edge in two different nontrivial components (why?).

Why the restriction of no loops? Because if there is an Eulerian cycle on a graph without loops, any number of loops can be added and there will still be an Eulerian cycle. And, vice versa, if there is an Eulerian cycle on a graph with loops, there will still be one if all the loops are removed. Not allowing loops makes the proof that follows a little easier.

Theorem 1 epitomizes a kind of mathematical theorem called an **existence theorem**. It postulates the existence of an entity (an Eulerian cycle) if a graph has a certain property (no vertices of odd degree). In addition, with the "if and only if", it claims that this property (no vertices of odd degree) is not only *sufficient* for there to be an Eulerian cycle but also *necessary*; there can be no Eulerian cycle in a graph which does not have this property (see Section 0.7 for a discussion of this terminology). Broadly speaking, there are two kinds of proofs of existence theorems:

1. Nonconstructive proofs, in which you do indeed prove the theorem, but don't, in so doing, produce any mechanism which enables you to *construct* the object whose existence is being proved.
2. Constructive proofs, in which the existence is proved (a) by exhibiting an object of the desired type or (b) by exhibiting an *algorithm* which can be shown to produce an object of the desired type.

For the past century, nonconstructive existence proofs have been far more common than constructive proofs; most mathematicians have preferred them esthetically. But it won't surprise you that we prefer constructive proofs whenever we can find them. We especially like type (b) constructive proofs, which we call *proofs by algorithm.*

To prove Theorem 1 we first consider the *sufficiency* or *if* part of the theorem, namely, that an Eulerian cycle exists if all vertices are of even degree. We give two proofs, the first a classical nonconstructive one and then a proof by algorithm. The nonconstructive proof, like many existence proofs, is a proof by contradiction.

PROOF 1 (nonconstructive) Suppose that having all vertices of even degree is not enough to ensure the existence of an Eulerian cycle. Then there must be at least one connected graph, all of whose vertices are of even degree, which does not have an Eulerian cycle. Among all graphs with even-degree vertices and no Eulerian cycle, let's choose a graph G with a minimum number of edges. Thus any connected graph with fewer edges, all of even degree, would have an Eulerian cycle. The graph G must have at least two edges because there is no one-edge graph whose vertices all have even degrees. (Remember: No loops are allowed.) Also G must have at least two vertices since there are no loops.

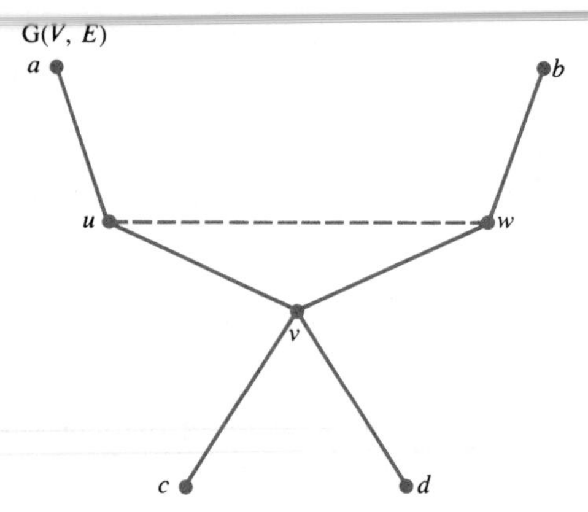

FIGURE 3.16

A Nonconstructive Proof of the Eulerian Cycle Theorem.

Since we are only considering connected graphs, there must be two edges of G which share a vertex (why?). Call these edges $\{u, v\}$ and $\{v, w\}$ (see Fig. 3.16). (The vertices u and w *could* be the same since we are allowing multigraphs. Make sure that you're convinced that the argument which follows—which assumes that G is simple—can be easily adapted to multigraphs.) Now consider the graph G' formed from G by deleting these two edges and replacing them by an edge $\{u, w\}$. If v is now isolated, delete it also. By this construction all vertices of G' also have even degree (why?).

Suppose first that G' is not connected (in which case v would not have been deleted). Then it consists of two components, each of which has fewer edges than G. So, each component has an Eulerian cycle because of our assumption that all graphs satisfying the hypotheses of the theorem, but which have fewer edges than G, have Eulerian cycles. One component must contain $\{u, w\}$ and the other v. We can now construct an Eulerian cycle on the original G (refer to Fig. 3.16) by replacing $\{u, w\}$ in the Eulerian cycle on its component with $\{u, v\}$ followed by the Eulerian cycle on the other component followed by $\{v, w\}$. Again, by referring to Fig. 3.16, convince yourself that this would give you an Eulerian cycle on G. Since we assumed no Eulerian cycle on G, this result is a contradiction.

Now suppose that G' is connected. As G' has one less edge than G, it must have an Eulerian cycle. Where $\{u, w\}$ appears in this Eulerian cycle, replace it by $\{u, v\}$ followed by $\{v, w\}$, which then gives an Eulerian cycle on G (why?). But this result is again a contradiction.

Thus whether G' has one or two components, we have obtained a contradiction of our assumption that G did not have an Eulerian cycle. This proves the sufficiency part of the theorem. ∎

Did you notice the induction-like flavor of this proof because we used an assumption about graphs with a smaller number of edges than the given graph to arrive at our contradiction?

PROOF 2 (constructive) We want to construct an algorithm which finds an

Eulerian cycle on a graph, all of whose vertices have even degree. Such an algorithm will certainly need some mechanism for "growing" paths which don't traverse any edge more than once. We begin therefore with procedure 3.3 Pathgrow (below) whose purpose is to grow paths. In this procedure V and E are the vertex and edge sets of a graph, v is any vertex in V, and $I(w)$ is the set of edges incident on vertex w. Therefore if w has degree d, the cardinality of $I(w)$ is d.

Procedure 3.3

Pathgrow

```
procedure PATHGROW(V, E, v)
    P ← ∅                                 [P is sequence of edges on path]
    U ← E                                 [U is set of edges not yet on path]
    w ← v                                 [Initialize w, the current end of P]
    repeat until I(w) ∩ U = ∅             [No untraversed edges from w]
        Pick e ∈ I(w) ∩ U                 [Choose an edge]
        x ← other end of e
        U ← U − {e}                       [Delete e from U]
        Add e to sequence of edges on P
        w ← x                             [Reset w]
    endrepeat
    return P                  [Path which is the output of the procedure]
endpro
```

At each step procedure Pathgrow picks an arbitrary edge, which has not been previously traversed, from among those incident on the current vertex and adds it to the path P. The procedure continues until it can go no further because there aren't any untraversed edges from the current vertex. Procedure Pathgrow is an example of a "follow-your-nose" algorithm in that it proceeds from one edge to the next by going in some available direction.

Now let's consider Algorithm 3.4 ECYCLE which uses Procedure Pathgrow to construct an Eulerian cycle on a connected graph, all of whose vertices are of even degree. In describing how Algorithm ECYCLE works, we will, in effect, give a proof that it constructs an Eulerian cycle. We'll use the graphs in Fig. 3.17 to illustrate the steps of the proof.

When we first call Procedure Euler, it calls Procedure Pathgrow, which starts traversing a path from v. Suppose that on the graph in Fig. 3.17(a) with $v = v_1$, we traverse the path

$$P = \underbrace{v_1, v_2, v_3, v_4, v_2, v_5, v_1}_{\text{Cycle in color in Fig. 3.17(a).}}$$

At this point we can't continue because there is no edge incident on v_1 which we have not traversed. It's important to realize that the only place we can get stuck is at the starting vertex v_1. The reason is that, when we reach any other vertex on the path, we have traversed an *odd* number of edges *at that vertex* and so, since all vertices are of even degree, there must be an unused edge on which to leave.

Algorithm 3.4

Ecycle

```
Input G(V, E)                                   [As a list of vertices and edges]
Algorithm ECYCLE
  procedure Euler(V', E', v')
    Pathgrow(V', E', v')                        [Finds a cycle P on G(V', E')]
    if P is not an Eulerian cycle on G(V', E')
       then Delete from E' the edges on the cycle P found
              and denote the nontrivial components of the
              subgraph which is left by G_1(V_1, E_1), G_2(V_2, E_2),
              ..., G_j(V_j, E_j); let v_i be some vertex in V_i on P.
         for i = 1 to j
           Euler(V_i, E_i, v_i)                 [Returns an Eulerian cycle
           Attach C to P at v_i                          on G_i named C]
         endfor
    endif
    C ← P
    return C                                    [C is Eulerian cycle on G(V', E')]
  endpro
  v ← Any vertex in V                           [Main algorithm]
  Euler(V, E, v)
Output C                                        [Eulerian cycle for G(V, E)]
```

The first time Euler is called, Pathgrow may find an Eulerian cycle on G, in which case P is assigned to C and we return to the main algorithm which then terminates. But if, as in the preceding paragraph, P is not an Eulerian cycle, we continue with the **then** portion of the algorithm and, for the graph of Fig. 3.17(a), delete edges e_1 through e_6. This leaves a graph with two nontrivial components, G_1 and G_2. The important thing for you to see here is that each of these components contains—and must contain—a vertex on the cycle already found; otherwise, the original graph would not have been connected. Also, each vertex in the resulting components still has even degree because the edges deleted each contributed an even number to the degree of each vertex on the cycle P.

Then Euler is called recursively for each component. In each recursive call the result of the call of Pathgrow may be an Eulerian cycle on the component. In this case the cycle C is attached to the cycle P found on the previous call of Pathgrow at a vertex they have in common. Otherwise, when Pathgrow finds a cycle which is not an Eulerian cycle, Euler is again called recursively. In Fig. 3.17(a), the call of Euler for component G_1 would certainly find an Eulerian cycle on this component (why?), which would then be attached at, say v_3 to the cycle found on the original call of Pathgrow. The cycle on G_1 might be

$$v_3, v_6, v_7, v_4, v_8, v_3,$$

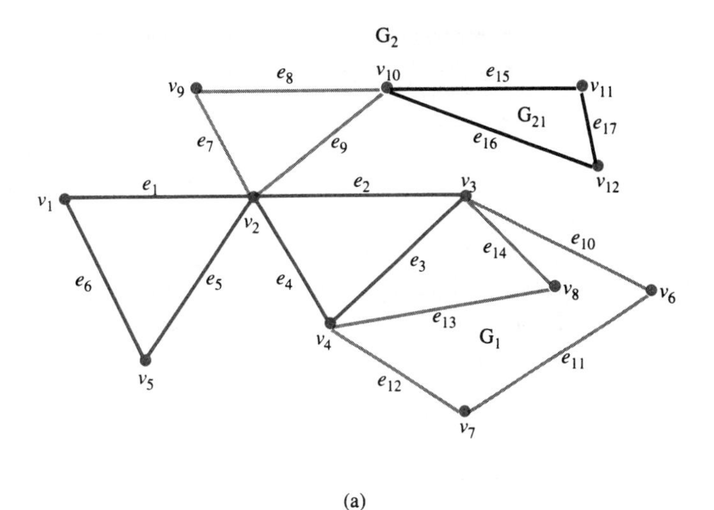

(a)

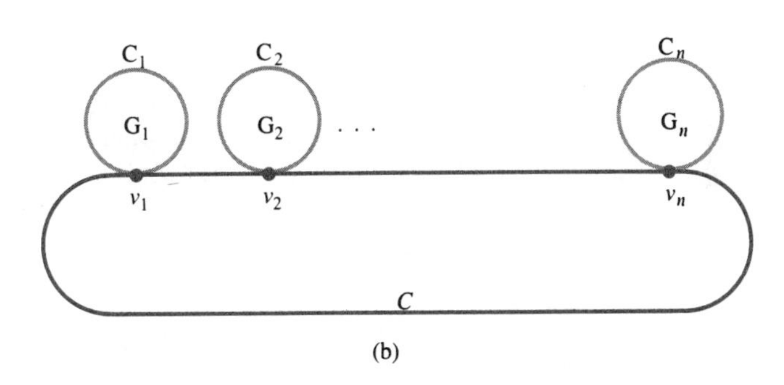

(b)

FIGURE 3.17 An Eulerian Cycle on a Connected Graph with All Vertices of Even Degree. (a) A connected graph with 12 vertices and 17 edges; the initial cycle $(v_1, v_2, v_3, v_4, v_2, v_5, v_1)$ is shown in color; a complete Eulerian cycle is v_1, v_2, v_3, v_6, v_7, v_4, v_8, v_3, v_4, v_2, v_9, v_{10}, v_{11}, v_{12}, v_{10}, v_2, v_5, v_1; (b) The general case.

or it might be

$$v_3, v_8, v_4, v_7, v_6, v_3,$$

depending on the choices made in the first line of the **repeat** loop in Pathgrow. Let's assume the former, so that P becomes

$$P = v_1, v_2, \underbrace{v_3, v_6, v_7, v_4, v_8, v_3,}_{\text{Eulerian cycle on } G_1} v_4, v_2, v_5, v_1, \tag{1}$$

as indicated in red and grey in Fig. 3.17(a). However, for component G_2, Pathgrow might just find the cycle

$$P = v_2, v_9, v_{10}, v_2, \tag{2}$$

which is not an Eulerian cycle on this component. The result would be another recursive call of Euler for the component labeled G_{21} in Fig. 3.17(a). This call would immediately find an Eulerian cycle on G_{21}, which would then be attached to P in Eq. (2) at vertex v_{10} to give

$$P = \underbrace{v_2, v_9, \overbrace{v_{10}, v_{11}, v_{12}, v_{10},}^{\text{Eulerian cycle on } G_{21}} v_2}_{\text{Eulerian cycle on } G_2}.$$

This would now be an Eulerian cycle on G_2 and would result in the end of the call to Euler for G_2. Control would then return to the original call and the Eulerian cycle found on G_2 would be attached to P in Eq. (1) at, say, the second occurrence of vertex v_2 obtaining finally an Eulerian cycle on the entire graph:

$$P = v_1, v_2, v_3, v_6, v_7, v_4, v_8, v_3, v_4, v_2, v_9, v_{10}, v_{11}, v_{12}, v_{10}, v_2, v_5, v_1.$$

Or we could have attached the cycle found on G_2 at the first occurrence of v_2 in Eq. (1).

Note that if a component G_i contains more than one vertex of the original cycle, as is the case with G_2 in Fig. 3.17(a), the new cycle can be attached at any of the vertices on the component. The simplest way to think about this process is that, while traversing the cycle P found on the initial call to Procedure Euler, you switch off to traversing the Eulerian cycle of the component when you come to some vertex on P to which the component is attached. Thus in Fig. 3.17(a), if we had arrived at v_3 from v_2 we could attach the Eulerian cycle on G_1 as

$$v_3, v_6, v_7, v_4, v_8, v_3,$$

although it could also be attached at v_4 as

$$v_4, v_8, v_3, v_6, v_7, v_4.$$

(Remember: Vertices can be repeated on an Eulerian cycle.) The complete Eulerian cycle in the general case shown in Fig. 3.17(b) is given by

$$v_1, C_1, v_2, C_2, \ldots, v_n, C_n, v_1.$$

But how do we know that the algorithm terminates? That is, could the recursion just get deeper and deeper without ever completing the initial call? Or even if it does terminate, might it fail to put the pieces just right to get an Eulerian cycle for the original graph? The answer is no, and the strong induction principle provides a simple proof. Let $P(n)$ be the proposition that the algorithm terminates and outputs an Eulerian cycle when G has n edges. The basis case, $P(2)$, requires the consideration of the only connected graph with two edges, no loops and all vertices of even degree, namely the one with two vertices and two edges between them. For this graph the first call to Euler results in Pathgrow finding an Eulerian cycle on this graph and returning immediately to the main algorithm which proves the basis case. Now suppose that the algorithm terminates for all graphs with $n-1$ or fewer edges. For a graph with n edges, after the first call of Pathgrow in which a cycle P is found, all the remaining nontrivial components, after the edges of P are deleted from G, have fewer than n edges (and at least two edges—why?). Therefore by the induction hypothesis, the algorithm terminates and provides an Eulerian cycle for each recursive call of Euler in the **for** loop. And thus it terminates for the original call for graph G, outputting an Eulerian cycle for G. ∎

Our constructive proof of the sufficiency part of Theorem 1 was somewhat longer than the nonconstructive one, and you may have found it more difficult to follow. But the important point is that, when you've found a constructive proof, you have not just a proof but a way of constructing—via an algorithm in this case—the object whose existence the theorem postulates.

Did you notice that, in effect, we also used induction in the portion of the proof preceding the termination argument? You should be able to restate this portion of the proof as a standard induction proof [12]. That our entire proof was really an induction proof should not surprise you because of the discussion in Chapter 2 of the close connection between induction and recursion.

We haven't yet finished the proof of Theorem 1 because all we've shown so far is that the even-degree vertex condition is sufficient to guarantee the existence of an Eulerian cycle. To complete the proof of the theorem, we must also show that this condition is necessary, namely, that any graph which has an Eulerian cycle must have all vertices of even degree. We'll also give a proof by algorithm of this, albeit an informal one.

Our algorithm is just to traverse the Eulerian cycle C on a graph G, starting at any vertex v and deleting each edge as we traverse it. What happens to the degree of each vertex as we do this? When we delete the first edge, we reduce the degree of v by 1 and the degree of the vertex at the other end of this edge by 1. For each subsequent edge, we also reduce the degrees of the vertices at each end by 1. For each vertex after v, this means that the degree is reduced by 2, since it is reduced by 1 when we arrive and by 1 more when we leave the vertex. As we traverse C, we may, of course, come back to a vertex (even v) more than once, but each time the degree is reduced by 2 as the edges are deleted. Finally, the last edge on C must be one that returns to v and the degree of v is reduced by 1 at this point. We are left with all the original vertices of G, each one now isolated because every edge has been deleted. (Recall the definition of an Eulerian cycle.) As the degree of each vertex has been reduced by an even number (one or more 2's), each vertex must originally have been of even degree.

Theorem 1 provides an immediate solution of the Königsberg bridge problem. As neither the graph in Fig. 3.15(b) nor that graph augmented by a single edge could have all vertices of even degree, there is no Eulerian path or cycle over the bridges of the river Pregel.

A consequence of Theorem 1 which you should be able to prove [13] is:

Corollary 2. A connected digraph with no loops has an Eulerian cycle if and only if every vertex has the same indegree as outdegree.

We may apply Corollary 2 to the de Bruijn cycle problem introduced in Example 2, Section 3.1. Figure 3.18 displays two examples of **de Bruijn graphs**, which are digraphs depending on a parameter n with the following properties:

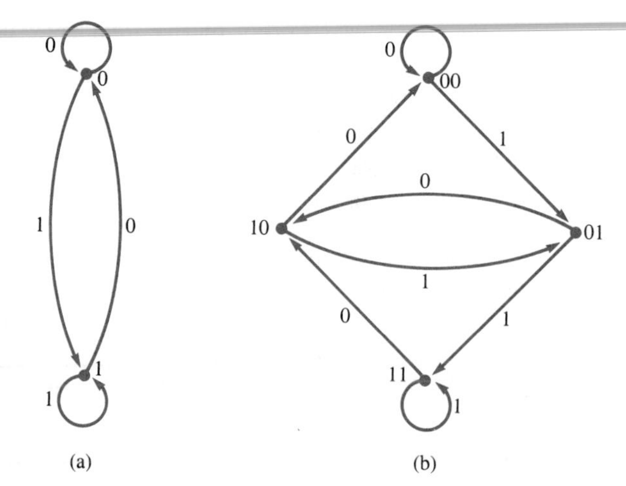

FIGURE 3.18

De Bruijn Graphs for (a) $n = 1$ and (b) $n = 2$. (For $n = 3$ see Fig. 3.2.)

a) A de Bruijn graph has 2^n vertices, each of which is labeled with one of the 2^n distinct sequences of n 0's and 1's.

b) Let a given vertex have the label

$$b_1b_2\ldots b_n, \qquad b_i = 0 \text{ or } 1. \tag{3}$$

Then from this vertex there are two outgoing edges labeled, respectively, 0 and 1 to the vertices with labels

$$b_2b_3\ldots b_n0 \qquad \text{and} \qquad b_2b_3\ldots b_n1. \tag{4}$$

From property b) and expression (4) it follows that each vertex (3) also has two incoming edges from the vertices with labels

$$0b_1b_2\ldots b_{n-1} \qquad \text{and} \qquad 1b_1b_2\ldots b_{n-1}. \tag{5}$$

The reason is that, whatever the value of b_n, each of the nodes with labels given by expression (5) will have one outgoing edge to the node with label (3).

Can we apply Corollary 2 to de Bruijn graphs? Clearly, every vertex has indegree and outdegree 2. But is a de Bruijn graph a connected graph? Yes, it is, but we leave a proof of this to [22]. So, from Corollary 2 we know that there is at least one Eulerian cycle on any de Bruijn graph (see Fig. 3.2 for the case $n = 3$). The length of this cycle (i.e., the number of edges on it) is 2^{n+1}, as each of the 2^n vertices has 2 edges emanating from it. If we traverse any such Eulerian cycle, writing down the label on each edge as we do so, the resulting sequence may be shown to include all possible subsequences of length $n+1$ if we allow wraparound from the end to the beginning of the sequence. For example, starting from the node labeled 11 in Fig. 3.18(b) and proceeding successively to nodes 10, 00, 00, 01, 10, 01, 11, and finally back to 11 again, we obtain the sequence

$$00010111, \tag{6}$$

which contains all possible subsequences of length 3 (find them!).

We conclude this section with one theorem about Hamiltonian paths:

Theorem 3. Every complete graph contains a Hamiltonian path.

PROOF This is an existence theorem, so we naturally prove it by using an algorithm. As with Pathgrow, this is a follow-your-nose algorithm.

Input $G(V, E)$ [A complete graph]
Algorithm Choose an arbitrary vertex v and traverse a path which never returns to the same vertex twice.
Output A Hamiltonian path

Not much of an algorithm, you may think. But it does the trick. Suppose you got stuck somewhere and couldn't proceed without returning to a vertex already on the path. But you can't be stuck at a vertex u if there is a vertex w not already on the path because, by the assumption of completeness, there is an edge $\{u, w\}$ in the graph and so we can go from u to w. Thus the process can be stopped only when all the vertices are on the path. ▮

A corollary whose proof you should be able to supply [23] is

Corollary 4. Every complete graph contains a Hamiltonian cycle.

A famous problem related to this corollary is the *traveling salesman problem*, which we will discuss in Section 4.2.

One final point. If you label the vertices of a graph 1, 2, ..., n, and then write these labels as you traverse a Hamiltonian graph, the result is a *permutation* of the integers 1, 2, ..., n. In Section 4.9, we will discuss ways of generating such permutations.

Problems: Section 3.4

1. What's the least number of new bridges which would have to be built in Königsberg for a cyclical walk without retracing one's steps? Using Fig. 3.15(b), is there more than one choice for where to build the bridge(s)?

2. Carry out Algorithm ECYCLE for the following graph twice, each time starting at 1.

a) First, whenever there is a choice of which edge to choose next, choose the permissible edge which goes to the lowest-numbered vertex. Whenever (in a recursive call) there is a choice of which vertex to start at, choose the lowest-numbered vertex. Show pictures of your partial results at enough stages so that a grader can be certain about what you have done and in what order.

b) Use the highest-numbered vertex throughout instead of the lowest-numbered.

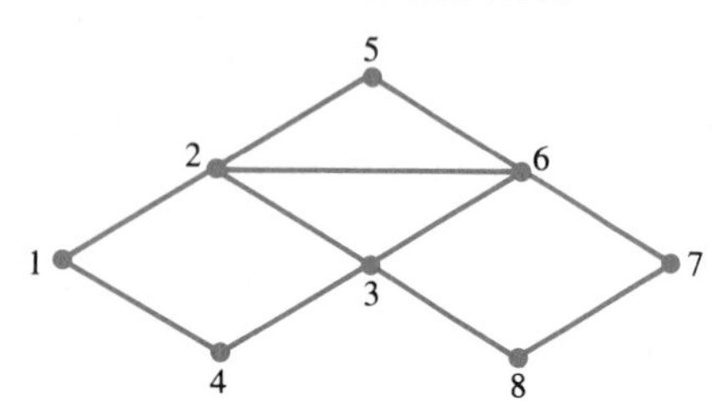

3. The Highway Department of Smallville is instructed to repaint the white lines which go down the middle of all roads in town. On the map of Smallville, the department's garage is marked with an X.

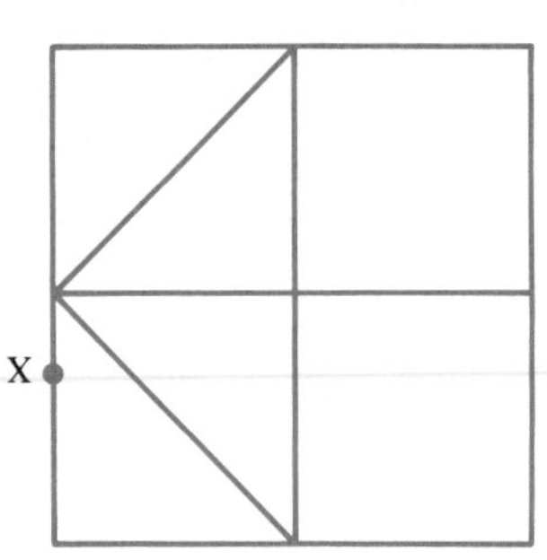

Is it possible for the Department to send out its paint truck so that it can paint all the lines without going over any stretch of road twice (and end up back at the garage)? Explain. If it is not possible, what is the smallest number of blocks (edges on the map) which the truck must travel twice?

4. Consider the de Bruijn graph $D(3,2)$, in which the 3 means that the vertices are *triples* and the 2 means that each of the three is a *binary* digit.

a) Draw this graph. It's already in the text, but draw it from scratch to show that you understand how to construct it.

b) Using our correspondence between paths in de Bruijn graphs and sequences of bits, find and highlight in $D(3,2)$ the cycle corresponding to the sequence 0101100 (*Hint*: At what vertex must the cycle end?)

c) Using a new copy of $D(3,2)$, pick a cycle of your choosing with six edges (none of which is a loop), highlight it, and then write the sequence S corresponding to it.

5. Consider the de Bruijn graph $D(1,3)$, in which the vertices are single "trits" (ternary digits, i.e., 0, 1, and 2).

a) Draw it.

b) Highlight the cycle corresponding to the sequence 01210211 starting at vertex 0. Note that this path overlaps itself. What caused that?

c) By eyeballing (or using our algorithm informally—you don't have to explain what you did), indicate a directed Eulerian cycle in $D(1,3)$. Indicate what your cycle is by labeling the edges 1, 2, 3, ... (in circles, so as not to confuse them with other numbers) in the order of traversal on the cycle.

d) Write a ternary "lock-picking" sequence which will open doors at a hotel whose room numbers are two trits.

6. In the actual Baltimore Hilton problem, each of the four digits can be anything from 0 to 9. Is there still a de Bruijn cycle for the burglar to use? (Obviously you aren't going to find out by drawing the graph!)

7. Generalize further. An *n-long, k-ary sequence* is a sequence of n symbols, each of which is chosen from a fixed set of k symbols. Thus the Baltimore Hilton problem is about 4-long, 10-ary sequences. For which n and k is there a de Bruijn cycle?

8. In the game of dominoes, two dominoes can be put end to end if the ends have the same number of dots on them. A standard set of dominoes has one piece for each (unordered) pair of distinct integers from 0 to 6 inclusive—for example $(0,3)$ and $(4,6)$—and one piece of the form (i,i) for each i from 0 to 6.

a) Is it possible to arrange all the dominoes into one big circle?

b) If all the pieces with no dots on one or both ends are excluded, is it possible to arrange the remaining pieces in a circle?

9. The following is a simplified version of problems that actually arise in DNA research. The graph theory that applies here also applies to the actual problems.

A chromosome is made up of building blocks A, B, and C, which are used repeatedly in various combinations. Biochemists believe that the chromosome is a single closed loop; that is, it is actually some one circular sequence of A's, B's, and C's. By a chemical technique, it is possible to break many copies of the chromosome randomly into pieces and then count the relative frequency of pieces which consist of exactly two blocks (e.g., pieces like AC or BB). When this has been done, the relative frequencies of the various two-block pieces are

AA 3, BB 0,
AB 2, BC 4,
AC 2, CC 1.

Assuming that each two-block sequence occurs in the data with the same relative frequency as in the chromosome, is this data consistent with the circular hypothesis? (Note that the subsequence BA isn't listed because the experiment doesn't preserve the order in which the subsequence occurs in the chromosome. Thus AB and BA are indistinguishable as types, and the occurrences of either are listed together under AB.)

10. Now suppose that the chemical technique is more precise. We get to pick a starting point on the chromosome and chop the whole thing from that point on into subsequences two blocks long; furthermore, we get an exact count of the number of each type of piece. Then we get to take another copy of the chromosome, shift over by one block from our previous starting point, and chop again. The first group of relative frequencies is

AA 2, BB 0,
AB 1, BC 2,
AC 0, CC 1,

and the second is

AA 1, BB 0,
AB 1, BC 2,
AC 2, CC 0.

(Again AB and BA, say, are indistinguishable as types.) Is this data still consistent with the hypothesis that the chromosome is a single loop?

11. Write an algorithm which, given as input an undirected graph in adjacency matrix form, determines whether all the vertices have even degree. It should print Yes or No; if no, it should name a vertex with odd degree.

12. Restate the proof-by-algorithm of Theorem 1 in a standard induction format.

13. Prove Corollary 2, the directed version of the Eulerian cycle theorem.

14. The Eulerian cycle theorem (Theorem 1) has a path version: A connected, undirected multigraph has an Eulerian path if and only if there are either zero or two vertices of odd degree. (If there are zero vertices of odd degree, every Eulerian path is a cycle.) Prove this version three ways:

a) Given an existence proof similar to Proof 1 of Theorem 1 in the text.

b) Give a proof by algorithm similar to Proof 2 of Theorem 1 in the text. (*Suggestion*: You need only consider the case of two odd-degree vertices. Why? Where should you start growing your path?)

c) Both a) and b) take as much work as the original Proofs 1 and 2, yet are very similar to them. Is there a way to avoid this repetition? Yes: Instead of slightly changing the proof, change the graph. If G has two vertices of odd degree, temporarily add a new edge between them. Call the resulting graph G'. Now apply Theorem 1 to G'. What does this tell you about G? Why?

This is a standard example of Reduce to the Previous Case. It is also sometimes called the **method of ideal elements**. The ideal element in this case is the new edge, which is not really in the graph. That is, it is ideal, not real.

15. Modify the algorithm [11] so that it prints Yes if either all vertices are of even degree or exactly two are of odd degree (and still prints No otherwise). When there are two vertices of odd degree, the algorithm should print the indices of those vertices.

16. The directed Eulerian cycle theorem has a path version. Here, for variety, we state the result to cover only paths which are not also cycles. (Compare with [15].)

A digraph has a directed Eulerian path (beginning and ending at different vertices) if and only if all vertices except two have indegree equal to outdegree, and one of those two has outdegree 1 greater than indegree while the other has indegree 1 greater than outdegree.

a) Prove this theorem by an existence argument.

b) Prove it by analyzing an algorithm.

c) Prove it by the method of ideal elements (see [14]).

17. Discover and prove a necessary and sufficient condition for a connected graph to be *traceable* (i.e., each edge appears once and only once on a path)

using exactly two paths. (The two paths can't be cycles or have an end in common.)

18. Suppose that you are given a connected digraph in which the indegree and outdegree are the same at every vertex. Obviously, if you strip off the arrows, you're left with a connected graph with even degree everywhere. The converse: Given a connected graph with even degree at every vertex, the edges that can be directed so that indegree = outdegree at each vertex—is not so obvious. Putting on an orientation which "balances" is more difficult than stripping it off. Nonetheless, the converse is true, and its proof by algorithm is easy. Explain. (*Hint*: Take a trip around the graph.)

19. Prove that the result of [18] is still true even in the unconnected case.

20. Prove: It is possible to direct the edges of any graph G so that, at each vertex v, the outdegree is either $\lfloor \deg(v)/2 \rfloor$ or $\lceil \deg(v)/2 \rceil$. (Note that this is a natural generalization of [18].)

21. Can the result of [20] be strengthened further? Is it always possible to choose in advance which vertices will have outdegree $\lfloor \deg(v)/2 \rfloor$?

22. In order to prove that a de Bruijn graph is connected, you need to show that there is a path from any vertex to any other. Do so by using the definition of a de Bruijn graph in Eqs. (3), (4), and (5) to find a sequence of vertex labels that defines the desired path.

23. Prove Corollary 4: Every complete graph has a Hamiltonian cycle.

24. How many distinct Hamiltonian cycles (i.e., cycles with the vertices reached in different orders) does the complete graph with n vertices have? (See also [12, Section 3.2])

25. Theorem: Every tournament contains a directed Hamiltonian path. We have already proved this theorem. Where?

26. Show that the following graph has no Hamiltonian path. (*Hint*: Label the vertices with A's and B's so that A vertices are adjacent only to B vertices and vice versa. What would happen if there were a Hamiltonian path?)

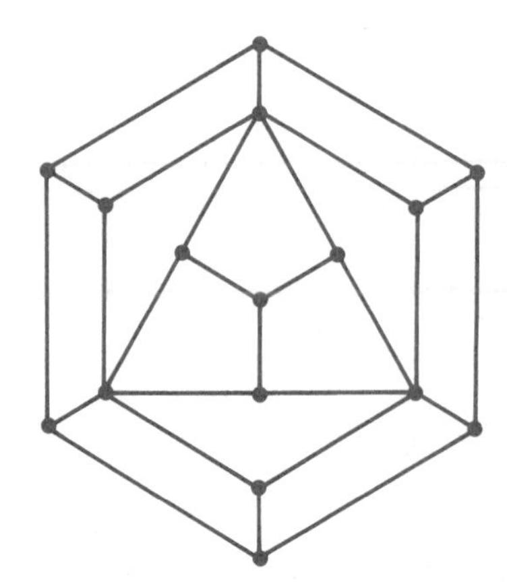

27. Suppose that you want to list all three-digit binary numbers in such an order that each number differs from the previous one in just one digit.

a) Turn this into a path problem about some graph.

b) Answer the question by drawing the graph and eyeballing a solution. (*Hint*: Try to draw the graph in a pleasing, symmetric way; the fact that each vertex has three digits may help.)

c) Is there still a solution if the first and last numbers must also differ in just one digit?

d) Can you generalize your answers to b) and c) to n-digit binary numbers?

e) Can you generalize to n-digit, k-ary numbers?

3.5 A Shortest Path Algorithm

Finding the shortest path between two vertices of a graph in which each edge has a *weight* associated with it is a common problem. Consider, for example, a road map of the United States such as the abbreviated one shown in Fig. 3.19. If you were going to drive from New York to Los Angeles and were constrained to follow a route using only the links in Fig. 3.19, you would want the shortest path (i.e., the path whose edge weights have minimum sum) between the two (unless, perhaps, you wanted to visit your grandparents in Minneapolis on the way).

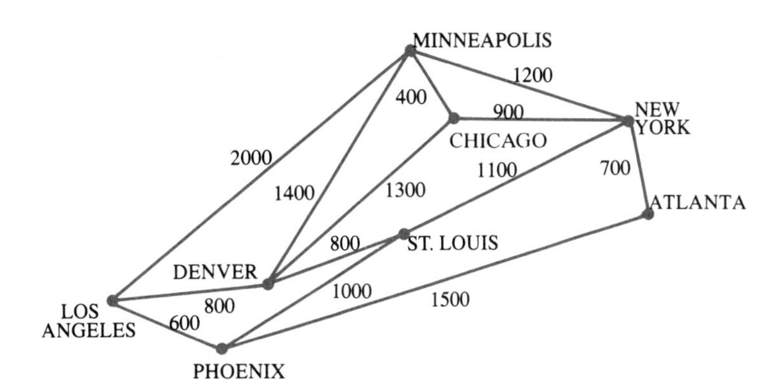

FIGURE 3.19

Portion of a Mileage Map.

Many other types of problems may also be interpreted as shortest path problems on graphs, even when the edge weights are not distances and the path cannot be interpreted as a trip. For example, we could reinterpret Fig. 3.19 as a chart in which the numbers represent the costs of sending messages between each pair of cities. In that case we could find the minimum-cost communication path from New York to Los Angeles. As another example, suppose that there were weights on the edges of the wolf-cabbage-goat-farmer graph (Fig. 3.1) which might represent the costs of trips with particular passengers. (It would be reasonable to charge more to transport a heavy wolf than a light—even if large—cabbage.) Then we might ask for the cheapest way to transport the wolf, the cabbage, and the goat across the river.

We will assume that shortest path problems are given as digraphs because, in fact, they often are. But even when they are not, as in the transportation and communications examples, we can convert the given graph into a digraph by replacing each edge by two edges, one in each direction, with equal weights. Consider, then, the digraph shown in Fig. 3.20. Our object is to find the shortest path from v_0 to v_7 and, in so doing, to develop a general shortest path algorithm for any digraph with nonnegative edge weights. We want this algorithm to be applicable to any digraph with a path between the two given vertices. (In Section 3.6 we'll consider the much easier special case when all the edge weights are 1, in which case the length of a path is just the number of edges on it.)

As with so many problems in discrete mathematics, the choice of a good approach rather than just any (brute force) approach can make a great difference in efficiency. One approach might be to compute first the *number of paths* from one vertex to the other by the technique already described in Section 3.3. Then we could search for each of these paths and calculate the length of each. Although possible, this approach would be quite slow and tedious because large digraphs will typically

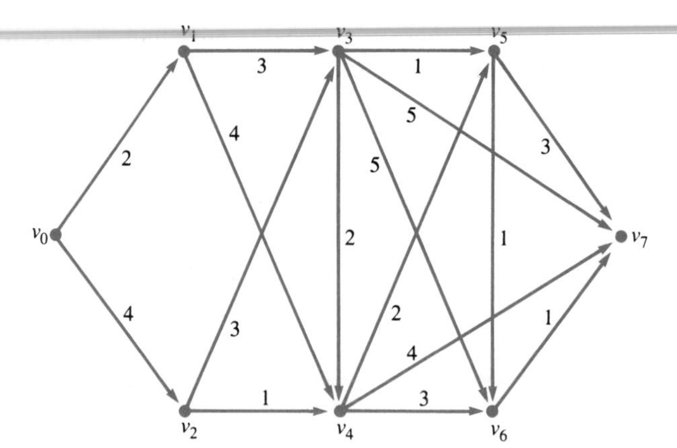

FIGURE 3.20

The Shortest Path Problem. Object: Find the shortest path from v_0 to v_7.

have many such paths. By contrast the method we are about to describe is not only quite elegant but also efficient.

Let our graph G have vertices $v_0, v_1, \ldots, v_n$ and suppose that we wish to find a shortest path from v_0 to v_n. We have seen (in the Prologue) that sometimes the best way to solve a problem is to generalize it. This is such a case. Our generalization is to find the shortest path from v_0 to *all* the other vertices. In order to do so, we first find the closest vertex to v_0 (i.e., the one with the shortest path from v_0), then the next closest, etc. Sooner or later our target vertex v_n will be the next closest and we'll have found the shortest path to it.

Now suppose that we have already found the $k-1$ vertices closest to v_0 (not counting v_0 itself; call them $u_1, u_2, \ldots, u_{k-1}$) and we want to find the next closest vertex u_k. We define

$$U_{k-1} = \{u_0(= v_0), u_1, u_2, \ldots, u_{k-1}\}.$$

Then the shortest path from v_0 to u_k must pass through only vertices in the set U_{k-1}. For if there were some vertex on the shortest path to u_k not in U_{k-1}, it would be closer to v_0 than u_k, which is a contradiction because we assumed that u_k is the closest vertex to v_0 not in U_{k-1}.

To find the shortest path to the vertices in V, we first define a *current distance function* from v_0 to each of the other vertices in V, which will be updated each time a new u_k is found. At the kth stage of our iteration, before we have found u_k, the current distance function will be the shortest path to each vertex of V from v_0, with the restriction *that all intermediate vertices be from* U_{k-1}. When there is no such path with this property, the distance will be set to ∞. At the beginning, when $k = 0$, the distance to all vertices other than v_0 is set equal to ∞. Once the shortest path to a vertex has been found and it becomes a member of the set U_k, its distance from v_0 will remain fixed from then on.

How may the distance to a vertex v not in U_k change from its previous value when u_k is found? The answer is that the distance to v can change it only if there is an edge from u_k to v. For the best path to v through vertices in U_k can only become shorter than it was if u_k is on it. And u_k must be the vertex *just before* v (i.e., there must be an edge from u_k to v); otherwise, there would be a shorter path to v bypassing u_k. (Why? See Fig. 3.21.) Using this observation, our definition of the distance function can be expressed in algorithmic form as follows, where $d(v)$ is the distance to vertex v and $\text{len}(u, v)$ is the length of the edge from u to v:

```
d(v_0) ← 0                                              [Initialization]
for i = 1 to n
    d(v_i) ← ∞
endfor

for i = 1 to n                                               [Updating]
    if v_i adjacent to u_k and not in U_k then   [Change d(v_i) only for vertices
(*)     d(v_i) ← min(d(v_i), d(u_k) + len(u_k, v_i))           adjacent to u_k]
    endif
endfor
```

The line labeled (*) needs some explanation. The new value of $d(v_i)$ is the minimum of the old value and the sum of the current distance function for the vertex u_k and the length of the edge from this vertex to v_i (if there is one). Note that the current distance function for u_k itself remains unchanged, as does the distance to any vertex not adjacent to u_k. You'll see shortly how these algorithm fragments become part of our complete shortest path algorithm.

Now, having found u_k in step k, how do we find u_{k+1} in step $k+1$? We need only find that vertex not in U_k for which the distance function is a minimum. This vertex becomes u_{k+1}, so we add it to U_k to get U_{k+1}, and then we update the current distance function. That's all there is to it.

A sketch of what we want the shortest path algorithm to be is then:

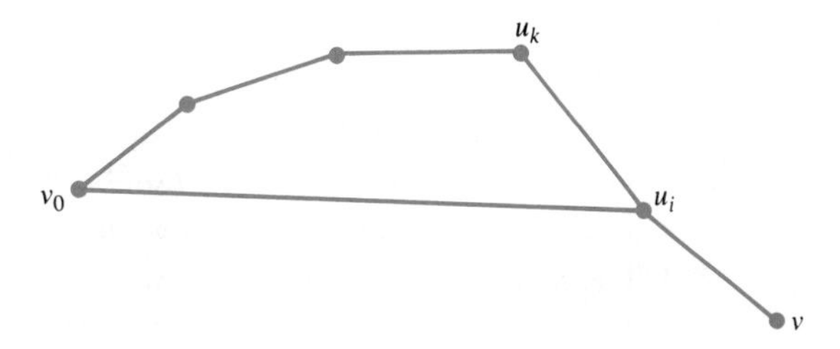

FIGURE 3.21 The Shortest Path to v. If u_k were not the last vertex on the path to v, then there would be a shorter path to v through u_i, as u_i is closer to v_0 than u_k for any $i < k$.

Initialize $d(v_0)$ to 0 and $d(v_i)$ to ∞, $i = 1, \ldots, n$
$U \leftarrow \{v_0\}$
repeat
  Update the current distance function.
  Find the vertex u not in U for which $d(u)$ is a minimum;
    if there is a tie choose u arbitrarily.
  $U \leftarrow U \cup \{u\}$
endrepeat when $u = v_n$

At the first entry into the loop, the current distance function is updated to give nonzero values for all vertices adjacent to v_0. (Do you see why?)

Before giving a more detailed statement of this algorithm in our notation, we'll show how it works on the graph in Fig. 3.20.

EXAMPLE 1 Find the shortest path from v_0 to v_7 in Fig. 3.20.

Solution The following table contains the complete answer, with the values at the end of each iteration listed. The values of k represent the successive iterations with $k = 0$, the state just before the first entry into the loop.

k	$d(v_1)$	$d(v_2)$	$d(v_3)$	$d(v_4)$	$d(v_5)$	$d(v_6)$	$d(v_7)$	Vertex Added to U
0	∞	∞	∞	∞	∞	∞	∞	v_0
1	2	4	∞	∞	∞	∞	∞	v_1
2	2	4	5	6	∞	∞	∞	v_2
3	2	4	5	5	∞	∞	∞	v_3
4	2	4	5	5	6	10	10	v_4
5	2	4	5	5	6	8	9	v_5
6	2	4	5	5	6	7	9	v_6
7	2	4	5	5	6	7	8	v_7

Remarks:

1. The vertex v_k which enters the set U on the kth step has the subscript k only because of the way we numbered the vertices.
2. In step 3, $d(v_3) = d(v_4)$, so we arbitrarily chose u_3 to be v_3.
3. Focusing on v_6, note that in step 4 the shortest path has length 10 (v_0, v_1, v_3, v_6) and is, in fact, the only path to v_6 using members of U. But in step 5, the path (v_0, v_2, v_4, v_6) with length 8 is now the shortest, and then in step 6 the path (v_0, v_1, v_3, v_5, v_6) with length 7 becomes the shortest.
4. Our method doesn't actually give the shortest path but only the *length* of the shortest path, namely, 8. To find the shortest path itself, we store with each new vertex that enters the set U the index of the *prior* vertex on the

shortest path to the new one. We can easily keep track of this prior vertex. Each time we change (i.e., decrease) the current distance function to a vertex v because of an edge (u, v), we just record the vertex u. Then, when v_n enters U, we only have to work our way back to v_0 by using these prior vertices. The following table records these prior vertices at the end of the distance calculation in each iteration. Thus in iteration 1 the shortest path to v_4 has prior vertex v_1, but in iteration 2 the shortest path has prior vertex v_2.

	Prior vertex on shortest path to						
k	v_1	v_2	v_3	v_4	v_5	v_6	v_7
0	v_0	v_0					
1			v_1	v_1			
2				v_2			
3					v_3	v_3	v_3
4						v_4	v_4
5						v_5	
6							v_6

Thus on the shortest path to v_7, the prior vertex is v_6; on the shortest path to v_6, the prior vertex is v_5, etc. Working our way back to v_0, we find that the shortest path from v_0 to v_7 is

$$(v_0, v_1, v_3, v_5, v_6, v_7). \blacksquare$$

Algorithm 3.5 SHORTESTPATH is a more complete statement of the shortest path algorithm. Remarks:

- Organizing the recording of the shortest path itself requires only a modest amount of additional effort [9].
- This algorithm works equally well for (undirected) graphs so long as the input includes, for each edge (u, v) another edge (v, u) with the same weight as (u, v).
- If all the lengths are 1, this algorithm will find the path with the least number of vertices on it (see also Algorithm PATHLENGTH in Section 3.6).
- Our representation of G as a list of edges might be unwieldy in practice because it requires many inefficient searches of the list during the algorithm. Would a representation of G by its adjacency matrix be an improvement [10]?

Algorithm SHORTESTPATH is certainly worth proving correct because, however well you have followed its derivation, it is sufficiently complex that its correctness is not obvious. As usual, our tool is induction using a loop invariant. This invariant represents the situation after the kth pass through the **repeat** ... **endrepeat** loop.

- For each vertex v not in U, $d(v)$ is the length of the shortest path from v_0 ending at v and otherwise passing through only vertices in $U - \{u\}$; if there is no such path, the length is infinite.
- for each v in U, $d(v)$ is the length of the shortest path from v_0 to v. (For v_0 itself, this length is 0.)

Algorithm 3.5
ShortestPath

```
Input G                                  [As a list of vertices, edges and weights]
    v0, vn                               [Starting and ending vertices]
Algorithm ShortestPath
  d(v0) ← 0                              [Initialize distance function]
  for i = 1 to n
      d(vi) ← ∞
  endfor
  U ← {v0}                               [Initialize U]
  u ← v0                                 [u is the most recent vertex entered into U]
  repeat                                 [Main loop]
      for i = 1 to n                     [Update distance function]
          if vi adjacent to u and not in U then
              d(vi) ← min(d(vi), d(u) + len(u, vi))
          endif
      endfor
      mindist ← ∞                        [Find minimum distance to vertex not in U]
      for i = 1 to n
          if vi not in U and d(vi) < mindist
              then mindist ← d(vi); u ← vi
      endfor
      U ← U ∪ {u}                        [u is next closest vertex]
  endrepeat when u = vn
Output d(vn)                             [Shortest path length]
```

- U consists of v_0 and the k closest vertices to v_0.

This is the most complicated loop invariant we have presented so far. Make sure you understand what it says before you go on to read the proof.

The basis case ($k = 0$) claims that the loop invariant is true before we enter the loop the first time. At this point, because of the initialization lines in the algorithm,

- since there are no vertices in $U - \{u\} = \{v_0\} - \{v_0\}$, the distances to all vertices other than v_0 are ∞ (i.e., the only path ending at any vertex is the null path at v_0);
- $U = \{v_0\}$ and $d(v_0) = 0$; and
- U does consist of v_0 and the 0 other closest vertices to v_0.

This completes the basis case.

Now we must show that, if the loop invariant is true after the $(k-1)$st pass through the **repeat** loop, it will still be true after the kth pass. At the end of pass $k-1$, we know U_{k-1} and, for all vertices v not in U_{k-1}, we know that $d(v)$ gives the shortest distance to v for paths passing through only $U_{k-1}-\{u_{k-1}\}$. We must show that the kth pass (1) correctly finds u_k and (2) updates $d(v)$ to give the shortest distance using $U_k-\{u_k\}$; that is, $d(v)$ must now take into account the possible use of paths including u_{k-1}. We have already argued item (2)—when we explained the algorithm fragment on p. 243. We do this update at the beginning of the **repeat** loop. Item (1) is taken care of in the remainder of the **repeat** loop. We argued earlier, using Fig. 3.21, that for the kth closest vertex u, the minimal distance from v_0 and the shortest distance using vertices in U_{k-1} are the same. Therefore $d(u)$ must be the minimum of all the $d(v)$ at this point. The second part of the **repeat** loop finds precisely this minimum.

Because the loop invariant is still true when we emerge from the loop the last time—at which point $u=v_n$ and $d(v_n)$ is the length of the shortest path to v_n—the algorithm does, indeed, do what it is supposed to do.

This is not an easy proof. But if you really understand what we are trying to do in SHORTESTPATH, we believe that you will be able to follow the proof. This algorithm, by the way, is often called **Dijkstra's algorithm**.

Let's now briefly analyze SHORTESTPATH. As the **repeat** ... **endrepeat** loop may have to be executed for all n vertices (as in Example 1) and each of the inner **for** ... **endfor** loops requires n executions of, at least, the test in their **if** statements, in the worst case the entire algorithm requires about n^2 steps. But since u will usually equal v_n before n traversals of the outer loop, the average case isn't nearly so bad as the worst case. Thus Dijkstra's algorithm is usually very efficient, even for large values of n.

If, instead of the shortest path between two specified vertices, you want to find the shortest path between all pairs of vertices in the graph, you can easily extend Dijkstra's algorithm to do so [11]. Alternatively, we can use Warshall's algorithm in the form described in [25, Section 3.3].

Problems: Section 3.5

1. Find the shortest path length from v_0 to v_5 in the following undirected graph, using Algorithm SHORTESTPATH.

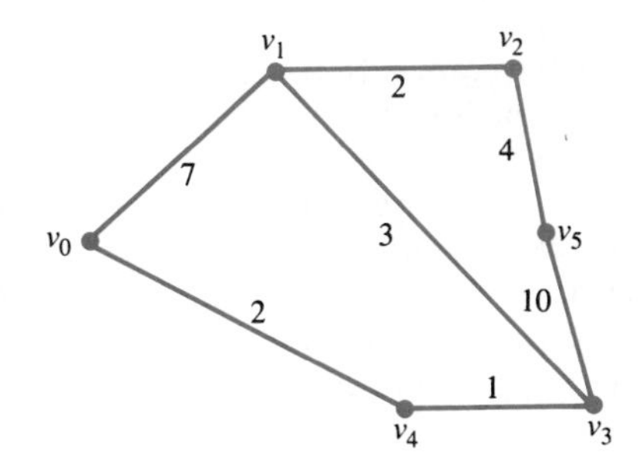

a) Write up your work in a table of the form shown on p. 244.

b) Write up your work on copies of the graph, one for each iteration of the algorithm. Highlight the current U with color and show the current value of $d(v)$ over each vertex.

2. Repeat [1], but now start from v_5 and go to v_0.

3. Find the shortest path from v_0 to v_1 in the digraphs on p. 248. Show your work in abbreviated form on one copy of each graph (otherwise it's too tedious).

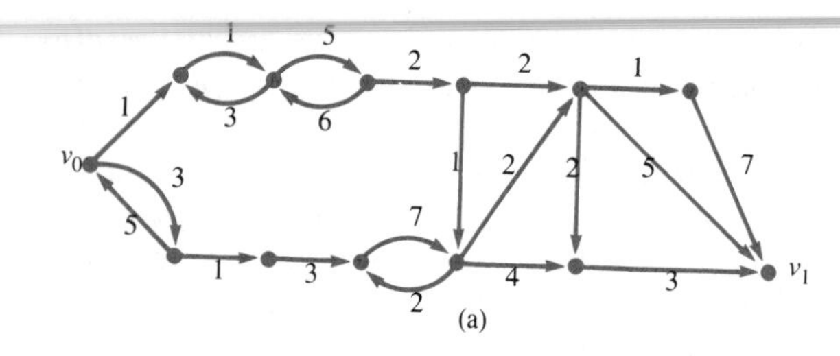

(a)

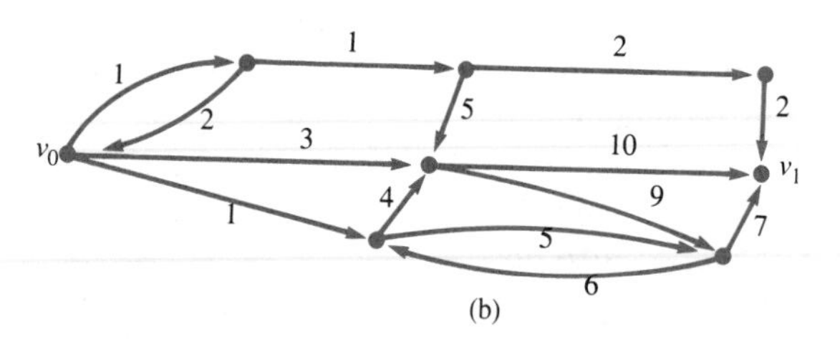

(b)

4. Consider the following **length matrix**, in which the (i, j) entry is the length of the directed edge from vertex i to vertex j. (If the (i, j) entry is "–", there is no edge from i to j.) Determine the length of the shortest path from v_1 to v_8 *without ever drawing any part of the graph.* Also determine which vertices are on this path and in which order, again without drawing any part of the graph. It is possible to show all your work right along the borders of the matrix; if you do so please explain what you are doing!

$$\begin{bmatrix} - & - & 5 & - & - & 2 & 3 & - \\ 4 & - & 6 & - & 7 & - & 5 & - \\ - & 3 & - & 9 & 2 & 6 & - & 7 \\ 3 & - & 2 & - & 1 & - & 7 & 6 \\ - & 5 & - & 1 & - & - & 4 & - \\ - & - & 2 & - & 8 & - & 9 & - \\ 1 & 2 & 3 & - & - & 6 & - & - \\ 5 & - & 8 & - & 2 & - & 9 & - \end{bmatrix}$$

5. For the graph in Fig. 3.20, find *all* paths from v_0 to v_7 and their lengths. Explain your method.

6. Let P be a 15-gon, with vertices labeled cyclically 1 through 15. Let G be the graph obtained from P by including the additional edges

$$\{1,5\}, \{4,8\}, \{7,9\}, \{8,12\}, \{11,15\}.$$

   a) What is the number of edges in the smallest cycle containing edge $\{1,15\}$? (*Hint*: Delete edge $\{1,15\}$ from the graph.)

   b) List the edges in a smallest cycle containing $\{1,15\}$.

   c) Is the answer to b) unique? Explain how you know.

7. Repeat [6], but use $\{5,6\}$ instead of $\{1,15\}$ in a) and b).

8. The **girth** of a graph is the length of its shortest simple cycle. Using SHORTESTPATH as a procedure, give an algorithm for finding the girth of a graph. In terms of the number n of vertices, analyze the complexity of your algorithm.

9. Modify SHORTESTPATH so that it outputs the shortest path (as a set of edges), as well as giving the shortest distances. You need only write you *additions* to the algorithm, and indicate where they go.

10. Algorithm SHORTESTPATH contains such English language statements as "**if** v_i adjacent to u **and not** in U". These statements can be eliminated by using G in adjacency matrix form as input and by defining a binary vector B to indicate when a vertex is in U. For instance, $B(u)$ could be set to 1 if u is in U, and set to 0 otherwise. Rewrite SHORTESTPATH so that it avoids all set notation and English language phrases and uses arrays instead. (This approach will make SHORTESTPATH much easier to implement in a standard computer language.)

11. Since Algorithm SHORTESTPATH finds the shortest path between any pair of vertices in a graph or digraph, you should not have difficulty adapting it to find the shortest paths between *all* pairs of vertices (in both directions for a digraph). Do so by showing only the statements that have to be added to SHORTESTPATH.

12. Suppose that you have a weighted **bigraph**: Each edge has two weights on it, one for each direction. (For instance, an edge might represent a stretch of three-lane road, with two northbound lanes and one southbound lane; the weights might represent the average time needed to traverse the road going both north and south.) Again, you need to find the shortest paths from v to all other vertices, where the length of a path is the sum of its weights in the directions the edges are traversed. Show how to solve this problem using Algorithm SHORTESTPATH.

13. Do the following inductive statements work? Here "work" means that, if you already knew the information in the previous case of the statements, it would be relatively easy to determine the information for the next case. Thus an efficient algorithm could be devised using iteration on this information, just as Algorithm SHORTESTPATH is obtained from iteration on knowledge of the shortest path lengths for the k closest vertices to v_0.

    a) The shortest paths from v_0 to $v_1, v_2, \ldots, v_k$ (where these are *not* necessarily the k closest vertices to v_0) are known. If you think that this doesn't work, explain why it *did* work in the Prologue.

    b) The shortest paths to the k farthest vertices (and which vertices these are) is known.

14. Suppose that you have a graph (or digraph) with a weight ≥ 1 on each edge, and you want to find the path from v_0 to v_n with the smallest *product* of its weights. Modify Algorithm SHORTESTPATH to find this path. Explain why your modification works. Why is the restriction that all weights be ≥ 1 necessary?

15. Nowhere in this section have we said that the paths we find must be simple paths. Yet they always are. Explain why.

16. Now assume that edges are allowed to have negative weights. The concept of shortest *simple* path still makes sense, but the concept of shortest path may not. Explain.

17. Returning to graphs with positive weights on the edges, suppose that you want to find the *longest* simple path from v_1 to other vertices. Will replacing min by max in the first **for**-loop inside the **repeat**-loop of Algorithm SHORTESTPATH and ∞ by 0 do the job? Justify your answer.

18. Let the weights on edges of graph G represent heights. Define the height of a path to be the maximum height on its edges. Suppose that you desire to find the path from v_0 to v_n with the minimum height. (This might be useful if the edges represent sections of pipe in a water distribution network; the work needed to pump water depends on the maximum elevation it will reach.) Modify Algorithm SHORTESTPATH to accomplish this task. Verify that your algorithm works.

3.6 Breadth First Search and Depth First Search

Finding paths between vertices, as we did in Sections 3.3–3.5, is just one aspect of a more general problem: that of *searching* a graph to find a vertex with some particular property or with some specific data associated with it. That searching a graph is an important task shouldn't be surprising. After all, it is plausible that, for many applications, we will want somehow to process the data associated with a vertex. To do so, we'll first have to search through the graph to find the data we want and usually we'll need assurance that we've processed the data at each vertex once and only once.

There are two general methods for searching a graph which occur so often in applications that we devote this section to them. They are called **breadth first search (BFS)** and **depth first search (DFS)**.

In breadth first search we do the searching by always visiting (i.e., processing the data associated with) the vertices nearest the one we start from before visiting any others. Then we apply this idea recursively, although—as you can see in Algorithm 3.6 BREADTHFIRSTSEARCH—the most convenient implementation of this idea is iterative. This algorithm introduces the idea of a **queue** (Fig. 3.22a), which is a linear structure in which items are inserted at one end and removed from the other. A queue is often contrasted with a **stack** (Fig. 3.22b), in which items are inserted

Algorithm 3.6

BreadthFirst-Search

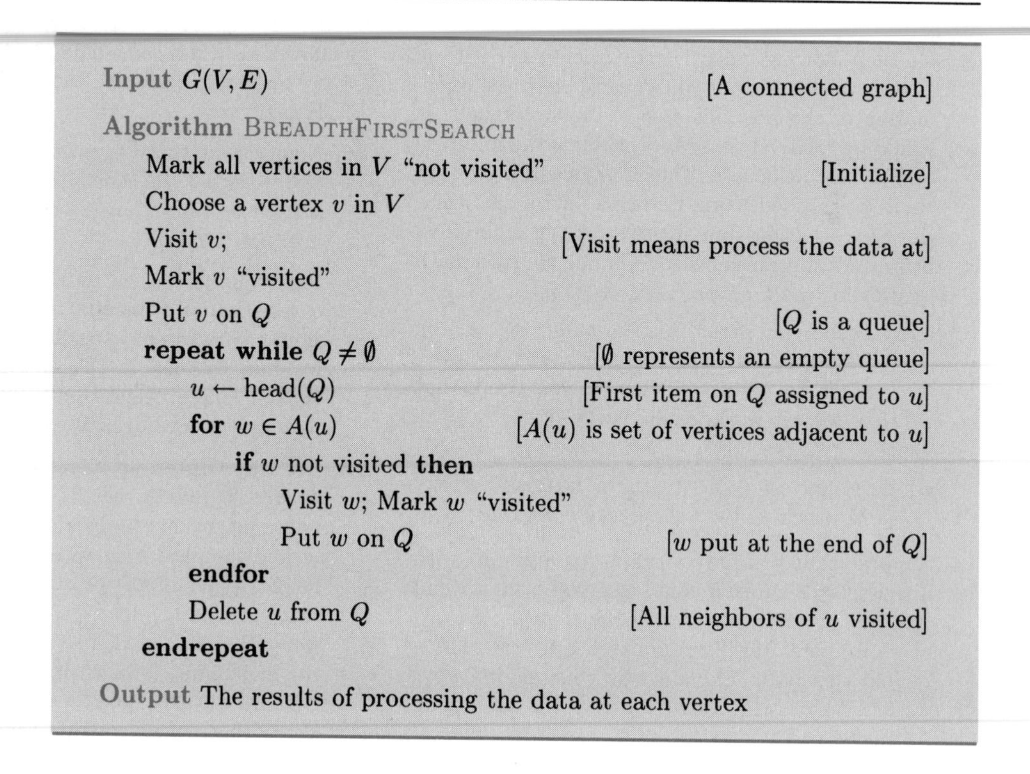

Input $G(V, E)$ [A connected graph]

Algorithm BREADTHFIRSTSEARCH

Mark all vertices in V "not visited" [Initialize]
Choose a vertex v in V
Visit v; [Visit means process the data at]
Mark v "visited"
Put v on Q [Q is a queue]
repeat while $Q \neq \emptyset$ [$\emptyset$ represents an empty queue]
    $u \leftarrow \text{head}(Q)$ [First item on Q assigned to u]
    for $w \in A(u)$ [$A(u)$ is set of vertices adjacent to u]
        if w not visited **then**
            Visit w; Mark w "visited"
            Put w on Q [w put at the end of Q]
    endfor
    Delete u from Q [All neighbors of u visited]
endrepeat

Output The results of processing the data at each vertex

and removed at the same end. If you know some computer science, you'll know that stacks are used in implementing recursive algorithms on computers.

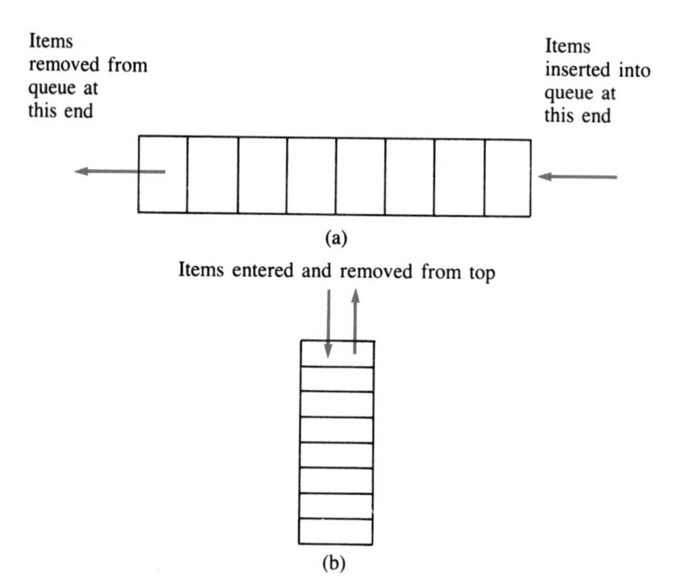

FIGURE 3.22

Queues and Stacks (a) A *queue* is a first-in-first-out (FIFO) structure. (b) A *stack* is a last-in-first-out (LIFO) structure.

Note that, if G is connected, then BREADTHFIRSTSEARCH visits all the vertices of G once and only once. We leave the proof to a problem [1].

Is breadth first search restricted to connected graphs? No, but for a graph which is not connected, we must have some criterion other than an empty queue for ending the algorithm. What should this criterion be? We leave the answer to this question and the design of a breadth first search algorithm which works for unconnected graphs to [1].

BREADTHFIRSTSEARCH does not impose any rule about the order in which vertices adjacent to the current vertex are visited. If desired, such an order could be imposed by, for example, numbering the vertices and entering the vertices adjacent to u into the queue in numerical order. If we do this for the graph in Fig. 3.23 and choose vertex v_1 as v, then the order in which the vertices are visited is

$$v_1, v_2, v_4, v_5, v_6, v_3, v_7, v_8, v_9.$$

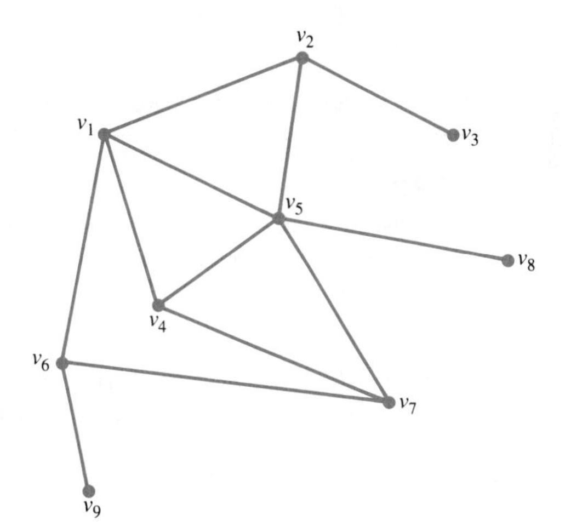

FIGURE 3.23

A Graph for Use with Breadth First Search and Depth First Search Examples.

You should not view BREADTHFIRSTSEARCH as a rigid model of how to do this search. Vertices could be visited as they are removed from the queue rather than just before they enter the queue. Or some vertices might not be visited at all, depending on some criterion. And it may not be necessary to initialize the vertices to "not visited" if the processing of a vertex marks it as "visited" in some unambiguous way. But BREADTHFIRSTSEARCH does provide an adaptable model for many specialized algorithms which are, in essence, based on breadth first search.

Suppose that we want to know the shortest path length from a vertex v to all the other vertices of a graph, where the length of a path is the number of edges on it. (This is just a special case of the shortest path problem considered in Section 3.5, where here the length of each edge is 1; this simplification makes the problem much easier.) For example, in a graph representing a grid of streets in a city where all the streets are of similar length, this method would tell you the shortest way to get from one point (say, your home) to any other.

Algorithm 3.7
PathLength

Input $G(V, E)$ [A connected graph]
v [A particular vertex in V]

Algorithm PATHLENGTH

Put v on Q [Q is a queue]
Mark v with length 0
repeat while $Q \neq \emptyset$
  $u \leftarrow \text{head}(Q)$
  $l \leftarrow$ length of path from v to u
  for $w \in A(u)$ [$A(u)$ is set of vertices adjacent to u]
    if w not marked with length **then**
      Mark w with $l+1$; Put w on Q
  endfor
  Delete u from Q
endrepeat

Output Each vertex marked with its length from v

Algorithm 3.7 PATHLENGTH fits quite closely the BREADTHFIRSTSEARCH paradigm. Marking each vertex with its length obviates the need to mark the vertices "visited".

EXAMPLE 1 Applying PATHLENGTH to the graph in Fig. 3.24(a), we obtain the path lengths from vertex v_1 which are shown in Fig. 3.24(b). ∎

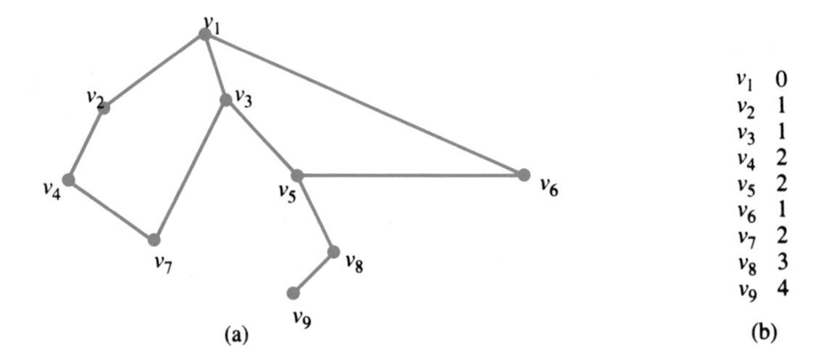

FIGURE 3.24

An Application of Algorithm PATHLENGTH: (a) graph and (b) path lengths.

PATHLENGTH results in a marking which is the *minimum* distance from v to each vertex. (Remember that there can be more than one path from v to another vertex and that these paths can have different lengths.) We leave the proof to [4]. Note also that PATHLENGTH doesn't generally find the shortest path when the edge weights are not all equal.

In effect, breadth first search (and depth first search) result in the construction of a spanning tree for any connected graph G. Recall that a spanning tree is a tree which includes all the vertices in G and only edges in G. We will consider spanning trees further in Section 3.8.

Now we turn to depth first search. Here the idea is to start at v and go as far as you can, following your nose, until you can't go any farther (i.e., until there are no unvisited vertices adjacent to the vertex you have arrived at). Then back up along the path just traversed until you find a vertex with an unvisited neighbor and start down a new path, again following your nose. Implementing this idea could be done using a sequence of applications of Procedure Pathgrow. But it's rather messy to keep track of where you have been on each path as you grow it, so that you can back up when you need to (in particular, see [24, Supplementary]). The way to avoid this mess is just to let our algorithmic language take care of the bookkeeping by making the algorithm recursive.

Algorithm 3.8

DepthFirst-Search

```
Input G(V,E)                                        [A connected graph]
Algorithm DEPTHFIRSTSEARCH
    procedure Depth(u)                                    [u is a vertex]
        Visit u
        Mark u "visited"
        for w in A(u)             [A(u) is set of vertices adjacent to u]
            if w is marked "not visited" then
                Depth(w)                                 [Recursive call]
        endfor
        return
    endpro
    Mark all vertices of V "not visited"               [Main algorithm]
    Choose some v arbitrarily
    Depth(v)
Output The results of processing the data at each vertex
```

Algorithm 3.8 DEPTHFIRSTSEARCH really just grows a path by, in effect, calling the path grower recursively every time you get to a new vertex. When you get to a point from which you can't continue—because there are no unvisited vertices adjacent to the vertex at which you've arrived—the recursion automatically goes back to a level at which there is a vertex with an unvisited neighbor.

If we apply DEPTHFIRSTSEARCH to Fig. 3.23 starting at vertex v_1 and if in the **for**-loop we choose the vertices to assign to w in numerical order, the order in which the vertices are visited is

$$v_1, v_2, v_3, v_5, v_4, v_7, v_6, v_9, v_8.$$

When a vertex has been visited, it never gets visited again, so no vertex is visited more than once. But does DEPTHFIRSTSEARCH visit all the vertices of a connected graph? Proving that it does is a nice example of how induction can be used to prove the correctness of recursive algorithms.

Theorem 1. DEPTHFIRSTSEARCH visits all the vertices of a connected graph G.

PROOF By strong induction on n, the number of vertices in G. When $n = 1$, the basis case, the result is immediate because the (solitary) vertex v gets visited when Procedure Depth is called from the main algorithm, and the algorithm terminates after this visit. Now assume that the theorem is true when G has fewer than n vertices. Then consider a connected graph G with n vertices. As in the algorithm, choose a vertex v at which to start and call Depth. The situation is as shown in Fig. 3.25, with $a, b, \ldots$ the vertices adjacent to v. Suppose that, without any loss of generality, a is chosen as the first vertex adjacent to v. Let G' be the graph consisting of G except for v and the edges which impinge on v. When Depth is called with argument a, the situation is the same as if the input had originally been the *component* C of G' containing a with a the vertex chosen in the main algorithm (why?).

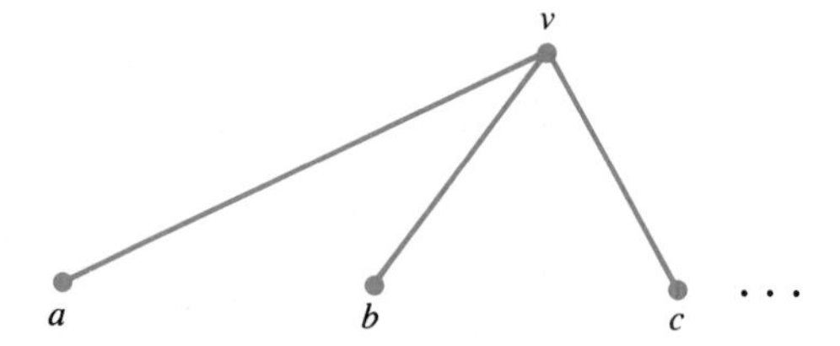

FIGURE 3.25

Induction Proof of Theorem 1.

Since this component has at least one less vertex than G, our induction hypothesis says that, when Depth is called with argument a, the algorithm visits all the vertices in C. After doing this, control goes back to the **for** statement within Depth(v), looking for any other vertices adjacent to v which are still unvisited. (There may be none because G' might be connected so that b, c, etc., would have been visited as a result of the call of Depth with argument a.) If there is an unvisited vertex adjacent to v, then Depth is called with this unvisited vertex as argument, and again, all vertices in one component of G' are visited. As there can be no more components of G' than there are vertices of G adjacent to v, successive calls of Depth will visit all the vertices of G. ∎

Depth first search embodies the important algorithmic idea of **backtracking**. The essence of any backtracking algorithm is that you search for a solution in a particular direction until either you find it or you reach some kind of dead end. If you reach a dead end, then you back up ("backtrack") to the closest previous point at which you had some decision to make about the direction in which to go. At this point, if there is a direction to go in and you haven't tried it before, do so; if not, backtrack further until you find such a direction. Either you will find a solution, if there is one, or you will end up back at the beginning with no untried directions. The crucial attributes of this idea are that you never try a direction more than once and that you never skip over any possible solution.

DEPTHFIRSTSEARCH is an instance of a backtracking algorithm because, starting from any vertex, we go down some path until we can go no farther (i.e., there is no unvisited vertex to go to). Then we back up on the path just traversed until a visited vertex is reached which has unvisited neighbors. We then apply this same idea repeatedly, each time going as deeply as possible and then backtracking until a new path is found on which there are unvisited vertices. Backtracking has many other applications, two of which are treated in [6–7].

Theorem 1 suggests a method of determining whether a given graph is connected. Each time a vertex is visited, you label it as visited and then, at the end, check to see whether all the vertices have been labeled [5a]. Note that breadth first search also can be used to determine whether a graph is connected [5b].

This completes our introduction to BFS and DFS. We will discuss various other applications of these algorithms in the remaining two sections of this chapter.

Problems: Section 3.6

1. a) Prove that Algorithm BREADTHFIRSTSEARCH visits all the vertices of a connected graph once and only once.

b) Show how to modify BREADTHFIRSTSEARCH for graphs which are not connected, so that it searches all components of the graph.

2. Let $G(V,E)$ be defined by $V = \{1,2,\ldots,15\}$ and let

$$E = \{\{u,v\} \mid u \equiv v\,(\mathrm{mod}\,3) \text{ or } u \equiv v\,(\mathrm{mod}\,5)\}.$$

Draw a BFS tree for G starting at vertex 1. Break ties by choosing the vertex with the lowest index. Show your work by drawing a picture of the graph with the final tree darkened and the tree edges labeled by their order of inclusion.

3. Apply Algorithm PATHLENGTH to the graph obtained in [2], starting at vertex 1.

4. Prove that Algorithm PATHLENGTH marks each vertex with the minimum distance from the starting vertex.

5. a) Write a DFS algorithm to test whether a given graph is connected.

b) Repeat for BFS.

6. The Eight Queens Problem. The object of this problem is to place eight queens on a chessboard so that no two queens are attacking each other (i.e., no two queens are in the same row, the same column, or on the same diagonal).

a) Describe a graph in which each vertex is one square of a chessboard and for which each solution to the eight queens problem is a set of mutually nonadjacent vertices. Don't try to draw the graph!

b) Actually, graphs are not a very useful tool in solving the eight queens problem, but backtracking is. Describe a backtracking algorithm for this problem which begins by putting a queen in the first column of the chessboard, then another which doesn't attack the first in the second column, a third which doesn't attack either of the first two in the third column, etc. When you can't find a place to put a queen in column i, backtrack to column $i-1$ and try again. In each column try the first row first, then the second row, etc., so that, when you have to backtrack, you can then try only rows which you haven't tried before.

c) Use your algorithm from b) to find one solution to the eight queens problem. It's a little tedious but shouldn't take you more than 10 minutes by hand. It's best to use an actual chessboard.

7. Consider the problem of forming sequences of the digits 1, 2, and 3 with the property that nowhere in the sequence are there two *adjacent* subsequences which are the same. Thus two adjacent digits can never be the same and, for example, 121323213231 does not satisfy the criterion because of the two adjacent subsequences 21323.

a) Use backtracking to devise a method for generating sequences of any length using the symbols 1, 2, and 3. (*Hint*: Each time you add a digit, how far back in the sequence do you have to look to make sure that there are no two identical adjacent subsequences?)

b) Display your method using our algorithmic language.

c) Use your algorithm to generate a sequence of length 10 having the desired property.

3.7 Coloring Problems

A **coloring** of a graph is any assignment of colors to vertices such that no two adjacent vertices have the same color. The idea of coloring graphs is an important one in graph theory because it has many applications. In this section we'll mention some of these applications, discuss some basic algorithms and theorems, and indicate how graph coloring is related to the problem of coloring a map.

A general class of problems to which graph coloring can be applied is *scheduling*. Suppose that you want to schedule the classes for a university. Because a professor normally teaches more than one course, you must make sure that each professor's courses are taught at a different time. One way to describe this problem is by a graph in which each vertex represents a class to be scheduled and an edge joining two vertices represents classes taught by the same instructor. If each color is then associated with a particular class time (red—8–8:50, MWF; blue—9–9:50, MWF; etc.), a coloring of the graph implies that no professor is scheduled for two classes at the same time. You can extend this idea to take into account other constraints on a class schedule. Suppose that Professor Sleeplate is not a "morning person" and thus should never be scheduled for an 8 A.M. class. Then if the graph also contains vertices corresponding to possible class times (8 A.M.—green; 9 A.M.—yellow; etc.), there would be an edge from the 8 A.M. vertex to each class vertex taught by Professor Sleeplate. A coloring of the graph would give a schedule in which all vertices representing classes colored the same as the 8 A.M. vertex could then naturally be scheduled at 8 A.M., but no class taught by Professor Sleeplate would be scheduled at 8 A.M.

Various aspects of computer scheduling may also be attacked using the graph coloring model. Suppose that we have a large database (e.g., student records at a university), which is accessed by many application programs (e.g., to update addresses, to send out tuition bills, to insert course grades, etc.). Suppose also that certain application programs should never access the database at the same time because a change made by one program might cause the other program to make an error. For example, the program which calculates a student's grade point average should never be accessing the database at the same time as the program which inserts course grades in the database. We can let the vertices in a graph represent the application programs, and an edge would join any two vertices corresponding to programs which cannot be active at the same time. Then a coloring of the graph provides a schedule for these application programs in which no two programs which interfere with each other can be active at the same time. Graph coloring can also be applied to other aspects of job scheduling on a computer, as well as to the scheduling of computer resources, such as the assigning of registers by a compiler to hold intermediate results of a computation.

We begin with the essential definition needed for a discussion of graph coloring problems.

Definition 1. The **chromatic number** $\chi(G)$ of a graph G is the minimum number of colors needed such that no two adjacent vertices have the same color. A graph G is said to be *k-colorable* for any $k \geq \chi(G)$.

Figure 3.26(a–c) displays a minimal coloring and thus the chromatic number of some graphs. For simplicity in what follows, we'll refer to the colors of vertices not as "red" or "green" or "blue" but rather as "1" or "2" or "3".

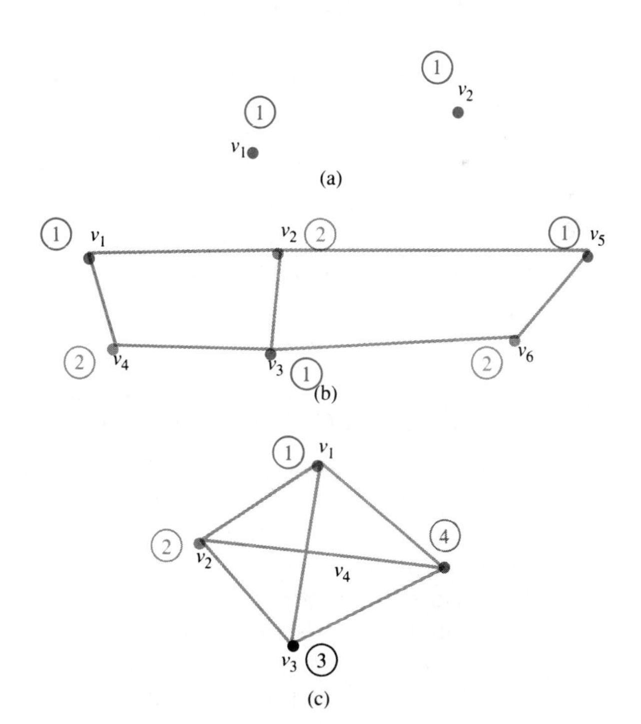

FIGURE 3.26

Examples of Chromatic Numbers (colors shown in circles): (a) $\chi(G) = 1$; (b) $\chi(G) = 2$; (c) $\chi(G) = 4$.

Note, in particular, that if a graph is 2-colorable, this means that its vertex set may be split into two disjoint subsets one subset colored 1 and the other colored 2, such that each edge in the edge set joins a vertex in one subset to a vertex in the other subset. Thus—recall Definition 5, Section 3.2—a graph is 2-colorable if and only if it is bipartite. For this reason, k-colorable graphs are sometimes referred to as k-partite graphs (see [12]).

Another property of bipartite graphs is expressed in the following theorem.

Theorem 1. A graph is 2-colorable (i.e., bipartite) if and only if it has no cycles of odd length.

PROOF We may assume that our graph G is connected because, if it isn't, we can consider each component separately (why?).

Proving the *necessity* part, namely, that a bipartite graph can have no odd-length cycles is easy. Suppose that a bipartite graph which has been 2-colored has a cycle. Choose any vertex v on the cycle, and suppose that it is colored 1. Then as you traverse the cycle starting at v, successive vertices are colored 1, 2, 1, 2, 1, ... When you get back to the vertex v, the previous vertex must be colored 2 if the coloring is to be valid. But this will be true only if the number of vertices on the cycle was even (i.e., if the vertices on the cycle can be divided into adjacent pairs, each colored 1, 2). Thus there can be no cycle of odd length.

Now we must prove the *sufficiency* part of the theorem that, if there are no odd cycles, the graph is bipartite. As this is really a statement about the existence of something, namely a 2-coloring of a graph, it cries out to be proved by presenting an algorithm. Algorithm 3.9 BIPARTITE is such an algorithm. ∎

Algorithm BIPARTITE actually does more than is required by the sufficiency part of Theorem 1. We'll show this by proving the following lemma and then returning to the proof of Theorem 1.

Lemma 2. BIPARTITE produces a two-coloring of a connected graph G if and only if G has no cycles of odd length.

PROOF OF LEMMA BIPARTITE either terminates by outputting a coloring because the queue is empty or else it halts with the message "Graph not bipartite". We begin with the first case. At termination, all vertices are colored. For suppose they weren't. Then some uncolored vertex must be adjacent to a colored vertex (why?). But if this were so, the uncolored vertex would have been colored and entered into the queue just before its colored neighbor was removed from the queue.

Algorithm 3.9
Bipartite

```
Input G(V,E)                                        [A connected graph]
Algorithm BIPARTITE
  Choose an arbitrary v ∈ V and color it 1
  Put v on Q                                                [Q is a queue]
  repeat while Q ≠ ∅               [That is, while queue is not empty]
    u ← head(Q)
    S ← all vertices adjacent to u                            [S is a set]
    if any vertex in S has the same color as u
      then print "Graph not bipartite"; stop
      else color all uncolored vertices in S opposite to u (i.e., 1 if
           u is colored 2 and 2 if u is colored 1) and put them on
           Q in arbitrary order
    endif
    Delete u from Q
  endrepeat
  print V                                [Including color of each vertex]
Output A 2-coloring of G (i.e., a list of members of V with their colors),
       or a message that G is not bipartite
```

This contradiction shows that the queue can be empty only when all vertices have been colored.

When BIPARTITE outputs a coloring of V, why must it be a two-coloring? Well, it must have been true each time through the main loop that no vertex in S had the same color as u. As each uncolored vertex in S is colored differently from u, this means that all neighbors of each vertex u have a color different from u. Therefore in this case the algorithm achieves a coloring and, as all vertices are colored either 1 or 2, it is a two-coloring.

Now, when can the algorithm stop with the message that G is not bipartite? Only if an already colored vertex u has a neighbor w (in the set S) which has the same color as u. Because each vertex that is colored in BIPARTITE is adjacent to another already colored vertex—which in turn is adjacent to another previously colored vertex, etc.—there must be a path of colored vertices from u back to the first colored vertex v and from w back to v. Thus there must be a path of colored vertices from u to w. As w is adjacent to u, this means that u and w are on a cycle. Alternate vertices on any path are colored differently by BIPARTITE. Thus the only way that adjacent vertices u and v on a cycle can have the same color is if there is an *odd* number of vertices on the cycle, which is the same as saying that the cycle is of odd length. Therefore the algorithm halts with the message that G is not bipartite only if there is a cycle of odd length. Otherwise we get a two-coloring. ∎

We have now proved three things:

1. BIPARTITE either produces a 2-coloring or it halts with the message "Graph not bipartite".
2. If it outputs the message, then there is an odd cycle.
3. If it produces a 2-coloring, then there is no odd cycle.

From (1) and (2) we get:

4. If there is no odd cycle, then BIPARTITE produces a two coloring.

That is, if there is no odd cycle, by (2) the message will not be output, and so by (1) the algorithm finds a 2-coloring. Statement (4) is the "if" part of the lemma and (3) is the "only if."

Now back to the proof of the sufficiency part of Theorem 1.

PROOF OF THEOREM 1 (CONTINUED) We need to show only that, if there are no odd cycles, the graph is bipartite. But the sufficiency part of Lemma 2 does so because it says that if there are no odd cycles, BIPARTITE produces a two-coloring, which is the same as saying that the graph is bipartite. So, as we started out to do, we have shown that, if there are no cycles of odd length, the graph is bipartite. ∎

The reason we proved a separate lemma about BIPARTITE is that this algorithm makes no assumption that the graph has no cycles of odd length. It runs in any event because it is constructed so that it halts with an appropriate message if the graph has an odd cycle.

Did you notice that BIPARTITE is really just an application of breadth first search? If not, go back and compare BIPARTITE and BREADTHFIRSTSEARCH. You'll see that the former is just an adaptation of the latter.

Now let's consider computing the chromatic number of a graph. We begin with a theorem which gives us some bounds on the chromatic number.

Theorem 3. Let $\chi(G)$ be the chromatic number of a graph $G(V, E)$, let $c(G)$ be the clique number of the graph (see Definition 6, Section 3.2), and let $\Delta(G)$ be the maximum degree of any vertex of G. Then

$$c(G) \leq \chi(G) \leq \Delta(G) + 1.$$

Algorithm 3.10
K-Colorable

Input $G(V, E)$ [A connected graph]

Algorithm K-COLORABLE

```
Choose an arbitrary vertex v in V and color it 1
V' ← V − {v}                                    [Initialize V']
repeat while V' ≠ ∅
    u ← an arbitrary vertex in V'
    S ← all vertices adjacent to u
    Color u the minimum color not yet assigned to a
        vertex in S
    V' ← V' − {u}
endrepeat
k ← maximum color assigned to a vertex
```

Output k and the color of each vertex in V

PROOF A clique is just a subgraph which, by itself, is a complete graph and $c(G)$ is the number of vertices in any largest clique of G. As there is an edge between any two vertices of a complete graph, each vertex of the clique must have a different color. Therefore

$$c(G) \leq \chi(G).$$

Now let's color G by choosing the vertices one at a time and coloring each vertex v with the smallest (i.e., lowest, number) color not used for any already colored vertex adjacent to it. Since no vertex has more than $\Delta(G)$ vertices adjacent to it, it will never be necessary to give v a color greater than $\Delta(G)+1$. Thus

$$\chi(G) \leq \Delta(G)+1. \blacksquare$$

Theorem 3 might be termed "half-useful". It's easy to determine the maximum degree of any vertex, so an algorithm to compute the chromatic number of a graph can stop whenever it determines that the chromatic number is at least $\Delta(G)+1$. On the other hand, the lower bound $c(G)$ isn't very useful because calculating the clique number is generally very difficult.

Now let's consider a graph coloring algorithm whose result will be an upper bound on $\chi(G)$. As before, we'll assume that our graph is connected since otherwise we could consider each component separately. A fairly obvious way to go about computing a bound on $\chi(G)$ is just to use the approach in Theorem 3 by choosing one vertex after another and giving it the lowest color possible. We illustrate this approach in Algorithm 3.10 K-COLORABLE. The output, k, of this algorithm shows only that G is k-colorable. However, k is not generally the chromatic number, as Fig. 3.27 illustrates. K-COLORABLE is an example of a greedy algorithm that doesn't always achieve the best possible result. Indeed, the computation of the

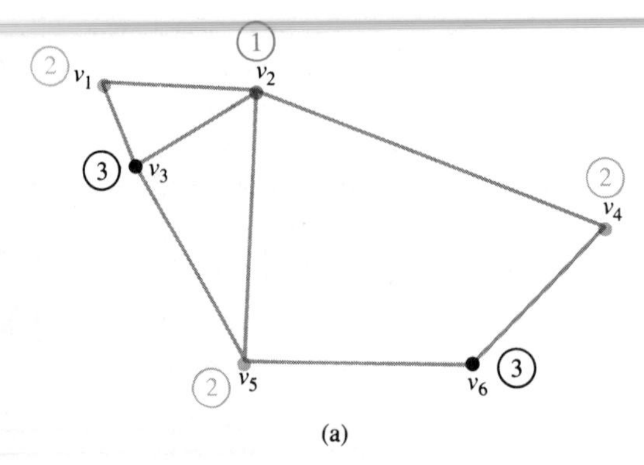

(a)

i) Color v_6 1
ii) Color v_1 1
iii) Color v_3 2 (since it is adjacent to v_1, which is colored 1)
iv) Color v_5 3 (since it is adjacent to vertices colored 1 and 2)
v) Color v_4 2 (since it is adjacent to a vertex colored 1)
vi) Color v_2 4 (since it is adjacent to vertices colored 1, 2 and 3)

(b)

FIGURE 3.27

Calculation of k-colorability. (a) A graph with $\chi(G) = 3$. A 3-coloring is denoted by the numbers in circles. (b) Application of Algorithm K-COLORABLE to the same graph. All we know from using the algorithm is that the graph is 4-colorable. For certain other orders of choosing the vertices, the algorithm would have 3-colored the graph.

chromatic number for general graphs is a very difficult problem, and we'll not pursue it further here.

Finally, suppose that you want to print a map of, say, the 48 contiguous states of the United States, using the minimum number of colors such that no two areas with a common boundary have the same color. (It's OK if two areas which meet at a single point have the same color.) This problem may be directly related to the graph coloring problem if you associate a vertex with each area of the map and have an edge between two vertices whenever the corresponding areas have a common boundary, as illustrated in Fig. 3.28. (This is essentially what we did in Section 3.4 when we derived a graph representation of the Königsberg bridge problem.) What is the minimum number of colors needed to print any map? The answer is given by the famous Four Color Theorem whose proof would be even longer than this chapter. We presume that you can guess what the answer is.

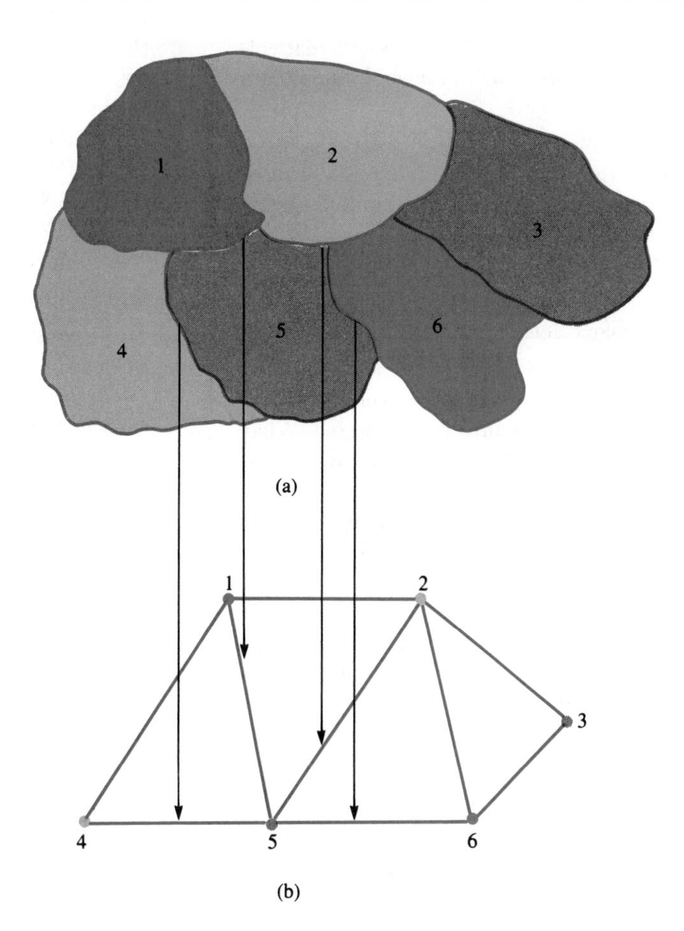

FIGURE 3.28

Maps and Graphs: (a) A map of six regions; (b) a graph corresponding to this map. The arrows indicate how the boundaries of region 5 become edges from vertex 5 to other vertices.

Problems: Section 3.7

1. What graphs are 1-colorable?

2. What are the chromatic numbers of the following graphs?

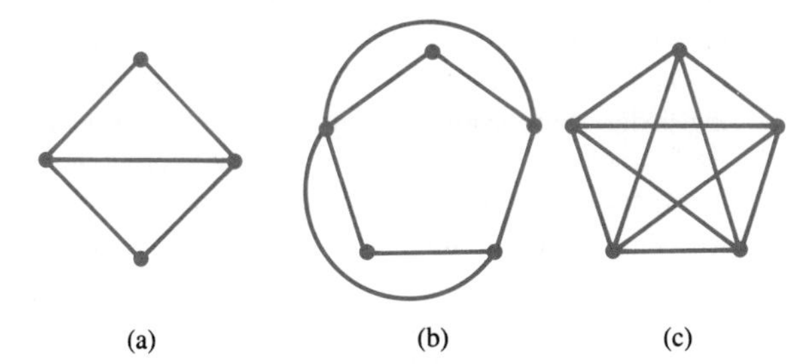

3. Determine the chromatic number of the following graph, called the Petersen Graph. Justify your answer.

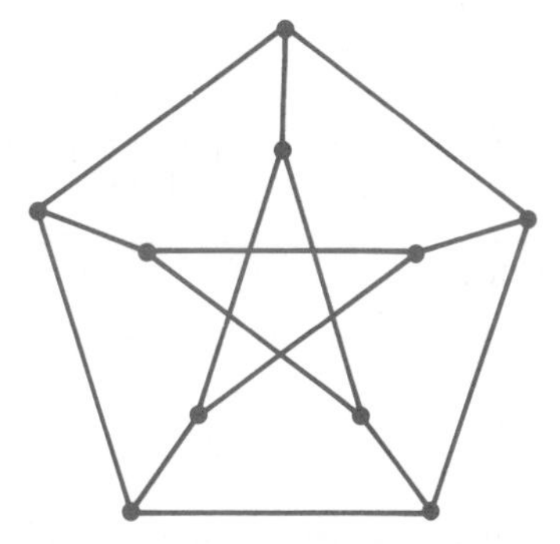

4. With as few colors as possible, color the vertices of

a cube so that adjacent vertices have different colors. How do you know that you have the minimum number?

5. With as few colors as possible, color the *faces* of a cube so that faces with an edge in common have different colors. How do you know that you have the minimum number? (Solve by turning this into a vertex coloring problem on an appropriate graph.)

6. How many colors does it take to color any tree? Why?

7. Suppose that you are given a tree on which two nonadjacent vertices are already colored red. What is the maximum number of colors needed to complete a coloring of the tree? Give a proof by algorithm: Exhibit an algorithm which you can show always uses at most the number of colors you have claimed.

8. Suppose that you are given a tree on which k mutually nonadjacent vertices have already been colored red. Determine the maximum number of colors needed to complete that coloring. Give a proof by algorithm.

9. A college registrar must schedule final exams so that no student is supposed to take two exams at the same time. Figure out how to turn this into a coloring problem. What do the vertices represent? When is there an edge between two vertices? What do the colors represent? Why might you wish to find a coloring with a minimum number of colors?

10. Use the same conditions as in [9], but now the registrar is also mindful of professors and doesn't want any professor to have to give two exams at the same time. How must the graph be changed so that the coloring still provides a solution?

11. Use the same conditions as in [10], except now there are further constraints. The Physics 1 exam cannot be scheduled for Monday morning, and the PoliSci 3 exam cannot be scheduled for Tuesday evening. Explain how to augment the graph to take care of this.

12. A graph is *k-partite* if the vertex set V can be partitioned into V_1, V_2, ..., V_k so that no edge has its two ends in the same set. Explain why a graph is k-colorable if and only if it is k-partite.

13. Determining whether a graph G has $\chi(G) = 3$ is not so tough. First, determine whether $\chi(G) < 3$, using Algorithm BIPARTITE. Next, try in turn each three-way partition of the vertices. If you find one for which there are no edges between any pair of vertices in the same set, then G is 3-partite, thus 3-colorable, and hence $\chi(G) = 3$ since we already verified that it isn't 2-colorable.) If no three-way partition works, then G is not 3-partite and hence not 3-colorable. Critique this claim.

14. The **generalized octahedral graphs** $\{O_n\}$ are defined recursively as follows:

   i) O_1 consists of 2 isolated vertices.
   ii) O_{n+1} is obtained from O_n by creating 2 new vertices and connecting each one to all the vertices in O_n.

   a) Draw O_1, O_2, and O_3.
   b) Why are these called octahedral graphs? (*Hint*: What is an octahedron?)
   c) Determine $\chi(O_1)$, $\chi(O_2)$, and $\chi(O_3)$.
   d) Conjecture and prove a formula for $\chi(O_n)$.

15. Consider the undirected graph given by the following adjacency matrix:

$$\begin{bmatrix} 0 & 0 & 1 & 1 & 0 & 1 \\ 0 & 0 & 1 & 0 & 1 & 1 \\ 1 & 1 & 0 & 0 & 1 & 0 \\ 1 & 0 & 0 & 0 & 1 & 0 \\ 0 & 1 & 1 & 1 & 0 & 1 \\ 1 & 1 & 0 & 0 & 1 & 0 \end{bmatrix}$$

With the aid of Algorithm BIPARTITE, not by inspection, either properly two-color this graph or display an odd elementary cycle. For consistency to ease grading, let the "arbitrary" start vertex be v_1 (represented by the first row). Show how you use the algorithm by drawing a picture for each iteration of the algorithm. Also, list all vertices in the order they appear in your queue. When you have a choice of which vertex to put in the queue next, put in first the one with the lowest index.

16. Construct a graph G as follows. Put vertices v_1 through v_9 evenly spaced along a circle, in increasing order. Now draw an edge from each vertex to its four closest neighbors (two on each side). For instance, v_1 is adjacent to v_8, v_9, v_2, and v_3.

   a) What are the bounds of $\chi(G)$ according to Theorem 3?
   b) What actually is $\chi(G)$? None of our theorems will answer this for you, so you must give a special justification for your answer, based on the specific nature of this graph.

17. Repeat [16], except now the circular graph has 10 vertices.

18. Repeat [16], except now the circular graph has 11 vertices.

19. Algorithm K-COLORABLE leaves the order in which vertices are colored up to you. What sort of vertices do you think it would be best to color first: high degree, low degree, some other condition? Give your reasons.

20. Consider the following two algorithms for coloring a graph. The first is a variant of Algorithm K-COLORABLE. In both, colors are described by numbers, not names. Thus the "lowest numbered" color is color 1.

Algorithm

```
Input G(V,E)                     [V = {v1, v2, ..., vn}]
      n                                        [n = |V|]
Algorithm COLOR-BY-VERTICES
  for i = 1 to n
     Color vi the lowest numbered color not
        already used for adjacent vertices
  endfor
Output A coloring of G
```

Algorithm

```
Input Same
Algorithm ONE-COLOR-AT-A-TIME
  c ← 1                          [Initialize to first color]
  U ← V
  repeat until U = ∅
     W ← ∅
     for i = 1 to n
        if vi ∈ U and is not adjacent
              to any vertex in W then
                 Color vi with color c
                 W ← W ∪ {vi}
                 U ← U − {vi}
     endfor
     c ← c+1
  endrepeat
Output A coloring of G
```

a) Apply both algorithms to the graph obtained in [16].

b) Apply both algorithms to the graph obtained in [17].

c) What do you observe about the relative behavior of the two algorithms?

d) Prove your observations in c).

21. *Edge Coloring.* Now let's color edges instead of vertices. An edge coloring is *proper* if no two edges incident on the same vertex have the same color. The **edge chromatic number** of G, written $\chi'(G)$, is the minimum number of colors needed to edge-color G properly.

a) Find χ' for the six-cycle (i.e., the graph with a cycle of six vertices and no other edges).

b) Find χ' for the seven-cycle.

c) Find χ' for the complete graphs having five and six vertices.

d) Generalize from c) and prove your result.

e) Prove that, for any G, $\chi'(G) \leq 2\Delta - 1$, where Δ is the maximum degree in G. Give a proof by algorithm. (In fact, $\chi'(G) \leq \Delta + m$, where m is the maximum edge multiplicity (number of edges between the same two vertices), but the proof is subtle. This result is known as Vizing's theorem.)

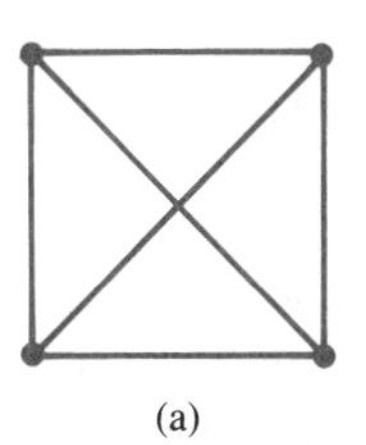
(a)

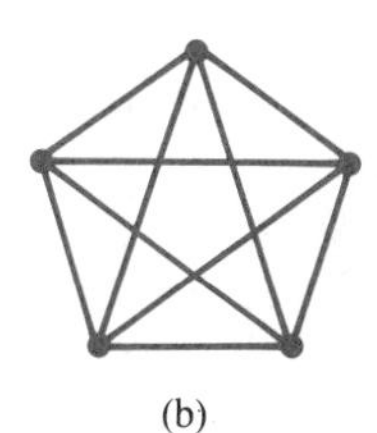
(b)

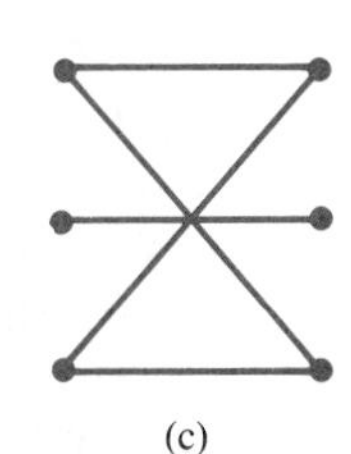
(c)

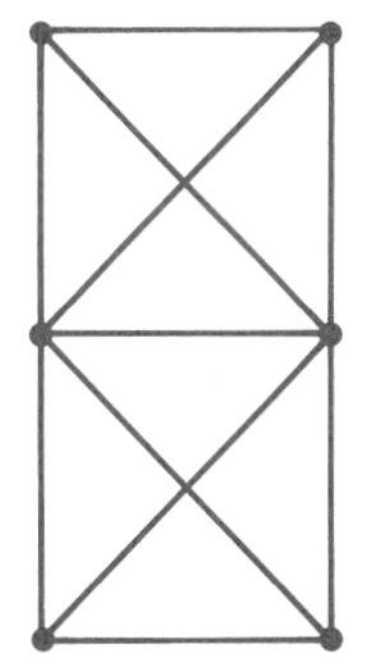
(d)

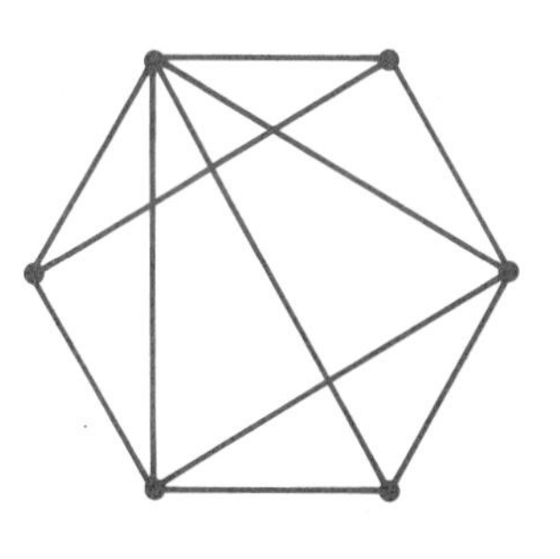
(e)

22. A graph is *planar* if it can be drawn on a plane surface with no edges intersecting except at vertices. Which of the graphs on p. 265, drawn with crossing edges, are planar (because they could be redrawn without crossing edges)? You needn't give a proof when you claim that the graph is not planar.

23. Let $G(V, E)$ be an acyclic graph with at least one edge. Give an algorithmic proof that there must be at least one vertex of degree 1. (*Hint*: Walk around the graph following your nose; if you are not at a vertex of degree 1, then either you can continue the walk or you have found a cycle.)

24. Use the result of [23] to show that a connected, acyclic graph with n vertices has $n-1$ edges. (*Hint*: Any degree-1 vertex and the edge incident on it can be removed from the graph without affecting the relative number of edges and vertices.)

25. An area surrounded by the edges of a planar graph and containing no smaller such area within it is called a **region**. The **infinite region** is the area not surrounded by edges of the graph (i.e., the area outside all the edges). Let G be a connected, planar graph with n vertices, m edges, and r regions where r includes the infinite region. Derive **Euler's formula**:

$$n - m + r = 2.$$

(*Hint*: Use induction on the number of regions with the result of the previous problem used for the basis case.)

26. For the complete bipartite graph K_{mn} (see [24, Section 3.2]),

a) show that K_{22} and K_{32} are planar; and

b) show that K_{33} is not planar. [*Hint*: By Euler's formula (see [25]), how many regions would a planar representation have? At the least, how many edges would surround each region? Now count the number of edges two ways and get a contradiction.]

27. The n-cube graph Q_n is the edge graph of the analog of a cube in n dimensions. Thus, Q_1 is an edge, Q_2 is a four-cycle, and Q_3 is the usual skeletal image of a cube.

a) Define Q_n for all n recursively.

b) What is the smallest positive integer n for which Q_n is not planar?

28. Let G be a simple, connected planar graph with no loops, n vertices, m edges, and $r > 1$ regions.

a) If each region has at least k edges bounding it, show that $kr \leq 2m$.

b) Show that $n \geq (1-2/k)m + 2$.

29. Suppose that G is a graph as in [28] in which every region is bounded by at least four edges. Show that G has a vertex of degree 4 or less.

30. In the spirit of [29], what can you prove about vertex degrees for a planar graph in which every face has at least five edges?

31. Draw maps which require exactly

a) two, **b)** three, and **c)** four

colors so that no two areas with a common boundary have the same color.

32. Suppose that we have six chemicals and, for each i from 1 to 4, chemical i cannot safely be stored in the same room as chemical $i+1$ or $i+2$. Determine how many rooms are needed for safe storage by turning this into a graph coloring problem.

33. Use the same conditions as in [32], but now the following additional pairs of chemicals cannot be stored together: $\{5, 6\}$, $\{5, 1\}$, $\{6, 1\}$, and $\{6, 2\}$.

3.8 Trees

We devote a section in this chapter to trees because they are by far the most important special case of a graph. Trees are useful in numerous applications, particularly in computer science. Sometimes a tree provides the most appropriate structuring of the data for *thinking* about a problem, and sometimes *using* a tree in an algorithm to solve the problem is most appropriate. In this section we'll discuss spanning trees, which we mentioned in Section 3.6, and binary trees, a special case useful in many applications.

Spanning Trees

We first encountered the idea of a spanning tree in Example 4, Section 1.2 about building a telephone network for Thirdworld. Formally, we define a spanning tree in the following manner.

Definition 1. A **spanning tree** of a graph $G(V, E)$ is a connected, acyclic subgraph of G, which includes all the vertices in V and only edges from E.

A spanning tree therefore is not a tree as in Definition 4, Section 3.2. According to that definition, a tree is a digraph. Instead, a spanning tree is a *free tree*, as defined in Section 3.2 after Definition 4. Does every graph have at least one spanning tree? No, because a free tree (we'll just say "tree" from now on when discussing spanning trees) is connected. Thus an unconnected graph cannot have a spanning tree. Is it sufficient for a graph to be connected for it to have a spanning tree? Yes, because of Algorithm 3.11 SPANTREE (in which we use the notation $E(C, U)$ to denote the set of edges which join a vertex in set C to a vertex in set U).

You may be satisfied that the result of SPANTREE is a spanning tree of G, but we think it's sufficiently unobvious to deserve a proof.

Theorem 1. If G is connected, SPANTREE terminates with the output T a spanning tree of G.

PROOF First, must the algorithm terminate? Yes, because V is finite and, as the subset U of V has one less element after each pass through the loop, U must eventually be empty. Could the algorithm fail before U is empty? That is, when U is not empty, might it not be possible to pick a c and u with an edge between them? No, because we have assumed that G is connected, so there must be a path

Algorithm 3.11

SpanTree

Input $G(V, E)$ [A connected graph]

Algorithm SPANTREE
  Choose a vertex v of G
  $T \leftarrow \{v\}$ [Put vertex on spanning tree]
  $C \leftarrow \{v\}$ [C contains the vertices on T]
  $U \leftarrow V - \{v\}$ [U contains vertices not on T]
  repeat until $U = \emptyset$
    Pick a $c \in C$ and a $u \in U$ such that $\{c, u\} \in E(C, U)$
    $T \leftarrow T \cup \{u, \{c, u\}\}$ [Add vertex and edge to T]
    $C \leftarrow C \cup \{u\}$
    $U \leftarrow U - \{u\}$
  endrepeat

Output T [All vertices in V and those edges in E, which together make a spanning tree]

from any vertex in U to any one in C. Some edge on this path will have one end in U and one in C.

But is the result really a spanning tree? We argue it is, using the loop invariant that, at each pass through the loop, T is a tree which spans a subgraph of G containing the vertices in C. This is certainly true before the first entry since the single vertex v is a tree and C contains precisely this vertex. Suppose (the induction hypothesis) that T has the desired property after k passes through the loop. At the subsequent pass, the edge added to T connects the vertex u added to T to a vertex c already in C. The induction hypothesis assures us that v is connected to c via a path in T, so the edge added extends this path to u. Thus the new T is connected (why?) and is still acyclic because there is only one edge from u to vertices on the previous T so u cannot be on a cycle. Therefore T is still a tree at each exit from the loop. Since U is empty at the final exit from the loop, C must equal V because the union of C and U is always V (why?). Thus at the end, T is a tree which contains all the vertices of G (and only edges from E) and is therefore, by definition, a spanning tree. ∎

Since SPANTREE does result in a spanning tree, Theorem 1 shows that every connected graph does have a spanning tree.

Suppose that G is not a connected graph. Then what is the result of SPANTREE? It is just a tree which spans that *component* of G containing the vertex chosen in the first step of the algorithm. Can you prove this assertion [5]?

One noteworthy aspect of SPANTREE is that it is not the first algorithm in this chapter which, in effect, finds spanning trees. What are the others? Two such are the algorithms for BFS and DFS. Think about it. BREADTHFIRSTSEARCH starts with a vertex v and then "visits " all the vertices adjacent to v. If, in so doing, we put these vertices *and* the edges joining them to v in a set T, we would have the start of a spanning tree. Then as subsequent vertices are visited, if we add them, too, to the set T, together with the edges joining them to their parent vertices, we would end up with a spanning tree. Figure 3.29 illustrates this process. Similarly, for Algorithm DEPTHFIRSTSEARCH, if we place the appropriate edge and vertex into T each time we visit a vertex, we get a spanning tree for G. This is illustrated in Fig 3.30. This means that our BFS and DFS algorithms are really nothing more than special cases of SPANTREE. In these cases, the rules to pick the vertices c and u are specified, so that vertices will be added to the spanning tree in BFS or DFS order. Our argument here doesn't constitute a proof of this fact, but it can be proved without too much trouble.

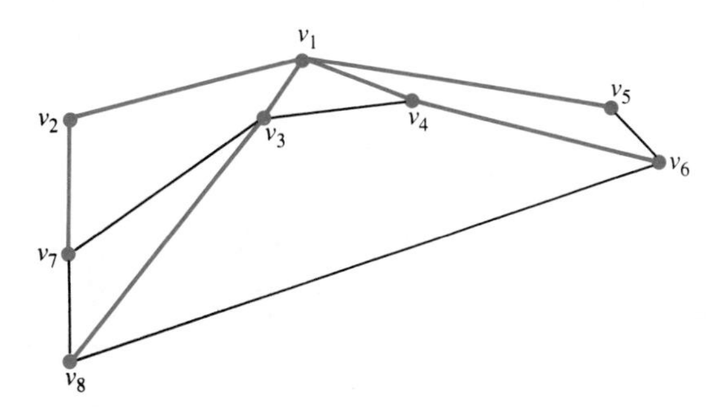

FIGURE 3.29

A Spanning Tree by Breadth First Search.

Spanning Tree: v_1, v_2, $\{v_1, v_2\}$, v_3, $\{v_1, v_3\}$, v_4, $\{v_1, v_4\}$, v_5, $\{v_1, v_5\}$, v_7, $\{v_2, v_7\}$, v_8, $\{v_3, v_8\}$, v_6, $\{v_4, v_6\}$

Another algorithm which, in effect, constructs a spanning tree is Algorithm SHORTESTPATH in Section 3.5. Here the search procedure to add vertices looks for the vertex closest to the initial vertex, which hasn't yet been added to the set U. Suppose each time we add a vertex u we also add the edge joining u to the vertex

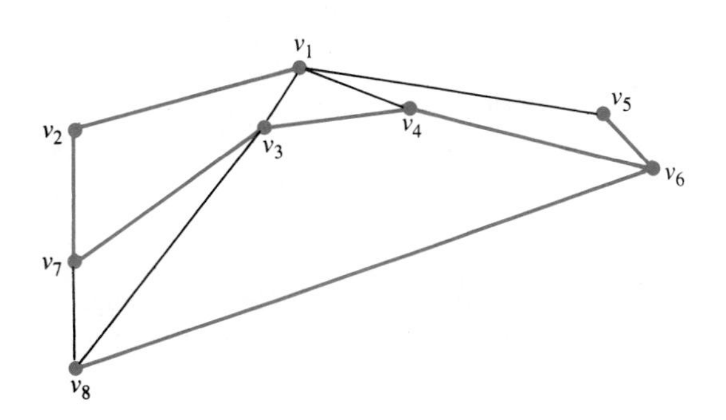

FIGURE 3.30

A Spanning Tree by Depth First Search.

Spanning Tree: v_1, v_2, $\{v_1, v_2\}$, v_7, $\{v_2, v_7\}$, v_3, $\{v_3, v_7\}$, v_4, $\{v_3, v_4\}$, v_6, $\{v_4, v_6\}$, v_5, $\{v_5, v_6\}$, v_8, $\{v_6, v_8\}$

in U on the shortest path from the initial vertex to u. Then we would again end up with a spanning tree (really a *directed* spanning tree since we defined this algorithm for digraphs) by the time all the vertices had been added to U [4]. (SHORTESTPATH may terminate *before* all the vertices have been added to u; if it doesn't, then we get, in effect, a spanning tree.) Thus the notion of spanning trees unifies a variety of different algorithms in graph theory.

Among the most interesting spanning trees are those that minimize the sum of weights associated with the edges of the graph. We first encountered such trees in Example 4, Section 1.2, where we considered finding a **minimum spanning tree**. In that example we presented a greedy algorithm and claimed that it finds *a* minimum spanning tree (*a* because, conceivably, more than one spanning tree could have the minimum weight). Now we'll prove that claim. Algorithm 3.12 MINSPANTREE is a generalization of Algorithm TELENET in Section 1.2 which works for any connected graph. An example of how MINSPANTREE works is given in Fig. 3.31.

Algorithm 3.12
MinSpanTree

Input $G(V, E)$ [A connected graph]
  $w(e)$ for all $e \in E$ [Weights of all edges]

Algorithm MINSPANTREE
  Find the edge $\{l, m\}$ with minimum weight among all edges; if
    there is a tie choose edge arbitrarily
  $T \leftarrow \{l, m, \{l, m\}\}$ [Initialize spanning tree]
  $U \leftarrow V - \{l, m\}$ [U is set of unused vertices]
  repeat until $U = \emptyset$ [That is, until all vertices are in T]
    Find the minimum weight edge $\{q, k\}$ among all edges from
      vertices q in T to vertices k not in T; if a tie,
      choose arbitrarily
    $T \leftarrow T \cup \{k, \{q, k\}\}$ [Add k and edge to T]
    $U \leftarrow U - \{k\}$
  endrepeat

Output T [The minimum weight spanning tree]

Theorem 2. If all the edge weights are distinct, MINSPANTREE finds a minimum weight spanning tree of a connected graph $G(V, E)$.

The restriction to distinct edge weights is not necessary but it does simplify the proof. Even so, to prove the theorem we first need the following lemma.

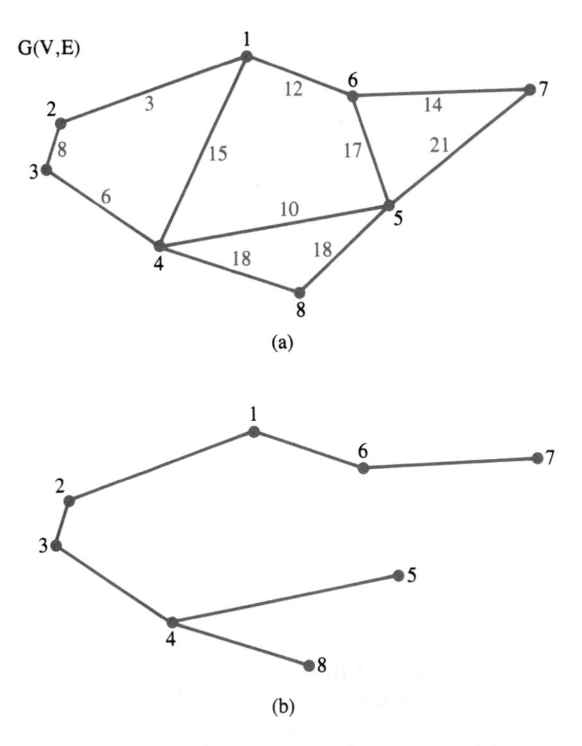

FIGURE 3.31

An Example of Algorithm MinSpanTree. (a) The given graph (b) The spanning tree that results after vertices and edges are added as follows: 1, 2 $\{1,2\}$; 3 $\{2,3\}$; 4 $\{3,4\}$; 5 $\{4,5\}$; 6 $\{1,6\}$; 7 $\{6,7\}$; 8 $\{4,8\}$. Note that at the last stage we chose $\{4,8\}$ arbitrarily instead of $\{5,8\}$.

Lemma 3. If the vertices of $G(V, E)$ are divided into any two disjoint nonempty sets, V_1 and V_2, then any minimum spanning tree of G contains the shortest edge connecting a vertex in V_1 to a vertex in V_2.

Figure 3.32 illustrates the meaning of this lemma. With V_1 and V_2 as shown, the shortest edge from a vertex in V_1 to a vertex in V_2 is D–F. If this edge is appended to the spanning tree in Fig. 3.32(b), then there would be a cycle A–D–F–C. But, if A–C, which also joins a vertex in V_1 to a vertex in V_2, is deleted, we again have a spanning tree, but a shorter one (length 29 versus 33). Is the new tree a minimum spanning tree?

PROOF OF LEMMA 3 Suppose the lemma were not true. Then there would be some minimum spanning tree T and two sets V_1 and V_2 such that the shortest

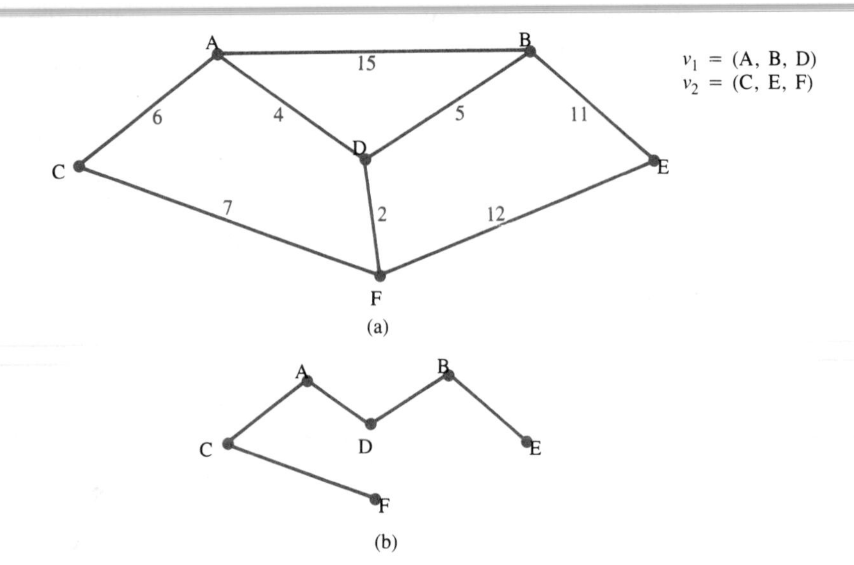

FIGURE 3.32

The Shortest Edge Between Disjoint Sets of Vertices: (a) Graph G; (b) a spanning tree of G.

edge—call it s—from a vertex in V_1 to a vertex in V_2 would not be T. Now add s to T to get T', which must create a cycle (why?). This cycle must have at least one other edge s' connecting a vertex in V_1 to a vertex in V_2 (why?). If we delete s' from T, we again have a spanning tree. But since s is shorter than s' (we assumed that s is the shortest edge joining the two sets), the new spanning tree is shorter (i.e., its edges have less total length) than the purported minimum spanning tree T. A contradiction! Therefore the assumption that led to the contradiction, namely, that s is not on T, must have been wrong—which proves the lemma. ∎

Now the theorem is easy to prove.

PROOF OF THEOREM 2 By the lemma, the shortest edge in G must be on any minimum spanning tree. (Just suppose that we divide the vertices of G into any two sets such that one end of the shortest edge is in one of the sets and the other end is in the other set.) Then at every stage of the algorithm, the edge s to be added by the algorithm is the shortest edge between a node on the tree and a node not yet on the tree. Again, we apply the lemma with one end of s in V_1, the set of all vertices already on the tree, and the other end in V_2, the set of all vertices not yet on the tree. This shows that s must be on any minimum spanning tree. When the algorithm ends, we have a spanning tree T for which, by Lemma 3, each edge is on any minimum spanning tree. But one spanning tree cannot be properly contained in another (why?). So T is a minimum spanning tree. (In fact, assuming that all edge weights are distinct, we have shown that T is the *unique* minimum spanning tree.) ∎

Can you spot where there would be a flaw in our proof if we had not made the assumption about distinct edge weights [10]?

Now that you have had more experience with the algorithmic notation than when we presented Example 4, Section 1.2, you should be able to write MINSPANTREE without resort to English [7]. MINSPANTREE is often called **Prim's algorithm**. A related algorithm, called **Kruskal's algorithm**, is considered in [8–9].

Binary Trees

A binary tree is a tree in which:

1. Each node n has at most two offspring. (The **offspring**, or **children**, of a node n are the nodes at the ends of branches leading from n.)
2. Any offspring of n must be designated as a left or a right offspring. Thus the following are distinct binary trees because one is a root with a single left branch and the other is a root with a single right branch.

Some further examples of binary trees are shown in Fig. 3.33.

Recall Example 3, Section 3.1, in which we showed how a list of words can be sorted using a tree. That tree was, in fact, a binary tree. Many sorting algorithms, as well as many other algorithms and problems related to trees, are naturally expressed

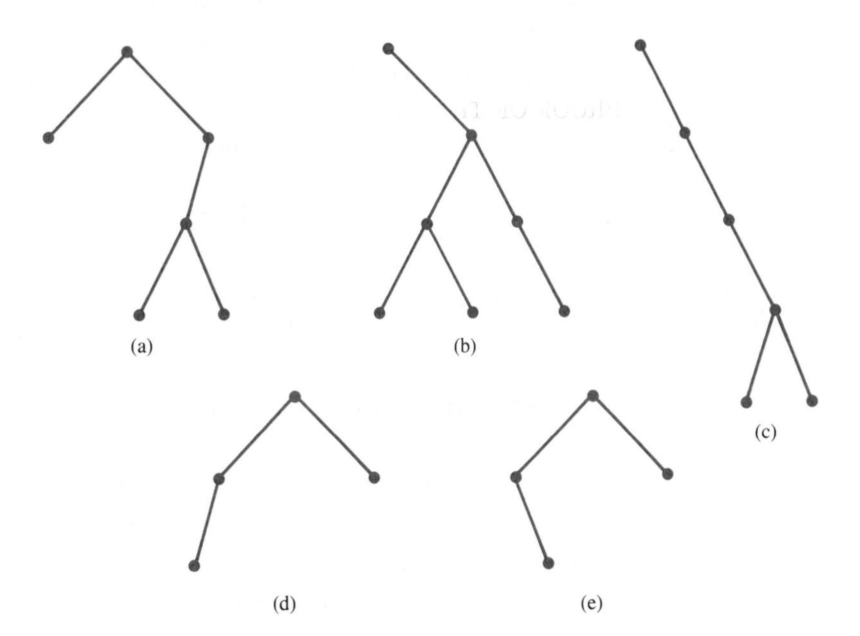

FIGURE 3.33

Some Examples of Binary Trees. (Note that (d) and (e) are isomorphic as *trees* but not as *binary* trees.)

in terms of binary trees. Another advantage of binary trees is that, because they are restricted to a maximum of two offspring per node, there are efficient computer representations of binary trees. In this section we consider only the problem of *traversing* binary trees. This is an important problem in applications of binary trees, and it also illustrates another application of depth first search.

In Example 3, Section 3.1, we gave a rule for alphabetizing the words stored in the binary tree of Fig. 3.3 (which is reproduced in Fig. 3.34) by just listing them "left to right". Now we'll develop a more formal and algorithmically practical method using binary tree traversal.

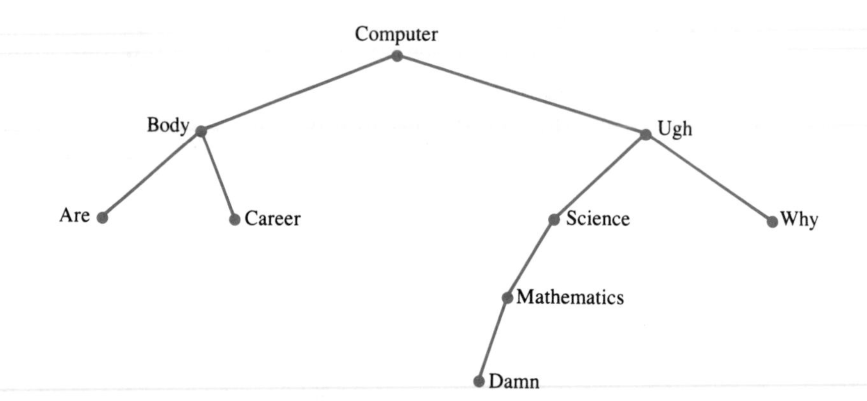

FIGURE 3.34

A Tree of Words to Be Sorted.

The general problem is this: We have a binary tree in which certain information is stored in (associated with) each node. This information might be words, as in Fig. 3.34, or much larger *records*, such as an employee's personnel record (name, address, date of birth, social security number, salary, etc.). We want to *process* the data associated with each node. Such processing might, for example, consist of

- alphabetizing the data (e.g., by surname) or numerically ordering it (e.g., by social security number);
- finding all data with a common characteristic (e.g., salary greater than $20,000); or
- printing mailing labels in order to send some document to all employees.

For any of these applications we need to process the data in each node. We naturally want some *systematic* way to do this which ensures that the data at each node is processed once and only once. For data stored in a binary tree there are three basic paradigms for doing this:

Preorder traversal

Process the root.

Traverse the left subtree (by Preorder).

Traverse the right subtree (by Preorder).

Inorder traversal

Traverse the left subtree (by Inorder).

Process the root.

Traverse the right subtree (by Inorder).

Postorder traversal

Traverse the left subtree (by Postorder).

Traverse the right subtree (by Postorder).

Process the root.

By this time you should have seen enough recursion and inductive definitions to be comfortable with descriptions like these, which define things in terms of themselves. There are, of course, three other traversal methods which we might define by interchanging left and right, but there is no good reason for doing so—except perhaps if your native language is Arabic or Hebrew.

Be sure you understand the meaning of "process the root" in each method of traversal. In order to traverse the tree we will have to look at some nodes more than once, in the sense of determining whether a node has a left or right child. But we'll only visit (i.e., process) the data associated with each node once.

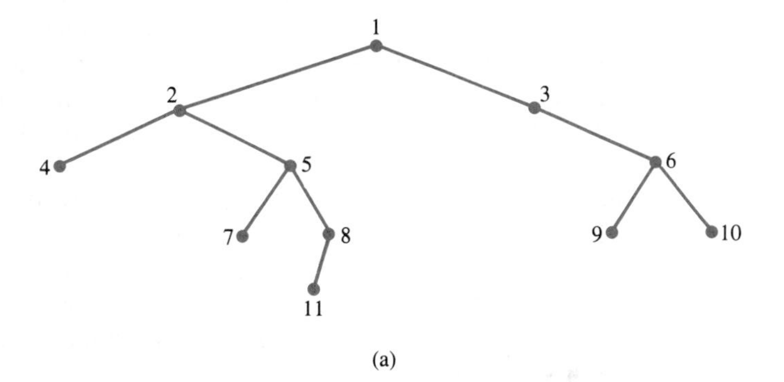

Preorder	Inorder	Postorder
1	4	4
2	2	7
4	7	11
5	5	8
7	11	5
8	8	2
11	1	9
3	3	10
6	9	6
9	6	3
10	10	1

(b)

FIGURE 3.35

Examples of Binary Tree Traversal: (a) diagram and (b) order of node processing.

An example of each of these methods of traversal is shown in Fig. 3.35, where for convenience, we have only numbered the nodes and haven't shown any data associated with them. Let's consider the Inorder traversal in detail to make sure that the definitions are really clear:

1. We first traverse the left subtree, that is, the subtree whose root is node 2.
2. Applying Inorder traversal to this subtree, we first traverse its left subtree, which is the single leaf node (i.e., one with no branches emanating from it) 4.
3. As node 4 is a leaf, its left subtree is empty, and so we "process the root", namely, node 4 itself.
4. As node 4 has no right subtree, this completes the traversal of the left subtree of node 2 and, according to Inorder traversal, we then process the root of this subtree, namely, node 2.
5. Now we traverse the right subtree of the subtree with root 2, which consists of the subtree with root node 5. Etc.

Which traversal method should we use to alphabetize the tree in Fig. 3.34? That is, which traversal method will process the nodes in alphabetical order? Inspection of Fig. 3.34 reveals that Inorder traversal has the desired property, which we can easily prove by strong induction on the number of nodes n of the tree. The basis case ($n = 1$) is immediate because there is only the item at the root. Now suppose (the induction hypothesis) that Inorder traversal processes in alphabetical order any tree with n or fewer nodes constructed as in Example 3, Section 3.1. Consider a tree with $n+1$ nodes. Inorder traversal first processes the left subtree. This subtree has n or fewer nodes so, by the induction hypothesis, it processes the left subtree in alphabetical order. Then it processes the root which is next in alphabetical order by the construction of the tree. Finally, Inorder traversal processes the right subtree, all of whose nodes follow the root in alphabetical order. By the induction hypothesis the right subtree is also processed in alphabetical order. Therefore the whole tree is processed in alphabetical order which completes the proof.

Because of their simple recursive form, we can readily express each of the three traversal methods in algorithmic form. Rather than present three separate algorithms, Algorithm 3.13 BinaryTreeTraversal is a three-in-one algorithm for all three traversals.

In actual algorithms for a particular form of tree traversal, only one of the three process(u) commands would appear. By comparing this algorithm with Algorithm DepthFirstSearch in Section 3.6, you can see that BinaryTreeTraversal is just a variation on DFS. In DFS we deal with a general graph. Because binary tree traversal deals with a special kind of digraph—namely, a binary tree—we are able to replace the **repeat** ... **endrepeat** loop in the DFS algorithm with the **if** statements in BinaryTreeTraversal. From the perspective of depth first search, the only difference between the three traversals is the place in the algorithm at which the processing of the data at the vertex is performed.

The algorithm given in Example 3, Section 3.1, for building the binary tree together with BinaryTreeTraversal used for inorder traversal, defines a method of **sorting** (e.g., alphabetizing) any set of data which has associated with it a *key* (e.g., social security number, surname). Methods of sorting have led to a large literature in computer science. **Tree sorting**, as presented here, is the basis of some efficient and useful methods of sorting. We'll discuss other methods of sorting in Section 7.7 and the Epilogue.

Algorithm 3.13

BinaryTree-Traversal

```
Input A binary tree
Algorithm BINARYTREETRAVERSAL
    procedure traverse(u)
        process(u)                          [For preorder traversal]
        if u has a left child lc then
            traverse(lc)
        process(u)                          [For inorder traversal]
        if u has a right child rc then
            traverse(rc)
        process(u)                          [For postorder traversal]
        return
    endpro
    v ← root of tree
    traverse(v)
Output The results of the processing of the data at each vertex
```

Problems: Section 3.8

1. a) Restate the algorithms for BFS and DFS so that the output of each is a spanning tree for the input graph.

   b) For graphs (i) and (ii), apply the algorithm for BFS from a) to find a spanning tree, starting at v_1. Where there is a choice of which vertex to visit next, visit the one with the lowest index. Show your work by drawing a picture of the graph with the final tree darkened and the tree edges labeled by their order of inclusion.

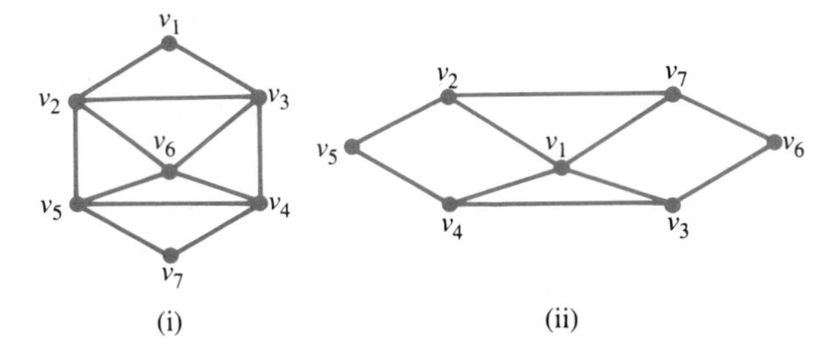

   c) Repeat b) for the algorithm for DFS found in a).

2. a) Redraw the graphs in [1b] with the trees dangling down from the root, each edge in the tree having the same length, and nontree edges fitting in where they must.

   b) Repeat for [1c].

3. a) Let G be a connected graph and let T be any spanning tree obtained by BFS applied to G. Let a "cross-edge" be any edge of G–T. Suppose that T has been drawn with the root v_0 at the top and all other vertices hanging down by levels (i.e., vertices two edges away from v_0 in T are drawn two units lower, and so on). If the cross-edges from G are now drawn in, argue that each cross-edge is either horizontal or goes between adjacent levels.

   b) Can you make the same argument for DFS applied to G.

4. Restate Algorithm SHORTESTPATH so that it does not terminate until all nodes have been added to the set U and so that it outputs a spanning tree for the graph.

5. For graphs which are not connected, prove that Algorithm SPANTREE results in a tree which spans that component of G in which the initial vertex chosen lies.

6. Perform Algorithm MINSPANTREE for the following graph G. Display your work by darkening the tree edges; also, to indicate the order in which you add edges to the tree, label those edges 1, 2, etc., in circles (to avoid confusion with the weights, which are uncircled).

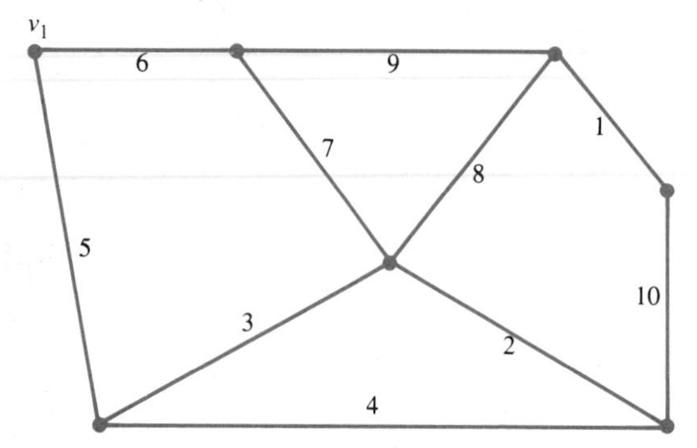

7. Rewrite Algorithm MINSPANTREE so that all the English language statements are replaced by statements in our algorithmic language.

8. For the graph of [6], apply the following variant of Algorithm MINSPANTREE: At each stage add the edge of minimum cost *anywhere* which does not form a cycle with any edges you have already chosen. (Thus the set of edges chosen at any point may be a disconnected acyclic set—a **forest**.) Label the edges with circled numbers in the order you add them.

9. Call the algorithm obtained in [8] MINSPANTREE$'$. (It is usually called Kruskal's algorithm.) Prove that MINSPANTREE$'$, like Algorithm MINSPANTREE, finds the MST if all edges have different weights. (Both algorithms are correct, even when some edge weights are the same, but the proofs are harder.) [*Hint*: Use Lemma 3, which we used to prove MINSPANTREE correct. When you add edge e using MINSPANTREE$'$, let v be either end node of e and let V_1 consist of v and all nodes connected to v by edges that you have already chosen (there may be none).]

10. Suppose that we don't require edge weights to be distinct. Why does the proof in the text that Algorithm MINSPANTREE works now fail? Why does your proof in [9] also fail?

11. There is a maximum weight spanning tree version of Prim's algorithm.
    a) State it and apply it to the graph in [6].
    b) Prove that it is correct when all edges have distinct weights.

12. There is also a maximum weight spanning tree version of Kruskal's algorithm (see [8–9]).
    a) State it and apply it to the graph in [6].
    b) Prove that it is correct when all edges have distinct weights.

13. Assertion: Algorithm MINSPANTREE is correct. What's wrong with the following proof of this assertion?

    "Proof" Induction on the number n of nodes. In the basis case, $n = 2$, there is only one edge, so the algorithm is clearly correct. Inductive step. Assume that the algorithm is correct when there are n nodes. We are given a graph G with $n+1$ nodes. Let v be the node which MINSPANTREE connects last. Suppose that we had run this algorithm on the subgraph G' connecting the other n nodes. Then the final tree from the run on G' would be the same as the tree obtained after n steps on G. Why? Because the subset of edges from which MINSPANTREE picks a minimum at each stage in G' is a subset from the set from which it picks in G. However, the subset contains the element which is picked in G, so the same edge is picked. Therefore we conclude that after n nodes are joined to the tree, when the algorithm works on G, it has picked the minimum tree connecting those nodes. Now MINSPANTREE goes on to connect the final node v. It picks the minimum length edge to v. As it has picked the minimum set of edges apart from v, and the minimum edge to v, it has picked the minimum tree connecting all $n+1$ nodes. ∎

    (*Note*: We don't know of a correct proof of the validity of MINSPANTREE using induction on the number of vertices.)

14. Although greed pays in the MST problem, in other problems it can backfire miserably. Consider the assignment problem: n people are to be assigned to n jobs. Each person is given a rating on each job, where higher is better. Assume that the goal is to maximize the sum of the ratings of the assignment (See [5, Supplementary, Chapter 1]).

a) Describe a greedy approach to this problem.

b) Show how your greedy approach would work for the situation depicted in the following graph.

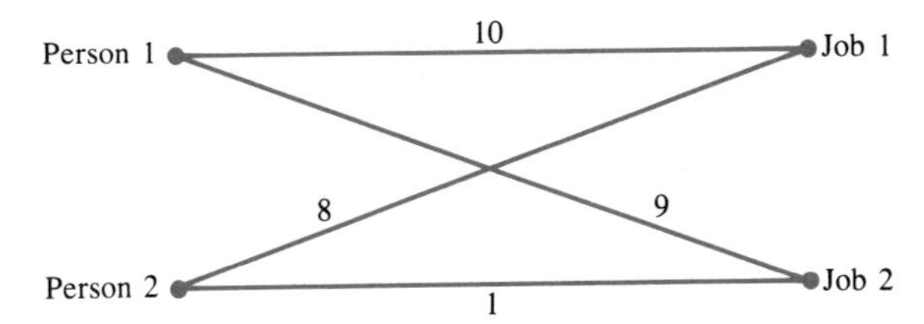

How good is your answer compared to the optimal solution?

15. Use Algorithm SPANTREE to explain why, if a tree has n nodes, it has exactly $n-1$ edges.

16. Show that any two of the following conditions imply the third and thus that any two may be used to define a free tree.

i) G is connected.

ii) G is acyclic.

iii) In G, $|E| = |V|-1$.

17. In chemistry, an **alkane** is a saturated acyclic hydrocarbon molecule. In graph theory terminology, this means that an alkane is a free tree with two types of nodes: C(arbon) nodes of degree 4 and H(ydrogen) nodes of degree 1.

a) Draw an alkane with three carbons. Draw three different alkanes with five carbons.

b) Chemistry books state without proof that every alkane is of the form C_nH_{2n+2}; that is, if there are n carbon atoms, there are $2n+2$ hydrogen atoms. Prove this assertion, using the result of [16] and Eq. (3), Section 3.2. Start by supposing that the alkane is C_nH_m and show that $m = 2n+2$.

18. Define a "hydronitron" to be an acyclic molecule made up of hydrogen atoms (one bond) and nitrogen atoms (three bonds). (Actually, very few such molecules exist naturally, but suppose they did.) Determine all mathematically possible formulas N_nH_m for hydronitrons.

19. Consider a family tree, that is, a tree diagram of some person and all his or her descendants. Describe in terms of this tree what it means for two people to be

a) siblings. b) cousins. c) second cousins.

d) second cousins once removed (if you don't know what this means, look it up in a dictionary or encyclopedia).

e) first cousins twice removed.

f) Based on these relations, write a general definition of "ith cousin jth removed".

20. Draw all distinct binary trees with

a) three nodes. b) four nodes.

21. Find and explain a recursion for B_n, the number of binary trees with n nodes. Use it to compute B_3, B_4, and B_5. (Compare your results with those of [20].)

22. A **ternary tree** is a tree in which each node has at most three offspring, which are the roots of left and/or center and/or right subtrees. Display all ternary trees with at most three vertices.

23. Derive a recursion, similar to that in [21], for the number of ternary trees with n nodes.

24. List the vertices of the binary tree shown in

a) Inorder. b) Preorder. c) Postorder.

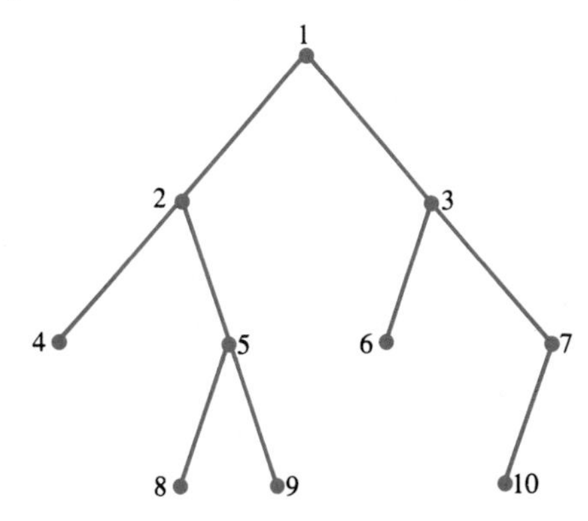

25. Suppose that you're given a list of words like that in Fig. 3.34 and want to build a binary tree as in Example 3, Section 3.1. Describe the trees for which the nodes will be visited in alphabetical order using

a) Preorder traversal. b) Postorder traversal.

26. Can you uniquely reconstruct a binary tree if you are told

a) the Preorder and Inorder traversals? Explain.

b) the Preorder and Postorder traversals? Explain.

c) the Inorder and Postorder traversals? Explain.

Problems 27–32 are about game trees. A **game tree** is a tree in which the root represents the position at the start of the game, the nodes at the first level down represent the positions after all possible first moves by player 1, the nodes at the next level down represent all possible responses by player 2 to each of player 1's moves, etc. The leaves of the tree represent the final positions of the game and are labeled W, D, or L, depending on whether the final position is a win, draw, or loss for player 1.

27. Consider the following *minimax* algorithm for labeling the nonleaf nodes of a game tree.

i) Suppose that all the children of a node have labels W, D, or L. If the node represents a position where it is player 1's move, label the node with the *best* result among the labels of its children's nodes. If the node represents a position where it is player 2's move, label the node with the *worst* result among the labels of its children's nodes. Thus in the tree below, the node A would be labeled D (since it is player 2's move) and node B would be labeled W (since it is player 1's move).

ii) Working up from the bottom, label all the nodes. The label of the root is called the *value* of the game.

a) Explain why step i) gives reasonable values to each of the nonleaf nodes and why, therefore, we call the root the value of the game.

b) Find the label of the root of the following tree.

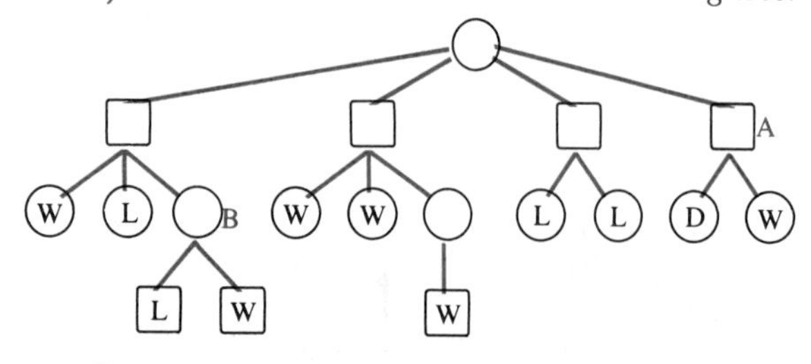

28. Consider 2×2 tic-tac-toe—not much of a game, but it serves a purpose. Draw the complete game tree. (For instance, the root has degree 4, with one edge for each corner in which player 1 can put an X.) Analyze the game; that is, for each node determine whether it should be labeled W or L.

29. There is a lot of symmetry in tic-tac-toe. For instance, in [28], each of the four possible moves is into a corner and thus they are all essentially the same. Using all such symmetries, draw and analyze the "reduced" game tree for 2×2 tic-tac-toe.

30. Using the symmetry described in [29], draw the reduced game tree for ordinary 3×3 tic-tac-toe, down through the point where each player has moved once. (The complete game tree is too much to ask for.)

31. In the preceding problems, the value assigned to a node (W, D, or L) is the value to player 1. There is another approach: Assign to each node the value of the game to the player who moves from that node should the game reach that point and play proceeds optimally. (This approach makes the analysis of some games easier as the next problem will illustrate.) Modify the minimax algorithm in [27] for evaluating a game so that it works under this alternative approach.

32. In the game of daisy, two players alternate picking petals from a daisy. Each time, each player gets to pick 1 or 2 petals. The person who picks the last petal wins. If the daisy has n petals, we say the game is n-petal daisy.

a) Draw and analyze the complete game tree for four-petal daisy.

b) Draw and analyze the complete game tree for five-petal daisy, working up from the bottom, as usual.

c) Determine the winner of five-petal daisy more quickly than in b) by making use of a) and labeling the nodes as described in [31]. That is, by looking at the four-petal tree, or just part of that tree, you can determine immediately who wins from the children nodes of the root of five-petal daisy. Explain, using the word "recursion".

Supplementary Problems: Chapter 3

1. a) Suppose that, after Algorithm BUILDTREE terminates (for whatever size list), a new word comes along. It's easy to add it to the tree. How?

b) Now, suppose that, after BUILDTREE terminates, you want to delete some word on the list and adjust the tree accordingly. (For example, the list might be names of students and one student may have dropped out.) You can't simply delete the corresponding vertex; the remaining vertices must be reconnected so that the tree looks just as it would if the deleted word had not been present when BUILDTREE built the tree. Of course, you could just run the algorithm again on the shorter list but there's a better way. Devise an appropriate algorithm to take the tree and the word to be deleted as input and produce the desired tree as output.

2. Prove that a graph without loops contains an elementary cycle of even length if the degree of each vertex is at least 3. (*Hint*: Show that, if there is a cycle of odd length, it must be possible to construct another cycle which intersects it to get a cycle of even length.)

3. Let G be a simple, non-complete graph such that every vertex has degree 3.

a) Explain why G has at least six vertices.

b) There are two nonisomorphic graphs of this sort having exactly six vertices. Find them. Can you explain why there are no others?

4. A **matching** in a graph G is a subgraph in which each vertex has degree 0 or 1. A **maximum matching** in G is a matching with as many edges as possible.

a) Find all nonisomorphic maximum matchings in the complete graph having four vertices.

b) Find all nonisomorphic matchings (maximum or not) in the complete graph having five vertices.

5. Given $G(V, E)$ and $U \subset V$, the **induced** subgraph of G on U, denoted by $\langle U \rangle$, is defined to have vertex set U and edge set $\{\{u,v\} \text{ in } E \mid u,v \in U\}$. That is, $\langle U \rangle$ consists of all edges of G between vertices of U. Now, let G be the graph

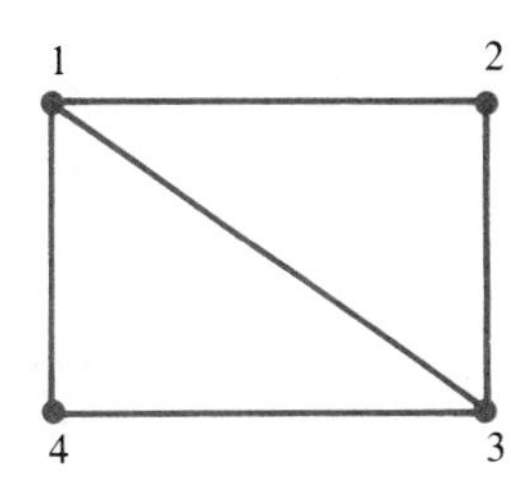

a) Draw $\langle\{1,2,3\}\rangle$ and $\langle\{1,3,4\}\rangle$.

b) Draw a subgraph of G which is *not* an induced subgraph. Explain why it isn't.

6. The idea of a component seems straightforward. It's obvious that every vertex is in *a* largest connected subgraph of the entire graph and that these subgraphs partition the whole graph. However, when looked at more generally, the idea of a component is not so straightforward.

Let's define a *P-component* of a graph or digraph to be a maximal subgraph with property P. For instance, if P is the property of being connected, then a P-component is just an ordinary component. To say that subgraph H of graph G is maximal means that H is not a proper subgraph of any other subgraph J of G which also satisfies property P.

a) Let P be the property of being weakly connected. Consider the accompanying digraph D. Show that there are two distinct maximal, weakly connected subgraphs containing vertex 2. Show that there is no way to partition the graph into maximal, weakly connected subgraphs.

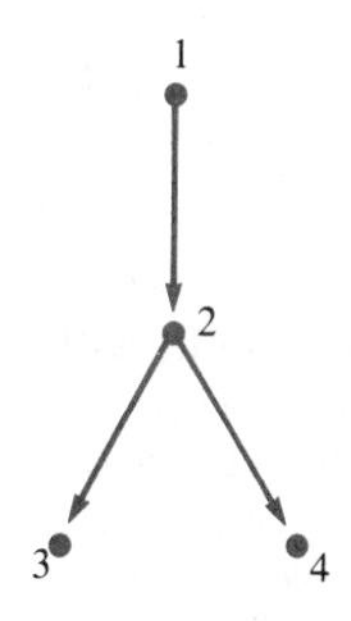

b) Now let P be the property of being strongly connected. For each vertex, find all the strongly connected components it belongs to in the accompanying digraph D.

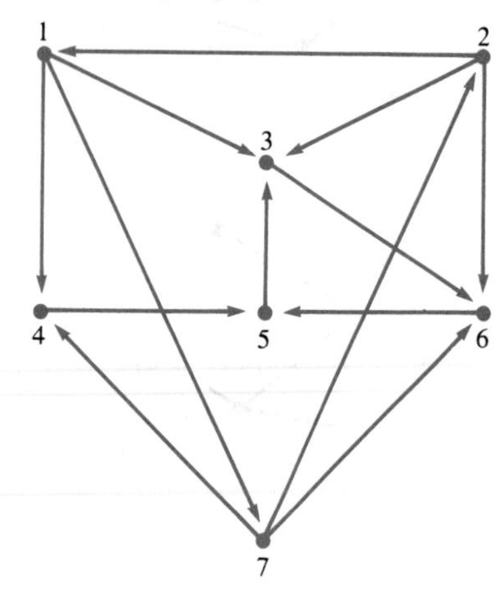

c) Show that if J and K are strongly connected subgraphs of some digraph, and $V(J)\cap V(K) \neq \emptyset$, then $J \cup K$ is also strongly connected.

d) Show that c) implies that each vertex of a digraph is in a unique maximal, strongly connected component and that these components partition the vertex set. (Note that edges need not be in any strongly connected component.)

7. Let B be a Boolean matrix (all entries 0 or 1), and let $B^{(k)}$ be the **Boolean product** (ordinary matrix product with all nonzero terms replaced by 1) of B with itself k times. As special cases, let $B^{(0)} = I$ and $B^{(1)} = B$. Now let A'' be the Boolean adjacency matrix of a graph: the (i,j) entry is 0 if there are no edges between i and j and 1 otherwise.

a) Prove that $(A''+I)^{(k)} = \sum_{i=0}^{k}(A'')^{(i)}$, where the summation is a Boolean sum—0 if the ordinary sum is 0 and 1 otherwise.

b) Using a) or otherwise, prove that the (i,j) entry of $(A''+I)^{(k)}$ is 1 if and only if there is a path of length at most k between i and j, including null paths.

8. Let D be a digraph with $V = \{v_1, v_2, \ldots, v_n\}$ and in which all edges point from lower to higher numbered vertices (as in the Prologue). There are no directed cycles in such a graph, so the total number of directed paths between any pair of vertices is finite. Therefore it makes sense to try to compute $N = [n_{ij}]$ where n_{ij} is the *total* number of directed paths from i to j. As usual, A is the adjacency matrix.

a) Prove that $N = (N+I)A$.

b) Part a) would seem to be of no help in obtaining N, because to obtain N on the left you already need to know it on the right. However, A is an **upper triangular matrix**, with all entries strictly above the main diagonal. Convince yourself that in order to determine the kth column of the product $(N+I)A$, you need only know the leftmost $k-1$ columns of N (and the kth column of A). Use this observation to develop an iterative algorithm to compute N.

c) Suppose that all you want to know is the number of directed paths from v_1 to each other vertex. There is a much simpler algorithm that can do this. Think recursively and figure it out. (In fact, this algorithm amounts to looking only at the first row of both sides of $N = (N+I)A$ and then using the ideas from b). It's simpler to think about this problem without referring to matrices.) Where have you seen this problem—and this algorithm—before?

9. Consider the graph with vertices labeled 1 through 50 and with vertex i adjacent to vertex j if either

i) the sum of the digits of i is the same as the sum of the digits of j (e.g., vertices 8, 17, and 35 are pairwise adjacent because $1+7 = 3+5 = 8$); or

ii) i is the square of j or vice versa (e.g., vertex 16 is adjacent to vertex 4).

Answer the following questions using Algorithm WARSHALL run on a computer.

a) Is vertex 25 connected by a path to vertex 36?

b) How many connected components does the graph have?

c) How many vertices are in the smallest component?

10. a) Answer the same questions as in [9], using Algorithm SHORTESTPATH. Show your work by drawing the shortest paths which the algorithm produces.

b) Do the same thing, using Algorithm BREADTHFIRSTSEARCH.

c) Do the same thing once more, using Algorithm DEPTHFIRSTSEARCH.

11. The surprising thing about Algorithm WARSHALL is what it does induction on: not the length of the path, nor the indices of the end vertices, but rather the indices of the intermediate vertices. Is such an unnatural choice necessary? To answer this question, determine whether the following seemingly more natural choices of inductive assumptions work. That is, find out whether determining the next case is easy if the previous cases are known, so that an efficient algorithm can be constructed.

a) An $n \times n$ matrix $M^{(k)}$ is known where the (i, j) entry is the length of the shortest path from v_i to v_j with at most k edges.

b) A $k \times k$ matrix $N^{(k)}$ is known where the (i, j) entry gives the length of the shortest path from v_i to v_j in the induced subgraph on $v_1, v_2, \ldots, v_k$. (See [5] for the definition of *induced*.)

12. Suppose that digraph D has some negative edge weights, but that no directed cycle has negative weight.

a) Explain why there is always a shortest path between any two points and why that path is simple.

b) Determine whether the shortest-path variant of WARSHALL works for such a path.

13. Here is a modification of the path-growing procedure which forms the first part of Algorithm ECYCLE. If $G(V, E)$ is the original graph, at each stage let G' be the graph with the original vertex set V and all so-far-unused edges. Instead of picking *any* unused edge from your current location, choose an edge whose removal does not increase the number of components of G' (where isolated vertices do count as components). If there are no such edges, pick any edge. The path-growing procedure with this modification is called **Fleury's algorithm**.

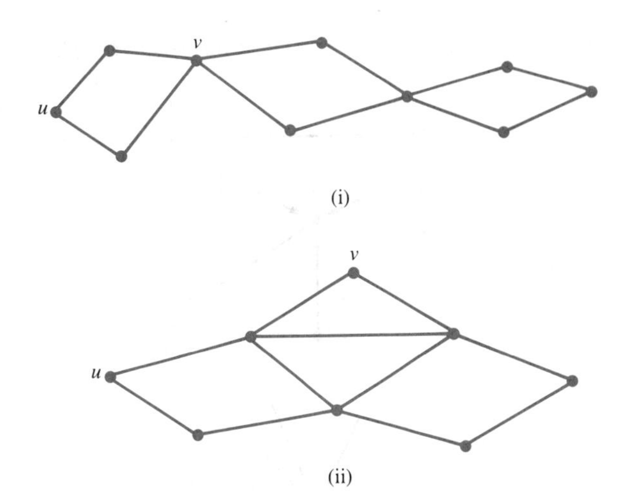

a) Apply Fleury's algorithm to graphs (i) and (ii) starting at u.

b) Apply it to the same graphs, starting at v.

c) There is a theorem that says: If G has an Eulerian cycle, then Fleury's algorithm finds it. That is, with Fleury, recursion is never necessary; you won't get stuck until every edge has been traversed. Prove this theorem. (*Hint*: Show that if G has an Eulerian cycle, then the only occasion in which there is not an edge from the current vertex v whose removal does not increase the number of components is when the degree of v in G' is 1. This means that, when there are no unused edges incident on the current vertex, there are no unused edges anywhere. Why? Why is this important?)

14. Prove that a tournament has a Hamiltonian cycle if and only if it is strongly connected.

15. Prove that a strongly connected tournament on n vertices has directed cycles of length k for every integer k from 3 to n. (*Hint*: Do induction on k.)

16. In the scheduling problem of the Prologue, the algorithm required the vertices of the directed graph D to be numbered so that all edges are directed from lower indices to higher. We did not explain there how such a numbering is supposed to be found, but we asserted that such a numbering exists if D has no directed cycles. It's possible to prove this assertion with a path-growing algorithm.

a) Explain why the nonexistence of directed cycles is necessary for the numbering to exist.

To prove sufficiency, assume that D has no such cycles and show how to construct a numbering:

b) Show that the only candidates for vertices to be numbered 1 are vertices with indegree $=0$.

c) Give a proof by algorithm that at least one such vertex exists.

d) Suppose that a vertex with indegree $=0$ has already been designated as v. Think iteratively and thus develop an algorithm to number all the vertices. Any such algorithm is called a **topological sort**.

17. Devise another topological sort (see [16]) which uses as input the adjacency matrix of D and never grows any paths.

18. [Computer Project] In Algorithm ECYCLE, splicing cycles into other cycles is more complicated than it appears—at least if a computer has to do it. Write a computer Program which actually runs ECYCLE. First, think hard about the data structure you want to use to represent paths.

19. Write a backtracking program, perhaps recursively, which determines conclusively whether a given (di)graph G has a Hamiltonian cycle. (The program will be terribly slow, even for moderate-sized graphs because, in effect, it tries every partial path emanating from your choice of start vertex.)

20. Algorithm SHORTESTPATH still runs even if the graph has some negative weights. Does it succeed in finding the shortest simple paths? If it does, explain why. If it doesn't, give an example and also give a reason why you might have expected it not to work even before you found a counterexample. (One way to do this would be to show that the recursive thinking which led to SHORTESTPATH breaks down—knowing the previous case does not make the current case easy to compute.)

21. Suppose that graph G is connected and has some negative weight edges, but no cycle has nonpositive weight.

a) Explain why there always is a shortest path between any two points, and why that path is simple.

b) Determine whether Algorithm SHORTESTPATH works for such a graph.

22. Devise an algorithm to find the *second* shortest simple path between two vertices in a graph.

23. The following algorithm is another way to find the shortest distances in a graph from some initial vertex v_0 to all others. It is called **Ford's Algorithm**.

a) Carry out Ford's algorithm for graph (i) at the right.

b) Carry out Ford's algorithm for the directed graph (ii). Edges without arrows can be traversed either way.

c) Compare the behavior of Ford's and Dijkstra's algorithms when all edges have positive weights.

d) Prove what you observed in c).

e) Show that Ford's algorithm is valid so long as there are no undirected edges with negative weights and no directed cycles with total

Algorithm

Input $G(V, E)$, v_0
$w(E)$ [A weight function on the edges]

Algorithm FORD
$U \leftarrow v_0$; $d(v_0) \leftarrow 0$
for $v \in V - \{v_0\}$
$d(v) \leftarrow \infty$
endfor
repeat until $U = \emptyset$
$v \leftarrow$ vertex in U with $d(v)$ minimum
$W \leftarrow$ all vertices adjacent to v
repeat while $W \neq \emptyset$
$u \leftarrow$ any vertex in W
if $d(v) + w(\{v, u\}) < d(u)$ **then**
$d(u) \leftarrow d(v) + w(\{v, u\})$
$U \leftarrow U \cup \{u\}$
endif
$W \leftarrow W - \{u\}$
endrepeat
$U \leftarrow U - v$
endrepeat

Output $d(v)$ for all $v \in V$ [Minimum distances from v_0]

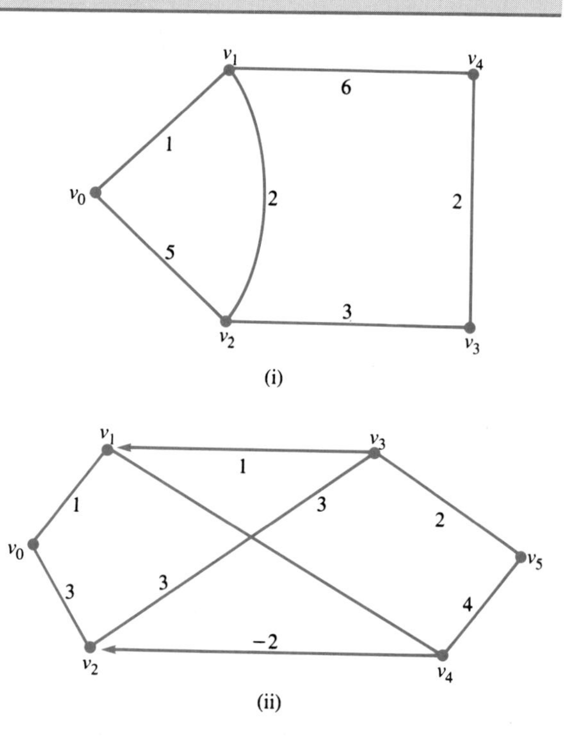

weight negative. (*Note*: An undirected negative edge amounts to a negative directed 2-cycle—go one way on the edge and then back. Thus the first condition is just a special case of the second.)

f) What does Ford's algorithm do if there is a negative weight directed cycle (or a negative undirected edge)?

g) Modify Ford's algorithm so that it outputs the shortest path from v_0 as well as the distances.

24. A century ago, Tarry proposed the following method for exploring a labyrinth and getting out alive. (The labyrinth consists of chambers (vertices) and passageways (edges). You are standing at the entrance and have a piece of chalk.) The method is:

- Step 1: If there is any unmarked passageway leading from the chamber (or entrance) where you are now, mark an x at its start and go to the other end, where you make another x. Then,

  i) if no other passageway into the chamber you just entered is marked, also put an E next to your x, but

  ii) if there are marks on other passageways into that chamber, immediately go back to the other end of the passageway you just used.

  and repeat this step.

- Step 2: If there are no unmarked passageways leading from where you are, then

  iii) if some passageway is marked with an E, move along that passage to its other end and go back to Step 1, but

  iv) if no passageway is marked with an E, stop.

a) Do a Tarry search of the following labyrinth starting at v. To help the grader, make a list of the edges (with direction traveled) in the order you traverse them. For instance, if you go from u to v, write (u, v). If later you go from v to u, write (v, u) at that point in your list.

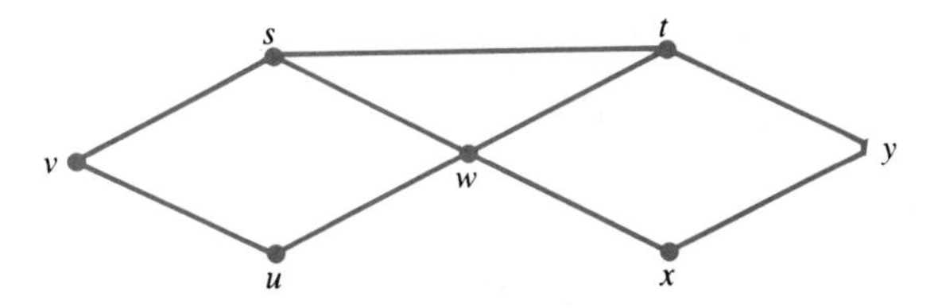

b) Tarry didn't call his method an algorithm because the word didn't exist then. But it is an algorithm. Write it up in our algorithmic language, preferably recursively.

c) It is a fact that, if the labyrinth is connected, Tarry's algorithm will cause you to explore every edge—in fact twice—and finish at the entrance. Prove this assertion.

d) How is Tarry's algorithm related to DFS?

25. As in [24], let a labyrinth consist of chambers (vertices) and passageways (edges). Assume that you are standing at the vertex labeled "Entrance" and you wish to find a path—any path—to the vertex labeled "Exit".

a) Describe what it means to reach a dead end in this labyrinth, that is, a place from which you must retrace your steps.

b) Devise a backtracking algorithm to find a way through the labyrinth which always takes the rightmost edge when leaving a vertex and suitably backtracks whenever a dead end is reached.

26. What is the least number m of vertices of degree 3 or more which a graph G must have before $\chi(G) = 4$? You need to exhibit a G with $\chi(G) = 4$ and with the number m of vertices of degree 3 or more, and you have to explain why any graph H with fewer vertices of degree ≥ 3 has $\chi(H) \leq 3$.

27. Generalize your answer in [26] to k-colorable graphs, and give a proof by algorithm.

28. [Computer project] Rewrite the algorithms of [20, Section 3.7], using arrays so that they can run in some actual computer language.

29. Prove: For any graph G, there is some way to number the vertices so that Algorithm ONE-COLOR-AT-A-TIME [20, Section 3.7] colors G with the minimum possible number of colors.

30. Consider the squares of a checkerboard as countries on a map. Then the usual coloring of a checkerboard proves that two colors suffice to color the graph derived from this map. Suppose that you have a small checkerboard (or large dominoes), so that a domino exactly covers two adjacent squares. Use the standard coloring of the board to prove that it is impossible to cover all but two opposing corner squares of a checkerboard using dominoes. (This problem shows that coloring is sometimes useful in contexts which don't seem to have anything to do with coloring.)

31. A graph is called **color-critical** if the removal of any vertex reduces the chromatic number.

a) Which cycles are color-critical?

b) Show that if $\chi(G) = n$ and G is color-critical, then every vertex of G has a degree at least $n-1$.

32. We call a graph **spherical** if we can draw it on the surface of a sphere without crossing edges. Explain why a graph is spherical if and only if it is planar. (*Hint*: To get from spherical to planar, puncture the sphere within one of the faces of the graph.)

33. A graph is **toroidal** if it can be drawn on a doughnut without crossing edges. (Mathematicians call the shape of a doughnut a **torus**.) Show by construction that the complete graph with five vertices is toroidal. (Note: The largest complete graph which is toroidal has seven vertices.)

34. When some edges have the same weight, a graph can have more than one minimum weight spanning tree.

a) Display all minimum weight spanning trees for the accompanying graph. (Find them by any method you please, including brute force.)

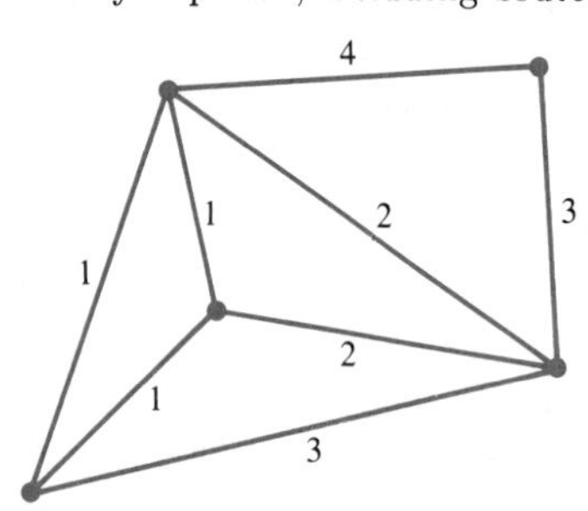

b) What do you observe about the weights on all the minimum weight spanning trees?

c) Prove your observation in b).

35. In cases of graphs with ties for one or more edge weights, Algorithm MINSPANTREE may be reinterpreted to say: When there is a tie for cheapest edge from the set of vertices connected so far to all other vertices, pick any one of the cheapest edges.

a) Prove: Every minimum weight spanning tree in such a graph can be obtained by some set of choices using this algorithm.

b) Prove: Every tree obtainable by making such choices is a minimum weight spanning tree.

36. Devise an algorithm for finding the *second* lowest weight spanning tree in a graph.

37. A **leaf** in a free tree is a vertex of degree 1. In [23, Section 3.7], we asked you to prove that an acyclic graph had at least one vertex of degree 1. That proof shows that any free tree also has a node of degree 1, but now we ask you to prove this result by induction. It's surprisingly tricky. (*Caution*: Don't do a faulty buildup proof. *Suggestion*: Strengthen the hypothesis.)

38. Given a spanning tree T in a graph G, and any edge $e \in G-T$, there is exactly one cycle in $T+e$. The set of cycles obtained in this way, as you consider each edge $e \in G-T$ in turn, is called the **fundamental system of cycles for** T.

a) Find the fundamental system of cycles in graph (i).

b) Find the fundamental system of cycles in graph (ii).

The edges in T are shown in color. Such fundamental systems form a basis in the sense of linear algebra and are important in algebraic graph theory.

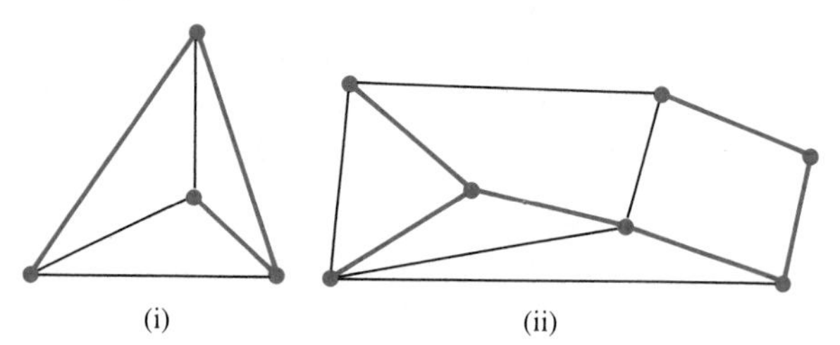

39. For a spanning tree T in a graph G, the deletion of any edge $e \in T$ divides the vertices of T into two disjoint sets. Consider the cut-set (see [14, Section 3.2]) for these two sets of vertices in $G-T+e$. The set of cut-sets obtained in this way as you consider each edge $e \in T$ in turn is called the **fundamental system of cut-sets for** T.

a) Find the fundamental system of cut-sets for the spanning tree in graph (i), [38].

b) Find the fundamental system of cut-sets for the spanning tree in graph (ii), [38].

40. Consider the expression

$$(4\times2) + [3 - ((4\times5)/(3+7))] \qquad \textbf{(1)}$$

We can represent expressions such as (1) by binary trees, since each stage of the calculation consists of doing a single arithmetic operation on two numbers. Because such expressions have an operator between the two numbers it operates on, the appropriate binary representation would have the given numbers on the leaves and the operations on the internal nodes. The left-to-right order of numbers and operators in (1) is exactly their order in an Inorder traversal of the tree. We now develop other ways to represent (1) by considering Preorder and Postorder traversals of the same tree.

a) Represent expression (1) with a binary tree.

b) Evaluate expression (1) using this tree.

c) Using the tree, write the numbers and operations in Postorder sequence. This representation is called **Reverse Polish Notation (RPN)**.

d) It is a fact that RPN expressions are unambiguous without any parentheses. (RPN is the system used by Hewlett-Packard calculators, which have no parenthesis buttons.) There is an unambiguous method for evaluating RPN expressions, as follows. Whenever an operator appears as you read from left to right, use that operator to combine the two immediately previous symbols in the current version of the sequence. (If the immediately previous two symbols are not both numbers, or there aren't two previous symbols, the algorithm aborts. But this can happen only if the Postorder order wasn't recorded properly.) As a result of the operation, the two numbers and the operator are now replaced by one symbol. Evaluate your expression from c) by this method, and check to see whether you get the same answer as in b).

e) Prove (induction!) the fact stated in d).

41. Consider again expression (1) in [40].

a) Write expression (1) according to Preorder. This representation is called **Polish Prefix Notation (PPN)**. (PPN like RPN is unambiguous and parenthesis-free.)

b) Devise a rule for computing expressions in PPN form, assuming that you are allowed to read from right to left. Apply your method to the expression obtained in a).

c) Assume that you must read from left to right. Think of another rule for computing PPN expressions. Apply your method to the expression obtained in a).

d) Prove that your method in b) works, using induction.

e) Prove that your method in c) works, using induction.

f) Why isn't your method in c) of much use on hand calculators?

The next five problems are about game trees and are a continuation of [27–32, Section 3.8].

42. The daisy game (for however many petals) is symmetric in that, wherever two nodes get label n (meaning n petals are left), the subtrees hanging from those nodes are the same. They are the same tree, and their nodes get assigned W and L the same as if we were to use the labeling scheme of [31, Section 3.8]. This approach allows us to condense the game tree into a smaller graph with one node for each number of remaining petals. In particular, this graph G for five-petal daisy has six nodes, labeled 5, 4, 3, 2, 1, and 0. Draw this graph. Assign W and L by looking at this graph (*not* by copying these labels from the original tree). Explain briefly what you are doing.

43. With the aid of [42], or otherwise, discover who wins n-petal daisy. Prove your conclusion by induction.

44. Can you give a simple rule for optimal play of n-petal daisy?

45. In the "misere" form of a game, the player who takes the last piece *loses*.

a) Draw and analyze the game tree for the misere form of four-petal daisy.

b) Discover who wins n-petal misere daisy. Prove it by induction.

c) Repeat the proof in b) by reference to the result of [44].

d) Give a memoryless algorithm for optimal play in n-petal misere daisy. (See [44].)

46. There are six sticks. A player must pick up 1, 2, or 3 sticks at each turn. The player who picks up the last stick wins. Determine who wins this game—if both players play optimally—by drawing the game tree and finding the value of the root.

CHAPTER 4

Fundamental Counting Methods

4.1 Introduction

You have already had several occasions in this book to count things. Indeed, number is one of humanity's fundamental concepts, and counting is one of the fundamental intellectual urges. Ever tried to count the stars in the sky — or just the students in your math class? Ever wondered how many matching outfits you could put together from your wardrobe?

We will concentrate in this chapter on *how* to count without worrying too much about *what* we count. We realize that this approach can be taken to an unfortunate extreme. Who cares how many ways a baseball manager can order nine players to form a lineup, or an anagrammist can rearrange the letters of "Mississippi"? The manager would never consider all possible lineups to be equally worthy, and the anagrammist would never need to know how many ways the reordering could take place.

However, there *is* something to say for doing such "toy" examples. First, they are easy to state clearly and quickly, and so we can get on with the main business of learning solution methods. Second, they remind us that, in many situations where there is a lot of choice, the real issue is to find an optimal solution. Often the first step in solving optimization problems is to get an idea of how many choices there are. In an age of computers, if the number of choices is not too large, it may be just as well to have a computer consider each of them in turn. More generally, given an algorithm to find the optimal solution, whether ingeniously or by brute force, we need to know how many steps the algorithm takes in order to know if it is fast enough for practical use. But determining the number of steps is just another counting problem. The subject of analysis of algorithms is full of counting problems. We mentioned earlier that the hardest part of an analysis is often the average case analysis. Average case analysis uses probability, and for the most part probability computations of this sort consist of doing two counting problems and taking the ratio. See Section 6.6.

The concepts introduced in this chapter occur throughout discrete mathematics, regardless of whether someone is explicitly trying to count something. We refer to such concepts as ordered and unordered arrangements (permutations and combinations) and one-to-one correspondence. We want you to gain confidence and dexterity in dealing with such concepts.

All the basic counting methods are treated at length in this chapter, except for recursive counting. Recursive counting is so important in relation to the algorithmic approach of this book that we devote all of the next chapter to it.

Specifically, Sections 4.2 and 4.3 treat the two fundamental rules on which other counting methods are based: the Sum and Product Rules. Section 4.4 introduces the permutations and combinations already alluded to. Many identities involve permutations and combinations, and Section 4.5 provides methods for deriving them. Section 4.6 is devoted to the single most important combinatorial identity: the Binomial Theorem. Section 4.7 explains how to count ways to place balls in urns and shows that many problems can be reformulated in these terms. The topic of Section 4.8 is Inclusion-Exclusion, a more advanced counting technique that is useful whenever there is a danger of overcounting because the same situation can be described in more than one way. Through 4.8 the material is fairly traditional, with the end product being a formula instead of an algorithm. In Section 4.9 we illustrate that counting formulas have interesting algorithmic questions associated with them. You may not need to know how many ways there are to do something but instead may need to list them all or list a random selection. Section 4.10 treats the celebrated Pigeonhole Principle—it's about what happens when there are more pigeons than holes—but with an algorithmic twist. Finally, we conclude with a chapter summary, in which we review the various methods discussed in the chapter and try to give some guidance as to which method to try when.

Mathematicians, like other people, like to dignify activities with fancy names. The study of counting is usually called **combinatorics**, from the verb "combine". Even longer phrases are sometimes used: **combinatorial analysis** or **combinatorial mathematics**. Actually, mathematicians do not quite agree about the bounds of combinatorics; many would include graph theory. Another name for counting, not subject to broader interpretations, is **enumeration**.

4.2 First Examples: The Sum and Product Rules

We introduce the Sum and Product Rules through examples. Try to work the examples yourself before reading our solutions.

EXAMPLE 1 In a certain computer system, a file name must be a string of letters and digits, 1 to 10 symbols long. The first symbol must be a letter and the computer makes no distinction between uppercase and lowercase letters in any position. Thus L3MAR and L3mar are both the same legitimate name, whereas 3Lmar is not legitimate. How many legitimate file names are there?

EXAMPLE 2 Until a few years ago, auto license plates in New Hampshire consisted of two (upper-case) letters followed by three digits, e.g., RD 357 and SK 092. How many different cars could be registered in New Hampshire?

EXAMPLE 3 A salesman living in Washington, D.C., sells supplies to state governments. He decides to make a sales trip to all 50 state capitals to sell his wares. Assuming that he goes to one capital after another, visiting them all before returning home, in how many different orders can he make his trip? (We introduced this problem earlier in Example 2, Section 3.1.)

EXAMPLE 4 How many additions and how many multiplications of individual numbers are involved in multiplying two $n \times n$ matrices? (See Section 0.6 for definitions.)

Here are our solutions.

Solution to Example 1 A file name either has one symbol, or two symbols, or ..., or 10 symbols. In each case, we have 26 choices for the first symbol (the 26 letters of the alphabet). However, for each later symbol we have 36 choices because the 10 digits are also allowed. So the number of one-symbol names is 26, and the number of two-symbol names is 26×36, because for each of the 26 possible first choices there are 36 second choices. Likewise, the number of 3-symbol names is $26 \times 36 \times 36$, because for each of the 26×36 choices of the first two symbols there are 36 choices for the third. Continuing in this fashion, we get that the total number of names is

$$T = \sum_{k=1}^{10} 26 \times 36^{k-1}. \tag{1}$$

Using the formula for summing a geometric series (see [8], Section 2.2) we find that

$$T = 26(36^{10} - 1)/35 \approx 2.7 \times 10^{15},$$

but Eq. (1) is sufficient for the purposes of this section. ∎

Solution to Example 2 All plates had exactly five symbols, so multiplying the number of allowed symbols for each position we obtain

$$26 \times 26 \times 10 \times 10 \times 10 = 676,000$$

possibilities. New Hampshire changed its system because, although for many years it had under 750,000 people (including children and others not owning cars), its population has now grown rapidly to over 1 million. First New Hampshire switched to four digits after the two letters and more recently further arrangements of digits and letters have been introduced. (Actually, New Hampshire tried to stave off a major change by allowing drivers to have however many letters and numbers they wanted if they would pay more for it. But apparently not enough drivers opted for this.) ∎

Solution to Example 3 Clearly, there are 50 choices for the first capital to visit. For each such choice there are 49 choices for the capital to visit next—any capital but the one already visited. Then there are 48 possibilities remaining for the third capital. This continues until just one choice remains—the last capital. Thus the total number of possible orders is

$$N = 50 \times 49 \times \cdots \times 3 \times 2 \times 1 = 50! \tag{2}$$

Now, this is just the sort of problem criticized in the introduction to this chapter. Even if there were such a salesperson, he or she wouldn't care how many possible orders there are for the sales trip. The significant question is: Which order is optimal? However, our counting problem is of use to this salesman in a negative sense. A simple mental calculation shows that the number of possible trip orders is astronomical: The first 41 factors of Eq. (2) are all ≥ 10, so $N \geq 10^{41}$. (Or, a scientific calculator shows that $50! \approx 3.0 \times 10^{64}$.) Even if we had a computer that could print out a *billion* visit orders a second, it would still take over 10^{32} seconds, or more than 10^{24} years, just to print out the list. For comparison, the entire universe is believed to be less than 10^{10} years old. And while computers may someday be that fast, the laws of physics put bounds on how fast a computer can be (for instance, information can't move around inside a computer faster than the speed of light). Thus we are *never* going to solve this problem by using brute force on a computer. We need a much more ingenious algorithm. In fact, although there *are* ingenious, fast algorithms that give near optimal answers to this problem (see Section 4.9 for one approach), there is no known algorithm that runs in a less than astronomical number of steps and that is guaranteed to give *the best* solution. This famous problem is called the Traveling Salesman Problem. (This name doesn't refer just to the problem of visiting the 50 U.S. state capitals, but to the general problem of finding the least distance order in which to visit n locations, starting and ending at the same one. Thus our state capitals example is an example with $n = 51$.) ∎

Solution to Example 4 Recall that if A and B are $n \times n$ matrices, then so is $C = AB$, and the ij entry of C is defined by

$$c_{ij} = \sum_{k=1}^{n} a_{ik} b_{kj} \,.$$

This formula involves n products and $n-1$ additions. (To sum n numbers you add $n-1$ times.) Since all this must be done for each of the n^2 c_{ij}'s, we obtain n^3 multiplications and $n^2(n-1)$ additions. ∎

Note exactly what we have shown: *If* we evaluate a matrix multiplication directly from the definition, then it takes the number of steps we have claimed. Could there be another, nonobvious way to do the multiplication that takes less steps? Until recent years, it was believed that the answer, for general matrices, was No.

However, it is now known that the answer is Yes. The improved methods involve Divide and Conquer recursions of the sort discussed in Section 5.9. No one knows what the optimal algorithm is for multiplying $n \times n$ matrices, but the best so far take under $n^{2.5}$ steps.

Let us state two general principles used in these examples:

The Sum Rule

If there are $n(A)$ ways to do A and, distinct from them, $n(B)$ ways to do B, then the number of ways to do A *or* B is $n(A) + n(B)$. Similarly, one adds three terms when one must do A or B or C; and so on.

The Product Rule

If there are $n(A)$ ways to do A and $n(B)$ ways to do B, then the number of ways to do A *and* B is $n(A) \times n(B)$. It is assumed that A and B are *independent*; that is, the number of choices in B is the same regardless of which choice in A is taken. Similarly, one multiplies three terms if A, B, and C are independent and one must do A and B and C; and so on.

The Product Rule is closely related to the fact that if you want to find the number of entries in a rectangular table, you multiply the number of rows by the number of columns. In order to make a chart of all possible pairs of an A choice and a B choice, we could make a matrix with one row for each A choice and one column for each B choice, as in Fig. 4.1.

FIGURE 4.1

A matrix showing the 12 possible pairs of choices for three A values and four B values

$A \backslash B$	1	2	3	4
p	p_1	p_2	p_3	p_4
q	q_1	q_2	q_3	q_4
r	r_1	r_2	r_3	r_4

How have we used these rules? In Example 1, our file name was either one symbol *or* two symbols *or* three symbols, ..., so we added the number of file names in each of these cases. But within each case we had *and* decisions. For instance, for the two-symbol names, we had to pick a first symbol *and* a second symbol, independently. Analyzing still more closely, in choosing the second symbol we had to make an *or* decision: We either picked a letter or a digit, for $26 + 10$ choices.

In Example 3, *which* choices are available to the traveling salesman for the second city depends on what is chosen first, but the *number* of second choices is independent of the first choice. Furthermore, the salesman has to make a first choice *and* a second choice, and so on. Thus we have used the product rule.

In the matrix multiplication example, we counted the number of operations by finding an "address" for each operation. For instance, each addition is addressed (or indexed) by which entry of the product matrix C it helps to compute, and then by its order of appearance in the formula for that entry. Since there are n^2 entries in C, and $n-1$ additions in the formula for each entry, the product rule gives us $n^2(n-1)$.

Although these rules should seem quite straightforward, it is easy to get confused in a problem with a lot of steps within steps. At each stage of analysis, consciously ask yourself if you are "and-ing" or "or-ing".

Problems: Section 4.2

1. How many entries are in a 7×3 matrix?

2. In a Chinese restaurant, you pick a soup, an appetizer, a main dish, and a dessert.
   - **a)** If you have a choice of 5 soups, 6 appetizers, 20 main dishes, and 3 desserts, how many different meals could you arrange?
   - **b)** Actually, you don't have to have anything except a main dish. Now how many meals can you arrange?

3. In another restaurant, for the main dish there are 10 meat, 5 fowl, and 8 fish choices. How many ways are there to pick a main dish?

4. In a certain summer school program, each student must take three courses. One must be in History, Philosophy, or Religion. One must be in Math or Science. One must be in Literature, Fine Arts, or Music. Each of these departments offers three courses. How many possible programs are there for a student?

5. In each of the following cases, how many 3-letter "words" are there, using lowercase letters, if
   - **a)** any letter can be used in any position;
   - **b)** the letters must be distinct;
   - **c)** the first and last letters must be consonants and the middle letter a vowel (a, e, i, o, u);
   - **d)** the first or last letter must be a consonant and the other two letters must be a double vowel (aa, ee, etc.)?

6. How many 3-letter "words" are there, using letters, if
   - **a)** the first letter may be uppercase or lowercase and the other two must be lowercase;
   - **b)** same as part (a), except the letters must be distinct (consider the uppercase and lowercase of the same letter to be distinct).

7. How many batting lineups of 9 baseball players are there,
   - **a)** given exactly 9 players to start with; or instead
   - **b)** given a roster of 20 players to start with?

8. How many integers are there from 1 to 999 inclusive? The answer is obvious, but practice using the Sum and Product Rules by solving as follows. Each integer has either 1, 2, or 3 digits. In each case, how many choices are there for the leftmost digit? For later digits? Finish the problem.

9. Do [8] again, devising a shorter method that avoids using the Sum Rule for the number of digits. *Hint*: When recording a number from 1 to 999 on a computer card, you are not allowed to leave any of the digits columns blank.

10. How many times does each of the following algorithm fragments "Do Something"? Assume that $n \geq 3$.
    - **a)**
    ```
    for i = 1 to n
        for j = 1 to n
            Do Something
        endfor
    endfor
    ```
    - **b)**
    ```
    for i = 1 to n
        for j = 2 to n
            Do Something
        endfor
    endfor
    ```

c) **for** $i = 1$ **to** n
    for $j = 1$ **to** n
        if $i \neq j$ **then** Do Something
    endfor
endfor

d) **for** $i = 1$ **to** n
    for $j = i{+}1$ **to** n
        Do Something
    endfor
endfor

11. In many states auto license plates consist of three capital letters followed by three numerals. Are there any states in which this is probably not enough (even if discarded plate numbers can be reused)? Are there any states in which using both three letters followed by three numerals and three numerals followed by three letters is probably not enough? How many license plates are possible where you live?

12. Before a company puts its computers to work solving an optimization problem, it wants to know whether the cost of obtaining the solution is worth it in terms of the potential savings from implementing the solution. Suppose the only way of solving the traveling salesman problem is brute force. That is, each of the $n!$ routes must be listed separately, the cost for each one must be computed separately, and the smallest cost among them must be singled out. Assume that a computer can compute and compare the total distances for a million traveling-salesman routes each second; that it costs 1 cent per second to do this, and that a company is prepared to spend $10,000 to find the optimal route for its salesman. What is the maximum number of cities the salesman can visit on his trip (in addition to his home town) in order for this budget to be sufficient to solve the problem on the computer by brute force?

13. The analysis in [12] is not very refined. How long it takes the computer to find the total distance for a proposed route depends, at the least, on how many links are in the route. Suppose each of the following steps is a basic computer operation: looking up one link distance, adding two numbers, and comparing two numbers to find which is larger. Suppose that anything else the computer has to do, such as figure out which route to analyze next, takes no time at all. If the computer can do a million basic operations per second, and the financial data of [12] still hold, now what is the maximum number of cities for a brute force solution? (*Note*: Every tour to n cities in addition to the salesman's home town involves $n+1$ links.)

14. It's intuitively clear that, having just visited an East Coast capital, we will not find an optimal Traveling Salesman tour by jumping immediately to a West Coast capital. On the contrary, we would expect to go to the capital of an adjacent state. Thus, after the first capital, we need not really consider 49 choices, but only a few. Assume that for each state, there are really only three other states that we should consider for the next visit. Do [12 and 13] again, using this (plausible but not always correct) assumption.

15. Telephone area codes consist of three digits. The first digit must be one of 2 through 9, the second must be 0 or 1, the third must be one of 1 through 9. How many area codes are there?

16. How many phone numbers are possible *within* an area code? A local phone number consists of seven digits, each 0 through 9, except that the first digit cannot be a 0 or 1.

17. Area codes cover the U.S., Canada, and parts of Mexico and the Caribbean. If every person in this coverage area had a personal phone, would there be enough numbers (including area codes) to go around?

18. The information in [16] is not quite right. In places where you don't dial 1 before the area code, the first three digits of the local number cannot be the same as an area code. (The reason is that current telephone equipment processes dialing sequentially. It has no way of waiting until you finish to tell how many digits you dialed and determining thereby whether the first three were an area code or an exchange.) How much local capacity is added by switching to dialing a 1 before an area code?

19. At Washington High School, 35 students play on the football team and 15 play on the basketball team. Can you determine how many ways there are to pick one person from the football and basketball team members? If you can't, explain why the Sum Rule does not apply.

4.3 Subtler Examples and the Division Rule

It may seem that the sum and product rules are so simple that nothing very interesting could be computed using them. We hope the examples in this section will convince you otherwise.

EXAMPLE 1 If there are n Senators on a committee, in how many ways can a subcommittee be formed? Also, in how many ways can the full committee split into two sides on an issue?

Solution The way most people naturally think of answering the first question is to count the number of subcommittees with one member (there are n), then the number of subcommittees with two members (it turns out there are $n(n-1)/2$ of them), then the number with three members (getting harder!) and so on, and then add up all these counts (Sum Rule). It *can* be done this way, and we'll do it this way in Section 4.4, but there is a much simpler way—using just the Product Rule. Don't first decide the size of the subcommittee; instead consider each Senator in turn and decide whether he or she will be on the subcommittee. Thus there are two possibilities, Yes or No, decided independently for each of the n Senators. This approach gives a product of 2 times itself n times, that is, 2^n subcommittees.

Figure 4.2 illustrates this approach when the full committee has three members. Yet another use for trees! Here they are called **decision trees**. Each path from the root to a leaf represents a sequence of possible Yes/No membership decisions for all the members of the full committee. The corresponding subcommittee is recorded at the leaf. Note that at each level (which corresponds to the membership decision for a particular full-committee member) the number of nodes doubles, just as the product rule says it should.

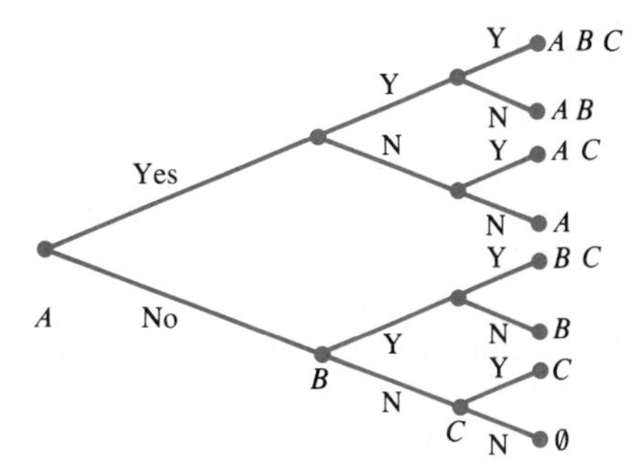

FIGURE 4.2 All possible subcommittees of a three-member committee

We have to be a little careful, though, about our simple solution. When we considered subcommittee size first, we started counting at size 1. However, the product method includes in its count one subcommittee with zero members (see Fig. 4.2). It also includes one subcommittee with all n members. Do we really want to include the "null" subcommittee and the subcommittee of the whole? This is a matter of definition, not mathematics. For instance, if Congress refuses to allow null subcommittees, but does allow committees of the whole, the correct answer to the first question in Example 1 is $2^n - 1$.

Note 1. Null committees do occur in the real world. If, say, Congress wants the President to set up a committee to investigate something and he doesn't want to, he might stall by "forgetting" to appoint the committee. That is, he appoints a null committee.

Note 2. If we allow both null and whole subcommittees, Example 1 is simply asking: How many sub*sets* does an n-set have? So this is not the first time we have discussed the problem; see Section 0.1 and [27], Section 2.2. However, the Product Rule solution given here is new.

The second question in Example 1, about splitting on an issue, is even more ambiguous. For instance, suppose the issue is a motion and the full committee consists of Senators A, B, C, D, and E. One way for them to split is to have A, B, and C vote Yes, and D and E vote No. Another way is to have D and E vote Yes and A, B, and C vote No. Are these different splits? The same people are in agreement and the same ones are in opposition in both cases. Indeed, the first split becomes the second if the motion is reworded in the negative.

Again, this is a matter of definition. If you are interested in who voted Yes, the two splits above are different; if you are interested only in the pattern of opposition, they are the same. Since we have two interpretations, let's answer both. Interpretation one: Who votes Yes makes a difference. Then each Senator either votes Yes or No, giving n 2-way choices and a product of 2^n splits.

With this interpretation, the answer is the same as the number of subcommittees. This can't just be happenstance. Indeed, for each Yes/No split, think of those who voted Yes as forming a subcommittee (the Yes-Subcommittee) and just don't think about the others. (*Note*: since everybody could vote Yes, or nobody could do so, we definitely must allow the full subcommittee and the null subcommittee in this context.) Conversely, for each subcommittee, imagine that all the subcommittee members have voted Yes on some issue, and all the others on the full committee have voted No. Thus we have a **one-to-one correspondence** between the objects counted in the first problem (subcommittees) and the objects counted in the second (Yes/No splits). Looking for such correspondences is a very powerful method in combinatorics. If you find a one-to-one correspondence between the problem before you and some other problem you have already solved, then you have the answer to the new problem without a single additional computation. Even if the second problem is not one you have already solved, it may be one you can solve more easily than the first.

But what is the number of split votes for the second interpretation? For each split in which the actual vote doesn't matter, there are two splits in which it does matter—one side votes Yes and the other No, or vice versa. Moreover, every Yes/No split corresponds to a vote-doesn't-matter split in this two-to-one fashion. Therefore the number of doesn't-matter splits is $2^n/2 = 2^{n-1}$. This is our first instance of the **Division Rule**. We'll do another example shortly before we try to state the rule more formally. ∎

Example 1 brings out an important point about counting problems. It is very hard in a brief description to make such a problem completely unambiguous. So what do you do when the meaning of a problem (on a test, say!) is unclear to you? Our suggestion:

- state that you think the problem is ambiguous;
- state unambiguously what you think is the most likely interpretation; and
- then answer that interpretation correctly.

Caution 1: If the interpretation you choose makes the problem trivial to answer, it is probably not the interpretation the professor intended, and you probably won't get much credit for solving it. *Caution 2*: Not all professors are willing to believe that their wonderfully worded questions are ambiguous. If the question uses only terms that the professor has taken great pains to define precisely in class, the professor may be right. But if you are honestly confused, we still think our approach is the best one. Even professors make mistakes!

EXAMPLE 2 Find a lower bound on the number of digits the thief in the Baltimore Hilton problem (Example 2, Section 3.1) must punch sequentially in order to try all 4-digit combinations.

Solution The thief tries a sequence of digits. Each digit tried becomes the first digit of at most one attempted combination. But there are 10^4 possible combinations (product rule!), so the thief must try at least 10^4 digits in a row.

Again we have used the idea of a correspondence, but here we have an inequality, not an equality. We said the first digit of *at most* one possible combination. After the kth attempted digit, if the thief does not try at least three more digits, then attempt k does not count as the first digit of any combination tried. However, the only digits that are not followed by three more digits are the last three. Therefore in order to cover all 10^4 combinations, the thief must try at least 10^4+3 digits. In fact, this number of digits is enough; see [6], Section 3.4. Therefore 10^4+3 is the optimal solution. ∎

EXAMPLE 3 Find an upper bound on the number of steps in Towers of Hanoi.

Solution Forget for the moment the main problem of how to get from one configuration of Towers of Hanoi to another. Just ask: How many legitimate configurations of rings on poles are there? If it turns out that there are, say, 37 configurations, then at most 36 steps are needed to solve the puzzle (if it can be solved at all). In other words, the worst that could happen is that you have to pass through every configuration to get to the desired end. To see this, you might think in terms of a state graph as described in Section 3.1. The different legitimate configurations are the vertices, and you seek a path between two specific vertices. The worst that could happen would be for the graph to be a path with those two vertices at the ends. You certainly never have to pass through the same configuration (vertex) more than once. (Why?)

So, how many configurations are there? First, you only have to worry about which rings are on which poles; the order of the rings on each pole is uniquely determined as largest to smallest. So you merely have to divide the n rings up among the poles. That is, each ring merely has to "vote", independently, for pole 1, 2, or 3. This gives 3^n configurations and an upper bound of $3^n - 1$ moves.

Not very impressive, you say, since the actual answer, $2^n - 1$, is so much less than this bound. But this bound applies to *any* version of TOH in which

- only one piece can move at a time;
- smaller pieces must always be on top of larger pieces; and
- any further restriction on moves depends solely on the current configuration, not on past moves. (Where have we used this last condition?)

Indeed, we have already discussed a natural modification of TOH in which the bound we have obtained is exact—Straightline TOH in Section 2.4. ∎

EXAMPLE 4 You have 13 identical-looking coins and a balance scale. You are told that one of the coins is fake and has a different weight from the others. You are asked to find which coin is fake, and whether it is light or heavy, using only the balance scale. Find a good lower bound on how many weighings this will take. (A weighing is one use of the scale, e.g., weighing coins 1 and 2 against 3 and 13 and finding which pair, if either, is heavier.)

Solution If you have tackled this coin problem before, you know that determining how to find the fake coin in a small number of steps is hard. But here we asked only for a lower bound on the number of steps, and a very good lower bound can be obtained by a simple product rule calculation.

The hard part of the complete problem is to devise a successful *strategy.* You must decide which coins to weigh first, and then, since you don't know what will happen, you must decide in advance which coins you will weigh second depending on the outcome of the first weighing. Such a strategy involves another decision tree. Figure 4.3 shows a successful strategy for three coins.

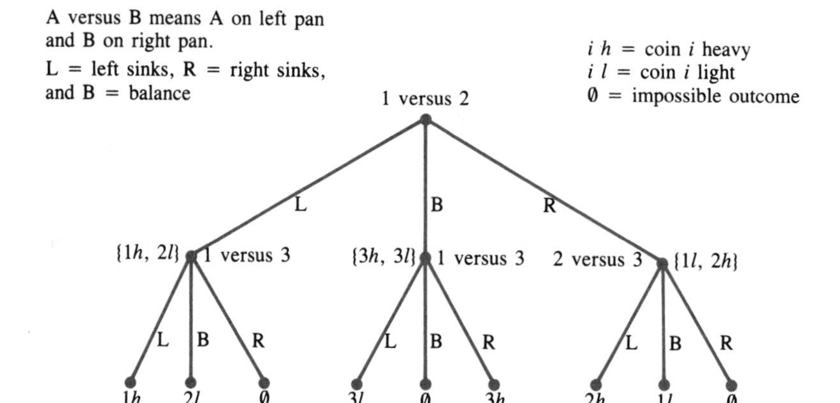

FIGURE 4.3

A decision tree for finding one fake coin from three with a pan balance. Each node is marked with the set of situations leading to it.

By "the situation", we mean the (initially unknown) information about which coin is fake and whether it is heavier or lighter. Each possible situation will cause you to go down one particular path in your strategy tree. Suppose that two different situations lead you down the same path. Then you won't be able to distinguish between them; they give you exactly the same results. In other words, a strategy is successful iff each leaf corresponds to at most one possible situation. A simple *necessary* condition, therefore, for a successful strategy is that there be at least as many leaves as situations. Also, the maximum number of edges leading from the root to a leaf in a successful strategy tree is the number of weighings that strategy requires in the worst case.

In Fig. 4.3, we have marked each node by the set of situations that lead to it. Since each leaf has at most one situation, the strategy is successful. However, if we lopped off the tree after one weighing, the strategy would not be successful. Also note that every leaf is at the same level; thus the worst case and the best case are the same.

So where's the product rule? Figuring out your strategy is hard, but the *shape* of the strategy tree is always the same: It's a **ternary** tree. Each weighing has three possible outcomes: The left pan sinks, the right pan sinks, or they balance. Thus a strategy with one weighing has 3 leaves, a strategy with at most two weighings has at most 9 leaves, and, in general, with n weighings there can be 3^n leaves. Think of it this way: Which leaf you reach is determined by the "history" of your weighing outcomes. Since there are three possible outcomes each time, the product rule gives 3^n histories and 3^n leaves.

In our specific case with 13 coins, there are 13×2 situations to distinguish; each coin may be heavy or light. Thus a successful strategy tree must have at least 26 leaves. Since $3^2 = 9$ and $3^3 = 27$, you should conclude that the very best strategy will need at least three weighings in the worst case. This is the bound you sought. (This bound is not exact. With more refined counting, you can show that at least four weighings are necessary. By actually devising strategies, you can then show that four are sufficient. See [13–16].) ∎

EXAMPLE 5 In how many distinguishable ways can the letters of "computer" be arranged? In how many distinguishable ways can the letters of "mathematics" be arranged?

Solution There are 8 letters in "computer", all distinct. Any one of them can go first, then any of the remaining 7, and so on. So the answer is $8! = 40,320$.

The letters of "mathematics", on the other hand, are not all distinct. There are two m's, two a's and two t's. But let us try to reduce this problem to the previous case. Suppose all the letters *were* different—say, the m's were labeled $\mathrm{m}_1, \mathrm{m}_2$, and so on. Then the answer would be 11!. Clearly we have overcounted, but how much? For each arrangement with copies of the same letter indistinguishable, there are 8 arrangements with the copies distinguished. There are 2×1 ways to put the subscripted m's in the two m slots, independently 2×1 ways to order the subscripted a's, and the same for the t's. See Fig. 4.4. The correct answer is $11!/8 = 4,989,600$. ∎

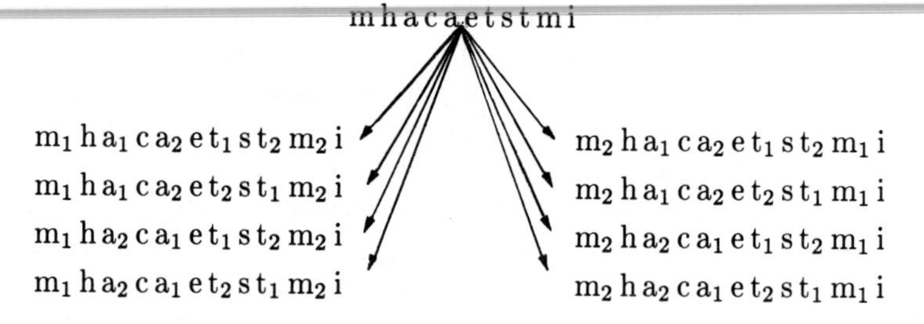

FIGURE 4.4

For each arrangement of the letters in "mathematics" with copies of the same letter indistinguishable, there are eight arrangements with distinguishable copies.

This then is the

Division Rule

If there is a k-to-1 correspondence of objects of type A with objects of type B, and there are $n(A)$ objects of type A, then the number of objects of type B is $n(A)/k$.

A **k-to-1 correspondence** from A-objects to B-objects is an onto mapping in which every B-object is the image of exactly k A-objects. See Fig. 4.5. This is a natural generalization of the concept of 1-to-1 correspondence discussed in Example 1.

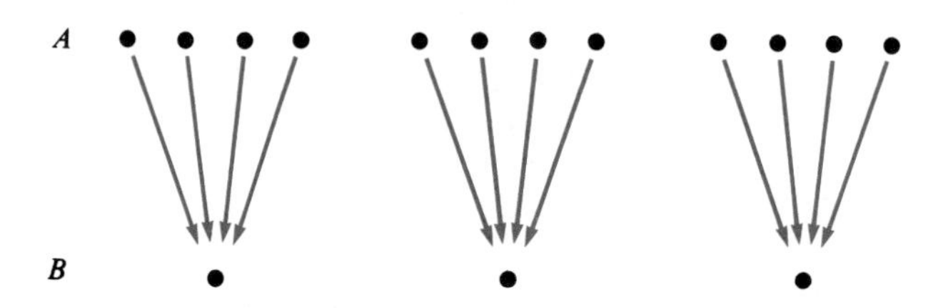

FIGURE 4.5

A 4-to-1 correspondence between A-objects and B-objects

The Division Rule is commonly used for counting ordered arrangements when some of the objects are indistinguishable (e.g., the m's in Example 5). It is also commonly used when order doesn't matter; often it's easiest to first suppose order does matter and then divide out [26-27].

We close this section with a word of caution about the division rule. It works only if every object you want to count corresponds to *exactly* k objects of the comparison set. It's easy to think that you have an exact k-to-1 correspondence when you don't, as the next example shows.

EXAMPLE 6 Count the number of 0–1 sequences of length n when a sequence and its reverse are considered to be the same. For instance, if $n = 4$, then 1010 and 0101 should together be counted just once.

(This problem is not so silly. Think about the Universal Product Code—the "bar graph" code on supermarket purchases. Electronic scanners can read this code backwards or forwards, so a code word and its reverse cannot represent different things.)

Solution Start It seems pretty clear that we can take the total number of binary sequences of length n and divide by 2, for an answer of $2^n/2 = 2^{n-1}$. However, this answer isn't right! To take a very simple example, when $n = 2$ this formula gives 2, yet the right answer is 3:

$$00, \quad 01, \quad 11.$$

What went wrong? What's the right formula? [19–20] ∎

In general, counting problems that involve a variable many-to-one correspondence are quite tricky. There is a general result called Polya's Theorem that helps when there is a lot of symmetry. This theorem is treated in a more advanced combinatorics course.

Problems: Section 4.3

Try to solve [1–7] using the Sum and Product Rules alone.

1. How many ways can a committee of n members form two disjoint subcommittees, if not everybody has to serve on either subcommittee?

2. How many ways can the two subcommittees in [1] be formed if they need not be disjoint and, again, not everybody has to serve?

3. In the game of Twenty Questions, one person (the answerer) thinks of something and the other person (the asker) tries to find out what it is by asking Yes/No questions. The answerer must reply truthfully and the asker is limited to 20 questions and a final claim ("I claim the thing is ..."). Suppose that the something must be an English word, of which there are about half a million. Can the asker guarantee a win? (Assume that the asker has a complete dictionary.) What is an optimal strategy?

4. Suppose that we change Towers of Hanoi to have four poles in a row, but otherwise legitimate configurations are unchanged. Give an upper bound on the number of steps to solve 4-pole TOH, a bound as general as the 3-pole bound we devised in Example 3. For instance, the bound should hold for the linear 4-pole version where the poles are in a row and rings must move to adjacent poles.

5. Returning to TOH with three poles, some sets are sold commercially with the three poles in a row and with the rings alternately colored red and blue from smallest to largest. Also, the floor under each pole is colored. The largest ring and the floor under the middle pole are the same color, and the floor under the left and right poles is the other color. The instructions say: Never put a ring directly on top of another ring of the same color or on a floor of the same color. With these additional requirements, how many legitimate configurations are there?

6. How many possible voting splits are there in a committee of n people if there is three-way voting: Yes, No, Abstain? Give several different interpretations of this question and therefore several

different answers.

7. Given a committee of n people, how many ways are there to pick a subcommittee and a chairman for that subcommittee?

8. How many ways are there to order the letters of
a) English? b) Economics? c) Mississippi?

9. How many nine-digit integers can be written using two 1's, four 2's and three 5's?

10. In a hypnosis experiment, a psychologist inflicts a sequence of flashing lights on a subject. The psychologist has a red light, a blue light, and a green light. How many different ways are there to inflict nine flashes, if two flashes are to be red, four blue and three green?

11. A family of five is to be seated at a round table. How many ways are there to do this? How many ways are there if only the relative positions of the people count? (The latter question provides an example where the Division Rule applies, but not because some individual objects are indistinguishable.)

12. Answer the same two questions as in [11], but now assume there are two extra place settings.

13. Show that no strategy for the coin problem of Example 4 is guaranteed to find the fake coin in at most three weighings. *Hint*: Suppose some strategy S did require only three weighings. Since two weighings can distinguish at most 9 situations, and there are 26 situations to be distinguished, the first weighing of S would have to divide the situations into three sets of exactly 9, 9 and 8 situations each. Does weighing one coin against one other in the first weighing accomplish this? No, because if the left pan sinks, you have 2 situations left (which ones?); if the right pan sinks, again you have 2 situations; if the pans balance, you have 22 left. What about two coins versus two, and so on?

14. Find a strategy for the coin problem that takes at most four weighings. Show that you are correct by drawing a tree diagram for your strategy, using the format of Fig. 4.3. Each leaf must correspond to at most one possibility.

15. Suppose that, in addition to the 13 coins, you have an additional "test coin" that you know is the right weight. Now there *is* a strategy that needs at most three weighings. Find it. Prove it correct.

16. Generalize the results of [13–15] to n coins. (This is tough.)

17. Again you are given 13 coins and one is fake. But now you are told the common weight of the good coins and the different weight of the fake one. Also, you are given a single-pan electronic scale that tells you the weight of what you put on it. Give a lower bound on the number of steps necessary to find the fake coin. Show that this lower bound is exact.

18. Generalize the result of [17] to n coins.

19. Define a sequence of digits of length n to be symmetric if the digit in position k is the same as the digit in position $n-k+1$ for all k. For the problem in Example 6, show that there is a 2-to-1 correspondence only after we (temporarily) exclude the symmetric binary sequences. Then solve the problem.

20. Solve the problem in Example 6 again for decimal sequences of length n.

21. A computer operating system requires file names to be exactly four letters long, with uppercase and lowercase letters distinct.
   a) How many file names are there?
   b) The operating system is changed so that uppercase and lowercase are no longer distinct. Figure out the new number of file names two ways: applying the Division Rule to (a) and directly.

22. Repeat [21] with "exactly" replaced by "up to". Can the Division Rule still be used?

23. A firm produces four-digit house-number signs. The first digit cannot be 0. The company need not produce every number from 1000 to 9999 because some signs can be gotten from others by rotating them 180 degrees; e.g., 6611 is just 1199 turned around. How many different signs must the company make?

24. If n players compete in a (single-elimination) tennis tournament, how many matches are there? *Hint*: If you draw the usual binary tree diagram for a tournament, it would seem that the answer is simplest to find when n is a power of 2. For other n, there have to be some byes (a player is advanced to the next round without having a match), and you might think this would make the answer dependent on how the byes are distributed. Actually, the answer is not dependent on that, nor even on

the usual binary tree set-up. This is shown by using a one-to-one correspondence. Think: What is the real purpose of each match?

25. If you multiply an nth-degree polynomial $P(x)$ by an mth-degree polynomial $Q(x)$ without combining any like terms as you go along, in general how many terms can you then eliminate by combining? For instance, multiplying $3x+2$ by $5x-7$, without combining, gives

$$15x^2 - 21x + 10x - 14.$$

Then you can eliminate one term to obtain the final answer: $15x^2-11x-14$. Why do we say "in general"?

26. How many ways are there to distribute four distinct balls evenly between two distinct boxes? *Hint*: Choose a first ball to put in the first box, then a second ball, then a first ball for the second box, then a second ball. But the order in which we put the balls in doesn't matter. What should we divide by?

27. Repeat [26], except now the boxes are indistinguishable. (See Section 4.7 for more on this sort of problem.)

4.4 Permutations and Combinations

Permutations and Combinations are the basic building blocks of combinatorial mathematics.

Definition 1. A **permutation of n things taken r at a time** is any arrangement in a row of r things from a set of n distinct things. Order in the row makes a difference: The same r things ordered differently are different permutations. $P(n,r)$ is the *number* of permutations of n things taken r at a time.

One also speaks of a "permutation of r things from n" or an "(n,r) permutation".

Definition 2. A **combination of n things taken r at a time** is any subset of r things from a set of n distinct things. Order makes no difference. The *number* of combinations of n things taken r at a time is denoted by $C(n,r)$ or

$$\binom{n}{r}$$

and pronounced "n choose r".

One also speaks of a "combination of r things from n" or an "(n,r) combination". We will tend to use the $C(n,k)$ notation within paragraphs and the $\binom{n}{k}$ notation within displays—the latter doesn't look so good in paragraphs.

For example, the permutations of the three integers 1, 2, and 3 taken two at a time are

$$(1,2) \qquad (1,3) \qquad (2,1) \qquad (2,3) \qquad (3,1) \qquad (3,2),$$

and the combinations of two things from these three integers are

$$\{1,2\} \qquad \{1,3\} \qquad \{2,3\}.$$

Recall that parentheses indicate an ordered pair and braces indicate a set, that is, an unordered pair.

The main difference between permutations and combinations is that, in a combination, order makes no difference. In particular, in computing $P(n,r)$, *which* r elements are chosen and *how* they are arranged both make a difference. In computing $C(n,r)$, only which elements are chosen makes a difference.

Let us find a formula for $P(n,r)$. We find it by counting the number of ways we can pick one permutation. Since order makes a difference, we organize our picking around position. We can begin by choosing the first entry in the permutation (say on the left), then the second (from the left), and so on. There are n choices for the first entry, $n-1$ for the second, $n-2$ for the third, and in general, $n-(k-1) = n-k+1$ for the kth. Thus by the Product Rule,

$$P(n,r) \;=\; \prod_{k=1}^{r}(n-k+1). \tag{1}$$

For instance,

$$P(5,2) = 5 \times 4 = 20 \quad \text{and} \quad P(6,4) = 6 \times 5 \times 4 \times 3 = 360.$$

There is a shorter way to write the general formula. Consider these numerical examples first:

$$P(5,2) = 5 \cdot 4 = 5 \cdot 4 \cdot \frac{3 \cdot 2 \cdot 1}{3 \cdot 2 \cdot 1} = \frac{5!}{3!}.$$

Similarly,

$$P(6,4) = \frac{6 \cdot 5 \cdot 4 \cdot 3 \cdot 2 \cdot 1}{2 \cdot 1} = \frac{6!}{2!}.$$

In general, multiplying $P(n,r)$ by 1 in the form $(n-r)!/(n-r)!$, we obtain

$$P(n,r) = \frac{n!}{(n-r)!}. \tag{2}$$

Note that Eq. (2) makes sense even when $r = 0$; it says that $P(n,0) = 1$. That is, the number of ways to line up no things from n is 1: There's just one way to do nothing! Since this interpretation is reasonable, and extends the domain of correctness of a simple formula, we adopt it.

For instance, according to Eq. (2), the number of permutations of the integers 1,2,3 taken two at a time should be

$$P(3,2) = \frac{3!}{1!} = 6,$$

just the number we got earlier.

Since we developed our formula for $P(n, r)$ through a stepwise analysis, it is quite easy to express the answer with an algorithm. Algorithm PERMVALUE, computes $P(n, r)$ using Eq. (1). Note that this algorithm works even when $r = 0$.

Algorithm 4.1

PermValue

```
Input n, r                                   [Nonnegative integers]
Algorithm PERMVALUE
    P ← 1
    for k = 1 to r
        P ← P × (n+1−k)
    endfor
Output P
```

Any combinatorial quantity can be described by giving an algorithm to compute it. Most of the quantities we develop in this chapter have simple formulas, so it would seem there is little point in providing a lengthier algorithmic description as well. However, sometimes formulas are not as useful as they first appear [15], and sometimes there aren't any good formulas.

To find a formula for $C(n, r)$, we reduce to the previous problem. Every combination can be made into a permutation by imposing an order. Indeed, given a combination of r things from n, there are $r!$ different orders that can be imposed (Product Rule). Moreover, every permutation of r things from n is obtained by imposing some order on some combination. Therefore we have an $r!$-to-1 correspondence between the set of all (n, r) permutations and all (n, r) combinations. Thus by the Division Rule,

$$\binom{n}{r} = \frac{P(n,r)}{r!} = \frac{n!}{r!(n-r)!}. \tag{3}$$

For instance,

$$C(3,2) = \frac{3!}{2!\,1!} = 3,$$

and

$$\binom{6}{4} = \frac{6!}{4!\,2!} = \frac{6 \cdot 5}{2 \cdot 1} = 15.$$

Notice that Eq. (3) makes sense even when $r = 0$, since $0! = 1$. Namely, $C(n, 0) = n!/(0!n!) = 1$. This value makes sense because there is one empty subset of any n-set.

Using Eq. (1) for $P(n, r)$, we may express $C(n, r)$ another way:

$$\binom{n}{r} = \frac{P(n,r)}{r!} = \frac{\prod_{k=1}^{r}(n-k+1)}{r!}. \tag{4}$$

EXAMPLE 1 How many different full houses are there in poker? A full house is a hand of 5 cards in which 3 of them are from one denomination and 2 from another. In a pack of cards there are 13 denominations (2, 3, ..., queen, king, ace) and 4 cards of each. The order in which cards are dealt makes no difference in counting different hands.

Solution 1 Since order makes no difference, we can specify a convenient order. Let us imagine the 3-of-a-kind are picked first, followed by the 2-of-a-kind. Then there are 52 choices for the first card, followed by 3 for the second (because it must be one of the remaining cards of the same denomination as the first), and 2 for the third. Then there are 48 choices for the fourth card (anything of some other denomination) and 3 for the fifth. We have thought in terms of permutations, albeit of a restricted sort, to eliminate differences in order. Have we completely succeeded? No! We have forced the 3-of-a-kind to be considered first, but we haven't eliminated the effects of order among them. Each 3-of-a-kind has therefore been counted 3! times. Likewise, each 2-of-a-kind has been counted 2! times. Therefore we have counted each hand 6×2 times. So, using the Division Rule, the correct answer is

$$\frac{52 \cdot 3 \cdot 2 \cdot 48 \cdot 3}{12} = 3744.$$

Solution 2 Since order makes no difference, we can think of this as a problem in choosing sets, i.e., combinations. First, we must pick a denomination for the 3-set and then a denomination for the 2-set. There are 13×12 ways to do this. Having chosen the denomination for the 3-set, there are $C(4,3) = 4$ ways to pick the 3-set. Similarly, there are $C(4,2) = 6$ ways to choose the 2-set. Therefore the answer is

$$13 \times 12 \times 4 \times 6 = 3744. \blacksquare$$

EXAMPLE 2 Diagonal chords are drawn between each pair of nonadjacent vertices of a convex polygon with n sides (see Fig. 4.6). How many chords are there?

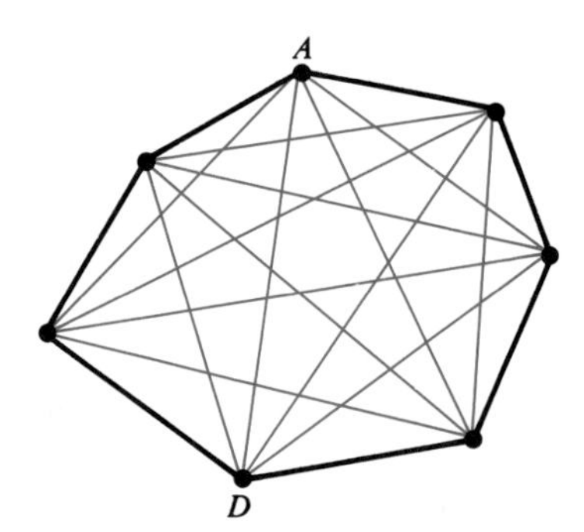

FIGURE 4.6

The chords of a convex polygon

Solution From each vertex, $n-3$ chords emanate. (Why?) Since there are n vertices, we get $n(n-3)$ chords total. However, we have counted every chord twice, once for each end. Put another way, we have counted directed chords, and now we must use the division rule to get undirected chords. The right answer is $n(n-3)/2$.

Alternative Solution A chord corresponds to a 2-subset of the vertices, of which there are $C(n, 2)$. However, not every 2-subset gives a chord. The n pairs of adjacent vertices give sides. Thus the answer is

$$C(n, 2) - n.$$

Please check that these answers are the same [12]. ∎

EXAMPLE 3 Considering the same n-gon, in how many interior points do its chords intersect? Assume that the n-gon is irregular, so that no three chords intersect at the same interior point.

Solution This is a rather hard problem. Most people first consider the number of intersection points on each chord. From Fig. 4.6, you can see that the number varies. However, you can try to figure out how many chords have a given number of intersections and proceed from there. This approach can work but it's tricky [5, Supplementary Problems]. Fortunately, there is an entirely different way to answer the question.

Namely, think as follows. What causes an intersection point in the first place? Answer: two criss-cross chords. What causes there to be two chords? Four vertices. Couldn't these two chords come from three vertices by sharing a vertex? Yes, but then they wouldn't intersect in the interior. Can any four vertices produce two intersecting chords (i.e., what happened to the observation that only nonadjacent vertices produce a chord)? Yes, by connecting the four vertices alternately to form a criss-cross of chords. Can the same four vertices produce two chords in more than one way? Yes, but only the criss-cross configuration already discussed results in two chords that intersect internally. See Fig. 4.7.

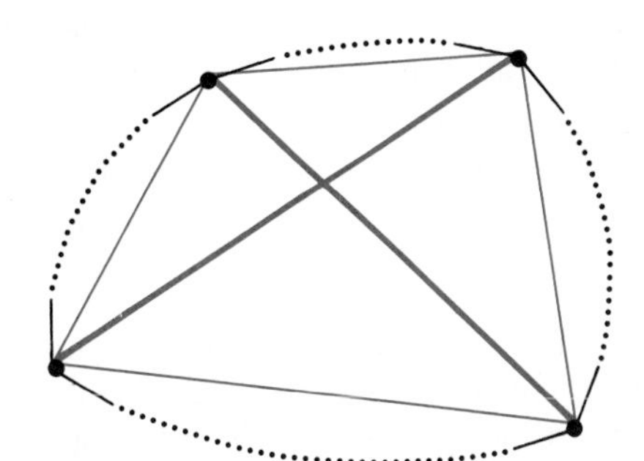

FIGURE 4.7

Any four vertices of an n-gon create just one pair of intersecting chords.

What have we shown? That there is a one-to-one correspondence between interior intersection points and combinations of the vertices taken four at a time. Therefore the number of intersections is simply $C(n, 4)$. ∎

EXAMPLE 4 Using the same n-gon as before, into how many line segments are the chords divided by their various intersections? By segments we mean nonoverlapping pieces. For instance, in Fig. 4.6, chord AD is divided into seven segments.

Partial Solution From the previous two examples, it should be clear that we will save ourselves much grief if we attack this problem using combinations and correspondences. *Hint*: For any given chord, how is the number of segments into which it is divided related to the number of intersection points on it? If we sum up this relation over all chords, we get a statement about the total number of segments. Finish the problem! [13] ∎

Permutations and Combinations with Repetitions

Suppose that we are allowed to use the same elements over again in making our permutations and combinations. (In probability and statistics, this is called sampling with replacement.) For instance, the collection of permutations taken two at a time from the set $\{1,2,3\}$ now becomes

$$(1,1) \quad (1,2) \quad (1,3) \quad (2,1) \quad (2,2) \quad (2,3) \quad (3,1) \quad (3,2) \quad (3,3).$$

The collection of combinations, again two at a time, becomes

$$\{1,1\} \quad \{1,2\} \quad \{1,3\} \quad \{2,2\} \quad \{2,3\} \quad \{3,3\}.$$

Let us call the number of these things $P^*(n,k)$ and $C^*(n,k)$. Are there formulas for these quantities?

(Unlike $P(n,k)$ and $C(n,k)$, which are rather standard notation, $P^*(n,k)$ and $C^*(n,k)$ is just temporary notation we have made up. There aren't universally accepted notations for these two quantities.)

Yes, there are formulas. For permutations with replacement the formula is very simple. At each of the k locations in the sequence, you get to choose any one of the n objects, whether or not it has been picked before. Therefore $P^*(n,k) = n^k$. For example, the number of ordered pairs of three things is $P^*(3,2) = 3^2 = 9$, in agreement with the list of ordered pairs displayed above.

The formula for $C^*(n,k)$ is harder to derive, though not much more complicated to write down. We will derive it one way in Section 4.7 and another way in Section 5.3.

Note. In talking about combinations with repetitions, we have abused the definition of set. For instance, we listed $\{2,2\}$ as a set of size two from $\{1,2,3\}$. We did indeed mean to list $\{2,2\}$ and we did mean it to have two elements. The abuse is in calling it a set. As we noted in Section 0.1, the thing we are talking about here is properly called a *multiset*. The *set* $\{2,2\}$ is the same as the set $\{2\}$ and has one element.

Problems: Section 4.4

1. Evaluate
   a) $P(5,1)$, $P(5,2)$ and $P(5,3)$,
   b) $C(5,1)$, $C(5,2)$ and $C(5,3)$.

2. Which is more work to compute, $P(100,3)$ or $P(100,97)$?

3. Which is more work to compute, $C(100,3)$ or $C(100,97)$?

4. When the expression

$$(A+a)(B+b)(C+c)(D+d)(E+e)$$

is multiplied out, how many terms will have three capital letters?

5. Show that the number of ways to pick a subset of one or two things from n distinct things is $n(n+1)/2$.

6. By algebra, show that

$$P(n,k) \times P(n-k,j) = P(n,k+j).$$

Can you also explain this equality without using algebra? (We discuss such non-computational proofs at length in the next section.)

7. How many ways are there to pick a combination of k things from $\{1,2,\ldots,n\}$ if the elements 1 and 2 cannot both be picked?

8. Three distinct balls are to be distributed among seven different urns so that each urn gets at most one ball.
   a) How many ways can this be done?
   b) What if the balls are indistinguishable?

9. How many ways are there to put eight rooks on a chessboard so that no one rook can capture another? (This means that no two are in the same row or same column.) How many ways are there to put five noncapturing rooks on a chessboard?

10. How many triangles can be formed from the edges and diagonals of an n-gon, if the vertices of each triangle must be vertices of the n-gon?

11. Same as [10], but now all three sides of the triangle must be diagonals of the n-gon. Be careful!

12. Check that the two solutions to Example 2 are the same.

13. Let $\mathcal{C}$ be a collection of lines, and let I be the set of their intersection points. Assume that no endpoint of any line is in I and that no three lines intersect at the same point. Let i_L be the number of intersection points on line L in $\mathcal{C}$.
    a) Explain why L is divided into i_L+1 segments.
    b) Explain why $\sum_{L\in\mathcal{C}} 1 = |\mathcal{C}|$.
    c) Explain why $\sum_{L\in\mathcal{C}} i_L = 2|I|$.
    d) Finish Example 4.

14. a) How many ways are there to permute a sequence of m identical a's and n identical b's?
    b) How many ways are there to choose a *set* of n objects from $m+n$ distinct objects?
    c) Why are the answers to (a) and (b), the first about permutations, the second about combinations, the same? Give a one-to-one correspondence.

15. Algorithm 4.2, PERMVALUE2, computes $P(n,r)$, based on the formula $n!/(n-r)!$. Although correct, PERMVALUE2 is not as good as PERMVALUE in the text. Why? If you're not sure, compute $P(25,3)$ with a hand calculator using both algorithms.

Algorithm 4.2. PERMVALUE2

```
Input n, r
Algorithm PERMVALUE2
   P ← 1
   for i = 2 to n
      P ← P * i
   endfor
   for i = 2 to n−r
      P ← P/i
   endfor
Output P
```

16. Come up with at least two algorithms to compute $C(n,k)$, of varying goodness in the sense of [15].

17. How many edges can a simple graph on n vertices have? Recall that a simple graph is one without multiple edges between any pair of vertices and without loops.

18. How many different simple, labeled graphs are there with n vertices? In a labeled graph (see [21,

Section 3.2]), the vertices are distinguished, say, by numbering them. A labeled graph with n vertices and exactly one edge, with the edge between vertices 1 and 2, is considered to be different from the labeled graph with n vertices and exactly one edge, with the edge between vertices 3 and 4.

19. How many edges can a simple digraph on n vertices have? For each ordered pair (u, v) of vertices, a simple digraph has at most one edge *from* u to v. Thus it may have two edges *between* u and v, one from u and one from v.

20. How many labeled simple digraphs are there on n vertices?

21. Recall that a tournament is a digraph such that, for each pair of vertices $\{u, v\}$, exactly one of (u, v) and (v, u) is an edge. A **partial tournament** is a tournament from which some of the edges may have been removed. How many labeled tournaments are there on n vertices? How many labeled partial tournaments?

4.5 Combinatorial Identities and Combinatorial Arguments

The numbers $C(n, r)$ are miraculous. They satisfy an immense number of identities, many very beautiful, many very surprising. We will introduce the few most fundamental ones. We will use their introduction as an opportunity to illustrate several methods of proof in combinatorics—algebra, induction, and a conceptual method often called **combinatorial argument**.

Theorem 1. For all nonnegative integers $k \leq n$,

$$\binom{n}{k} = \binom{n}{n-k}. \qquad (1)$$

PROOF 1 We have a formula for $C(n, r)$: Eq. (2) of Section 4.4. Let's write it down for $r = k$ and $r = n-k$ and see if we can make them look the same.

$$\binom{n}{k} = \frac{n!}{k!(n-k)!},$$

$$\binom{n}{n-k} = \frac{n!}{(n-k)!(n-(n-k))!} = \frac{n!}{(n-k)!k!} = \frac{n!}{k!(n-k)!}.$$

PROOF 2 Let us think about what $C(n, k)$ and $C(n, n-k)$ count and look for a one-to-one correspondence between them. $C(n, k)$ counts the number of k-element subsets of an n-set. $C(n, n-k)$ counts the number of $(n-k)$-element subsets. But in picking a k-element subset we inescapably pick an $(n-k)$-element subset as well, namely, all the elements we left out of the k-subset. In short, within the n-set, complementation (see Section 0.1) provides a one-to-one correspondence between k-subsets and $(n-k)$-subsets. ∎

Theorem 2. For all positive integers $k < n$,

$$\binom{n}{k} = \binom{n-1}{k} + \binom{n-1}{k-1}. \tag{2}$$

Note that Eq. (2) is recursive. If you know the combinations of $n-1$ things, you need only a single addition to find $C(n,k)$. Equation (2) leads to a simple algorithm for computing combinations [10].

Traditionally, Theorem 2 and its algorithm are depicted graphically, as in Fig. 4.8. This is a row-by-row chart of the combinations $C(n,k)$, where n indexes the row and k varies across the row. The top row is row 0 and the leftmost entry in any row is the 0th entry. With the rows centered as shown, Theorem 2 says that each entry is the sum of the two entries immediately above it. For instance,

$$\binom{4}{1} = 4 = 1+3 = \binom{3}{0} + \binom{3}{1},$$

as highlighted by the colored arrows. You should be able to fill in the next row by sight [9].

```
            1
          1   1
        1   2   1
      1   3   3   1
    1   4   6   4   1
  .   .   .   .   .   .
```

FIGURE 4.8

Pascal's Triangle

This famous figure is called **Pascal's Triangle**, after the French mathematician and philosopher Blaise Pascal (1623–1662). Note that the entries are symmetric around the center, which is exactly what Theorem 1 says.

PROOF 1 (ALGEBRAIC) We begin with the RHS (right-hand side) of Eq. (2), since it is more complicated. We substitute in the factorial formula for combinations, Eq. (2), Section 4.4, and massage the expression until it looks like the LHS (left-hand side):

$$\begin{aligned}
\binom{n-1}{k} + \binom{n-1}{k-1} &= \frac{(n-1)!}{k!(n-1-k)!} + \frac{(n-1)!}{(k-1)!\big((n-1)-(k-1)!\big)} \\
&= \frac{(n-1)!(n-k)}{k!(n-1-k)!(n-k)} + \frac{k(n-1)!}{k(k-1)!(n-k)!} \\
&= \frac{(n-1)!(n-k)}{k!(n-k)!} + \frac{(n-1)!k}{k!(n-k)!} \\
&= \frac{(n-1)!}{k!(n-k)!}\big((n-k)+k\big) = \frac{n!}{k!(n-k)!} = \binom{n}{k}. \quad \blacksquare
\end{aligned}$$

[whew!]

PROOF 2 We look for interpretations of the two sides of Eq. (2) that make it clear they are counting the same thing. The LHS is about picking k-subsets from an n-set. The RHS is about picking k-subsets and $(k-1)$-subsets from an $(n-1)$-set. How can choosing from an $(n-1)$-set be about the same thing as choosing from an n-set? Perhaps by singling out one element first, making our decision about that, and then dealing with the rest. That's it! Both sides are about, say, fielding a set of k players from a team roster of n players. On the LHS, we simply choose a set of k players all at once. On the RHS, we have singled out a certain player, call him Joe, and decide about him first. Either we don't field Joe, in which case we have to pick a full set of k players from the remaining $n-1$ players on the roster, or else we do field Joe, in which case we must pick only $k-1$ players from the remaining $n-1$. ∎

For both Theorems 1 and 2, the second proof is a combinatorial argument. There wasn't any computation. Instead, we showed that both sides of the equation count the same thing. In one case we showed this by exhibiting a one-to-one correspondence. In the other, we made up a story about two different approaches to accomplishing the same choice. Combinatorial arguments are harder to think up than algebraic proofs—there is no formula for finding an appropriate correspondence or story, whereas in the algebraic proof you just grind it out (assuming your algebra is good enough!). But with practice you can get good at finding combinatorial arguments. In the process, you will be developing mathematical insight. It may also become a source of satisfaction to you to solve a problem while avoiding computation. Mathematicians love to do this.

Note. We have not defined $C(n,k)$ for either $k < 0$ or $k > n$. Because the previous theorem involves formulas where n and k change by 1, it would be very convenient to let k "slip over" the 0 and n bounds. Precisely because $C(n,k)$ hasn't been defined beyond these bounds, we can define it any way we wish. Mathematicians always strive to extend definitions so that theorems which already hold for the restricted definitions will continue to hold more generally. It turns out in this case that the "right" extension is

---

Definition 1. Let n and k be integers with $n \geq 0$. Then $C(n,k) = 0$ for $k < 0$ and $k > n$.

By plugging into this definition [14], it is easy to check

Theorems 1 and 2 Extended. Let n and k be integers. Then Eq. (1) is true whenever $n \geq 0$ and Eq. (2) is true whenever $n > 0$.

We are now ready for a more substantial result:

Theorem 3. For n any nonnegative integer,

$$\sum_{k=0}^{n} \binom{n}{k} = 2^n. \tag{3}$$

PROOF 1 Actually, we've already proved this equation, and by a combinatorial argument! We just didn't have the notation then to write the result down. To wit, we noted in Section 4.3 two different ways to go about choosing a subcommittee from a committee of n. Either you first choose the size of the subcommittee and then choose a subset with that size, or else you go through the full committee one by one and decide for each member whether that member goes on the subcommittee. The first method gives the LHS (Sum Rule); the second, the RHS (Product Rule). ∎

PROOF 2 We can't give an algebraic proof of the sort we gave for the previous two theorems because Eq. (3) involves a variable number of terms, depending on n. This sort of algebraic situation is ripe for mathematical induction, especially since we already have a result, Theorem 2, that ties combinations of n things to combinations of $n-1$. Let $P(n)$ be the assertion that Eq. (3) is true for the particular value n.

$P(0)$. The RHS is $2^0 = 1$; the LHS is $0!/0!0! = 1$ also.

$P(n-1) \Longrightarrow P(n)$. Take the LHS of Eq. (3), break it down using Theorem 2, and substitute into Eq. (3) itself in the case $n-1$ (the inductive hypothesis). Let's hope this gives us the RHS of Eq. (3). Indeed, here are the computations:

$$\begin{aligned}
\sum_{k=0}^{n} \binom{n}{k} &= \sum_{k=0}^{n} \left[\binom{n-1}{k} + \binom{n-1}{k-1} \right] \\
&= \sum_{k=0}^{n} \binom{n-1}{k} + \sum_{k=0}^{n} \binom{n-1}{k-1} \\
&= \sum_{k=0}^{n-1} \binom{n-1}{k} + \sum_{j=-1}^{n-1} \binom{n-1}{j} \qquad [j = k-1 \text{ in second sum}] \\
&= 2^{n-1} + \sum_{j=0}^{n-1} \binom{n-1}{j} \\
&= 2^{n-1} + 2^{n-1} = 2^n. \ \blacksquare
\end{aligned}$$

Note that we used the extended form of Theorem 2 in the first equation. What allowed us to replace n on top of the first $\sum$ in the second line with $n-1$ on top of the first $\sum$ in the third line? What allowed us to replace $j = -1$ below the second $\sum$ in the third line with $j = 0$ below the $\sum$ in the fourth line?

Our final theorem of this section is not anywhere near as important as the previous ones, but it is a good challenge for you to see if you can now anticipate the combinatorial argument we give to prove it.

Theorem 4. For any nonnegative integer n,

$$\sum_{k=0}^{n} k\binom{n}{k} = n2^{n-1}. \tag{4}$$

Let's think it through. The LHS tells us that we can think in terms of subcommittees again, where we decide on subcommittee size first. But what is this extra factor k? Since it multiplies the number of size k subcommittees, it must be an extra choice for each of these subcommittees. But k is the number of members of such a subcommittee, so we must be singling one of them out for special attention—the subcommittee chair! So the appropriate story is about how to pick a subcommittee and a leader for it. The RHS must give another way to accomplish the same task. All those 2's say that once again, we are making Yes/No subcommittee appointment decisions for committee members. But this time we are doing this for only $n-1$ members. Before that there is one n-way choice. Ah, picking the chair for the subcommittee before making other choices. Now we are ready to write down a more formal argument.

PROOF Both sides of Eq. (4) count the number of subsets of an n-set when each subset also has a distinguished element. We can either first decide the subset size k, then pick the subset, then decide which of the k elements in it is distinguished (LHS); or we can first decide which of the original n elements is to be the distinguished subset element and then, for each of the remaining $n-1$ elements, decide whether or not it is in the subset too (RHS). ∎

Problems: Section 4.5

1. Verify the following equalities by both algebraic and combinatorial arguments:

$$(n-k)\binom{n}{k} = (k+1)\binom{n}{k+1} = n\binom{n-1}{k}.$$

2. Verify that

$$\binom{n+1}{k+1} = \frac{n+1}{k+1}\binom{n}{k}.$$

3. Verify two ways that

   a) $C(n,k)C(n-k,j) = C(n,j)C(n-j,k)$,

   b) $C(n,k)C(n-k,j) = C(n,k+j)C(k+j,k)$.

   Then verify, directly from a) and by a combinatorial argument, that

   c) $C(n,k)P(n-k,j) = P(n,j)C(n-j,k)$.

4. Show that

$$C(n,k) = C(n-2,k-2) + 2C(n-2,k-1) + C(n-2,k).$$

5. Show two ways that

$$P(n,k) = n\,P(n-1,k-1).$$

6. Show two ways that

$$P(n,k) \;=\; P(n-1,k) + k\,P(n-1,k-1).$$

7. Show two ways that

$$P(n,k)\,P(n-k,j) \;=\; P(n,j)P(n-j,k).$$

8. Verify both by induction and by a combinatorial argument:

$$\sum_{k=m}^{n} \binom{k}{m} = \binom{n+1}{m+1}.$$

What does this equality say about Pascal's Triangle?

9. Fill in the next two rows of Fig. 4.8.

10. Use Theorem 2 to write an iterative algorithm that computes Pascal's Triangle down through the nth row.

11. Theorem 2 can also be used to write a recursive algorithm which, given (n,k), outputs $C(n,k)$. The recursive algorithm doesn't compute every number in the nth row to get $C(n,k)$. It doesn't compute all entries in earlier rows either. However, it computes certain previous entries many times (unless a lot of "flags" are included to keep it from doing computations more than once).

a) Write this recursive algorithm (don't worry about flags).

b) What previous combinations does your algorithm compute to get $C(n,k)$? Answer by shading in part of Pascal's Triangle.

c) Conjecture how many times your algorithm computes various entries when invoked to compute $C(n,k)$. (Proving the answer requires the methods of Chapter 5, and even then its tricky.) Which algorithm, the one for this problem or the one in the previous problem, appears to be more efficient?

12. If p is a prime, show that all numbers in row p of Pascal's Triangle are divisible by p (except the 1's at the ends).

13. Show the converse of [12]: If every entry in row n of Pascal's Triangle (except the ends) is divisible by n, then n is a prime.

14. Check that Theorems 1 and 2 are true for all nonnegative integers n and all integers k, if $C(n,k)$ is defined to be 0 for $k<0$ and $k>n$.

15. Use the second equality in [1] to give an algebraic proof of Theorem 4.

16. Give an algebraic, inductive proof of Theorem 4.

17. Prove two ways that

$$\sum_{i=0}^{k} \binom{n}{i}\binom{m}{k-i} = \binom{n+m}{k}.$$

18. For n odd, show that

$$\sum_{k=0}^{\lfloor n/2 \rfloor} \binom{n}{k} = \sum_{k=\lceil n/2 \rceil}^{n} \binom{n}{k}.$$

Therefore, what are both sums equal to?

19. Do [18] again for n even.

4.6 The Binomial Theorem

Multiplying binomials was probably the bane of your high school existence. How fast can you still do $(3x-7)(2x+6)$? A special case was taking a power of a single binomial. Squares like $(2x+3)^2$ probably didn't cause you too much trouble, but higher powers, like $(x+2)^4$, may have been a pain. The Binomial Theorem gives a general formula for all such binomial powers. Actually, now that it is possible to buy a hand calculator which can do all these powers for you, the fact that the Binomial Theorem can do them too may not be so impressive. However, because of the coefficients in its formula, the Binomial Theorem also has many applications to counting, and thus it is still a very important theorem.

First we use the inductive paradigm to guess the statement of the theorem. Then we give some examples. Then we prove it.

To guess the theorem we start by expanding the first few positive integer powers of $x+y$. The results are

$$\begin{aligned}(x+y)^1 &= x+y,\\ (x+y)^2 &= x^2+2xy+y^2,\\ (x+y)^3 &= x^3+3x^2y+3xy^2+y^3,\\ (x+y)^4 &= x^4+4x^3y+6x^2y^2+4xy^3+y^4.\end{aligned}$$

The powers of x and y on the RHS form an easy pattern: it's always x^iy^j, where i goes down by 1 for each new term and j goes up by 1. The sum of i and j is always n, the power on the LHS.

What about the coefficients? These should look familiar, especially if we rewrite $1x$ for x and $1y$ for y: They are just the entries of Pascal's Triangle. We are missing the top of the triangle, but that's easily fixed. It's reasonable to define $(x+y)^0$ to be 1, which gives us the top.

Therefore, with considerable confidence we assert:

Theorem 1, The Binomial Theorem. For all nonnegative integers n,

$$(x+y)^n = \sum_{k=0}^{n} \binom{n}{k} x^{n-k}y^k. \tag{1}$$

Because of this theorem, the $C(n,k)$'s are also called **binomial coefficients**. (Do you see [14] that we could also write the RHS as

$$\sum_{k=0}^{n}\binom{n}{k}x^ky^{n-k} \;=\; y^n+\binom{n}{1}xy^{n-1}+\binom{n}{2}x^2y^{n-2}+\cdots+\binom{n}{n-1}x^{n-1}y+x^n\,?$$

Sometimes it is convenient to write a polynomial with descending powers of x, as in Eq. (1), and other times with ascending powers, as in this alternative version.)

EXAMPLE 1 Expand $(x+2)^4$.

Solution Setting $y=2$ and $n=4$ in the Binomial Theorem, we have

$$\begin{aligned}(x+2)^4 &= 1x^42^0+4x^32^1+6x^22^2+4x^12^3+x^02^4\\ &= x^4+8x^3+24x^2+32x+16. \;∎\end{aligned}$$

EXAMPLE 2 Set $x=y=1$ in the Binomial Theorem and see what happens.

Solution The LHS of the Binomial Theorem is easy: $(1+1)^n = 2^n$. The RHS simplifies nicely too: All the powers of x and y go away. Therefore we obtain

$$2^n = \sum_{k=0}^{n}\binom{n}{k}.$$

We got this same result two other ways in Section 4.5. ∎

EXAMPLE 3 Derive a general expansion for $(x-y)^n$ and then apply the special case $x = y = 1$ to it.

Solution Since $x - y = x + (-y)$, we can reduce to the previous case by substituting $-y$ for y in the Binomial Theorem. This substitution is perfectly legitimate: y represented any number, and so does $-y$. We get

$$\begin{aligned}(x-y)^n &= \sum_{k=0}^{n} \binom{n}{k} x^{n-k}(-y)^k \\ &= \sum_{k=0}^{n} (-1)^k \binom{n}{k} x^{n-k} y^k.\end{aligned}$$

Now, when $x = y = 1$, the LHS is even simpler than before: It's 0, at least when $n \geq 1$. So we get

$$0 = \sum_{k=0}^{n} (-1)^k \binom{n}{k}. \quad \blacksquare \tag{2}$$

Can you think of a nice way to put in words what this last identity says? Can you think of a combinatorial argument for it? [7]

EXAMPLE 4 Compute $(1.02)^3$ to four decimal places.

Solution We haven't talked about approximate computations before, so let's evaluate this expression exactly, using the Binomial Theorem, and see what we get. Plugging into the theorem with $x = 1$ and $y = .02$ we get

$$(1.02)^3 = 1 + 3(.02) + 3(.0004) + 1(.000008).$$

Generalizing, we see that whenever y is close to zero, the terms involving higher and higher powers of y will contribute less and less to the answer. In this case, we now realize that it would have sufficed to quit after the y^2 term. In particular, to 4 decimal places,

$$(1.02)^3 \approx 1 + .06 + .0012 = 1.0612. \quad \blacksquare$$

EXAMPLE 5 Compute $\sqrt{1.02}$.

Solution $\sqrt{1.02} = (1.02)^{1/2}$, so again set $x = 1$ and $y = .02$ and plug in:

$$\sqrt{1.02} = \sum_{k=0}^{1/2} \binom{\frac{1}{2}}{k} 1^{\frac{1}{2}-k} (.02)^k. \tag{3}$$

The factor $1^{(\frac{1}{2}-k)}$ equals 1 for any k. The factor $C(\frac{1}{2}, k)$ is a bit of a problem—you can't choose k things from $\frac{1}{2}$ thing! But recall from Eq. (4), Section 4.4, that one formula for $C(\alpha, k)$ is

$$\binom{\alpha}{k} = \frac{1}{k!}\prod_{j=1}^{k}(\alpha+1-j) = \frac{\alpha(\alpha-1)(\alpha-2)\cdots(\alpha-k+1)}{k!}, \tag{4}$$

which makes sense for any *real number* α. (We have switched from n to α to emphasize this fact; traditionally n refers to integers only.) *Note*: Since there are k factors in the product, in both numerator and denominator, k must still be a nonnegative integer. ($k = 0$ is all right because empty products are defined to be 1, as discussed in Section 2.7. See also [22].) Thus

$$\binom{\frac12}{0} = 1, \quad \binom{\frac12}{1} = \frac{1}{2}, \quad \binom{\frac12}{2} = \frac{(\frac12)(-\frac12)}{2} = -\frac{1}{8}, \quad \binom{\frac12}{3} = \frac{(\frac12)(-\frac12)(-\frac32)}{6} = \frac{1}{16},$$

and so on.

There's one more novelty in Eq. (3): a noninteger on top of the summation sign. Recall that $\sum_{k=a}^{b}$ means to start the sum at $k = a$ and go up by 1 until you hit b. But starting at 0, k will never hit $\frac12$. So let's just keep going:

$$\begin{aligned}\sqrt{1.02} &= 1 + (\frac{1}{2})(.02) + (-\frac{1}{8})(.0004) + (\frac{1}{16})(.000008) + \cdots \\ &\approx 1.00995005\,. \ ∎\end{aligned}$$

Is this a joke? We take the Binomial Theorem, which we have asserted for nonnegative integers n, and blithely assume that it is also true for all real α. This is pretty outrageous, considering that the only evidence we have for the theorem are the calculations we made at the beginning of the section for $n = 1, 2, 3, 4$, and the only general proof technique we have is induction, which only works for integers. Yet we have just charged ahead, giving $C(\alpha, k)$ and $\sum_{k=0}^{\alpha}$ the first plausible interpretation that occurs to us. This is the sort of unthinking plug-in that math teachers attribute to their most mechanically minded students!

Yes, it is a joke. But like all good jokes, there's something serious underneath. The method works! 1.00995005 is extremely close to $\sqrt{1.02}$. In fact, it is correct to seven decimal places. Furthermore, if we added all those other terms for $k = 4, 5, \ldots$, it turns out we would get $\sqrt{1.02}$ *exactly.*

The point is, it's a rather marvelous fact about mathematics that results which are proved in limited domains often hold much more generally, if we can only figure out how to state them. Therefore, sometimes it's worth our while to let our minds wander expansively and test if our formulas still work.

The infinite sum binomial theorem we have hit upon is called the Binomial *Series* Theorem. We are in no position to prove it yet, because we won't even give a precise definition of infinite sum until Chapter 9. (You can't actually do an infinite number of additions, so infinite sum has to be defined in terms of "limits" of finite sums.) However, we will state the series theorem precisely, right after we prove the ordinary (finite sum) Binomial Theorem. We do that now. We give two proofs: the first, an algebraic proof by induction; the second, a combinatorial argument.

FIRST PROOF OF THEOREM 1 The case $n = 1$ is an easy check. Assume that the formula is true for n and now extend it to $n + 1$. Since

$$(x+y)^{n+1} = (x+y)^n(x+y),$$

we may assume that

$$(x+y)^{n+1} = \left(\sum_{k=0}^{n} \binom{n}{k} x^{n-k}y^k\right)(x+y). \tag{5}$$

Now all we have to do is multiply out the two factors on the right, combine like terms, and see if it's what we want. Since the y's have simpler exponents than the x's, let's concentrate on combining terms with the same power of y. After we expand the RHS of Eq. (5), y appears to each integer power from 0 to $n+1$. Let j be any such power. It is helpful to restate the RHS of Eq. (5) as

$$\left(\cdots + \binom{n}{j-1}x^{n-(j-1)}y^{j-1} + \binom{n}{j}x^{n-j}y^j + \cdots\right)(x+y).$$

The only way to get y^j when we multiply out is to multiply $\binom{n}{j}x^{n-j}y^j$ from the first factor by x from the second, or multiply $\binom{n}{j-1}x^{n-(j-1)}y^{j-1}$ from the first factor by y from the second. Combining the two products, we get

$$\left[\binom{n}{j-1} + \binom{n}{j}\right] x^{n+1-j}y^j.$$

But, by Theorem 2, Section 4.5, this is just

$$\binom{n+1}{j}x^{n+1-j}y^j.$$

Since this argument holds for any j (see [13]), we conclude that the RHS of Eq. (5) is

$$\sum_{k=0}^{n+1}\binom{n+1}{k}x^{n+1-k}y^k.$$

But this is just what the case $n+1$ of the Binomial Theorem states. The inductive step is complete. ∎

SECOND PROOF OF THEOREM 1 We are looking for a formula for the expansion of

$$(x+y)(x+y)\cdots(x+y)$$

when there are n factors. Terms in the expansion arise by (and only by) our selecting exactly one of x and y from each of the n factors and multiplying our selections together. For instance, if we pick x from the second and fourth factors and y from all the others, we get a contribution of x^2y^{n-2} to the expansion. The expansion can be simplified by combining like terms. For instance, choosing x from the first and second factors and y from the others also leads to x^2y^{n-2}. So we merely need to decide what sorts of terms of the form x^iy^j *can* occur and how many of them *do* occur.

Since we get one power of either x or y from each factor $(x+y)$, and there are n factors, we conclude that $i+j=n$ and both i and j are nonnegative. That is, all terms are of the form $x^{n-k}y^k$, where k goes from 0 to n. Now, how many ways can we pick exactly k y's from the n factors $(x+y)$? We are simply picking a subset of k y's from a set of n y's. The number of ways to do this is $C(n,k)$. Thus the expansion is $\sum_{k=0}^{n} C(n,k)x^{n-k}y^k$. ∎

We hope you liked the second proof better—it reduces the Binomial Theorem to recognizing the definition of $C(n,k)$. Surprisingly, it doesn't seem to involve induction at all. However, that's a misleading perception. When you find two proofs of something, one clearly using induction and the other seemingly not, what has usually happened in the second proof is that the induction has been pushed back to something so well accepted that you don't immediately recognize it as an induction. Where is the induction in our second proof? In the claim that to multiply a string of binomial factors, you add all the terms you can create by taking one summand from each of the factors. For instance, for two factors you get

$$(a_1+a_2)(b_1+b_2) = a_1b_1 + a_1b_2 + a_2b_1 + a_2b_2.$$

This fact may be very familiar, but a careful proof for arbitrarily many factors requires induction and a firm command of algebra. To get from n factors to $n+1$, you use the distributive law. See [28].

We promised a statement of the Binomial Series Theorem.

Theorem 2, Binomial Series Theorem. If α is any *real* number, and $|y| < |x|$, then

$$(x+y)^\alpha = \sum_{k=0}^{\infty} \binom{\alpha}{k} x^{\alpha-k}y^k, \tag{6}$$

where $\binom{\alpha}{k}$ is defined as in Eq. (4).

Actually, what one usually sees is the special case

$$(1+x)^\alpha = \sum_{k=0}^{\infty} \binom{\alpha}{k} x^k \qquad \text{when } |x| < 1. \tag{7}$$

(The number 1 has been substituted for x in Eq. (6) and then y has been renamed x.) This case isn't really any less general, because for any x and y we may write

$$(x+y)^\alpha = x^\alpha\Big(1+\frac{y}{x}\Big)^\alpha.$$

Thus instead of using x and y in Eq. (6), we may use Eq. (7) with y/x in place of x and then multiply the result by x^α. Also, while the condition $|y| < |x|$ in the general theorem may seem mysterious, the equivalent condition $|x| < 1$ in the special case is quite plausible: If $|x| < 1$, then the factors x^k go to 0 as k increases, so it is credible that the sum effectively stops growing after some number of terms. Conversely, if $|x| \geq 1$, then x^n does not go to 0, so it is credible that there is no limiting value for the sum as we add more terms. But as we've said, we are in no position to give careful proofs of any of these claims.

The Multinomial Theorem

Why stop at a formula for $(x+y)^n$? How about $(x+y+z)^n$, or even $(\sum_{k=1}^m x_k)^n$. The second proof of the Binomial Theorem suggests how to proceed. For instance, in the case of $(x+y+z)^n$, every term in the expansion should be of the form $C(n; i, j, k)x^i y^j z^k$, where i, j, and k are nonnegative integers summing to n, and $C(n; i, j, k)$ is the number of ways to pick x from i of the n factors, y from j of them, and z from the remaining k of them. The only remaining step is to find a formula for $C(n; i, j, k)$, which is called a trinomial coefficient. Let's begin with a general definition.

Definition 1. The **trinomial coefficient** $C(n; i, j, k)$ is the number of ways to partition n distinct objects into three distinct sets, the first set with i elements, the second with j, and the third with the remaining $k = n-i-j$. It is often written

$$\binom{n}{i, j, k}.$$

Similarly, the **multinomial coefficient** $C(n; k_1, k_2, \ldots, k_m)$, often written

$$\binom{n}{k_1, k_2, \ldots, k_m},$$

is the number of ways to partition n distinct objects into m distinct sets, with k_i elements in the ith set.

Note. The word "partition" means that each object appears in exactly one of the sets. Therefore, although we didn't say it explicitly in the definition, the sum of the k_i's must be n.

Theorem 3. For all nonnegative integers n,

$$(x+y+z)^n = \sum_{i+j+k=n} \binom{n}{i, j, k} x^i y^j z^k, \tag{8}$$

where the sum is understood to range over all triples of nonnegative integers i, j, k that add to n. More generally,

$$(x_1+x_2+\cdots+x_m)^n = \sum_{k_1+k_2+\cdots+k_m=n} \binom{n}{k_1, k_2, \ldots, k_m} x_1^{k_1} x_2^{k_2} \cdots x_m^{k_m}, \tag{9}$$

where each k_i must be a nonnegative integer.

Equation (9) is called the **Multinomial Theorem**. The special case, Equation (8), is sometimes called the **Trinomial Theorem**.

PROOF It suffices to prove the general case, Eq. (9). When we multiply n sums together, as on the LHS of Eq. (9), the result is the sum of all terms obtained by multiplying together one term from each sum. For instance,

$$(a+b+c)(p+q+r)(x+y+z) \;=\; apx + apy + apz + aqx + \cdots + cry + crz.$$

In Eq. (9), all sums on the LHS are the same. Therefore the result contains like terms that can be combined. Each term in the result is the product of a certain number of x_1's, x_2's, ..., x_m's, with n of these factors altogether (counting multiplicities). Thus the result will contain the term ${x_1}^{k_1} x_2^{k_2} \cdots {x_m}^{k_m}$ as many times as we can pick x_1 from k_1 of the copies of $S = x_1 + x_2 + \cdots + x_m$, pick x_2 from k_2 other copies of S, and so on. But this is just the number of ways to partition the set of n copies of S into m distinct sets, of sizes $k_1, k_2, \ldots, k_m$, respectively. By definition, this is $\binom{n}{k_1, k_2, \ldots, k_m}$. Thus, after combining like terms, we obtain the RHS of Eq. (9). ∎

Theorem 3 isn't of much help unless we can evaluate multinomial coefficients. The next theorem shows how.

Theorem 4.

$$\binom{n}{i, j, k} = \frac{n!}{i!j!k!}. \tag{10}$$

More generally,

$$\binom{n}{k_1, k_2, \ldots, k_m} = \frac{n!}{k_1!k_2!\cdots k_m!}. \tag{11}$$

PROOF We outline an inductive proof. For a one-to-one correspondence, which gets the multinomial case all at once, see [30]. To verify Eq. (10), think in terms of picking the three subsets sequentially. For the first subset, we must pick i of the n elements. There are $C(n, i)$ ways to do this (an ordinary binomial coefficient). Now, whatever $n{-}i$ elements are left, there are $C(n{-}i, j)$ ways to pick the second subset. At this point, no choices are left: The third subset must consist of all the remaining elements. By the Product Rule, we conclude that

$$\binom{n}{i, j, k} = \binom{n}{i}\binom{n-i}{j} = \frac{n!}{i!(n-i)!}\,\frac{(n-i)!}{j!(n-i-j)!} = \frac{n!}{i!j!(n-i-j)!}.$$

Since $k = n{-}i{-}j$, we have Eq. (10). Then we may prove Eq. (11) by induction with Eq. (10) as the basis case. The key idea of the inductive step is: Partitioning n into $m+1$ sets can be accomplished by first picking the set of size k_1, and then partitioning the remaining $n{-}k_1$ elements into m sets [29]. ∎

Problems: Section 4.6

1. Evaluate:
a) $(x+y)^5$ **b)** $(x-1)^4$ **c)** $(2x-1)^3$

2. Approximate to four decimal places:
a) $(1.01)^4$ **b)** $(.98)^3$ **c)** $(2.01)^3$

3. Show that $\sum_{k=0}^{n} C(n,k)2^k = 3^n$.

4. Simplify
a) $\sum_{k=0}^{n}(-1)^k C(n,k)2^k$
b) $\sum_{k=0}^{n}(-1)^k C(n,k)4^k$
c) $\sum_{k=0}^{n}(-1)^k C(n,k)2^{n-k}$
d) $\sum_{k=0}^{n}(-1)^k C(n,k)2^{n-k}3^k$
e) $\sum_{k=0}^{n}(-1)^k C(n,k)2^n 3^k$

5. What is the effective rate of interest per year if the "nominal" rate is 12% and compounding is monthly? (If a nominal rate of r is compounded k times a year, that means the principal is multiplied by $(1+\frac{r}{k})$ a total of k times. Thus a nominal rate of 20% compounded semiannually means that the principal is multiplied by $(1+.1)^2 = 1.21$, so the effective (or "annualized") interest rate is 21%.

6. Show that $(a+\epsilon)^n \approx a^n + n\epsilon a^{n-1}$. Here ϵ stands for a number small enough relative to a that ϵ^2 and higher powers of ϵ are negligible. How is this problem relevant to the previous problem?

7. **a)** Eq. (2) can be put nicely into words as a statement about the number of odd and even subsets. Make such a statement.
b) What does Eq. (2) tell you about the size of $\sum_{k \text{ even}}^{n} C(n,k)$?

8. Use the Binomial Theorem to prove that

$$(1+x)^n \geq 1+nx$$

for any $x \geq 0$ and n nonnegative. This was proved directly by induction in Example 1, Section 2.2.

9. Use the Binomial Theorem to prove that

$$(1+x)^n > 1+nx$$

for any $x > 0$ and $n > 1$.

10. Approximate $\sqrt{1.5}$. Let $x = 1$ and $y = .5$ in the Binomial Series and evaluate terms until it seems there is no further change in the first three decimal places. Compare with the answer from a calculator.

11. If ϵ is much smaller than a, a good approximation of $\sqrt{a^2+\epsilon}$ is

$$S = a + \frac{\epsilon}{2a}.$$

Justify this claim by simply squaring S. Now explain how you might *discover* this fact by using the Binomial Series Theorem.

12. Approximate $\sqrt[3]{9}$ using the Binomial Series Theorem. Compute four terms. Let $x = 8$ and $y = 1$, or equivalently, note that $\sqrt[3]{9} = 2\sqrt[3]{1+\frac{1}{8}}$ and use the special case of the Binomial Series Theorem. Compare your answer with that from a calculator.

13. In the first proof of Theorem 1, we hid some things under the rug. We said our j could be any integer from 0 to $n+1$ but, in fact, if $j = 0$ or $j = n+1$, then some of the things we did need more explanation. What things? What's the explanation?

14. Prove the version of the Binomial Theorem displayed right after Eq. (1)—the version with ascending powers of x—four ways:
a) Substitute y for x and x for y in Eq. (1).
b) Substitute $n-k$ for k in Eq. (1).
c) Do an induction similar to the first proof of the Binomial Theorem in the text.
d) Give a combinatorial argument parallel to the second proof in the text.

15. Compute

$$\binom{5}{2,2,1} \quad \text{and} \quad \binom{6}{1,2,3}.$$

16. Are $C(6;1,2,3)$ and $C(6;3,2,1)$ the same number? Do they count the same thing?

17. Simplify:
a) $\sum_{i+j+k=n} C(n;i,j,k)$,
b) $\sum_{i+j+k=n} (-1)^k C(n;i,j,k)2^j/3^{i+j}$

18. Express $C(n,k)\,C(n-k,j)$ as a trinomial. Do the same for $C(n,k)\,C(k,j)$.

19. Find a simple formula for $C(-1,k)$.

20. Verify that

$$\binom{-n}{k} = (-1)^k\binom{n+k-1}{k}.$$

21. **a)** Verify that

$$\binom{-1/2}{k} = \frac{(-1)^k(2k)!}{2^{2k}(k!)^2}.$$

b) Find a similar formula for $C(1/2, k)$.

22. $C(n, k)$ was initially defined for integers $0 \le k \le n$ (Def. 2, Section 4.4). We've extended that definition twice, in Def. 1, Section 4.5 and in Eq. (4). Are there any pairs (n, k) to which the two extensions assign conflicting values?

23. **a)** When $|x| < 1$, the infinite geometric series $\sum_{k=0}^{\infty} x^k$ sums to $1/(1-x)$. Show that this fact is a special case of the Binomial Series Theorem.

b) Find an infinite series which sums to $1/(1-x)^2$ when $|x| < 1$.

c) Generalize.

24. Since $x^{\alpha-k}y^k = x^\alpha(y/x)^k$, the Binomial Series Theorem may be rewritten as: If $|y| < |x|$ then

$$(x+y)^\alpha = x^\alpha \sum_{k=0}^{\infty} \binom{\alpha}{k}\left(\frac{y}{x}\right)^k.$$

This form of the theorem makes the condition $|y| < |x|$ less mysterious. Why?

25. Implicit in our introduction of multinomial coefficients is a new meaning for binomial coefficients. Specifically, there ought to be a binomial coefficient $C(n; i, j)$, where $j = n-i$, and it ought to be the number of ways to partition an n-set into two subsets of size i and j. But the binomial coefficient we talked of previously, and wrote as $C(n, i)$, was defined as the number of ways to pick out *one* subset of size i from an n-set. Is there a contradiction between the two uses of "binomial"?

26. Prove the Trinomial Theorem by noting that

$$(x+y+z)^n = [x + (y+z)]^n$$

and applying the Binomial Theorem twice.

27. Think up and prove analogs of Theorems 1 and 2, Section 4.5, for trinomial coefficients. Ever heard of Pascal's pyramid?

28. In the second proof of the Binomial Theorem, we hid the induction in the claim that to multiply a string of binomial factors, you add all the terms you can create by taking one summand from each of the factors. Make the induction explicit by proving this claim inductively. Start by stating the claim as an equation; the LHS should be $\prod_{k=1}^{n}(a_k + b_k)$.

29. Complete the proof of Theorem 4.

30. We claim there is a one-to-one correspondence between the number of ways to partition an n-set of distinct elements into m distinct subsets of sizes $k_1, \ldots, k_m$ and the number of ways to permute the letters of a word of length n, when there are only m distinct letters in the word and the ith letter occurs k_i times. Justify this claim and use it to prove Eq. (11).

The following problems are for students who know calculus.

Derive Theorem 4, Section 4.5, again as follows. Take the Binomial Theorem, regard x as a constant, and differentiate with respect to y. Now make an appropriate substitution for x and y.

32. Derive a new theorem, similar to Theorem 4, Section 4.5, by differentiating the Binomial Theorem with respect to x. Can you also prove your new theorem directly from the Binomial Theorem without calculus?

33. Evaluate

a) $\sum_{k=0}^{n} k(k-1)\binom{n}{k}$ **b)** $\sum_{k=0}^{n} k^2\binom{n}{k}$

c) $\sum_{k=0}^{n} \binom{n}{k}/(k+1)$. (Careful!)

34. Assuming that $(fg)' = f'g + fg'$, prove Leibniz' formula for the nth derivative of a product:

$$(fg)^{(n)} = \sum_{k=1}^{n} \binom{n}{k} f^{(k)} g^{(n-k)}.$$

4.7 Four Common Problems with Balls and Urns

How can *any* problems with balls and urns be common? The answer is that they form paradigms. Many problems that don't on the face of it have anything to do with balls or urns are equivalent to problems of putting balls in urns. So if we can count the number of ways to place balls in urns, we can solve a lot of other counting problems.

We could have introduced these paradigms anywhere in this chapter. Introduced earlier, they could have provided our first examples of one-to-one correspondences. However, we would not have had the techniques to solve most of the counting problems associated with the paradigms. We do now, or will shortly.

We assume throughout this section that there are b balls and u urns. We also assume that each urn can hold any number of balls and that there is no order to the balls within an urn. These assumptions leave us with four situations: The balls can be distinguishable or indistinguishable, and independently the urns can be distinguishable or indistinguishable.

"Distinguishable" and "indistinguishable" are long words; sometimes "labeled" and "unlabeled" are used instead. In any event, Fig. 4.9 indicates what these words mean. The balls are numbered (distinguishing them), and the urns have different colors (distinguishing them). Thus Fig. 4.9(a), (b), and (c) represent different ways to put three balls into two urns. However, if we ignore the numbers on the balls (making them *in*distinguishable), then (a) and (b) are the same, but (c) is still different. If instead we ignore the colors (making the *urns* indistinguishable), then (a) and (c) are the same and (b) is different. If we ignore both numbers and colors, then all three parts are the same.

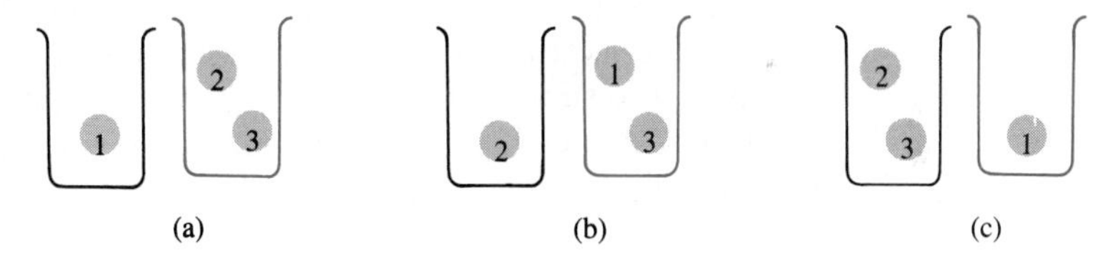

FIGURE 4.9 Three ways to put 3 balls into 2 urns. How many of them are different depends on what is distinguishable.

There are, to be sure, further possibilities. For instance, some of the balls may be distinguishable and others not. Some urns may hold only a limited number of balls. The urns might instead be stacks, which means the balls are necessarily ordered within them, and this order might make a difference. The urns might be in a row, in which case *their* order might make a difference. But the four basic problems above will be enough to keep us busy for now.

Before we try to find counting formulas for these four paradigms, let's consider several problems that reduce to them. For now we only do the reductions. Later in the section we will solve most of the problems.

EXAMPLE 1 How many different configurations are there in Towers of Hanoi with n rings?

Putting a ring on a pole is like putting a ball in an urn. The rings are distinguishable by size, so we must use distinguishable balls. The poles are distinguishable too—it's not enough, for instance, to get all the rings on *some* pole; it's got to be the target pole. True, the rings are ordered on the pole, whereas the balls are unordered in the urn. However, the rules of TOH result in a unique ordering. That is, given any specific set of rings to put on a given pole, there is just one way to do it, just as there is only one way to put balls as an unordered set in an urn. In conclusion, we wish to count the number of ways to put n distinguishable balls in three distinguishable urns.

EXAMPLE 2 How many solutions are there in nonnegative integers to the equation

$$x + y + z + w = 73\,?$$

Think of building up a solution unit by unit, that is, by starting with 73 1's and assigning them one at a time to the variables. So the 1's are like balls and the variables are like urns. What's distinguishable? In the end, if three 1's get assigned to x, it makes no difference if they are the first three 1's or the last three. But it does makes a difference if they are assigned to x or to y. For example, (3,40,20,10) and (40,20,10,3) are considered different solutions. Thus we are asking: How many ways can 73 indistinguishable balls be put in 4 distinguishable urns?

EXAMPLE 3 In chemistry, the electrons in a given ring of an atom can each be in one of several energy states. How many ways can six electrons be assigned to five energy states?

It's easy enough to think of electrons as balls and the states as urns. The urns are distinguishable, because we can measure different energy levels. The harder question is: Do electrons act distinguishably? We think of them as distinct entities, so it is natural to suppose they will act distinguishably. However, it is an empirical fact that they don't. The appropriate model is putting *in*distinguishable balls in distinguishable urns, as in Example 2.

EXAMPLE 4 How many ways can $4n$ bridge players be split into $2n$ teams for a round-robin tournament?

Putting a person on a team is like putting a ball in an urn. People are certainly distinguishable, so we need distinguishable balls. What about the teams (urns)? We can, and probably do, give them different names, but does it really make a difference if Jones and Smith are called Team A and White and Roberts are called Team B—instead of vice versa? No. Therefore we are putting $4n$ distinguishable balls into $2n$ indistinguishable urns. Since a bridge team always has two members, we must further insist that every urn gets exactly two balls. (We will not include such restrictions when we try to solve the balls and urns problems later, but in fact, for distinguishable balls in indistinguishable urns, such restrictions make the problem easier. [18])

EXAMPLE 5 A **partition** of a positive integer n is any set of positive integers which sum to n. Thus $\{5,1,1\}$ and $\{4,3\}$ are partitions of 7. So is $\{3,4\}$, but it's not a different partition from $\{4,3\}$. How many partitions are there of n?

This is like Example 2, except that

- we are not restricted to a fixed number of variables;
- the variables are not distinguished, e.g., $x = 3, y = 4$ is not different from $x = 4, y = 3$; and
- each variable must be positive.

At most there are n variables (since the longest way to partition n into positive integers is to divide it into n 1's), but there might be only one. For instance, there are 7 partitions of 5:

$$\begin{array}{lllll} 5 \\ 4 & 1 \\ 3 & 2 \\ 3 & 1 & 1 \\ 2 & 2 & 1 \\ 2 & 1 & 1 & 1 \\ 1 & 1 & 1 & 1 & 1 \end{array}$$

(We have written the partitions in the traditional form: in descending order of size of the parts. Partitions are unordered, but writing them down imposes an order, so the best way to avoid unintended repetitions is to use a standard order.)

We can make Example 5 look much more like Example 2 by insisting that there be exactly n integers in the partition, but allowing 0's. Thus the list above becomes

$$\begin{array}{lllll} 5 & 0 & 0 & 0 & 0 \\ 4 & 1 & 0 & 0 & 0 \\ 3 & 2 & 0 & 0 & 0 \\ 3 & 1 & 1 & 0 & 0 \\ 2 & 2 & 1 & 0 & 0 \\ 2 & 1 & 1 & 1 & 0 \\ 1 & 1 & 1 & 1 & 1 \end{array}$$

Now Example 5 is the same as Example 2, except that the variables here are indistinguishable. In short, we are asking for the number of ways to put n indistinguishable balls (1's) into n indistinguishable urns (the variables).

What we have called partitions are sometimes called **unordered partitions** to make clear the distinction from the situation in Example 2. However, the study of unordered partitions is such an old and venerable part of mathematics (it's considered part of number theory) that usually they are just called partitions.

EXAMPLE 6 How many ways are there to pick a combination of k things from a set of n distinct things if repetitions are allowed? (This is just the number $C^*(n,k)$ of multisets discussed at the end of Section 4.4.)

We turn this problem into balls and urns as follows. The urns are the n distinct types of objects. The balls are the k individual selections of objects. In other words, each time we put a ball into urn j, that means we select distinct object j another time. Since we are counting combinations, not permutations, the order of selection makes no difference. For instance, if two balls go in urn j, we don't care if they are the first two or the last two. Thus the balls are indistinguishable. In summary, $C^*(n,k)$ is just the number of ways to put k indistinguishable balls into n distinguishable urns.

Alternatively, it suffices to note that $C^*(n,k)$ equals the number of nonnegative integer solutions to

$$x_1 + x_2 + \cdots + x_n = k. \tag{1}$$

The point is: To uniquely determine a k-multiset from n distinct objects, we merely need to determine how many objects we will choose of each type. Let x_j be the number we choose of type j. Then our decision amounts to a solution to Eq. (1). Conversely, every solution to Eq. (1) defines a multiset. Finally, from Example 2, we already know that the number of solutions is the number of ways to put k indistinguishable balls in n distinguishable urns.

These correspondences are illustrated in Fig. 4.10 for a few of the possible arrangements with $b = 6$ and $u = 3$.

Enough examples. Let's now stick to balls and urns and try to solve the four problems. At the end of this section we will summarize the results.

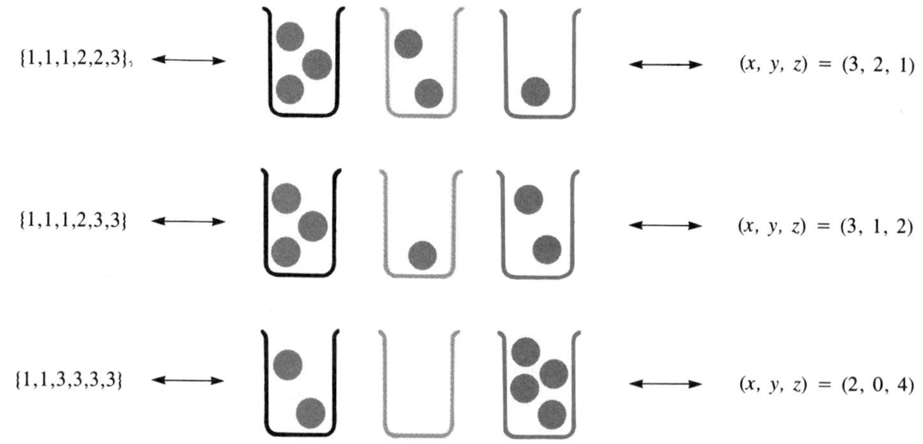

{1,1,1,2,2,3} ⟷ ⟷ (x, y, z) = (3, 2, 1)

{1,1,1,2,3,3} ⟷ ⟷ (x, y, z) = (3, 1, 2)

{1,1,3,3,3,3} ⟷ ⟷ (x, y, z) = (2, 0, 4)

FIGURE 4.10 Illustration of the one-to-one correspondence from 6-multisets with 3 distinct objects, to ways to put 6 indistinguishable balls into 3 distinguishable urns, to the positive integer solutions to $x + y + z = 6$.

Question 1: In how many ways can b distinguishable balls be put in u distinguishable urns?

Since the balls are distinguishable, we can talk of a first ball, a second ball, and so on; furthermore, we can decide independently for each ball which urn to put it in. Since the urns are distinguishable, there are u choices per ball. By the product rule, the total number of choices is u^b.

Solution to Example 1 The number of TOH configurations is 3^n. ∎

Question 2: In how many ways can b indistinguishable balls be put in u distinguishable urns?

We are going to solve this by an ingenious one-to-one correspondence. In Chapter 5, we will show a more systematic way to discover this formula. Also see [19] for another ingenious method. As always, the value of developing more advanced techniques is that more problems become solvable by standard applications of those techniques rather than requiring a flash of insight. But the insights are still pretty!

Since the urns are distinguishable, we can imagine them numbered from 1 to u and lined up from left to right by number. Drawn schematically, with sides of consecutive urns flush with each other, they look like

Now let's imagine an arrangement of the balls in the urns. As the balls are indistinguishable, we might just as well line up the balls in each urn from left to right. Our picture becomes, for example,

We can simplify this picture still further to

What do we see? A collection of balls and *walls*. We must begin and end with walls, we must have b balls, and we must have $u-1$ internal walls. (Why?) But the order of the b balls and $u-1$ internal walls is arbitrary. Every sequence of balls and internal walls is a schematic picture of a way of putting indistinguishable balls into distinguishable urns (distinguished by their order from left to right). Conversely, each way of putting indistinguishable balls into distinguishable urns reduces to a balls and walls picture by the process of pictorial abstraction that we've just gone through. So we merely have to count the number of ways to line up b balls and $u-1$ internal walls. There are $b+u-1$ positions in the line-up, and b of them must be balls. Therefore the answer to Question 2 is $C(b+u-1, b)$. That's it! No calculations at all!

Solution to Example 2 The number of solutions is

$$\binom{73+4-1}{73} = \frac{76!}{73!3!} = \frac{76 \cdot 75 \cdot 74}{3 \cdot 2 \cdot 1} = 70,300. \blacksquare$$

Solution to Example 6 $C^*(n,k) = C(n+k-1,k)$. Combinations both with and without repetitions are counted by binomial coefficients. ∎

> **Question 3:** In how many ways can b distinguishable balls be put in u indistinguishable urns?

Watch closely. We will set up a many-to-one correspondence between this problem and the problem in Question 1. Namely, for each way to put the balls in indistinguishable urns, we can distinguish these urns in $u!$ ways (just number the urns every possible way). Moreover, every arrangement of b distinguishable balls in u *dis*tinguishable urns is associated in this way with some arrangement of b distinguishable balls in u *in*distinguishable urns—just take the numbers off the urns. Thus we have a $u!$-to-1 correspondence. So the answer to Question 3 is the answer to Question 1 divided by $u!$, or $u^b/u!$.

Convinced? We've warned you to be careful with the Division Rule (end of Section 4.3). Is our answer plausible? Every natural number up to u is a factor of its denominator, but only factors of u appear in the numerator. For instance, with $u = 3$ and $b = 2$,

$$\frac{u^b}{u!} = \frac{3 \cdot 3}{3 \cdot 2 \cdot 1} = \frac{3}{2}.$$

This isn't an integer. It can't be right!

The error is: Our correspondence is not always $u!$-to-1. Some configurations with indistinguishable urns match up with fewer than $u!$ configurations with distinguishable urns. For instance, suppose that there are four balls, 1,2,3,4, and three indistinguishable urns. One arrangement is

|_①_| |_② ④_| |_③_| (2)

If we now label the urns by their positions, then arrangement (2) certainly does correspond to $3! = 6$ different arrangements with distinguishable urns: Just consider each permutation of the urns. For instance, switching the last two urns, we obtain

|_①_| |_③_| |_② ④_|

This is a different arrangement of labeled urns than arrangement (2), because now balls 2 and 4 are in the third urn.

But suppose that our initial arrangement, before we give the urns labels by position, is

|_① ② ③ ④_| |___| |___| (3)

Now if we label the urns by position, and then switch the last two urns, we are left with exactly the same labeled arrangement: four balls in urn one and no balls in either urn two or urn three. Arrangement (3), viewed as having indistinguishable urns, corresponds to only three arrangements with distinguishable urns—there are only three choices for where the set of balls $\{1,2,3,4\}$ can be put.

Our $u!$-to-1 correspondence breaks down only when two or more urns are empty. So for the moment, let's just worry about arrangements in which *no* urn is empty. Specifically, define

$$T(b,u) = \text{the number of ways to put } b \text{ distinguishable balls into } u \text{ distinguishable urns, with no empty urns.}$$

Also define

$$S(b,u) = \text{the number of ways to put } b \text{ distinguishable balls into } u \text{ } \textit{in}\text{distinguishable urns, with no empty urns.}$$

Then we really do have a $u!$–to-1 correspondence between the things counted by $T(b,u)$ and $S(b,u)$, so

$$S(b,u) = \frac{1}{u!}T(b,u). \tag{4}$$

Next, let

$$B(b,u) = \text{the number of ways to put } b \text{ distinguishable balls into } u \text{ indistinguishable urns, with empty urns allowed.}$$

Thus $B(b,u)$ is just a name for the answer to Question 3. Because the urns are indistinguishable, if k of them are empty, it doesn't matter which k. It is as if we only had $u-k$ urns to start with. Thus

$$B(b,u) \;=\; \sum_{j=0}^{u} S(b,j) \;=\; \sum_{j=0}^{u} \frac{T(b,j)}{j!}\,. \tag{5}$$

So what? We still haven't solved the problem. But we have reduced it to another problem—putting balls in urns with required "occupancies"—a problem worth tackling for itself. (It shows up frequently in probability.) We too will tackle and solve it, but it requires an additional counting technique: Inclusion-Exclusion. That is the subject of the next section. The answer to Question 3 is completed in Example 6 of that section.

In the preceding equations, S stands for the mathematician James Stirling (1692-1770), and $S(b,u)$ is known as a **Stirling number of the second kind.** (Stirling numbers of the first kind are mentioned in the analysis of Algorithm MAXNUMBER in Section 6.6.) The B above stands for E.T. Bell, a 20th century mathematician.

Question 4: In how many ways can b indistinguishable balls be put in u indistinguishable urns?

Let us call the answer $p(b,u)$. Note that the answer to Example 5 is $p(n,n)$. Thus $p(b,u)$ is called a partition function.

We are tempted to tell you: To compute $p(b,u)$ take the answer to Question 2 and de-distinguish the urns by dividing by $u!$—but we wouldn't fool you this time. In fact, this $u!$-to-1 correspondence breaks down even more than the previous one [30].

There are many beautiful theorems about such partition functions. There are algorithms that allow you to list all the partitions of a given size (if you have the time and paper) [23] and recursions which allow you to compute any particular values [24]. There are also ***asymptotic results***; these describe the rate of growth of partition functions very precisely. Many of these results are very deep, and they depend on both continuous and discrete methods. But there are very few closed-form formulas in partition theory. In particular, there is no nice, closed-form solution to Question 4.

We summarize the results about balls and urns (including the result we won't get until the next section) in Fig. 4.11.

Urns \ Balls	Distinguishable	Indistinguishable
Distinguishable	u^b	$\binom{b+u-1}{b}$
Indistinguishable	$\sum_{j=0}^{u}\sum_{k=0}^{j}\frac{(-1)^k}{j!}\binom{j}{k}(j-k)^b$ (See Example 6, Section 4.8)	No closed-form formula

FIGURE 4.11

Summary of solutions for b balls in u urns

Problems: Section 4.7

1. Finish Example 3.

2. How many legitimate configurations are there for TOH with n rings and four poles?

3. How many possible 4-letter "words" are there using the roman alphabet? What type of ball and urn problem is this?

4. Biochemists have found that, in cell replication, all amino acids are coded for using just four nucleotide bases, known as U, C, A and G. Each string of three nucleotide bases codes for a specific amino acid. At most how many amino acids are there? (In actuality there are fewer, because several base triples code for the same acid. Also, some triples code for "end of chain".)

5. For each of our four models, in how many ways can four balls be put in three urns? (For the two models with indistinguishable urns, answer by actually listing the possibilities. Be systematic!)

6. Do [5] again for six balls and four urns. The case of distinguishable balls in indistinguishable urns will now have too many cases to list out. So analyze indistinguishable balls and indistinguishable urns first and, for each possibility, figure out how many cases this would split into if the balls were distinguishable.

7. How many ways can you put $b = b_1+b_2+\cdots+b_u$ indistinguishable balls into u distinguishable urns

if each urn i must be given b_i balls? (This is easy!)

8. How many ways are there to put b indistinguishable balls into u distinguishable urns if each urn may hold at most one ball? *Hint*: This is actually much easier than Question 2, where urn capacity is unrestricted.

9. How many different "code words" of length $D+d$ can be made using D dots and d dashes? Answer this question by first interpreting it as a problem about putting balls in urns. (This is a hard way to answer a simple question, but we want you to practice changing problems from one representation to another.)

10. How many solutions are there to $x+y+z+w = 10$ if

a) the unknowns must be nonnegative integers?

b) the unknowns must be positive integers?

c) the same as (a), except that $x + y$ must also equal 6? *Hint*: Temporarily think of $x + y$ as a single variable.

11. Make up a ball and urn problem for which $\binom{n}{i,j,k}$ is the answer. Be sure to state what is distinguishable. Generalize to arbitrary multinomials.

12. Do permutations with repetitions correspond to one of our models? Explain.

13. Make a chart illustrating the one-to-one correspondence between indistinguishable balls in distinguishable urns and "balls and walls" in the case $b = 3, u = 3$. That is, draw pictures of all 10 arrangements for each problem and indicate by arrows which pairs correspond.

14. In Example 5, there was a pattern to the order in which we listed the partitions. List the partitions of the number 4 following the same pattern. Do the same for the partitions of the number 6.

15. Rubik's Cube is a $3 \times 3 \times 3$ cube constructed from 27 unit cubes so that each face of 9 unit squares can rotate freely. Initially, the cube is painted with a different color on each face, i.e., the 9 unit squares on each face get the same color. But then, as you rotate different faces, things get all mixed up!

a) Considering just one face, how many different color arrangements are possible? That is, one arrangement has green in the top left, blue to its right, and so on. Don't worry about whether the arrangement can actually be obtained by rotating the faces.

b) Considering just one face, how many different color frequencies are possible? For instance, one frequency is to have five red squares, four blue ones and no other colors.

c) What do (a) and (b) have to do with balls and urns?

16. How many ways are there to take four distinguishable balls and put two in one distinguishable urn and two in another if

a) the order of balls within each urn makes a difference?

b) the order does not make a difference?

17. Repeat [16] for indistinguishable urns.

18. Answer Example 4.

19. Here is another way to find the number of combinations with repetitions. As shown on the left side of Fig. 4.10, such a combination may be depicted as a sequence that is increasing but not necessarily *strictly* increasing. The set of all (n, k)-combinations with repetitions is thus the set of all k-long increasing sequences whose elements are taken from $1, 2, \ldots, n$.

We claim there is a one-to-one correspondence between the set of such sequences and the set of all k-long *strictly* increasing sequences whose elements are taken from $1, 2, \ldots, n, n+1, \ldots, n+k-1$. Namely, for each sequence of the first set, and for each position i from 1 to k in that sequence, add $i-1$ to the number in the ith position. The result is the corresponding sequence in the second set. For instance, for the sequences in Fig. 4.10, the correspondence gives

$$\begin{array}{ccc} 1,\ 1,\ 1,\ 2,\ 2,\ 3 & \longleftrightarrow & 1,\ 2,\ 3,\ 5,\ 6,\ 8 \\ 1,\ 1,\ 1,\ 2,\ 3,\ 3 & \longleftrightarrow & 1,\ 2,\ 3,\ 5,\ 7,\ 8 \\ 1,\ 1,\ 3,\ 3,\ 3,\ 3 & \longleftrightarrow & 1,\ 2,\ 5,\ 6,\ 7,\ 8 \end{array}$$

Conversely, for each strictly increasing sequence of length k using 1 through $n+k-1$, subtracting $i - 1$ from the ith term gives a merely increasing sequence with elements taken from 1 through n. The number of such strictly increasing sequences is easy to count. How many are there? Why?

20. Combinations with gaps. Count the number of k-subsets of $\{1, 2, \ldots, n\}$ such that no two consecutive integers occur in the set. For example, $\{1, 3, 9\}$ is such a 3-subset (assuming $n \geq 9$) but $\{1, 8, 9\}$ is not. *Hint*: Reverse the idea presented in the previous problem.

21. Suppose that balls and urns are distinguishable, that urns are stacks (so the order of balls within them matters) and that even the order of the urns makes a difference. Again we ask: In how many ways can b balls be put in u urns? Try to solve this problem by a method similar to the one we used for counting indistinguishable balls in distinguishable (unstacked) urns.

22. Consider the question: How many ways are there to put b different flags on u different flagpoles? Answer it using the following steps to get started.

a) How many ways are there to put the first flag on some pole?

b) How many ways are there to put up the second flag on some pole? *Hint*: The pole with the first flag on it has been divided by that flag into two parts, the part above that flag and the part below.

c) How many ways are there to put up the third flag?

d) Now answer the original question. The correct expression is often called the **rising factorial**.

e) Reframe the original question as one about putting balls into urns. How is it related to [21]?

23. How many ways can b indistinguishable balls be placed in u distinguishable urns if each urn must contain at least k balls?

24. Write an algorithm for the procedure you followed in solving [14].

25. Define $p^*(b,u)$ to be the number of ways to put b indistinguishable balls into u indistinguishable urns so that no urn is empty. Recall from the discussion after Question 4 that $p(b,u)$, with no star, is the number of ways to put b indistinguishable balls into u indistinguishable urns with no restrictions.

a) For what i and j is $p^*(b,u) = p(i,j)$?

b) Express $p(i,j)$ as a sum of p^* terms.

c) Deduce a recursion for $p^*(b,u)$.

d) Deduce a recursion for $p(i,j)$.

26. Let $p(n)$ be the number of partitions of n, i.e., the quantity sought in Example 5. Use the results of [25d] to help you compute $p(1)$ through $p(6)$. Check your answers against Example 5 and [14].

27. Write an algorithm to automate your method from the previous problem.

28. With $p(b,u)$ as in Question 4 of the text, find closed-form formulas for

a) $p(b,1)$ **b)** $p(b,2)$ **c)** $p(b,3)$

That is, the formulas should be in terms of b, not in terms of other values of $p(b,u)$.

29. Let $p'(b,2)$ be the number of (unordered) partitions of b into any number of parts, none of which is greater than 2. Show that $p'(b,2) = p(b,2)$. Does this relation generalize?

30. Here is an argument that $T(b,u) = P(b,u)u^{b-u}$, where $P(n,k)$ is the number of (n,k)-permutations and $T(b,u)$ is defined as in this section. Since no urn can be empty, each urn must have at least one ball. So let's pick a sequence of u balls to be the initial balls in urns $1, 2, \ldots, u$, respectively. That is, we want a permutation of u of the b balls, of which there are $P(b,u)$. Now, the remaining balls can be put in the urns without restriction. Since the balls and urns are distinguishable, there are u^{b-u} ways to do so.

Is this argument right? If not, where does it break down?

31. In the specific case $b = 5, u = 3$, write out the correspondence between the ways to put indistinguishable balls in distinguishable urns and the ways to put indistinguishable balls in indistinguishable urns. Verify the statement in the text that this correspondence fails to be an exact many-to-one correspondence on even more occasions than the correspondence discussed for Question 3.

32. In Eq. (5), why did we start j at 0? Isn't this unnecessary because $S(b,0)$ and $T(b,0)$ will always be 0? Aren't there always no ways to put b balls into 0 urns?

4.8 Inclusion-Exclusion

In this section we develop a different sort of counting method, one of particular use when we deal with overlapping sets.

EXAMPLE 1 At Washington High School, 35 students play on the football team, 15 students play on the basketball team, and 4 students play on both teams. How many students play on at least one of the two teams?

EXAMPLE 2 At Washington High School, 35 of the 1543 enrolled students play on the football team, 15 play on the basketball team, and 4 play on both teams. How many students don't play on either team?

Both examples can be illustrated by Fig. 4.12. Think of the set of students at the school as represented by points within the box, and the set of students on a team as represented by points within the appropriate circle. This is a Venn diagram, which we introduced in Section 0.1.

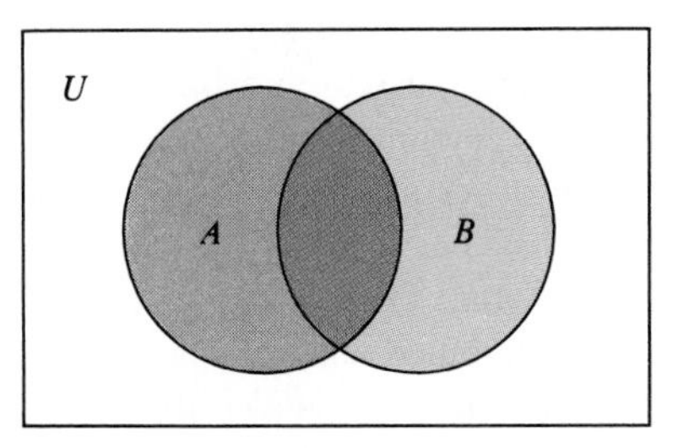

FIGURE 4.12 A Venn diagram of the Washington High School student body: U = students; A = football team; B = basketball team.

The point is: You can't just add the cardinalities (sizes) of the football and basketball teams, because then the students who play on both teams will be counted twice. You have to make an adjustment. After *including* players once for each team, you have to *exclude* the students who play on both teams.

Solution to Example 1 Based on the preceding explanation, the answer to the question in Example 1 is

$$35 + 15 - 4 = 46. \blacksquare \tag{1}$$

Solution to Example 2 The students who don't play either sport are everybody else, so the answer to the question in Example 2 is

$$1543 - 46 = 1497. \blacksquare \tag{2}$$

The general principle behind Eq. (1) is: If A and B are any two sets, then

$$|A \cup B| = |A| + |B| - |A \cap B|. \tag{3}$$

To express the principle behind Eq. (2), note that we want the number of things not in A and not in B. That is, we want $|\overline{A} \cap \overline{B}|$, where complements are relative to the appropriate universal set U (in this case, the set of students at Washington High). By one of DeMorgan's Laws [7, Section 2.3], $\overline{A} \cap \overline{B} = \overline{A \cup B}$, and, by definition, this in turn equals $U - (A \cup B)$. So the general principle behind Eq. (2) is

$$\begin{aligned} |\overline{A} \cap \overline{B}| &= |U| - |A \cup B| \\ &= |U| - |A| - |B| + |A \cap B|. \end{aligned} \tag{4}$$

So much for two sets. Using the inductive paradigm, let's figure out what happens with three sets and hope we see a pattern.

EXAMPLE 3 At Washington High School, of the 1543 students, 35 play football, 15 play basketball, and 30 play baseball; 4 play both football and basketball, 8 play both football and baseball, 7 play both basketball and baseball; 1 plays all three. How many students play at least one of these sports? How many play none of them?

Note. When we say 35 play football, we mean *at least* football. When we say 7 play both basketball and baseball, we mean at least these two sports. If we want to say that a certain number play basketball and baseball only, we will include "only" or some other special indicator word. This ambiguity in English (does "x" mean "at least x" or "only x"?) is avoided with set notation. When we define A to be the set of football team members, we understand that, as always, membership in the set it determined solely by whether the defining property (playing football) is met. Whether a member is also in the set B of basketball players is irrelevant. Similarly, $A \cap B$ unambiguously means the set of students who play at least football and basketball.

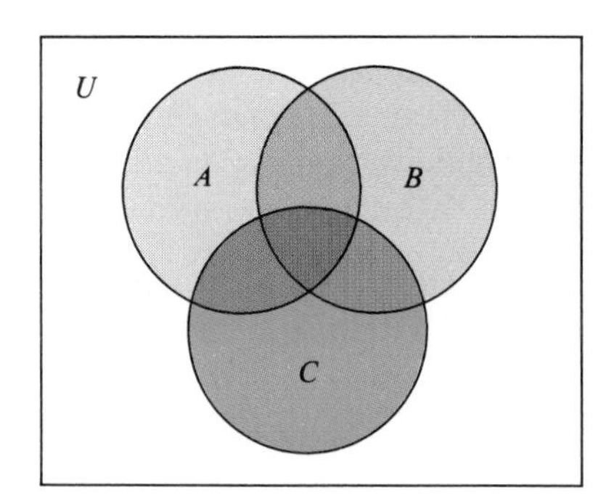

FIGURE 4.13 Another Venn diagram of the Washington High School student body: U = students; A = football team; B = basketball team; C = baseball team.

Solution to Example 3 Now the appropriate diagram is Fig. 4.13, where C is the set of baseball players. To answer the first question in Example 3, we need the total number of students within the three circles. Let's begin by adding $|A|, |B|$ and $|C|$. This counts the students who play exactly one sport just once, as we want, but those who play two or more sports are counted twice or more. (Who's counted exactly twice? Who more?) So let's subtract $|A \cap B|$, $|A \cap C|$, and $|B \cap C|$ once each. Now the students who play exactly two of these sports are also counted just once, but the students who play all three sports have been counted $3-3=0$ times. (Why?). So let's add back $|A \cap B \cap C|$. We get

$$\begin{aligned} |A \cup B \cup C| &= |A| + |B| + |C| - |A \cap B| \\ &\quad - |A \cap C| - |B \cap C| + |A \cap B \cap C|. \end{aligned} \tag{5}$$

The general formula to answer the second question in Example 3 is now obtained by subtracting the RHS of Eq. (5) from $|U|$:

$$\begin{aligned} |\overline{A} \cap \overline{B} \cap \overline{C}| &= |U| - |A \cup B \cup C| \\ &= |U| - |A| - |B| - |C| + |A \cap B| \\ &\quad + |A \cap C| + |B \cap C| - |A \cap B \cap C|. \end{aligned} \tag{6}$$

In particular, using Eqs. (5) and (6), we find that the specific numerical answers to the questions in Example 3 are

$$35 + 15 + 30 - 4 - 8 - 7 + 1 = 62 \text{ students play on at least one of the teams,}$$

and

$$1543 - 62 = 1481 \text{ students play on none of the three teams.}$$ ∎

The situation for n sets, $A_1, A_2, \ldots, A_n$, should now be clear: You alternately add and subtract the cardinalities of the intersections of the A's, changing sign each time you intersect more sets. When you compute $|\bigcup A_i|$—this generalizes Eq. (5)—you start the alternating sum with the $|A_i|$. When you compute $|\bigcap \overline{A}_i|$—this generalizes Eq. (6)—you start with $|U|$.

We will state and prove the generalization of Eq. (6). Just as Eq. (6) is derivable from Eq. (5) by subtracting both sides of Eq. (5) from $|U|$, so is Eq. (5) derivable from Eq. (6) by subtracting both sides of Eq. (6) from $|U|$. (Verify!). This relationship is also true of their generalizations. Thus it suffices to obtain either one. Let us start by defining

$$\begin{aligned} N &= |U|, \\ N_0 &= |\bigcap_{i=1}^{n} \overline{A}_i|, \\ N_i &= |A_i|, \\ N_{ij} &= |A_i \cap A_j|, \\ N_{ijk} &= |A_i \cap A_j \cap A_k|, \\ &\quad \cdots \end{aligned}$$

$$N_{123\dots n} = |\bigcap_{i=1}^{n} A_i|.$$

Then the generalization of Eq. (6) is

Theorem 1, Inclusion-Exclusion. With notation as above,

$$N_0 = N - \sum_i N_i + \sum_{i,j} N_{ij} - \sum_{i,j,k} N_{ijk} + \cdots + (-1)^n N_{123\dots n}. \quad (7)$$

Here $\sum_i$ means the sum over all i from 1 to n, $\sum_{i,j}$ means the sum over all $C(n,2)$ combinations of two distinct indices, and so on. The last term is positive if n is even, negative if n is odd.

In many cases (see Examples 5 and 6), all the numbers N_i are equal to each other, all the N_{ij} are equal to each other, and so forth. In such cases, there is a simpler way to write Eq. (7).

Corollary 2, Symmetric Inclusion-Exclusion. Suppose for each k from 1 to n, the number of elements in the intersection of any k of the sets A_1 through A_n is the same. Call this value n_k. Also let $n_0 = |U|$. Then

$$N_0 = \sum_{k=0}^{n} (-1)^k \binom{n}{k} n_k. \quad (8)$$

The proof is simple (assuming Theorem 1): The $\sum$ in (7) with k distinct subscripts has $\binom{n}{k}$ terms; if each is n_k, then the total is $\binom{n}{k} n_k$.

Before proving Theorem 1, we give several more examples to show its power.

EXAMPLE 4 Two integers are relatively prime if they have no factors in common, other than 1. How many positive integers ≤ 100 are relatively prime to 100?

Solution An integer is relatively prime to $100 = 2^2 5^2$ if it doesn't have 2 or 5 as a factor. Therefore set

$$\begin{aligned} U &= \{1, 2, \dots, 100\}, \\ A &= \{n \in U \mid 2 \text{ divides } n\}, \\ B &= \{n \in U \mid 5 \text{ divides } n\}. \end{aligned}$$

We seek $|\overline{A} \cap \overline{B}|$, so we can use Eq. (4). Since 2 divides a number iff the number is a multiple of 2, $|A| = 100/2$. Likewise, $|B| = 100/5$. Finally, $A \cap B$ is the set of integers ≤ 100 that are divisible by both 2 and 5, i.e., divisible by 10. Therefore $|A \cap B| = 100/10$. Thus the answer is

$$100 - 50 - 20 + 10 = 40. \blacksquare$$

This problem was simple enough that you might do it by brute force. However, the Inclusion-Exclusion method works on divisibility problems no matter how large the initial number is (here it was 100) or how many prime factors it has [5, 6, 9].

EXAMPLE 5 How many m-digit integers contain each of the digits 0–9 at least once? (Allow an integer to start with 0, e.g., 013 is a 3-digit integer in this problem.)

Solution This is a very hard problem to do using the methods of earlier sections, except for special values of m. (What's the answer when $m = 9$? When $m = 10$?) It would be an even harder problem if we didn't allow 0 to be the first digit [21]. For general m, you might try to solve the problem by breaking it into many cases and using the sum rule. For instance, the digit 9 can go in exactly one of the m places, or in more than one. If it goes in exactly one, then the digit 8 can go in exactly one of the remaining $m-1$ places or more than one. If the digit 9 goes in $k > 1$ places, then the digit 8 can go in exactly one of the remaining $m-k$ places or more than one. We have already broken the problem into four cases, and we have yet to consider any digits less than 8. If you try to pursue this sort of approach, we bet almost anything you will end up overcounting some things and undercounting others, all in rather haphazard ways. Beware of breaking into many cases unless you have a very systematic method.

Inclusion-Exclusion is such a method. Let U be the set of all m-digit positive integers. A natural first choice for A_i is the set of m-digit integers in which digit i appears at least once. But let's call that set B_i instead until we determine if it's a helpful choice. (Only if it is helpful should we call it A_i and use Theorem 1 on it.) Is either $\bigcup_{i=0}^{9} B_i$ or $\bigcap_{i=0}^{9} \overline{B_i}$ what we want to count? Recall, the latter expression is what Theorem 1 is about (if we use A_i instead of B_i) and the former expression may be obtained from the latter by subtraction.

The answer is No. The former is the set of m-digit integers in which at least one of 0–9 appears (an easy club to join!). The latter is the set of m-digit integers in which none of 0–9 appears (a tough club to join!!). We want *all* of 0–9 to appear, just the opposite. In short, we want $\bigcap B_i$. This suggests that maybe we should start with

$$A_i = \overline{B_i} = m\text{-digit integers in which digit } i \text{ does } \textit{not} \text{ appear.}$$

Indeed, this works, because now $\bigcap B_i = \bigcap \overline{A_i}$, and so Theorem 1 applies.

Using the notation of Theorem 1, we find, for instance, that $N_{2,5,9}$ is just the number of m-digit integers in which neither 2 nor 5 nor 9 appears (but any other digit may). Thus for each of the m positions there are $10-3 = 7$ remaining digits to choose from. So $N_{2,5,9} = 7^m$. Furthermore, the specific values 2,5,9 are irrelevant. For any three digits (i.e., for any $S = \{i, j, k\}$), $N_S = 7^m$. Likewise, for any k digits,

$N_S = (10-k)^m$. Therefore, substituting $(10-k)^m$ for n_k in Corollary 2, and noting that there are 10 different digits, we find that the answer is

$$\sum_{k=0}^{10}(-1)^k\binom{10}{k}(10-k)^m\,. \quad \blacksquare \tag{9}$$

Moral: If you want $|\bigcap B_i|$, start with $A_i = \overline{B}_i$. It seems backwards, but it works.

EXAMPLE 6 Finish answering Question 3 from Section 4.7: How many ways can b distinguishable balls be put in u indistinguishable urns?

Solution Recall that we found a formula for the answer, Eq. (5), Section 4.7, in terms of $T(b,u)$, the number of ways to put b distinguishable balls into u distinguishable urns so that every urn is occupied. All (!!) that remains is to find a formula for $T(b,u)$.

We just did, in Example 5—except we were short-sighted and only wrote down the formula for base 10. There is a one-to-one correspondence between the number of b-digit integers in base u, when every "digit" from 0 to $u-1$ must show up at least once, and the number of ways to put b distinguishable balls into u distinguishable urns, when every urn must contain at least one ball. Namely, think of putting ball i into urn j as a way of declaring that the ith digit from the left must be j.

Therefore, substituting in Eq. (9), we obtain

$$T(b,u) = \sum_{k=0}^{u}(-1)^k\binom{u}{k}(u-k)^b\,. \tag{10}$$

So, from Eq. (5), Section 4.7, the answer to Question 3 is

Theorem 2. The number of ways to put b distinguishable balls into u indistinguishable urns is

$$B(b,u) = \sum_{j=0}^{u}\sum_{k=0}^{j}\frac{(-1)^k}{j!}\binom{j}{k}(j-k)^b\,. \tag{11}$$

Pretty complicated! It's hard to imagine how we might have discovered this formula without Inclusion-Exclusion. (Incidentally, such complicated formulas can lead to intriguing difficulties when you try to evaluate them; see [25].) ∎

PROOF OF THEOREM 1 Let e be any object in the universal set U. We need to show two things: (1) if e is in none of the sets A_i, then it is counted once on the RHS of Eq. (7); and (2) if it is in one or more of the A_i, then it is counted zero times.

Suppose e is in none of the A_i. Then it is counted once in N, the first term on the right in Eq. (7), and it is not counted in any of the others. So (1) is proven.

Suppose e is in exactly one of the A-sets, say A_3. Then it is counted once in N, once in N_i for $i = 3$, and zero times elsewhere. But N is added to the RHS and N_3 is subtracted (in the second term), so we are all right.

Suppose e is in exactly two of the A-sets, say A_4 and A_7. Then e is counted $+1$ time in N, -1 time in each of N_4 and N_7, $+1$ time in $N_{4,7}$, and nowhere else, for a net count of 0, as required.

In general, if e is in exactly k of the A_i, it will be counted $+1$ time in N, -1 time in k of the terms N_i, $+1$ time in $C(k,2)$ of the terms N_{ij} (because there are $C(k,2)$ ways to pick two of the k sets e is in), and so on. That is, e will be given a net count of

$$\sum_{j=0}^{k} (-1)^j \binom{k}{j}.$$

But as we showed in Example 3, Section 4.6, this sum is 0: Apply the Binomial Theorem to $(1-1)^k$. ∎

Problems: Section 4.8

1. At Jefferson High School, 40 students play field hockey, 30 run track, and 1335 of the 1400 enrolled students do neither. How many do both? How many play field hockey but don't run track? How many only run track?

2. For [1], let F and T be the set of students on the field hockey and track teams, respectively. Let U be the entire student body.
   a) Translate the information given in [1] into set notation.
   b) Take all the formulas you used or devised to answer all three questions in [1] and translate them into set notation too. For instance, for the second question you need a formula for $|F \cap \overline{T}|$.

3. Jefferson High School also has a swimming team with 50 members. One student is on both the swimming and field hockey teams, five are on both the swimming and track teams, and none is on all three teams.
   a) How many students play none of these sports?
   b) How many play at least field hockey but don't swim?
   c) How many only run track?
   d) How many are on at most the track and swimming teams?

4. For [3], let F, T, and S be the set of students on the field hockey, track, and swimming teams, respectively. Let U be the entire student body.
   a) Translate the information given in [3] into set notation.
   b) Write in set notation the formula you devised to answer [3c].
   c) Write in set notation your formula for [3d].

5. How many positive integers less than 105 are relatively prime to 105?

6. How many of the integers from 1 to 1000 are divisible by at least one of 3, 5, and 7?

7. How many positive integers less than 100 have no perfect squares (other than 1) as a factor?

8. How many integers from 1 to 1000 have a perfect cube (other than 1) as a factor?

9. Explain why an integer between 2 and 100 is prime iff either (a) it is a prime less than 10, or (b) no prime less than 10 divides it. Then use Inclusion-Exclusion to compute the number of primes less than 100. (You may know other methods that will

answer this question faster, but if we replaced 100 by 1 million, this method would be a winner.)

10. What is the value of

$$\sum_{k=0}^{b}(-1)^k C(b,k)(b-k)^m$$

when

a) $m<b$, **b)** $m=b$?

Don't try to calculate; reread Example 5 and think.

11. A **derangement** of size n is a permutation of the numbers 1 to n such that no number k is in the kth place. Thus 4321 and 2413 are derangements, but 41$\underline{3}$2 and $\underline{1}$32$\underline{4}$ are not, because of the numbers underlined. Let D_n be the number of derangements of size n. Show that

$$D_n = \sum_{k=0}^{n}(-1)^k\binom{n}{k}(n-k)!\,.$$

Hint: Let U be the set of all permutations of 1 through n and let A_i be the set of permutations in U that *do* keep number i in the ith place.

12. How many permutations of *aabcccdeeee* have the two a's together? How many don't?

13. Find the number of permutations of *aabbcc* in which no identical letters are next to each other.

14. Find the number of 6-long sequences of a's, b's and c's so that no identical letters are next to each other. *Warning*: The small difference in wording from the previous problem makes a big difference in the easiest general approach.

15. A group of n male–female couples go to a dance. How many ways can all $2n$ people dance at once if each male dances with a female other than the one he came with? Come up with a general formula and then evaluate it in the case $n=4$.

16. Show that the answer to Example 5 can also be written

$$\sum_{k=0}^{10}(-1)^k\binom{10}{k}k^m. \qquad \textbf{(12)}$$

Now change the base for writing integers from 10 to b. Is formula (9) still correct? Is formula (12) still correct? For which bases b?

17. Verify the following closed-form formulas for the Stirling numbers. See Eq. (5), Section 4.7, as well as Example 6 in this section.

$$\begin{aligned} S(b,u) &= \sum_{k=0}^{u}\frac{(-1)^k}{u!}\binom{u}{k}(u-k)^b \\ &= \frac{1}{u!}\sum_{k=0}^{u}(-1)^{u-k}\binom{u}{k}k^b. \end{aligned}$$

18. **a)** How many positive integers $\le$ a million are divisible by 5^3 but not 5^4?

b) How many are divisible by 5^3 but not 5^4, and are also divisible by 2^3 but not 2^4?

c) How many are divisible by 5^3 but not 5^4, or are divisible by 2^3 but not 2^4?

d) How many are divisible by 5^3 or 2^3 but are not divisible by either 5^4 or 2^4?

19. Write down the Inclusion-Exclusion formula that generalizes Eq. (5) the same way that Theorem 1 generalizes Eq. (6). Use N' for the number of objects in U that are in at least one of the A_i's.

20. Write down the symmetric Inclusion-Exclusion formula that generalizes Eq. (5) the way that Corollary 2 generalizes Eq. (6) in the symmetric case.

21. Solve Example 5 again if an m-digit integer is *not* allowed to start with 0.

22. Although Eq. (7) is the standard way to state Inclusion-Exclusion, we are not very happy with it. It relies too much on ellipses, both in the equation itself and in the definition of the N's just before the theorem. One of the purposes of set notation and Σ notation is to avoid such potentially misleading use of dots.

We can avoid elapses in Theorem 1 by using the following idea. First, define all the N's at once. Namely, for each $S\subset[n]=\{1,2,\ldots,n\}$, define

$$N_S = \Big|\bigcap_{i\in S}A_i\Big|.$$

As a special case, define $N_\emptyset=|U|$. (This is consistent with the discussion of empty products in Section 0.6.) Using this notation, restate Eq. (7) in the form $N_0=\sum_{S\subset[n]}??$, where you have to figure out what ?? is. *Note*: N_0 and $N_\emptyset$ are not the same. $N_\emptyset$ has just been defined; N_0 was defined just before Eq. (7).

23. Let P be a finite set of properties and let U be some set.

a) There is a theorem which counts the number of elements of U that satisfy *none* of the properties in terms of the number of elements that satisfy *at least* various subsets of the properties. State the theorem. (The only work here is recognizing that this is something you have already seen. Explain the disguise.)

b) There is also a theorem which counts the number of elements of U that satisfy *all* of the properties in terms of the number of elements that satisfy *at most* various subsets of the properties. State this theorem. Explain. (This is a little trickier, but Example 5 should help.)

24. In Eq. (9), why do we bother with the term when $k = 10$? Won't it always be 0? Similarly, in Theorem 2, isn't it enough to let k in the inner sum stop at $j - 1$?

25. It is quite possible for the value of a sum to be small even though the individual terms are large. For instance, we know from [10] that

$$\sum_{k=0}^{15}(-1)^k\binom{15}{k}(15-k)^{14} = 0,$$

but most of the terms are quite large.

a) Use a computer or hand calculator to evaluate the LHS of the preceding equation. Why don't you get the right answer? It has to do with the fact that you subtract as well as add. (With a hand calculator, it's less work to use 10 and 9 instead of 15 and 14, and you will probably still get a wrong answer. Why?)

b) Now consider the Bell numbers $B(b, u)$ which answer Question 3 from the previous section. You'd have the same computational problem if you tried to evaluate them using the RHS of Eq. (11), and for the Bell numbers you don't know the value in advance. There was a similar problem computing permutations and combinations from the original formulas, but there the problem involved the fact that you divide as well as multiply; see problems [15 and 16, Section 4.4]. We got around that problem by finding a recursion for $C(n, k)$ and computing with it. Try to find a recursion for the Bell numbers. If you find one, use it to compute B(5,4) exactly.

4.9 Combinatorial Algorithms

You've learned about such basic building blocks as permutations and combinations, and you've learned simple formulas for counting how many there are. But suppose that you're asked to produce a specific permutation of some six things or to produce all permutations of these six things. How would you do that? These are intrinsically algorithmic questions. They are also very useful questions. Consider again the Traveling Salesman Problem of Section 4.2. The only solution method we proposed was to compute the length for each possible tour. If the cities (other than the salesman's home city) are labeled 1 through n, then there is a one-to-one correspondence between tours and permutations of 1 to n. For instance, if $n = 5$, the tour

home, city 2, city 4, city 3, city 5, city 1, home

corresponds to the permutation 24351. So the first step in the brute force approach of Section 4.2 is to generate all the permutations.

We pointed out that, for $n = 50$, this approach is well beyond the capacity of any conceivable computer. So what do you do? One alternative is to look for algorithms which run quickly and find near-optimal solutions. One such algorithm is to take a *random* selection of tours and find the shortest among them. (Random means that each possible tour is equally likely.) This tour probably won't be the best, but in general it will be close to best. To see this, first suppose you found just one random tour. On average, in the ranking of all tours from best to worst it would be right in the middle. (For instance, if there were just three tours and you picked one at random, 1/3 of the time you would pick the best tour, 1/3 of the time you would pick the second best, and 1/3 of the time you would pick the worst, so on average you would pick the second best.) If you found exactly two tours at random, it should be plausible that the better of them would be, on average, 1/3 the way down the ranking from best to worst. Indeed, it's a theorem that if you choose n things at random from a collection of N things, the average rank of your best choice is $1/(n+1)$ of the way down the ranking from rank 1 (best) to rank N (worst). Specifically, if you take 99 random tours, only 1/100 of the remaining $N-99$ tours are likely to be shorter than the shortest of your 99.

The point is, it is very valuable to know how to produce a collection of random permutations. It suffices to know how to produce one random permutation, because as soon as you know how to produce one, you just repeat the process as many times as you want (say, 99 times).

Furthermore, you don't need to take such complicated problems as the Traveling Salesman Problem to see the usefulness of random combinatorial arrangements. For example, suppose that 100 students have signed up for a course and there is room for only 50. One fair way to choose the class is to make a random selection. This means a random combination of 50 things from the total of 100. So you need an algorithm to produce such random combinations.

So much for motivation. The rest of this section is devoted to algorithms themselves. To keep things simple, we discuss algorithms for permutations only and only permutations of n things taken n at a time (n-permutations). The problems let you try your hand at algorithms for other combinatorial objects and suggest a few more applications.

We break this section into two subsections. In the first we devise three algorithms for generating a single random permutation. We try to motivate how to devise such algorithms and show how to analyze them. In the second subsection we discuss an algorithm for generating all n-permutations. We also show that such generating algorithms have applications to seemingly nonalgorithmic questions.

To talk about random objects involves talking about probability. It takes time to develop probability ideas carefully. All the statements we make about probabilities and "average values" in this section are correct, but we won't be able to justify them fully until Chapter 6. For now we will explain them with plausibility arguments. Do not hesitate to make similar "leaps of intuition" in doing the problems for this section. You may be wrong occasionally, but mathematics would not get very far if people never tentatively proposed solutions they couldn't completely justify.

In order to generate random permutations (or other random objects), we need some sort of randomizing ability. Since permutations are made out of numbers, it would be nice to have a **random number generator**. We will assume that our algorithmic language contains a function RAND$[m, n]$ which, on each call, picks an integer between m and n, inclusive, with each choice equally likely and independent of all previous choices.

(How can such a function be implemented on a computer, a deterministic device? It can't! What computers actually have are **pseudorandom number generators**, i.e., programs that produce numbers which look random for all essential statistical purposes but which aren't really random. The mathematics that explains how this is done is very interesting, but it is beyond the scope of this book.)

Random Permutations

All right, let's start finding algorithms to produce random permutations. For each algorithm we will ask two questions:

1. Is it really a solution? That is, does it really produce a permutation of n things and is each of the $n!$ permutations equally likely?
2. Is it an efficient solution? That is, can we think of another approach that takes fewer steps and/or is easier to state in algorithmic language?

Here is a simple approach that we suspect occurs to most people first. (What would you try?) Since we want to produce the numbers from 1 to n in random order, it seems reasonable to invoke RAND[1,n]. Can we just invoke it n times?

No. Since each choice is independent of previous choices, we might repeat one number and leave out another.

The simplest way around this problem is to keep a record of which numbers have already been picked, and each time we get a repetition, throw it away. Then keep going until all numbers have been picked once. This approach is easy to illustrate by putting slashes through numbers which get discarded. For instance, with $n = 5$ the sequence

$$3 \quad 1 \quad 4 \quad \not{1} \quad 5 \quad \not{4} \quad \not{3} \quad 2$$

gives the permutation 31452.

Let's write this method up as Algorithm 4.3, PERMUTE-1. Let

$$\text{Perm} = [\text{Perm}(1),\ \text{Perm}(2), \ldots,\ \text{Perm}(n)]$$

be a sequence that will hold the permutation when the algorithm terminates. Let Flag be another n-sequence. (That is, Flag(1) is the first entry in the sequence, Flag(2) the second, and so on. We may also call such sequences vectors or arrays.) Flag(i) $= 0$ will mean that, so far, the integer i has not been placed in the permutation; Flag(i) $= 1$ will mean that it has.

It should be clear that when this algorithm terminates, the output is a permutation: The loop invariant for the outer loop is that, for each i, the numbers Perm(1) through Perm(i) are distinct integers in the interval $[1, n]$. Since the algorithm terminates after n iterations of the outer loop, the final result is a permutation.

It may not be so obvious that each of the $n!$ permutations is equally likely to be chosen. But we can argue informally as follows. After Perm(1) through Perm(k) have been chosen, only the $n - k$ unchosen numbers are eligible, and the algorithm plays no favorites among them. Thus any one of them will be picked with probability

Algorithm 4.3

Permute-1

```
Input n                                              [Length of permutation]
Algorithm PERMUTE-1
    Flag ← 0                              [Set all components of Flag to 0]
    for i = 1 to n                      [ith pass will determine Perm(i)]
        repeat
            r ← RAND[1, n]                         [Pick a random number
        endrepeat when Flag(r) = 0               until an unused one found]
        Perm(i) ← r
        Flag(r) ← 1                                      [r has been used]
    endfor
Output Perm                   [A random permutation given as a sequence]
```

$1/(n-k)$. Thus the probability that any given number will be the first number picked (when $k = 0$) is $1/n$; the probability that any given number among the remaining ones will be the second number picked is $1/(n-1)$, and so on. Thus the probability that we will get any particular permutation is $1/n!$, just what we want.

So PERMUTE-1 is correct. But it's not efficient. Towards the end it spends a lot of time picking things that get discarded. How much time?

Let us find the expected number of RAND calls in each traverse of the outer loop. When k values of Perm have already been chosen (i.e., when the **for** loop variable i has just become $k+1$) the next call of RAND has probability $(n-k)/n$ of giving an unchosen number. In general, if an action has probability p of success each time it is tried, we hope it seems reasonable that, on average, it will take $1/p$ repetitions to achieve the first success (see Section 6.6). Thus the expected number of RAND calls for the entire algorithm is

$$\sum_{k=0}^{n-1} \frac{n}{n-k} = n \sum_{j=1}^{n} \frac{1}{j}. \tag{1}$$

The harmonic series $\sum 1/j$ is one that you have met several times before (e.g., Example 3 of Section 1.6). It turns out that $H(n) = \sum_{j=1}^{n} 1/j$ grows as $\log n$. Therefore we conclude from Eq. (1) that Algorithm PERMUTE-1 requires on the order of $n \log n$ calls of RAND to generate one permutation.

Could we hope for better? Sure, we could hope for just n calls of RAND, with none wasted. So PERMUTE-1 is inefficient by a factor of $\log n$—not bad for small n but considerable as n gets large. So let's try to do better.

Here's one way. After k numbers have already been chosen, we have only $n - k$ to choose from. So let's invoke RAND$[1,n-k]$ instead of RAND$[1,n]$. Unfortunately, it's highly unlikely that the unchosen numbers are just $\{1, 2, \ldots, n-k-1, n-k\}$. That's all right. Keep a list of the unchosen numbers, let $r =$ RAND$[1, n-k]$, and pick the rth unchosen integer (in increasing order). For instance, suppose that

$n = 5$ and suppose further that the results of calls to RAND are as shown in the left column on the left in Table 4.1. Then the columns on the right show the corresponding status of our choices, where slashed numbers are numbers picked previously and circled numbers are numbers picked now. For instance, in the fourth row of Table 4.1, the value of RAND is 2. Since the numbers 1, 3, and 5 have already been picked, the second number still available is 4. Therefore we pick 4 by circling it. The net result of all the RAND choices in the Table is the permutation 35142.

TABLE 4.1
Typical Output of Algorithm PERMUTE-2 **for** $n = 5$

RAND[1, 5] = 3	1	2	③	4	5
RAND[1, 4] = 4	1	2	3̸	4	⑤
RAND[1, 3] = 1	①	2	3̸	4	5̸
RAND[1, 2] = 2	1̸	2	3̸	④	5̸
RAND[1, 1] = 1	1̸	②	3̸	4̸	5̸

This approach gives us Algorithm 4.4, PERMUTE-2. The only hard part is determining which is the rth unchosen number, so that it can be picked next. (Humans can eyeball this, as we did in Table 4.1.) The variables c and j are used for this purpose. The algorithm marches up from 0 (by incrementing j), and c counts how many of the numbers from 1 on have not been picked previously. When c equals r, the current j is the rth unchosen number.

The proof that PERMUTE-2 gives a random permutation is similar to the proof for PERMUTE-1. Each number from 1 to n is chosen once, and after r numbers have been chosen, each of the yet unchosen numbers has probability $1/(n-r)$ of being chosen.

What about efficiency? RAND is invoked just n times, but this is not the right measure for PERMUTE-2. Whereas Flag was checked each time RAND was invoked in PERMUTE-1 (except when Flag was initialized), here Flag is checked much more often. On a computer, each RAND call will be more time-consuming than each Flag check, but if there are many more flag checks, then the number of Flag checks will be the overriding factor. So let's count the number of times Flag is checked in the main loop.

Flag is checked once in the *inner* loop for each value of j up through the value that gets assigned to Perm(i). Since the previously assigned numbers are randomly distributed, and the next assigned number is randomly distributed among the unassigned, it seems reasonable that the next assigned number is randomly distributed from 1 to n. Therefore it should seem reasonable that it's average position is $(n+1)/2$. (Why is $n/2$ wrong?) Therefore, on average, $(n+1)/2$ checks of Flag are made in the inner loop for *each* value of i. Thus the average number of checks

Algorithm 4.4

Permute-2

```
Input n
Algorithm PERMUTE-2
    Flag ← 0
    for i = 1 to n                                  [Pick ith entry in Perm]
        r ← RAND[1, n−i+1]
        c ← 0                      [Number of unchosen values passed over]
        j ← 0                    [Initializes number considered for Perm(i)]
        repeat while c < r
            j ← j + 1
            if Flag(j) = 0 then c ← c + 1
                                          [Another unchosen number found]
        endrepeat
        Perm(i) ← j
        Flag(j) ← 1
    endfor
Output Perm
```

in the main loop is $n(n+1)/2$. In short, even though Algorithm PERMUTE-2 saves on RAND calls, it is *worse* than Algorithm PERMUTE-1 because $n(n+1)/2$ grows faster than $n \log n$—the ratio of $n(n+1)/2$ to $n \log n$ gets arbitrarily large.

The reason PERMUTE-2 didn't work well is that the yet unchosen numbers got spread out, and so we had to keep checking all the numbers to see whether they had been chosen. The next algorithm, PERMUTE-3, avoids this problem by keeping all the yet unchosen numbers together.

Here's how. Suppose, as in Table 4.1, that we choose Perm(1) = 3. In that table we kept 3 in its place after it was chosen. This time, we *exchange* the positions of 1 and 3. See Table 4.2. This exchange will keep the unchosen numbers $\{1, 2, 4, 5\}$ contiguous. Now, to determine Perm(2), we need merely call RAND[2,5], and whatever number is in the *position* that RAND picks becomes Perm(2). Say RAND[2, 5] = 5. Then we exchange the numbers in the fifth and second positions. The unchosen numbers are again together, in positions $\{3, 4, 5\}$. Table 4.2 gives a complete example. In the column on the left we show the values for RAND. In the columns on the right we show in each row the ordering of 1 to $n = 5$ just *before* making the exchange called for by RAND in that row. The two numbers with asterisks are the ones about to be switched. (If a number is switched with itself, it gets two asterisks.) The numbers underlined are the ones which are already known to be in their Perm positions. Note that this method has the added benefit that the last row *is* the permutation.

The proof that Algorithm PERMUTE-3 gives a permutation is easy. Perm *starts* as a permutation, and switching two elements can never destroy that. In short, the

Algorithm 4.5

Permute-3

```
Input n
Algorithm PERMUTE-3
    for i = 1 to n
        Perm(i) ← i                                [Initialize Perm]
    endfor
    for i = 1 to n
        r ← RAND[i, n]
        Perm(i) ↔ Perm(r)                          [Exchange values]
    endfor
Output Perm
```

loop invariant is that Perm is a permutation. The proof that the permutation is random is the same as before.

How efficient is Algorithm PERMUTE-3? As in PERMUTE-2, RAND gets called n times. So let's see if anything else gets exercised more. There isn't any Flag. The only thing left to look at is assignments of entries of Perm. We find that there are n initially, plus one exchange for each of the n times through the second loop. If we think of an exchange as composed of ordinary one-way assignments, each exchange requires 3 assignments. (Why?) Thus there are $4n$ assignments altogether. In short, PERMUTE-3 is *linear* in n, the "size" of the problem. For this reason it is the best of the three algorithms. Note that it is also the shortest in algorithmic language.

All Permutations

Now we turn to the problem of listing all permutations. In order to be correct and efficient, we must make sure to list each n-permutation once—but only once. A list inescapably comes out in some order, and we can turn this fact to our advantage.

TABLE 4.2
Typical Output of Algorithm PERMUTE-3 for $n = 5$

RAND[1, 5] = 3	1*	2	3*	4	5
RAND[2, 5] = 5	$\underline{3}$	2*	1	4	5*
RAND[3, 5] = 3	$\underline{3}$	$\underline{5}$	1**	4	2
RAND[4, 5] = 5	$\underline{3}$	$\underline{5}$	$\underline{1}$	4*	2*
RAND[5, 5] = 5	$\underline{3}$	$\underline{5}$	$\underline{1}$	$\underline{2}$	4**
	$\underline{3}$	$\underline{5}$	$\underline{1}$	$\underline{2}$	$\underline{4}$

That is, one way to develop a listing algorithm is to start by deciding what order you would like and then attempting to devise an algorithm that achieves this order.

A very natural order for permutations is "odometer" order: We view each n-permutation as an n-digit number (we've already been doing this), and then we list them in the order they would appear on an odometer. For instance, if $n = 3$, the order would be

123
132
213
231
312
321

The more general name for this sort of order is **lexical** or **lexicographic** order. Lexical order is odometer (i.e., numerical) order when the individual symbols are digits and alphabetical order when the symbols are letters.

There is a simple recursive way to describe an algorithm which produces permutations in lexical order. Notice that all the permutations with first symbol 1 precede all those with first symbol 2, and so on. Furthermore, if we take all the permutations that start with 1 and strip off the 1, we are left with the lexical ordering of all $(n-1)$-permutations on the symbols 2 through n. Similarly, if we take all those n-permutations that start with 2 and strip off the 2, we are left with the lexical ordering of all $(n-1)$-permutations on the symbols $1, 3, 4, \ldots, n$. And so on. Thus we can describe the algorithm as n recursive calls to itself, one for each possible starting symbol.

Although this recursive approach is simple in principle, like most recursions it is hard for humans to carry out. This one is also hard to implement in most current computer languages; the difficulty is the need to handle permutations of nonconsecutive integers. (The algorithm is not so hard to write in our algorithmic language because our language allows **for** loops over sets [19].) Therefore it is worth our while to seek an iterative algorithm for producing lexical order, even if the iterative algorithm is a bit complicated.

An iterative algorithm has other advantages here, too. First, it forces us to understand carefully which digits get changed going from one permutation to the next. (Just as on an odometer, sometimes only a few digits change, sometimes many.) This is significant because the real efficiency question for an all-permutations algorithm is: How much work does the algorithm take on average to generate *one* permutation? (The total number of permutations is the same for all such algorithms, $n!$.) Second, when an iterative and a recursive algorithm produce the same output, the iterative one is generally much faster since there is less computer overhead. Finally, the algorithm we are about to develop is related to an algorithm for generating "truth tables" in Chapter 7.

It wasn't a waste to think recursively first, however. You'll see how this helps a little later.

The key to developing an iterative algorithm for lexical order is to determine how to generate the next permutation from any given permutation. Then we just feed in the starter permutation $123\ldots n$ and let things run until we reach $n\ldots 321$. So suppose we have a typical permutation

$$[\text{Perm}(1),\ \text{Perm}(2),\ \ldots,\ \text{Perm}(n)]. \tag{2}$$

To find the next permutation we reason as follows.

1. If we start from the *right* in permutation (2) and move to the left as long as the digits are *increasing*, then we have a sequence of digits that is the largest possible formed from the digits in it. (Example: If $n = 7$ and permutation (2) is [4,2,5,7,6,3,1], then we stop at 7 and 7631 is the largest possible number that can be formed from 1, 3, 6, and 7.) No change in just these digits can result in a larger permutation.
2. The digit immediately to the left of the leftmost digit considered in the previous paragraph is the first one (from the right) which can be exchanged with something to its right to get a permutation later in the order. We exchange it with the smallest digit to its right that is greater than it, since this gives the minimum possible increase in this digit. (In the example, we exchange 5 and 6, obtaining [4,2,6,7,5,3,1].)
3. After this exchange, we still have a sequence increasing to the right where we had one before. (Now we have 7531 instead of 7631.) To get the smallest permutation greater than the previous one, we must arrange the digits in this sequence in their minimum odometer order. To do this we need only to reverse them. (In our example, the new permutation is [4,2,6,1,3,5,7].)

This reasoning is embodied in Algorithm 4.6, AllPerm. As an example, when $n = 4$, AllPerm generates the permutations shown in Table 4.3. Read down the left half, then down the right half of the table.

The operations in AllPerm are comparisons of Perm entries, switches of entries, and reassignments of counters. In each passage through the main loop, the number of such steps varies from permutation to permutation. In fact, the analysis of AllPerm is fairly difficult and hard to make intuitive. We will take it up in Section 6.6.

We've said all we want to say on combinatorial algorithms per se. We got onto these algorithms because counting combinatorial objects is not always enough. However, it is an intriguing fact that such algorithms can shed light back on counting problems and other "structural" combinatorial problems that don't appear to be related to algorithms. We want to stimulate you with an example of this idea.

We know that there are $n!$ permutations of the symbols 1 to n. When we devised Algorithm AllPerm to generate them all without repetitions, we were in effect constructing a one-to-one correspondence between the set $[n!] = \{1, 2, \ldots, n!\}$ and the set $\mathcal{P}$ of permutations. A different algorithm to generate all permutations (there are many, and AllPerm isn't necessarily the best [36]) would involve a different one-to-one correspondence.

If we view the correspondence as going from $\mathcal{P}$ to $[n!]$, we have what is known as the **rank function**. For instance, in Table 4.3, Rank(1324) = 3, because 1324 is the third permutation in lexical order. If we view the correspondence as going

Algorithm 4.6

AllPerm

```
Input n
Algorithm AllPerm
    for i = 1 to n                              [Generate first permutation]
        Perm(i) ← i
    endfor
    repeat                                      [Main permutation loop]
        print Perm
        b ← n−1      [b will be position of leftmost digit to be changed]
        repeat until b = 0 or Perm(b) < Perm(b+1)
            b ← b−1
        endrepeat
        if b = 0 then stop                      [All permutations found]
        c ← n            [c will be position to be exchanged with b]
        repeat until Perm(c) > Perm(b)
            c ← c−1
        endrepeat
        Perm(b) ↔ Perm(c)                       [Exchange digits]
        d ← b + 1; f ← n                        [Initialize for reversal]
        repeat until d ≥ f
            Perm(d) ↔ Perm(f)                   [Reverse by exchanging]
            d ← d+1;  f ← f−1
        endrepeat
    endrepeat
Output  All n! permutations of 1, 2, ..., n
```

from $[n!]$ to $\mathcal{P}$, we have what is known as the **unrank function**. For instance, for lexical order Unrank(3) = 1324.

We don't have to look at a table to determine the rank function. This is where the recursive analysis of lexical order is quite handy. Given a permutation Perm, look at Perm(1). Suppose that it is k. For each j from 1 to $k-1$, all $(n-1)!$ permutations beginning with j come before Perm in lexical order. Then, if we strip off the first digit, we can apply the same analysis recursively to determine how many permutations beginning with k also come before Perm.

For example, consider $n = 4$ and let Perm = 3241. Then all the permutations that begin with 1 or 2 come before Perm; there are $2 \times 3! = 12$ of these. Now strip off the 3 from Perm and consider 241 as a permutation on the set $\{1, 2, 4\}$. The $1 \times 2! = 2$ permutations on this set that begin with 1 all come before 241. Now strip

TABLE 4.3
Output for Algorithm ALLPERM **when** $n = 4$

Permutation	b	c	Permutation	b	c
1 2 3 4	3	4	3 1 2 4	3	4
1 2 4 3	2	4	3 1 4 2	2	4
1 3 2 4	3	4	3 2 1 4	3	4
1 3 4 2	2	3	3 2 4 1	2	3
1 4 2 3	3	4	3 4 1 2	3	4
1 4 3 2	1	4	3 4 2 1	1	2
2 1 3 4	3	4	4 1 2 3	3	4
2 1 4 3	2	4	4 1 3 2	2	4
2 3 1 4	3	4	4 2 1 3	3	4
2 3 4 1	2	4	4 2 3 1	2	3
2 4 1 3	3	4	4 3 1 2	3	4
2 4 3 1	1	3	4 3 2 1	0	

off the 2 and consider 41 as a permutation on $\{1, 4\}$. Then $1 \times 1! = 1$ permutations on this set start with a smaller number than 4 (namely, the permutation 14). Finally, strip off the 1, leaving the only 1-permutation on the symbol 4. So we conclude that

$$2 \times 3! + 1 \times 2! + 1 \times 1! + 0 \times 0! = 15$$

permutations come *before* 3241 in lexical order. Therefore, Rank(3241) = 16. (Some authors define the rank to be the number of things which come before, in which case 15 itself is the answer.)

So what? Well, there was nothing special about 3142. For any n-permutation, it has a specific rank, and we have illustrated how (one less than) that rank can be computed in the form

$$\sum_{k=1}^{n} a_k (n-k)!\,, \qquad 0 \le a_k \le n-k. \tag{3}$$

For instance, when Perm = 3241 we have obtained $a_1 = 2$, $a_2 = 1$, $a_3 = 1$ and $a_4 = 0$. More generally, a_k is the number of digits to the right of the kth digit of Perm that are numerically smaller than the kth digit.

Furthermore, if Perm and Perm$'$ are two different n-permutations, then the sequences $a_1, a_2, \ldots, a_n$ and $a'_1, a'_2, \ldots, a'_n$ of coefficients they generate must be different. (Why?) Therefore we have the following result: Every number from 0 to $n! - 1$ is *uniquely* representable in the form of Eq. (3).

This little result may not be so important (we haven't even labeled it as a theorem), but the point is: Each algorithm that generates all the permutations leads to a different theorem of this sort. Each algorithm that generates all the (n, k)-combinations leads to another theorem of this sort, and so on. The idea of relating combinatorial algorithms to combinatorial identities (and also to various combinatorial "structure" theorems) has proved very fruitful in recent research.

Problems: Section 4.9

1. It is a misnomer to call RAND a function. Why? (Nonetheless, it is common practice to do so.)

2. Most computer languages do not actually have our RAND as their basic (pseudo)random number generator. More often, they have a command, call it RAN, which generates a decimal number strictly between 0 and 1, where every decimal number is equally likely. How could you define our RAND$[n, m]$ in terms of RAN?

3. Actually, Algorithm PERMUTE-1 isn't an algorithm at all, according to the precise definition in Chapter 1. It is logically possible for it to go on running forever. How? Yet we claim it is "essentially" an algorithm. Explain.

4. Let $n = 5$.

   a) Given

   $$2\ 4\ 3\ 4\ 1\ 2\ 4\ 3\ 5\ 4\ 2\ldots$$

   as a stream of outputs from RAND[1,5], what permutation would Algorithm PERMUTE-1 produce?

   b) Given the outputs

RAND$[1, 5] = 2$	RAND$[1, 2] = 1$
RAND$[1, 4] = 4$	RAND$[1, 1] = 1$
RAND$[1, 3] = 2$	

   what permutation does PERMUTE-2 produce?

   c) Given the outputs

RAND$[1, 5] = 2$	RAND$[4, 5] = 4$
RAND$[2, 5] = 4$	RAND$[5, 5] = 5$
RAND$[3, 5] = 4$	

   what permutation does PERMUTE-3 produce?

5. There is a variant of Algorithm PERMUTE-3 that calls

   RAND$[1,n]$, RAND$[1,n-1]$, RAND$[1,n-2]$, ...

   instead of

   RAND$[1,n]$, RAND$[2,n]$, RAND$[3,n]$, ...

   (It fills in the permutation from the right.) Figure it out and write it down.

6. Consider Table 4.2.

   a) The last line is not necessary; the permutation does not change from the previous line. Why not?

   b) How can Algorithm PERMUTE-3 be changed to take advantage of this fact?

   c) How is its efficiency analysis affected?

7. Argue as carefully as you can that the permutation produced by Algorithm PERMUTE-3 is random.

8. Explain why the following algorithm does or does not generate a random permutation of $1, 2, \ldots, n$.

```
Algorithm PERMUTE-4
  for i = 1 to n
    Perm(i) ← i                          [Initialize]
  endfor
  for k = n to 2 step −1
    Perm(1) ↔ Perm(Rand[1, k])
  endfor
Output Perm(1),..., Perm(n)
```

9. We sluffed over one subtlety in arguing that Algorithm PERMUTE-2 is correct. It is clear that Perm(1), Perm(2), etc., are all different integers, but how do we know that they are all between 1 and n? In PERMUTE-2, Perm$(i) \leftarrow j$, and j is incremented several times before this assignment takes place, without any explicit instruction to keep j from going above n. Explain why it doesn't.

10. For $n = 5$ what are the first 10 permutations output by Algorithm ALLPERM? What are the last 10?

11. Step through Algorithm ALLPERM when $n = 3$. Make a table like Table 4.3, but also depict the process of creating each permutation from the previous one by showing the result of exchanging two entries and then showing the result of reversing part of the sequence. That is, show steps 2 and 3 in the text description of ALLPERM separately.

12. Consider the rank and unrank functions for Algorithm ALLPERM.

a) What is Rank(462135)?

b) When $n = 5$, what is Unrank(90)?

c) Devise a general procedure for computing the unrank function.

13. **Reverse lexical** order is just lexical order run backwards. For instance, if the symbols are digits, reverse lexical order lists numbers from largest to smallest. Do [12] again for reverse lexical order. Also come up with a general procedure (not using the lexical rank) for computing the reverse lexical rank function.

14. **Colexical** order is lexical order, but with the rightmost symbol the most significant instead of the leftmost. For instance, if the symbols are letters, colexical order would be alphabetical order if we read English from right to left. Answer [12] again for colexical order. Also come up with a general procedure for computing the colexical rank function.

15. The coefficients a_k in Eq. (3) that arise out of Algorithm ALLPERM are intimately related to Algorithm PERMUTE-2. Explain how.

16. In Eq. (3) it suffices to sum from $k = 1$ to $n-1$ instead of 1 to n. Explain.

17. We didn't define lexical order precisely in the text but instead relied on your prior understanding of odometer order. Give a precise definition. A recursive definition is one possibility.

18. One way to generate all the n-permutations is to take any algorithm for generating single random permutations and run it until all permutations have been found, discarding repeats along the way. If we count each call of the random permutation algorithm as one step, on average how many steps does this approach take?

19. Write a recursive version of Algorithm ALLPERM. Make the recursive procedure AllPerm(S), where S is the set of symbols to be permuted. The body of AllPerm(S) will consist of a loop that starts **for** $s \in S$. Assume this loop chooses the elements s in lexical order. You may also use English to instruct the algorithm to insert symbols in front of the output of subcalls.

20. What sort of permutation corresponds to the Traveling Salesman Problem on n cities when there isn't a designated "home city"?

21. In the text we indicated how to find an approximate solution to the Traveling Salesman Problem by using random permutations. Another plausible approach is to try a greedy algorithm: Start with the shortest edge and grow a path, at each step adding the shortest edge from either end of the path to vertices not on the path. When all vertices have been reached, close the path by adding the edge between the ends.

Unfortunately, this need not work well. Consider the complete graph in Fig. 4.14. The shortest cycle happens to be 1–2–3–4–5–1, with length 900. What cycle, with what length, does the greedy algorithm obtain?

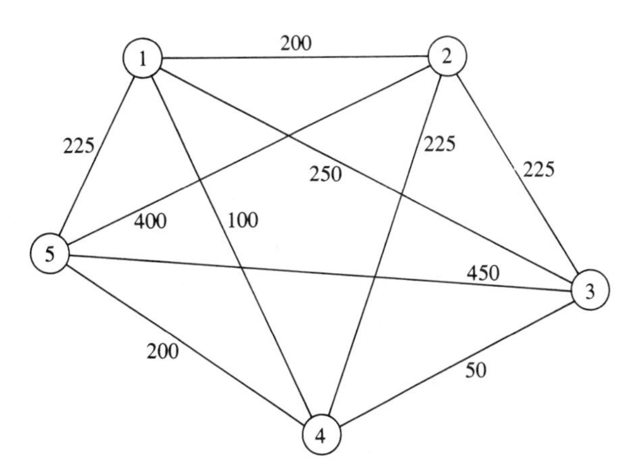

FIGURE 4.14
Greed need not pay

22. For [21], verify that the cycle claimed to be shortest is the shortest. This is bearable by hand (24 tours), but use a computer, basing your program on Algorithm ALLPERM.

23. Make up a traveling salesman problem with six cities. Make up the distances between each pair of cities. Write a computer program similar to

ALLPERM to test the length of each tour and thus find the shortest one.

24. Make up a traveling salesman problem with 20 cities. Make up the distances between each pair of cities. (This is a lot of pairs—let the computer generate the distances.) Write a computer program similar to PERMUTE-3 which tests N random tours and outputs the shortest tour found. Run this program for $N = 10, 100, 1000$. Report on your results.

25. Run your program from [24] again, but now run it on the six-cities of [23], for which you know the exact answer. Do some experimenting. How many random tours do you need to pick before you get within 10% of the optimal answer?

26. Write an algorithm to output a random ordered pair (i, j) where each entry is between 1 and n, inclusive.

27. Write an algorithm to output all ordered pairs (i, j) with entries between 1 and n, inclusive.

28. Here is an algorithm which purports to output a random increasing ordered pair (i, j), where $1 \le i \le j \le n$.

Input n
Algorithm RISING-PAIR
  $i \leftarrow \text{RAND}[1, n]$
  $j \leftarrow \text{RAND}[i, n]$
Output (i, j)

This outputs an increasing ordered pair, all right. What's wrong with the algorithm?

29. Write an algorithm which outputs a truly random, *strictly* increasing ordered pair with entries between 1 and n. That is, it should output (i, j) where $1 \le i < j \le n$.

30. Write an algorithm which outputs a truly random, increasing ordered pair with entries between 1 and n. That is, it should output (i, j) where $1 \le i \le j \le n$.

31. Modify our three random permutation algorithms to produce random permutations of n things taken k at a time.

32. Devise an algorithm to construct all permutations of n things taken k at a time.

33. Develop an algorithm to find a random combination of n things taken k at a time. *Hint*: You don't have to wander very far from a permutation algorithm. On the printed page, a combination has to appear in some order, but nobody said anything about caring what that order is.

34. Develop an algorithm to output all combinations of n things taken k at a time, without listing any combination twice. *Suggestion*: If for no other reason than to keep from repeating things, it is now important to output each combination in a standard way, say, in increasing order.

35. Suppose that 100 students sign up for course CS101 and 90 are male. Also suppose that there is room for only 30 students in the course, and the CS Department decides to make a random selection.

a) The CS Department Chair is worried that the random selection might result in all 10 females being excluded and that there will be claims of discrimination. He would like to see what typical random selections look like with regard to gender. Provide him with 10 such random selections.

b) Generate 100 random selections and use these to estimate the probability that there will be no females in the class; at most 1 female in the class. (If you know how to compute the probability exactly, do so and compare your answers with your experimental results.)

36. Algorithm ALLPERM sometimes requires a lot of reassignments of the permutation entries to get between the current permutation and the next one. It turns out there is another algorithm which generates the permutations in such an order (*not* lexical order) that each permutation differs from the previous one by the minimum possible reassignment—the exchange of two adjacent entries. For instance, here is a listing of the six permutations of $\{1, 2, 3\}$ with this "minimal exchange" property:

$$321 \quad \underline{23}1 \quad 2\underline{13} \quad \underline{12}3 \quad 1\underline{32} \quad \underline{31}2.$$

(Underlining indicates those two digits that were just exchanged.)

a) Think recursively to devise such an algorithm. *Hint*: Suppose that you already have a minimal exchange listing of the permutations on the set $\{1, 2, \ldots, n-1\}$. For each permutation in the list, consider all the ways to insert the symbol n. Do it so as to maintain the minimal exchange property.

b) Find a recursive formula for the rank function for your algorithm.

c) Use (one less than) the rank function to devise a representation theorem for the numbers from 0 to $n! - 1$ that is different from the theorem in the text.

d) Find an iterative algorithm which outputs the n-permutations in the same order as your algorithm in part a).

4.10 Algorithmic Pigeonholes

In addition to putting balls into urns, mathematicians like to put pigeons into holes. This hobby is based on a very simple observation.

The Pigeonhole Principle

If $n + 1$ or more pigeons are put into n holes, at least two pigeons are put in the same hole.

This principle has many applications, some quite surprising. Most of the applications are "structure" results. They tend to say: No matter how disordered you try to be, there is some structure in what you create. This vague statement will make more sense after some examples.

What's "algorithmic" got to do with it? You'll see.

EXAMPLE 1 You have five pairs of socks, all different colors, jumbled together as ten individual socks in your dresser drawer. You go into your room before sunrise to get a pair and discover the light is out. How many socks must you pull out to be certain to have a matched pair among them?

Solution The socks are pigeons and the colors are holes. There are five holes and you want to have two pigeons in some hole. Thus to be certain, you have to take six socks. ∎

EXAMPLE 2 You and a friend play the following game. The friend gets to pick any ten integers from 1 to 40. You win if you can find two different sets of three from those ten that have the same sum. Prove that you can always win this game.

Solution Let the pigeons be the possible 3-sets. Let the holes be the possible sums. There are $C(10,3) = 120$ pigeons. The smallest possible sum is $1+2+3 = 6$ (and that's possible only if your friend picked 1, 2, and 3). Likewise, the largest possible sum is $38+39+40 = 117$. Therefore there are at most $117-5 = 112$ holes. So at least two pigeons (3-sets) go in the same hole (have the same sum). ∎

We claimed that the Pigeonhole Principle is about disorder and structure. Now we can explain better what we meant. In Example 1, the socks were disordered, but you couldn't help but find order (a matching pair!) if you took six. In Example 2, your friend can pick the ten numbers from 1 to 40 any way at all, with irregular gaps between their values. But this can't keep you from finding a certain order to things, specifically, two triples that have the same sum.

Note that the Pigeonhole Principle does not say *which* pigeons share a hole, or *how* to find them. For this reason, the principle has usually been regarded as a purely existential result (see the discussion of existential and constructive proofs in Section 3.4). However, it doesn't have to be existential. Instead of asking

> Given anything of type A, prove that it contains something of type B,

instead ask

> Devise an algorithm that, given anything of type A, outputs something of type B from it if it's there.

The second form is the reason for our section title. Moreover, the second is easier to get started on, and if you are successful, as a by-product you will usually have an answer for the first form as well.

We illustrate this with a problem that is very hard when treated the existential way. It concerns subsequences of a sequence. Given a sequence of numbers, a **subsequence** is a subset considered in the same order as the numbers appear in the original. Thus $1, 5, 2$ is a subsequence of $S = 1, 3, 5, 7, 6, 4, 2,$ but $1, 2, 5$ is not. Given any sequence of *distinct* numbers, a subsequence is **monotonic** if it is always increasing or always decreasing. Thus $1, 3, 5, 7$ and $6, 4, 2$ are both monotonic subsequences of S, but $1, 5, 2$ is not.

The problem is:

> Given a sequence of distinct real numbers, what can you say about the longest monotonic subsequence?

We have stated this so open-endedly because, in fact, there are two quite different forms in which this problem has circulated. The existential version goes:

> Show that any sequence of length such-and-such has a monotonic subsequence of length so-and-so.

("Such-and-such" and "so-and-so" are specific expressions, but we don't want to give them away. That's the problem with the existential approach; you are presented with the results as a fait accompli and have no idea how you could find them yourself.) The algorithmic version goes:

> Devise an algorithm that takes any sequence and finds the length of its longest monotonic subsequence.

(In both versions, and henceforth in this section, it is understood that we are talking about sequences of distinct real numbers.)

Let's try to solve the algorithmic problem. (In so doing, our goal is to gain insight about the existential problem as well.) Use the recursive paradigm! Suppose the given sequence is $a_1, a_2, \ldots, a_m$. Suppose you already knew how to find a longest monotonic subsequence for sequences of length k. (We say "a longest", not "the longest", because perhaps several subsequences tie for being longest.) How would you handle a sequence $a_1, a_2, \ldots, a_k, a_{k+1}$? Well, either some longest monotonic subsequence is a subsequence of the first k terms, in which case you already know how to find it, or else all longest subsequences end with a_{k+1}. In the latter case, let S be a longest monotonic subsequence. The term in S just before a_{k+1} is some a_j with $j \leq k$. Furthermore, S with a_{k+1} deleted (call this subsequence S') itself must be a longest monotonic subsequence in the following sense: If S is increasing (respectively decreasing) then S' has to be a longest increasing (respectively decreasing) subsequence ending at a_j. (Why? Figure 4.15 should help.)

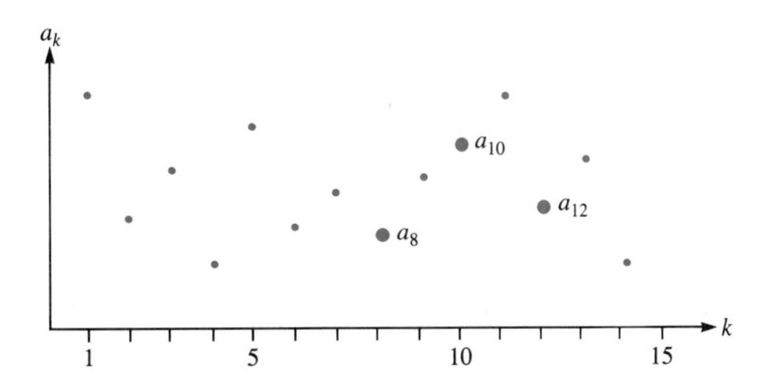

FIGURE 4.15

The graph of part of a sequence. The value of a term is indicated by its height. The term a_{12} may be appended to any decreasing subsequence ending at a_{10} or to any increasing subsequence ending at a_8.

This suggests that to succeed at recursion it's not enough to know the length of the longest monotonic subsequence *within* the initial sequence $a_1, \ldots, a_k$. We need to know the length of the longest increasing and the longest decreasing sequence *ending* at a_j for each $j < k+1$.

So let's try the recursive paradigm again with a stronger inductive hypothesis. Suppose you already knew the lengths u_j and d_j of the longest up and the longest down subsequences ending at a_j, for each j from 1 to k. How could you then determine u_{k+1} and d_{k+1}? Once you answer this question, you can figure out all the pairs (u_k, d_k) for k from 1 to m (the length of the given sequence), and then the maximum of all these numbers will be the answer to the original question.

We've already made the key observation for an algorithm to find u_{k+1} and d_{k+1}. The longest monotonic sequence ending at a_{k+1} must have *some* a_j as the previous term. Therefore we look at *each* previous term a_j, take the longest increasing sequence ending there (if $a_j < a_{k+1}$) or the longest decreasing sequence ending there (if $a_j > a_{k+1}$) and tack a_{k+1} on the end. Then we maximize over all these a_j. Algorithm 4.7, MONOTONIC, carries this out, and Table 4.4 gives an example. (In the algorithm, u_k and d_k don't become the lengths of the longest up and down sequences ending at a_k until the end of the inner loop.)

Algorithm 4.7

Monotonic

Input $a_1, a_2, \ldots, a_m$ [Distinct numbers]

Algorithm MONOTONIC

$u_1 \leftarrow 1;\ d_1 \leftarrow 1$

for $k = 2$ **to** m

  $u_k \leftarrow 1;\ d_k \leftarrow 1$ [Initialize]

  for $j = 1$ **to** $k-1$

    if $a_j < a_k$ **then** $u_k \leftarrow \max\{u_k, u_j + 1\}$

        else $d_k \leftarrow \max\{d_k, d_j + 1\}$ [$a_j > a_k$]

  endfor

endfor

length $\leftarrow \max\{u_1, u_2, \ldots, u_m, d_1, d_2, \ldots, d_m\}$

Output *length*

TABLE 4.4
Trace of Algorithm MONOTONIC **for the sequence** $7, 3, 8, 4, 6$

k	u_k	d_k
1	1	1
2	1	2
3	2	1
4	2	2
5	3	2

j	u_4	d_4
initial	1	1
1	1	2
2	2	2
3	2	2

for $k = 4$

The box on the left shows values at the end of the outer loop; the box on the right shows values at the end of the inner loop for $k = 4$. In this inner loop, u_4 and d_4 start at 1, representing the sequence consisting of $a_4 = 4$ alone. Then sequences ending at a_4 with $a_1 = 7$ as the previous term are considered. Since these are monotonic down, and $u_1 = 1$, d_4 increases to 2. Next, $a_2{=}3 < a_4$, so sequences ending at a_4 with a_2 as the previous term must be monotonic up, and u_4 increases to $u_2{+}1 = 2$. Finally, $a_3 > a_4$, so $d_3{+}1 = 2$ is the length of a down sequence ending at a_4. Since d_4 was already 2, it does not change.

All right, we can solve the algorithmic problem, but so far our method doesn't seem to suggest any patterns we can use for the existential version. That is, it hasn't shown us any nice relationship between m, the length of the original sequence, and n, the length of the longest monotonic subsequence. Besides, where are the pigeons and where are the holes?

Answering the last question will lead us to a pattern. Since an assignment of pigeons to holes is a function mapping one set (pigeons) to another (holes), look to see if Algorithm MONOTONIC sets up any functions. Yes, for each k it assigns the couplet (u_k, d_k). Furthermore, this function is one-to-one. To see why, look at the two lines inside the inner loop; they say that, for each k, (u_k, d_k) is different from all previous (u_j, d_j)—either $u_k \geq u_j + 1$ or $d_k \geq d_j + 1$. So let the pigeons be the k's (the indices of the sequence), and let the holes be ordered pairs of positive integers. The length of the longest monotonic subsequence is the maximum value of all the u_j's and d_j's for the holes actually occupied by pigeons. If this maximum is n, then at most the n^2 holes from (1,1) to (n, n) are occupied. Since each hole has at most one pigeon, there are at most n^2 pigeons. In other words, we have solved the existential problem by proving the following "structure" theorem of numerical sequences.

Theorem 1. Let a sequence consist of distinct numbers. If the longest monotonic subsequence has length n, then the sequence has at most n^2 terms. Put differently, a sequence of $n^2 + 1$ distinct numbers necessarily has a monotonic subsequence of at least $n + 1$ terms.

Now that you know how to obtain this existential theorem via the algorithmic approach, you too can impress and mystify your friends by telling them only the theorem and the following short proof. (This is all that is typically said if one presents the problem existentially.)

PROOF Let the sequence be $a_1, \ldots, a_{n^2+1}$. For each k, let u_k (respectively d_k) be the length of the longest increasing (respectively decreasing) subsequence ending at a_k. Note that u_k and d_k are both positive integers. The mapping $k \to (u_k, d_k)$ is an injection, because for any pair $j < k$, either the longest up sequence or the longest down sequence to a_j can be continued on to a_k (depending on whether $a_j < a_k$ or vice versa). Therefore there are $n^2 + 1$ distinct pairs (u_k, d_k), so some u_k or d_k must be at least $n + 1$. ∎

Sometimes the Pigeonhole Principle is useful because we *don't* want to put more than one pigeon in a hole, or we know for a fact that we haven't. For instance, the solution of the monotonic subsequence problem hinged on realizing that each pigeon was necessarily in a different hole. In such cases, it's the contrapositive form of the principle that we use:

If n pigeons fit in different holes, then there are at least n holes.

We conclude with an easy example.

EXAMPLE 3 In order for a keyboard to send its output to a computer, there must be a code for each character—uppercase and lowercase letters, numerals, punctuation marks and such commands as "carriage return". Each of these characters must be converted into a string of 0's and 1's. Are binary strings of length six enough to encode all characters? If not, what's the minimum length needed?

Solution There are about 100 actual symbols and a variable number of function keys. These letters are the pigeons and the 6-bit strings are the holes. It is essential that each pigeon go in a different hole. (Why?) Unfortunately, there are only $2^6 = 64$ 6-bit strings. Therefore 6 bits are not enough. On the other hand, unless you have a keyboard with a lot of special function keys, 7 bits are enough. And sure enough, American Standard Code for Information Interchange (ASCII) is 7 bits. ∎

Problems: Section 4.10

1. The **Extended Pigeonhole Principle** says: If ______ pigeons are put in n holes, then some hole contains at least $k+1$ pigeons. Fill in the blank. Explain.

2. In Example 1, instead of one pair of each of five colors, suppose that you have seven identical pairs of each color. Does the answer change?

3. Repeat [2], except that you want to pull out three pairs all of the same color. How many socks must you take to be certain of success.

4. Repeat [3], except that the three pairs need not be the same color.

5. Same as Example 1, except that you want the red pair. How many socks must you draw to be certain? (This problem shows that if you start specifying *which* pigeonhole you want, the Pigeonhole Principle doesn't apply.)

6. Consider Example 2 (equal sums of 3-sets when ten numbers are picked between 1 and 40).

   a) Suppose your friend picks ten numbers between 1 and <u>50</u>. Show that the sort of pigeonhole analysis in the solution is now insufficient to prove that you can always win. That is, show that the number of 3-sets is not greater than the number of possible sums. *Note*: Showing this does not prove that you *can't* force a win. A much more subtle analysis might still show that you can always beat your friend; see part g) for an example. If you can't always beat him, a proof of that would exhibit a particular ten-set of $\{1, 2, \ldots, 50\}$ such that all 120 of its 3-subsets have different sums.

   b) Show that the pigeonhole analysis of the solution is insufficient to show you can always win if your friend picks nine numbers between 1 and 40.

   c) Suppose we want to show that, when any 10 integers are picked from 1 to m, then some two 3-subsets of the 10-set necessarily have the same sum. What is the largest m for which the sort of analysis in Example 2 shows this?

   d) Suppose we want to show that, when any n integers are picked from 1 to 100, then some two 3-subsets of the n-set necessarily have the same sum. What is the smallest n for which the sort of analysis in Example 2 shows this?

   e) Same as part c), except now consider 4-subsets instead of 3-subsets.

   f) Same as part d), but consider 4-subsets.

   g) Suppose your friend now picks ten integers between 1 and 24, and you try to find two *pairs* of his numbers with the same sum. There are $C(10, 2) = 45$ pairs and 45 possible sums (from 1+2=3 to 23+24=47). So the pigeonhole analysis alone does not prove that you can always win. Nonetheless, you can. Prove it. *Hint*: Show that your friend, to have a chance to force a win, must include in his ten-set 1, 2, 23, and 24. So what?

7. Example 2 was presented in the existential form. An algorithmic form would ask for an algorithm which, given n distinct numbers chosen from 1 to m, finds all 3-subsets with the same sum (if there are any).
   a) Outline such an algorithm.
   b) Explain how finding such an algorithm might lead someone to solve the existential version, that is, to identify the pigeons and the holes and to find m and n (like $m = 40$ and $n = 10$) which guarantee the existence of 3-subsets with equal sums.

8. Show that, given *any* $n + 1$ integers from the set $\{1, 2, \ldots, 2n\}$, one of the integers is a multiple of another. (In fact, one of them is a power of 2 times another.) *Hint*: Look at the largest odd factor of each number.

9. If computers used trits instead of bits, how long would strings have to be to transfer keyboard input? (See Example 3. Trits are the "digits" 0, 1, and 2.)

10. Why isn't every sequence of distinct terms monotonic? After all, for any sequence of distinct terms, each term is either $>$ or $<$ the previous term.

11. We devised Algorithm MONOTONIC recursively, yet we wrote it iteratively. Why? When we need information about the jth case, why not obtain it by a recursive procedure call?

12. In Algorithm MONOTONIC, why did we initialize u_k and d_k to 1? Find a situation in which any other initialization would lead to a wrong answer.

13. Algorithm MONOTONIC can be refined.
   a) The line

$$\textit{length} \leftarrow \max\{u_1, \ldots, u_m, d_1, \ldots, d_m\}$$

   cannot be implemented as a single line in most current computer languages. Rewrite MONOTONIC so that length is computed iteratively as part of the main loop.
   b) Modify MONOTONIC further so that it returns the terms of a longest monotonic subsequence (instead of just its length).
   c) If there are several longest monotonic subsequences, you could return all of them, but it's better to return one that is uniquely determined by further properties. Think of at least one property that would uniquely determine one longest subsequence and modify MONOTONIC further to output that subsequence.

14. Theorem 1 didn't really finish the job on monotonic subsequences. It shows that n^2+1 terms force a monotonic subsequence of length $n+1$, but it doesn't show that some lengths less than n^2+1 don't also force the same conclusion. Maybe some even more subtle pigeonhole argument replaces n^2+1 by a smaller value. Show that Theorem 1 is the best possible by exhibiting a sequence of n^2 distinct numbers with no monotonic subsequence of length more than n. For your sequence, verify that the pairs (u_k, d_k) take on every value from (1,1) to (n, n).

15. How do the analysis and conclusions of the monotonic subsequence problem change if the terms of the original sequence need not be distinct?

16. Consider a sequence

$$(a_1, b_1),\ (a_2, b_2),\ \ldots,\ (a_m, b_m)$$

of ordered pairs of real numbers. Such a "vector sequence" is monotonic if the a entries are monotonic and the b entries are monotonic (it's okay if, say, the a's are increasing and the b's are decreasing). Solve the longest monotonic subsequence problem again for such vector sequences. Devise both an algorithmic and an existential solution.

17. The Pigeonhole Principle hardly requires a formal proof, but it is interesting to note that one can give a formal proof (of the contrapositive form), using the Sum Rule. Do so.

18. Give a formal proof of the Extended Pigeonhole Principle of [1].

Summary: Chapter 4

You've seen a lot of counting formulas and methods in this chapter. When a problem appears on a test (say) instead of at the end of a section, is there any easy way to tell which method to try on it?

The bad news is this: There's no simple algorithm for choosing the right approach. First of all, there are a lot of methods you haven't learned yet—they come later in this book or in a combinatorics course. But even if you were already an expert, you would find that a lot of easily stated problems are difficult or impossible to attack successfully.

Nonetheless, a few words of review might be helpful. Given a problem, first ask yourself: is it about ordered arrangements or sets? If so, it might be doable directly with permutations or combinations. If the problem isn't about permutations or combinations, but you can describe the situation in a reasonable number of and/or steps, you might get an answer straight from the Sum and Product Rules. If this doesn't work, then ask: Can I make it into a problem of balls and urns? If it's a standard type, you're home free.

If none of these approaches works, try to find a one-to-one correspondence or a combinatorial argument that reduces the problem to something more tractable. Or maybe a k-to-1 correspondence obtains, in which case the Division Rule will help.

Solving a counting problem is not just a matter of finding *any* correct formula. Usually, if you are willing to be careful and sum a great many terms, you can express the answer. But this is not a good solution if there is a much simpler expression for the same number. If you have summed a lot of terms with combinations in them, the Binomial Theorem may help to simplify the sum.

If your problem involves overlapping sets, then you probably should skip over all the suggestions above and first try Inclusion-Exclusion.

Finally, there were some things in this chapter that weren't problems with numerical answers. You might have to generate arrangements rather than count them. Section 4.9 gave some examples of how to approach this sort of algorithmic question. Or you might want to know whether some correspondence is one-to-one or not, or want to ensure that it is. Then try thinking in terms of pigeons and holes.

That's about all we can say at this level of generality. Remember, there are many methods we haven't discussed yet. For instance, we have made much less use of recursive methods than in earlier chapters. We have rarely said: relate the answer you want to the answer to a smaller version of the same problem. This method will often work, once you learn the techniques in the next chapter.

Supplementary Problems: Chapter 4

1. Investigate the pattern of odd and even entries in Pascal's Triangle. It will help to replace each even entry by 0 and each odd entry by 1.

2. Twenty scientists are working on a secret project. The government trusts the scientists collectively, but not individually. An administrator decides that any group of five or more of the scientists should be allowed to enter the documents room together, but that no smaller group should be allowed to do so. The idea is to put a number of locks on the door and distribute keys to individual scientists so that each group of five will have all the keys among them, but no smaller group will.

   Is this a workable decision? At the minimum, how many locks are necessary? How many keys must be made for each lock? How many keys must each scientist carry? *Hint*: Show that there must be a one-to-one correspondence between 4-sets of scientists and locks.

3. How many triangles are formed using chords and sides of an n-gon, where the vertices of the triangle need not be vertices of the n-gon. For instance, one such triangle is highlighted in Figure 4.16. Assume that no three chords meet at the same interior point. *Hint*: Relate different sorts of triangles to different-sized sets of vertices of the *polygon*.

4. Repeat [3], but now only chords of the n-gon can be sides of the triangle.

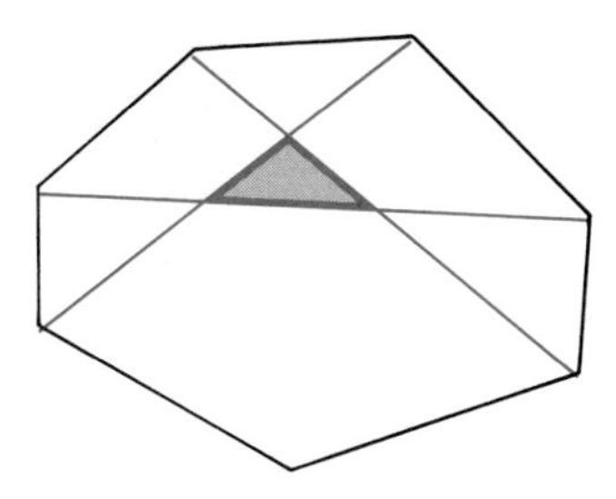

FIGURE 4.16

5. Solve Example 3, Section 4.4 (number of intersection points of the chords of a polygon) by the first method outlined in the text. Let c_k be the number of chords which have exactly k intersection points with other chords of the polygon. Use the Division Rule to express the answer in terms of $S = \sum kc_k$. Now, if you can deduce a formula for c_k and then figure out how to simplify S, you're in business.

6. As Eq. (2), Section 4.6 shows, the number of odd subsets of an n-set equals the number of even subsets.
   a) When n is odd, this fact has a particularly simple one-to-one correspondence proof using complementation. Give it. Why does it break down when n is even?
   b) Here is a one-to-one correspondence between the odd subsets of an n-set and the even subsets—a correspondence which works whether n is odd or even. Let x be any one element in the n-set, N. Then for any $S \subset N$, define

$$f(S) = \begin{cases} S \cup \{x\} & \text{if } x \notin S, \\ S - \{x\} & \text{if } x \in S. \end{cases}$$

   Show that f is a correspondence between odd and even sets.

7. In how many ways can five indistinguishable balls be put in three distinguishable urns if:
   a) The first urn must hold exactly three balls.
   b) The first urn must hold at least three balls.
   c) The first two urns each must hold at least one ball.
   d) The first two urns together must hold at least one ball.
   e) No urn may be empty.

8. Do [7] again for distinguishable balls. Which parts get much harder?

9. We outline another proof of Inclusion-Exclusion. Whereas the proof given in Section 4.8 merely verifies the formula, the following method provides another way to discover it.

   For each $A \subset U$, define the function

$$A(x): U \to \{0, 1\}$$

by

$$A(x) = \begin{cases} 1 & \text{if } x \in A, \\ 0 & \text{if } x \notin A. \end{cases}$$

$A(x)$ is called the **characteristic function** of set A.

   a) Explain why $\sum_{x \in U} A(x) = |A|$.
   b) Show that $A(x)B(x) = (A \cap B)(x)$. In words, the product of the characteristic functions of two sets is the characteristic function of their intersection.
   c) Show that $1 - A(x) = \overline{A}(x)$. ($\overline{A}$ is the complement of A in U.) Also, restate this equation in words.
   d) Show that $\prod_{i=1}^{n} \big(1 - A_i(x)\big)$ is the characteristic function of $\bigcap_{i=1}^{n} \overline{A}_i$.
   e) Derive the Inclusion-Exclusion formula by evaluating

$$\sum_{x \in U} \Big[\prod_{i=0}^{n} \big(1 - A_i(x)\big)\Big]$$

   two ways: directly and by first multiplying out the product.

10. Find an inductive proof of the Inclusion-Exclusion theorem. (You should expect this theorem to have an inductive proof, but the only inductive proof we know is quite intricate.)

11. How many nonnegative integral solutions are there to $x+y+z+w = 20$ if no unknown can be 10 or greater? *Hint*: The number of solutions in which x, say, *is* 10 or greater is the same as the number of all nonnegative solutions to $x' + y + z + w = 10$. Now use Inclusion-Exclusion.

CHAPTER 5

Difference Equations

5.1 Introduction

Conditions that define a sequence recursively are called **difference equations**. Another name, aptly enough, is **recurrence relations**. We've presented several examples of recursively defined sequences already. For instance, when we first analyzed Towers of Hanoi in Section 1.6, we considered the sequence $\{h_n\}$, where h_n is the number of moves necessary to solve the puzzle with n rings by our algorithm. We showed that this sequence is specified by the conditions

$$h_{n+1} = 2h_n + 1 \quad \text{for } n \geq 1, \qquad h_1 = 1. \tag{1}$$

Also, in Example 1 of Section 2.6 we let C_n be the worst case number of steps taken by Algorithm BINSEARCH, and we showed that

$$C_n = 1 + \max\{C_{\lfloor \frac{n-1}{2} \rfloor}, C_{\lceil \frac{n-1}{2} \rceil}\} \quad \text{for } n > 1, \qquad C_1 = 1. \tag{2}$$

Technically speaking, the first equation in each display above is the difference equation and the second is an **initial condition**, but we will often regard such pairs as a unit and refer to that unit as a difference equation. This same situation arose with induction: There are two parts, the basis and the inductive step, but they are referred to together as induction.

We wish to *solve* difference equations, that is, find an explicit form for the nth term of the sequence. For instance, in Section 2.2 we proved that the solution to Eq. (1) is $h_n = 2^n - 1$. Similarly, in Section 2.6 we proved that the solution to Eq. (2) is $C_n = \lfloor \log_2 n \rfloor + 1$. However, in both cases we succeeded only because we were able to *guess* the explicit form; from that we could prove it by induction. What we want are systematic methods that *produce* the right formula for us.

Systematic methods are one of the things this chapter is about. If you are eager to get to them, you can turn to Section 5.4 and start right in. However, we are going to devote two substantial sections before that doing more of what we have already done: treat difference equations by ad hoc methods.

Why? There are several reasons. First, most equations discussed in this chapter can be solved by computer algebra systems. That is, if an explicit solution is known for a standard difference equation, any good computer algebra system will come up with it in a flash. Soon hand calculators will too. So it's not so important that you learn how to solve these equations by hand as it once was. Instead it's important that you understand the mathematical ideas that lead to these solutions (since these ideas, for instance, "linearity", also show up in other parts of mathematics). Also, it is important to learn about certain patterns that arise in the solutions to difference equations and to understand why they arise (for instance, solutions to linear difference equations are usually geometric sequences or sums of geometric sequences).

Second, there are many types of difference equations for which explicit solutions are not known. But you can still generate as many terms as you want, using a computer if necessary. Often you can find what you need (the value of a specific term, the rate of growth of the sequence, etc.) by looking at such output. So a very important skill is to be able to take a problem and determine the difference equation associated with it. This is called *modeling* a situation with a difference equation and is what Section 5.2 is about. Almost as important is to be able to prove patterns you generate from such output. This aspect is treated in Section 5.3, using the difference equations set up in Section 5.2.

Finally, the examples of difference equations in earlier chapters have been from mathematics and computer science, and mostly concern algorithms. However, difference equations are useful in an immense range of disciplines. Therefore another purpose of the next two sections is to show you the variety of applications.

In Sections 5.4 through 5.8 we develop systematic methods for solving various sorts of linear difference equations. (We will explain the meaning of linear in Section 5.4.) Then in Section 5.9, we use this knowledge to look more deeply into the use of difference equations in the analysis of algorithms.

There is a very strong parallel between the theory of differ*ence* equations and the theory of differ*ential* equations, and we have arranged our material to emphasize this parallel. Those of you who have already studied differential equations, perhaps as a chapter in a calculus course, should find these parallels striking. If you study differential equations later, you should find the parallels striking then.

There are some advanced methods for solving difference equations that we won't discuss at all in this chapter. The most important is the method of "generating functions". We will introduce that method in Chapter 9.

One more point: What we are solving for in this chapter are *sequences*, not individual numbers. If there weren't a sequence, there wouldn't be terms to relate via a difference equation. But since the sort of solution we are looking for is a formula that holds for all terms, we often talk about solving for a typical term, say C_n, instead of for the sequence $\{C_n\}$. For instance, if you look back at the first paragraph of this section, you will see that we treated TOH "properly" by referring to a sequence, but for Algorithm BINSEARCH we lapsed into "term-talk" right away. Such term-talk is quite common—usually it reads better and causes no confusion. But never forget that we are really talking about sequences and that C_n, say, is referring to a *generic* term in the sequence.

5.2 Modeling with Difference Equations

The purpose of this section is to take you through the process of setting up difference equations. We won't solve the equations we set up, but defer that until Section 5.3 or later. We have set up a few difference equations before, so in this section we'll emphasize examples that are harder or that illustrate applications not touched on earlier. However, it's a good idea to start with something simple and familiar.

EXAMPLE 1 Describe arithmetic and geometric sequences using difference equations.

Solution Probably you have seen arithmetic sequences defined as sequences of the form

$$a, \quad a+d, \quad a+2d, \ \ldots, \ a+nd, \ \ldots$$

and geometric sequences as sequences of the form

$$a, \quad ar, \quad ar^2, \ \ldots, \ ar^n, \ \ldots .$$

An arithmetic sequence is one that starts at some value a and goes up by some fixed difference d each time. Well, this last sentence is just a difference equation in words. So $s_0, s_1, \ldots$ is an arithmetic sequence if and only if

$$s_n = s_{n-1} + d, \quad n \geq 1; \qquad s_0 = a.$$

Similarly, a geometric sequence is one that starts at some value a and is changed by a fixed multiplicative ratio r each time. So $t_0, t_1, \ldots$ is a geometric sequence if and only if

$$t_n = rt_{n-1}, \quad n \geq 1; \qquad t_0 = a. \ ∎$$

EXAMPLE 2 Describe arithmetic and geometric *series* using difference equations.

Solution A **series** is the sum of terms from a sequence. So an arithmetic series is

$$a + (a+d) + (a+2d) + \cdots + (a+nd),$$

and a geometric series is

$$a + ar + ar^2 + \cdots + ar^n.$$

Even if we are only interested in these sums for one specific value of n, it is worthwhile to consider all the values $n = 0, 1, 2, \ldots$. This gives us a new sequence, the

sequence of partial sums, and so we can look for a difference equation to describe the typical term in that sequence. Specifically, for $n \geq 0$ define

$$S_n = a + (a+d) + (a+2d) + \cdots + (a+nd) = \sum_{k=0}^{n}(a + kd),$$

$$T_n = a + ar + ar^2 + \cdots + ar^n = \sum_{k=0}^{n} ar^k.$$

We want difference equations for the sequences $\{S_n\}$ and $\{T_n\}$.

Now, the very definition of $\sum$ is recursive: S_n is just S_{n-1} with one more term, $s_n = a + nd$, added to it. Therefore we may describe $\{S_n\}$ by

$$S_n = S_{n-1} + (a+nd), \quad n \geq 1; \qquad S_0 = a.$$

Similarly,

$$T_n = T_{n-1} + ar^n, \quad n \geq 1; \qquad T_0 = a. \quad ■$$

This same technique specifies the partial sums for *any* sequence. Hence, for any sequence $\{a_n\}$,

$$A_n = \sum_{k=0}^{n} a_k \iff \big[A_0 = a_0 \quad \text{and} \quad A_n = A_{n-1} + a_n,\ n \geq 1\big]. \tag{1}$$

But so what? The goal is presumably to find a closed form for these sums, and all we've obtained is difference equations. True, but that will change in Section 5.7, which covers solutions for recursions like Eq. (1). The method of difference equations is not the only way to find formulas for series, nor is it the simplest, but probably it is the most systematic.

EXAMPLE 3 Devise a difference equation for the expected (average) number of comparisons in sequential search (Algorithm SEQSEARCH from Section 1.6) when the new word w is known to be on the list.

Solution Recall that the algorithm simply marched through the list until it found w. Let E_n be the expected (i.e., average) number of comparisons when there are n words in the list. Let k be any integer from 1 to $n-1$. We may now think of sequential search as a two-stage, recursive process. The first stage is sequential search applied to the first k words. If w is found, the process ends. Otherwise, we do stage two, sequential search on the remaining $n-k$ words.

What is the expected number of comparisons in the first stage? If w is among the first k words, the answer by definition is E_k. If it is not, the answer is k: We check all k first-stage words. Furthermore, the probability is k/n that w is in the first k, since it is assumed that w is equally likely to be any one of the n words. Thus

$$\text{Average no. of comparisons for stage 1} = \frac{k}{n}E_k + \frac{n-k}{n}k\,.$$

What about the second stage? The average number of comparisons *during* stage two is E_{n-k}, but we only get there with probability $(n-k)/n$. In short,

$$\text{Average no. of comparisons for stage 2} = \frac{k}{n}0 + \frac{n-k}{n}E_{n-k}.$$

Adding, we conclude that

$$E_n = \frac{k}{n}E_k + \frac{n-k}{n}(k + E_{n-k}). \tag{2}$$

Finally, since this equation holds for any k less than n, let's pick one that's easy to compute with, namely, $k = 1$. It's easy to see that $E_1 = 1$. (Why?) Substituting 1 for k and simplifying, we find

$$E_n = \frac{n-1}{n}E_{n-1} + 1. \tag{3}$$

Compute E_2 and E_3 from Eq. (3) and you'll see what the explicit formula for E_n is. Sure enough, it checks with the formula we obtained by completely different methods in Chapter 1.

(If our probability manipulations disturb you, e.g., multiplying an average by a probability and adding it to other things to get another average, don't be upset. We haven't formally analyzed the concept of averages, and we won't until Chapter 6. But we hope that such manipulations will seem at least plausible for the time being.) ∎

EXAMPLE 4 Is an IRA (Individual Retirement Arrangement) a good investment? Set up some difference equations that can help answer this question.

IRAs are an option available to US taxpayers that shields investment income from taxes until retirement. For instance, if you put money in an interest-bearing IRA account, the interest is not taxed until it is withdrawn many years later, whereas in an ordinary account the interest is taxed each year. It used to be that tax on the principal put down to open an IRA account was also deferred, but the 1986 tax reform eliminated that feature for many people.

This tax deferral is supposed to save you money. Since IRAs have the disadvantage that your money is "locked up" for a long time, the saving should be substantial to be worth it. Is it?

Solution Individual circumstances vary, so let's make some simplifying assumptions. (Other assumptions may appeal to you more; see [19].) Assume the choice is between an ordinary interest-bearing account and an interest-bearing IRA. (A person can also have IRA stock funds, etc.) Assume both accounts earn interest at a fixed rate of r a year (e.g., $r = .1$ means 10%). Assume you are not allowed to defer the tax on your principal. Thus the choice of an ordinary account or an IRA has no effect on your taxes this year, so you would have the same amount of after-tax money to start either account. Call this amount P. Assume you open the account at time 0 and plan to withdraw all the proceeds after exactly N years. Finally, assume your tax rate is t, now and in the future, under either option. For each option, you want to compute how much money you get after paying taxes in year N.

So far we have no sequence. Let P_k be the amount in the account exactly k years from the start if you set up an ordinary account. Let $\hat{P}_k$ be the amount after exactly k years if you set up an IRA. ($\hat{P}$ is read "P hat". Remember our choice this way: Hats make things bigger, as IRAs are supposed to do.) For the IRA, the compounding of interest is uninterrupted, so

$$\hat{P}_{k+1} = \hat{P}_k(1+r), \quad k \geq 0; \qquad \hat{P}_0 = P. \tag{4}$$

For the ordinary account, t times the interest is taxed away each year, leaving less money to draw interest the next year. That is,

$$P_{k+1} = P_k(1+r) - P_k rt, \quad k \geq 0; \qquad P_0 = P. \tag{5}$$

These are the difference equations. Let V and $\hat{V}$ be the after-tax values upon withdrawal at year N. For the ordinary account, $V = P_N$, because all taxes, including taxes in year N, have already been factored in. For the IRA,

$$\hat{V} = \hat{P}_N - (\hat{P}_N - P)t,$$

because all tax on the interest, $P_N - P$, is paid after withdrawal. So the analysis will be easy to complete as soon as we solve the difference equations. Do you recognize them as a type you already know how to solve? ∎

The next example isn't so much a single example as an explanation of how a whole subfield of economics can be treated with difference equations.

EXAMPLE 5 **National Income Accounting**

Every student of introductory economics learns these equations:

$$Y = C + I + G,$$
$$C = cY. \qquad (0 < c < 1)$$

What do they mean? First, that national income Y (the sum of all personal incomes) equals national consumption C (the sum of all personal spending), plus investment I, plus government spending G. Second, that consumption is a fraction c of national income.

Why are these equations right? The second is pretty clear; the only question is whether c is a constant, and if not, what it depends on. For simplicity, in introductory economics c is assumed to be a constant; we will assume the same. As for the first equation, a brief explanation goes like this (economists please forgive us!). All money that is spent eventually goes into the pockets of *people*. For instance, money invested in municipal bonds might be spent to build a bridge, in which case all the money eventually gets to the contractors and their employees, to the workers who make the steel girders, etc.

One trouble with this model (there are lots of troubles) is that it assumes all these transfers take place instantaneously. In fact, if you buy a bridge bond, it may be a year before a steelworker takes some of that money home. A better model builds in such time lags. To model this with discrete mathematics, we divide up time into periods, say, years or quarters. Let Y_n be the national income in period n, and likewise for C, I and G. Then a slightly more reasonable model is

$$\begin{aligned} Y_n &= C_{n-1} + I_{n-1} + G_{n-1}, \\ C_n &= cY_{n-1}. \end{aligned} \tag{6}$$

Just as intro econ students solve the simpler model for Y, let us try to solve for Y_n. First let us take the case that I and G do not vary from period to period. Then substituting the second equation above into the first (with n replaced in the second equation by $n-1$), we get

$$Y_n = cY_{n-2} + (I + G),$$

where $I + G$ is constant. So, once we know Y for two consecutive periods, we can figure it out for all future periods.

To take a slightly more complicated version, let's continue to assume that government spending is constant, but now let

$$I_n = A + dC_{n-1}, \tag{7}$$

where A and d are constants. That is, we now assume there is a rock-bottom amount A of investment even if the economy is bust ($C = 0$), but that the more people have spent in the previous period, the more confidence people and business will have for investing in the current period.

Substituting Eq. (7) into the first line of Eq. (6), and then substituting the second line of Eq. (6) into the first twice, we find that

$$Y_n = cY_{n-2} + cdY_{n-3} + (A + G).$$

The actual economic models on which public policy is based are much more complicated than these, but these illustrate the basic technique. Economic forecasting involves lots of difference equations.

EXAMPLE 6 **Business inventory**

Trusty Rent-A-Car has offices in New York City and Los Angeles. It allows customers to make local rentals or one-way rentals to the other location. Each month, it finds that half the cars that start the month in NYC end it in LA, and one third of the cars that start the month in LA end it in NYC. (See Fig. 5.1.) If at the start of operations Trusty has 1000 cars in each city, what can it expect n months later? Model this situation with a difference equation.

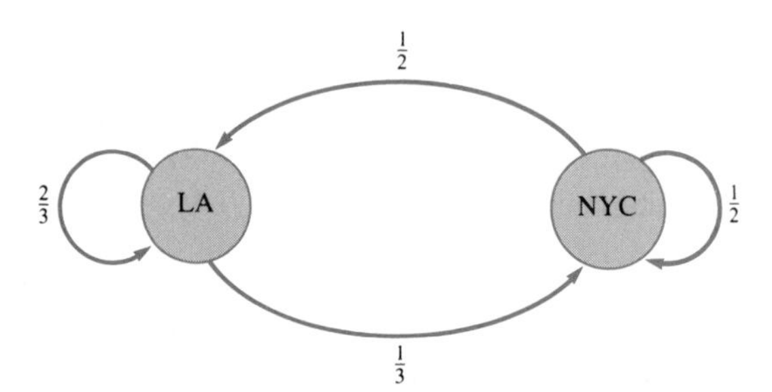

FIGURE 5.1

A digraph of monthly car movements for Trusty Rent-A-Car.

Solution Let N_n be the number of cars in New York at the beginning of month n, with the start of operations considered to be the beginning of month 0. Define L_n similarly. Then we have the linked difference equations

$$N_n = \tfrac{1}{2}N_{n-1} + \tfrac{1}{3}L_{n-1},$$
$$L_n = \tfrac{1}{2}N_{n-1} + \tfrac{2}{3}L_{n-1},$$
$$L_0 = N_0 = 1000.$$

We say "linked" because each variable depends on previous values of the other variable as well as of itself. It's considerably harder to separate the variables here than in Example 5. Try it! It can be done [29], but it's easier to study this sort of problem by matrix methods. (See Section 8.9.) However, it's not hard at all to compute some terms of $\{N_n\}$ and $\{L_n\}$ and look for a pattern. In fact, do you have any intuition about what will happen? Will Trusty suffer wild oscillations in the locations of its cars or will things "settle down"? ∎

EXAMPLE 7 Find a recursion for the number of indistinguishable balls in distinguishable urns.

We solved this problem in Section 4.7, but we promised another approach and this is it. Let us denote the answer by $C^*(u,b)$. (We could use double-index notation like $c_{u,b}$ instead. You should get accustomed to "function notation" and "index notation" for both one-variable and two-variable difference equations.) We use C^* and put u first because we showed in Section 4.7 that the answer we seek now is the number of combinations of b things from u allowing repetitions, which we have already called $C^*(u,b)$.

Solution We devise a difference equation and from that compute a chart of values to look at for conjectures. Think recursively. Let's pick one of the urns; since they're distinguishable we can give it a number, say, 1. Either some balls go in urn 1 or none do. If none do, we must put all b balls in the other $u-1$ urns. There are $C^*(u-1,b)$ ways to do this. If some balls go in urn 1, let's begin by dropping exactly one ball in—any one. There's only one way to do this, not b ways, since balls are indistinguishable. Now what? We have $b-1$ balls to put in how many urns? The answer is u, not $u-1$, because now we are allowed to use urn 1 again. Thus there are $C^*(u,b-1)$ ways to finish up. By the Sum Rule we have the recurrence

$$C^*(u,b) = C^*(u,b-1) + C^*(u-1,b). \tag{8}$$

This argument makes sense so long as $u \geq 2$ and $b \geq 1$. (Why?) So we must compute the base cases $u=1$ or $b=0$ directly. This is easy. $C^*(1,b)=1$ because there is exactly one way to "distribute" b balls into one urn, even if $b=0$. Also, $C^*(u,0)=1$ for any $u \geq 1$ because there is exactly one way to distribute no balls into u urns: Do nothing!

These base cases give us the first column and row of Fig. 5.2. Now we use Eq. (8) to fill in the rest; it says that each entry is the sum of the entry directly below it and the entry directly to the left. Note that the first variable indexes columns, and the second indexes rows, with the row number increasing as you go *up* the page, just as in the standard (x,y) coordinate system.

b \ u	1	2	3	4	5
4	1	5	15	35	70
3	1	4	10	20	35
2	1	3	6	10	15
1	1	2	3	4	5
0	1	1	1	1	1

FIGURE 5.2

The number of ways to put b indistinguishable balls in u distinguishable urns, computed from Eq. (8)

In any event, can you look at this chart, use the inductive paradigm, and complete the solution yourself? ∎

EXAMPLE 8 Discuss the use of difference equations in population models.

Solution This is an enormous field. We will sketch a few ideas for setting up models and let you do some more in the problems.

The simplest population assumption is constant growth rate. Let P_n be the numerical size of some population at time period n (measured in some convenient units, maybe thousands or millions). If the population grows by 2% over each time period, then we have

$$P_{n+1} = (1.02)P_n.$$

This, of course, gives a geometric sequence (the discrete analog of exponential functions in calculus).

However, there may be a crowding effect. A better equation might be

$$P_{n+1} = P_n\left[1 + r\left(1 - \frac{P_n}{C}\right)\right]. \tag{9}$$

where the positive constant r is the idealized growth rate and $1 - (P_n/C)$ models the crowding. So long as $P_n < C$, the population increases from P_n to P_{n+1}, but the closer P_n is to C, the smaller the amount of increase. Similarly, if $P_n > C$, then $P_{n+1} < P_n$, but the closer P_n is to C the smaller the decline. Finally, if $P_n = C$, then $P_{n+1} = P_n$; the population doesn't change. See Fig. 5.3, which illustrates all three situations. Do you see why C is called the **carrying capacity** of the system? Also, look particularly at the plot that starts at $P_0 = .2$ (dark color). Notice how the population rises more and more steeply for a while, and then less and less steeply as it approaches the carrying capacity. This is very common population behavior. Data like this is said to trace a "logistic" curve (or S-curve).

There is no reason to suppose a real-world population follows exactly this model of crowding. Instead of the term $r\big(1-(P_n/C)\big)$ there is probably some more general function $f(P_n)$. The key requirement is that f be decreasing with a zero at some number C. However, even with the simple form of Eq. (9), lots can happen; see [22–24]. Eq. (9) is a **quadratic** difference equation: if you multiply it out, you get a term P_n^2. This is the simplest sort of **nonlinear difference equation**; if we used $f(P_n)$ instead we would get a more general nonlinear equation.

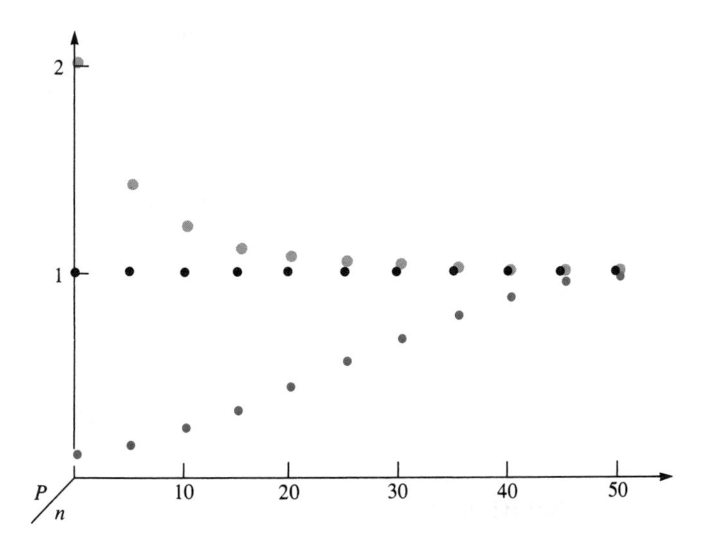

FIGURE 5.3

Graphical solution to population model (9) with $r = .1$, $C = 1$ (in thousands, say) and initial conditions $P_0 = .2$ (dark color), $P_0 = 1$ (black) and $P_0 = 2$ (light color). Every fifth term is shown.

For population studies it is a crude approximation to assume that births are some fraction of the whole population, as we have assumed above. Except for single-cell organisms, only part of the population is of "child bearing" age. Thus, to refine the model a bit, let C_n and A_n be, respectively, the number of children and adults at time n. Then you might have a set of difference equations like this:

$$\begin{aligned} C_n &= aC_{n-1} + bA_{n-1}, \\ A_n &= cC_{n-1} + dA_{n-1}. \end{aligned} \tag{10}$$

Here, a would take into account that a few children die, and some graduate into adulthood. The constant b would take into account that adults give birth to children. And so on. Even this is crude. For human populations, demographers break things down, at the least, into age groups like 5–9, 10–14, etc.

Let us briefly mention the case of competing populations. Let W_n and R_n be the number of wolves and rabbits on a certain range during time period n. It seems reasonable to suppose that

$$R_n = aR_{n-1} - bW_{n-1},$$

because rabbits produce rabbits and wolves eat rabbits. Indeed, this again leads to a linked linear system as in Eqs. (10)—what's a plausible equation for W_n?—but now at least one coefficient is negative. In any event, matrix methods can be applied. Interesting work has been done in just the last ten years on what can be determined about matrix difference equations when all that is certain about the coefficients is their signs. ∎

Finally, we can't leave the subject of difference equations in population models without introducing the most famous population problem ever: Fibonacci's rabbit problem. In 1202, the Italian mathematician Fibonacci posed the following:

EXAMPLE 9 Suppose a pair of rabbits give birth at the end of every month to one baby pair of rabbits, and that these baby pairs in turn give birth to a pair at the end of every month, starting at the end of their second month, and so on with each generation. If you start with one pair of newborn rabbits, how many pairs will you have exactly n months later? Assume that rabbits never die (not an unreasonable simplifying assumption if the values of n that really interest you are less than the lifespan of rabbits).

Solution Probably you've already guessed that the answer is the Fibonacci sequence we defined in Section 2.7. But let's apply the recursive paradigm from scratch. Let r_n be the number of pairs alive exactly n months after you start with the newborn pair. Thus $r_0 = 1$. Can we express r_n in terms of earlier r's? Of course. Those alive exactly n months after the start are just those alive $n-1$ months after the start plus those born exactly n months after the start. We can write this as

$$r_n = r_{n-1} + b_n,$$

but this gives us two variables, r and b. The key additional observation is that the number born n months after the start (b_n) equals the number alive $n-2$ months after the start—because there is one newborn pair for each previous pair that is at least two months old, and a pair is at least two months old at time n iff it was alive at time $n-2$. Thus

$$r_n = r_{n-1} + r_{n-2}.$$

This is the Fibonacci recursion from Chapter 2. Using the initial conditions $r_0 = 1$ and $r_1 = 1$ (where does the latter come from?) we can easily compute as many terms as we want:

$$1,\ 1,\ 2,\ 3,\ 5,\ 8,\ 13,\ 21,\ 34,\ 55,\ \ldots$$ ∎

This sequence shows up in an amazing number of places having nothing to do with rabbits. For instance, it shows up in the complete answer to a question touched on in Example 5 of Section 1.2: how efficient is Euclid's gcd algorithm? See [12], which gives just a hint of the connection to Fibonacci numbers.

Problems: Section 5.2

The purpose of this exercise set is to have you practice setting up difference equations. Even where a problem seems to ask for an explicit solution (e.g., [13]), it suffices to set up the appropriate difference equation and initial conditions. If you can guess an explicit formula from your equation, great; if you can prove it, super; but neither is part of the assignment—yet.

1. Devise a difference equation for the series $1 + 4 + 9 + \cdots + n^2$.

2. Consider the sequence 1, 3, 7, 13, 21, 31,... . Devise a difference equation for the sequence. Devise a difference equation for the associated series.

3. An **arithmetic-geometric** sequence is one that starts at some value a, and thereafter each term is obtained from the previous one by first multiplying by a constant ratio r and then adding a constant difference d. Find a difference equation for this sort of sequence.

4. In Example 3, what recursion do you get if you substitute $n-1$ for k in Eq. (2)? Also, find a recursion relating E_{2n} to E_n by replacing n by $2n$ and k by n.

5. Modify the second equation of the national income model (6) so that $C_n = cY_n$. That is, we now think of personal income as being consumed instantaneously, at least relative to the rate at which $C+I+G$ turns into people's income.

a) Find the difference equation for Y_n (involving no other variables) assuming that $I+G$ is constant.

b) Find the difference equation for Y_n assuming that G is constant but I satisfies Eq. (7).

6. Modify the national income model so that G is no longer constant. Specifically, assume that $G_n = \overline{G} - kY_n$, where $\overline{G}$ and k are constants.

a) Explain why this difference equation for G_n is plausible if the government has the philosophy that it should try to counter economic cycles. What do $\overline{G}$ and k represent?

b) Find the difference equation for Y_n (involving no other variables) assuming I is constant.

c) Find the difference equation for Y_n if I satisfies Eq. (7).

7. Los Angeles suffers an earthquake and becomes less popular. Trusty Rent-a-Car now finds that only 1/4 the cars in NYC end up in LA the next month, whereas 4/5 the cars rented in LA end up in NYC. Revise the linked difference equations of Example 6 to reflect this change.

8. Developing countries have a problem with too many people moving to the cities. Country X does a study and finds that, in a given year, 10% of the rural population moves to a city, but only 1% of the urban population goes back to the country. Describe this situation with difference equations.

9. Change Fibonacci's problem slightly. Once a pair of rabbits starts giving birth, assume they give birth to two new pairs every month. Also, assume that you start at $n=0$ with three newborn pairs. What is the recursion now? What are the first ten terms. Do you see a pattern?

10. Change Fibonacci's problem again. At age two months, a pair gives birth to one new pair; every month after that it gives birth to two pairs. What is the recursion now?

11. Change Fibonacci's problem once more. As in the original problem (Example 9), rabbits give birth at the start of every month beginning with their second, but now each pair dies—just before the start of its 12th month. Now find a recursion for r_n. Caution: It's tempting to say that

$$r_n = r_{n-1} + r_{n-2} - r_{n-12}$$

because r_{n-2} counts the new births at month n, and $r_{n-1} - r_{n-12}$ counts the pairs which continue living from the previous month—but this analysis is wrong!

12. Let F_n be the nth Fibonacci number. How many iterations does Algorithm EUCLID (Example 5, Section 1.2) use to compute $\gcd(F_n, F_{n-1})$? Let a_n be the answer, and find a difference equation for it.

13. How many n-symbol code words of dots and dashes are there if dashes may not be consecutive?

14. How many monotonic increasing sequences of positive integers are there which end at n and do not contain any consecutive integers? For instance, when $n=1$ there is just one, the singleton sequence (1). When $n=4$ there are three: (1,4), (2,4) and (4).

For [15–17], you might review Example 1 from Section 2.4, where a difference equation was devised for counting the number of regions into which n lines divide the plane.

15. A set of ovals are drawn on the plane, so that each pair intersect in two points and no three ovals intersect at the same point. Into how many regions do n ovals divide the plane?

16. Great circles are drawn on a sphere so that no three intersect at the same point. Into how many regions do n of them divide the sphere?

17. n planes float in space. They are in **general position**—every two intersect in a line, every three intersect in a single point, but no four have a point in common. Into how many solid regions do the planes divide space? *Hint*: Look at the intersection of the nth plane with the others; this consists of lines dividing the nth plane into regions.

18. If your income is low enough, you may defer taxes on the principal of an IRA as well as on the investment income. In this case, you can put more principal into an IRA than into an ordinary account (contrary to the assumption in the text) because you need to put less money aside to pay taxes. Redo the difference equation analysis in the text for this case. Find difference equations for P_k and $\hat{P}_k$ and formulas for V and $\hat{V}$.

19. We don't claim that the IRA analysis in the text and in [18] cover all possibilities. For instance, there is a cap of $2000 on IRA contributions, regardless of how much money P you have available. Also, our analysis in [18] seems to assume that people spend a fixed amount and save whatever is left over; other assumptions may be more reasonable. Another point is that we assumed a fixed tax rate. Many people expect to have a lower tax rate during retirement. Come up with at least one other IRA model, incorporating whatever changes you think are most needed.

20. Actually, you don't just start an IRA one year and let it sit. You are allowed to put $2000 in every year. Do Example 4 again under the assumption that you have P available for augmenting your principal every year. As in the text, assume that P is the same whether you use an IRA or an ordinary account.

21. Let $c_{u,b}$ be the number of ways to put b indistinguishable balls into u distinguishable urns if each urn can hold at most one ball. Find a difference equation and initial conditions for $c_{u,b}$. Create a chart of values like Fig. 5.2. Do you recognize these numbers?

22. Consider Eq. (9), the model of population growth with crowding. Compute several terms of the sequence for $C = 1$, $P_0 = .5$ and $r =$

a) .5 b) 1 c) 1.5
d) 2 e) 2.5 f) 3.

23. Do [22] again, except use $P_0 = 1$. (You should be able to solve all the parts at once.)

24. In light of [22, 23], what do you think it means for a difference equation to have an **equilibrium point**? To be **stable**? For what values of r does it seem that the Eq. (9) is stable? (The surprisingly complicated behavior one gets from even simple nonlinear difference equations like Eq. (9) is a hot topic of current research, involving such things as "chaos" and "fractals".)

25. Consumer loans work as follows. The lender gives the consumer a certain amount P, the principal. At the end of each payment period (usually each month) the consumer pays the lender a fixed amount p, called the payment. Payments continue for a prearranged number N of periods, say 60 months = 5 years. The value of p is calculated so that, with the final payment, the principal and all interest due have been paid off exactly. During each payment period, the amount that the consumer still owes increases by a factor of r, the period interest rate; but it also decreases at the end of the period by p, since the consumer has just made a payment.

Let P_n be the amount the consumer owes just after making the nth payment. Find a difference equation and boundary conditions for P_n. (Boundary conditions are initial conditions that aren't consecutive terms.)

26. The payment p in [25] is considered to have two parts: payment of interest and return of principal. (This distinction is important for tax purposes.) The interest is the increase in the amount due during the period (from multiplying by r); the return of principal is the rest. These amounts vary from period to period, but their sum is always p.

a) Let I_n and R_n be the interest and return in period n. Express these in terms of P_n (or P_{n-1} if that's simpler).

b) Consumers are often surprised when the lender shows them a table of values of I_n and R_n. For the longest time, it seems that you mainly pay interest, but towards the end R_n shoots up rapidly. This suggests that R_n grows exponentially. Show that this is correct by finding a difference equation for R_n.

c) Find a difference equation for I_n.

27. In demography (human population studies) it is

sufficient to keep track of the number of females. Models used for serious projections usually divide females into age groups year by year. Let $P(n,a)$ be the number of females of age a years in the United States at the start of calendar year n. For each age a (say, from 1 to 100), such a model would express $P(n{+}1,a)$ as a function of $P(n,b)$ for various b. What is a reasonable difference equation for $P(n{+}1,1)$? For $P(n{+}1,a)$ where $a>1$?

28. Using the notation $[n]=\{1,2,\ldots,n\}$, define $a_n = \sum_{P\subset[n]} |P|$. Find a difference equation relating a_{n+1} and a_n. *Hint*: for every subset P of $[n]$ you get two subsets of $[n{+}1]$, namely P and $P\cup\{n{+}1\}$.

29. Unlink the difference equations in Example 6. Find an expression for N_n in terms of N_{n-1} and N_{n-2}. No values of L should appear.

30. Even when a quantity has an explicit formula, numerical calculations may be more accurate if you use difference equations. This is most likely to be true when the quantity is doubly indexed, such as $C^*(u,b)$. Develop difference equations for

a) $P(n,k)$ and $P^*(n,k)$ from Section 4.4 (permutations and permutations with repetitions).

b) $B(b,u)$, the number of ways to put b distinguishable balls into u indistinguishable urns; see Eq. (5) in Section 4.7.

5.3 Getting Information from Difference Equations

This section is about the use of the inductive paradigm in the analysis of difference equations. Continuing with several examples from the previous section, we will guess or otherwise concoct closed forms and then prove them correct by induction. We also show that you can often establish very useful information even when you can't find a closed form. As we explained in Section 5.1, the approaches illustrated in this section are important because many difference equations cannot be solved by the mechanical methods of later sections. Even when those mechanical methods apply, the approaches of this section are sometimes faster.

Throughout this section, example numbers refer to examples from the previous section. We start with Example 3 because it's the easiest, but mostly we go in order.

EXAMPLE 3 Continued

Conjecture and prove an explicit formula for the expected length of sequential search.

Solution Section 5.2 we showed that $E_1 = 1$ and

$$E_n = \frac{n-1}{n}E_{n-1} + 1. \tag{1}$$

From this equation it is easy to compute

$$E_2 = \frac{3}{2}, \qquad E_3 = 2, \qquad \text{and} \qquad E_4 = \frac{5}{2},$$

from which we guess

$$E_n = \frac{n+1}{2}.$$

We will now prove this equality by induction. Let $P(n)$ be the assertion that the guess is correct for the particular value n. We already know that $P(1)$ is true. Assuming $P(n{-}1)$ and substituting in Eq. (1), we have

$$E_n = \frac{n-1}{n}\left(\frac{(n-1)+1}{2}\right) + 1 = \frac{n-1}{2} + 1 = \frac{n+1}{2}. \quad ∎$$

Even when you can't guess an explicit formula for a series by computing values from its difference equation, you can still use the difference equation to prove a formula if you discover it by some other means. For instance, for the geometric series $T_n = \sum_{k=0}^{n} ar^k$ we showed that

$$T_n = T_{n-1} + ar^n, \qquad T_0 = a. \tag{2}$$

Perhaps you seem to recall that

$$T_n = \frac{a(r^{n+1} - 1)}{r - 1}, \tag{3}$$

but you are not sure it's correct. If we can prove Eq. (3) inductively using (2), it's correct.

EXAMPLE 2 Continued Prove Eq. (3) by induction using Eq. (2).

Solution The initial case of Eq. (3), $n = 0$, is easy: does $a = a(r-1)/r-1$? Yes, at least when $r \neq 1$, which reminds us that Eq. (3) *is only correct when* $r \neq 1$. The inductive step is the following calculation:

$$\begin{aligned} T_n &= \frac{a(r^{(n-1)+1} - 1)}{r - 1} + ar^n = \frac{ar^n - a}{r - 1} + \frac{ar^n(r - 1)}{r - 1} \\ &= \frac{(ar^n - a + ar^{n+1} - ar^n)}{r - 1} = \frac{a(r^{n+1} - 1)}{r - 1}. \quad \blacksquare \end{aligned}$$

Even if you *mis*remember a formula, induction can help you recover. For instance, if you try to prove a wrong explicit formula for T_n using Eq. (2), it won't work, and *why* it doesn't work may suggest what you are misremembering [7].

EXAMPLE 4 Continued Complete the analysis of IRAs.

Solution Using P for the original principal, P_k for the money in the account at time k, V for the value after taxes upon withdrawal at time N, and hats for the IRA option, we found that

$$\begin{aligned} P_{k+1} &= P_k(1 + r) - P_k rt, \qquad (4) \\ \hat{P}_{k+1} &= \hat{P}_k(1 + r), \\ V &= P_N, \\ \hat{V} &= \hat{P}_N - (\hat{P}_N - P)t = \hat{P}_N(1 - t) + Pt. \end{aligned}$$

Clearly $\{\hat{P}_n\}$ is a geometric sequence, so

$$\begin{aligned} \hat{P}_N &= P(1 + r)^N, \\ \hat{V} &= P\big[(1 + r)^N(1 - t) + t\big]. \end{aligned}$$

Now the key observation: the sequence $\{P_k\}$ is also geometric, because Eq. (4) may be rewritten as

$$P_{k+1} = P_k[1 + r(1 - t)].$$

Thus

$$V = P_N = P[1 + r(1-t)]^N.$$

So which investment is better? It makes sense to look at $\hat{V}/V$. We find that

$$\frac{\hat{V}}{V} = \frac{(1+r)^N(1-t)+t}{[1+r(1-t)]^N}. \tag{5}$$

This equation doesn't simplify readily, but if we delete the positive term t from the numerator we get

$$\begin{aligned} \frac{\hat{V}}{V} &> \frac{(1+r)^N(1-t)}{[1+r(1-t)]^N} \\ &= \left[\frac{1+r}{1+r(1-t)}\right]^N (1-t). \end{aligned} \tag{6}$$

Consider the bottom line in display (6). The factor

$$\frac{1+r}{1+r(1-t)}$$

is a constant greater than 1, so if the power N is large enough,

$$\left[\frac{1+r}{1+r(1-t)}\right]^N (1-t)$$

will be greater than 1 even though the final factor $1-t$ is less than 1. In fact, the larger N is, the more $1-t$ will be overwhelmed.

For example, consider the typical values $r = .1$ and $t = .25$. If you start an IRA at age 35 and retire upon reaching 65, then $N = 65 - 35 = 30$ and inequality (6) becomes

$$\frac{\hat{V}}{V} > \left[\frac{1.1}{1.075}\right]^{30}(.75) \approx 1.495.$$

Thus an IRA gains you about 50% more after-tax income. (Using Eq. (5) rather than inequality (6), we find that $\hat{V}/V$ is 1.523.) And had you been perspicacious enough to start your IRA at age 30 instead of 35, then from (6) we would find $\hat{V}/V \approx 1.677$. (The precise value from Eq. (5) is 1.697.) From either age, the gain over a non-IRA investment is substantial enough that we think it's worth your while to lock up some of your money this way. (If you are allowed to defer taxes on principal as well, the advantage of an IRA is even greater [11]. Even if you are not allowed to do so, $\hat{V}/V \geq 1$ for all positive integers N [10].) ∎

EXAMPLE 6 Continued

Rent-A-Car Inventory

In Section 5.2 we determined that

$$\begin{aligned} N_n &= \tfrac{1}{2}N_{n-1} + \tfrac{1}{3}L_{n-1}, \\ L_n &= \tfrac{1}{2}N_{n-1} + \tfrac{2}{3}L_{n-1}, \\ L_0 &= N_0 = 1000, \end{aligned}$$

where N_k and L_k are the number of cars in NYC and LA, respectively, at time k. Although individual cars will keep moving from NYC to LA and back, the *number* of cars in each place might stabilize. If we owned the rental agency, that would certainly make life simpler! If stability occurred, we would say that the agency was in **equilibrium**. What would this equilibrium look like?

Solution If an equilibrium is achieved, then for all n after some point, $N_{n-1} = N_n = N$ and $L_{n-1} = L_n = L$. Let $x = N/2000$ be the fraction of rental cars in New York after equilibrium is achieved. Then substituting N and L into the first two lines of the display above and dividing by 2000, we obtain

$$\begin{aligned} x &= \tfrac{1}{2}x + \tfrac{1}{3}(1-x), \\ 1-x &= \tfrac{1}{2}x + \tfrac{2}{3}(1-x). \end{aligned}$$

Although these are two equations in one unknown, they do have a solution: $x = 2/5$. Since Trusty has 2000 cars, the one equilibrium solution is 800 in NYC and 1200 in LA.

What we now know with certainty is this: If Trusty Rent-A-Car starts with 2/5 of its cars in NYC, it will continue to have 2/5 of its cars there at the end of each period. It is also true, but we aren't prepared to prove it to you here, that no matter how the cars are distributed initially, in time the distribution converges on this (2/5, 3/5) equilibrium. ∎

In Example 6, the total number of cars was not changing. Difference equations for such "closed" systems usually result in sequences which head towards equilibrium. Such closed difference equation systems are called **Markov Chains**. We give a precise definition and study them by matrix methods in Section 8.10.

EXAMPLE 7 Continued

Determine the number of ways to put b indistinguishable balls in u distinguishable urns.

Solution In Section 5.2 we obtained the recursion

$$C^*(u,b) = C^*(u,b-1) + C^*(u-1,b) \tag{7}$$

and computed the values shown in Fig. 5.2, repeated here as Figure 5.4. Voilà, this is Pascal's Triangle, only arranged funny. Thus $C^*(u,b) = C(n,k)$, where n,k are the "Pascal coordinates" of the position (u,b) in rectilinear coordinates. For each

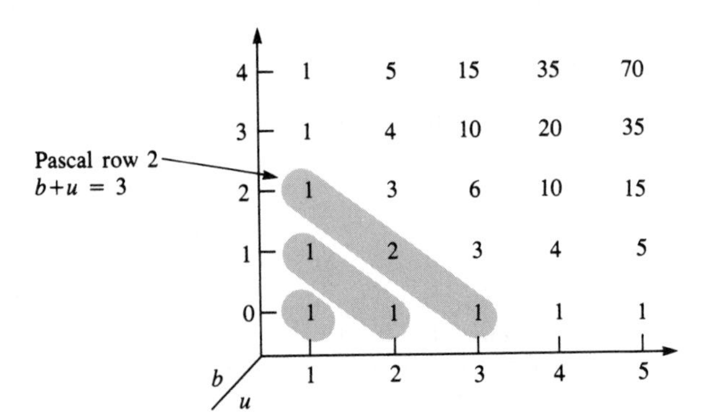

FIGURE 5.4

Ways to put b indistinguishable balls into u distinguishable urns, with Pascal's triangle noted in color

Pascal row (marked in color), $u + b$ is constant and 1 greater than the Pascal row number—recall that the top row of the triangle is row 0. Furthermore, if we think of each Pascal row as starting at the lower right, then the kth entry in the row (starting with $k = 0$) is also in the kth horizontal row. Thus

$$C^*(u,b) = \binom{u+b-1}{b} = \frac{(u+b-1)!}{b!(u-1)!}. \tag{8}$$

Having discovered Eq. (8), we prove it by induction using Eq. (7). Let $P(u,b)$ be the statement that Eq. (8) holds for the particular pair (u,b). (*Note*: $P(u,b)$ is a *proposition*, not a count of permutations.) Equation (7) suggests that we should assume the previous cases $P(u,b-1)$ and $P(u-1,b)$. Then

$$\begin{aligned} C^*(u,b) &= C^*(u,b-1) + C^*(u-1,b) && \text{[Eq. (8)]} \\ &= \binom{b+u-2}{b-1} + \binom{b+u-2}{b} && [P(u-1,b),\ P(u,b-1)] \\ &= \binom{b+u-1}{b}, && \text{[The binomial recursion]} \end{aligned}$$

so $P(u,b)$ is true.

Is this legit? This **double induction** involves two variables that are both changing, and we didn't treat that in Chapter 2. Yes, it is legitimate—once we verify an appropriate basis. Look at Figure 5.5. The square in column u and row b represents $P(u,b)$. Think of $P(u,b)$ as a domino with square cross section as seen from the top, so that we can use the language of Chapter 2. The three-line displayed equation above shows that $P(u,b)$ falls down (is true) if the two dominoes immediately below it and to its left fall down. So suppose we can show as a basis that all dominoes along the bottom and left sides of the infinite array fall down. These are shaded in Figure 5.5. The arrows illustrate why all the dominoes fall. Thus $P(2,1)$ falls down because $P(2,0)$ and $P(1,1)$ are in the basis. Then $P(3,1)$ falls down because $P(2,1)$ just fell over and $P(3,0)$ is in the basis. Continuing this way (a single-variable induction!), we see that the whole row $P(m,1)$ falls down. Now we apply the same reasoning to see that the whole row $P(m,2)$ falls down, and then row $P(m,3)$, and so on (another single-variable induction, this time on the second variable). We say that this proof reduces double induction to **iterated induction**.

So, to finish this example, we need to verify the basis: is $P(u,0)$ true for all $u \geq 1$ and is $P(1,b)$ true for all $b \geq 0$? We need to show that the right-hand side of Eq. (8) equals 1 in all these cases, since we showed in Section 5.2 that all the border entries of Figure 5.4 are 1's. Well, $P(u,0)$ claims that $\binom{u-1}{0} = 1$, and that's always true; $P(1,b)$ claims that $\binom{b}{b} = 1$, and that's always true. ∎

EXAMPLE 9 Continued Analyze the Fibonacci numbers further.

Solution It's practically impossible to guess a formula for the Fibonacci numbers. But since we can compute them so easily using the associated difference equation, $F_n = F_{n-1} + F_{n-2}$, we can analyze them numerically as far as we want. It is a semimystical principle in mathematics that it is always revealing to look at rates

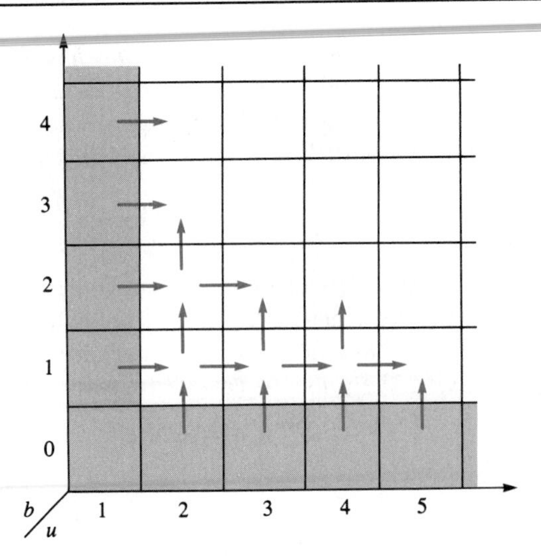

FIGURE 5.5

A valid double induction principle

of growth. To do that, we look at ratios as in Section 2.4. Table 5.1 shows the first ten ratios F_{n+1}/F_n. Clearly the ratio seems to approach a limit. Therefore we conclude that the rabbit population grows quite regularly at about 62% a month. This knowledge, so easy to comprehend, may be much more useful than a closed-form expression for the exact number.

Have you seen this ratio before? The ancient Greeks called it the **Golden Ratio** [21] over 1500 years before Fibonacci. ∎

TABLE 5.1
Fibonacci numbers and their ratios

n	F_n	F_{n+1}/F_n
0	1	1
1	1	2
2	2	1.5
3	3	1.667
4	5	1.6
5	8	1.625
6	13	1.615
7	21	1.619
8	34	1.618
9	55	1.618

Problems: Section 5.3

1. Consider the difference equation

$$a_n = a_{n-1} - a_{n-2} \quad \text{for } n > 1,$$
$$a_0 = a_1 = 1.$$

Compute several terms. Do you see a pattern? Can you prove it?

2. Consider the same recursion as in [1], but now let a_0 and a_1 be arbitrary. What happens? Can you prove it?

3. Consider the difference equation $a_{n+2} = 5a_{n+1} - 6a_n$. Compute the first 6 terms if:

a) $a_0 = 1, a_1 = 2$ b) $a_0 = 5, a_1 = 10$
c) $a_0 = 1, a_1 = 3$ d) $a_0 = 2, a_1 = 5$
e) $a_0 = 0, a_1 = 1$.

Do you see a pattern? Can you prove it?

4. Discover and prove a closed form formula for the modification of Example 3 Continued in which the word w is on the list with probability 1/2. Assume that, if w is not on the list, it is equally likely to be in any of the gaps: before w_1, between w_1 and w_2, and so on.

5. In [4, Section 5.2], you were asked to find a formula relating E_{2n} and E_n for Algorithm SEQSEARCH. Use that formula and the condition $E_1 = 1$ to find and prove an explicit formula for E_n when n is a power of 2. (This may seem silly, since we already found a formula for E_n for *any* n, but it provides more practice, and it previews a "powers of 2" approach that is very important in Section 5.9.)

6. Use Eq. (2), Section 5.2, to find a recursion relating E_{n+2} to E_n. Use this recursion to find and prove an explicit formula for E_{2n}.

7. Suppose that you remembered Eq. (3) to be the negative of what it really is. Try doing the inductive proof using Eq. (2). Things won't cancel right. Do you see how this might suggest to you where your error is?

8. Using the IRA analysis of Example 4 Continued, find $\hat{V}/V$ for

a) $r = .1, t = .25, N = 5$
b) $r = .11, t = .25, N = 30$
c) $r = .1, t = .3, N = 30$.

Notice we have changed just one of the three parameters r, t, N at a time from the example in the text. From this you can make general conjectures about, say, the effect on the advantage of IRAs when t alone changes.

9. You may withdraw money from an IRA early, but under most circumstances you must then pay a penalty of 10% of all the interest withdrawn, in addition to paying all the regular taxes on that interest. A regular interest-bearing account can be withdrawn without penalty at any time. For this reason people are wary of putting money in an IRA if they think they may need it for an emergency before retirement.

a) Let $r = .1$, $t = .25$. Show that if you withdraw all the money after $N = 20$ years, an IRA is still advantageous, despite the penalty. (Assume as in the text that you cannot defer tax on the principal.)

b) With $r = .1$ and $t = .25$, what is the first year N for which an IRA is advantageous despite the penalty?

10. Using the IRA analysis of Example 4 Continued, verify that $\hat{V}/V = 1$ for $N = 1$ and $\hat{V}/V > 1$ for $N > 1$. *Hint*: expand Eq. (5) by the Binomial Theorem.

11. Redo the analysis of Example 4 Continued if you are permitted to defer tax on your IRA principal until withdrawal. (This builds on [18, Section 5.2]; the additional work is actually slightly simpler than what we had to do in the text.) Include numerical evaluations for the same values of r, t, and N considered in the text.

12. The analysis in [18, Section 5.2] assumes that a person who can defer tax on IRA principal can deposit more money to start an IRA than to start an ordinary account, because the extra money from reduced current taxes is saved. But some people might spend the "windfall", or bump up against the legal limit on yearly IRA contributions, and have the same amount P to invest despite the tax deferral. Model this situation. Show by example that if N is not large enough, $\hat{V}/V$ can be less than 1. Remember: If you can and do defer taxes on the principle, you must pay these taxes when the account is closed.

13. Consider an IRA to which you add P each year,

but otherwise the hypotheses of Example 4 are unchanged. (This builds on [20, Section 5.2].) You are not yet prepared to find closed forms for V and $\hat{V}$ in this model, but there is no reason why you can't do some iterative computer evaluations. Use $r = .1$, $t = .25$, and $N = 30$ and 35 as before. Do you expect $\hat{V}/V$ to be more or less than in the one-time deposit model of the text?

14. Show that there is a unique equilibrium point for Trusty Rent-A-Car for the post-earthquake conditions described in [7, Section 5.2].

15. Consider underdeveloped Country X from [8, Section 5.2]. If migration continues the same way and equilibrium is reached, what fraction of the population will live in the cities?

16. Figure 5.6 displays values of a function $q = f(m,n)$ defined for pairs of positive integers with $n \le m$. Figure out a formula for f.

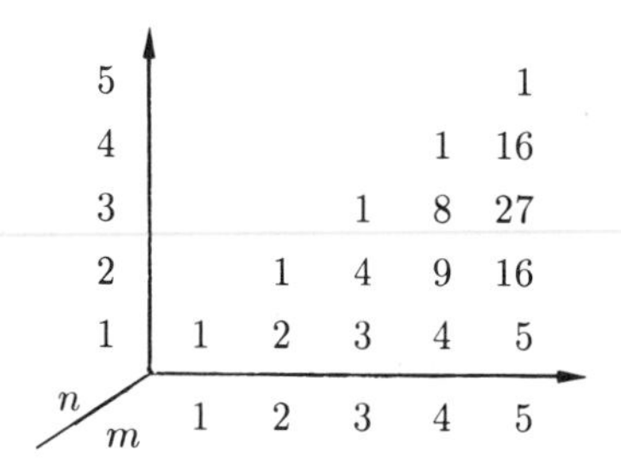

FIGURE 5.6

17. Here is a double induction principle different from the one used in Example 7 Continued. To prove $P(m,n)$ for all natural numbers $m \ge m_0$ and $n \ge n_0$, it suffices to prove

i) Basis: $P(m, n_0)$ for all $m \ge m_0$ and $P(m_0, n)$ for all $n \ge n_0$;

ii) Inductive Step: For all $n > n_0$ and $m > m_0$, $[P(m-1, n)$ and $P(m-1, n-1)] \Longrightarrow P(m,n)$.

Justify this principle by a figure and an argument similar to that used for the double induction principle in the text.

18. State the double induction principle from Example 7 Continued in a general way. See [17].

19. Set up a value and ratio table like Table 5.1 for the following modifications of the Fibonacci numbers.

a) The **Lucas numbers** $\{L_n\}$: these numbers satisfy the same recursion as the Fibonaccis, but the initial conditions are $L_0 = 1$ and $L_1 = 3$.

b) The sequence in [9, Section 5.2].

c) The sequence in [10, Section 5.2]. (Start with one pair of rabbits.)

20. Let $r_n = F_{n+1}/F_n$, the ratio of consecutive Fibonacci numbers.

a) Find a difference equation relating r_n and r_{n-1}. Use it to check the ratios in Table 5.1 without actually computing any Fibonacci numbers. You also need the initial condition $r_0 = 1$.

b) If r_n really does go to a limit, call it r, then r_{n-1} goes to r as well. Thus rewrite the recursion in (a) as an equation in r and solve. You should get two roots, but one of them could not possibly arise from the Fibonacci numbers. Why? Does the remaining root agree with the data in Table 5.1?

21. According to the ancient Greeks, the most pleasing rectangle is one where the ratio of length to width is the same as the ratio of the sum of length and width to length alone. The Greeks knew that all rectangles with this property have the same ratio of length to width, so they called it the Golden Ratio. Find a formula for this ratio.

22. Let $G_n = \sum_{i=0}^{n} F_i$, where F_i is again the ith Fibonacci number. Compute the first several terms of $\{G_n\}$. Find a pattern. Prove it.

23. Consumer loans continued. In [25, Section 5.2] you were asked to develop the equation $P_{n+1} = P_n(1+r) - p$ for the amount of principal owed at the end of period n+1. Payment p is initially an unknown; you know how much you want to borrow, $P = P_0$; you know the period interest rate, r, the lender will charge; and you know the number of periods, N, after which the loan must be completely repaid. Compute P_1, P_2, and P_3 in terms of P_0, p, and r by repeatedly substituting into the difference equation. At this point you should see a pattern. Now express P_N in terms of P_0, p and r and solve for p.

24. Consider the difference equation

$$A_0 = 0$$

and for $n > 0$ and c constant,

$$A_n = A_{n-1} + \left(\frac{1}{n+1}\right)^2 (c - A_{n-1}).$$

Find and prove a formula for A_n.

25. Consider the difference equation

$$a_{n+1} = \frac{1}{2}\left[a_n + \frac{4}{a_n}\right], \quad n \geq 0,$$

Compute the first ten terms (use a hand calculator) for any $a_0 > 0$ and then any $a_0 < 0$. What happens?

5.4 Solving Difference Equations: Preliminaries

We have two matters to discuss: (1) terminology, and (2) a general existence and uniqueness theorem that underlies most of what we will do in this chapter.

Before we can give general methods for solving classes of difference equations, we need some words to describe the classes we are talking about. The key phrases we need to define are kth order, linear, constant coefficient, and homogeneous.

The **order** of a difference equation is the difference between the highest and lowest indices (subscripts) in the equation. Consider

$$\begin{aligned}
&\text{(A)} \quad a_{n+1} = \tfrac{1}{2}\left(a_n + \frac{4}{a_n}\right)\\
&\text{(B)} \quad a_n = a_{n-1}a_{n-2},\\
&\text{(C)} \quad a_n = a_{n-10},\\
&\text{(D)} \quad w_n = 1 + w_{\lfloor n/2 \rfloor}.
\end{aligned}$$

Equation (A) is first order because $(n+1) - n = 1$. Equation (B) is second order. Equation (C) is tenth order (even though there is only one previous term). Finally, (D) has no order, at least no constant order. The difference between the highest and lowest index is $n - \lfloor n/2 \rfloor = \lceil n/2 \rceil$, which varies with n.

A difference equation is **linear** if the highest indexed term equals a sum of multiples of previous terms. The multipliers can either be constants or functions of the index variable. (In all the equations above, the index variable is n.) Optionally, there can be additional terms, which must also be either constants or functions of the index variable. Of the equations above, only (C) and (D) are linear. Equation (C) does not have any additional terms; (D) does—the constant 1. The following equations are linear also:

$$\begin{aligned}
&\text{(E)} \quad a_{n+1} = 3a_n + na_{n-1} + 2^n,\\
&\text{(F)} \quad S_n = S_{n-1} + n^3 - 2n^2 - 3n + 2,\\
&\text{(G)} \quad P_{n+1} = P_n(1+r) - p.
\end{aligned}$$

All of these have additional terms. For instance, in (F) the additional terms are the cubic polynomial in n.

A linear difference equation has **constant coefficients** if all the multipliers of subscripted terms are constants; that is, they do not depend on the index variable. Thus, (C), (D) and (F) are all constant coefficient equations. Equation (E) is not, because of the n in front of a_{n-1}. Equation (G) is a constant coefficient equation (r is a constant).

A linear difference equation is **homogeneous** if there are no terms other than subscripted terms. Thus (C) is the only homogeneous recurrence above. Additional terms, when they occur in a linear recursion, are called **nonhomogeneous terms** or the **nonhomogeneous part**.

(These linearity ideas are not special to difference equations. They are general ideas from "linear algebra", which we discuss in Chapter 8.)

In the following sections we develop complete solution methods for kth-order constant coefficient linear difference equations (first the homogeneous ones, then nonhomogeneous ones) and the general first-order linear equation (variable coefficients allowed).

But before we look for methods, it's good to know in advance that there *are* solutions and to know to what extent they are unique. This leads to the second topic of this section.

The reason that difference equations come with initial conditions is because initial conditions are intended to specify a unique solution. For instance, many sequences satisfy the recurrence $h_{n+1} = 2h_n + 1$, but one and only one sequence (the TOH sequence) also satisfies $h_1 = 1$. If we are given only the recurrence, we speak of finding the **general solution**, that is, the set of all sequences satisfying the recurrence. If we are also given initial conditions which specify a unique sequence, we talk about finding the **particular solution**.

The following is a fairly general theorem about when a recurrence and (one or more) initial conditions define a unique solution. Loosely stated, it says that a kth-order recurrence and k consecutive initial conditions uniquely determine a sequence.

Theorem 1. Suppose that we are given a function $f(x_1, x_2, \ldots, x_k)$ defined for all k-tuples of real numbers. Suppose that we are also given k real numbers $c_0, c_1, \ldots, c_{k-1}$. Then there exists one and only one sequence $a_0, a_1, a_2, \ldots$ satisfying

$$a_n = f(a_{n-1}, a_{n-2}, \ldots, a_{n-k}), \quad n \geq k \tag{1}$$

and

$$a_0 = c_0,\ a_1 = c_1,\ \ldots,\ a_{k-1} = c_{k-1}. \tag{2}$$

This theorem is completely parallel to the Principle of Mathematical Induction. Induction says: Once the initial case or cases are checked, the inductive step settles the claim for all remaining cases. This theorem says: Once the initial terms are specified as in Eqs. (2), then Eq. (1) uniquely specifies all the remaining terms. It should be no surprise, then, that the proof of this theorem is by induction.

PROOF Let $P(n)$ be the statement that there is one and only one choice for a_n. We do strong induction in the form that uses k base cases, $P(0)$ through $P(k-1)$. These base cases are true by Eqs. (2). Now consider any $n \geq k$ and suppose that $P(j)$ is true for all $j < n$. In particular, $a_{n-1}, a_{n-2}, \ldots, a_{n-k}$ are all uniquely determined. Since these are all real numbers, they may be plugged into Eq. (1), and since f is a function, there is a unique output. This output is the one and only choice for a_n. ∎

Why does the theorem make a fuss about f having *all* real k-tuples in its domain? Because otherwise [12] there may be *no* sequence $\{a_n\}$ satisfying Eqs. (1) and (2). Why is Eq. (1) given in a form with a_n by itself on one side? Because otherwise there may be *many* sequences satisfying these equations [13]. We will mostly study linear recursions (as defined previously), and for linear recursions Theorem 1 always applies.

As we mentioned earlier, Theorem 1 is an "existence and uniqueness" theorem. It says that one sequence exists (existence) and only one (uniqueness). The existence part will serve us only as a form of inspiration; since we know there must be a solution, there might be a closed form and it's worth looking for. The uniqueness part will serve us more directly. At several points in the next few sections we will invoke it to show that we have finished solving a problem. The next example illustrates how.

EXAMPLE 1 Use Theorem 1 to prove the geometric series formula:

$$\sum_{k=0}^{n} ar^k = \frac{a(r^{n+1}-1)}{r-1}, \qquad r \neq 1. \tag{3}$$

Solution We need a difference equation. If we set $T_n = \sum_{k=0}^{n} ar^k$, then we get

$$T_n = T_{n-1} + ar^n \quad \text{for } n > 0, \qquad T_0 = a. \tag{4}$$

Theorem 1 tells us that Eq. (4) has a unique solution—k is 1, the sequence is $\{T_n\}$, and $f(T_{n-1}) = T_{n-1} + ar^n$. Therefore, to prove Eq. (3) it suffices to check that the sequence with $T_n = a(r^{n+1}-1)/(r-1)$ is *one* solution to Eq. (4). Checking the initial condition is easy:

$$\frac{a(r^{0+1}-1)}{r-1} = a. \quad \checkmark$$

The recurrence is not much harder:

$$\begin{aligned}
\frac{a(r^{n+1}-1)}{r-1} &\stackrel{?}{=} \frac{a(r^n-1)}{r-1} + ar^n \\
&\stackrel{?}{=} \frac{ar^n - a}{r-1} + \frac{ar^{n+1} - ar^n}{r-1} \\
&= \frac{a(r^{n+1}-1)}{r-1}. \quad \checkmark
\end{aligned}$$

We are done. ∎

Problems: Section 5.4

For each difference equation in [1–10], state its order and whether it is linear. If it is linear, state if it is homogeneous; if it has constant coefficients.

1. $a_n = 2/a_{n-1}$.
2. $w_n = \max\{w_{n-1}, w_{n-3}\}$.
3. $h_{n+1} = 2h_n + 1$.
4. $a_{n-2} = 2^n a_{n-6} + (-3)^n$.
5. $F_n = F_{n-1} + F_{n-2}$.
6. $P_{k+1} = P_k(1+r) - P_k rt$.
7. $d_n = (n-1)(d_{n-1} + d_{n-2})$.
8. $E_n = \dfrac{3}{n} + \dfrac{n-2}{n}(2 + E_{n-2})$.
9. $p_{n+1} = \sum_{k=0}^{n} p_k$.
10. $t_n = \sum_{k=1}^{n-1} t_k t_{n-k}$.
11. The text defined classifications only for difference equations for a single sequence involving a single index variable. Nonetheless, we think you can figure out reasonable classifications for the following. Do so.

    a) $\left\{\begin{array}{l} N_n = \frac{1}{2}N_{n-1} + \frac{1}{3}L_{n-1} \\ L_n = \frac{1}{2}N_{n-1} + \frac{2}{3}L_{n-1} \end{array}\right\}$

    b) $C^*(u,b) = C^*(u-1,b) + C^*(u,b-1)$.

12. Consider the difference equation
$$r_n = \frac{r_{n-2}}{r_{n-1} - 2}, \qquad n > 1.$$
Suppose that we set initial conditions $r_0 = 1$, $r_1 = 2$. Show that there is no infinite sequence satisfying these conditions—because one can't even define r_2. Why isn't Theorem 1 violated?

13. Consider the difference equation
$$a_n^2 = a_{n-1}^2 \quad \text{for } n > 1, \qquad a_1 = 1.$$
Find all infinite sequences that satisfy this difference equation. Why isn't Theorem 1 violated?

14. Consider the difference equation
$$a_n = 2a_{n-2} \quad \text{for } n > 2, \qquad a_1 = 2,\ a_3 = 3.$$
Show that no sequence satisfies these conditions. Why isn't Theorem 1 violated?

15. Theorem 1 shows that, in general, k consecutive terms define the rest of a kth-order difference equation uniquely. Show that fewer than k consecutive terms never do. *Hint*: Fewer than k terms are given, so you can make up the kth term.

16. Use Theorem 1 to prove the formula for the sum of an arithmetic series:
$$\sum_{k=1}^{n} a + (k-1)d \;=\; an + \frac{n(n-1)}{2}d.$$

17. **We've proved the formula for a geometric series twice, in Example 2 Continued, Section 5.3, and in Example 1 in this section. The first proof used induction, the second didn't. Or did it?**

5.5 Second-Order, Constant Coefficient, Homogeneous Difference Equations

In this section, "difference equation" means the second-order, constant coefficient, homogeneous linear difference equation
$$a_n = ca_{n-1} + da_{n-2}.$$

We will show how to solve all such difference equations, obtaining the particular solution if initial conditions are given and the general solution otherwise. We develop the solution method through examples and then summarize it as Theorem 2. We conclude with a subsection on the numerical behavior of solutions. The fact that solutions always have a certain algebraic form makes them easy to analyze.

The advantage of considering only second-order problems is that all the behavior of higher-order problems already occurs in the second-order case, but we don't get bogged down in the notation needed for the general case. In Section 5.6 we use more notation and state the general results.

To conjecture what second-order solutions might look like, it is actually best to review the first-order problem:

$$a_n = ra_{n-1}. \tag{1}$$

This equation—it says that the ratio of consecutive terms is constant—defines a geometric sequence, so the solution is $a_n = ar^n$, where $a = a_0$. Have we also seen geometric growth associated with second-order difference equations? Yes, for instance, the Fibonacci numbers grow with a ratio closer and closer to a constant as n increases. Also, the difference equation in [3, Section 5.3] was second order, and the solutions to parts a)–c) were exactly geometric. Let's guess, therefore, that *every* difference equation (linear, second-order, homogeneous, constant-coefficient) has a geometric solution. If this turns out to be correct, then we'll worry about combining geometric solutions to get other solutions.

EXAMPLE 1 Find all *geometric* solutions to

$$a_n = a_{n-1} + 2a_{n-2}, \quad n > 1. \tag{2}$$

This difference equation arose in [9, Section 5.2], a problem about rabbits.

Solution If $a_n = r^n$ is a solution, then

$$r^n = r^{n-1} + 2r^{n-2}, \quad n > 1.$$

We need to find all ratios r for which this is true. One immediate solution is $r = 0$, but this is the trivial solution—all terms are 0. Excluding this solution as uninteresting, we can divide out the common factor r^{n-2} to obtain

$$r^2 = r + 2 \qquad \text{or} \qquad r^2 - r - 2 = (r-2)(r+1) = 0.$$

This polynomial equation is called the **characteristic equation** of the difference equation. Its solutions are $r = -1, 2,$ so $a_n = (-1)^n$ and $a_n = 2^n$ are solutions to Eq. (2). Indeed, since every step of this algebraic derivation reverses, the sequence $\{r^n\}$ is a nontrivial geometric solution $\iff r = -1, 2.$ ∎

Note. We use "r" in the characteristic equation because it suggests both "ratio" and "root".

EXAMPLE 2 Find all geometric solutions to the Fibonacci recursion.

Solution Proceeding as before, from $F_n = F_{n-1} + F_{n-2}$ we obtain

$$\begin{aligned} r^n &= r^{n-1} + r^{n-2}, \\ r^2 &= r + 1, \\ r^2 - r - 1 &= 0. \end{aligned}$$

This characteristic equation doesn't factor, but the quadratic formula gives

$$r = \frac{1 \pm \sqrt{5}}{2} = 1.618\ldots \text{ and } -.618\ldots.$$

So there are two geometric solutions $\{r^n\}$, one for each value of r just displayed. ∎

For neither Example 1 nor Example 2 is either of the geometric solutions precisely the solution we looked at in Section 5.2. As for [9, Section 5.2], the solution was a sum involving both $(-1)^n$ and 2^n. As for the Fibonacci numbers, 1.618 seems to be just the ratio we observed at the end of Section 5.3. Moreover, $(-.618)^n$ gets closer and closer to zero. Could F_n just be a multiple of 1.618^n perturbed by a multiple of $(-.618)^n$? The answer is Yes. This is a special case of a general result.

Theorem 1. If $\{b_n\}$ and $\{b'_n\}$ are both solutions to the difference equation

$$a_n = ca_{n-1} + da_{n-2}, \tag{3}$$

then $\{Bb_n + B'b'_n\}$ is also a solution for arbitrary constants B and B'.

The sequence $\{Bb_n + B'b'_n\}$ is called a **linear combination** of the sequences $\{b_n\}$ and $\{b'_n\}$. So Theorem 1 may be summarized by saying that the set of solutions to Eq. (3) is **closed** under linear combinations.

PROOF To say that $\{b_n\}$ is a solution is to say that

$$b_n = cb_{n-1} + db_{n-2}$$

for all $n \geq 2$ (we assume b_0, b_1 are the first two terms). To say that $\{b'_n\}$ is a solution is to say that

$$b'_n = cb'_{n-1} + db'_{n-2}$$

for all $n \geq 2$. Multiplying the last equation by B', the previous one by B, and adding, we get

$$\begin{aligned} Bb_n + B'b'_n &= B(cb_{n-1}+db_{n-2}) + B'(cb'_{n-1}+db'_{n-2}) \\ &= c(Bb_{n-1}+B'b'_{n-1}) + d(Bb_{n-2}+B'b'_{n-2}). \end{aligned}$$

But this equation says that $\{Bb_n + B'b'_n\}$ satisfies Eq. (3). ∎

We now know how to parlay a small number of geometric solutions into an infinite set of solutions. But is this set large enough to include a particular solution for any given initial conditions? To get at the answer, let's extend the previous examples.

EXAMPLE 3 In Example 1, the two geometric solutions to Eq. (2) were $(-1)^n$ and 2^n. Is there a linear combination of these two solutions that also satisfies the initial conditions $a_0 = a_1 = 3$? (These were the initial conditions of [9, Section 5.2].)

Solution We are asking if there are specific constants A and B for which

$$\begin{aligned} A(-1)^0 + B2^0 &= \quad A + \quad B = a_0 = 3, \\ A(-1)^1 + B2^1 &= -A + 2B = a_1 = 3. \end{aligned} \tag{4}$$

Solving these equations, we find that $A = 1$, $B = 2$, so the answer is Yes, there is exactly one such linear combination, $a_n = (-1)^n + 2^{n+1}$. ∎

Notice that by Theorem 1, Section 5.4, we can say that this linear combination is in fact the *only* solution *of any form* to Eq. (2) and the initial conditions $a_0 = a_1 = 3$. And there is nothing special about these initial conditions. No matter what a_0 and a_1 are (and for each solution to Eq. (2) they have to be something), it is easy to check that Eqs. (4) can be solved for A and B and the solutions are unique. This check shows that $A(-1)^n + B2^n$ is the general solution to Eq. (2), because Theorem 1, Section 5.4, says that there are no other solutions. For this reason, $\{(-1)^n\}$ and $\{2^n\}$ are said to be a **basis** of the solutions to Eq. (2).

EXAMPLE 4 Find a closed-form formula for the nth Fibonacci number.

Solution From Example 2 and Theorem 1, every sequence of the form

$$F_n = A\left(\frac{1+\sqrt{5}}{2}\right)^n + B\left(\frac{1-\sqrt{5}}{2}\right)^n$$

is a solution to $F_n = F_{n-1}+F_{n-2}$. Since the initial conditions for the Fibonaccis are $F_0 = F_1 = 1$, it is sufficient to find A and B that satisfy

$$\begin{aligned} A\left(\frac{1+\sqrt{5}}{2}\right)^0 + B\left(\frac{1-\sqrt{5}}{2}\right)^0 &= A + B = F_0 = 1, \\ A\left(\frac{1+\sqrt{5}}{2}\right)^1 + B\left(\frac{1-\sqrt{5}}{2}\right)^1 &= A\left(\frac{1+\sqrt{5}}{2}\right) + B\left(\frac{1-\sqrt{5}}{2}\right) = F_1 = 1. \end{aligned}$$

After considerable algebra, we find that

$$A = \frac{1}{\sqrt{5}}\left(\frac{1+\sqrt{5}}{2}\right), \qquad \text{and} \qquad B = -\frac{1}{\sqrt{5}}\left(\frac{1-\sqrt{5}}{2}\right).$$

Therefore

$$F_n = \frac{1}{\sqrt{5}}\left[\left(\frac{1+\sqrt{5}}{2}\right)^{n+1} - \left(\frac{1-\sqrt{5}}{2}\right)^{n+1}\right]. \tag{5}$$ ∎

Fibonacci never discovered Eq. (5); it was first obtained 650 years later! It's certainly not very guessable. In fact, it's hard to believe the right-hand side is always an integer, but it must be since all Fibonacci numbers are integers. This matter is pursued further in [8–9, Supplement].

So far the examples in this section have had characteristic equations with distinct roots. A new wrinkle arises when there is a multiple root.

EXAMPLE 5 Find the general solution to

$$a_n = 4a_{n-1} - 4a_{n-2}, \tag{6}$$

and the particular solution when $a_0 = 5$, $a_1 = 4$.

Solution Substituting r^n for a_n and simplifying, we obtain

$$r^n = 4r^{n-1} - 4r^{n-2},$$
$$r^2 - 4r + 4 = (r-2)^2 = 0.$$

There is only one root, $r = 2$, so we get that 2^n is one basic solution. However, we need two basic solutions if we are going to satisfy two initial conditions by taking a linear combination. What's the other?

To find it, one approach is to pick two initial conditions repeatedly, generate terms in the numerical solution by brute force, and use the methods of Section 2.4 to guess a closed form each time, in hopes that we see a general pattern. Indeed, if you do this [9], you will see that each time you get a solution of the form $A2^n + Bn2^n$. Therefore we guess that this is the general solution. We can verify this guess by doing two things. First, we check by direct substitution [10] that $n2^n$ solves Eq. (6). Thus by Theorem 1 everything of the form $A2^n + Bn2^n$ is a solution. To show that it is the general solution, we must show that for any initial conditions a_0 and a_1, there are values for the constants A, B that work. Substituting the cases $n = 0$ and $n = 1$ into the putative general solution, we obtain

$$\begin{aligned} A \qquad &= a_0, \\ 2A + 2B &= a_1. \end{aligned}$$

Direct calculation shows that this pair of equations has a unique solution for A and B. For instance, when $a_0 = 5$ and $a_1 = 4$ as in this example, $A = 5$ and $B = -3$. In other words, the particular solution is $a_n = 5 \cdot 2^n - 3n2^n$. ∎

These examples illustrate the general pattern, which we now confirm with a definition and a theorem.

Definition 1. Let c and d be given constants. The characteristic equation of the difference equation

$$a_n = ca_{n-1} + da_{n-2}$$

is

$$r^2 - cr - d = 0.$$

The polynomial on the left side is called the **characteristic polynomial**.

Theorem 2. If the characteristic equation of difference equation

$$a_n = ca_{n-1} + da_{n-2} \tag{7}$$

has distinct roots r_1, r_2, then the general solution is

$$a_n = Ar_1^n + Br_2^n, \tag{8}$$

where A and B are arbitrary constants. If the characteristic equation has a double root r_1, then the general solution is

$$a_n = Ar_1^n + Bnr_1^n. \tag{9}$$

In either case, for any specific initial conditions a_0 and a_1, the difference equation has a unique particular solution with unique values for A and B. These values are found by substituting the general solution into the initial conditions and solving the resulting linear equations.

PROOF $a_n = r^n$ is a solution to Eq. (7) iff

$$r^n = cr^{n-1} + dr^{n-2}. \tag{10}$$

Excluding the trivial case $r = 0$, Eq. (10) is satisfied iff

$$r^2 = cr + d \qquad \text{or} \qquad r^2 - cr - d = 0,$$

that is, iff r is a root of the characteristic equation.

Now a quadratic has either two distinct roots r_1, r_2 or one double root r_1. In the former case, every sequence of the form in Eq. (8) is a solution to the difference equation by Theorem 1. In the latter case, we know that $a_n = r_1^n$ is a solution; as soon as we check directly that $a_n = nr_1^n$ is another solution, it follows from Theorem 1 that every sequence of the form in Eq. (9) is a solution.

To check that nr_1^n is a solution, we first note that r_1 is a double root of $r^2 - cr - d = 0$ iff

$$r^2 - cr - d = (r - r_1)^2 = r^2 - 2r_1r + r_1^2.$$

In other words,

$$c = 2r_1, \qquad \text{and} \qquad d = -r_1^2.$$

Therefore we must check that $a_n = nr_1^n$ satisfies the difference equation

$$a_n = 2r_1a_{n-1} - r_1^2a_{n-2}.$$

The check is a plug-in:

$$\begin{aligned} nr_1^n &\stackrel{?}{=} 2r_1[(n-1)r_1^{n-1}] - r_1^2[(n-2)r_1^{n-2}] \\ &\stackrel{?}{=} 2(n-1)r_1^n - (n-2)r_1^n \\ &\stackrel{?}{=} [(2n-2) - (n-2)]r_1^n \\ &\stackrel{\checkmark}{=} nr_1^n. \end{aligned}$$

Finally, we must show that *every* solution to Eq. (7) is in the form of either Eq. (8) or Eq. (9). It suffices to show that, for any initial values a_0 and a_1, there is at least one solution in the form of Eq. (8) if the roots are distinct and at least one in the form of Eq. (9) if they are not. By Theorem 1, Section 5.4, it will then follow that there are no other solutions.

So in the case of distinct roots, it suffices to show that there are constants A and B such that

$$\begin{aligned} Ar_1^0 + Br_2^0 &= A \quad + B \quad = a_0, \\ Ar_1^1 + Br_2^1 &= Ar_1 + Br_2 = a_1. \end{aligned} \tag{11}$$

In the case of a double root, we need to show the same for

$$\begin{aligned} Ar_1^0 + B \cdot 0 \cdot r_1^0 &= A \qquad\qquad = a_0, \\ Ar_1^1 + B \cdot 1 \cdot r_1^1 &= Ar_1 + Br_1 = a_1. \end{aligned} \tag{12}$$

We can use straightforward algebra to find (unique) solutions in both cases. (The result is expressions for A and B in terms of the already-known values a_0, a_1, r_1 and r_2.) For instance, in Eq. (11) we begin by multiplying the first equation by r_1 and then subtracting the second equation from the first [20]. ∎

Note 1. The roots r_1, r_2 of the characteristic equation $r^2 - cr - d = 0$ can never be 0, because $d \neq 0$ (otherwise the difference equation would not really be second-order). We used the fact that the roots are nonzero in Eqs. (11) and (12). (How?)

Note 2. Although we have always used the terms a_0 and a_1 for the initial conditions, any two consecutive terms will do. There will still be a unique solution expressed with unique values of A and B. However, sometimes you get a difference equation for which the given terms are something like $a_0 = 1$ and $a_{100} = 0$. (Such "initial" conditions are more properly called **boundary conditions.**) The method of solution is still the same: Write down the general solution from Eqs. (8) or (9) and then solve for the constants using the boundary conditions. However, for boundary conditions instead of initial conditions, Theorem 1, Section 5.4, does not guarantee that there is a unique solution. In some cases there will be infinitely many solutions [25], in some other cases there will be none [26]. The same things can happen if you have the "wrong" number of initial conditions, that is, more than two or less than two.

Note 3. Nothing in the proof of Theorem 2 requires the roots to be real numbers. Some examples with complex roots are treated in the next subsection.

Qualitative Analysis

Sometimes it's not important to have an exact value for a_n. Sometimes it's even misleading: If the model that leads to a difference equation is not very exact, it would be wrong to believe in the exactness of the answer that comes from that difference equation. In such cases one is more interested in qualitative results. For instance, what happens to the sequence in the long run? Does a_n go to 0, head to some nonzero limit, oscillate indefinitely, go to plus or minus infinity, or what? We will analyze long-term behavior in this subsection. Our analysis will be informal, since we do not treat limits carefully until Chapter 9. Theorem 2 says that every solution to a second-order difference equation has a simple formula built around geometric sequences. Therefore it's not hard to analyze the behavior of second-order difference equations.

EXAMPLE 6 Analyze the long-term behavior of the following sequences:

$$\begin{aligned} a_n &= 2(\tfrac{1}{3})^n + 4(\tfrac{1}{2})^n, \\ b_n &= 3 - 5(\tfrac{1}{2})^n, \\ c_n &= (-2)^n + 5(\tfrac{1}{2})^n, \\ d_n &= 2^n - 5(1.5)^n. \end{aligned}$$

Solution For $\{a_n\}$, the two geometric sequences are $\{2(\frac{1}{3})^n\}$ and $\{4(\frac{1}{2})^n\}$. Since the ratio in both cases is between 0 and 1, both sequences head towards 0. Therefore their sum goes to 0 as well. This is shown graphically in Fig. 5.7(a).

For $\{b_n\}$, the first geometric sequence is the constant sequence with nth term $3 = 3 \cdot 1^n$. Since the second geometric sequence goes to 0, we see that $b_n \to 3$. The sequence $\{b_n\}$ is shown in color in Fig. 5.7(a).

For the third sequence, the second term again dies out. The first term has ratio -2, and therefore alternates positive and negative, growing without bound in absolute value. This result is sometimes called **explosive oscillation**. Therefore the combined effect of both terms is also explosive oscillation.

For d_n, both ratios are greater than 1. The first term, 2^n, is always positive, whereas the second term, $-5(1.5)^n$, is always negative. The first term has the slightly greater ratio, but the second has a larger multiplicative constant (5 as opposed to 1). Which wins out in the long run? The larger ratio. To show this result, we rewrite the formula for d_n as follows:

$$d_n = 2^n\left[1 - 5(\tfrac{1.5}{2})^n\right].$$

Now note that $(1.5/2)^n$ goes to 0 as n gets large, so the quantity inside the brackets approaches 1. Therefore in the long run d_n looks essentially like 2^n. This result is illustrated in Fig. 5.7(b). Note how initially d_n is negative, that is, the term $-5(1.5)^n$ is initially more significant. But then the sum turns positive and starts to grow like 2^n. ∎

Example 6 involved sequences that arise from difference equations with distinct roots. Now let us look at sequences arising from multiple roots.

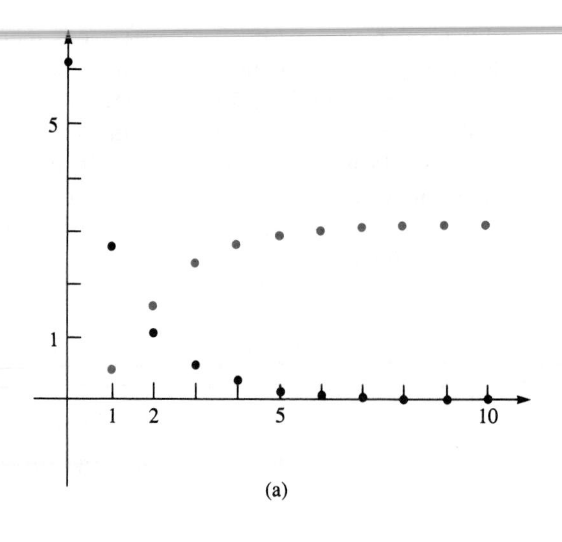

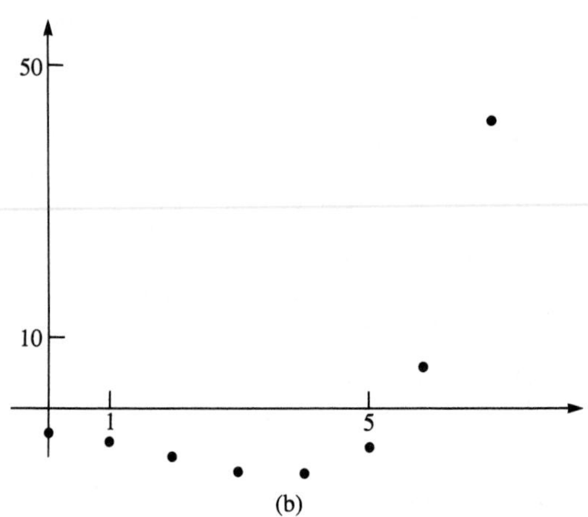

FIGURE 5.7 (a) Graph of $a_n = 2(\frac{1}{3})^n + 4(\frac{1}{2})^n$ (black) and $b_n = 3 - 5(\frac{1}{2})^n$ (color). (b) Graph of $d_n = 2^n - 5(1.5)^n$.

EXAMPLE 7 Analyze the long-term behavior of the following sequences:

$$e_n = -3 \cdot 2^n + .7n2^n,$$
$$f_n = 3(.6)^n + 4n(.6)^n.$$

Solution We may rewrite the first formula as $(.7n-3)2^n$. As n gets large, both factors go to positive infinity, so their product does also. (Note that e_n is initially negative, since $3 > .7n$ for small n, but eventually $.7n$ is much larger than 3.)

As for f_n, the same sort of factoring yields $(4n+3)(.6)^n$. Now the first factor goes to infinity as n increases, but the second goes to 0. Which wins out? In Section 0.5 we said that the exponential function $g(n) = a^n$ grows much faster than any polynomial, so long as $a > 1$. It turns out that exponentials are also "more powerful" when $0<a<1$; as $n \to \infty$, these exponentials go to 0 so much faster than polynomials increase that the product goes to 0. For the example at hand, think of it intuitively this way: When n is large, each increase by 1 in n reduces $(.6)^n$ by 60%, but the relative increase in $(4n+3)$ is very small. In any event, the long-run behavior is that f_n goes to 0. This result is illustrated graphically in Fig. 5.8. Note that f_n increases at first—the relative growth in $(4n+3)$ is substantial when n is still small. ∎

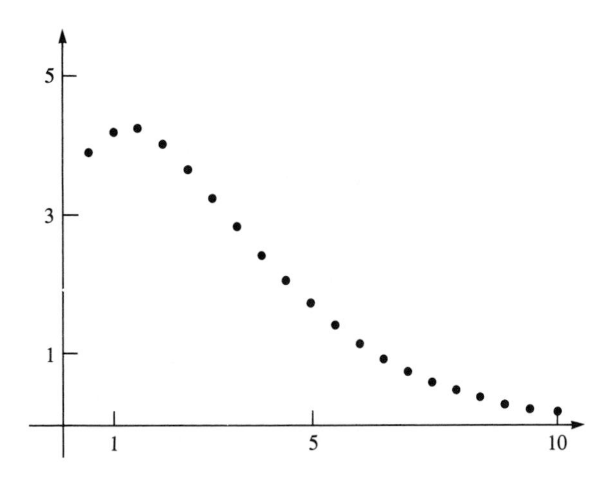

FIGURE 5.8 Graph of $f_n = 3(.6)^n + 4n(.6)^n$

The basic pattern is clear: The significant term in the long run involves r_1^n if $|r_1| > |r_2|$ or involves nr_1^n if there is a double root. But not so fast. This explanation doesn't cover the whole story. What happens, for instance, if $r_1 \neq r_2$ but $|r_1| = |r_2|$?

Consider the simple case $a_n = 1^n + (-1)^n$. Then the sequence alternates 2, 0, 2, 0, ... Another way to say the same thing is that we have a sequence with **period** 2. In this case the geometric ratios both have absolute value 1. Related behavior occurs for

$$b_n = 3(1.2)^n + 2(-1.2)^n.$$

Here both ratios again have the same absolute value, but it's not 1. When n is even, the negative sign drops out and we have

$$b_n = 5(1.2)^n, \qquad n \text{ even}.$$

When n is odd, the negative sign stays and we get

$$b_n = (1.2)^n, \qquad n \text{ odd}.$$

The graph of the sequence $\{b_n\}$ is shown in Fig. 5.9. Instead of alternating between

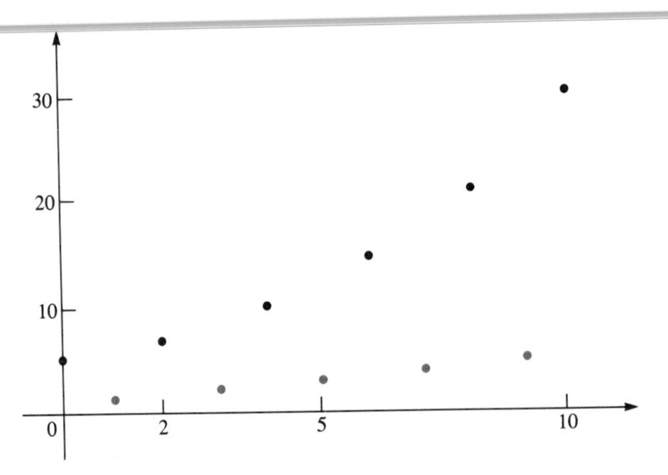

FIGURE 5.9

The sequence $b_n = 3(1.2)^n + 2(-1.2)^n$, alternates between $b_n = 5(1.2)^n$ for n even (black) and $b_n = (1.2)^n$ for n odd (color).

two numbers, this sequence alternates between two simple patterns, both of which are geometric, differing only by the constant.

The story is still not over. The geometric ratios in solutions to difference equations are roots of polynomials; what if the roots are complex numbers? (Readers with little knowledge of complex numbers should read this paragraph and the next lightly.) Figure 5.10 shows two examples, for the sequences

$$c_n = i^n + (-i)^n \qquad \text{and} \qquad d_n = (.8i)^n + (-.8i)^n.$$

First note that the c's and d's are real numbers, despite being expressed using complex numbers, because the sum of two complex conjugates raised to the same power is real. Next notice that the c sequence is periodic of period four, and that the d sequence also has a pattern of length four, but each set of four has values closer to 0 than does the previous set. In fact, these sequences are just discrete analogs of the "harmonic" and "damped harmonic" motion one studies in calculus. To show this analogy, we have superimposed the appropriate cosine curves over the discrete graph.

There is nothing special about period four. With appropriate complex roots, it is possible to get any period and certain more general types of "cyclic" behavior. If you know well the algebra of complex numbers, including the concepts of modulus and argument and the Theorem of DeMoivre, the analysis of geometric series with complex ratios is no harder than for real ratios. The same basic result applies: If $|r| > 1$, then ar^n is unbounded; if $|r| < 1$, then $ar^n \to 0$; and if $|r| = 1$, then ar^n is cyclic. The complex case is explored in some problems [8, 24–29] for those with appropriate background.

Let us get back to the essential point. The qualitative behavior of the solution

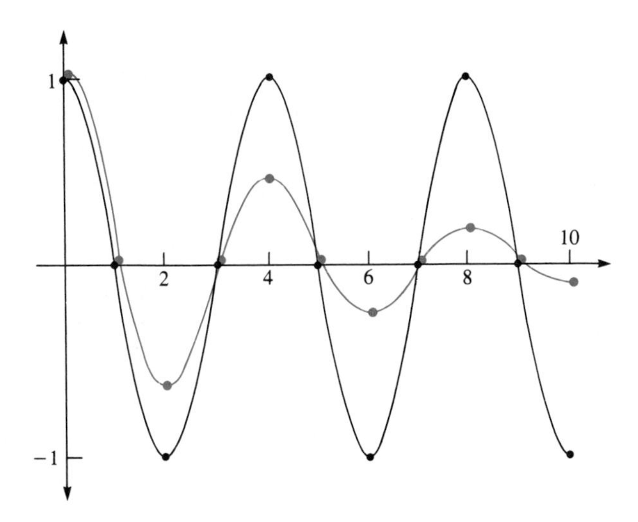

FIGURE 5.10

The sequences $c_n = i^n + (-i)^n$ (black) and $d_n = (.8i)^n + (-.8i)^n$ (color). These sequences are discrete analogs of the continuous circular functions $y = \cos(\pi x/2)$ and $y = (.8)^x \cos(\pi x/2)$.

to a linear difference equation depends on the geometric ratios. Furthermore, it is very easy to find this maximum ratio *without* solving the difference equation completely. All that matters is the characteristic polynomial.

EXAMPLE 8 Give a qualitative analysis of the general solution to

$$a_n = a_{n-1} + 6a_{n-2}.$$

Solution The characteristic polynomial is $r^2 - r - 6 = (r+2)(r-3)$. Thus the general solution is

$$A3^n + B(-2)^n.$$

The first term will predominate, except in the rare case that initial conditions force $A = 0$. Even if A is tiny and B is huge, eventually the term $A3^n$ will predominate. Therefore the sequence may show oscillating behavior at first (if B is large and thus $B(-2)^n$ is more significant for small n) but eventually the sequence will stay positive and growing (if $A > 0$) or stay negative and growing (if $A < 0$). ∎

Problems: Section 5.5

In [1–8], find the particular solutions by the method of this section.

1. $a_n = a_{n-1} + 2a_{n-2}$, $a_1 = 2$, $a_2 = 3$.
2. $a_n = a_{n-1} + a_{n-2}$, $a_0 = 1$, $a_1 = 3$.
3. $a_{n+1} = 5a_n - 6a_{n-1}$, $a_1 = -1$, $a_2 = 1$.
4. $a_n = -2a_{n-1} - a_{n-2}$, $a_0 = 1$, $a_1 = 2$.
5. Same as [3] except $a_1 = 5$, $a_2 = 7$.
6. Same as [4], but $a_0 = 1$, $a_1 = -2$.
7. Solve [3acd, Section 5.3].
8. (Requires complex numbers) Solve [1, Section 5.3].
9. Pick four different pairs of values for a_0, a_1, compute ten terms of the solution to Eq. (6) for each pair, and then use the methods of Section 2.4 to guess a closed form for the solution in each case.
10. Verify that $\{n2^n\}$ is indeed a solution to Eq. (6).
11. In the solution to Example 1, and again in the proof of Theorem 2, we dismissed the case $r = 0$ in order to divide by r and get a quadratic. However, the sequence $\{0^n\}$ *is* a solution to the original difference equation, and if we claim in Theorem 2 that we have all solutions, we better have that trivial solution too. Do we? How?
12. Ever wonder why we usually start our sequences at a_0 instead of a_1? One reason is because it makes the algebra less tedious when you have to find a particular solution from initial conditions *by hand*. Check this out with Examples 3 and 5. In each case, do the algebra to find A and B using a_0 and a_1 as the initial values. Then find the same A and B again using a_1 and a_2.
13. We now have two formulas for the Fibonacci numbers: the recursion

$$F_n = F_{n-1} + F_{n-2}, \qquad F_0 = F_1 = 1,$$

and Eq. (5). Which do you feel is the more efficient formula to use, and why, if you want to

a) compute the first 25 Fibonacci numbers;
b) analyze the long term growth rate of the Fibonacci sequence;
c) know the order of magnitude of F_{100};
d) know the exact value of F_{100}.

14. Determine the long-range behavior of the sequence with nth term

a) $2^n + 3^n$ b) $3 + (.5)^n$
c) $3(1.1)^n - (1.2)^n$ d) $3(1.1)^n + (-1.2)^n$
e) $2(1.2)^n + 3(-1.2)^n$ f) $(.9)^n + 2(-.9)^n$
g) $3(\frac{1}{2})^n + 5n(\frac{1}{2})^n$ h) $2(1.2)^n - 3n(1.2)^n$

15. Consider the directed graph in Fig. 5.11. Count the number of directed paths from v_0 to v_{10}. *Hint*: Generalize the problem to ask for a sequence of counts and find a difference equation for the sequence.

FIGURE 5.11

16. Suppose national income in each period is the average of national income in the preceding and succeeding periods.

a) Model this situation with a difference equation and find the general solution.
b) Find the particular solution if U.S. national income is \$1 trillion in period 0 and \$1.05 trillion in period 1.

17. **Consider binary sequences of length n in which 1's may not appear consecutively except in the rightmost two positions. In Example 4, Section 2.5, we incorrectly proved that there are 2^n such sequences. Find the right number. *Hint*: Consider how to get sequences of length $n+1$ by putting 0 or 1 to the left of sequences of length n.**
18. Repeat [17], except now consecutive 1's may not appear anywhere.
19. Generalize the argument at the end of the solution to Example 6: Show that if $|r_1| > |r_2|$ and $A \neq 0$, then in the long run $Ar_1^n + Br_2^n$ behaves like Ar_1^n.
20. Finish the verification from the proof of Theorem 2 that the linear systems (11) and (12) have solutions.
21. Show: For any second-order linear difference equation, there is always a unique solution given any two consecutive values a_k and a_{k+1}, if

a) the characteristic equation has distinct roots;
b) it has a double root.

22. Give an example of two linear equations in two unknowns that have no solution. By changing the constants on the right-hand side only, obtain a system that has infinitely many solutions. (We made a fuss in Theorem 2 and in [20, 21] that certain types of linear systems always have a unique solution. This problem shows that the fuss is called for; linear systems do not automatically have unique solutions.)

23. Prove that F_n is the closest integer to

$$\frac{1}{\sqrt{5}}\left(\frac{1+\sqrt{5}}{2}\right)^{n+1}$$

for all $n \geq 0$.

For the rest of the problems in this set, solid knowledge of complex numbers is very useful or even necessary.

24. Consider the recursion $a_{n+1} = a_n - a_{n-1}$, where a_0 and a_1 are anything you please. Show that $\{a_n\}$ is periodic—starting with a certain a_n, it simply repeats itself every 6 terms. If you already showed this result in [2, Section 5.3] by induction, this time do it using the theory of this section.

25. Let $\{a_n\}$ satisfy the recursion of [24] *and* satisfy the boundary conditions $a_0 = 1, a_6 = 1$. Show that $\{a_n\}$ is not unique; in fact, there are infinitely many sequences meeting these conditions.

26. Again let $\{a_n\}$ satisfy the recursion of [24], but now let the boundary conditions be $a_0 = 1, a_6 = 2$. Show that *no* sequences satisfy these conditions. (Assume Theorem 2.)

27. For second-order difference equations, qualitative information about the long-term behavior of solutions can be read directly from the *coefficients* of the quadratic characteristic polynomial. That is, you don't have to solve the quadratic. Let $ax^2 + bx + c$ be an arbitrary quadratic with real coefficients. Show that:

a) If the roots are complex, then both roots have modulus $< 1 \iff |c/a| < 1$.

b) If the roots are real, then both have absolute value $< 1 \iff |b/a| < 1+(c/a)$ and $|b/a| < 2$.

c) What conditions on the coefficients ensure that both roots are greater than 1 in absolute value?

28. Find a linear difference equation for which $\{\sin n\theta\}$ and $\{\cos n\theta\}$ are solutions. *Hint*:

$$\cos n\theta = (e^{in\theta}+e^{-in\theta})/2 = \tfrac{1}{2}r_1^n + \tfrac{1}{2}r_2^n,$$

where $r_1 = e^{i\theta}$ and $r_2 = e^{-i\theta}$.

29. Use your solution to [28] to write a computer program which plots a circle by plotting the 360 points $(\sin\theta, \cos\theta)$ for integer degrees θ from 1 to 360. Whatever machine you do this on, compare the running time for this program and for the more straightforward program of the form

```
for θ = 1 to 360
    plot (sin θ°, cos θ°)
endfor
```

30. The simplest correct recursions for [28] are second order. Come up with a system of linked first-order difference equations relating $\sin(n+1)\theta$ and $\cos(n+1)\theta$ to $\sin n\theta$, $\cos n\theta$ and some initial values. Write a program for plotting a 360-point circle using this linked recursion. Compare its speed to that of the two programs in [29].

5.6 Difference Equations of Arbitrary Order

As in Section 5.5, all difference equations in this section are constant-coefficient, linear, and homogeneous, but now we allow arbitrary order instead of just second-order. The approach used in Section 5.5 still works. We illustrate this with two examples and then state a general theorem.

EXAMPLE 1 What is the general solution to

$$a_n = 2a_{n-1} + 5a_{n-2} - 6a_{n-3}\,?$$

Solution Guessing a geometric solution $a_n = r^n$, substituting, dividing by the highest power of r we can, and factoring, we obtain the characteristic equation

$$r^3 - 2r^2 - 5r + 6 \;=\; (r-1)(r+2)(r-3) \;=\; 0.$$

(How are you supposed to find this factorization? See the following paragraph.) Therefore the sequences $\{1^n = 1\}$, $\{(-2)^n\}$, and $\{3^n\}$ are solutions. Assuming the methods of Section 5.5 still apply, we conclude that the general solution is

$$a_n \;=\; A + B(-2)^n + C3^n. \;\blacksquare$$

(Factoring polynomials. There is a theorem—perhaps you learned it in high school—that says: If $(r-c)$ is a factor of a polynomial with integer coefficients, and c is an integer, then c must divide the constant term. Since the only divisors of 6 are ± 1, ± 2, ± 3, and ± 6, it's easy enough to find the factors here by brute force. Or let a machine do it for you. Actually, difference equations that arise from real-world data are not likely to have integer roots, so what you really want to use is a root approximation algorithm such as the bisection method that we discuss in Section 9.7.)

EXAMPLE 2 Solve

$$b_{n+1} = 5b_n - 3b_{n-1} - 9b_{n-2}; \qquad b_0 = 6,\; b_1 = -8,\; b_2 = -22. \tag{1}$$

Solution The characteristic polynomial is

$$r^3 - 5r^2 + 3r + 9 = (r+1)(r-3)^2.$$

Thus with one polynomial we get both situations that we treated in Section 5.5: more than one root and multiple roots. The upshot is: we get as basic solutions every sequence that we would get from $(r+1)$ alone and from $(r-3)^2$ alone. That is, the basic solutions are $(-1)^n$, 3^n, and $n3^n$. Therefore there are constants A, B, and C so that the general solution is

$$b_n = A(-1)^n + B3^n + Cn3^n.$$

Plugging into the initial conditions by setting $n = 0, 1, 2$, we obtain three simultaneous linear equations for the three unknown constants:

$$\begin{aligned} A + \;\; B \qquad\;\; &= \;\; 6,\\ -A + 3B + \;\; 3C &= -\,8,\\ A + 9B + 18C &= -22. \end{aligned} \tag{2}$$

Solving these (tedious by hand) we find that $A = 5$, $B = 1$, and $C = -2$. Thus the particular solution to Eqs. (1) is

$$b_n \;=\; 5(-1)^n + 3^n - 2n3^n. \;\blacksquare$$

What if the characteristic polynomial has a factor $(r-c)^3$. Then you get basic solutions c^n, nc^n and n^2c^n. If instead $(r-c)^4$ is the highest power of $(r-c)$, then n^3c^n is another basic solution. And so on. We summarize all this in the next definition and theorem.

Definition 1. Let $c_1, c_2, \ldots, c_k$ be given constants. The characteristic equation of the difference equation

$$a_n = \sum_{i=1}^{k} c_i a_{n-i}$$

is

$$r^k - \sum_{i=1}^{k} c_i r^{k-i} = 0.$$

This is the polynomial equation obtained by replacing each a_m in the difference equation with r^m, dividing by the highest common power of r, and putting all terms on the left. The polynomial on the left is called the characteristic polynomial.

Theorem 1. Suppose that the characteristic equation of a linear, homogeneous, constant-coefficient, difference equation is

$$\prod_{i=1}^{k} (r - r_i)^{m_i} = 0.$$

That is, the equation has distinct roots $r_1, r_2, \ldots, r_k$, with multiplicities $m_1, m_2, \ldots, m_k$, respectively. Then every solution $\{a_n\}$ to the difference equation can be expressed uniquely as a linear combination of the following solutions:

$$\begin{gathered}\{r_1^n\},\quad \{nr_1^n\},\ \ldots\ ,\ \{n^{m_1-1}r_1^n\},\\ \{r_2^n\},\ \ldots\ ,\ \{n^{m_2-1}r_2^n\},\\ \cdots\\ \{r_k^n\},\ \ldots\ ,\ \{n^{m_k-1}r_k^n\}.\end{gathered}$$

Why is Theorem 1 true? Unfortunately, this is not easy to explain. We have proved some parts but are merely stating others. We proved the fact that

$$r^n \text{ is a (nonzero) solution } \iff r \text{ is a root of the characteristic equation}$$

in Example 1 of Section 5.5. (We showed it there for a particular second-order difference equation, but the method of proof was general.) We also proved the fact that linear combinations of solutions are solutions in Theorem 1 of Section 5.5. (As stated, that theorem is only about second-order problems and linear combinations of just two solutions. To make it apply to arbitrary order requires only the careful use of Sigma signs or dot-dot-dots [13], and induction shows that the same result holds for linear combinations of any number of solutions [14].) Consequently, we really have shown that everything of the form $a_n = \sum_{i=1}^{k} A_i r_i^n$. is a solution.

What we have left out is an explanation of where the additional basic solutions come from in the multiple-root case, and a proof in either case that linear combinations of the basic solutions give us *all* solutions. As for the first gap, a complete proof would be too much for this first course. As for the second gap, Theorem 1, Section 5.4, says that, if every set of initial values can be matched using a linear combination of these basic solutions, then there are no other solutions. So the second gap boils down to a question about whether a certain type of linear equations, such as those in system (2), always has a unique solution. We treat this sort of question in Chapter 8, but a full answer requires an even deeper treatment than we provide there. A proof of the existence of a unique solution to the linear equations in the distinct root case appears as one of the harder problems in that chapter [44, Section 8.8].

Qualitative Analysis

Using Theorem 1, it is possible to do a complete analysis of the long-run behavior of solutions to linear difference equations. However, we won't attempt to be complete, for two reasons. First, there are many more cases than for second-order problems. Second, it's not wise to attempt long-run analysis in complicated cases without a careful grounding in limits. Instead, we illustrate and then state a theorem that covers the most important case: distinct roots of different magnitudes.

EXAMPLE 3 Suppose that the general solution to a difference equation is

$$A3^n + B(-2)^n + C(1.4)^n.$$

What is the long-run behavior?

Solution Factor out the term involving the geometric ratio with the largest magnitude, in this case 3^n:

$$A3^n\left(1 + \tfrac{B}{A}(-\tfrac{2}{3})^n + \tfrac{C}{A}(\tfrac{1.4}{3})^n\right). \tag{3}$$

Since $|-\frac{2}{3}| < 1$ and $|\frac{1.4}{3}| < 1$, the expression in large parentheses goes to

$$1 + \tfrac{B}{A}\cdot 0 + \tfrac{C}{A}\cdot 0 = 1$$

in the long run. Therefore in the long run the solution behaves like $A3^n$. (More precisely, the ratio of $A3^n$ to the exact solution goes to 1.) The only exception to this analysis is the unlikely event that you are interested in a particular solution where $A = 0$ (in which case the division by A to get Eq. (3) is invalid). Then, if $B \neq 0$, a similar analysis shows that the solution behaves like $B(-2)^n$. ∎

This example illustrates the following theorem.

Theorem 2. Suppose that $\{a_n\}$ is the solution to a linear, constant-coefficient, homogeneous difference equation whose characteristic equation has distinct roots. Assume that r_1 is the root with the largest absolute value and that no other root has the same absolute value. Assume further that the coefficient of r_1^n in the explicit formula for a_n is not 0. Then in the long run, $\{a_n\}$ behaves like a multiple of the geometric sequence $\{r_1^n\}$.

Problems: Section 5.6

For [1–9], use the method of this section to find the particular solution if initial conditions are given and the general solution otherwise. The solutions for first-order problems are pretty obvious, but using a general method on a simple case is always a good test of your understanding of that method.

1. $a_n = a_{n-1}, \quad a_0 = 1$
2. $d_n = -2d_{n-1}, \quad d_1 = 3$
3. $e_{n+1} = 3e_n, \quad e_2 = 2$
4. $f_n = f_{n-1} + 8f_{n-2} - 12f_{n-3}$.
5. $c_n = -3c_{n-1} - 3c_{n-2} - c_{n-3}$
6. $e_n = 3e_{n-2} - 2e_{n-3}$
7. $a_n - 2a_{n-1} - a_{n-2} + 2a_{n-3} = 0$, $a_0 = -1,\ a_1 = 6,\ a_2 = 2$.
8. For the recursion in [5] find the particular solution for
   a) $c_0 = 1, \quad c_1 = -1, \quad c_2 = 1$
   b) $c_0 = 0, \quad c_1 = -1, \quad c_2 = 2$
   c) $c_0 = 0, \quad c_1 = -1, \quad c_2 = 2$
   d) $c_0 = c_1 = c_2 = 1$
9. For the recursion in [6] find the particular solution for
   a) $e_0 = e_1 = e_2 = 1$
   b) $e_0 = 1, \quad e_1 = 2, \quad e_2 = 3$
10. Suppose that a difference equation has order k. Explain why the degree of its characteristic polynomial is also k.
11. Explain why a monic polynomial (leading coefficient 1) is the characteristic polynomial of some linear difference equation $\Longleftrightarrow$ the constant term of the polynomial is nonzero.
12. What is the degree of the polynomial in Theorem 1?
13. Prove: If $\{b_n\}$ and $\{b'_n\}$ are both solutions to
$$a_n = \sum_{i=1}^{k} c_i a_{n-i} \quad (c_1, \ldots, c_k \text{ constants}),$$
then so is every linear combination $\{Bb_n + B'b'_n\}$. This is the generalization to nth-order difference equations of Theorem 1, Section 5.5.
14. Prove: If any linear combination of two solutions to a homogeneous linear difference equation is a solution, then any linear combination of any number of solutions is a solution.
15. Find a difference equation whose characteristic polynomial is $(r-3)(r+4)$.
16. Find a difference equation whose characteristic polynomial is $(r-2)^5$.

In the text we worked from difference equations to polynomials to explicit solutions, but the process can be reversed. Problems [15, 16] asked you to reverse this partly, from polynomial to difference equation; [17–23] ask you to go all the way. Each problem gives you a sequence. Name the simplest difference equation you can (including initial conditions) for which it is a solution.

17. $a_n = 2^n$
18. $b_n = 4 \cdot 3^n$
19. $c_n = 2^n + 3^n$
20. $d_n = 7 + (-1)^n$
21. $e_n = 2^n - (-2)^n + 3$
22. $f_n = r_1^n + r_2^n$, where r_1 and r_2 are the zeros of $x^2 - 5x + 3$. You can answer this problem without ever computing r_1 and r_2. Finding the recursion is easy; finding the initial conditions is a little harder.

23. $g_n = r_1^n + r_2^n + r_3^n$, where r_1, r_2 and r_3 are the roots of $x^3 - x^2 + 2x - 3$. (The idea behind [22, 23] is known as the method of Newton sums.)

24. Consider the difference equation with characteristic equation $(r-c)^2(r-d) = 0$, where $c \neq d$. Verify that the sequences $\{c^n\}$, $\{nc^n\}$, and $\{d^n\}$ are solutions.

5.7 Nonhomogeneous Difference Equations

We have now solved all linear, constant-coefficient, homogeneous difference equations. It's time to tackle the next level, the nonhomogeneous case. As you will see, it is possible in part to reduce to the previous case.

Throughout this section, "difference equation" means "constant-coefficient, linear difference equation". The only variation will be whether or not the equation is homogeneous or nonhomogeneous.

The first thing to ask ourselves about nonhomogeneous equations is whether Theorem 1, Section 5.5, still holds. Is it enough to find some basis of solutions and take linear combinations? We find out by trying to mimic the proof of that theorem.

Suppose, for example, that both $\{a_n\}$ and $\{b_n\}$ satisfy

$$s_n = s_{n-1} + 2s_{n-2} + n^2. \tag{1}$$

Substituting first a and then b for s, and then adding, we obtain

$$a_n + b_n \;=\; (a_{n-1} + b_{n-1}) \;+\; 2(a_{n-2} + b_{n-2}) \;+\; 2n^2.$$

Thus $\{a_n + b_n\}$ satisfies an equation with the same linear part, but the nonhomogeneous term, $2n^2$, is different. This is enough to show that solutions to a nonhomogeneous difference equation are *not* closed under linear combinations.

However, if adding makes the nonhomogeneous part worse, then subtracting will make it better. In fact, it will go away:

$$b_n - a_n \;=\; (b_{n-1} - a_{n-1}) \;+\; 2(b_{n-2} - a_{n-2}). \tag{2}$$

That is, if we define $h_n = b_n - a_n$ (h for homogeneous), Eq. (2) is

$$h_n = h_{n-1} + 2h_{n-2}, \tag{3}$$

which is just Eq. (1) with the nonhomogeneous term deleted. Since Eq. (2) is homogeneous, we know how to solve it using the method of Section 5.5. Since $b_n = a_n + h_n$, this means that, if we know *any one* solution of Eq. (1), call it a_n, then any *other* solution b_n can be found by adding to a_n some solution of the homogeneous equation which results when the nonhomogeneous term is deleted. This is best made clear by an example.

EXAMPLE 1 Find the general solution to

$$R_n = R_{n-1} + 2R_{n-2} + 1. \tag{4}$$

Solution It happens that the constant sequence $a_n = -1/2$ is one solution—try it! (We'll explain how to find such "starter" solutions later.) Since Eq. (4) without the nonhomogeneous term is the same as Eq. (1) without its nonhomogeneous term, it follows from the discussion preceding this example that any other solution b_n of Eq. (4) is given by $b_n = a_n + h_n$, where h_n is some solution of Eq. (3). By the characteristic polynomial method (Section 5.5), we know that the general solution for Eq. (3) is $A2^n + B(-1)^n$. Therefore every solution b_n to Eq. (4) is of the form

$$b_n = -\tfrac{1}{2} + A2^n + B(-1)^n.$$

But is the converse true? Is every sequence of the form $b_n = -\frac{1}{2}+A2^n+B(-1)^n$ a solution to Eq. (4)? Yes, because the part $a_n = -1/2$ satisfies $a_n = a_{n-1} + 2a_{n-2} + 1$, the part $h_n = A2^n + B(-1)^n$ satisfies Eq. (3), and so adding we find that

$$b_n = a_n + h_n = (a_{n-1} + h_{n-1}) \ + \ 2(a_{n-2} + h_{n-2}) + 1.$$

That is, $\{b_n\}$ satisfies Eq. (4). ∎

For any nonhomogeneous difference equation, the homogeneous equation obtained by deleting the nonhomogeneous part is called the **associated** (homogeneous) **difference equation**. The reasoning in Example 1 generalizes to the following theorem.

Theorem 1. Let $\{a_n\}$ be any particular solution to a nonhomogeneous, constant-coefficient linear difference equation. Then $\{b_n\}$ is the general solution if and only if

$$b_n = a_n + h_n,$$

where $\{h_n\}$ is the general solution to the associated homogeneous difference equation.

PROOF Repeat the argument in Example 1, except in place of the particular difference equation (4), use the generic equation

$$a_n = \sum_{i=1}^{k} c_i a_{n-i} \ + \ f(n),$$

($f(n)$ is the nonhomogeneous part), and instead of using the specific solution $a_n = -1/2$, just use a generic particular solution a_n. Note that the first paragraph of the argument in Example 1 proves "only if" and the second paragraph proves "if". Since the if argument uses subtraction, you should not be surprised that the reverse only-if argument uses addition. ∎

The significance of Theorem 1 is that it reduces the problem of finding all nonhomogeneous solutions to that of finding just one. We already know how to find a basis of solutions for the associated homogeneous problem, and we so we simply add on all linear combinations of that basis to our one nonhomogeneous solution.

The coefficients in this linear combination may be used to satisfy initial conditions, as the next example shows.

EXAMPLE 2 Find all solutions to

$$a_n - 2a_{n-1} = 3^n. \tag{5}$$

Then find the particular solution for which $a_0 = 4$.

Solution First we need to find some particular solution; it need not satisfy $a_0 = 4$. Because we don't yet have a systematic way to produce such a solution, let's guess (intelligently!). Unless a_n involves 3^n in some way, it seems unlikely that subtracting $2a_{n-1}$ is going to create 3^n. So let's guess $a_n = A3^n$, where A is to be determined. Substituting in the LHS of Eq. (5) gives

$$A3^n - 2(A3^{n-1}) = A3^n - \frac{2}{3}A3^n = \frac{A}{3}3^n.$$

Therefore

$$\frac{A}{3}3^n = 3^n \qquad \text{or} \qquad A = 3.$$

Thus $a_n = 3(3^n) = 3^{n+1}$ is a particular solution. Since the general solution of the associated homogeneous equation is $B2^n$, the general solution to Eq. (5) is

$$a_n = 3^{n+1} + B2^n.$$

Now, to satisfy the initial condition, we proceed as in the previous sections. Since a_0 is given, we plug in $n = 0$ and find

$$4 = a_0 = 3 + B \qquad \text{or} \qquad B = 1.$$

Therefore the particular solution for the initial condition is

$$a_n = 3^{n+1} + 2^n. \blacksquare$$

This approach is great if you are good at guessing a starter solution, but that's a skill which people new to a subject rarely have. For instance, you might think from Example 2 that you should always guess a multiple of the nonhomogeneous part. Alas, that doesn't always work (although it does work often [17]).

Fortunately, you don't have to learn how to guess here, because there is a general procedure that is guaranteed to pick a good guess for you! (Actually, it's only guaranteed if the nonhomogeneous part $f(n)$ is of a certain form, but the form is general enough to cover most cases that arise in practice [10, 18].)

We first state the procedure below and then give two examples of its use. We won't prove it: All the proofs we know are sophisticated, and the amount of effort needed for you to absorb a proof is out of proportion to the number of times you will need the procedure. We do need to solve nonhomogeneous difference equations at various places in this book, but they are usually simple ones that we can handle in other ways. For those who are curious, we present the key idea of one proof in [20].

General Procedure for Solving
Constant Coefficient, Nonhomogeneous Linear Difference Equations

1. Write the equation in the form
$$a_n - \sum_{i=1}^{k} c_i a_{n-i} = f(n),$$
where $f(n)$ is the nonhomogeneous part.
2. From step 1, write down and factor $p(r)$, the characteristic polynomial of the associated homogeneous difference equation.
3. Find and factor $q(r)$, the characteristic polynomial of the (simplest) homogeneous equation for which $f(n)$ is itself a solution.
4. Write down the general solution for the homogeneous equation associated with $P(r) = p(r)q(r)$. Call it $H(n)$.
5. Temporarily delete that part of $H(n)$ which is the general solution of the associated homogeneous form of the original equation. Call this deleted part $D(n)$ and the remaining part $R(n)$. $R(n)$ is the proper guess for finding one solution of the original equation.
6. Plug $R(n)$ into the original equation, simplify, and solve for the constants by matching coefficients. For instance, if 2^n appears on either side, it must appear on both sides with the same coefficient. Likewise, if $n^2 3^n$ appears on either side, it must appear on both sides with the same coefficient.
7. Write down $R(n)$ with the specific constant values just computed. This is a particular solution to the original difference equation (but not, generally, to the initial conditions).
8. Append $D(n)$ to this particular solution. This gives the general solution.
9. If there are initial conditions, substitute them into the general solution to solve for the remaining constants (as with homogeneous problems). This gives the particular solution to the problem.

EXAMPLE 3 Using the general procedure, find all solutions to
$$a_n + a_{n-1} - 6a_{n-2} = 2^n - 1. \tag{6}$$

Solution

Step 1. This was done for us in the problem statement: $f(n) = 2^n - 1$.

Step 2. The characteristic polynomial of the associated homogeneous equation is
$$p(r) = r^2 + r - 6 = (r+3)(r-2).$$

Step 3. The nonhomogeneous part of Eq. (6) is a linear combination of 2^n and 1^n. Therefore that part is a solution of the homogeneous difference equation with polynomial

$$q(r) = (r-2)(r-1).$$

Step 4. $P(r)$ is

$$p(r)q(r) = [(r+3)(r-2)][(r-2)(r-1)] = (r+3)(r-2)^2(r-1).$$

The general solution for the associated difference equation is

$$H(n) = A(-3)^n + B2^n + Cn2^n + D.$$

Step 5. The part to be deleted temporarily is

$$D(n) = A(-3)^n + B2^n,$$

since the terms $(-3)^n$ and 2^n correspond to $p(r)$. The remaining part, the guess, is

$$R(n) = Cn2^n + D.$$

Note that this guess is *not* a multiple of the original nonhomogeneous part.

Step 6. Plugging $R(n)$ into the original difference equation, we obtain

$$(Cn2^n + D) + [C(n-1)2^{n-1} + D] - 6[C(n-2)2^{n-2} + D] = 2^n - 1.$$

After considerable simplification—begin by writing 2^{n-1} as $\frac{1}{2}2^n$ and 2^{n-2} as $\frac{1}{4}2^n$—all the n's drop out except those in the exponents, and we find that

$$\frac{5}{2}C2^n - 4D = 2^n - 1.$$

For any solution this is an identity for all nonnegative integers n, so the coefficients of 2^n on both sides must be equal, and the constant terms must be equal. That is,

$$\frac{5}{2}C = 1 \qquad \text{and} \qquad -4D = -1.$$

Thus $C = 2/5$ and $D = 1/4$.

Step 7. The particular solution obtained in step 6 is

$$\frac{2}{5}2^n + \frac{1}{4}.$$

Step 8. Appending the deleted part to the particular solution just obtained, we find that the general solution is

$$A(-3)^n + B2^n + \frac{2}{5}n2^n + \frac{1}{4}.$$

Step 9. No initial conditions were given, so there is no step 9 in this case. ∎

EXAMPLE 4 Find a closed formula for $\sum_{k=1}^{n} k^2$.

Solution We explained a general method for representing a series by a difference equation in Example 2, Section 5.2. For the current example, that method says to consider the sequence whose typical term is $s_n = \sum_{k=1}^{n} k^2$. Then the difference equation is

$$s_n = s_{n-1} + n^2, \qquad s_1 = 1. \tag{7}$$

This is a first-order constant-coefficient nonhomogeneous problem—for series problems it always is—so we can apply the general procedure.

Step 1. The equation is $s_n - s_{n-1} = n^2$.

Step 2. The characteristic polynomial is $p(r) = r - 1$.

Step 3. The nonhomogeneous part is $n^2 = n^2 1^n$, so the simplest $q(r)$ is $(r-1)^3$. (We have used Theorem 1, Section 5.6.)

Step 4. $P(r) = (r-1)^4$, so

$$H(n) = An^3 + Bn^2 + Cn + D.$$

(We used Theorem 1 from Section 5.6 again.)

Step 5. $D(n) = D$ and $R(n) = An^3 + Bn^2 + Cn$.

Step 6. Substituting $R(n)$ in the original recursion, Eq. (7), we find that

$$\begin{aligned}
[An^3+Bn^2+Cn] - [A(n-1)^3 + B(n-1)^2 + C(n-1)] &= n^2, \\
A(3n^2-3n+1) + B(2n-1) + C &= n^2, \\
&\text{[Binomial Theorem]} \\
(3A)n^2 + (2B-3A)n + (C-B+A) &= 1n^2 + 0n + 0.
\end{aligned}$$

Thus

$$3A = 1, \qquad 2B - 3A = 0, \qquad C - B + A = 0,$$

which results in

$$A = \frac{1}{3}, \qquad B = \frac{1}{2}, \qquad C = \frac{1}{6}.$$

Step 7. A particular solution is

$$\frac{2n^3 + 3n^2 + n}{6}. \tag{8}$$

Step 8. The general solution is

$$\frac{2n^3 + 3n^2 + n}{6} + D.$$

Step 9. Substituting $s_1 = 1$ in the general solution, we find that $D = 0$. Thus in this case $R(n)$ happens to be *the* particular solution as well, and so Eq. (8) itself is a closed formula for $\sum_{k=1}^{n} k^2$. Factoring it, we obtain the more commonly used formula

$$\sum_{k=1}^{n} k^2 = \frac{1}{6}(2n + 1)(n + 1)n. \ \blacksquare$$

Note how mechanical the general procedure is. Furthermore, with just a slight modification (mostly an option to skip a number of steps), this procedure subsumes the procedures covered previously for homogeneous problems. No wonder computer algebra systems can solve standard difference equations easily.

Problems: Section 5.7

1. Find the general solution to $b_n = 3b_{n-1} + f(n)$, where $f(n)$ is

   a) 2^n b) $n2^n$ c) 3^n
   d) 1 e) $(-2)^n$ f) $5 \cdot 3^n$

2. Find the general solution to

$$a_n = 3a_{n-1} - 2a_{n-2} + f(n),$$

   for each choice of $f(n)$ in [1].

3. Find the general solution to $c_{n+1} = 4c_n - 4c_{n-1} + f(n)$ for each choice of $f(n)$ in [1].

4. Solve the TOH recursion using the method of this section.

5. Solve $s_n = 3s_{n-1} + 2$, $s_1 = 2$ (the Straightline TOH recursion) using the method of this section.

6. Solve Example 1 again, this time using the general procedure.

7. Solve Example 2 again, using the general procedure.

8. In Theorem 1, is it necessary that a_n and b_n satisfy a *constant-coefficient* nonhomogeneous linear difference equation? Is it necessary that they satisfy a *linear* difference equation? Why?

9. If $\{a_n\}$ satisfies $s_n = s_{n-1} + 2s_{n-2} + n^2$, show that $\{3a_n\}$ does not satisfy this same equation. What difference equation does $\{3a_n\}$ satisfy? (This problem shows that solutions to nonhomogeneous linear difference equations are not closed under multiples.)

10. The general procedure works only if you can find a constant-coefficient, linear homogeneous difference equation which $f(n)$ satisfies. But it's not hard to tell when $f(n)$ has this property because you already know all possible solutions to such equations. Describe them.

11. If $f(n)$ does satisfy some homogeneous difference equation, it is not necessary to write down such an equation in order to use the general procedure. All that is necessary is to write down the characteristic polynomial in factored form. This is easy. What is the factored characteristic polynomial for the following f's ?

    a) 2^n b) $n2^n$
    c) $2^n + 3^n$ d) $2^n + 1$
    e) $n^2 + 3n - 2$ f) $n^3 + (-1)^n n^2$

12. Solve [23, Section 5.3] (consumer loans) by the methods of this section.

13. Using the methods of this section, you can now find closed-form formulas for all sorts of sums, using the recursive formulation of sums in Section 5.2. Find formulas for the following sums.

    a) $\sum_{k=1}^n k$ b) $\sum_{k=1}^n k^3$
    c) $\sum_{k=1}^n k2^k$ d) $\sum_{k=1}^n ka^k, \quad a \neq 1$
    e) $\sum_{k=1}^n k^2 2^k$ f) $\sum_{k=1}^n (-1)^k k^2$
    g) $\sum_{k=1}^n k^2/2^k$ h) $\sum_{\substack{0 \le k \le n \\ k \text{ even}}} k^2 2^k$

14. For what functions $f(k)$ can this section be used to find closed form formulas for $\sum_{k=0}^n f(k)$?

15. Suppose that you follow the general procedure but forget to delete all of $D(n)$ before plugging into the original equation in step 6. What will happen?

16. Explain how to modify the general procedure to cover the homogeneous case too.

17. Under what circumstances will the proper guess for a particular solution to a nonhomogeneous difference equation simply be a multiple of the nonhomogeneous part? (This pleasant situation arose in Example 2 but not in Example 3.)

18. If $f(n)$ is of a form to which the general procedure does not apply (see [10]), that doesn't mean you have to give up. Theorem 1 still applies. So if by hook or by crook you can find one solution to your nonhomogeneous difference equation, you automatically get them all. For instance, consider

$$a_n = 2a_{n-1} + \frac{1}{n}. \tag{9}$$

    By playing around—setting $a_0 = 0$ and using the inductive paradigm to compute terms and look for a pattern—you might soon conjecture that one solution is

$$a_n = \sum_{k=1}^{n} \frac{2^{n-k}}{k}.$$

    Prove this and then find all solutions to Eq. (9).

19. Find all solutions to $b_n = 3b_{n-1} + \log n$. See [18].

20. Suppose $\{f_n\}$ satisfies

$$f_n - 3f_{n-1} = 0. \tag{10}$$

(Of course, f_n must be a multiple of 3^n, but thinking about a closed form would simply obscure the point in what follows.) Suppose further that $\{a_n\}$ satisfies

$$a_n - 2a_{n-1} - 3a_{n-2} = f_n. \tag{11}$$

Substitute Eq. (11) into Eq. (10) to obtain a *homogeneous* equation for $\{a_n\}$. Without regrouping this equation in standard form, determine its characteristic polynomial. How is it related to the polynomial for Eq. (10) and the polynomial for the homogeneous difference equation associated with Eq. (11)? What part of the general procedure does this example help explain?

21. Except for the nonhomogeneous part, Eq. (4) is the difference equation for the rabbit problem [9, Section 5.2]. Make up a plausible story about rabbits that fits Eq. (4), including the nonhomogeneous part.

5.8 The General First-Order Linear Difference Equation

In this section we derive a formula for the general term of any sequence $a_0, a_1, \ldots$ satisfying a recursion of the form

$$a_n = c_n a_{n-1} + d_n, \quad n \geq 1. \tag{1}$$

Here c_n and d_n can both vary with n. Heretofore, only d_n was allowed to vary (as the nonhomogeneous part in Section 5.7.) In other words, we generalize to nonconstant coefficients but we restrict to first-order—there is no general formula for variable coefficient, kth-order linear difference equations. Many recursions fit into the general first-order mold—many that we've solved by other means and many that we couldn't. Thus the formula we will obtain has considerable unifying value.

Our plan of attack is this: First we give some examples of recursions that fit this mold, some old, some new. Then we derive the general solution formula. Finally we apply the formula to the same examples.

EXAMPLE 1 If $c_n = 1$ and $d_n = d$ for all $n \geq 1$, then

$$a_n = a_{n-1} + d.$$

That is, we are talking about the general arithmetic sequence.

EXAMPLE 2 If $c_n = c$ and $d_n = 0$ for all $n \geq 1$, what sequence are we talking about?

EXAMPLE 3 Let $c_n = 1$ for all $n \geq 1$, let $a_0 = 0$, and let d_n be arbitrary. The resulting first-order equation is simply the recursive way to define

$$a_n = \sum_{k=1}^{n} d_k.$$

For instance, if $d_n = n$, then a_n is the sum of the first n integers. If $d_n = n^2$, then a_n is the sum of the first n perfect squares.

EXAMPLE 4 What choices of c_n, d_n, and a_0 yield the recursive way to define the general sequence of **partial products**

$$a_n = \prod_{i=1}^{n} p_i\,?$$

EXAMPLE 5 Let $c_n = c$ and $d_n = d$. This gives us the general arithmetic-geometric sequence. For instance, if $c = 2$ and $d = 1$, we get the TOH recursion.

Example 4 is one we haven't dealt with before. Another is

$$a_n = na_{n-1} + n!, \qquad n \geq 1; \qquad a_0 = 1. \tag{2}$$

The nonhomogeneous part cannot be handled by the general procedure in Section 5.7, and neither can the linear part because the coefficient of a_{n-1} is not constant.

To find the promised general formula, let's apply the inductive paradigm to Eq. (1): Expand the first few terms of $\{a_n\}$ and look for a pattern.

$$\begin{aligned}
a_0 &= a_0,\\
a_1 &= c_1a_0 + d_1,\\
a_2 &= c_2a_1 + d_2 = c_2c_1a_0 + c_2d_1 + d_2,\\
a_3 &= c_3a_2 + d_3 = c_3(c_2c_1a_0 + c_2d_1 + d_2) + d_3\\
&\qquad\qquad\quad = c_3c_2c_1a_0 + c_3c_2d_1 + c_3d_2 + d_3.
\end{aligned}$$

Do you see a pattern in all this? Look at the subscripts in each term. Using $\sum$ and $\prod$ notation, we see the pattern

$$\begin{aligned}
a_n &= \Big(\prod_{i=1}^{n} c_i\Big)a_0 + \Big(\prod_{i=2}^{n} c_i\Big)d_1 + \cdots + \Big(\prod_{i=n}^{n} c_i\Big)d_{n-1} + d_n\\
&= \Big(\prod_{i=1}^{n} c_i\Big)a_0 + \sum_{j=1}^{n}\Big(\prod_{i=j+1}^{n} c_i\Big)d_j\,.
\end{aligned}$$

Note that when $j = n$, the product $\prod_{i=j+1}^{n} c_i$ is the empty product, which we long ago declared equal to 1. Thus the last term on the second line is d_n, just as it is on the first.

The first product in the last line of the preceding display is just crying out to be grouped along with the others. If we define

$$d_0 = a_0$$

(which we can do because we have not previously used d_0), we get

$$a_n = \sum_{j=0}^{n}\Big(\prod_{i=j+1}^{n} c_i\Big)d_j\,. \tag{3}$$

Clearly Eq. (3) is a notational triumph, but does it really tell us anything? Does it help us do computations, or does it just look pretty? When we continue with our examples, you'll find out. But first, since our fast footwork with $\sum$ and $\prod$ may have thrown you for a loop, let's clarify some things by proving the formula.

Theorem 1. If the sequence $\{a_n\}$ satisfies

$$a_n = c_n a_{n-1} + d_n, \quad n \geq 1; \qquad a_0 = d_0,$$

then a_n satisfies Eq. (3).

PROOF Use induction on n. For $n = 0$, Eq. (3) becomes

$$a_0 = \sum_{j=0}^{0} \Bigl(\prod_{i=j+1}^{0} c_i \Bigr) d_j.$$

Now, a sum from $j = 0$ to 0 is just a single term with $j = 0$, so the above is just

$$\Bigl(\prod_{i=1}^{0} c_i \Bigr) d_0 = d_0.$$

(An empty product = 1.) Thus the basis case $n = 0$ is true.

Now we prove the formula for $n + 1$, assuming it is correct for n.

$$\begin{aligned}
a_{n+1} &= c_{n+1} a_n + d_{n+1} \\
&= c_{n+1} \Bigl[\sum_{j=0}^{n} \Bigl(\prod_{i=j+1}^{n} c_i \Bigr) d_j \Bigr] + d_{n+1} \\
&= \Bigl[\sum_{j=0}^{n} \Bigl(\prod_{i=j+1}^{\boxed{n+1}} c_i \Bigr) d_j \Bigr] + (1) d_{n+1} \\
&= \sum_{j=0}^{n} \Bigl(\prod_{i=j+1}^{n+1} c_i \Bigr) d_j + \Bigl(\prod_{i=n+2}^{n+1} c_i \Bigr) d_{n+1} \\
&= \sum_{j=0}^{\boxed{n+1}} \Bigl(\prod_{i=j+1}^{n+1} c_i \Bigr) d_j. \blacksquare
\end{aligned}$$

It is very important in this proof to note where the bounds on $\sum$ and $\prod$ change. We have boxed the key places for your attention.

We now apply Theorem 1 to all the examples in this section and see if we obtain anything new.

EXAMPLE 1 Continued **Arithmetic sequence,** $a_n = a_{n-1} + d$

For $n \geq 1$, substitute 1 for c_n and d for d_n in Eq. (3). Recall that $d_0 = a_0$ in every use of Eq. (3) by definition. We find that the inside product on the right in Eq. (3) is always 1 and thus

$$a_n = \sum_{j=0}^{n} d_j = a_0 + \sum_{j=1}^{n} d = a_0 + nd, \tag{4}$$

just as it should be. ∎

EXAMPLE 2 Continued

Geometric sequence, $a_n = ca_{n-1}$
Substituting c for c_n, and 0 for each d_n with $n \geq 1$, we find that every term of the sum in Eq. (3) is 0, except the term containing $d_0 = a_0$. Thus

$$a_n = \Big(\prod_{i=0+1}^{n} c\Big)a_0 = a_0 c^n,$$

again, just as it should be. ∎

EXAMPLE 3 Continued

General series, $a_n = a_{n-1} + d_n,\ a_0 = 0$
The calculation is similar to that in Eq. (4). However, we don't get the rightmost equality as the d_n's are not all the same. We obtain

$$a_n = a_0 + \sum_{j=1}^{n} d_j = \sum_{j=1}^{n} d_j, \tag{5}$$

because we defined $a_0 = 0$. Sure enough, Eq. (5) describes the sequence of partial sums of the series $\sum d_j$, but is it a *formula* in the sense we had hoped for? Unfortunately it isn't; it's just the Sigma notation statement of the definition of a general series. It doesn't automatically tell us, for instance, what $\sum k$ or $\sum k^2$ is. Here is one case, then, where Theorem 1 isn't all that helpful. In general, formulas containing lots of Sigmas and/or Pi's, such as Eq. (3), give useful information only if the terms have enough structure so that we already know how the sums and products can be rewritten without the $\sum$ and $\prod$ signs. ∎

EXAMPLE 4 Continued

Partial product $\prod_{i=1}^{n} p_i$
The right choice for the c's and d's are:

$$c_n = p_n, \quad \text{and} \quad d_n = 0, \quad \text{for } n \geq 1; \qquad d_0 = a_0 = 1.$$

Sure enough, plugging into Theorem 1 gives

$$a_n = \prod_{j=1}^{n} p_j,$$

but no more. ∎

EXAMPLE 5 Continued

Arithmetic-geometric sequence, $a_n = ca_{n-1} + d$
In Eq. (3) set $d_j = d$ for $n > 0$ and $d_0 = a_0$. Then

$$a_n = \sum_{j=0}^{n}\Big(\prod_{i=j+1}^{n} c\Big)d_j = \Big(\prod_{i=1}^{n} c\Big)a_0 + \sum_{j=1}^{n}\Big(\prod_{i=j+1}^{n} c\Big)d$$
$$= a_0c^n + \sum_{j=1}^{n} c^{n-j}d = a_0c^n + d\Big(\sum_{k=0}^{n-1} c^k\Big).$$

Using our knowledge of geometric series on the parenthesized expression and assuming $c \neq 1$ (the case $c = 1$ is handled by Example 2), we obtain

$$a_n = a_0c^n + d\Big(\frac{c^n - 1}{c-1}\Big) = c^n\Big(a_0 + \frac{d}{c-1}\Big) - \frac{d}{c-1}. \tag{6}$$

Equation (6) is a new formula. We could also obtain it by the method of Section 5.7; indeed, if you thoroughly understood that section, it should have been obvious when we started Example 5 that $a_n = ca_{n-1} + d$ must have its solution in the form $Ac^n + B$. However, that method requires solving for A and B as a second stage. Here their values fall out as specific expressions involving a_0. Thus Theorem 1 can do some good. ∎

EXAMPLE 6 Solve Eq. (2), which we noted is not of a form handled by any previous systematic methods.

Solution Setting $c_n = n$ and $d_n = n!$ in Theorem 1, we get

$$a_n = \sum_{j=0}^{n}\Big(\prod_{i=j+1}^{n} i\Big)j! = \sum_{j=0}^{n}\Big(\frac{n!}{j!}\Big)j!$$
$$= \sum_{j=0}^{n} n! = (n+1)n! \qquad [n! \text{ is a constant relative to } j]$$
$$= (n+1)!.$$

Again we see that Theorem 1 can do some good. ∎

Problems: Section 5.8

Solve [1–10] by the method of this section, even though the first few are easier to do by other means.

1. $a_n = a_{n-1} + 1, \quad a_0 = 0$
2. $b_n = 2b_{n-1} + 2^n, \quad b_0 = 1$
3. $c_n = 3c_{n-1}, \quad c_0 = 1$
4. $d_n = nd_{n-1}, \quad d_0 = 1$
5. $e_n = ne_{n-1} + d_n$, where $e_0 = d_1 = d_2 = 1$ and all other $d_k = 0$.
6. $f_n = [(n-1)/n]\, f_{n-1}, \quad f_0 = 1$
7. $g_n = [(n+1)/n]\, g_{n-1}, \quad g_0 = 1$
8. $h_n = nh_{n-1} - n!, \quad h_0 = 1$
9. $m_n = 2nm_{n-1} + n!2^n, \quad m_0 = 1$
10. Solve $a_n = c_na_{n-1} + d_n$ if
    a) $c_n = c^n, d_n = 0, a_0 = 1$
    b) $c_n = c^n, d_n = a_0 = 1$
11. Solve for the average time of sequential search (Eq. (3) of Section 5.2) using the method of this section. Note that $E_0 = 0$.
12. Even though it isn't of any type you have solved systematically before this section, you ought to be

able to solve Eq. (2) without the theory of this section. Do it.

13. Solve the TOH recursion by the method of this section.

14. Solve the Straightline TOH recursion by the method of this section. See [41, Section 2.2] for the definition of this version of TOH.

15. Obtain Eq. (6) by the theory of Section 5.7.

16. Solve

$$S_1 = 1;$$
$$S_n = \frac{n-1}{n} S_{n-1} - \frac{1}{n^2(n-1)}, \qquad n \geq 2.$$

17. Solve $t_n = \frac{1}{n} t_{n-1} + \frac{1}{n!}, \quad t_0 = 1$

18. Solve $r_n = [(n-1)/n]\, r_{n-1} + d_n$ if

a) $r_0 = d_1 = 1$, all other $d_k = 0$

b) $r_0 = 1,\ d_1 = 2,\ d_2 = 3$, all other $d_k = 0$

c) $r_0 = 1,\ d_n = n + 1$ for all n

19. Here is a trick for solving $a_n = ca_{n-1} + d$ without using the theory of this section *or* the last. Namely, there is a constant k (you find it), so that, after adding k to both sides and rewriting the RHS, the given equation becomes

$$a_n + k = c(a_{n-1} + k).$$

Now define $b_n = a_n + k$ and rewrite the recursion in terms of b's. A closed-form formula for b_n in terms of b_0 should now be clear. Work backwards to a formula for a_n. Check that the formula is equivalent to the formula obtained in Example 5 Continued.

20. In [20, Section 5.2] we asked you to come up with a difference equation for both an IRA and an ordinary interest-bearing account if you deposit principal P each year. (For the IRA, it was assumed you could defer taxes on interest only.) At that point you couldn't solve the equations to obtain explicit formulas for the final values of the accounts, but now you can.

a) Do so.

b) Using your formulas in (a), find the numerical value of the amounts at withdrawal in both the ordinary and IRA accounts if $r = .1$, $t = .25$, $P = \$2000$, and $N = 30$.

21. Redo [20] under the assumption that you are allowed to defer all taxes on principal as well as interest.

22. Find and simplify a closed-form formula for a_n, if $a_n = ra_{n-1} + s^n$ and $a_0 = 1$, where r and s are constants and

a) $r \neq s$ **b)** $r = s$

23. Solve the following problems from Section 5.3 by the method of the current section.

a) [23] **b)** [24]

24. The nth partial sum of the general geometric series, $\sum_{k=0}^{n} ar^k$, is also the nth *term* of a certain arithmetic-geometric *sequence*. Describe that sequence recursively, find a closed form for the nth term using the methods of this section, and verify that this closed form is equivalent to the usual formula for the sum of a geometric series.

5.9 Applications to Algorithms

When you want to determine how many steps an algorithm will take, difference equations occur naturally. They are especially natural if the algorithm is recursive.

We warm up with two simple, iterative algorithms for evaluating polynomials. Then we move on to several recursive algorithms. One of the points we will demonstrate is that a recursive algorithm, though easy to understand, is often *not* the most efficient way to do a computation. Often it is horrendously inefficient. We will demonstrate this with the Fibonacci numbers.

Finally, we give several examples of a special class of recursive algorithms that are often surprisingly efficient—**Divide and Conquer** algorithms. The recursive counting method that goes with Divide and Conquer is also special, and we give several examples.

OK, let's evaluate polynomials. The first algorithm, STRAIGHTFORWARD, needs no introduction because it corresponds exactly to the way most people think to evaluate a polynomial: You compute each term in order and add them up.

Algorithm 5.1
Straightforward

Input $a_0, a_1, \ldots, a_n, x$
Algorithm STRAIGHTFORWARD
$S \leftarrow 0$
for $k = 0$ to n
$S \leftarrow S + a_k x^k$
endfor
Output $S = \sum_{k=0}^{n} a_k x^k$

The next algorithm needs some explanation. Think recursively. Suppose we already know how to evaluate polynomials of degree $n-1$ and now wish to evaluate

$$P(x) = a_0 + a_1 x + \cdots + a_{n-1}x^{n-1} + a_n x^n. \tag{1}$$

The obvious way is

$$\Bigl(a_0 + a_1 x + \cdots + a_{n-1}x^{n-1}\Bigr) + a_n x^n.$$

That is, first evaluate the $(n-1)$st degree polynomial in parentheses and then add $a_n x^n$. This method of evaluation is easy to implement iteratively, and it gives us Algorithm STRAIGHTFORWARD. But it's also possible to "grow $P(x)$ from the right". Namely, consider the $(n-1)$st degree polynomial that uses the rightmost n coefficients from Eq. (1):

$$a_1 + a_2 x + a_3 x^2 + \cdots + a_{n-1}x^{n-2} + a_n x^{n-1}.$$

(Note that the subscripts on a and the powers on x differ by 1.) Then

$$P(x) = a_0 + \Bigl(a_1 + a_2 x + \cdots + a_{n-1}x^{n-2} + a_n x^{n-1}\Bigr)x.$$

(Note the x at the end.) This method of evaluation can be implemented iteratively, too, and is known as **Horner's Algorithm.***

---

* William Horner, 1786–1837, was an English mathematician and school headmaster. He used this algorithm as part of "Horner's Method", a now obsolete paper-and-pencil procedure for computing roots of polynomials to several decimal places without going completely mad. Horner's *algorithm*, however, is far from obsolete, as you are about to see. The traditional hardcopy format can be found in most elementary algebra books under the name "synthetic division" or "synthetic substitution".

Algorithm 5.2

Horner

Input and **Output** same as that for STRAIGHTFORWARD

Algorithm HORNER

```
    S ← a_n
    for k = n − 1 downto 0
        S ← Sx + a_k
    endfor
```

EXAMPLE 1 Compare the efficiency of Algorithms STRAIGHTFORWARD and HORNER by counting the number of multiplications.

Solution It's easy to eyeball the answers here, but let's practice thinking systematically. Let M_k be the number of multiplications in STRAIGHTFORWARD after k passes through the loop, and let M_k^* be the corresponding number for HORNER. In STRAIGHTFORWARD the kth pass through the loop requires k multiplications (except for $k = 0$, where the computer takes one multiplication to figure out a_0 as $a_0 \cdot 1$). In HORNER each pass through the loop requires just one multiplication. Therefore we have the difference equations

$$M_k = M_{k-1} + k, \quad k \geq 1; \qquad M_0 = 1;$$
$$M_k^* = M_{k-1}^* + 1, \quad k \geq 1; \qquad M_0^* = 0.$$

Since both algorithms pass through their loops n times, we want M_n and M_n^*. It's easy to see from what we just said that

$$M_n^* = n \qquad \text{and} \qquad M_n = 1 + \sum_{i=1}^{n} i = 1 + \frac{n(n+1)}{2}.$$

However, if for some reason we didn't see these closed forms right away, we could obtain them by the methods for nonhomogeneous linear difference equations. In any event, note how much more efficient HORNER is: It's an *order n* algorithm (i.e., linear) while STRAIGHTFORWARD is order n^2 (quadratic). In fact, it is known that HORNER is best possible in terms of the number of multiplications. ∎

EXAMPLE 2 Consider the following two, slightly different recursive procedures for Towers of Hanoi. Call them Procedures TOH and TOH*. (We omit the rest of the algorithms in which these procedures reside.) Is one more efficient than the other?

```
procedure TOH(n, i, j)                [Move n rings from pole i to pole j]
    if n > 1 then TOH(n−1, i, 6−i−j) endif
    move Ring n from Pole i to Pole j
    if n > 1 then TOH(n−1, 6−i−j, j) endif
    return
endpro
```

```
procedure TOH*(n, i, j)
    if n = 1 then move Ring n from Pole i to Pole j
             else
                TOH*(n−1, i, 6−i−j)
                move Ring n from Pole i to Pole j
                TOH*(n−1, 6−i−j, j)
    endif
    return
endpro
```

Both procedures are correct (for $n \geq 1$ rings); indeed, both perform exactly the same minimal sequence of moves. This result is easy to show by induction [7]. The question is whether one requires more peripheral activity than the other. The first has fewer lines. Does that make it better?

Solution Let us use as our measure of efficiency the number of times n is compared to 1. In Procedure TOH, these comparisons take the form "if $n > 1 \ldots$", and in TOH*, the form "if $n = 1 \ldots$". This count is a reasonable measure because, other than invoking itself and moving rings, these comparisons are all that either procedure does. Let C_n be the number of comparisons made by TOH(n, i, j), including comparisons in subcalls. Define C_n^* analogously. Then we have

$$\begin{aligned} C_1 &= 2 \qquad \text{and} \qquad C_n = 2 + 2C_{n-1} \quad \text{for } n > 1, \\ C_1^* &= 1 \qquad \text{and} \qquad C_n^* = 1 + 2C_{n-1}^* \quad \text{for } n > 1. \end{aligned}$$

For instance, the first of these lines is correct because Procedure TOH has two explicit comparisons of n and 1, and for $n > 1$ it also results in two calls of TOH with $n{-}1$ rings.

From our knowledge of nonhomogeneous recursions, we conclude that

$$C_n = A2^n + B, \qquad C_n^* = A^*2^n + B^*,$$

where A, B, A^*, B^* are constants to be determined. Substituting into the recursions to determine B and B^*, and substituting the initial values to determine A and A^*, we find that

$$C_n = 2 \cdot 2^n - 2, \qquad C_n^* = 2^n - 1.$$

In short, Procedure TOH makes twice as many comparisons, which is significant because the number of comparisons grows exponentially. ∎

EXAMPLE 3 Analyze Algorithms FIBA and FIBB, which find the nth Fibonacci number, F_n.

Solution Let us use as our measure the number of assignments of values. Let A_n be the number of assignments if Algorithm FIBA is run, and B_n the number for FIBB. In both algorithms there are two initial cases:

$$A_0 = A_1 = 2, \qquad B_0 = B_1 = 1.$$

So FIBB is more efficient for the initial cases. For $n > 1$, in Algorithm FIBA we

Algorithms 5.3 and 5.4

FibA and FibB

```
Input n
Algorithm FibA
    F_0 ← 1;  F_1 ← 1
    for k = 2 to n
        F_k ← F_{k-1} + F_{k-2}
    endfor
Output F_n
```

```
Input and Output same
Algorithm FibB
    Function F(k)
        if k = 0 or 1 then F ← 1
                      else F ← F(k-1) + F(k-2)
        endif
        return F
    endfunc
    F_n ← F(n)
```

simply make one more assignment for each pass through the loop. Thus

$$A_n = A_{n-1} + 1, \qquad n > 1.$$

It follows that

$$A_n = n + 1, \qquad n > 1.$$

In Algorithm FibB, when $n > 1$ we make the one assignment on the **else** line *plus* all assignments incurred by calling the function for $n-1$ and $n-2$. That is,

$$B_n = 1 + B_{n-1} + B_{n-2}, \qquad n > 1. \tag{2}$$

Thus the recursion for $\{B_n\}$ is identical to that for $\{F_n\}$, the Fibonacci numbers themselves, *except* that the B-recursion has an additional, nonhomogeneous constant 1.

We can solve Eq. (2) by our standard methods, but it's a mess—$\sqrt{5}$ appears many times, even more than in solving the ordinary Fibonacci recursion (Example 4, Section 5.5). Fortunately, all we need is qualitative information: How do B_n and $A_n = n+1$ compare? We answer this question by showing that $B_n \geq F_n$, which we know is approximately $(1.618)^n$ and thus much larger than $n+1$.

Why is $B_n \geq F_n$? Because $B_0 = F_0$, $B_1 = F_1$, and the B-recursion has an additional positive term (the 1) on its right-hand side. It follows by induction [12] that $B_n > F_n$ for $n > 1$. (To obtain a formula for B_n without all the messy algebra alluded to, see [13,14].)

Conclusion: Algorithm FIBB is lousy compared to FIBA. In fact, it's lousy period. It would take fewer steps if we simply started at 0 and added 1 over and over until we reached F_n that way! ∎

Why is this recursion so slow? Because the algorithm repeats itself flagrantly. For instance, suppose we use Algorithm FIBB to compute $F(10)$. It first must compute $F(8)$ and $F(9)$. In computing $F(8)$ it will first compute $F(7)$. Later, in computing $F(9)$, it will compute $F(7)$ and $F(8)$ again, both *from scratch.* The smaller k is, the more times FIBB recomputes $F(k)$.

In general, if a recursive function or procedure calls on itself more than once in its definition, the total amount of work to run it will be exponential. Sometimes this can't be avoided—every algorithm for Towers of Hanoi takes at least $2^n - 1$ moves. But often it can be avoided: Don't use recursion (or, if you are a knowledgeable programmer, use recursion but save the results of calls and use flags to avoid making the same call twice).

All this is too bad. Recursive algorithms are relatively easy to prove correct and to analyze for the number of steps. Iterative algorithms are not always so easy to analyze as Algorithm FIBA. So you might think of finding a recursive algorithm for a problem as a first step. A recursive algorithm, though often inefficient, proves that a problem can be constructively solved and thus gives us the fortitude to search further for a more efficient algorithm.

Divide and Conquer

The Divide and Conquer approach is a special case of the recursive paradigm. Instead of supposing that you already know how to do the immediately preceding case, or all preceding cases, you assume that you know how to do the cases approximately half the size of the current case. Recursive algorithms built on this assumption often run much faster than more straightforward algorithms. In this subsection we give several examples of the Divide and Conquer approach and show you how to analyze their run time.

First, here are some sketches of Divide and Conquer solutions.

EXAMPLE 4 Find if a word w is on an alphabetical list of n words.

Solution We already gave a Divide and Conquer solution to this problem with Algorithm BINSEARCH of Section 1.6. Recall the idea: Check if w is the middle word. If not, and it comes earlier in the dictionary, search for it on the first half of the list. (We assume you already know how to do this.) If w is later in the dictionary, search for it on the second half. Also recall the analysis: Whereas the more obvious sequential search took n comparisons in the worst case, BINSEARCH at worst took about $\log n$ comparisons, a very substantial saving for large n. (We gave an intuitive argument for $\log n$ in Section 1.6 and then proved an exact formula by induction in 2.6.) ∎

EXAMPLE 5 Find the maximum of n numbers.

Solution Our previous solution, Algorithm MAXNUMBER of Section 1.6, was iterative, but a Divide and Conquer solution is equally easy to devise. Suppose we already know how to find the maximum of a list of $n/2$ numbers. We then divide the list of n in half, find the maximum of each recursively, and compare the two winners. To be more precise, that's what we do if n is even. If n is odd, we divide the list into lists of sizes $\lfloor n/2 \rfloor$ and $\lceil n/2 \rceil$. This illustrates why we used the phrase "approximately half the size" in the first paragraph of this subsection. ∎

EXAMPLE 6 Given a real number x and a positive integer n, compute x^n.

Solution The obvious approach is to start with x and keep multiplying by x a total of $n-1$ times; e.g., $x^3 = x \cdot x \cdot x$, which involves two multiplications. But think recursively. If n is even, find $x^{n/2}$ and do just one more multiplication to square it. If n is odd, find $y = x^{\lfloor n/2 \rfloor}$ and then do just two more multiplications to compute $x^n = y^2 x$. ∎

EXAMPLE 7 Arrange n numbers in increasing order.

Solution This **sorting problem**, a classic of computer science, will be discussed at some length later in the book. But it's not hard now to outline a Divide and Conquer approach to the solution. Divide the list in half and sort each half by recursion. Combine half lists, maintaining the increasing order. The smallest number of all is the smaller of the smallest numbers in the two half lists. Having picked this number off, the next smallest of all is the smaller of the smallest numbers *remaining* in the half lists. Think of the half lists as numbers held in chutes (larger numbers higher up) above a narrow "bucket" (See Figure 5.12). One at a time, the bottom numbers in the chutes are compared and the smaller one drops into the bucket. Numbers in the bucket are always in increasing order, so when all have dropped into the bucket the algorithm is finished. This method is known as **Merge Sort**. We show later in this section that this method takes around $n \log n$ comparisons. The methods many people learn first (e.g., Insertion Sort or Bubble Sort; see Section 7.7 and the Epilogue) take n^2. ∎

Now let's write down the difference equations for these Divide and Conquer methods. As we argued in Section 2.6, the recursion for the worst case of BINSEARCH is

$$A_n = 1 + \max\{A_{\lfloor \frac{n-1}{2} \rfloor}, A_{\lceil \frac{n-1}{2} \rceil}\}, \qquad A_1 = 1. \tag{3}$$

The argument was: You compare the search word w with the middle word in any case (that gives the 1); in the worst case, w is on whichever side is itself worse.

As for Example 5, finding the maximum takes just one more comparison after solving the half problems, but unlike for binary search both half problems must be solved. Also, the case where $n = 1$ takes no work. So the recursion is

$$B_n = 1 + B_{\lfloor \frac{n}{2} \rfloor} + B_{\lceil \frac{n}{2} \rceil}, \qquad B_1 = 0. \tag{4}$$

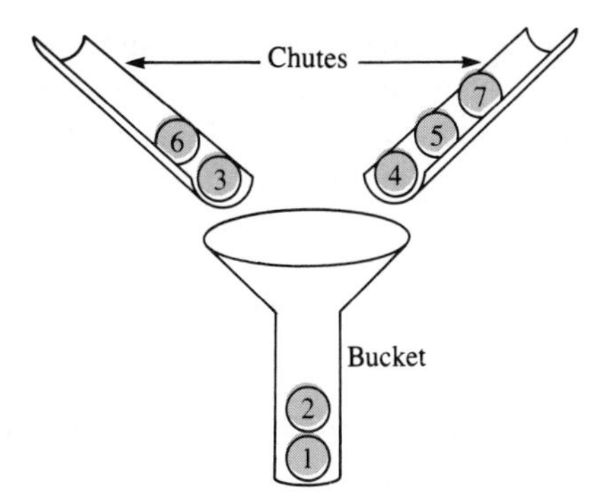

FIGURE 5.12

The merge phase of Merge Sort

Turning to the computation of x^n, both the size of the subproblem and the number of multiplications after the subproblem seem to depend on whether n is odd or even. Actually, there is a single expression for the exponent of the subproblem, namely, $\lfloor n/2 \rfloor$. So we may write the recursion this way:

$$C_n = C_{\lfloor n/2 \rfloor} + t, \qquad \text{where } t = \begin{cases} 1 & \text{if } n \text{ even} \\ 2 & \text{if } n \text{ odd} \end{cases} \quad \text{and} \quad C_1 = 0. \tag{5}$$

Finally, for Merge Sort, the two subproblems are as close to $n/2$ as possible and the number of comparisons in the merge phase is at most $n-1$. (Why? How could it be less? [35]) Therefore it suffices to solve the recursion

$$D_n = D_{\lfloor n/2 \rfloor} + D_{\lceil n/2 \rceil} + n-1, \qquad D_1 = 0. \tag{6}$$

Fine and dandy. We have recursions. Unfortunately, they are very elaborate—full of floors and ceilings and tiny little symbols you can hardly read. More to the point, they don't seem to be the sort that you have learned how to solve in this chapter. The only one we have already solved (binary search, in Section 2.6) required guessing the solution and then performing a technically tricky induction.

Fortunately, there is a simple procedure that converts all these recursions to sorts that we do know how to solve. We introduce it with an example.

EXAMPLE 8 Find a solution to Eq. (5) for computing x^n.

Solution We need to simplify the recursion. Before treating the main culprits—the floors and ceilings—let's see if anything else is unwieldy. Yes, t varies between 1 and 2. Let's replace it with its greatest value, 2. That will cause our answer to err on the high side, but that's okay. In analysis of algorithms, erring on the low side is the thing to avoid. Erring on the high side will only cause things to be better than you predict but erring on the low side could cause disasters like missile failures or nuclear plant explosions.

As for those floors and ceilings, we would like them to go away. If n is divisible by 2, they do go away, and we get

$$C_n = C_{n/2} + 2, \qquad C_1 = 0. \tag{7}$$

Now, here's the main idea. Suppose we could find a subsequence of indices $\{n_k\}$ such that for each k, $n_{k-1} = n_k/2$ exactly. Then setting $c_k = C_{n_k}$, Eq. (7) would reduce immediately to a first-order difference equation in c_k. But of course we do know such a subsequence: $n_k = 2^k$. Therefore Eq. (7) becomes

$$c_k = c_{k-1} + 2, \qquad c_0 = 0.$$

This is so simple we can solve it by sight:

$$c_k = 2k.$$

Finally, substitute C's and n's back in. Since $n = 2^k$, then $k = \log n$ (all our logs are base 2) and

$$C_{2^k} = C_n = 2\log n\,. \quad ∎$$

Warning. That's the solution method but we don't claim the answer is exactly correct, even as an upper bound. The argument presented shows only that $2\log n$ is an upper bound when n is a power of 2. In fact, whenever an answer involves logs, it *can't* be exactly correct for other n because then it's not even an integer!

Nonetheless, it is a pleasant fact that the answer obtained by this method generally *is* correct, either in the sense that it is a valid upper bound or in the sense that its order of magnitude is correct. But one needs to do a little more work to show this. There are two main ways. One is to regard the answer you obtain as a conjecture and then prove it correct for all n by strong induction. The other is to argue that it is correct for powers of 2 and then show that it is therefore off by at most a multiplicative factor for other n. We illustrate both methods below, the first in Example 9 and the second in Example 10.

EXAMPLE 9 Verify that the solution to Example 8 is a valid upper bound. That is, if C_n is the number of multiplications needed to compute x^n by the Divide and Conquer method of Example 6, then verify that $C_n \le 2\log n$ for *all* positive integers n.

Solution By strong induction, with $P(n)$ the statement $C_n \le 2\log n$. The basis case is easy: We know that $C_1 = 0$, and sure enough, $2\log 1 = 0$, too. As for the inductive step, since we are assuming the conclusion for all $i < n$, assume it in particular for $\lfloor n/2 \rfloor$. Then by Eq. (5)

$$\begin{aligned} C_n &\le 2\log\lfloor n/2\rfloor + 2 = 2(\log\lfloor n/2\rfloor + 1) \\ &\le 2[\log(n/2) + 1] && [\,\lfloor n/2\rfloor \le n/2] \\ &= 2\log n. \quad ∎ && [1 = \log 2] \end{aligned}$$

With all the attention given to the solution method, you may have lost track of the bottom line: Whereas the straightforward method for finding x^n takes $n-1$ multiplications, Divide and Conquer takes at most $2\log n$.

EXAMPLE 10 Solve the Merge Sort recursion, Eq. (6), for n a power of 2, and discuss the validity of your answer for general n.

Solution First, replace $n-1$ in Eq. (6) by n. This means at best you're aiming for an upper bound, but that was true anyway: $n-1$ was a worst case for the merge phase already. Now let $2^k = n$, let $d_k = D_{2^k}$, and rewrite the difference equation in terms of k:

$$d_k = 2d_{k-1} + 2^k, \qquad d_0 = 0. \tag{8}$$

This one is hard to eyeball, but the step-by-step procedure of Section 5.7 applies. The right guess for the initial particular solution is $Ak2^k$ (not $A2^k$, which is a solution of the associated homogeneous equation). Plugging this guess into Eq. (8) we find that $A = 1$. So the general solution is $B2^k + k2^k$. Using the initial condition, we find that $B = 0$. Thus the particular solution to Eq. (8) is simply $k2^k$. Substituting n back in, we find that $D_n = (\log n)n$. Finally, recalling that at best we have an inequality, and writing the right-hand side in the more usual order, we assert that

$$D_n \leq n \log n. \tag{9}$$

The argument so far proves inequality (9) for n a power of 2, because the translation of Eq. (6) to Eq. (8) is exact for such n (except for increasing $n-1$ to n, which causes the inequality). Is inequality (9) correct for all n? It turns out it is, but we don't know any easy proof (try induction and see where you bog down). We don't stop to give a proof because it's easy to show in any event that $n \log n$ is off by at most a small constant factor. The exact solution D_n to Eq. (6) is clearly an increasing sequence (do an induction!), so if $2^{k-1} < n < 2^k$, the largest that D_n could be is $2^k \log 2^k$, the right-hand side of inequality (9) when $n = 2^k$. How far off are we if we assume that $D_n \leq n \log n$ for all n? Let's first answer for a specific n. Consider $n = 2^5 + 1$. Our assumption tells us to assert $D_n \leq 33 \log 33$. The best bound we have proved says that D_n could be as large as $64 \log 64$, but that is bigger than $33 \log 33$ by about 2. In general, at worst we would be off by a factor of

$$\frac{2^k \log 2^k}{n \log n} > \frac{2^k \log 2^k}{2^{k-1} \log 2^{k-1}} = 2\frac{k}{k-1},$$

which is little more than 2 for any n large enough to be of interest. So we say that Merge Sort is (at most) an $n \log n$ algorithm. ∎

One final point. We have used n of the form 2^k because that makes all the floors and ceilings go away. More precisely, it makes them go away if they are all of the form $\lfloor n/2 \rfloor$ and/or $\lceil n/2 \rceil$; both these expressions become $n/2 = 2^{k-1}$. However, if you have something like $\lfloor (n-1)/2 \rfloor$, as in Eq. (3), then substituting 2^k does not give you exactly $n/2$. In this case there are two ways to proceed. The simplest (and it's the one given in the procedure box at the end of this section) is to replace the floor-ceiling expressions by $n/2$ anyway. This imprecise simplification adds yet another reason why your solution need not be exact, but in general the method still works.

The second approach is to find a subsequence of n's other than 2^k for which the floors and ceilings you have vanish exactly. For instance, for Eq. (3) we want a subsequence $\{n_k\}$ where, for each k, n_k-1 is divisible by 2. In fact, we want $n_{k-1} = (n_k-1)/2$, for then Eq. (3) reduces exactly to a first-order recursion. Restated, we want $n_k = 2n_{k-1} + 1$. Also, we want $n_1 = 1$. (Why?) But this is just the TOH recursion! So if you use the "TOH sequence" $n_k = 2^k-1$, you can solve Eq. (3) exactly. Both approaches to Eq. (3) are pursued in the problems.

We conclude with a summary of the solution procedure for Divide and Conquer recursions.

Procedure for Solving Divide and Conquer Recursions

Assume the recursion uses the letter A, that is, it expresses A_n in terms of things like $A_{\lfloor n/2 \rfloor}$.

1. Replace all expressions for approximately half of n by exactly $n/2$. Perhaps make other simplifying changes, but only if they would increase the value of A_n.
2. Substitute 2^k for n throughout.
3. Define $a_k = A_{2^k}$ and rewrite the equation as a first-order difference equation for a_k.
4. Solve this difference equation. Substitute n back in.
5. See if your solution is valid for general n. Argue by induction or by comparing A_n with A_{2^k} where $2^{k-1} < n < 2^k$.

Problems: Section 5.9

Note: all logs in this problem set are base 2.

1. Compute the number of additions in the two polynomial evaluation algorithms, STRAIGHTFORWARD and HORNER. Incidentally, HORNER has appeared earlier in this book, without being named or analyzed. Where?

2. Count the number of assignments of variables in STRAIGHTFORWARD and HORNER. Consider each quantity in the Input to be one assignment.

3. It seems that HORNER does much better than STRAIGHTFORWARD—the former is as good as the latter on assignments and additions and does much better on multiplications. Why, then, do we use STRAIGHTFORWARD in practice? "That's what our silly teachers taught us" is not a good enough answer.

4. Prove that HORNER is correct, that is, it really does output the value of the polynomial at x.

5. Consider Algorithm 5.5, POWERPOLY, which evaluates polynomials and attempts to save work on evaluating x^k by computing lower powers first and saving the results. Compare this algorithm for efficiency with the two in Example 1.

6. Algorithms can be implemented on hand calculators as well as on computers, but the appropriate measures of efficiency are different. For hand calculators the obvious measure is the number of button pushes. For your favorite hand calculator, compute the number of button pushes for Algorithms

a) STRAIGHTFORWARD b) HORNER

Algorithm 5.5. POWERPOLY

```
Input and Output same as in Example 1
Algorithm PowerPoly
  P ← x
  S ← a_0 + a_1 x
  for k = 2 to n
    P ← Px
    S ← S + a_k P
  endfor
```

c) POWERPOLY from [5]

Since calculators differ, you must describe in words or "button pictures" how your calculator would handle these algorithms, as well as show your counting calculations. For instance, the answers will be very different for calculators that have an exponentiation key and those that don't, and for calculators that have memories and those that don't. Also, the number of pushes depends on the number of digits in your input data. For simplicity, assume each input quantity requires P button pushes each time you punch it in.

7. Prove that Procedures TOH and TOH* perform the same sequence of moves, given the same input (n, i, j).

8. Measure the relative efficiency of Procedures TOH and TOH* by the number of procedure calls needed to run them with n rings. For instance, running TOH(n, i, j) requires the initial call plus two subcalls with $n-1$ rings, plus whatever calls those subcalls make.

9. Explain intuitively why Procedure TOH requires exactly twice as many comparisons of n and 1 as Procedure TOH*.

10. Compare Algorithms FIBA and FIBB for the Fibonacci numbers on the basis of how many additions they each require.

11. Compare FIBA and FIBB on the basis of the number of additions and subtractions they perform. In FIBB, when function $F(k)$ calls $F(k-1)$, a subtraction is performed in figuring out the calling value. Similarly, subtractions are performed in FIBA to evaluate subscripts.

12. Prove that $B_n > F_n$ for all $n > 1$, where B_n is as in Eq. (2). Use induction and the recursive definitions of B_n and F_n. Do not use closed-form formulas.

13. Solve for B_n in Eq. (2) by the general method of Section 5.7. It's actually not so much work if you make use of work already done in Example 4, Section 5.5. At some point you will have to solve equations of the form

$$Cg^0 + D\bar{g}^0 = e_1,$$
$$Cg^1 + D\bar{g}^1 = e_2,$$

for C and D, where

$$g = (1+\sqrt{5})/2, \quad \bar{g} = (1-\sqrt{5})/2,$$

and e_1, e_2 will be known constants. If you now compare your equations with the similar equations solved in Example 4, Section 5.5, you should be able to simply write down the answer.

14. Here is another trick for solving for B_n. Note that

$$B_n = 1 + B_{n-1} + B_{n-2}$$

may be rewritten as

$$(B_n + 1) = (B_{n-1} + 1) + (B_{n-2} + 1).$$

Now set $G_n = B_n + 1$. What are the recursion and initial condition for G_n? Solve for B_n by solving for G_n.

15. Show that if $n = 2^k$, then $3^k = n^{\log 3}$. Generalize. *Hint*: $3 = 2^{\log 3}$.

Problems 16–27 present Divide and Conquer recursions. Solve them for powers of 2 but express your answers in terms of general n. However, you need not determine the validity of your answers for general n.

16. $A_n = A_{\lfloor n/2 \rfloor} + A_{\lceil n/2 \rceil}, \quad A_1 = 1$

17. $B_n = 3B_{n/2}, \quad B_1 = 1$ (See [15].)

18. $C_n = C_{\lfloor (n-1)/2 \rfloor} + C_{\lfloor n/2 \rfloor} + C_{\lceil n/2 \rceil} + C_{\lceil (n+1)/2 \rceil}$, $C_1 = 1$

19. $D_n = 3D_{n/2} + n, \quad D_1 = 1$

20. $E_n = 3E_{n/2} + n^2, \quad E_1 = 1$

21. $F_n = 4F_{n/2} + n^2, \quad F_1 = 1$

22. $G_n = G_{\lfloor n/2 \rfloor} + G_{\lceil n/2 \rceil} + n, \quad G_1 = 1$

23. $H_n = 5H_{n/2} + \log n, \quad H_1 = 1$

24. $I_n = I_{\lfloor (n-1)/2 \rfloor} + I_{\lceil (n-1)/2 \rceil} + n \log n, \quad I_1 = 1$

25. $J_n = 3J_{n/2} + n \log n, \quad J_1 = 1$

26. $K_n = 2K_{n/2} + \frac{1}{n}, \quad K_1 = 1$

27. $L_n = 7L_{n/2} + (4.5)n^2$, $L_1 = 1$. This recursion may seem completely cooked up, but it's not. It counts the number of operations in a Divide and Conquer method for multiplying $n \times n$ matrices discovered about 20 years ago. Does it beat the order n^3 standard algorithm?

28. Use our algorithmic language to write up the Divide and Conquer recursion discussed in

a) Example 4. *Suggestion*: For the recursive procedure, use as parameters the index of the first and last word in the subsequence to be searched.

b) Example 5

c) Example 6

d) Example 7.

29. If n is input in binary form, then our Divide and Conquer algorithm to compute x^n can be rewritten as a very simple iterative loop. Do so. By "input in binary" we mean that you are given n as a sequence of bits $a_m, a_{m-1}, \ldots, a_0$.

30. Consider Eq. (4) for the Divide and Conquer maximum algorithm.

a) Solve it exactly for powers of 2 by the method of this section.

b) In this case, it just so happens that the answer to (a) is exact for all values of n. Show this.

31. Theorem. Any algorithm that correctly finds the maximum of n numbers, using comparisons of two of the numbers at a time, must use at least $n-1$ comparisons. Proof: Let G be a graph whose vertices are the numbers and which has an edge between two numbers if they get compared at any point in the algorithm. We claim that if the algorithm is correct, the graph must be connected. Why? How does this prove the theorem?

32. Prove or disprove: The solution for *all* $n \geq 1$ to

$$\begin{aligned} r_n &= 2 + r_{\lfloor n/2 \rfloor} + r_{\lceil n/2 \rceil}, \quad n > 2, \\ r_1 &= 0, \qquad r_2 = 1, \end{aligned}$$

is $r_n = \lceil 3n/2 \rceil - 2$. (There are two initial conditions because the recursion doesn't take hold until $n > 2$.)

33. Consider Eq. (3) for Algorithm BINSEARCH.

a) Solve it by the powers of 2 method. (You will *not* get the same expression obtained in Section 2.6.) Verify that your answer is a valid upper bound for the exact answer for all n.

b) As discussed just before the summary of the Divide and Conquer method, there is another approach to Eq (3), using the TOH sequence. Use this approach to find A_n exactly for $n = 2^k - 1$. How does this answer compare to the answer in Section 2.6?

34. Let D_n be the exact solution to Eq. (6). Prove that $\{D_n\}$ is an increasing sequence.

35. If ordered lists of size p and q are merged as in Merge Sort, what is the least number of comparisons necessary before the order of the entire list is known? Give an example with $p = 2$ and $q = 4$ to show that this result could actually happen.

36. Solve the Merge Sort recursion Eq. (6) for powers of 2 without simplifying $n-1$ to n. The algebra is more complicated, but the approach is exactly the same. (It happens the answer is *not* correct as an upper bound for n not a power of 2; check it out by hand for $n = 3$.)

37. Consider Eq. (5) for finding x^n. By replacing t in that equation with 1, find a lower bound on C_n. Verify that this bound is correct for all n.

38. Here is another Divide and Conquer approach to finding x^n. If n is even, do the same as before. But if n is odd, assume that you know both $x^{\lfloor n/2 \rfloor}$ and $x^{\lceil n/2 \rceil}$ and do just *one* more multiplication. Analyze this approach. Is it better or worse than the one in the text?

39. Describe sequential search (Example 1 of Section 1.6) in Divide and Conquer form. (A word description, as in Examples 4–7, is fine.) What is the associated difference equation for the average number of comparisons?

40. Consider the problem of finding both the maximum and the minimum from n numbers.

a) Devise an iterative solution, based on Algorithm MAXNUMBER from Section 1.6.

b) Devise a Divide and Conquer solution. *Suggestion*: You'll get a better algorithm if you have two base cases, $n = 1$ and $n = 2$.

c) Determine and compare the efficiency of the two approaches.

41. The **median** of a set of numbers is the number such that half the other numbers are smaller. Is Divide and Conquer a good approach for finding medians? Explain.

42. Although dividing in two is usually very good, it is not always best. Sometimes dividing in three is better.

a) Determine the exact number of multiplications to obtain x^{27} by the method of Example 6. (The upper bound solution we obtained is of no use here. Instead, actually carry out the recursion.)

b) Figure out a 3-way division recursion for x^{27} and count the number of multiplications.

43. To our knowledge, there is still no known formula or efficient algorithm for the *exact* minimum number of multiplications to find x^n for arbitrary n. Call this minimum number $M(n)$.

a) Show that $M(2^k) = k$.

b) Show that $\log n \le M(n) \le 2\log n$ for all n.

Summary: Chapter 5

The most important thing about difference equations is knowing how to set them up. This skill will be valuable even when we all have pocket calculators that solve standard difference equations immediately. The only way to learn how to set equations up is to practice. If you have only one number you want (the population in the year 2000, or the number of regions created when 10 lines cross on a plane), embed that number in a sequence (the population in year t, or the number of regions when there are n lines). Given a sequence of numbers, think recursively. (Suppose you already knew the population in past years; how is that related to current population?)

The second most valuable skill is knowing how to get information from your difference equation even when you can't obtain an explicit formula by a mechanical method. Generate terms, display and manipulate them in various ways (tables, graphs, ratios, etc.), and look for patterns; then try to prove your conjectures by mathematical induction. Or try to find a substitution that reduces your difference equation to a simpler one. Failing this, replace your equation by a simpler inequality and try to solve it for an upper bound on your sequence.

Finally, learn to recognize the standard types of *linear* recurrences and know the solution forms. In simple cases, even in the future it will probably be easier to figure these out by hand or in your head than to use that pocket calculator. (Doing it in your head is often quite possible if all you want is the general solution and you don't need to solve for the constants.)

If you have a constant coefficient kth order homogeneous difference equation:

$$a_n = \sum_{i=1}^{k} c_i a_{n-i}, \tag{1}$$

then the general solution is a sum of geometric sequences:

$$a_n = \sum_{i=1}^{k} A_i r_i^n,$$

where $r_1, \ldots, r_k$ are the roots of the characteristic polynomial (assuming they are distinct). You solve for the A's from the initial conditions. If the characteristic polynomial has multiple roots, you also get terms such as Anr^n and Bn^2r^n. See Sections 5.5 and 5.6.

In the nonhomogeneous case,

$$a_n = f(n) + \sum_{i=1}^{k} c_i a_{n-i},$$

you first have to find one particular solution, and then you add on all the solutions to the associated homogeneous problem. Often a multiple of $f(n)$ is a particular solution; in any event, the procedure at the end of Section 5.7 gives a general method for solving the whole problem.

If your recurrence has nonconstant coefficients, but is first-order,

$$a_n = c_n a_{n-1} + d_n,$$

there is another type of formula:

$$a_n = \sum_{j=0}^{n} \Big(\prod_{i=j+1}^{n} c_i \Big) d_j .$$

If you can simplify the product and then the sum, the result will be a useful explicit formula. See Section 5.8.

If you have a Divide and Conquer recurrence—A_n is expressed in terms such as $A_{\lfloor n/2 \rfloor}$ and $A_{\lceil n/2 \rceil}$—then substituting $a_k = A_{2^k}$ reduces the problem to a first-order difference equation. See the procedure at the end of Section 5.9.

Supplementary Problems: Chapter 5

1. The Fibonacci recursion can be used to work backwards to define F_{-1}, F_{-2}, etc., by writing the recursion as
$$F_{n-2} = F_n - F_{n-1}.$$
Thus $F_{-1} = 1-1 = 0$, $F_{-2} = 1-0 = 1$, and so on. Compute some more terms. Prove what you find.

2. Consider the recursion of Example 3, Section 5.5:
$$a_n = a_{n-1} + 2a_{n-2}, \qquad a_0 = a_1 = 3.$$
Show that this recursion can be worked backwards in a unique way and compute the terms a_{-1} to a_{-10}. Check that the explicit formula obtained in Section 5.5 works for these negatively-indexed terms as well. Why was this bound to happen?

3. Suppose the sequence $\{a_n\}$ repeats itself twice, i.e., for some $p > 1$, $a_p = a_0$ and $a_{p+1} = a_1$. Suppose further that $\{a_n\}$ satisfies some second-order linear, constant-coefficient homogeneous recursion. Show that $\{a_n\}$ is periodic. What can you say about the characteristic polynomial of such a recursion?

4. Show that for any second-order linear difference equation whose characteristic equation has distinct roots, there is a unique solution for any two initial values a_k and a_{k+2}, so long as one root is not the negative of the other.

5. In [11, Section 4.5], you were asked to write an algorithm to compute $C(n,k)$ by recursively computing earlier entries in Pascal's triangle. Assume that this algorithm is set up using a recursive function named, not surprisingly, $C(n,k)$. You were asked to conjecture how many times this function is called (including the main call) when the main call is $C(n,k)$. You are now in a position to find and prove an answer by setting up a double-indexed difference equation and solving it. Do so.

6. Recall that a spanning tree of an undirected graph is a subgraph which is a tree and which includes all the vertices.

   a) The double cycle C_n^2 is a cycle of length n except that every edge is doubled; consider Fig. 5.13(a). How many spanning trees does C_n^2 have? Assume that each edge and vertex is labeled, so that trees that look the same but don't use exactly the same edge or vertex sets are different.

   b) Let G_n be the graph with vertex and edge sets
$$V = \{x, y, 1, 2, \ldots, n\},$$
$$E = \{\, \{k,x\}, \{k,y\} \mid k = 1, 2, \ldots, n \,\}.$$
See Fig. 5.13(b). How many spanning trees does G_n have? Again assume that everything is labeled.

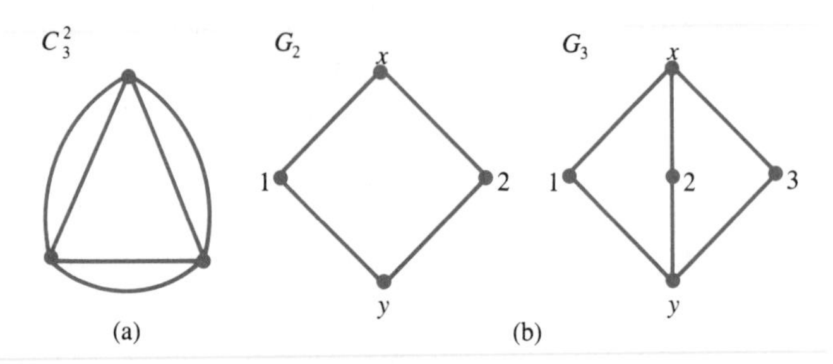

FIGURE 5.13

7. **Annuities** are loans in reverse. You pay a lump sum P to an insurance company and the insurance company pays you back with interest in the form of a fixed payment p at the end of each period, usually a month. With "annuities certain" the payment is made for a fixed number of periods (even if you die in the meantime), at the end of which your "loan" is completely paid back. But more common is an annuity in which you are paid p per period until you die. This adds an element of chance—for the insurance company. It has to figure out how much p should be so that it has just enough money to pay everybody. (For simplicity, we've ignored profit.) It figures this out by grouping annuitants by age at the beginning of the contract, and by statistics on mortality.

   Consider a group of people who buy annuities (payments until death) when they turn 65. Let m_n be the probability that a person at age 65 will live at least n more payment periods (months). We claim that an appropriate recursion for computing the annuity payment p to persons in this group is $P_n = P_{n-1}(1+r) - m_n p$. Explain. (The first part involves figuring out just what P_n means—it's not the actual amount each person is owed, because some annuitants will have died already.)

Assume (as insurance companies do) that everybody dies by age 100, so that all the money should be paid back by then. Find a formula involving a summation for the value of p in terms of P, r, and the various m_n, where p is the period payment to annuitants who pay in P the day they turn 65 and r is the monthly interest rate.

8. It seems hard to believe that the RHS of Eq. (4), Section 5.5, will always be an integer, let alone a Fibonacci number. Prove that, for any integer a and any positive integers b and n, both

$$\left(\frac{a+\sqrt{b}}{2}\right)^n + \left(\frac{a-\sqrt{b}}{2}\right)^n \qquad (1)$$

and

$$\frac{1}{\sqrt{b}}\left[\left(\frac{a+\sqrt{b}}{2}\right)^n - \left(\frac{a-\sqrt{b}}{2}\right)^n\right] \qquad (2)$$

are integers—so long as $a^2 - b$ is divisible by 4.

Note: This problem is quite a challenge if you attack it directly, but if you think of expressions (1) and (2) as terms of two sequences and work backwards to a difference equation for the sequences, then the problem is not very hard at all. This problem illustrates an important point. We are all trained to seek closed forms, such as those above. The underlying reason for this training is the assumption that, whatever knowledge we seek, it will be easier to obtain from a closed form. However, sometimes this is simply false! Sometimes a recursive equation is not only sufficient to yield the information, but it also yields the information more easily.

9. For given integers a and b, let $P(a, b)$ be the proposition that expressions (1) and (2) in [8] are integers for *any* integer n, not just for n a positive integer.

a) Prove $P(1, 5)$. (*Note*: when $a = 1$ and $b = 5$, expression (2) is the $(n-1)$st Fibonacci number.)

b) Find general conditions relating a and b for which $P(a, b)$ is true.

10. Let C_n be the number of calls of the function F needed to compute $F(n)$ in Algorithm FIBB, Section 5.9. For instance, $C_0 = 1$, since there are no calls except the initial call, $F(0)$. Find a formula for C_n.

11. Refine [10] as follows. Let $C_{n,k}$ be the number of calls of $F(k)$ invoked by Algorithm FIBB when you initially set out to compute $F(n)$. For instance, $C_{0,0} = 1$ and $C_{2,0} = 1$. Find a formula for $C_{n,k}$.

12. From the definitions in [10, 11] we have

$$C_n = \sum_{k=0}^{n} C_{n,k}.$$

In light of the results of those two problems, this identity becomes an interesting formula concerning the Fibonacci numbers. Can you prove this Fibonacci formula directly?

13. Let F_k be the kth Fibonacci number. It is a fact that

$$F_n = F_k F_{n-k} + F_{k-1} F_{n-k-1}.$$

a) Try to prove this identity by induction on n.

b) Try to prove it by induction on k.

14. For the Fibonacci numbers, prove that

$$F_{2n} = F_n{}^2 + F_{n-1}^2.$$

15. Suppose that national income in any period is the average of national income in the previous two periods and the income in the following one period. Model this relationship with a difference equation and find the general solution. Discuss the qualitative behavior of solutions. For instance, is explosive growth possible? Is oscillation towards a limit possible? Is constant income possible? Are any of these likely?

16. Repeat [15], but make national income the average of national income in the previous two periods and the following *two* periods. (*Note*: the characteristic polynomial can be factored.)

17. Consider the recursion

$$r_n = r_{n-1} + r_{n-2} + \cdots + r_1 + r_0 \quad \text{for } n > 0,$$
$$r_0 = 1.$$

This is an example of a **variable order** linear difference equation, because the number of terms varies with n. In many cases it is possible to reduce such problems to fixed order linear difference equations by a substitution. For instance, in this example,

$$r_n = r_{n-1} + (r_{n-2} + \cdots + r_0)$$
$$= r_{n-1} + r_{n-1}$$

So we are now down to a first-order equation. Note, however, that the first-order recurrence is only valid for $n \geq 2$. Why? What, therefore, is the solution to the original problem?

Note: With this method, the simpler recurrence typically applies only after a few terms.

18. Solve the following variable order recurrences (see [17]).

a) $r_n = 1 + \sum_{k=1}^{n-1} r_k$ for $n > 1$, $r_1 = 1$.

b) $r_n = r_{n-2} + r_{n-3} + \cdots + r_0$ for $n \geq 2$, $r_1 = r_0 = 1$.

c) $r_0 = 1$ and for $n > 0$,

$$r_n = r_{n-1} + r_{n-3} + r_{n-5} + \cdots + r_\alpha,$$

where $\alpha = 0$ if n is odd and $\alpha = 1$ if n is even.

d) $r_n = \sum_{k=1}^{n} k r_{n-k}$ for $n > 0$, $r_0 = 1$.

19. Solve the following variable order nonhomogeneous recursions. See [17]

a) $a_n = 1 + \sum_{k=1}^{n} a_{n-k}$ for $n > 0$, $a_0 = 1$.

b) $b_n = n + \sum_{k=1}^{n} b_{n-k}$ for $n > 0$, $b_0 = 1$.

c) $c_n = 2^n + \sum_{k=1}^{n} c_{n-k}$ for $n > 0$, $c_0 = 1$.

20. Here is a device that usually works for solving

$$a_n = \sum_{j=1}^{k} c_j a_{n-j} + d$$

using only the homogeneous theory. Namely, there is usually a constant c so that, after adding c to both sides and rewriting the RHS, the equation becomes

$$a_n + c = \sum_{j=1}^{k} c_j (a_{n-j} + c).$$

Define $b_n = a_n + c$ and rewrite the recursion in terms of b's. The general solution for b_n may now be obtained by the homogeneous theory. Then subtract c to get the general solution for a_n.

a) Apply this trick to solve

$$a_n = a_{n-1} + 2a_{n-2} + 4.$$

b) When is it impossible to solve for c?

c) This trick works for difference equations in several variables as well. You happen to know the solution to

$$C(n,k) = C(n-1,k) + C(n-1,k-1),$$
$$C(n,0) = C(n,n) = 1.$$

Now find the solution to

$$D(n,k) = D(n-1,k) + D(n-1,k-1) + 1,$$
$$D(n,0) = D(n,n) = 1.$$

21. Here is a Divide and Conquer method for evaluating a polynomial, described through an example. To evaluate

$$a_3 x^3 + a_2 x^2 + a_1 x + a_0,$$

rewrite it as

$$(a_3 x + a_2) x^2 + (a_1 x + a_0).$$

That is, reduce evaluating a four-term polynomial to evaluating two-term polynomials, and continue recursively. Compute the number of multiplications needed to evaluate a polynomial with 2^k terms by this method

a) If the multiplying factors such as x^2 outside the parentheses are computed by brute force,

b) If they too are computed using Divide and Conquer.

22. Here is another Divide and Conquer method for polynomial evaluation. To evaluate

$$a_3 x^3 + a_2 x^2 + a_1 x + a_0,$$

rewrite it as

$$(a_3 x^2 + a_1) x + (a_2 x^2 + a_0). \qquad (3)$$

(Note that this is not the same regrouping as in [21]. Here we have factored a single power of x out of half the terms, leaving two polynomials in x^2.)

a) Compute the number of multiplications needed to evaluate a polynomial with 2^k terms by this method.

b) Suppose that you want to evaluate a polynomial with 2^k terms at $x = \pm 1$. Do you see that Eq. (3) alone (that is, the first level of the recursion, without going further) already saves work? Explain. (This is the key idea of the **Fast Fourier Transform**. If the polynomial is to be evaluated at complex roots of unity, the saving at the first-level breakdown can be repeated at later levels as well.)

23. Consider the recursive function $F(k)$ in Algorithm FIBB, Section 5.9. Suppose that it is modified so that the first time F is called for any particular k it stores the answer, and for each later call it simply looks up that answer instead of computing it again.

a) Make a tree of the calls when the main call is $F(4)$. When it is $F(6)$.

b) Actually, there are two trees in each case, depending on whether, on the line

$$F \leftarrow F(k-1) + F(k-2),$$

Algorithm FIBB first calls $F(k-1)$ or first calls $F(k-2)$. Whatever assumption you made in (a) about the order of calls, now draw the trees for the other assumption. (Represent the first call by the edge to the left child.)

c) Count the total number of calls in each tree and make a conjecture. Prove it.

24. Suppose a binary tree has the property that at each internal (non-leaf) node, the longest path going left from that node is one edge longer than the longest path going right. For instance, if a node has one child, it must be a left child and a leaf.

a) Give two examples of such binary trees.

b) Show that for each $n \geq 0$, there is a unique binary tree with this property and such that the longest path down from the root contains n edges.

c) Let a_n be the number of nodes in the unique tree in (b). Find a formula for a_n.

25. Consider the two-index recursion

$$a_{0,0} = 1,$$
$$a_{n,k} = 0 \quad \text{if } k < 0 \text{ or } k > n,$$
$$a_{n,k} = \tfrac{1}{2}(a_{n-1,k-1} + a_{n-1,k}) \quad \text{otherwise.}$$

Guess a formula for $a_{n,k}$ by computing some values. Prove it by induction.

26. Define the sequence of functions $\{F_n(k)\}$, where F_n has domain the integers from 0 to n, by

$$F_n(n) = p^n \quad \text{for } n \geq 0,$$
$$F_n(0) = q^n \quad \text{for } n \geq 0,$$
$$F_n(k) = pF_{n-1}(k-1) + qF_{n-1}(k) \quad \text{otherwise,}$$

where p and q are constants. Discover an explicit formula for $F_n(k)$. Prove it by induction.

27. Let M be a two-by-two matrix you pick at random. Let $\mathbf{v}_0$ be an ordered pair (vector in R^2) you also pick at random. For any ordered pair $\mathbf{v} = (a, b)$, define $||\mathbf{v}|| = \max\{|a|, |b|\}$. Finally, define

$$\mathbf{v}_{n+1} = M\frac{\mathbf{v}_n}{||\mathbf{v}_n||}, \qquad n > 0.$$

Compute $\mathbf{v}_1$ through $\mathbf{v}_{10}$. (Use a calculator or computer.) What do you find?

28. Let

$$w = \frac{-1 + i\sqrt{3}}{2}, \qquad \overline{w} = \frac{-1 - i\sqrt{3}}{2}.$$

(These are the two complex cube roots of 1.)

a) Compute the first several points in the sequence defined by

$$a_n = 1 + i\sqrt{3}w^n - i\sqrt{3}\overline{w}^n. \tag{4}$$

What pattern do you observe?

b) Show how to reach the same conclusion without messy computations by first working backwards from Eq. (4) to the difference equation for a_n.

29. Suppose $\{a_n\}$ is a sequence of real numbers that satisfy $a_n + ba_{n-1} + ca_{n-2} = 0$, where b and c are real numbers and $c > b^2/4$.

a) Show there are real constants A, B, and θ such that

$$a_n = c^{n/2}(A \sin n\theta + B \cos n\theta).$$

b) Show there are also real constants D and α so that a_n has the alternative formula

$$a_n = c^{n/2}\big(D \sin(n\theta + \alpha)\big).$$

This equation says every solution is a discrete sinusoidal motion with a "phase shift" (α) and a multiplier $c^{n/2}$ that causes explosive, steady, or damped behavior.

CHAPTER 6

Probability

6.1 Introduction

Probability theory is one of the most broadly useful parts of mathematics. To give you a sense of its breadth, we start by stating several problems, each of which we solve at some point in the chapter. After stating them, we will discuss the role of this chapter.

EXAMPLE 1 A 30-question multiple choice test (say, a section from the College Boards) gives five choices for each answer. You lose 1/4 as much credit for each wrong answer as you gain for each right answer. (Problems left unanswered neither gain nor lose you credit.) If you guess at random on questions for which you don't know the answer, what is the probability that you will raise your total score?

EXAMPLE 2 Medical tests are not perfect. The standard tine test for tuberculosis attempts to identify *carriers*—people who have been infected by the tuberculin bacteria. However, 8% of carriers test negative (a "false negative") and 4% of noncarriers test positive (a "false positive"). Suppose you take the tine test and get a positive result. What is the probability that you are a carrier? (A carrier need not be sick. The bacteria may have been killed off naturally, or they may be in a latent state, capable of causing serious illness later. However, it is very difficult to distinguish between these two cases, so identified carriers are usually treated.)

Tuberculosis is no longer a leading threat in advanced nations, so maybe this example doesn't grab you. But the issue of imperfect tests applies to almost any disease. Substitute AIDS for tuberculosis. If you tested positive, wouldn't you want to know exactly the probability that you have AIDS and what this probability means? (We use tuberculosis instead of AIDS because the data for tuberculosis are well established. The AIDS tests and the data for them are changing rapidly.)

EXAMPLE 3 In Section 4.9, we gave several algorithms for constructing permutations of n objects. In the first algorithm, at each stage we simply picked one of the objects at random, rejecting it if it had already been picked. If k of the n objects have already been placed in the permutation, how many more picks should we expect (on average) before obtaining another previously unpicked object?

EXAMPLE 4 You receive a form letter in the mail saying, "Congratulations, you may have already won an all-expenses-paid trip to Europe worth $10,000." A lottery number is printed on an entry form with your letter. If you return this form to the sponsors—and if the number is the one which has already been chosen by their computer—you win the prize. Should you return the form? (Remember, you will have to pay the postage.)

EXAMPLE 5 Many genetic characteristics are determined by a single gene. Let's say eye color is such a characteristic, and that in order to have blue eyes, you must have received the "b" version of the eye gene from both your parents (in genetic terminology, blue is "recessive"). If 20% of the population have blue eyes, what is the probability that a child will have blue eyes if *neither* parent does?

EXAMPLE 6 Two people play the following version of "Gambler's Ruin". They each start with a certain number of dollars and repeatedly flip a fair coin. Each time the coin comes up heads, person A gives person B $1. Each time it comes up tails, B gives A $1. This continues until one of them is broke (ruined). If A starts with $3 and B starts with $5, what is the probability that A goes broke?

EXAMPLE 7 In Section 1.6 we considered Algorithm MaxNumber for finding the largest number on a list $x_1, \ldots, x_n$. For k from 2 to n, that algorithm compares x_k to the largest number so far, called m; initially $m \leftarrow x_1$ and if x_k is larger than m, then $m \leftarrow x_k$. Since the number of comparisons is always $n-1$, analyzing this algorithm means answering the question: On average, how many times does m get assigned a new value?

Think about these examples. Only Example 6 was fun and games, and even it can be restated as a serious problem about "random walks" of gas molecules in long tubes, or about competition of biological species for fixed resources [3]. (If you are an addicted gambler, the problem is serious already!) Basically, life is full of situations where information is uncertain and various outcomes are possible. Probability theory is the means to cope with such uncertainty and thus all informed citizens ought to know something about it.

But why treat probability in *this* book? Only two of these examples, Examples 3 and 7, deal with issues we have considered previously.

There are two reasons. First, those two examples concern the core subject in this book, algorithms. For the analysis of algorithm efficiency, we have said that the average case is most important but also the most difficult. That analysis depends on probability. This chapter gives you the tools to do the job.

The second reason for including probability is this: Everyone needs to know something about it, it's not included elsewhere in the first two years of core college math, and the discrete approach provides the easiest access to the subject.

We should make clear, though, that probability theory is not a subset of discrete math. Like most branches of mathematics, it has both continuous and discrete aspects, and their interplay is often more powerful than both used separately. For instance, in Section 6.4 we will discuss the "binomial distribution"; it models the outcome of repeated coin tossing. As the number of tosses increases, this distribution approaches the famous "normal" distribution, which is continuous. It is usually easier, and sufficient, to model a problem approximately using the normal distribution than exactly with the binomial distribution. Alas, due to the length of this book we include hardly any discussion of continuous probability theory.

There is a funny thing about probability. Everybody has an intuitive idea of what it means, but when it comes to getting quantitative, it is easy to go wrong. In many problems, two different arguments lead to very different answers, yet both arguments seem right. (We give one famous example, the second child problem, in [19, Section 6.3].)

When a subject is slippery in this way, a formal approach is called for. Thus in this chapter more than others, we state definitions precisely in displays and solve problems by carrying out calculations in detail using the definitions. (For discrete probability theory, all we need for such precision is the knowledge of sets and functions from Chapter 0.) This formal approach won't guarantee that you won't make mistakes (one of us once gave an incorrect analysis of the second child problem in class!) but it helps a lot.

This chapter is organized as follows. In Sections 6.2 through 6.5, we introduce the formulation that mathematicians have devised for discussing probability. In Section 6.2 we define "probability space". Section 6.3 concerns "conditional probability", where we show how to modify probabilities when additional information becomes available. In Section 6.4 we introduce "random variables"; these are functions that allow one to reduce a lot of different-looking probability spaces to a few standard spaces. In Section 6.5 we treat "expected value", the generalization of average value.

In the remaining two sections of this chapter we get some payoffs from all that concept building. In Section 6.6 we treat average case analysis of algorithms by taking examples done intuitively in earlier chapters—we had no choice but to do them intuitively—and doing them right. Thus in terms of feedback to the core topics of this book, Section 6.6 is the centerpiece of the chapter. We hasten to say, however, that in terms of the general usefulness of probability payoffs appear as early as Section 6.3—we solve Example 2 there. Finally, in Section 6.7 we show that the connection between probability and our core topics is not all one way. The recursive paradigm, one of our core methods, turns out to be just the thing to solve some otherwise very hard types of probability problems, and in Section 6.7 we show how.

Two final points. First, if it struck you that some of the questions in Examples 1–7 are not well defined—you may feel they need some more information or at least a clearer specification of the assumptions before they can be solved—you are absolutely right. But then, specifying a question clearly and determining the information needed to get started are usually the greater part of the battle in applying mathematics. We will define these questions better at appropriate places.

Second, in most textbooks most probability problems are about tossing coins, throwing dice, or picking balls from urns. These activities are less significant than those in our examples. But there is a reason for such textbook choices: Such situations are simpler to understand. Also, most people find them fun. We too will couch most of our examples in these terms. However, occasional examples like those we've introduced will remind you that probability is useful as well as fun.

Problems: Section 6.1

1. Examples 2 and 4 are among those for which not enough information is given.

a) For Example 2, suppose one person out of 100 is a tuberculosis carrier. Can you answer the question now?

b) For Example 4, suppose the number on your entry form is 7-493-206. Can you answer the question now?

2. In Example 1, suppose you guess on exactly one problem. What is the probability that you increase your score? What's the probability if you guess on exactly two problems?

3. Suppose two species, say birds, occupy the same "ecological niche". This means they roost in the same areas, compete for the same food, etc. It is reasonable therefore to suppose that the total number of birds of both species that can be supported in a given locale is fixed. That is, if one more bird of the first species survives in the next generation, then one less of the second species survives. Do you see that we can model the situation with gambler's ruin? If one species is "fitter" than the other, what simple change in the model accommodates this fact?

It is a theorem that, with probability 1, eventually one of the gamblers is ruined—even if the coin is fair. Thus if the gambling model is appropriate for biology, we conclude that we should never find two species occupying the same niche. (Why?) This conclusion has been confirmed by observation. Biologists refer to it as Gause's Law.

6.2 Probability Space

In this section we will set up the basic mathematical formulation for probability problems. In any such problem, there are various possible **outcomes**, also called **events**. The key step is to represent these events as *sets*. The set of all possible outcomes for a problem is called the **sample space**. All other events are subsets of the sample space. Probabilities are numbers between 0 and 1, inclusive, assigned to these sets; the closer the probability of an event is to 1, the more certain that the event will happen. All this is formalized in Definition 1, but we wish to give some examples first. One of the lessons of these first examples is that there isn't always just one right sample space for a problem.

EXAMPLE 1 A regular die is to be tossed once. What is an appropriate sample space?

Solution Presumably the only thing of interest is which face comes up, in which case the natural sample space is $\{1,2,3,4,5,6\}$. Here, $\{2\}$ is the event that the face with two dots comes up. The set $\{1,3,5\}$ is the event that an odd number comes up. ∎

An event consisting of exactly one element from the sample space is called an **atomic event**. All other events are **nonatomic** or **compound**. In the above example, "two dots" is atomic and "odd number" is compound. For a fair die, all the atomic events are equally likely; that is, each has probability $1/6$.

EXAMPLE 2 A regular die is to be tossed twice. What is an appropriate sample space?

Solution There is no one right answer to this. If the only question we will ever ask is, "How many dots come up total?", then we might as well choose our sample space to be $\{2,3,\ldots,11,12\}$. However, if we are going to ask about the probabilities of more complicated situations:

Is the first toss a 2?

Is one of the tosses a 2?

Are both faces the same?

then this sample space isn't refined enough. A better choice would be

$$\{\ (1,1),\ (1,2),\ (2,1),\ \ldots\ ,\ (5,6),\ (6,5),\ (6,6)\ \}.$$

Here, the atomic event $\{(2,5)\}$ is the event that the first toss results in a 2, the second in a 5. ∎

The more refined sample space has the additional advantage that, if the die is fair, each element of the sample space is equally likely. Thus to determine the probability that both faces are the same, we merely have to divide the *number* of sample space elements in which both faces are the same by the total number of elements [2]. On the other hand, with the first sample space, the probability that should be assigned to each atomic event is not immediately obvious.

To figure it out, you would probably resort to the second model anyway [8].

EXAMPLE 3 Two different but indistinguishable dice are tossed together. What is an appropriate sample space?

Solution Although we can't distinguish them, they are different dice. We might imagine that one is painted red, the other blue, and we just happen to be completely color blind. In this case, either sample space of Example 2 can be used; in the second space, we may interpret the element $(2,5)$ to mean that the red die shows up 2 and the blue die 5.

However, the indistinguishability suggests a third sample space. The ordered pairs $(2,5)$ and $(5,2)$ could be replaced by a single element, the *set* $\{2,5\}$. Thus the sample space would be

$$\{\{1,1\}, \{1,2\}, \ldots, \{1,6\}, \{2,2\}, \{2,3\}, \ldots, \{2,6\}, \{3,3\}, \ldots, \{6,6\}\}.$$

Which sample space is best? The last sample space suggests (but does not require) that getting two 1's should have the same probability as getting a 1 and a 2. However, to the contrary we feel that there is only one way to get two 1's, but two ways to get a 1 and a 2 (depending on whether the 1 is on the red die or the blue). This is borne out by empirical experiments: When two dice are tossed, $\{1,2\}$ shows up about twice as often as $\{1,1\}$. Thus the space of ordered pairs is preferred.

However, if we were dealing with certain atomic particles (atomic as in physics), then the space of unordered pairs is better. If you ask certain questions about which excitation state various electrons are in, then having one electron in state 1 and a second in state 2 is *not* twice as likely as having both in state 1. In short, not only can't we humans distinguish these particles, but it seems nature can't either! In this case the unordered sets are equiprobable and thus form the best sample space.

The moral is: The best sample space may not be obvious at the outset. ∎

In Examples 1–3, we preferred the sample space in which atomic events are equally likely. However, it's not always possible to concoct such a sample space. For instance, suppose we are tossing a biased coin. The natural sample space is still $\{H, T\}$, but H and T are not equally likely. So it is time to formalize the general framework for assigning probabilities to events.

Definition 1. A **probability measure** on a set S is a real valued function, Pr, with domain the set of subsets of S, such that

1. for all subsets A of S, $0 \le \Pr(A) \le 1$;
2. $\Pr(\emptyset) = 0$, $\Pr(S) = 1$;
3. if subsets A and B of S are disjoint, then

$$\Pr(A \cup B) = \Pr(A) + \Pr(B).$$

The value $\Pr(A)$ is called the probability of event A. The set S is called the sample space. A sample space, together with a probability measure, is called a **probability space**.

Note that a probability is a number assigned to an individual event. A probability measure is a function from the set of all events to numbers. Thus a probability is a particular value of a probability measure.

Definition 1 explains what probability *is*, in the sense that it gives all the rules necessary to *use* probabilities; however, it does not say what probability *means*. But this is true of all formal definitions—and it's a good thing too, because it separates the issue of use from the often stickier issue of meaning. In fact, when examined closely, the meaning of probability is quite sticky. We return to this point briefly at the end of this section.

We can give a first illustration of the use of Definition 1 by returning to Example 1 (tossing a die). We have $S = \{1, 2, \ldots, 6\}$ and for $A \subset S$, $\Pr(A) = |A|/6$. (Recall that $A \subset S$ means A is a subset of S and $|A|$ is the number of elements in A.) You should check that conditions 1)–3) are satisfied. Also check [6] that the probability of the event "an odd number is tossed" is 1/2.

In general, for any finite sample space S, the function defined by

$$\Pr(A) = \frac{|A|}{|S|}$$

is called the **equiprobable measure**, for it makes all atomic events equally likely. As we have already noted, the advantage of this measure is that computing a probability reduces to counting the number of elements in a set—something you've done a lot! Also, in the absence of any information to the contrary, it is usually reasonable to assume that atomic events are equally likely (but reread Example 3).

For any probability measure, it follows by induction from 3) that if $A_1, \ldots, A_k$ are disjoint, then [10]

$$\Pr\Big(\bigcup_{i=1}^{k} A_i\Big) = \sum_{i=1}^{k} \Pr(A_i). \tag{1}$$

As a special case, if $A = \{e_1, e_2, \ldots, e_k\}$, then setting $A_i = \{e_i\}$ and writing $\Pr(e_i)$ instead of $\Pr(\{e_i\})$ to avoid eye strain, we have

$$\Pr(A) = \sum_{i=1}^{k} \Pr(e_i).$$

In other words, for any finite subset A of any sample space S,

$$\Pr(A) = \sum_{e \in A} \Pr(e). \tag{2}$$

Thus a probability measure on a finite sample space is completely determined by addition from the probability of the atoms.

EXAMPLE 2 Continued

What is the right probability measure for the first sample space in Example 2, in the sense that it corresponds to the equiprobable measure on the second sample space?

Solution It suffices to determine probabilities for the atomic events in the first sample space. These atomic events correspond to events in the second space as follows:

$$\begin{aligned}
\{2\} &\longleftrightarrow \{(1,1)\},\\
\{3\} &\longleftrightarrow \{(1,2),(2,1)\},\\
\{4\} &\longleftrightarrow \{(1,3),(2,2),(3,1)\},\\
&\vdots\\
\{7\} &\longleftrightarrow \{(1,6),(2,5),(3,4),(4,3),(5,2),(6,1)\},\\
\{8\} &\longleftrightarrow \{(2,6),(3,5),(4,4),(5,3),(6,2)\},\\
&\vdots\\
\{12\} &\longleftrightarrow \{(6,6)\}.
\end{aligned}$$

Since atomic events in the second space are equiprobable, and there are 36 atoms, we conclude that, in the first space,

$$\Pr(2) = \frac{1}{36},\ \Pr(3) = \frac{2}{36},\ \ldots,\ \Pr(7) = \frac{6}{36},\ \ldots,\ \Pr(12) = \frac{1}{36}.$$

This may be stated completely and succinctly as

$$\Pr(n) = \begin{cases} \dfrac{n-1}{36} & \text{for } 2 \le n \le 7, \\ \dfrac{13-n}{36} & \text{for } 8 \le n \le 12. \end{cases}$$ ∎

There are all sorts of little rules for computing the probability of one event in terms of related events. The next two theorems are examples. When you need such a rule, it's usually best to redevelop it by intuition or by drawing a picture, as we illustrate below. However, all such rules follow logically from conditions 1)–3) in Definition 1.

Definition 2. Let $\sim E$ be the event "not E". That is, if S is the sample space,

$$\sim E = S - E.$$

For instance, in Example 1, if E is the event "an odd number shows up", then $\sim E$ is the event "an even number shows up". Note that $\sim E$ is just the set complement of E (see Section 0.1), so we could also write $\overline{E}$ as we did there.

How is $\Pr(\sim E)$ related to $\Pr(E)$? Think intuitively: Since $\sim E$ is everything other than E, it must get all the remaining probability. Since the total probability is 1, this means that $\Pr(\sim E) = 1 - \Pr(E)$.

To illustrate that such facts follow from Definition 1, we give this simple fact a formal statement and proof:

Theorem 1. For any event E in a sample space,

$$\Pr(\sim E) = 1 - \Pr(E).$$

PROOF S is the disjoint union of E and $\sim E$. Therefore by 2) and 3),

$$1 = \Pr(S) = \Pr(E) + \Pr(\sim E).$$

Now subtract $\Pr(E)$ from both sides. ∎

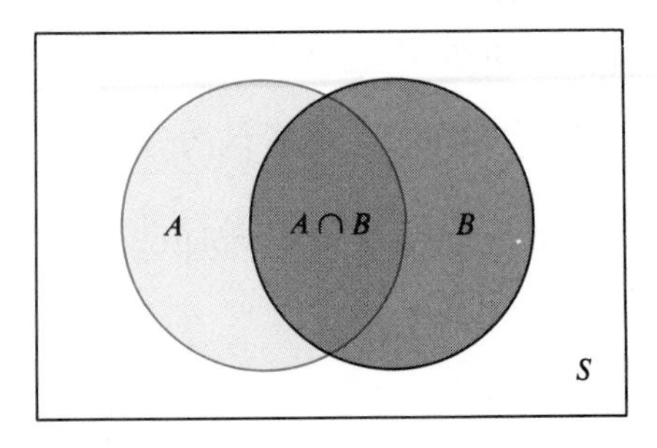

FIGURE 6.1 $\Pr(A \cup B) = \Pr(A) + \Pr(B) - \Pr(A \cap B)$

What can be said about $\Pr(A \cup B)$ when A and B are not disjoint? A Venn diagram tells it all (see Fig. 6.1). Think of the probability of a set as its area in the diagram. If we simply add Pr(A) and Pr(B), we count $\Pr(A \cap B)$ twice. We should subtract $\Pr(A \cap B)$ once to compensate. Therefore we believe:

Theorem 2. For arbitrary events $A, B \subset S$,

$$\Pr(A \cup B) = \Pr(A) + \Pr(B) - \Pr(A \cap B).$$

PROOF At this point, about the only thing available for starting the proof is Definition 1. The only part of that definition which looks close to what we want is 3), which is about disjoint sets. Thus we must restate the situation of interest, namely, the union of two arbitrary sets, in terms of disjoint sets. Note that

$$A = (A-B) \cup (A \cap B), \qquad B = (B-A) \cup (A \cap B),$$

and

$$A \cup B = (A-B) \cup (A \cap B) \cup (B-A),$$

where $A-B$, $A \cap B$ and $B-A$ are disjoint. Thus using condition 3) twice in the first equality below, and once in the last, we obtain

$$\begin{aligned}\Pr(A)+\Pr(B) &= \big[\Pr(A{-}B)+\Pr(A{\cap}B)\big] \;+\; \big[\Pr(B{-}A)+\Pr(A{\cap}B)\big]\\ &= \big[\Pr(A{-}B)+\Pr(A{\cap}B)+\Pr(B{-}A)\big] \;+\; \Pr(A{\cap}B)\\ &= \Pr(A{\cup}B)+\Pr(A{\cap}B).\end{aligned}$$

Subtracting $A \cap B$ from both sides gives us the theorem. ∎

Does this theorem remind you of Inclusion-Exclusion (Section 4.8)? See [16, 17].

Types of Probability Spaces

All our examples so far have been of finite probability spaces. That is, S has been finite. There are, however, many types of infinite probability spaces. The two simplest types are discrete infinite spaces and continuous spaces. For instance, suppose we wish to study the number of telephone calls passing through a phone company switching center each day. In principle, it could be any nonnegative integer, so the sample space is the set of natural numbers. (The phone company needs to know the probability measure on this space when building or expanding a switching center. Why?) This is a discrete infinite probability space.

Now suppose we pick a number from 0 to 1 at random. Since every number is eligible (theoretically), the sample space is the interval $0 \leq x \leq 1$. This is a continuous space. Spaces can also be partly discrete and partly continuous.

Definition 1 must be modified for infinite spaces. For instance, even with induction, condition 3) is not sufficient for deducing the probabilities of all events just from the probabilities of the atomic events—you can get only to finite sums by induction and now some sets will have infinitely many atoms [21]. It turns out one must assume the following infinite version of Eq. (1): If $A_1, A_2, \ldots$ is an infinite sequence of disjoint events in S, then

$$\Pr\Big(\bigcup_{i=1}^{\infty} A_i\Big) \;=\; \sum_{i=1}^{\infty} \Pr(A_i)\,. \tag{3}$$

We hope this sounds reasonable, but we won't discuss infinite sums carefully until Chapter 9.

Going to infinite sample spaces also imposes some surprising restrictions. You can't always assign probabilities the way you want. In Example 4, we give a simple situation in which there *cannot* be an equiprobable measure. In fact, for many sample spaces, some subsets cannot be assigned *any* probability value. But this is a very subtle matter, studied in the advanced subject called measure theory. Fortunately, it need not concern us further. All "nice" sets (events people typically consider) turn out to be "measurable" (probabilities can be assigned to them in reasonable ways), and the few times we slip over into infinite probability spaces we will always look at nice sets.

EXAMPLE 4 Show that there is no equiprobable measure on the positive integers.

Solution An equiprobable measure would require each integer to have the same probability. What could this probability be? It can't be positive, no matter how small. For suppose that $\Pr(1) = \epsilon$. (ϵ, the Greek lowercase epsilon, is the traditional symbol in mathematics for a very small but arbitrary positive number.) Then for some positive integer n, $n\epsilon > 1$. Thus by equiprobability and condition 3), $\Pr(\{1,2,\ldots,n\}) = n\epsilon > 1$, contradicting condition 1) of Definition 1. On the other hand, we can't have $\Pr(1) = 0$ for, by Eq. (3),

$$\Pr(S) = \sum_{k=1}^{\infty} \Pr(k) = 0 + 0 + \cdots = 0,$$

when $\Pr(S)$ should be 1. ∎

Example 4 should bother you because, offhand, the notion of picking a positive integer at random makes sense. But think about it. If you were told to pick a positive integer at random, what would you actually do? [22]

Defining Probability Spaces

Picking an appropriate probability space is the key to a good start on a probability problem. The examples we have given are reasonably easy. In a real-world example it's a lot harder; you need the sort of judgment that one develops with experience. Here are two more (still not very hard) examples.

EXAMPLE 1, Section 6.1, Continued

Guessing on the College Boards

Devise a probability space for this problem. Recall that there are five choices for each question and you want to know whether to guess.

Solution What hasn't been said is how many of the 30 questions you guess on randomly. Let's call that number n. Then we can make the elements of the sample space the n-tuples of possible guesses. For example, the n-tuple $(\mathrm{A}, \mathrm{A}, \mathrm{B}, \mathrm{E}, \mathrm{D}, \ldots)$ represents the atomic event that you guess answer A on the first guessed-at question, A again on the second, B on the third, and so on. Because you guess randomly, the appropriate probability measure is the equiprobable measure. However, all that really matters is whether you guess right or wrong. So a simpler space is the set of "words" of length n on the two symbols R and W. Thus the word RWWRW... represents the event that you guess right on the first guessed-at question, wrong on the second, and so on. This reduction causes no complication in figuring out the probability measure: Because you guess randomly, each R has probability $1/5$ and each W probability $4/5$. Each of these repeated, "independent" choices is called a "Bernoulli trial". We'll discuss Bernoulli trials in Section 6.4. ∎

EXAMPLE 4, Section 6.1, Continued

Mail-in contest

Devise a probability space. Recall that you might win a \$10,000 trip, but you have to pay for a stamp to mail in your lottery number.

Solution We assume that the winning number is chosen randomly (i.e., with equiprobable measure); that's what the cover letter usually says. So a natural choice of S is the set of all ticket numbers. Unfortunately, you don't know how many there are. Also, in most cases the contest promises that all prizes will be awarded. In other words, S really consists only of *returned* ticket numbers. Again, it's best to simplify. Let $S = \{\text{Win, Lose}\}$, and evaluate the probability that you Win by estimating the number of returned tickets. If your estimate is N tickets, then your estimate of your chance of winning is $\Pr(\text{Win}) = 1/N$. ∎

The last example shows that determining the probability space can be hard—how do you obtain N? At least once you've done that, clearly you should use the equiprobable measure. What about problems such as tossing a warped coin, where you don't have equiprobability (or some other form of symmetry) to guide you? The standard approach is to assign probabilities by the **ratio interpretation**. The probability of an event E is defined as the fraction of the time E happens if we repeat the situation being modeled without limit. For instance, E might be the event that a head shows up if we toss a particular warped coin once. Imagine repeatedly tossing the coin forever. We expect that the fraction of the times that E occurs would approach a limit. We define $\Pr(E)$ to be that limit.

This approach is fraught with difficulties. First, you can't repeat an experiment forever. Consequently, even if the fraction of times E occurs appears to settle down very nicely after a large number of repetitions, you can't be absolutely certain what the limit at infinity is, or even that there *is* a limit. Second, some events are repeatable, but it's hard to get the data. For instance, there are lots of mail-in contests, but it may be hard to find out from the various sponsors what fraction of the tickets are returned (and hardly worth the effort when your maximum loss is the cost of postage!). Finally, some events are not repeatable even once. What sense, for instance, does the ratio interpretation make for getting even a rough idea of the probability that the world will end in nuclear war?

These remarks lead us into the *philosophy* of probability, a fascinating but difficult subject. Fortunately, we can pull back. Mathematically, we do not need to justify where a probability measure comes from. So long as the measure satisfies Definition 1, the results in this chapter will hold. Also, so long as we are dealing with a repeatable situation, then in some sense the ratio interpretation is forced on us, at least as a conceptual way of understanding what probability means. For so long as we intend the statement "$\Pr(E) = 1$" to mean that E surely happens, there is a theorem which says: If an experiment is repeated independently and infinitely many times, it surely happens that the fraction of times that any event A occurs goes to a limit, and the limit is $\Pr(A)$. In other words, if we are talking about a repeatable event, the ratio interpretation must be right—either that or our whole mathematical formulation of probability is not a good model of reality. The theorem just referred to is called the **Law of Large Numbers**.

Problems: Section 6.2

1. Using the sample space of Example 1, describe in set notation the event that the number which comes up is
   a) Even
   b) A perfect square
   c) Divisible by 2 or 3
2. Consider the three events listed in the second paragraph of Example 2.
   a) Using the preferred sample space for that example, describe each of these events as a set of elements. Also describe the event "the sum of the dots is 5" and "the difference of the dots is 3".
   b) Using the equiprobable measure on this sample space, determine the probability for each event in a).
3. For each atomic event in the first sample space suggested in Example 2, express that event as a subset of the preferred sample space.
4. Suppose that a penny is tossed three times. Let S be the sample space. The event $E =$ "the second toss is an H" is an event in the space, but it is not an atomic event. Why not? Express E as a set of atomic events.
5. Let A, B, C be events in some sample space S. Using set notation, express the event
   a) At least one of A, B, and C happens.
   b) At least two of A, B, and C happen.
   c) At most two of A, B, and C happen.
   d) At most one of them happens.
   e) Exactly one of them happens.
6. For Example 1, verify that $\Pr(A) = |A|/6$ meets the definition of a probability measure. Verify that the probability of $\{1,3,5\}$ is 1/2.
7. Prove that the equiprobable measure really is a probability measure. That is, prove that the function $\Pr(A) = |A|/|S|$ satisfies Definition 1.
8. In the definition of probability measure (Definition 1), the requirement that $\Pr(\emptyset) = 0$ is redundant: It follows from condition 3. Show this. *Hint*: $\emptyset$ is disjoint from every set.
9. How many atomic events are there in the third sample space of Example 3? This was the space in which $(2,5)$ and $(5,2)$ are the same. What type of ball and urn problem (Section 4.7) is this?
10. Prove Eq. (1) by induction from condition 3 in Definition 1.
11. Prove that Eq. (2) is equivalent to condition 3 of Definition 1. That is, in any finite sample space, any function Pr which satisfies condition 3 satisfies Eq. (2), and vice versa.
12. Prove that $\Pr(A{-}B) = \Pr(A) - \Pr(A{\cap}B)$.
13. Prove that
$$\Pr(\sim\!A \cap \sim\!B) = 1 - \Pr(A) - \Pr(B) + \Pr(A{\cap}B).$$
14. Show that if $A_1, A_2, \ldots, A_n$ are disjoint and A is their union, then
$$\Pr(A \cap B) = \sum_{i=1}^{n} \Pr(A_i \cap B).$$
15. Use [12] to give another proof of Theorem 2. *Hint*: Write $A \cup B$ using $A - B$.
16. Consider Theorem 2, which is about $\Pr(A{\cup}B)$. We hinted that this theorem is related to Inclusion-Exclusion.
    a) Generalize the statement of the theorem.
    b) Assume you have an equiprobable measure. Prove your generalization in a few lines as a corollary of Inclusion-Exclusion (Theorem 1, Section 4.8, or the variant discussed in [19, Section 4.8], which is the exact analog of the probability result you want).
    c) Prove your generalization for arbitrary probability measures. Now you can't simply quote the Inclusion-Exclusion *result*. However, the *proof* from Section 4.8 can be modified to give you a probability proof.
17. Theorem 2 is the probability analog of the first case of Inclusion-Exclusion:
$$|A \cup B| = |A| + |B| - |A \cap B|.$$
    We proved that case in a line or two in Section 4.8 but proving Theorem 1 took half a page. Can you explain why?
18. Prove: if $A \subset B$ then $\Pr(A) \le \Pr(B)$.
19. Odds. When one says that the odds against event E are $s{:}t$ (pronounced "s to t"), that means E

will fail to occur with probability $s/(s+t)$. In this problem, assume that exactly one of three events, A, B, and C, must happen.

a) If the odds against A are 1:1 and the odds against B are 2:1, what are the odds against C?

b) If the odds against A are instead 3:2 and against B 3:1, what are the odds against C?

c) If the odds against A are $p\colon q$ and those against B are $r\colon s$, what are the odds against C?

20. Prove an inequality relating

$$\Pr\Big(\bigcup A_i\Big) \quad \text{and} \quad \sum \Pr(A_i).$$

21. Let function F be defined on all subsets A of R by

$$F(A) = \begin{cases} 0 & \text{if } |A| \text{ is finite,} \\ 1 & \text{if } |A| \text{ is infinite.} \end{cases}$$

a) Prove: For all $A, B \subset R$, $F(A \cup B) \leq F(A) + F(B)$.

b) By induction, prove for all positive integers n that

$$F\Big(\bigcup_{i=1}^{n} A_i\Big) \leq \sum_{i=1}^{n} F(A_i).$$

c) Nonetheless, find an infinite sequence of sets $A_1, A_2, \ldots$ such that

$$F\Big(\bigcup_{i=1}^{\infty} A_i\Big) \not\leq \sum_{i=1}^{\infty} F(A_i).$$

Moral: Proving something for the infinitely many positive integers is not the same as proving it for infinity.

22. (Example 4 continued.) To pick a positive integer at random, we think you would do the following. You would pick some large positive integer N and pick an integer at random *between* 1 and N. (This is what a computer must do, since each computer has a largest integer known to it.) Or you might pick a string of *digits* at random, continuing until you got tired. (But does this make each integer equally likely?) In any event, so long as there is some largest N that you will consider, you have a finite space and there can be an equiprobable measure on it.

But in some ways this begs the question. It seems perfectly reasonable, for instance, to contend that the probability of picking an even integer by random choice from among *all* positive integers is $1/2$. Fortunately, we can justify such statements by using the idea in the preceding paragraph, if we also use limits. Namely, let P_N be the probability that, when an integer from 1 to N is picked at random, it is even. Then our contention may be restated this way: The limit of P_N as $N \to \infty$ is $1/2$.

For those who already have studied limits, show that this contention is correct.

23. (Philosophy question). Consider Example 2, Section 6.1 but forget about the tine test. Just consider the question, "What is the probability I have tuberculosis?" One argument goes: The answer is 0 or 1; either I have it or I don't. Claims that the probability should be some other number, say .001, are nonsense, because my tuberculosis status is not a repeatable experiment. It's a fact.

Argue against this claim. Come up with an interpretation of the statement "I have tuberculosis with probability .001" which allows it to have a ratio interpretation.

6.3 Conditional Probability, Independence, and Bayes' Theorem

Since probabilities measure the likelihood of events, they change as new information is discovered. Example 2, Section 6.1 (tuberculosis), provides a good example. Once you are tested, the probability that you are a carrier changes; it goes up if you test positive, down if you test negative. On the other hand, sometimes new information is irrelevant. If you find out that the previous toss of a coin was heads, that doesn't affect the probability that the next toss will be heads.

In this section we show how to fit such considerations into the probability space formulation of the previous section. We do so using Definitions 1 and 2. First we state these definitions, then we motivate them and show how to use them. We conclude this section with a result called Bayes' Theorem. It is just a restatement of Definition 1, but a particularly useful one.

In many places in this section we divide by a probability. Needless to say, that probability must be nonzero. We make this explicit in Definitions 1 and 2, but thereafter it will be implicit.

Definition 1. Suppose that $\Pr(A) \neq 0$. The **conditional probability of B given A**, denoted $\Pr(B|A)$, is defined to be

$$\Pr(B|A) = \frac{\Pr(A \cap B)}{\Pr(A)}. \tag{1}$$

The quantity $\Pr(B|A)$ is also called the **probability of B conditioned on A** or **the probability of B given A.**

Definition 2. Events A and B are **independent** if

$$\Pr(A \cap B) = \Pr(A) \cdot \Pr(B), \tag{2}$$

or equivalently (when $\Pr(A) \neq 0$), if

$$\Pr(B|A) = \Pr(B). \tag{3}$$

We motivate Definition 1 by continuing with the tuberculosis example. Consider Fig. 6.2. The universe S is the set of all people to whom the statistics apply (the U.S. population, say), T is the set of all tuberculosis carriers, and P represents the set of people who test positive (or would if they were tested). We were asked: If all we know about you is that you are in P, what is the probability that you are also in T? This is what $\Pr(T|P)$ is supposed to mean.

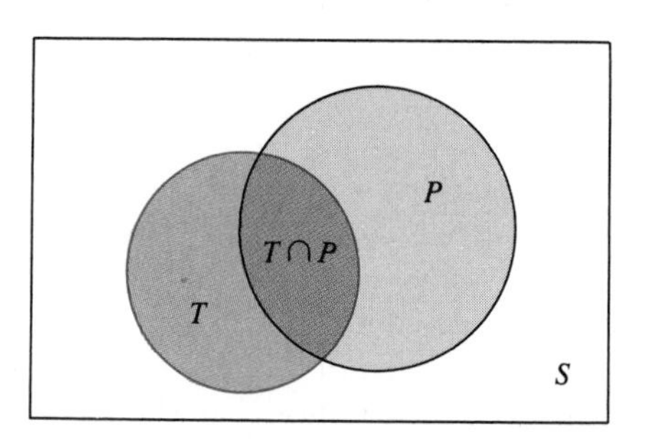

FIGURE 6.2

Tuberculosis testing

We need to show that Eq. (1) captures that intended meaning. The example asks "How probable is $T \cap P$ relative to P?" This relative probability might be close to 1 even if $\Pr(T \cap P)$ is small. So what we need is a *change of scale.* Since only P "counts" anymore, we want P to have probability 1. So for every $X \subset P$, we multiply $\Pr(X)$ by $1/\Pr(P)$. In short,

$$\Pr(T|P) = \frac{\Pr(T \cap P)}{\Pr(P)}.$$

This is precisely what Eq. (1) says, with $A = P$ and $B = T$.

We now give some numerical examples of the use of Definition 1. The key to a correct application of the definition is a careful identification of the sample space and the sets A and B.

EXAMPLE 1 An urn contained two black balls and two white balls. Balls were removed one at a time, randomly, and not replaced. You were not present. (a) What is the probability (from your viewpoint, based on just this information) that the first ball removed was black? (b) You are now informed that the second ball removed was black. How should you change your answer?

Solution Surely your answer to (a) should be 1/2, since half the balls available for the first draw were black. You do not need to think consciously in terms of a sample space to arrive at this conclusion. However, in order to answer (b), conscious thought about a sample space is helpful. Let F (for first) be the event that the first ball was black; N (for next) the event that the second ball was black. The sample space may be taken to be the 4×3 different ways that the first two balls may be chosen. The measure is equiprobable; i.e., each element is an ordered pair, such as (b_2, w_1), where $\{b_1, b_2, w_1, w_2\}$ are the four balls. Then F is the subset of all couplets with the first entry black. There are 2×3 of these, since there are two black balls for the first draw and three remaining balls for the arbitrary second draw:

$$F = \{(b_1, b_2), (b_1, w_1), (b_1, w_2), (b_2, b_1), (b_2, w_1), (b_2, w_2)\}.$$

Similarly, N is a subset with six couplets [1]. Clearly, $F \cap N = \{(b_1, b_2), (b_2, b_1)\}$. Consequently,

$$\Pr(F \cap N) = \frac{2}{12}, \qquad \Pr(N) = \frac{6}{12}, \qquad \text{and} \qquad \Pr(F|N) = \frac{2/12}{6/12} = \frac{1}{3}.$$

Thus the information that the second ball was black *lowers* the probability that the first ball was black. Is this intuitively reasonable? [20] ∎

EXAMPLE 2 A fair coin was tossed twice while you weren't present. (a) What is the probability that the first toss was a head? (b) What is the probability that the first toss was a head, given that the second toss was a head?

Solution Again, the first answer is 1/2. We hope that you feel very strongly that the second answer is also 1/2, but let's check it out as carefully as in the previous example. We may take S to be the four-element set $\{(\mathrm{H}, \mathrm{H}), (\mathrm{H}, \mathrm{T}), (\mathrm{T}, \mathrm{H}), (\mathrm{T}, \mathrm{T})\}$.

Taking F to be the event that the first toss is a head and N that the second is a head, we find

$$F \cap N = \{(H,H)\}, \qquad N = \{(H,H),\,(T,H)\},$$
$$\Pr(F \cap N) = \frac{1}{4}, \qquad \Pr(N) = \frac{1}{2},$$

and thus

$$\Pr(F|N) = \frac{1/4}{2/4} = \frac{1}{2}. \quad \blacksquare$$

If you felt at the outset that both answers in Example 2 should be the same, probably you knew intuitively that one coin toss is "independent" of another in the sense that knowledge of one shouldn't affect probabilities for the other. This notion of independence is easy to put in symbols: Events A and B are independent if $\Pr(B|A) = \Pr(B)$. This is exactly what Eq. (3) in Definition 2 says. However, Eq. (3) is not the primary display in the definition. That honor goes to Eq. (2), which says: A and B are independent if $\Pr(A \cap B) = \Pr(A) \cdot \Pr(B)$. Equation (2) is first for two reasons: it is slightly more general since $\Pr(A) = 0$ is allowed, and the roles of A and B in it are symmetric—mathematicians love symmetry.

In any event, it's easy to show that the two equations are equivalent when $\Pr(A) \neq 0$. Assuming Eq. (3) and substituting Eq. (1) into it, we get

$$\frac{\Pr(A \cap B)}{\Pr(A)} = \Pr(B).$$

Multiplying by $\Pr(A)$ gives Eq. (2). Reversing all the steps gets us back to Eq. (3).

Caution. Many people think that Eq. (2) is true for *any* events A and B. It's not. It is true if and only if the events are independent. Thus $\Pr(A \cap B)$ can be very different from $\Pr(A) \cdot \Pr(B)$; see [18].

Another caution. Many people (often the same ones) also have the exact opposite problem: They don't believe events are independent even when they are! Some years ago, one of us volunteered for an experiment at a leading business school. The "experiment" consisted of our being shown a penny (we were allowed to examine it to see that it was ordinary), being told that in a sequence of N tosses it had come up heads m times, and being asked what we then expected the next toss would be. The past results were always very unusual, for instance, in 10 tosses there had been 8 heads. Every time, we shrugged in disbelief at the question and said "It's 50–50." Finally, the experimenter said, "Why don't you think the past history has an effect?" We explained our mathematical background and that we knew coin tosses are independent events. "Well, in that case we won't count your answers in our data; you'll bias our study." We still wonder sometimes what this study claimed to prove.

We now consider how to compute $\Pr(B|A)$ when A occurs *before* B. This problem is usually easier to think through than conditioning on later events, which is what we have done so far. (On the other hand, conditioning on later events is done more often in scientific investigations. Scientists usually want to study something which is not directly observable but leaves observable traces. For instance, a tuberculin infection in the lungs in not directly observable, but usually it leaves an observable trace in the form of an allergic reaction to the tine test.)

EXAMPLE 3 Returning to the urns in Example 1, what is the probability that the second ball is black, given that the first one is?

Solution Surely the answer is 1/3, because there are just three balls left, one of them black. ∎

Strictly speaking, this solution is too easy! We haven't used Definition 1, yet that's the definition which applies whether A or B comes first. To apply that definition, we must reckon in terms of subsets of the original sample space. In this case, that sample space consists of all pairs of the four balls. Yet the easy solution we just gave dealt with picking a single ball from three. Was our answer right?

Yes it was, and so was the method. In general, when conditioning on prior events, you may simply "throw away" what has already happened and you will get the right answer. The reason is that the quick method implicitly involves a probability ratio, and the sets in that ratio are "equivalent" to the ones you would use when following Definition 1. In the next two paragraphs we explain this reason in detail for this particular example. Hereafter, we will stick to the easy solution. You may, too.

What have we really done in our quick solution? The additional information that the first ball was black told us that we could limit our space to $\{b, w_1, w_2\}$, where b is the name of whichever black ball is left. Furthermore, we are interested in the outcome event $\{b\}$. Therefore the answer is $\Pr(b)/\Pr(b, w_1, w_2)$, which is 1/3 since atomic events are equiprobable.

Had we solved the problem using Definition 1, the answer would be

$$\frac{\Pr(N \cap F)}{\Pr(F)},$$

where N and F are as in Example 1. However, we make two observations. First, the couplets in F come in pairs—one couplet with first entry b_1, the other couplet with first entry b_2. For instance, (b_1, w_1) and (b_2, w_1) form such a pair. So do (b_1, b_2) and (b_2, b_1). Replacing both b_1 and b_2 by b, and writing just one couplet for each such pair of couplets in F, we get

$$F' = \{(b, b), (b, w_1), (b, w_2)\}.$$

Now, since the first entry of each remaining couplet is b, we can just ignore the first entries, obtaining

$$F'' = \{b, w_1, w_2\}.$$

Similarly, $N \cap F$ reduces to $\{b\}$, so we are indeed just considering picking one ball out of three. Furthermore, in the process of reducing F to F'' and $F \cap N$ to $\{b\}$, we did not change the ratio of their probabilities. The ratio stayed the same because all atomic events were equiprobable, and we did just as much amalgamating on top of the fraction as on the bottom—there was a two-to-one correspondence.

Sequential Decisions

Heretofore, we have used knowledge of $\Pr(A)$ and $\Pr(A \cap B)$ to compute $\Pr(B|A)$, using Eq. (1). But when $\Pr(B|A)$ is already known, we can turn things around and use knowledge of $\Pr(A)$ and $\Pr(B|A)$ to compute $\Pr(A \cap B)$. That is, multiplying both sides of Eq. (1) by $\Pr(A)$, we obtain

---

Theorem 1. For any events A and B in a sample space,

$$\Pr(A \cap B) = \Pr(A) \cdot \Pr(B|A). \tag{4}$$

Caution. Eq. (4) is true for all events A and B, but Eq. (2) is true only when A and B are independent.

Theorem 1 is very handy for a common type of problem: computing the probability of a string of sequential events.

EXAMPLE 4 No Nonsense College requires each freshman to either take mathematics or start a foreign language. If a student chooses mathematics, she can take either continuous or discrete. If she takes a language, it can be French, German, or Russian. For French there is only one starting course, but in German and Russian one may choose between regular and scientific versions. A student decides to make each decision in turn, randomly! What is the probability she takes discrete math? Scientific German?

A problem like this is best summarized by a decision tree (Fig. 6.3). Formally, the sample space consists of all sequences of decisions which branch from the root of the tree to one of the leaves. Each such sequence is an atomic event. For instance, the sequence (Math, Discrete) is one such atom; (Language, German, Scientific) is another. The *event* "Math" is the set of all sequences which involve the *decision* to take math, namely, the set

$$\{\,(\text{Math}, \text{Continuous}),\ (\text{Math}, \text{Discrete})\,\}.$$

The question is: What is the probability measure for the space?

Solution to Example 4 The student does not decide everything at once. For instance, she does not contemplate taking discrete math, but rather contemplates taking some math, and *given* that she randomly chooses that, then she makes a random choice between continuous and discrete mathematics. That is, the student does not deal with $\Pr(\text{Math} \cap \text{Discrete})$ directly, but rather deals with $\Pr(\text{Math})$ and

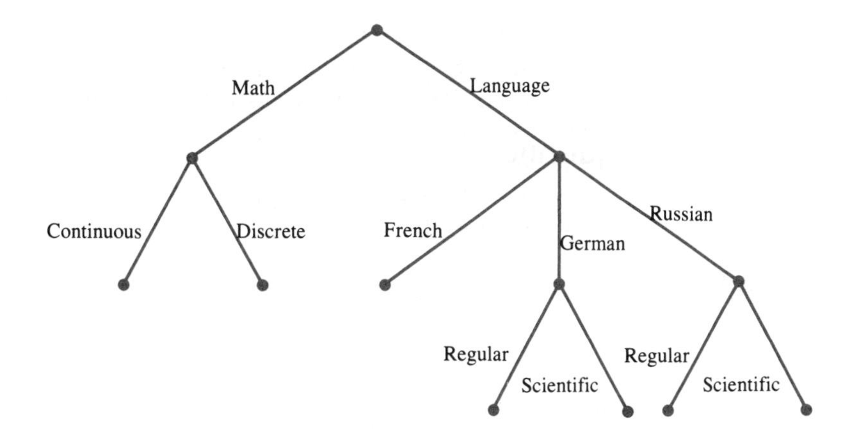

FIGURE 6.3

A decision tree for college courses

then with Pr(Discrete|Math). Moreover, we know what these last two probabilities are, since each decision is random:

$$\Pr(\text{Math}) = \frac{1}{2}, \qquad \Pr(\text{Discrete}\,|\,\text{Math}) = \frac{1}{2}.$$

Thus $\Pr(\text{Math} \cap \text{Discrete}) = 1/4$.

Similarly, let L be the event that the student takes a language, G that she takes German, and Sc that she takes the scientific version of some course. Then

$$\begin{aligned}\Pr(L \cap G \cap Sc) &= \Pr(L \cap G) \cdot \Pr(Sc\,|\,L{\cap}G)\\ &= \Pr(L) \cdot \Pr(G|L) \cdot \Pr(Sc\,|\,L{\cap}G).\end{aligned}$$

Furthermore, we know what these last three probabilities must be: $\Pr(L) = 1/2$, because there was a two-way choice between Math and Language; $\Pr(G|L) = 1/3$, because once the student chooses to take some language, three languages are available; and $\Pr(Sc\,|\,G{\cap}L) = 1/2$, because once she chooses German, there is merely a two-way choice between scientific and regular. Thus $\Pr(L{\cap}G{\cap}Sc) = 1/12$.

Note that there are seven courses the freshman can take, and making sequential random decisions about them does *not* lead to each course being equally likely. ∎

Bayes' Theorem

Let $A_1, A_2, \ldots, A_n$ be a set of **mutually exclusive and exhaustive** events. That is, every element of the sample space S must be in exactly one of the A_i. ("Mutually exclusive" means disjoint; "exhaustive" means their union is S.) Let B be any event in S. Bayes' Theorem is a formula for $\Pr(A_i|B)$, one that's especially useful when B is observable and the A's are not. Before stating the theorem, we illustrate the sort of situation in which it can be used.

EXAMPLE 5

In a certain county, 60% of the registered voters are Republicans, 30% are Democrats, and 10% are Independents. 40% of the Republicans oppose increased military spending, while 65% of the Democrats and 55% of the Independents oppose it. A voter writes a letter to the county paper, arguing against increased military spending. What is the probability that this voter is a Democrat?

Let A_1 be the event "the voter is a Republican". Similarly, for A_2 substitute Democrat and for A_3 Independent. Clearly these are mutually exclusive events; according to the data given they are also exhaustive. Let B be the event "the voter is against increased military spending". Note that, if we read the paper but don't know the writer personally, B is observable but the A's are not. Bayes' Theorem will tell us how to assign revised probabilities to the A's in light of B.

Now let's state Bayes' Theorem, then prove it and apply it to Example 5.

Theorem 2, Bayes' Theorem. Let $A_1, \ldots, A_n$ be mutually exclusive and exhaustive events in sample space S. Let B be any event in S. Then for each i from 1 to n,

$$\Pr(A_i|B) = \frac{\Pr(A_i)\Pr(B|A_i)}{\sum_{j=1}^{n} \Pr(A_j)\Pr(B|A_j)}. \tag{5}$$

PROOF The set manipulations of this proof are best understood by looking at Fig. 6.4. Since $S = \bigcup A_j$, we have

$$B = \bigcup_{j=1}^{n} (B \cap A_j).$$

As all the sets $B \cap A_1, \ldots, B \cap A_n$ are disjoint,

$$\Pr(B) = \sum_{j=1}^{n} \Pr(B \cap A_j).$$

Furthermore, by Theorem 1, for any particular i,

$$\Pr(A_i \cap B) = \Pr(A_i)\Pr(B|A_i).$$

Thus

$$\begin{aligned}
\Pr(A_i|B) &= \frac{\Pr(A_i \cap B)}{\Pr(B)} \\
&= \frac{\Pr(A_i)\Pr(B|A_i)}{\sum_{j=1}^{n} \Pr(A_j \cap B)} \\
&= \frac{\Pr(A_i)\Pr(B|A_i)}{\sum_{j=1}^{n} \Pr(A_j)\Pr(B|A_j)}. \quad \blacksquare
\end{aligned}$$

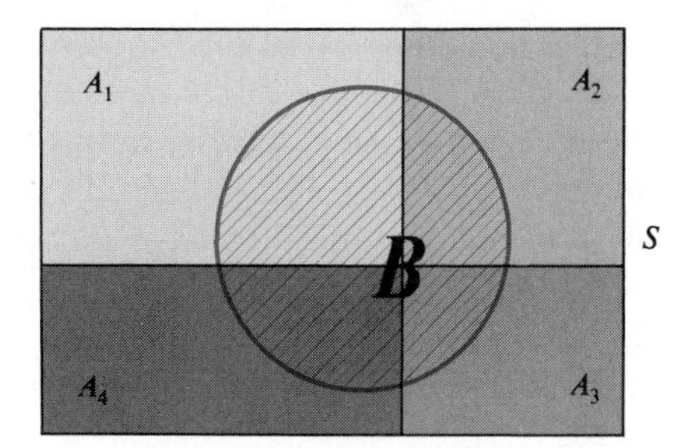

FIGURE 6.4

Diagram for Proof of Bayes' Theorem

Solution to Example 5 Using the notation as set up in the paragraph after the example, we were given the following information:

$$\Pr(A_1) = .6, \qquad \Pr(A_2) = .3, \qquad \Pr(A_3) = .1$$
$$\Pr(B|A_1) = .4, \qquad \Pr(B|A_2) = .65, \qquad \Pr(B|A_3) = .55.$$

Plugging into Bayes' formula, we find that

$$\begin{aligned}\Pr(A_2|B) &= \frac{(.3)(.65)}{(.6)(.4) + (.3)(.65) + (.1)(.55)} \\ &= \frac{.195}{.49} \approx .398\,.\end{aligned}$$

In short, knowledge of the letter increases the probability that the voter is a Democrat from 30% to about 40%. ∎

Solution to the tuberculosis problem (finally!) This problem (Example 2, Section 6.1) asked: What is $\Pr(T|P)$, the probability you are a carrier given that you have tested positive? Plugging into Bayes' Theorem, using T for A_1, $\sim T$ for A_2 and P for B, we get

$$\Pr(T|P) = \frac{\Pr(T)\Pr(P|T)}{\Pr(T)\Pr(P|T) + \Pr(\sim T)\Pr(P|\sim T)}\,.$$

Using the data on false positives and false negatives from the original statement of the problem, this equation becomes

$$\Pr(T|P) = \frac{.92\Pr(T)}{.92\Pr(T) + .04\big(1-\Pr(T)\big)}\,.$$

Now, at least, it is clear what we still need to know. We need to know what fraction of the total population are tuberculosis carriers. For the United States the answer seems to be about $\frac{3}{4}$%. Thus

$$\Pr(T|P) = \frac{(.92)(.0075)}{(.92)(.0075) + (.04)(.9925)} \approx .148.$$ ∎

Does it surprise you that the answer is so low? One's reaction to such a fact might be to question the value of the tine test and to demand a better test. But this reaction shows a lack of understanding of what the tine test is all about. Its real purpose is to show that a large number of people *don't* have TB. As [37] shows, it does this very well: Anyone from the general population who tests negative has probability .9994 of not having TB. Those people who test positive are probably okay too, but they need to have another, more precise test (the Mantoux test or a chest X-ray). Why don't doctors simply give better tests to begin with? Because they are much more time-consuming and costly. The tine test drastically reduces the number of people who need more elaborate tests. For more on tuberculosis, see [38–40].

Problems: Section 6.3

1. Suppose that

$$\Pr(A) = .6,\ \Pr(B) = .5,\ \text{and}\ \Pr(A \cap B) = .4.$$

   a) Are A and B independent?

   b) Determine $\Pr(A|B)$ and $\Pr(B|A)$.

2. Repeat [1], except $\Pr(A \cap B) = .3$.

3. Repeat [1], except instead of information about $A \cap B$, you are given that $\Pr(A \cup B) = .8$.

4. What is $\Pr(A|B)$ if $\Pr(B|A) = .8$, $\Pr(B|{\sim}A) = .3$, and $\Pr(A) = .2$?

5. What is $\Pr(A|Z)$ if events A, B, C are mutually exclusive and

$$\begin{array}{ccc} \Pr(A) = .2, & \Pr(B) = .3, & \Pr(C) = .5 \\ \Pr(Z|A) = .9, & \Pr(Z|B) = .5, & \Pr(Z|C) = .1\,? \end{array}$$

6. At one Chinese restaurant, you have a choice of two soups—eggdrop or Wonton—and then you have a choice of three main dishes—pork, chicken, or beef. Assume that there's nothing else on the menu, that you make each choice randomly, and that you don't skip any choices.

   a) Draw a decision tree and determine the probability of your choosing each possible meal.

   b) What is the probability that you'll eat beef?

   c) What is the conditional probability you chose Wonton soup given that you choose pork?

7. At another Chinese restaurant, you have the same choice of soups, but if you choose eggdrop soup, then you must choose pork or beef, and if you choose Wonton soup, then you must choose either beef, chicken, or Peking Duck. Again assume that you choose randomly between the soups and then randomly among the main course entrees available to you.

   a) Draw a decision tree and determine the probability of your choosing each possible meal.

   b) What is the probability that you'll eat beef?

   c) What is the conditional probability that you chose Wonton soup given that you choose pork?

   d) What is the conditional probability that you chose Wonton soup given that you choose beef?

8. In Example 1, name the elements of event N.

9. Weather on consecutive days is not independent. Let us suppose that all days can be classified as Good or Bad as far as weather is concerned. Suppose also that the probability of weather today conditioned on weather yesterday is as follows:

	Good yesterday	Bad yesterday
Good today	.7	.4
Bad today	.3	.6

Finally, suppose the probability that any given day is Good is 4/7.

   a) If today is Good, what is the probability that yesterday was Good?

   b) If today is Bad, what is the probability that yesterday was Good?

Note: It turns out that the probability that any given day is Good is redundant; it can be deduced from the table by a clever argument. Can you figure it out? (If not, wait until you study Markov Chains in Section 8.10.)

10. From the definition of conditional probability, check that

$$\Pr(B|A) = P(B) \iff \Pr(A|B) = \Pr(A).$$

11. You are given that $\Pr(B) = .5$ and $\Pr(A|B) = .4$. For each of the following probabilities, either determine its value or show that its value is not unique. (Do the latter by giving two different "weighted" Venn diagrams, each of which is consistent with the given information. By a weighted Venn Diagram involving sets A and B, we mean one in which all four events, $A \cap B$, $A \cap \sim B$, $\sim A \cap B$, and $\sim A \cap \sim B$, have been assigned probabilities which add to 1.)

a) $\Pr(A \cap B)$ **b)** $\Pr(\sim A|B)$
c) $\Pr(A|\sim B)$ **d)** $\Pr(\sim A)$

12. In Example 5, what percent of the registered voters are against increased military spending?

13. A penny is tossed until either a head comes up or three tosses have been made, whichever comes first. Assume the coin is fair.

a) List all the atomic events in the sample space and give the probability for each.

b) What is the probability that the game ends with a head? With the third toss? With the second toss?

14. One penny in a million has two heads. A friend tosses the same penny 10 times and reports that he got 10 heads. Based on this information alone, what probability should you give to the possibility that his penny has two heads? (Assume that all pennies other than two-headed pennies are fair coins with heads and tails.)

15. How does the answer to [14] change if your friend got all heads on 20 tosses? How does it change if you find out that there are also two-tailed pennies, again one in a million?

16. The following list gives some information about students who take introductory calculus at Podunk U. Each row lists a department, then the percentage of calculus students who end up majoring in that department, and finally the percentage of those majors who got an A in calculus. For instance, the first row says that 25% of calculus students eventually major in math, and of those math majors, 40% got an A in calculus.

Math	25	40
Science	20	30
Engineering	40	25
Social science	10	20
Humanities	5	15

a) What fraction of the students enrolled in calculus get A's?

b) John Doe gets an A. What is the probability he will major in Math? In Engineering?

17. A mediocre student gets a perfect paper on a 20-question multiple choice test. The teacher is suspicious and speaks to the student. The student acknowledges that he didn't know the answers to the last 10 questions, but claims that he simply guessed at random and was lucky. Each question had 5 answers listed. The teacher estimates that 1 student in 1000 cheats. Should she believe this student? What if her estimate is wrong and only 1 student in 10,000 cheats?

18. For Example 1, here are three other sample spaces we might have chosen. (They are stated in the parts below.) Which ones are appropriate with an equiprobable measure? What are F and N in each? Check that you get the same conditional probability $\Pr(F|N)$ as in the text for each sample space that you declare appropriate.

a) The set of all 4! permutations of $\{b_1, b_2, w_1, w_2\}$.

b) The set of all 4!/2!2! permutations of the multiset $\{b, b, w, w\}$.

c) The set of all ordered pairs each of whose elements is either b or w.

19. The second-child problem. In the following parts, you are asked probability questions about the sex of children. In each case, assume that any child is equally likely to be male or female and that the sex of any one child in a family is independent of the sex of any other children in that family. The questions are quite tricky, but only because your intuition may mislead you; no linguistic or other cheap tricks are intended. The way to keep from going wrong is to be very careful about defining your probability space and the particular events that you consider in it. State your space and your events carefully.

a) You visit the home of an acquaintance, who says, "I have two kids." A boy walks into the room. The acquaintance says, "That's my older child." What is the probability that the younger one is a boy?

b) You visit the home of an acquaintance, who says, "I have two kids." A boy walks into the room. The acquaintance says, "That's one of my kids." What is the probability that the other one is a boy?

c) You live in a culture where, when children are introduced, male children are always introduced first, in descending order of age, and then female children, also in descending age order. You visit the home of an acquaintance, who says, "I have two kids, let me introduce them." He yells, "John come here." (John is a boy's name.) What is the probability that the other child is a boy?

d) You go to a parent-teacher meeting. The principal is sitting in the first row. You've heard that the principal has two children. The teacher in charge asks everyone who has a son (meaning at least one) to raise a hand. The principal raises her hand. What is the probability that the principal has two sons?

20. (Philosophical question) In Example 1, we said the first draw depended on the second. But this is absurd. Something can never depend on something else that happens later. Which ball shows up second can never affect which ball showed up first. Thus the conditional probability should equal the unconditional probability (as in Example 2). Respond to this objection.

21. Let $\Pr(A) = 2/3$, $\Pr(B) = 3/4$. What is the largest that $\Pr(A \cap B)$ can be? The smallest? Is every number in between also possible?

22. If $\Pr(A) = 2/3$ and $\Pr(B) = 3/4$, what is the largest and smallest $\Pr(B|A)$ can be? Answer the same question for $\Pr(A|B)$.

23. Do [21] again in general. If $\Pr(A) = p$ and $\Pr(B) = q$, what is the range of possible values for $\Pr(A \cap B)$?

24. "Being independent is like being perpendicular." This admittedly fuzzy assertion can be made precise in a surprising number of situations in mathematics—situations that on the face of it have nothing to do with either probabilistic independence or geometric perpendicularity! Here we give one simple way to connect these two notions: rectangular Venn diagrams. Represent sample space S by a rectangle. For all events $E \subset S$, let $\Pr(E)$ be represented by the relative area we assign to E in the diagram, i.e,

$$\Pr(E) = \frac{\text{Area } E}{\text{Area } S}.$$

Now, if independence is like perpendicularity, let's see if we can make events A and B independent by representing them as perpendicular rectangles. That is, let A be horizontal, stretching all the way across S. Let B be vertical, stretching from top to bottom of S. Now verify that

$$\Pr(A \cap B) = \Pr(A) \cdot \Pr(B).$$

25. It is a fact that

$$\Pr(B \cap C \,|\, A) = \Pr(C \,|\, A \cap B) \Pr(B|A).$$

a) Give an intuitive explanation. *Hint*: Imagine that A, B, C are events which, if they happen at all, must happen in that order.

b) Of course, the events don't have to happen in the order A, B, C, so if you used the hint in a) your proof in not really general. Give a "real" proof, using Definition 1.

26. For Example 4, compute the probabilities of all the atomic events. Use them to compute:

a) The probability that the student takes a regular language course.

b) The probability she takes a scientific language course.

c) The probability that she takes a scientific language course, given that she takes a language.

d) The probability that she takes Russian, given that she takes a language. The answer to this part should be obvious, but compute it by the method requested. Then see the comments following part e).

e) The probability that she takes Russian, given that she takes a scientific language course.

We were not entirely honest about what we were doing in Example 4. We made it look like we were deciding the value of $\Pr(L \cap G \cap Sc)$ using knowledge of various conditional probabilities. But this is backwards! Conditional probabilities are defined only in terms of unconditioned probabilities such as $\Pr(L \cap G \cap Sc)$. That is, the probability measure must come first.

What we were really doing in Example 4 was *defining* the probability measure, using what we *wanted* the conditional probabilities to be as a guide in choosing our definitions. In the previous parts of this problem, you were asked to finish defining that probability measure by computing the probabilities of all atomic events.

f) Use what you found in a)–e) to verify that all the conditional probabilities, blithely treated as known in Example 4, are in fact exactly what we assumed.

We didn't mention all these subtleties in the text because the main thing in this first course is to learn the technique presented in Example 4 for doing sequential decision problems. It's easy and it always works. (The probability measure obtained will always imply that the conditional probabilities really are what they were assumed to be.) We omit both a precise statement and a proof of this theorem, because both are a bit tricky to set up. (If you'd like to give it a try, we suggest proof by induction on the length of the decision tree.)

27. Actuaries (life insurance mathematicians) have collected various sorts of statistics about mortality. One important statistic is p_n, the probability that a person who lives to age n years will continue to live until at least age $n+1$. (Actually, there are separate statistics for men, women, smokers, etc., but let's ignore that.) Note that the p_n are actually conditional probabilities; in what probability space? Express the following probabilities in terms of the p_n's. The answers may be complicated expressions involving $\sum$ and $\prod$ notation.

a) The probability that a person age 30 lives to at least age 50

b) The probability that a person age 30 dies before age 31

c) The probability that a person age 30 dies at age 65

d) The probability that a person age 30 dies before age 65

e) The probability that a person age 60 dies between 65 and 70

f) The probability that a newborn baby lives to age 65

g) The probability that a person who died before age 40 died after age 30

28. Express your answers to [27] using algorithms rather than formulas (which is what actuaries do today anyway—computers have liberated them from many tedious calculations).

29. Another important actuarial statistic is l_n, the probability that someone just born will live to at least age n years. Express all the parts of [27] again, using the l_n's.

30. Express the answers to [29] using algorithms.

31. Prove: If A and B are independent, then so are A and $\sim B$, so are $\sim A$ and B, and so are $\sim A$ and $\sim B$.

32. Definition. $A_1, \ldots, A_n$ are **mutually independent** if for all $I \subset \{1, 2, \ldots, n\}$,

$$\Pr\Big(\bigcap_{i \in I} A_i\Big) = \prod_{i \in I} \Pr(A_i).$$

In other words, the "product rule" for probabilities holds for any subset of these events.

a) Show that if A_1, A_2, A_3 are mutually independent, then

$$\Pr(A_1 \cap A_2 \cap {\sim}A_3) = \Pr(A_1)\Pr(A_2)\Pr({\sim}A_3).$$

In other words, A_1, A_2, and $\sim A_3$ are also independent.

b) Generalize part a).

c) If every *pair* of A_1, A_2, A_3 are independent (Definition 2), does this imply that the three events are *mutually* independent?

33. Definition. Let S be a sample space with probability measure Pr. Let A be any subset with $\Pr(A) \neq 0$. Then the **induced probability measure** $\Pr_A$ on A is defined on subsets C of A by

$$\Pr_A(C) = \frac{\Pr(C)}{\Pr(A)}.$$

Show that this definition does indeed define a probability measure with A as its sample space. (That is, show that $\Pr_A$ meets the conditions in Definition 1, Section 6.2.)

34. Show that if S is a finite set and Pr is the equiprobable measure on S, then any induced measure on a subset of S is also an equiprobable measure.

35. The fact that we have called $\Pr(B|A)$ a conditional *probability*, not just a conditional number, suggests that it really ought to satisfy the requirements for probability in its own right. Show that this is correct.

36. In the proof of Bayes' Theorem, we expanded $\Pr(A_i \cap B)$ as $\Pr(A_i)\Pr(B|A_i)$. Why didn't we expand it as $\Pr(B)\Pr(A_i|B)$? Isn't the second way correct, too? Wouldn't it give an equally true theorem?

37. (Tuberculosis testing) Using $\Pr(T) = .0075$ and the data of Example 2, Section 6.1, verify the claim on page 460 that $\Pr(\sim T|\sim P) \approx .9994$.

38. The ideal in medical tests is to find a cheap, quick test which is so precise that no further tests are needed. However, it is unlikely one can find such tests when the condition being tested for is rare—because the test must be *extremely* precise, more precise than most things in medicine are. Suppose we found a test such that every TB carrier tests positive. Let p be the probability that someone who is not a carrier also tests positive on this test. How small does p have to be in order for $\Pr(T|P)$ to equal .95? Assume that $\Pr(T) = .0075$. (*Note*: For a test to be completely accurate for those people who really are carriers, the test probably has to err in favor of positive results. Is a p this low likely, given such "bias"?)

39. The prevalence of tuberculosis infection in the United States is generally low, but it is far from uniform. In certain immigrant or minority neighborhoods it is as high as 15%. That is, if S is restricted to certain subpopulations, $\Pr(T) = .15$. Using this value for $\Pr(T)$, and otherwise continuing to use the probabilities given in Example 2, Section 6.1, recompute $\Pr(T|P)$ and $\Pr(\sim T|\sim P)$. (Doctors call $\Pr(T|P)$ the "sensitivity" of the test and $\Pr(\sim T|\sim P)$ its "specificity".)

40. For various good reasons, the only group in the general U.S. population that currently receives widespread TB testing is preschool children. For this group $\Pr(T) = .003$. Do [39] again using this value.

41. In Example 5, not only do we know that the letter-writing voter is opposed to increased military spending, but we also know that he or she feels strongly enough about it to write to the newspaper. Suppose we somehow know that 5% of Democrats in that county oppose military spending strongly enough to write (at some point) to the newspaper about it and that 2% of Republicans and 3% of Independents also feel that strongly. Does this change the answer to the question?

42. As in Example 5, Section 6.1, assume that blue eyes are recessive. But make the assumption that 20% of eye genes are type b, as opposed to assuming that 20% of people have blue eyes. (This new assumption is easier to work with mathematically, but it is not directly observable.) Assume that each person's two eye genes are independent, i.e., each gene is type b with independent probability .2. Also assume that the eye genes of married couples are independent. (This assumption is questionable; for instance, blue-eyed people might tend to select each other as mates.) With all these assumptions, answer the question of Example 5, Section 6.1.

43. Solve Example 5, Section 6.1, itself, with the same assumptions about gene independence as in [42].

44. The following questions may shed some light on why many people believe that past performance of coin-tossing (or repeatable random events in general) affects future performance. In your answers, be explicit about your sample space and probability measure.

a) An ordinary coin was tossed 10 times and heads came up 8 times. If it is tossed again, what is the probability of getting a head?

b) An ordinary coin was tossed 10 times and heads came up 8 times. What is the probability that the tenth toss was a head?

c) An ordinary coin was tossed 10 times and heads came up 6 times. However, 4 of the first 5 tosses were heads. What is the probability that the tenth toss was a head?

45. We defined $\Pr(B|A)$ only when $\Pr(A) \neq 0$, but in fact it is often possible to give $\Pr(B|A)$ a useful definition even when $\Pr(A) = 0$. (This happens most often for continuous probability spaces. In any event, this fact is one reason why we rarely state $\Pr(A) = 0$.) Show that Theorem 1 is true when $\Pr(A) = 0$ no matter how $\Pr(B|A)$ is defined.

6.4 Random Variables and Probability Distributions

Random variables are a device for transferring probabilities from complicated sample spaces to simple spaces whose elements are numbers. Once such a transfer is made, the probability measure is usually described as a "distribution"—intuitively the probability is distributed on the real line. More important, once such transfers are made, different-looking probabilities spaces often turn out to be the same. In fact, a small number of distributions cover most situations. We explain these things in this section and introduce the most important discrete distributions.

The definition of a random variable is quite simple; only the name is mystifying. So we go right to the definition and then explain the name.

Definition 1. A **random variable** is any real-valued function on a sample space.

EXAMPLE 1 Consider the usual sample space for tossing two dice:

$$\{ (i,j) \mid 1 \le i,\ j \le 6 \}.$$

Let $X(i,j)$ be the sum of the dots when toss (i,j) occurs. That is, $X(i,j) = i+j$. Then X is a random variable. ∎

EXAMPLE 2 Let Y be a man's shoe size (length only, American sizes). So Y is really a function whose domain S is the set of American men. For instance, Y(Stephen Maurer) = $12\frac{1}{2}$. When S is viewed as a sample space, Y is a random variable. ∎

The reason for the name "variable" is simple: The quantity varies. It's just like calling y both a variable and a function of x when we consider $y = x^2$. The name "random", though unnecessary, reminds us of the probability context. If we are viewing the set of men as a sample space, presumably we have some chance events to consider. Perhaps we run a shoe store and want to know how many shoes of different sizes to stock to meet customer demand. The variable is random in that we don't know what its value will be when the next atomic event occurs, i.e., another man enters our store.

The set of all American men is an unwieldy sample space. The elements have long names, they can't all be represented easily in a graph, etc. Besides, we don't particularly care *who* buys our shoes; we care about their shoe sizes. The set of shoe sizes is finite and has a natural order as points on the x-axis. It is much simpler to associate a probability with each shoe size than with each person. For instance, if size 10 has the highest probability, then we will want to stock more size 10 shoes than other sizes.

We started with the idea of a random variable, but we immediately went on to a second important idea: probability *on* a random variable. In effect, a random variable from S to R transfers the probabilities from S to a new sample space consisting of a subset of R. For instance, in Example 2 the new sample space consists of all those integers and half integers that are American men's shoe sizes. The number 10 in the new space gets assigned the probability of the event in the original space consisting of all men who wear size 10 shoes. Put another way, the probability of size 10 is the probability that the next man who enters the store wears size 10.

We may formalize this transfer with the following two definitions.

Definition 2. Let X be a random variable on sample space S with probability measure Pr. Then the probability that X takes on the value c, written $\Pr(X{=}c)$ or $\Pr_X(c)$, is defined by

$$\Pr(X{=}c) \;=\; \Pr\big(\{s{\in}S \,|\, X(s){=}c\}\big).$$

Similarly,

$$\Pr(X{\leq}c) \;=\; \Pr\big(\{s{\in}S \,|\, X(s){\leq}c\}\big),$$

and more generally, if C is a set of real numbers, then

$$\Pr(X{\in}C) \;=\; \Pr_X(C) \;=\; \Pr\big(\{s{\in}S \,|\, X(s){\in}C\}\big).$$

In the spirit of this definition, we may also use, for instance, the notation $\{X = c\}$ to represent the event $\{s{\in}S \,|\, X(s){=}c\}$. See Example 3.

We have used X to transfer probability from S to values $c \in \text{Range}(X)$. If we now think of c as varying, we do indeed get a probability measure on Range(X). When this range is a discrete set of points, it is customary to describe this new measure using the terminology and notation of the next definition. We use x instead of c to emphasize that we have a variable.

Definition 3. Let X be a random variable on a sample space S with probability measure Pr. Assume that the range of X is a discrete subset of the real numbers. Then the function f defined on Range(X) by

$$f(x) \;=\; \Pr(X{=}x) \tag{1}$$

is called the **probability distribution** or **point distribution** associated with X. The function

$$F(x) \;=\; \Pr(X{\leq}x) \tag{2}$$

is called the **cumulative distribution** function for X (or for f).

The letters f and F in Eqs. (1) and (2) are customary, to distinguish these probability functions on the real numbers from the underlying probability measure on S. Actually, though it is natural for probability distributions to arise from transferring probability to the real numbers R from some more general sample space, the original space could itself be a discrete subset of R, in which case a probability measure on it is already a distribution. In other words, a probability distribution is just the restriction to the atoms of *any* probability measure on a discrete subset of R. The cumulative distribution is just the "running sum" of the point distribution.

EXAMPLE 3 Let S be the usual equiprobable, ordered-pair sample space for tossing two dice, and let $X(i,j) = i+j$. (See Example 1.) Determine the real-number sample space associated to S by X and determine its point distribution and its cumulative distribution.

Solution The new sample space is the range of X, namely, $\{2,3,\ldots,12\}$. The point distribution is

$$\begin{aligned}
f(2) &= \Pr(X{=}2) = \Pr(\{(1,1)\}) = \frac{1}{36},\\
f(3) &= \Pr(X{=}3) = \Pr(\{(1,2),(2,1)\}) = \frac{2}{36},\\
&\vdots\\
f(7) &= \Pr(X{=}7) = \Pr(\{(1,6),(2,5),\ldots,(6,1)\}) = \frac{6}{36},\\
&\vdots\\
f(12) &= \Pr(X{=}12) = \Pr(\{(6,6)\}) = \frac{1}{36}.
\end{aligned} \tag{3}$$

Thus the cumulative distribution is

$$\begin{aligned}
F(2) &= f(2) = \frac{1}{36},\\
F(3) &= f(1) + f(2) = \frac{3}{36},\\
&\vdots\\
F(12) &= f(1) + \cdots + f(12) = 1.
\end{aligned}$$

Go back to Example 2 Continued, Section 6.2. Do you see that, in computing $f(x)$ in display (3), we have repeated exactly what we did there? Thus the explicit formula for $\Pr(n)$ there is also a formula for $f(n)$ here. By using formulas for summing arithmetic series, we could also obtain [6] an explicit two-case formula for $F(n)$. ∎

Logically speaking, it is not necessary to define both a probability distribution and a cumulative distribution; each can be derived from the other. For instance, if the elements of the sample space are $x_1 < x_2 < \cdots < x_n$, then for each k

$$F(x_k) = \sum_{i=1}^{k} f(x_i)$$

and

$$f(x_k) = \sum_{i=1}^{k} f(x_i) - \sum_{i=1}^{k-1} f(x_i) = F(x_k) - F(x_{k-1}).$$

However, we have defined both concepts because they are both convenient to use. For instance, probabilities that are very natural to ask about, e.g., $\Pr(a<X<b)$, are easily described in terms of F; see [17].

Note. When we call something simply a "distribution" (e.g., "the uniform distribution" below), we mean the probability and cumulative distributions as a pair. Unfortunately, another common usage is for "distribution" to mean "probability distribution" alone. Even more unfortunately, in books that discuss continuous distributions, "distribution" sometimes means "cumulative distribution", because the analog of the point distribution in that context has a very different name: the density function. Such is life.

We now turn to four important (classes of) distributions. All four show up frequently in applications, and the first three show up repeatedly in other sections of this chapter.

EXAMPLE 4

The Finite Uniform Distribution

For this distribution, if $S \subset R$ and $n = |S|$, then for all $x \in S$,

$$f(x) = \frac{1}{n}.$$

In other words, the finite uniform distributions are precisely the distributions in which atomic events are equiprobable.

As for the cumulative distribution function, if we again label the elements of S as $x_1 < x_2 < \ldots < x_n$, then

$$F(x_k) = \frac{k}{n}.$$

In some cases, the notation is even simpler. For instance, if $S = \{1, 2, \ldots, n\}$, then $F(k) = k/n$.

One advantage of distributions over arbitrary probability measures is that the order on the real line makes a graphical display easy and informative. In Fig. 6.5, we show the point and cumulative distributions for the uniform measure on $\{1, 2, 3, 4, 5\}$. ∎

EXAMPLE 5

The Binomial Distribution

An experiment, with probability p of success and thus $1-p$ of failure, is repeated independently n times. Such an experiment is called a sequence of **Bernoulli trials**. We seek the point distribution, $B_{n,p}(k)$, which gives the probability that exactly k trials are successes.

The natural sample space for Bernoulli trials consists of n-tuples like SSFSFFF..., where S means success and F means failure. We map this space over to the real numbers $\{0, 1, 2, \ldots, n\}$ by grouping together all n-tuples that have the same number of successes. Since the trials are assumed to be independent, the appropriate probability model involves multiplication. That is, if an n-tuple contains

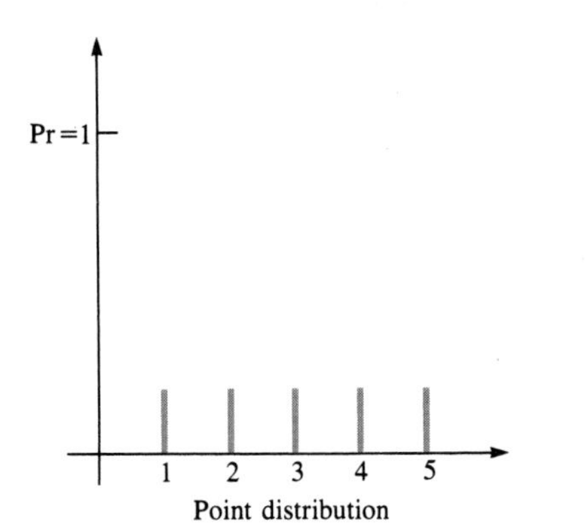

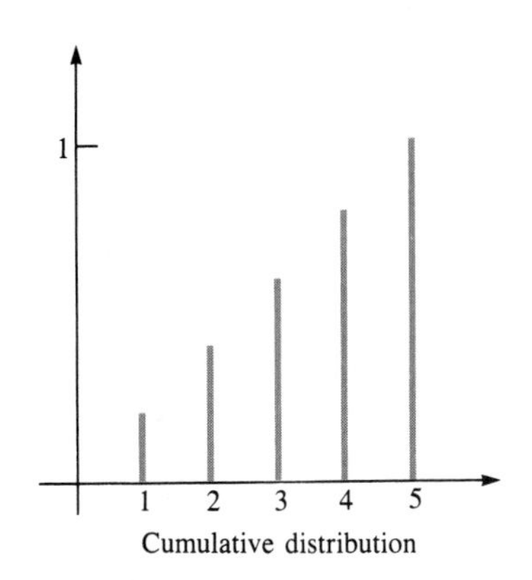

FIGURE 6.5

The Uniform Distribution on $\{1,2,3,4,5\}$, shown in point and cumulative form with bar graphs

k S's and $(n-k)$ F's, then the probability of the tuple is $p^k(1-p)^{n-k}$. Finally, there are $C(n,k)$ n-tuples with k S's (why?), and they represent disjoint events. Thus the point distribution is

$$B_{n,p}(k) = \binom{n}{k} p^k (1-p)^{n-k}. \tag{4}$$

The distribution for Bernoulli trials is called the binomial distribution because of Eq. (4), which looks like a term out of the Binomial Theorem. Only occasionally is it called the Bernoulli distribution.

There's no simple exact formula for the cumulative binomial distribution; from the definition, we simply have to say

$$F(k) \;=\; \sum_{j=0}^{k} B_{n,p}(j) \;=\; \sum_{j=0}^{k} \binom{n}{j} p^j (1-p)^{n-j}.$$

Figure 6.6 shows the binomial distribution for $n = 10$ and $p = .5$. Figure 6.7 shows it for $n = 20$ and $p = .2$. Note that the vertical scales on the point distributions have been stretched to improve readability. From these point distributions you can easily see which values k are most likely and which are very unlikely. Are the results what you would expect? ∎

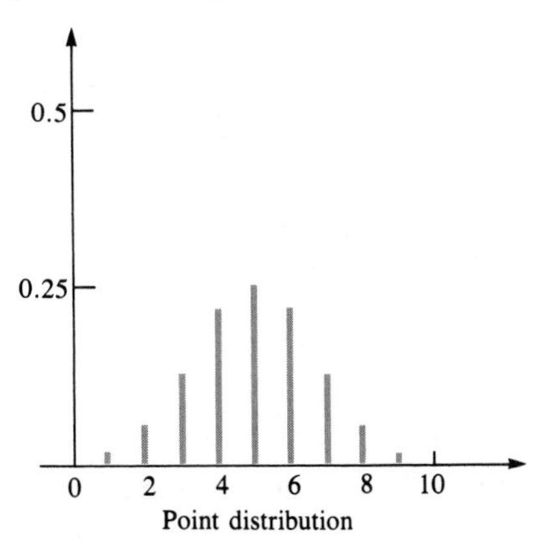

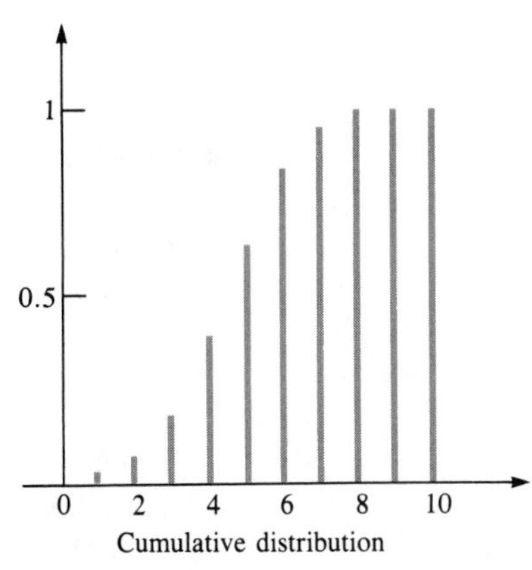

FIGURE 6.6

The Binomial Distribution for $n = 10$ and $p = .5$. Note the different vertical scales.

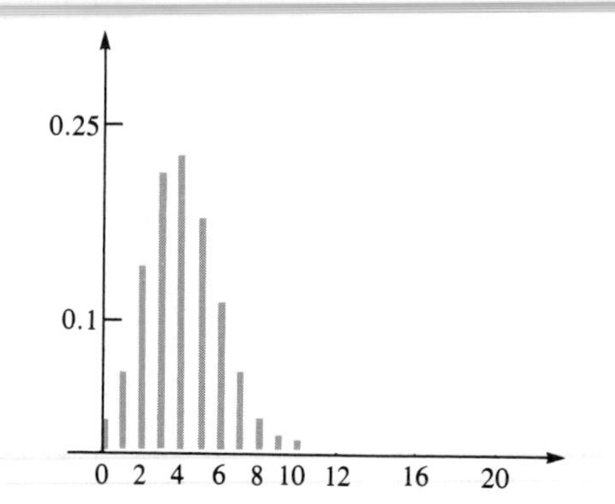

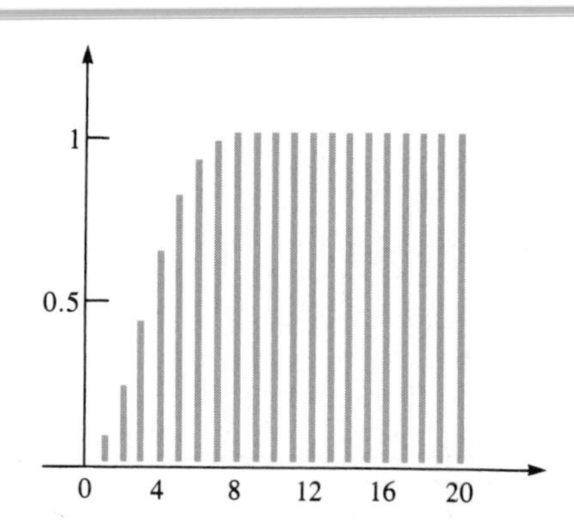

FIGURE 6.7

The Binomial Distribution for $n = 20$ and $p = .2$

Note. Throughout the study of probability, many formulas involve both p and $(1-p)$. To shorten things, the following notation has become the

$$\text{Probabilist's tradition:} \quad q = 1 - p.$$

For instance, the binomial distribution may be written as

$$B_{n,p}(k) = \binom{n}{k} p^k q^{n-k}.$$

Solution to Example 1, Section 6.1 (Guessing on College Boards) The question was: What is the probability that you will improve your score? The example didn't say how many problems you guess on. Let's say it's 10. Your score on these problems is $r - (w/4)$, where r is your number of right answers and $w = 10 - r$ is your number wrong. Thus in order to come out ahead, you must have

$$r - \frac{10 - r}{4} > 0 \iff r > 2.$$

Since there are 10 independent guesses, each with probability .2 of being right, we are talking about the binomial distribution $B_{10,.2}(r)$. Therefore the probability of improving your score is

$$\sum_{r=3}^{10} B_{10,.2}(r).$$

This expression involves eight terms, and they are "going the wrong way" insofar as the cumulative distribution is concerned. So we apply a little algebra:

$$\begin{aligned}
\sum_{r=3}^{10} B_{10,.2}(r) &= \sum_{r=0}^{10} B_{10,.2}(r) - \sum_{r=0}^{2} B_{10,.2}(r) \\
&= 1 - \sum_{r=0}^{2} B_{10,.2}(r) \qquad (5) \\
&= 1 - F(2). \qquad (6)
\end{aligned}$$

Tables have been compiled for all the common cumulative distributions. If we have such a table, Eq. (6) will be easy to look up. Even if we don't, the right-hand side of Eq. (5) requires summing only three terms instead of eight. Specifically,

$$F(2) = \left(\frac{4}{5}\right)^{10} + 10\left(\frac{1}{5}\right)\left(\frac{4}{5}\right)^{9} + 45\left(\frac{1}{5}\right)^{2}\left(\frac{4}{5}\right)^{8} \approx .6778.$$

Thus the probability of improving your score by pure guessing on 10 questions is

$$1 - F(2) \approx 1 - .6778 = .3222.$$

This seems like a pretty good chance of success for no mental effort, but before you go out and try it, also compute the probability of decreasing your score [25]. In fact, for an informed judgment, you really need to take into account the probabilities of various *amounts* of increase and decrease. In the next section, we show how to summarize such proliferating information. ▮

EXAMPLE 6 **The Negative Binomial Distribution**

Again suppose you run an experiment repeatedly and independently, but this time you keep going until you obtain your nth success. What is the probability that you need exactly k trials?

Solution The underlying sample space now consists of S/F-tuples of varying length. For instance, if $n = 2$, some of them are:

SS, FSFFFS, and FFFFSFFS.

There are also infinite-tuples in the space—all F's and all F's except for one S anywhere. (What's the probability of any one of the infinite-tuples? [28]) Any one sequence which takes k trials to obtain n successes will have probability $p^n q^{k-n}$. There will be many such sequences; the final S must go at the end, but the $n-1$ others may go in any of the other $k-1$ slots. Thus

$$f(k) = \binom{k-1}{n-1} p^n q^{k-n}, \qquad k \geq n.$$

This is the probability distribution for the negative binomial of **order** n. (If one refers simply to the negative binomial, that means the negative binomial of order 1.) Note that the domain of this distribution is an infinite discrete set—all integers $k \geq n$. Note also that $k \geq n$ is the reverse inequality from that in the binomial distribution; hence n and k switch roles in all the formulas. (Why don't we switch the meanings of n and k in the definition of the negative binomial and thus make the roles in the formulas the same in both distributions? Because the usage we have chosen is consistent in a more important way: In both distributions n is a *parameter*, that is, an index which indicates which of several related distributions we are talking about; and in both k is the random variable.)

Again, the cumulative distribution is obtained by summing.

Figure 6.8 shows the negative binomial of order 3 with $p = .5$. We can't show the whole domain, so we have cut it off at $k = 20$.

Why is this distribution called the negative binomial? Because its terms arise in the Binomial Series Theorem (Theorem 2, Section 4.6) when $-n$ is the exponent [32].

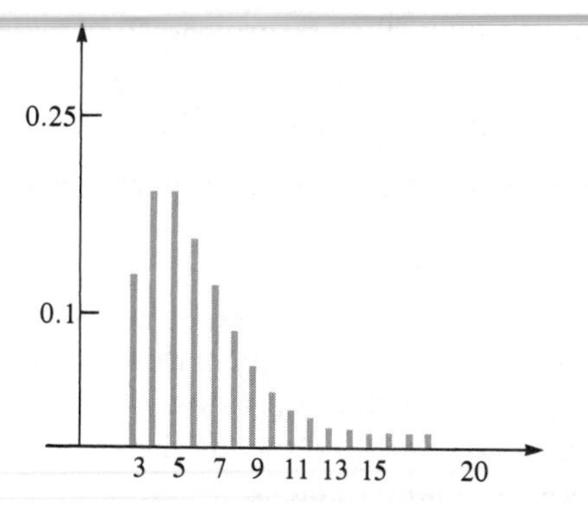

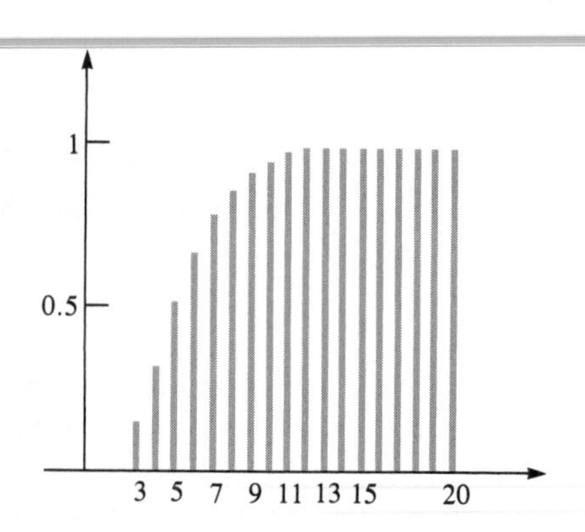

FIGURE 6.8

The Negative Binomial of Order 3 with $p = .5$

Note that Example 3, Section 6.1 (constructing a permutation by random selections), is really about the negative binomial distribution. We repeatedly search for another unused number, until the first success. Once the kth new number has been selected, the probability in each succeeding trial of successfully picking a $(k+1)$st unused number is $p = (n-k)/n$. We are interested in the total number of trials before all numbers from 1 to n have been used. ∎

EXAMPLE 7 **The Poisson Distribution**

A large computer receives, on average, λ new jobs a minute. What is the probability that exactly k new jobs will come in during a given minute?

Solution There are many problems of this sort. Another is: If the telephone company handles, on average, λ calls a minute, what is the probability that, in the course of a day, at some point they will need k circuits (if no caller is to receive an "all circuits busy" signal)?

Problems of this sort are usually modeled very well by the following distribution:

$$f_\lambda(k) \;=\; e^{-\lambda}\frac{\lambda^k}{k!}, \qquad k = 0, 1, 2, \ldots. \tag{7}$$

The function f_λ is called the **Poisson distribution**. Sorry about the λ (a Greek lambda); a roman letter would do fine (so long as it does not make you think the value must be an integer, or that it varies), but λ is traditional. Also in Eq. (6), e is the famous base of the natural logarithms, $2.71828\ldots$.

Where does Eq. (7) come from? It is the limit of a Bernoulli trials model. To see this, assume that at most one new computer job comes in per second. If, on average, λ new jobs come in per minute, that suggests that the probability of receiving a new job in any given second is $\lambda/60$. From this we can calculate the probability that new jobs begin during exactly k of the 60 seconds—it's just the probability of k successes (job starts) in 60 Bernoulli trials (the seconds).

The reason this Bernoulli model isn't good enough is that it doesn't allow more than one new job to come in each second. So, let's divide seconds into tenths, and model the situation again with 600 segments in a minute instead of 60. If we keep increasing the number of segments, in the limit there is no bound on the number of jobs that could come in per minute (or per any nonzero time period). The formula we get in the limit is Eq. (7).

To show that we get Eq. (7), we need a fact from calculus about e and some familiarity with limit calculations. The fact from calculus is:

$$e^x = \lim_{n\to\infty} \left(1 + \frac{x}{n}\right)^n.$$

We use it as follows. Fix λ and let n be the number of intervals into which the single Poisson period is to be divided in a Bernoulli-trials approximation. Then the probability of "success" in each interval is $p = \lambda/n$. Hence, the probability of k successes in the binomial approximation is

$$\begin{aligned}\binom{n}{k} p^k q^{n-k} &= \binom{n}{k}\left(\frac{\lambda}{n}\right)^k\left(1-\frac{\lambda}{n}\right)^{n-k} \\ &= \binom{n}{k}\left(\frac{\lambda/n}{1-\frac{\lambda}{n}}\right)^k\left(1-\frac{\lambda}{n}\right)^n.\end{aligned}$$

Now we let n go to infinity, use the fact from calculus, and simplify—in the limit many of the numbers in the first two factors cancel [29].

Figure 6.9 shows the Poisson distribution for $\lambda = 4$. If we were to approximate this by a binomial distribution with $n = 20$, then p would be $4/20 = .2$. But $B_{20,.2}$ is exactly the binomial distribution in Fig. 6.7. Compare! ∎

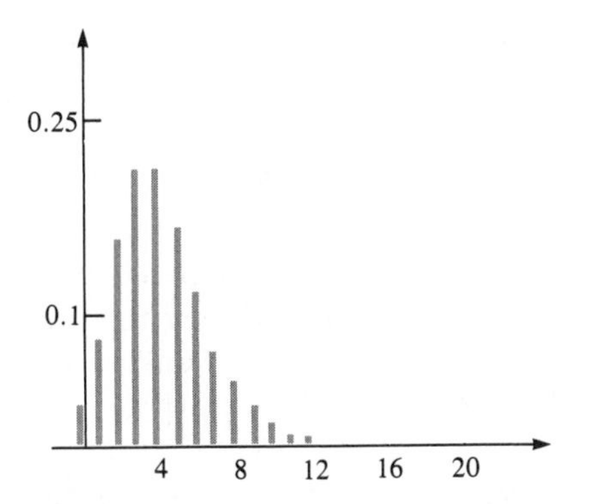

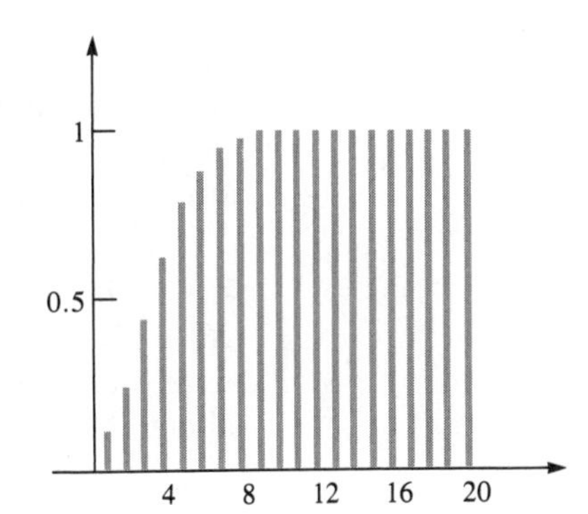

FIGURE 6.9

The Poisson Distribution for $\lambda = 4$ (k shown up to 20)

New Distributions from Old

There are lots of ways to make new distributions directly from previous distributions. We close this section with a discussion of the most useful way, addition. If X and Y are random variables on the same space S, then so is $X + Y$, for is it not a real-valued function? Therefore it has a distribution.

But why should this interest us? Because mathematicians prefer thinking to remembering. There are a lot of distributions out there. We can either try to catalog and remember them all (ugh!), or we can seek methods of generating many from a few and remember only those few (yeah!). Reducing something to a small number of simple base cases is a general principle of good mathematics.

In fact, we will use our ability to relate random variables through addition in two ways. One is to add the variables we have already discussed and see what we get. The other is to see if we can represent the variables we have already discussed as sums of still simpler variables. The latter approach will be very helpful in Sections 6.5 and 6.6.

EXAMPLE 8 Two distinct dice are tossed. Let X be the number of dots which come up on the first die and Y the number of dots on the second. Then $X+Y$ is the total number of dots. We found the distribution for this in Example 3. It is not one of our basic distributions—it would be Bernoulli trials with $n = 2$ but every trial has six possible outcomes instead of two. On the other hand, the distributions for X and Y are standard distributions, namely, uniform distributions. Thus, once we have a general method for calculating sum distributions from the summands (see the analysis after Example 10), and once we apply this method to standard distributions, then we will not need to remember the distribution of $X+Y$ as an isolated case. ∎

EXAMPLE 9 An experiment is run n times. Let X_k be a random variable which can take on just two values as follows:

$$X_k = \begin{cases} 1 & \text{if experiment } k \text{ is a success,} \\ 0 & \text{if experiment } k \text{ is a failure.} \end{cases}$$

Then $\sum_{k=1}^{n} X_k$ is the total number of successes. We already know the distribution for this sum—the binomial distribution. In other words, we have shown how to represent every binomial distribution as a sum of extremely simple distributions. A random variable which equals 1 or 0, such as X_k, is called a **characteristic function** or **binary random variable**. ∎

EXAMPLE 10 Let X_n be the number of calls entering a central phone office during minute n. Then $X_n + X_{n+1}$ is the number of calls entering during the two-minute stretch starting with minute n. If each X_i is Poisson with parameter λ, do you have any hunch about the distribution $X_n + X_{n+1}$ will have? [37] ∎

We now develop the general formula for the distribution of the sum of two random variables X and Y. (For sums of more variables, one proceeds by induction.) We assume the variables are discrete and that X can take on values $x_1, \ldots, x_n$. Then in order for $X+Y$ to take on value z,

$$\begin{array}{llll} \text{either} & X = x_1 & \text{and} & Y = z - x_1, \\ \text{or} & X = x_2 & \text{and} & Y = z - x_2, \\ & \vdots & & \vdots \\ \text{or} & X = x_n & \text{and} & Y = z - x_n. \end{array}$$

Since each pair of X and Y values are a disjoint event from any other pair, we can add probabilities and obtain

$$\Pr(X+Y=z) \;=\; \sum_{i=1}^{n} \Pr(X=x_i \text{ and } Y=z-x_i). \tag{8}$$

This is as far as we can go, unless we assume that X and Y are independent random variables [39], in which case we get

$$\Pr(X+Y=z) \;=\; \sum_{i=1}^{n} \Pr(X=x_i)\Pr(Y=z-x_i). \tag{9}$$

Finally, it's a nuisance to have all these subscripts, and to have to declare how many points are in the range of X. So we simply write

$$\Pr(X+Y=z) \;=\; \sum_{x} \Pr(X=x)\Pr(Y=z-x). \tag{10}$$

We can think of this sum as being over all real numbers x; if $x \notin \text{Range}(X)$, then $\Pr(X=x) = 0$, so the sum reduces to the same finite sum as in Eq. (9). Similarly, even when $\Pr(X=x) \neq 0$, it may be that $\Pr(Y=z-x) = 0$, which reduces the number of terms even further.

EXAMPLE 11 Find the distribution of $X+Y$ if X has distribution $B_{n,p}$, Y has distribution $B_{m,p}$, and X and Y are independent.

Solution From Eqs. (4) and (10), we obtain

$$\begin{aligned}\Pr(X+Y=k) &= \sum_j \left[\binom{n}{j} p^j q^{n-j}\right]\left[\binom{m}{k-j} p^{k-j} q^{m-(k-j)}\right] \\ &= \left[\sum_j \binom{n}{j}\binom{m}{k-j}\right] p^k q^{m+n-k}.\end{aligned}$$

The sum in the last line runs over integers j from 0 to n. (Why?) We saw in Chapter 4 that

$$\sum_j \binom{n}{j}\binom{m}{k-j} = \binom{m+n}{k}.$$

We conclude that

$$\Pr(X+Y=k) \;=\; \binom{m+n}{k} p^k q^{m+n-k}.$$

In short, $X+Y$ also has a binomial distribution, namely, $B_{m+n,p}$. With hindsight, we can justify this conclusion without a single computation. Let X and Y be defined on the same sample space of $m+n$ Bernoulli trials with success probability p. Let X be the number of successes in the first n trials and Y the number in the last m. They are certainly independent. In this context, what does $X+Y$ measure? ∎

Problems: Section 6.4

1. For the uniform distribution on

$$S = \{10, 11, \ldots, 99\},$$

what is

a) $\Pr(25)$? b) $\Pr(30 \le x \le 40)$?

2. Consider the binomial distribution with $n = 3$ and $p = .4$. Make a complete table of values for its point and cumulative distribution functions.

3. Consider the Poisson distribution with $\lambda = 2$. What is the probability that the random variable of this distribution, call it k, satisfies

a) $k = 2$? b) $k \le 2$? c) $k \ge 2$?

4. What is the best probability distribution to use to study each of the questions a)–g)? Be as precise as you can; if the distribution you choose has parameters, and you can give their specific values, do so. On the other hand, some of these situations are vaguely stated, and there isn't necessarily one right distribution to use. Maybe no distribution we have studied is appropriate. Be prepared to defend your choice.

a) How many 6's will show up in ten throws of a fair die?

b) How many 7's will show up in ten throws of two fair dice?

c) How many customers will arrive at the First National Bank between 10 and 11 A.M.?

d) How many alpha particles will be emitted in the next minute from a 1-kg piece of radioactive uranium?

e) How many throws of two fair dice will be necessary to get a 7?

f) How many throws of one fair die will be necessary to get 6 twice?

5. Suppose that our sample space is the set of all possible infinite sequences of tosses of a fair coin. Let the random variable X give, for each infinite sequence, the number of the trial when the first H appears. What distribution does X have?

6. Find an explicit formula for $F(n)$ in Example 3. Your formula will have two cases: $1 \le n \le 7$ and $7 < n \le 12$.

7. Two fair dice are tossed repeatedly.

a) Let $\Pr(m, n)$ be the probability that the first 2 or 11 appears on toss m and the first 7 appears on toss n. Find a formula for $\Pr(m, n)$. *Hint*: Consider three cases, $m > n$, $m = n$ and $m < n$.

Let $f(k)$ be the probability that the first 7 shows up on the kth toss, without either 2 or 11 appearing earlier.

b) Find a formula for $f(k)$.

c) Why is $f(k)$ *not* a point distribution? This is a conceptual question; it has nothing to do with the form of the answer to part b).

8. Maximum likelihood value. For the uniform distribution, clearly each value of the variable is equally likely. Which value is most likely for each of the other discrete distributions we have introduced? Try to guess the answer; then use the method described next to find out for sure.

Let $a_1, a_2, \ldots$ be a sequence of positive numbers. Define the ratio $r_k = a_{k+1}/a_k$. Suppose that the sequence $\{r_k\}$ is decreasing and that n is the first index for which $r_n < 1$.

a) Show that a_n is the maximum term in $\{a_k\}$.

b) Suppose, in addition, that $r_{n-1} = 1$. What does this say about the maximum of $\{a_k\}$?

c) Apply your results in parts a) and b) to the binomial distribution $B_{n,p}$.

d) Apply it to the negative binomial.

e) Apply it to the Poisson distribution.

Problems 9–15 are best done with a computer having graphics capabilities, or a hand calculator having a graphics screen. At the very least, use a calculator with a few scientific function keys. *Caution*: For some of the problems, you will have to think hard about how to compute the binomial coefficients. If you just use the definition directly, the factorials will cause overflows or underflows.

9. Compute, or plot, the Poisson distribution, for $k = 0, 1, \ldots, 10$, in the cases $\lambda = 1, 2, 3, 4$.

10. Compute, or plot, the negative binomial distribution (order 1) for $k = 0, 1, \ldots, 10$, and $p = .3, .5$, and $.7$.

11. Repeat [10], but for the negative binomial of order 2.

12. Repeat [10], but for order 3.

13. People often believe that if you toss a coin 100 times, the most likely outcome is 50 heads. That's true, but they often believe even more: That this event is *quite* likely, or at least, almost certainly the number of heads will fall within a few of 50. Furthermore, they believe that if 100 is replaced by, say, 1000, and 50 by 500, then the certainty of getting 500 heads, or being within a few of 500, is even greater. This problem asks you to check out these beliefs.

a) With p fixed and n increasing, what happens to $B_{n,p}(k)$, where k is the most likely value? Base your answer on computations, using $p = .5$ and $n = 10, 20, 50$.

b) Repeat part a) with $p = .7$.

c) With p fixed and n increasing, what happens to the probability that k is within 2 of the maximum likelihood value? That is, if the maximum likelihood value is 10, k should be from 8 to 12. Use the same values for p and n as in parts a) and b), but if you aren't certain of the pattern, try some larger n's.

d) With p fixed and n increasing, what happens to the probability that k is within $.1n$ of the maximum likelihood value? That is, if that value is 10, k should be within $10 \pm .1n$. The advice about p and n given in part c) holds.

14. Repeat [13] for the Poisson distribution: that is, check if the most likely value gets more likely. You might think this problem doesn't make any sense: There is no n, representing a number of trials, to let grow. But reason as follows. If λ is a measure of the number of phone calls expected per time period, then if we take longer and longer time periods, we should use larger and larger values for λ. Furthermore, since a longer time period should mean a more definitive sample, for large λ we should expect the probability to be concentrated around the most likely value. So when we say repeat [13], we mean use increasing values of λ. Find out what happens to the probability of the most likely value—by itself, $\pm$ a constant amount, and $\pm$ a percentage amount.

15. Do the analysis of [13] for the negative binomial distribution.

16. [13] and its variants can, of course, be answered analytically, but you need some knowledge of limits of binomial coefficients to do it. The fundamental tool for this sort of thing is **Stirling's Formula**:

$$n! \approx \sqrt{2\pi}\, n^{n+\frac{1}{2}} e^{-n}. \tag{11}$$

The $\approx$ means that, as $n \to \infty$ the ratio of the right side to the left side goes to 1. This means the right side can be substituted for the left side in approximating probabilities.

Use Eq. (11) to show that the probability of the most likely value of $B_{n,p}$ is approximately $1/\sqrt{2\pi np(1-p)}$. So what is the answer to [13], parts a) and b)?

17. The cumulative distribution function tells us $\Pr(X \le x)$, but this may not seem very interesting: It's more likely that you would want to know the probability that X is in some interval. However, every such question can be answered by combining values of the cumulative function—hence it's importance. Let $x_1 < x_2 < \ldots < x_n$ be the values of random variable X. Express each of the following using the cumulative function $F(x) = \Pr(X \le x)$.

a) $\Pr(X > x)$

b) $\Pr(X \ge x_i)$

c) $\Pr(x_i < X \le x_j)$

d) $\Pr(x_i < X < x_j)$

e) $\Pr(x_i \le X \le x_j)$

f) $\Pr(x_i \le X < x_j)$

18. A probability distribution is **symmetric** if $f(x) = f(-x)$ for all real numbers x. If $F(x)$ is the cumulative function for a symmetric distribution, what is the probabilistic interpretation of $G(x) = 2F(x) - 1$? Over what domain for x does this interpretation hold?

19. An urn contains r red balls and b black balls. A ball is removed at random, the color is noted, and the ball is returned to the urn. This procedure is called **selection with replacement**. Suppose that it is repeated n times. Let X be the number of times a red ball is selected. Determine the probability distribution of the random variable X. It's one you've seen before.

20. An urn contains r red balls and b black balls. A ball is removed at random, the color is noted, and the ball is thrown away. This procedure is called **selection without replacement**. Suppose that it is repeated n times. (Note that $n \le r+b$.) Let Y be the number of times a red ball is selected. Determine the probability distribution of the random variable Y. This is called the **hypergeometric distribution**.

21. A widgit manufacturer has trouble with her production line. A certain machine malfunctions one quarter of the time, but in a way that cannot be directly detected. She finds out about it only because more widgits than usual are defective when tested. If the machine is running properly, only 1 widgit in each batch of 10 is defective. If the machine malfunctions, 3 out of 10 are defective. A batch of 10 is produced, and the manufacturer has her quality control specialist choose two of the widgits (without replacement) and test them. One of the two is defective.

a) What is the probability of this outcome (one of two defective) if the machine is running properly?

b) What is the probability of this outcome if the machine is malfunctioning?

c) What is the probability that the machine is malfunctioning?

22. Do [21] again, but now the troublesome machine is running properly if, for each widgit independently, the probability that it is defective is .1; and the machine is malfunctioning if, for each widgit independently, the probability that it is defective is .3.

23. The Trinomial Distribution. An experiment has three possible outcomes, a, b, and c, with probabilities p, q, and r. The experiment is repeated independently n times. Let $f_n(k, j)$ be the probability that event a happens k times and b happens j times. Find a formula for f_n. *Note*: f_n is not a distribution in the sense of Definition 3 because its domain is a subset of $R \times R$, not R. In many books the definition of distribution is broader to include functions like f_n.

24. The Multinomial Distribution. Figure out what this is—what situation it models and what the formula for it must be. See Theorem 3, Section 4.6.

25. Consider Example 1, Section 6.1 (College Boards) again.

a) As in the solution in the text, assume that you guess on 10 questions. What is the probability that your score goes down? That it stays the same?

b) Now assume that you guess on 20 questions. Determine the probabilities that your score goes up, stays the same, or goes down. Compare these numbers to those in part a) and the text. Do you have an intuitive explanation for what's going on?

26. Some multiple choice exams with five answers per question only take off 1/5 point for a wrong answer. If you guess at random on 10 questions on such an exam, what is the probability that you improve your score?

27. Back to the College Boards, with 1/4 off for each wrong answer. Suppose that you can eliminate one answer on each of 10 questions, and guess randomly among the remaining answers. Now what is the probability that you increase your score?

28. In the negative binomial distribution of order n, so long as $0 < p < 1$, the probability of getting any one of the infinite-tuples is 0. Why? What is the probability of the event consisting of all infinite tuples in the space; that is, what is the probability that you never obtain an nth success?

29. Complete the derivation of the Poisson distribution from the binomial distribution, as outlined at the end of Example 7.

30. We have used the fact that the binomial distribution approaches the Poisson distribution to motivate the claim that the Poisson distribution is appropriate for situations like telephone calls. But as with all limit relationships, it cuts both ways. Suppose that a problem satisfies the binomial distribution exactly, but n is large and p is small. Then it is easier (and usually sufficient) to find approximate probabilities using the Poisson distribution than to get exact values using the binomial distribution

a) Approximate $B_{100,.01}(k)$ for $k = 0, 1, 2$. Compare your Poisson approximation with the actual binomial value.

b) Approximate $B_{50,.98}(k)$ for $k = 48, 49, 50$. *Caution*: p is not small, so you have to rewrite the binomial probabilities before you can approximate them with the Poisson distribution.

31. The negative binomial distribution is often useful with the roles of success and failure exchanged. We ask for the probability $f_n(k)$ that the nth failure happens on the kth trial. For instance, if each trial consists of seeing whether a light bulb lasts for another day, then to say that the first failure occurs on day k is to say that the light bulb lasts k days.

The negative binomial, viewed as a model of failure probabilities, has an interesting property: The chance of future failure is independent of the past record. We illustrate this broad claim with a specific instance. Let X be the trial in which the first failure occurs. Consider the statement

$$\Pr(X{=}k \mid X{>}k_0) \;=\; \Pr(X{=}k{-}k_0). \qquad (12)$$

In what sense does this statement say that the distribution of X forgets its past? Prove that the negative binomial (with $n = 1$) satisfies Eq. (12).

Do you think the negative binomial is a good model for light bulbs? (Don't be too sure; get some data!) What about for human mortality, where success in a trial means living another year? What about for having an accident with your car, where success in a trial means making it through another month without having an accident?

32. Use the Binomial Series Theorem (Theorem 2, Section 4.6) to show that

$$(1-q)^{-n} = \sum_{k=n}^{\infty} \binom{k-1}{n-1} q^{k-n}$$

and conclude that

$$\sum_{k=n}^{\infty} \binom{k-1}{n-1} p^n q^{k-n} = 1.$$

What is the probabilistic interpretation of this equality using the negative binomial distribution?

33. Let X be a random variable with domain some sample space S and range some discrete set R' within the real numbers R. We said in the text that X transfers probabilities from S to R' and makes R' into a sample space. Verify this assertion. That is, show that R' meets the definition of sample space in Section 6.2 when you use Definition 2 in this section to define the probabilities on R'. We say that X **induces** a probability measure on R' from the original measure on S.

34. Compute the distribution for the number of dots on two dice using Eq. (10) and Example 8.

35. Generalize [34]. If X_k is the uniform random variable with values $1, 2, \ldots, k$, what is the distribution of $X_m + X_n$? (Assume the variables are independent.)

36. Let $X_{n,p}$ be the number of the trial in which the nth success occurs for Bernoulli trials with probability of success p. It is a fact that, if $X_{n,1}$ and $X_{m,p}$ are independent, then their sum has the distribution of $X_{m+n,p}$.

a) Justify this fact conceptually as follows. Let the sample space for both $X_{n,p}$ and $X_{m,p}$ be the set of all possible infinite Bernoulli trials. Interpret $X_{n,p}$ in the usual way, but reinterpret $X_{m,p}$ to be the number of additional trials needed to obtain m more successes after the nth success. This reinterpretation doesn't change the distribution of $X_{m,p}$. Why? Now, what does $X_{n,p} + X_{m,p}$ count?

b) Prove the fact algebraically in the special case $n = m = 1$, using Eq. (9).

c) Prove it algebraically in general. *Hint*: Mimic Example 11. As in that example, the powers of p and q will combine nicely, and you will be left with a sum of products of binomial coefficients—but a different product than before. Use a combinatorial argument to reduce that sum to a single binomial; you already know what binomial coefficient that must be, if the fact is to be true.

37. Let X and Y be independent variables, both with the Poisson distribution with parameter λ. Guess and then prove algebraically a fact about the distribution of $X+Y$. *Hint*: Let X and Y count the number of successes in consecutive periods. Now double the length of a period.

38. Let X_λ have the Poisson distribution with parameter λ. If X_λ and X_μ are independent, guess and prove the distribution of their sum.

39. Independent variables. We talked about independent variables, and multiplied their probabilities together, as if we had defined what we were talking about long ago. However, we have defined independence only for *events*, never for *variables*. To put our talk on a firm foundation, we have to explain independence of variables in terms of independence of events.

Definition 4. Let X and Y be random variables on space S. Then X and Y are **independent** if for any $A \subset \text{Range}(X)$ and $B \subset \text{Range}(Y)$ the events $\{X{\in}A\}$ and $\{Y{\in}B\}$ are independent; in other words, if

$$\Pr(X{\in}A \text{ and } Y{\in}B) \;=\; \Pr(X{\in}A)\Pr(Y{\in}B). \quad (13)$$

Recall that $\{X{\in}A\}$ really is an event since it is shorthand for $\{s{\in}S \mid X(s){\in}A\}$.

The most useful case of Eq. (13) is when $A = \{x\}$ and $B = \{y\}$. We get

$$\Pr(X{=}x \text{ and } Y{=}y) \;=\; \Pr(X{=}x)\Pr(Y{=}y), \quad (14)$$

which is what we used in the text.

Prove: For random variables with finite ranges, Eqs. (13) and (14) are equivalent. That is, Eq. (13) is true for all A and B iff Eq. (14) is true for all x and y.

40. Derive a formula for the distribution of $X - Y$,

a) without assuming X and Y are independent;

b) assuming they are independent.

You may assume in both cases that both variables have finite range.

6.5 Expected Value and Variance

Now that we can associate numbers with a probability space (via random variables or distributions), we can combine those numbers into a single summarizing number. Such a number is called a **statistic**. The most important statistic is the average or expected value. Recall that the average of the numbers $x_1, x_2, \ldots, x_n$ (no probability space yet) is defined by

$$\text{average} \;=\; \frac{1}{n}\sum_{i=1}^{n} x_i. \quad (1)$$

Formula (1) also applies to an equiprobable sample space, but when different numbers have different probabilities a modified definition is necessary.

In this section we define and motivate the general definition of average and develop its key mathematical properties. In a brief subsection at the end, we also introduce the second most important statistic, variance. Then in the next section we use our knowledge of expected value to determine the average run time of various algorithms.

Average value is important because it serves as a *predictor*. Suppose you have two different algorithms for solving the same problem. They may take different numbers of steps, depending on the data; sometimes the first algorithm may be faster, at other times the second. How do you choose between them? Generally you should pick the one with the smaller expected value. On average it will finish the job sooner.

The usefulness of expected value is not limited to analyzing algorithms. Suppose you have the choice of several investments, each of which involves some risk. You can choose among them using their expected values [17, 20].

Even when you have no choices, an expected value can be useful as a *descriptor*. Suppose there is only one section of the math course you plan to take next term. Suppose the average grade given last year by the professor who teaches that course was low. You better plan to work hard! Or consider that average from the professor's point of view. Suppose it is lower than the average grade for the same course two years ago. Then the professor can conclude that last year's class wasn't so smart—either that or she's getting tougher.

To see the need to generalize formula (1), imagine you are in a contest that offers a very large first prize, to be given to one person, and a modest second prize, to be given to 100 people. Clearly, if you want to summarize your expectations, you should take a weighted average, with the weights determined by the different probabilities of winning first prize, second prize, or nothing. This leads to the following definition. It has two parts because we use both random variables and distributions.

Definition 1. The **expected value** of a discrete random variable X, written $E(X)$, is

$$E(X) \;=\; \sum_x x\,\mathrm{Pr}(X{=}x), \tag{2}$$

where x runs over all values in Range(X). Expected value is also called **expectation, mean** or **average** and is often denoted by $\overline{X}$, or even $\overline{x}$. Similarly, the expected value of a discrete probability distribution f is

$$E(f) \;=\; \sum_x x f(x), \tag{3}$$

where x runs over the domain of f.

The two parts of the definition are two sides of the same coin. For a random variable X, the associated distribution, $f(x) = \mathrm{Pr}(X{=}x)$, has the same expectation. When discussing expectations, we will move freely between variable and distribution language, using whatever seems most natural at the time.

In Definition 1, x in the sums may take on a finite or at most a countably infinite number of values. ("Countably infinite" means that these values can be labeled $x_1, x_2, x_3, \ldots$.) In the finite case, Eqs. (2) and (3) are ordinary finite sums. In the countably infinite case, they are the limit of finite sums—something we must continue to deal with informally until Chapter 9.

EXAMPLE 1 Show that the expectation of an equiprobable variable (i.e., uniform distribution) is the same as the average of its values.

Solution Let $x_1, \ldots, x_n$ be the values of X. We are given that $\mathrm{Pr}(X{=}x_i) = 1/n$ for each i. Hence

$$E(X) \;=\; \sum_{i=1}^{n} x_i \frac{1}{n} \;=\; \frac{1}{n}\sum_{i=1}^{n} x_i\,.$$

But the last expression is just the average. ∎

Solution to Example 4, Section 6.1 (should you mail in your lottery number?) Recall that there is one prize, worth \$10,000, but you have to pay the postage (currently \$.25) if you participate. Suppose you do participate. Let X be your net gain. X is a random variable which takes on one of two values: \$9999.75 if you win, $-\$.25$ if you lose. Let p be the probability of winning. Thus

$$E(X) \;=\; p(9999.75) + (1-p)(-.25) \;=\; 10000p - .25.$$

Since the net gain if you don't participate is 0, it is worth participating if

$$10000p - .25 > 0, \qquad \text{that is,} \qquad p > .000025.$$

Presumably, every participant has an equal chance of winning, so we set $p = 1/n$, where n is the number of participants. Thus we conclude that

$$E(X) > 0 \iff n < \frac{1}{.000025} = 40,000.$$

Of course, you don't know what n is! But you can estimate it. Presumably, the lottery company doesn't waste numbers. They probably sent out at least as many invitations as the number on your ticket. The lottery tickets of this sort that we have received have numbers in the tens of millions. Even if only 1 person in a 100 replies (a low estimate, we think), it's not worth it to play; but see [1, Supplementary]. ∎

EXAMPLE 2 What is the expected count when two dice are tossed?

Solution Let X be the count. Then

$$E(X) \;=\; \sum_{i=2}^{12} i\,\Pr(X{=}i). \tag{4}$$

Assuming as always that the 36 pairs (i,j) are equiprobable, Eq. (4) becomes

$$\begin{aligned} E(X) &= 2\cdot\frac{1}{36} + 3\cdot\frac{2}{36} + 4\cdot\frac{3}{36} + \cdots + 7\cdot\frac{6}{36} + 8\cdot\frac{5}{36} + \cdots + 11\cdot\frac{2}{36} + 12\cdot\frac{1}{36} \\ &= \frac{252}{36} \;=\; 7. \end{aligned}$$

Such a simple answer (7 is halfway between the minimum number of dots, 2, and the maximum, 12) should have a simpler computation. It does. You will be able to do it the simpler way later in this section, once you learn the various methods for simplifying expectation computations [4]. ∎

We now compute the expectation for some of the standard distributions.

EXAMPLE 3 What is $E(B_{n,p})$, the expectation for the binomial distribution $B_{n,p}(k)$?

Solution By definition, the answer is

$$\sum_{k=0}^{n} k\binom{n}{k}p^k q^{n-k} \qquad (q = 1-p). \tag{5}$$

Can this be simplified? It looks tough, but there are lots of successful ways to attack it. We will show two in the text and leave others for the problems [23, 24].

To begin with, let's conjecture what the answer is on purely conceptual grounds. There are n trials and each has probability p of success. Therefore we should expect np successes. The only question is: How do we make Eq. (5) simplify to this?

Method 1: Direct algebraic attack. Does Eq. (5) look like anything you've seen before? Yes: If we throw out the initial factor k, we have the expression from the Binomial Theorem with $x = p$ and $y = q$. This realization suggests that our attack may be more successful if we generalize; That is, we replace p by x and q by y (where y varies independently of x) and try to simplify

$$\sum_{k=0}^{n} k \binom{n}{k} x^k y^{n-k}. \tag{6}$$

In the Binomial Theorem, the sum without the initial k simplified to $(x+y)^n$, so we might expect a power of $(x+y)$ in the simplification of expression (6) too. The fact that our special case should simplify to np suggests the general case should be $nx(x+y)^n$, except that the power of $(x+y)$ might be different. Evidence for a different power is provided by another special case of expression (6). Namely, Theorem 4, Section 4.5, stated that the case $x{=}y{=}1$ reduces to $n2^{n-1} = n(x+y)^{n-1}$. (The proof was combinatorial and involved picking a team and a captain.)

It turns out that $n-1$ is the right exponent in general, as the following approach shows. The trick is to get rid of the initial factor k in expression (6), thus reducing it to the Binomial Theorem. We use the identity

$$k\binom{n}{k} = n\binom{n-1}{k-1},$$

which is easily proved either algebraically or combinatorially; see [1, Section 4.5], where $k+1$ is used instead of k. The appearance of $k-1$ in the identity suggests making $k-1$ the basic variable and then giving it a new, single-letter name. Then expression (6) becomes

$$\sum_{k=0}^{n} n\binom{n-1}{k-1} x^{1+(k-1)} y^{(n-1)-(k-1)} = nx \sum_{j=-1}^{n-1} \binom{n-1}{j} x^j y^{(n-1)-j}. \quad [j = k-1]$$

In the last expression, we may delete the term $j = -1$. (Why?) Thus by the Binomial Theorem, we get

$$\sum_{k=0}^{n} k\binom{n}{k} x^k y^{n-k} = nx(x+y)^{n-1}.$$

Finally, setting $x = p$ and $y = q$, we get that $E(B_{n,p}) = np$.

Method 2: Recursion. As usual, recursion is shorter and sweeter, but more subtle. Let $E_n = E(B_{n,p})$. The key idea is: If we are running n trials, then after we run the first we are back to the situation for $B_{n-1,p}$ and can use knowledge of E_{n-1}. Specifically, with probability q, the first trial is a failure. Hence the only successes will be those in the remaining trials, so we can expect E_{n-1} successes.

On the other hand, if the first trial is a success, then all told we expect $1 + E_{n-1}$ successes. In short,

$$E_n = q\,E_{n-1} + p\,(1+E_{n-1}) = E_{n-1} + p. \tag{7}$$

Clearly this holds for $n > 1$. Furthermore, it's easy to compute E_1 directly from the definition:

$$E_1 = 0\,q + 1\,p = p.$$

The explicit solution to this recursion and initial condition is immediate:

$$E_n = np.$$

Note how the algebra in the recursive solution is much closer to our intuition—each trial increases the expected number of successes by p—than the direct algebraic attack. ▮

EXAMPLE 4 Find the mean of the negative binomial of order 1. That is, find the expected number of Bernoulli trials through the first success, given that the chance of success each time is p.

Solution The negative binomial distribution of order 1 is $f(k) = q^{k-1}p$. So by Definition 1, the answer is

$$\sum_{k\geq 1} kq^{k-1}p = p\sum_{k\geq 1} kq^{k-1}. \tag{8}$$

As usual let's guess the simplification first. Since each trial achieves on average the fraction p of successes, it should take $1/p$ trials to get 1 full success. Of course, successes don't come in fractional parts, so this is just a plausibility argument. In fact, in some similar problems this argument is simply wrong [13]. But here it will stand us in good stead.

Equation (8) is tricky for us to attack directly, because it involves an infinite sum. Nevertheless, let's plunge right in.

Method 1: Direct attack. We compute the sum from $k = 1$ to an arbitrary n, look for a formula, and then decide intuitively whether the formula approaches a limit as $n \to \infty$. The summand is of the form Akq^{k-1}, so we can find a formula by the methods of Sections 5.7 and 5.8. (See Sections 9.4 and 9.5 for still further methods.) One finds after some simplification [25, 26] that

$$p\sum_{k=1}^{n} kq^{k-1} = \frac{1}{p} - \Big(n + \frac{1}{p}\Big)q^n. \tag{9}$$

Equation (9) is of the form $A-(Bn+C)q^n$, where $0 \leq q = 1-p \leq 1$. So long as $q < 1$ (i.e., $p > 0$), in the limit everything vanishes except $A = 1/p$. The term Cq^n goes to 0 since q^n goes to 0. The tricky term is Bnq^n, as the growth in n might conceivably compensate for the decline in q^n; however, exponentials (such as q^n) always overpower polynomials (like Bn). In conclusion, $E(X) = 1/p$, as anticipated.

Method 2: Recursion. Let X be the random variable for the number of the trial with the first success. We separate the initial trial from the rest. With probability p we succeed in the first trial and quit, so $X = 1$. With probability q we fail and must go on to further trials. That is, it is just as if we are starting from scratch, so we can expect $E(X)$ more trials. But we have already done one trial. Thus

$$E(X) = p1 + q(E(X)+1).$$

Solving for $E(X)$, we find again that $E(X) = 1/p$. Note that this time there is no subscript on E; there is no parameter n because the sequence of trials has no predetermined stopping point. When a random variable (here the number of trials) has an infinite range, the direct calculation of E from the definition gets harder. However recursion gets easier! (We will give more examples of this type in Section 6.7.) ∎

Solution to Example 3, Section 6.1 (picking the next item for a permutation) Recall the issue: Given n objects, and having already picked k for a permutation, how long will it take to find a new $(k+1)$st object if we repeatedly pick from all n at random.

If there are $n - k$ objects left, the chance of success on any one random pick is $p = (n-k)/n$. Hence by Example 4, the expected number of additional trials to get the $(k+1)$st object is $1/p = n/(n-k)$. ∎

Expectations of Composite Variables

At the end of Section 6.4, we showed that random variables can often be expressed in terms of two or more simpler random variables. We now develop some theorems which show how to compute the expectation of such composite variables in terms of the expectations of the simpler variables. This process often provides a particularly easy way to compute expectations of complicated variables.

Theorem 1. Let X and Y be random variables on sample space S. (X and Y need *not* be independent.) Then

$$E(X+Y) = E(X) + E(Y). \tag{10}$$

Similarly,

$$E\left(\sum X_i\right) = \sum E(X_i). \tag{11}$$

Before proving this theorem, we give several examples of its use.

EXAMPLE 3 Continued

Expected Number of Successes in n Bernoulli Trials
Let X be the total number of successes. Let X_i be a binary random variable defined by

$$X_i = \begin{cases} 1 & \text{if trial } i \text{ is a success,} \\ 0 & \text{if trial } i \text{ is a failure.} \end{cases}$$

Note that

$$X = \sum_{i=1}^{n} X_i.$$

The whole point of introducing the X_i is that they are simple to analyze yet they can be summed to get X. For each i, $E(X_i) = p$. (We computed this before, after Eq. (7).) Thus, by Eq. (11), $E(X) = np$. ∎

EXAMPLE 4 Continued

Expected Value of the Negative Binomial of Order 1

Let X be the number of trials through the first success. Define X_i by

$$X_i = \begin{cases} 1 & \text{if the } i\text{th trial takes place,} \\ 0 & \text{if the } i\text{th trial does not take place.} \end{cases}$$

Then $X = \sum_{i=1}^{\infty} X_i$. (Why?) Furthermore, $\Pr(X_i{=}1) = q^{i-1}$. (Why?) Thus

$$E(X_i) = 1\Pr(X_i{=}1) + 0\Pr(X_i{=}0) = q^{i-1}$$

and

$$E(X) = \sum_{i\geq 1} q^{i-1} = \sum_{j=0}^{\infty} q^j = \frac{1}{1-q} = \frac{1}{p},$$

as anticipated. ∎

Note. The X_i variables are definitely not independent [35].

PROOF OF THEOREM 1 Let z run over Range$(X{+}Y)$ and let x and y run over all pairs with $x \in$ Range(X) and $y \in$ Range(Y). By definition,

$$\begin{aligned} E(X{+}Y) &= \sum_z z\Pr(X{+}Y{=}z) \\ &= \sum_{x,y} (x{+}y)\Pr(X{=}x, Y{=}y) \\ &= \sum_{x,y} x\Pr(X{=}x, Y{=}y) + \sum_{x,y} y\Pr(X{=}x, Y{=}y). \end{aligned} \tag{12}$$

We now analyze the first sum in line (12). We rewrite it using a double sum:

$$\sum_x \sum_y x\Pr(X{=}x, Y{=}y) = \sum_x \Big(x \sum_y \Pr(X{=}x, Y{=}y)\Big). \tag{13}$$

For any particular x_0 the inner sum on the right divides the ways X can equal x_0 into mutually exclusive and exhaustive cases, depending on what value Y takes on at the same time. Thus

$$\sum_y \Pr(X{=}x_0, Y{=}y) = \Pr(X{=}x_0).$$

Therefore, the RHS of Eq. (13) becomes

$$\sum_x x \Pr(X{=}x) \;=\; E(X).$$

Similarly, $\sum_{x,y} y \Pr(X{=}x, Y{=}y)$ in line (12) is $E(Y)$, and we have proved Eq. (10). The generalized form, Eq. (11), follows for finite sums by induction [34]. ∎

By essentially the same method, one proves

Theorem 2. Let X and Y be random variables, and let a and b be constants. Then

$$E(aX + bY) \;=\; aE(X) + bE(Y).$$

For instance,

$$E(2X) \;=\; 2E(X) \qquad (\text{set } a = 2,\, b = 0).$$

Also,

$$E(X{-}Y) \;=\; E(X) - E(Y) \qquad (\text{set } a = 1,\, b = -1).$$

There is also a theorem about $E(XY)$. Sure enough, it equals $E(X)\,E(Y)$, but only if X and Y are *independent* [44].

Variance and Standard Deviation

We said the purpose of expected value is to summarize a lot of information about a random variable in a single number. But no one number can tell it all. Ever hear about the statistician who drowned trying to wade across a stream whose average depth was known to be 2 feet? If you don't get the joke, consider the following two random variables: X such that

$$X = 1,\, 2,\, 3, \qquad \text{each with probability } 1/3, \tag{14}$$

and Y such that

$$Y = 0,\, 6 \qquad \text{with} \quad \Pr(0) = 2/3, \quad \Pr(6) = 1/3. \tag{15}$$

You can easily see that both X and Y have expected value 2. Look at Fig. 6.10. However, the values of X are much closer to 2 than the values of Y are. (And 6 feet is deeper than the statistician could wade through!)

In short, expected value is a measure of the *center*. We also need a measure of *dispersion* around the center. Here's how to get one. The variable $Z = \big(X{-}E(X)\big)^2$ measures the difference of X from it's center—the square makes everything non-negative so that differences on opposite sides don't cancel out [22, 53]—and to get a single number we just take $E(Z)$, which we denote by $\mathrm{Var}(X)$:

Definition 2. Let X be a random variable on some probability space. The **variance** of X, denoted $\mathrm{Var}(X)$, is defined by

$$\mathrm{Var}(X) = E\big([X-E(X)]^2\big).$$

The **standard deviation** of X, denoted σ_X, is defined by

$$\sigma_X = \sqrt{\mathrm{Var}(X)}\,.$$

Consequently, $\mathrm{Var}(X)$ is also denoted by σ_X^2.

We've written this definition in terms of a random variable, but just as for expectations there is a distribution version as well.

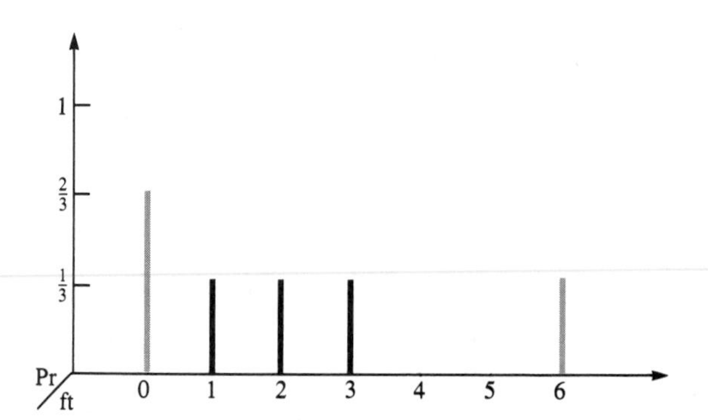

FIGURE 6.10 Two distributions (one black, one in color) with the same mean 2 but different variance

EXAMPLE 5 Compute the variance and standard deviation for X and Y as defined in lines (14) and (15).

Solution We have

$$\begin{aligned}\mathrm{Var}(X) &= \sum_k \Pr(X{=}k)[k-E(X)]^2 \\ &= \frac{1}{3}(1-2)^2 + \frac{1}{3}(2-2)^2 + \frac{1}{3}(3-2)^2 = \frac{2}{3}.\end{aligned}$$

Similarly,

$$\begin{aligned}\mathrm{Var}(Y) &= \sum_k \Pr(Y{=}k)[k-E(Y)]^2 \\ &= \frac{2}{3}(0-2)^2 + \frac{1}{3}(6-2)^2 = \frac{24}{3} = 8,\end{aligned}$$

which is much bigger. Taking positive square roots, we obtain the standard deviations:

$$\sigma_X = \sqrt{2/3} \approx .816, \qquad \text{and} \qquad \sigma_Y = \sqrt{8} \approx 2.828.$$

Again the value for Y is much bigger. ∎

Based on Definition 2 and Example 5, the following fact should be intuitively reasonable: If the variance of a random variable is very small, then with near certainty a random value of that variable will be very close to the mean. It further turns out that, for any random variable, if values are sampled repeatedly and independently, then the variance of the average of the sample values is very small. (More formally, if we start with a random variable X, then even if the variance of X is large, the variance of the new variable $(1/n)\sum_{i=1}^{n} X$ will be small, if n is large enough.) Thus the *average* of a *sample* is a good estimate of the *mean* of the random variable *as a whole*. This fact is of fundamental importance in statistics, the science of determining underlying structure from sample data.

Problems: Section 6.5

1. Let X_A be the characteristic function for event A: $X_A(s) = 1$ for $s \in A$ and $X_A(s) = 0$ otherwise. Show that $E(X_A) = \Pr(A)$.

2. From the definition of mean value, find the mean of the discrete uniform distribution with values $1, 2, \ldots, n$.

3. Find the mean of the discrete uniform distribution whose n values are evenly spaced from a to b.

4. We claimed in Example 2 (average number of dots when two dice are thrown) that there are easier ways to get the answer. Using Theorem 1, show an easier way.

5. You receive a lottery ticket in the mail. It says that you may already have won the first prize of $10,000 or one of the 1000 runner-up prizes of a $25 bond—the winning tickets have already been selected, but only those who mail back their stubs will collect their prizes. There are indications that 10 million people were sent tickets. If it costs you 25 cents to mail back your ticket to participate, what is your expected net gain

a) assuming one person in 10 mails back the stub;

b) assuming one person in 100 mails it back?

6. A common sort of slot machine has three wheels which spin, each showing a single entry when it comes to rest. If two or more of the entries at rest are the same, you win. Suppose that each wheel has the same 10 distinct entries. Suppose that it costs 25¢ to play, you win 50¢ if exactly two entries are the same, and you win $10 if all three entries are the same. Assume that all possible combinations are equally likely. What is your expected net gain?

7. A horse has probability p of winning a race. If you bet $1 on the horse and it wins, you get $\$w$ back; if it loses you get 0. At what value should racetrack officials set w so that your expected return is 0?

8. Show that, if racetrack officials set the winning amounts according to [7], then no matter how many people bet whatever amounts on whichever horses, the expected profit to the track (the amount the track keeps after paying off bets) is 0. Assume that the probabilities on the horses add to 1. Give simple examples (say with just two horses in the race) to show that the actual profit can be positive; that it can be negative.

9. Four horses, A,B, C and D, are in a race. The probabilities that each horse will win are .4, .3, .2 and .1, respectively. The race track has announced that if you bet $\$x$ on a given horse and it wins, you will receive

$$\begin{array}{rl} \$\ 2x & \text{if it is horse } A, \\ \$\ 3x & \text{if it is horse } B, \\ \$\ 4x & \text{if it is horse } C, \\ \$9.5x & \text{if it is horse } D. \end{array}$$

(If your horse loses, you get nothing back.)

a) For each horse, compute the expected monetary outcome from betting $1 on that horse.

b) Suppose you spread your risk by betting 25¢ on each horse. Now what is your expected outcome?

c) There is a way to split up your $1, betting different amounts on each horse, so that you achieve *exactly* your expected outcome no matter which horse wins. What is this split? Show that you are right.

Note: This part assumes the race track accepts fractional bets. Even if the track only accepts bets in multiples of $2, as is customary, you can approximate the needed fractions very closely if your total bet is large. Also, there is a method behind this example. No matter what probabilities the horses have of winning, and no matter how much the track will pay for winning tickets, you can always split your bet so that you know the outcome in advance. Unfortunately, tracks arrange things so that this outcome is negative!

10. Racetracks don't like to run the risk of losing money on a race. They also have the ability to adjust the odds on horses as betting proceeds, in light of the amounts bet. (That is, they keep real-time records of how much has been bet.) As they adjust the odds, they adjust the amount you win from a $1 bet if your horse wins. Show how a racetrack can set the winning amounts when the betting closes so that the track makes a profit of *exactly* 0 no matter which horse wins. (Tracks use this system, but they modify it slightly so that they are guaranteed a profit of, say, 5% of all money bet.)

11. A betting game is said to be **fair** if the expected monetary return to each player is 0. For instance, suppose Ann and Bill agree to flip a good penny, with Ann paying Bill $1 if they get heads, and Bill paying Ann $1 if they get tails. This game is fair.

a) Suppose instead they agree to throw a good die, with Ann paying Bill $1 if 1 dot comes up, and Bill paying Ann $1 otherwise. Obviously, this is not fair. What are the expected payoffs to each player?

b) Unfair games can be made fair by having initial payments. In this case, Ann should pay Bill a certain amount at the start of the game to make it fair. How much?

c) Yvonne and Zack agree to play the following game. They will throw two fair dice once. If 2 or 11 comes up, Yvonne pays Zack $2. If 7 comes up, Zack pays Yvonne $1. Otherwise the game is a draw. Who should pay whom how much at the start to make the game fair?

12. Solve [2] again, but this time do it recursively. (In this case, a recursive solution is much more work. Even finding the recursion is work. However, it's a good review.)

13. A friend picks an integer from 1 to 3 at random, and you get three guesses to figure out which one she picked. We know you can do it, but think about the average number of guesses you will need. One argument goes: On average, you will have 1/3 of a success on the first guess, 1/2 of a success on the second (if you get to a second guess), and a full success on the third. Since $\frac{1}{2}+\frac{1}{3} < 1$, it will take on average more than two guesses and less than three to have one full success, i.e., to find the number.

This answer is wrong. Find and justify the right answer using the definition of expected value.

14. College Boards again. You guess at random on n questions. Recall that there are five answers per question and you get 1/4 off for each wrong answer.

a) For each individual question on which you guess, what is the expected difference in your score from what it would be if you had left that question blank? (Here the difference is a signed number; if your score goes down, the difference is negative.)

b) What is the expected difference in your score for guessing on all n questions?

c) For each individual question, what is the expected absolute value of the difference in your score from guessing?

d) Consider the question: What is the expected absolute value of the difference in your score from guessing on all n problems? This is much harder to answer than part b). Why? Answer it in the specific case $n = 5$. (*Note*: Absolute difference is what the College Board is talking about when it makes statements like "If the same student takes an equivalent test on the next day, his or her score will typically differ by as much as 30 points from the previous score.")

e) Answer parts a) and b) again on the assumption that you were able to eliminate one answer on each question before guessing randomly among the remaining answers.

f) With the assumption in part e) and $n = 5$, determine the probability that guessing increases your score. (This is [27, Section 6.4] with $n = 5$ instead of $n = 10$.)

g) How do you reconcile the fact that the answer to part f) was less than .5 with the fact that the expected difference from part e) is positive? Does this same situation arise when $n = 10$?

15. (Example 3, Section 6.1, extended) What is the expected number of attempts until one finally gets a complete permutation of the n objects?

16. The average of $x_1, x_2, \ldots, x_n$ has the property that, when it is added to itself n times, the result is the sum of the numbers. If we replace "added to" by "multiplied by" and "sum" by "product", the expression with this property is called the **geometric mean**. Derive a formula for the geometric mean. (In light of this definition, the average of n numbers is sometimes called their **arithmetic mean**.)

17. You want to invest some money in the stock market for a year. A broker suggests the stock of XYZ Company. He believes there is a 40% chance the price will double in a year, a 40% chance it will halve, and a 20% chance it will stay the same. Assuming he's right, what is your expected gain if you buy $5000 of XYZ stock now and sell it in a year?

18. The answer to [17] is positive, but that doesn't mean you should buy the stock, even assuming the broker is right and even if there are no other options. It all depends on what the money is worth to you. Suppose you are just making ends meet. While it would be nice to gain $5000, it would be a *disaster* to lose $2500. Economists like to measure the worth of money (and everything else) to individuals in hypothetical units called "utils". Suppose gaining $5000 is worth 1000 utils to you, but losing $2500 is worth −3000 utils. Assume that just breaking even on the stock purchase is worth 0 utils. Your gain in utils is a random variable. It has an expected value. Should you buy the stock?

19. You have $10,000 to invest. You can either put it in a bank certificate of deposit, at 10% interest per year, or you can buy a bond which returns 12% a year. Whereas the face value of the certificate never varies from $10,000, the selling price of the bond can go up or down, depending on prevailing interest rates. (This has no effect on the interest you receive, which is fixed.) Your broker estimates that, one year from now, there is a 30% chance the bond will be worth 10% more, a 30% chance that its price will be unchanged, and a 40% chance its price will be 10% less. You intend to cash in your investment in one year. Which investment has the greater expected return?

20. It may be that expected monetary value is not the right measure in [19]. Just as in [18], you should probably consider the utility of money. Assume that simply preserving your $10,000 is worth 0 additional utils to you; that each dollar lost at the end of the year is worth −1 util; and that each additional dollar gained at the end of the year is worth +1 util, up to $1000, but that thereafter each additional dollar is worth only +1/2 util. Which investment strategy, bank certificate or bond, has the greater expected utility?

21. Although it is usually wise to choose between options using the expected value, sometimes that's not a sure thing. Some years ago one of us commuted to New York City by train to work for a foundation. To get from the foundation's office to the train station we could either walk or take the subway. A few weeks' experience showed that, from office desk to train: (1) walking took an average of 25 minutes with a standard deviation of 1 minute; and (2) the subway took an average of 20 minutes with a standard deviation of 10 minutes. The disutility of missing the train was extremely high and the disutility of waiting around at the station was also high. Which do you think was better, walking or subway? (You don't have enough information to do a thorough analysis, but give your intuitive reasons.)

22. Let X be any random variable. What is

$$E\big(X-E(X)\big)\,?$$

23. Again let $B_{n,p}(k)$ be the point distribution for the number of successes in n Bernoulli trials with probability of success p on each trial. Prove that $E(B_{n,p}) = np$ by induction. *Hint*: Start with a recursion for combinations of n things in terms of combinations of $n-1$ things.

24. (Assumes calculus) Prove that $E(B_{n,p}) = np$ by starting with the Binomial Theorem, regarding x as the variable and y as a constant, and differentiating.

25. Verify Eq. (9) by the methods of Section 5.7 and 5.8.

26. Here is a special, shorter method for verifying Eq. (9). Note that

$$\sum_{k=1}^{n} kf(k) = \sum_{j=1}^{n}\sum_{k=j}^{n} f(k).$$

Now substitute q^{k-1} for $f(k)$ and on the right use twice the formula for the sum of a geometric series.

27. (Assumes calculus) Derive Eq. (9) by differentiating something.

28. Let $N_{n,p}$ be the negative binomial distribution of order n. Thus $N_{n,p}(k)$ is the probability of obtaining the nth success on the kth trial, if each trial has success probability p. Let X_i be the number of additional trials after the $(i-1)$st success up to and including the ith success. Use the X_i's to derive $E(N_{n,p})$.

29. Derive $E(N_{n,p})$ directly from the definition of expected value. *Hint*: Rewrite $k\,C(k-1, n-1)$ so that there is no k in front; then rewrite it further so that the Binomial Series Theorem applies [Eq. (7), Section 4.6].

30. Find $E(N_{n,p})$ by recursion on n.

31. Let X_λ be the Poisson random variable. Find $E(X_\lambda)$ directly from the definition of expected value. You need just one fact from calculus:

$$\sum_{k\geq 0} \frac{x^k}{k!} = e^x.$$

If you regroup and make the right change of index, the answer falls out.

32. (Assumes calculus) Find $E(X_\lambda)$ using differentiation.

33. Balls labeled 1 through n are put in an urn. A ball is picked at random, its number is recorded, and it is put back in the urn. Let X_k be the sum of the numbers recorded if this is done k times. Now suppose a ball is picked at random, its number is recorded, and it is thrown away. Let Y_k be the sum of the numbers recorded if this is done k times. Compare $E(X_k)$ and $E(Y_k)$.

34. Verify that Eq. (11) follows by induction from Eq. (10) if the number of variables is finite. (In Example 4 Continued there are infinitely many variables. Equation (11) is *usually* true for infinitely many variables; it is always true if the variables only take on nonnegative values, as in Example 4. However, this infinite case cannot be proved by induction. It's proof requires the careful use of limits.)

35. Verify that X_i and X_j of Example 4 Continued are not independent.

36. Prove Theorem 2. *Hint*: Mimic the proof of Theorem 1.

37. We defined $E(X)$ in terms of probabilities of *range* values x of X, but these in turn are defined in terms of probabilities of events in S. Thus it is possible to express $E(X)$ directly in terms of such sample space probabilities. Figure 6.11 illustrates what is going on: The probability $\Pr(X{=}x)$ which "weights" x is merely the sum of the probabilities of the elements s which X maps to x. It follows that

$$E(X) = \sum_{e\in S} X(e)\Pr(e). \qquad (16)$$

Now let S be the sample space whose atoms are all triple tosses of a fair coin. Let X be the total number of heads obtained. Compute $E(X)$ using Definition 1. Compute it again using Eq. (16)

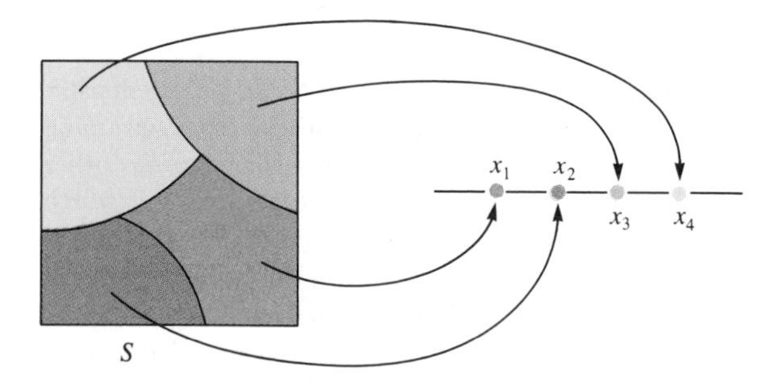

FIGURE 6.11

38. Let f be any real-valued function (f does not mean distribution here). Let X be any random variable on some sample space S. Then $f(X)$ is another random variable—for any $s \in S$, $f(X(s))$ is a real number, which is all that is required of a random variable. Thus by definition,

$$E\big(f(X)\big) = \sum_{z\in \mathrm{Range}(f)} z\Pr\big(f(X){=}z\big). \qquad (17)$$

Explain why this is equivalent to the following, often more convenient formula:

$$E\big(f(X)\big) = \sum_{x\in \text{Range}(X)} f(x)\Pr(X{=}x). \qquad (18)$$

39. Let X be the number of dots when you toss one die. Use Eq. (17) to compute $E\big((X-4)^2\big)$. Check your work by using Eq. (18).

40. Let X be a random variable and let a and b be constants. Show that $E(aX+b) = aE(x)+b$. Use this to answer [3] again.

41. Let X be a random variable on S, and let $A \subset S$. Then $E(X|A)$, the **expectation of X given A**, is the expectation of X on the induced probability space A. (That is, A is the sample space and $\Pr_A(C) = \Pr(C)/\Pr(A)$. See [33, Section 6.3].) Find a formula for $E(X|A)$ in terms of probabilities on the original space S.

42. Prove that

$$E(X) = \Pr(A)E(X|A) + \Pr(\sim\!A)E(X|\sim\!A).$$

We have already used this fact many times in the text; where?

43. You toss two dice. Let X be the number of dots showing on the first and Y the number on the second. Using the definition of expectation, compute
a) $E(X)$, b) $E(X^2)$, c) $E(XY)$.
Confirm that your results are consistent with the remark after Theorem 2 about expectations of products.

44. Prove: If X and Y are independent random variables, then $E(XY) = E(X)\,E(Y)$.

45. Let X be a binary random variable with

$$\Pr(X{=}1) = p.$$

Find $\text{Var}(X)$ and σ_X.

46. Let Y be the number of dots that come up when a fair die is tossed once. Find $\text{Var}(Y)$ and σ_Y.

47. Let Z be the number of heads when a fair coin is tossed twice. Find $\text{Var}(Z)$ and σ_Z.

48. Show that

$$\text{Var}(cX) = c^2\text{Var}(X) \quad \text{and} \quad \sigma_{cX} = |c|\sigma_X.$$

49. Prove: If X and Y are independent, then

$$\text{Var}(X+Y) = \text{Var}(X) + \text{Var}(Y).$$

You will need the result from [44].

50. Use [45] and [49] to find the variance of the binomial distribution $B_{n,p}$.

51. Let $X_1, \ldots, X_n$ be independent copies of a random variable X (such as you get by sampling from X n times). Suppose X has variance V and standard deviation σ. What are the variance and standard deviation of $(1/n)\sum_{i=1}^n X_i$?

52. If X and Y are independent, show that

$$\sigma_{X+Y} \le \sigma_X + \sigma_Y.$$

One says therefore that standard deviation is **subadditive**.

53. If variance is supposed to measure dispersion, why isn't it defined as $E\big(|X-E(X)|\big)$ instead of $E\big([X-E(X)]^2\big)$? The former expression keeps positive and negative values of $X - E(X)$ from canceling without introducing any distortion from squaring. What's the advantage of the latter expression?

6.6 Applications to Algorithms: Proofs of Prior Claims

At various points earlier in this book, we gave intuitive probabilistic arguments about algorithms. In many cases we made claims we said we would justify in Chapter 6. Now's the time!

It's time because this chapter has built a lot of muscles, and you should be ready to use your new strength. We don't need any new example algorithms because the old ones provide quite enough opportunity for a workout. In each of the following subsections, we take an algorithm discussed earlier, review how we argued its analysis before, and show how this argument is justified with our new concepts and theorems (i.e., muscles). The subsections are ordered from easier to harder, but none is trivial.

Average Case Analysis of Algorithm MAXNUMBER

MAXNUMBER, which we introduced in Example 3, Section 1.6, is repeated as Algorithm 6.1. It finds the maximum of distinct numbers $x_1, x_2, \ldots, x_n$ by comparing each element x_k in turn with m, the maximum of the previous elements, and updating m if x_k is larger. The hard question is: On average, how many times is m assigned? (Actually, the question asked in Section 1.6 was: How many times is m *re*assigned? The difference in the number of assignments is 1, as there is one initial assignment, $m \leftarrow x_1$. Henceforth we discuss the number of assignments, because the results are a little prettier.)

Algorithm 6.1
MaxNumber

```
Input n, x_1, x_2, ..., x_n
Algorithm MAXNUMBER
    m ← x_1
    for k = 2 to n
        if x_k > m then m ← x_k
    endfor
Output m
```

We argued in Section 1.6 that it is sufficient to consider the case where the x's are a random permutation of $[n] = \{1, \ldots, n\}$. For a given permutation, the number of assignments of m equals the number of times, as k goes from 1 to n, that x_k is the largest x so far. However, instead of looking at the number of assignments per permutation, we looked at *positions*. We computed the probability, over all permutations of $[n]$, that the kth entry is the largest so far. This probability is clearly $1/k$, because for each group of permutations using the same k integers for its first k entries, $1/k$ of them have the largest of these k integers in the kth spot. Thus we claimed that the expected number of assignments was $\sum_{k=1}^{n} 1/k$. (For reassignments the sum starts at $k = 2$.) Why is this right?

First of all, you can now recognize that we are talking about random variables. Let Y be the random variable (with domain the set of permutations of $[n]$) which gives for each permutation the number of times the variable m is assigned by the algorithm. We are doing an average case analysis, so we want $E(Y)$. We obtain it by writing Y as a sum of simpler variables. Specifically, let X_k be the binary random variable which is 1 when a permutation causes MAXNUMBER to make an assignment of m at the kth spot, and 0 if it doesn't. Clearly,

$$Y = \sum_{k=1}^{n} X_k. \tag{1}$$

By the sum theorem for expectations (Theorem 1, Section 6.5), it suffices to find $E(X_k)$ for all k. Since each X_k is a 0–1 variable, we know that

$$E(X_k) = \Pr(X_k{=}1) = \frac{1}{k}.$$

(This equality is correct even for X_1, for a special reason. Why?) Thus

$$E(Y) = \sum_{k=1}^{n} E(X_k) = \sum_{k=1}^{n} \frac{1}{k},$$

and the justification is done.

Notice that we did *not* compute $E(Y)$ directly from its definition:

$$E(Y) = \sum_{k} k \Pr(Y{=}k). \tag{2}$$

We didn't because it looked hard. If we could count the *number* of n-permutations that require k assignments of m, then we would know $\Pr(Y{=}k)$: Just divide the count by $n!$, the total number of permutations. But short of brute force, how to do this count isn't obvious—and after doing it, we would still have to evaluate the sum in Eq. (2). That's why we turned from looking at assignments per permutation (i.e., Y) to assignments per position (X_k). This problem provides an excellent example of why decompositions like Eq. (1), coupled with the sum theorem for expectations, are so valuable.

Actually, the number of n-permutations requiring k assignments of m *can* be counted, and Eq. (2) *can* be evaluated directly. One way is to use doubly indexed recursions [29]. Indeed, the counts of permutations turn out to be the absolute values of the **Stirling numbers of the first kind**, important numbers that appear in other contexts as well [32].

One caution. Although decompositions like Eq. (1) provide a powerful solution technique, they can't do everything. At the end of Section 6.5 we mentioned that expected value is not the whole story and that you ought at least to know the variance, too. Unfortunately, the sum theorem for variances requires independent variables. For Algorithm MaxNumber, we luck out: The X_k's are independent and so decomposition leads to an easy computation of the variance [2]. In general it is not possible to obtain an independent decomposition and the variance must be obtained by more advanced methods.

Average Case Analysis of Sequential Search

We considered this average case problem in Sections 1.6, Example 1, and again in Section 5.2, Example 3. It is the more sophisticated recursive approach in Section 5.2 that we return to now.

With E_n as the average number of comparisons (when the search word is guaranteed to be on the list), we concluded in Section 5.2 that

$$E_n = \frac{k}{n} E_k + \frac{n-k}{n}(k + E_{n-k}). \tag{3}$$

We obtained Eq. (3) by first claiming that the number of comparisons with the first

k elements of the list is, on average,

$$\frac{k}{n}E_k + \frac{n-k}{n}k. \tag{4}$$

Then we claimed that the average number of comparisons with the remaining $n-k$ words is

$$\frac{n-k}{n}E_{n-k}. \tag{5}$$

Finally, we added these expressions to get Eq. (3). Why is this right?

First of all, once again we are dealing with a random variable, namely, the number of comparisons during the search, and we are seeking its expected value. Call that variable X, so that $E_n = E(X)$. Once again we use a sum decomposition: $X = Y+Z$, where Y is the number of comparisons with the first k words, and Z is the number with the remaining $n-k$ words. Thus so long as expression (4) is $E(Y)$ and expression (5) is $E(Z)$, then adding them to obtain Eq. (3) is correct by the sum theorem for expectations.

So why is expression (4) equal to $E(Y)$? The argument we gave at the time broke the analysis into two cases: Either we found the search word among the first k words, for an average of E_k over these cases only, or we didn't find the search word among the first k, in which case the average value of Y is k because the actual value of Y is always k. Finally, we weighted these two cases according to their probabilities (k/n for the first case and $(n-k)/n$ for the second).

In short, we computed $E(Y)$ by computing Y's expected values when restricted to two complementary events and then taking a weighted average of these values. That is, define

$$E(Y|A) = \text{Expected value of } Y \text{ given that event } A \text{ occurs,}$$

and let A be the specific event "search word found among the first k words". Then we are asserting $E(Y)$ equals expression (4) because this assertion is of the form

$$E(Y) = \Pr(A)E(Y|A) + \Pr(\sim A)E(Y|\sim A). \tag{6}$$

(That is, $\Pr(A) = k/n$, $E(Y|A) = E_k$, and so on.) However, Eq. (6) is not one that we've discussed before, so it demands a proof.

Theorem 1. For any event A and random variable Y, Eq. (6) is true.

PROOF First, you must understand what the **conditional expectation** $E(Y|A)$ means. It means that we restrict our attention to event A, making it the whole sample space, and weight the values of Y by their conditional probabilities in this space. Thus

$$E(Y|A) = \sum_y y\Pr(\{Y{=}y\}|A) = \sum_y y\,\frac{\Pr(\{Y{=}y\} \cap A)}{\Pr(A)}.$$

Now to the proof proper. By definition

$$E(Y) = \sum_y y \Pr(Y{=}y).$$

So

$$\begin{aligned}
\Pr(A)E(Y|A) &+ \Pr(\sim A)E(Y|\sim A) \\
&= \Pr(A)\sum_y y\,\frac{\Pr(\{Y{=}y\}\cap A)}{\Pr(A)} + \Pr(\sim A)\sum_y \frac{\Pr(\{Y{=}y\}\cap \sim A)}{\Pr(\sim A)} \\
&= \sum_y y\Pr(\{Y{=}y\}\cap A) + \sum_y y\Pr(\{Y{=}y\}\cap \sim A) \\
&= \sum_y y\big[\Pr(\{Y{=}y\}\cap A) + \Pr(\{Y{=}y\}\cap \sim A)\big] \\
&= \sum_y y\Pr(Y{=}y) = E(Y). \quad \blacksquare
\end{aligned}$$

Returning to sequential search, we have now justified that $E(Y)$ is given by expression (4). How come $E(Z)$ is given by expression (5)? For the same reason—just substitute Z for Y in the theorem. Recall that Z is the number of comparisons on the back $n-k$ words, and A is the event that we succeed in finding the search word before getting there. Thus $E(Z|A) = 0$, because Z is constantly 0 for event A, and $E(Z|\sim A) = E_{n-k}$, because once the search of the first k words fails, we are back to the case of searching $n-k$ words from scratch (with success guaranteed since success on the whole list of n was assumed).

This finishes the justification of Eq. (3). For an alternative derivation, without recourse to summing expressions (4) and (5), see [9]. ∎

Analysis of Algorithm PERMUTE-1

We reprint PERMUTE-1 (from Section 4.9) as Algorithm 6.2. It repeatedly picks a random integer from 1 to n, putting them into the permutation if they haven't appeared before and throwing them out otherwise.

In Section 4.9 we had to answer several questions about PERMUTE-1 informally. The first was: Does PERMUTE-1 really pick a *random* permutation, that is, does each permutation of $1, \ldots, n$ have an *equal* chance of being selected?

This would not be a question if Algorithm PERMUTE-1, like the later permutation algorithms,

- immediately picked any integer from 1 to n with probability $1/n$;
- then immediately picked any unpicked integer with probability $1/(n-1)$;
- then immediately picked a third unpicked integer with probability $1/(n-2)$;
- and so on.

For then each permutation would have probability $1/n!$. However, at each stage PERMUTE-1 picks *any* integer from 1 to n with probability $1/n$, but then throws it out if it's been picked before. Let k be the number of entries *already* put into

Algorithm 6.2

Permute-1

```
Input n                                          [Length of permutation]
Algorithm PERMUTE-1
    Flag ← 0                              [Set all components of Flag to 0]
    for i = 1 to n                        [ith pass will determine Perm(i)]
        repeat
            r ← RAND[1, n]                   [Pick a random number
        endrepeat when Flag(r) = 0             until an unused one found]
        Perm(i) ← r
        Flag(r) ← 1                                      [r has been used]
    endfor
Output Perm                  [A random permutation given as a sequence]
```

the permutation. (In terms of the variable i in the algorithm, $k = i-1$.) We must show that, for each k from 0 to $n-1$, each of the $n-k$ so-far unpicked integers has probability $1/(n-k)$ of being the next integer to be picked and retained. (We introduce k because we often need to refer to the number of so-far unpicked integers, and $n-k$ is more pleasant than the equivalent expression $n-i+1$.)

Actually, it is sufficient to show that each so-far unpicked integer has an *equal* chance of being the next new number picked, for then each of the $n-k$ possible outcomes must have probability $1/(n-k)$. (The only other possible outcome—previously picked integers keep coming up forever—has probability 0.) Thus it suffices to show that, for each so-far unpicked integer, the probability that it is picked can be described by the same expression.

Let j be any one of the $n-k$ so-far unpicked integers. The event that j is the next to be picked breaks into infinitely many disjoint cases: It is picked on the first try; the first try results in a previously picked integer but the second gets j; the first two tries result in previously picked integers but the third gets j; and so on. On each trial the probability of getting a previous pick is k/n, and the probability of getting j is $1/n$. Thus the probability that j is the next pick is

$$\sum_{m=0}^{\infty} \left(\frac{k}{n}\right)^m \left(\frac{1}{n}\right). \tag{7}$$

Whatever expression (7) equals—we don't treat infinite sums until Chapter 9—it's the same for each j, so we are done. Algorithm PERMUTE-1 is validated as a random algorithm. (Actually, expression (7) is a geometric series—k and n are constants—so you can probably evaluate it right now [17].)

The other thing we claimed in a hand-waving way about PERMUTE-1 was that the average number of RAND calls is

$$n \sum_{j=1}^{n} \frac{1}{j},$$

i.e., $nH(n)$ where $H(n)$ is the nth harmonic number. But this expression is easy to justify: From the solution in Section 6.5 to Example 3, Section 6.1, we know that the expected number of RAND calls to pick the $(k+1)$st number is $n/(n-k)$, and thus by the additivity of expectations, the expected number of calls to pick all n numbers is

$$\sum_{k=0}^{n-1} \frac{n}{n-k} = n \sum_{j=1}^{n} \frac{1}{j}.$$

Average Case Analysis of Algorithm Permute-2

Permute-2 (from Section 4.9) is repeated as Algorithm 6.3. When k numbers have already been picked (again $k = i-1$), the algorithm picks a random integer r from 1 to $n-k$ and then finds the value j of the rth so-far unused integer by checking the Flags. So there is no problem about the permutation being random. The hard question was: What is the expected number of Flag checks?

Algorithm 6.3
Permute-2

```
Input n
Algorithm PERMUTE-2
    Flag ← 0
    for i = 1 to n                           [Pick ith entry in Perm]
        r ← RAND[1, n−i+1]
        c ← 0              [Number of unchosen values passed over]
        j ← 0            [Initializes number considered for Perm(i)]
        repeat while c < r
            j ← j + 1
            if Flag(j) = 0 then c ← c + 1
                                     [Another unchosen number found]
        endrepeat
        Perm(i) ← j
        Flag(j) ← 1
    endfor
Output Perm
```

In each iteration of the i loop, there is one Flag check for each value of j, from 1 to the final value assigned to Perm(i). In Section 4.9 we said it was plausible that this final value of j is uniformly distributed from 1 to n, because the numbers previously put in the permutation are randomly distributed, and the next assigned number is randomly distributed among the unassigned (via r being random).

Let's firm this up. Let J be a random variable for the final value of j in the ith pass of the outer loop. We want to show that, for each integer $m \in [n]$, $\Pr(J{=}m) = 1/n$. The previous paragraph breaks the evaluation of J into two parts—what happened in previous passes of the outer loop and what happens in the current pass. Since the second part depends on the first, we are talking about the product of probabilities where one is conditional on the other. Specifically, let U_m be the event that integer m is not picked in previous passes of the outer loop (U for unpicked). Then

$$\Pr(J{=}m) \;=\; \Pr(U_m)\Pr(J{=}m\,|\,U_m).$$

The value of $\Pr(J{=}m\,|\,U_m)$ is definitely $1/(n{-}k)$, because m is previously unchosen, r is chosen uniformly from set $[n{-}k]$, and *some one* value of r results in $J = m$. (We don't know what that value of r is—it depends in a complicated way on the numbers already chosen for Perm—but we don't need to know it.)

Next, $\Pr(U_m)$ is not hard to determine. We already know that PERMUTE-2 picks permutations randomly, so $\Pr(U_m)$ is the fraction of permutations of length k chosen from $[n]$ that don't use the integer m. There are $n!/(n{-}k)!$ permutations of length k chosen from $[n]$ without restrictions. Of these, $(n{-}1)!/(n{-}1{-}k)!$ don't use m. Dividing and simplifying, we obtain $\Pr(U_m) = (n{-}k)/n$. Therefore Eq. (8) gives

$$\Pr(J{=}m) \;=\; \frac{n-k}{n}\,\frac{1}{n-k} \;=\; \frac{1}{n}.$$

Finishing up the average case analysis of PERMUTE-2 is easy. First,

$$E(J) \;=\; \sum_{m=1}^{n} m\Pr(J{=}m) \;=\; \sum_{m=1}^{n}\frac{1}{n}\,m \;=\; \frac{1}{n}\frac{n(n{+}1)}{2} \;=\; \frac{n+1}{2}.$$

This equality gives the expected number of Flag checks per pass. Since PERMUTE-2 consists of n passes, and since expectations add, we get that the total number of expected Flag checks is $n(n{+}1)/2$, as claimed in Section 4.9.

Notice that we did not proceed as in previous subsections; we did not find $E(J)$ by decomposing J into a sum. In a way that may be surprising, as the algorithm essentially computes j as a sum. Specifically, the final j equals $r+y$, where y is the number of previously picked integers that occur before the rth unpicked integer. Let R and Y be the random variables of which r and y are values. Then with J as before, we have $J = R+Y$. So to show that J is uniform, it seems we only have to find the distributions of R and Y and combine them. Furthermore, it's easy to get started: R is uniform on $[n{-}k]$. Unfortunately, Y is not independent of R: The bigger r is, the larger y is likely to be (why?). So determining $\Pr(J{=}m)$ by the summation route requires the general formula for the distribution of a sum of two variables, Eq. (8) of Section 6.4, rather than the simpler Eq. (9) there. It can be done, but it's tricky [19, 20].

Actually, we don't need to find the distribution of J. All we want is $E(J)$, which equals $E(R) + E(Y)$ no matter how dependent Y is on R. But now we need to find $E(Y)$, for which we need the distribution of Y. Finding this is tricky too, even though the answer is simple [21].

Analysis of AllPerm

AllPerm (from Section 4.9), reprinted as Algorithm 6.4, lists all the permutations of $[n]$, in order of increasing size if they are regarded as n-digit numbers base n. For example, when n=5, the first permutation is 12345, and the permutation immediately after 32541 is 34125. AllPerm obtains the listing as follows. After printing the current permutation, it finds the longest decreasing subsequence in it on the right (that's 541 in 32541). Then AllPerm interchanges the digit just to the left of this subsequence (2 in 32541) with the first digit larger than it in the subsequence (4, obtaining 34521). Finally, it reverses the modified subsequence (obtaining 34125).

What is the complexity of AllPerm? Note that it is a deterministic algorithm, so the average case, worst case and best case are the same. It might seem therefore

Algorithm 6.4
AllPerm

```
Input n
Algorithm AllPerm
    for i = 1 to n                              [Generate first permutation]
        Perm(i) ← i
    endfor
    repeat                                          [Main permutation loop]
        print Perm
        b ← n−1        [b will be position of leftmost digit to be changed]
        repeat until b = 0 or Perm(b) < Perm(b+1)
            b ← b−1
        endrepeat
        if b = 0 then stop                          [All permutations found]
        c ← n               [c will be position to be exchanged with b]
        repeat until Perm(c) > Perm(b)
            c ← c − 1
        endrepeat
        Perm(b) ↔ Perm(c)                                  [Exchange digits]
        d ← b + 1; f ← n                          [Initialize for reversal]
        repeat until d ≥ f
            Perm(d) ↔ Perm(f)                       [Reverse by exchanging]
            d ← d + 1; f ← f − 1
        endrepeat
    endrepeat
Output  All n! permutations of 1, 2, ..., n
```

that probability theory has nothing to offer, but it does. The amount of work needed to produce the next permutation varies in a random-looking way, so to attempt an exact calculation of the work in ALLPERM is not advisable. Instead, because each permutation is created equally often (1 time out of $n!$), we can treat each permutation as random, compute the average amount of work per permutation by probabilistic methods, and multiply the result by $n!$.

What shall we count as a step of work? Unlike previous algorithms, ALLPERM does a lot of printing, so let each symbol printed be one step of work. Right away (no probability yet, and no decision about what else counts as work) we note that ALLPERM has order at least $n(n!)$: All $n!$ permutations are printed, and each printing consists of n symbols. This is already an extreme rate of growth (greater than 2^n), but you have to expect high complexity for algorithms which list all possibilities of something. (Just don't run any algorithm for printing all permutations when n is greater than 5 or 6.) Since $n(n!)$ printing steps is a given for any permutation listing algorithm, the relevant question is: Are there other items we should count as work and what is the complexity of those items? For instance, suppose we found two algorithms to print all permutations, and the first had $5n!$ steps of nonprinting work while the second had $2n(n!)$ steps of nonprinting work. Then *including* the printing, both algorithms have complexity of order $n(n!)$, but the first algorithm has lower nonprinting complexity and is preferable for $n \geq 3$.

In order for ALLPERM to produce the next permutation from the previous one (one pass of the main loop), it goes through three inner loops, and these loops do different amounts of work for different permutations. When the work in some part of an algorithm is variable, we have always tried to measure that work. For ALLPERM, let us therefore declare that each iteration of each inner loop is also one step of work. One iteration of the first inner loop is not equal to one iteration of the third inner loop, if you count both comparisons and assignments, but these "inequities" do not invalidate the claim in the next sentence, as you should check.

With iterations of inner loops counted as steps, the nonprinting complexity of ALLPERM is at most order $n(n!)$. To see this, let's follow the algorithm through one pass of the main loop. The first inner loop finds the length of the longest decreasing sequence on the right of the current permutation by working backwards from $b = n-1$ until $\text{Perm}(b) < \text{Perm}(b+1)$. This work takes at most n executions of the **repeat until** line that starts the inner loop, and thus at most $n-1$ iterations of the whole inner loop. Since we are just trying to find a simple bound, let's say there are n iterations. The second inner loop determines the smallest digit in the decreasing sequence just found that is bigger than the digit $\text{Perm}(b)$ immediately to the left of the sequence; it too works backwards, this time from $c = n$, until $\text{Perm}(c) > \text{Perm}(b)$. This work takes at most n iterations (actually $n-2$). Finally, the third inner loop reverses the decreasing sequence by interchanging symmetrically placed pairs. This reversal takes at most $n/2$ iterations. Thus we have at most $n+n+\frac{n}{2}$ nonprinting steps per permutation, or $\frac{5}{2}n(n!)$ in the whole algorithm.

We just did a worst-case analysis of each pass through the main loop of ALLPERM. Since not all passes are worst case, we have overstated the nonprinting complexity of the whole algorithm. For instance, we said that finding the length of the longest decreasing sequence on the right (henceforth abbreviated LDSR) could take up to n iterations, but in general it will take many fewer. (Here's where probability starts to come in.) For instance, half the time the LDSR will have length 1. (Why?) In fact, it turns out that all the work needed to find the next permutation takes, on average, a *constant* amount of work (in particular, the work is not a function of n). Thus for the whole algorithm the number of nonprinting steps is not of the form $Cn(n!)$ but rather just $C(n!)$. Put another way, as n increases, the nonprinting work becomes negligible relative to the printing work, and thus the total complexity of ALLPERM is $(1+\epsilon)n(n!)$, where $\epsilon = C/n$.

Our claim that the nonprinting work is of the form $C(n!)$ is hard to show completely with what we assume you know, but we can take you most of the way. Let us restrict our attention to the work involved in the first inner loop, finding the length of the LDSR. The work in the second inner loop is less, since c is decreased fewer times than b is, and the work in the third inner loop is also less [26–28]. So it suffices to show that the work in the first inner loop is bounded by a constant.

We will find a bound on the expected number of comparisons between Perm(b) and Perm(b+1) needed to determine the length of the LDSR. (With one exception—what?—this is one more than the number of complete iterations of the first inner loop.) If the LDSR has length k, then the number of comparisons is k. Therefore, if p_k is the probability that the length of the LDSR in a random permutation is k, we seek $\sum_{k=1}^{n} kp_k$. To determine p_k, we group permutations of n things according to the set of things which appear in the last (rightmost) $k+1$ places; within each group the numbers that appear earlier (farther left) are fixed in place. For instance, if $n = 5$ and $k+1 = 3$, the group containing 14325 is

$$14235, \quad 14253, \quad 14325, \quad 14352, \quad 14523, \quad 14532.$$

Note that each group has $(k+1)!$ permutations in it, and each permutation belongs to exactly one group.

Now observe that, within any group, there are exactly k ways to arrange the last $k+1$ integers to make the LDSR of the permutation have length k. Namely, there are k among these integers which are not the largest, and any of these may go in the leftmost of the $k+1$ positions; thereafter, the others must go in decreasing order. For instance, for the group of permutations displayed above, only the $k = 2$ permutations 14253 and 14352 have LDSRs of length $k = 2$. Therefore, within each group, and thus within the set of all $n!$ permutations, the probability that the LDSR has length k is $k/(k+1)!$. (See [22] to test your understanding of this assertion.) It follows that the average number of comparisons is

$$\sum_{k=1}^{n} k \, \frac{k}{(k+1)!} \;=\; \sum_{k=1}^{n} \frac{k^2}{(k+1)!}\,. \tag{9}$$

The sum in Eq. (9) does not have a closed-form equivalent, but that isn't really what concerns us. The question is: Does it grow without bound as n increases (perhaps it is approximately $n/2$, say) or is it bounded by some constant C? It should be intuitive that it is bounded—the fraction $k/(k+1)!$ gets small very fast. Indeed, a few minutes playing with a hand calculator will convince you that the sum has a limit somewhere between 1.7 and 1.8. But can you prove this?

Yes, if we tell you one famous fact and allow you to manipulate infinite sums by the rules which work for finite sums (a process that is sometimes legitimate, including here). The fact is

$$\sum_{k=0}^{\infty} \frac{1}{k!} = e \approx 2.71828. \tag{10}$$

Clearly, Eq. (9) is related to the left-hand side of Eq. (10), so you should not be surprised that the limit of Eq. (9) as $n \to \infty$ involves e. The first step in making the connection is to replace $k+1$ by k and n by ∞, turning the right side of Eq. (9) into

$$\sum_{k=2}^{\infty} \frac{(k-1)^2}{k!}.$$

(The change in k is merely a substitution and doesn't change the value. The replacement of n *does* change the value, but the way to show that Eq. (9) is bounded for all n is to show that the infinite sum is finite. The infinite sum *is* a bound on the sum up to n for all n.)

Next note that we can let k start at 1 again instead of 2. (Why?) Then, expanding the square and regrouping, we obtain

$$\sum_{k=1}^{\infty} \frac{(k-1)^2}{k!} = \sum_{k=1}^{\infty} \frac{k^2}{k!} - 2\sum_{k=1}^{\infty} \frac{k}{k!} + \sum_{k=1}^{\infty} \frac{1}{k!}. \tag{11}$$

From Eq. (10) we already know the value of the last sum: $e-1$. (Where does the -1 come from?) By further, similar manipulations, you can evaluate the first two sums as well [23, 24]. The result is that the left side of Eq. (11) is a constant. Thus for every n, the exact average work per permutation of n things, given by Eq. (9), is bounded by this constant, and we are done.

Problems: Section 6.6

1. In the analysis of Algorithm MAXNUMBER, what is the special reason that makes $E(X_1) = 1$?

2. Consider the variables $X_1, X_2, \ldots, X_n$ in the analysis of Algorithm MAXNUMBER.

   a) Show that they are independent. It suffices to show (See [39, Section 6.4]) that for $i \neq j$

   $$\Pr(X_i{=}a, X_j{=}b) = \Pr(X_i{=}a)\Pr(X_j{=}b)$$

   for $a = 0, 1$ and $b = 0, 1$.

   b) Recall that $Y = \sum_{k=1}^{n} X_k$ is the random variable for the number of assignments in MAXNUMBER. Using the fact that variances add for independent variables [49, Section 6.6], show that

   $$\mathrm{Var}(Y) = \sum_{k=1}^{n} \frac{1}{k} - \sum_{k=1}^{n} \frac{1}{k^2}.$$

3. Write an algorithm MINNUMBER, similar to MAXNUMBER, to find the minimum of n inputted numbers. Determine the expected number of assignments. (The answer is the same as for MAXNUMBER. The point is for you to go through the steps yourself.)

4. Consider Algorithm 6.5, BIGNEXT, which finds the Biggest and Next-biggest of n distinct numbers. Do a best, worst, and average case analysis for the

   a) number of assignments;

   b) number of comparisons.

Algorithm 6.5. BIGNEXT

```
Input n, x_1, ..., x_n                              [n ≥ 2]
Algorithm BIGNEXT
   if x_1 > x_2 then B ← x_1; N ← x_2
                else  B ← x_2; N ← x_1
   endif                                  [Initialize B and N]
   for k = 3 to n
      if x_k > B then N ← B; B ← x_k
         x_k > N then N ← x_k
      endif            [3 cases; if x_k < N do nothing]
   endfor
Output B, N
```

5. In Algorithm BIGNEXT of [4], suppose we replace the for-loop with

```
for k = 3 to n
   if x_k ≤ N then do nothing
      x_k ≤ B then N ← x_k
              else  N ← B; B ← x_k
   endif
endfor
```

Now what is the average number of assignments? The average number of comparisons?

6. On a given day in Miami, the probability that it will rain is 1/3. If the average amount of rain on a rainy day is 1 cm, what is the expected amount of rain per day? What is the expected amount of rain per year?

7. On a typical January day in Buffalo, the probability of no precipitation is 20%, the probability of rain is 10%, and the probability of snow is 70%. When it rains, the average amount is .5 cm. When it snows, the average amount (in melted equivalent) is 1.5 cm. What is the expected amount of precipitation (in liquid measure) on a typical January day?

8. Redo the proof of Theorem 1 using the alternative definition of $E(X)$ in Eq. (16) of [37, Section 6.5], where $E(X)$ is expressed as a weighted sum over events, not variable values. (The algebra of the proof is unchanged, but the notation is perhaps a little simpler.)

9. Justify Eq. (3) as follows. Let X be the number of comparisons with the whole list of n words. Let A be the event "search word is among the first k words on the list". Apply Theorem 1 directly to X. You must show that $E(X|{\sim}A) = k + E_{n-k}$.

10. For the proof of Theorem 1, give the reason for each equality in the final, long display.

Let X be a random variable over a probability space and let $A_1, A_2, \ldots, A_n$ be a set of mutually exclusive and exhaustive events in the space. Prove

$$E(X) \;=\; \sum_{i=1}^{n} \Pr(A_i)\, E(X|A_i).$$

12. Justify Eq. (7), Section 6.5, which gave a recursive relation among expected values for the binomial distribution.

13. Example 4, Section 6.5, gave two methods for finding the expected value of the negative binomial distribution with $n = 1$. The second method depended on the recursive equation

$$E(X) \;=\; p\cdot 1 + (1-p)(E(X)+1).$$

Justify this equation.

14. Consider Algorithm BINSEARCH of Example 2, Section 1.6. Let A_n be the average number of comparisons when there are n words on the list and the search word is known to be on the list. Argue carefully that

$$\begin{aligned} A_n &= 1 + \frac{\lfloor \frac{n-1}{2} \rfloor}{n} A_{\lfloor \frac{n-1}{2} \rfloor} + \\ &\qquad \frac{\lceil \frac{n-1}{2} \rceil}{n} A_{\lceil \frac{n-1}{2} \rceil}, \qquad (12)\\ A_1 &= 1. \end{aligned}$$

Hint: Use the three-term generalization of Theorem 1.

15. Using the procedure for solving Divide and Conquer recurrences in Section 5.9, you can make some progress towards solving Eq. (12).

a) What difference equation do you get if you use that procedure. What is its solution? Is that solution the exact solution to Eq. (12) for any values of n?

b) Following the statement of the procedure in Section 5.9 there was a brief discussion of a more refined approach. Using that approach, obtain a slightly different difference equation from Eq. (12).

c) Solve the difference equation from part b). *Caution*: This is tough.

16. We wish to produce a random multiset of size two from the integers 1 to 3. What's wrong with the following algorithm?

Algorithm MULTISET
$x \leftarrow$ RAND[1,3]
$y \leftarrow$ RAND[1,3]
Output x, y

17. Use the formula for the sum of an infinite geometric series to show that

$$\sum_{m=0}^{\infty} \left(\frac{k}{n}\right)^m \frac{1}{n} = \frac{1}{n-k}.$$

18. Suppose that we do Bernoulli trials until one of the mutually exclusive events $A_1, A_2, \ldots, A_n$ occurs. Let $\Pr(A_i)$ be the probability that A_i occurs on a given trial. It is not assumed that all the $\Pr(A_i)$ are the same. Let $p(A_i)$ be the probability that A_i is the event that finally occurs and thus causes the trials to stop.

a) Assuming that $\Pr(A_j) > 0$, show that

$$\frac{p(A_i)}{p(A_j)} = \frac{\Pr(A_i)}{\Pr(A_j)}.$$

b) Express $p(A_i)$ in terms of the various Pr's.

c) Check that $\sum_{k=1}^{n} p(A_k) = 1$, thus verifying that failure to ever terminate has probability 0. *Note*: Do not assume that the A_i's are exhaustive, but only that at least one of them has a positive probability.

19. For Algorithm PERMUTE-2, consider the alternative approach using $J = R{+}Y$ mentioned on page 500. Consider the case $n{=}5$, $i{=}3$. Show by direct calculation using Eq. (7), Section 6.4, that $\Pr(J{=}3) = 1/5$. *Hint*: $J{=}3$ can be accomplished three ways: $R{=}1$ and $Y{=}2$; $R{=}2$ and $Y{=}1$; and $R{=}3$ and $Y{=}0$. Now determine, for instance, in what ways it is possible for Y to be 2 when $R = 1$.

20. Despite the large amount of arithmetic in [19], it is possible to carry out that approach in general. Consider the ith pass of Algorithm PERMUTE-2 when there are n integers. Using the notation in the text, show that

$$\Pr(Y{=}y \,|\, R{=}r) = \frac{\binom{y+r-1}{y}\binom{n-y-r}{k-y}}{\binom{n}{k}}.$$

From this equality, find $\Pr(R{=}r, Y{=}y)$ and then find and prove a summation formula to verify that $\Pr(J{=}m) = 1/n$.

21. Using the notation at the end of the discussion of Algorithm PERMUTE-2 on page 500, find $E(J)$ again by finding $E(R) + E(Y)$. We noted that the distribution of Y given R is complicated, but surprisingly, the distribution of Y by itself (that is, averaged over all possibilities for R) is simple—it's uniform.

To see this, assume again that there are n integers and that we are picking the ith for our permutation. As in the text, let $m = r{+}y$ be the integer picked. Then y is the number of integers smaller than m in the new partial permutation; said another way, m is the $(y{+}1)$st smallest. For instance, if $i{=}4$, if the partial permutation picked previously is 4,2,7, and if now $m{=}5$, then m is the third smallest in the new partial permutation 4,2,7,5.

Now, recall that at each stage of PERMUTE-2, the partial permutation obtained so far is truly random. Therefore when the ith integer is picked, it must be equally likely to be the smallest, the second smallest, or whatever, of those picked so far.

a) What is $E(Y)$?

b) What is $E(J)$?

22. Here is an argument that, in a random permutation of n integers, the longest decreasing sequence on the right has length k with probability $1/(k!)$. What's wrong with the argument?

Whatever the k integers on the right are, there is only one way to put them in decreasing order. With no restrictions, there are $k!$ ways to arrange them. Therefore the probability requested is $1/(k!)$.

23. Show that the first two terms of Eq. (11) cancel out, so that the sum equals the value of the last term, namely, $e - 1$. *Hint*:

$$\sum_{k\ge1} \frac{k^2}{k!} = \sum_{k\ge1} \frac{(k-1)+1}{(k-1)!}.$$

24. (Assumes calculus) Evaluate Eq. (9) starting with

$$\sum_{k\ge0} \frac{x^k}{k!} = e^x$$

and differentiating (among other things).

25. There is an error in Eq. (9): It is not right for the term with $k = n$, because this term does not follow the general pattern. Why not? What is the correct value for this term? Does the correction make any difference in the calculation in [23]?

26. In Algorithm ALLPERM, the inner repeat loop that begins

$$\textbf{repeat until } \mathrm{Perm}(c) > \mathrm{Perm}(b) \qquad (13)$$

determines the position of that number in the longest decreasing sequence on the right (LDSR) which is the smallest number in that sequence larger than the first number in the permutation to the left of that sequence. Determine the average number of times this line is executed per permutation. *Hint*: Divide the permutations into groups by the length, k, of their LDSR. Argue that, within each group, each number of executions from 1 to k is equally likely. Now invoke the general form of Theorem 1 and then do the sort of analysis done in [23].

27. How many permutations of $[n]$ have length k for their LDSR and cause the instruction of Algorithm ALLPERM shown in [26], display (13), to be executed exactly j times? Use your answer to derive a summation formula for the average number of executions of instruction (13).

28. In the text, we used probability to analyze the expected number of iterations within the first inner loop of Algorithm ALLPERM. In this problem you are asked to extend that probabilistic analysis.

a) Find the constant C so that, as $n \to \infty$, the total number of iterations within all three inner loops is $C(n!)$.

b) Now count each comparison and each assignment as a step instead of each iteration. Find D so that, as $n \to \infty$, the total number of nonprinting steps of ALLPERM is $D(n!)$.

In problems 29–32 below, let $P_{n,k}$ be the number of permutations of $[n]$ in which exactly k of the numbers $x_1, \ldots, x_n$ are the largest of the x's up to that point. Let us say that such a permutation has k *temporary maxes*.

29. We seek a recursion for $P_{n,k}$. A natural way to start is with a recursive procedure for generating the n-permutations from the $(n-1)$-permutations. We need a method which generates n different n-permutations from each $(n-1)$-permutation, since there are $n!$ n-permutations but only $(n-1)!$ $(n-1)$-permutations. Perhaps the most natural method is: For each permutation P of set $[n-1]$, create n n-permutations by inserting the number n at each of the n possible slots before, in between, and after the $n-1$ numbers in P. Call these n n-permutations the permutations *derived* from P.

a) Illustrate this method by using it to generate all the permutations of $\{1,2,3\}$ from the two permutations of $\{1,2\}$. Indicate which 3-permutations are derived from which 2-permutation.

b) This method will lead easily to a recursion for $P_{n,k}$ if there is a simple relationship between the number of temporary maxes of the derived permutations and the number of temporary maxes of the starting permutation. Alas, there is not a simple relationship. Why?

Here is a perhaps less natural method for generating n-permutations from $(n-1)$-permutations. For a given $(n-1)$-permutation P of $[n-1]$, create n sequences by appending at the end, in turn, $1, 2, \ldots, n$. Only the last of these is an n-permutation. (Why?) So modify the others as follows: If the number appended at the end is k, then for all j from k to $n-1$, replace j in P by $j+1$.

c) Illustrate this method by using it to generate all the permutations of $\{1,2,3\}$ from the two permutations of $\{1,2\}$. Indicate which 3-permutations are derived from which 2-permutation.

d) Justify this method in general. Each $(n-1)$-permutation P leads to n different sequences of length n. Show that they are all permutations, and that each n-permutation is generated in this way exactly once.

e) There *is* a simple relationship between the number of temporary maxes of P and the number in the n-permutations derived from it in this second way. The reason is that the only change in order is at the end. Find this relationship and use it to write a recurrence for $P_{n,k}$.

The numbers $s_{n,k} = (-1)^{n+k} P_{n,k}$ are the Stirling numbers of the first kind. They satisfy a recursion very similar to the recursion of part e). The "temporary max" numbers $P_{n,k}$ are thus the absolute values of the Stirling numbers of the first kind; sometimes the temporary max numbers are themselves called Stirling numbers.

30. Let A_n be the expected number of assignments of m in Algorithm MAXNUMBER when n numbers are input.

a) Show that A_n satisfies the recursion

$$A_n = A_{n-1} + \frac{1}{n}.$$

Hint: Plug the recursion from [29e] into the definition of $E(A_n)$. You will then have to do a fair amount of juggling of sums.

b) To verify that $A_n = \sum_{k=1}^{n} 1/k$, it now suffices to check that the sequence $\{A_n\}$ satisfies the right initial condition. Check this.

31. Let $p_{n,k} = P_{n,k}/(n!)$ be the probability that a random n-permutation has k temporary maxes. Develop a recursion for $p_{n,k}$ two ways:

a) By substituting $(n!)p_{n,k}$ for $P_{n,k}$ in the recursion of [29e].

b) By arguing directly using the idea behind the permutation generation scheme preceding [29c].

32. Permutations in cycle form. We have thought of permutations as a reordering of set $[n]$ or perhaps as a one-to-one function from $[n]$ to itself. For instance, the permutation 45132 is the function P which maps 1 to 4 (i.e., the first position holds 4), 2 to 5, 3 to 1, 4 to 3 and 5 to 2.

There is another representation which is often useful. Starting with 1, look at $P(1)$, $P(P(1))$, and so on, until you get back to 1. For the permutation 45132,

$$P(1) = 4, \quad P(P(1)) = P(4) = 3, \quad \text{and} \quad P(3) = 1.$$

Thus 1 generates the cycle 143. Similarly, 2 generates the cycle 25, and in this case, every number in the permutation is now accounted for. The permutation as a whole is represented as (143)(25). (Does the order in which we consider numbers make any difference? Not really. Had we started with 5 and then gone to 4, we would have come up with the representation (52)(431), which is considered the same, since each parenthesized string represents a cycle with no natural beginning and end.)

Let $C_{n,k}$ be the number of n-permutations whose cycle representation involves k cycles. Show two ways that $C_{n,k} = P_{n,k}$:

a) By a one-to-one correspondence. That is, show that for each n-permutation with k cycles there is a (usually different) n-permutation with k temporary maxes.

b) By showing that $C_{n,k}$ satisfies the same recurrence and same initial conditions as $P_{n,k}$.

6.7 Recursive Methods in Probability

From time to time we have shown that recursive techniques often make the solution of a probability problem much easier. In this section we give several more examples, all couched in terms of various games. Among the examples is a complete solution to the game Gambler's Ruin, a particular case of which is the one remaining unanswered problem from Section 6.1 (Example 6). Gambler's Ruin is next to impossible to solve directly, but thinking recursively leads to a second-order difference equation and the solution falls out.

EXAMPLE 1 Two people play the following game. They alternate tossing a single fair die. The first person to throw a 6 wins. What is the probability that the first player wins?

Solution This problem is not that hard to solve directly, using an infinite sum [3]. But it is even easier using recursion. In fact, we show two different solutions using recursion.

Method 1. Let p_1 denote the probability of a win by player 1. Either she wins on the first toss (probability $\frac{1}{6}$), or winning takes more tosses. If it takes more, this means that neither player tosses a 6 on the first try (probability $\frac{5}{6} \cdot \frac{5}{6}$), after which it is as if the game is starting over. Thus

$$p_1 = \frac{1}{6} + \left(\frac{5}{6}\right)\left(\frac{5}{6}\right)p_1. \tag{1}$$

Solving Eq. (1), we find that $p_1 = 6/11$.

Method 2. Let p_2 be the probability that player 2 wins. In order for him to win, it is necessary that player 1 not toss a 6 on her first throw. Thereafter, it is as if player 2 were the first player. Thus

$$p_2 = \frac{5}{6}\, p_1.$$

Combining this equality with

$$p_1 + p_2 = 1,$$

we readily solve and find again that $p_1 = 6/11$. ∎

Example 1 is typical of a type of problem where, after certain sequences of events, it is as if you have started all over again. You can always solve such a problem recursively, by assuming you know the answer in the case that the problem starts over again inside itself. You don't even get a difference equation to solve—just an ordinary algebraic equation in the unknown probability.

EXAMPLE 2 Find the expected number of tosses in Example 1 (a) in general; (b) in the case that player 1 wins; (c) in the case that player 2 wins.

Solution We solved (a) in Section 6.5. In fact, the one simple solution there was by recursion. If we don't care who wins, this game is just Bernoulli trials until the first 6 is tossed, so the answer, call it E, is $1/p = 6$.

To solve (b) and (c), let E_i be the expected number of tosses if player i wins. From Theorem 1, Section 6.6, we know that

$$E \;=\; p_1E_1 + p_2E_2 \;=\; \frac{6}{11}E_1 + \frac{5}{11}E_2. \tag{2}$$

Now, since every win by player 2 looks just like a win by player 1 with an extra toss at the front,

$$E_2 = 1 + E_1.$$

Thus Eq. (2) becomes $6 = E_1 + (5/11)$, from which it follows that

$$E_1 = \frac{61}{11} \qquad \text{and} \qquad E_2 = \frac{72}{11}. \;∎$$

EXAMPLE 3 A and B play tennis. If A has probability p of winning any given point, what is the probability that A will win a game? (A game of tennis is won by the player who first gets 4 points, except that one must always win by 2 points. By tradition, the score is recorded in a funny way, but we won't worry about that and will use the obvious recording method, e.g., 4–2. Also, a game is just part of a set, which in turn is part of a match, but we won't worry about that either, yet.)

Solution If we temporarily ignore cases where the game goes into "overtime" due to the win-by-2 rule, the problem is not hard. Player A can win 4–0, 4–1, or 4–2. The probability of winning 4–0 is p^4. The probability of winning 4–1 is the number of sequences of 5 points ending in a win times the probability of any such sequence, which is p^4q. (Recall that $q = 1-p$.) The number of such 5-point sequences is $C(4,1)$, because we can allow the one lost point to be any one of the first four. Similarly,

$$\Pr(A \text{ wins } 4\text{–}2) = \binom{5}{2}p^4q^2.$$

So far there's no recursion in sight, but let's get back to the win-by-2 rule. This rule makes a difference only if play reaches 3–3. To concentrate on the essentials, let's figure out the probability that A wins *given* that 3–3 has been reached. Call this p^*. To win, A must either win two points in a row, or else A and B must split points and then A has probability p^* of winning from there—because of the win-by-2 rule it is as if they were back at 3–3. Therefore

$$p^* = p^2 + (2pq)p^*,$$

from which simple algebra gives

$$p^* = \frac{p^2}{1-2pq}.$$

(Why is it $2pq$ and not just pq for the probability of winning 1 point each?) Thus the probability that A wins in overtime, starting from the beginning, is p^* times the probability of reaching 3–3, that is, $C(6,3)p^3q^3p^*$. In conclusion, if p_G is the total probability that A wins, then

$$p_G = p^4 + \binom{4}{1}p^4q + \binom{5}{2}p^4q^2 + \binom{6}{3}\frac{p^5q^3}{1-2pq}. \quad ∎ \qquad (3)$$

EXAMPLE 4 **Winning a Point in Volleyball or Racquetball**

In either of these games, in order to win a *point* you must have the serve. If the other side has the serve, your winning the *rally* merely transfers the serve to you. Suppose that you have probability p of winning a rally. What is the probability that you will win the next point if (a) you have the serve; (b) the other side has the serve?

Solution

(a) You can win the next point by either winning a rally right away or by trading rally wins back and forth with your opponent for a while before winning a rally on your serve. Thinking recursively, either you win a point right away, or your opponent wins one rally, you win the next, and then it is as if the game is just starting with you serving. So, letting p_a be the desired probability, we have

$$p_a = p + (pq)p_a.$$

Thus

$$p_a = \frac{p}{1-pq}.$$

(b) Think recursively again. Your opponent is serving, so you must win the first rally. But then it is exactly as if you were starting the game by serving, case (a). Thus

$$p_b = \frac{p^2}{1-pq}. \quad \blacksquare$$

A more natural question about volleyball or racquetball is: Given the probability of winning a rally, what's the probability of winning a game? (Both games are of the first-to-21 type.) We return to this question later. But there are already interesting special cases you can answer [17].

EXAMPLE 5 **Gambler's Ruin**

The general situation is this. Player A starts with $\$m$ and player B with $\$n$. They repeatedly play some game where player A has probability p of winning each time. At the end of each game, the loser pays the winner \$1. Play continues until one of the players is broke (ruined). The question is: What is the probability that A will go broke?

Solution Your first thought might be to approach this question directly, counting the number of ways A could go broke and multiplying each by its probability. By a "way" we mean a sequence of games, for instance, $BBABB$, meaning five games were played and B won all but game 3. If A starts with \$3 and B with \$5, as in Example 6, Section 6.1, then $BBABB$ is one way for A to go broke. Its probability is pq^4.

Of course, there are infinitely many ways to go broke because sequences can go on arbitrarily long, with A and B trading back and forth before B pulls ahead. So we would have to break this approach into pieces: For each k, find the number of sequences of length k which result in A's ruin. A necessary condition on such a k-sequence is that A lose m more times than B does. Furthermore, it is not hard to count the number of such sequences using combinations [21]. However, this does not help us. Again considering Example 6, Section 6.1, look at the sequence of length 13 in which A wins the first 5 and loses the remaining 8. This does not result in A's ruin, because B is ruined after the fifth game and play stops.

In other words, the sequences we want to count must not only have a certain overall frequency of losses by A, but also the losses must not bunch up towards the end. Whenever there are such internal restrictions, counting sequences or probabilities by direct methods is usually very hard. With experience, you would know right away to try recursion.

So, let $P_{m,n}$ be the probability that A goes broke, given that A starts with $\$m$ and B with $\$n$, and that A wins any particular game with probability p. Assume that we already know the ruin probabilities for other starting amounts of money. With probability p, after one game the distribution of wealth is $(m+1, n-1)$; otherwise it is $(m-1, n+1)$. Thus we already have our recursion:

$$P_{m,n} = p\,P_{m+1,n-1} + q\,P_{m-1,n+1}. \tag{4}$$

The initial conditions are easy because generalizing the problem to all starting amounts allows us to consider the degenerate case when one player starts with 0: If A starts with 0, she's already ruined; if B starts with 0, he's already ruined, so A will never be. In other words,

$$P_{m,0} = 0, \quad m > 0, \qquad \text{and} \qquad P_{0,n} = 1, \quad n > 0. \tag{5}$$

Unfortunately, we have a recursion with two subscripts. Although we have treated some **partial difference equations** before, we have done so by guessing. We haven't introduced any general methods. (There are some but they're difficult.) If we can reduce the two subscripts to a single index, we'll be much better off.

We can! The total amount of money in the game is fixed at $m+n$, so the second subscript is unnecessary—it is always $m+n$ minus the first subscript. Therefore our problem becomes

$$P_m = p\,P_{m+1} + q\,P_{m-1}, \qquad P_0 = 1, \quad P_{m+n} = 0. \tag{6}$$

(The recursion is just Eq. (4) with the second subscripts removed. The initial conditions correspond, respectively, to A having none of the $m+n$ dollars and A having them all.) Eq. (6) is easy to solve by the methods of Sections 5.5. Rewrite the recursion as

$$p\,P_{m+1} - P_m + q\,P_{m-1} = 0.$$

The characteristic equation is

$$p\,r^2 - r + q = 0.$$

(Remember: p and $q = 1-p$ are constants; r is the unknown.) By the quadratic formula, the roots are

$$\frac{1 \pm \sqrt{1-4p(1-p)}}{2p} = \frac{1 \pm |1-2p|}{2p} = \left(\frac{1-p}{p}, \frac{p}{p}\right) = \left(\frac{q}{p}, 1\right).$$

Thus the general solution is

$$P_m = C\,(q/p)^m + D,$$

except when $q = p$, which gives a double root. (This important special case of "fair-per-game Gambler's Ruin" is left to the problems [22, 25, 26].) To find C and D, solve from the boundary conditions:

$$C + D = 1, \qquad C(q/p)^{m+n} + D = 0.$$

We find

$$C = \frac{1}{1-(q/p)^{m+n}}, \qquad D = 1 - C.$$

For convenience in writing, set

$$r_* = (q/p).$$

Then, after a little more algebra, we have

$$P_m = \frac{r_*^m - r_*^{m+n}}{1 - r_*^{m+n}}. \tag{7}$$

We are done. We have the probability of A's ruin in explicit form, and it's even fairly simple.

Of course, we are not really done until we interpret what this formula means. For instance, how do the sizes of the initial "pots", m and n, affect the outcome? Such questions are usually answered by considering what happens as m or n gets large. These quantities appear as powers in Eq. (7), and thus that equation is easiest to interpret when $r_* < 1$, because

$$\text{if} \quad 0 < r_* < 1, \quad \text{then} \quad r_*^t \to 0 \quad \text{as} \quad t \to \infty,$$

even if r_* is just slightly less than 1. So, to make $r_* < 1$, consider the case $p > q$. That is, A has the advantage play by play. If m is large, it follows that Eq. (7) is essentially $(0-0)/(1-0) = 0$. In other words, player A has very little chance of being ruined. This conclusion is true even if n is much larger than m and if p is only slightly greater than 1/2. For instance, if $p = .51$ and $m = 100$, A's chance of ruin is 1.8% (to one decimal place), whether B has \$200 or \$200 million [23]. If $m = 200$, A's chance of ruin is merely .03%.

What if n is large but m is small? Then Eq. (7) is essentially $(r_*^m - 0)/(1-0) = r_*^m$. In other words, A's chance of ruin goes down geometrically as m increases.

Now, what if A has the disadvantage play by play? There are two ways to study this situation. One is to rewrite Eq. (7) in terms of $\bar{r} = 1/r_*$, (since now $\bar{r} < 1$) and consider large m and n again [24]. However, it is simpler to use symmetry. To find out what happens when a player has the disadvantage, simply restate the previous results from B's point of view.

The result is: Any player with even a slight disadvantage play by play is in big trouble. For instance, suppose B has probability .49 (almost even) of winning any given play. Then if A starts with a mere \$100, the probability that B will be ruined is *at least* 98.2% no matter how much money B starts with. And if A starts with \$200, B will go broke at least 99.97% of the time!

This analysis should tell you something about betting against the house on casino trips. Set B = you, A = the casino, and consider that A starts with a lot of money and p is always greater than .5 (though only slightly, so as to entice people who haven't read our book). ∎

EXAMPLE 6 **First-to-N Games**

Suppose A and B play a game where A has probability p of winning the next point, whatever has happened so far. Suppose the winner of the game is the player who gets N points first. What is the probability that A wins the game?

There are many games of this sort. For instance, except for win-by-2 rules, in tennis the winner of a set is the first to 6 games; in racquetball, the first to 21 points; and in American squash, the first to 15 points. To keep things simple, this example does not include a win-by-2 rule, so it provides only a first pass at a model for these games. Also, in some of these games the probability of winning a point is heavily dependent on who serves and the serve alternates. See [36, 37] for further discussion.

Solution In one sense we have already solved this problem. Generalizing the analysis in Example 3, the probability A wins is

$$\sum_{i=0}^{N-1} \binom{N-1+i}{i} p^N q^i \tag{8}$$

However, expression (8) can be unwieldy to evaluate. Is there an easier equivalent, either in the form of a more efficient algorithmic evaluation or another formula?

The answer is Yes to both. We will show you a recursion for the probability of winning the game (this time with two indices that cannot be reduced to one). Like other recursions, this one leads to an algorithm. You can then compute with the algorithm, look at the output, and conjecture a simple formula. In fact, this is the way the authors (re)discovered a simpler formula for this problem. You can then use mathematical induction to prove the formula. Or, you can look for a simpler, entirely different combinatorial argument to justify the formula. Such a combinatorial argument happens to exist—it almost always does if the formula is simple enough. Progress would be faster if we could hit upon combinatorial arguments in the first place (before trying recursion), but life rarely works that way.

On to a recursion! Suppose we already knew the probability of winning smaller games, whatever "smaller" should turn out to mean. If A wins the first point, then thereafter she has to win $N-1$ points, whereas B has to win N; if B wins the first point, it's the reverse. It is now clear that "smaller" does not only mean "first to m, where $m < N$", but also includes the possibility that the two players must obtain different numbers of points. Therefore we generalize the problem to ask: What is $p_{m,n}$, the probability that A gets m points before B gets n points? The recursion is:

$$p_{m,n} = p\, p_{m-1,n} + q\, p_{m,n-1}. \tag{9}$$

The initial conditions are

$$p_{m,0} = 0, \quad m > 0; \qquad \text{and} \qquad p_{0,n} = 1, \quad n > 0. \tag{10}$$

(Why?) Anyway, algorithmically speaking, the problem is done. The answer to the original question is $p_{N,N}$, and you can write a program [32] to compute this for any initial p and N. However, much more can be done; see [33–35]. ∎

Problems: Section 6.7

1. Two people alternate flipping a fair coin until it comes up heads. Player 1 goes first, and the person who gets the head is the winner. What is the probability that player 1 wins?

2. Modify Example 1 by letting the second player win if he tosses either a 6 or a 5. Again find the probabilities p_1 and p_2

a) directly via infinite sums,

b) via recursion.

3. Solve Example 1 directly by summing an infinite geometric series.

4. If a problem can be solved by a recursion involving a single unknown instead of a sequence, e.g., Eq. (1), then the underlying process must have the possibility of going on forever. Explain why. For instance, in Example 1, it is logically possible that a 6 might never show up (although the probability is 0).

5. Three players rotate tossing a fair die until one of them wins by throwing a 6. ("Rotate" means the order in which they toss is always the same: player 1, player 2, player 3.) Use recursion to determine the probabilities of a win by each of the players.

6. Change [5] so that, after each toss, which of the other two players tosses next is chosen at random. Determine the winning probabilities.

7. For Example 2, here is another recursive approach to finding E_1, the expected number of tosses if player 1 wins. If she wins, she either wins in one toss, or each player tosses once and then it is as if the game starts over. Use Theorem 1, Section 6.6, to obtain an equation for E_1. *Caution*: When using this theorem, you should weight the event that player 1 wins on the first toss by 11/36, not 1/6. Why?

8. For the game with three players in [5], determine E_i, the expected number of tosses if player i wins

a) using the method presented in Example 2,

b) using the method of [7].

9. Players 1 and 2 alternate flipping a fair coin until heads comes up twice in a row. The person who flipped the second of these heads is the winner. Determine the probability that player 1 wins. What if the coin is not fair?

10. In Example 3 (tennis), if A wins 4–2 there were 6 points. So why isn't the probability of this sort of win $C(6,2)p^4q^2$?

11. The scoring system of a game is "good" if it magnifies the difference in ability between players. That is, if player A is only slightly better than player B on a per-point basis, player A should nonetheless have a strong likelihood of winning the game.

a) Show that tennis (Example 3) is good by making a table of values of p and p_G from Eq. (3).

b) Continuing with tennis, also verify that

$$p{=}.5 \implies p_G{=}.5.$$

(Since tennis treats the players symmetrically, if p_G did not turn out to be .5, we would rightly suspect that we made a mistake in our analysis.)

You might also like to make point/game probability tables for other games discussed in this section. As for tennis, it should be no surprise that the goodness is magnified further once one goes on to consider sets and matches.

12. In many women's tennis tournaments, a match consists of two out of three sets. That is, as soon as one player wins two sets, the match is over. If player A has probability p of winning any given set against B, what is the probability that A wins the match?

13. In many men's tennis tournaments, a match consists of three out of five sets. If player A has probability p of winning any given set against B, what is the probability that A wins the match?

14. In the old days, to win a set in tennis a player had to reach six games first, except that one always had to win by two games. (Now, at a certain point there is usually a tie-breaker.) Assume that A has probability p of winning any given game against B. Under the old rules, what is the probability that A wins the set?

15. Players A and B reach a 3–3 tie in a tennis game ("game" is defined in Example 3).

a) Find the expected number of additional points to end the game.

b) Find the expected number of additional points to end the game with A the winner.

c) Answer b) again with B the winner.

16. For Example 3 (tennis) compute the expected number of points per game.

17. Players A and B play racquetball. Player A has probability p of winning any given rally. What is the probability that A shuts out player B? This means A gets 21 points and B gets 0. Recall that a player wins a point only by winning a rally while serving. Who serves first is chosen at random.

18. In Example 4 (racquetball), assume that player A is about to serve.

a) What is the expected number of rallies for someone to win a point?

b) What is the expected number of rallies for A to win the first point?

c) What is the expected number of rallies for B to win the first point?

19. Modify Example 4 by assuming that player A has probability p of winning a rally if A serves and probability p' of winning if player B serves. Determine the probability that A wins the next point if:

a) A has the serve, b) B has the serve.

20. Do [18] again under the revised probability assumptions of [19].

21. Assuming that players A and B play k games, what is the number of ways B could win m more games than A does? What is the probability that B wins m more games than A, if A has probability p of winning any given game?

22. Solve Gambler's Ruin when $p = .5$. In particular, answer the question in Example 6, Section 6.1 ($m = 3$, $n = 5$).

23. Use Eq. (7) to show that if B has $n > n_0$ dollars and $p > q$, then A's chance of ruin is between

$$r_*^m - r_*^{m+n_0} \quad \text{and} \quad \frac{r_*^m}{1 - r_*^{m+n_0}}.$$

Use this range to verify the claim in the text that "if $p = .51$ and $m = 100$, A's chance of ruin is 1.8% (to one decimal place) whether B has \$200 or \$200 million".

24. Analyze Gambler's Ruin in the case $p < q$ by rewriting Eq. (7) with $\bar{r} = 1/r_* = p/q$.

a) Show that Eq. (7) becomes

$$P_m = \frac{1 - \bar{r}^n}{1 - \bar{r}^{m+n}}. \tag{11}$$

b) What does Eq. (11) say if m is large but n is not?

c) What does Eq. (11) say if n is large?

d) Use Eq. (11) to show that if B has $n \geq n_0$ dollars then player A's chance of ruin is at least $1-\bar{r}^{n_0}$. Hence show that if $p = .49$ and B starts with a mere \$100, then the probability that A will be ruined is *at least* 98.2%, no matter how much she starts with.

e) Suppose that player A has .499 probability of winning an individual game. What is the least amount of money that player B needs so that A's chance of ruin is at least .99, no matter how much money A starts with?

25. In Gambler's Ruin, suppose one player has the edge per game, even just a slight edge. The analysis in the text (and also in [23–24]) shows that there is some fixed amount of money which protects that player against the other, no matter how big the other's pot, in the sense that the first player has, say, at most a 1% chance of ruin. Is the same true when each game is 50–50?

26. In Gambler's Ruin with $p = q$, what is the expected number of games until somebody is ruined?

27. In Gambler's Ruin with $p \neq q$, what is the expected number of games until somebody is ruined? As usual, player A starts with \$$m$ and B starts with \$$n$.

28. Show that winning by 2 (e.g., winning a tennis game after a 3–3 tie) is a special case of Gambler's Ruin. Reconcile the formulas we obtained for p^* in Example 3 with Eq. (7) in Example 5.

29. In Gambler's Ruin, suppose both players start with \$$n$.

a) Show from one of the formulas in this section that the probability player 1 gets ruined is

$$\frac{q^n}{p^n + q^n}.$$

b) A result this simple deserves a simple, direct derivation. With hindsight, can you find one? That is, don't use recursion or difference equations; argue from first principles. If you don't see a general argument, try to give an argument in the case $n = 2$.

30. Solve Gambler's Ruin without a difference equation in the case both players start with \$2. That is, relate $P_{2,2}$ to other $P_{i,j}$ just as before, but now there are so few equations that you can solve them directly. Verify that your answer agrees with the answer we got in Example 5.

31. In Gambler's Ruin, let $N_{k,m,n}$ be the number of sequences of exactly k games that begin with player A having \$$m$ and player B having \$$n$, and end with A going broke.

a) Write a recursion and boundary conditions for $N_{k,m,n}$.

b) Check your recursion by using it to compute $N_{5,3,5}$, a value that is easy to determine in your head.

c) Compute $\sum_{k=3}^{10} N_{k,3,5}$ (use a computer).

d) If $m = 3$, $n = 5$ and each game is fair ($p = q = .5$), what is the probability that A is ruined in 10 games or less?

32. Solve the first-to-N game by writing a program to compute $p_{m,n}$ using Eqs. (9) and (10). An iterative algorithm is much wiser than a recursive one. (Why?) Run the program for squash (N=15), for $p = .51$, .55 and .60.

33. This problem is about a simpler formula than Eq. (8) for the probability that player A gets N points first.

a) Start by picking the simplest interesting value for p, namely, 1/2, and compute some values of $p_{m,n}$ using Eqs. (9) and (10). You should recognize a pattern. The formula we have in mind still involves a sum, but it is more familiar than Eq. (8) and easy to approximate using the continuous normal distribution. *Hint*: Except for the boundary conditions and the division by 2, Eq. (9) with $p = 1/2$ is the recursion for combinations with repetitions, the table of which contains Pascal's triangle along the diagonals. See Example 7 Continued, Section 5.3.

b) Now guess a formula for arbitrary p.

c) Prove your general formula by double induction.

34. Here is a combinatorial argument for the formula we had in mind in [33]. Consider the following two games, G and G'. In G, player A wins if she gets m points before player B gets n points. In G', $m+n-1$ points are played no matter what, and at the end A is declared the winner if she got m or more; otherwise B is the winner. Assume that A has probability p of winning any given point.

a) Show that A has the same probability of winning G as G'. Show this with a many-to-one correspondence between possible scorecards for G' and scorecards for G. By a scorecard we mean a listing $AABABBA\ldots$ that indicates who won each point.

b) It's easy to write a binomial sum for the probability A wins G'. Do it. By part a), this is also a formula for G.

35. Many first-to-N games have an additional win-by-2 rule. You already know how to compute the probability of a win by player A, given that the game has reached a stage at which the latter rule is invoked.

a) Modify a single case of the recursion in Eq. (9) so that the values $p_{m,n}$ are correct for such games.

b) Modify the argument and the summation in [34] to accomplish the same.

36. The first-to-N model developed so far is inappropriate for *sets* of tennis, because serving makes a tremendous difference in tennis, and the server changes with each game. (The model *is* appropriate for *games* of tennis, for the same person serves the entire game. It is also appropriate in squash, because having the serve makes little difference in that game, at least between good players.) A more appropriate model for tennis sets is: Player A has probability p of winning a game when she serves and p' when her opponent serves.

Assume that a set of tennis is a first-to-6 game without a win-by-2 or a tie-breaker rule.

a) Revise the recursive method of Eq. (9), assuming that A serves first.

b) Revise it again, assuming that B serves first.

c) The first serve is determined at random. In terms of the probabilities determined in parts a) and b), what is the probability that A wins the set?

37. The model developed in [32–35] is also inappropriate in any game in which a player must have the serve to win a point—even if the serve itself has no effect on the probability of winning a rally. The reason is that the probability of winning the next point (as opposed to the next rally) is not independent of who won the previous point.

a) Illustrate this dependence. Example 4 should help.

b) Explain how both the recursive and the direct approaches we have taken to our previous model assume independence.

c) Modify the recursion Eq. (8), which is about winning points, to account for this dependence. Assume as the basic building block that player A has probability p of winning any given rally and that rallies are independent. *Hint*: You will now need *three* indices—the number of points A must win, the number B must win, and a 0–1 index indicating who serves first (0 means A serves first). We do not know any combinatorial argument, or similarly direct approach, for treating this revised model.

38. A drunkard leaves an inn located at position m on a straight line between the police station, at 0, and his home, at n. Each minute, he randomly staggers either one unit towards home or one unit towards the police station. Determine the probability he gets home first.

39. Consider another drunkard who does his drinking at home. He staggers out his front door and, each second, randomly goes one unit to the left or one unit to the right.

a) Show that he returns home at some point with probability 1.

b) Determine the expected number of seconds until he first returns home.

40. Take the model of a tennis game in Example 3 and corroborate the results by a simulation. That is, write the scoring rules into a computer program that "plays" a tennis game with player A having some probability p of winning any given point. Then run the program many times and compute the fraction of games that A won. Compare this to the probability of winning computed in Example 3. You wouldn't expect the two numbers to be identical, but they should be close. Specifically, one shows in statistics that, if you run the game 100 times, with 95% certainty your fraction of wins should differ from the answer in Example 3 by at most .1. If you run your simulation 1000 times, your fraction should differ from the theoretical result by at most .03.

41. We attempted to make our models more and more realistic as we went along in this problem set, but surely you can think of ways in which we could go further. However, we have gone about as far as we think we should by theoretical methods. Make up a more detailed model for one of the examples which interests you, and use simulation to estimate the probability of winning the game.

Supplementary Problems: Chapter 6

1. In [18, 19], Section 6.5, we introduced utility as a preferred measure to money. We also followed the assumption about utility usually made by economists: The more money you have, the fewer utils the next dollar is worth. (Another way to say this is: The utility function, the function $U(x)$ that tells you how many utils $\$x$ is worth to you, is concave down.) After all, at some point there is little more you will want to buy—how much caviar can you eat?

However, you can argue that this assumption is not always correct. Sure, you get tired of little things like caviar after a while, but maybe what you really want is the sense of power one gets from owning General Motors, or the instant fame from winning a million-dollar state lottery. In other words, the utility of more money may be small until you get to very large amounts of money, at which point the utility may suddenly become large. We can summarize this situation by saying the utility function is concave up.

A third possibility is that your utility function is linear: $U(x) = ax + b$, with $a > 0$.

a) Show that if you have a linear utility function, you can work with money and forget utility. That is, you will make the same decisions by choosing the actions which maximize your expected monetary return as you will if you maximize your expected util return.

b) In the solution to Example 4, Section 6.1 (which appears in Section 6.5), would the answer change if we analyzed the problem using a concave-down utility function in which \$1 extra is worth at most 1 util?

c) Complete Example 4, Section 6.1, using the following concave-up utility function: Additional dollars are worth 1 util up to \$5000 (and losses are worth −1 util each), but gains above \$5000 are worth 10 utils per dollar. Assume that 250,000 people return their tickets and are thus eligible for prizes.

2. Review the conditional life probabilities p_n of [27, Section 6.3]. Using mathematical expressions, and then using algorithmic language, find expressions for

a) the expected age at which a newborn infant will die;

b) the expected age at which a person who has just turned 50 will die.

Note: For the purposes of such computations, actuaries assume that no one lives beyond 100, i.e., $p_{100} = 0$.

3. Life annuities. Insurance companies offer a contract where you pay them a lump sum, and then they pay you a fixed amount at regular intervals until you die. (Usually the interval is monthly, but for the purposes of this problem we assume that it is yearly.) The lump sum depends on the amount you want to receive in each payment, and how old you are when you start. Suppose that you want $5000 a year, that you just turned 50, and that you want your first payment when you turn 51. Using the p_n's (see [27, Section 6.3]), what is the expected amount the company will pay you during your lifetime? Answer using formulas and algorithms.

4. The answer to [3] is *not* the amount insurance companies charge you for the annuity. They charge less, because they can earn money on the part of the lump sum they haven't paid you back yet. Assume that they can earn 10% per year. How much money do they need now to cover the expected amount they will pay you during your lifetime? Answer using formulas and algorithms.

   Note: The answer to this problem is a good first approximation of what insurance companies will charge, but there is much more to it. For instance, they must charge some extra in order to set up reserves in case they have bad luck—you and the other people they sell annuities to may live longer than you're supposed to! To plan for this type of situation and others (e.g., there is the little matter of profit), more complicated probability analysis is called for.

5. Suppose a discrete probability distribution f is symmetric around c. This condition means that $f(c+w) = f(c-w)$ for all numbers $c+w$ in the domain of the distribution. Show that c is the mean of the distribution. This fact should be obvious; the question is how to prove it. *Hint*: Show that $\sum_x xf(x) = \sum_x cf(x)$ by regrouping the terms in pairs.

6. Assume that X and Y are independent variables and that Y is symmetric around 0 (see [5]). Show that $X-Y$ and $X+Y$ have the same distribution.

Problems 7–10 ask you to find several ways to compute $E(B^2_{n,p})$, the expectation of the square of the Binomial distribution.

7. Let X_i be the characteristic function of success on the ith trial from a sequence of Bernoulli trials. Then $B_{n,p} = \sum_i X_i$, and thus

$$B^2_{n,p} = \sum_i X_i^2 + \sum_{i \neq j} X_i X_j.$$

   Find $E(B^2_{n,p})$ from this equality.

8. Find $E(B^2_{n,p})$ directly from the definition. Method 1 of Example 3, Section 6.5, might help.

9. Find $E(B^2_{n,p})$ by recursion.

10. (Assumes calculus) Find $E(B^2_{n,p})$ by differentiating something.

11. A fair coin is tossed four times.

    a) What is the probability that the third toss is a head?

    b) What is the probability that the third toss is the last head?

    c) What is the probability that the third toss is a head, given that exactly two of the tosses are heads?

    d) What is the probability that the third toss is the last head, given that exactly two of the tosses are heads?

12. What is the range of values for $\Pr(A \cap B \cap C)$ if $\Pr(A) = p$, $\Pr(B) = q$ and $\Pr(C) = r$?

13. A simple card game is played as follows. There are two cards, High and Low. One of the cards is dealt, face down, to the first player, who then gets to look at it. At this point, he can either *fold*, paying the second player $1 and ending the game, or *play* (by simply announcing "I play"). If he plays, the second player must either *fold*, paying the first player $1, or *see*. If she sees, then the first player reveals his card; if it is High, he wins $2 from the second player; if it is Low, he pays $2 to the second player.

    It turns out that the optimal strategies for this game are the following. For the first player: If he is dealt High, he should always play; if he gets Low, he should fold with probability 2/3 and play the rest of the time. (In other words, he should bluff 1/3 the time.) For the second player: In those games where the first player plays, she should fold 1/3 the time and see the rest. (How do you determine that these are the best strategies? Take a mathematics course called "Game Theory".)

Assuming that the card is dealt randomly and both players play optimally, find the probability that, in a given game:

a) The first player folds.

b) The second player folds.

c) A bluff is called.

d) The first player makes money.

e) The second player sees but she would have been better off to fold.

f) The first player folded, given that you know he lost.

g) If the first player loses, it's because his bluff is called.

14. Refer to Example 6, Section 6.1, and suppose you are told that player A got ruined. With this additional information, what is the probability that A lost the first game?

15. Suppose that players A and B play a first-to-21 game in which one must be serving in order to receive a point for winning a rally, and in which the game must be won by 2 points. Player A has probability p of winning any given rally, and A and B have battled to a 20–20 tie. Let p_A^* be the probability that A goes on to win the game if it is now her serve. Let p_B^* be the probability that A goes on to win if it is now B's serve. Determine p_A^* and p_B^* by "co-recursion", that is, by establishing and solving two simultaneous equations in these two unknowns.

Suggestion: Life will be simpler if you first define p_A (respectively p_B) as the probability A wins a point given that A (respectively B) now has the serve.

16. Suppose that we redefine the rules of Gambler's Ruin as follows. In each round, the loser gives \$1 to a referee, instead of directly to the winner. When someone goes broke, the referee gives everything he has collected to the remaining player. Write a recursion and boundary conditions for this modified game. What is the solution?

17. A bug moves from vertex to vertex of square $ABCD$ at random. That is, whatever vertex it is currently at, in one step it moves with equal probability to either adjacent vertex. Assume that it starts at A.

a) What is the probability that the bug reaches C before it visits B?

b) What is the expected number of steps for it to reach C (whether or not it passes through B)?

c) What is the expected number of steps to reach B?

18. The eight vertices of a cube are at the points (a_1, a_2, a_3), where each a_i is either 0 or 1. A bug moves at random from vertex to adjacent vertex. Assume it starts at (0,0,0). For each other vertex, what is the expected number of steps for the bug to reach that vertex for the first time?

19. Suppose that the rules of a tiebreaker are that the winner is the first player to win 2 points in a row.

a) Show by example that this is different from winning by 2 points, as discussed in Section 6.7.

b) If the first player has probability p of winning a point, find the probability that she wins the tiebreaker. *Hint*: Write down an infinite sequence which represents all possible winning sequences for player 1.

c) Answer part b) again, recursively. We do not see how to do this directly, but it can be done if you consider two cases and use co-recursion. Namely, let P_p be the probability that player 1 wins starting with winning the first point of the tiebreaker. Let P_q be the probability that player 1 wins starting with losing the first point.

d) Which tiebreaker rule, win-by-2 or 2-in-a-row, is better in the sense that it more greatly magnifies small differences in players on a point-by-point basis? See the discussion in [11, Section 6.7].

20. Consider the general class of difference equations

$$a_k = 1 + c_k a_{k-1}, \qquad a_1 = 1,$$

where the constants c_k are all close to 1. (For instance, c_k might be $(k-1)/k$.) Intuitively, the solution $\{a_k\}$ should be very close to the sequence $\{k\}$, which is the solution to

$$a_k = 1 + 1\,a_{k-1}, \qquad a_1 = 1.$$

Although this intuition is often true (and hence useful), it can be false.

a) Take the specific case $c_k = .99$, that is,

$$a_k = 1 + .99a_{k-1}, \qquad a_1 = 1.$$

Show that $\{a_k\}$ is bounded—no term is greater than 100—so that $\{a_k\}$ is not close to $\{k\}$ in the long run.

b) Now consider the case $c_k = .999$. How do $\{a_k\}$ and $\{k\}$ compare?

c) Suppose that each c_k satisfies $0 \le c_k \le 1$ and $c_k \to 1$ as $k \to \infty$. Show that $\{a_k\}$ is at least like $\{k\}$ in the sense that $\{a_k\}$ is unbounded. *Hint*: Starting at some index k_0, $c_k \ge .99$ for all $k \ge k_0$. Show that $a_{k_0} \ge 1$ and therefore eventually a_k becomes large enough to approach or exceed the bound in part a). Generalize.

d) Despite the result in c), even the condition that $\{c_k\}$ approaches 1 is not enough to make $\{a_k\}$ approach $\{k\}$. Consider the specific case $c_k = (k-1)/k$. Obtain an explicit formula for a_k. What happens to the difference $k - a_k$? What happens to the ratio a_k/k?

e) Consider the specific case

$$c_k = \frac{\log \frac{k+1}{2}}{\log k},$$

where all logs are base 2. Show that $\{c_k\}$ approaches 1 from below. Obtain an explicit formula for a_k. What happens to the ratio a_k/k?

f) Now consider the specific difference equation

$$\begin{aligned} a_k &= 1 + \left(\frac{2^k - 2}{2^k - 1}\right) a_{k-1}, \\ a_1 &= 1. \end{aligned}$$

(This equality is a Divide and Conquer recursion for the average complexity of binary search. Compare with Eq. (12) in [14, Section 6.6].) Find an explicit formula for a_k. Now what happens to the difference $k - a_k$? *Hint*: The general formula from Section 5.8 will lead to a sum which you can evaluate by methods earlier in Chapter 5.

21. Consider Algorithm BINSEARCH from Example 2, Section 1.6. Let A_n be the average number of comparisons when there are n words on the list and the search word is known to be on the list. There is a way to find A_n other than the complicated route through a Divide and Conquer recurrence in [20f]. Namely, consider the binary tree of possible searches through the list. If the list has $n_k = 2^k - 1$ words, the tree is a "complete" binary tree with k levels, with a word at every vertex. (The root is at level 1, the children of the root at level 2, and so on.) If the search word w is on the list, the number of comparisons done to find w is the level of w in the tree. Therefore, to determine the average amount of work, you merely have to evaluate $\sum_{i=1}^{k} i n_i$, where n_i is the number of words at level i, and then divide by the total number of words, n. (Why?) Do this for a complete binary tree.

22. The method of [21], unlike the method of [20f], works even when the number of words, n, is not of the form $2^k - 1$.

a) For any positive integer n, let

$$k = \lfloor \log_2(n+1) \rfloor .$$

Thus n satisfies

$$2^k - 1 \le n < 2^{k+1} - 1.$$

Show that the binary search tree for n words has $k+1$ levels (unless $n = 2^k - 1$), and every level except for the bottom level is complete. (Use induction on n.)

b) Find a sum similar to that set up in [21], evaluate it, and divide by n. You may leave your answer in terms of both n and k.

23. In analyzing a game of tennis (Example 3, Section 6.7), we noted that the win-by-2 rule goes into effect only if the game reaches 3–3. That's the way tennis players think of it (3–3 is the first score called "deuce"), but in fact the win-by-2 rule goes into effect at 2–2 as well, because you can't win 4–3.

Once again, suppose player A beats player B on a given point with probability p and let p_G be the probability A wins the game. Use the observation about win-by-2 at 2–2 to come up with a different-looking formula for p_G than in Example 3, Section 6.7. Verify that these two formulas are equal.

CHAPTER 7

An Introduction to Mathematical Logic

7.1 Introduction, Terminology, and Notation

Logic is one of the oldest branches of mathematics. For example, the notion of a *syllogism* was important to the ancient Greeks, and the phrase "Aristotelian logic" is still commonly used. But the flowering of logic in modern mathematics did not begin until the mid-nineteenth century, particularly with the work of George Boole (1815–1864). In the twentieth century, particularly with the growing importance of computers, logic has become one of the most important and useful branches of modern mathematics.

Twice earlier we've made brief forays into mathematical logic. In Section 0.7 we introduced much of the terminology of mathematical proofs, which we then used in later chapters. Then in Chapter 2, we used *propositions*, a key concept in logic. One of the purposes of this chapter is to recapitulate somewhat more formally what we've said earlier about proofs in order to stress the importance of logical thinking in mathematics. Another purpose is to introduce you to a variety of areas of mathematical logic which are finding increasing application in various disciplines, including computer science. In particular, since the study of algorithms has been an important theme throughout this book, we want to show you how to use logic in the formal study of algorithms as mathematical objects. And last—but certainly not least—we want to provide you with some tools for organizing your thoughts about mathematics and about reasoning generally.

We'll begin by introducing the **propositional calculus**, which we'll then use to discuss some formal aspects of mathematical proofs and to introduce the notion of algorithm verification. Then we'll discuss **Boolean algebra**—an adaptation of the propositional calculus—which has many applications, particularly in the design of digital circuits. Next we'll introduce some of the basic ideas of the **predicate calculus**—an extension of the propositional calculus. We'll then use the predicate calculus to discuss some more aspects of algorithm verification. Finally, we'll consider an algorithm directly related to mathematical logic but having wider ramifications.

Let's begin as usual with some examples illustrating the variety of problems to which the subject of this chapter can be applied.

EXAMPLE 1 Arguments expressed in ordinary English can be expressed in the notation of mathematical logic and then studied. Consider the following argument:

> If it is cloudy in the morning, then it is going to rain in the afternoon.
>
> If it is going to rain in the afternoon, I should take my raincoat with me.
>
> Therefore, if it is cloudy in the morning, I should take my raincoat with me.

This argument involves various propositions, which we may represent by letters:

> P: It is cloudy in the morning.
>
> Q: It is going to rain in the afternoon.
>
> R: I should take my raincoat with me.

Just like the propositions we used in discussing mathematical induction, these propositions can be *true* or *false*. Unlike the propositions in Chapter 2, however, these propositions do not contain a variable n. In Sections 7.2 and 7.6 (but mainly the latter), we'll consider propositions called *predicates*, which do contain variables.

We may formalize the preceding argument as

$$\begin{array}{c} P \Longrightarrow Q \\ Q \Longrightarrow R \\ \hdashline P \Longrightarrow R, \end{array}$$

where, as we discussed in Section 0.7, the arrow ($\Longrightarrow$) should be read as "implies". The two propositions above the dashed line are the **premises** and the one below the dotted line is the **conclusion**. Do you find this kind of argument convincing? Suppose that "cloudy" were changed to "sunny" everywhere in the preceding argument. How would you feel about the argument then?

Now consider the following.

> If astrology is a true science, then the economy is improving.
>
> The economy is improving.
>
> Therefore, astrology is a true science.

Again, we can represent propositions by letters:

> P: Astrology is a true science.
>
> Q: The economy is improving.

The argument may be represented as

$$\begin{array}{l} P \Longrightarrow Q \\ Q \\ \text{-------} \\ P. \end{array}$$

How do you feel about this argument (whatever your opinions about astrology!)? In this chapter we will consider interpretations of arguments like those above. ∎

EXAMPLE 2 You are designing a computer and want to build a circuit whose input will be a decimal digit expressed in binary form and whose output will be the **complement** of this digit (where the complement of a digit d is defined as $9-d$). Table 7.1 expresses what you want to accomplish. The circuits available to you and their properties are shown in Fig. 7.1. How should you design the circuit? We'll solve this problem in Section 7.5. ∎

TABLE 7.1
Complementing a Decimal Digit

Input Digit		Output Digit	
Decimal	Binary	Binary	Decimal
0	0000	1001	9
1	0001	1000	8
2	0010	0111	7
3	0011	0110	6
4	0100	0101	5
5	0101	0100	4
6	0110	0011	3
7	0111	0010	2
8	1000	0001	1
9	1001	0000	0

EXAMPLE 3 Our first algorithm in Section 1.2 (Algorithm MULT reproduced on the next page as Algorithm 7.1) was an algorithm to multiply two digits by repeated addition. The argument we gave in Section 1.2 probably convinced you that this algorithm does what it is supposed to do. (See also [16, Section 2.6].) But is that the same as *proving* that this algorithm is correct? We'll return to this question in Section 7.4. ∎

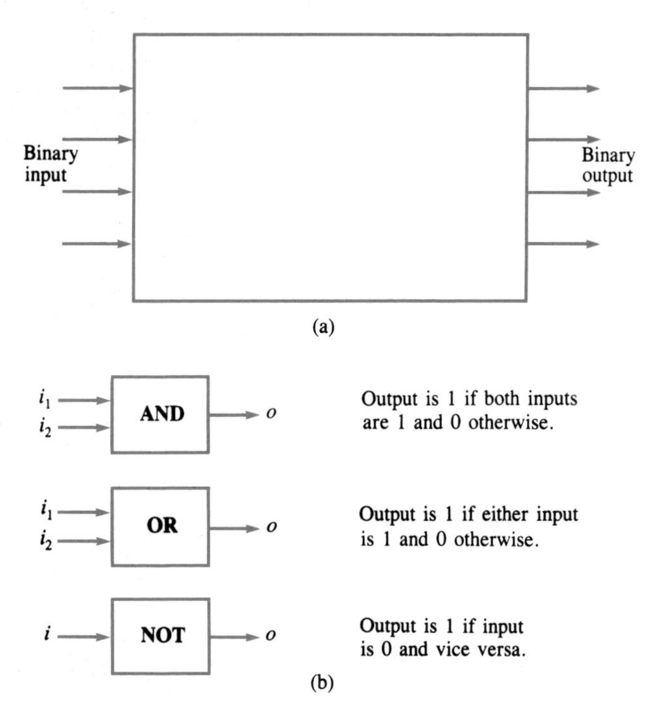

FIGURE 7.1

(a) Designing a Complementer; (b) Available Circuit Elements

As in most of the previous chapters in this book, we must first introduce some of the terminology and notation which we shall use in the remainder of the chapter. Since logic is concerned with correct reasoning, what we will do, in general, is to begin with some propositions called premises and then show that these lead to a correct conclusion. Our propositions will be statements, like those in Example 1, which can be either true or false. No "maybes" are allowed.† The reasoning process

Algorithm 7.1

Mult

```
Input x                                        [Integer ≥ 0]
      y                                        [Integer]
Algorithm MULT
   prod ← 0; u ← 0
   repeat until u = x                          [When u = x, skip rest of loop]
      prod ← prod + y
      u ← u + 1
   endrepeat
Output prod                                    [= xy]
```

† Not all systems of logic are two-valued. In addition to *multiple-valued logics*, there is a system of logic, called fuzzy logic, which in effect allows "maybes".

which leads from premises to a conclusion is called a **deductive process** or just a **deduction**.

It is very important that you understand the distinction between the two words **validity** and **truth** as used in logic. We say that an argument or a reasoning process or a deduction is valid if the truth of the conclusion would follow from the truth of the premises. Consider the argument:

> No even number is a prime.
>
> 2 is an even number.
>
> Therefore 2 is not a prime.

This argument is valid, even though its conclusion is false. The reason is that *if* the premises ("No even number is a prime" and "2 is an even number") were both true (of course, the first one isn't), then the conclusion would be also. So with one or more *factually incorrect* premises, an argument may still be valid (i.e., *logically correct*) although its conclusion may be false. Similarly, the argument in Example 1 is valid as given and would still be valid if "cloudy" were replaced by "sunny".

It's not true, however, that a valid argument based on false premises necessarily leads to a false conclusion, as in

> No even number is a prime.
>
> 4 is an even number.
>
> Therefore 4 is not a prime.

The two preceding arguments recall the computer aphorism, garbage-in, garbage-out (GIGO). Of course, bad input *may* nevertheless yield good output, but this result would be unexpected and you should never count on it. So, also, false premises may lead to a correct conclusion, but you should never depend on such an argument.

Have you noticed that we haven't yet defined what we mean by an **argument**? For example, is the following an argument?

> February is the second month of the year.
>
> Sand contains silicon.
>
> Therefore the sum of the integers from 1 to n is $n(n+1)/2$.

The premises are factually correct and the conclusion is true but we hope you don't find this reasoning characteristic of the kind we've stressed in this book. The problem, of course, is that the premises are *irrelevant* to the conclusion, something that was not true in the previous arguments even when the premises were false. So, at the risk of some subjectivity, we allow something to be called an argument only if the premises are related to the conclusion.

If an argument is valid (i.e., logically correct) and if the supposed facts on which it is based are true (i.e., the premises are true), then the argument is sometimes called **sound**. In mathematical logic we focus on the validity of arguments rather than on their soundness because we often do not know whether the premises are true. This may seem strange at first. But consider the argument:

Premise: n is an even number.

Conclusion: n^2 is an even number.

You should recognize this as a valid argument but, of course, the premise may or may not be true for a given n. The important thing about a valid argument is that, whenever the premises are true, the conclusion *must* be true. Conversely, an invalid argument is one in which the conclusion may be false even if the premises are true. Mathematical proofs are valid arguments because the truth of the premises must imply the truth of the conclusion, but such proofs are often not sound arguments because the premises may not be true in certain cases (e.g., if n above is odd). What can you say about that argument when n is odd? You may remember the old saw about being able to prove anything from false premises. We'll return to this question in Section 7.2.

EXAMPLE 4 Both of the previous arguments about prime numbers are valid but unsound. Here are some examples of arguments which illustrate other aspects of soundness and validity.

- A valid and sound argument:

  All even numbers greater than 2 are composite (i.e., can be factored).

  12 is an even number greater than 2.

  Therefore 12 is composite.

- An invalid and unsound argument:

  All primes except 2 are odd.

  All products of two odd numbers are odd.

  Therefore all products of two odd numbers are prime.

  (Here the argument is invalid and unsound because, while the premises are correct, the conclusion is false.) ∎

We can't give an example of an invalid but sound argument. Why?

In all the foregoing examples, we gave only the premises and conclusions of arguments. Usually, however, we begin a valid argument with its premises and follow a reasoning process to deduce a conclusion which is true if the premises are true. (For example, although it is correct that the square of an even number is even, the conclusion does not follow immediately from the premises. Can you supply a proof?) Much of this chapter is about how to find the reasoning process (i.e., the deduction) which leads from premises to a conclusion.

You may have noticed that our description of a valid argument has a flavor much like that of algorithmic thinking. We begin an algorithm with a specification of the allowable input (in a sense, the premises of the algorithm) and proceed through a series of steps to obtain some desired output (in a similar sense, a true conclusion). Thus the structures of valid arguments and correct algorithms are quite similar. One result of this similarity is that computers can be programmed to produce valid arguments. Another is that valid arguments can be used to explore whether an algorithm is correct. In the remainder of this chapter we will keep reminding you of this connection between logic and algorithms.

7.2 The Propositional Calculus

As its name implies, this section is about the propositions we introduced in Section 7.1—how to manipulate them and how to calculate with them. Hence the use of "calculus". As in the previous section, for now we'll use capital letters, P, Q, R, ..., to denote propositions. (Later on we'll also use A, B, C, ... to denote propositions.) These are really **propositional variables** since their domain is a set of propositions, that is, a set of declarative statements. These statements might be ordinary sentences such as "I am not a mathematician yet but I will be soon" or "The moon is blue"; or mathematical statements such as "$5+3 \geq 7$". When the domain of a propositional variable is not intended to be *all* propositions, we will indicate this. Because propositions may be just English sentences—as in some of our examples—the propositional calculus is sometimes called the *sentential calculus.*

A proposition may be viewed as an assertion that something is true. Since, in fact, a particular proposition may not be true (i.e., we may have made an incorrect assertion), a proposition is said to have a **truth value**, which is either True (T) or False (F). Thus the assertion that the moon is blue is true or it isn't. Except that it might be partly blue and partly not. Then what would be the truth value? Although, as we noted in Section 7.1, some systems of logic allow more than two truth values, we'll restrict the discussion in this chapter to propositions whose truth values are either true or false, without any ambiguity. Sometimes we'll say that a proposition is true (or false) meaning that its truth value is T (or F).

Most of the propositions in this section will not contain variables, but occasionally you'll encounter expressions such as

$$x > 2, \qquad prod = uy, \qquad \text{and} \qquad p_n < \frac{r^2}{4}. \tag{1}$$

We introduced the second expression in (1) in Section 1.2, where we called it a *loop invariant* for Algorithm MULT (reproduced in Example 3, Section 7.1).

Propositions whose truth value depends on the values of the variables in them are called *predicates*, a term we first introduced in Section 0.5. In this section we will be interested only in the truth value of propositions like those in (1) when each variable in them is given a specific value. Viewed this way such propositions are called Boolean expressions† (which is consistent with the use of this term in programming languages). In Section 7.6 we'll consider how to deal with predicates

† Boolean expressions are just what we called logical expressions in discussing our algorithmic language in Section 1.3.

when we know only that the variables may take on values from some set (of, say, integers or real numbers).

Propositions like those in (1) are the **atoms** of the propositional calculus. They cannot be subdivided into two or more simpler propositions. To use such atoms to form more complicated entities, **molecules** or **molecular sentences**, if you will, we must define ways of combining them. For example, given the propositions

P: Jill is 13,

and

Q: Andrew is 12,

how could we combine these into the single proposition

R: Jill is 13 and Andrew is 12?

The question answers itself: We use **connectives** such as "and". (Careful, though. Just because there is an "and" in a proposition doesn't mean that it's not atomic. For example, "Jill and Andrew are siblings" is an atomic proposition because it expresses a single thought.)

What connectives should we allow? The answer depends upon the kinds of propositions which we want to be able to express and the ease with which we wish to be able to express them. The three connectives

and
or
not

are of fundamental importance because they allow expression of the kinds of propositions familiar in everyday discourse as well as in mathematics. Indeed, they are used in almost all programming languages to form compound (or molecular) Boolean expressions from simple (or atomic) ones [e.g., $(x<5)$ and $(y^2 \geq 8)$]. In our algorithms we have already used these connectives without calling special attention to them, as in

check map **and** board first correct train

in the Washington METRO algorithm (Example 2, Section 1.2). Also without calling special attention to them, we have used these connectives in logical expressions in our algorithms, as in

if v_i **not** in U **and** $d(v_i) < \mathit{mindist}$

in SHORTESTPATH (Section 3.5). We counted on your intuitive understanding of words like "and", "or", and "not" when we used them previously without formal definition. Here, however, we will formalize their meanings in a way which should conform precisely to your intuition. Such formalization is necessary in order to achieve a precision of language and notation which is crucial in introducing mathematical logic.

If P and Q are two propositions, then

$$P \textbf{ and } Q \tag{2}$$

means just what you think it does: It is the assertion of both P and Q. Thus the truth value of P **and** Q is determined by the truth values of P and of Q, as shown in Table 7.2(a). This table gives the truth value of (2) for all possible truth values (i.e., T or F) of P and Q. Such a table—which gives the truth values of a molecular or compound proposition for all possible values of its atoms—is called a **truth table**.

TABLE 7.2
Truth Tables for (a) "and", (b) "or", and (c) "not"

(a)

P	Q	P and Q
T	T	T
T	F	F
F	T	F
F	F	F

(b)

P	Q	P or Q
T	T	T
T	F	T
F	T	T
F	F	F

(c)

P	not P
T	F
F	T

We will consistently use the word "and" to describe the compound proposition (2). But you should be aware that in ordinary English we have a variety of ways of expressing the *thought* "and" without necessarily using the *word* "and". For example:

Mathematics is hard and computer science is useful.

Mathematics is hard; computer science is useful.

Whereas mathematics is hard, computer science is useful.

Mathematics is hard, but computer science is useful.

Although mathematics is hard, computer science is useful.

Or, using examples containing mathematical notation:

$x < 0$ and $y > 14$.

$x < 0$ whereas $y > 14$.

$x < 0$ but $y > 14$.

Although $x < 0$, $y > 14$.

Indeed, in both of the preceding groups, each sentence has the same meaning (i.e., the same truth value), except perhaps for nuance. However, in each group you would probably prefer one (which?) rather than the others in ordinary discourse. In our algorithmic notation, we have allowed more than one way of expressing the same thing (can you think of some examples?) because, when thinking algorithmically, expressing a thought in one way rather than another is often more conducive to accuracy. But in writing mathematical expressions there must be some very good reason to allow more than one way to express the same thing because precision is so important. Put another way, if you consistently say the same thing the same way in a poem, it might be boring, but in mathematics consistency of language is a good way to avoid ambiguity. Therefore we allow only the "and" form in the mathematical expressions above.

EXAMPLE 1 Use "and" and propositional variables to express the sentence:

I am going to the store; then I am going to the movies.

Solution With

P: I am going to the store

Q: I am going to the movies

the sentence becomes just P **and** Q. Note two things:

1. The propositional notation does not capture the temporal aspect of the English sentence, namely, that first you are going to the store and then you are going to the movies. Although there are systems of logic which can handle temporal ideas, propositional logic cannot.
2. If someone tells you that "I am going to the store", you tend to accept the statement as implicitly true. But in the propositional calculus you must remember that any proposition can be true or false; in other words, there is no implicit bias to telling the truth. ∎

Table 7.2(b) and (c) give the truth tables for "or" and "not". Thus P **or** Q is the assertion of either P or Q, and **not** P is the assertion of the negation of P (compare to the discussion of negation in Section 0.7). You should be able to think of other ways to express the concept of "or" or "not" in English [1, 2]. Note that "or" is *not* the assertion that either P or Q but not both is true because, when both are true, P **or** Q *is* true according to the truth table. Because the latter assertion for "or" is also a possible meaning of "or" in English, the "or" described in Table 7.2(b) is called the **inclusive or** while the other is called the **exclusive or**. (What would be the truth table for the "exclusive or" [3a]?)

EXAMPLE 2 Use **and**, **or**, and **not** and propositional variables to express each of the following:

a) If my mother doesn't meet my plane, my father will.

b) If I go to eat at The Fisherman tonight, I'll eat lobster, but if I go to another restaurant, I'll have steak.

Solution

a) With

P: My mother will meet my plane

Q: My father will meet my plane

this sentence becomes P **or** Q. Or does it? Should it be the "exclusive or" instead of the "inclusive or"? Does the English sentence allow the possibility that both parents will meet the plane? We think the answer is No so that, in fact, P **or** Q is not a very good answer. We'll return below to the question of how to express the "exclusive or" (see also [3]).

b) With

P: I am going to eat at The Fisherman

Q: I'm going to eat lobster

R: I'm going to eat steak

this sentence becomes

$$(P \textbf{ and } Q) \textbf{ or } ((\textbf{ not } P) \textbf{ and } R).$$

Note the use of parentheses to associate operands (i.e., propositions) with operators (i.e., **and**, **or**, and **not**). We'll return below to how and when to use parentheses in forming compound propositions. ∎

"And", "or", and "not" are called **logical connectives**, or **logical operators**. We will use the former term in this chapter. The "and" and "or" connectives are called **binary connectives** because they have two operands, whereas "not" is a **unary connective** because it has only one operand.

FIGURE 7.2

Notation for Logical Connectives

English Word	Notation	Mathematical Term
and	$\wedge$	conjunction
or	$\vee$	disjunction
not	$\neg$	negation

Just about now we've gotten tired of typing all those "and"s and "or"s and "not"s. This tedium is a sure sign that we should introduce some mathematical notation as a shorthand for these three words. Be warned that there are a variety of such notations in the mathematical literature. The notation displayed in Fig. 7.2 is, however, quite common. It bears a striking resemblance to the notation we introduced in Section 0.1 for set intersection, union, and complement. This similarity should not greatly surprise you. There is a close analogy between the intersection of two sets (which is the set with elements in one set *and* the other) and the "and" of two propositions, and likewise for union and "or". The analogy is not quite so close between set complement and "not" because the "universe" in the proposition case consists only of the proposition and its negation. Nevertheless, you will sometimes find an overbar ($\overline{P}$) used for "not" P instead of $\neg P$. We have chosen to use the latter in this section to emphasize the notion of an operator ($\neg$) operating on an operand (P). However, in Section 7.5 we'll use the overbar notation. Also shown in Fig. 7.2 are mathematical terms for the three connectives which we shall use from time to time in what follows.

EXAMPLE 3 How many different binary connectives which operate on two propositions can be defined?

Solution Since the truth table for any binary connective has four entries (one for each possible pair of values of true and false) and since each entry can have one of two values (T or F), the Product Rule says that there are $2^4 = 16$ possible binary truth tables or **truth functions**. These are shown in Fig. 7.3. The notation in the various columns is that commonly used for these connectives. Only 10 of the 16 columns correspond to useful, non-trivial binary connectives. In the other six, those headed "True", P, Q, $\neg Q$, $\neg P$, and "False", the value of the connective depends upon neither or on only one of the propositions P and Q. ∎

Is Example 3 just a game? Are any of the other connectives as useful as "and", "or", and "not" or even useful at all? Yes, indeed, some are very useful, as we will now discuss.

The two most important connectives besides the three already introduced are those shown in the fifth and seventh columns of Fig. 7.3. The first is the **implication connective** in column 5, which is often called the **conditional connective** because it corresponds to the idea in English:

if P, then Q.

We discussed implication in Section 0.7. So, too, this construct has appeared as the conditional or selective expression in our algorithmic language:

if P **then** Q.

The implication connective asserts that: The truth of P implies the truth of Q. It is important in mathematical logic and mathematics generally because we are often interested in determining when the truth of one proposition does—or does not—imply the truth of another. Recall that all our examples in Section 7.1 had this characteristic if you interpret P as the compound expression consisting of all the premises "anded" together.

		1	2	3	4	5	6	7	8
P	Q	"True"	$P \vee Q$	$P \Longleftarrow Q$	P	$P \Longrightarrow Q$	Q	$P \Longleftrightarrow Q$	$P \wedge Q$
T	T	T	T	T	T	T	T	T	T
T	F	T	T	T	T	F	F	F	F
F	T	T	T	F	F	T	T	F	F
F	F	T	F	T	F	T	F	T	F

		9	10	11	12	13	14	15	16
P	Q	$P \mid Q$	$P \not\Longleftrightarrow Q$	$\neg Q$	$P \not\Longrightarrow Q$	$\neg P$	$P \not\Longleftarrow Q$	$P \downarrow Q$	"False"
T	T	F	F	F	F	F	F	F	F
T	F	T	T	T	T	F	F	F	F
F	T	T	T	F	F	T	T	F	F
F	F	T	F	T	F	T	F	T	F

Names of Connectives
2—inclusive or, disjunction
3, 5—conditional, implication
7—equivalence
8—and, conjunction
9—Sheffer stroke, nand (not and)
10—nonequivalence, exclusive or
11, 13—not, negation
12, 14—nonimplication
15—Peirce's arrow, nor (not or)

FIGURE 7.3

The 16 Binary Truth Functions

Does it bother you that $P \Longrightarrow Q$ is defined in Fig. 7.3 to be true when P is false? Many people find this objectionable at first. You might want to argue that $P \Longrightarrow Q$ should be false when P is false because the hypothesis on which it is based is false. But to so argue would be to confuse the conditional statement, $P \Longrightarrow Q$, with its hypothesis, P. They are *not* the same thing. Why? Because $P \Longrightarrow Q$ doesn't purport to say anything about P but only that something else–Q–is true *if* P is true.

Why not just give $P \Longrightarrow Q$ no truth value when P is false? After all, it's implicit when you say **if** P **then** Q that you are not making a claim about Q when P is false. So why assign a truth value to a nonclaim? Well, in mathematics, leaving things undefined is not just untidy; it often makes subsequent analysis and reasoning much more difficult. If we want—and we do—to use $P \Longrightarrow Q$ as a part of larger logical constructs, we would quickly find ourselves drowning in a sea of complication if we had to take care of undefined cases. So even though we may not care about $P \Longrightarrow Q$ when P is false, giving it a value is still useful in this case.

But which value? Why did we choose to define $P \Longrightarrow Q$ as true in the "don't care" cases when P is false? The answer is related to the fact that many mathematical theorems are stated in the form "if P, then Q". For example, in Section 4.5 we stated the theorem: For any nonnegative integer n

$$\sum_{k=0}^{n} kC(n,k) = n2^{n-1} \tag{3}$$

which has the "if P, then Q" form with P: $n \geq 0$ and Q the assertion in Eq. (3).

Now, as with this theorem, if we prove Q assuming that P is true, we should be allowed to assert that $P \Longrightarrow Q$. If we define the implication connective to be such that $P \Longrightarrow Q$ is true when P is false (i.e., $n < 0$), the proof of the theorem is, indeed, the same as asserting $P \Longrightarrow Q$.

The convention that $P \Longrightarrow Q$ is true when P is false has another advantage. If P is always false, we can say that $P \Longrightarrow Q$ is *vacuously* true. Vacuously true statements are often useful. For instance, in Chapter 2 we sometimes took as the basis case for an induction proof a smaller n than you might have expected because then the basis case turned out to be trivial. For example, in TOH, $P(n)$ can be restated as "If n rings have to be moved, then TOH can be done." When $n = 0$ this is vacuously true because P, namely, that zero rings have to be moved, is false because nothing "has to be moved". When we start with $P(1)$ instead, we actually have to move a ring to prove the basis case.

EXAMPLE 4 State the theorem, "If n is an even integer, then n^2 is also an even integer" as an implication.

Solution With

P: n is an even integer

Q: n^2 is an even integer

this theorem becomes $P \Longrightarrow Q$. Do you see that $P \Longrightarrow Q$ is true even if n is odd? ∎

The connective in column 7 of Fig. 7.3 is the **equivalence connective**; it is sometimes called the **biconditional**. It was also discussed in Section 0.7. A notation sometimes used instead of $\Longleftrightarrow$ is $\equiv$. The notation $P \Longleftrightarrow Q$ asserts that P and Q have the same truth value. For this reason it is sometimes called the "if and only if" or the "necessary and sufficient" connective (compare to Section 0.7). The latter makes sense because, if $P \Longleftrightarrow Q$ is to be true, then when Q is true, it is not just sufficient that P be true (as with $P \Longrightarrow Q$); it is also necessary that P be true. If P and Q are propositions such that $P \Longleftrightarrow Q$ has the truth value T for all possible combinations of values of the atoms in P and Q, we say that P and Q are **logically equivalent**. Analogously, if $P \Longrightarrow Q$ has the truth value T for all possible combinations of the atoms in P and Q, we say that P **logically implies** Q.

The connectives in Fig. 7.3 which we haven't discussed can be expressed in terms of those we have discussed as in the following example.

EXAMPLE 5 Use the connectives we've discussed to express the proposition:

P "exclusive or" Q.

Solution You can find the answer in Fig. 7.3 because column 10 gives just the truth values of P "exclusive or" Q. As the symbol in the column heading indicates, the "exclusive or" idea is the same as nonequivalence. Another way of writing the idea of the "exclusive-or" is

$$\neg(P \Longleftrightarrow Q).$$

If this isn't quite clear, just construct a truth table for $\neg(P \Longleftrightarrow Q)$ and you'll see that this is just the truth table for the "exclusive or" [3b]. ∎

Constructing and Evaluating Compound Propositions

Now that we have atoms and connectives, it's time to consider the rules which must be followed to form legal compound propositions. Can we just write strings such as

$$P \vee Q \wedge R \Longrightarrow S \Longleftrightarrow T$$

or

$$P \wedge Q \vee \neg\neg R \tag{4}$$

or

$$\neg PQ \wedge R?$$

As with arithmetic expressions, we need rules about forming compound propositions from atomic propositions and connectives. Let's first agree to use P, Q, R, ... only for atomic propositions and A, B, C, ... for general propositions which may be atomic or compound. Thus we might denote $P \wedge \neg Q \vee R$ by A. A **logical expression** (or just an *expression*) is any string of letters representing atomic or compound propositions together with connectives. Thus the strings in (4) are all expressions as is

$$P\neg QA \wedge B\neg C \vee R.$$

A legal expression is called a **well-formed formula**†, commonly abbreviated **wff** and pronounced "woof" (or, in some precincts, "wiff").

The rules for forming wffs from atomic propositions and connectives are simple and intuitive:

1. An individual atomic proposition P, Q, R, ..., is a wff.
2. If A is a wff, then the negation of A — $\neg A$ — is a wff.
3. If A and B are wffs, then $A \circ B$ is a wff, where $\circ$ is any of the binary connectives that we have defined ($\wedge$, $\vee$, $\Longrightarrow$, $\Longleftrightarrow$).

You should recognize these as *syntactic rules*, which define the *structure* of wffs. They enable you to determine whether a particular string is a wff.

EXAMPLE 6 Are the strings (a) $P \wedge \vee Q$ and (b) $P \wedge \neg\neg\neg Q$ wffs?

Solution For (a) the answer is No because the rules above do not allow any two binary connectives to be adjacent (do you see why [12]?). For (b), however, the answer is Yes because:

† Often called by logicians just a formula.

TABLE 7.3
Truth Tables for $(P \wedge Q) \Longrightarrow R$ and $P \wedge (Q \Longrightarrow R)$

P	Q	R	$P \wedge Q$	$(P \wedge Q) \Longrightarrow R$	$Q \Longrightarrow R$	$P \wedge (Q \Longrightarrow R)$
T	T	T	T	T	T	T
T	T	F	T	F	F	F
T	F	T	F	T	T	T
T	F	F	F	T	T	T
F	T	T	F	T	T	F
F	T	F	F	T	F	F
F	F	T	F	T	T	F
F	F	F	F	T	T	F

P is a wff by rule 1.
Q is a wff by rule 1.
$\neg Q$ is a wff by rule 2.
$\neg\neg Q$ is a wff by rule 2.
$\neg\neg\neg Q$ is a wff by rule 2.
$P \wedge \neg\neg\neg Q$ is a wff by rule 3. ∎

What we are really interested in, however, is the **semantics**, or meaning, of wffs. That is, given any truth values of the atomic propositions in a wff, we'd like to know the truth value of the wff. But now we have some problems. For example, what is the interpretation of

$$P \wedge Q \implies R? \tag{5}$$

If we first "and" P and Q and then apply the implication, we get a result different from that if we first evaluate $Q \Longrightarrow R$ and then "and" this with P. (See Table 7.3, where we have used parentheses to indicate the order in which the connectives are applied.)

Now consider the interpretation of

$$P \implies Q \implies R. \tag{6}$$

This time, as Table 7.4 illustrates, the interpretation $(P \Longrightarrow Q) \Longrightarrow R$ is different from $P \Longrightarrow (Q \Longrightarrow R)$, where again we have used parentheses to indicate the order of application of the connectives. Can you explain in words why expression (6) has two possible interpretations [15b]?

One way to avoid ambiguities like those in expressions (5) and (6) is to use parentheses. This gives one last rule for forming wffs:

4. Any wff formed by rules 1–3 or by this rule may have parentheses placed around it.

TABLE 7.4
Truth Table for $P \Longrightarrow Q \Longrightarrow R$

P	Q	R	$P \Longrightarrow Q$	$(P \Longrightarrow Q) \Longrightarrow R$	$Q \Longrightarrow R$	$P \Longrightarrow (Q \Longrightarrow R)$
T	T	T	T	T	T	T
T	T	F	T	F	F	F
T	F	T	F	T	T	T
T	F	F	F	T	T	T
F	T	T	T	T	T	T
F	T	F	T	F	F	T
F	F	T	T	T	T	T
F	F	F	T	F	T	T

Thus, for example, (P), $(P \vee Q) \Longrightarrow R$, $P \wedge \neg(\neg Q)$, and $((P \vee (\neg Q)) \Longleftrightarrow R)$ are wffs. As in ordinary algebra, we always evaluate (i.e., calculate truth values of) the parts of wffs within parentheses before evaluating these portions outside parentheses. Thus the parentheses we used in Tables 7.3 and 7.4 resulted in interpretations in accordance with this rule.

Rule 4 allows but does *not* require the use of parentheses so that ambiguous wffs are still possible. Therefore we need two additional rules in order to be able to disambiguate any wff:

I. Whenever there are two or more binary connectives not separated by parentheses, the wff is evaluated using the precedence rule for connectives:

$$\begin{array}{cl} \neg & \text{High} \\ \wedge & \big\downarrow \\ \vee & \\ \Longrightarrow & \\ \Longleftrightarrow & \text{Low} \end{array} \tag{7}$$

When a wff without parentheses is evaluated or when we evaluate a portion of a wff within a given pair of parentheses, we must first handle all negations (do you see why the unary connective $\neg$ must have the highest precedence?), then all $\wedge$s, then $\vee$s, then implications, and, finally, equivalences. Therefore if we don't use parentheses in expression (5), its correct interpretation is as if we had written

$$(P \wedge Q) \Longrightarrow R$$

because $\wedge$ has a higher precedence than $\Longrightarrow$, so that $P \wedge Q$ is evaluated first. Thus if we wanted the second interpretation of Table 7.3, we would be *forced* to use parentheses as in $P \wedge (Q \Longrightarrow R)$.

II. If two or more instances of the same binary connective are not separated by parentheses, evaluate them from left to right. Thus with no parentheses in expression (6), the correct interpretation is as if we had written

$$(P \Longrightarrow Q) \implies R.$$

This rule is analogous to the rule in algebra which says that the correct evaluation of $16/8/2$ is 1 (i.e., $16/8=2$ and $2/2=1$) and not 4 (i.e., $8/2=4$ and $16/4=4$). If you want the interpretation in the last column in Table 7.4, then you must use parentheses as in $P \Longrightarrow (Q \Longrightarrow R)$.

Summarizing, the rules for evaluating wffs are:

- Evaluate connectives not separated by parentheses in the order shown by the precedence (7).
- When two or more identical connectives appear consecutively, not separated by parentheses, evaluate them left to right.

EXAMPLE 7 What are the correct interpretations of the following wffs?

a) $P \wedge Q \wedge \neg R \Longleftrightarrow Q \wedge R$.

b) $P \wedge Q \wedge (R \Longrightarrow P \wedge R) \wedge (\neg R)$.

Solution

a) Using the rules above and inserting parentheses to avoid visual ambiguity, we evaluate this expression in the order:

$$\neg R; \qquad \text{[Rule I]}$$

$$P \wedge Q \qquad \text{and} \qquad Q \wedge R; \qquad \text{[Rules I and II]}$$

$$(P \wedge Q) \wedge \neg R; \qquad \text{[Rule I]}$$

$$((P \wedge Q) \wedge \neg R) \iff (Q \wedge R). \qquad \text{[Rule I]}$$

Table 7.5(a) gives the truth table for this wff.

b) Working first inside parentheses and using Rule I we evaluate

$$P \wedge R \qquad \text{and} \qquad \neg R$$

and then

$$R \implies (P \wedge R).$$

Then, using rule II, we evaluate:

$$P \wedge Q;$$

$$(P \wedge Q) \wedge (R \Longrightarrow (P \wedge R));$$

$$((P \wedge Q) \wedge (R \Longrightarrow (P \wedge R))) \wedge (\neg R).$$

Table 7.5(b) gives the truth table for this wff. ∎

TABLE 7.5
Truth Tables for (a) $P \wedge Q \wedge \neg R \Longleftrightarrow Q \wedge R$ and (b) for $P \wedge Q \wedge (R \Longrightarrow P \wedge R) \wedge (\neg R)$

(a)

P	Q	R	$\neg R$	$P \wedge Q$	$Q \wedge R$	$P \wedge Q \wedge \neg R$	$P \wedge Q \wedge \neg R \Longleftrightarrow Q \wedge R$
T	T	T	F	T	T	F	F
T	T	F	T	T	F	T	F
T	F	T	F	F	F	F	T
T	F	F	T	F	F	F	T
F	T	T	F	F	T	F	F
F	T	F	T	F	F	F	T
F	F	T	F	F	F	F	T
F	F	F	T	F	F	F	T

(b)

P	Q	R	$P \wedge R$	$\neg R$	$R \Longrightarrow (P \wedge R)$	$P \wedge Q$	$(P \wedge Q) \wedge (R \Longrightarrow (P \wedge R))$	$(P \wedge Q) \wedge (R \Longrightarrow (P \wedge R)) \wedge (\neg R)$
T	T	T	T	F	T	T	T	F
T	T	F	F	T	T	T	T	T
T	F	T	T	F	T	F	F	F
T	F	F	F	T	T	F	F	F
F	T	T	F	F	F	F	F	F
F	T	F	F	T	T	F	F	F
F	F	T	F	F	F	F	F	F
F	F	F	F	T	T	F	F	F

Rules 1–4 and I and II enable us to construct and evaluate any wff, a fact we won't prove but which we'll discuss further in Section 7.8. We'll explore in a problem some other examples of when wffs would be ambiguous without rules I and II [16].

Two natural algorithmic questions raised by this discussion are:

1. Given a string consisting of atomic propositions, connectives, and parentheses, can we devise an algorithm to determine whether it is a wff?
2. Given a wff, can we generate its truth table, as in Tables 7.4 and 7.5?

The first question is quite difficult to answer in general, but in Section 7.8 we will give a recursive algorithm for a simplified version of it. Since the algorithm for evaluating a wff is quite similar to that for determining whether a string is a wff, the second question is also quite difficult to answer. However, one aspect of generating a truth table, namely, generating the values of the atomic propositions in each row, is relatively easy. So let's just assume that we can use our algorithmic language to evaluate wffs, given a truth value assignment to each of its atoms. Using this assumption, let's design an algorithm to generate a truth table. The key to calculating the values in each row may be inferred from Table 7.5. Looking at the first three columns in (a) and (b) and working from the right to the left, you can see that each row after the first, which is all T's, can be found from its predecessor by

- finding the first T from the right and changing it to an F, and
- changing all columns to the right of the changed column to T.

This idea works in general for any number of atomic propositions. Do you see why? We leave the answer to [19]. (*Hint*: Consider successive binary integers and how the 0's and 1's change from one integer to the next.)

Algorithm 7.2 TRUTHTABLE does the job for us. Do you see the connection of TRUTHTABLE to the algorithm used for our robot adder at the beginning of Chapter 1? Also compare TRUTHTABLE to ALLPERM in Section 4.9. Do you see that both are based on the same idea of working leftward to find the first position where a change is needed? Why is TRUTHTABLE somewhat easier to design than ALLPERM was [20]?

You should be able to apply TRUTHTABLE to the wffs of Example 7 to obtain the truth tables (without the intermediate columns) of Table 7.5.

Tautologies

So far the material in this section may have seemed like no more than a mathematical game—a fun game perhaps—but does it relate to anything practical? One quite practical thing it relates to is testing the validity of arguments expressed in ordinary English. In Example 7.1 an argument about when to wear a raincoat was expressed in words as

> If (P) it is cloudy in the morning, then (Q) it is going to rain in the afternoon.
>
> If (Q) it is going to rain in the afternoon, then (R) I should take my raincoat.
>
> Therefore, if (P) it is cloudy in the morning, (Q) I should take my raincoat with me.

More formally, we then expressed this argument as

$$P \Longrightarrow Q$$
$$Q \Longrightarrow R$$
$$\text{-------}$$
$$P \Longrightarrow R.$$

Our discussion in this section implies that this argument is valid if and only if the wff

$$((P \Longrightarrow Q) \wedge (Q \Longrightarrow R)) \implies (P \Longrightarrow R) \tag{8}$$

is true for all possible values of its atoms. Note that both "if ... then" and "therefore" in the English formulation of the argument have been replaced by $\Longrightarrow$ in

Algorithm 7.2

TruthTable

```
Input n        [Number of distinct atomic propositions in wff: P1 to Pn]
      S                                         [String of symbols in wff]

Algorithm TRUTHTABLE
   procedure newrowatoms                         [Next row of values for
      j ← n                                        atomic propositions]
      repeat while j > 0 and Pj = F              [Change F to T from
         Pj ← T; j ← j − 1                                right to left]
      endrepeat
      if j = 0 then end ← true            [If j = 0, previous row was last]
               else Pj ← F                     [Rightmost T changed to F]
      endif
      return
   endpro
   procedure evaluate
      print the row of current values of the atoms
      evaluate the wff
      print the evaluation
      return
   endpro
   end ← false                          [Main algorithm; initialize end]
   for i = 1 to n
      Pi ← T                                           [First row all T's]
   endfor
   repeat while end = false
      evaluate
      newrowatoms
   endrepeat

Output Truth table with n + 1 columns, the last being the values
       of the wff
```

the wff. Actually, "therefore" in English connotes something stronger than "if ... then", but in translating to wffs we use $\Longrightarrow$ for both meanings. You should also be able to see that the example in Section 7.1 about astrology can be expressed in wff form as

$$((P \Longrightarrow Q) \wedge Q) \Longrightarrow P \tag{9}$$

Note that some of the parentheses in expression (9) aren't necessary (which ones?) in order to express the arguments as wffs. But we will continue to use more parentheses than required if doing so makes the wffs more readable. You should make a habit of doing so, too.

TABLE 7.6
Truth Tables for the Arguments of Example 1, Section 7.1:
(a) wff1 = $((P \Longrightarrow Q) \wedge (Q \Longrightarrow R)) \Longrightarrow (P \Longrightarrow R)$;
(b) wff2 = $((P \Longrightarrow Q) \wedge Q) \Longrightarrow P$

(a)

P	Q	R	$P \Longrightarrow Q$	$Q \Longrightarrow R$	$(P \Longrightarrow Q) \wedge (Q \Longrightarrow R)$	$P \Longrightarrow R$	wff1
T	T	T	T	T	T	T	T
T	T	F	T	F	F	F	T
T	F	T	F	T	F	T	T
T	F	F	F	T	F	F	T
F	T	T	T	T	T	T	T
F	T	F	T	F	F	T	T
F	F	T	T	T	T	T	T
F	F	F	T	T	T	T	T

(b)

P	Q	$P \Longrightarrow Q$	$(P \Longrightarrow Q) \wedge Q$	wff2
T	T	T	T	T
T	F	F	F	T
F	T	T	T	F
F	F	T	F	T

Were our arguments in Section 7.1 examples of valid reasoning? That is, do the conclusions [$P \Longrightarrow R$ in expression (8) and P in expression (9)] follow from the premises [$P \Longrightarrow Q$, $Q \Longrightarrow R$ in (8) and $P \Longrightarrow Q$, Q in (9)]? To answer these questions, we need only construct the two truth tables. In Table 7.6(a) we see that expression (8) is true for all values of P, Q, and R. Although we are probably interested only (why?) in whether the conclusion $P \Longrightarrow R$ is true when both premises are true (rows 1, 5, 7, and 8), this will be the case if and only if the last column is T in *every* row. The reason is that any time either of the premises is F, the conclusion will always be true. (Recall the properties of implication.) In Table 7.6(b), one of the entries in the last column is false, and so our argument about astrology was not valid. (Bet you knew that anyhow!) The interpretation of the F in the last column is that, so far as any information contained in the example is concerned, the economy could be improving even if astrology is false (whew!).

A wff which has the property, as in expression (8), that it is true for all possible values of its atomic propositions, is called a **tautology**. (In ordinary English usage a "tautology" is a statement devoid of information. What's the use of making a statement that is true independent of the facts? If I ask you, "Were you there?" and you answer, "Either I was or I wasn't", you haven't been helpful. However, in mathematical logic, tautologies are often highly nonobvious and no stigma is attached to this word.)

We noted earlier that when a wff of the form $A \Longrightarrow B$ is a tautology, we say that A logically implies B[†]. It's important that you understand the distinction between A *implies* B ($A \Longrightarrow B$), which is a wff whose truth value may be true or false depending upon the truth values of its atoms, and A *logically implies* B in which the truth value of $A \Longrightarrow B$ is always true whatever the truth values of the atoms. (Remember: A and B are typically compound propositions and $A \Longrightarrow B$ may therefore be true for all values of the atoms in it.)

The cloudy-rain-raincoat example suggests that testing an argument for validity is the same as determining whether a certain wff is a tautology. We can state this assertion as a theorem.

Theorem 1. Let $A_1, A_2, \ldots, A_n$ be wffs which are the premises of an argument and let B be a wff which is the conclusion. Then the argument

$$\begin{array}{c} A_1 \\ A_2 \\ \vdots \\ A_n \\ \text{-----} \\ B \end{array}$$

is valid if and only if

$$(A_1 \wedge A_2 \wedge \cdots \wedge A_n) \Longrightarrow B \tag{10}$$

is a tautology.

PROOF Recall that an argument is valid if, when the premises are *assumed* true, the conclusion must be true. If expression (10) is a tautology, then B must be true for all truth values of the atomic propositions which make the premises true and so the argument is valid. Conversely, if the argument is valid, then B is true when all the premises are true. But, if any of the premises is false, then wff (10) is true from the definition of the implication connective. Therefore the wff is true for any values of its atoms and so is a tautology. (Do you see why it was necessary to define implication when the hypothesis is false as we did in order for this theorem to be true?) ∎

[†] Some authors, who use a single-shaft arrow ($\rightarrow$) to represent ordinary implication, use a double-shaft arrow ($\Longrightarrow$) to represent logical implication.

Besides being a useful notion when we want to consider the validity of an argument, the idea of a tautology is also valuable when we want to determine whether two wffs are equivalent. Suppose, as we will discuss further in Section 7.5, that we are designing an integrated circuit chip for a computer or some other electronic device. The output function of such a circuit is often expressed as a wff of its inputs. The form of this wff will determine just how physical circuitry must be used to realize the desired output. If the wff can be simplified, it may be possible to save on circuitry and, therefore, on cost. The situation is:

- You are given wff1 describing the output in terms of the inputs.
- Someone presents you with a simpler wff2, purporting to give the same output for all values of the input.

Question: Are they really logically equivalent? That is, is

$$\text{wff1} \iff \text{wff2} \tag{11}$$

a tautology? [Remember that, if wff1 and wff2 are wffs, then so is expression (11).]

How would you answer the question above? The obvious way at this point is just to calculate the truth table, as in the following example.

EXAMPLE 8 Are the two wffs

$$(P \wedge Q) \vee (P \wedge R) \vee (\neg Q \wedge R) \qquad \text{and} \qquad (P \wedge Q) \vee (\neg Q \wedge R)$$

logically equivalent? That is, can we just leave out the middle term in the first wff?

Solution The relevant truth table, shown in Table 7.7, indicates that the two wffs are equivalent because the last column is all T's; that is, the second term in wff1, $P \wedge R$, is superfluous because the truth value of the wff is the same whether it is present or absent. ∎

TABLE 7.7
Truth Table for the Equivalence of Two wffs:
wff1 = $(P \wedge Q) \vee (P \wedge R) \vee (\neg Q \wedge R)$;
wff2 = $(P \wedge Q) \vee (\neg Q \wedge R)$

P	Q	R	$P \wedge Q$	$P \wedge R$	$\neg Q \wedge R$	wff1	wff2	wff1 $\iff$ wff2
T	T	T	T	T	F	T	T	T
T	T	F	T	F	F	T	T	T
T	F	T	F	T	T	T	T	T
T	F	F	F	F	F	F	F	T
F	T	T	F	F	F	F	F	T
F	T	F	F	F	F	F	F	T
F	F	T	F	F	T	T	T	T
F	F	F	F	F	F	F	F	T

EXAMPLE 9 Prove that the following expressions are tautologies.

a) $\underbrace{\neg\neg\cdots\neg}_{2n} P \Longleftrightarrow P.$

b) $\underbrace{\neg\neg\cdots\neg}_{2n-1} P \Longleftrightarrow \neg P.$

Solution

a) When $n=1$ the wff is $\neg\neg P \Longleftrightarrow P$ for which the truth table is

P	$\neg P$	$\neg\neg P$
T	F	T
F	T	F

Thus the wff is a tautology. Now how would you prove this for any n? By induction, of course! We leave the proof to a problem [27a].

b) The proof for b) is essentially the same as that for a). Alternatively, define Q to be the proposition $\neg P$ and use a) immediately [27b]. ∎

In the propositional calculus as in other branches of mathematics we are interested in proving theorems. Such theorems may take the form of expression (10): Show that the conjunction of some premises $(A_1, A_2, \ldots, A_n)$, which serve as the *hypotheses* of the theorem, logically imply the desired *conclusion* B. Proving such a theorem is therefore equivalent to showing that an expression of the form in (10) is a tautology. Theorem 1 shows that proving a theorem of this form is also equivalent to showing that a particular argument is valid.

Is expression (10) the only type of theorem we want to prove? How about trying to prove that two wffs (such as wff1 and wff2 in Example 8) are equivalent. Doesn't that really have the flavor of a theorem? Yes, but can we fit it into the paradigm of expression (10)? We can by allowing n in expression (10) to be 0, that is by allowing a logical implication to have no premises which we could write as

$$\Longrightarrow B, \tag{12}$$

where B is a wff which could (but need not) have the form wff1 $\Longleftrightarrow$ wff2. The argument corresponding to (12) has the form

$$\frac{\quad}{B.}$$

The notation (12) means only that we must prove B true under all circumstances, which is the same as saying that B must be true for *all* values of the atomic propositions in the wffs, that is that B must be a tautology. Thus, as before, proving a theorem is equivalent to showing that some wff is a tautology. Indeed, you may have noticed that expression (10) itself can be rewritten as a theorem with no premises in the form $\Longrightarrow ((A_1 \wedge A_2 \wedge \cdots \wedge A_n) \Longrightarrow B)$ [32].

If you find the form of (12) somewhat disturbing, the following may help. It is easy to show with a truth table [24] that

$$(\text{wff1} \Longleftrightarrow \text{wff2}) \iff ((\text{wff1} \Longrightarrow \text{wff2}) \wedge (\text{wff2} \Longrightarrow \text{wff1}))$$

is a tautology, that is, that $\text{wff1} \Longleftrightarrow \text{wff2}$ is logically equivalent to $(\text{wff1} \Longrightarrow \text{wff2}) \wedge (\text{wff2} \Longrightarrow \text{wff1})$. The interpretation of this expression is that, instead of stating the equivalence of two wffs, wff1 and wff2, we can state that wff1 implies wff2 *and* that wff2 implies wff1. Therefore the theorem

$$\Longrightarrow (\text{wff1} \Longleftrightarrow \text{wff2})$$

can be proved by proving the two theorems

$$\text{wff1} \Longrightarrow \text{wff2} \quad \text{and} \quad \text{wff2} \Longrightarrow \text{wff1}, \tag{13}$$

where the premise in the first is wff1 and the conclusion is wff2; in the second theorem the roles of the two wffs are reversed.

Well, then, how do we prove theorems in the propositional calculus? We know at least one answer to this question, namely, that we just use a truth table or part of one. That is, we

- use all sets of values of the atoms in the premises, compute the value of each premise A_i, and
- then for each row where all the premises are true, compute the value of B.

If and only if B is T for all such values, the theorem is true. What we have shown, therefore, is that there is a truth table proof for each theorem in the propositional calculus.

EXAMPLE 10 Prove (or disprove) the theorem

$$((P \Longrightarrow (Q \wedge R)) \wedge (R \Longrightarrow S) \wedge (\neg(Q \wedge S))) \Longrightarrow \neg P. \tag{14}$$

Solution The appropriate truth table is shown in Table 7.8. As the conclusion is true for the four rows of the table where all three premises are true, the theorem is proved; that is, expression (14) is a tautology. ∎

That *should* cover proving theorems in the propositional calculus. We've outlined a nice mechanical procedure which will always work. But it isn't enough. Why not?

- One answer might be that, when the number of atomic propositions gets fairly large, the truth table becomes long and unwieldy. True, but with computers around, this reason won't wash unless the number of variables is very large, indeed.
- A better answer is that proof by truth table does *not* lead to insight about a theorem or to possible generalizations of it. Another method might do so. A related answer—in the case of a theorem that two wffs are equivalent—is that we usually know only one of the wffs and need a method to derive another (simpler) wff. Another proof method might be more helpful in doing this than truth tables.

TABLE 7.8
Truth Table for $((P \Longrightarrow Q \wedge R) \wedge (R \Longrightarrow S) \wedge \neg(Q \wedge S)) \Longrightarrow \neg P$

P	Q	R	S	$P \Longrightarrow (Q \wedge R)$	$R \Longrightarrow S$	$\neg(Q \wedge S)$	$\neg P$
T	T	T	T	T	T	F	
T	T	T	F	T	F	T	
T	T	F	T	F	T	F	
T	T	F	F	F	T	T	
T	F	T	T	F	T	T	
T	F	T	F	F	F	T	
T	F	F	T	F	T	T	
T	F	F	F	F	T	T	
F	T	T	T	T	T	F	
F	T	T	F	T	F	T	
F	T	F	T	T	T	F	
F	T	F	F	T	T	T	T
F	F	T	T	T	T	T	T
F	F	T	F	T	F	T	
F	F	F	T	T	T	T	T
F	F	F	F	T	T	T	T

- A still better answer is that very few portions of mathematics allow a completely mechanical proof procedure like the truth table technique of the propositional calculus. (Imagine how much easier high school geometry would have been if there were a crank you could have turned to generate proofs!) Because one reason for studying the propositional calculus is to provide a simple subsystem of mathematics from which to get a better feel for the logical methods of mathematics in general, developing a deductive method of proving theorems in the propositional calculus, as well, would be wise.

Thus pursuing another method of proving things in the propositional calculus is important. We'll do precisely that in Section 7.3.

Problems: Section 7.2

1. Give some words or phrases which, as far as their truth value is concerned, are equivalent to "or". Use them in some sentences to show what you mean.

2. Give some words or phrases which, as far as their truth value is concerned, are equivalent to "not". Use them in some sentences to show what you mean.

3. **a)** Give the truth table for the "exclusive or".
   b) By constructing the truth table for
   $$\neg(P \Longleftrightarrow Q)$$
   show that the solution to Example 5 is correct.

4. Express in words the meaning of each of the connectives in columns 9, 10, 12, 14, and 15 of Fig. 7.3.

5. Each of the binary connectives in columns 9, 10, 12, 14, and 15 of Fig. 7.3 can be expressed in terms of the logical connectives, $\wedge$, $\vee$, $\neg$, $\Longrightarrow$, and $\Longleftrightarrow$. For example, $P \not\Longleftrightarrow Q$ can be expressed as $\neg(P \Longleftrightarrow Q)$ because both have the same truth tables. Show how each of the other four can be similarly expressed.

6. Which of the following implications are vacuously true and which are not? Explain.

   a) If x is an even prime greater than 2, then x^2 is even.

   b) If Ms. Jones is my grandfather, then I am Ms. Jones' grandson.

   c) If x_1, x_2, and x_3 are solutions of $ax^2+bx+c = 0$, then $x - x_1$ is a factor of ax^2+bx+c.

   d) If Jane and Jack are my parents, then I am the daughter of Jane, Jack, and Jill.

7. Using J for "John is here" and M for "Mary is here", write the following sentences as wffs. (Sometimes even simple English sentences can be ambiguous, especially if they involve "not". If you think one of these sentences is ambiguous, state in your own words the different things you think it can mean and translate each meaning into symbols.)

   a) John is not here.

   b) John isn't here.

   c) John is here and Mary is here.

   d) John and Mary are here.

   e) John and Mary are not here.

   f) John and Mary are not both here.

   g) Neither John nor Mary is here.

   h) John or Mary is here.

   i) John is here but Mary isn't.

   j) One of John or Mary isn't here.

   k) At least one of John or Mary isn't here.

8. Let R stand for "It is raining" and C stand for "It is cloudy". Write each of the following sentences as a wff. The instructions about ambiguity in [7] still apply.

   a) If it's raining, then it's cloudy.

   b) If it is raining, there are clouds.

   c) Only if it's raining are there clouds.

   d) For rain it is necessary that there be clouds.

   e) For rain it is sufficient that there are clouds.

   f) It isn't raining unless it is cloudy.

   g) It rains just in case it's cloudy.

9. Write each of the following sentences using logic and math notation. For instance, "if x is 2, then x squared is 4" becomes $(x = 2) \Longrightarrow (x^2 = 4)$. Also, indicate whether each assertion is True or False.

   a) Whenever x is 2, then x squared is 4.

   b) Given that x squared is 4, it follows that x is 2.

   c) Assuming that x is 2, we have x cubed is 8.

   d) If x squared is 4 and x is positive, then we can conclude that x is 2.

   e) If x is positive, then x squared being 4 implies that x is 2.

   f) From the assumption that x cubed is 8, it is necessary that x be 2.

   g) It is not true that x squared being 4 forces x to be 2.

   h) It is not true that if x squared is 4, then x is 2.

   i) If x squared is not 4, then x is 2.

   j) If x squared is 4, then x is 2 or x is -2.

   k) Either x squared is 4 implies that x is 2, or x squared is 4 implies that x is -2.

   l) 4 is greater than 6 just in case 0 is greater than 2.

10. Write each of the following sentences using math and logic notation. Use x and y for arbitrary numbers. For instance, the statement "the sum of two numbers is 6" should be translated as "$x+y=6$".

    a) If the sum of two numbers is negative, then at least one of the numbers is negative.

    b) The product of two numbers is 0 if and only if at least one of them is 0.

    c) The sum of two positive numbers is positive.

    d) A number isn't 3 unless it's positive.

    e) When the absolute values of two numbers are the same, the numbers are themselves the same or one is the negative of the other.

11. Find all wffs without parentheses, using just the atom P and the operands $\neg$ and $\wedge$, that can be obtained by at most four applications of Rules 1–3 defining wffs. For instance, $P \wedge \neg P$ is one such wff. The first rule (atoms are wffs) was used to get P.

Then one application of the rule about "nots" gives us $\neg P$. Then an application of the rule about $A \circ B$ gives us $P \wedge \neg P$. Thus we used three applications all told.

12. Use Rules 1–3 to prove that no wff has two consecutive binary connectives.

13. Use Rules 1–4 to determine whether each of the following is a wff. That is, it is not enough to state that a string is or is not a wff. You must show how application of the rules leads to your conclusion.

a) $P \vee Q \wedge \neg R \neg\neg P$.

b) $(\neg\neg P) \wedge Q \vee \neg\neg R$.

c) $((P \vee Q)) \wedge R \wedge \neg(Q \Longrightarrow P)$.

d) $(P \Longleftrightarrow Q) \neg R \vee S \wedge (\neg T)$.

e) $(Q \wedge R) \vee \neg Q \wedge \neg\neg Q$.

f) $\neg P(\vee(\neg Q)) \wedge R$.

g) $((P \vee Q) \wedge (Q \wedge \neg R) \Longrightarrow Q) \wedge S$.

14. Evaluate the following expressions when all the atoms are True. Use the precedence rules.

a) $P \wedge \neg Q \Longrightarrow R$.

b) $P \Longrightarrow \neg P \vee Q \wedge \neg R$.

c) $P \Longrightarrow Q \Longrightarrow \neg R \Longrightarrow P$.

15. a) For each of the connectives $\wedge$, $\vee$, $\Longrightarrow$, and $\Longleftrightarrow$, determine whether order matters. For instance, for $\Longrightarrow$, you determine whether

$$(P \Longrightarrow Q) \Longrightarrow R \quad \text{and} \quad P \Longrightarrow (Q \Longrightarrow R)$$

are equivalent. Another way is to answer the question: Which of these four connectives satisfy an associative law?

b) For each case in a) where the connective does not obey the associative law, state in words the two possible interpretations of $P \circ Q \circ R$ where $\circ$ represents the connective.

16. Which of the following would be ambiguous without Rules I and II?

a) $P \vee Q \Longrightarrow R$. b) $P \Longrightarrow Q \Longleftrightarrow R$.
c) $P \Longleftrightarrow Q \Longrightarrow Q$. d) $P \Longleftrightarrow Q \vee R$.
e) $P \Longrightarrow Q \wedge R$. f) $P \Longrightarrow Q \wedge R \Longrightarrow S$.

17. Construct the following truth tables.

a) For those strings of [13] that are wffs.

b) For the wffs of [14].

c) For the wffs of [16] (using Rules I and II).

d) For the wffs of Example 7.

18. Display the first n columns of the result of Algorithm TRUTHTABLE when

a) $n = 4$. b) $n = 5$.

19. Give a proof that TRUTHTABLE generates a truth table for any n.

20. Compare TRUTHTABLE and ALLPERM, Section 4.9. What characteristics of generating truth tables makes this an easier problem than generating all the permutations of $1, 2, \ldots, n$.

21. Suppose that you need to find the truth table for a logical expression that involves just three atoms, P, Q, and R. Write an algorithm using nested loops to list all triples of truth values for these three atoms. They should be listed in the same order as in the text. (This algorithm requires less ingenuity than TRUTHTABLE, but it is less flexible.)

22. Suppose that a wff is a tautology. Explain in words why it is still a tautology if one or more of its atoms are replaced by other wffs. If more than one instance of a given atom is replaced, it must be replaced by the *same* wff each time. Why?

23. Let A be any set of premises and let B be a desired conclusion. We want to determine whether it is a theorem that A implies B. Using [22] and a tautology, explain why it is just as good to prove that $\neg B$ implies $\neg A$. What did we call the implication $\neg B \Longrightarrow \neg A$ in Section 0.7?

24. Show by a truth table that, for any wffs A and B,

$$[A \Longleftrightarrow B] \Longleftrightarrow [(A \Longrightarrow B) \wedge (B \Longrightarrow A)].$$

25. For each of the following arguments in words, first explain informally in words why you think it is valid or invalid. Then translate it into symbols and produce a truth table to determine its validity.

a) Either a Republican is President or a Democrat is President.
If a Democrat is President, then the Democrats control the Senate.
The Democrats control the Senate.
\-\-\-\-\-\-\-\-\-\-\-\-
Therefore a Democrat is President.

b) If Bob lives in Chicago, then he lives in Illinois.
If Bob lives in Chicago, then he lives in the USA.
Bob lives in Illinois.
\-\-\-\-\-\-\-\-\-\-\-\-

Therefore Bob lives in the USA.

c) If you have a headache, then you feel nervous. If you have a headache, then you need aspirin. You feel nervous.

Therefore you need aspirin.

26. Determine whether each of the following symbolic arguments is valid.

a) $A \Longrightarrow B$
$\neg A \Longrightarrow C$
$\neg B$

C.

b) $(A \vee C) \Longrightarrow (B \vee C)$

$A \Longrightarrow B$.

c) $(A \Longrightarrow B) \vee C$
$\neg B$

$A \Longrightarrow C$.

27. Complete the proof of Example 9 by:

a) Induction for both parts of Example 9.

b) Using the result of [22] for the second part of Example 9.

28. Which of the following wffs are tautologies? (Use Fig. 7.3 for the definitions of the connectives.)

a) $(P \mid Q) \vee (P \Longleftrightarrow Q)$.

b) $(P \mid Q) \vee (P \downarrow Q)$.

c) $(P \not\Longleftrightarrow Q) \vee (P \vee Q)$.

d) $(P \Longrightarrow Q) \vee (Q \Longrightarrow P)$.

e) $((P \wedge Q) \vee R) \vee ((P \vee Q) \wedge R)$.

f) $((P \vee Q) \Longrightarrow R) \vee (\neg(P \vee Q) \Longrightarrow R)$.

g) $(P \Longrightarrow Q) \Longleftrightarrow (\neg Q \Longrightarrow P)$.

29. Which of the following pairs of wffs are logically equivalent?

a) $(P \wedge Q) \Longrightarrow R$ and $(P \Longleftrightarrow R) \vee (Q \Longleftrightarrow R)$.

b) $(P \Longleftrightarrow Q) \wedge (Q \Longleftrightarrow R)$ and $P \Longleftrightarrow R$.

c) $(P \wedge (Q \Longrightarrow R)) \vee (\neg P \wedge Q)$ and $(\neg P \vee \neg Q \vee R) \wedge (P \vee Q)$.

d) $(\neg P \wedge (R \Longrightarrow Q)) \vee (P \wedge \neg(Q \Longrightarrow R))$ and $P \wedge Q \wedge \neg R \vee \neg P \wedge Q \vee \neg P \wedge \neg R$.

30. For which of the following pairs of wffs does the first logically imply the second? (Use Fig. 7.3 for the definitions of the connectives.)

a) $P \mid Q$ and $P \downarrow Q$.

b) $P \downarrow Q$ and $P \mid Q$.

c) $P \wedge (P \wedge \neg(Q \wedge \neg R))$ and R.

d) $P \wedge (P \wedge \neg(Q \wedge \neg Q))$ and Q.

e) $P \vee Q \vee R$ and $(Q \vee (P \vee R)) \vee P$.

f) $(((P \Longrightarrow Q) \Longrightarrow (\neg R \Longrightarrow \neg S)) \Longrightarrow R) \Longrightarrow T$ and $(T \Longrightarrow P) \Longrightarrow (S \Longrightarrow P)$.

g) $P \Longrightarrow Q \Longrightarrow R$ and $P \Longrightarrow R$.

h) $P \Longleftrightarrow Q \Longleftrightarrow R$ and $P \Longleftrightarrow R$.

i) $(P \Longrightarrow Q) \Longleftrightarrow R$ and $(P \Longleftrightarrow Q) \Longrightarrow (P \Longleftrightarrow R)$.

31. Which of the following sentences are tautologies?

a) Without food, people die.

b) If John said that he would do it and he did it, then he did it.

c) If 2 is not equal to 2, then the moon is made of blue cheese.

d) Canadians are North Americans.

32. a) Can a theorem without premises always be reformulated as a theorem with premises? Explain.

b) Can a theorem with premises always be reformulated as a theorem without premises? Explain.

7.3 Natural Deduction

This section is an introduction to proving theorems in the propositional calculus by deductive reasoning rather than, as in Section 7.2, by using truth tables. Be forewarned: This is a very large topic, whose surface we'll only be able to scratch. Nevertheless, we hope to give you a feel for what it means to prove a theorem formally. We'll return to the ideas of this section at various places later in this chapter.

We begin with three examples which illustrate how, instead of using truth tables, we might reason about theorems in the propositional calculus.

EXAMPLE 1 In Section 7.2 (see Table 7.6(a)) we used a truth table to prove the theorem

$$((P \Longrightarrow Q) \wedge (Q \Longrightarrow R)) \implies (P \Longrightarrow R).$$

Perhaps it occurred to you then that, instead of using a truth table to prove this theorem, we might have reasoned as follows:

> If the truth of one proposition implies the truth of a second and if the truth of that second implies the truth of a third, then by an application of the notion of transitivity (see Section 0.2), the truth of the first implies the truth of the third. That is, if $P \Longrightarrow Q$ and $Q \Longrightarrow R$ are true, then $P \Longrightarrow R$ is also true. ▮

EXAMPLE 2 Consider the theorem

$$\neg(P \wedge Q) \iff (\neg P \vee \neg Q), \tag{1}$$

that is, the claim that expression (1) is a tautology. You could easily prove this theorem using a truth table but, instead, you might reason as follows:

> If $\neg(P \wedge Q)$ is true, then P or Q or both must be false, in which case $\neg P \vee \neg Q$ is also true. If $\neg(P \wedge Q)$ is false, then both P and Q must be true, in which case $\neg P \vee \neg Q$ is also false. ▮

EXAMPLE 3 Consider the theorem

$$\neg P \vee Q \vee (P \wedge \neg Q),$$

where, as in Example 2, this theorem claims that the expression is a tautology. Note that this is our first example of a theorem without premises which is not expressed in the form

$$\text{wff1} \iff \text{wff2}.$$

To prove this theorem we could again use a truth table. But also, we could reason as follows:

$\neg(P \wedge \neg Q) \iff (\neg P \vee Q)$.	[Using expression (1) with Q replaced by $\neg Q$ and deleting the double negation; see Example 9, Section 7.2]
$\neg(P \wedge \neg Q) \vee (P \wedge \neg Q)$.	[This is a tautology because a wff or its negation must be true (why?)]
$\neg P \vee Q \vee (P \wedge \neg Q)$.	[This, too, is a tautology, using the equivalence in the first line to substitute $\neg P \vee Q$ for $\neg(P \wedge \neg Q)$ in the second line]

The format of this proof should remind you of two-column proofs in geometry. ▮

The proofs in Examples 1 and 2 are informal with the arguments given in discursive English. The proof in Example 3 is more formal; each step is justified by appeals to intuitive **rules of inference**, such as:

- For any wff you can substitute a logically equivalent wff (why does this make intuitive sense?)—see [22], Section 7.2.
- A wff or its negation must be true (why?).

In the remainder of this section we will develop further the notion of a formal deductive proof, particularly the idea of rules of inference. In so doing we shall develop a **formal deduction system** which consists of:

- A set of symbols together with rules specifying how legal strings of these symbols may be formed. (In the propositional calculus the symbols are the atomic propositions, the connectives, parentheses, T, and F, and the rules are those we gave in Section 7.2 for constructing wffs.)
- An initial set of legal strings which are given to you. These are called **axioms**. (More below on axioms in the propositional calculus.)
- A set of ways of constructing new legal strings from others. These are what we called rules of inference above, but now they will be more formal. In the propositional calculus many of the rules of inference have the form

$$\begin{array}{l} \text{wff1} \\ \text{wff2} \\ \quad\vdots \\ \text{wff}(n-1) \\ \text{-------} \\ \text{wff}n. \end{array}$$

That is, such a rule of inference is a statement of the argument that wff1, wff2, ..., wff($n-1$) imply that wff$n$ is valid or, equivalently, that

$$(\text{wff1} \wedge \text{wff2} \wedge \cdots \wedge \text{wff}(n-1)) \implies \text{wff}n$$

is a tautology. Two of the rules of inference that we'll present shortly have the preceding form.

One way of looking at our definition of a formal deduction system is that it defines a game. You start with symbols and rules for their formation into strings. At the beginning, you are given certain "winning" strings, the axioms. The object of the game is to construct *new* "winning" strings using the rules of inference.

In plane geometry, the axioms—for example, that through every pair of points, one and only one line can be drawn—are supposed to be self-evident, nonprovable truths which must be used as a starting point from which to derive the theorems of plane geometry. Do we need such a set of axioms as part of a formal deduction system for the propositional calculus? Although various versions of the propositional calculus are defined with sets of axioms, axioms are not, in fact, necessary. Moreover, in order to model patterns of reasoning like those used in human reasoning, we prefer not to be encumbered with axioms which are seldom an explicit part of human reasoning. A system with rules of inference but no axioms, such as we will discuss here, is called a **natural deduction system.**

But back to our game as it applies in the propositional calculus. Using our language of propositions and connectives, we want to use rules of inference—we'll give some examples just below—to derive (or deduce) conclusions from premises. A **derivation** (or deduction) of a conclusion B from a set of premises $A_1, A_2, \ldots, A_n$ uses the rules of inference to derive a sequence of conclusions $B_1, B_2, \ldots, B_n = B$ ending with the desired conclusion B. Thus the notion of a derivation is essentially the same as the idea of proving a theorem. That is, the steps in the proof of the theorem [see expression (10), Section 7.2]

$$(A_1 \wedge A_2 \wedge \cdots \wedge A_n) \implies B$$

might consist of the proof that each B_i is a valid conclusion from the premises A_i, $i = 1, \ldots, n$ and already verified conclusions B_j, $j = 1, \ldots, i-1$.

Now what about rules of inference? In fact, many such rules can be stated. But as our only purpose here is to introduce you to the notion of a formal proof and give some examples, we'll content ourselves with stating only a few rules of inference and then show how they can be used to prove some simple theorems in the propositional calculus. We'll begin with just two rules of inference.

Rule 1: *Modus Ponens*†

If A and B are two wffs such that $A \Longrightarrow B$ is true and A is true, then B is true. More formally,

$$\begin{array}{l} A \implies B \\ A \\ \text{-------} \\ B, \end{array}$$

which we can express as the tautology

$$((A \Longrightarrow B) \wedge A) \implies B. \qquad (2)$$

Thus this rule of inference says that any time you can assert the truth of the two premises $(A \Longrightarrow B)$ and A, you may also assert the truth of the conclusion B. (Does this make intuitive sense to you?) An argument in English which takes the form of modus ponens is

If you are reading this book, you must be crazy.	$A \Longrightarrow B$
You are reading this book. (Aren't you?)	A
Therefore you must be crazy.	B

† Latin for, roughly, a method of placing.

Here "you are reading this book" plays the role of A and "you must be crazy" plays the role of B. Please remember that this argument has the form: *If* the implication is true and A is true, *then* B is true. We don't, of course, believe that $A \Longrightarrow B$ is true!

Rule 2: *Conditional Proof*

If, by temporarily ***assuming*** that A is true, it is possible to derive that B is true (in any number of steps using any inference rules of natural deduction), then we may derive $A \Longrightarrow B$.

Thus this rule says that, if there is a valid argument beginning with A and ending with the conclusion B,

$$\begin{array}{c} A \\ \vdots \\ \hline B, \end{array}$$

then we may assert that $A \Longrightarrow B$. Rule 2, therefore, states a condition for determining the validity of an argument but not its soundness (why?). When we use Rule 2 in a deductive proof, we indent the **conditional assumption** A and all the subsequent steps until the conclusion B. Then we write $A \Longrightarrow B$ unindented because Rule 2 says that $A \Longrightarrow B$ is true independently of A. (Do you see why, after assuming A and deriving B from it, it is then sensible to conclude $A \Longrightarrow B$ whether A is true or not?)

EXAMPLE 4 Prove the theorem

$$((P \Longrightarrow Q) \wedge (Q \Longrightarrow R)) \implies (P \Longrightarrow R).$$

(See Table 7.6(a), Section 7.2, for a truth table proof of this theorem.)

Solution In two-column format, we proceed as follows:

Derivation	Justification
1. $P \Longrightarrow Q$.	Premise
2. $Q \Longrightarrow R$.	Premise
3. P.	Assumption
4. Q.	Rule 1 (modus ponens)—lines 1 and 3
5. R.	Modus ponens, 2 and 4
6. $P \Longrightarrow R$.	Rule 2, 3 and 5

The numbers in each justification show the line numbers of the wffs used in the inference rule. The little "subroutine" 3–4–5 enables us to show that there is an argument beginning with the assumption P and concluding with the assertion of R. This argument makes use of lines 1 and 2 which, in this sense, act like "global" data to the subroutine. ∎

It's important that you understand why the proof in Example 1 is a *formal* proof. Each step is

- a restatement of a premise,
- a conditional assumption, or
- a new wff which is the result of a purely formal manipulation of the symbols involved using one of the rules of inference; this wff will be a valid conclusion of an argument beginning with the premises of the theorem.

We used the term "subroutine" above because, in fact, the notion of assuming that something A is true and then using it to conclude that B is true corresponds very closely to the idea of a subroutine in a computer program in which some input parameters (the assumption A) are provided and some output (the conclusion B) is computed.

To conclude this section, we give one additional rule of inference and then present two examples which use this rule.

Rule 3: *Modus Tollens*†

If A and B are two wffs such that $A \Longrightarrow B$ and $\neg B$ are true, then we may derive the truth of $\neg A$. In our argument schema,

$$A \Longrightarrow B$$

$$\neg B$$

$$\text{-------}$$

$$\neg A.$$

As with the other rules, you should convince yourself that this one is reasonable. Thus if you know that the truth of A implies the truth of B and you also know that B is false, is it reasonable to conclude that A is false [1]?

EXAMPLE 5 Prove the theorem

$$((P \Longrightarrow Q) \wedge (Q \Longrightarrow R) \wedge \neg R) \Longrightarrow \neg P$$

---

† Latin for, roughly, a method of removing.

Solution

Derivation	Justification
1. $P \Longrightarrow Q$.	Premise
2. $Q \Longrightarrow R$.	Premise
3. $\neg R$.	Premise
4. $\neg Q$.	Modus tollens, 2 and 3
5. $\neg P$.	Modus tollens, 1 and 4 ∎

Our last example uses Rules 2 and 3.

EXAMPLE 6 Prove that

$$(P \Longrightarrow Q) \implies (\neg Q \Longrightarrow \neg P).$$

Solution Before reading the following proof, satisfy yourself by reasoning informally that this theorem is true.

Derivation	Justification
1. $P \Longrightarrow Q$.	Premise
2. $\neg Q$.	Assumption
3. $\neg P$.	Modus tollens, 1 and 2
4. $\neg Q \Longrightarrow \neg P$.	Rule 2, 2 and 3

Do you see that you could prove that $(\neg Q \Longrightarrow \neg P) \Longrightarrow (P \Longrightarrow Q)$ in just the same way? Putting these two results together is a formal justification that proving the contrapositive of a theorem is equivalent to proving the theorem itself. ∎

There are a number of other possible rules of inference in addition to the three we have introduced. Some are given in [4]. Using them, we could prove more extensive theorems (most of which, by the way, would make very tedious reading). But since our aim has been only to introduce you to the ideas of formal deduction, we'll stop here. We emphasize again that this section has been only a very brief introduction to a large subject. However, the terminology and methodology introduced should be useful if you ever encounter natural deduction again. And the formality of this proof technique should help in Section 7.4 on the verification of algorithms.

Problems: Section 7.3

1. a) Express rules of inference 1 and 3 as theorems.
   b) Prove each of these theorems using truth tables.

2. Show that Rule 2, the rule of conditional proof, is true. *Hint*: The first step is to state an equivalent wff.

3. The Game of Stars and Slashes (Schagrin, Rapaport, and Dipert) is played with four symbols, A, B, $*$, and $/$. Strings are formed using these four symbols, where each may be repeated as often as desired. The rules for forming "winning" strings are:

   - A or B by itself is a winning string.
   - If S_1 and S_2 are winning strings, then $/S_1 * S_2/$ is a winning string.

   a) Construct all winning strings with nine or fewer symbols.
   b) From the result of a) can you infer a formula for the length of possible winning strings? Can you prove that your formula is correct?
   c) Using the rules for winning strings, determine whether each of the following is a winning string.

   i) $//A * A/ * //B * A/ * A//$
   ii) $/A * /A * B/$
   iii) $/////B * B/ * A/ * B/ * B * B//$

4. Consider the following additional rules of inference, each stated in the form of an argument

$$\begin{array}{ccc} A \wedge B & A \wedge B & A \\ ---- & ---- & ---- \\ B \wedge A. & A. & A \vee B. \end{array}$$

   a) Explain each of these rules in words.
   b) Prove that each is true using a truth table.

5. Use the rules of inference in the text and those of [4] to come up with natural deductions for the following arguments, where A and B are arbitrary wffs. That is, provide the necessary intermediate steps and the justification for each step.

   a) $A \wedge B$
   $----$
   $B.$

   (*Note*: This is *not* a case of the 2nd rule of [4]. You must take two steps and combine two rules to get it. In a formal system, you may not make without comment the obvious jumps which are standard in informal proofs.)

   b) A
   $A \Longrightarrow B \wedge C$
   $C \Longrightarrow D$
   $--------$
   $D.$

   c) $A \Longrightarrow B \wedge C$
   $B \Longrightarrow D$
   $--------$
   $A \Longrightarrow D.$ (*Hint*: Use conditional proof.)

   d) $A \Longrightarrow B$
   $B \vee C \Longrightarrow D$
   $--------$
   $A \Longrightarrow D.$

   e) $\neg\neg\neg P \Longrightarrow P$
   $--------$
   $\neg\neg P.$ (Use method of [6].)

6. Suppose that the assumption of P allows you to prove both Q and $\neg Q$ where Q is any wff. State in words why you can then assert $\neg P$.

7. Suppose you are in a subproof of a natural deduction as in lines 3–5 of Example 4. Explain why (at least) the line before the subproof (e.g., line 2 in Example 4) could be, if desired, inserted as a line of the subproof.

8. Use the ideas in [6–7] to prove:
   a) $\neg\neg P \Longrightarrow P.$ b) $P \Longrightarrow \neg\neg P.$
   (*Hint*: Q in [6] can itself be P.)

7.4 Algorithm Verification

We touched on proving algorithms correct at various times in Chapters 2–6. The proofs of algorithm correctness given in those chapters were all informal in that they consisted of arguments in discursive English. In this section we will discuss a much more formal approach to proofs of algorithm correctness. We believe that our previous proofs of algorithm correctness were convincing, so why is a more formal approach useful? Here are three answers:

1. Algorithms themselves are rather formal structures which are closely related to those still more formal structures, computer programs. Not only would it be nice if we could prove the correctness of programs but, if we had a formal technique for doing so, then we *might* be able to mechanize (i.e., have a computer do) these proofs. Thus being able to prove algorithms correct formally is closely related to the goal of being able to prove programs correct formally. We must admit, however, that research in this direction has proceeded quite slowly because achieving such mechanical proofs is inherently very difficult.
2. Nevertheless, the techniques developed to *try* to prove programs correct have proved very useful in the writing of algorithms and programs. One aspect of this approach, mentioned several times already in this book, is the importance of thinking about loop invariants when writing algorithms and proving them correct. Loop invariants are an important part of the technique for proving algorithms correct which we will discuss in this section.
3. Finally, the formal discipline to be introduced in this section breaks the proof of an algorithm up into a number of subproofs. This approach is useful in thinking about proofs of long, complicated algorithms even when each subproof will be done informally.

We'll describe the ideas involved in proving algorithms correct by carrying through an example. The example we will use is Example 3, Section 7.1, where we posed the question of how to prove that the multiplication algorithm reproduced in Fig. 7.4 is correct.

The crucial step in proving an algorithm correct is to add to the algorithm **assertions** about the variables in the algorithm. These assertions take the form of one or more Boolean expressions as in expression (1), Section 7.2, joined by the propositional connectives $\wedge$, $\vee$, and $\neg$. Two of the assertions in Fig. 7.4—at the beginning and end of the algorithm—are called the **input specification (IS)** and **output specification (OS)**, respectively. What we really want to prove is that, if the input specification is true, the execution of the algorithm allows us to assert the truth of the output specification. That is, we want to prove the theorem

$$\text{IS} \Longrightarrow \text{OS}, \tag{1}$$

which, because of the temporal aspect of executing an algorithm, should be read:

> **if** IS is true at the beginning of execution of the algorithm,
>
> **then** OS is true when the algorithm terminates.

Input x [Integer ≥ 0]
y [Integer]

Algorithm MULT
$\{x \geq 0\}$ [Input specification]
$prod \leftarrow 0;\ u \leftarrow 0$
$\{prod = uy\}$ [Loop invariant]
repeat until $u = x$
$prod \leftarrow prod + y$
$u \leftarrow u + 1$
endrepeat
$\{prod = uy \wedge u = x\}$ [Loop termination condition]
$\{prod = xy\}$ [Output specification]

Output $prod$ [$= xy$]

FIGURE 7.4

Algorithm MULT with Assertions

The use of "assertion" is standard in the language of algorithm verification. Except for the output specification, however, assertions are just hypotheses or premises, as we have used these terms earlier in this chapter.

Now let's use Fig. 7.4 to explain what we mean by the terms in the preceding paragraph. We distinguish assertions *about* an algorithm from the statements of the algorithm itself by placing the assertions in braces ({ and }). The input specification

$$\{x \geq 0\}$$

provides a constraint on the input which must be observed if the algorithm is to produce the desired output. This assertion is a predicate, since x in this example can be any nonnegative integer. The truth values of this assertion and the ones that follow will *depend on the current values of the variables in the algorithm.* Thus we view the input specification as asserting that *the particular value of* x input to the program will be nonnegative. (You should be able to see why this algorithm won't work if x is negative.)

To be completely precise and comprehensive, we should have stated the input specification as

$$\{x, y \text{ integers}, x \geq 0, -\infty < y < \infty\}. \tag{2}$$

But we want to avoid being quite so pedantic because

- we'll take the *type* of the variables (i.e., integer) to be understood; whenever this is not obvious, the type should be given as in expression (2);
- when no constraint on a variable, such as y in Fig. 7.4 is asserted, it is implicit that any value of the type is allowable; thus, the $-\infty<y<\infty$ in expression

(2) is really redundant because, once y is specified to be an integer, these limits are implicit.

Note, by the way, that the use of the word "specification" is natural because the IS *specifies* the character of the input.

The assertions within the algorithm itself are intended to express truths about the values of the variables contained in them whenever control reaches that particular place in the algorithm. But how do we know where to put assertions in the algorithm (other than the input and output specifications, which must always be at the beginning and end, respectively)? A general rule is that the assertions should be inserted so as to break up the proof of the correctness of the entire algorithm into manageable (i.e., relatively simple) subproofs. A specific rule which follows from this is

> Each loop should be preceded and followed by assertions. The former is the loop invariant; the latter, the **loop termination condition**, combines the loop invariant with the Boolean expression that is tested each time through the loop.

This rule is reasonable because, since loops are crucial portions of all significant algorithms, we should want to prove that each loop does what it is supposed to do. In the example in this section—and in the example that we will present in Section 7.7—all assertions except for the input and output specifications will satisfy this rule. However, in longer algorithms, or in those which contain complicated **if** ... **endif** structures, it is often useful to add further assertions which break up the proof of the correctness of the algorithm into more manageable subproofs.

Now let's get to the assertions in Algorithm MULT. The first of these

$$\{prod = uy\}$$

was our first illustration of a loop invariant way back in Example 1, Section 1.2. It asserts that each time the **repeat** ... **endrepeat** loop is entered, the current value of *prod* must equal uy, using the current values of these variables. How do you know that this assertion is, in fact, intended to say something about the variables each time the loop is entered, rather than just the first time? Actually, this is just a convention of the notation shown in Fig. 7.4, which we have emphasized by drawing the arrow from **endrepeat** to the assertion, not to **repeat**.

The loop termination condition

$$\{prod = uy \wedge u = x\} \tag{3}$$

is an assertion about the variables which should be true when the loop is exited. We could have written this assertion as

$$\{prod = xy\} \tag{4}$$

because the truth of proposition (3) implies the truth of proposition (4) (why?). We prefer (3) because the proof that the algorithm is correct is more straightforward if the loop termination condition consists of the conjunction of the loop invariant and the Boolean expression, which is tested each time through the loop. We do, however, use proposition (4) as the output specification.

The output specification (4) is, of course, just the assertion of the desired output of the algorithm. Getting the output specification right is relatively easy; you usually know pretty well what you want your algorithm to do. On the other hand, getting the input specification right is often much harder. Specifying the conditions on the input variables that must be satisfied in order for the output to be correct may be very difficult. (Think, for example, of an algorithm to compute a payroll whose input must include all the constraints imposed by the tax laws.)

To prove the correctness of MULT, let's first give the assertions the names of propositional variables:

$$\begin{aligned} &P_1 : x \geq 0. && \text{[Input specification]} \\ &P_2 : \mathit{prod} = uy \\ &P_3 : \mathit{prod} = uy \wedge u = x \\ &P_4 : \mathit{prod} = xy && \text{[Output specification]} \end{aligned}$$

What we would like to prove is that

$$P_1 \implies P_4.$$

To prove this it would be sufficient (why?) to show that

$$\begin{aligned} &P_1 \implies P_2, \\ &P_2(\text{before}) \implies P_2(\text{after}), \\ &P_2 \implies P_3, \\ &P_3 \implies P_4, \end{aligned}$$

where each of these is a theorem in the propositional calculus. The second theorem is necessary because we must prove that, if the loop invariant is true at any entry into the loop, then it remains true when we enter the loop the next time.

These theorems have quite a different flavor from those considered in the propositional calculus because the propositions (e.g., $\mathit{prod} = uy$) contain numerical, not propositional variables as atoms. These theorems cannot therefore be proved using truth tables or by the natural deduction method of Section 7.3. Our approach to each proof will be to give an (informal) argument which attempts to show that each assertion in the algorithm follows from its predecessor:

1. The first assertion after the input specification is the loop invariant assertion. When we first come to this assertion ($\mathit{prod} = uy$), it is correct because both *prod* and u have just been set to 0. Thus $P_1 \Longrightarrow P_2$.
2. Is P_2 still true when we return to it from the end of the loop? That is, assuming that the loop invariant is true when we enter the loop, is it still true when we reenter it the next time? Each "next time" involves a value of u that is 1 greater than the previous one. So, if we define $P_2(k)$ to be the proposition

   $$P_2 \text{ holds for } u = k,$$

   the theorem $P_2(\text{before}) \Longrightarrow P_2(\text{after})$ becomes the theorem

   $$P_2(k) \implies P_2(k+1) \tag{5}$$

for values of k from 0 to $x-1$. Does expression (5) look familiar? What we have done is recast this theorem in terms of a *finite induction* with basis case $k=0$. (The induction is finite because $P_2(k)$ need only be true for $k=0, 1, \ldots, x$; finite inductions are usually what are needed to prove algorithms and programs correct.) If the IS is true, the basis case is true because $P_1 \Longrightarrow P_2$ is really $P_1 \Longrightarrow P_2(0)$. So all we need to do to prove the theorem (5) is the inductive step. Let $prod_k$ denote the value of *prod* before execution of the loop for a particular value $u=k$. Then the induction hypothesis is

$$prod_k = ky.$$

Because of the statements in the loop itself, the value of *prod* after this execution (and before the next execution) is

$$\begin{aligned} prod_{k+1} &= prod_k + y \\ &= ky + y \quad \text{[By the induction hypothesis]} \\ &= (k+1)y, \end{aligned} \tag{6}$$

which is just $P_2(k+1)$. This completes the proof of the theorem (5).

3. The loop termination condition must be true when we reach it because we have just shown that $prod = uy$ each time we reach the end of the loop and $u=x$ is just the condition which causes us to exit from the loop. Therefore $P_2 \Longrightarrow P_3$.
4. Finally, $P_3 \Longrightarrow P_4$ is true because, as we argued above, the truth of the loop termination condition implies the truth of the output specification.

Does this complete our proof? Not quite. A proof of algorithm correctness requires not only that we show that expression (1) is a theorem, but also that the algorithm actually does terminate if the input specification is true. In particular, to prove the correctness of any loop, we must not only prove that the loop invariant is, indeed, invariant but we must also prove that the loop terminates. We prove that Algorithm MULT terminates as follows:

1. The only thing that could prevent its termination would be that the **repeat** ... **endrepeat** loop is executed endlessly.
2. Because u is initialized to 0 and x is nonnegative if the IS is true, $u \leq x$ when the loop is first entered.
3. Each time the loop is executed, u is increased by 1. Therefore, sooner or later, u will have the same value as x and the loop will be exited, which completes the proof.

Do you see why this proof is invalid if the IS is not true [4b]? Did you notice that this is just another induction proof [4c]?

Algorithm MULT is a very simple algorithm, so you shouldn't be misled into thinking that proofs of the correctness of algorithms are always as easy and straightforward as this one. They're not. But even when they are much more difficult, the discipline of trying to prove an algorithm correct by inserting assertions in it *as you develop it* can be a valuable aid in constructing correct algorithms. In Section 7.7 we'll give a somewhat more difficult example of proving an algorithm correct.

Problems: Section 7.4

1. In this section, "assertion" and "hypothesis" mean the same thing. As used in most writing, how do they differ?

2. Write an algorithm which inputs a real number x and a positive integer n and computes $x+n$ by repeated addition of 1. Insert appropriate assertions and use them to prove that the algorithm is correct and terminates.

3. Here is a trivial algorithm to compute the reciprocal of a nonzero number.

Algorithm RECIPROCAL

```
Input x                              [x ≠ 0]
Algorithm RECIPROCAL
  y ← 1/x
Output y                    [Reciprocal of x]
```

Obviously the correctness of this algorithm depends on the input meeting the input specifications. Use assertions to prove the algorithm correct. (The assertions and the proof are quite simple. The point of the exercise is to show how the input specifications are used in the verification process.)

4. **a)** Suppose that the input specification for MULT is changed to assert only that x is an integer. Where does the proof of correctness break down?
   b) Why is the proof of the termination of MULT invalid if the input specification is false?
   c) Restate the proof of the termination of MULT as an induction.

5. Write input specifications for the following algorithms.
   a) An algorithm to determine whether three positive integers satisfy the Pythagorean theorem.
   b) Algorithm BINSEARCH (Section 1.6).
   c) Algorithm PERMUTE-1 (Section 4.9).

6. Write output specifications for each of the algorithms in [5]. You'll have to use ellipsis for c). In Section 7.6 we'll develop an easier way to express this OS.

7. For each of the following write the algorithm requested, insert appropriate assertions in it, and then prove that your algorithm is correct.
   a) Determine whether a nonnegative integer n is even by successive subtraction of 2 from n.
   b) Sum n given numbers.
   c) Find the greatest power of 2 less than or equal to a given positive integer n.
   d) Given an array of numbers $a_1, a_2, \ldots, a_n$, find the subscript of the largest number in the array.

7.5 Boolean Algebra

It is remarkable that the subject of this chapter—originally developed a century and a half ago by George Boole to provide a formal way to describe and manipulate logical propositions in an algebraic framework†—has so many practical applications

† Actually the term "Boolean algebra" refers to a whole class of algebras, of which the one we discuss here is the simplest and most practically useful.

today. Probably the most important application is in the design of digital circuits. At the end of this section we'll solve the problem introduced in Example 2, Section 7.1, concerning the design of a complementer.

Algebra, as used in the title of this section, has a more general meaning than what you know as high school algebra. In mathematics, an algebra consists of

- a set S, and
- the definition of operators which map tuples (see Section 0.1)

$$(s_1, s_2, \ldots, s_p), \quad s_i \in S,$$

  into elements of S.

Thus, for example, if $p=2$, we would have a binary operator which maps (s_1, s_2) into s, which is an element of S. (Does this remind you of the propositional calculus?—see below) The index p may be different for different operators. An operator on a p-tuple is called a ***p*-ary operator**. As in high school algebra, we may then define variables whose values are in S and manipulate them using the operators.

Boolean algebra is very closely related to the propositional calculus, as we'll demonstrate below. Why then devote a section to Boolean algebra? One reason is that the notation and perspective on truth values and variables are different in Boolean algebra from those in the propositional calculus. Another reason is to discuss how this formulation of the propositional calculus can be conveniently used for circuit design applications.

Computers normally operate on binary digits (bits), so it is convenient in Boolean algebra to replace the truth values, T and F by 1 and 0, respectively. And instead of using five operators corresponding to the five logical connectives we previously emphasized, we will (at first anyway), restrict ourselves to three operators (a term we'll use in place of connective in this section) corresponding to $\wedge$, $\vee$, and $\neg$. But we'll represent these operators in this section by $\cdot$, $+$, and $^-$ (overbar), respectively, because this notation is traditional in Boolean algebra. Nonetheless, we will continue to refer to these three operators as the **and**, **or**, and **not** operators, respectively.

EXAMPLE 1 Show how Boolean algebra fits into the paradigm for an algebra, as we have just defined it.

Solution The set $S = \{0, 1\}$, there are two binary operators, $\cdot$ and $+$, and one unary operator $^-$. Thus, for example, $+$ maps the tuple $(0, 1)$ into 1, since $0+1=1$ in Boolean algebra (i.e., this corresponds to the truth value T for F $\vee$ T in the propositional calculus). This example illustrates an important distinction between the propositional calculus and Boolean algebra, namely, that the truth values of propositions in the former become the values of variables in the latter. ∎

Corresponding to the atoms which were the variables of the propositional calculus, we have Boolean variables, which are the atoms of Boolean algebra, and

TABLE 7.9
Truth Tables for $+$, $\cdot$, and $^{-}$

$p+q$

$p \backslash q$	0	1
0	0	1
1	1	1

$p \cdot q$

$p \backslash q$	0	1
0	0	0
1	0	1

p	$\bar{p}$
0	1
1	0

which we will denote by p, q, r, ... These variables can take on the values 0 or 1. Table 7.9 gives the truth tables for the three Boolean operators. This table is precisely analogous to Table 7.2, Section 7.2. Note, in particular, that $1+1$ in Boolean algebra is 1. It couldn't be 2 because there is no value 2 in Boolean algebra. It is defined to be 1 because $1+1=1$ in Boolean algebra corresponds to T as the truth value of $\mathrm{T} \vee \mathrm{T}$ in the propositional calculus.

We will denote expressions in Boolean algebra, which may be atomic or compound, by a, b, c, ... The rules for forming valid expressions in Boolean algebra are precisely the same as those for forming wffs in the propositional calculus:

- An atom is a Boolean expression.
- If a is a Boolean expression, then so is $\bar{a}$.
- If a and b are Boolean expressions, then so is $a \cdot b$ and $a+b$.
- If a is a Boolean expression, then so is (a).

The left-to-right and precedence rules for evaluating wffs in Section 7.2 also apply to Boolean expressions with $^{-}$ highest in precedence followed by $\cdot$ and $+$ [compare to (7), Section 7.2].

Boolean expressions may be regarded as **Boolean functions** which map k-tuples of 0's and 1's, where k is the number of variables in the expression, into a 1-tuple which is either 0 or 1. Thus the expression

$$\bar{p} \cdot q + p \cdot \bar{q} \cdot r + p \cdot \bar{r},$$

with $p=1$, $q=0$, and $r=1$, maps $\{1,0,1\}$ into $\{1\}$, since this is the value of the expression for these values of the variables (why?).

You may have noted that our definition of Boolean algebra is closely akin to the definition of set algebra in Section 0.1 (although we didn't call it an algebra there). In set algebra

- the objects, that is the elements of S, are just sets, such as sets of integers, and
- the operations of intersection, union and complement correspond to $\cdot$ (**and**),

+ (**or**), and $^-$ (**not**). Note, however, the unfortunate confusion of notation: For set intersection and union, we use a notation, $\cap$ and $\cup$, akin to that in the propositional calculus, whereas for set complement we use the same notation as in Boolean algebra.

The rules which govern the manipulation of expressions in Boolean algebra are the same as those governing the manipulation of wffs, so we don't need to repeat them. For example, corresponding to the wff

$$(P \wedge Q) \vee (Q \wedge R)$$

there is the Boolean expression

$$(p \cdot q) + (q \cdot r),$$

whose values can be tabulated in a truth table, as in Table 7.10.

TABLE 7.10
Truth Table for a Boolean Expression

p	q	r	$p \cdot q$	$q \cdot r$	$(p \cdot q) + (q \cdot r)$
0	0	0	0	0	0
0	0	1	0	0	0
0	1	0	0	0	0
0	1	1	0	1	1
1	0	0	0	0	0
1	0	1	0	0	0
1	1	0	1	0	1
1	1	1	1	1	1

Note that we list the values of the atoms in their natural binary order, whereas in the truth tables of Section 7.2 we always started with a row of T's (=1's). This is merely conventional and has no other significance.

A typical theorem in Boolean algebra is a statement that two Boolean expressions are equivalent using "equivalent" in the same sense as in the propositional calculus. However, instead of using the equivalence operator $\Longleftrightarrow$, in Boolean algebra we just use an equal sign. Thus

$$pq + pr + \bar{q}r \quad = \quad pq + \bar{q}r \tag{1}$$

is a theorem which states that the values of both sides of Eq. (1) are the same for all possible values of the variables. Thus in Boolean algebra two expressions are equal if, for all possible choices of values from $\{0,1\}$, the two expressions are both 0 or both 1. Thus it should not surprise you that expressions in Boolean algebra are equal if and only if the corresponding expressions in the propositional calculus are logically equivalent. In common with the language of ordinary algebra, we shall call entities like (1) *equations*.

Note that in Eq. (1) we dropped the $\cdot$ and simply used the convention that the $\cdot$ is implied when two Boolean variables are written next to each other. We'll almost always use this convention in this chapter. This convention is, of course, the same as that used with ordinary multiplication. Indeed, the $\cdot$ operator in Boolean algebra is sufficiently akin to multiplication in ordinary algebra and the $+$ is sufficiently akin to addition, that $\cdot$ is sometimes called the "logical multiplication operator" and $+$ the "logical addition operator". We will therefore often refer to $a+b$ as a sum and to ab as a product.

The following theorem states many of the important algebraic properties of expressions in Boolean algebra:

Theorem 1. If a, b, and c are Boolean expressions, the following equations are Boolean theorems:

(i) $a+0=a.$ [Additive identity]

(ii) $a\cdot 1=a.$ [Multiplicative identity]

(iii) $a+b=b+a.$
(iv) $ab=ba.$ [Commutative laws]

(v) $(a+b)+c=a+(b+c).$
(vi) $(ab)c=a(bc).$ [Associative laws]

(vii) $a(b+c)=ab+ac.$
(viii) $a+bc=(a+b)(a+c).$ [Distributive laws]

(ix) $a+\bar{a}=1.$
(x) $a\bar{a}=0.$

(xi) $\overline{a+b}=\bar{a}\bar{b}.$
(xii) $\overline{ab}=\bar{a}+\bar{b}.$ [De Morgan's laws]

In the remainder of this section we'll refer to these equations by using the lowercase Roman numerals.

We can easily prove each of these identities using truth tables [2]. Equations (ix) and (x) have no analogs in ordinary algebra but are true in Boolean algebra because one of a and $\bar{a}$ is 1 and the other is 0. The last two identities, De Morgan's laws, are given a place equal in prominence to the others because they are so often useful in Boolean algebra. You may recognize (xii) as analogous to the propositional calculus theorem (1), Section 7.3 [3].

Of the first eight identities in Theorem 1, the only one which should not have been familiar to you from high school algebra is the second distributive law (viii), which has no analog in ordinary algebra. Whereas the first distributive law (vii), like its counterpart in ordinary algebra, says that multiplication distributes over addition, (viii) says that addition distributes over multiplication. Does it disturb you that in Boolean algebra

$$a + bc = (a{+}b)(a{+}c),$$

whereas in ordinary algebra

$$a^2 + ab + ac + bc = (a{+}b)(a{+}c)? \tag{2}$$

It shouldn't because, in any mathematical system, of which ordinary algebra and Boolean algebra are two examples, we are free to define the rules of the game as we see fit. In Boolean algebra the definitions of the **and** and **or** operators determine that (viii) is a theorem, while in ordinary algebra the definitions of addition and multiplication determine that Eq. (2) is a theorem. Of course, if the "games" we play in mathematics are to be useful practically, then the rules we define must correspond to some sphere of application.

The distributive law (viii) is not the only "weird" fact in Boolean algebra. For instance,

$$a + a = a \qquad \text{and} \qquad a \cdot a = a. \tag{3}$$

You should be able to prove Eqs. (3) and explain their meaning [4].

The six pairs of identities in Theorem 1 exemplify another fact about Boolean algebra. In any Boolean theorem, if $+$ is replaced everywhere by $\cdot$ and $\cdot$ by $+$ and if 0 is replaced by 1 and 1 by 0, the result is still a theorem. We consider this **principle of duality** further in a problem [6].

We can prove theorems in Boolean algebra just as we did in the propositional calculus by using truth tables or by using natural deduction. We can also prove all Boolean theorems by using the identities of Theorem 1 as axioms. Here is an example of how Theorem 1 can be used to prove a Boolean theorem.

EXAMPLE 2 Prove that $a = \bar{\bar{a}}$.

Solution On the right in brackets we indicate which portions of Theorem 1 are used in each line.

$$\begin{aligned}
a &= a \cdot 1 && [\text{(ii)}]\\
&= a(\bar{\bar{a}}{+}\bar{a}) && [\text{(ix) with } a \text{ replaced by } \bar{a} \text{ and then (iii)}]\\
&= a\bar{\bar{a}} && [\text{(vii) and (x)}]\\
&= \bar{\bar{a}}a + 0 && [\text{(iv) and (i)}]\\
&= \bar{\bar{a}}a + \bar{\bar{a}}\bar{a} && [\text{(x), with } a \text{ replaced by } \bar{a}\text{, and (iv)}]\\
&= \bar{\bar{a}}(a{+}\bar{a}) && [\text{(vii)}]
\end{aligned}$$

$$= \bar{a} \cdot 1 \qquad [(ix)]$$
$$= \bar{a}. \ \blacksquare \qquad [(ii)]$$

The problems provide some practice with using truth tables to prove theorems, as well as with using the method of Example 2 [5, 7].

We noted in Section 7.2 that one purpose of deductions in the propositional calculus is to take a given wff and simplify it. Actually, such simplification is more common in Boolean algebra because the Boolean algebra formulation is normally used in designing electronic circuits. But what do we mean by *simplification*, a term we have thus far left to your intuition? For our purposes here we define the *complexity of an expression* as the number of Boolean operators in it. (Don't forget the implied "ands" in Boolean expressions.) Thus to simplify an expression is to reduce the number of operators in it.

For example, consider the Boolean theorem

$$\bar{p} + q + p\bar{q} = 1,$$

which states that the expression on the left is a tautology (i.e., it is equal to 1 (true) for all values of p and q). This is the theorem in Boolean algebra which corresponds to the theorem in the propositional calculus in Example 3, Section 7.3. To prove this Boolean theorem deductively as in Example 2, we would write:

Derivation	Justification
$\bar{p} + q + p\bar{q}$	
$= \overline{p\bar{q}} + p\bar{q}$	$\bar{p} + q = \overline{p\bar{q}}$ by (xii) with b replaced by $\bar{b}$ and using $\bar{\bar{b}} = b$ from Example 2
$= 1$	By (ix) with $p\bar{q}$ for a

This derivation takes an expression with five operators ($^-$, $+$, $+$, $\cdot$, and $^-$) and simplifies it to an expression with none. You may find this format, in which each line contains a transformation of the expression on the previous line, somewhat more congenial than the format used in Section 7.3.

Another method of simplifying Boolean expressions is through the use of **Karnaugh maps**. The idea of a Karnaugh map is connected to the idea of writing a Boolean expression in **disjunctive normal form (DNF)**. An expression in DNF has the form of a sum of products in which each product contains *once each* the variables in the entire expression, with each variable complemented or uncomplemented. For example, here is a three-variable expression in disjunctive normal form:

$$\bar{p}qr + \bar{p}q\bar{r} + p\bar{q}r.$$

Each term in a DNF expression corresponds to one row of a truth table for the entire expression and makes the value of the expression 1 for that row. For example, $\bar{p}q\bar{r}$ corresponds to the row in which $p = 0$, $q = 1$, and $r = 0$.

Note that when we apply (vii) of Theorem 1 to the preceding expression, it becomes

$$\bar{p}q(r+\bar{r}) + p\bar{q}r = \bar{p}q + p\bar{q}r \qquad [\text{Since } r+\bar{r}=1]$$

Thus the DNF version of a Boolean expression is usually not its simplest form but, as you will see, we can use it in an intermediate stage when simplifying Boolean expressions.

EXAMPLE 3 Convert the Boolean expression

$$pq + pr + \bar{q}r$$

to DNF.

Solution We will, in effect, present an algorithm for converting any Boolean expression to DNF although, for the sake of brevity, we won't state this algorithm in our usual notation.

1. In the first step we insert, for each variable missing from any term, the sum of that variable and its complement. For this example we get

   $$pq(r+\bar{r}) + p(q+\bar{q})r + (p+\bar{p})\bar{q}r.$$

   This expression must be equivalent to the original one because the sum of any variable and its complement is 1 and the product of 1 and any expression is the expression.
2. Then we apply the distributive law (vii) and delete any terms which appear more than once (because $p+p=p$). In our example we obtain

   $$\begin{aligned} &pqr + pq\bar{r} + pqr + p\bar{q}r + p\bar{q}r + \bar{p}\bar{q}r \\ = \; &pqr + pq\bar{r} + p\bar{q}r + \bar{p}\bar{q}r \end{aligned} \tag{4}$$

   The resulting expression is still equivalent to the original one and is now in DNF. ∎

An alternative algorithm would be to write the truth table for the expression and then use the rule stated preceding Example 3 to write the DNF [13, 14]. For expressions with relatively few terms but many variables, this approach would be quite tedious.

Our first step in "simplifying" the expression in Example 3 has certainly made things worse. But bear with us. The next step is to convert the expression in DNF to a Karnaugh map. One-, two-, three-, and four-variable Karnaugh map templates are shown in Fig. 7.5. The key to the construction of these maps is that each box should correspond to a different possible term in a DNF expression. If there are n variables in a Boolean expression, there are 2^n possible terms in the DNF (why?), which explains the number of boxes in each part of Fig. 7.5. Note that the labels of each box in each part of the figure are chosen so that their product is one of the possible terms in a DNF expression. For example, in the three-variable template, the upper right-hand box corresponds to the term $p\bar{q}\bar{r}$.

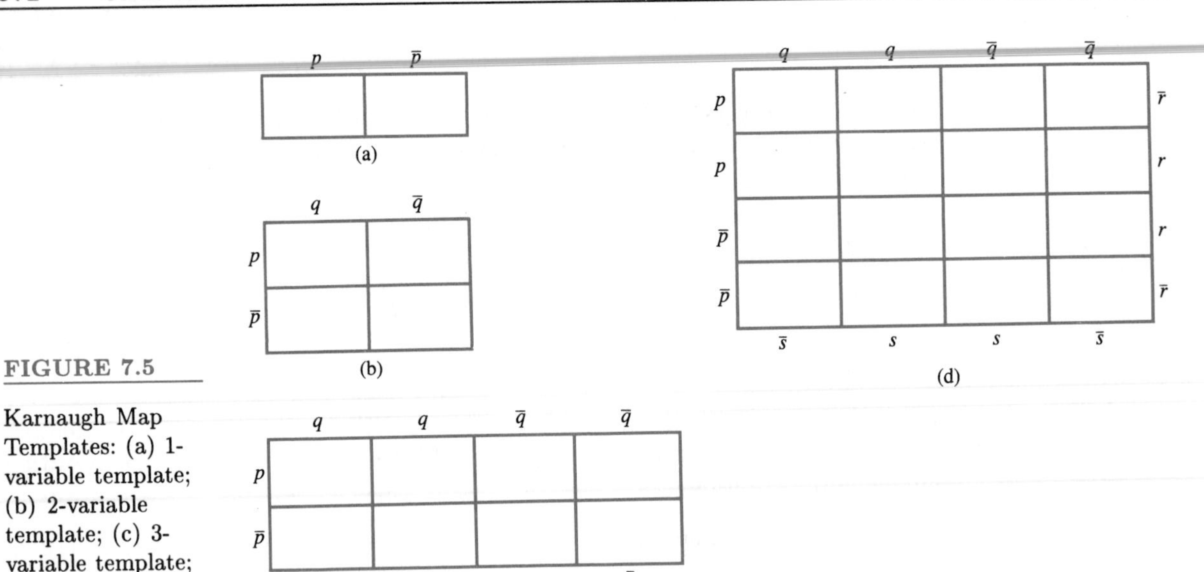

FIGURE 7.5

Karnaugh Map Templates: (a) 1-variable template; (b) 2-variable template; (c) 3-variable template; (d) 4-variable template

A Karnaugh map template is converted into a Karnaugh map for a particular Boolean expression by inserting 1's in each box corresponding to a term of the DNF for this expression. This step is shown in Fig. 7.6 for expression (4).

The pattern of *regions* in the Karnaugh map can be used to simplify Boolean expressions. For example, Fig. 7.6 shows two 2×1 regions outlined in color. The terms corresponding to the upper left-hand 2×1 region are $pq\bar{r}$ and pqr. As two of the three parts of each term are the same (pq) and the other appears uncomplemented in one term and complemented in the other (r and $\bar{r}$), the sum of these two terms is

$$pqr + pq\bar{r} \;=\; pq(r+\bar{r}) \;=\; pq. \tag{5}$$

Similarly for the 2×1 region at the right the sum of the two terms is

$$p\bar{q}r + \bar{p}\bar{q}r \;=\; (p+\bar{p})\bar{q}r \;=\; \bar{q}r. \tag{6}$$

When we put Eqs. (5) and (6) together, the four-term DNF expression in Eq. (4) simplifies to:

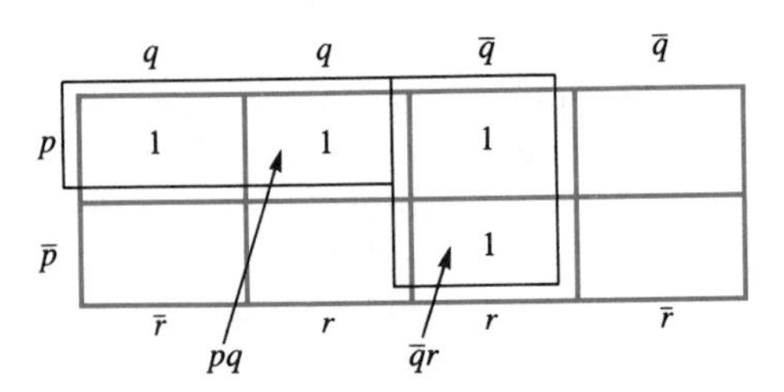

FIGURE 7.6

Karnaugh Map for Expression (4), Example 3

$$\begin{aligned} pqr + pq\bar{r} + p\bar{q}r + \bar{p}\bar{q}r \\ &= pq(r+\bar{r}) + (p+\bar{p})\bar{q}r \\ &= pq + \bar{q}r. \end{aligned}$$

This simplification is the one we obtained in Example 8, Section 7.2, using a truth table. The Karnaugh map approach provides an approach to simplification of Boolean expressions which, at first glance, may seem to rely on human visual pattern recognition of map regions. However, it isn't hard to imagine designing an algorithm which, given a Boolean expression as input, forms its DNF equivalent, then searches for patterns in the corresponding Karnaugh map, and, finally, simplifies the original expression.

Here is a second example of the use of Karnaugh maps, which illustrates some other aspects of these maps.

EXAMPLE 4 Simplify the DNF expression whose Karnaugh map is shown in Fig. 7.7(a).

Solution The 2×2 region corresponds to four terms, all of which contain q and r and which otherwise contain the four possible combinations of p and s, complemented and uncomplemented:

$$pqrs, \qquad pqr\bar{s}, \qquad \bar{p}qrs, \qquad \text{and} \qquad \bar{p}qr\bar{s}.$$

Therefore, the sum of these four terms can be simplified to qr. If you don't see this, do the calculation. You'll see the general rule that, in any 2×2 region, the

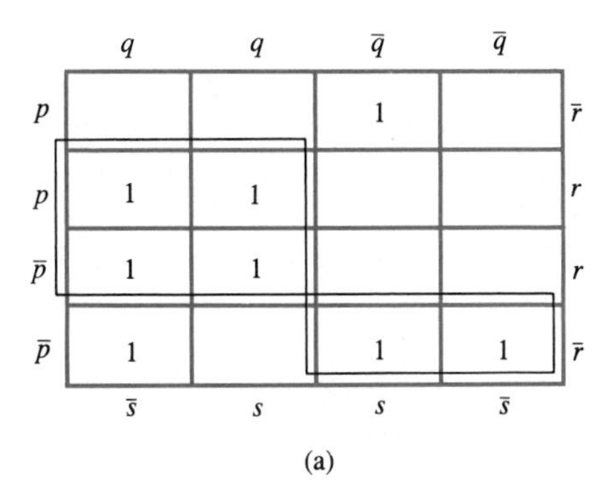

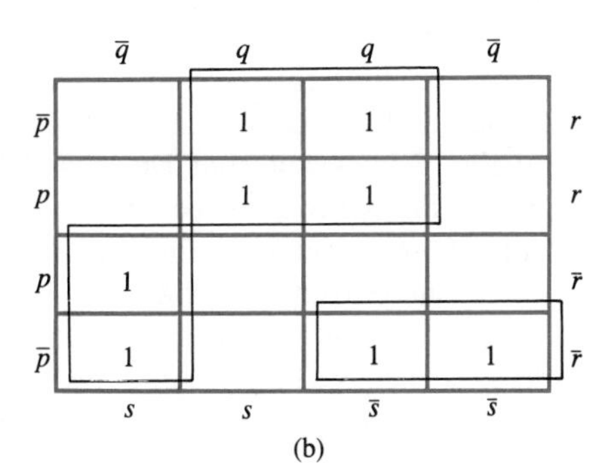

FIGURE 7.7

Karnaugh Maps for Example 4

variables which are not constant throughout the region drop out [16]. In the 2×1 region in Fig. 7.7(a), the s part of the two terms drops out, leaving $\bar{p}\bar{q}\bar{r}$. For the remaining two boxes, no simplification seems possible, so we get the expression

$$qr + \bar{p}\bar{q}\bar{r} + p\bar{q}\bar{r}s + \bar{p}q\bar{r}\bar{s}.$$

Is this the simplest possible expression represented by the Karnaugh map in Fig. 7.7(a)? No, it's not, and the reason is that there are two other 2×1 regions that we could have used instead of the one shown in Fig. 7.7(a). The way the regions appear in the Karnaugh map depends on which of the many possible ways to label the boxes we choose. Figure 7.7(b) shows another labeling, in which once again the 16 boxes are labeled by the 16 possible combinations of four variables. The 1's in Fig. 7.7(b) correspond to precisely the same terms as the 1's in Fig. 7.7(a). The 2×2 square corresponds, as before, to qr. But now there are two 2×1 rectangles, which correspond to $\bar{q}\bar{r}s$ and $\bar{p}\bar{r}\bar{s}$, both different from the 2×1 term we obtained using the map in Fig. 7.7(a). Thus we may write the complete expression as

$$qr + \bar{q}\bar{r}s + \bar{p}\bar{r}\bar{s},$$

which is certainly simpler than the preceding one. ∎

It's a bit unfortunate if the expression we get depends on how we label the boxes in the Karnaugh map. Could we have generated the simpler expression using Fig. 7.7(a)? Yes, we could have, simply by recognizing that a 2×1 rectangle which enables two four-variable terms to be combined into one three-variable term need not necessarily be formed from two adjacent squares. The pattern can also be formed by a *wraparound* from the bottom to the top or the right to the left or vice versa. Another way of saying the same thing is that we can consider opposite borders of the map to be contiguous. Then, referring to Fig. 7.7(a) and ignoring the 2×1 rectangle of actually contiguous squares, we have two other 2×1 rectangles: The first contains the $p\bar{q}\bar{r}s$ and $\bar{p}\bar{q}\bar{r}s$ squares; the second contains the $\bar{p}q\bar{r}\bar{s}$ and $\bar{p}\bar{q}\bar{r}\bar{s}$ squares. Using these rectangles leads to the expression obtained directly from Fig. 7.7(b).

How do we know which rectangles and squares to choose to get the simplest possible expressions? This isn't an easy question, and we won't attempt to answer it here. Dealing with Karnaugh maps and simplifying Boolean expressions involves a great deal more than we can possibly present here. However, some further topics are covered in [20, 21].

To conclude this section we return to Example 2, Section 7.1, which concerned the design of a complementer. In Table 7.11 and Fig. 7.8, we have repeated Table 7.1 and Fig. 7.1, except that we have added some labels and displayed the standard symbols for the **logic circuits** which implement the "and", "or", and "not" operations. Our object is to determine what needs to go in the rectangle in Fig. 7.8, so that the outputs will be correct for any inputs which represent the digits 0–9. The question we need to answer is: What Boolean function of the inputs b_8, b_4, b_2, and b_1 is each of the outputs? If we knew the answer to this question, we could—as you'll soon see—quite easily design the desired logic circuit.

TABLE 7.11
Complementing a Decimal Digit

Input Digit					Output Digit				
Decimal	Binary				Binary				Decimal
	b_8	b_4	b_2	b_1	c_8	c_4	c_2	c_1	
0	0	0	0	0	1	0	0	1	9
1	0	0	0	1	1	0	0	0	8
2	0	0	1	0	0	1	1	1	7
3	0	0	1	1	0	1	1	0	6
4	0	1	0	0	0	1	0	1	5
5	0	1	0	1	0	1	0	0	4
6	0	1	1	0	0	0	1	1	3
7	0	1	1	1	0	0	1	0	2
8	1	0	0	0	0	0	0	1	1
9	1	0	0	1	0	0	0	0	0

Note that the subscript associated with each binary digit is the value of the power of 2 associated with that place in positional notation.

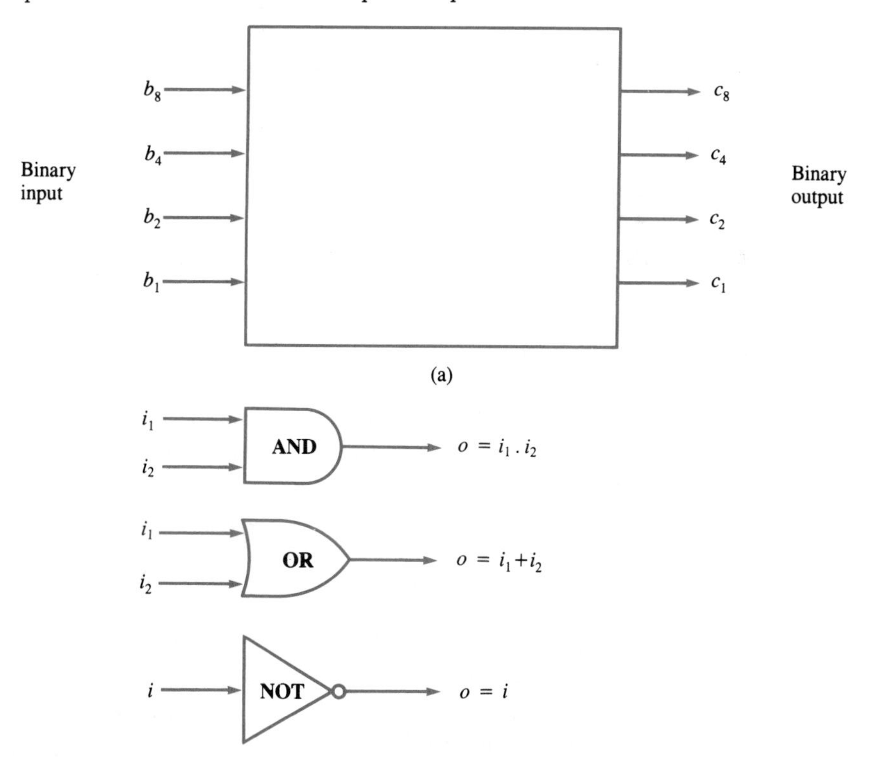

FIGURE 7.8

(a) Designing a Complementer; (b) Available Logic Circuits

We'll solve this problem two ways. First we'll use Karnaugh maps and, as an example, we'll focus on c_4. Referring to Table 7.11, we observe that c_4 equals 1 in only four positions. For example, the 1 in the c_4 column in the third row (input 2, output 7) corresponds to the input values

$$b_8 = 0, \qquad b_4 = 0, \qquad b_2 = 1, \qquad \text{and} \qquad b_1 = 0,$$

for which the product

$$\bar{b}_8\bar{b}_4 b_2 \bar{b}_1 = 1.$$

Corresponding to each of the other three 1's in the c_4 column is a similar product of the b_i's that equals 1. Now suppose that we logically add these four products together to obtain

$$\bar{b}_8\bar{b}_4 b_2\bar{b}_1 + \bar{b}_8\bar{b}_4 b_2 b_1 + \bar{b}_8 b_4\bar{b}_2\bar{b}_1 + \bar{b}_8 b_4\bar{b}_2 b_1. \tag{7}$$

The crucial point is that the value of this sum will be 1 if and only if the b_i's have just the values of one of the rows in Table 7.11 in which $c_4 = 1$. The b_i's in each of these rows are such that one and only one of the terms in expression (7) is 1 and all the rest are 0. And for all other rows, the b_i's are such that all four terms are 0. Therefore for any row in Table 7.11 the sum in expression (7) gives the value of c_4.

	b_2	b_2	$\bar{b}_2$	$\bar{b}_2$	
b_1	1				$\bar{b}_4$
b_1				1	b_4
$\bar{b}_1$				1	b_4
$\bar{b}_1$	1				$\bar{b}_4$
	$\bar{b}_8$	b_8	b_8	$\bar{b}_8$	

FIGURE 7.9

Karnaugh Map for Expression (7)

Now the sum in expression (7) is in DNF. Therefore we may try to simplify it using the Karnaugh map shown in Fig. 7.9. This Karnaugh map has two 2×1 rectangles, the one on the left resulting from a top-to-bottom wraparound. The two terms corresponding to these two rectangles are $b_2\bar{b}_4\bar{b}_8$ and $\bar{b}_2 b_4\bar{b}_8$, so the equation for c_4 is

$$c_4 = b_2\bar{b}_4\bar{b}_8 + \bar{b}_2 b_4\bar{b}_8.$$

Similarly, we could generate equations for c_1, c_2, and c_8, but we'll leave that to a problem [34]. Before showing you how the equation for c_4 might be converted into a logic circuit, we'll present our other approach to this problem.

In this approach, we simply look at Table 7.11 and—using visual pattern recognition but rather differently from that used with Karnaugh maps—try to infer from it the equation for each c_i in terms of the b_i's. We reason as follows:

- As the b_2 and c_2 columns are identical,

$$c_2 = b_2. \tag{8}$$

- Each entry in the c_1 column is the complement of the corresponding entry in the b_1 column; therefore

$$c_1 = \bar{b}_1. \tag{9}$$

- There is a 1 in the c_4 column if there is a 1 in the b_2 column or the b_4 column but not both (what connective discussed in Section 7.2 does this correspond to?); therefore c_4 is 1 if and only if both $b_2 + b_4$ and $\bar{b}_2 + \bar{b}_4$ are 1. From this it follows that

$$c_4 = (b_2+b_4)(\bar{b}_2+\bar{b}_4). \tag{10}$$

- There is a 1 in the c_8 column only when there are 0's in the b_2, b_4, and b_8 columns; therefore

$$c_8 = \overline{b_2 + b_4 + b_8}. \tag{11}$$

But wait a minute. Equation (10) for c_4 is different from the one we obtained earlier. How can this be? Could it be that they are really the same but just expressed in a different form? If we use the distributive law (vii) on Eq. (10), we obtain

$$\begin{aligned} c_4 &= (b_2+b_4)(\bar{b}_2+\bar{b}_4) \\ &= b_2\bar{b}_2 + b_2\bar{b}_4 + \bar{b}_2 b_4 + b_4\bar{b}_4 && (12) \\ &= b_2\bar{b}_4 + \bar{b}_2 b_4. && [\text{Since } p\bar{p} = 0] \end{aligned}$$

This result looks a little better, but there still is no $\bar{b}_8$, as in the earlier equation. Actually, both equations are correct because $\bar{b}_8 = 1$ in all rows in which $c_4 = 1$. That is to say, if we multiply $b_2\bar{b}_4+\bar{b}_2 b_4$ by $\bar{b}_8$, its value remains the same for all rows of Table 7.11. The Karnaugh map approach, as we have described it, is not capable of determining this result and thereby enabling deletion of the $\bar{b}_8$. The cause of this failure is that only 10 of the 16 possible combinations of the b_i's are relevant in this example. By recognizing that we don't care what the other six combinations give, we can in fact use the Karnaugh map approach to derive Eq. (12) [20]. Thus both of our approaches will give the same result if properly used.

Which, then, is to be preferred? We like the Karnaugh map approach better because of its inherently algorithmic, constructive flavor. The second approach, while quicker here, smacks too much of ad hoc cleverness. It would likely fail in more complex circuit design than the simple one we've considered. To be fair, we should admit that the Karnaugh map approach is not of much use when there are more than 5 or 6 variables because the maps and their labeling become unwieldy. But there are systematic approaches to the design of logic circuits which take an approach similar to that of the Karnaugh map idea and are applicable to circuits of any complexity.

Finally, what does the logic circuit which implements Eqs. (8), (9), (11), and (12) look like? It is shown in Fig. 7.10 and is actually nothing more than a pictorial representation of these equations. For example, c_4 as given by Eq. (12) is the logical sum of two logical products or, stated another way, the "or" of two "and"s.

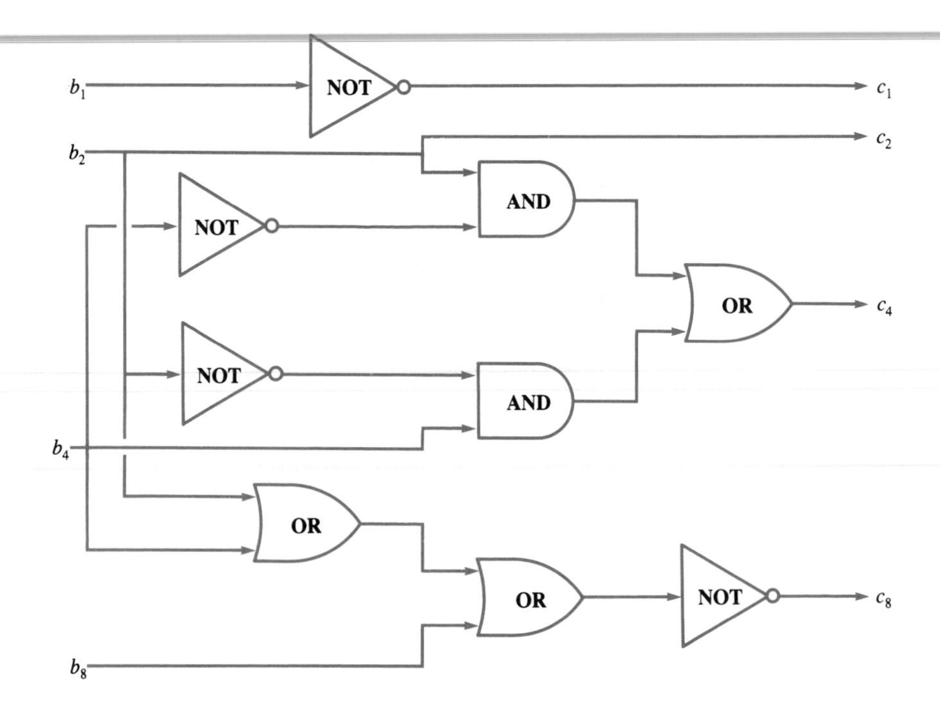

FIGURE 7.10

Logical Circuit for a Complementer

Therefore we "and" b_2 and $\bar{b}_4$, as well as $\bar{b}_2$ and b_4, and then "or" the two results together. The rest of Fig. 7.10 may be interpreted similarly.

In the actual design of logic circuits, it turns out that circuits which implement the "nand" and "nor" connectives (see Fig. 7.3, Section 7.2) are more common than those that use "and", "or", and "not". We'll explore some of the reasons for this usage in [4–8, Supplementary].

Problems: Section 7.5

1. Simplify $1+a$ and $0 \cdot a$. Prove that your simplifications are correct.

2. Verify all 12 pairs of Theorem 1 using truth tables.

3. Show that (xii) of Theorem 1 is the Boolean equivalent of expression (1), Section 7.3.

4. **a)** Explain the meaning of both parts of Eq. (3) in words.

   b) Give deductive proofs of both parts of Eq. (3) in words. *Hint*: Use (ix) and (x) in Theorem 1.

5. Prove the following Boolean theorems using truth tables.

   a) $p\bar{q}+\bar{p}q = (p+q)(\bar{p}+\bar{q})$.

   b) $\overline{pr+q\bar{r}} = \bar{p}r+\bar{q}\bar{r}$.

   c) $p\bar{q}+q\bar{r}+\bar{q}r+\bar{p}q = p\bar{q}+q\bar{r}+\bar{p}r$.

   d) $p\bar{q}+q\bar{r}+\bar{q}r+\bar{p}q = \bar{q}r+p\bar{r}+\bar{p}q$.

6. Parts (i)–(x) of Theorem 1 could be taken to be the axioms of Boolean algebra in that all possible theorems of Boolean algebra can be deduced from these 10 identities.

   a) Verify that in each pair of identities in Theorem 1 one part can be obtained from the other by exchanging $\cdot$ for $+$ and 0 for 1. For this reason each member of a pair is said to be the **dual** of the other.

b) Why therefore is it true that, from every theorem in Boolean algebra, you can derive a **dual theorem** by exchanging $\cdot$ for $+$ and 0 for 1?

c) Write the dual of each of the Boolean theorems in [5].

7. Prove the following Boolean theorems deductively.

a) $p + pq = p$. [*Hint*: Use (ii) of Theorem 1]

b) $p(p+q) = p$.

c) $p+(\bar{p}+q) = 1$.

d) $p(\bar{p}q) = 0$.

8. Both of de Morgan's laws [(xi) and (xii) of Theorem 1] can be proved deductively using (i)–(x). Do it. *Hint*: Use (vii) and (viii), c) and d) of [7], and the fact that if $a+b=1$ and $ab=0$, then a and b are complements of each other.

9. a) Prove the generalized distributive law for Boolean algebra:

$$p(q_1+q_2+\cdots+q_n) \;=\; pq_1 + pq_2 + \cdots + pq_n.$$

b) Prove deductively that

$$(p+q)(p+r)(p+s) \;=\; p + qrs.$$

10. Using DNF, show that $(p+q)(p+r) = p+qr$. Which side represents the simpler circuit to build?

11. Use Karnaugh maps to simplify the following expressions.

a) $pr+(pq+\bar{q})\bar{r}$

b) $(p+q)r + \bar{p}r + p\bar{r}$

c) $p\bar{q}+q\bar{r}+\bar{q}r+\bar{p}q$
[Find two distinct simplifications both with the same number of terms (compare to [5c,d]).]

12. Diagram circuits for the Boolean expressions in [11] using AND, OR, and NOT gates as the basic components.

13. a) Describe how the truth table of a Boolean expression can be used to derive the DNF for that expression.

b) Apply the result of a) to $pq+pr+\bar{q}r$ to get Eq. (4).

14. Write DNF expressions for the Boolean expressions represented by the last columns of the following truth tables.

a)

p	q	
0	0	0
0	1	1
1	0	0
1	1	1

b)

p	q	
0	0	1
0	1	1
1	0	1
1	1	0

c)

p	q	r	
0	0	0	1
0	0	1	0
0	1	0	0
0	1	1	1
1	0	0	1
1	0	1	0
1	1	0	1
1	1	1	1

d)

p	q	r	
0	0	0	1
0	0	1	1
0	1	0	0
0	1	1	0
1	0	0	0
1	0	1	0
1	1	0	1
1	1	1	0

15. Apply the Karnaugh map method to rewrite $p(q+r)$ in another form. What do you end up with? Does the result represent a simpler circuit than the original?

16. Show by truth tables that the DNF for the 2×2 region discussed in the second paragraph of Example 4 can be expressed as qr.

17. Write the DNF for each of the following.

a) $pq+\bar{p}\bar{q}$.

b) $(\bar{p}+\bar{q})(p+q+r)$.

c) $pq\bar{r}+\bar{q}\bar{r}+\bar{p}r$.

d) $pqrs+\bar{p}\bar{q}+p\bar{r}\bar{s}+q\bar{r}\bar{s}$.

18. Use the results of [17] and Karnaugh maps to simplify each of the expressions in [17].

19. a) Design an array for Karnaugh maps with five variables. *Hint*: Since there are only four sides to an array but five variables, some side must have more than one set of labels.

b) If we could draw them in three dimensions, what would be the best way to arrange boxes for Karnaugh maps of three variables? What would be the advantage of this over the two-dimensional arrangement we have shown?

20. If we *don't* care what the output is for a certain set of inputs, we can put a "d" in the Karnaugh map box for those inputs and include that box in a group or not include it, whichever will help the simplification most. Use this observation to obtain Eq. (12).

21. In this problem we refer to the rows of a 4-variable truth table by their decimal equivalents. Thus 9 refers to the row whose first four columns are 1 0 0 1 since 1001 in binary is equivalent to 9 in decimal.

Simplify the following Boolean expressions where "don't care" refers to rows for which the value of the expression is irrelevant. That is, the original expression may evaluate to 1 in a "don't care" row, but the simplified expression can evaluate to 0 for such a row, or vice versa. This allows you to put 1's in any "don't care" box in a Karnaugh map if doing so will make a useful pattern of 1's which can be simplified.

a) $\bar{p}qr\bar{s}$ [Don't care: 10, 11, 12, 13, 14, 15]

b) $\bar{p}\bar{r}+pq\bar{s}+\bar{p}\bar{q}rs+pqrs$ [Don't care: 2, 6, 7]

c) $\bar{p}q\bar{r}+\bar{p}\bar{r}s+\bar{q}\bar{s}$ [Don't care: 0, 2, 3, 7, 9, 10, 11, 13, 14, 15]

22. Show that the Boolean expression defined by any possible truth table (i.e., a table with one more column than the number of atoms), regardless of how many atoms are involved, can be expressed using just **and**, **or**, and **not**. *Hint*: Consider DNF.

23. Show that the expression defined by any possible truth table can be expressed using just **and** and **not**.

24. Show that the expression defined by any possible truth table can be expressed using just **or** and **not**.

25. Consider an n-variable truth table. Let's call the value of the expression in the ith row and last column f_i and denote the DNF product corresponding to row i by m_i. Show that the DNF for the expression defined by the truth table is given by

$$\sum_{i=0}^{2^n-1} f_i m_i. \tag{13}$$

26. Use de Morgan's laws to give an informal proof of the following rule: To find the complement of a Boolean expression (i.e., an expression whose value is the complement of the original expression, whatever the values of the atoms), change all $\cdot$ to $+$ and all 0's to 1's, and vice versa, and replace each variable by its complement. *Hint*: Use the fact that any Boolean expression which contains $+$ or $\cdot$ can be written as $a+b$ or $a\cdot b$ for suitable Boolean expressions a and b.

27. As an alternative to the rule given in [26], argue that the complement of the expression in (13) is given by

$$\sum_{i=0}^{2^n-1} \bar{f}_i m_i. \tag{14}$$

28. As in [25] let m_i represent the DNF product of the variables in the ith row of a truth table. Similarly, let M_i represent the sum of the same variables. Prove the correctness of the following identities.

a) $\bar{m}_i = M_{2^n-1-i}$. **b)** $\overline{M}_i = m_{2^n-1-i}$.

Hint: Use the rule in [26].

29. **Conjunctive Normal Form (CNF)** is any expression of the form:

$$(p+q+r)(p+q+r)\cdots(p+q+r),$$

where each variable under consideration appears once in each factor, either plain or complemented, and no factor appears more than once. Otherwise, the number of factors is not fixed. Do you see that CNF is just like DNF with the roles of $\cdot$ and $+$ interchanged? Therefore CNF is the dual of DNF. Just as every wff can be written in DNF, so can every wff be written in CNF. Use the rule in [26] and the result of [28] applied to expression (14) in [27] to obtain the CNF form of any expression defined by a truth table as

$$\prod_{i=0}^{2^n-1} (f_i+M_{2^n-1-i}).$$

If you were given a truth table and asked to derive the CNF representation of this expression, which rows of the truth table would you have to pay attention to? Why?

30. Figure out the CNF for the truth tables in [14].

31. Karnaugh maps can also be used to simplify CNF, with each box now representing a sum rather than a product. Use the following CNF Karnaugh map to answer a) and b).

	q	$\bar{q}$	$\bar{q}$	q
p		1	1	
$\bar{p}$	1			
	r	r	$\bar{r}$	$\bar{r}$

a) To what three terms in a CNF expression do the three 1's correspond?

b) Show that the CNF corresponding to the 2×1 rectangle can be simplified to $p+\bar{q}$.

32. Rewrite each of the wffs in [17] in CNF.

33. Simplify each result in [32] using a Karnaugh map. Show that each result is equivalent to the result of [18].

34. Use Karnaugh maps to find expressions for c_1, c_2, and c_8 in the complementer example.

35. Use the same technique for designing a complementer to design a circuit to add 1 to a decimal digit, where $9+1$ equals 0.

7.6 The Predicate Calculus

Recall from Section 7.2 that a predicate is a propositional function, which contains one or more variables such as

$$\lfloor j/3 \rfloor * 3 \;=\; j.$$

For each assignment of specific values to the variables in a predicate the truth value of the predicate can be calculated. Thus the predicate above will be true whenever j is a multiple of 3 and false otherwise.

The **predicate calculus** is an extension of the propositional calculus, which, as its name implies, lets us deal with predicates without requiring, as in the propositional calculus, that each variable be given a value in order to convert it to a proposition. Thus because it enables us to handle predicates as well as propositions, the predicate calculus is a much more powerful language than the propositional calculus for expressing complicated mathematical thoughts. In this section we will present a brief introduction to the predicate calculus, just enough so that when you meet it again, most likely in a mathematics or computer science course, you'll understand the basic concepts and notation.

As an instance where the propositional calculus fails us, consider the following common type of argument:

All prime numbers are integers.

7 is a prime number.

Therefore 7 is an integer.

This argument *seems* like some of those we considered back in Section 7.1 having the form:

$$\begin{array}{c} P \\ Q \\ \hline R, \end{array}$$

where P and Q represent the first two sentences in the argument and R is the conclusion. This argument certainly seems to be valid but how would you prove it using the propositional calculus? In fact, you can't, because the first proposition is really a shorthand for an infinite number of propositions:

2 is an integer.

3 is an integer.

5 is an integer.

7 is an integer.

$\vdots$

Indeed, the conclusion R is but one of the infinite number of propositions P. But the propositional calculus can't handle propositions which, instead of being single, distinct propositions, actually stand simultaneously for many or an infinite number of propositions.

Does this example also have something to do with predicates? Although perhaps not obvious, we'll show that it does.

Another way in which the propositional calculus fails us is illustrated in the following algorithm verification example. Suppose that you are given a sequence of numbers, $n_1, n_2, \ldots, n_n$, and you want to *sort* them; that is, you want to transform the sequence $\{n_i\}$ into a new sequence $\{N_i\}$ such that the N_i's will be in numerical order. There are many algorithms for doing this. Suppose that you had designed or had been given such an algorithm and you wanted to insert assertions into it, as we discussed in Section 7.4. The assertion for the input specification might be

$$\{1 \leq m \wedge n[1:m]\},$$

where the bracket notation means that n is a list of m numbers.

Now what would the assertion for the output specification look like? You want to say that there is a list $N[1:m]$ with the property that $N_1 \leq N_2$, $N_2 \leq N_3$, $N_3 \leq N_4$, etc. So you would have to write something like†

$$\{N_1 \leq N_2 \wedge N_2 \leq N_3 \wedge \cdots \wedge N_{n-1} \leq N_n\}.$$

Not very pretty. It would be nice to have a notation which enables us to avoid the ellipsis.

Now consider a constraint on the input which states that all the numbers in the list $n[1:m]$ must be distinct. How would you express that in an assertion [1]?

Problems like these, which involve statements of infinite or arbitrary length, are common in mathematical logic and its applications. Therefore an extension of the propositional calculus—which allows us to deal with these kinds of problems in a convenient and efficient way—would be valuable. The predicate calculus does just this.

Recall first the distinction we made in Section 7.2 between atomic propositions, P, Q, R, etc., which could have only two truth values, T and F, and predicates which contain one or more variables whose values must come from some underlying domain such as N or R. For each substitution of a possible value (or set of values) of the variable(s), the predicate becomes a proposition which is true or false. For example, the predicate

x is a feminist,

where x can be any living animal (i.e., this is the underlying domain), is true if x is Gloria Steinem, false if x is Phyllis Schafly, and (presumably) false if x is your

† Alternatively, you might wish to write $N_1 \leq N_2 \leq N_3 \leq \cdots \leq N_m$ but this expression isn't very pretty either.

pet cat Rusty. Thus a predicate may be thought of as a **propositional function** of its variables. A common—and natural—notation for a predicate with variable x is $P(x)$, with x the argument of the predicate P. If the predicate has two variables, x and y, we would write $P(x,y)$.

Our predicate variables, unless otherwise stated, will have N or N^+ as their domains throughout this section. Instead of writing $P(i)$ for such predicates, we'll use subscripts, so that our notation becomes P_i for a predicate with these domains. Examples of such predicates are

$$P_i\colon\ 1+2+\cdots+i = 1(i+1)/2;$$
$$P_j\colon\ \lfloor j/2 \rfloor * 2 = j.$$

Note that the first predicate is true for all values of i in N^+, but the second is true only for certain values of j. (Which ones?)

A wff in the propositional calculus becomes a predicate if any of its atomic propositions is replaced by a proposition depending on some variable. For example,

$$P \wedge Q \implies R$$

becomes a predicate

$$((x>7) \wedge Q) \implies (y<5),$$

if P and R are replaced as above. Note that a predicate may contain some atomic propositions, such as Q, in addition to propositions with variables.

EXAMPLE 1 The entities called Boolean expressions in our algorithmic language or in programming languages are just predicates. For instance,

$$i < j; \qquad \text{[Two-variable predicate } P_{ij}\text{]}$$
$$\lfloor(\sqrt{i})\rfloor^2 = i; \qquad \text{[Is } i \text{ a perfect square?]}$$
$$(i<10) \vee (j>20);$$
$$((x>10) \wedge (x<20)) \implies (x=15). \qquad \text{[Compound predicates]}$$ ■

For each value (or set of values) of the variable(s) in a predicate, the predicate will have the value T or F [2]. The number of distinct variables in a predicate is not restricted, but all our examples will have either one or two variables.

Quantifiers

If you deal with predicates—that is, with propositions which depend on variables—you may want to consider more, perhaps many more, than one value of the variable(s) in the predicate. For example, if you have the predicate

$$P_i\colon\ i^2 > 8i+7,$$

you might want to consider all $i \in N$ or some range of values of i. The mechanism in the predicate calculus for doing this kind of thing is called **quantification**. A **quantifier** (see Section 0.7) enables us to gather many propositions under a single rubric. The first of the two most important quantifiers used in mathematical logic is defined as follows:

Definition 1. The expression

$$(\forall i)P_i \tag{1}$$

represents the infinite conjunction

$$P_1 \wedge P_2 \wedge P_3 \wedge \cdots.$$

The symbol $\forall$ (an upside down A, where "A" is short for "all") is called the **universal quantifier**.

The notation (1) should be read, "For all i, P_i is true", which is the same as saying that P_1 **and** P_2 **and** P_3, etc., are true. In expression (1), P_i is any predicate with domain N or N^+. Indeed, as you can now see, a predicate is just a shorthand for writing an infinite number of propositions, instead of just listing some of them and using an ellipsis to indicate the rest.

EXAMPLE 2 Write the proposition:

All odd integers are prime

using quantifier notation.

Solution Using P_i to represent the proposition that i is odd and Q_i to represent the proposition that i is prime, this proposition becomes

$$(\forall i)(P_i \Longrightarrow Q_i) \quad \blacksquare$$

A quantified predicate like that in Example 2 has a truth value. That is, the expression is true or it is false. Either $P_i \Longrightarrow Q_i$ is true for all i or it's not. Therefore using propositional connectives with quantified predicates, as we'll start to do shortly, shouldn't cause any problems. (Note, however, that we've oversimplified things a bit here. Must a quantified predicate have a truth value? We'll discuss this later under "bound and free variables", but you might try and figure out the answer now.)

If the propositions in Definition 1 were ones which depended on real numbers rather than on integers such as

$$|x| \; < \; |2x|,$$

then, instead of expression (1), we would write

$$(\forall x)P(x).$$

Although the idea of a quantifier sometimes seems strange or difficult when first encountered, you actually have met the essential idea before. Summation and product notation, which we first discussed in Section 0.5, are both notations which *summarize* something unwieldy or impossible to write out completely, as in

$$\sum_{i=1}^{n} a_i \;=\; a_1 + a_2 + \cdots + a_n, \tag{2}$$

and

$$\prod_{i=1}^{\infty} b_i \;=\; b_1 \cdot b_2 \cdot b_3 \cdots. \tag{3}$$

Equation (2) is analogous to stating a finite, but perhaps very large, number of propositions. Equation (3) is analogous to induction or the case in Example 2 where we wished to state an infinite number of propositions. Quantifiers enable us to summarize conjunctions (and also disjunctions, as we'll show) of many propositions in much the same way that sum and product notation allow us to summarize the addition and multiplication of mathematical expressions.

Although logicians and mathematicians almost always apply the universal quantifier to infinite sets of propositions, computer scientists often apply it to finite sets of propositions, as in the algorithm verification example at the beginning of this section. There the output specification that the numbers be in numerical order could be stated

$$(\forall i : 1{\le}i{<}m) N_i \;\le\; N_{i+1}. \tag{4}$$

In expression (4), the portion after the colon indicates the range of values of i to which the quantifier applies. Here,

$$N_i \;\le\; N_{i+1}$$

is a predicate whose underlying universe is the finite number of values of i from 1 to $m-1$. [Make sure, by the way, that you understand why "$<m$" rather than "$\le m$" appears in expression (4).] If you want to make the domain of quantification, such as the real numbers, explicit for a predicate $P(x)$, you could write

$$(\forall x)(R(x) \implies P(x)),$$

where $R(x)$ is the predicate "x belongs to R (the set of real numbers)". Or, analogously to Eq. (4), you might write

$$(\forall x\colon R(x))P(x).$$

The analogy of quantifier notation with summation or product notation becomes even more evident when you realize that something like Eq. (2) could be written in a "linear" notation as

$$\textstyle\sum i\colon 1 \;\le\; i \;\le\; n\colon a_i.$$

Or, going in the other direction, we could have written Eq. (4) as

$$\underset{1\le i<m}{(\forall i)}\ N_i \le N_{i+1},$$

which is like the notation we used in Section 0.5 [11].

Universal quantifier notation allows us to express the principles of mathematical induction compactly. Using this notation we can express the First (or weak) Principle as follows: In order to prove

$$(\forall n{:}\, n{\ge}n_0)P(n), \tag{5}$$

it suffices to prove

$$P(n_0) \;\wedge\; (\forall n{:}\, n{>}n_0)[P(n{-}1) \Longrightarrow P(n)]. \tag{6}$$

Using this notation, we can give a better answer than before to the common question: Why isn't it enough to do just the inductive step? That is, isn't the proof finished as soon as $P(n)$ is proved from $P(n{-}1)$? The answer is: We are trying to prove $(\forall n{:}\, n{\ge}n_0)P(n)$, which is not what we get in the inductive step because the $(\forall n)$ applies to the implication $P(n{-}1) \Longrightarrow P(n)$, not to $P(n)$ alone. In other words, the question above arises from the failure to appreciate the implicit quantifiers within the induction principle. This sort of confusion is not surprising because quantifiers are rarely made explicit in informal mathematical writing.

Quantifiers also allow us to rebut the claim that induction is valid if we prove $P(n{-}1) \Longrightarrow P(n)$ but invalid if we prove $P(n) \Longrightarrow P(n{+}1)$. In the latter case (so the argument goes), we are assuming what we want to prove rather than concluding with it. Again, we are trying to prove $(\forall n)P(n)$, not just $P(n)$. Within the inductive step, the n is a dummy variable, so whether we use $n-1$ and n, or n and $n+1$ is irrelevant. (But how would we have to change expression (6) if we wanted $P(n) \Longrightarrow P(n{+}1)$ in it [8]?)

For strong induction, again wanting to prove expression (5), it is enough to prove

$$P(n_0) \;\wedge\; (\forall n{:}\, n{>}n_0)\left\{[(\forall j{:}\, n_0{\le}j{<}n)P(j)] \;\Longrightarrow\; P(n)\right\}. \tag{7}$$

Note the use of the universal quantifier a second time to express that the truth of all the propositions $P(j)$ for $n_0{\le}j{<}n$ may be used to prove that $P(n)$ is true.

As noted just above, the universal quantifier is used implicitly rather than explicitly in informal mathematical writing. For example, the commutative law for the sum of two integers,

$$m+n \;=\; n+m,$$

is really the statement

$$(\forall m)(\forall n)\;\; m+n \;=\; n+m, \tag{8}$$

where $m{+}n = n{+}m$ is a two-variable predicate. Here, we used double quantification just as we previously used double summation. You should read expression (8) as "For all m and for all n, $m{+}n = n{+}m$."

Indeed, in all mathematical identities the universal quantifier is implicit because such identities apply to all values of their variables in the underlying universe. This is really the distinction between an identity and an equation. The former is true for many (usually an infinite number of) values of its variables, whereas the latter is true for only one or a few values.

As we implied at the beginning of this section with "All prime numbers are integers", we often use quantifiers in ordinary speech. Sometimes we make them explicit, using words like "all" or "every", but sometimes they are implicit, as in "Math majors do well on tests", which would usually be taken to mean "All math majors do well on tests".

The other essential quantifier used in mathematical logic is defined as follows:

Definition 2. The expression

$$(\exists i)P_i \tag{9}$$

represents the infinite disjunction

$$P_1 \vee P_2 \vee P_3 \cdots.$$

The symbol $\exists$ (a backward E, where "E" is short for "exists") is called the **existential quantifier**.

The notation (9) should be read, "There exists an i for which the predicate P_i is true", which is the same as saying that P_1 **or** P_2 **or** P_3, etc., is true. Corresponding to Eq. (4) there is a similar form for the existential quantifier:

$$(\exists i\colon m \le i \le n)P_i, \tag{10}$$

which is true if $P_m \vee P_{m+1} \vee \cdots \vee P_n$ is true and false otherwise. As with the universal quantifier, expression (9) defines a proposition which is true or false. Either there exists an i for which P_i is true or there doesn't.

When we make statements in ordinary English which correspond conceptually to the existential quantifier, we usually use words like "one" or "some", as in

One of these days I'll...

Some people really like mathematics.

In either case the implication is that at least one day or person satisfies whatever condition follows. That is, in either case there *may* be more than one day or person for which the statements are true. In general, there *may* be many P_i which are true, perhaps even all of them, but there *must* be at least one if expression (9) is to be true.

As with the universal quantifier, we may use multiple existential quantifiers, as in

$$(\exists x)(\exists y)[x^2 + y^2 = 1].$$

You should be able to see from this example and the one above for the universal quantifier that it doesn't matter how you order the variables when a number of quantifiers of the same kind are concatenated. That is, just as you can change the order of summation when all the limits of summation are constants, so you can change the order of repeated universal or repeated existential quantifiers. We leave the proof to a problem [10], and we also leave to a problem [11] the situation in which the universe of values of an inner quantification depends on a value of the variable in an outer quantification.

Just as we can have summation and product notation in the same expression, we can have both existential and universal quantifiers in the same expression. For example,

$$(\forall i)(\exists j)2^j \leq i < 2^{j+1}$$

says that any integer i lies between two successive powers of 2. But what about the order in which you write the quantifiers when there is a mixture of universal and existential quantifiers? For example, is

$$(\forall i)(\exists j)P_{ij} \iff (\exists j)(\forall i)P_{ij} \tag{11}$$

valid? That is, is the truth value of the two sides of expression (11) always the same, independent of P_{ij}? The answer is No, as the following example illustrates.

EXAMPLE 3 Consider the predicate

$$P_{ij}: i = j.$$

Then

$$(\forall i)(\exists j)P_{ij}$$

is true because for any i there exists a j such that $i = j$. On the other hand

$$(\exists j)(\forall i)P_{ij}$$

is false because there is no j such that $i = j$ for all i. Thus the two sides of expression (11) can have different truth values. ∎

What is true, however, is that

$$(\exists j)(\forall i)P_{ij} \implies (\forall i)(\exists j)P_{ij} \tag{12}$$

always has truth value T independent of P_{ij} because, whenever there exists a j (say, j_0) such that P_{ij} is true for all i, then for all i there must exist an j (namely, j_0) for which P_{ij} is true.

Is expression (12) a tautology as we defined that term in Section 7.2? Yes, but "tautology" is normally reserved only for use in the propositional calculus. So, instead of calling expression (12) a tautology, we say that a predicate calculus expression which is always true is *logically valid.*

EXAMPLE 4 The following are logically valid formulas in the predicate calculus, where P_i is assumed to be a valid predicate.

$$(\forall i)P_i \implies P_j.$$ [If the left-hand side is true, all P_j are true]

$$(\exists i)P_i \lor (\forall i)\neg P_i.$$ [Since, if no P_i is true, all P_i must be false] ■

The second formula in Example 4 suggests an answer to the question: Can a universally quantified predicate be replaced by the existential quantification of some other predicate and vice versa? Thus is it possible, for example, to express

$$(\forall i)P_i$$

using the existential instead of the universal quantifier? Well, P_i is true for all i if and only if there is no i for which P_i is false, that is, if and only if there is no i for which $\neg P_i$ is true (so that $\neg(\exists i)\neg P_i$) is true). Conversely, it is true that there is no P_i such that $\neg P_i$ is true (i.e. $\neg(\exists i)\neg P_i$ is true) if and only if P_i is true for all i. Therefore

$$(\forall i)P_i \iff \neg(\exists i)\neg P_i \tag{13}$$

is logically valid.

Thus the two quantifiers are related through the idea of negation. By negating both sides of expression (13) and then by substituting $\neg P_i$ for P_i, we may generate the following three related logically valid formulas [12]:

$$\neg(\forall i)P_i \iff (\exists i)\neg P_i, \tag{14}$$

$$(\exists i)P_i \iff \neg(\forall i)\neg P_i, \tag{15}$$

$$\neg(\exists i)P_i \iff (\forall i)\neg P_i. \tag{16}$$

Negation causes some of the most difficult problems with quantifiers. The following example illustrates the difficulties.

EXAMPLE 5 Consider the proposition:

> For every positive integer k there is an N such that for all $n > N$,
>
> $$\frac{n^k}{1.01^n}$$
>
> is as close to 0 as you wish.

State the negation of this proposition.

Solution You first need to recognize this as a quadruply quantified proposition, which can be stated as

$$(\forall k)\big[(\forall \epsilon\colon \epsilon>0)\,\{(\exists N)[(\forall n\colon n>N)(n^k/1.01^n<\epsilon)]\}\big]$$

where k, n, and N are positive integers and ϵ is a positive real number. The negation of this proposition is

$$\neg(\forall k)\big[(\forall \epsilon\colon \epsilon>0)\,\{(\exists N)[(\forall n\colon n>N)n^k/1.01^n < \epsilon]\}\big] \tag{17}$$

Then, applying expressions (14) and (16) successively, we obtain propositions equivalent to expression (17):

$$(\exists k)\big[\neg(\forall \epsilon\colon \epsilon>0)\,\{(\exists N)[(\forall n\colon n>N)(n^k/1.01^n<\epsilon)]\}\big];$$
$$(\exists k)\big[(\exists \epsilon\colon \epsilon>0)\,\{\neg(\exists N)[(\forall n\colon n>N)(n^k/1.01^n<\epsilon)]\}\big];$$
$$(\exists k)\big[(\exists \epsilon\colon \epsilon>0)\,\{(\forall N)\neg[(\forall n\colon n>N)(n^k/1.01^n<\epsilon)]\}\big];$$
$$(\exists k)\big[(\exists \epsilon\colon \epsilon>0)\,\{(\forall N)[(\exists n\colon n>N)(n^k/1.01^n\geq\epsilon)]\}\big].$$

In the last line we applied the negation by changing $<$ to $\geq$. In each step we changed a universal quantifier to an existential quantifier or vice versa and applied the negation to the remaining portion of the proposition [13]. Thus we have "pushed through" the $\neg$ from left to right, changing quantifiers as we went. Now what does the final proposition say in mathematical English? It says that there exists a k and an $\epsilon > 0$ such that for all N there is an $n > N$ such that

$$\frac{n^k}{1.01^n} \geq \epsilon.$$

Caution: We have not proved that this negation is true; we have merely figured out what it is. The original proposition was true because the exponential of any number greater than 1 ultimately grows faster than n raised to any power. Therefore the negation of this statement is in fact false; no matter which k and ϵ we choose, there will be some N such that $n^k/1.01^n$ is less than ϵ for all $n > N$. ∎

Bound and Free Variables

Is

$$(\forall i)(i^2>i) \tag{18}$$

a predicate with variable i? No, because the variable i has been "quantified out". It makes no sense to talk about giving i a value in expression (18) because the quantifier encompasses *all* values of i. What we have in expression (18) is not a propositional function but, as noted earlier, just a proposition itself.

What about

$$(\exists i)(i^2 = j)? \tag{19}$$

In expression (19) we are concerned with whether j is a perfect square. For each j the proposition is true or false but expression (19) itself is still a predicate whose value depends on the value assigned to j. In an expression such as (19) the variable j is said to be **free** because its value is unconstrained (except by the underlying universe), and i is said to be **bound** because the quantifier does not allow us to give specific values to i. By contrast, in

$$(\forall j)(\exists i)(i^2 = j), \tag{20}$$

both variables are bound—j by the universal and i by the existential quantifier. Thus expression (20) is a proposition with truth value T or F (which is it?). Did you recognize that expression (19) answers the earlier question about when a quantified predicate does not have a truth value? Only when each variable in a predicate is quantified, as in expression (20), do we get a proposition rather than a predicate.

Theorems and Proofs

In the predicate calculus, as in the propositional calculus, there are rules for forming wffs. We won't discuss them here, however, because they are rather complicated. Instead, we'll rely on your intuitive feel for appropriate formulas whenever necessary below and in the problems for this section.

We'll conclude this section with some brief remarks on theorems and rules of inference in the predicate calculus. A theorem in the predicate calculus, as in the propositional calculus, begins with some (zero or more) wffs, which are premises. We then use rules of inference to derive a conclusion, which must be true if the premises are true. To prove theorems in the predicate calculus, we need some rules of inference involving quantifiers in addition to those in Section 7.3 for the propositional calculus. Then, as in the propositional calculus, we apply natural deduction. (Note that the truth table method of proof is generally impossible in the predicate calculus. Even when the domain of a predicate variable is finite, as in Eq. (4), there is normally a variable [m in Eq. (4)] which can assume values from an infinite set.)

We will content ourselves with a single example of a proof in the predicate calculus, as any further discussion would soon get considerably more intricate than in the case of the propositional calculus.

EXAMPLE 6 Prove: If all primes are integers and all integers are rational, then all primes are rational. (Are the premises true?)

Solution First we state the argument corresponding to this theorem formally:

$$(\forall x)(PM(x) \implies I(x))$$
$$(\forall x)(I(x) \implies R(x))$$
$$\overline{\hspace{8em}}$$
$$(\forall x)(PM(x) \implies R(x)),$$

where $PM(x)$ is the predicate that x is a prime, $I(x)$ is the predicate that x is an integer, and $R(x)$ is the predicate that x is rational. Our proof will use two rules of inference for the predicate calculus (which we number 4 and 5 following rules 1–3 in Section 7.3):

- *Rule 4—Universal instantiation.* If $(\forall x)P(x)$ is true, we may assert $P(x)$ for any x in the domain of the quantification.
- *Rule 5—Universal generalization.* If $P(x)$ is true for an x chosen *arbitrarily* in a domain, we may assert $(\forall x)P(x)$ where the quantification is over the same domain. (To prove $P(x)$ for an arbitrary x means that in the proof we cannot use any special properties of the x chosen. Thus to prove a theorem about quadrilaterals, we couldn't consider a square and then use some property of "squareness" in the proof.)

Now here's the proof of the validity of the above argument:

	Derivation	Justification
1.	$(\forall x)(PM(x) \Longrightarrow I(x))$.	Premise
2.	$(\forall x)(I(x) \Longrightarrow R(x))$.	Premise
3.	$PM(x) \Longrightarrow I(x)$.	Rule 4, Universal Instantiation
4.	$I(x) \Longrightarrow R(x)$.	Rule 4
5.	$PM(x)$.	Assumption with x arbitrary
6.	$I(x)$.	Modus ponens, 5, 3
7.	$R(x)$.	Modus ponens, 6, 4
8.	$PM(x) \Longrightarrow R(x)$.	Rule 2, Conditional Proof
9.	$(\forall x)(PM(x) \Longrightarrow R(x))$.	Rule 5, Universal Generalization

Make sure that you understand each step of this proof. If you do, you probably understand the basic ideas of quantification. One caution, though: It is *not* true that everything in the predicate calculus can be proved by stripping the quantifiers, doing some propositional calculus, and then putting the quantifiers back. You always can strip them, but whether you can put them back depends on the order in which they were stripped, the order in which you want them put back, and what was done to the variables in between. But we won't pursue this topic further here. ∎

Problems: Section 7.6

1. a) Without using quantifiers, write an input specification for an algorithm to sort a list $n[1, m]$ of m *distinct* numbers.

b) Rewrite this IS using quantifiers.

2. This problem refers to the four predicates of Example 1.

a) For $j = 17$, for what values of i is $P(i, j)$: $i < j$ true?

b) For what values of $i > 0$ is the predicate "i is a perfect square" true?

c) Describe or draw a diagram to illustrate the pairs (i, j) such that $(i<10) \vee (j>20)$ is true.

d) For what real numbers x is the predicate $[(x>10) \wedge (x<20)] \Longrightarrow (x = 15)$ true?

3. a) Use quantifiers to write an output specification for an algorithm to sort a list $n[1, m]$ of m *distinct* numbers.

b) Use quantifiers to write an output specification for PERMUTE-1 (Section 4.9).

4. For the following statements, let the domain be the set R of real numbers. For each statement, try to say it in words. If it is false, give an example to show why. Example:

$$(\forall x)(\exists y)(x<y)$$

says "for every number there is a bigger number". This is true.

a) $(\forall x)(\exists y)[x \neq 0 \Longrightarrow xy = a]$.

b) $(\exists y)(\forall x)[x \neq 0 \Longrightarrow xy = 1]$.

c) $(\exists x)(\forall y)[y \leq x]$.

d) $(\forall x)(\exists y)(x+y = x)$.

e) $(\exists y)(\forall x)(x+y = x)$.

f) $(\forall x)(\forall y)(\exists z)[x < z < y]$.

g) $(\forall x)(\forall y)[x \neq y \Longrightarrow (\exists z)(x<z<y \vee x>z>y)]$.

h) $(\forall x)(\forall y)(\forall z)[(x>y \wedge y>z) \Longrightarrow x > z]$.

5. Let the domain be R again. Translate each of the following sentences into a formula of the predicate calculus.

a) Every number has an additive inverse.

b) There is a smallest number.

c) Every positive number has a square root.

d) Every positive number has a positive square root.

e) There is a multiplicative identity (1, but don't say so!).

f) Any two numbers have a sum (that is, the sum is another real number).

g) No number equals twice itself.

6. Now let the domain be the integers Z. Try to say each of the following statements in words. If it is false, give an example to show why.

a) $(\forall i)(\exists j)(j = i{+}1)$.

b) $(\forall i)(\exists j)[Z(i) \Longrightarrow (Z(j) \wedge j = i + 1)]$.
[$Z(i)$ is the predicate "i is an integer"]

c) $(\forall i)(\forall j)[(i^2 = j) \Longrightarrow ((-i)^2 = j)]$.

d) $(\forall i)(\exists j)(j = i^2)$.

e) $(\forall i)(\forall j)(\forall k)(i \mid j \wedge j \mid k) \Longrightarrow (i \mid k)$.
[($\mid$ means "divides without remainder")]

f) $(\exists j)(\forall i) ij = i$.

g) $(\forall i)(\forall j)(\forall k)[(i \mid j \wedge j \mid k) \Longrightarrow \lfloor\lfloor k/j \rfloor/i\rfloor = \lfloor k/(\lfloor j/i \rfloor)\rfloor]$

7. With the domain the natural numbers N, translate the following sentences into formulas of the predicate calculus.

a) There is a smallest nonnegative integer.

b) For each number not a prime, there is another integer which divides it without remainder.

c) All numbers less than 100 are less than 2^7.

d) All numbers less than 100 lie between two successive powers of 2.

e) Every number is even or odd.

f) The product of two odd numbers is odd.

g) The product of an odd and an even number is even.

h) If one number divides a second and the second divides a third, then there is an integer which, when multiplied by the first number, gives the third.

8. a) Rewrite expression (6), replacing $P(n{-}1) \Longrightarrow P(n)$ with $P(n) \Longrightarrow P(n{+}1)$.

b) Rewrite expression (7), replacing $P(n)$ with $P(n{+}1)$.

9. In the following expressions, identify free variables and bound variables.

a) $x{+}y = y{+}x$.

b) $(\forall x)(x{+}y = y{+}x)$.

c) $(\forall x)(\forall y)(x{+}y = y{+}x)$.

d) $(\forall x)[(x{>}0 \Longrightarrow (\exists y)(x = 2y)]$.

e) $(\forall x)P(x) \Longrightarrow (x{>}0 \wedge (\exists y)Q(y))$.

f) $g(n) = \sum_{i=1}^{n}(2i{+}1)$

10. Explain in words why the order of two universal quantifiers can be switched but $(\forall i)(\exists j)$ cannot be switched. What about the order of two existential quantifiers?

11. Switch the order of quantifiers in each of the following. In each case the domain is N.

a) $(\forall n{:}\, 1{<}n{<}10)(\forall m{:}\, n{<}m{<}10)P(m,n)$.

b) $(\forall n{:}\, 0{\le}n{\le}5)(\forall m{:}\, 0{\le}m{\le}n)P(m,n)$.

c) $(\forall n{:}\, 1{\le}n{\le}5)(\forall m{:}\, 0{\le}m{\le}n)P(m,n)$.

d) $(\forall n{:}\, 0{\le}n{<}10)(\forall m{:}\, n{\le}m{\le}n^2)P(m,n)$.

12. Argue in words that each of the formulas (14)–(16) is logically valid.

13. In the text we showed that, to negate an expression with a single quantifier, we can replace it with the other quantifier and negate the predicate inside. This generalizes to arbitrary strings of quantifiers. For instance,

$$\neg(\forall x)(\exists y)(\exists z)(\forall w)P(x,y,z,w) \Longleftrightarrow (\exists x)(\forall y)(\forall z)(\exists w)\neg P(x,y,z,w).$$

Prove this generalization by induction.

14. Apply the result of [13] to:

a) The formulas of [4].

b) The formulas of [6].

In each case state the resulting formula in words and, if it is false, give an example to show why.

7.7 Algorithm Verification Using the Predicate Calculus

At the beginning of Section 7.6 we considered briefly the problem of taking a list $n[1\colon m]$ of numbers and *sorting* it into a list $N[1\colon m]$ of numbers in numerical sequence. We also considered this problem in Section 5.9, where we analyzed the merge sort method. (See also Section 3.8.) Here we'll consider another algorithm for sorting, which illustrates nicely how quantification plays a role in verifying algorithms.

The method we'll consider here is called **insertion sort**. The idea is illustrated in Fig. 7.11. We begin with the list $n[1\colon m]$ and apply the recursive paradigm as follows. Suppose we had a list whose first $i-1$ elements were in numerical order (Fig. 7.11(b)), where the primes indicate that the element in positions $1, \ldots, i-1$, will normally not be the elements that were there originally. In order to insert the ith element in its proper place we

- first take n_i and put it in a temporary location (temp in Fig. 7.11(c) and (d));
- then compare n_i with n'_{i-1}; **if** $n_i \geq n'_{i-1}$ **then stop** (because n_i is already in its proper place—not necessarily its *final* place but its correct place among the first i items) **else** move n'_{i-1} into the slot previously occupied by n_i; the result if $n_i < n'_{i-1}$ is shown in Fig. 7.11(c);
- continue comparing and moving until a j is found such that

$$n_i > n'_j,$$

  as shown in Fig. 7.11(d);
- *insert* n_i into slot $j+1$, as shown in Fig. 7.11(e), where $''$ indicates that the items in these slots are not the ones that were there at the start of ith step.

Of course, if $n_i < n'_1$, no j will be found for which $n_i \geq n'_j$; this situation is easily recognized when j gets to 0, in which case n_i gets inserted into slot 1. Summarizing, insertion sort works by inserting the ith element in its proper place in the already sorted list of $i-1$ elements. As we start with a sorted list of one element (namely, n_1), the result of applying the steps above for $i = 2, 3, \ldots, m$ should result in a completely sorted list.

The entire process is illustrated in Fig. 7.12 and is translated into our algorithmic language in Algorithm 7.3, INSERTION. Algorithm INSERTION also includes assertions, as discussed in Section 7.4. Look first at the algorithm, ignoring the assertions, to make sure that you understand how it implements sorting by insertion. Although we have described sorting by insertion using the recursive paradigm, as with binary search implementing the algorithm using iteration rather than recursion is easier and more straightforward.

Now let's see whether we can use the assertions to *prove* the algorithm correct. Remember: Informal reasoning that sorting by insertion works is one thing; proving that an algorithm to implement the insertion idea is correct is something quite different. We reason as follows:

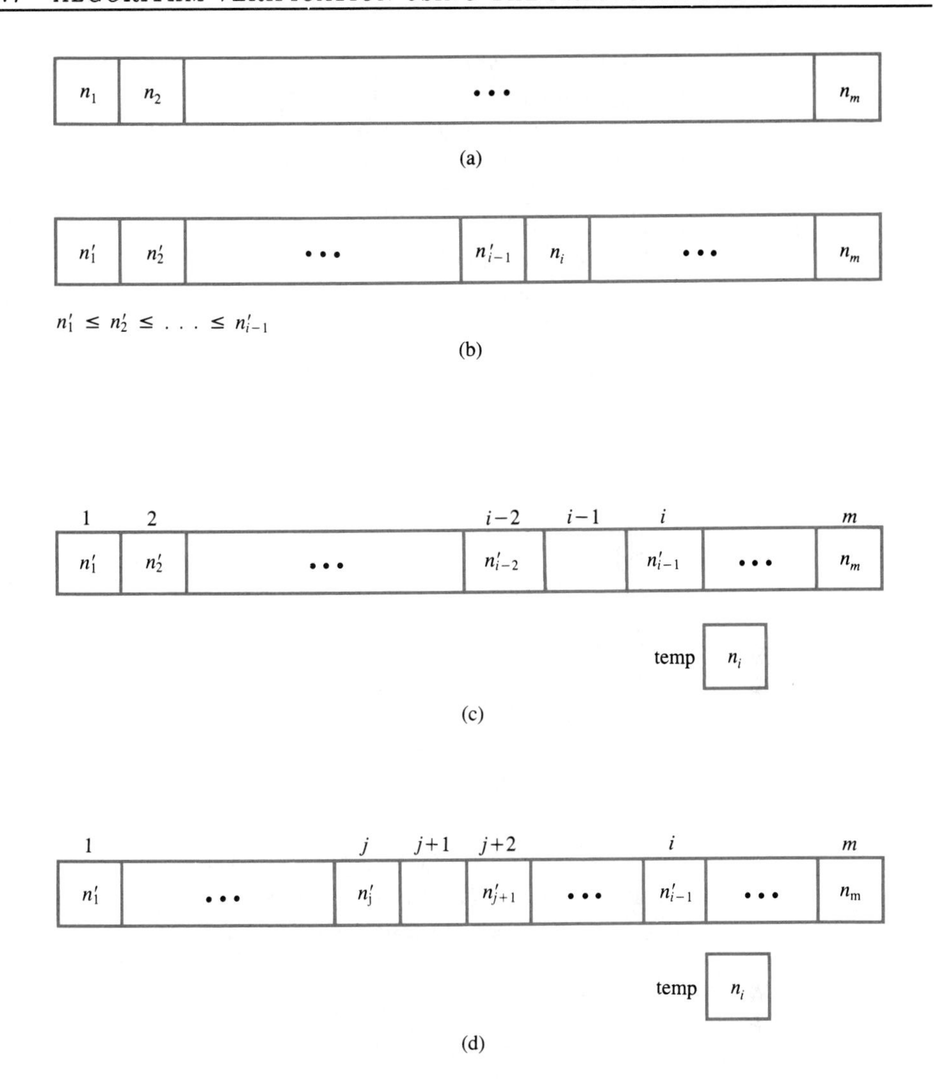

FIGURE 7.11 Insertion Sort (a) A list to be sorted; (b) the list after the first $i-1$ elements are in numerical order; (c) the list after the first comparison of temp with n'_{i-1}, assuming that $n_i < n'_{i-1}$; (d) the list with n_i ready to be inserted in its proper place; (e) the list with the first i elements in numerical order.

$n[1:9]$:	[15]	12	9	17	6	12	2	8	16
End of loop with $i=2$:	[$\underline{12}$	15]	9	17	6	12	2	8	16
End of loop with $i=3$:	[$\underline{9}$	12	15]	17	6	12	2	8	16
End of loop with $i=4$:	[9	12	15	$\underline{17}$]	6	12	2	8	16
End of loop with $i=5$:	[$\underline{6}$	9	12	15	17]	12	2	8	16
End of loop with $i=6$:	[6	9	12	$\underline{12}$	15	17]	2	8	16
End of loop with $i=7$:	[$\underline{2}$	6	9	12	12	15	17]	8	16
End of loop with $i=8$:	[2	6	$\underline{8}$	9	12	12	15	17]	16
End of loop with $i=9$:	[2	6	8	9	12	12	15	$\underline{16}$	17]

FIGURE 7.12 An Example of Insertion Sort (Item inserted is underlined; brackets enclose sorted portion of list.)

Algorithm 7.3

Insertion

Input m [Integer > 0]
$n[1:m]$ [List to be sorted]

Algorithm INSERTION

$\{m>0\}$ [Input specification]

$\{(\forall k: 1 \le k \le i-1) n_k \le n_{k+1}\}$ [Outer loop invariant]

for $i = 2$ **to** m

$temp \leftarrow n_i$

$j \leftarrow i-1$

$\{(\forall k: 1 \le k \le j-1) n_k \le n_{k+1} \wedge (\forall k: j+2 \le k \le i-1)$
$n_k \le n_{k+1} \wedge (\forall k: j+2 \le k \le i) temp < n_k\}$ [Inner loop invariant]

repeat until $j = 0$ **or** $temp \ge n_j$

$n_{j+1} \leftarrow n_j$

$j \leftarrow j-1$

endrepeat

$\{(\forall k: 1 \le k \le j-1) n_k \le n_{k+1} \wedge (\forall k: j+2 \le k \le i-1)$
$n_k \le n_{k+1} \wedge (\forall k: j+2 \le k \le i) temp < n_k$
$\wedge (j = 0 \vee n_j \le temp)\}$ [Loop termination condition]

$n_{j+1} \leftarrow temp$

endfor

$\{(\forall k: 1 \le k \le i-1) n_k \le n_{k+1} \wedge i = m\}$ [Loop termination condition]

$\{(\forall k: 1 \le k \le m-1) n_k \le n_{k+1}\}$ [Output specification]

Output $n[1:m]$ [In numerical order]

1. The input specification—$m>0$—states only the assumption that the list is not empty.
2. The loop invariant for the outer loop states that, before entering this loop with a new value of i, the first i items in the list are sorted. When the loop is entered for the first time with (implicitly) $i=1$, this assertion is vacuously true because there are no values of k between 1 and $i-1$ $(=0)$.
3. The purpose of the inner loop is to insert the ith element in its proper position with respect to the prior $i-1$ elements.

   a) The first part of the loop invariant says that the first j elements are properly ordered. When the loop is entered the first time with $j=i-1$, this says that the first $i-1$ elements are properly ordered. But this is precisely what the outer loop invariant says on entry into the outer loop. (Remember: As the outer loop is entered, i is increased by 1; thus in terms of this new i, the first $i-1$ elements were in order when the loop was entered.) Nothing between the beginning of the outer loop and entry into the inner loop changes any element of the list, so this condition must still be true at the first entry into the inner loop.

   b) The second part of the inner loop invariant says that the elements from $j+2$ to i are in order. At the first entry into the inner loop, this is vacuously true because there are no values of k between $j+2$ $(=i+1)$ and $i-1$.

   c) The third part of the inner loop invariant says that n_i is less than all values it has been compared with thus far. This condition is also vacuously true the first time the loop is entered.

4. Does the truth of the inner loop invariant on one entry into this loop imply its truth on the next entry? What happens in the loop (assuming that the condition for exit is not satisfied)? The value that was in the j position is moved to the $j+1$ position and j is reduced by 1. The first part of the inner loop invariant is still true because it now refers to one less element, as the new j is one less than the old j and these elements are unchanged.

   Is the second portion of the loop invariant now true for

$$j_{\text{new}} = j_{\text{old}} - 1?$$

   The only new value of k is

$$j_{\text{new}} + 2 = j_{\text{old}} + 1. \quad (1)$$

   So, is

$$n_k \le n_{k+1}$$

   for this value of k? Yes, because these two values are both from the $i-1$ elements of the list which were sorted on entry into the outer loop.

   For the third part of the loop invariant again the only value of k is $j_{\text{new}}+2 = j_{\text{old}}+1$. On entry to the loop $temp<n_{j_{\text{old}}}$ or the exit would have

been taken. Since $n_{j_{\text{old}}}$ becomes $n_{j_{\text{old}}+1}$ in the loop, on exit $temp < n_{j_{\text{old}}+1} = n_{j_{\text{new}}+2}$ which is what we want.

This is the only hard part of the proof of correctness. The rest is all downhill.

5. When the inner loop is exited, either $temp \geq n_j$ or $j = 0$. The loop termination condition is the conjunction of the exit condition with the loop invariant.
6. Now, how about the outer loop? If its loop invariant is true on one entry, is it true on the next entry? When the inner loop is exited elements 1 through j and $j+2$ through i are in order. Suppose that the inner loop was exited with $j \neq 0$. We need then only show that the elements j, $j+1$, and $j+2$ are in order. What the outer loop does, after the inner loop is exited, is to put *temp* (the old n_i) in the $j+1$ position. Now *temp* is greater than or equal to the element in the j position because that's what caused the exit from the inner loop. And it's less than the element in the $j+2$ position because, if it weren't, the inner loop would have been exited earlier. So the three elements are in order.

   Suppose that $j = 0$ on exit from the inner loop. We'll let you provide the appropriate argument [1].

   Therefore the outer loop invariant really is invariant from one pass through the loop to the next.
7. The outer loop termination condition is just the conjunction of the loop invariant and the exit condition from the **for** loop ($i = m$). Is the output specification true? Yes, because it simply is the result of replacing i by m in the outer loop termination condition.

This completes our verification of Algorithm INSERTION except that we haven't shown that it terminates. Does it always? We leave this to a problem [2].

How important was the use of the universal quantifier in the assertions for Algorithm INSERTION? Well, we could have done without it if we had been willing to use a lot of ellipsis, which would have made the assertions considerably less readable. For more complicated algorithms with more complex loop invariants, quantifier notation would be even more desirable.

The verification of INSERTION was not simple. But, if you understood it and—even better—understand how to construct proofs for your own algorithms, you will have developed an important ability. As with all informal mathematical proofs, you should convince yourself that our proof really is a proof.

We didn't use induction explicitly in the verification of INSERTION. But you may have seen or, at least, had a sneaking suspicion that verification of an algorithm with one loop nested inside another, where both loops are controlled by a counter, was likely to have involved some kind of induction. Indeed, the verification that each of the loop invariants really was invariant was really an induction proof in each case [4].

In the Epilogue we'll analyze insertion sort and discuss it in relation to other methods of sorting.

Problems: Section 7.7

1. Argue that the outer loop invariant in INSERTION is satisfied on reentry to the outer loop when the condition which causes exit from the inner loop was $j = 0$.

2. Give a proof that INSERTION always terminates if the input specification is satisfied. What would happen if the input specification is not satisfied?

3. When the list to be sorted by INSERTION has a length of 1, so that sorting it means doing nothing, the input specification is satisfied. But does the algorithm work properly? Why?

4. Restate as proofs by induction the proofs that the inner and outer loop invariants hold on initial entry to the loop and at each reentry.

5. Bubble sort. Here the idea is to compare successively adjacent items on a list and interchange them if they aren't in numerical order.
   a) Suppose that you start at the beginning of a list and proceed to the end, first comparing the elements in the first two positions, interchanging them if they are out of order, then comparing the elements in positions 2 and 3 and interchanging, if necessary, and so on. Show that, after comparing and, if necessary, interchanging the elements in the last two positions, the largest element will be at the end of the list.
   b) Then, going back to the beginning again and proceeding as before, show that the next-to-largest element ends up in the next-to-last position. You don't need to compare all successive pairs of elements this time. Where can you stop?
   c) Argue that by continuing this procedure, you will eventually get the list in sorted order.

6. a) Write Algorithm BUBBLE to implement the notion in [5].
   b) Show how your algorithm works on the data in Fig. 7.12.

7. a) How many passes are required (i.e., the number of times you went back to the start of the list) in Algorithm BUBBLE?
   b) If the list were initially ordered, how could you modify your algorithm to prevent unnecessary passes? Do it.
   c) Apply your modified algorithm to the data in Fig. 7.12.

8. In bubble sort, as described in [5], larger items are "bubbled" (if you visualize the list as going from top to bottom, really they sink) to the end of the list. Modify BUBBLE so that small items bubble to the beginning.

9. Your Algorithm BUBBLE should have two nested loops.
   a) Determine a loop invariant for each. *Hint*: The outer loop invariant must state not only which items are already sorted but also the relation of these items to the unsorted elements. The inner loop invariant only has to state the relationship between the last item on a sublist and the ones which precede it.
   b) For both loop invariants argue that they are true on entry to their respective loops and on each subsequent reentry.
   c) State appropriate loop termination conditions and use the outer loop termination condition to show that the appropriate output specification is satisfied.

10. We can do still better than the improvement suggested in [7]. Sometimes a pass results in putting more than one number in its correct final place. How could you discover this and thereby improve your algorithm? *Hint*: Each time you interchange two values, record the position of the interchange. Rewrite the modified algorithm from [7] to exploit your answer to the preceding question.

11. [Computer project] Convert INSERTION and your BUBBLE from [5], [7], or [10] to computer programs. Compare their running times for a variety of data to try to determine which executes most rapidly.

7.8 Wffs and Algorithms

Although algorithms have played an important role in this chapter—because of our emphasis on algorithm verification—we've only presented one algorithm (Algorithm TRUTHTABLE) related to mathematical logic itself. In this section we'll present another, to introduce you to some ideas which are important not just in mathematical logic but also in a variety of other applications (such as in writing compilers for programming languages).

Consider this question: If you were given a string of atomic propositions, connectives, and parentheses, how could you determine whether it is a wff in the propositional calculus? As it stands, this is a rather difficult question. To make it easier to answer, we'll add one constraint to the rules for forming wffs stated in Section 7.2. It is that any binary connective with its left and right operands *must* be enclosed in parentheses. Thus we do not allow

$$A \vee B \wedge (C \vee D),$$

but insist that it be written

$$((A \vee B) \wedge (C \vee D)).$$

Wffs written in this way are called **fully parenthesized expressions.**

The requirement that a wff be a fully parenthesized expression means that we must modify the definition of a wff given in Section 7.2 as follows:

1. An individual atomic proposition P, Q, R, ... is a wff.
2. **if** A is a wff, **then** $\neg A$ is a wff.
3. **if** A and B are wffs, **then** $(A \circ B)$ is a wff, where $\circ$ is any of the binary connectives we have defined.

According to this definition, the following strings are legal wffs:

$(((P \vee Q) \wedge R) \wedge (P \vee S))$	[Nothing restricts P from appearing more than once]
$((P \vee (Q \wedge \neg P)) \wedge R)$	[A $\neg$ may follow another connective directly]
$((P \vee Q) \vee \neg\neg R)$	[Any number of negations in a row may appear]
P	[An atom by itself is a legal wff]

However, the following strings are not legal wffs:

$((P \vee Q) \wedge R))$	[Too many right parentheses]
$(P \vee (Q \wedge \vee P))$	[Two consecutive binary connectives]
$P \vee (Q \wedge Q)$	[$(Q \wedge Q)$ is legal; the problem here is the missing outer parentheses]

Before reading further, think about how you would design an algorithm to take a string consisting of single letters representing atomic propositions, connectives, and parentheses (*no spaces allowed*) and determine whether this string is a wff. Because of the requirement that expressions be fully parenthesized, you can do this in a single *scan* across the string; that is, start at the first character and examine one character at a time until you either determine that the string cannot be a wff or you get to the end.

First, we define some conventions for the algorithm which follows:

- There is a function read(ch) which, each time it is invoked, reads the next symbol in the string, starting at the left, and assigns this symbol to the variable ch (for *ch*aracter).
- Logical expressions of the form

$$\begin{aligned} ch &= \text{letter} \\ ch &= \text{connective} \\ ch &= (\\ &\vdots \end{aligned}$$

  are true or false depending on the last symbol scanned (i.e., read).

Whenever a letter can appear in a string, it can be preceded by any number of negations. For convenience, therefore, we provide Procedure Notscan which, any time a $\neg$ is scanned, reads subsequent characters until something other than a $\neg$ is found.

At the start of Algorithm 7.4, WFF we need to determine whether the string begins

- with a letter or some negations followed by a letter (in which case there must be no other characters if it is to be a wff), or
- with a left parenthesis (.

In the latter case, the "(" signals the start of the wff that is the left operand of a connective. It should be followed by the connective itself and then by the wff that is the right operand. These operands can be atomic propositions (i.e., single letters) or they can be arbitrarily more complicated wffs. The procedure Seekwffpair looks for two wff operands separated by a connective and surrounded by parentheses.

Any fully parenthesized wff which is not an atom or some number of negations of an atom, whether part of a larger wff or not, must begin with a new left parenthesis. Thus it is natural (do you see why?) to use a recursive approach to look for correct operands. Each time we make a recursive call, we may succeed or fail in finding a legal part of a wff. When we fail we will set a variable—naturally called *wff*—to false so that, when we return from each level of recursion, we can see whether continuing at the next higher level is worthwhile.

Each of the two procedures, Nextchar and Seekwffpair, calls the other. We say, therefore, that they are **mutually recursive**.

With regard to Algorithm WFF you need to understand that, whenever a symbol appears which could not be correct (e.g., anything but a left parenthesis, a $\neg$, or a letter after a connective), *wff* is set to F. The recursion is then sent to

Algorithm 7.4

WFF

```
Input S                                                        [String of characters]
Algorithm WFF
   procedure Notscan                                         [Scan over a string of ¬s]
      repeat until ch ≠ ¬
         read(ch)
      endrepeat
      return
   endpro
   procedure Nextchar                      [Reads next character and decides
      read(ch)                                                   what to do next]
      if ch = ¬ then Notscan                                          [Scan over ¬s]
      if ch = ( then Seekwffpair                  [Call procedure to look for wffs]
         ch ≠ letter then wff ← F                [If ch not (, it must be letter
      endif                                                       to be legal wff]
      return
   endpro
   procedure Seekwffpair  [Determine whether string starting with (
                 is two wffs separated by a connective followed by a )]
      Nextchar                              [Start search for left operand; may
                                         result in recursive call to Seekwffpair]
      if wff = F then return                                [Left operand not valid]
      read(ch)                                           [Connective only legal ch]
      if ch ≠ connective then wff ← F; return
      Nextchar                             [Start search for right operand; may
                                         result in recursive call to Seekwffpair]
      if wff = F then return                               [Right operand not valid]
      read(ch)
      if ch≠) then wff ← F                                        [ ) only valid ch]
      return                                           [If wff still T, then success]
   endpro
   wff ← T                                         [Main algorithm; initialize wff]
   Nextchar                                                         [First symbol]
   if wff = T then                                             [String legal so far]
      read(ch)
      if ch ≠ blank then wff ← F                                [Blank only legal ch]
   endif
   if wff = T then print "String is a wff"
             else  print "String is not a wff"
   endif
Output "String is a wff"     or     "String is not a wff"
```

the next higher level which, because Procedure Seekwffpair always checks to see if it is worth continuing (i.e., whether *wff* is still true), will result in a return to the main algorithm. If *wff* is still true when the blank after the last symbol is encountered, then the entire string is a *wff*. You may be surprised that Algorithm WFF does not explicitly test whether or not there are the same number of left and right parentheses. Do you see why this isn't necessary [2]?

EXAMPLE 1 Apply Algorithm WFF to the strings

a) $(\neg\neg P \vee (Q \wedge \neg\neg(R \Longleftrightarrow S)))$.

b) $((P \wedge \neg\neg(Q \vee R))$.

Solution For a) we may describe the action of the algorithm as follows:

Symbol Read	Read by Procedure	Action Taken
(	Nextchar	Call Seekwffpair—level 1
$\neg$	Nextchar	Call Notscan
$\neg$	Notscan	Continue in Notscan
P	Notscan	Return to Nextchar and then to Seekwffpair (left operand found)
$\vee$	Seekwffpair	
(	Nextchar	Call Seekwffpair—level 2
Q	Nextchar	Return to Seekwffpair—left operand found
$\wedge$	Seekwffpair	
$\neg$	Nextchar	Call Notscan
$\neg$	Notscan	Continue in Notscan
(	Notscan	Return to Nextchar and then call Seekwffpair—level 3
R	Nextchar	Return to Seekwffpair—left operand found
$\Longleftrightarrow$	Seekwffpair	
S	Nextchar	Return to Seekwffpair
)	Seekwffpair	Right operand found—return to level 2
)	Seekwffpair	Right operand found—return to level 1
)	Seekwffpair	Right operand found—return to main algorithm and output that string is a wff

Similarly for b), but somewhat abbreviated:

Symbol(s) Read	Action Taken
(	Call Seekwffpair—level 1
(	Call Seekwffpair—level 2
$P\wedge\neg\neg($	Call Notscan, then Seekwffpair—level 3
$Q\vee R)$	Return to level 2
)	Return to level 1
blank	) paren expected, so set *wff* to F and return to main algorithm and output that string is not a wff

---

The string is $(\neg\neg P\vee(Q\wedge\neg\neg(R\Longleftrightarrow S)))$. Applying the rules in the text we get:

Root of tree:	$\vee$	[Rule 3]
Left operand: $\neg\neg P$ which give the subtree:		
Root:	$\neg\neg P$	[Rule 1]
Right operand: $Q\wedge\neg\neg(R\Longleftrightarrow S)$ which give the subtree:		
Root:	$\wedge$	[Rule 3]
Left child:	Q	[Rule 1]
Right subtree:		
Root:	$\neg$	[Rule 2]
Right child:	$\neg$	[Rule 2]
Right subtree:		
Root:	$\Longleftrightarrow$	[Rule 3]
Left child:	R	[Rule 1]
Right child:	S	[Rule 1]

which gives the tree:

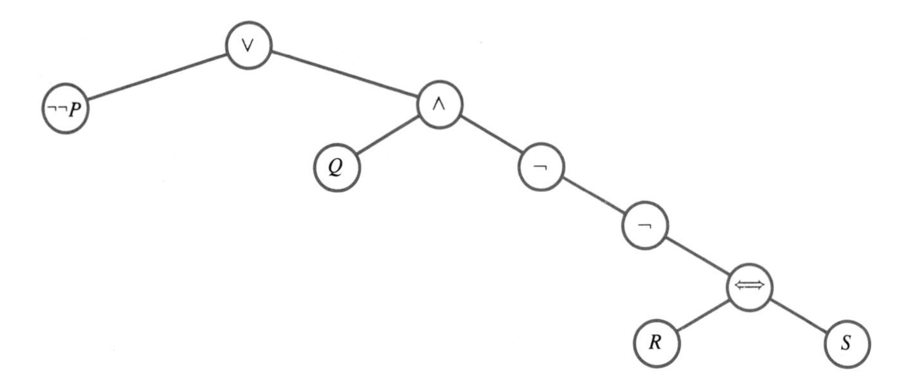

FIGURE 7.13

The Binary Tree Corresponding to the String in Example 1(a)

In [3] we ask you to trace this algorithm for several other strings. ∎

We can display any legal string, such as that in Example 1(a), as a binary tree as follows:

1. If the string has no binary connectives, the tree has only a root which contains the entire string.
2. If there are any leading (i.e., not enclosed in parentheses) negations, we place them, successively, at the root of the tree and at the roots of successive right subtrees until all are used up.
3. We place the main binary connective (the one enclosed in only a single pair of parentheses) if there is one, at the root of the entire tree or at the root of the next right subtree if there are leading negations.
4. We then consider the two operands of the main binary connective after stripping off the outer pair of parentheses; we then find the left and right subtrees of the node for the main binary connective by applying rules 1–3 recursively.

For the string in Example 1(a), this tree is displayed in Fig. 7.13. An inorder traversal of this tree (see Section 3.8) visits the nodes in the order of the symbols in the original string [4]. Note that the tree is a parenthesis-free representation of the string. This observation lies at the heart of an algorithm to determine whether a string is a wff when wffs are not required to be fully parenthesized.

Problems: Section 7.8

1. Use the rules for constructing fully parenthesized wffs to prove that they have as many left parentheses as right parentheses.

2. Explain how WFF checks indirectly whether a wff candidate has as many left parentheses as right parentheses. If a candidate didn't, but otherwise was OK, where would WFF set *wff* to False?

3. Trace WFF for each of the following strings.

a) $P \vee Q$. **b)** $(P \vee Q)$.
c) $(P \wedge (Q \Longrightarrow \neg R))$. **d)** $(P \vee Q \vee R)$.
e) $((P \wedge (Q \wedge (RS))) \Longrightarrow \neg T)$.
f) $(((P \Longleftrightarrow Q) \Longrightarrow R) \vee (\neg(S \wedge T) \vee P))$.

4. a) Show that an inorder traversal of the tree in Fig. 7.13 visits the symbols in the order in which they appear in the string of Example 1(a).

b) Construct the corresponding binary tree for the string of Example 1(b) after adding a ")" at the end of this string.

c) Show that an inorder traversal of the binary tree corresponding to any wff which is constructed using rules 1–4 will visit the symbols in their order in the wff. *Hint*: Use induction on the number of binary connectives.

5. We noted in the text that any wff in the propositional calculus may be represented as a binary tree. However, not all binary trees are trees for wffs. What additional condition must wff trees satisfy?

6. Draw the binary trees for the following wffs.

a) $(P \vee Q)$.
b) $((P \wedge Q) \Longrightarrow (R \Longrightarrow (S \vee \neg T))$.
c) $(((P \wedge Q) \wedge R) \wedge \neg\neg S)$.
d) $((P \wedge Q) \wedge (R \wedge P))$.
e) $\neg((P \vee Q) \Longrightarrow \neg(P \wedge \neg Q))$.

7. For each of the wffs in [6], compute the truth value

a) when every atom is True.
b) when every atom is False

Do you see how you could make use of your tree to evaluate a wff using any assignment of truth values to its atoms? (This gives a hint for an algorithm which checks a string for wff-hood and determines its truth value simultaneously.)

8. Suppose that the rules for fully parenthesized wffs were changed so that, if A is a wff, then $(\neg A)$ is a wff instead of $\neg A$. That is, negated statements must be enclosed in parentheses just as compound statements must be. Rewrite WFF to be correct for the new rules.

9. Suppose that the rules in the text for fully parenthesized wffs were changed so that, if A is a wff, then $\neg(A)$ is a wff instead of $\neg A$. That is, negated statements must be enclosed in parentheses before they are negated. Rewrite WFF to be correct for the new rules.

10. **a)** Do a postorder traversal of the tree in Fig. 7.13, writing the symbols left to right as you visit the nodes (with $\neg\neg P$ considered to be a single symbol).

b) Reassemble the postorder string into a fully parenthesized string as follows:

> Scan the string left to right, writing down each operand as you come to it (i.e., $\neg\neg P$, Q, R, etc.).
>
> Each time you come to a binary operator place it between the *two most recently written operands*, enclose the result in parentheses, and thereafter consider the result to be a new single operand.
>
> Each time you come to a $\neg$, place it before the most recently written operand.

You should obtain precisely the original fully parenthesized string. Explain why this happens. (See [40, Supplementary, Chapter 3]).

11. Repeat both parts of [10] with "postorder" replaced by "preorder" and "left to right" replaced by "right to left".

Supplementary Problems: Chapter 7

1. The five major logical connectives, $\wedge$, $\vee$, $\neg$, $\Longrightarrow$, and $\Longleftrightarrow$, are not independent. For instance, if somebody doesn't understand $\Longleftrightarrow$, you can explain that $P \Longleftrightarrow Q$ means $[(P \Longrightarrow Q) \wedge (Q \Longrightarrow P)]$.

a) Using only $\wedge$ and $\neg$, can you explain any of the other three connectives? Which ones? How?

b) Same as a), but start with $\vee$ and $\neg$.

c) Same as a), but start with $\Longrightarrow$ and $\neg$.

d) Show that, using $\Longleftrightarrow$ only, none of the other four connectives can be explained. *Hint*: First show that truth tables for expressions with P, Q and $\Longleftrightarrow$ must have 0, 2, or 4 T values.

Note: More problems concerning expressibility of compound statements in terms of a limited number of logical operators appear below.

2. Suppose you want to show that a whole series of statements $S_1, S_2, \ldots, S_n$ are equivalent, that is, that $S_i \Longleftrightarrow S_j$ is a tautology for each i and j. Typically, such a mutual equivalence is proved by proving a number of implications; e.g., $S_1 \Longrightarrow S_2$, $S_2 \Longrightarrow S_3$, and so on. For instance, once these two implications are proved, it isn't necessary to prove $S_1 \Longrightarrow S_3$ directly, so half of $S_1 \Longleftrightarrow S_3$ has been done. What is the least number of implications necessary to prove the mutual equivalence of all the S_i's? Prove that you are right. *Hint*: Use graph theory.

3. You are kidnapped by a gang of logic-fiends and placed in a room with two doors. The leader tells you that the red door leads to freedom and the green door to death but you can't see the colors. He also tells you that you may ask his assistant one Yes–No question and that his assistant understands logic perfectly, but either is always truthful or is a complete liar. Use the methods of this chapter to figure out a question to ask. *Hint*: There are only two variables—which door leads to freedom and whether the assistant lies or tells the truth. Let's say that you want an answer of Yes to your question to mean take the red door. For all four pairs of values for the variables, figure out what the truthful answer to your question would have to be. From this result you can construct the question.

4. Suppose you found a tribe whose language included a word for | (the Sheffer stroke—see Fig. 7.3), but which didn't have words for any of the other logical connectives.
   a) How could you explain "not" to these people?
   b) How could you explain "and"?
   c) How could you explain "or"?

   It follows that you can express every wff to these people; see [22], Section 7.5.

5. Suppose that you found another tribe whose language included a word for ↓ (Peirce's arrow—see Fig. 7.3), but which didn't have equivalents for any of the other logical connectives.
   a) How could you explain "not" to these people?
   b) How could you explain "and"?
   c) How could you explain "or"?

   It follows that you can express every wff to these people; see [22], Section 7.5.

6. Show how to construct circuits for all five standard logical operators ($\neg$, $\wedge$, $\vee$, $\Longrightarrow$, and $\Longleftrightarrow$) using NAND ($=|$) gates. *Hint*: See [1] and [4].

7. Show how to construct circuits for all five standard logical operators using NOR ($=\downarrow$) gates. *Hint*: See [5].

8. Why are most actual logic circuits constructed with just NAND and NOR gates rather than AND, OR, and NOT gates?

9. [13, Section 7.6], doesn't give a complete description of how to negate a formula in the predicate calculus. Why? Because "inside" may consist of several pieces joined by operators like $\wedge$, and the pieces themselves may have quantifiers. For instance, the formula might be

$$(\forall i)[((\exists j)P_j \;\wedge\; (\forall k)Q_{ik}) \Longrightarrow \neg(R_i \wedge \neg S)].$$

Come up with a recursive description of how to negate all predicate calculus formulas by changing the quantifiers. *Suggestion*: All valid formulas in the predicate calculus may be viewed as having a binary tree structure, with operators and quantifiers on the internal nodes and predicates and atoms on the leaves. For instance, the formula above can be drawn as

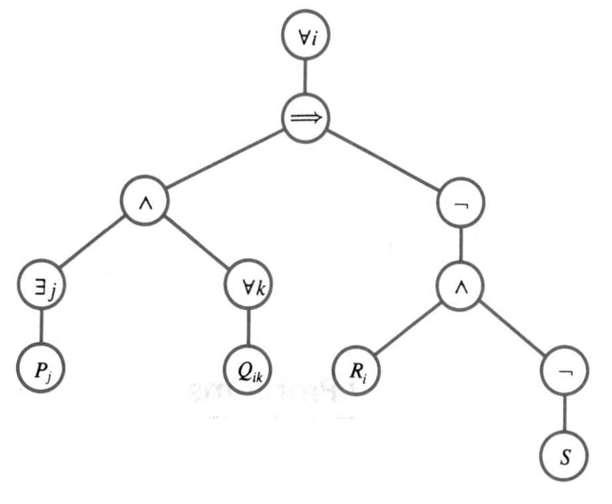

Figure out a rule for negation which explains what to do to the root and to the subtree hanging from the root. For the given formula, give both the negation tree and the negation formula. Describe in general terms how to negate any valid formula using its tree.

CHAPTER 8

Algorithmic Linear Algebra

8.1 Introduction

Whenever you consider sets of linear equations or sets of linear functions, and seek systematic ways of dealing with them, you are doing **linear algebra**. Thus we have already brushed against this subject several times—we just haven't tried to be systematic yet. Here are reminders of some of those earlier meetings with the subject.

EXAMPLE 1 In Section 0.6, we considered a set of variables y_1, y_2, y_3 defined in terms of x_1, x_2, and then another set of variables z_1, z_2 defined in terms of the y's, and then determined how to express the z's in terms of the x's. The y's were linear functions of the x's, and the z's were linear functions of the y's. We showed that the z's were also linear functions of the x's. ∎

EXAMPLE 2 In [5, Section 2.2], we used the method of undetermined coefficients for guessing the closed formula for $\sum_{k=1}^{n} k$. This method consisted of guessing that the answer was a quadratic $an^2 + bn + c$, where the coefficients a, b, c were unknown. Plugging in the easily computed value of the sum for $n = 1, 2, 3$ gives three *linear* equations in a, b, c (no powers of the variable n remain), which we solved ad hoc. ∎

EXAMPLE 3 In Section 5.5 you learned that to solve a homogeneous nth-order linear difference equation you first ignore the initial conditions and find the general solution. The general solution contains unknown constants—as many as there are initial conditions. Then you plug in the initial conditions and are left with n linear equations. You solve these—we did it then by ad hoc methods—to find the values of the constants. ∎

EXAMPLE 4 In Example 7, Section 5.2, we showed an example of linked difference equations. (Trusty Rent-A-Car with locations in NYC and LA.) That is, the problem statement itself involved a set of two linear equations, where the unknowns are terms in the two sequences. In this chapter we explain a method for solving such linked recurrences using matrices and "eigenvalues". Indeed, this matrix method can also be used to solve linear difference equations in one variable and many probability problems as well. ∎

Finally, if all this doesn't convince you that it is worth developing systematic methods for solving linear equations, surely all the time you spent on such equations in high school should be a motivation!

There are at least two aspects to linear algebra problems. First, one would like to have efficient, correct algorithms for solving the problems, regardless of how many variables are involved. Second, one would like to have general theorems that say what sort of problems have what sort of solutions. For instance, concerning Example 3, we stated in Chapter 5 that if the initial conditions for an nth order linear recurrence are n consecutive terms of the sequence, then there is always a unique solution, but if the initial conditions are not consecutive, there may be many solutions or none. Linear algebra is the theory within which such statements are proved.

Traditionally, linear algebra theory is presented abstractly, using nonconstructive methods, and the algorithms are presented separately. You should not be surprised to learn that it doesn't have to be this way and, in this book, it won't be. Specifically, the first goal of this chapter is to thoroughly acquaint you with **Gaussian elimination**. This is a very efficient and widely used general algorithm for solving systems of linear equations. We will use properties of Gaussian elimination to prove various key theorems as well.

Linear algebra is usually the subject of an entire course, often given in the second half of the sophomore year. Our coverage is not the same as in that course. However, we do cover the linear algebra we believe most students really need during their first two college years. Linear algebra *is* discrete mathematics, it's what we need to tie up a number of loose ends, and it's chock full of interesting algorithms and tractable, interesting efficiency questions. When you need more linear algebra, you can take an upper level course.

We have talked about linear algebra as though its full purpose is to provide a more general and theoretical context for solving linear equations. Solving equations is the fundamental issue with which linear algebra starts, all right, but there's lots more. In fact, linear algebra is about any mathematical structure in which there is a reasonable concept of addition and "scalar multiplication" (multiplication by numbers, in this context called **scalars**). When variables are added and multiplied by scalars, you get your familiar linear equations, but we've already shown that sequences can be added and multiplied by scalars too. So can arbitrary functions. So can points in space (if you think of them in terms of their coordinates). The list goes on and on. Linear algebra can be studied as a theory which abstracts out the common elements in all such situations. We will lead you up to, but not really into, abstract linear algebra.

To outline this chapter, the next two sections develop Gaussian elimination, first in Section 8.2 for systems where there are as many unknowns as equations, then in Section 8.3 for more general systems. In Section 8.4, we take a deeper look at the algorithm, deducing some general consequences about solving equations. Section 8.5 explains some refinements of the algorithm, related to the so-called *LU* factorization, that are used in efficient implementations on computers. In Section 8.6, we begin to develop the algebraic theory, by tying Gaussian elimination to the matrix algebra introduced in Section 0.6. Section 8.7 introduces the fundamental concept of "vector space" and gives examples arising from linear equations and matrices. Section 8.8 presents the basic theory of vector spaces. In Section 8.9, we introduce the second half of linear algebra, the theory of "eigenvalues" and "eigenvectors". This theory has many applications to iterative processes, as indicated by Example 4 above. We devote Section 8.10 to one particularly important application of eigenvalue theory, "Markov chains".

8.2 Gaussian Elimination: Square Systems

All right, we are going to reinvent Gaussian elimination—the standard method for solving systems of linear equations. The complete algorithm needs a lot of bells and whistles, so in this section we concentrate on the most familiar case: when there are just as many equations as unknowns. Past experience should tell you that, generally, such a system has a unique solution.

By an $m \times n$ **system** of linear equations, we mean a set (system) of m equations involving n unknowns. If $m = n$, we typically refer to an $n \times n$ or **square system**. For instance, a typical 2×2 system is

$$\begin{aligned} 2x + 3y &= 7 \\ 3x - 2y &= 4. \end{aligned}$$

Armed with the recursive paradigm, we can jump in right away with $n \times n$ systems:

$$\begin{array}{ccccccccc} a_{11}x_1 & + & a_{12}x_2 & + & \cdots & + & a_{1n}x_n & = & b_1 \\ a_{21}x_1 & + & a_{22}x_2 & + & \cdots & + & a_{2n}x_n & = & b_2 \\ \vdots & & \vdots & & & & \vdots & & \vdots \\ a_{n1}x_1 & + & a_{n2}x_2 & + & \cdots & + & a_{nn}x_n & = & b_n. \end{array} \tag{1}$$

How can we solve such a system? Well, suppose we already knew how to solve $(n-1) \times (n-1)$ systems. Then to reduce to the previous case we have to eliminate one variable. There is a standard way to do this: Subtract the first equation enough times from the others to eliminate x_1 from them. (Alternatively, use the first equation to solve for x_1 in terms of $b, x_2, x_3, \ldots$ and then substitute the result in the other equations.) We get something of the form

$$
\begin{aligned}
a_{11}x_1 + a_{12}x_2 + a_{13}x_3 + \cdots + a_{1n}x_n &= b_1 \\
a'_{22}x_2 + a'_{23}x_3 + \cdots + a'_{2n}x_n &= b'_2 \\
a'_{32}x_2 + a'_{33}x_3 + \cdots + a'_{3n}x_n &= b'_3 \\
\vdots \qquad \vdots \qquad\quad \vdots \quad &\quad \vdots \\
a'_{n2}x_2 + a'_{n3}x_3 + \cdots + a'_{nn}x_n &= b'_n.
\end{aligned}
\tag{2}
$$

Now by assumption we solve the $(n-1)\times(n-1)$ subsystem, leaving us with:

$$
\begin{aligned}
a_{11}x_1 + a_{12}x_2 + a_{13}x_3 + \cdots + a_{1n}x_n &= b_1 \\
x_2 \qquad\qquad\qquad\qquad &= c_2 \\
x_3 \qquad\qquad\quad &= c_3 \\
\ddots \qquad\quad &\quad \vdots \\
x_n &= c_n.
\end{aligned}
$$

Finally, we plug the solutions for x_2 through x_n into the top equation and solve it for x_1: Subtract all the terms $a_{1i}c_i$ to the other side and divide by a_{11}. This method is the basic form of Gaussian elimination.

Plugging-in and subtracting terms can be done as a single step by subtracting *equations.* That is, subtract $x_2 = c_2$ from the top equation a_{12} times, subtract $x_3 = c_3$ from the top a_{13} times, and so on. This equation subtracting approach is the convenient way to proceed when we get to the format of Example 2 later. We already mentioned the equivalence of equation subtracting and substitution when we went from system (1) to system (2).

EXAMPLE 1 Solve the system

$$
\begin{aligned}
2x - 2y + 4z &= -6 \\
2x - \ \ y + 6z &= -6 \\
-x + 2y - 3z &= \ \ 6.
\end{aligned}
$$

Solution To begin, subtract the first row from the second:

$$
\begin{aligned}
2x - 2y + 4z &= -6 \\
y + 2z &= \ \ 0 \\
-x + 2y - 3z &= \ \ 6.
\end{aligned}
$$

Now multiply the first row by $1/2$ and add the result to the third row. Actually, it's better to say: Multiply the first row by $-1/2$ and *subtract* the result from the third—we will eventually want to write the method in our algorithmic language, so it's best if we always use subtraction instead of switching between subtraction and addition. Anyway, we get

$$
\begin{aligned}
2x - 2y + 4z &= -6 \\
y + 2z &= \ \ 0 \\
y - \ \ z &= \ \ 3.
\end{aligned}
$$

Now we are down to the form of display (2) above. By recursion, we invoke the procedure all over again, applying it to the 2×2 system in the bottom rows. We eliminate y from the lower equations of that subsystem (all 1 of them) by subtracting:

$$\begin{aligned} 2x - 2y + 4z &= -6 \\ y + 2z &= 0 \\ -3z &= 3. \end{aligned}$$

We have reduced to the previous case again, which is now a 1×1 system. Since this subsystem has only one row, there are no further subcalls. It is time to solve for the last variable and work our way back up. From here on, therefore, the process is known as **back-substitution**. (In some books Gaussian elimination is considered to end at this point, and back-substitution is a separate process. We consider back-substitution to be a subprocedure of Gaussian elimination.)

Specifically, we begin the back-substitution by dividing the last row by -3:

$$\begin{aligned} 2x - 2y + 4z &= -6 \\ y + 2z &= 0 \\ z &= -1. \end{aligned}$$

Now we return to the second call of the solve procedure. To substitute the solution of the bottom 1×1 system into the 2×2 system, subtract the third equation twice from the second:

$$\begin{aligned} 2x - 2y + 4z &= -6 \\ y &= 2 \\ z &= -1. \end{aligned}$$

At this point we are finished with the 2×2 subsystem; the coefficient of y is already 1. So we return to the main call on the whole system. We subtract the second row -2 times from the first and subtract the third row 4 times:

$$\begin{aligned} 2x &= 2 \\ y &= 2 \\ z &= -1. \end{aligned}$$

Finally, we divide the top equation by 2 and we are done:

$$\begin{aligned} x &= 1 \\ y &= 2 \\ z &= -1. \ \blacksquare \end{aligned}$$

Did you notice that the actual letters x, y, z were just dead weights? All they did is act as place holders. The same can be said of the equals signs. Only the numbers changed. This suggests a format improvement: Forget the variables and equalities and write all the numbers in matrix form. This format holds their places just as well and takes less writing. It is also better for computers, since they know all about matrix arrays.

For our 3×3 system, the initial matrix is 3×4:

$$\left[\begin{array}{rrr|r} 2 & -2 & 4 & -6 \\ 2 & -1 & 6 & -6 \\ -1 & 2 & -3 & 6 \end{array}\right].$$

The leftmost three columns are the coefficients of the variables. The last (rightmost) column consists of the constants on the right-hand side of the equations. It is traditionally marked off by a vertical line. A matrix such as this is called an **augmented** matrix.

EXAMPLE 2 Carry through the work of Example 1 again but in matrix format.

Solution First we need to eliminate x from all but the first equation. In matrix terms, this means we need to use the top left entry to drive all the entries directly below it to 0. Such an entry is called a **pivot** and is marked by an asterisk:

$$\left[\begin{array}{rrr|r} 2^* & -2 & 4 & -6 \\ 2 & -1 & 6 & -6 \\ -1 & 2 & -3 & 6 \end{array}\right].$$

Notice that we repeated the initial matrix, for the usual convention is to star the pivot for the operation you are *about to do* rather than the operation you *just did.*

The process of driving all the entries below a pivot to 0 is called a **downsweep**. In equation terms, we subtract one equation the right number of times from each equation below it. In matrix terms, we subtract the entries of one row the right number of times from the entries of the lower rows, column by column. In this example we subtract the first row 1 time from the second row and $-1/2$ times from row 3. The result is

$$\left[\begin{array}{rrr|r} 2 & -2 & 4 & -6 \\ 0 & 1^* & 2 & 0 \\ 0 & 1 & -1 & 3 \end{array}\right].$$

Notice that where we left blanks in the equations, we now write a 0. This is commonly done, but a blank would be OK too.

We now proceed with the work on the lower 2×2 submatrix. The asterisk in the previous display indicates the pivot for the next step. After downsweeping with it we have

$$\left[\begin{array}{rrr|r} 2 & -2 & 4 & -6 \\ 0 & 1 & 2 & 0 \\ 0 & 0 & -3^* & 3 \end{array}\right].$$

The remaining 1×1 matrix needs no downsweep so we are ready to begin the back-substitution. (*Note*: Many books revert to equations at this point and continue as in Example 1, but we continue with matrices. We will also do the back-substitution in a different order than in Example 1; see [13]. We will discuss the relative merits of the matrix and nonmatrix approaches after we present Gaussian elimination in algorithmic language.) The matrix analog of solving for z is to make the nonzero entry on the left in the third row equal to 1. Therefore we **scale** that row by dividing by -3:

$$\left[\begin{array}{ccc|c} 2 & -2 & 4 & -6 \\ 0 & 1 & 2 & 0 \\ 0 & 0 & 1^* & -1 \end{array}\right].$$

We now subtract the bottom row just the right number of times (determined by the starred pivot) from all the rows above it to zero-out all the other entries in the third column. This process is called an **upsweep**. In this case we subtract the bottom row 2 times from the second row and 4 times from row 1.

$$\left[\begin{array}{ccc|c} 2 & -2 & 0 & -2 \\ 0 & 1^* & 0 & 2 \\ 0 & 0 & 1 & -1 \end{array}\right].$$

The second row is already scaled, so next we subtract that row enough times from all the rows above it to zero-out all the other entries in the second column. There is only one row above it, so we subtract -2 times and get

$$\left[\begin{array}{ccc|c} 2^* & 0 & 0 & 2 \\ 0 & 1 & 0 & 2 \\ 0 & 0 & 1 & -1 \end{array}\right].$$

Finally, we scale the first row:

$$\left[\begin{array}{ccc|c} 1 & 0 & 0 & 1 \\ 0 & 1 & 0 & 2 \\ 0 & 0 & 1 & -1 \end{array}\right].$$

We're done: This matrix corresponds to the final display in Example 1: $x = 1$, $y = 2$, and $z = -1$. Notice that the square matrix on the left is just an identity matrix. See Section 0.6. ∎

To carry out Gaussian elimination in matrix form above, we repeatedly applied just two **row operations**: Subtracting a multiple of one row from another and scaling a row by a nonzero constant.

We mean to illustrate a general algorithm for square systems, but one of the requirements for an algorithm is that it work for all allowed inputs. Does this algorithm meet that requirement? Well, suppose that we started with

$$\left[\begin{array}{ccc|c} 1^* & 2 & 3 & 1 \\ 4 & 5 & 6 & 2 \\ 7 & 8 & 9 & 1 \end{array}\right]. \tag{3}$$

Then after one downsweep we have

$$\left[\begin{array}{ccc|c} 1 & 2 & 3 & 1 \\ 0 & -3^* & -6 & -2 \\ 0 & -6 & -12 & -6 \end{array}\right]$$

and after the second

$$\left[\begin{array}{ccc|c} 1 & 2 & 3 & 1 \\ 0 & -3 & -6 & -2 \\ 0 & 0 & 0 & -2 \end{array}\right].$$

Now it's impossible to proceed: The entry in the (3,3) position is 0 so we can't scale by it. However, do you see that we don't have to proceed? At this point we know there is *no* solution, because the third line says that $0x + 0y + 0z = -2$, that is, that $0 = -2$. Since no triple (x, y, z) can make this possible, there are no solutions.

A complete algorithm would have to identify this sort of case, but it is not the only special case. Suppose we changed the input of matrix (3) in the (3,4) entry only:

$$\left[\begin{array}{ccc|c} 1^* & 2 & 3 & 1 \\ 4 & 5 & 6 & 2 \\ 7 & 8 & 9 & 3 \end{array}\right].$$

Then after two downsweeps the result is

$$\left[\begin{array}{ccc|c} 1 & 2 & 3 & 1 \\ 0 & -3 & -6 & -2 \\ 0 & 0 & 0 & 0 \end{array}\right].$$

Now there is no longer a contradiction: Any triple (x, y, z) satisfies $0 = 0$. It is as if there weren't any third equation at all. We are left with two equations in three unknowns, the sort of situation we said we would discuss in the next section. However, we hope your experience tells you that this is an **underdetermined** system and thus is likely to have many solutions.

We are still not done with the list of possible special cases. Fortunately, they all have one thing in common. The only way we will be interrupted from reducing the original matrix to an identity matrix is if some entry we wish to pivot on is 0. In the two preceding cases, such a 0 appeared at the very last step of downsweeping, but it could appear much earlier. For example, if the initial augmented matrix is

$$\left[\begin{array}{ccc|c} 0 & 1 & 2 & 3 \\ 4 & 5 & 6 & 7 \\ 4 & 3 & 2 & 1 \end{array}\right],$$

then we are interrupted by a 0 right at the start.

Think about how you might handle situations like this. In the next section we will tell you what is done. For now—because we have more important conceptual matters to deal with—we are content to state a version of the algorithm which works only when no 0's appear in pivot positions. Note: although our algorithm forces many entries to 0, it never intentionally forces a 0 in a position where we later want to pivot.

Our algorithm is Algorithm 8.1, GAUSS-SQUARE. The input is the augmented matrix. Henceforth this augmented matrix is called $\hat{A}$, the square submatrix of *coefficients* is called A, and the rightmost column of *constants* is called **b**. In symbols, $\hat{A} = [A|\mathbf{b}]$. Calling the whole matrix $\hat{A}$ allows us to use the letter a for entries in either part. For instance, $a_{i,n+1} = b_i$.

A few comments are in order. We set up three procedures because they correspond to the three different types of steps we had already identified and because there are variants of Gaussian elimination in which the order and nature of these parts get changed. For instance, Scale(i) is so short that it is hardly worth setting up as a separate procedure, except that scaling is a bit more involved in some other variants.

In Downsweep we use an inner loop to subtract one row from another, term by term, except that we assign a_{ki} to 0 instead of subtracting to make it 0; this saves one computation. In Scale and Upsweep an inner loop is unnecessary because there are only two nonzero entries to worry about—in the pivot column and in the rightmost column. Furthermore, we choose the scaling and subtracting factors to make the entry in the pivot column simple (1 when scaling and 0 when upsweeping), so here too we make that change by an assignment.

Finally, we mentioned earlier that many books stop using matrices once the final pivot is found. In other words, were they to use our algorithmic language, their version of GAUSS-SQUARE would contain our Downsweep but would replace Scale and Upsweep with a procedure which worked on the unknowns directly rather than on the matrix. You are asked to write such an algorithm in the problems [22].

One final point about the nature of the algorithm. So far, every number in our examples has been an integer. The reason is that we picked our numbers very carefully. The fact is, Gaussian elimination does not depend in any way on the form of the numbers—they can be integers, rationals, reals or even complex numbers [14]. The only operations performed are $+$, $-$, $\times$, and $\div$, which make sense for all these types of numbers. To be sure, every example in this chapter will involve only integers and modest fractions, but we arranged that only to keep ugly numbers from getting in the way of the ideas. In actual practice (on computers), the coefficients of linear equations are usually read in and out in decimal form.

Algorithm Analysis

Let us analyze the running time of GAUSS-SQUARE. We will assume that we somehow pick matrices which don't lead to 0's for pivots. (Although it is impossible just to look at a large square matrix and tell whether it leads to 0's for pivots, if you pick the matrix entries at random, a matrix almost never results in 0's.) We will count each numerical multiplication or division as one step and ignore assignments, additions and subtractions. As usual, these last three items do not affect the order of magnitude of the run time [20].

Let us compute the work for each call of the three procedures in the algorithm separately. In fact, we will compute the work done on the **b** column separately from

Algorithm 8.1
Gauss-Square

```
Input Â              [n×(n+1) augmented matrix [A|b], with entries a_ij]
Algorithm GAUSS-SQUARE                          [Works if no 0 pivots]
    procedure Downsweep(i)
                              [Subtract multiples of row i from lower rows]
        for k = i+1 to n                    [k varies over rows below i]
            m_k ← a_ki/a_ii
            a_ki ← 0              [Assigning saves a step over computing]
            for l = i+1 to n+1          [l varies over entries in row k]
                a_kl ← a_kl − m_k a_il
            endfor
        endfor
    endpro
    procedure Scale(i)
        a_i,n+1 ← a_i,n+1/a_ii
        a_ii ← 1
    endpro
    procedure Upsweep(i)
                             [Subtract multiples of row i from higher rows]
        for k = i − 1 downto 1
            a_k,n+1 ← a_k,n+1 − a_ki a_i,n+1
            a_ki ← 0
        endfor
    endpro
    for i = 1 to n                                     [Main algorithm]
        Downsweep(i)
    endfor
    for i = n downto 1
        Scale(i)
        Upsweep(i)
    endfor
Output Â                                     [Last column is solution]
```

the work on the coefficient matrix A. Separating the analysis this way will help in Section 8.5. Then we sum over all the calls and finally sum over all the procedures. It may be helpful to look at the summary in Table 8.1 as you read along.

Scale(i): For each call of this procedure, the only change to the coefficient matrix is an assignment, so there are zero steps. For the **b** column there is one division, so one step.

Upsweep(i): In the coefficient matrix all the entries above the pivot are reset to 0 and once again this takes zero steps. In the last column, each entry in rows $i-1$ to 1 is recomputed using a formula involving one multiplication. Therefore, there are $i-1$ steps.

Downsweep(i): First, m_k is computed by a division for every row below the pivot row. This takes $n-i$ steps. Then, every entry simultaneously below and to the right of the pivot entry is recomputed using a formula involving one multiplication. There are $n-i$ rows so affected, and $n-i$ entries affected in each such row (not counting the entries in the **b** column yet), for a total of $(n-i)^2$ steps. Adding everything so far, we get $(n-i)^2+(n-i)$ steps. As for the **b** column, each entry in the last $n-i$ rows gets recomputed by the same formula, so there are $n-i$ steps.

TABLE 8.1
Step Count of Algorithm GAUSS-SQUARE

	Coefficient Matrix A	Constant Column **b**
Scale(i)	0	1
All Scaling	0	n
Upsweep(i)	0	$i-1$
All Upsweeping	0	$\frac{n^2-n}{2}$
Downsweep(i)	$(n-i)^2+(n-i)$	$n-i$
All Downsweeping	$\frac{n^3-n}{3}$	$\frac{n^2-n}{2}$
Total	$\frac{n^3-n}{3}$	n^2
Combined Total	$\frac{n^3+3n^2-n}{3}$	

Now, each of these procedures is iterated between $i=1$ and n. (Downsweep could just as well end at $n-1$, since the nth downsweep is vacuous.) For Scale the result is easy: For A and **b** the sums are 0 and n, respectively. For Upsweep the sum of $i-1$ for i from 1 to n is $\sum_{s=0}^{n-1} s = (n-1)n/2$. (We made the substitution

$s = i-1$ so that we could use the formula for $\sum s$ directly.) Thus the total amount of work for Upsweep is

$$0 \text{ for } A \qquad \text{and} \qquad \frac{n^2-n}{2} \text{ for } \mathbf{b}.$$

Finally, for Downsweep it is simplest to make the substitution $t = (n-i)$. Thus t ranges from 0 to $n-1$ and the work for $\mathbf{b}$ is $\sum_{t=0}^{n-1} t = (n-1)n/2$. The work for A is $\sum_{t=0}^{n-1}(t^2+t)$. This quantity can be computed by several methods you learned in Chapter 5, or simply by the formulas for $\sum t^2$ and $\sum t$ proved by induction in Chapter 2. (For $\sum t^2$ see [2, Section 2.4].) In any event, the resulting closed form is

$$\frac{n^3-n}{3}.$$

Now let's sum over all three procedures. For the A part, there is only one nonzero summand, the display just above. For the $\mathbf{b}$ part the sum leads to a pleasant simplification:

$$n + \frac{(n-1)n}{2} + \frac{(n-1)n}{2} = n^2.$$

Thus the total operation count for the augmented matrix $\hat{A}$ is

$$\frac{n^3+3n^2-n}{3}.$$

It's rarely important to remember the exact formulas for operation counts of algorithms. What is important to remember is the highest order term (see Section 9.3). In this case that's $n^3/3$. It also turns out to be important to remember the work for the $\mathbf{b}$ column: that's n^2.

Finally, is there anything to be gained in efficiency on a computer if we replace Scale and Upsweep by a back-substitution algorithm? Since we have left writing the latter to you, we can't answer this by pointing to specific lines and doing a count. But assuming you write Algorithm BACK-SUB along the lines we intend, we can say the following. There is *no* saving in multiplications and divisions, because the only work BACK-SUB eliminates is some assignments of 1's and 0's. Nonetheless, there is a saving. Although assignments themselves take very little computer time, *finding* the matrix location in which to put the assignments does take some time. That is, evaluating double indices (as in a_{ki}) and figuring out where in memory this refers to, is not free. By not making these 0–1 assignments, and in one other way, the Algorithm BACK-SUB we have in mind does less work with double indices [23].

In summary, given the conventions we follow throughout this book about what to count in algorithm analysis, there is no difference between our Algorithm GAUSS-SQUARE and one with back-substitution. However, on actual computers, there is some difference. Consequently, all efficient real-world computer codes use back-substitution. We will too, for this and several other reasons, when we get to our final version of Gaussian elimination in Section 8.5.

Validity

The rest of this section deals with why Gaussian elimination works. We will continue to assume that we somehow avoid the special problems with 0's in pivot positions. Our concern now is with the logic behind the procedure: Why is the solution to the last matrix also the solution to the first?

To answer this question, we need to be absolutely clear about what it means to "solve" an equation or system of equations. It means to come up with a final system whose solutions are obvious (e.g., $x = 2$, or $x = 2$, $y = 3$) and which has exactly the same solutions as the original system. That is, a set of values for the variables should satisfy the original system *if and only if* it satisfies the obvious system.

It is a theorem (proved below) that if one matrix is obtained from another by row operations, then the systems of equations they represent are *equivalent*: Each system has the same solutions. No solutions are ever lost and no "extraneous" solutions are introduced.

It is important to understand that this assertion is not a triviality. It is *not* true that every type of manipulation you have used to solve algebra problems preserves solutions. For instance, to try to solve

$$\sqrt{x} = -2, \tag{4}$$

you might square both sides to obtain

$$x = 4, \tag{5}$$

but Eq. (4) has no solutions and Eq. (5) has one. All that we have shown (and all that most computations show) is that *if* a set of values for the variables satisfies the first equation *then* it satisfies the last; but we want *if and only if*. In other words, sequential computations show that the solutions to the first equation are a *sub*set of the solutions to the last; we need to test whether the two solution sets are equal by demonstrating containment the other way.

One common test is to plug the solutions to the final equation (or system) back into the original. For instance, if we plug the only solution to Eq. (5) back into Eq. (4), we see right away that it doesn't work. Another general method is to see if all the steps in the computation reverse. Above, we went from Eq. (4) to Eq. (5) by squaring, but this operation does not reverse completely. Indeed, the left-hand side of Eq. (4) is the *positive* square root of the left hand side of Eq. (5), but the right-hand side of Eq. (4) is the *negative* square root of the right-hand side of Eq. (5). Thus Eq. (4) does not follow from Eq. (5).

The pleasant thing about row reduction is that all operations do reverse.

Theorem 1. Let $\mathcal{S}$ be a system of linear equations, and let $\mathcal{T}$ be any other system obtained from it by Gaussian elimination steps. ($\mathcal{T}$ need not be the result of a complete Gaussian elimination; in fact, the steps need not be carried out in the specific order of the Gaussian elimination algorithm.) Then $\mathcal{S}$ and $\mathcal{T}$ have the same solutions.

PROOF It suffices to show that every type of computation carried out in Gaussian elimination reverses. Every operation is a row operation, and there are two types:

1. Dividing a row i by the nonzero coefficient a of one of its terms.
2. Subtracting a row i some number of times m from another row j.

In the first type, all rows but i are left alone. Thus 1) is reversed by multiplying the new row i by a. In the second type, all rows but j are left alone. 2) is reversed by doing 2) again, but this time subtracting row i from row j a total of $-m$ times.

There is yet another row operation allowed in Gaussian elimination: switching the position of rows i and j. (See Section 8.3.) But since this does not change the *set* of equations, it has no effect on the set of solutions, so we don't even have to verify that it reverses. Of course it does reverse: Just switch the rows back. ∎

Problems: Section 8.2

1. Solve the system

$$\begin{aligned} x + 2y &= 3 \\ 2x + 3y &= 4. \end{aligned}$$

 a) by the method of Example 1,
 b) by the matrix method of Example 2.

2. Solve the system

$$\begin{aligned} 2x + 3y &= 4 \\ x + 2y &= 3. \end{aligned}$$

 a) by the method of Example 1,
 b) by the matrix method of Example 2.

 Note that this is the same *set* of equations as in [1], so the solution will be the same. However, the work will be different.

3. Solve the system

$$\begin{aligned} x + 2y + 3z &= 1 \\ 2x + 3y + 4z &= 2 \\ -x + y + 2z &= 1. \end{aligned}$$

 a) as in Example 1; **b)** as in Example 2.

4. Generalized Motors produces one type of car and one type of truck. Each car requires 500 kg of steel and 250 kg of aluminum. Each truck requires 1500 kg of steel and 400 kg of aluminum.

 a) If the company wants to produce 20 cars and 10 trucks, how much steel and aluminum must it purchase?

 b) If the company has on hand 15,500 kg of steel and 5,300 kg of aluminum, how many cars and trucks can it produce if all the steel and aluminum is to be used?

5. Verify that there is no solution for the system

$$\left[\begin{array}{ccc|c} 1 & 0 & -1 & 1 \\ 2 & 1 & 0 & 1 \\ 3 & 2 & 1 & a \end{array}\right],$$

 if $a = 2$. For what value of a does this system have solutions?

6. Assume that the following matrix shows the number of grams of nutrients per ounce of food indicated:

$$\begin{array}{l} \\ \text{protein} \\ \text{fat} \\ \text{carbohydrates} \end{array} \begin{array}{c} \begin{array}{ccc} \text{meat} & \text{potato} & \text{cabbage} \end{array} \\ \begin{bmatrix} 20 & 5 & 1 \\ 30 & 3 & 1 \\ 15 & 20 & 5 \end{bmatrix} \end{array}.$$

 The Army desires to use these delectable foods to feed new recruits a dinner providing 305 grams of protein, 365 grams of fat, and 575 grams of carbohydrates. How much of each food should be prepared for each recruit?

7. Trusty Rent-A-Car (Example 6, Section 5.2) has 2000 cars and two locations, NYC and LA. In any given month, half the cars rented in NYC are driven to LA and 1/3 of those rented in LA are driven to NYC. The rest are returned to the city

where they are rented. Assume that Trusty is in a steady state—the number of cars in each location remains the same from month to month. Use the methods of this section to determine the number of cars in each city.

8. Solve the system given by the following augmented matrix. We highly recommend programming Algorithm GAUSS-SQUARE and using a computer.

$$\left[\begin{array}{cccc|c} 2 & 3 & 4 & 5 & 6 \\ 1 & 2 & 3 & 4 & 5 \\ 4 & 2 & 5 & 3 & 1 \\ 5 & 3 & 1 & 4 & 2 \end{array}\right].$$

9. Use the method of undetermined coefficients to rederive the formula for $\sum_{k=0}^{n} 3^n$ as follows. Based on intuition or the general theory of Section 5.7, we guess that the answer is of the form $A3^n + B$, where A and B are undetermined constants. So evaluate $\sum_{k=0}^{n} 3^n$ by hand for the two simplest cases, $n = 0, 1$, and solve for A and B. Reconcile the form of your answer with the form you get by using the general formula for the sum of a geometric series.

10. Use the method of undetermined coefficients to rederive the formula for $\sum_{k=0}^{n} r^n$, where r is an arbitrary constant. Solve for A and B as in [9], but now the solution will be in terms of r. Is there any value of the constant r for which the two linear equations in A and B cannot be solved? Why?

11. Based on the theory of Section 5.7, there ought to be a formula of the form

$$\sum_{k=0}^{n} k2^k = An2^n + B2^n + C.$$

Find A, B, and C by the method of undetermined coefficients.

12. There is a unique cubic polynomial $P(x)$ such that

$$P(-1) = -1, \quad P(0) = 3,$$
$$P(1) = 5, \quad \text{and} \quad P(2) = 11.$$

Find it by setting up and solving a system of linear equations. (The polynomial is said to **interpolate** the data. There are several methods for finding interpolating polynomials, and the method in this problem is far from the easiest when the number of data is large.)

13. When we came to the upsweep in Example 2, we did not actually do the subtractions in the same order as in Example 1.

a) Indicate the order in which entries would have been zeroed out had we followed the exact same order. Explain in general terms which apply to $n \times n$ problems as well as this 3×3.

b) Explain why there is no harm in changing the order.

c) Why might the order we showed in Example 2 be better for people or for computers? In any event, it *is* the standard order.

14. a) Suppose that the initial augmented matrix for a linear system contains integers only. Explain why the solution will consist, at worst, of rational numbers.

b) What, at worst, will the solution consist of if the initial matrix entries are rational numbers?

c) What, at worst, will happen if the initial matrix entries are real numbers?

15. In Algorithm GAUSS-SQUARE we treated the square coefficient matrix together with the constant column as a single $n \times (n+1)$ matrix $\hat{A}$, with all its entries referred to in the form a_{ij}. Although this approach made the algorithm easier to write and read, conceptually the constants play a different role than the coefficients, and in some variants of the algorithm they are manipulated separately. To get used to this separate treatment, rewrite the algorithm with input $[A|\mathbf{b}]$, where A is $n \times n$ and its entries are called a_{ij}, and $\mathbf{b}$ is $n \times 1$ and its entries are called b_i.

16. In procedure Downsweep, why doesn't the inner loop extend from $l = 1$ to $n + 1$? Isn't the whole row i subtracted from row k?

17. What goes wrong in GAUSS-SQUARE if a pivot entry is 0? What lines would provoke error messages?

18. In procedure Scale(i) we changed the rightmost entry first, but by hand it is more natural to scale from left to right. Can we switch the order of the two lines in the body of the procedure?

19. In Downsweep, we could eliminate the line in which m_k is defined if we rewrote the instruction three lines below it as $a_{kl} \leftarrow a_{kl} - a_{ki}a_{il}/a_{ii}$. Would this make the algorithm longer or shorter to run? By how many steps?

20. In this problem we consider the number of assignments, additions and subtractions in Algorithm GAUSS-SQUARE.

a) Check that, when computing matrix entries, GAUSS-SQUARE does no additions and does one subtraction per multiplication. Argue that, were you to count the number of subtractions exactly (don't!), the highest power of n in the operation count would still be n^3.

b) Check that GAUSS-SQUARE has more assignments of matrix entries than multiplications and divisions. Count the assignments that are not associated with multiplications or divisions. From this show that, even counting assignments, GAUSS-SQUARE is a cubic algorithm.

c) GAUSS-SQUARE also has additions and assignments of index variables. Count the exact number of these operations and show that the algorithm is still cubic.

21. Some people like to make the leading coefficient of an equation 1 before subtracting that equation from others. For instance, had we followed this approach in Example 1, the first display after the start would have been

$$\begin{aligned} x - y + 2z &= -3 \\ 2x - y + 6z &= -6 \\ -x + 2y - 3z &= \ \ 6. \end{aligned}$$

Then we would have subtracted the first row twice from the second. Carry out this approach in matrix form for this example. What's your gut reaction: Does this method seem easier or harder than the method in the text? This question is analyzed quantitatively in [27].

22. Write an Algorithm BACK-SUB to replace Scale and Upsweep. Don't modify the matrix $\hat{A}$ after Downsweep finishes, but instead set up an array $(x_1, x_2, \ldots, x_n)$ of the unknowns and solve them iteratively starting with x_n.

23. Compare the relative efficiency of Scale and Upsweep to your Algorithm BACK-SUB in [22] as follows:

a) Count the number of multiplications/divisions for an $n \times n$ system with BACK-SUB and thus verify that it is the same as for Scale and Upsweep. (Here and in later parts, we refer to the sum total of all calls of Scale(i) and Upsweep(i) in GAUSS-SQUARE.)

b) Count the number of lookups of doubly subscripted variables in BACK-SUB and compare it to the number in Scale and Upsweep.

c) Count the number of assignments of singly subscripted x's in BACK-SUB. Compare your answer to the number of assignments of doubly subscripted variables in Scale and Upsweep. (Computers can find singly subscripted variables in memory faster than they can find doubly subscripted ones.)

24. In applying the recursive paradigm to reinvent Gaussian elimination, we deleted the first variable to break down to the previous case. Although that's the tradition in this subject, it doesn't have to be done that way. Indeed, in most places in this book, we've deleted the last variable. Redo the arithmetic of Example 2 by deleting the last variable every time a new subcall is made.

25. In applying the recursive paradigm to reinvent Gaussian elimination, we modified the last $n-1$ equations to break down to the previous case. That is, we subtracted out x_1 from all but the first equation. It doesn't have to be that way. We could, for instance, use the last original equation to subtract out x_1 from the first $n-1$ equations. Redo the arithmetic of Example 2 using this approach.

26. Redo Example 2 using the changes from both [24] and [25].

27. In the following parts, you are asked to consider variants of Gaussian elimination. Each variant involves the same input and output as the original (not shown). Each variant involves a Downsweep, an Upsweep, and a Scale, but they may have to be somewhat different from those in regular Gaussian elimination. In every case, Scale(i) means whatever needs to be done at that point so that the pivot entry in row i gets value 1 (and so that the solution set is unchanged). Downsweep means whatever needs to be done so that the entries below the pivot in column i are turned into 0's. Upsweep is whatever needs to be done so that the entries above the pivot in column i are turned into 0's. In each problem part, state your version of these procedures precisely.

a) In this variant of Gaussian elimination the scaling is done early.

```
Algorithm Gauss-A
  for i = 1 to n
    Scale(i)
    Downsweep(i)
  endfor
  for i = n downto 1
    Upsweep(i)
  endfor
```

How many multiplication/division operations are required now? (At least determine the highest power term.) *Note*: In this version of Downsweep, no multiplier needs to be computed before subtracting. Why?

b) Here is a variant where all the scaling is done together in the middle.

```
Algorithm Gauss-B
  for i = 1 to n
    Downsweep(i)
  endfor
  for i = 1 to n
    Scale(i)
  endfor
  for i = n downto 1
    Upsweep(i)
  endfor
```

How many multiplication/division operations are required now?

c) Now consider a variant where scaling is postponed as long as possible.

```
Algorithm Gauss-C
  for i = 1 to n
    Downsweep(i)
  endfor
  for i = n downto 1
    Upsweep(i)
  endfor
  for i = 1 to n
    Scale(i)
  endfor
```

How many multiplication/division operations does this require?

d) Now let's do both down and up-sweeps together, with scaling first.

```
Algorithm Gauss-D
  for i = 1 to n
    Scale(i)
    Upsweep(i)
    Downsweep(i)
  endfor
```

How many multiplication/division operations are required?

This variant goes under the name **Gauss-Jordan reduction.**

28. Show that equivalence of systems of equations as defined in this section (each set has the same solutions) is an equivalence relation as defined in Section 0.2.

29. Consider the following attempt to solve $x = 2 + \sqrt{x}$.

$$x = 2 + \sqrt{x} \qquad \text{(i)}$$
$$x - 2 = \sqrt{x} \qquad \text{(ii)}$$
$$x^2 - 4x + 4 = x \qquad \text{(iii)}$$
$$x^2 - 5x + 4 = 0 \qquad \text{(iv)}$$
$$(x-4)(x-1) = 0 \qquad \text{(v)}$$
$$x = 1, 4 \qquad \text{(vi)}$$

a) Is each step legitimate? That is, does each line follow correctly from the previous line?

b) Do lines (i) and (vi) have the same solutions?

c) If your answer to **b**) is No, which line or lines don't reverse? That is, which lines don't imply the previous line?

30. Problems with radicals in them are not the only problems in elementary algebra where extraneous roots show up. Consider

$$\frac{2x}{x-1} + \frac{1}{x-1} = \frac{3}{x-1}.$$

"Solve" in the usual way. What fails to reverse?

8.3 Gaussian Elimination: General Case

We have two issues to resolve to obtain the general algorithm: how to deal with nonsquare matrices and how to proceed when we hit a zero pivot. Resolving the first issue will suggest how to resolve the second.

There is also another set of questions about interpreting the results of the algorithm. A system of linear equations can have zero, one, or infinitely many solutions. How do we identify which we have? In Section 8.2, it was easy. When we had just one solution, it stared us right in the face at the end of the algorithm with a set of equations of the form

$$x_1 = c_1, \qquad x_2 = c_2, \quad \ldots, \quad x_n = c_n.$$

Similarly, when there was no solution, we knew it because we obtained an obviously impossible equation, $0 =$ nonzero. Will we be so fortunate with the general algorithm? The answer is Yes.

The case of infinitely many solutions is even more intriguing. What does it even mean to solve a problem for which it's impossible to finish listing the solutions? One answer is: If there is a nice way to describe the infinite set so that we feel we have a good grasp of it, then we consider that to be a solution. But offhand, it's not obvious what such a nice representation for linear systems might be, or that any such representation exists. By Theorem 1 of the previous section, we do know that every system of equations we obtain from the original set expresses implicitly the exact solution set we want. The question is: Can we get to a system which makes an infinite solution set easy to comprehend?

It turns out that if we generalize Gaussian elimination in a fairly natural way, the algorithm itself will suggest what the nice representation is, and will find it.

We will generalize Gaussian elimination through a series of examples, state the result in our algorithmic language, and then recapitulate through two more examples. Let's start. A good strategy is to try to isolate different issues. To confront the issue of multiple solutions, here is a system with more variables than equations. (Warning: More variables than equations is neither necessary nor sufficient for multiple solutions, but it is the most common case.)

EXAMPLE 1 Solve

$$\begin{aligned} 2x - 2y + 4z + \phantom{\tfrac{3}{2}}w &= -6 \\ 2x - y + 6z \phantom{{}+\tfrac{3}{2}w} &= -6 \\ -x + 2y - 3z + \tfrac{3}{2}w &= 6. \end{aligned} \tag{1}$$

Solution This is the same system as in Examples 1 and 2, Section 8.2, except we have added a new variable w. Switching to augmented matrix form, and doing the first downsweep, we obtain

$$\left[\begin{array}{cccc|c} 2 & -2 & 4 & 1 & -6 \\ 0 & 1^* & 2 & -1 & 0 \\ 0 & 1 & -1 & 2 & 3 \end{array}\right].$$

The second downsweep (the pivot is marked above) gives us

$$\left[\begin{array}{rrrr|r} 2 & -2 & 4 & 1 & -6 \\ 0 & 1 & 2 & -1 & 0 \\ 0 & 0 & -3^* & 3 & 3 \end{array}\right].$$

We have marked the third pivot but as always there is no downsweep to do from the bottom row. Heretofore we would have gone to the fourth column to look for another pivot, but there are no more rows in which to place a pivot. That suggests we should turn to scale and upsweep, using the pivot entries we have already attained. Scaling from the (3,3) entry and then upsweeping gives us in turn

$$\left[\begin{array}{rrrr|r} 2 & -2 & 4 & 1 & -6 \\ 0 & 1 & 2 & -1 & 0 \\ 0 & 0 & 1^* & -1 & -1 \end{array}\right] \quad \text{and} \quad \left[\begin{array}{rrrr|r} 2 & -2 & 0 & 5 & -2 \\ 0 & 1^* & 0 & 1 & 2 \\ 0 & 0 & 1 & -1 & -1 \end{array}\right].$$

(Notice that scaling and upsweeping have more work to do than before; in Section 8.2 they affected only the current pivot column and the **b** column.) We now return to the second pivot, at (2,2). It is already scaled, so we upsweep with it by subtracting the second row -2 times from the first row:

$$\left[\begin{array}{rrrr|r} 2^* & 0 & 0 & 7 & 2 \\ 0 & 1 & 0 & 1 & 2 \\ 0 & 0 & 1 & -1 & -1 \end{array}\right].$$

Finally we return to the first pivot. Scaling, we arrive at

$$\left[\begin{array}{rrrr|r} 1 & 0 & 0 & 7/2 & 1 \\ 0 & 1 & 0 & 1 & 2 \\ 0 & 0 & 1 & -1 & -1 \end{array}\right].$$

Since we did exactly the same operations as in Example 2, Section 8.2, every column of the present result is the same as in the final matrix there, except of course for the new column.

At this point in the original problem we were done, so now let's switch back to variables and equations to see what they suggest:

$$\begin{aligned} x \qquad\quad + \tfrac{7}{2}w &= 1 \\ y \quad + \; w &= 2 \\ z - \; w &= -1. \end{aligned}$$

The unknowns x, y, and z have been isolated just as well as before, except that there are w's on the left-hand side. To finish isolating x, y, z, subtract w to the other side:

$$\begin{aligned} x &= 1 - \tfrac{7}{2}w \\ y &= 2 - \; w \qquad\qquad (2) \\ z &= -1 + w. \end{aligned}$$

Now here's the important thing to notice: *We may let w take any value we please,* *and then x, y, z are completely determined.* No value of w can lead to any contradictions because w appears only on the right side of the equations. Because w has this property we say that system (1) has one **degree of freedom.**

To make this single degree of freedom more apparent, we rewrite system (2) using column vectors, a special type of matrix explained in Section 0.6:

$$\begin{pmatrix} x \\ y \\ z \end{pmatrix} = \begin{pmatrix} 1 \\ 2 \\ -1 \end{pmatrix} + w \begin{pmatrix} -7/2 \\ -1 \\ 1 \end{pmatrix}.$$

In other words, Gaussian elimination has partitioned the variables into two sets, a set of **free** variables, which can take on any values (in this case there is just one free variable, w), and a set of **bound** variables, which are completely determined in terms of the free variables. Free variables are also called **nonpivot** variables, and bound variables are also called **pivot** variables, because the bound variables are exactly the ones having a pivot in their columns.

One more point and we are done with this first effort. Typically, one wants to read off all the variables in a solution in one fell swoop, so it is convenient to have each variable appear on the left on a line by itself. Currently w does not. We need to add a new equation to achieve this result, and this equation must not change the set of solutions. The equation $w = w$ will accomplish this: It is satisfied by every w. So our final system of equations is

$$\begin{aligned} x \qquad\qquad\qquad &= 1 - \tfrac{7}{2}w \\ y \qquad\qquad &= 2 - \ \ w \\ z \qquad &= -1 + \ w \\ w &= \qquad\quad w, \end{aligned}$$

or in vector form,

$$\begin{pmatrix} x \\ y \\ z \\ w \end{pmatrix} = \begin{pmatrix} 1 \\ 2 \\ -1 \\ 0 \end{pmatrix} + w \begin{pmatrix} 7/2 \\ -1 \\ 1 \\ 1 \end{pmatrix}.$$

We will refer to a solutions display like this as **standard vector form.** It is a particularly good way to represent solutions to a linear system when there are infinitely many vectors in the solution set. ■

Now let's see how to do Gaussian elimination when there are more rows than columns.

EXAMPLE 2 Solve the system

$$\begin{aligned} 2x - 2y &= -6 \\ 2x - \ \ y &= -6 \\ -x + 2y &= \ \ 6. \end{aligned}$$

Solution This is the same as system (1), except we deleted the variables z and w. After two downsweeps we obtain

$$\left[\begin{array}{cc|c} 2 & -2 & -6 \\ 0 & 1^* & 0 \\ 0 & 0 & 3 \end{array}\right].$$

Because of the third line, this system is inconsistent. There are no solutions and we can stop. ∎

EXAMPLE 3 Solve the system

$$\begin{aligned} 2x - 2y &= -6 \\ 2x - y &= -6 \\ -x + 2y &= 3. \end{aligned}$$

Solution This is the same system as in Example 2, except we have decreased the last constant by 3. Thus after two downsweeps we get

$$\left[\begin{array}{cc|c} 2 & -2 & -6 \\ 0 & 1^* & 0 \\ 0 & 0 & 0 \end{array}\right].$$

Now the third row does not represent an impossible equation. Rather it represents the equation $0 = 0$, which is vacuously satisfied by every pair of values (x, y). Thus we can ignore the third row and continue. Scaling and upsweeping with the last pivot entry, we obtain

$$\left[\begin{array}{cc|c} 2^* & 0 & -6 \\ 0 & 1 & 0 \\ 0 & 0 & 0 \end{array}\right].$$

Finally, we scale on the previous pivot position (as marked) and get

$$\left[\begin{array}{cc|c} 1 & 0 & -3 \\ 0 & 1 & 0 \\ 0 & 0 & 0 \end{array}\right], \qquad \text{that is,} \qquad \begin{pmatrix} x \\ y \end{pmatrix} = \begin{pmatrix} -3 \\ 0 \end{pmatrix}. \ ∎$$

Now let's face the music on the second problem identified at the beginning of this section: 0's in pivot positions.

EXAMPLE 4 Solve the system of equations represented by

$$\left[\begin{array}{ccc|c} 1^* & 2 & 3 & 1 \\ 2 & 4 & 5 & 1 \\ -1 & 1 & 2 & 4 \end{array}\right]. \tag{3}$$

Solution After downsweeping from the pivot we have

$$\left[\begin{array}{ccc|c} 1 & 2 & 3 & 1 \\ 0 & 0^* & -1 & -1 \\ 0 & 3 & 5 & 5 \end{array}\right].$$

We are blocked from the natural pivot by a 0. However, there is an easy out. There is another row farther down with a nonzero entry in the same column. We can *switch* these two rows, obtaining

$$\left[\begin{array}{ccc|c} 1 & 2 & 3 & 1 \\ 0 & 3^* & 5 & 5 \\ 0 & 0 & -1 & -1 \end{array}\right],$$

and proceed normally. Indeed, the remaining downsweeps change nothing, so we may proceed with scale and upsweep. The final matrix is

$$\left[\begin{array}{ccc|c} 1 & 0 & 0 & -2 \\ 0 & 1 & 0 & 0 \\ 0 & 0 & 1 & 1 \end{array}\right],$$

and so the unique solution is $(x, y, z) = (-2, 0, 1)$. This example shows why we included switching as a row operation in the proof of Theorem 1, Section 8.2. ∎

EXAMPLE 5 Solve the system of equations represented by

$$\left[\begin{array}{ccc|c} 1^* & 2 & 3 & 1 \\ 2 & 4 & 5 & 1 \\ -1 & -2 & 2 & 4 \end{array}\right].$$

Solution All we have done is change the (3,2) entry from the previous start matrix (3), but it makes a big difference. After downsweeping the first column we get

$$\left[\begin{array}{ccc|c} 1 & 2 & 3 & 1 \\ 0 & 0^* & -1 & -1 \\ 0 & 0 & 5 & 5 \end{array}\right], \tag{4}$$

and there is nothing farther down in the second column to save us.

Well, in all previous cases we moved both down and right when looking for new pivots. If we can't move down, let's move right:

$$\left[\begin{array}{ccc|c} 1 & 2 & 3 & 1 \\ 0 & 0 & -1^* & -1 \\ 0 & 0 & 5 & 5 \end{array}\right].$$

Continuing as if nothing unusual has happened, we downsweep column 3:

$$\left[\begin{array}{ccc|c} 1 & 2 & 3 & 1 \\ 0 & 0 & -1 & -1 \\ 0 & 0 & 0 & 0 \end{array}\right].$$

As we have noted before, the thing to do now is to ignore the bottom row. It is time to start scaling and upsweeps. Previously, we always scaled and upswept using the same pivot positions as for downsweeps, but in the reverse order. Let's try the same approach again. So scaling and then upsweeping with the (2,3) entry as the pivot, we obtain

$$\left[\begin{array}{ccc|c} 1^* & 2 & 0 & -2 \\ 0 & 0 & 1 & 1 \\ 0 & 0 & 0 & 0 \end{array}\right].$$

Now we are back to considering the original pivot position. Since it is already scaled, and since there is never any upsweep from the first row, we are done—if this matrix provides a good form for reading off all the solutions.

It does. Look at the last matrix turned into equation form:

$$\begin{aligned} x + 2y \qquad &= -2 \\ z &= \ \ 1. \end{aligned}$$

As in earlier examples, we have isolated certain variables in the sense that each appears in just one equation and with a coefficient of 1. But this time the variables are the first and third—they are not consecutive from the left. There's nothing wrong with that! We can still transfer the free variable to the right:

$$\begin{aligned} x \qquad &= -2 - 2y \\ z &= \ \ 1. \end{aligned}$$

We can still get the free variable alone on the left of a line by including $y = y$:

$$\begin{aligned} x \qquad\quad &= -2 - 2y \\ y \quad &= \qquad\ \ y \\ z &= \ \ 1. \end{aligned}$$

We can still rewrite the result in column vector form:

$$\begin{pmatrix} x \\ y \\ z \end{pmatrix} = \begin{pmatrix} -2 \\ 0 \\ 1 \end{pmatrix} + y \begin{pmatrix} -2 \\ 1 \\ 0 \end{pmatrix}. \ \blacksquare$$

When there are infinitely many solutions, why does the final matrix of Gaussian elimination give us a set of equations from which we can read off the solutions easily? Think of it this way. For those columns where Gaussian elimination finds a nonzero pivot, the combined effect of downsweep, scale, and upsweep is to create a so-called **unit column vector**: all 0's except one 1. That means the variable associated with that column has been isolated; it appears in just one equation. Furthermore, no two pivot variables are isolated in the same equation, because different pivots are in different rows as well as different columns. Thus all the nonpivot variables may be moved to the right and given any values we please; this will uniquely determine the pivot variables without any chance of the assigned values being in contradiction.

All right, let us state the algorithm formally. We'll call it GAUSS. It's quite long, so first here is an outline:

```
repeat while maybe-more-pivots
     ScanSwitch                    [Scan for pivot; maybe switch rows]
     if another-pivot then Downsweep
endrepeat
for each pivot found, in reverse order
     Scale
     Upsweep
endfor
```

Downsweep, Scale, and Upsweep are basically unchanged from GAUSS-SQUARE except, as noted earlier, Scale and Upsweep have to attend to more columns than before. The new procedure ScanSwitch looks for the next pivot. If there is one, but it's in the "wrong" row, ScanSwitch also makes the appropriate row switch. In any event, it either reports back the position of the next pivot or reports that there is none.

Because the number of pivots need no longer equal the total number of rows, GAUSS needs a counter for the number of pivots found so far. We will use r, since the rth pivot will still be in the rth row. Also, since pivots need no longer be on the main diagonal, GAUSS needs to store the column index of the rth pivot for use in Scale and Upsweep; we'll call that value c_r. GAUSS determines when to stop searching for new pivots using the values of r and c_r, rather than additional variables with names like another-pivot as in the outline. The complete version appears as Algorithm 8.2.

Now that we have the complete algorithm, we apply it in a large example. This will allow us to recapitulate all the maneuvers that can occur and to step through the algorithmic language to check that it accomplishes what we said it should. Our example is cooked up to put the algorithm through all its paces; it's highly unlikely an example like this would occur in practice.

EXAMPLE 6 Apply Algorithm GAUSS to the augmented matrix

$$\hat{A} = \left[\begin{array}{cccccc|c} 1^* & 2 & 1 & 1 & 4 & 1 & 2 \\ 2 & 4 & 2 & 2 & 8 & 5 & 10 \\ -1 & -2 & 1 & -3 & 2 & 1 & 0 \\ 1 & 2 & 4 & -2 & 13 & 8 & 13 \end{array}\right].$$

Solution The inputs to GAUSS are $m = 4$, $n = 6$, and the 4×7 matrix $\hat{A} = [a_{ij}]$. It's clear by eyeballing that the first "real" work should be downsweeping the first

Algorithm 8.2

Gauss

```
Input m, n, Â                                   [Â = [A | b] is m×(n+1)]
Algorithm GAUSS
    procedure ScanSwitch(r, j)          [Look for pivot starting at (r, j)]
        i ← r                                  [New row variable so r saved]
        repeat while j ≤ n and a_ij = 0
                                     [Until j out of range or pivot found]
            if i < m then i ← i+1                   [Continue down column]
                      else i ← r, j ← j+1
             [Return to topmost candidate row and try next column]
            endif
        endrepeat
        if j ≤ n and i > r           [Another pivot found but not in row r]
            then row i ↔ row r                    [Move pivot row to row r]
        c_r ← j                               [Pivot position saved as (r, c_r)]
        return c_r                                    [c_r needed elsewhere]
    endpro
    procedure Downsweep(i, j)
        for k = i+1 to m
            m_k ← a_kj/a_ij
            a_kj ← 0
            for l = j+1 to n+1
                a_kl ← a_kl − m_k a_il
            endfor
        endfor
    endpro
    procedure Scale(i, j)
        for l = j+1 to n+1
            a_il ← a_il/a_ij
        endfor
        a_ij ← 1
    endpro
```

(Continues on next page.)

Algorithm 8.2 Continued

Gauss

```
procedure Upsweep(i, j)
    for k = i−1 downto 1
        for l = j+1 to n+1
            a_kl ← a_kl − a_kj a_il
        endfor
        a_kj ← 0
    endfor
endpro

r ← 1, j ← 1,          [Main algorithm; initialize pivot search position]
repeat while r ≤ m
    ScanSwitch(r, j)          [Finds next pivot and puts it at (r, c_r)]
    if c_r > n then exit                              [No more pivots]
    Downsweep(r, c_r)
    r ← r+1, j ← c_r+1                       [Next pivot search position]
endrepeat
for i = r−1 downto 1                       [r−1 is last row with a pivot]
    Scale(i, c_i)
    Upsweep(i, c_i)
endfor
Output Â
```

column with the pivot at (1,1), but we want to step through GAUSS and check that this happens. The main algorithm (henceforth called Main) begins by setting $(r, j) = (1, 1)$. That is, GAUSS does indeed first look for a pivot at (1,1). As $r \leq m$, the **repeat** loop in Main is activated, and we call ScanSwitch(1,1).

In ScanSwitch we set the temporary row variable i to 1. (The value of r is never changed by ScanSwitch.) Since $a_{ij} = a_{11} \neq 0$, we skip over the **repeat** loop. In the **if** line after that loop, we find that $i \not> r$, so there is no row switch. The current value 1 of j is assigned to c_1 and returned to Main. (The value c_1 will be used to locate the first pivot entry again when we scale and upsweep.) In short, we correctly found a pivot in the first place we looked.

Returning to Main, $c_1 \not> n$, so we proceed to Downsweep(1,1). We obtain

$$\left[\begin{array}{cccccc|c} 1 & 2 & 1 & 1 & 4 & 1 & 2 \\ 0 & 0 & 0 & 0 & 0 & 3 & 6 \\ 0 & 0 & 2 & -2 & 6 & 2 & 2 \\ 0 & 0 & 3 & -3 & 9 & 7 & 11 \end{array}\right].$$

(The row operations in Downsweep, Scale, and Upsweep should be quite familiar to you by now, so we'll only go step by step in Main and in procedure ScanSwitch.) Now we increment r by 1 and set j to $c_1+1 = 2$. Thus the next place we will begin to look for a pivot is at (2,2), one row down and one column over from the previous success.

Eyeballing the preceding display, it's clear that we won't find a pivot in column 2, and even in column 3 we will have to do a row switch to get the pivot into the second row. But how does GAUSS discover all this? Since $2 = r \le m = 4$, we begin the **repeat** loop of Main again. It calls ScanSwitch(2,2). Setting $i = 2$, we find that $j \le n$ and $a_{22} = 0$, so we start executing the loop in ScanSwitch. Since $i < m$ (we are not yet looking at the bottom row), we activate the **then** clause and increase i by 1. That is, we look in the next row down (but same column). Again, $a_{ij} = 0$, so we increase i to 4. Once more $a_{ij} = 0$, but this time $i = m$, so the **else** is activated. We set i back to 2 and up j to 3. In other words, we go back to the first row we looked at in this scan, but move right one column. Now $a_{23} = 0$, so we do another **repeat** loop, which increases i to 3. We return to the **repeat** line, but now $a_{33} \ne 0$, so we exit to the **if** below **endrepeat**. As $3 = i > r = 2$, we switch rows 2 and 3:

$$\left[\begin{array}{cccccc|c} 1 & 2 & 1 & 1 & 4 & 1 & 2 \\ 0 & 0 & 2^* & -2 & 6 & 2 & 2 \\ 0 & 0 & 0 & 0 & 0 & 3 & 6 \\ 0 & 0 & 3 & -3 & 9 & 7 & 11 \end{array}\right].$$

At this point, we return to Main with the pivot position $(r, c_r) = (2,3)$. This pivot is not out of bounds ($c_r \not> n$) so we downsweep from it, obtaining

$$\left[\begin{array}{cccccc|c} 1 & 2 & 1 & 1 & 4 & 1 & 2 \\ 0 & 0 & 2 & -2 & 6 & 2 & 2 \\ 0 & 0 & 0 & 0 & 0 & 3 & 6 \\ 0 & 0 & 0 & 0 & 0 & 4 & 8 \end{array}\right].$$

Preparing for the next pivot search, we update (r, j) to (3,4), reenter Main's **repeat** loop, and call ScanSwitch.

Looking at the preceding display, we see that we won't find a pivot in either column 4 or 5, and that the first entry to consider in column 6 is 3. The algorithm discovers all this as follows. We pass through the **repeat** loop in ScanSwitch four times, looking at the 0 values a_{34}, a_{44}, a_{35}, and a_{45}. Then we reach the nonzero a_{36}, which kicks us out of the **repeat**. Because $3 = i = r$, the row-switch command is not activated and we return to Main with $c_3 = 6$ and do Downsweep(3,6):

$$\left[\begin{array}{cccccc|c} 1 & 2 & 1 & 1 & 4 & 1 & 2 \\ 0 & 0 & 2 & -2 & 6 & 2 & 2 \\ 0 & 0 & 0 & 0 & 0 & 3^* & 6 \\ 0 & 0 & 0 & 0 & 0 & 0 & 0 \end{array}\right]. \tag{5}$$

At this point we humans can see that the first half of Gaussian elimination is over, but in the algorithm we are still in Main and update (r, j) to (4,7). Main starts its **repeat** loop again since the row index is in bounds. We call ScanSwitch(4,7).

But the **repeat** loop in ScanSwitch is never activated because $7 = j > n = 6$. The **if** structure below **endrepeat** is not activated either, for the same reason. In other words, ScanSwitch knows to do nothing once j goes out of bounds. Furthermore, we return to Main at the line **if** $c_r > n$ **then exit**. So Main does nothing more in its **repeat** loop either, thus ending the downsweep stage. We are left with a value of (r, c_r) that doesn't represent a pivot.

That's why the **for** loop in Main starts at $r - 1$. It proceeds to Scale and Upsweep from the pivot positions in reverse order. From $(i, c_i) = (3, 6)$ these two procedures together give

$$\left[\begin{array}{cccccc|c} 1 & 2 & 1 & 1 & 4 & 0 & 0 \\ 0 & 0 & 2^* & -2 & 6 & 0 & -2 \\ 0 & 0 & 0 & 0 & 0 & 1 & 2 \\ 0 & 0 & 0 & 0 & 0 & 0 & 0 \end{array}\right].$$

Returning to the second pivot, they give

$$\left[\begin{array}{cccccc|c} 1^* & 2 & 0 & 2 & 1 & 0 & 1 \\ 0 & 0 & 1 & -1 & 3 & 0 & -1 \\ 0 & 0 & 0 & 0 & 0 & 1 & 2 \\ 0 & 0 & 0 & 0 & 0 & 0 & 0 \end{array}\right]. \tag{6}$$

Finally, on the first pivot Scale(1,1) doesn't change anything because $a_{11} = 1$, and Upsweep(1,1) is always vacuous. So the previous matrix is the output matrix. This completes the trace of Algorithm GAUSS.

We still have to write the solutions in standard vector form. Translating matrix (6) into equations directly, we obtain

$$\begin{aligned} u + 2v \qquad + 2x + \ y \qquad &= \ \ 1 \\ w - \ \ x + 3y \qquad &= -1 \\ z &= \ \ 2. \end{aligned}$$

The free variables are v, x, y, so moving them to the other side we obtain

$$\begin{aligned} u &= \ \ 1 - 2v - 2x - \ y \\ w &= -1 \qquad + \ x - 3y \\ z &= \ \ 2. \end{aligned}$$

Adding the redundant equations $v = v$, $x = x$, and $y = y$ and rewriting the system as vectors, we finally obtain

$$\begin{pmatrix} u \\ v \\ w \\ x \\ y \\ z \end{pmatrix} = \begin{pmatrix} 1 \\ 0 \\ -1 \\ 0 \\ 0 \\ 2 \end{pmatrix} + v \begin{pmatrix} -2 \\ 1 \\ 0 \\ 0 \\ 0 \\ 0 \end{pmatrix} + x \begin{pmatrix} -2 \\ 0 \\ 1 \\ 1 \\ 0 \\ 0 \end{pmatrix} + y \begin{pmatrix} -1 \\ 0 \\ -3 \\ 0 \\ 1 \\ 0 \end{pmatrix}. \blacksquare$$

Note. Had the bottom entry of the **b** column in matrix (5) been nonzero, the equations associated with the original matrix would have had no solutions. Yet Algorithm GAUSS would have just kept going. We ask you to fix this, and consider various other improvements, in the problems.

EXAMPLE 7 Solve the system

$$\begin{aligned} u + 2v + w + x + 4y + z &= 0 \\ 2u + 4v + 2w + 2x + 8y + 5z &= 0 \\ -1u - 2v + w - 3x + 2y + z &= 0 \\ u + 2v + 4w - 2x + 13y + 8z &= 0. \end{aligned}$$

Solution Notice that all the constants on the right are 0's. From Chapter 5 it should not surprise you to hear that such a system of linear equations is called **homogeneous**. When we put a homogeneous system into matrix form, there is no need to include the **b** column. No matter what row operations we do in what order, we will still have $\mathbf{b} = \mathbf{0}$. (Boldface 0 means a vector in which each entry is the number 0.) Homogeneous systems play a very important role in linear algebra theory, as you will see at the start of Section 8.7.

In any event, except for the **b** column, this problem is the same as Example 6. Since the values of entries in the **b** column never affect either what GAUSS does or the values computed elsewhere, we conclude that the final matrix is

$$\begin{bmatrix} 1^* & 2 & 0 & 2 & 1 & 0 \\ 0 & 0 & 1 & -1 & 3 & 0 \\ 0 & 0 & 0 & 0 & 0 & 1 \\ 0 & 0 & 0 & 0 & 0 & 0 \end{bmatrix},$$

where we haven't bothered to write the all-zero **b** column. Similarly, the standard vector form for the solutions is

$$\begin{pmatrix} u \\ v \\ w \\ x \\ y \\ z \end{pmatrix} = v \begin{pmatrix} -2 \\ 1 \\ 0 \\ 0 \\ 0 \\ 0 \end{pmatrix} + x \begin{pmatrix} -2 \\ 0 \\ 1 \\ 1 \\ 0 \\ 0 \end{pmatrix} + y \begin{pmatrix} -1 \\ 0 \\ -3 \\ 0 \\ 1 \\ 0 \end{pmatrix},$$

where again it is customary in a homogeneous system to leave out the constant vector $\mathbf{0}$ on the right. ∎

Problems: Section 8.3

Throughout this problem set, solve all systems using Algorithm GAUSS except where explicitly told otherwise. If a problem is stated using variables, give your final answer in standard vector form. If the problem is stated using a matrix, stop at the final matrix obtained by GAUSS. If a matrix has no **b** column, a column of zeros is understood.

1. Solve the system $\left\{\begin{aligned} x+y+\ z &= \ \ 0 \\ -x+y+3z &= -2 \\ 2x-y+4z &= \ \ 4 \end{aligned}\right\}.$

2. Solve the system $\left\{\begin{aligned} x+\ y+z+\ w &= 3 \\ x+2y-z+3w &= 5 \end{aligned}\right\}.$

3. Solve $\left\{\begin{aligned} 2x-2y+3z &= 1 \\ -x+\ y \qquad &= 1 \\ 3x-3y+5z &= 2 \end{aligned}\right\}.$

4. In this problem you are asked to solve [1–3] again, but with the order of rows changed. Of course, the solutions should be the same, but the steps in the algorithm will differ.
   - a) Solve [1] with rows 1 and 2 interchanged.
   - b) Solve [1] with rows 2 and 3 interchanged.
   - c) Solve [2] with the rows switched.
   - d) Solve [3] with rows 1 and 2 switched.
   - e) Solve [3] with the constant on the right of the third equation changed to 3.
   - f) Solve [1] again with the variables in the reverse order.

5. Apply Algorithm GAUSS to

a) $\left[\begin{array}{ccc|c} 1 & 2 & 3 & 4 \\ 2 & 3 & 4 & 5 \end{array}\right]$ b) $\begin{bmatrix} 1 & 2 & 3 & 4 \\ 0 & 2 & 3 & 4 \\ 0 & 0 & 0 & 4 \end{bmatrix}$

c) $\begin{bmatrix} 1 & 2 & 3 & 4 \\ 0 & 0 & 0 & 4 \\ 0 & 2 & 3 & 4 \end{bmatrix}$ d) $\begin{bmatrix} 0 & 0 & 3 \\ 0 & 1 & 4 \\ 0 & 2 & 5 \end{bmatrix}$

e) $\left[\begin{array}{cccc|c} 1 & 3 & 3 & 2 & 1 \\ 2 & 6 & 9 & 5 & 4 \\ -1 & -3 & 3 & 0 & 2 \end{array}\right]$

f) $\begin{bmatrix} 1 & 2 & 2 \\ 2 & 1 & -2 \\ 2 & -2 & 1 \end{bmatrix}$ g) $\begin{bmatrix} 0 & 0 & 0 \\ 0 & 0 & 0 \end{bmatrix}$

Note: A zero matrix as in g) would rarely come up in practice, but it's always good to test an algorithm on extreme cases. This helps you to understand the algorithm and/or catch errors.

6. (parts **a** through **g**) Assume that each matrix in [5] came from a homogeneous system of linear equations using variables $x_1, x_2, \ldots$, appearing from left to right. (Ignore the **b** columns in parts **a** and **e**.) Continue on from the final matrix you obtained to write the solutions in vector form.

7. Suppose in solving some problem we reach a situation like the following, where the 2 in the second row was the last pivot.

$$\left[\begin{array}{cccc|c} 2 & 1 & 3 & -1 & 4 \\ 0 & 0 & 2 & 7 & 3 \\ 0 & 0 & 0 & 0 & 5 \\ 0 & 0 & 0 & 0 & 2 \end{array}\right]$$

According to our algorithm, we should stop looking for new pivots. But why didn't we write the algorithm to ignore the vertical line and continue on to pivot on the 5?

8. Suppose Gaussian elimination reduces A to

$$R = \begin{bmatrix} 1 & 2 & 0 & -3 \\ 0 & 0 & 1 & 1 \\ 0 & 0 & 0 & 0 \end{bmatrix}.$$

   - a) Name all solutions to $A\mathbf{x} = \mathbf{0}$.
   - b) But you were never told the entries of A! What theorem justifies your answer to part a)?
   - c) Suppose further that Gaussian elimination required no row switches to reach R. Find a vector **b** such that $A\mathbf{x} = \mathbf{b}$ has no solutions. *Caution*: If all you show about your **b** is that $R\mathbf{x} = \mathbf{b}$ has no solutions, you're not done.

9. Consider a homogeneous system $A\mathbf{x} = \mathbf{0}$.
   - a) Explain why there is always at least one solution.
   - b) Show that if A is $m \times n$ with $n > m$, then there are infinitely many solutions. (Consider the form of the final matrix obtained by GAUSS.)
   - c) Make up an example where $n < m$ and yet there are still infinitely many solutions. (What has to happen in the matrix at the end of GAUSS?)

10. Modify Algorithm GAUSS so that it stops and reports "no solutions" when and if this is obvious from the matrix.

11. In Example 6, the column index j was already out of bounds when ScanSwitch was called for the last time. Give an example where j goes out of bounds *during* ScanSwitch.

12. True or false: In GAUSS, ScanSwitch is always called one more time than the number of pivots. Explain.

13. True or false: In GAUSS, the row index r never goes out of bounds. Explain.

14. Algorithm GAUSS ends with the final augmented matrix, but you aren't really finished until you write the solutions in standard vector form. Write an add-on to GAUSS which outputs each vector on the right in this form, along with the index of the free variable which multiplies it. (That is, if a certain vector is multiplied by x_3, it is sufficient for your algorithm to output "x3" before outputting that vector.)

15. In Algorithm GAUSS the line in ScanSwitch that switches rows i and r is not written in as much detail as the other lines. Flesh it out.

16. The procedure Downsweep in GAUSS misses a trick. If there was a row switch just before Downsweep is called, then Downsweep does a lot of unnecessary work. What work? Can you make a simple change to correct this?

17. In GAUSS suppose that $\hat{A}$ is input in two parts: the coefficient matrix A and the constant column $\mathbf{b}$. Rewrite the algorithm in this form. Within your algorithm refer to entries of A as a_{ij} and entries of $\mathbf{b}$ as b_i.

Problems 18–24 introduce the subject of **network flows**, which involves both linear algebra and graph theory.

18. Consider a directed graph with an incompressible material flowing through it. You can think of the material as a fluid or an electric current, but it really doesn't matter what it is. The important point is the incompressibility: The total amount flowing into each vertex equals the amount flowing out of that vertex. We define a **flow** to be any vector $\mathbf{x} = (x_1, x_2, \ldots)$ such that, if x_i is the amount of material flowing along edge e_i in the direction of its arrow, then the incompressibility condition is met at every vertex. (If the flow is opposite to the arrow, x_i is negative.) Using the methods of this section, determine all flows for the graph in the Fig. 8.1

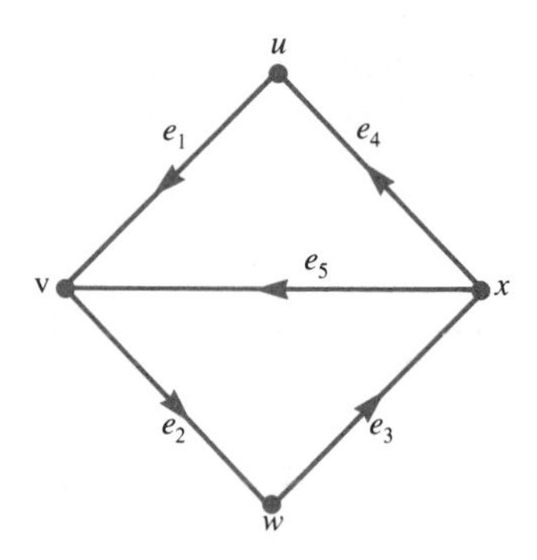

FIGURE 8.1

19. Have you seen before the sort of matrix obtained in [18]? *Hint*: Look over the first few problems for Section 3.3.

20. Find all flows for the graph of [18] with the added restriction that $x_1 = 1$. (You can find them by adding the equation $x_1 = 1$ to the set of equations you started with in [18] and solving from scratch, but if x_1 was a free variable in your solution to [18], there is a much easier way. What? If x_1 wasn't a free variable, could you modify your solution algorithm slightly so that it is?)

21. A **potential difference graph** is a digraph with a number y_i on each edge e_i and such that, for each cycle in the digraph, the "signed sum" of the y's around the cycle is 0. (In a signed sum, y_i is added if the arrow of e_i faces in the direction of the cycle, and y_i is subtracted if the arrow faces the other way.) Each y_i is called the potential difference of e_i and $\mathbf{y} = (y_1, y_2, \ldots)$ is called the **potential difference vector**. Each y_i might be the differences in elevation between the tail of e_i and the head of e_i if the digraph represents a road system, or y_i might represent drop in voltage if the digraph represents an electrical circuit; but it really doesn't matter. The thing to understand is that y_i represents the change in the potential (whatever potential means) in the direction of the arrow. That's why, if you traverse an edge in the opposite direction to its arrow, the change you add to the sum is $-y_i$.

For the digraph of [18], one cycle is e_1, e_5 (against the arrow), and e_4. Another is e_5, e_2, e_3. Assuming these are the only cycles in the graph, determine all potential difference vectors. (But see [22].)

22. There is another cycle in Fig. 8.1—e_1, e_2, e_3, e_4— that we told you not to use when computing all the potential differences in [21]. Solve for all potential differences again, using this cycle as well. You will find that the answer is the same. (In general, it is only necessary to require that the sums be 0 around *some* cycles to ensure that the sum is 0 around all cycles. We leave to another course the problem of characterizing *which* sets of cycles have this property.)

23. For the graph of [18], find all potential differences in which:

a) $y_1 = 1$ **b)** $y_1 = 1$ and $y_5 = -2$

24. In the graph of [18], on how many edges must the potential differences be specified before there is a unique potential difference vector? Can any set of this many edges have their differences specified with any values we please, or will such conditions sometimes be inconsistent or redundant?

25. A resistive electric circuit is a connected set of wires with resistances in them. Such a circuit is modeled by a digraph. On each edge e_k, let i_k be the current flowing through it in the direction of the arrow and let r_k be the resistance of that edge to current. Usually the resistances are given and the currents are unknown. To find the currents, electrical engineers use three principles:

- The currents form a flow as defined in [18]. This is called **Kirchhoff's First Law**.
- On each edge there is a "voltage drop", and the set of voltage drops form a potential difference vector, as defined in [21]. This principle is called **Kirchhoff's Second Law**.
- For each edge e_k, there is a component v_k of the voltage drop that satisfies

$$v_k = r_k i_k.$$

This is known as **Ohm's Law**. (On most edges v_k is the total voltage drop, but batteries and current sources add further terms; see [26].)

Using these principles, find all possible currents in the graph shown in Fig. 8.2. (The edge marked with r_k is e_k and has current i_k. Assume there are no batteries or current sources.) Suggestion: Let your unknowns be i_1, i_2, i_3, i_4, i_5. That is, eliminate the v_k's immediately using Ohm's Law and the resistances shown on the graph. (This still leaves a lot of work if you do Gaussian elimination by hand. For this graph and many others, but not all, hand work can be speeded-up by "series-parallel decomposition", which is taught in physics and engineering courses.)

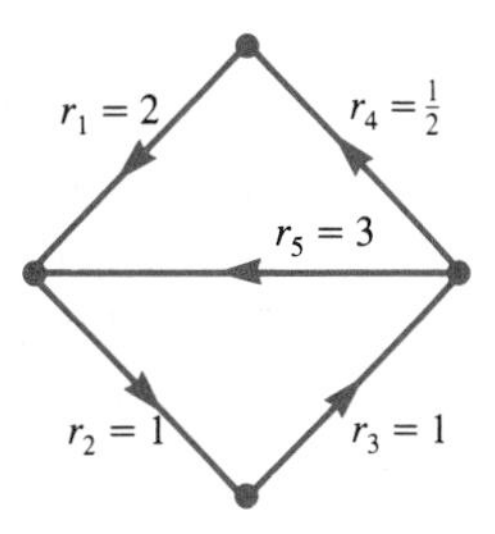

FIGURE 8.2

26. Solve for the currents in [25] again with the additional assumption that:

a) There is a 2-volt battery on e_2. This raises the voltage by 2 volts, so the voltage drop on e_2 is the drop caused by Ohm's Law *minus* 2.

b) There is a current source of 1 amp on e_1. This source provides whatever voltage boost is necessary to make $i_1 = 1$. Since you don't know what that voltage is, delete any equation that makes an assertion about the voltage drop around a loop containing e_1.

8.4 Gaussian Elimination: A Closer Look

In the previous section, you learned how to do Gaussian elimination for any system of linear equations. This remarkable algorithm can also be used to develop a theoretical understanding of linear systems. In this section we take a closer look. We introduce the concept of echelon form and then present some general results about when a linear system has solutions. We also discuss the validity and computational complexity of the algorithm.

Definition 1. A matrix is in **echelon form** if

1. Any all-zero rows are below all other rows; and
2. The first nonzero entry of any row (called the **leading entry** or a **pivot entry**) is strictly farther right than the first nonzero entry of any row above it.

A matrix is in **reduced echelon form** if all the following are true:

1. It is in echelon form.
2. The first nonzero entry of any row is a 1. This is called a **leading 1** or a **pivotal 1**.
3. All entries above a pivotal 1 are 0.

Echelon form is also called **row-echelon form**, and reduced echelon form is also called **row-reduced echelon form** or **reduced row-echelon form**.

EXAMPLE 1 Consider

$$\left[\begin{array}{cccccc|c} 1 & 2 & 1 & 1 & 4 & 1 & 2 \\ 0 & 0 & 2 & -2 & 6 & 2 & 2 \\ 0 & 0 & 0 & 0 & 0 & 3 & 6 \\ 0 & 0 & 0 & 0 & 0 & 0 & 0 \end{array}\right].$$

This is the augmented matrix at the end of the downsweep stage from Example 6, Section 8.3. Call it $[A' \mid \mathbf{b}']$. The coefficient part A' is in echelon form. We've added shading to highlight the echelon form. (The entire augmented matrix is in echelon form as well, but this isn't the important point.)

EXAMPLE 2 Consider

$$\left[\begin{array}{cccccc|c} 1 & 2 & 0 & 2 & 1 & 0 & 1 \\ 0 & 0 & 1 & -1 & 3 & 0 & -1 \\ 0 & 0 & 0 & 0 & 0 & 1 & 2 \\ 0 & 0 & 0 & 0 & 0 & 0 & 0 \end{array}\right].$$

This is the final augmented matrix from Example 6, Section 8.3. Call it $[A'' \mid \mathbf{b}'']$.

The coefficient part A'' is in reduced echelon form. Again we've highlighted this with shading.

It's no accident that our examples of echelon form came from Gaussian elimination. (That's why the leading entries are also called pivotal entries.) If you thoroughly understand what Algorithm GAUSS does, your reaction to the next theorem should be "Of course!".

Theorem 1. Suppose that Gaussian elimination (in the form of Algorithm GAUSS) is applied to $[A \mid \mathbf{b}]$. Let $[A' \mid \mathbf{b}']$ be the matrix obtained just after the last downsweep, and let $[A'' \mid \mathbf{b}'']$ be the final matrix. Then A' is in echelon form and A'' is in reduced echelon form.

The importance of this theorem is that it summarizes why Gaussian elimination allows us to read off solutions in a useful way. If $[A'' \mid \mathbf{b}'']$ is in reduced echelon form, then when we translate back to equations, each equation involves exactly one pivot variable and expresses that variable solely in terms of free variables. The reason is that, for each pivotal 1 in a reduced echelon matrix, every other entry in its column is 0; conditions 1 and 3 in the definition of reduced echelon form ensure this. If $[A' \mid \mathbf{b}']$ is in echelon form, each equation involves exactly one *leading* pivot variable and expresses that variable solely in terms of free variables and those pivot variables farther to the right. Thus if we solve the equations from the bottom, by the time we get to any pivot variable, we already know formulas for all the other pivot variables in its equation and thus can express the new pivot variable solely in terms of free variables. In short, echelon form is what makes back-substitution work.

PROOF OF THEOREM 1 A' is in echelon form because of procedures Downsweep and ScanSwitch: Downsweep makes all entries below a pivot zero. After that, ScanSwitch assures that the next pivot (if any) is in the next row, that it is strictly to the right of the previous pivot, and that all entries left and down from the new pivot are 0. ScanSwitch also assures that any zero row appearing above a nonzero row gets moved down.

A'' is in reduced echelon form because of Scale and Upsweep. First, these procedures only modify entries to the right or above pivots, so in going from A' to A'', echelon form is maintained. Second, Scale makes each pivot a 1, and Upsweep makes each entry above a pivot a 0. ∎

When we say Gaussian elimination reduces A to B, we mean that B is the final, reduced echelon matrix obtained by GAUSS. When we say B is the reduced echelon form of A, we again mean that it is the reduced echelon form obtained using GAUSS. (In fact, even if you use some other sequence of elementary row operations, this is the only reduced echelon matrix you can obtain from A; see [2, Supplementary Problems].)

All right, the above explains why Gaussian elimination works for any given linear system, but a lot of questions remain. Can we make any general observations about when we get no solutions, exactly one solution, or many solutions? Yes we can.

Below, by the system $[A \,|\, \mathbf{b}]$ we mean the system of linear equations for which $[A \,|\, \mathbf{b}]$ is the matrix shorthand. (We'll develop more standard notation in Section 8.6.)

Theorem 2. Consider the system $[A \,|\, \mathbf{b}]$, where A is $m \times n$.

1) This system has at least one solution $\Longleftrightarrow$ when Gaussian elimination is applied, no row appears with a nonzero in the $\mathbf{b}$ column only.

Now fix A and let $\mathbf{b}$ vary. That is, consider any one $m \times n$ matrix A and all column m-tuples $\mathbf{b}$. Then

2) The system $[A \,|\, \mathbf{b}]$ has at *least* one solution for every $\mathbf{b} \Longleftrightarrow$ Gaussian elimination finds a pivot in every row of A.
3) The system $[A \,|\, \mathbf{b}]$ has at *most* one solution for every $\mathbf{b} \Longleftrightarrow$ Gaussian elimination finds a pivot in every column of A.
4) The system $[A \,|\, \mathbf{b}]$ has *exactly* one solution for every $\mathbf{b} \Longleftrightarrow$ Gaussian elimination reduces A to the identity matrix I.

PROOF All four numbered statements are of the form $P \Longleftrightarrow Q$. We will prove the first three by proving $Q \Longrightarrow P$ and $\neg Q \Longrightarrow \neg P$ (where $\neg$ is "not" as in Section 0.7 or Chapter 7).

As for statement 1), if some row is all 0's except in the $\mathbf{b}$ column (i.e., $\neg Q$), that means one equation in the system is $0 =$ nonzero, so the system has no solutions. (Furthermore, the system at this point has the same solutions as the original system, by Theorem 1, Section 8.2.) On the other hand, if Gaussian elimination does not produce a row that is all 0's except in the $\mathbf{b}$ column, then we read off one or more solutions from the final reduced echelon form.

As for statement 2), first recall that the entries in $\mathbf{b}$ have no effect on what Gaussian elimination does to A, so it makes sense to talk about Gaussian elimination working on A alone. If there is a pivot in every row of A, Gaussian elimination never creates an all-zero row in A, so in reducing $[A \,|\, \mathbf{b}]$ there can never be a row that is zero except in $\mathbf{b}$. Thus there is at least one solution no matter what $\mathbf{b}$ is. On the other hand, suppose Gaussian elimination did produce a zero row. Take this matrix with a zero row and augment it with a $\mathbf{b}'$ that has 1 in that row and anything you please elsewhere. Now run Gaussian elimination backwards from this augmented matrix, obtaining $[A \,|\, \mathbf{b}]$ for some $\mathbf{b}$. Then $[A \,|\, \mathbf{b}]$ has no solution.

Now consider statement 3). If there is a pivot in every column, then there are no free variables, so there is at most one solution for any $\mathbf{b}$. On the other hand, if some column does not have a pivot, then there are free variables, so depending on $\mathbf{b}$ there is either no solution or infinitely many. Furthermore, for some $\mathbf{b}$'s, there are surely solutions. Specifically, for $\mathbf{b} = \mathbf{0}$ there is always the trivial solution (all variables 0), so for $\mathbf{b} = \mathbf{0}$ the free variables result in infinitely many solutions. This contradicts the claim that every $\mathbf{b}$ has at most one solution.

Finally, statement 4) is proved from the logical conjunction of 2) and 3). $[A \,|\, \mathbf{b}]$ has exactly one solution (for every $\mathbf{b}$) $\iff$ it has both at least one and at most one $\iff$ A has a pivot in every row and every column. That means A is square, that the pivotal 1's are the only nonzero entries, and that these pivotal 1's run down the main diagonal (since there are no extra columns for "gaps"). Thus A has a pivot in every row and column $\iff$ the reduced echelon form of A is I. ∎

Using Theorem 2 we can analyze other cases of interest. For instance, what Property of A will guarantee that the system $[A \,|\, \mathbf{b}]$ will have infinitely many solutions for every $\mathbf{b}$? Answer: that there be a pivot in every row but not in every column. This follows in two steps. First, from 2) we know that a pivot in every row ensures at least one solution for every $\mathbf{b}$. Second, if some column has no pivot, then there is a free variable, which means infinitely many solutions whenever there is at least one. Other cases of interest are treated in the problems [10–12].

We will now restate Theorem 2 in a more standard form. By the **rank** of a matrix A we mean the number of pivots produced when A is reduced to reduced echelon form using Algorithm GAUSS. Notice that this definition seems to depend on the algorithm as well as the matrix. Conceivably, A could have a different rank if the rows or columns were arranged in a different order, or if a different variant of Gaussian elimination were used. (We will pursue this issue further in Section 8.8.)

By the rank of an augmented matrix $[A \,|\, \mathbf{b}]$ we mean the number of pivots when $[A \,|\, \mathbf{b}]$ is row-reduced as if it were an *un*augmented matrix, that is, as if $\mathbf{b}$ were an ordinary column. In computational practice, one never treats $[A \,|\, \mathbf{b}]$ this way, but this definition is convenient for our theorem.

Theorem 3. Let A be $m \times n$. Then:

1) For any particular m-tuple $\mathbf{b}$, $[A \,|\, \mathbf{b}]$ has at least one solution $\iff$ $\text{rank}(A) = \text{rank}([A \,|\, \mathbf{b}])$.
2) The system $[A \,|\, \mathbf{b}]$ has at least one solution for every m-tuple $\mathbf{b}$ $\iff$ $\text{rank}(A) = m$.
3) The system $[A \,|\, \mathbf{b}]$ has at most one solution for every $\mathbf{b}$ $\iff$ $\text{rank}(A) = n$.
4) The system $[A \,|\, \mathbf{b}]$ has exactly one solution for every $\mathbf{b}$ $\iff$ $\text{rank}(A) = m = n$.

PROOF Claims 2) and 3) are direct translations of claims 2) and 3) in Theorem 2 into rank language. Claim 4) is simply the conjunction of 2) and 3). (We leave out the part from Theorem 2 about getting I because we want to concentrate on what can be said directly using the rank concept.) Claim 1) is equivalent to 1), as we now show. Let us write $[A\,\mathbf{b}]$ when we don't want to consider column $\mathbf{b}$ as augmentation. (Note the lack of a vertical bar.) If Algorithm GAUSS applied normally to $[A\,|\,\mathbf{b}]$ does not lead to a row with a nonzero in the $\mathbf{b}$ column only, then when GAUSS is applied to $[A\,\mathbf{b}]$ exactly the same calculations are performed on A and $\mathbf{b}$ as before. (Why?) In particular, the pivots are the same and so $\text{rank}(A) = \text{rank}(\,[A\,|\,\mathbf{b}]\,)$. On the other hand, if Gaussian elimination applied to $[A\,|\,\mathbf{b}]$ does lead to a row with a nonzero only in $\mathbf{b}$, then the downsweep will stop at this point; but in $[A\,\mathbf{b}]$ the algorithm will go on and pivot on that nonzero $\mathbf{b}$ entry. Hence $\text{rank}(A) < \text{rank}(\,[A\,|\,\mathbf{b}]\,)$. ∎

Analysis and Verification

We will be brief and informal. As for the correctness of Algorithm GAUSS, there are three things to show: 1) the algorithm terminates properly (neither going on forever nor aborting because we ask it to do something impossible); 2) when it terminates it has obtained a reduced echelon form; and 3) the reduced echelon form has the interpretation we have claimed for the solutions of the original problem. We have already argued 2) and 3), so let us turn to 1).

The only way GAUSS could abort is if we asked it to divide by 0 or look up a matrix entry for a nonexistent row or column. GAUSS never tries to divide by 0 because the only entries it tries to divide by are those a_{ij} that ScanSwitch has determined are valid pivots.

GAUSS would try to look up a nonexistent matrix entry only if some variable that is incremented by assignment—i in ScanSwitch, r in Main, or j in both Main and ScanSwitch—gets out of bounds. The **if** clause in the **repeat** loop of ScanSwitch keeps i in bounds. As for r, it is changed only in the last line of the **repeat** loop of Main, and as soon as that line makes $r > m$ (out of bounds), the **while** condition at the top of the loop kicks us down to the **for** loop, where r is immediately lowered. As for j, the crucial case is when it becomes $n{+}1$; the value $j = n{+}1$ is in bounds because of the $\mathbf{b}$ column, but it is not a value that should produce a pivot. As noted in Example 6, Section 8.3, as soon as j becomes $n{+}1$ (in either ScanSwitch or Main), various lines in ScanSwitch and Main move us along to set $c_r = n{+}1$ and then the **if** clause of Main pushes out right into the **for** loop of Main.

This finishes the validation. Of course, a really complete argument would have to do a lot more. In Chapter 7 we indicated the sorts of things that need to be done to complete the argument. Also, although we haven't used the word "induction" or the phrase "loop invariant", we hope you felt their presence.

As for a complexity analysis of GAUSS, for the average time analysis we would need a probabilistic model of the occurrence of 0's in potential pivot positions, which is tricky. However, it turns out that the average case is essentially the same as the worst case, which occurs when no 0's appear in pivot positions. As usual, assume that we are given $[A\,|\,\mathbf{b}]$, where A is $m \times n$. The computation for the worst case itself breaks into two cases, $m \leq n$ and $m \geq n$. The first is quite easy if you note that, as there are m pivots and they are assumed to occur in the first m columns, GAUSS treats the remaining $n - m$ columns of A as if they are additional

b columns. Furthermore, when no 0's get in the way, GAUSS does computations on a square system matrix the same way GAUSS-SQUARE does. Therefore, GAUSS does $(m^3-m)/3$ computations on the leftmost square matrix (from Section 8.2) and m^2 computations on each of the $m-n+1$ "**b**" columns. Summing and dropping all but the highest order terms, we find that the complexity is

$$\text{order} \quad nm^2 - \tfrac{2}{3}m^3. \tag{1}$$

We ask you to verify this in more detail, and do the case $m \geq n$, in [17,18].

Problems: Section 8.4

1. Determine whether each of the following matrices is in echelon form or reduced echelon form.

a) $\begin{bmatrix} 1 & 2 & 4 \\ 0 & 3 & 5 \\ 0 & 0 & 6 \end{bmatrix}$ b) $\begin{bmatrix} 1 & 2 & 3 & 4 \\ 0 & 5 & 0 & 6 \\ 0 & 0 & 0 & 1 \end{bmatrix}$

c) $\begin{bmatrix} 1 & 0 & 0 \\ 0 & 0 & 1 \\ 0 & 1 & 0 \end{bmatrix}$ d) $\begin{bmatrix} 2 & 1 & 0 & 0 \\ 0 & 0 & 1 & 0 \\ 0 & 0 & 0 & 1 \end{bmatrix}$

2. Is $\begin{bmatrix} 1 & 1 & 0 \\ 0 & 0 & 1 \end{bmatrix}$ in reduced echelon form? Aren't there too many pivotal 1's?

3. Determine the rank for each matrix in [1].

4. Determine the rank for each matrix in [5] of Section 8.3.

5. In a 6×7 reduced echelon matrix with five pivotal 1's, at most how many entries are nonzero? How about in an $m \times n$ reduced echelon matrix with p pivots?

6. In a 6×7 echelon matrix with five pivotal entries, at most how many entries are nonzero? How about in an $m \times n$ echelon matrix with p pivots?

7. It may seem pretty silly to row reduce the following matrices—they are already reduced—but a good test of your understanding of an algorithm is to apply it to cases where there seems to be nothing to do. So step through Algorithm GAUSS for the following matrices, and explain why the algorithm changes nothing.

a) $\begin{bmatrix} 0 & 0 & 0 \\ 0 & 0 & 0 \\ 0 & 0 & 0 \end{bmatrix}$ b) $\left[\begin{array}{ccc|c} 1 & 0 & 0 & 1 \\ 0 & 1 & 0 & 2 \\ 0 & 0 & 1 & -1 \end{array}\right]$

8. True or false: When Algorithm GAUSS is applied to $[A \mid \mathbf{b}]$, and the downsweep half is completed, the entire augmented matrix (not just A) is in echelon form. If false, when is it false?

9. True or false: When Algorithm GAUSS is applied to an augmented matrix, in the end the entire augmented matrix is in reduced echelon form. If false, when is it false?

10. Characterize those matrices A for which, whatever $\mathbf{b}$ is, if $[A \mid \mathbf{b}]$ has at least one solution, then it has infinitely many solutions. Provide a characterization in terms of Gaussian elimination (as in Theorem 2) and then in terms of rank (as in Theorem 3).

11. Repeat [10] for those matrices A for which there are $\mathbf{b}$'s for which $[A \mid \mathbf{b}]$ has no solution, and there are no $\mathbf{b}$'s for which it has infinitely many solutions.

12. Are there any matrices A for which $[A \mid \mathbf{b}]$ never has a solution except for $\mathbf{b} = \mathbf{0}$? Explain.

13. Recall that the rth pivot in GAUSS is always in position (r, c_r). Explain why $c_r \geq r$.

14. In the main part of GAUSS, can the order of the lines "Scale(i, c_i)" and "Upsweep(i, c_i)" be switched? Why? If not, can one or the other of these procedures be modified slightly so that the switch is OK?

15. There is a variant of GAUSS in which Scale(i, j) and Upsweep(i, j) are replaced by the following single procedure:

```
procedure Scale-Upsweep(i, j)
  for l = j+1 to n+1
    if a_il ≠ 0 then
      a_il ← a_il/a_ij
      for k = i-1 downto 1
        a_kl ← a_kl - a_kj a_il
      endfor
    endif
  endfor
  for k = i-1 downto 1
    a_kj ← 0
  endfor
  a_ij ← 1
endpro
```

This procedure does several things differently than in GAUSS, but it's still correct. Explain what it does differently and why it's advantageous to do it this way.

16. In GAUSS suppose we replace the innermost line of Scale(i, j) with

if $a_{il} \neq 0$ **then** $a_{il} \leftarrow a_{il}/a_{ij}$

Suppose we also replace the innermost line of Upsweep(i, j) with

if $a_{il} \neq 0$ **then** $a_{kl} \leftarrow a_{kl} - a_{kj}a_{il}$

Do we gain any advantage? Are these changes as advantageous as those in [15]?

17. Using the detailed method of Section 8.2, determine the step count for Algorithm GAUSS when $[A\,|\,\mathbf{b}]$ is $m\times(n+1)$ and $m \leq n$. Assume that no 0's appear where you want to pivot. (No pivot 0's is the worst case.) Make a chart like Table 8.1 (Section 8.2). Verify that expression (1) gives the right order of magnitude.

18. Same as [17] except now $m \geq n$.

19. Gauss-Jordan reduction was defined as a variant of GAUSS-SQUARE in [27d, Section 8.2].

a) Modify Gauss-Jordan reduction so that it succeeds for nonsquare matrices as well. That is, make it a variant of GAUSS.

b) For an $m \times n$ matrix A, determine the number of multiplications and divisions needed to perform Gauss-Jordan. Assume no 0's arise where you want to pivot. (The answer will depend on which is bigger, m or n.)

8.5 Algorithm LU-GAUSS

In the preceding sections, we developed a general and powerful algorithm for solving linear equations, but for the real world this is usually only the start. Pretty soon somebody comes up with a practical situation which the algorithm doesn't treat well enough. You should look at such setbacks positively. They are invitations to understand your algorithm better and to devise even more useful variants.

In this section we find and fix a difficulty with Algorithm GAUSS-SQUARE. (For simplicity, we assume throughout that we are dealing with square systems and that no 0's appear where we want to pivot. In the problems you will be asked to treat nonsquare systems and thus fix the corresponding difficulty with Algorithm GAUSS.) The result is a variant algorithm, LU-GAUSS. It is essentially the algorithm actually used in commercial computer codes. The letters L and U stand for "lower triangular" and "upper triangular" matrices, which play an important role, as you'll see.

The difficulty is: It often happens that one has to solve several linear systems which differ only in the constants on the right-hand side. That is, one has several problems $[A\,|\,\mathbf{b}]$ with the *same* A. For instance, problems [25, 26, Section 8.3] dealt with finding the current in an electrical network. As indicated there, so long as the wire connections remain the same, different specifications about current and voltage result only in different **b**'s. The coefficient matrix depends solely on the graph representing the network.

So suppose your boss gives you a sequence of problems

$$[A\,|\,\mathbf{b}_1],\quad [A\,|\,\mathbf{b}_2],\quad \ldots,\quad [A\,|\,\mathbf{b}_t]. \tag{1}$$

How can you solve them? What's the difficulty?

One way to solve sequence (1) is obvious: Use Algorithm GAUSS-SQUARE over and over. But think of all the redundancy. Everything to the left of the augmentation line is the same each time. This is the sort of situation we meant when we said in the first paragraph that your general algorithm might not work well enough.

So, do all the problems at the same time: Write down A just once and make each **b** another augmentation column.

EXAMPLE 1 Solve the two systems

$$\begin{aligned} 2x - 2y + 4z &= -6 \\ 2x - y + 6z &= -6 \\ -x + 2y - 3z &= 6 \end{aligned} \qquad \text{and} \qquad \begin{aligned} 2x - 2y + 4z &= -2 \\ 2x - y + 6z &= -3 \\ -x + 2y - 3z &= 3. \end{aligned}$$

Solution We set up the doubly augmented matrix

$$\left[\begin{array}{ccc|cc} 2^* & -2 & 4 & -6 & -2 \\ 2 & -1 & 6 & -6 & -3 \\ -1 & 2 & -3 & 6 & 3 \end{array}\right].$$

We now proceed with downsweeps as usual. (Since the first set of equations happens to be that in Example 1, Section 8.2, the only new work is in the final column and we go quickly.) When the downsweeps are done we have

$$\left[\begin{array}{ccc|cc} 2 & -2 & 4 & -6 & -2 \\ 0 & 1 & 2 & 0 & -1 \\ 0 & 0 & -3^* & 3 & 3 \end{array}\right]. \tag{2}$$

Continuing with scale and upsweep, we finally arrive at

$$\left[\begin{array}{ccc|cc} 1 & 0 & 0 & 1 & 2 \\ 0 & 1 & 0 & 2 & 1 \\ 0 & 0 & 1 & -1 & -1 \end{array}\right].$$

Thus the solution to the first system is $x = 1$, $y = 2$, and $z = -1$ (as we already knew) and the solution to the second is $x = 2$, $y = 1$, and $z = -1$. ∎

We hope it is clear why this method works. Each row is no longer an abbreviation for one equation but rather for two equations, one from each system. Since we perform the same operations in both systems, we can write them both together.

There is just one hitch: This approach requires that you wait until you have all the problems before you solve any of them. But the first and last problems may arrive days apart, and your boss always wants each answer right away. A practical algorithm must be one which can be run sequentially.

We can't just forget the row reduction of A and work only with the new **b**'s, because what happens to A determines what to do to **b**. So we ask: What is the minimum amount of information from the reduction of A which we need to save from the first problem in order to do later ones?

Answer: We need to know the scaling factors and the number of times each row is subtracted from each other row. This suggests creating and saving a second matrix. The (k, i) entry of this matrix will be the number of times row i is subtracted from row k, except if $i = k$ it will be the factor by which row i is scaled. (We are using index letters exactly the way we used them in GAUSS-SQUARE, which you should consult.) Note that we scale each row just once and subtract each row i from each other row k just once. (If $i < k$ the subtraction occurs in the downsweep; if $i > k$ it occurs in the upsweep.) Thus there will be no ambiguity as to which operation an entry in this new matrix refers to.

Now the surprising part: We don't need to declare a new matrix at all. All this information can be—and to a large extent already is—stored in A during the operation of GAUSS-SQUARE.

Consider the echelon matrix U obtained after all downsweeps. The scaling factors are the diagonal entries. (Remember, we've assumed that no 0's appear in pivot positions during downsweep.) Furthermore, the multiplicative factors for upsweep are the entries above the diagonal. For instance, in matrix (2), the scaling factor for row 3 is -3 and the number of times row 3 is subtracted from row 2 (once row 3 is scaled) is 2—remember, our goal is to zero out the entries above the pivot. Similarly, the scaling factor for row 2 is 1, and after scaling we will subtract row 2 from the first row -2 times.

What about the multiples for downsweep? They never appear in A as it is reduced. All that ever happens to a_{ki} for $k > i$ during Downsweep(i) is that it becomes 0. In fact, a_{ki} becomes 0 just after we determine the multiple (called m_k in GAUSS-SQUARE) by which row i should be subtracted from row k. But that's great. Who needs the 0? It was just a place holder. So let's put the multiplier in there.

EXAMPLE 2 For the linear systems in Example 1, compute the matrix which stores all the scale and multiplier information as just described. Then use this information to solve the second system separately from the first.

Solution We go step by step, since we are doing something new. The initial matrix (we don't need the constant column for either system yet) is

$$\begin{bmatrix} 2^* & -2 & 4 \\ 2 & -1 & 6 \\ -1 & 2 & -3 \end{bmatrix}.$$

Downsweeping the first column corresponds to subtracting the first equation $2/2 = 1$ times from the second and $-1/2$ times from the third. So the matrix we write down this time is

$$\left[\begin{array}{c|cc} 2 & -2 & 4 \\ \hline 1 & 1^* & 2 \\ -1/2 & 1 & -1 \end{array}\right].$$

Note that we boxed off the entries which appear where we used to put 0's. They do not correspond to coefficients in any equations.

Now pivot in the second column. We subtract the second row 1 time from the third and write

$$A' = \left[\begin{array}{c|c|c} 2 & -2 & 4 \\ \hline 1 & 1 & 2 \\ \cline{2-2} -1/2 & 1 & -3^* \end{array}\right].$$

We have completed the downsweep and have what we need. We give this matrix its own name, A', because we later refer to it repeatedly. We may now use A' to solve the second system (actually, either system). The **b** vector for the second system is

$$\begin{pmatrix} -2 \\ -3 \\ 3 \end{pmatrix}.$$

We know the first thing that should be done to **b** is the downsweep corresponding to the first column of the original A. The subtraction multiples are sitting in the first column of A', below the line. So we subtract the first entry of **b** once from the second entry and $-1/2$ times from the third:

$$\begin{pmatrix} -2 \\ -1 \\ 2 \end{pmatrix}.$$

The next thing we do to **b** is the downsweep generated by the second pivot. The entry in the second column of A' below the line tells us to subtract the second entry of **b** once from the third:

$$\begin{pmatrix} -2 \\ -1 \\ 3 \end{pmatrix}.$$

Now we are ready for scale and upsweep from each pivot in reverse order. The (3,3) entry of A' tells us that the third scaling factor is -3, so we divide the third entry of **b** by that:

$$\begin{pmatrix} -2 \\ -1 \\ -1 \end{pmatrix}.$$

Next, we subtract multiples of the third row from higher rows. Matrix A' tells us to subtract 4 times from the first row and 2 times from the second. Once again, carrying this out on **b** only gives

$$\begin{pmatrix} 2 \\ 1 \\ -1 \end{pmatrix}.$$

Then, we scale the second entry (no work since A' tells us the scaling factor is 1) and subtract the result -2 times from the first entry:

$$\begin{pmatrix} 4 \\ 1 \\ -1 \end{pmatrix}.$$

Finally, we scale the top entry by the scaling factor in the (1,1) position of A':

$$\begin{pmatrix} 2 \\ 1 \\ -1 \end{pmatrix}.$$

Thus the solution to the second problem in Example 1 is $x = 2$, $y = 1$, and $z = -1$. Of course, this is the same answer we got before, but the point is that we did it without further row operations on the original A or any other coefficient matrix. We just read off entries from A' and did things to **b** with them. ∎

This revised method appears as Algorithm 8.3, LU-Gauss. We conclude with a few more general remarks about the method.

First, if we are going to use LU-Gauss every time after the first that we get a problem $[A\,|\,\mathbf{b}]$ with the same A, then we might as well use it the first time too and have just one algorithm. For instance, in Example 2, after getting A' we would apply exactly the same steps to $\mathbf{b} = (-6,-6,6)^{\mathrm{T}}$ that we applied to $\mathbf{b} = (-2,-3,3)^{\mathrm{T}}$. Indeed, we would work on $(-6,-6,6)^{\mathrm{T}}$ first.

Second, although we have put all the coefficient data we need in one matrix A', conceptually it has two separate parts, which we have indicated by the vertical and horizontal lines within the matrix. Each part can be written as a matrix by itself. The part below and left of the lines is called L because it is **lower triangular**, that is, all entries are on or below the main diagonal. The other part is called U because it is **upper triangular**. More precisely, L and U are names for whole matrices which contain these entries. For instance, for Example 2,

$$L = \begin{bmatrix} 1 & 0 & 0 \\ 1 & 1 & 0 \\ -1/2 & 1 & 1 \end{bmatrix}, \qquad U = \begin{bmatrix} 2 & -2 & 4 \\ 0 & 1 & 2 \\ 0 & 0 & -3^* \end{bmatrix}.$$

For U we have simply filled in 0's below the main diagonal. This is precisely the matrix we called U early in this section, that is, the echelon matrix which Gauss-Square obtains when downsweep is finished. For L we have filled in 0's above the main diagonal, but on the main diagonal itself we have placed 1's. For now, simply regard this as a convention; we give a better reason at the end of the next section. Anyway, the method of Algorithm LU-Gauss is often called **LU Decomposition**.

Algorithm 8.3

LU-Gauss

```
Input A, b                                          [A is n×n, b is n×1]
Algorithm LU-GAUSS                                  [Works if no zero pivots]
    procedure L-Downsweep(i)
        for k = i+1 to n
            a_ki ← a_ki/a_ii                        [a_ki is now the multiplier m_ki]
            for l = i+1 to n
                a_kl ← a_kl − a_ki a_il [Normal row reduction on U part]
            endfor
        endfor
    endpro
    procedure B-Sweep
        for i = 1 to n                              [Downsweep on b]
            for k = i+1 to n
                b_k ← b_k − a_ki b_i
            endfor
        endfor
        for i = n downto 1
            b_i ← b_i/a_ii                          [Scale b]
            for k = i−1 downto 1                    [Upsweep on b]
                b_k ← b_k − a_ki b_i
            endfor
        endfor
    endpro
    for i = 1 to n                                  [Main algorithm]
        L-Downsweep(i)
    endfor
    B-Sweep
Output A, b      [b = solutions, A = pivot instructions, called A′ in text]
```

Our final remark concerns efficiency. As noted in Section 8.2, scale/upsweep and back-substitution are not quite the same. The numbers of multiplications and divisions are the same, but in back-substitution you compute only the values of the variables; you don't continue to update the coefficient matrix (by assigning 0's and 1's). Well, LU-GAUSS makes the same savings as back-substitution. If you did [22, Section 8.2], which asked you to rewrite GAUSS-SQUARE to do back-substitution directly, then what you came up with probably looks very much like the second for-loop of the B-Sweep procedure in LU-GAUSS. But LU-GAUSS does even more.

It doesn't assign 0's in the downsweep either, but rather assigns to those spots the multiples m_{ij} which were computed but then discarded in GAUSS-SQUARE. So LU-GAUSS is not only more practical because it saves work in later problems, but it is slightly more efficient even for one problem. For this reason, commercial codes for solving linear systems are built around the LU idea.

The existence of LU-GAUSS is one of the reasons we separated the step count for the **b** column from the rest. The analysis of LU-GAUSS is a triviality given the analysis we did of GAUSS-SQUARE. The output of the two algorithms looks different, but that's because they make different assignments, and we've counted each assignment as zero steps. Thus for an $n \times n$ system, LU-GAUSS takes exactly $(n^3-n)/3$ steps on the coefficient matrix and n^2 on each **b** column. Therefore it takes about $n^3/3$ work for the first problem it is given with a particular matrix A, and n^2 work for each succeeding problem. In other words, by cleverly filling in what used to be 0 entries, we reduce our work for future problems with A by a factor of $n/3$.

Finally, we must inform you that we have barely scratched the surface with regard to smart ideas for improving Gaussian elimination, and furthermore, additional improvements *are* necessary. Although our version of Gaussian elimination works fine assuming "exact arithmetic", with real-world data computers introduce roundoff error. If you pivot on a small entry such as .00001, you can introduce a lot of error. Even if you always pivot on medium-sized entries, in problems requiring hundreds of pivots a good deal of error can accumulate. So clever codes stay as far away as possible from pivoting on small entries.

The most widely used approach is called **partial pivoting**. If there has just been a pivot in position (i, j), then *every* entry in column $j+1$ below the pivot is considered for the next pivot, even if $a_{i+1,j+1} \neq 0$. The entry with the largest absolute value is chosen; its row is exchanged with row $i+1$, and it becomes the next pivot. An alternative is **complete pivoting**. In this version, if there was just a pivot in position (i, j), then every a_{kl} with $k > i$ and $l > j$ is considered, and the next pivot is again the one with $|a_{kl}|$ largest. If this pivot is not in position $(i+1, j+1)$, then both a row switch and a column switch are performed to move it to the $(i+1, j+1)$ position. Note: column switches, like row switches, do not change the solution set, but they do require you to update the order of the variables.

Finally, there are methods for solving linear systems that aren't based on Gaussian elimination at all, notably the "QR" method. Although these methods work much better for certain types of problems, nothing is as efficient as Gaussian elimination as a general method. But the situation may change because linear algebra algorithms is currently a very active research area—nobody is certain yet what sort of equation solving algorithm is best for parallel-processor computers.

If you are interested in learning more about algorithms for solving linear equations, you might look at some of the programs in the LINPACK package of linear algebra routines, available commercially in the IMSL library and in such matrix algebra software as MATLAB.

Problems: Section 8.5

1. Solve the systems

$$\begin{aligned} x + 2y &= 3 \\ 2x + 3y &= 4 \end{aligned} \qquad \text{and} \qquad \begin{aligned} x + 2y &= -1 \\ 2x + 3y &= \ \ 0 \end{aligned}$$

a) by the method of Example 1;
b) by the LU method of Example 2.

In part **b**), as you move from one column **b** to another, indicate what you are doing. For instance, if you subtract the third row 6 times from the second row, you might write $6_{2,3}$ next to it. Explain whatever shorthand you use.

2. Solve the systems

$$\begin{aligned} 2x + 3y &= 4 \\ x + 2y &= 3 \end{aligned} \qquad \text{and} \qquad \begin{aligned} 2x + 3y &= \ \ 0 \\ x + 2y &= -1 \end{aligned}$$

a) by the method of Example 1;
b) by the LU method of Example 2.

Note that this is the same set of equations as in [1], so the solution will be the same. However, the work will be different.

3. Use Algorithm LU-GAUSS to solve the system

$$\begin{aligned} x + 2y + 3z &= 1 \\ 2x + 3y + 4z &= 2 \\ -x + y + 2z &= 1. \end{aligned}$$

4. We've indicated that LU-GAUSS is better for computers than GAUSS, even when A won't be used again. In your opinion, which algorithm is better for people, especially people who do this sort of thing only occasionally for homework? Explain.

5. In [6, Section 8.2], you were asked to help the Army feed its recruits. Now, after a Congressional investigation, there is so much clamor that the Army feels compelled to provide more nutriment. It decides to provide recruits a dinner containing 390 g of protein, 510 g of fat, and 600 g of carbohydrates. However, the Army does not change from serving meat, potatoes, and cabbage.

a) You are hired to tell the Army how many ounces of each type of food to serve at dinner. Do so. See the data in [6, Section 8.2].

b) Congress is not satisfied with this change and some time later passes a law requiring 360 g of protein, 504 g of fat, and 400 g of carbohydrate. Assuming that the Army still insists on meat, potatoes, and cabbage, how much should it serve now?

c) Congress could have mandated 440 g of each of these three nutrients. That sounds good (more protein, less fat), but what's wrong? Assume that the Army is committed to meat, potatoes, and cabbage.

6. Solve $[A \mid \mathbf{e}_1]$, $[A \mid \mathbf{e}_2]$, and $[A \mid \mathbf{e}_3]$ for

$$A = \begin{bmatrix} 1 & 1 & 0 \\ 3 & 1 & 1 \\ 2 & 1 & 1 \end{bmatrix},$$

$$\mathbf{e}_1 = \begin{pmatrix} 1 \\ 0 \\ 0 \end{pmatrix}, \quad \mathbf{e}_2 = \begin{pmatrix} 0 \\ 1 \\ 0 \end{pmatrix}, \quad \mathbf{e}_3 = \begin{pmatrix} 0 \\ 0 \\ 1 \end{pmatrix}.$$

Solve two ways: by the method of Example 1 and by the method of Example 2.

7. Use LU-GAUSS to solve the equations associated with

$$\left[\begin{array}{rrr|r} 1 & 1 & 1 & 0 \\ -1 & 1 & 3 & -2 \\ 2 & -1 & 4 & 4 \end{array}\right].$$

8. In Algorithm GAUSS-SQUARE, let $V = [v_{ij}]$ be what A has been reduced to just before we are ready to scale and upsweep from pivot i. It is clear that the scaling factor is v_{ii} and that the number of times row i will be subtracted from row k for $k < i$ is v_{ki}. However, what we claim in the text is that these factors are u_{ii} and u_{ki}. That is, we claim we don't have to wait until we get to V in order to read off these factors; they are already present in U, the echelon matrix after downsweep. Justify this claim. (If U weren't in echelon form, the claim would be false.)

9. Suppose we decide to perform the variant of GAUSS-SQUARE in which all upsweeping is done before all scaling. [See 27b, Section 8.2.] It is still possible to determine all the scale factors and upsweep multiples from U. How?

10. Suppose we decide to perform the variant of GAUSS-SQUARE in which all scaling is done before all upsweeping. [See 27c, Section 8.2.] It is

still possible to determine all the scale factors and upsweep multiples from U. How?

11. Devise a variant of Algorithm GAUSS-SQUARE that implements partial pivoting. (Use GAUSS-SQUARE, not LU-GAUSS, because row switches make the latter quite tricky; see [15].)

When A is not square, or there are 0's where you want to pivot, implementing an LU method gets tricky. In [12–15], you are asked to extend LU-GAUSS to progressively more general matrices, starting in each problem with specific matrices.

12. Consider matrices where there are more rows than columns, but 0's don't arise where you want to pivot.
    a) Devise an LU-type algorithm for solving
$$\left[\begin{array}{cc|c} 1 & 2 & 1 \\ 3 & 4 & 2 \\ 5 & 6 & 3 \end{array}\right] \quad \text{and} \quad \left[\begin{array}{cc|c} 1 & 2 & 3 \\ 3 & 4 & 2 \\ 5 & 6 & 1 \end{array}\right].$$
    b) Write up what you did in algorithmic language. Your algorithm has to work only for $m \times n$ matrices with $m > n$ and in which no 0's arise where you want to pivot.

13. Now consider problems with more columns than rows but in which no 0's arise where you want to pivot.
    a) Devise an LU-type algorithm to solve
$$\begin{aligned} x + y + z + w &= 3 \\ x + 2y - z + 3w &= 5 \end{aligned}$$
    and
$$\left[\begin{array}{ccc|c} 1 & 2 & 3 & 4 \\ 2 & 3 & 4 & 5 \end{array}\right].$$
    Hint: Treat the nonpivot columns as if they were additional **b** columns.
    b) Write up what you did in algorithmic language. Your algorithm has to work only for $m \times n$ matrices with $m < n$ and in which no 0's arise where you want to pivot.

14. Now consider problems where any of the pivot-finding steps of Algorithm GAUSS can arise except row switches.
    a) Devise an LU-type algorithm to solve
$$\left[\begin{array}{cccc|c} 1 & 3 & 3 & 2 & 1 \\ 2 & 6 & 9 & 5 & 4 \\ -1 & -3 & 3 & 0 & 3 \end{array}\right]$$
    and
$$\begin{bmatrix} 1 & 2 & 3 & 4 \\ 1 & 4 & 6 & 8 \\ 1 & 0 & 0 & 1 \end{bmatrix} \quad \text{(homogeneous)}.$$
    b) Write up what you did in algorithmic language.

15. The hardest part of making an LU-type algorithm work for general matrices is dealing with row switches. In LU-GAUSS, and probably in all your algorithms for the past few problems, the algorithm knew what to do with an entry in the L-and-U matrix by its position—on the main diagonal, below it, in a column with a higher index than the last row, etc. But if there are row switches, most of this positional information no longer means anything.
    a) Devise an LU-type algorithm to solve
$$\left[\begin{array}{cccc|c} 0 & 0 & 1 & -1 & 7 \\ 1 & 2 & 0 & 0 & 5 \\ 0 & 3 & 0 & 1 & 10 \\ 1 & 5 & 1 & 0 & 14 \end{array}\right].$$
    b) Write up what you did in algorithmic language.

16. Let A be an $n \times n$ matrix which does not lead to any 0's in pivot positions. Count the number of assignments to subscripted variables when $[A \mid \mathbf{b}]$ is reduced by GAUSS-SQUARE, by GAUSS and by LU-GAUSS.

17. A **tridiagonal matrix** is a square matrix in which all entries are necessarily 0 except on the main diagonal and the diagonals directly above and below it. If A is tridiagonal, Gaussian elimination (either GAUSS-SQUARE or LU-GAUSS) can be speeded up considerably. Write an improved algorithm and analyze its efficiency. (This is not an idle effort, for many applied problems lead to tridiagonal matrices or other **band matrices**—matrices with 0's outside some set of diagonals.)

18. There is a modification of Gauss-Jordan reduction [27d, Section 8.2] which relates to Gauss-Jordan reduction the same way that LU-GAUSS relates to GAUSS. Figure it out and write it up.

8.6 Gaussian Elimination and Matrix Algebra

So far we have treated the solution of linear systems as a purely numerical endeavor. We even stripped off the letters x, y, z and the equals signs and dealt solely with matrices. But in Section 0.6 we showed that such matrices have an algebraic structure: One can add and multiply them. In this section we put this matrix algebra to use, first as a way to write the statements of linear algebra problems concisely, and then as a way to express and understand what Gaussian elimination does to solve these problems.

If Section 0.6 was your first acquaintance with matrix algebra, we highly recommend you review it before reading on. The key ideas that you will need for the rest of this chapter are the definition of matrix multiplication, the concept of row and column vectors as special types of matrices, and the idea of an identity matrix (I).

Our first goal is to use matrices to rewrite the general linear system

$$\begin{aligned} a_{11}x_1 + a_{12}x_2 + \cdots + a_{1n}x_n &= b_1 \\ a_{21}x_1 + a_{22}x_2 + \cdots + a_{2n}x_n &= b_2 \\ \vdots \qquad \vdots \qquad\qquad \vdots \quad &\quad \vdots \\ a_{m1}x_1 + a_{m2}x_2 + \cdots + a_{mn}x_n &= b_m. \end{aligned} \tag{1}$$

Recall that when a matrix product AB is formed, the rows of A are multiplied term by term against the columns of B. Each row on the left above is in this same form—a sum of products of two factors. This suggests making the coefficients of each equation into a row of a matrix, and making the unknowns the column of another matrix:

$$\begin{bmatrix} a_{11} & a_{12} & \cdots & a_{1n} \\ a_{21} & a_{22} & \cdots & a_{2n} \\ \vdots & \vdots & & \vdots \\ a_{m1} & a_{m2} & \cdots & a_{mn} \end{bmatrix} \begin{pmatrix} x_1 \\ x_2 \\ \vdots \\ x_n \end{pmatrix}.$$

This product of a matrix and a column vector is itself a column vector whose entries are just the left-hand side of system (1). Thus to assert system (1) is simply to assert the single equation

$$\begin{bmatrix} a_{11} & a_{12} & \cdots & a_{1n} \\ a_{21} & a_{22} & \cdots & a_{2n} \\ \vdots & \vdots & & \vdots \\ a_{m1} & a_{m2} & \cdots & a_{mn} \end{bmatrix} \begin{pmatrix} x_1 \\ x_2 \\ \vdots \\ x_n \end{pmatrix} = \begin{pmatrix} b_1 \\ b_2 \\ \vdots \\ b_m \end{pmatrix}.$$

This still takes a lot of space, so let's define

$$A = \begin{bmatrix} a_{11} & a_{12} & \cdots & a_{1n} \\ a_{21} & a_{22} & \cdots & a_{2n} \\ \vdots & \vdots & & \vdots \\ a_{m1} & a_{m2} & \cdots & a_{mn} \end{bmatrix}, \qquad \mathbf{x} = \begin{pmatrix} x_1 \\ x_2 \\ \vdots \\ x_n \end{pmatrix}, \qquad \text{and} \qquad \mathbf{b} = \begin{pmatrix} b_1 \\ b_2 \\ \vdots \\ b_m \end{pmatrix}.$$

Voilà, our problem can now be stated short and sweet:

$$A\mathbf{x} = \mathbf{b}.$$

Furthermore, notice that $[A \,|\, \mathbf{b}]$ is precisely the augmented matrix we have been using all along, and A and $\mathbf{b}$ are the names we have already used for these parts.

Concise notation is only part of what we gain. The fact that a problem with any number of variables now looks like a problem with just one variable is very suggestive.

Think of it this way. Suppose that a high school Algebra I student walked into your class and saw written on the board that you were studying how to solve $A\mathbf{x} = \mathbf{b}$. She might say "Hey, that's easy; just divide by A to get $x = b/A$." In other words, since she wouldn't realize that capital A means something different from what she's used to (she's thinking of a number), and similarly with $\mathbf{b}$, she would wonder what the fuss is about.

Maybe she's right! We haven't defined division for matrices yet, but remember that ordinary division is just multiplication by the "multiplicative inverse". That is, b/a means $a^{-1}b$, where a^{-1} is the number which satisfies $a^{-1}a = aa^{-1} = 1$.

Therefore, let us define the (multiplicative) **inverse** of A, written A^{-1}, to be that matrix (if it exists) that satisfies

$$AA^{-1} = A^{-1}A = I.$$

We say "if it exists" because, just as the number 0 doesn't have an inverse, some matrices don't have inverses. For instance, no nonsquare matrix has an inverse. This follows by a simple "shape" argument. Suppose A is $m \times n$ with $m \neq n$. If A^{-1} exists, then for both products AA^{-1} and $A^{-1}A$ to make sense, A^{-1} must be $n \times m$. But then AA^{-1} is $m \times m$ and $A^{-1}A$ is $n \times n$, so they can't possibly be equal. Thus A^{-1} can't exist.

We insist that both AA^{-1} and $A^{-1}A$ equal I because, in general, matrix multiplication is not commutative. (For real numbers the condition $a^{-1}a = aa^{-1}$ above is redundant.) If A^{-1} exists, we say that A is **invertible**. A synonym is **nonsingular**.

The main purpose of this section is to prove the following.

Theorem 1. If A is an $n \times n$ matrix, then it has an inverse A^{-1} if and only if Gaussian elimination reduces A to I. In this case, for every column n-tuple $\mathbf{b}$ the unique solution to $A\mathbf{x} = \mathbf{b}$ found by Gaussian elimination is just $A^{-1}\mathbf{b}$.

This theorem does *not* say that we can now forget about Gaussian elimination and solve linear systems by dividing by A. In fact, the best general method for computing A^{-1} uses Gaussian elimination (see Example 3). Even using Gaussian elimination, it takes more work to find A^{-1} and then multiply by $\mathbf{b}$ than it takes simply to solve $A\mathbf{x} = \mathbf{b}$ as we have already done in Section 8.2. Thus the point of Theorem 1 is conceptual. It is interesting to know that there is a shorthand notation for what we are doing with Gaussian elimination, and that this notation reveals Gaussian elimination as a generalization of dividing out by a constant—only now the constant is a matrix. This viewpoint will help you think about what you are doing, for instance, when you solve one linear system and then chain the results into another [27, 28].

The proof of Theorem 1 is not hard, but you need familiarity with a few more ideas before it will go down smoothly. The first idea is that of a **linear combination**. This means "sum of multiples". For instance,

$$\begin{pmatrix}5\\7\\9\end{pmatrix} \quad \text{is a linear combination of} \quad \begin{pmatrix}1\\2\\3\end{pmatrix} \quad \text{and} \quad \begin{pmatrix}1\\1\\1\end{pmatrix}$$

because

$$\begin{pmatrix}5\\7\\9\end{pmatrix} = 2\begin{pmatrix}1\\2\\3\end{pmatrix} + 3\begin{pmatrix}1\\1\\1\end{pmatrix}.$$

That is, we take the multiples 2 and 3 of the vectors on the right and add them to get the vector on the left. Similarly, the row vector $(-3, 2, 0)$ is a linear combination of $(1, 0, 2)$ and $(2, -1, 1)$ because

$$(-3, 2, 0) = (1, 0, 2) - 2(2, -1, 1).$$

Here the multiple of the first vector on the right is 1 and the multiple of the second is -2. Going a step further, consider infinite sequences. If

$$\{a_n\} = \{1, 2, 4, 8, \ldots\} \qquad \text{and} \qquad \{b_n\} = \{1, 1, 1, 1, \ldots\},$$

then

$$\{c_n\} = \{3, 4, 6, 10, \ldots\}$$

is a linear combination of $\{a_n\}$ and $\{b_n\}$ because $\{c_n\} = \{a_n\} + 2\{b_n\}$, that is, $c_n = a_n + 2b_n$ for each n.

In general, if $x_1, x_2, \ldots, x_n$ are objects that can be multiplied by numbers and then added, and if $c_1, c_2, \ldots, c_n$ are numbers, then $\sum c_i x_i$ is a linear combination of the x's.

We can now discuss two important results (Lemmas 2 and 3) about how to interpret matrix multiplication in terms of rows and columns. These results are used in the proof of Theorem 1.

Lemma 2. Each row of the matrix product AB is a linear combination of the rows of B, with coefficients taken from the corresponding row of A. Similarly, each column of AB is a linear combination of the columns of A, with the coefficients taken from the corresponding column of B. Specifically, let $A = [a_{ij}]$ be $m \times n$, let $B = [b_{jk}]$ be $n \times p$, let $\mathbf{a}_j$ be the jth column of A, and let $\mathbf{b}_j$ be the jth row of B. Then the ith row of AB is $\sum_{j=1}^{n} a_{ij}\mathbf{b}_j$ and the kth column of AB is $\sum_{j=1}^{n} \mathbf{a}_j b_{jk}$.

EXAMPLE 1 Consider the product

$$C = \begin{bmatrix} 1 & 2 & 3 \\ 4 & 5 & 6 \end{bmatrix} \begin{bmatrix} 10 & 11 \\ 12 & 13 \\ 14 & 15 \end{bmatrix}.$$

The first row of C is

$$1(10,11) + 2(12,13) + 3(14,15) \;=\; (76,82).$$

(The entries of the first row of A, namely, $1,2,3$, have been used as coefficients for the rows of B.) The second row of C is

$$4(10,11) + 5(12,13) + 6(14,15) \;=\; (184,199).$$

Similarly, the first column of C is a linear combination of the columns of A:

$$\begin{pmatrix} 1 \\ 4 \end{pmatrix} 10 + \begin{pmatrix} 2 \\ 5 \end{pmatrix} 12 + \begin{pmatrix} 3 \\ 6 \end{pmatrix} 14 \;=\; \begin{pmatrix} 76 \\ 184 \end{pmatrix}.$$

The coefficients $10, 12, 14$ are obtained by going down the first column of B. ∎

EXAMPLE 2 Do the multiplication

$$\begin{bmatrix} 1 & 3 \\ 2 & 2 \\ 3 & 1 \end{bmatrix} \begin{bmatrix} 1 & -1 \\ 1 & 1 \end{bmatrix}$$

in your head.

Solution Call the matrices A and B. We use the interpretation of AB in terms of columns. The first column of AB is a linear combination of the columns of A, with coefficients 1 and 1 (taken from the first column of B). Therefore the first column of AB is the sum of the columns of A. Similarly, the second column of AB is the difference when the first column of A is subtracted from the second column of A, because the coefficients are -1 and 1 (taken from the second column of B). In conclusion,

$$AB = \begin{bmatrix} 4 & 2 \\ 4 & 0 \\ 4 & -2 \end{bmatrix}.$$

Given that this example was to be done your head, do you see why we chose to use the column interpretation rather than the row interpretation [8]? ∎

PROOF OF LEMMA 2 Let $[c_{ik}] = AB$. By definition, $c_{ik} = \sum_{j=1}^{n} a_{ij}b_{jk}$. We illustrate this formula for several values of k in Fig. 8.3. We show just the ith row of A and several columns of B. For instance, c_{i1} is obtained using row i of A and column 1 of B, as shown. The computations of $c_{i1}, c_{i2}, \ldots, c_{ik}$ are parallel: a_{i1} is multiplied by the first entry of each column from B, a_{i2} is multiplied by the second entry of each column, and so on. Therefore we may write the parallel processes as a single process using row vectors.

$$(c_{i1}, \ldots, c_{ip}) = a_{i1}(b_{11}, \ldots, b_{1p}) + a_{i2}(b_{21}, \ldots, b_{2p}) + \cdots + a_{in}(b_{n1}, \ldots, b_{np}).$$

See Fig. 8.3, where the second term in this sum is indicated by colored shading. Using the notation from the lemma, the previous display says that the ith row of AB is $\sum_{j=1}^{n} a_{ij}\mathbf{b}_j$, as claimed. The proof of the interpretation of columns of AB is the same, except that the role of rows and columns is interchanged. Start with a figure showing the typical jth column of B being multiplied against all the rows of A. [13] ∎

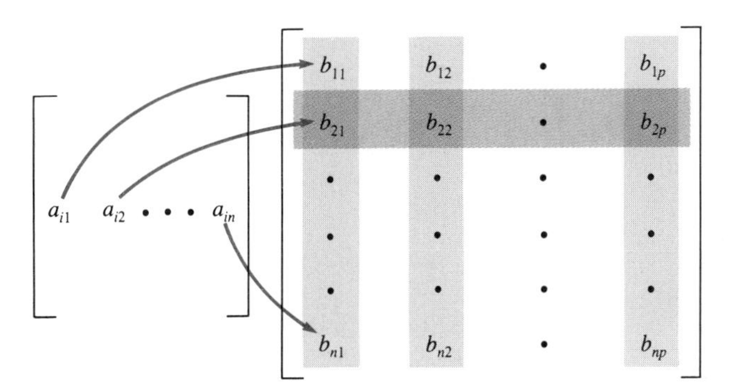

FIGURE 8.3

Row i of AB is obtained by multiplying the entries of the ith row of A against the rows of B.

Lemma 2 has a converse, which provides an important way to interpret linear combinations of vectors in terms of matrix multiplication.

Lemma 3. If the rows of matrix C are linear combinations of the rows of B, then there is a matrix A such that $C = AB$. Similarly, if the columns of matrix P are linear combinations of the columns of Q, then there is a matrix R such that $P = QR$.

We say that C is obtained by **premultiplying** B by A, and that P is obtained by **postmultiplying** Q by R. With this terminology we may summarize the lemmas by saying that row operations on a matrix amount to premultiplication and column operations amount to postmultiplication.

PROOF OF LEMMA 3 We prove the first sentence and leave the second to you [14]. Suppose that row i of C equals $\sum_{j=1}^{n} a_{ij}\mathbf{b}_j$, where $\mathbf{b}_j$ is the jth row of B. Then simply define row i of A to be $[a_{i1}, a_{i2}, \ldots, a_{in}]$. As shown in the proof of Lemma 2, this definition is just what is needed to make $AB = C$. ∎

The first part of Lemma 3 has an immediate application to Gaussian elimination. If A' is obtained from A by a single row operation, then every row of A' is a simple linear combination of the rows of A. Thus you can think of every step of Gaussian elimination as obtained by premultiplying the current matrix. For instance, suppose A' is obtained by subtracting the first row of A from the third row of A six times. Letting $\mathbf{r}_j$ and $\mathbf{r}'_j$ be the jth rows of A and A', we can say that

$$\mathbf{r}'_3 = -6\mathbf{r}_1 + \mathbf{r}_3 \qquad \text{and} \qquad \mathbf{r}'_j = \mathbf{r}_j \quad \text{for } j \neq 3.$$

Thus, if A has four rows, then

$$A' = \begin{bmatrix} 1 & 0 & 0 & 0 \\ 0 & 1 & 0 & 0 \\ -6 & 0 & 1 & 0 \\ 0 & 0 & 0 & 1 \end{bmatrix} A.$$

It follows that *the entire Gaussian elimination process amounts to a single premultiplication.* For instance, if A, A', and A'' are successive steps in an elimination, then we have just shown that there are matrices E and E' such that

$$A' = EA \qquad \text{and} \qquad A'' = E'A'.$$

Thus

$$A'' = E'(EA) = (E'E)A.$$

The last equality follows from the associativity of matrix multiplication. So, if we obtain A'' from A by two row operations, then we may express A'' as a premultiplication of A. The general result follows by induction [16].

We are finally ready to start on the proof of Theorem 1. First we will show necessity: If A^{-1} exists, then Gaussian elimination reduces A to I. Then we will show sufficiency.

PROOF OF THEOREM 1 We must show that the existence of an algebraic entity, A^{-1}, forces an algorithm to terminate a certain way (with I). How do we make that bridge from algebra to algorithm? Well, the reduced echelon form at termination is intimately tied to the number and type of solutions to $A\mathbf{x} = \mathbf{b}$, whatever $\mathbf{b}$ is. (Review Theorem 2, Section 8.4.) Therefore, let us seek to tie the existence of A^{-1} to information about solutions to $A\mathbf{x} = \mathbf{b}$.

We can do so with a little simple algebra. Given that A^{-1} exists, we may write

$$\begin{aligned} A\mathbf{x} &= \mathbf{b}, \\ A^{-1}A\mathbf{x} &= A^{-1}\mathbf{b}, \\ I\mathbf{x} = \mathbf{x} &= A^{-1}\mathbf{b}. \end{aligned}$$

This calculation shows that if $\mathbf{x}$ is to be a solution to the original problem, then $\mathbf{x} = A^{-1}\mathbf{b}$ is the only candidate. Of course, it may be extraneous. (See the subsection on validity at the end of Section 8.2.) So we plug in $A^{-1}\mathbf{b}$ for $\mathbf{x}$ in the first line to see if it works:

$$A(A^{-1}\mathbf{b}) = (AA^{-1})\mathbf{b} = I\mathbf{b} = \mathbf{b}.$$

It checks. (Notice that we used both parts of the definition of matrix inverse, the fact that $A^{-1}A = I$ in the "solve" and the fact that $AA^{-1} = I$ in the "check".)

What we have shown is: If A^{-1} exists, then for every $\mathbf{b}$, there exists exactly one solution to $A\mathbf{x} = \mathbf{b}$. By Theorem 2, Section 8.4, the reduced echelon form of A must be I.

Now for sufficiency. If Gaussian elimination reduces A to I, we must show that there is some matrix X such that $XA = AX = I$. Let's start by finding an X that satisfies $AX = I$, since except for the fact that X and I are square matrices instead of column vectors, this is a familiar problem. In fact, this equation breaks up into several equations of the column vector sort. Let $\mathbf{x}_1, \mathbf{x}_2, ..., \mathbf{x}_n$ be the columns of X and let

$$\mathbf{e}_1 = \begin{pmatrix} 1 \\ 0 \\ \vdots \\ 0 \end{pmatrix}, \qquad \mathbf{e}_2 = \begin{pmatrix} 0 \\ 1 \\ \vdots \\ 0 \end{pmatrix}, \qquad \cdots, \qquad \mathbf{e}_n = \begin{pmatrix} 0 \\ 0 \\ \vdots \\ 1 \end{pmatrix}$$

be the columns of I. These $\mathbf{e}$'s are called the **standard unit vectors**. Because of the way matrix multiplication is defined, $AX = I$ is equivalent to the n equations

$$A\mathbf{x}_1 = \mathbf{e}_1, \quad A\mathbf{x}_2 = \mathbf{e}_2, \quad \ldots, \quad A\mathbf{x}_n = \mathbf{e}_n. \tag{2}$$

Now, because we are assuming that A reduces to I, we know that each of these equations will have a (unique) solution. Put all the solutions $\mathbf{x}_1, \ldots, \mathbf{x}_n$ side by side and we have an X satisfying $AX = I$.

What about $XA = I$? Since Gaussian elimination turns A into I via row operations, we know from Lemma 3 that there is a matrix Y so that $YA = I$. Indeed, this sequence of row operations turns $[A\,|\,I]$ into $[I\,|\,X]$—this is simply the result of solving all equations in (4) together, as in Example 1, Section 8.5. Therefore $YI = X$ as well as $YA = I$. Thus $Y = X$ and so $XA = I$ as needed. ∎

One final remark. We have just proved a theorem which says in part that if a certain thing happens (Gaussian elimination reduces A to I) then something else exists (A^{-1}). Naturally, we prefer a proof which is constructive; it should show *how* to construct the second thing when the first thing happens. Well, our proof *is* constructive. Do you see where it shows how to compute A^{-1}? Right at Eqs. (4). Solve these equations simultaneously, by starting with $[A\,|\,I]$, and you get A^{-1}.

EXAMPLE 3 Find the inverse of

$$A = \begin{bmatrix} 2 & -2 & 4 \\ 2 & -1 & 6 \\ -1 & 2 & -3 \end{bmatrix},$$

or show that it doesn't exist.

Solution This matrix is the coefficient matrix of Example 1, Section 8.2, so we already know that it reduces to I. Therefore A^{-1} exists, and we can find it by reducing

$$\left[\begin{array}{rrr|rrr} 2 & -2 & 4 & 1 & 0 & 0 \\ 2 & -1 & 6 & 0 & 1 & 0 \\ -1 & 2 & -3 & 0 & 0 & 1 \end{array}\right].$$

After completing the downsweep we have

$$\left[\begin{array}{rrr|rrr} 2 & -2 & 4 & 1 & 1 & 1 \\ 0 & 1 & 2 & -1 & 1 & 0 \\ 0 & 0 & -3 & 3/2 & -1 & 1 \end{array}\right],$$

and after the Scale and Upsweep we obtain

$$\left[\begin{array}{rrr|rrr} 1 & 0 & 0 & 3/2 & -1/3 & 4/3 \\ 0 & 1 & 0 & 0 & 1/3 & 2/3 \\ 0 & 0 & 1 & -1/2 & 1/3 & -1/3 \end{array}\right].$$

Therefore

$$A^{-1} = \begin{bmatrix} 3/2 & -1/3 & 4/3 \\ 0 & 1/3 & 2/3 \\ -1/2 & 1/3 & -1/3 \end{bmatrix}. \blacksquare$$

We said at the beginning of this section that you shouldn't actually solve linear systems using the inverse, but there are other occasions where it is useful to know what it is. Therefore it behooves us to ask for an analysis of the matrix inversion algorithm we have just illustrated. Assume that A is $n \times n$. The algorithm is to apply Gaussian elimination to $[A \,|\, I]$. We know that Gaussian elimination takes about $n^3/3$ steps on the coefficient matrix and n^2 on each column of constants. As there are n columns of constants, the total work has magnitude

$$\tfrac{1}{3}n^3 + n(n^2) = \tfrac{4}{3}n^3.$$

However, I is not just *any* collection of n columns. It is especially simple. Conceivably, at many points the values of various entries can be assigned instead of computed. Indeed, a considerable saving *is* possible. You are asked to explore this in [29–33].

In this section we have just scratched the surface concerning the relation between solving linear systems and matrix algebra. We close by mentioning one more important relationship. As discussed in Section 8.5, for computation it is best to store two aspects of the Gaussian elimination procedure: the echelon matrix U produced from downsweeps and the lower triangular set of downsweep subtraction multiples. We stored all this information in a single array, but we said there were reasons to write them as two separate matrices, L and U. For instance, for the matrix A in Example 3,

$$L = \begin{bmatrix} 1 & 0 & 0 \\ 1 & 1 & 0 \\ -1/2 & 1 & 1 \end{bmatrix}, \qquad U = \begin{bmatrix} 2 & -2 & 4 \\ 0 & 1 & 2 \\ 0 & 0 & -3 \end{bmatrix}.$$

In L we included a main diagonal of 1's and promised that we would explain why later. The reason for the 1's is: When L is defined this way, *it's a theorem that*

$A = LU$. See [34]. That is, in the process of solving $A\mathbf{x} = \mathbf{b}$ Gaussian elimination factors A into a special form. More than that, Gaussian elimination solves $A\mathbf{x} = \mathbf{b}$ by first inverting L and then inverting U. To see this, note that this theorem says we start with $LU\mathbf{x} = \mathbf{b}$ (except we don't know yet what L and U are!) and thus, after the downsweep, Gaussian elimination has reduced LU to U. Since this reduction is accomplished by premultiplication, we must have premultiplied by the inverse of L. That is, we went from $LU\mathbf{x} = \mathbf{b}$ to

$$U\mathbf{x} = L^{-1}LU\mathbf{x} = L^{-1}\mathbf{b}.$$

Then Gaussian elimination goes on and reduces U to I, so the premultiplication for this part inverted U. That is, premultiplying the previous display by U^{-1} we obtain

$$U^{-1}U\mathbf{x} = I\mathbf{x} = \mathbf{x} = U^{-1}L^{-1}\mathbf{b}.$$

In other words, Gaussian elimination doesn't invert A directly, but rather (1) factors it; (2) inverts one factor; and (3) inverts the other factor. Stated this way, it sounds like a silly, roundabout approach. But the remarkable fact is that it is much shorter than direct inversion. The special triangular nature of L and U make the combined work much less than the direct approach. The delight in occasionally finding such an effective indirect attack is one of the joys of the careful study of algorithms.

Problems: Section 8.6

1. A good algorithm to construct something constructs it when it exists and gives you a clear message when it doesn't. In Example 3 we showed how Gaussian elimination constructs A^{-1} when it exists. How does it tell you if A^{-1} doesn't exist?

2. For each matrix below, find its inverse or show that it has none. (These all arise in graph theory. They are incidence matrices of cycles.)

a) $\begin{bmatrix} 1 & 1 & 0 \\ 0 & 1 & 1 \\ 1 & 0 & 1 \end{bmatrix}$ b) $\begin{bmatrix} 1 & 1 & 0 & 0 \\ 0 & 1 & 1 & 0 \\ 0 & 0 & 1 & 1 \\ 1 & 0 & 0 & 1 \end{bmatrix}$

c) $\begin{bmatrix} 1 & 1 & 0 & 0 & 0 \\ 0 & 1 & 1 & 0 & 0 \\ 0 & 0 & 1 & 1 & 0 \\ 0 & 0 & 0 & 1 & 1 \\ 1 & 0 & 0 & 0 & 1 \end{bmatrix}$

3. For each matrix below, find its inverse or show that it has none. (These are all incidence matrices of directed cycles.)

a) $\begin{bmatrix} 1 & -1 & 0 \\ 0 & 1 & -1 \\ -1 & 0 & 1 \end{bmatrix}$ b) $\begin{bmatrix} 1 & -1 & 0 & 0 \\ 0 & 1 & -1 & 0 \\ 0 & 0 & 1 & -1 \\ -1 & 0 & 0 & 1 \end{bmatrix}$

c) $\begin{bmatrix} 1 & -1 & 0 & 0 & 0 \\ 0 & 1 & -1 & 0 & 0 \\ 0 & 0 & 1 & -1 & 0 \\ 0 & 0 & 0 & 1 & -1 \\ -1 & 0 & 0 & 0 & 1 \end{bmatrix}$

4. Invert the following matrices.

a) $\begin{bmatrix} 1 & 2 \\ 3 & 4 \end{bmatrix}$ b) $\begin{bmatrix} a & b \\ c & d \end{bmatrix}$

c) $\begin{bmatrix} 1 & 2 & 2 \\ 2 & 1 & -2 \\ 2 & -2 & 1 \end{bmatrix}$

Note that **b**) gives you a handy formula for inverting any 2×2 matrix. (There isn't such a simple formula for larger matrices.)

5. In [4b], what assumption did you make about the values of a, b, c, d in order to reduce the matrix to

I? You should be able to express this assumption as a single equation or inequality.

6. Every real number except 0 has a multiplicative inverse. Does every square matrix which is not all 0's have an inverse? Does every square matrix which has no 0 entries have an inverse?

7. Let the rows of B be $\mathbf{r}_1, \mathbf{r}_2, \mathbf{r}_3$. You want to premultiply B to obtain

$$\begin{bmatrix} \mathbf{r}_1+\mathbf{r}_2 \\ \mathbf{r}_2-2\mathbf{r}_3 \\ \mathbf{r}_1+\mathbf{r}_3-2\mathbf{r}_2 \end{bmatrix}$$

What is the premultiplier matrix?

8. Answer the question in the last sentence of Example 2.

9. In your head, figure out the entries in the following products.

a) $\begin{bmatrix} 1 & 1 & 0 \\ 0 & 1 & 1 \\ 1 & 0 & -1 \end{bmatrix} \begin{bmatrix} 1 & 2 & 3 \\ 4 & 5 & 4 \\ 3 & 2 & 1 \end{bmatrix}$

b) $\begin{bmatrix} 1 & 2 & 3 \\ 4 & 5 & 4 \\ 3 & 2 & 1 \end{bmatrix} \begin{bmatrix} 1 & 1 & 0 \\ 0 & 1 & 1 \\ 1 & 0 & -1 \end{bmatrix}$

10. Let

$$P = \begin{bmatrix} 1 & 1 & 0 \\ 1 & 0 & 1 \\ 0 & 1 & 1 \end{bmatrix}, \quad Q = \begin{bmatrix} 2 & 0 & 2 \\ 0 & 1 & 1 \\ -1 & -1 & 0 \end{bmatrix}.$$

a) From inspection, write a B such that $BP = Q$.

b) It is not so easy to write from inspection a C such that $PC = Q$. But use some method from this section to compute C (if it exists).

11. There are three types of elementary row operations, as described in the proof of Theorem 1, Section 8.2. From Lemma 3, each corresponds to premultiplication by a certain sort of **elementary matrix**. Describe these three sorts of elementary matrices.

12. There are three types of elementary column operations, exactly parallel to the row situation. Describe the three sorts of elementary matrices which arise from them. (Except for column switches in complete pivoting, these column operations are *not* used in solving systems of equations, because they change the solutions.)

13. Prove that the kth column of AB is a linear combination of the columns of A with the coefficients taken from the kth column of B. (This is the second half of the proof of Lemma 2.) Start with a figure analogous to Fig. 8.3.

14. Finish the proof of Lemma 3.

15. Suppose that you have three sets of variables, $\{x_1, x_2, \ldots, x_m\}$, $\{y_1, \ldots, y_n\}$, and $\{z_1, \ldots, z_p\}$. Suppose also that each z is a linear combination of the y's and that each y is a linear combination of the x's. Prove that each z is a linear combination of the x's. *Hint*: Turn this into a matrix problem. (*Note*: This generalizes the example discussed in Section 0.6 and again in Example 1, Section 8.1.)

16. Prove by induction that the effect of doing Gaussian elimination on any matrix is to premultiply it by something. Do variants of Gaussian elimination have this same premultiplication property? What is essential to make an algorithm that modifies a matrix be equivalent to a premultiplication?

17. Suppose we change A to B by a sequence of row operations. We know there is a premultiplier P such that $PA = B$. We could find it by writing the elementary matrix for each row operation, as in [11], and multiplying them all together; but that would be very tedious. Find a better way. *Hint*: Start with $[A \,|\, I]$.

18. In ordinary algebra, it is traditional to write x divided by a as x/a. We have avoided this notation, and even the word division, because for matrices they are ambiguous. Should X/A mean $A^{-1}X$ or XA^{-1}? These products are different for matrices, and either one may be appropriate depending on the circumstances.

a) Divide $\begin{bmatrix} 2 & 3 \\ 4 & 5 \end{bmatrix}$ by $\begin{bmatrix} 1 & 2 \\ 3 & 4 \end{bmatrix}$ two ways.

b) Divide $\begin{pmatrix} 2 \\ 4 \end{pmatrix}$ by $\begin{bmatrix} 1 & 2 \\ 3 & 4 \end{bmatrix}$. Only one way makes sense. Why?

c) Divide (2,3) by $\begin{bmatrix} 1 & 2 \\ 3 & 4 \end{bmatrix}$. Only one way makes sense. Why?

19. Where in our proof do we justify the last sentence of Theorem 1?

20. Here is another proof that [Gaussian elimination reduces A to I] $\Longrightarrow$ [A^{-1}exists]. It doesn't require

Lemmas 2 and 3. (They are very important anyway.)

The first part is the same as in the text: Gaussian elimination reduces $[A\,|\,I]$ to $[I\,|\,X]$, where $AX = I$. Now argue that there must be a sequence of elementary row operations that turns $[X\,|\,I]$ into $[I\,|\,A]$ and from there complete the proof.

21. A **diagonal** matrix is a square matrix with all entries 0 except on the diagonal, where they can be anything. For a diagonal matrix, let $d_1, d_2, \ldots, d_n$ be the diagonal entries.

a) When is a diagonal matrix invertible?

b) When it is invertible, show that the inverse is also diagonal, and name the diagonal entries.

22. Prove two ways that any square matrix A has at most one inverse.

a) Use a little bit of algebra. *Hint*: Suppose that A had two inverses, X and X'. Simplify XAX'.

b) Use Theorem 1. *Hint*: If A has at least one inverse then Gaussian elimination reduces A to I, so what can you say about the number of solutions to $AX = I$?

23. Let A be invertible and $n \times n$. How much work does it take to solve $A\mathbf{x} = \mathbf{b}$ by first finding A^{-1} and then premultiplying $\mathbf{b}$ by it? How much more work is this than doing Gaussian elimination? Assume that finding A^{-1} takes $\frac{4}{3}n^3$ steps, as in the text; but see [29–33].

24. Your boss gives you a huge square matrix A and tells you to solve $A\mathbf{x} = \mathbf{b}$. Just as you are about to start, your fairy godmother arrives and hands you A^{-1} on a silver platter. To solve the problem most efficiently, what should you do?

25. Repeat [24], but this time it so happens that yesterday you solved $A\mathbf{x} = \mathbf{c}$ by Algorithm LU-GAUSS and you've saved the final matrix. Now what should you do?

26. Your boss gives you two square matrices A and B and asks you to verify that $B = A^{-1}$. Which is smarter, checking directly whether B meets the definition, or solving for A^{-1} from scratch using Gaussian elimination?

27. Let

$$A = \begin{bmatrix} 1 & 2 \\ 3 & 4 \end{bmatrix}, \quad B = \begin{bmatrix} 1 & 2 \\ 2 & 3 \end{bmatrix}, \quad \mathbf{c} = \begin{pmatrix} 3 \\ 8 \end{pmatrix}.$$

Solve $AB\mathbf{x} = \mathbf{c}$ two ways:

a) By multiplying to obtain AB and then using Gaussian elimination.

b) By solving $A(B\mathbf{x}) = \mathbf{c}$ for $\mathbf{d} = B\mathbf{x}$ and then solving this last equation for $\mathbf{x}$.

28. Suppose that A and B are general $n \times n$ matrices. Compare the amount of work involved in the two methods of solving $AB\mathbf{x} = \mathbf{b}$ discussed in [27].

29. Fill in the missing steps in the matrix inversion of Example 3.

30. If you did [29], you found that a lot of entries which change during Gaussian elimination on an arbitrary set of augmentation columns do not change during a matrix inversion. This allows you to write a modification of Algorithm GAUSS for matrix inversion which takes fewer steps. Write such an algorithm.

31. How many steps does your algorithm in [30] take to invert an $n \times n$ matrix?

32. Do [30] again, but this time modify LU-GAUSS instead of GAUSS.

33. Many books discuss matrix inversion using Gauss-Jordan reduction instead of Gaussian elimination. (See [27d, Section 8.2] for the definition of Gauss-Jordan reduction.) Although Gauss-Jordan reduction is slower for solving $A\mathbf{x} = \mathbf{b}$, maybe it pulls ahead in the case of matrix inversion.

a) Do Example 3 again using Gauss-Jordan reduction.

b) Write a special version of Gauss-Jordan reduction for inverting matrices. As in [30], there may be savings from replacing computations with assignments.

c) Analyze the efficiency of your algorithm from b) and compare it to your answer to [31].

34. Suppose A is square and Gaussian elimination produces L and U when reducing A. That is, L and U are the lower and upper triangular matrices associated with LU-GAUSS. Prove: $A = LU$. Assume no 0 pivots or row switches occur. *Hint*: Think of L as a premultiplier of U affecting its rows.

35. In addition to the LU decomposition of A, there is also an LDU decomposition. Here D is defined to be the diagonal matrix whose diagonal entries are the pivots when A is reduced to I by GAUSS or LU-GAUSS. Then U is redefined to be whatever matrix makes DU equal the previous U. Thus we now have the theorem $A = LDU$.

a) Find the LDU decomposition for A from Example 3.

b) Explain why U in the LDU decomposition always has all 1's on its main diagonal. (This makes the fact that we defined L to have all 1's on its diagonal seem more natural—there is now a symmetry.)

c) Find an interpretation in terms of some version of Gaussian elimination for the off-diagonal entries of the new U.

36. Suppose A and B are conformable, that is, AB exists. Prove: If A and B are invertible, so is AB. *Hint*: Guess a formula for $(AB)^{-1}$ in terms of A^{-1} and B^{-1} and verify that your formula meets the definition. (*Caution*: You will probably guess the wrong formula at first.)

37. In the morning you put on your socks and then you put on your shoes.

a) In the evening, most people want to invert this process. How do you do it?

b) What does this have to do with [36]?

38. The result in [36] is often stated as

$$(AB)^{-1} = B^{-1}A^{-1}.$$

However, this equality is misleading because it suggests a bi-implication—that the existence of both inverses on the right implies the existence of the inverse on the left (this is [36]) and vice versa. However, vice versa is false: A and B don't even have to be square (hence certainly not invertible) and yet AB can be square and invertible.

a) Come up with a specific example confirming the previous sentence.

b) Prove: If AB is invertible and A is square, then both A and B are invertible and $(AB)^{-1} = B^{-1}A^{-1}$.

39. We argued that a nonsquare matrix A can't have an inverse by showing that $A^{-1}A$ and AA^{-1} wouldn't even have the same shape, let alone be equal. But what if we change the definition of inverse and ask only that $A^{-1}A$ and AA^{-1} both be identity matrices, but maybe with different sizes? Prove or disprove: Now some nonsquare matrices have inverses. *Hint*: Look carefully at the proof of Theorem 1.

8.7 Vector Spaces: Definition and Examples

In this section and the next we present one of the most pervasive and applicable ideas in mathematics: the construct called a vector space. It will give us the power to understand better many situations we have already confronted.

For instance, we have found that the solution set to a set of linear equations has a certain degree of freedom—the number of free variables found by Gaussian elimination. But suppose we had listed the variables in a different order. The solution set doesn't change, but the final form found by Gaussian elimination does change. (See [25, Section 8.2] and [4].) Different variables can become the free variables. Can the *number* of free variables change? The answer is No: That number is intrinsic to the system of equations. To show this, we will show that this number is the "dimension" of a certain vector space defined independently of Gaussian elimination.

In this section we define vector space, and give several important examples. In Section 8.8 we introduce key attendant concepts, such as dimension, and prove their key properties.

The power of vector space theory is two-fold. The concepts alone allow a precision and conciseness of thought without which further work would be very difficult. For instance, we will use these concepts repeatedly in Section 8.9. Second, specific theorems will answer natural questions, e.g., the question above about whether the number of free variables can change.

Actually, we are not going to discuss all vector spaces, but only those that reside in R^n, the set of all n-tuple vectors of real numbers:

$$R^n = \{(x_1, x_2, \ldots, x_n) \mid x_i \in R\}.$$

Thus when we say "let V be a vector space", we mean a vector space in R^n. We indicate an element of R^n with boldface type, e.g., $\mathbf{x} \in R^n$. The display above represents the elements of R^n as row vectors, but we allow ourselves to write them as column vectors, too. We allow $n = \infty$ as well as $n = 1, 2, 3, \ldots$. That is, one R^n is R^∞, the set of all infinite sequences. Thus the theory we develop in this section will apply to everything in Chapters 5 and 9 as well as to the solutions of linear systems you've studied so far in this chapter.

Before giving the definition of vector space (Definition 1), let us give two examples that motivate it.

EXAMPLE 1 **Homogeneous Linear Equations**

Consider the set of solutions to any system of homogeneous linear equations, for instance, the system in Example 7, Section 8.3. We claim that any multiple of a solution is a solution and any sum of solutions is a solution. For instance, in Example 7 the solutions were precisely those vectors of the form

$$\begin{pmatrix} u \\ v \\ w \\ x \\ y \\ z \end{pmatrix} = v \begin{pmatrix} -2 \\ 1 \\ 0 \\ 0 \\ 0 \\ 0 \end{pmatrix} + x \begin{pmatrix} -1 \\ 0 \\ 1 \\ 1 \\ 0 \\ 0 \end{pmatrix} + y \begin{pmatrix} -1 \\ 0 \\ -3 \\ 0 \\ 1 \\ 0 \end{pmatrix},$$

where v, x, y could take on any values. So take any particular solution, say,

$$\mathbf{x} = \begin{pmatrix} -3 \\ 1 \\ -11 \\ -2 \\ 3 \\ 0 \end{pmatrix} = 1 \begin{pmatrix} -2 \\ 1 \\ 0 \\ 0 \\ 0 \\ 0 \end{pmatrix} + (-2) \begin{pmatrix} -1 \\ 0 \\ 1 \\ 1 \\ 0 \\ 0 \end{pmatrix} + 3 \begin{pmatrix} -1 \\ 0 \\ -3 \\ 0 \\ 1 \\ 0 \end{pmatrix}.$$

Now take any multiple, say, 10. Then $10\mathbf{x}$ is a solution because

$$10\mathbf{x} = \begin{pmatrix} -30 \\ 10 \\ -110 \\ -20 \\ 30 \\ 0 \end{pmatrix} = 10 \begin{pmatrix} -2 \\ 1 \\ 0 \\ 0 \\ 0 \\ 0 \end{pmatrix} + (-20) \begin{pmatrix} -1 \\ 0 \\ 1 \\ 1 \\ 0 \\ 0 \end{pmatrix} + 30 \begin{pmatrix} -1 \\ 0 \\ -3 \\ 0 \\ 1 \\ 0 \end{pmatrix}.$$

That is, we simply choose the coefficients v, x, y to be 10 times what they were before. Similarly, take any other solution, say,

$$\mathbf{y} = \begin{pmatrix} -18-\pi \\ 5 \\ \pi-24 \\ \pi \\ 8 \\ 0 \end{pmatrix} = 5\begin{pmatrix} -2 \\ 1 \\ 0 \\ 0 \\ 0 \\ 0 \end{pmatrix} + \pi\begin{pmatrix} -1 \\ 0 \\ 1 \\ 1 \\ 0 \\ 0 \end{pmatrix} + 8\begin{pmatrix} -1 \\ 0 \\ -3 \\ 0 \\ 1 \\ 0 \end{pmatrix}.$$

Then $\mathbf{x}+\mathbf{y}$ is also a solution because

$$\mathbf{x}+\mathbf{y} = \begin{pmatrix} -21-\pi \\ 6 \\ \pi-35 \\ \pi-2 \\ 11 \\ 0 \end{pmatrix} = 6\begin{pmatrix} -2 \\ 1 \\ 0 \\ 0 \\ 0 \\ 0 \end{pmatrix} + (\pi-2)\begin{pmatrix} -1 \\ 0 \\ 1 \\ 1 \\ 0 \\ 0 \end{pmatrix} + 11\begin{pmatrix} -1 \\ 0 \\ -3 \\ 0 \\ 1 \\ 0 \end{pmatrix}.$$

We say that the solutions are **closed** under addition and scalar multiplication. ∎

EXAMPLE 2 **Linear Homogeneous Difference Equations**

Consider the general solution to any linear homogeneous difference equation; that is, consider the set of sequences which satisfy the recurrence when no initial conditions are given. In Chapter 5 we showed that the general solution is of the form

$$a_k = \sum_i A_i r_i^k,$$

where r_i is the ith root of the characteristic equation and the A_i's are arbitrary constants. (Recall: If there are multiple roots, there are also terms like kr_i^k and $k^2 r_i^k$; but this doesn't spoil what we are about to say.) For exactly the same reasons as in Example 1, the set of solutions is closed under addition and scalar multiplication. Imagine the same sort of displays as in Example 1, except instead of column 6-tuples, the vectors are row ∞-tuples. The A_i's play the role that v, x, and y did in Example 1. ∎

The closure property illustrated in Examples 1 and 2 is exactly what makes a vector space:

Definition 1. A **vector space** in R^n $(n = 1, 2, 3, \ldots \text{ or } \infty)$ is any nonempty subset V of R^n that satisfies

1. If $\mathbf{x} \in V$, then $c\mathbf{x} \in V$ for any $c \in R$ (Closure under scalar multiplication).
2. If $\mathbf{x}, \mathbf{y} \in V$, then $\mathbf{x}+\mathbf{y} \in V$ (Closure under vector addition).

Make sure you understand what we mean by $c\mathbf{x}$ and $\mathbf{x}+\mathbf{y}$. If $\mathbf{x} = (x_1, x_2, \ldots, x_n)$ and $\mathbf{y} = (y_1, y_2, \ldots, y_n)$, we mean that

$$c\mathbf{x} = (cx_1, cx_2, \ldots, cx_n), \qquad \text{and} \qquad \mathbf{x}+\mathbf{y} = (x_1+y_1,\, x_2+y_2, \ldots, x_n+y_n).$$

(We don't have to state this as a definition because vectors are special cases of matrices and we already defined addition and scalar multiplication of matrices in Section 0.6. The above equations use those definitions.)

Recall from Section 8.6 that a linear combination of vectors is any sum of scalar multiples of those vectors. If we apply (1) and (2) from Definition 1 repeatedly, we see that a vector space is closed under arbitrary linear combinations. In fact, closure under linear combinations is really the important property conceptually. However, *verifying* that something is a vector space is easier if we merely have to check single additions and single scalar multiplications. Hence we use Definition 1.

In light of this definition and Examples 1 and 2, the set of solutions to a homogeneous system of linear equations is a vector space, and so is the general solution to a homogeneous linear difference equation. The latter is a vector space within R^∞. If the former comes from a system of m equations in n unknowns, is it a vector space in R^m or R^n [2]?

EXAMPLE 3 R^n itself is a vector space. This is easy to check. Clearly, R^n is closed under scalar multiplication and addition. ∎

Note. Because R^n is a vector space itself, we say that any vector space V within R^n is a **subspace** of R^n. Sub*space* means more than sub*set*. In general, to say that A is a subwhatsit of B is to say that B is a whatsit, $A \subset B$, *and* A is a whatsit, too. Thus, to say that A is a (vector) subspace of B is to say that B is a (vector) space, $A \subset B$, and A is a (vector) space too.

EXAMPLE 4 Is the First Quadrant of the coordinate plane a vector space?

Solution We are talking about the subset $Q = \{\,(x, y) \mid x, y \geq 0\,\}$ of R^2. It fails condition (1) of Definition 1: It's not closed under multiplication by negative scalars. For instance, $(1, 1) \in Q$ but $(-1)(1, 1) = (-1, -1) \notin Q$. So the answer is No.

Note that Q does satisfy condition (2), but that's not enough. ∎

EXAMPLE 5 Determine if $S = \{\,(x, y) \mid x, y \geq 0 \text{ or } x, y \leq 0\,\}$ is a vector space.

Solution S satisfies condition (1) OK. For instance, if $x, y \geq 0$, then either $cx, cy \geq 0$ (if $c \geq 0$) or $cx, cy \leq 0$ (if $c \leq 0$). However, S fails condition (2). For instance, both $(4, 4)$ and $(-2, -6)$ are in S, but their sum $(2, -2)$ is not. So S is not a vector space. ∎

EXAMPLE 6 **Geometric Vector Spaces**

Consider a plane through the origin in space (Fig. 8.4). Any point $\mathbf{p}$ on the plane can be identified with its x, y, z coordinates; that is, it is a vector in R^3. Point **p** can also be associated with an arrow from the origin to that point. (Such arrows are often called **geometric vectors** and are used heavily in physics.) Thinking in terms of arrows, it should be clear that $2\mathbf{p}$ is just the point at the end of the arrow obtained by stretching the arrow for **p** by 2, and thus also sits on the plane. Furthermore, there is nothing special about the scalar 2. Second, if **q** is another point on the plane, the arrow for $\mathbf{p}+\mathbf{q}$ is obtained by putting the arrows for **p** and **q** end to end, as in Fig. 8.4. (This figure illustrates the so-called parallelogram rule for adding geometric vectors. To see why it's called that, see Fig. 8.5, which shows geometric vectors in two dimensions for simplicity.) Thus $\mathbf{p}+\mathbf{q}$ is also on the plane. In short, a plane through the origin is a vector space. ∎

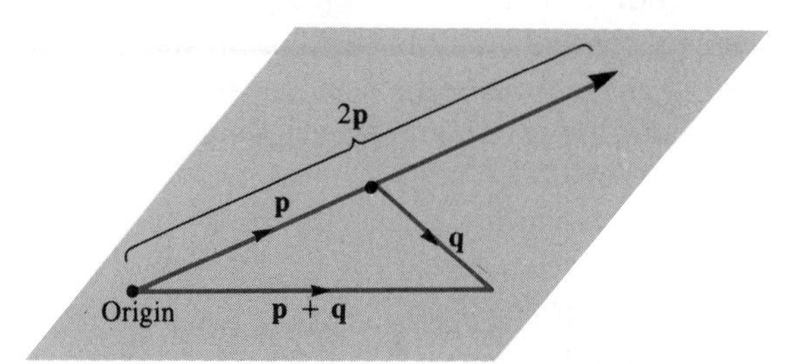

FIGURE 8.4

A plane through the origin is a vector space

We have gone over the idea of geometric vectors very hastily because we won't make further direct use of them. But even if you have never seen them before, Example 6 should suggest to you that vector space theory is just the thing for understanding our 3-dimensional world. Indeed, it's just the thing for understanding the 4th dimension as well, or for that matter, any number of geometric dimensions. When viewed algebraically using vector spaces, n-dimensional space is no harder to cope with than 3 dimensions! Many geometric ideas—for instance perpendicularity—have elegant linear algebra translations, and we regret that we can't go into them.

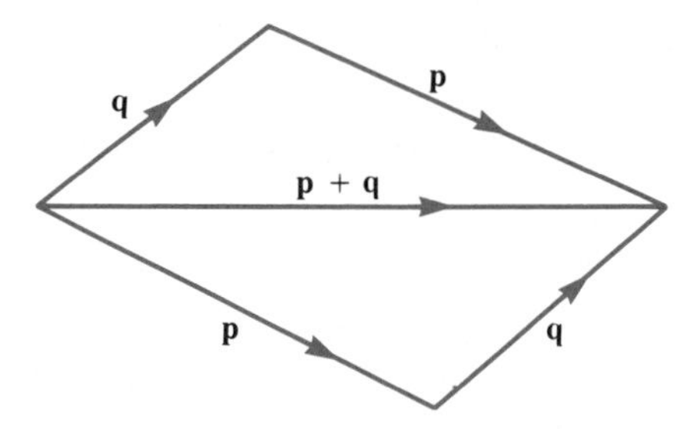

FIGURE 8.5

The parallelogram rule of vector addition

Mostly, we will be concerned with vector spaces associated with systems of linear equations, or equivalently, with matrices. Example 1 shows one way to make an association, but there are several others. In particular, we are going to associate one vector space with the rows of a matrix and another with the columns. We can't say that the rows themselves form a vector space, for they aren't closed under either scalar multiplication or addition. Our next concept gets at how to augment a *set* of vectors to get a *space* of vectors.

Definition 2. Let $S = \{\mathbf{v}_1, \mathbf{v}_2, \ldots, \mathbf{v}_k\}$ be a set of vectors in R^n. Then the **span** of S is the set of all linear combinations of vectors from S:

$$\text{Span}(S) = \Big\{\sum_{i=1}^{k} c_i\mathbf{v}_i \mid c_i \in R\Big\}.$$

(Here we mean k to be an integer and not ∞, so S is a finite set. This restriction ensures that $\sum_{i=1}^{k} c_i\mathbf{v}_i$ makes sense. Actually, this restriction that S be finite is not essential; see [24, Supplement]. If S is the empty set $\emptyset$ in R^n, we define $\text{Span}(S) = \{\mathbf{0}\}$, the *non*empty set containing the single vector $\mathbf{0}$. Think of it this way: The only linear combination of vectors in $\emptyset$ is the empty sum, and just as we defined the empty sum of numbers to be the number 0, we define the empty sum of vectors to be $\mathbf{0}$. We wouldn't mention the case $S = \emptyset$ at all—it's quite unimportant—except that unless we treat this case, $\{\mathbf{0}\}$ would become an annoying exception to several results we state later.)

Theorem 1. $\text{Span}(S)$ is a vector space.

PROOF We must use Definition 1. If $S = \{\mathbf{v}_1, \ldots, \mathbf{v}_t\}$ is in R^n, then clearly $\text{Span}(S)$ is in R^n, so we merely need to check closure. Pick any two vectors from $\text{Span}(S)$, say,

$$\mathbf{x} = \sum_{i=1}^{k} a_i\mathbf{v}_i, \qquad \mathbf{y} = \sum_{i=1}^{k} b_i\mathbf{v}_i.$$

Then $c\mathbf{x} \in \text{Span}(S)$ because

$$c\mathbf{x} = c\sum_{i=1}^{k} a_i\mathbf{v}_i = \sum_{i=1}^{k} (ca_i)\mathbf{v}_i.$$

(In other words, $c\mathbf{x}$ is a linear combination of the $\mathbf{v}$'s with coefficients ca_i.) Also, $\mathbf{x} + \mathbf{y} \in \text{Span}(S)$ because

$$\mathbf{x}+\mathbf{y} = \sum_{i=1}^{k} a_i\mathbf{v}_i + \sum_{i=1}^{k} b_i\mathbf{v}_i = \sum_{i=1}^{k} (a_i+b_i)\mathbf{v}_i. \quad \blacksquare$$

In light of Theorem 1, Span(S) is also called the space **generated** by S. (If it weren't for Theorem 1, we couldn't have said "space".) If we start with a space

V and somehow show that $V = \text{Span}(S)$ for some set of vectors S, we say that S **spans** V.

Definition 3. Let A be an $m \times n$ matrix. The **row space** of A, denoted $\mathcal{R}_A$, is the span of the rows. The **column space**, denoted $\mathcal{C}_A$, is the span of the columns.

It is correct to call $\mathcal{R}_A$ and $\mathcal{C}_A$ *spaces* because of Theorem 1.

EXAMPLE 7 Determine if $(1, 2, 3, 4)^{\mathrm{T}}$ is in the column space of

$$B = \begin{bmatrix} 1 & 0 & 0 \\ 1 & 1 & 0 \\ 0 & 1 & 1 \\ 0 & 0 & 1 \end{bmatrix}.$$

(Recall that A^{T} is the transpose of A, in which columns and rows are switched. Here we transpose $(1, 2, 3, 4)$ just to save space: Rows fit nicely in paragraphs but columns don't, and transposing allows us to write a column as a row.)

Solution By definition, the question is: Are there coefficients c_1, c_2, c_3 so that

$$\begin{pmatrix} 1 \\ 2 \\ 3 \\ 4 \end{pmatrix} = c_1 \begin{pmatrix} 1 \\ 1 \\ 0 \\ 0 \end{pmatrix} + c_2 \begin{pmatrix} 0 \\ 1 \\ 1 \\ 0 \end{pmatrix} + c_3 \begin{pmatrix} 0 \\ 0 \\ 1 \\ 1 \end{pmatrix}?$$

In other words, is there a solution to

$$\begin{bmatrix} 1 & 0 & 0 \\ 1 & 1 & 0 \\ 0 & 1 & 1 \\ 0 & 0 & 1 \end{bmatrix} \begin{pmatrix} c_1 \\ c_2 \\ c_3 \end{pmatrix} = \begin{pmatrix} 1 \\ 2 \\ 3 \\ 4 \end{pmatrix}?$$

Use Gaussian elimination! Doing the downsweep we find that

$$\left[\begin{array}{ccc|c} 1 & 0 & 0 & 1 \\ 1 & 1 & 0 & 2 \\ 0 & 1 & 1 & 3 \\ 0 & 0 & 1 & 4 \end{array}\right] \longrightarrow \left[\begin{array}{ccc|c} 1 & 0 & 0 & 1 \\ 0 & 1 & 0 & 1 \\ 0 & 0 & 1 & 2 \\ 0 & 0 & 0 & 2 \end{array}\right],$$

so from the final row on the right we see that there is no solution. Thus $(1, 2, 3, 4)^{\mathrm{T}} \notin \mathcal{C}_A$. ∎

EXAMPLE 8 Is $(1,2,3,4)$ in the row space of

$$A = \begin{bmatrix} 1 & 1 & 0 & 0 \\ 0 & 1 & 1 & 0 \\ 0 & 0 & 1 & 1 \end{bmatrix}?$$

Solution Transpose everything in sight. More precisely, it is equivalent to ask: Is $(1,2,3,4)^{\mathrm{T}}$ in the column space of A^{T}? Since A^{T} is just B from Example 7, the answer is No. ∎

Suppose one day somebody gives you a matrix A and asks if $\mathbf{v}$ is in its row space *and* if $\mathbf{u}$ is in the column space. The method of Examples 7 and 8 would require two Gaussian eliminations, one to seek a solution to $A\mathbf{x} = \mathbf{u}$ and another to seek a solution for $A^{\mathrm{T}}\mathbf{y} = \mathbf{v}^{\mathrm{T}}$. It would be nice if you could answer row and column questions about A simultaneously, just using A. After reading the next section you will be able to.

Actually, Example 7 has its priorities backwards. One is rarely interested per se in whether some vector is in some column space. But one is often interested in whether a matrix equation has a solution. Example 7 actually shows that the two questions are equivalent. In other words, we now have a general criterion for when $A\mathbf{x} = \mathbf{b}$ will have a solution: precisely when $\mathbf{b} \in \mathcal{C}_A$. Operationally this criterion doesn't change things; you still answer the question by doing Gaussian elimination. But conceptually the criterion helps because it ties the question of solving equations to an important theoretical construct. You are being asked about membership in a vector space.

Here is one more definition about special vector spaces from matrices. This one simply codifies what we already did in Example 1.

Definition 4. The **null space** of matrix A, denoted $\mathcal{N}_A$, is the set of solutions to the homogeneous system $A\mathbf{x} = \mathbf{0}$.

$\mathcal{N}_A$ is called a *null* space because A "annihilates" all the vectors in it. And $\mathcal{N}_A$ is a *space* because, from Gaussian elimination, we know that the solutions to $A\mathbf{x} = \mathbf{0}$ are precisely the span of certain vectors. For an alternative proof, see [13].

Problems: Section 8.7

1. Which of the following subsets of R^2 are vector spaces? Why?

a) $\{(x,0) \mid x \in R\}$

b) $\{(x,x) \mid x \in R\}$

c) All solutions to $x + 2y = 0$

d) All solutions to $x + 2y = 3$

e) All points (x,y) on the straight line through $(1,-3)$ and $(-2,6)$

f) All points (x, y) on the straight line through $(1, -3)$ and $(2, 4)$
g) $\{(x, y) \mid x, y > 0\}$
h) The union of the x- and y-axes in R^2
i) $\{(m, n) \mid m, n \in N\}$

2. Let A be $m \times n$. Is the set of solutions to $A\mathbf{x} = \mathbf{0}$ a vector space in R^m or R^n?

3. Which of the following are subspaces of R^∞?
a) Those sequences satisfying $x_n = 3x_{n-1}$
b) Those satisfying $x_n = x_{n-1} + 2x_{n-2}$
c) Those satisfying $x_{n+1} = x_n^2$
d) Those satisfying $x_n = 2x_{n-1} + 1$
e) Those with $x_1 = x_3$
f) Those with all terms equal
g) Those with only a finite number of terms nonzero
h) Those which are bounded, i.e., for some B, $|x_i| \le B$ for all i

4. Consider the system

$$\begin{aligned} x + y + 2z &= 1 \\ 2x + 2y + 3z &= 3. \end{aligned}$$

Find all vectors which satisfy this system, using Gaussian elimination in the following ways. Note whether the degrees of freedom remain the same.
a) With the variables and equations in the order given
b) With the equations rewritten so that the variables appear from left to right in the order y, z, x
c) Same as b) but with the variables in the order z, x, y
d) Same order of variables, but reverse the order of the equations

5. The set $\mathcal{M}_{m\times n}$ of all $m \times n$ matrices can be viewed as R^{mn}: The mn-tuple is simply broken into m rows of length n rather than stretched out in one row of length mn. Therefore it makes sense to ask if various subsets of $\mathcal{M}_{m\times n}$ are vector spaces. Which of the following are? Why?
a) Those matrices with all diagonal entries 0
b) Those matrices with all diagonal entries 1
c) Diagonal matrices (all entries off the main diagonal 0)
d) **Symmetric** matrices: For all i, j, $a_{ij} = a_{ji}$
e) Upper triangular matrices
f) All matrices in which no 0's appear in pivot positions when row-reduced.
g) Matrices in reduced echelon form
h) Echelon matrices

6. Let $A = \begin{bmatrix} 1 & 3 & 5 \\ 2 & 2 & 6 \\ 3 & 1 & 7 \end{bmatrix}$.
a) Determine whether $(2, 3, 4)^{\mathrm{T}} \in \mathcal{C}_A$.
b) Determine whether $(1, 1, 1) \in \mathcal{R}_A$.

7. Show that $\{\mathbf{0}\}$ is a vector space. More precisely, for any vector space V, let $\mathbf{0}$ be the zero of that space. Show that $\{\mathbf{0}\}$ is a subspace of V.

8. Show that $\{\mathbf{0}\}$ is the only nonempty vector space with only a finite number of vectors in it.

9. Is (1,1,1) in the span of (1,2,0), (0,1,2), (2,0,1)?

10. Is (1,1,1,1) in the row space of $\begin{bmatrix} 1 & 1 & 0 & -1 \\ 0 & 0 & 2 & 1 \end{bmatrix}$?

11. A set U in R^n is a **translate of a vector space** if there is a vector space $V \in R^n$ and a fixed vector $\mathbf{t} \in R^n$ such that

$$U = \{\mathbf{t} + \mathbf{v} \mid \mathbf{v} \in V\}.$$

Use Gaussian elimination to explain why, for any linear system $A\mathbf{x} = \mathbf{b}$, the solution set is a translate of a vector space.

12. Consider Example 3 again. Was it necessary for the plane to go through the origin in order for the plane to be a vector space?

13. Suppose that A is $m \times n$.
a) If $\mathcal{C}_A = R^m$, what does that say about the reduced echelon form of A? What does it say about the rank of A (defined in Section 8.4).
b) If $\mathcal{R}_A = R^n$, what does that say about the reduced echelon form of A? About the rank of A?

14. Here is the start of an alternative proof, using matrix notation, that $\mathcal{N}_A$ is a vector space. By definition $\mathcal{N}_A = \{\mathbf{x} \mid A\mathbf{x} = \mathbf{0}\}$. Using properties of matrix multiplication, show that if $\mathbf{x}$ satisfies $A\mathbf{x} = \mathbf{0}$ then so does $c\mathbf{x}$. Keep going.

15. Explain why a set V in R^n is a vector space if and only if it is closed under linear combinations.

16. Show that $S \subset T \implies \mathrm{Span}(S) \subset \mathrm{Span}(T)$. If S is a proper subset of T, must $\mathrm{Span}(S)$ be a proper subset of $\mathrm{Span}(T)$?

17. If the point of the definition of Span(S) is to augment S just enough to get a vector space, why didn't we define Span S to be

$$\{c\mathbf{v} \mid c \in R, \mathbf{v} \in S\} \cup \{\mathbf{u}+\mathbf{v} \mid \mathbf{u}, \mathbf{v} \in S\}\ ?$$

If that's no good, why not

$$\{c\mathbf{u} + d\mathbf{v} \mid c, d \in R,\ \mathbf{u}, \mathbf{v} \in S\}\ ?$$

18. Let $\mathbf{u}$ and $\mathbf{v}$ be any vectors in the same R^n. Let $\mathbf{w} = 2\mathbf{u} + 3\mathbf{v}$ and $\mathbf{x} = 3\mathbf{u} - 4\mathbf{v}$. Finally, define $\mathbf{y} = 5\mathbf{w} + \mathbf{x}$. Show that $\mathbf{y}$ is a linear combination of $\mathbf{u}$ and $\mathbf{v}$.

19. [18] generalizes: Linear combinations of linear combinations are linear combinations. State this assertion more carefully and then prove it. *Hint*: You can do this nicely using matrices. If you make your initial vectors into the rows of a matrix, what piece of algebra represents the first linear combinations?

20. The **left null space** of matrix A, denoted $\mathcal{N}_{A^{\mathrm{T}}}$, is defined to be $\{\mathbf{y} \mid \mathbf{y}A = \mathbf{0}\}$.

a) Prove that $\mathcal{N}_{A^{\mathrm{T}}}$ is a vector space.

b) Find a basis for $\mathcal{N}_{A^{\mathrm{T}}}$ if $A = \begin{bmatrix} 1 & 2 \\ 2 & 2 \\ 2 & 1 \\ 1 & 1 \end{bmatrix}$.

21. Apply Gaussian elimination to

$$\left[\begin{array}{ccc|c} 1 & 4 & 7 & b_1 \\ 2 & 5 & 8 & b_2 \\ 3 & 6 & 9 & b_3 \end{array}\right].$$

a) Show that $(b_1, b_2, b_3)^{\mathrm{T}} \in \mathcal{C}_A$ (where A is the square matrix to the left of the augmentation line) iff $b_1 - 2b_2 + b_3 = 0$.

b) Define $Q = [1, -2, 1]$. Why is $\mathcal{C}_A = \mathcal{N}_Q$?

22. Pick any set $S = \{\mathbf{v}_1, \ldots, \mathbf{v}_k\}$ in R^n. Let $V = \mathrm{Span}(S)$. Prove: $V = \mathcal{N}_Q$ for some matrix Q. *Note*: This result generalizes [21].

8.8 Vector Spaces: Basic Theory

In this section we study three key concepts in linear algebra: dependence, basis and dimension. We define these terms and prove key theorems about them. The glory of these definitions is that, by reducing previously intuitive concepts to precise algebraic formulations, they allow solid proofs of intuitive beliefs (like every space has a specific dimension).

Towards the end of this section, we will apply our results to show that the degrees of freedom in a system of linear equations, and the rank of the associated matrix, are independent of the algorithm used to solve the system. Some other specific consequences of the theorems in this section appear in the problems [44], [18, Section 8.10]. But the main payoff to you will be in a framework for better understanding anything involving linear algebra.

This payoff has a price. At first most people find the definitions and proofs in this section hard to grasp. We'll do our best to help. We'll display definitions, state them precisely, and call attention to their subtleties. We'll motivate the proofs as best we can. It will also help if you keep in mind that learning how to do linear algebra proofs is *not* the primary goal of this chapter. Rather, the primary goal is to understand the key concepts of linear algebra. A proof can sometimes elucidate a concept as well as an example can. So read the proofs for understanding and constantly look at how they amplify the key concepts. Also, note how frequently proofs hinge on knowledge of Gaussian elimination. That's why this is *algorithmic* linear algebra.

Dependence and Independence

We start with a definition of "depends on", a concept not defined in most books, but one which we think helps to motivate the definition of "dependent", which *is* defined in every book.

Definition 1. Vector $\mathbf{u}$ is said to **depend on** vectors $\mathbf{v}_1, \mathbf{v}_2, \ldots, \mathbf{v}_n$ if $\mathbf{u}$ is a linear combination of the $\mathbf{v}$'s.

EXAMPLE 1 $(2,3)$ depends on $(1,0)$ and $(0,1)$ because

$$(2,3) = 2(1,0) + 3(0,1). \blacksquare$$

EXAMPLE 2 Determine if $(1,2,3,4)$ depends $(1,1,0,0)$, $(0,1,1,0)$ and $(0,0,1,1)$.

Writing all the vectors as columns, we are simply asking: Is $\mathbf{b} = (1,2,3,4)^{\mathrm{T}}$ in the column space of

$$A = \begin{bmatrix} 1 & 0 & 0 \\ 1 & 1 & 0 \\ 0 & 1 & 1 \\ 0 & 0 & 1 \end{bmatrix} ?$$

In other words, does $A\mathbf{x} = \mathbf{b}$ have a solution? In Example 7 of the previous section we determined by Gaussian elimination that the answer is No. $\blacksquare$

EXAMPLE 3 Show that $\mathbf{0}$ depends on any set of vectors.

Solution Actually, we need to be more precise, since there is a different $\mathbf{0}$, i.e., n-tuple of 0's, for each R^n. What we mean is: For any n and any set $S = \{\mathbf{v}_1, \mathbf{v}_2, \ldots, \mathbf{v}_k\} \subset R^n$, show that the n-tuple $\mathbf{0}$ depends on S. The solution is simple:

$$\mathbf{0} = \sum_{i=0}^{k} 0\mathbf{v}_i\,.$$

That is, take all the multipliers in the linear combination to be 0. $\blacksquare$

Definition 2. A set of vectors $S = \{\mathbf{v}_1, \mathbf{v}_2, \ldots, \mathbf{v}_k\} \subset R^n$ is **dependent** if there exists some set of constants c_i, *not all* 0, so that

$$\sum_{i=0}^{k} c_i\mathbf{v}_i = \mathbf{0}.$$

Note two things especially about this definition. In ordinary English something always depends *on* something else, but here that's not true. The set is simply dependent (if you will, on itself). Second, the restriction that not all coefficients be 0 is crucial. Indeed, there is a shorthand name for this restriction: S is dependent if there exists a **nontrivial solution** to $\sum_{i=0}^{k} c_i\mathbf{v}_i = \mathbf{0}$. Without this restriction, every set of vectors would be dependent: Choose all $c_i = 0$. (That's the **trivial solution.**) What is probably not at all obvious at this point is why, with just this one restriction on the c's, the definition is useful. If there has to be a restriction, why isn't it that *no* c_i be 0? We begin to explain with the following theorem.

Theorem 1. A set $S = \{\mathbf{v}_1, \mathbf{v}_2, \ldots, \mathbf{v}_k\} \subset R^n$ is dependent $\iff$ some $\mathbf{v}_j \in S$ depends on the others.

In other words, a set is dependent if not every vector in it gives "new" information; at least one vector can be obtained by combining the others.

PROOF Sufficiency ($\Longleftarrow$). Almost all proofs in elementary linear algebra amount to translating both the hypothesis and the conclusion into simple algebraic statements, and then doing straightforward algebra to get from the former to the latter. The hard part is figuring out the translation. However, here there is very little choice, because we have essentially nothing to work with so far but a few definitions. If you trace back from the definition of "depends on" to the definition of "linear combination" (Section 8.6), you will see that the hypothesis says there exist constants c_i such that

$$\mathbf{v}_j = \sum_{\substack{i=1 \\ i\neq j}}^{k} c_i\mathbf{v}_i. \tag{1}$$

We don't know what these constants are, but we are told they exist, so we may use them in this generic form. As for the conclusion, it says there are constants (perhaps different from the c's, so let's call them d's) such that

$$\sum_{i=0}^{k} d_i\mathbf{v}_i = \mathbf{0}. \tag{2}$$

In this case we are not told such d's exist but rather are challenged to find them. Thus if we can find any d's for which Eq. (2) holds, we are done. We are left to devise some algebra that starts with Eq. (1) and gets all the $\mathbf{v}$'s on one side and 0 on the other, as in Eq. (2). So subtract $\mathbf{v}_j$ from both sides:

$$\mathbf{0} \;=\; c_1\mathbf{v}_1 + \cdots + c_{j-1}\mathbf{v}_{j-1} + (-1)\mathbf{v}_j + c_{j+1}\mathbf{v}_{j+1} + \cdots + c_k\mathbf{v}_k\,.$$

In other words, Eq. (2) does hold, with

$$d_i = \begin{cases} c_i & \text{if } i \neq j, \\ -1 & \text{if } i = j. \end{cases}$$

Note that at least one d_i is nonzero, because $d_j = -1 \neq 0$.

Necessity ($\Longrightarrow$). Now we need to go from Eq. (2) to Eq. (1); i.e., this time we are told the d's in Eq. (2) exist and we are challenged to make up c's for Eq. (1). By the definition of dependence, at least one $d_i \neq 0$. Let j index such a d. Subtract the jth term from Eq. (2):

$$-d_j\mathbf{v}_j = \sum_{\substack{i=1 \\ i\neq j}}^{k} d_i\mathbf{v}_i \, .$$

Now divide by $-d_j$ to obtain

$$\mathbf{v}_j = \sum_{\substack{i=1 \\ i\neq j}}^{k} \left(-\frac{d_i}{d_j}\right)\mathbf{v}_i \, .$$

In other words, Eq. (1) does hold with $c_i = -(d_i/d_j)$ for all $i \neq j$. ∎

EXAMPLE 4 Show that $\mathbf{e}_1, \mathbf{e}_2, \mathbf{e}_3$ and $(3,0,7)$ are dependent. (Recall that the $\mathbf{e}$'s are the standard unit vectors: $\mathbf{e}_1 = (1,0,0,\ldots)$, $\mathbf{e}_2 = (0,1,0,0,\ldots)$, and so on. From the context, it should be clear that in this case we mean the $\mathbf{e}$'s to be in R^3.)

Solution By inspection, we see that

$$3\begin{pmatrix}1\\0\\0\end{pmatrix} + 0\begin{pmatrix}0\\1\\0\end{pmatrix} + 7\begin{pmatrix}0\\0\\1\end{pmatrix} - 1\begin{pmatrix}3\\0\\7\end{pmatrix} = \begin{pmatrix}0\\0\\0\end{pmatrix}.$$

Indeed, no matter what fourth vector we picked, the set would be dependent because the $\mathbf{e}$'s span R^3. ∎

EXAMPLE 5 Are the vectors $(1,1,0,0)$, $(0,1,1,0)$, $(0,0,1,1)$ and $(1,2,3,4)$ dependent?

Solution We are asking whether there is a nontrivial linear combination of them (that is, not all coefficients 0) which equals $\mathbf{0}$. Once again, if we write the vectors as the columns of a matrix, the question becomes one about a linear system: Are there nontrivial solutions to

$$\begin{bmatrix} 1 & 0 & 0 & 1 \\ 1 & 1 & 0 & 2 \\ 0 & 1 & 1 & 3 \\ 0 & 0 & 1 & 4 \end{bmatrix} \begin{pmatrix}c_1\\c_2\\c_3\\c_4\end{pmatrix} = \begin{pmatrix}0\\0\\0\\0\end{pmatrix} ?$$

From our knowledge of Gaussian elimination, we know this translates into yet another question: Do we get any nonpivot columns (and thus free variables)? This same matrix appeared in Example 7, Section 8.7, except there the fourth column was augmentation. The result of downsweep there was

$$\left[\begin{array}{ccc|c} 1 & 0 & 0 & 1 \\ 0 & 1 & 0 & 1 \\ 0 & 0 & 1 & 2 \\ 0 & 0 & 0 & 2 \end{array}\right].$$

Deleting the augmentation line, we see that we can go on to pivot on the 2 in position $(4,4)$, and that the final result is I; in particular, every column has a pivot. By Theorem 2 part (4), Section 8.4, the equation two displays above has exactly one solution, which is the trivial solution. Therefore the four vectors we started with are not dependent. ∎

We hope you are beginning to see a pattern in answering questions about vectors. We attempt to turn the question into one about a linear system by putting the vectors in a matrix. So far, we've made the vectors into columns, but sometimes rows work better. Then we keep translating the question until it becomes one about solving the linear system associated with the matrix.

The observations we made in solving Example 5 lead to an important general result.

Theorem 2. If A is an $m \times n$ matrix with $m < n$, then its columns are dependent.

PROOF Since there is at most one pivot in each row, and there are more columns than rows, there are some nonpivot variables. Thus $A\mathbf{x} = \mathbf{0}$ has nontrivial solutions; i.e., the columns are dependent. ∎

Since being not dependent is at least as important as being dependent, it is given its own, not very surprising name:

Definition 3. A set of vectors that is not dependent is **independent**.

This definition is important enough to restate it in several equivalent ways. We have

$\{\mathbf{v}_1, \ldots, \mathbf{v}_k\}$ is independent

$\iff$ It is not the case that there exists a nontrivial solution to $\sum_{i=1}^{k} c_i\mathbf{v}_i = \mathbf{0}$

$\iff$ The only solution to $\sum_{i=1}^{k} c_i\mathbf{v}_i = \mathbf{0}$ is the trivial solution

$\iff$ $\sum_{i=1}^{k} c_i\mathbf{v}_i = \mathbf{0}$ implies $c_i = 0$ for all i.

EXAMPLE 6 Show that the vectors $\mathbf{v}_1 = (1,1,0,0)$, $\mathbf{v}_2 = (0,1,1,0)$, $\mathbf{v}_3 = (0,0,1,1)$ and $\mathbf{v}_4 = (1,2,3,4)$ are independent.

Solution We show that the only solution to $\sum_{i=1}^{4} c_i\mathbf{v}_i = \mathbf{0}$ is $(c_1, c_2, c_3, c_4) = \mathbf{0}$. Writing the $\mathbf{v}$'s as columns, we must solve

$$\begin{bmatrix} 1 & 0 & 0 & 1 \\ 1 & 1 & 0 & 2 \\ 0 & 1 & 1 & 3 \\ 0 & 0 & 1 & 4 \end{bmatrix} \begin{pmatrix} c_1 \\ c_2 \\ c_3 \\ c_4 \end{pmatrix} = \begin{pmatrix} 0 \\ 0 \\ 0 \\ 0 \end{pmatrix}?$$

This is the same matrix equation as in Example 5, because both examples started with the same vectors. We found that the matrix reduces to I, so the trivial solution is the only solution for (c_1, c_2, c_3, c_4). The $\mathbf{v}$'s are independent. ∎

Basis

On many occasions in mathematics, we wish to study an infinite set. The way to do this is to find a finite subset that somehow represents the whole set. For instance, when a homogeneous linear system has infinitely many solutions, Gaussian elimination picks out a finite set of vectors that spans the whole solution set. In fact, Gaussian elimination picks out a rather special sort of "lean" spanning set.

Definition 4. A subset $B = \{\mathbf{v}_1, \mathbf{v}_2, \ldots, \mathbf{v}_k\}$ of vector space V is a **basis** of V if B satisfies these two conditions:

1. B spans V;
2. B is independent.

Thus a basis is a spanning set which is lean in that there are no dependencies among its vectors.

EXAMPLE 7 Show that $\{\mathbf{e}_1, \mathbf{e}_2, \ldots, \mathbf{e}_n\}$ is a basis of R^n.

Solution The $\mathbf{e}$'s span R^n because any vector $\mathbf{b} = (b_1, \ldots, b_n) \in R^n$ satisfies $\mathbf{b} = \sum_{i=0}^{n} b_i\mathbf{e}_i$. In other words, to get $\mathbf{b}$ as a linear combination of the $\mathbf{e}$'s, just use the *coordinates* of $\mathbf{b}$ as the *coefficients* of the $\mathbf{e}$'s. The $\mathbf{e}$'s are independent because the only way to get $\sum_{i=0}^{n} c_i \mathbf{e}_i = \mathbf{0}$ is to set all the c's to 0. ∎

Standard unit vectors are a delight to work with because most facts about them are clear from inspection. Here's a harder example.

EXAMPLE 8 Determine if $B = \{(1,2,3), (4,5,6), (7,8,9)\}$ is a basis of R^3.

Solution Consider the matrix with the given vectors as columns:

$$A = \begin{bmatrix} 1 & 4 & 7 \\ 2 & 5 & 8 \\ 3 & 6 & 9 \end{bmatrix}$$

The requirement that B span R^3 means that $A\mathbf{x} = \mathbf{b}$ must have a solution for every $\mathbf{b} \in R^3$. Thus Gaussian elimination must find a pivot in every row of A. The requirement that B be independent means that $A\mathbf{x} = \mathbf{0}$ must have exactly one solution, $\mathbf{x} = \mathbf{0}$. Thus Gaussian elimination must find a pivot in every column. But there is only one reduced echelon form with a pivot in every row and every column: I. Therefore, we row-reduce A to see if we get I. After two downsweeps we obtain

$$\begin{bmatrix} 1 & 4 & 7 \\ 0 & -3 & -6 \\ 0 & 0 & 0 \end{bmatrix},$$

so we already know that the reduced echelon form is not I. We conclude that B is not a basis. ∎

Note that we have actually proved a theorem. In fact, combine the above argument with Theorem 1, Section 8.6, and we have proved

---

Theorem 3. Let B be the column set of the $n \times n$ matrix A. Then the following conditions are equivalent:

1. B is a basis of R^n;
2. Gaussian elimination reduces A to I;
3. A is invertible.

This theorem tells us how to check if a set of vectors is a basis of R^n (put the vectors in the columns of a matrix and row reduce), but most of the time we need to find bases of subspaces of R^n. We return to the general problem of finding bases shortly, but now that you have some feel for the definition, we'll indicate what bases are good for. The next theorem gives their most important property.

Theorem 4. If $B = \{\mathbf{v}_i, \ldots, \mathbf{v}_k\}$ is a basis of V, then every $\mathbf{u} \in V$ can be expressed as a linear combination of the $\mathbf{v}$'s in a unique way.

In other words, a basis B is a lean spanning set in yet another way; there is no redundancy in expressing vectors using B.

PROOF Consider any $\mathbf{u} \in V$. We already know that $\mathbf{u}$ is some linear combination of the $\mathbf{v}$'s since B spans V. We must show that $\mathbf{u}$ is a linear combination in just one way. That is, we must show that

$$\text{if} \quad \sum_{i=0}^{k} d_i \mathbf{v}_i = \mathbf{u} \quad \text{and} \quad \sum_{i=0}^{k} d_i' \mathbf{v}_i = \mathbf{u}, \qquad \text{then} \quad d_i = d_i' \quad \text{for } i = 1, 2, \ldots, k.$$

Now, we are assuming that B is a basis; we haven't yet used that B is therefore independent. That B is independent reduces to the algebraic statement that $\sum 0\mathbf{v}_i$ is the only linear combination which produces $\mathbf{0}$. So we have to go from the two sums above equaling $\mathbf{u}$ to one sum equaling $\mathbf{0}$. Only then can we conclude anything about the coefficients. Once again, we are compelled to subtract:

$$\sum_{i=0}^{k} (d_i - d_i')\mathbf{v}_i = \mathbf{0}.$$

Thus for each i, $d_i - d_i' = 0$. Therefore, $d_i = d_i'$. ∎

The converse of Theorem 4 is also true [22]. In other words, the unique representation property in the theorem *characterizes* bases. (Mathematicians say that property P characterizes structure S if something is of type S if and only if it satisfies property P.)

Think of what Theorem 4 means this way. Let B be a basis of V and let $\mathbf{v}$ be a vector in V. Suppose you and a friend are each told to express $\mathbf{v}$ as a linear combination using B. It does not matter how you go about it—maybe you use Gaussian elimination and your friend uses divine inspiration—if neither of you makes mistakes you will get the same answer. Had you been given a spanning set that is not a basis, you two could well come up with different answers and both be right.

Now we return to finding bases. How to find them depends on how a space is presented. The most common ways are as the span of a set of vectors or as the solutions to a set of homogeneous equations. If a space is given in the latter way, it is a null space. If it is given in the former way, we can describe it as either a row space or a column space, depending on whether we write the set of generating vectors as the rows of a matrix or the columns. The following theorem gives a complete recipe for finding bases in all three cases. Recall from Section 8.3 that "standard vector form" is our way of writing solutions to linear systems in terms of linear combinations of vectors: We apply Gaussian elimination to the original matrix, rewrite the reduced echelon form in terms of variables, move the free variables to the right, add a redundant equation $x = x$ for each free variable x, and then regroup using column vectors. See Eq. (1) in Example 10.

Theorem 5. Suppose that Gaussian elimination is applied to A and that U is the echelon form obtained from the downsweep. Then

1. the nonzero rows of U are a basis for $\mathcal{R}_A$;
2. the columns of A in the same positions as the pivot columns of U form a basis of $\mathcal{C}_A$; and
3. the column vectors on the right in the standard vector form of the solutions to $A\mathbf{x} = \mathbf{0}$ are a basis for $\mathcal{N}_A$.

This takes a little proving, so let's give some examples first.

EXAMPLE 9 Find a basis for $V = \text{Span}\{(1, 2, 3), (4, 5, 6), (7, 8, 9)\}$.

Solution We answer this two ways. First let's write these vectors as the rows of a matrix:

$$A = \begin{bmatrix} 1 & 2 & 3 \\ 4 & 5 & 6 \\ 7 & 8 & 9 \end{bmatrix}.$$

Downsweeping we obtain

$$U = \begin{bmatrix} 1 & 2 & 3 \\ 0 & -3 & -6 \\ 0 & 0 & 0 \end{bmatrix}.$$

Therefore by conclusion (1) of Theorem 5, a basis for V is

$$\{(1, 2, 3), (0, -3, -6)\}.$$

Now write the original vectors as the columns of a matrix:

$$A' = \begin{bmatrix} 1 & 4 & 7 \\ 2 & 5 & 8 \\ 3 & 6 & 9 \end{bmatrix}.$$

As we saw in Example 8, after downsweeping we have

$$U' = \begin{bmatrix} 1 & 4 & 7 \\ 0 & -3 & -6 \\ 0 & 0 & 0 \end{bmatrix}.$$

The shape tells us that there were pivots in the first and second columns. Therefore conclusion (2) of Theorem 5 says that the corresponding columns of A' are a basis of V. These columns are

$$\begin{pmatrix}1\\2\\3\end{pmatrix}, \begin{pmatrix}4\\5\\6\end{pmatrix}. \blacksquare$$

Note the asymmetry in the methods. Conclusion (2) of Theorem 5 always gives a subset of the original generating set as the basis. Conclusion (1) does not. Also, when using (1) you work with U. When using (2) you must go back to the original matrix. For instance, can you see an easy reason why the first two columns of U' could not be a basis of V? [24]

EXAMPLE 10 Find a basis for the null space of

$$\begin{bmatrix} 1 & 2 & 1 & 1 & 4 & 1 \\ 2 & 4 & 2 & 2 & 8 & 5 \\ -1 & -2 & 1 & -3 & 2 & 1 \\ 1 & 2 & 4 & -2 & 13 & 8 \end{bmatrix}.$$

Solution This is the same matrix we worked with in Example 7, Section 8.3. We determined that the null space (at that time called the set of solutions of the homogeneous equations) was all vectors of the form

$$\begin{pmatrix}u\\v\\w\\x\\y\\z\end{pmatrix} = v\begin{pmatrix}-2\\1\\0\\0\\0\\0\end{pmatrix} + x\begin{pmatrix}-2\\0\\1\\1\\0\\0\end{pmatrix} + y\begin{pmatrix}-1\\0\\-3\\0\\1\\0\end{pmatrix}. \tag{3}$$

In other words, the vectors on the right span the space. In light of Theorem 5, they are also a basis. ∎

We start the proof of Theorem 5 with a lemma of independent interest.

Lemma 6. Suppose that matrix B can be obtained from matrix A by elementary row operations. Then

$$\mathcal{N}_A = \mathcal{N}_B \quad \text{and} \quad \mathcal{R}_A = \mathcal{R}_B.$$

PROOF The first equality states that $A\mathbf{x} = \mathbf{0}$ and $B\mathbf{x} = \mathbf{0}$ have the same solutions. This is true by Theorem 1, Section 8.2.

The second equality takes more work. Since B can be obtained by premultiplying A, all rows of B are linear combinations of rows of A. That means the rows of B are in any vector space containing the rows of A, because vector spaces are closed under scalar multiplication and addition. In particular, the rows of B are in $\mathcal{R}_A$. Likewise, all linear combinations of the rows of B are in any space containing the rows of B. Therefore

$$\mathcal{R}_B \subset \mathcal{R}_A.$$

To get inclusion the other way, recall that all elementary row operations are reversible. That is, A can be obtained from B by some sequence of row operations. Hence $\{\text{rows of } A\} \subset \mathcal{R}_B$ and so

$$\mathcal{R}_A \subset \mathcal{R}_B.$$

The last two displays combined give $\mathcal{R}_A = \mathcal{R}_B$. ∎

PROOF OF THEOREM 5 Item 1 (nonzero rows of U are a basis of $\mathcal{R}_A$). In light of Lemma 6 ($\mathcal{R}_A = \mathcal{R}_U$), it suffices to show that the nonzero rows of U are a basis of $\mathcal{R}_U$. Call these rows $\mathbf{u}_1, \mathbf{u}_2, \ldots, \mathbf{u}_k$. Clearly, they span $\mathcal{R}_U$, as the zero vectors below them contribute nothing. We must show that the u's are independent, i.e., the only solution to

$$\sum_{i=1}^{k} c_i \mathbf{u}_i = \mathbf{0} \tag{4}$$

is to have all $c_i = 0$. Concentrate on the columns of U containing pivots. Figure 8.6 shows this schematically, with p_1, p_2, p_3 being the first three pivots, and $\times$'s representing the arbitrary values that can go above pivots; nonpivot columns and any additional pivot columns are represented by the blank space to the left and right of each column shown. When restricted to the first pivot column, Eq. (4) says

$$c_1 p_1 + c_2 0 + c_3 0 + \cdots \ = \ 0.$$

Since $p_1 \neq 0$, we conclude that $c_1 = 0$. In light of this conclusion, Eq. (4) when restricted to the second pivot column says

$$0 + c_2 p_2 + c_3 0 + \cdots \ = \ c_2 p_2 \ = \ 0.$$

And so on. By induction, all the c's are 0.

FIGURE 8.6

The Pivot Columns of an Echelon Form Matrix

$$\begin{bmatrix} & p_1 & & \times & & \times & \\ & 0 & & p_2 & & \times & \\ & 0 & & 0 & & p_3 & \\ & 0 & & 0 & & 0 & \end{bmatrix}$$

Item 2 (pivot columns of A are a basis of $\mathcal{C}_A$). First, we show that the pivot columns of A span $\mathcal{C}_A$ by showing that, for any $\mathbf{b}$, if $A\mathbf{x} = \mathbf{b}$ has a solution (that is, if $\mathbf{b} \in \mathcal{C}_A$), then there is a solution in which all the free variables are 0 (that is,

b can be written as a linear combination of the pivot columns alone). This claim follows from the form in which Gaussian elimination provides the solutions for **x**: a fixed vector plus a sum of other vectors, each multiplied by a different free variable. If we set all the free variables to 0, then the fixed vector is a solution (in fact, the only solution). Second, we show that the pivot columns of A are independent. This is equivalent to proving that the trivial solution is the only solution to $A\mathbf{x} = \mathbf{0}$ in which all the free variables are 0. But there is always just one solution to any $A\mathbf{x} = \mathbf{b}$ in which all the free variables are set to 0. Finally, since the pivot columns span and are independent, they are a basis.

Item 3 (standard vector form for solutions to $A\mathbf{x} = \mathbf{0}$ gives a basis for $\mathcal{N}_A$). Since the standard form vectors spans $\mathcal{N}_A$, we must show only that they are independent, that is, the only linear combination of them equaling **0** is the trivial combination. The key is to look just at the rows of the free variables in the standard form solution. We claim these rows always form an identity matrix. (For instance, consider Fig. 8.7, which shows the solutions to Example 7, Section 8.3, again, this time with the rows of free variables marked with arrows.) The rows of free variables form an identity matrix because they correspond to equations $x = x$, $y = y$, etc., where $x, y, \ldots$ are the free variables. The columns of an identity matrix are independent, so the longer columns in which these columns sit must also be independent. (If a linear combination of these longer columns equaled **0**, that is, if across every row the combination equaled the number 0, then the combination would equal 0 across the subset of rows making up the identity matrix, contradicting the independence of the columns of I.) ∎

$$\begin{matrix} \\ \longrightarrow \\ \\ \longrightarrow \\ \longrightarrow \\ \\ \end{matrix} \begin{pmatrix} u \\ v \\ w \\ x \\ y \\ z \end{pmatrix} = v \begin{pmatrix} -2 \\ 1 \\ 0 \\ 0 \\ 0 \\ 0 \end{pmatrix} + x \begin{pmatrix} -2 \\ 0 \\ 1 \\ 1 \\ 0 \\ 0 \end{pmatrix} + y \begin{pmatrix} -1 \\ 0 \\ -3 \\ 0 \\ 1 \\ 0 \end{pmatrix}$$

FIGURE 8.7

The rows of free variables form an identity matrix.

Corollary 7. Every finite set of vectors contains a basis of the space it spans.

PROOF Write the vectors as the columns of a matrix A. By Theorem 5 part (2), Gaussian elimination finds an independent subset with the same span, i.e., a basis of the span. ∎

Dimension

We now come to the final definition in the basic theory.

Definition 5. The **dimension** of a vector space V, denoted $\dim(V)$, is the cardinality of any basis.

For instance, $\dim(R^n)$ is n, because $\mathbf{e}_1, \mathbf{e}_2, \ldots, \mathbf{e}_n$ is a basis. (If the dimension of R^n hadn't been n, this concept would not have been named dimension!) Also, for A in Example 9, $\dim(\mathcal{R}_A) = \dim(\mathcal{C}_A) = 2$, and for A in Example 10 $\dim(\mathcal{N}_A) = 3$.

From this definition and Theorem 5, we can immediately conclude the following.

Theorem 8. For any matrix A, $\dim(\mathcal{R}_A) = \dim(\mathcal{C}_A)$.

PROOF By Theorem 5 part (1), $\mathcal{R}_A$ has a basis with one vector for each pivot and, by Theorem 5 part (2), so does $\mathcal{C}_A$. Thus both dimensions equal the number of pivots when A is row-reduced. ∎

This is a very surprising theorem. If A is not square, $\mathcal{R}_A$ and $\mathcal{C}_A$ aren't even made up of the same length tuples, so why on earth should one expect their dimensions to be the same? If you weren't surprised, chalk that up to the simple proof that Gaussian elimination permits. In any event, Theorem 8 is a powerful theorem, with deep consequences. For instance, see [44] for a proof of a key claim from Chapter 5 about the satisfiability of difference equation initial conditions. (Mathematicians say that a theorem is powerful if many general results are easy consequences of it. These consequences are deep if they are very hard to prove directly from basic principles.)

There is a quick way to state Theorem 8: *row-rank* = *column-rank*. That is, the **row-rank** of A is defined to be $\dim(\mathcal{R}_A)$ and **column-rank** is defined to be $\dim(\mathcal{C}_A)$. Now recall from Section 8.3 that we defined the rank of A (no qualifiers like "row") to be the number of pivots when A is reduced using Algorithm GAUSS. As the proof of Theorem 8 shows, we can make an even stronger statement:

$$\text{row-rank} = \text{column-rank} = \text{rank}.$$

A quantity with a purely computational definition (rank) has been shown to equal two other quantities with conceptual definitions—nothing in the definitions of row-rank and column-rank refers to any particular algorithm. Thus it follows that the rank is independent of the Gaussian elimination variant we use. For instance, we could reduce from the bottom right instead of the top left and we would still get the same number of pivots. Equivalently, we could rearrange the rows or columns before we start [14, 15].

However, there is a serious gap in our development of dimension. Perhaps you caught it. In the definition, we said *the* dimension. But a vector space has lots of bases [26]. Do they really all have the same number of vectors? If they don't, then, for instance, we have *not* proved that the number of pivots is independent of the algorithm.

Obvious as the uniqueness of dimension may seem (based on intuitive geometric ideas), nothing on the face of the definition forces uniqueness. Indeed, there are mathematical "independence structures" more general than vector spaces where bases are defined but they can have different sizes [43]. Thus a proof of uniqueness is called for. Indeed, logically this proof should precede Definition 5. Again, we first prove a lemma of independent interest.

Lemma 9. Let $\mathbf{u}_1, \ldots, \mathbf{u}_n$ be vectors in a vector space W with a basis $\mathbf{v}_1, \ldots, \mathbf{v}_m$. If $n > m$, then the $\mathbf{u}$'s are dependent.

PROOF If $W = R^m$, we have already proved this lemma (Theorem 2), for then the $\mathbf{u}$'s may be thought of as columns in a matrix that is wider than it is deep. So the whole issue in the current proof is to reduce the general case to the $m \times n$ matrix case. Well, W is some tuple space (maybe not R^m or even inside R^m), so we may still write the $\mathbf{u}$'s as the columns of a matrix U. Likewise, we write the $\mathbf{v}$'s as the columns of a matrix V. Using the column interpretation of multiplication, we must show there is a column vector $\mathbf{x} \neq \mathbf{0}$ such that $U\mathbf{x} = \mathbf{0}$. Since the $\mathbf{v}$'s are a basis of W, each $\mathbf{u}$ is a linear combination of the $\mathbf{v}$'s. That is, there is a coefficient matrix C such that

$$VC = U.$$

(The jth column of C gives the coefficients of the $\mathbf{v}$'s in the linear combination resulting in $\mathbf{u}_j$. See Example 11 below.) Each column of C has m entries and there are n columns, so C is wider than it is deep. Thus $C\mathbf{x} = \mathbf{0}$ has many solutions (there are at least $n - m$ nonpivot columns). To lift this conclusion back to U (that is, to show that $U\mathbf{x} = \mathbf{0}$ has many solutions), multiply $\mathbf{0} = C\mathbf{x}$ by V:

$$\mathbf{0} = V\mathbf{0} = V(C\mathbf{x}) = (VC)\mathbf{x} = U\mathbf{x}. \quad \blacksquare \tag{5}$$

EXAMPLE 11 Let $\mathbf{v}_1, \mathbf{v}_2$ be a basis of some vector space and suppose

$$\begin{aligned} \mathbf{u}_1 &= \mathbf{v}_1 + 2\mathbf{v}_2, \\ \mathbf{u}_2 &= 3\mathbf{v}_1 + 4\mathbf{v}_2, \\ \mathbf{u}_3 &= 5\mathbf{v}_1 + 6\mathbf{v}_2. \end{aligned}$$

Show that $\mathbf{u}_1, \mathbf{u}_2, \mathbf{u}_3$ are dependent by exhibiting a nontrivial linear combination that equals $\mathbf{0}$.

Solution In matrix form (with the $\mathbf{u}$'s and $\mathbf{v}$'s viewed as columns), the given information is

$$\begin{bmatrix} \mathbf{v}_1 & \mathbf{v}_2 \end{bmatrix} \begin{bmatrix} 1 & 3 & 5 \\ 2 & 4 & 6 \end{bmatrix} = \begin{bmatrix} \mathbf{u}_1 & \mathbf{u}_2 & \mathbf{u}_3 \end{bmatrix}.$$

This is the equation $VC = U$ from the proof of Lemma 9. The proof then calls for a nontrivial solution of $C\mathbf{x} = \mathbf{0}$. From Gaussian elimination, one solution is $(1, -2, 1)^{\mathrm{T}}$. (In fact, this is the basis of $\mathcal{N}_C$ that Gaussian elimination finds.) Therefore the **u**'s are dependent because

$$\mathbf{u}_1 - 2\mathbf{u}_2 + \mathbf{u}_3 = \mathbf{0}.$$

If you are incredulous (after all, how can we know this when we've never told you exactly what the **u**'s and **v**'s are!), substitute the definitions of the **u**'s in terms of the **v**'s into the left-hand side. You will find that it all cancels out. This substitution is exactly what is going on in Eq. (5). ∎

Theorem 10. Any two bases of the same vector space have the same number of vectors. Thus dimension is well defined.

PROOF Suppose not. Then some vector space W has two bases $\mathbf{u}_1, \ldots, \mathbf{u}_n$ and $\mathbf{v}_1, \ldots, \mathbf{v}_m$ with $n > m$. By Lemma 9, the **u**'s are dependent—contradiction. ∎

In light of Theorem 10, we say that dimension is **well defined**, or that it is an **invariant** of a space.

Briefly, here are two important applications of this theorem. First, our proof that rank is intrinsic to a matrix is complete. Since rank = row-rank, and the latter is well defined and depends only on the matrix, the former depends only on the matrix too. In this regard, please turn back to Theorem 3, Section 8.4. You can now see that it provides an intrinsic characterization of when $A\mathbf{x} = \mathbf{b}$ will have various sorts of solution sets. At the time you first read it, the theorem only *appeared* intrinsic; the dependence on the algorithm was hidden in the definition of rank. But now we know it really *is* intrinsic. Indeed, except for the use of "system $[A \mid \mathbf{b}]$" instead of the equation $A\mathbf{x} = \mathbf{b}$, the statement is the standard one.

Second, we wondered at the beginning of Section 8.7 whether the number of free variables obtained in solving a homogeneous linear system is an artifact of Gaussian elimination (that is, of the order in which we arranged the variables or the equations) or whether it too is intrinsic. Answer: It too is intrinsic. If you understand the relationship between the number of pivots and the number of free variables [16], this answer follows immediately from the fact that rank is intrinsic. But here is a direct one-sentence proof: There is one free variable (degree of freedom) for each vector in a basis of the null space (Theorem 5), and the size of a basis is invariant. Indeed, most mathematicians don't talk in terms of degrees of freedom. Why use that vague phrase when the precise phrase "dimension of the null space" applies?

One closing comment. There's still one gap in our discussion of dimension. We've proved that any two bases of a space are the same size, but maybe some vector space has no bases at all! For instance, our main tool for finding bases, Theorem 5, requires a (finite) matrix to start with. Thus Theorem 5 doesn't help us to find bases of subspaces of R^∞. Even for subspaces of R^n we've still got problems.

Theorem 5 does fine if the subspace has some finite spanning set, or is the solution set to some finite set of linear equations. But maybe some large subspace of some large R^n is so peculiar that no finite set of vectors or equations can describe it.

Fortunately, it is a theorem that every vector space has a basis. For subspaces of R^n this theorem is treated in the problems [26–31]. For subspaces of R^∞ (and other large vector spaces such as discussed in Section 8.10), very different "infinitary" methods are needed.

Problems: Section 8.8

1. Consider the vectors

$$\mathbf{v}_1 = \begin{pmatrix}1\\1\\0\\0\end{pmatrix},\quad \mathbf{v}_2 = \begin{pmatrix}0\\1\\1\\0\end{pmatrix},$$

$$\mathbf{v}_3 = \begin{pmatrix}0\\0\\1\\1\end{pmatrix},\quad \mathbf{v}_4 = \begin{pmatrix}1\\1\\1\\1\end{pmatrix}.$$

a) Does $\mathbf{v}_4$ depend on the others?
b) Does $\mathbf{v}_1$ depend on the others?
c) Does $\mathbf{v}_2$ depend on the others?
d) Are $\mathbf{v}_1, \mathbf{v}_2, \mathbf{v}_3, \mathbf{v}_4$ dependent?

2. For each set of vectors below, determine if it is independent or dependent.

a) $(1,1,0,0)$, $(0,1,1,0)$, $(0,0,1,1)$, $(1,0,0,1)$.
b) $(1,2,0,0)$, $(0,1,2,0)$, $(0,0,1,2)$, $(2,0,0,1)$.
c) $\begin{pmatrix}1\\-1\\0\\0\end{pmatrix}, \begin{pmatrix}0\\1\\-1\\0\end{pmatrix}, \begin{pmatrix}0\\0\\1\\-1\end{pmatrix}, \begin{pmatrix}-1\\0\\0\\1\end{pmatrix}.$

3. Determine if the columns of the following matrix are independent.

$$\begin{bmatrix} 3.7 & \pi^2 & 10^6 \\ 4/3 & \log 10 & \sin 10^\circ \end{bmatrix}$$

4. Let

$$A = \begin{bmatrix} 1 & 3 & 5 \\ 2 & 2 & 6 \\ 3 & 1 & 7 \end{bmatrix}.$$

Do all the following. Think ahead; they can all be done with a single set of computations from a single matrix.

a) Find a basis for $\mathcal{R}_A$.
b) Find a basis for $\mathcal{C}_A$.
c) Determine the dimension of $\mathcal{N}_A$.
d) Determine whether $(2,3,4)^\mathrm{T} \in \mathcal{C}_A$.
e) Determine whether $(1,1,1) \in \mathcal{R}_A$.

5. Do [4] again for $A = \begin{bmatrix} 1 & 2 & 3 \\ 4 & 5 & 6 \\ 7 & 8 & 9 \end{bmatrix}$.

6. Let $A = \begin{bmatrix} 1 & 1 & 0 & 0 \\ 2 & 2 & 1 & 0 \\ 3 & 3 & 2 & 1 \\ 1 & 1 & 2 & 3 \end{bmatrix}$. Find

a) a basis for $\mathcal{C}_A$;
b) a basis for $\mathcal{N}_A$;
c) a basis for $\mathcal{R}_A$ which is a subset of the rows of A;
d) another basis for $\mathcal{R}_A$.

7. Do [6] again for

$$A = \begin{bmatrix} 1 & 1 & 1 & 1 \\ 2 & 2 & 2 & 2 \\ 1 & 2 & 3 & 4 \\ 2 & 3 & 4 & 5 \end{bmatrix}.$$

8. Let A be $m \times n$ and let p be the number of pivots when A is row-reduced. What is the dimension of

a) $\mathcal{C}_A$ b) $\mathcal{R}_A$ c) $\mathcal{N}_A$ d) $\mathcal{N}_{A^\mathrm{T}}$

9. Show that any set of vectors containing $\mathbf{0}$ is dependent

a) From the definition;
b) From Example 3 and Theorem 1.

10. Explain why subsets of independent sets are independent. Explain why supersets of dependent sets are dependent. (A is a superset of B if $B \subset A$.)

11. In the proof of necessity in Theorem 1, there appears to be a lot of freedom in the choice of the index j. For most sets of vectors this is true. Come up with a set of two or more vectors for which there is only one choice for j.

12. Where would the proof of Theorem 1 break down if we didn't include in the definition of dependence that not all the c's can be 0?

13. Suppose that $\mathbf{u}$ depends on S. Show that

$$\text{Span}(S \cup \{\mathbf{u}\}) = \text{Span}(S).$$

14. Suppose that augmented matrices $\hat{A}$ and $\hat{B}$ differ only in the order of rows. Use results in this section to show quickly that Algorithm GAUSS finds the same number of pivots for both.

15. Suppose that augmented matrices $\hat{A}$ and $\hat{B}$ differ only in the order of columns and that the $\mathbf{b}$ column is the same in both. Prove that Algorithm GAUSS finds the same number of pivots for both.

16. Starting with the fact that the rank of a matrix A is independent of the Gaussian elimination algorithm, explain why the degrees of freedom of the solutions to $A\mathbf{x} = \mathbf{b}$ is also invariant. See [8].

17. Let $\mathbf{v}_1, \mathbf{v}_2, \ldots, \mathbf{v}_m$ be any m vectors in some vector space. Explain why any $m+1$ vectors in the span of the $\mathbf{v}$'s are dependent.

18. In the text, we showed that the number of degrees of freedom is invariant for any *homogeneous* system of linear equations. Explain why the word "homogeneous" may be dropped.

19. Find the dimension of the space of 3×3 symmetric matrices.

20. What is the dimension of the set of solutions to the difference equation $a_n = a_{n-1} + 2a_{n-2}$? Why?

21. $\{\mathbf{e}_1, \mathbf{e}_2, \ldots, \mathbf{e}_n\}$ is such a nice basis. Why should we consider any others?

22. Prove the converse of Theorem 4.

23. Show that Theorem 5 is still correct if you substitute reduced echelon form for echelon form.

24. Give a simple reason (evident from inspection) why the first two columns of U' are not a basis of V in Example 9.

25. Prove that the rows of a square triangular matrix with nonzeros on the main diagonal are independent.

26. Given a nonempty basis $\mathbf{v}_1, \mathbf{v}_2, \ldots, \mathbf{v}_k$ of an arbitrary vector space V, show how to generate infinitely many different bases for V.

27. Suppose $S = \{\mathbf{v}_1, \mathbf{v}_2, \ldots, \mathbf{v}_k\}$ is an independent set in vector space V and $\mathbf{v}_{k+1} \in V - \text{Span}(S)$. Show that $\{\mathbf{v}_1, \ldots, \mathbf{v}_k, \mathbf{v}_{k+1}\}$ is an independent set.

28. Prove: If $S = \{\mathbf{v}_1, \mathbf{v}_2, \ldots, \mathbf{v}_k\}$ is dependent, then at least one $\mathbf{v}_j \in S$ has the property that $\text{Span}(S - \{\mathbf{v}_j\}) = \text{Span}(S)$.

29. Consider the following algorithm.

```
Input V                              [A vector space]
Algorithm GROW-BASIS
  S ← ∅
  repeat until Span(S) = V
    Pick v ∈ V−Span(S)
    S ← S ∪ {v}
  endrepeat
Output S
```

a) Let V be any subspace of R^n with n finite. Show that this algorithm finds a basis for V. Thus this algorithm proves that every vector space within any R^n has a basis.

b) Is this algorithm really an algorithm? Why or why not?

30. Consider the following algorithm.

```
Input S, V     [S a finite spanning set
                          of vector space V]
Algorithm SHRINK-TO-BASIS
  repeat until S is independent
    Pick v ∈ S such that v depends on
       S−{v}
    S ← S−{v}
  endrepeat
Output S
```

a) Let V be any subspace of R^n with n finite. Given S, show that this algorithm finds a basis for V.

b) Is this really an algorithm? Why or why not?

31. Theorem: Any n independent vectors in an n-dimensional space are a basis.

a) Prove this theorem using Algorithm GROW-BASIS from [29].

b) Prove it using Gaussian elimination. *Hint*: What does the hypothesis of independence imply about the reduced echelon form? What does this say about solutions to $A\mathbf{x} = \mathbf{b}$ for all $\mathbf{b}$?

32. Let A be $m \times n$. Prove

a) The columns of A span $R^m \iff$ the rows of A are independent.

b) The columns of A are independent $\iff$ the rows of A span R^n

This problem gives a typical **duality** result; a fact about the row space (respectively, the column space) is equivalent to a related result about the column space (respectively, row space).

33. A basis is a lean spanning set in even more ways than we have indicated so far. Theorem: In any finite-dimensional vector space, a basis is a *minimum* spanning set. That is, a basis has strictly fewer vectors than any nonbasis spanning set. Justify this assertion. *Note*: If the invariance of dimension doesn't show up in your justification, you haven't been complete.

34. Let

$$M = \begin{bmatrix} 1 & 1 & 0 & 0 \\ 0 & 1 & 1 & 1 \\ 1 & 0 & 1 & 0 \end{bmatrix}, \quad N = \begin{bmatrix} 3 & 1 & 0 & -1 \\ 3 & -1 & 0 & 1 \end{bmatrix}.$$

Determine if $\mathcal{R}_N \subset \mathcal{R}_M$.

35. Devise and justify an algorithm for determining if the span of one set of vectors is contained in the span of another.

36. Prove: If U and V are both subspaces of R^n, so is $U \cap V$.

37. Let U be the span of $\{(1,2,3),\ (2,3,1)\}$. Let V be the span of $\{(3,1,2),\ (1,0,-1)\}$. Find a basis for $U \cap V$.

38. Suppose that the columns of matrix A are independent and so are the columns of B. Prove that the columns of AB are independent.

39. Show how to determine simultaneously if $\mathbf{v} \in \mathcal{R}_A$ and $\mathbf{u} \in \mathcal{C}_A$.

40. By the **truncation** of a vector in R^n to R^m, where $m < n$, we mean lopping off the last $n - m$ coordinates. Show that the truncation of the vectors in a dependent set always leads to a dependent set, but that the truncation of the vectors in an independent set need not lead to an independent set.

41. By an **extension** of a vector in R^n to R^m, where $m > n$, we mean adding on $m - n$ coordinates arbitrarily. Show that any extension of the vectors in an independent set results in an independent set, but that the extension of the vectors in a dependent set need not be dependent.

42. An independent set S in vector space V is **maximal** if there is no proper superset of S that is also independent. That is, there does not exist T such that $S \subset T$, $S \neq T$ and T is independent. Explain why every maximal independent set is a basis.

43. Call a set U of vertices in a graph G **independent** if no two of the vertices in U are incident to a common edge. Call a set of vertices a **vertex basis** if it is a maximal independent set. (This is a reasonable definition of basis because it corresponds to one of the ways basis can be defined in a vector space; see [42].)

a) Not all bases in a graph have the same size. Let G be a square (4-cycle) with one diagonal edge. G has three vertex bases. What are they?

b) Find one graph that has a vertex basis consisting of a single vertex and another vertex basis consisting of ten vertices.

c) A **vertex cover** in a graph is a set of vertices that is incident to every edge. Prove or disprove: In every graph, some vertex basis is a vertex cover.

44. In Chapter 5 we showed that the general solution to the typical kth-order homogeneous linear difference equation is $a_n = \sum_{i=1}^{k} A_i r_i^n$, where $r_1, r_2, \ldots, r_k$ are distinct constants. We claimed there is a unique particular solution (choice of A's) for each possible set of k consecutive initial conditions

$$a_0 = b_0, \quad a_1 = b_1, \quad \ldots, \quad a_{k-1} = b_{k-1}.$$

You will now prove this. Let

$$M = \begin{bmatrix} 1 & 1 & \cdots & 1 \\ r_1 & r_2 & \cdots & r_k \\ r_1^2 & r_2^2 & \cdots & r_k^2 \\ \vdots & \vdots & & \vdots \\ r_1^{k-1} & r_2^{k-1} & \cdots & r_k^{k-1} \end{bmatrix}.$$

M is called the **Vandermonde matrix**.

a) Explain why the claim from Chapter 5 is equivalent to the claim that $M\mathbf{x} = \mathbf{b}$ has a unique solution for every $\mathbf{b}$.

b) Suppose that $M^{\mathrm{T}}\mathbf{c} = \mathbf{0}$ had a nontrivial solution. Explain why this is equivalent to saying that there is a polynomial $P(x) = \sum_{i=0}^{k-1} c_i x^i$ for which

$$P(r_1) = 0, \quad P(r_2) = 0, \quad \ldots, \quad P(r_k) = 0.$$

In other words, P would be a polynomial of degree at most $k-1$, yet it would have at least k roots. This is impossible by the **Fundamental Theorem of Algebra**.

c) Assuming the Fundamental Theorem of Algebra, prove the claim from Chapter 5. *Hint*: Translate both the claim in part a) and the conclusion of part b) into statements about the rank of M.

8.9 Eigenvalues

So far, all our applications of matrices have been to solving linear systems. There are many other applications. In particular, another major use of matrices is to model transitions—situations where the current values in a set of data depend on previous values. To use matrices to analyze transitions, we will introduce two important concepts, eigenvalues and eigenvectors.

Matrix transition models encompass difference equations in several variables, as our first example shows.

EXAMPLE 1 In Section 5.2, Example 7 we considered a rent-a-car company with offices in NYC and LA. Letting N_k and L_k be the number of cars in New York and Los Angeles, respectively, in month k, the company found that

$$\begin{aligned} N_{k+1} &= \tfrac{1}{2}N_k + \tfrac{1}{3}L_k, \\ L_{k+1} &= \tfrac{1}{2}N_k + \tfrac{2}{3}L_k. \end{aligned} \tag{1}$$

The question was: If $N_0 = L_0 = 1000$, then what happens in the long run? Show how to set this up as a matrix problem.

Solution Eq. (1) is a linear system, so we may express it as

$$\begin{pmatrix} N_{k+1} \\ L_{k+1} \end{pmatrix} = \begin{bmatrix} 1/2 & 1/3 \\ 1/2 & 2/3 \end{bmatrix} \begin{pmatrix} N_k \\ L_k \end{pmatrix}.$$

If we now set

$$A = \begin{bmatrix} 1/2 & 1/3 \\ 1/2 & 2/3 \end{bmatrix} \quad \text{and} \quad \mathbf{u}_k = \begin{pmatrix} N_k \\ L_k \end{pmatrix},$$

we have

$$\mathbf{u}_1 = A\mathbf{u}_0, \quad \mathbf{u}_2 = A\mathbf{u}_1 = A^2\mathbf{u}_0,$$

and in general

$$\mathbf{u}_n = A^n\mathbf{u}_0.$$

Thus the rental car problem is now a simple geometric sequence problem, except that the "base" is a matrix A instead of a number r. While it is easy to determine how r^n behaves in the long run solely on the basis of the value of r, it's not evident—yet—how to analyze A^n. ▮

Example 1 is typical of all the problems in this section and the next. We want to analyze a set of values that change over time. In other words, we have a vector sequence $\{\mathbf{u}_n\}$, where $\mathbf{u}_n$ is the vector of values at time n. Then we find that the transition from time n to $n+1$ can be expressed as a matrix difference equation, $\mathbf{u}_{n+1} = A\mathbf{u}_n$, where the matrix A is independent of n. Therefore $\mathbf{u}_n = A^n\mathbf{u}_0$. From this equation we can determine $\mathbf{u}_n$ for any particular n by simply multiplying out. However, we also want to analyze the long-term behavior by finding a general pattern. This is where eigenvalues come in.

Example 1 shows how a first-order linear difference equation in several real variables can be restated as a first-order linear difference equation in a single vector variable. Using a clever idea, any higher order difference equation in a single real variable can also be restated as a first-order matrix difference equation.

EXAMPLE 2 In Example 1, Section 5.5, we considered the recursion

$$r_{n+1} = r_n + 2r_{n-1}, \qquad r_0 = r_1 = 3. \tag{2}$$

(As we pointed out in Section 5.2, r_n can be interpreted as the number of rabbit pairs if each pair gives birth to two pairs starting when the parents are two months old.) Set up this second-order difference equation as a first-order matrix recursion.

Solution The clever idea is to set

$$\mathbf{u}_n = \begin{pmatrix} r_{n+1} \\ r_n \end{pmatrix}.$$

Then

$$\mathbf{u}_n = \begin{pmatrix} r_{n+1} \\ r_n \end{pmatrix} = \begin{pmatrix} r_n + 2r_{n-1} \\ r_n \end{pmatrix} = \begin{bmatrix} 1 & 2 \\ 1 & 0 \end{bmatrix} \begin{pmatrix} r_n \\ r_{n-1} \end{pmatrix} = \begin{bmatrix} 1 & 2 \\ 1 & 0 \end{bmatrix} \mathbf{u}_{n-1}.$$

Thus

$$\mathbf{u}_n = \begin{bmatrix} 1 & 2 \\ 1 & 0 \end{bmatrix}^n \begin{pmatrix} 3 \\ 3 \end{pmatrix}. \quad ▮$$

Matrices that generate one vector in a sequence from another are called **transition matrices**. Note that transition matrices are square. In this section, when we want to refer to an arbitrary square matrix, we will say that it is $k \times k$. We reserve n for the general term in the sequence $\mathbf{u}_0, \mathbf{u}_1, \ldots$.

Let us turn to the long-run behavior of A^n. There is one type of matrix A for which A^n is easy to analyze: diagonal matrices. We have

$$\begin{bmatrix} d_1 & & & \\ & d_2 & & \\ & & \ddots & \\ & & & d_k \end{bmatrix}^n = \begin{bmatrix} d_1^n & & & \\ & d_2^n & & \\ & & \ddots & \\ & & & d_k^n \end{bmatrix}.$$

Thus for diagonal matrices, $\{A^n\}$ involves k (usually different) numerical geometric sequences $\{d_1^n\}$, $\{d_2^n\}, \ldots, \{d_k^n\}$. Each numerical sequence can be made to appear explicitly in $A^n\mathbf{u}_0$ by the right choice of $\mathbf{u}_0$, because

$$A^n\mathbf{e}_1 = d_1^n\mathbf{e}_1, \quad A^n\mathbf{e}_2 = d_2^n\mathbf{e}_2, \quad \ldots, \quad A^n\mathbf{e}_k = d_k^n\mathbf{e}_k.$$

Furthermore, by choosing $\mathbf{u}_0$ to be a linear combination of the $\mathbf{e}$'s, $A^n\mathbf{u}_0$ acts like a sum of geometric terms. For instance,

$$A^n(3\mathbf{e}_1+2\mathbf{e}_2) = 3A^n\mathbf{e}_1 + 2A^n\mathbf{e}_2 = 3d_1^n\mathbf{e}_1 + 2d_2^n\mathbf{e}_2.$$

The pleasant fact is that this is basically as complicated as the behavior of any A^n gets. For most $k\times k$ matrices A, there is a set of k vectors $\mathbf{v}_1,\ldots,\mathbf{v}_k$, in fact, *independent* vectors which thus form a basis of k-dimensional space, such that for each i, $\{A^n\mathbf{v}_i\}$ grows geometrically. (In general, the $\mathbf{v}$'s will *not* be the standard unit $\mathbf{e}$'s.) Consequently, to analyze $\{A^n\mathbf{u}_0\}$ for any given $\mathbf{u}_0$, it suffices to decompose $\mathbf{u}_0$ into a linear combination of the $\mathbf{v}_i$'s and then sum the geometric behavior on the pieces.

That was pretty fast! In the rest of the section we flesh out this analysis more slowly. We begin with a definition.

Definition 1. Let A be a $k\times k$ matrix. The number λ is said to be an **eigenvalue** of A if there is some nonzero vector $\mathbf{v}$ such that

$$A\mathbf{v} = \lambda\mathbf{v}. \tag{3}$$

If λ is an eigenvalue, then *any* $\mathbf{v}$ such that Eq. (3) holds is said to be an **eigenvector** of A.

Note 1. Eigenvalues are sometimes called **characteristic values**—"characteristic" is what "eigen" means in German. Similarly, eigenvectors are sometimes called **characteristic vectors**. The symbol λ is the Greek letter lambda and is traditional in this business.

Note 2. For all the examples in the text, λ will be a *real* number. However, in general some of the eigenvalues of a matrix will be *complex* numbers (even if all the entries in the matrix are real) [36]. When an eigenvalue is real, there is always an associated real eigenvector (all entries real), but when an eigenvalue is complex, the eigenvectors will generally have complex entries. In particular, if all the eigenvalues of a $k\times k$ matrix happen to be real, we can hope to find k independent real eigenvectors, thus obtaining a basis of R^n; but more generally the most one can hope for is a basis of C^n, the space of n-tuples of complex numbers. In any event, to fully understand the behavior of vector sequences $\{A^n\mathbf{u}_0\}$—more fully than we expect in this book—you must first understand the behavior of complex geometric sequences $\{(a+bi)^n\}$ as well as real geometric sequences. This same knowledge is needed for a full understanding of solutions to linear difference equations; see [27, Section 5.5] and [30, Supplement, Chapter 5]. *Caution:* R and C as used here are different from script $\mathcal{R}$ and $\mathcal{C}$, which stand for row and column spaces.

Note 3. For λ real and $k = 2$, there is a nice geometric interpretation of Definition 1. If we represent the vector $\mathbf{v} = (x, y)$ on the coordinate plane as an arrow from the origin to the point (x, y), then $\lambda\mathbf{v} = (\lambda x, \lambda y)$ is parallel to (x, y). (If $\lambda > 1$, $\mathbf{v}$ is stretched to $\lambda\mathbf{v}$; if $0 \leq \lambda \leq 1$, $\mathbf{v}$ is shrunk to $\lambda\mathbf{v}$; if $\lambda < 0$, $\mathbf{v}$ is turned 180°.) If $\mathbf{u}$ is not an eigenvector of A, then $A\mathbf{u}$ is at some skew angle to $\mathbf{u}$. See Fig. 8.8.

Query. Why do we insist that there be a nonzero vector satisfying Eq. (3) before we call λ an eigenvalue [11]?

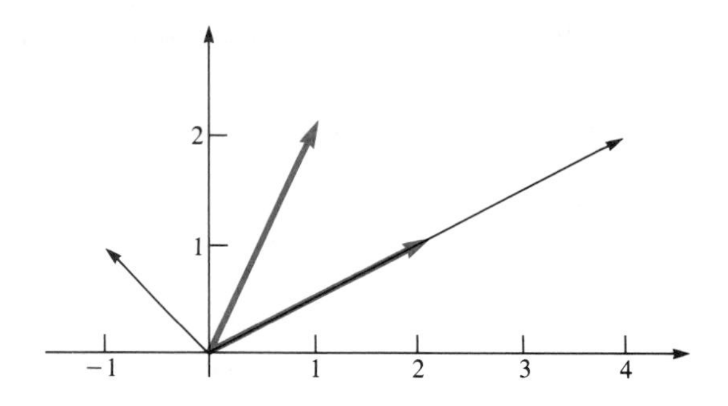

FIGURE 8.8 For $A = \begin{bmatrix} 3 & -2 \\ 1 & 0 \end{bmatrix}$, $\begin{bmatrix} 2 \\ 1 \end{bmatrix}$ is an eigenvector because $A\begin{bmatrix} 2 \\ 1 \end{bmatrix} = \begin{bmatrix} 4 \\ 2 \end{bmatrix}$ is parallel to $\begin{bmatrix} 2 \\ 1 \end{bmatrix}$. But $\begin{bmatrix} 1 \\ 2 \end{bmatrix}$ is not an eigenvector because $A\begin{bmatrix} 1 \\ 2 \end{bmatrix} = \begin{bmatrix} -1 \\ 1 \end{bmatrix}$ is not parallel to $\begin{bmatrix} 1 \\ 2 \end{bmatrix}$

EXAMPLE 3 If $\mathbf{u}$ is an eigenvector of A for eigenvalue λ, and $\mathbf{v}$ is an eigenvector of A for eigenvalue μ, then what are $A(\mathbf{u}+\mathbf{v})$ and $A^2(3\mathbf{u}+5\mathbf{v})$?

Solution

$$A(\mathbf{u}+\mathbf{v}) = A\mathbf{u} + A\mathbf{v} = \lambda\mathbf{u} + \mu\mathbf{v},$$

and

$$\begin{aligned} A^2(3\mathbf{u}+5\mathbf{v}) &= A\big[A(3\mathbf{u}+5\mathbf{v})\big] = A\big[A(3\mathbf{u})+A(5\mathbf{v})\big] = A\big[3A\mathbf{u}+5A\mathbf{v}\big] \\ &= A[3\lambda\mathbf{u} + 5\mu\mathbf{v}] = 3\lambda A\mathbf{u} + 5\mu A\mathbf{v} = 3\lambda^2\mathbf{u} + 5\mu^2\mathbf{v}. \ \blacksquare \end{aligned}$$

The point is that, by the distributive law of matrix multiplication, A acts on each eigenvector separately, and each additional power of A brings out another power of the eigenvalue. In general, if $\mathbf{v}_1, \mathbf{v}_2, \ldots, \mathbf{v}_k$ are eigenvectors of A corresponding to eigenvalues $\lambda_1, \lambda_2, \ldots, \lambda_k$, then by induction we have

$$A^n\Big(\sum_{i=1}^{k} c_i\mathbf{v}_i\Big) = \sum_{i=1}^{k} c_i\lambda_i^n\mathbf{v}_i\,. \tag{4}$$

We now address how to find eigenvalues and eigenvectors. Here is the first step.

Theorem 1. λ is an eigenvalue of A iff the null space of matrix $A-\lambda I$ is nontrivial, in which case that null space is precisely the set of all eigenvectors for λ.

PROOF λ is an eigenvalue if there exists $\mathbf{v} \neq \mathbf{0}$ such that $A\mathbf{v} = \lambda\mathbf{v}$. Subtracting $\lambda\mathbf{v}$ from both sides, we obtain

$$A\mathbf{v} - \lambda\mathbf{v} = A\mathbf{v} - \lambda I\mathbf{v} = (A - \lambda I)\mathbf{v} = \mathbf{0}.$$

That is, $\mathbf{v} \in \mathcal{N}_{A-\lambda I}$. Since $\mathbf{v} \neq \mathbf{0}$, we conclude that the null space is nontrivial. Furthermore, all the steps in this argument reverse: If $\mathcal{N}_{A-\lambda I}$ is nontrivial, then there is a $\mathbf{v} \neq \mathbf{0}$ so that $(A - \lambda I)\mathbf{v} = \mathbf{0}$. Thus $A\mathbf{v} = \lambda\mathbf{v}$. Finally, for any λ and any $\mathbf{v}$, $A\mathbf{v} = \lambda\mathbf{v} \iff \mathbf{v} \in \mathcal{N}_{A-\lambda I}$. ∎

Because of this theorem, the set of eigenvectors for a given λ is called its **eigenspace**. It follows from Theorem 1 that there are infinitely many eigenvectors for each eigenvalue. Therefore, for each eigenspace we will look for a minimal set of vectors that can represent it, i.e., a basis.

EXAMPLE 4 For the matrix of Example 2, determine if $\lambda = 2$ and $\lambda = 1$ are eigenvalues. If either is, determine its eigenspace by finding a basis.

Solution The matrix is

$$A = \begin{bmatrix} 1 & 2 \\ 1 & 0 \end{bmatrix}, \tag{5}$$

so

$$A - 2I = \begin{bmatrix} -1 & 2 \\ 1 & -2 \end{bmatrix}.$$

In general, to determine the null space of a matrix, use Gaussian elimination. However, here it is easy to see by inspection that the second column will not have a pivot and that $(2, 1)$ is a basis of the null space. Therefore, $\lambda = 2$ *is* an eigenvalue of A and $(2, 1)$ is a basis of its eigenspace. (More precisely, we should say that the column vector $(2, 1)^{\mathrm{T}}$ is a basis of the eigenspace, but we will write numerical vectors in row form when it fits better on the page to do so.)

Similarly, inspecting

$$A - 1I = \begin{bmatrix} 0 & 2 \\ 1 & -1 \end{bmatrix}$$

it is easy to see that $A-1I$ reduces to I, so the null space is trivial. (The first row operation is a row switch.) Therefore $\lambda = 1$ is *not* an eigenvalue of A. ∎

Although the method of Example 4 is fine for checking proposed eigenvalues, it is no good for finding them—there are infinitely many numbers λ that might work.

So we need to work with the symbol λ. For instance, continuing with matrix (5) from Example 4, we need to determine in general when the null space of

$$A - \lambda I = \begin{bmatrix} 1-\lambda & 2 \\ 1 & -\lambda \end{bmatrix} \tag{6}$$

is nontrivial. So let us use the downsweep part of Gaussian elimination to determine when $A-\lambda I$ has a nonpivot column. We obtain

$$\begin{bmatrix} 1-\lambda^* & 2 \\ 1 & -\lambda \end{bmatrix} \longrightarrow \begin{bmatrix} 1-\lambda & 2 \\ 0 & -\lambda - \frac{2}{1-\lambda} \end{bmatrix} = \begin{bmatrix} 1-\lambda & 2 \\ 0 & \frac{\lambda^2-\lambda-2}{1-\lambda} \end{bmatrix}. \tag{7}$$

This has a nontrivial nullspace just if the lower right entry in the last matrix is 0, that is, if

$$\lambda^2 - \lambda - 2 = (\lambda-2)(\lambda+1) = 0. \tag{8}$$

Therefore A has exactly two eigenvalues, $\lambda = 2$ and $\lambda = -1$. Notice that the quadratic in Eq. (8) is precisely the characteristic polynomial we obtained (in Chapter 5) for the recursion (2). Also note that this quadratic is the product of the two pivots in display (7). Neither fact is an accident.

The presence of a variable in Eq. (6) made Gaussian elimination messier than heretofore. In fact, it's worse than that: Our computation was incomplete. In order to pivot on the $(1,1)$ entry, we tacitly assumed that $1-\lambda \neq 0$. This assumption effectively excluded the case $\lambda = 1$ from our calculations, so we have to check separately that $\lambda = 1$ is not also an eigenvalue. As it happens, we already did this in Example 4.

By finding a nonzero eigenvector for each eigenvalue of matrix (5), we can obtain by matrix methods an exact formula for $A^n\mathbf{u}_0$:

EXAMPLE 5 Use matrix methods to find a formula for r_n from Example 2.

Solution Since $\mathbf{u}_n = (r_{n+1}, r_n)^{\mathrm{T}}$, we seek a formula for the second coordinate of $\mathbf{u}_n = A^n\mathbf{u}_0$, with A from Eq. (5). From the discussion around Eq. (8), we know that A has two eigenvalues, $\lambda_1 = 2$ and $\lambda_2 = -1$. From Example 4 we already know that $\mathbf{v}_1 = (2,1)$ is an eigenvector for λ_1. For λ_2 we find that

$$A - \lambda_2 I = \begin{bmatrix} 2 & 2 \\ 1 & 1 \end{bmatrix},$$

so by inspection a nonzero eigenvector is $\mathbf{v}_2 = (1,-1)$. (Indeed, the space of eigenvectors for λ_2 is one-dimensional and $(1,-1)$ is a basis.)

Now note: The two eigenvectors we have chosen, one for each eigenvalue, form a basis of R^2. Therefore we may express $\mathbf{u}_0 = (3,3)$ as a linear combination $c_1\mathbf{v}_1 + c_2\mathbf{v}_2$. To solve for $\mathbf{c} = (c_1, c_2)$, we do Gaussian elimination on the augmented matrix $[\mathbf{v}_1, \mathbf{v}_2 \,|\, \mathbf{c}]$:

$$\left[\begin{array}{cc|c} 2 & 1 & 3 \\ 1 & -1 & 3 \end{array}\right] \longrightarrow \left[\begin{array}{cc|c} 1 & 0 & 2 \\ 0 & 1 & -1 \end{array}\right].$$

Thus $\mathbf{c} = (2,-1)$ and

$$\mathbf{u}_0 = 2\mathbf{v}_1 - \mathbf{v}_2 = 2\begin{pmatrix}2\\1\end{pmatrix} - \begin{pmatrix}1\\-1\end{pmatrix}.$$

Therefore, by Eq. (4),

$$\begin{aligned}\begin{pmatrix}r_{n+1}\\r_n\end{pmatrix} = \mathbf{u}_n = A^n\mathbf{u}_0 &= A^n[2\mathbf{v}_1+(-1)\mathbf{v}_2]\\ &= 2\cdot 2^n\mathbf{v}_1 + (-1)(-1)^n\mathbf{v}_2\\ &= 2\cdot 2^n\begin{pmatrix}2\\1\end{pmatrix} - (-1)^n\begin{pmatrix}1\\-1\end{pmatrix},\end{aligned}$$

and so

$$r_n = 2\cdot 2^n - (-1)^n(-1) = 2^{n+1} + (-1)^n.$$

Sure enough, this is the same formula we obtained in Section 5.5. ∎

For a 2×2 matrix, we have illustrated that the eigenvalues are the roots of a quadratic polynomial. In Example 5 the roots were distinct, each eigenspace had dimension 1, and the union of the two eigenspace bases formed a basis of R^2. The situation in Example 5 is the most common case, but for an arbitrary 2×2 matrix A other things can happen. The quadratic might have only one (double) root. If that happens, the one eigenspace might have dimension 2, that is, it might equal R^2. Or it might still have dimension 1, in which case there is no basis of R^2 consisting of eigenvectors for A. See [6] for examples of each case.

The same sorts of things can happen for $k\times k$ matrices, but there are more cases. The most common case is k distinct eigenvalues, each with an eigenspace of dimension 1, but there can be fewer eigenvalues and the dimension of an eigenspace can be any integer from 1 to k.

In the $k\times k$ case the eigenvalues are the roots of a certain kth degree polynomial, called the "characteristic polynomial" of A. For $k=2$ we have shown how to get this polynomial by Gaussian elimination. However, Gaussian elimination is not the best tool in general—as k increases, reducing a matrix containing a variable gets messier and messier, and involves more and more special cases that have to be considered separately. (The reason is that, if you pivot on an entry which involves λ, you are assuming λ does not have a value which makes that expression 0). The right tool for theoretical purposes is the so-called "determinant", which we have not introduced. For instance, using determinants, one can give a simple, precise definition of the characteristic polynomial and prove that it is the product of the pivots of $A-\lambda I$. Thus, if you simply ignore all the special cases and take the product of the pivot entries you obtain on the main diagonal, all fractions will cancel, leaving just the right polynomial.

We don't introduce determinants because they would require at least another section and because they are *only* a theoretical tool. For computing eigenvalues, determinants and Gaussian elimination are both terribly inefficient. The best, complete computational methods are subtle, but shortly we'll give a good method for finding the largest eigenvalue.

Let us summarize the key facts:

1. The eigenvalues of a $k \times k$ matrix A are the roots of a certain kth degree characteristic polynomial, obtained by taking the product of the pivots of $A - \lambda I$. Thus A has at least 1 and at most k distinct eigenvalues.
2. For most matrices A, there is a basis of R^k (or of the complex number space C^k) consisting of eigenvectors of A; this is always the case if the characteristic polynomial has k distinct roots.
3. If R^k (or C^k) has a basis $\mathbf{v}_1, \ldots, \mathbf{v}_k$ of eigenvectors of A, corresponding respectively to (not necessarily distinct) eigenvalues $\lambda_1, \ldots, \lambda_k$, then for any $\mathbf{u} \in R^k$ (or C^k) there exist constants $c_1, \ldots, c_k$ such that

$$\mathbf{u} = \sum_{i=0}^{k} c_i \mathbf{v}_i \qquad \text{and thus} \qquad A^n \mathbf{u} = \sum_{i=0}^{k} c_i \lambda_i^n \mathbf{v}_i\,.$$

We won't prove (1). The one precise claim in (2), concerning the case of distinct roots, has a nice recursive proof [40]. (We won't even attempt to make precise, let alone prove, the broad claim about "most" matrices.) The truth of claim (3) is immediate from the definitions of basis and eigenvectors. The key thing to remember is that, for most matrices A (the ones with a basis of eigenvectors), the sequence $\{A^n\}$ behaves just like a set of ordinary geometric sequences, with a (possibly different) ratio λ_i for each vector $\mathbf{v}_i$ in a certain basis.

Finding Eigenvalues and Eigenvectors

We now discuss an algorithm to find the largest eigenvalue of a matrix and an associated eigenvector. In general, this is all you need if you want the long-term rate of growth of $A^n \mathbf{u}_0$. See [16, 17] for some approaches to finding other eigenvalues.

The algorithm we will develop is our first explicit example of an **approximation algorithm**. That is, it does not come up with the exact answer in a finite number of steps. Rather, it iterates for as long as we choose, getting closer and closer to what we want. Such algorithms are particularly useful in continuous mathematics. See Section 9.7.

So suppose we are interested in the sequence $\{\mathbf{u}_n = A^n \mathbf{u}_0\}$, where A is $k \times k$. Suppose further that there is a basis $\mathbf{v}_1, \ldots, \mathbf{v}_k$ of R^k (or C^k) consisting of eigenvectors of A, where $\mathbf{v}_i$ corresponds to λ_i. (At this point, we don't yet know the $\mathbf{v}_i$'s or the λ_i's.) Since the $\mathbf{v}$'s form a basis, we may write

$$\mathbf{u}_0 = \sum_{i=1}^{k} c_i \mathbf{v}_i. \tag{9}$$

Finally, let the eigenvectors be indexed so that

$$|\lambda_1| > |\lambda_i| \quad \text{for } i \geq 2.$$

In other words, we consider only the case where there is a unique largest eigenvalue. (Among other things, this ensures that λ_1 is a real number [42].) Then

$$\mathbf{u}_n = \sum_{i=1}^{k} c_i \lambda_i^n \mathbf{v}_i = \lambda_1^n \Big(c_1 \mathbf{v}_1 + \sum_{i=2}^{k} c_i \Big(\frac{\lambda_i}{\lambda_1}\Big)^n \mathbf{v}_i\Big). \tag{10}$$

As n gets large, $(\lambda_i/\lambda_1)^n$ goes to 0, so the only significant term in $\mathbf{u}_n$ is $c_1\lambda_1^n\mathbf{v}_1$. (This conclusion assumes $c_1 \neq 0$, but the probability that $c_1 = 0$ is very small. In fact, in computational practice c_1 is never 0; this is an interesting consequence of roundoff error [38].) In other words, if our goal is to find the long-range behavior of a sequence, all we need to find is the largest eigenvalue and one eigenvector for it; $\mathbf{u}_n = A^n\mathbf{u}_0$ grows like $c_1\lambda_1^n\mathbf{v}_1$.

Equation (10) not only shows that the largest eigenvalue and an eigenvector for it are the important items for long-term behavior, but it also suggests an iterative procedure for finding them. Namely, compute the vectors $\mathbf{u}_1, \mathbf{u}_2, \ldots$ by matrix multiplication. If we could divide the results by $\lambda_1, \lambda_1^2, \ldots,$ we would get $c_1\mathbf{v}_1 + \sum_{i=2}^{k} c_i(\lambda_i/\lambda_1)^n\mathbf{v}_i$, which converges to (approaches, stabilizes at) $c_1\mathbf{v}_1$, itself an eigenvector for λ_1. But we don't know λ_1, you say. True, but the point of dividing by the powers of λ_1 is simply to scale our results to get something which (in the limit) is a fixed size. There are lots of other ways to do this. For instance, we could scale each vector so that its first coordinate is $+1$.

Specifically, consider Algorithm 8.4, MAX-EIGEN. It iteratively computes $A\mathbf{x}$ and scales the result. Since the scaling keeps the size fixed (in the sense that the first coordinate is kept at 1), the iterations should stabilize on an eigenvector for λ_1. Moreover, given that the final $\mathbf{x}$ is an eigenvector for λ_1, it's easy to compute λ_1: It is the "ratio" of $\mathbf{x}$ and $\mathbf{y} = A\mathbf{x} = \lambda_1\mathbf{x}$. That is, $\lambda_1 = \mathbf{y}(i)/\mathbf{x}(i)$ for any coordinate i. The ratio is easiest to compute at $i = 1$, because $\mathbf{x}(1) = 1$. In short, $\lambda_1 = \mathbf{y}(1)$.

Note. This is a bare bones algorithm. We leave it to you to include bells and whistles which make it stop as soon as a limit is detected [39].

Algorithm 8.4
Max-Eigen

Input $A, \mathbf{x}$ [Transition matrix and initial vector]
Algorithm MAX-EIGEN

```
    for k = 1 to 100
        y ← Ax
        x ← y/y(1)                [Scale to make first coordinate 1]
    endfor
```

Output $\mathbf{x}$, $\mathbf{y}(1)$ [Eigenvector and eigenvalue]

We use Algorithm MAX-EIGEN to analyze the following example, which is typical of population models. The same idea is used to model human populations, but with much bigger matrices.

EXAMPLE 6

A certain animal has a maximum life span of four years—it dies by its fourth birthday. On average, 90% of newborns (animals under age 1) live another year, as do 80% of those age 1 and 70% of those age 2. Newborns are too young to have offspring during the coming year, but on average during that period each animal age 1 has .5 offspring, each animal age 2 has .6 offspring, and each animal age 3 has .2 offspring. The current population of this animal (in thousands) is $(.5, 1, 2, 1)$; that is, there are 500 newborns, 1000 age 1, and so on. What can we say about the long-term population prospects for this animal?

Solution Let the 4×1 column vector $\mathbf{u}_n = (u_{n0}, u_{n1}, u_{n2}, u_{n3})^{\mathrm{T}}$ represent the population (in thousands) of this animal n years from now. That is, u_{n0} is the number of newborns n years from now, u_{n1} is the number of 1-year-olds, etc. Then all the words in the example statement boil down to this: Let

$$\mathbf{u}_0 = \begin{pmatrix} .5 \\ 1 \\ 2 \\ 1 \end{pmatrix}, \qquad A = \begin{bmatrix} & .5 & .6 & .2 \\ .9 & & & \\ & .8 & & \\ & & .7 & \end{bmatrix},$$

where the blank spaces in A are 0's; then $\mathbf{u}_{n+1} = A\mathbf{u}_n$, so $\mathbf{u}_n = A^n\mathbf{u}_0$. One way to see this is to multiply the rows of A by the column $\mathbf{u}_n$, that is, write out the product $A\mathbf{u}_n$, and interpret. For instance, the first entry in $A\mathbf{u}_n$ is

$$0u_{n0} + .5u_{n1} + .6u_{n2} + .2u_{n3}.$$

If $A\mathbf{u}_n = \mathbf{u}_{n+1}$, then this sum should be the number of newborns in year $n+1$. Is it? Yes, because the problem said newborns in year n produce no newborns by year $n+1$, but those animals that were age 1 produce .5 newborns, those that were age 2 produce .6 newborns, and those that were age 3 produce .2 newborns. We leave the interpretation of the three other entries in $A\mathbf{u}_n$ to a problem [27].

We now compute and scale $A^n\mathbf{u}_0$ using Algorithm MAX-EIGEN, looking to see if the results settle down. The results—we obtained them with a computer but a hand calculator with matrix functions would do as well—are that by the 25th iteration the results have stabilized to four decimal places. Specifically, letting $\mathbf{x}_n$ be the nth value of $\mathbf{x}$ computed by the algorithm, we have

$$\mathbf{x}_{24} = \mathbf{x}_{25} = \begin{pmatrix} 1.0000 \\ 0.9059 \\ 0.7295 \\ 0.5140 \end{pmatrix}, \qquad \mathbf{y}_{24}(1) = \mathbf{y}_{25}(1) = 0.9935\,.$$

Now remember, the coordinates of $\mathbf{x}$ are *not* the actual age populations at time n, because of the scaling. Rather, $\mathbf{x}$ tells us the *ratio* of the age populations. For instance, for every 10,000 newborns at time period 24 (or 25), there are 9059 one-year-old animals. The fact that $\mathbf{x}$ has stabilized tells us that these ratios have become fixed for all later time periods. (We have hit upon an important fact about populations that can be modeled this way; they may not become fixed in size, but they do become fixed in age *distribution*.)

From $\mathbf{y}_{24}(1)$ we find that $\lambda_1 = 0.9935$. Because this λ is the yearly overall rate of growth of the population, we conclude that this animal is dying out! To be sure, its demise is very slow—by less than 1% a year—but demise it is. ∎

EXAMPLE 7 Do Example 6 over again, changing the initial population vector to $(1,1,1,1)$.

Solution Using a computer again, we find that by the 25th iteration we get the same $\mathbf{x}$ and $\mathbf{y}$ as before. (The actual population vectors $\mathbf{u}_n$ are not the same as before [31].) This calculation illustrates that the long-term age distribution and overall rate of growth are independent of the initial population. This is another important fact in population theory and is a consequence of Eq. (10).

Actually, we could have predicted this independence from our theory. Assuming as in Eq. (9) that there exists a basis of eigenvectors, the only thing that changes in representing the new $\mathbf{u}_0$ is the scalars:

$$\mathbf{u}_0' = \sum_{i=1}^{k} d_i \mathbf{v}_i.$$

Thus in the long run, the scaled values of $A^n \mathbf{u}_0'$ settle down to some multiple of $\mathbf{v}_1$, as in Example 6. The rule we have used to scale is the same in both cases (first coordinate = 1), so MAX-EIGEN produces the same multiple of $\mathbf{v}_1$. ∎

We don't want to mislead you. We can't guarantee that Algorithm MAX-EIGEN will come up with the same eigenvector all the time. As illustrated in the problems, any of the following could go wrong. Matrix A might have a basis of eigenvectors, but two or more vectors in that basis might correspond to the same largest eigenvalue [34]. Or there might be two different eigenvalues which tie for largest magnitude [35, 36]. Or we might pick a $\mathbf{u}_0$ which happens to have $c_1 = 0$; see [37], but then see [38], which shows that roundoff error is actually a boon when you want to find the largest eigenvalue.

For most types of real-world problems, none of these complications is likely. Indeed, for population matrices of the form shown in Example 6 there is a theorem that there is a unique largest eigenvalue and so long as you start with a nonnegative $\mathbf{u}_0$ you will converge to it. However, in Section 8.10, we discuss Markov chains, an important type of eigenvalue problem for which some of these complications do arise. Fortunately, for Markov chains there are methods that bypass the use of MAX-EIGEN.

In any event, if you find that $\mathbf{x}$ remains fixed after a while in MAX-EIGEN, you *have* found an eigenvector, because $\mathbf{y} = A\mathbf{x}$ is a multiple of $\mathbf{x}$.

The method behind MAX-EIGEN is often called the **power method**. Except for scaling, we have computed $A^n \mathbf{u}_0$ for the first so many powers n.

Eigenvalues and Differential Equations

(assumes calculus)

Eigentheory is tremendously important in science and engineering, because it can be used to solve differential equations as well as difference equations. The continuous analog of the difference equation $u_n = au_{n-1}$ is the differential equation $u'(t) = au(t)$. Just as there are higher order linear difference equations and linked linear difference equations, there are higher order linear differential equations, such as

$$u''(t) = u'(t) + 2u(t),$$

and linked linear differential equations such as

$$u'(t) = \tfrac{1}{2}u(t) + \tfrac{1}{3}v(t),$$
$$v'(t) = \tfrac{1}{2}u(t) + \tfrac{2}{3}v(t).$$

Just as both sorts of difference equations can be put into the matrix form $\mathbf{u}_n = A\mathbf{u}_{n-1}$, both sorts of differential equations can be expressed, using matrices and vector derivatives, as $\mathbf{u}' = A\mathbf{u}$. Whereas the solution to the matrix difference equation is

$$\mathbf{u}_n = A^n\mathbf{u}_0, \tag{11}$$

the solution to the matrix differential equation turns out to be

$$\mathbf{u}(t) = e^{At}\mathbf{u}(0). \tag{12}$$

(Yes, the constant e raised to a matrix power!) Just as eigenvalues and eigenvectors are crucial to interpreting Eq. (11), so are they crucial to interpreting Eq. (12). You have learned that Eq. (11) may usually be rewritten as

$$\mathbf{u}_n \;=\; \sum_{i=1}^{k} c_i\lambda_i^n\mathbf{v}_i.$$

It turns out that Eq. (12) may usually be rewritten as

$$\mathbf{u}(t) \;=\; \sum_{i=1}^{k} c_i e^{\lambda_i t}\mathbf{v}_i.$$

Problems: Section 8.9

1. Use matrix methods to determine the distribution of rental cars in Example 1 exactly three months from the start.

2. If $\mathbf{u}$ and $\mathbf{v}$ are eigenvectors of A for eigenvalues 2 and 7, respectively, what are

   a) $A\mathbf{v}$? b) $A(3\mathbf{u}-4\mathbf{v})$? c) $A^3(\mathbf{u}+3\mathbf{v})$?

3. Explain why, if a nonzero vector $\mathbf{v}$ is an eigenvector for A, then so is $c\mathbf{v}$ for any scalar c. Why, therefore, is any vector in the span of some eigenvectors actually a sum of eigenvectors?

4. For the following matrices, determine if $\lambda = 2$ is an eigenvalue. If it is, find an eigenvector.

   a) $\begin{bmatrix} 5 & -1 \\ -1 & 5 \end{bmatrix}$ b) $\begin{bmatrix} 7 & 8 \\ -4 & -5 \end{bmatrix}$

5. Repeat [4] for $\lambda = -1$.

6. For the following matrices, find all the eigenvalues and a basis of the eigenspace for each eigenvalue.

   a) $\begin{bmatrix} 1 & 1 \\ 1 & 1 \end{bmatrix}$ b) $\begin{bmatrix} 2 & 0 \\ 0 & 2 \end{bmatrix}$

   c) $\begin{bmatrix} 1 & 2 \\ 0 & 1 \end{bmatrix}$ d) $\begin{bmatrix} 1 & 0 & 0 \\ 0 & -1 & 0 \\ 0 & 0 & -1 \end{bmatrix}$

7. Repeat [6] for the matrices in [4].

8. Explain why a square matrix A is invertible iff 0 is not an eigenvalue. More generally, when is $A - kI$ invertible?

9. If 7 is an eigenvalue of A, name an eigenvalue of $A - 3I$. Generalize.

10. If 3 is an eigenvalue of A, name an eigenvalue of A^2. Generalize.

11. Answer the Query after Definition 1.

12. Show that two vectors **u** and **v** in R^n are dependent iff there exists some constant c so that one of them is c times the other. (Geometrically, they are dependent iff they are parallel.) Show that one cannot always insist that $\mathbf{v} = c\mathbf{u}$ (that is, that **v** be the "one" and that **u** be the "other").

13. Use [12] to show, without using either Gaussian elimination or determinants, that λ is an eigenvalue of $A = \left[\begin{smallmatrix} a & b \\ c & d \end{smallmatrix}\right]$ if and only if $(a-\lambda)(d-\lambda) - bc = 0$.

14. Use Gaussian elimination to reprove the result in [13]. Be sure to consider any special cases, for instance, the possibility that the top left entry of $A - \lambda I$ is 0.

15. Suppose A is invertible. Show that λ is an eigenvalue of $A \iff \lambda^{-1}$ is an eigenvalue of A^{-1}.

16. Use [15] to devise a method for finding the eigenvalue of A with the smallest absolute value. Assume that this eigenvalue is unique.

17. Suppose you have reason to believe there is an eigenvalue of A close to 3. Assume it's neither the smallest nor the largest in absolute value. Devise a method for finding it. *Hint*: Modify A so that you are looking for the smallest eigenvalue.

18. Consider $A = \begin{bmatrix} 0 & 1 & 1 \\ 1 & 0 & 1 \\ 1 & 1 & 0 \end{bmatrix}$.

a) Find the eigenvalues by doing Gaussian elimination on $A-\lambda I$, being very careful to consider all special cases. That is, if you pivot on an expression involving λ, you are assuming that it is nonzero, so you must consider the case that it *is* zero separately.

b) Find the eigenvalues using the theorem that the eigenvalues are the roots of the characteristic polynomial, where that polynomial is the product of the pivots upon reducing $A-\lambda I$ and you don't worry about special cases.

19. For the matrix in [18], find the eigenspaces. If there is a basis of R^3 consisting of eigenvectors, give such a basis.

20. How can it possibly be more work to do Gaussian elimination on a square matrix with a variable in it than on a square matrix with numbers only? Didn't we prove in Section 8.2 that Gaussian elimination takes about $n^3/3$ steps, and didn't that proof make no assumptions about the nature of the matrix entries? Explain.

21. Carry out the eigenvalue analysis of Example 1. That is:

a) Find the eigenvalues and an eigenvector for each eigenvalue.

b) Verify that the eigenvectors you found form a basis of R^2.

c) Find an exact expression for $\mathbf{u}_n = A\mathbf{u}_0$ in terms of these eigenvalues and eigenvectors.

22. Consider the recursion $r_n = 5r_{n-1} - 6r_{n-2}$, with $r_0 = 4$, $r_1 = 7$. Solve this recursion using the matrix method illustrated in Examples 2 and 5. Check your answer by solving the recursion again by the methods of Chapter 5.

23. Show that

$$\begin{bmatrix} 2 & 1 & 0 \\ 0 & 2 & 1 \\ 0 & 0 & 2 \end{bmatrix}$$

does *not* have a basis of eigenvectors.

24. Consider the linked third-order difference equations

$$s_n = 2s_{n-1} + t_{n-1} - 3t_{n-2},$$
$$t_n = t_{n-1} + 3s_{n-2} - t_{n-3}.$$

Write them as a first-order matrix difference equation. How many different numerical geometric sequences would you expect as terms in the solution?

25. Solve the linked difference equations

$$v(n+1) = 3v(n) - w(n),$$
$$w(n+1) = 4v(n) - 2w(n),$$
$$v(0) = 1, \quad w(0) = -2.$$

26. In Example 6, suppose we rescaled the eigenvector we found so that the *sum* of the coordinates is 1. In terms of the application, what useful interpretation would the coordinates now have?

27. Finish the verification from Example 6 that $A\mathbf{u}_n = \mathbf{u}_{n+1}$.

28. Suppose in Example 6 that the animal improves its survival rate from age 2 to age 3. Specifically, change the $(4,3)$ entry of A to .75. Using a computer, determine the long range population behavior. That is, determine whether the relative sizes of the age groups stabilize in the long run and, if so, to what. Also determine whether the population decreases, holds steady, or increases.

29. Repeat [28], supposing that the survival rates are unchanged from Example 6 but that the 3-year-olds are more prolific at child bearing. Specifically, make $A(1,4) = .25$.

30. To model human population by the matrix technique of Example 6, we might want to use ages 0 to 99. In any event, where would the nonzero entries of the matrix be?

31. For Examples 6 and 7, compute the exact population vectors after 25 years.

32. In Algorithm MAX-EIGEN, we computed scaled values of $A^n\mathbf{u}_0$, but we didn't compute values of A^n, scaled or otherwise. Do so, for A in Example 6. What happened? Can you explain what happened in terms of the theoretical results we have stated?

33. Look again at [27, Supplement, Chapter 5]. In terms of what you now know, what was going on?

34. Consider

$$A = \begin{bmatrix} 2 & 0 & 1 \\ 0 & 2 & 1 \\ 0 & 0 & 1 \end{bmatrix}.$$

Run Algorithm MAX-EIGEN on A, starting with $\mathbf{x} = (1,1,1)$ and then with $\mathbf{x} = (1,2,3)$. You should discover that the algorithm finds an eigenvector for the maximum eigenvalue all right, but even after scaling it finds a different eigenvector each time. (The difficulty here is that the eigenspace for this eigenvalue is not one-dimensional.)

35. Consider

$$A = \begin{bmatrix} 1 & 0 \\ 0 & -1 \end{bmatrix}.$$

Run Algorithm MAX-EIGEN on A, starting with any vector $\mathbf{x}$ other than $\mathbf{e}_1$ or $\mathbf{e}_2$ and printing intermediate values of $\mathbf{x}$. Does $\mathbf{x}$ converge? If so, does it converge to an eigenvector? (The difficulty here is that there are two different eigenvalues, both of which are largest in magnitude. Things could be even worse: You could get even more than two eigenvalues tied for largest in magnitude if some of them are complex numbers.)

36. Consider $A = \begin{bmatrix} 0 & -1 \\ 1 & 0 \end{bmatrix}$.

a) Find its eigenvalues and two independent eigenvectors.

b) Show that Algorithm MAX-EIGEN doesn't work for A. What goes wrong? Why?

37. Consider

$$A = \begin{bmatrix} 2 & 1 \\ 0 & 1 \end{bmatrix}.$$

The eigenvalues are 2 and 1. Show that if you run MAX-EIGEN starting with $(1,-1)$ or any integer multiple thereof, you converge to an eigenvector for the smaller eigenvalue but that if you start with any other integer vector, you converge to an eigenvector for the larger eigenvalue. (The difficulty here is that, with the notation of Eq. (9), if we start with $\mathbf{u}_0 = k(1,-1)$ we have $c_1 = 0$. We ask for integer vectors to avoid roundoff errors; see [38c].)

38. Consider

$$A = \begin{bmatrix} 1 & 1 \\ 1 & 0 \end{bmatrix}.$$

(If we had used the Fibonacci recursion in Example 2, this is the matrix we would have gotten.)

a) Show that the eigenvalues are $(1\pm\sqrt{5})/2$.

b) Show that a corresponding basis of eigenvectors is

$$\mathbf{v}_1 = \begin{pmatrix} 1 \\ -\frac{1-\sqrt{5}}{2} \end{pmatrix}, \quad \mathbf{v}_2 = \begin{pmatrix} 1 \\ -\frac{1+\sqrt{5}}{2} \end{pmatrix}.$$

c) In [37] we were able to pick a starting vector so that Algorithm MAX-EIGEN converged to an eigenvector for the *smaller* eigenvalue "by mistake". Try to mimic that here. Start with a good numerical approximation to $\mathbf{v}_2$, say $(1,-1.618)$. Carry out MAX-EIGEN and see what happens. The point is: Even though picking a starting vector which avoids the largest eigenvalue is theoretically possible, because of roundoffs in real computations the largest eigenvalue will usually appear anyway, and so MAX-EIGEN will converge to an eigenvector for the largest eigenvalue even when it shouldn't!

39. Modify Algorithm MAX-EIGEN so that it stops early if $\mathbf{x}$ has already stabilized and so that it reports "no eigenvector found" if $\mathbf{x}$ never stabilizes. "Stabilizes" should mean that the current and previous values of $\mathbf{x}$ are within some reasonable tolerance of each other. (If you insist on exact equality or choose too small a tolerance, due to roundoff the algorithm might never report that it has found an eigenvector.) If you can think of other improvements, state what they are and write them into the algorithm as well.

40. Theorem: Suppose $\mathbf{v}_1, \ldots, \mathbf{v}_k$ are nonzero eigenvectors for matrix A, corresponding to eigenvalues $\lambda_1, \ldots, \lambda_k$. If the λ_i are distinct, then the $\mathbf{v}_i$ are independent.

Prove this theorem. *Hint*: Use induction on k. For the inductive step, note that if

$$\sum_{i=1}^{k} c_i \mathbf{v}_i = \mathbf{0},$$

then

$$A\Big(\sum_{i=1}^{k} c_i \mathbf{v}_i\Big) = \sum_{i=1}^{k} c_i \lambda_i \mathbf{v}_i = \mathbf{0}.$$

Subtracting the first equation λ_1 times from the second reduces the theorem to the previous case.

41. Explain why a $k \times k$ matrix has at most k distinct eigenvalues. This is easy to show from [40]; no knowledge of characteristic polynomials is necessary.

42. Assume as in the text that A is a $k \times k$ matrix with real entries, that it has eigenvectors $\mathbf{v}_1, \ldots, \mathbf{v}_k$ with eigenvalues $\lambda_1, \ldots, \lambda_k$, and that $|\lambda_1| > |\lambda_i|$ for $i \geq 2$. Using Eq. (10), we argued that Algorithm MAX-EIGEN converges to $\mathbf{v}_1$. Explain why it follows that λ_1 is a real number.

8.10 Markov Chains

A square matrix is **stochastic** if all its entries are nonnegative and every column sums to 1. A **Markov chain** is a matrix transition problem ($\mathbf{u}_{n+1} = A\mathbf{u}_n$) where the transition matrix is stochastic.

EXAMPLE 1 Consider Trusty Rent-A-Car again. The transition matrix (found in Example 1 of the previous section) is

$$A = \begin{bmatrix} 1/2 & 1/3 \\ 1/2 & 2/3 \end{bmatrix}. \tag{1}$$

This matrix is stochastic, so Trusty's inventory problem is a Markov chain. ∎

Markov chains (named after the Russian mathematician A. A. Markov, 1856–1922) have a great many applications. They arise naturally in recursive probability problems and in "closed systems". Probability applications form the main part of this section and we address them shortly. A **closed system** is any situation in which the objects under study may change or move around but the total number of objects is constant. For instance, Trusty's world is a closed system because the number of cars is fixed (only the distribution changes). On the other hand, population models are rarely closed systems because the total population changes (as well as the distribution).

In this section, we tie together three previously discussed topics. First, because Markov chains involve transition matrices, they involve eigenvalues. Second, we use Markov chains to provide a systematic method (but somewhat tedious by hand) to solve the sort of probability problems we solved by insightful recursive thinking in Section 6.7. Third, we point out an interplay between Markov chains and directed graph theory.

Stochastic matrices are a special type of square matrices, so it shouldn't surprise you that certain results hold for Markov chains that don't hold for all transition problems. For instance, the largest eigenvalue of a stochastic matrix is always 1. (This makes intuitive sense for closed systems. For such systems, $A^n\mathbf{u}_0$ cannot change in total size over time; thus $|\lambda_1| \not> 1$ or else $A^n\mathbf{u}_0$ would blow up, and $|\lambda_1| \not< 1$ or else $A^n\mathbf{u}_0$ would shrink to $\mathbf{0}$.) Moreover, certain situations that rarely happen for general square matrices (e.g., λ_1 through λ_k are *not* all distinct) are natural in the context of Markov chains and we need to consider them.

We now turn to probability. Markov chains arise when, at each time $t = 0, 1, 2, \ldots$, some phenomenon must be in one of k mutually exclusive states, and the probability of moving to state i at time $n+1$ depends only on the previous state j at time n (recursion!). Note that this probability, called p_{ij}, is independent of n. The matrix $A = [p_{ij}]$ is called the **transition matrix** of the Markov chain. Matrix A is indeed stochastic: The entries are nonnegative because they represent probabilities. Each column sums to 1 because the columns represent all possible states at time n, and rows represent all possible states at time $n+1$. Thus, the sum down column j is the total probability of everything that could happen next time after state j; by definition, that's 1. (Note that the columns represent where you're coming *from* and the rows represent where you're going *to*.)

For instance, we can reinterpret Example 1 as a probability Markov chain. The "phenomenon" is the location of a given rental car. The "states" are "being in NY" and "being in LA". The matrix A in Eq. (1) is still the transition matrix, but now we interpret the (i, j) entry as the probability of a car moving from state j to state i instead of the fraction of cars that move. This new interpretation is an improvement. It's highly unlikely that *exactly* 1/2 the cars in NY at time n would end up in LA at time $n+1$. But the *probability* of any given car traveling from NY to LA might be 1/2.

Here are two more examples of probability Markov chains.

EXAMPLE 2 **Continued Crossings with Hybrids**

Following Mendel, let us consider some experiments with pea plants. The trait we'll consider is whether they have green pods or yellow pods. Suppose we start with a plant of unknown gene type. We cross-pollinate it with a known hybrid, that is, a plant known to have one copy of the dominant green-pod gene and one copy of the recessive yellow-pod gene. We take an offspring of this cross breeding and again cross it with a known hybrid, and so on. Set this up as a Markov chain.

Solution Every plant carries two pod-color genes, each either G (green, dominant) or y (yellow, recessive). The states are the possible genotypes (gene-pair types) of the current descendant of the original plant. There are three possibilities: GG (pure green), Gy (hybrid), and yy (pure yellow).

To see what is going on, suppose the original plant is GG and mate it with a hybrid. We get one gene from each parent. If we get the y from the hybrid, the offspring is type Gy. Then we mate this plant with another hybrid. If we get a y from both, the new offspring will be yy. But this "history" is just one of many possibilities. The point of setting up a Markov chain is to find the probabilities for all genotypes of offspring in all generations and to determine any limiting behavior.

To set this problem up as a Markov chain is to determine the transition matrix. In other words, we need to ascertain all possible transition probabilities for one generation. First let's find the probabilities for the genotype of the offspring if the parent of unknown type is actually GG. When this GG is mated with a known hybrid, one of two things happens: Either with probability 1/2 we get a G from both plants, resulting in a GG offspring; or with probability 1/2 we get G from the plant of unknown type and y from the hybrid, resulting in a Gy offspring. A yy offspring is impossible. This gives the first column of the following matrix. To fill in the second column, find the probabilities when the untyped parent is Gy. For the third column, the untyped parent is yy. The result is

$$\begin{array}{c} \\ \text{GG} \\ \text{Gy} \\ \text{yy} \end{array} \begin{array}{c} \begin{array}{ccc} \text{GG} & \text{Gy} & \text{yy} \end{array} \\ \begin{bmatrix} 1/2 & 1/4 & 0 \\ 1/2 & 1/2 & 1/2 \\ 0 & 1/4 & 1/2 \end{bmatrix} \end{array}.$$

Remember, the columns index the possible gene types of the parent plant ("from") and the rows index the possible gene types of the offspring ("to"). If the probabilities don't make sense, ask a friend who is more familiar with genetics about them. ∎

EXAMPLE 3 People A and B alternately toss a die until a 6 comes up. The person who tosses the 6 wins. (This was Example 1, Section 6.7.) Set this up as a Markov chain.

Solution The phenomenon is the status of the game. Let time n be the moment just *before* the nth toss is made. Two of the states are "A is about to toss" and "B is about to toss". However, these cannot be the only states because in a Markov chain the phenomenon must always be in one of the states. So we add "A has won" and "B has won". Also, the definition of Markov chain requires the process to go on forever, so we think of the game this way: Rounds do go on forever, but once A or B has won, there is no further action in any round. The game simply remains in whichever winning state it reached.

As for the transition matrix, to conform with a convention useful later, let us index the four states in the order (1) A has won, (2) B has won, (3) A is about to toss, and (4) B is about to toss. Then we have

$$A = \begin{bmatrix} 1 & & 1/6 & \\ & 1 & & 1/6 \\ & & & 5/6 \\ & & 5/6 & \end{bmatrix}. \qquad (2)$$

For instance, the entries in column 3 say that, if A is about to toss, with probability 1/6 the transition is to "A has won" (because A tosses a 6) and with probability 5/6 the transition is to "B is about to toss" (because A doesn't toss a 6). ∎

Unlike the previous examples, Example 3 has **absorbing states**, that is, states from which there is no exit. The existence of absorbing states makes a big difference in the types of long-term behavior of the chain. For instance, both absorbing states are eigenvectors for $\lambda = 1$. (More precisely, $\mathbf{e}_1$ and $\mathbf{e}_2$ are eigenvectors for $\lambda = 1$.) Furthermore, the long-run behavior of the game, that is, the probability of who wins, depends on who starts. In particular, Algorithm MAX-EIGEN (previous section) converges to different eigenvectors for different choices of the initial vector.

The transition matrix for a Markov chain can be multiplied by various vectors to obtain useful information. Whereas in the rent-a-car example we took the initial vector to represent numbers of cars, now it is more natural to let the vectors represent probabilities for all the states. So let us define $p_i^{(n)}$ as the probability that the chain is in state i at time n. Let $\mathbf{p}_n$ be the column vector $(p_1^{(n)}, p_2^{(n)}, \ldots, p_k^{(n)})^{\mathrm{T}}$. In other words, $\mathbf{p}_n$ is the probability distribution vector (or just "probability vector") at time n. We can now show that probability Markov chains are matrix transition problems in the sense that the $\mathbf{p}$'s can be computed by matrix multiplication:

Theorem 1. In a Markov chain, for every n,

$$\mathbf{p}_{n+1} = A\mathbf{p}_n \qquad \text{and so} \qquad \mathbf{p}_n = A^n\mathbf{p}_0.$$

PROOF To get to state i at time $n+1$, the phenomenon must have been in some state at time n. If we let $E_i^{(n)}$ be the event that the phenomenon is in state i at time n, we have

$$p_i^{(n+1)} = \Pr(E_i^{(n+1)}) = \sum_j \Pr\big(E_i^{(n+1)} \cap E_j^{(n)}\big).$$

By definition of conditional probability,

$$\begin{aligned} \Pr\big(E_i^{(n+1)} \cap E_j^{(n)}\big) &= \Pr\big(E_i^{(n+1)} \mid E_j^{(n)}\big)\Pr\big(E_j^{(n)}\big) \\ &= p_{ij}p_j^{(n)}. \end{aligned}$$

Thus

$$p_i^{(n+1)} = \sum_j p_{ij}p_j^{(n)},$$

which by the definition of matrix multiplication is the product of the ith row of A and the column $\mathbf{p}_n$. Therefore $\mathbf{p}_{n+1} = A\mathbf{p}_n$. ∎

Not surprisingly, the individual entries of A^n have a probability interpretation. Theorem 2 states this interpretation. It can be proved by calculations similar to those in Theorem 1 [15].

Theorem 2. Let A be the transition matrix of a Markov chain. The (i, j) entry of A^n is the conditional probability that the chain is in state i at time n, given that it was in state j at time 0.

Theorem 1 says we may use Algorithm MAX-EIGEN to determine the long-range probabilities in a Markov chain. Indeed, we can delete the scaling step because, if $\mathbf{x}$ is a probability vector, so is $A\mathbf{x}$ and there is nothing to scale. ($A\mathbf{x}$ is a probability vector by Theorem 1: if $\mathbf{x}$ is the probability distribution at some time, $A\mathbf{x}$ is the distribution one time period later. Or, see [20].) So replace the two lines

$$\mathbf{y} \leftarrow A\mathbf{x}$$
$$\mathbf{x} \leftarrow \mathbf{y}/\mathbf{y}(1)$$

with

$$\mathbf{x} \leftarrow A\mathbf{x}.$$

For instance, Trusty Rent-A-Car started with 1000 cars in each city, so its initial probability vector is $\mathbf{x} = (1/2, 1/2)$. Applying MAX-EIGEN, we find that after only five iterations $\mathbf{x}$ has stabilized to four decimal places at $(.4, .6)$. Similarly, to use MAX-EIGEN on the coin-tossing problem, we start with $\mathbf{x} = (0, 0, 1, 0)$, because we said in Chapter 6 that player A tosses first. We find after 49 steps that $\mathbf{x}$ has stabilized to four decimal places at $(.5454, .4545, 0, 0)$, agreeing with the conclusion in Chapter 6 that A wins with probability $6/11 = .5454\ldots$. If we start with $\mathbf{x} = (0, 0, 0, 1)$, though, we find after 49 steps that $\mathbf{x}$ stabilized to $(.4545, .5454, 0, 0)$.

Of course, we can also use Theorem 1 to compute probabilities after a specific number of steps instead of in the long run. After all, the rental company is not just interested in the long run. It also needs to know how many cars it will have in each place next month and the month after.

We need two definitions and then we can state the two main theorems of Markov chains. A Markov chain is **absorbing** if it contains at least one absorbing state and if from every state it is possible to get to some absorbing state. A Markov chain is **regular** if for some n it is possible to get from any state to any other in exactly n transitions. To say that it is possible to get from j to i in n steps means that the probability is positive. Thus in light of Theorem 2, a Markov chain is regular if some specific power of its transition matrix has all entries positive. For instance, the rent-a-car chain is regular because the transition matrix (1) itself has all positive entries.

For moderate-sized problems, the best way to see whether a chain is regular or absorbing is to look at a picture. The **state diagram** of a Markov chain (also called the **transition diagram**) is the digraph defined as follows. There is a vertex for every state and there is a directed edge (j, i) if it is possible to get directly from j to i, i.e., $p_{ij} > 0$. (You have already seen such state diagrams in Chapter 3. Where?) For instance, the state diagrams for Examples 1, 2, and 3 are shown in Fig. 8.9.

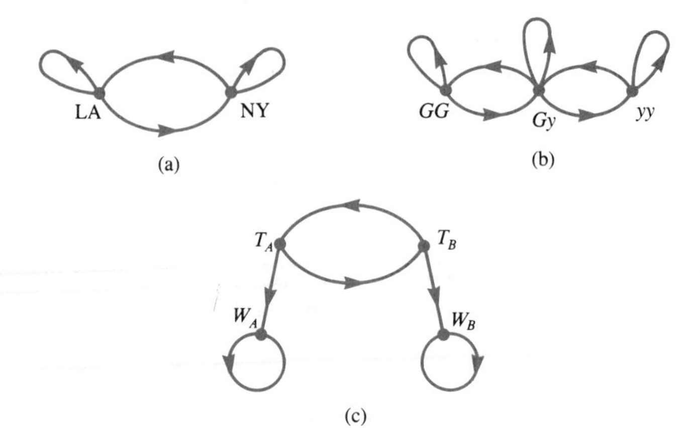

FIGURE 8.9

State Diagrams for (a) Rental Cars, (b) Peas and (c) Coin Tosses

From these diagrams, it is clear that the rental car problem is regular because its digraph is a complete directed graph. As for the pea problem, visual inspection shows that every state can be reached from every other state in exactly two steps (as well as exactly three or any larger number). Finally, for the coin toss problem it's clear that states 3 and 4 are absorbing—no edges go out—and that from each of the other two states at least one absorbing state can be reached eventually (indeed, both can).

A Markov chain need not be either regular or absorbing. The digraphs in Fig. 8.10 represent Markov chains which are neither. First note that neither digraph has any absorbing states. As for digraph (a), it always takes an odd number of steps to get from either state to the other, and it always takes an even number of steps to get from a state to itself. Thus there is no one n for which *all* transitions are possible. Figure 8.10(a) is an example of a **periodic chain**; here the period is 2. As for digraph (b), states $\{3, 4, 5\}$ are an absorbing *set*, but there are no individual absorbing states.

Thus the following two theorems do not cover all cases. However, most other

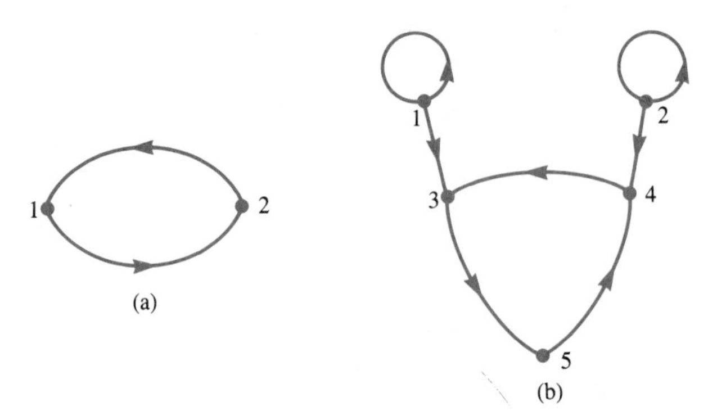

FIGURE 8.10

Markov Chains That Are Neither Absorbing nor Regular

cases can be reduced to them. For instance, in Fig. 8.10(b) we could temporarily declare $\{3,4,5\}$ to be a single state, ignoring the transitions between them, and then we have an absorbing chain. Once we have analyzed that, we can look at $\{3,4,5\}$ as a chain by itself, and it happens to be a regular chain. With ideas like this, and some results about periodic chains, one can (in another course) parlay Theorems 3 and 4 into an analysis of all Markov chains.

Theorem 3. Suppose that a regular Markov chain has transition matrix A. Then

1. $\lambda = 1$ is an eigenvalue and its eigenspace has dimension 1. There is exactly one probability vector $\mathbf{w}$ in this space (hence $\mathbf{w}$ is a basis), and all its coordinates are positive. All other eigenvalues of A have $|\lambda| < 1$.
2. For any initial probability vector $\mathbf{p}$, the sequence $\{A^n\mathbf{p}\}$ converges to $\mathbf{w}$.

We won't prove this theorem, but the usual method of doing so is to use the ideas in [19] to show that A^n converges to a matrix in which every column is the same. That common column is $\mathbf{w}$.

Theorem 3 tells us that regular Markov chains do indeed forget their past. Whatever the initial $\mathbf{p}$, in the long run the probability distribution is essentially $\mathbf{w}$. It also tells us that $\mathbf{w}$ is very easy to find: By (1) it is the unique probability vector satisfying $(A-I)\mathbf{w} = \mathbf{0}$. Please note that this equation captures the recursive idea with which we solved our first Markov chain problem (rental cars) back in Chapter 5. Namely, suppose we already knew that the chain settled down to some distribution $\mathbf{w}$. Then going through one more iteration wouldn't change things. That is, $A\mathbf{w} = \mathbf{w}$. But this is an equation we can use to solve for $\mathbf{w}$. It's equivalent to $(A-I)\mathbf{w} = \mathbf{0}$.

EXAMPLE 4 Show that there is a steady state for the cross-breeding problem in Example 2.

Solution We already saw from the state diagram that the Markov chain for this problem is regular. Therefore we need merely to solve

$$(A-I)\mathbf{w} = \begin{bmatrix} -1/2 & 1/4 & 0 \\ 1/2 & -1/2 & 1/2 \\ 0 & 1/4 & -1/2 \end{bmatrix} \mathbf{w} = \mathbf{0}.$$

By Gaussian elimination, by inspection, or by guess and check, we find that $\mathbf{w} = (1/4, 1/2, 1/4)$ is a solution and therefore the only probability vector solution. In the long run, a hybrid descendant is twice as likely as either pure type. ∎

Note that we did *not* solve this problem using any version of Algorithm MAX-EIGEN. Why use an approximation method (and one where most of the work is directed at finding the eigenvalue) when Theorem 3 tells us the eigenvalue and allows us to use an exact method? But see [16].

As for absorbing chains, we need a different theorem because in general Theorem 3 does not hold for them. Consider Example 3 (tossing for a 6). If you start in the state "A has won", then you stay there, and similarly for "B has won". Therefore there is no one vector $\mathbf{w}$ to which all initial probability vectors converge. Technically, the problem is that in Example 3 the eigenspace for $\lambda = 1$ has dimension > 1. This happens whenever a Markov chain has at least two absorbing states. In Section 8.9 we paid little attention to eigenspaces with dimension > 1, but now we need to consider them.

The transition matrix A for an absorbing chain can always be put in a special form by listing the absorbing states first. Specifically, if the absorbing states are numbered 1 through ℓ and the other states (called **transient states**) are numbered $\ell + 1$ through k, then

$$A = \begin{bmatrix} I & R \\ 0 & Q \end{bmatrix}, \tag{3}$$

where I is $\ell \times \ell$, Q is $(k-\ell) \times (k-\ell)$, and R is $\ell \times (k-\ell)$. For instance, the matrix in Eq. (2) is already in this form. The reason for the identity matrix is that an absorbing state can only lead to itself. The reason for the zero submatrix below the identity is that moving from an absorbing state to a transient state is not possible. No special form is claimed for the submatrices R and Q.

Theorem 4. Suppose we have an absorbing Markov chain with ℓ absorbing states and $k - \ell$ transient states, so that we may write its transition matrix as in Eq. (3). Let B be the $\ell \times (k-\ell)$ matrix whose (i, j) entry is the probability that the chain ends up in absorbing state i if it starts in the jth transient state (that is, state $\ell + j$). Then $I - Q$ is invertible and

$$B = R(I - Q)^{-1}. \tag{4}$$

The matrix $(I-Q)^{-1}$ is called the **fundamental matrix** of the Markov chain and is usually denoted by N. Note that the I in $I - Q$ must be the same size as Q, namely, $(k-\ell) \times (k-\ell)$, so it need not be the same size as the I in Eq. (3).

EXAMPLE 5 Find the probability that player A wins the die tossing game in Example 3 if she goes first.

Solution In matrix A of Eq. (2) we find that

$$Q = \begin{bmatrix} 0 & 5/6 \\ 5/6 & 0 \end{bmatrix}, \qquad R = \begin{bmatrix} 1/6 & 0 \\ 0 & 1/6 \end{bmatrix}.$$

Using Gaussian elimination we compute that

$$(I - Q)^{-1} = \begin{bmatrix} 36/11 & 30/11 \\ 30/11 & 36/11 \end{bmatrix},$$

so

$$B = R(I - Q)^{-1} = \begin{bmatrix} 6/11 & 5/11 \\ 5/11 & 6/11 \end{bmatrix}.$$

Because player A tosses first, that is, we begin in the first transient state. (The ordering of the states was given just before Eq. (2).) We want to know the probability of finishing in the first absorbing state. Therefore we want the $(1,1)$ entry of B, which is $6/11$. At no extra cost, B also gave us the answer to the three other questions we might have asked about how the game ends. ∎

Proof sketch for Theorem 4. Our goal is to show you just enough so that you see that the Eq. (4) for B is based on the same recursive idea we used in Section 6.7. First, one may show from the definition of matrix multiplication that A^n has the same form as A, namely,

$$A^n = \begin{bmatrix} I & R_n \\ 0 & Q^n \end{bmatrix}.$$

From $A^{n+1} = A^n A = AA^n$ one obtains the two relations

$$R_{n+1} = R + R_n Q \qquad \text{and} \qquad R_{n+1} = R_n + RQ^n.$$

Then one shows that Q^n approaches the 0 matrix. From this one can show that $(I-Q)^{-1}$ exists. Using the second relation for R_{n+1} it also follows from $Q^n \to 0$ that R_n approaches a limiting matrix. Now, the (i,j) entry of R_n is the probability that the process has reached absorbing state i by the nth step, given that it starts in the jth transient state. So the limit of this term is the probability that the process *ever* ends up in absorbing state i, having started in transient state j. (In particular, this probability is well defined.) That is, the limit of R_n is what we called B (and we have shown that B exists). Now, taking the first relation for R_{n+1} to the limit, we get

$$B = R + BQ \implies B(I - Q) = R \implies B = R(I - Q)^{-1}. \blacksquare$$

The equation $B = R+BQ$ is essentially what we used in Chapter 6. We argued this way: Suppose we already knew the probabilities of ending up in various absorbing states. Either we end up there right away (R), or else we make one transition to another transient state (Q) and then it is as if we are starting from the beginning again (B). Therefore $B = R+BQ$. What we have done in this section is to represent these ideas by matrix addition and multiplication.

A Warning about Rows and Columns

There are two conflicting conventions concerning products of matrices and vectors. In the convention we have followed, vectors are written as columns and are premultiplied by matrices. In the other convention, vectors are written as rows and are postmultiplied by matrices. For instance, we describe a matrix transition problem as

$$\mathbf{u}_n = A^n \mathbf{u}_0, \tag{5}$$

whereas in the other convention it is

$$\mathbf{u}_n = \mathbf{u}_0 A^n. \tag{6}$$

Since an equation in one form can be turned into an equation in the other by transposing, the conventions are entirely equivalent. It's just that the roles of rows and columns are interchanged. For instance, A in Eq. (6) is the transpose of A in Eq. (5).

We know of no linear algebra book that uses the row convention. Such a book would have to present and explain elementary column operations instead of elementary row operations, and when it expanded out an augmented matrix into a system of equations, the individual equations would run down the page instead of across. However, there are quite a few abstract algebra books and probability books that use the row convention. In such a book, the transition matrix of a Markov chain has *rows* summing to 1 and the (i, j) entry is the probability of going *from* state i *to* state j, the reverse of our convention.

Why would anyone follow the row convention when it conflicts with linear algebra usage? Because in certain contexts it is more natural. It's traditional in mathematics to name rows first, then columns; it's traditional in ordinary discourse to mention where you're coming from before you say where you are going to. (People say "I'm going from NY to LA", not "I'm going to LA from NY".) With the row convention, you can follow both traditions.

Moral: When you look up Markov chains in another book, determine immediately whether equations look like (5) or (6). If (6), the conventions are backwards from ours.

Problems: Section 8.10

1. Verify that $\lambda = 1$ is an eigenvalue for the transition matrices in Examples 1, 2, and 3.

2. Use Theorem 1 to determine the probability in Example 3 that player B has won by the fourth toss. Recall that player A goes first.

3. A bug meanders from vertex to vertex of a tetrahedron in a math professor's office. From any vertex the bug is equally likely to go to any *other* vertex. Set this up as a Markov chain and determine the probabilities for the long-term location of the bug.

4. In [9, Section 6.3], we considered a model for weather in which one day's weather depends on the previous day's. Represent the model as a Markov chain. Show that the long-range probabilities asserted there for the weather on a given day are consistent with the transition probabilities.

5. A sociologist finds that a certain society has three work classes: professional, skilled, and unskilled. Furthermore, the children of professionals become professional 80% of the time, skilled 10%, and unskilled 10%. For the children of skilled workers the percentages (in the same order) are 20%, 60%, and 20%. Finally, for unskilled workers the percentages are 25%, 50%, and 25%.

   a) Set this up as a Markov chain. Describe the states and give the transition matrix.

b) Suppose in the current generation the distribution of workers is 10%, 50%, and 40%. What will be the distribution two generations from now?

c) What happens in the long run?

6. A tennis game has reached "deuce" (3 points each) and the server has probability p of winning each additional point. Use a Markov chain to determine the probability that the server wins the game. (For the rules of tennis, see Section 6.7, where we solved this same problem by recursion.)

7. Set up (don't solve) Gambler's ruin (Section 6.7) as a Markov chain. Assume player A has $\$m$, B has $\$n$, and A wins each toss with probability p. Name the states and give the transition matrix. Which theorem, 3 or 4, would be relevant if you were asked to solve Gambler's ruin?

8. Consider the following **discrete random walk** with reflecting barriers. A particle must be at positions 1, 2, 3, or 4 on the number line. If it is at 2 or 3 at time n, at $n+1$ it moves one position right or left with equal probability. If it is at either end position, with equal probability it either stays there or moves one unit back from the end. Set this up as a Markov chain and determine the long-run behavior. (By considering a larger and larger number of positions, one approximates the movement of a gas particle in a closed tube.)

9. Model the effects of inbreeding on a dominant-recessive gene pair using a Markov chain. By inbreeding, we mean that we mate two members of the species, then mate two of their offspring, then two of the offspring of that pair, and so on. Thus at each generation n a state consists of a pair of genotypes, i.e., four genes total.

10. There is a unique stochastic matrix for which the digraph in Fig. 8.11 is the state diagram. What is that matrix? Why? Are there any other digraphs for which the stochastic matrix is unique?

11. As originally stated, the Trusty Rent-A-Car problem is **deterministic**: We know exactly how many cars go where. When a regular Markov chain is deterministic instead of probabilistic, Theorem 3 doesn't say quite what we need. Fortunately, we may replace statement (2) with (2′): For any initial distribution vector $\mathbf{u}$, the sequence $\{A^n\mathbf{u}\}$ converges to $c\mathbf{w}$, where c is the sum of the entries in $\mathbf{u}$. ($\mathbf{w}$ is defined in statement (1), which doesn't change.) Use this modified Theorem 3 to verify the results from Chapter 5 about the long-run distribution of Trusty's cars.

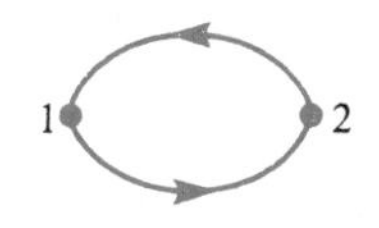

FIGURE 8.11

12. Draw state diagrams for the Markov chains of problems [3–6, 8].

13. For the state diagrams in Fig. 8.12, determine whether the associated Markov chains are regular, absorbing or neither.

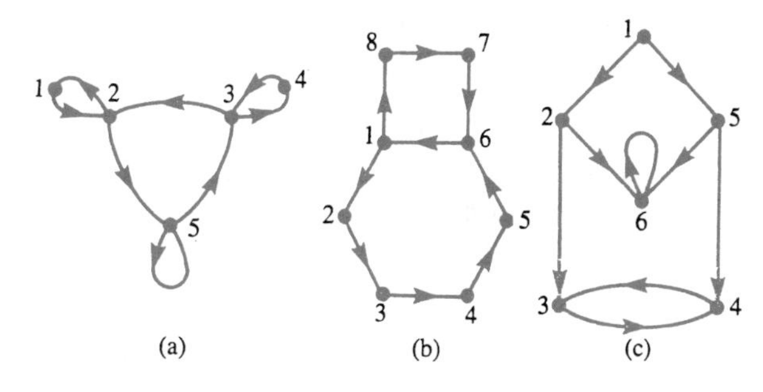

FIGURE 8.12

14. Suppose that the state diagram of a Markov chain is strongly connected and at least one vertex has a loop. Show that the chain is regular.

15. a) Prove Theorem 2. Both the statement and the proof of this result are very much like something in Chapter 3. What?

b) Use Theorem 2 to give an alternative proof that $\mathbf{p}_n = A^n\mathbf{p}_0$.

16. Suppose you need to find the limit probability vector $\mathbf{w}$ for a regular Markov chain with 1000 states. Which algorithm would be more efficient, MAXEIGEN or Gaussian elimination? Explain your reasoning.

17. Let $\mathbf{p}_0$ be the initial probability distribution in an absorbing Markov chain. Show that the limit distribution is

$$\begin{bmatrix} I & B \\ 0 & 0 \end{bmatrix} \mathbf{p}_0,$$

where I and B are as in Theorem 4.

18. Show that λ is an eigenvalue of $A \iff \lambda$ is an eigenvalue of A^{T}. *Hint*: Show that $A^{\mathrm{T}} - \lambda I$ and $A - \lambda I$ have the same rank by considering the row space of one and the column space of the other.

19. Let A be a stochastic matrix.

a) Show that the n-tuple $(1, 1, \ldots, 1)$ is an eigenvector for A^{T}. What is its eigenvalue?

b) For any vector $\mathbf{x}$, let $m(\mathbf{x})$ be its smallest entry and $M(\mathbf{x})$ its largest. Show that

$$m(\mathbf{x}) \le m(\mathbf{x}A) \le M(\mathbf{x}A) \le M(\mathbf{x}).$$

What does this say about the eigenvalues of A^{T}?

c) Using [18], what can you conclude about A?

20. If $\mathbf{x}$ is a vector, define $S(\mathbf{x})$ to be the sum of its coordinates.

a) Fact: For any $k \times k$ stochastic matrix A, $S(A\mathbf{x}) = S(\mathbf{x})$. Prove this fact two ways. First, rewrite the LHS as a double sum and change the order of summation. Second, let $\mathbf{y}$ be the row k-tuple $(1, 1, \ldots, 1)$. Then for any column k-tuple $\mathbf{w}$, $S(\mathbf{w})$ equals the matrix product $\mathbf{y}\mathbf{w}$. Now use [19a].

b) Show: If A is a stochastic matrix and $\mathbf{v}$ is an eigenvector for $\lambda \neq 1$, then $S(\mathbf{v}) = 0$.

21. Suppose we have a phenomenon where the probability of being in state i at time n depends on the states at time $n-1$ *and* $n-2$. That is, the quantities

$$p_{ijk} = \Pr\big(E_i^{(n+1)} \,|\, E_j^{(n)} \cap E_k^{(n-1)}\big)$$

are independent of n. Offhand, this does not appear to be a Markov chain, but by a device similar to that used in Example 2, Section 8.9, it may be turned into a Markov chain. (Indeed, there is nothing special about two past states; the phenomenon can be set up as a Markov chain so long as there is some fixed finite number of past states which count.) The device is this: Redefine what you mean by a state. Define the new states to be all ordered pairs of old states. Being in the new state (i, j) at time n is defined to mean being in old state i at time n and old state j at time $n-1$.

Suppose there are exactly two old states. Using the p_{ijk} notation of the preceding equation, write the transition matrix which shows that the phenomenon is a Markov chain. Label the rows and columns with the names of the new states to make clear what you're doing.

22. There is an extension of Theorem 4. Using the same notation as there, this extension states: (1) The (i, j) entry of $N = (I - Q)^{-1}$ is the expected number of times the chain passes through transient state i before reaching an absorbing state, if it starts in transient state j; and (2) the sum of column j of N is the expected number of steps before absorption if the process starts in transient state j.

Use this extension to determine the average number of tosses until the game in Example 3 is over. Compare your answer and your work with the solution in Example 2, Section 6.7.

Supplementary Problems: Chapter 8

1. Two matrices have the same **echelon shape** if they both have the same $m \times n$ size, they are both in echelon form, and they both have their pivots in the same positions.

a) Draw all echelon shapes for 2×3 matrices. (Represent pivots by $*$'s and other entries which may be nonzero by $\times$'s. The actual values of the nonzero entries are irrelevant.)

b) Draw all echelon shapes for 3×2 matrices.

c) How many echelon shapes are there for $n \times n$ matrices?

d) How many echelon shapes are there for $m \times n$ matrices where $m > n$?

e) How many echelon shapes are there for $m \times n$ matrices where $m < n$?

2. Sometimes we want to know if two different looking systems of homogeneous equations, $A\mathbf{x} = \mathbf{0}$ and $B\mathbf{x} = \mathbf{0}$, have identical solution sets. (The two systems must have the same variables $x_1, x_2, \ldots$, but perhaps one system has more equations than the other.) If we do Gaussian elimination on A and B and find that they have the same reduced echelon form (except perhaps for different numbers of zero rows), then surely these two systems have the same solutions. Happily, the converse is also true, so *two systems* $A\mathbf{x} = \mathbf{0}$ *and* $B\mathbf{x} = \mathbf{0}$ *have identical solution sets* $\iff$ *A and B have the same reduced echelon form (except perhaps for different numbers of zero rows)*. Therefore we have an algorithm for determining if two homogeneous systems have the same solution: Compute the reduced echelon forms of their initial matrices and check whether these reduced echelon forms are the same. We say that reduced echelon form is a **canonical form**.

Actually, for $A\mathbf{x} = \mathbf{0}$ and $B\mathbf{x} = \mathbf{0}$ to have the same solutions, it is *sufficient* that A and B reduce to the same echelon matrix after the downsweep part of Gaussian elimination. However, here the converse is *false*. The two systems can look different after the downsweep, yet have the same solutions. Thus echelon form is *not* canonical.

a) For each pair of matrices below, use the preceding facts to determine if the matrices represent homogeneous systems with the same solutions. If they don't, give a set of values of the variables which satisfy one system but not the other.

i) $$A = \begin{bmatrix} 1 & 1 & 0 & 2 \\ 1 & 1 & 1 & 1 \\ 2 & 2 & 3 & 1 \end{bmatrix},$$

$$B = \begin{bmatrix} 2 & 2 & -1 & 3 \\ -1 & -1 & 3 & -4 \\ 4 & 4 & 5 & -1 \end{bmatrix}.$$

ii) A as above and

$$C = \begin{bmatrix} 2 & 2 & 1 & 3 \\ -2 & -2 & 2 & -6 \\ 5 & 5 & 4 & 6 \end{bmatrix}.$$

iii) A as above and

$$D = \begin{bmatrix} 0 & 1 & 0 & -1 \\ 3 & 2 & 3 & 4 \\ -1 & 3 & -1 & -5 \end{bmatrix}.$$

b) Exhibit two different matrices in echelon form which represent homogeneous systems with the same solutions. *Hint*: Work backwards. Start with one reduced echelon form.

c) Prove that two systems with different reduced echelon forms have different solution sets. *Hint*: If you found a systematic way to answer part a), you're on the way. *Second hint*: If there is a variable which is free in one system and bound in the other, use it to build a vector that satisfies one system and not the other.

3. How much work does it take to solve $L\mathbf{x} = \mathbf{b}$ by Gaussian elimination if L is an $n \times n$ lower triangular matrix with no 0's on the main diagonal? As always, where computations can be replaced by assignments due to the special form of L, do so.

4. How much work does it take to solve $U\mathbf{x} = \mathbf{b}$ by Gaussian elimination if U is an $n \times n$ upper triangular matrix with no 0's on the main diagonal?

5. Let L be an $n \times n$ lower triangular matrix with no 0's on the main diagonal. Determine how much work is necessary to invert it. (Whenever work can be saved by assigning a matrix entry a value rather than computing it, do so.)

6. Do [5] for an upper triangular matrix.

7. Determine whether $\{(1,2,3,2), (2,0,1,-1)\}$ and $\{(4,4,7,3), (5,-2,0,-5)\}$ are both bases of the same space.

8. Devise and justify a general algorithm for determining whether two sets of vectors are both bases for the same space.

9. Determine if the span of $\{(1,-2,1), (1,1,2)\}$ is contained in the null space of

$$\begin{bmatrix} 1 & 2 & 3 \\ 2 & 1 & 0 \end{bmatrix}.$$

10. Give and justify a general method for determining if

    a) The space generated by a given set of vectors is contained in the null space of a given matrix.

    b) The space generated by a given set of vectors equals the null space of a given matrix.

11. Let

$$M = \begin{bmatrix} 1 & -1 & -2 & 2 \\ 2 & -2 & 2 & -2 \end{bmatrix},$$

and

$$N = \begin{bmatrix} 1 & 2 & 3 & -6 \\ 2 & -3 & 4 & -3 \\ -1 & 2 & 3 & -4 \end{bmatrix}.$$

Determine if $\mathcal{N}_N \subset \mathcal{N}_M$.

12. Give a general method for determining if two matrices have the same null space.

13. It is a theorem that [$A\mathbf{x} = \mathbf{b}$ has more than one solution for some $\mathbf{b}$] $\iff$ [$A\mathbf{x} = \mathbf{0}$ has more

than one solution]. This theorem illustrates an important principle in linear algebra: If a notable phenomenon happens somewhere, it (usually) happens at the origin. Anyway, prove this theorem two ways:

a) Using Gaussian elimination. What can you say about the reduced echelon form of A if $A\mathbf{x} = \mathbf{b}$ has more than one solution for some $\mathbf{b}$? For $\mathbf{b}=\mathbf{0}$?

b) Using matrix algebra. If $A\mathbf{x} = \mathbf{b}$ and $A\mathbf{x}' = \mathbf{b}$, what can you do to get an equation involving A and $\mathbf{0}$? Which half of the $\Longleftrightarrow$ does this prove? Does the algebra reverse?

14. In Section 8.2 we said that Algorithm GAUSS-SQUARE can be written so as to do back-substitution directly on the variables, without further manipulation of the coefficient matrix. See [22, Section 8.2]. Rewrite GAUSS so that it too does back-substitution directly. The hard part is getting straight exactly what you are trying to do. You can't just work with the variables numerically because they won't have unique values if there are any free variables.

15. Write a variant of Algorithm GAUSS which keeps all the pivots on the main diagonal by allowing column switches. Be sure to keep track of which variable is associated with which column.

16. The matrix A_R is said to be a **right inverse** of A if $AA_R = I$.

a) Verify that

$$\begin{bmatrix} -5/3 & 2/3 \\ 4/3 & -1/3 \\ 0 & 0 \end{bmatrix}$$

is a right inverse of

$$A = \begin{bmatrix} 1 & 2 & 3 \\ 4 & 5 & 6 \end{bmatrix}.$$

b) If A is $m \times n$, what size must A_R be?

c) Whereas a matrix can have at most one inverse, it can have infinitely many right inverses. Find all right inverses for A from part a).

17. The matrix A_L is said to be a **left inverse** of A if $A_L A = I$.

a) Verify that

$$\begin{bmatrix} -2 & 1 & 0 \\ 3/2 & -1/2 & 0 \end{bmatrix}$$

is a left inverse of

$$A = \begin{bmatrix} 1 & 2 \\ 3 & 4 \\ 5 & 6 \end{bmatrix}.$$

b) If A is $m \times n$, what size must A_L be?

c) For the A in part a), find some other left inverses. In fact, there are infinitely many.

18. Find a matrix which has neither a right inverse nor a left inverse. Explain why not.

19. Show that if A has both a right inverse and a left inverse, then in fact they are identical.

20. Suppose A has at least one right inverse. Show that for any column $\mathbf{b}$, $A\mathbf{x} = \mathbf{b}$ has at *least* one solution. We therefore say that a right inverse is an existence condition for solutions of linear systems.

21. Suppose A has at least one left inverse. Show that for any column $\mathbf{b}$, $A\mathbf{x} = \mathbf{b}$ has at *most* one solution. We therefore say that a left inverse is a uniqueness condition for solutions of linear systems. (Uniqueness here means that *if* there is a solution, there is at most one. It does not guarantee a solution. If we want to say that we are guaranteed *exactly* one solution, we say we know both existence and uniqueness.)

22. Prove: If A is square and has a right inverse, then it has an inverse. *Hint*: What conclusion does the hypothesis force concerning the reduced echelon form of A?

23. You may know from analytic geometry that a plane in space can be represented as the solutions to a single linear equation $ax + by + cz = d$.

a) If the plane goes through the origin, what is d?

b) Give an algebraic proof that a plane through the origin is a vector space.

24. If $S \subset R^n$ and $|S| = \infty$, we can't define Span(S) to be everything of the form $\sum_{\mathbf{v}\in S} c_\mathbf{v}\mathbf{v}$ because infinite sums don't always make sense (they don't always have a limit). The way around this is to restrict linear combinations to those sums in which only a finite number of the coefficients $c_\mathbf{v}$ are nonzero. With this definition, prove that Span(S) is a vector space regardless of the cardinality of S.

25. As in [24], redo the definition of "$\mathbf{v}$ depends on S" and "S is dependent" so that they make sense even if S is infinite. Prove that it is still true that S is dependent $\Longleftrightarrow$ some $\mathbf{v} \in S$ depends on $S - \{\mathbf{v}\}$.

26. Show that with the definition of Span and independence for infinite sets from [24], it is still true that B is a basis of $V \iff$ every $\mathbf{v} \in V$ has a unique representation as a linear combination from B.

27. Let $A = [a_{ij}]$ be $m \times n$ and suppose $a_{ij} = ij$. What is the maximum value of n if all the columns are independent? (The answer may depend on m.)

28. Let $A = [a_{ij}]$ be $m \times n$ and suppose $a_{ij} = i + j$. What is the maximum value of n if all the columns are independent? (The answer may depend on m.)

29. Suppose $\mathbf{u}, \mathbf{v}, \mathbf{w}$ are independent. Prove: $\mathbf{u}$, $\mathbf{u}+\mathbf{v}$ and $\mathbf{u}+\mathbf{v}+\mathbf{w}$ are independent.

30. Prove or disprove: Let $\mathbf{u}$,$\mathbf{v}$,$\mathbf{w}$ be vectors in some space V. If the sets $\{\mathbf{u}, \mathbf{v}\}$, $\{\mathbf{u}, \mathbf{w}\}$, and $\{\mathbf{v}, \mathbf{w}\}$ are each independent, then the single set $\{\mathbf{u}, \mathbf{v}, \mathbf{w}\}$ is independent.

31. Call a set $\{\mathbf{v}_1, \ldots, \mathbf{v}_k\} \subset R^n$ **strongly dependent** if there exists some set of constants, *none* 0, so that $\sum_{i=0}^{k} c_i \mathbf{v}_i = \mathbf{0}$.

a) How does strong dependence differ from dependence?

b) Prove that if $S = \{\mathbf{v}_1, \ldots, \mathbf{v}_k\}$ is strongly dependent, then every $\mathbf{v}_i \in S$ depends on $S - \{\mathbf{v}_i\}$.

c) Is the converse of part b) true? (Give a proof or counterexample.)

We made up this definition of strong dependence. It doesn't have much use except to help you understand (ordinary) dependence better.

32. Suppose Gaussian elimination requires no row switches to reduce A, and there are p pivots. Show that the first p rows of A are a basis of $\mathcal{R}_A$. Show that this conclusion can be false if there are row switches.

33. Find some statements in the text or problems of Section 8.7 and Section 8.8 that are false if we define $\text{Span}(\emptyset) = \emptyset$ instead of $\text{Span}(\emptyset) = \{\mathbf{0}\}$. Explain.

34. A *non*homogeneous matrix difference equation is anything of the form

$$\mathbf{u}_n = A\mathbf{u}_{n-1} + \mathbf{c}_n,$$

where $\mathbf{u}_0$ and the sequence of $\mathbf{c}_n$'s are given.

a) Prove that Theorem 1, Section 5.7, holds for matrix difference equations.

b) In Chapter 5, when the nonhomogeneous part was constant, in most cases the right guess for a particular solution was some other constant. Now consider $\mathbf{u}_n = A\mathbf{u}_{n-1} + \mathbf{c}$, that is, the nonhomogeneous part of the matrix recursion is a constant vector. Show that if 1 is not an eigenvalue of A, then there is always a constant particular solution.

35. Figure out how to use Algorithm 8.4, MAX-EIGEN, to find the largest *positive* eigenvalue of a square matrix A. *Hint*: modify A first so that it has no negative eigenvalues.

CHAPTER 9

Infinite Processes in Discrete Mathematics

9.1 Introduction

Discrete mathematics does not mean *finite* mathematics, something we stressed right at the beginning of this book. Indeed, we've already discussed infinite or potentially infinite processes at several places in this book:

- Mathematical induction (Chapters 2 and 7) is an explicitly infinite process because the proposition $P(n)$ almost always allows n to be any member of N or N^+.
- The recurrence relations presented in Chapter 5 almost always allow the parameter in the relation to take on any positive value; in fact, we are often interested in what happens to the solution of such recurrence relations as this parameter gets arbitrarily large, a point we'll return to shortly.
- In Chapter 7 we sometimes restricted the quantification parameter to a finite number of values, but we also allowed it to take on all values from an infinite set such as N.

In fact, infinite processes in discrete mathematics form the natural bridge from discrete to continuous mathematics in two ways. The first is that the *limiting behavior* of processes in discrete mathematics—as a discrete parameter gets arbitrarily large—often leads naturally to concepts of importance in continuous mathematics. (We made this point in Section 6.4 where we noted that the binomial distribution becomes the Poisson distribution in the limit as the binomial parameter gets arbitrarily large.) At various places in this chapter we will point out how the topic we are discussing forms a bridge to topics normally studied in calculus.

The other kind of bridge from discrete to continuous mathematics pertains to the use of discrete mathematics to solve approximately problems which cannot be solved directly using continuous mathematical techniques. Most often this involves finding approximations to the real number which is the solution of a continuous mathematical problem. We will discuss approximation algorithms briefly in Section 9.5, and they are the subject of Section 9.7.

As usual, we will begin with several examples which illustrate the topics that we will discuss in this chapter.

EXAMPLE 1 **Fibonacci Numbers**

In Section 5.5, we said that the behavior of

$$Ar_1^n + Br_2^n$$

depends on the r_i with the largest absolute value. (In Section 8.9 we claimed the same thing when there are more than two terms.) For example, for the Fibonacci sequence, this claim says that in the long run

$$f_n = \left(\frac{1}{\sqrt{5}}\right)\left[\left(\frac{1+\sqrt{5}}{2}\right)^{n+1} - \left(\frac{1-\sqrt{5}}{2}\right)^{n+1}\right] \tag{1}$$

behaves like

$$\left(\frac{1}{\sqrt{5}}\right)\left(\frac{1+\sqrt{5}}{2}\right)^{n+1}.$$

Can we make this more precise and justify it? We'll answer this question in Section 9.2. And in Section 9.6 we'll show you a way different from that in Section 5.5 to solve difference equations, such as the one for the Fibonacci numbers. ∎

EXAMPLE 2 **Continuous Interest**

Here is a very practical problem concerning something you may have thought about yourself. If you invest \$1000 at an annual simple (i.e., no compounding) interest rate of 10%, at the end of one year you will have

$$\$1000(1+.1) = \$1100.$$

If, however, the interest is compounded quarterly, the interest is $10/4 = 2.5\%$ each quarter, and so you would calculate as follows:

$$\begin{aligned} \text{End of first quarter:} &\quad \$1000.00(1+0.025) = \$1025.00; \\ \text{End of second quarter:} &\quad \$1025.00(1+0.025) = \$1050.63; \\ \text{End of third quarter:} &\quad \$1050.63(1+0.025) = \$1076.89; \\ \text{End of fourth quarter:} &\quad \$1076.89(1+0.025) = \$1103.81. \end{aligned}$$

Thus you have made \$103.81 and the *effective* interest rate is 10.381%. More compactly, this calculation is equivalent to

$$\$1000(1+0.1/4)^4 = \$1103.81 = \$1000(1+0.10381).$$

If, instead, the interest were compounded monthly, you would calculate it as

$$\$1000(1{+}0.1/12)^{12} \;=\; \$1104.71 \;=\; \$1000(1{+}0.10471),$$

and if compounded daily, as

$$\$1000(1{+}0.1/365)^{365} \;=\; \$1105.11 \;=\; 1000(1{+}0.10511). \qquad (2)$$

But what if, as banks sometimes advertise, the interest is compounded continuously? What is the effective rate of interest? And what does "compounding continuously" mean anyway? In this example, the parameter which is getting large is the number of times the interest is compounded each year. (At the same time another parameter—the time between each compounding—is getting very small.) We'll answer these questions, too, in Section 9.2. ▮

EXAMPLE 3 **Solving Algebraic Equations**

In high school algebra you solved a lot of equations of the form

$$f(x) \;=\; 0, \qquad (3)$$

including linear equations $[f(x) = ax + b]$ and quadratic equations $[f(x) = ax^2 + bx + c]$. In both of these special cases, there was a formula for finding the value or values of x which satisfy Eq. (3). But for almost all $f(x)$ there is no formula. Are there, however, efficient algorithms which achieve accurate numerical solutions for arbitrary $f(x)$? We'll return to this question in Section 9.7. ▮

Each of these examples involves explicitly or, in the case of Example 3, implicitly an infinite process. In the rest of this chapter we'll discuss various processes in discrete mathematics which involve infinity in one way or another.

Problems: Section 9.1

1. Look again at Eq. (1). The right-hand side is an integer, as we discussed in Section 5.5. You could verify this for any particular value of n by performing the tedious multiplications using all those square roots of 5. Let's make it a little easier and see what we can learn. Calculate $1/\sqrt{5}$, $(1{+}\sqrt{5})/2$, and $(1{-}\sqrt{5})/2$ to three decimal places and use the results to calculate f_n for $n = 2, 3, \ldots, 10$. Is it always easy to tell what the true integer value is? In particular, do the results suggest that some of the calculation you did was superfluous; see also [27, Section 9.2].

2. Do the following interest calculations, where P is the principal, r is the yearly interest rate, t is the period of compounding, and y is the number of years over which you want to calculate the interest. In each case, calculate the value of the principal and the *effective* yearly rate of interest.

a) $P = \$1000$, $r = 8\%$, $t =$ bimonthly, and $y = 2$.

b) $P = \$500$, $r = 10\%$, $t =$ semiannually, and $y = 3$.

c) $P = \$5000$, $r = 10\%$, $t =$ daily, and $y = 3$.

d) $P = \$10000$, $r = 15\%$, $t =$ monthly, and $y = 2$.

3. When the yearly interest rate is 10% and there is quarterly compounding, the result is clearly different from that obtained by compounding just once a year (see Example 2). Repeat the calculation in Example 2 for an interest rate of 1% a year and quarterly compounding. How different is the result from that obtained by compounding just once a year? How do you explain the difference in the results for 1% and 10%? *Hint*: Use the Binomial Theorem.

4. A question commonly asked when someone puts money in an interest bearing account is: How long will it take my money to double? Specifically: If interest is credited only at the end of each year, display a function of r, the yearly interest rate, whose value is the year in which your money has at least doubled in value.

5. You deposit $10 every week in a bank account which offers 6% yearly interest compounded weekly. (Assume that a year has exactly 52 weeks.) Write an algorithm to compute the amount of money in the account at the end of the year assuming that you
   a) make the deposit at the beginning of each week; or
   b) make the deposit at the end of each week.

   If you have a computer available, convert this algorithm into a program and print out the value of your money week by week in each case.

6. Consider the quadratic equation
$$x^2 - x - 1 = 0.$$
   a) Solve this equation using the quadratic formula. If we call the two solutions x_1 and x_2, show how you can express the nth Fibonacci number in terms of x_1 and x_2. [See Eq. (1).]
   b) Find two nonnegative values of x, call them x_1 and x_2, at which the quadratic on the left has different signs.
   c) Look at the point halfway between x_1 and x_2. The value of the quadratic at this point will be different from the value at either x_1 or x_2. Use this fact to derive a method for finding an arbitrarily good approximation to a zero of the quadratic. Use this method to calculate a root of the given equation correct to two decimal places. The result should be consistent with the result of a). (We will discuss this method further in Section 9.7.)
   d) Now find two nonpositive values at which the signs are different and repeat c). Again, the result should be consistent with that in a).

7. Suppose that we want to solve
$$x = \frac{1}{2}x + 1. \tag{4}$$
This is simple to do directly, but note that (4) is of the form
$$x = f(x). \tag{5}$$
Let's change this slightly to
$$x_{n+1} = f(x_n).$$
Now we can plug in any initial guess x_0 and iterate. Such a sequence often has a limit, and if it does, that limit value satisfies (5). Solve (4) by this method using a) $x_0 = 1$; b) $x_0 = 3$.

8. Suppose that we want to find the positive root of $x^2 - x - 1 = 0$.
   a) Rewrite this equation as $x = x^2 - 1$ and use the method of [7]. Try various values of x_0 between 0 and 3. What happens?
   b) Here is an uglier way to rewrite the original equation: Subtract (x^2-1) from both sides, add $3x$ to both sides, and divide both sides by 2. The result is
$$x = \frac{3x - x^2 + 1}{2}.$$
   Now try the method of [7] again for the same range of values of x as in a).

   Any explanations for what happened in this problem?

9.2 Sequences and Their Limits

We first introduced sequences in Section 0.1. You'll recall that an infinite sequence
$$\{a_n\} \tag{1}$$
is an ordered list of real values $a_0, a_1, a_2, \ldots$. Equivalently, we can define a sequence as a function $f(n)$ whose domain is the nonnegative integers N. For convenience, we often start the sequence with a_1 or something else other than a_0.

In denoting sequences we will often use k instead of n in (1), and we will sometimes use letters other than a. For example, it is natural to denote the sequence of Fibonacci numbers by $\{f_n\}$. Our main interest in this section is in what happens to the values of a_n in (2) as n approaches ∞. Do they get large in magnitude (that is, in absolute value) without bound? Or do they *converge* to some stable value or *limit*? Or neither? Note that we say "approaches ∞". As we can't actually carry out an infinite computation, we will be interested in (1) what would happen if we could and (2) how "close" we have to get to ∞—that is, how large n has to be—before the value of a_n gets as close as we would like to the value it would have at ∞.

Examples of sequences with the three different types of behavior mentioned in the preceding paragraph are:

- $\{n\} = N^+ = 1, 2, 3, \ldots$, where the terms of the sequence get large without bound, as do the terms of $\{2^n\}$.
- $\{1/n\}$ converges to 0 as n gets large; $\{(2n+3)/(3n+5)\}$ converges to $2/3$ (why?).
- 0, 1, 0, 1, 0, 1, 0, 1,..., converges to no limit; neither does the sequence $\{(-1)^n\}$.

But all this is just intuitive. We need an explicit definition of what we mean by words such as "converge" and "limit". The essential question about convergence is: As n gets larger and larger, is there some value, called the limit, around which the values of the sequence *cluster*? Put another way, do the values of the sequence *get* close and *stay* close to some limiting value? In the case of $\{1/n\}$, as n gets large the values get closer and closer to 0 and no value is farther from 0 than its predecessor. But consider this sequence:

$$a_n = \begin{cases} 1 & \text{if } n \text{ is a power of 10;} \\ 1/n & \text{otherwise.} \end{cases} \tag{2}$$

What happens as n increases from, say, 1 million to 10 million? The values of a_n get closer and closer to 0 but, then suddenly at 10 million ($= 10^7$), the value is 1 again. As n increases further, the values of the sequence get closer and closer to 0, but they don't stay close for *all* values of n. How can we capture the notion of getting close and staying close? We want to say something like this: If a sequence gets closer and closer to some value—call it L for limit—and stays close, then no matter how small a distance from L we choose, all the values of the sequence beyond a certain point will be within this distance of L. We illustrate this notion in Fig. 9.1 and formalize it in the following definition.

Definition 1. Let $\{a_n\}$ be a sequence. We say that the sequence **converges to a limit** L if, given any positive number $\epsilon > 0$, there exists an N such that

$$|a_n - L| < \epsilon$$

for all $n \geq N$.

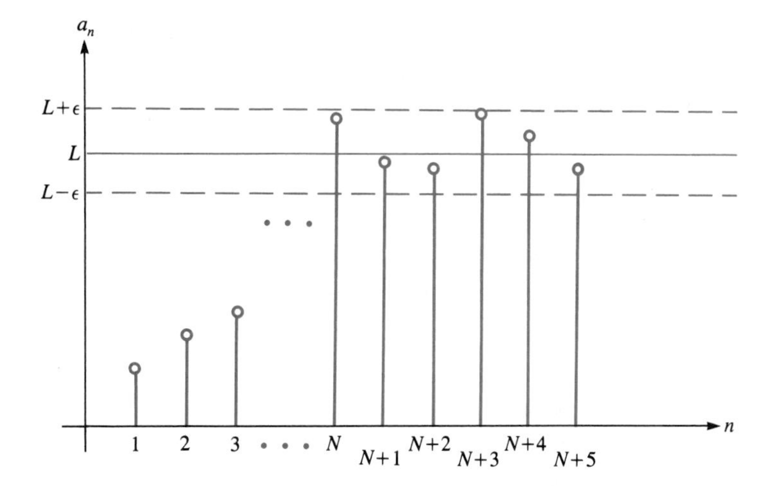

FIGURE 9.1

Convergence of a Sequence

As Fig. 9.1 illustrates, ϵ defines a distance from L but, no matter how small ϵ is, if you go far enough out in the sequence (i.e., choose N sufficiently large), all terms in a convergent sequence after the one with subscript N will be within ϵ of L.

The concept of a limit is one of the most important in all mathematics. It is crucial to the study of calculus, which is another reason why the material in this chapter serves as a bridge from discrete to continuous mathematics. Whereas here we discuss limits of sequences only—that is, limits of functions whose domain is a set of nonnegative integers— in calculus the essential interest is in limits of functions whose domain is an interval of the real number line. An understanding of limits in the sense in which we discuss them here should make it easier to understand their use in calculus.

Showing that a sequence converges is like playing a game. Suppose I claim that a sequence converges to L. Then you get to choose any positive number ϵ, and I have to find an N such that $|a_n - L| < \epsilon$ for any $n \geq N$. If I can do this, no matter how tiny the ϵ you choose, then I win the game; that is, the sequence converges. However, if you can find an ϵ for which I can't find an N, you win because the sequence does not converge. Do you see why I can win the game for the sequence $\{1/n\}$ but not for the sequence defined by Eq. (2)?

Because you can choose as small an ϵ as you want—10^{-56}, for example—convergence of a sequence to L means that the terms get and stay *arbitrarily* close to L. If ϵ is 10^{-56}, then N might have to be 10^{100}, but so what? You're allowed to go as far out as you have to in the sequence to get within ϵ of the limit. This illustrates that the value of N in Definition 1 depends on the value of ϵ. The smaller the ϵ chosen, generally the larger N must be. Since the value of N depends on the value of ϵ, it's reasonable to say that N is a function of ϵ. For this reason, in Definition 1 we could have written $N(\epsilon)$ instead of just N.

Another way of writing Definition 1, using the quantifier notation of Section 7.6, is to say that convergence of a sequence to a limit L occurs if

$$(\forall \epsilon : \epsilon > 0)(\exists N)(\forall n : n \geq N)(|a_n - L| < \epsilon).$$

Make sure that you understand how these three quantifiers capture the constraints of Definition 1.

When a sequence does converge, we say the limit *exists* and use the notation

$$\lim_{n \to \infty} a_n = L.$$

We have used this notation without explanation earlier in this book (e.g., see Sections 0.5 and 6.4). When, as is often the case, the $n \to \infty$ is obvious, we will just write

$$\lim a_n = L.$$

If a sequence does not converge, we say that it **diverges**. Two ways for a sequence to diverge (there also are other, more complicated ways) are for the magnitudes of the terms to grow without bound or for the terms to oscillate as in

$$\{0, 1, 0, 1, 0, 1, \ldots\}.$$

EXAMPLE 1 Use Definition 1 to prove that $\{1/n\}$ converges to 0.

Solution The game approach alluded to earlier works nicely here. Suppose that you choose $\epsilon = 0.1$. Then I can choose $N = 11$ because

$$\frac{1}{n} \leq \frac{1}{11} < 0.1, \qquad \text{for all } n \geq 11.$$

(Note that I could also choose any $N > 11$.) If you choose $\epsilon = 10^{-100}$, I would just have to choose $N = 10^{100}+1$, since

$$\frac{1}{n} \leq \frac{1}{10^{100}+1} < 10^{-100}, \qquad \text{for all } n \geq 10^{100}+1.$$

More generally, for any $\epsilon > 0$ you choose, the N that I choose would be one such that

$$\frac{1}{N} < \epsilon. \tag{3}$$

Then

$$\left|0 - \frac{1}{n}\right| \leq \frac{1}{N} < \epsilon, \qquad \text{for all } n \geq N,$$

so, according to Definition 1, 0 is the limit of the sequence. ∎

EXAMPLE 2 Does the sequence $\{1+(-1)^n/n\}$ converge and, if so, what is its limit?

Solution If the answer doesn't jump right out at you, use the inductive paradigm and compute some values of a_n. If you do, you'll quickly conjecture that

$$\lim_{n \to \infty} a_n = 1.$$

To prove that this equality is correct, just compute

$$|1 - a_n| = \left|\frac{-(-1)^n}{n}\right| = \frac{1}{n}.$$

As in Example 1, if you choose any ϵ, then all I have to do is choose an N such that $1/N < \epsilon$. I can always do so, which is sufficient to prove that $\{a_n\}$ converges to 1. ∎

Mathematicians state formal definitions, such as Definition 1, so that they can then build a structure of theorems based on the definitions. Thus we should be interested in theorems which enable us to build on results in simple cases such as those in Examples 1 and 2 in order to determine whether more complicated sequences converge and, if so, to what. We state only one of the many such theorems in order to give you their flavor.

Theorem 1. If $\{a_n\}$ and $\{b_n\}$ are two sequences which converge to a and b, respectively, then

(i) $\lim(a_n+b_n) = a + b$; $\lim(a_n-b_n) = a - b$.
(ii) $\lim ca_n = ca$, where c is any real number.
(iii) $\lim a_n b_n = ab$.
(iv) $\lim a_n/b_n = a/b$, if $b \neq 0$.
(v) $\lim |a_n| = |a|$.
(vi) $\lim(a_n)^x = a^x$ for positive x and, if $a \neq 0$, for negative x.

We'll prove only part (i) of this theorem and leave some of the other parts to a problem [9b–d].

PROOF The proof of (i) involves only one step which is not straightforward, so we consider the mathematics needed in this step first. Definition 1 implies that, given a sequence $\{a_n\}$ which converges to a limit L and any fixed positive constant d, for any $\epsilon > 0$ there exists an N such that

$$|a_n - L| < d\epsilon,$$

for all $n \geq N$. Why? What gives us the right to replace ϵ in the inequality in Definition 1 by $d\epsilon$? The answer is that $d\epsilon$ is just some other positive number, like ϵ itself. Definition 1 says only that, if a sequence converges, then for any given positive number, if we go far enough out in the sequence (i.e., N large enough), the terms of the sequence get within L of that positive number. Whether we call that positive number ϵ or $d\epsilon$ makes no essential difference.

This argument allows us to say that, if $\{a_n\}$ converges to a and $\{b_n\}$ converges to b, then for any $\epsilon > 0$, there exist N_1 and N_2 such that

$$\begin{aligned} |a_n - a| &< \epsilon/2, \quad \text{for all } n \geq N_1; \\ |b_n - b| &< \epsilon/2, \quad \text{for all } n \geq N_2. \end{aligned} \tag{4}$$

All we have done in (4) is set d in the preceding paragraph to $\frac{1}{2}$.

One last point before getting to the proof itself. We needed an N_1 and an N_2 in (4) because, for any given ϵ, one of the sequences may converge faster (i.e., get close to its limit quicker) than the other. But suppose that we choose $N = \max(N_1, N_2)$. then we may rewrite (4) as

$$|a_n - a| < \frac{\epsilon}{2} \quad \text{and} \quad |b_n - b| < \frac{\epsilon}{2}, \quad \text{for all } n \geq N, \tag{5}$$

because N is at least as large as either N_1 or N_2.

Now we're ready to prove (i). (We'll prove only the + case; the proof of the − case is almost identical and we'll leave it to a problem [9a].) The hypotheses are that $\{a_n\}$ converges to a and $\{b_n\}$ converges to b. These hypotheses assure us that

$$|a_n - a| \quad \text{and} \quad |b_n - b| \tag{6}$$

get arbitrarily small as n gets large. But, from Definition 1, the conclusion asks us to show that

$$|a_n + b_n - (a+b)| = |(a_n - a) + (b_n - b)| \tag{7}$$

gets arbitrarily small. Comparing (6) with Eq. (7) suggests using the triangle inequality to enable us to use the hypotheses. So, given any $\epsilon > 0$, we proceed as follows:

$$\begin{aligned} |a_n + b_n - (a+b)| &\leq |a_n - a| + |b_n - b| && \text{[Triangle inequality; see Section 0.4]} \\ &< \frac{\epsilon}{2} + \frac{\epsilon}{2} && \text{[Using (5) for any } n \geq N\text{]} \\ &= \epsilon && (8) \end{aligned}$$

Since (8) is true for any $\epsilon > 0$, it follows from Definition 1 that $\{a_n + b_n\}$ converges to $a + b$, which completes the proof. ∎

Example 3 applies Theorem 1 to the problem discussed in Example 1, Section 9.1.

EXAMPLE 3 Given a linear homogeneous difference equation with constant coefficients

$$a_n = \sum_{i=1}^{k} c_i a_{n-i},$$

describe the limiting behavior of the solution as n gets large, assuming that the roots of the characteristic equation are real and distinct in magnitude.

Solution We stated in Theorem 1, Section 5.5 that the solution of the difference equation is given by

$$\sum_{i=1}^{k} A_i r_i^n, \tag{9}$$

where the A_i are constants depending on the initial conditions, and the r_i's are the roots of the characteristic equation

$$r^k - \sum_{i=1}^{k} c_i r^{k-i} = 0.$$

We've assumed that the roots are real and distinct in magnitude. Therefore we can also assume that they're ordered in magnitude, so that

$$|r_1| > |r_2| > \cdots > |r_n|. \tag{10}$$

Now let's further assume that $A_1 \neq 0$, and let's rewrite (9) in the form

$$r_1^n \sum_{i=1}^{k} A_i \left(\frac{r_i}{r_1}\right)^n \tag{11}$$

Considering the summation portion only, what is the limit as $n \to \infty$? This limit involves the sum of k terms. Theorem 1 tells us that the limit of the sum of two terms is the sum of the limits. It's easy to prove for any k that the limit of the sum of k terms is the sum of the limits. (How would you prove this? We hope that by now the answer is clear [10].) Therefore

$$\lim \sum_{i=1}^{k} A_i \left(\frac{r_i}{r_1}\right)^n = \sum_{i=1}^{k} \lim A_i \left(\frac{r_i}{r_1}\right)^n, \tag{12}$$

with all limits as $n \to \infty$. Now, how about each limit on the right-hand side of Eq. (12)? Each is a product of a constant (A_i) and a term which depends on n $[(r_i/r_1)^n]$. Therefore, by (ii) of Theorem 1,

$$\lim A_i \left(\frac{r_i}{r_1}\right)^n = A_i \lim \left(\frac{r_i}{r_1}\right)^n \tag{13}$$

Now, how about $\lim(r_i/r_1)^n$? When $i = 1$ the limit of this ratio is just 1 (why?). When $i > 1$, $|r_i/r_1| < 1$ from (10), and the limit is 0 (why?) [11]. Therefore, using Eqs. (12) and (13), we find that the limit of the summation portion of (11) is A_1, which we've assumed is nonzero. Therefore, for large n, the term

$$A_1 r_1^n$$

determines the behavior of the solution of the difference equation, as we claimed in Example 1, Section 9.1. (In Section 9.3 We'll introduce a more precise way of expressing the dominance of the term above.) If r_1 itself has magnitude less than 1, then [again using (ii) of Theorem 1] the entire solution approaches 0 as $n \to \infty$; if the magnitude is greater than 1, then the solution gets arbitrarily large in magnitude as n gets large. And if $|r_1| = 1$, the magnitude of the solution approaches $|A_1|$ as n approaches infinity. ∎

What changes must we make in the solution of Example 3 if our assumption that $A_1 \neq 0$ is not correct [13a]?

When the roots of the characteristic equation are real but not distinct in magnitude [i.e., Eq. (10) is not satisfied], we can still use Theorem 1 to find the limiting behavior [13b]. When the roots are not all real, the situation is rather more complicated.

For the specific case of the Fibonacci sequence, Example 3 justifies our claim in Example 1, Section 9.1, that the dominant term for large n is

$$\left(\frac{1}{\sqrt{5}}\right)\left(\frac{1+\sqrt{5}}{2}\right)^{n+1}.$$

Indeed, we can show [27] that, for *any* n, this value *rounded* to the nearest integer is the nth Fibonacci number.

Theorem 1 is very useful for deriving general results like that in Example 3. But in order to use it to find the limit of a particular sequence you need to know the limits of the component sequences $\{a_n\}$ and $\{b_n\}$. Of course, you can always try to find such limits using Definition 1 directly, as in Examples 1 and 2. But it would be nice to have some theorems which tell you when sequences of certain kinds converge and also help you determine what they converge to. Theorem 2 exemplifies this kind of result. But first we need a definition.

Definition 2. A sequence $\{a_n\}$ is **monotone** (constantly) **increasing** if, for all n,

$$a_n \leq a_{n+1}. \tag{14}$$

If the $\leq$ in (14) can be replaced by $<$, then the sequence is **strictly monotone increasing**. And if $\leq$ and $<$ are replaced by $\geq$ and $>$, respectively, we get the corresponding definitions of a **monotone decreasing sequence** and a **strictly monotone decreasing sequence**.

Sometimes (14) is true for all n only after some value N, in which case we speak of an **eventually monotone** sequence.

Monotonicity in a sequence (or, for that matter, in any function) refers therefore to a tendency to keep going in one direction after a certain point. If we're interested in whether a sequence has a limit, determining that for monotonic sequences is likely to be easier than for those which may oscillate between increasing and decreasing behavior. Now here's the theorem:

Theorem 2. Let $\{a_k\}$ be a monotone increasing sequence and let B be such that

$$a_k < B \qquad \text{for all } k.$$

Then the sequence must converge to a limit.

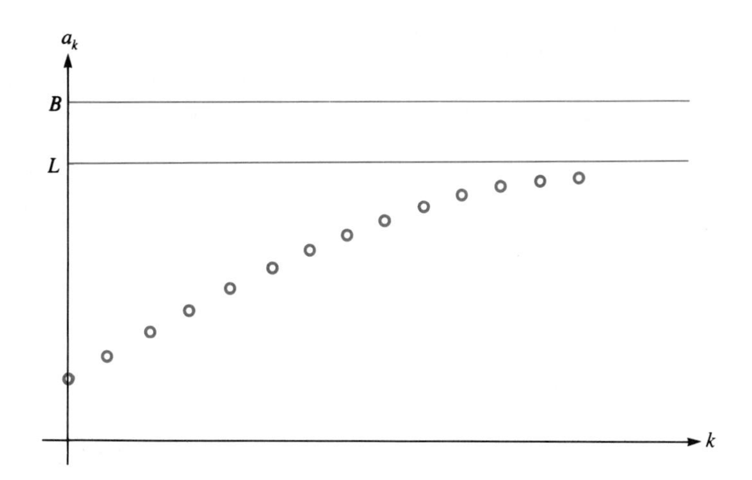

FIGURE 9.2

Convergence of a Monotonic Sequence

We call B, which need not itself be the limit, an **upper bound** of the sequence. There is an obvious analog to this theorem for monotone decreasing sequences [16].

The proof of Theorem 2 relies on understanding of the real numbers. We won't give it here, but see [17]. However, Figure 9.2 should suggest the truth of Theorem 2 to you. Because the terms of the sequence increase but never get to B, intuitively they have to bunch up at (i.e., get arbitrarily close to) some value L less than or equal to B.

Theorem 2 seems to suggest an algorithm for finding as good an approximation as you want to the limit of a monotonically increasing sequence which is bounded above. Just compute successive values until they bunch up sufficiently. But this is trickier than it may seem; we'll consider such an algorithm in a problem [19].

Example 4 illustrates the usefulness of Theorem 2.

EXAMPLE 4 Does the sequence $\{(1+1/n)^n\}$ converge and, if so, what is its limit?

Solution Let's begin by applying the inductive paradigm and computing a few terms of the sequence:

n	a_n
1	2.000
2	2.250
3	2.370...
4	2.441...
5	2.488...
6	2.521...
7	2.546...
8	2.565...
9	2.581...
10	2.593...

Each ellipsis indicates that we have given only the first three figures of an infinite decimal. This table at least suggests that the sequence converges, although it may give you no hint as to what it converges to.

Let's see if we can apply Theorem 2. Is the sequence monotone increasing and bounded above? Yes, on both counts but the algebra to show this is quite tedious, so we'll leave it to a problem [20]. So by Theorem 2 the sequence converges. To seven decimal places, the value of L is

$$2.7182818\ldots.$$

As you may know, this number is called e (after Euler) and is the base of **natural logarithms**. To summarize:

$$e = \lim_{n\to\infty} \left(1+\frac{1}{n}\right)^n.$$

Euler's number e arises in many places in mathematics (see, for example, Section 6.4). ∎

Although the mathematics is beyond us here, we can generalize the result of Example 4 to

$$\lim \left(1+\frac{r}{n}\right)^n = e^r. \tag{15}$$

In Example 2, Section 9.1 we showed, in effect, that the value of principal P invested at an interest rate r compounded every $1/n$ of a year is

$$P\left(1+\frac{r}{n}\right)^n.$$

This expression defines a sequence $\{P_n\}$. The result of compounding continuously is defined to be the limit of this sequence, if it exists. Why is this reasonable? Because as $n \to \infty$, the compounding is done over smaller and smaller fractions of a year (i.e., $1/n$); as this fraction approaches 0, the compounding becomes, in effect, continuous. But does $\{P_n\}$ converge? Yes, from Eq. (15) and (ii) of Theorem 1. The limit is

$$Pe^r = P[1+(e^r-1)].$$

Therefore the effective yearly rate of interest is $e^r - 1$. Table 9.1 lists some values of $e^r - 1$ for various values of r. It's interesting, by the way, to compare the line for $r = 0.1$ (10.52) with the corresponding value (10.511) from Eq. (2), Example 2, Section 9.1. As you can see, the advantage of continuous compounding over daily compounding is very small.

TABLE 9.1
Effective Rates of Interest for Continuous Compounding

Rate of Interest r	$100(e^r - 1)$
$5\% = 0.05$	5.13%
0.06	6.18
0.07	7.25
0.08	8.33
0.09	9.42
0.10	10.52
0.11	11.63
0.12	12.75
0.13	13.88
0.14	15.03
0.15	16.18

Problems: Section 9.2

1. What are the limits as $n \to \infty$ of the sequences whose nth term is given in a)–e)? Use Definition 1 to guide your thinking, but don't try to prove your answers correct.

a) $n/(n+1)$ **d)** $(n^2-1)/(n^2+1)$
b) $(2n+3)/(n+500)$ **e)** a^n, $|a| < 1$
c) $n/2^n$

2. Now use Definition 1 to *prove* whether each of the sequences whose nth term is given in a)–e) converges or diverges.

a) $1/n^2$ **d)** a^n, $|a| > 1$
b) $1/n + (-1)^n$ **e)** a^n, $|a| = 1$
c) $(-1)^n(1/n)$

3. Consider the sequence whose first three terms are

$$1/2,\ (1/2)/(3/4),\ \text{and}\ [(1/2)/(3/4)]/[(5/6)/(7/8)]$$

a) Find a way to define this sequence precisely.

b) Calculate enough terms (perhaps with the help of a calculator or computer) to make a conjecture about the limit of the sequence.

4. Prove that the limit of a sequence, if it has one, must be *unique* (i.e., a sequence cannot converge to two separate limits). *Hint*: Assume that there are two unequal limits L_1 and L_2 and derive a contradiction by picking an ϵ smaller than $|L_1 - L_2|$.

5. Books suffer from typographical errors (typos), although the situation usually improves with later editions. Assume that in the first edition of a book, one symbol in 25,000 is a typo. Assume further that 80% of the typos are caught and changed in each later edition. Unfortunately, with each edition one symbol in 500,000 becomes a new typo (due to resetting of pages, computer glitches, or whatever).

Assume that a book contains 3 million symbols. Let t_n be the number of typos in the nth edition. Based on the assumptions that we have made, find a closed form expression for t_n. Do typos die out?

6. Justify the following statements or give a counterexample.

a) If $\lim_{n\to\infty} a_n$ exists, then $\lim_{n\to\infty} |a_n|$ exists.

b) If $\lim_{n\to\infty} |a_n|$ exists, then $\lim_{n\to\infty} a_n$ exists.

7. Can a sequence diverge without diverging to infinity? Explain.

8. Using a computer or calculator, approximate

$$\lim_{n\to\infty}\left(1+\frac{2}{n}\right)^n.$$

9. **a)** Make the necessary changes in the proof of (i) of Theorem 1 so that it is correct for $a_n - b_n$.

b) Prove (ii) of Theorem 1.

c) Prove (v) of Theorem 1.

d) Prove (iii) of Theorem 1. *Hint*: Write $a_n b_n - ab$ as $a_n b_n - a_n b + a_n b - ab$ and then use an approach like that used to prove (i).

10. Prove that the limit of the sum of k terms is the sum of the limits.

11. Use Definition 1 and Theorem 2 to prove that $\{r^n\}$ converges to 0, if $|r| < 1$.

12. Suppose that

$$P(x) = \sum_{k=0}^{r} a_k x^k.$$

What is

$$\lim_{n\to\infty}\frac{P(n)}{n^r}?$$

Justify your answer.

13. **a)** Suppose that A_1 in (9) is 0. How must the solution of Example 3 be modified?

b) Suppose that the two roots of the characteristic equation of greatest magnitude are real and unequal but have equal magnitude, and assume that the coefficients of both in (9) are not 0. How must the solution of Example 3 be modified?

14. In the text we described as a game proving that a sequence $\{a_n\}$ has a limit L. The first player names an ϵ and the second has to counter with an N such that for all $n \geq N$, $|a_n - L| < \epsilon$. If the second player can *always* counter the first, the second player wins and L is the limit.

Let's consider some variations of this game. For each variation try to describe those sequences for which the second player is the winner. (These variations are not themselves important; their purpose is to help you understand the original version of the game better.)

a) Player 2 has to announce her N before player 1 announces his ϵ; for her to win, N has to counter *any* ϵ the first player chooses.

b) The original game is played but for only one round. That is, player 1 gets to choose ϵ just once and player 2 has to counter with an N so that $|a_n - L| < \epsilon$ for *that* ϵ.

c) Player 1 chooses N and player 2 must counter with an $\epsilon > 0$ so that for all $n \geq N$, $|a_n - L| < \epsilon$. Player 2 wins if she can counter every choice of N.

15. If the choice of N in Definition 1 need not depend on ϵ, then it must be true that all a_k in the convergent sequence $\{a_k\}$ must be equal for $k >$ some K. True or false? Prove it.

16. State the analog of Theorem 2 for monotone decreasing sequences. Does your theorem or Theorem 2 need to be restated at all for strictly monotone decreasing or increasing sequences?

17. Let $\{a_k\}$ be a sequence satisfying the conditions of Theorem 2. Let $\{A_k\}$ be the sequence of integer parts of a_k. Thus if $a_k = 4.721$, $A_k = 4$.

a) Prove that the sequence $\{A_k\}$ must reach some maximum value and then stay there forever. *Hint*: Why must there be some maximum value for this sequence? What would it mean if $\{A_k\}$ reached this value for some k and then later slipped back to a smaller value?

b) By generalizing a) to each decimal place of $\{a_k\}$ sketch a proof of Theorem 2. *Hint*: Show that the limit is $L = A.D_1D_2D_3\ldots$, where D_k is the limit of the sequence $\{d_{ik}\}$, with d_{ik} the kth decimal place of a_i.

18. Consider the sequence $\{a_n\}$ whose domain is N^+ and where

$$a_n = \frac{n^2 - n}{2n^2 + n}.$$

a) Show that this sequence is monotonic.

b) Calculate values of a_n for $n = 10, 20, 30, \ldots$, until the first two decimal places seem to have reached their maximum values. Argue plausibly—rather than proving—that the values you have found are, indeed, the maximum values of these two decimals.

19. Derive an algorithm for computing an approximation to the limit of a sequence which satisfies the conditions of Theorem 2. Your algorithm should input a value ϵ such that, when two successive approximations computed by your algorithm differ in magnitude by less than ϵ, the algorithm halts.

What can you say about how close your last approximation is to the true limit? Be careful. This is tricky. Could two successive approximations be arbitrarily close without the sequence converging? (See Example 3, Section 9.5.)

20. To show that the sequence $\{(1+1/n)^n\}$ is monotonic and bounded above:

a) Expand the nth and $(n+1)$st terms using the binomial theorem.

b) Show that each term of the second expansion is greater than a corresponding term of the first expansion. This result shows that the sequence is monotonic increasing.

c) In the expansion of $(1+1/n)^n$, show that the ith term is less than or equal to $1/2^{i-2}$ for $i > 2$.

d) Use the result in c) to show that the sequence is bounded above.

21. a) Complete this sentence: A sequence $\{a_n\}$ of positive numbers is monotonic increasing if and only if $a_{n+1}/a_n \ldots$.

b) What does the value of $\lim a_{n+1}/a_n$ tell you about the convergence or divergence of a monotonic sequence?

c) Repeat a) for monotonic decreasing sequences.

d) Now what does $\lim a_{n+1}/a_n$ tell you about the convergence or divergence of the sequence?

22. Use the results in [21] to test the convergence of each of the following sequences $\{a_n\}$. For those that converge, reason informally to find the limit. In each case the value of a_n is given.

a) $n^3/3^n$ **b)** $n!/2^n$
c) $(n^2+1)/n^2$ **d)** $n^n/n!$

23. a) Let the terms of the sequence $\{a_n\}$ be given by

$$a_0 = 1 \text{ and } a_n = a_{n-1} + 1/2^{n-1};\ n \geq 1.$$

Conjecture and prove a closed form for a_n.

b) Now let $a_0 = 1$, $a_n = a_{n-1}+1/n!$, $n \geq 1$. Use a) to prove that this sequence converges.

c) Calculate the terms of the sequence in b) for $n = 1, 2, \ldots, 6$. Compare these results with those of Example 4 and with the value of e given in this section. What do you surmise about the limit of this sequence?

24. Suppose that two sequences $\{a_n\}$ and $\{b_n\}$ both approach ∞ as $n \to \infty$. Then the ratio a_n/b_n approaches ∞/∞. This form is said to be **indeterminate** because, for any number L, there are sequences $\{a_n\}$ and $\{b_n\}$ which both approach ∞ and for which

$$\lim_{n\to\infty} \frac{a_n}{b_n} = L \qquad (16)$$

For example, if $a_n = 2n$ and $b_n = n$, then $L = 2$. Find $\{a_n\}$ and $\{b_n\}$ so that L in Eq. (16) is

a) 0 **b)** 1 **c)** -83.2 **d)** ∞

25. Similarly, $0/0$ is an indeterminate form. Replace ∞ by 0 everywhere in [24] (except in $n \to \infty$) and repeat all four parts.

26. Find (if it exists):

a) $\lim_{n\to\infty}(2n^3+3/n)/(n^2-100n)$

b) $\lim_{n\to\infty}(2n^2+3/n)/(n^3-100n)$

27. Fibonacci numbers again. Refer back to [1, Section 9.1]. Show how the results of this section enable you to claim that the nth Fibonacci number can be found by rounding

$$\left(\frac{1}{\sqrt{5}}\right)\left(\frac{1+\sqrt{5}}{2}\right)^{n+1}.$$

28. Argue that $\sqrt[n]{2} \to 1$. Begin by showing that the sequence is decreasing.

29. Consider the sequence with nth term

$$a_n = \left(1+\frac{1}{n^2}\right)^n.$$

a) Do some numerical investigation and conjecture the limit.

b) Try to prove your conjecture. (*Hint*: Put a_n in a form in which the result of Example 4 can be applied.)

9.3 Growth Rates and Order Notation

Mathematicians are interested mainly in convergent sequences. Not surprising, you may think. Why should anyone be interested in sequences whose values go off to ∞ as $n \to \infty$, or which don't converge to any value at all? Well, it turns out that if you're interested in algorithms and their analysis, your focus will be on divergent sequences. In this section, in addition to discussing why this is so, we'll consider how we can compare the rates of divergence of different sequences.

In all the algorithms we have analyzed thus far, there was at least one positive integer parameter—usually we called it n—which determined the behavior of he algorithm. The speed of the algorithm (in either the worst or average case) was always expressed as some function of n, that is, as a *sequence*. Here are some examples from previous chapters:

- The number of moves in Algorithm HANOI (Section 1.4) is expressed by the sequence $\{2^n - 1\}$
- The average number of comparisons in Algorithm SEQSEARCH (Section 1.6) is $\{(n+1)/2)\}$
- The number of comparisons in the worst case of Algorithm BINSEARCH (Section 2.6) is $\{\lfloor \log_2 n \rfloor + 1\}$
- The number of calls of RAND in Algorithm PERMUTE-3 (Section 4.8) is $\{n-1\}$

In all these examples the sequences are divergent because any power of n or the logarithm of n grows without bound as n gets large. It's important to understand that, if a problem can be solved by an algorithm whose time of execution sequence does *not* diverge to ∞, looking for a better algorithm isn't likely to be worthwhile. True, if the time of execution of an algorithm did not depend on n but was very large, we would be interested in looking for a better algorithm. Almost always, however, algorithms whose time of execution sequence does not diverge to ∞ are very easy to execute for all n.

For algorithms whose time of execution becomes infinite as n becomes infinite, it is certainly important to look for the most efficient formulation of the algorithm or to look for better algorithms for the same task. Searching for an item on an ordered list is a good example. In SEQSEARCH (see Section 1.6), the sequence representing the average number of comparisons grows as the first power of n. Because we sometimes need to search through very long lists (i.e., very large values of n) on computers, it is important to have a better method if one exists. BINSEARCH is such a better method. Its execution sequence in the worst case, although it also diverges, does so more slowly, since $\lceil \log_2(n+1) \rceil$ is less than n for all $n > 2$ (see the discussion at the end of this section). It is only a modest oversimplification to say that the analysis of algorithms is concerned mainly with finding and comparing the execution sequences of different algorithms for the same task.

In earlier chapters, when we analyzed algorithms, we typically showed that the number of operations in the worst or average case was proportional to n, some power of n, or perhaps to $\log n$. It would be nice to have some way of expressing rather more precisely this notion of "proportional growth". It would be particularly nice to be able to do this when we are trying to compare two or more algorithms for the same task. Thus the main purpose of this section is to develop and then use a notation for expressing algorithmic growth rates. This notation also has many uses in areas of mathematics other than the analysis of algorithms.

Definition 1. Let $\{a_n\}$ and $\{b_n\}$ be two sequences. Then we say that

$$a_n = O(b_n), \tag{1}$$

if $|a_n| \leq c|b_n|$ for some constant c for all except perhaps a finite number of values of n.

When talking about the notation in Eq. (1), we say that "a_n is 'big Oh' of b_n". Note that the "=" is stated as "is". Big Oh notation is traditional in mathematics and is widely used by writers about algorithms. However, it doesn't quite do the job when our purpose is to compare two algorithms with different execution sequences. Thus, for example, according to Definition 1, $n^2 = O(n^3)$ because $n^2 \leq n^3$ for all (nonnegative) values of n. A more useful definition is the following one.

Definition 2. Let $\{a_n\}$ and $\{b_n\}$ be two sequences. Then we say that

$$a_n = \text{Ord}(b_n),$$

if

$$\lim_{n\to\infty} \frac{a_n}{b_n} = c \neq 0. \tag{2}$$

Note that if $a_n = \text{Ord}(b_n)$, then also $b_n = \text{Ord}(a_n)$ (why?). If two sequences satisfy Definition 2, the growth of one is proportional to the other. When two sequences satisfy Definition 2, we say that one is of the **order** of the other. A notation sometimes used instead of Ord in Definition 2 is $a_n = \Theta(b_n)$ [2].

Note also that Definition 2 says nothing about the convergence or divergence of either sequence although, as we said earlier, we will almost always be concerned with sequences which diverge to ∞. Suppose that $\{a_n\}$ converges. Can you name a (trivial) sequence $\{b_n\}$ such that $a_n = \text{Ord}(b_n)$ [4]?

When all the terms in the sequences $\{a_n\}$ and $\{b_n\}$ in Definitions 1 and 2 are positive, as they will be for the execution sequences of algorithms, we can dispense with the absolute values in Definition 1 and we'll have $c > 0$ in Definition 2. However, for general results, such as in Example 1 below, we need to allow the possibility of negative terms.

We will use the notation introduced in Definition 2 exclusively in this section. What we really want to do is to relate a sequence $\{a_n\}$, which represents the speed of execution of an algorithm, to a *standard sequence* $\{b_n\}$. Then to compare two algorithms for the same task, we only have to compare the two corresponding standard sequences. If we had an appropriate *family* of standard sequences, we could compare the execution speed of almost any algorithm with a member of that family. Such a family of standard sequences should, for convenience, be restricted to a relatively small number of characteristic sequences. Our standard sequences are

$$\{a^n\}, \qquad \{n^r\}, \qquad \text{and} \qquad \{n^r \log n\}, \tag{3}$$

where a is any positive real constant greater than 1, r can be any nonnegative integer, and the base of the logarithm is unspecified because the order of the logarithm function is unaffected by the base [6]. Note that when $r = 0$ in the last item of (3), the sequence is just $\{\log n\}$. Note also that each of the sequences (3) is itself a family because of the set of values which can be assumed by a or r.

Our justification for choosing the functions in (3) is that the order of sequences which arise in the analysis of algorithms is usually one of those in (3).† An algorithm whose execution sequence has an order given by the second or third member of (3) is called a **polynomial time** algorithm because it can be executed in a time proportional to some power of n (or some power of n multiplied by $\log n$) In contrast, an algorithm whose execution sequence has the same order as $\{a^n\}$ for some value of a is said to exhibit **exponential growth**. Algorithm HANOI, whose execution sequence is $\{2^n - 1\}$, is an example of such a sequence with $a = 2$. Such algorithms will be very difficult or even impossible to execute (because no computer is fast enough) for even moderate values of n. If no nonexponential algorithm is known for a task, we can use the exponential algorithm for practical values of n. For larger values of n, we can try to find a polynomial time algorithm which will give a good approximation to the true result (see also Section 9.7).

EXAMPLE 1 What is the order of the sequence $\{6n^2 - 3n + 4\}$?

Solution We claim that the order of any polynomial in n of degree i is just n^i. To show this, we let the polynomial be

$$P_i(n) = c_i n^i + c_{i-1} n^{i-1} + \cdots + c_1 n + c_0.$$

Once again we use the fact (see Example 3, Section 9.2) that the limit of a sum is the sum of the limits:

---

† Sometimes sequences of the form $\{n^r \log\log n\}$ or $\{n!\}$, as well as various others, also arise.

$$\begin{aligned}\lim_{n\to\infty}\frac{P_i(n)}{n^i} &= \lim_{n\to\infty}\left(c_i+\frac{c_{i-1}}{n}+\cdots+\frac{c_1}{n^{i-1}}+\frac{c_0}{n^i}\right)\\ &= \lim_{n\to\infty} c_i + \lim_{n\to\infty}\frac{c_{i-1}}{n}+\cdots+\lim_{n\to\infty}\frac{c_1}{n^{i-1}}+\lim_{n\to\infty}\frac{c_0}{n^i}\\ &= c_i\end{aligned}$$

since the limit of a constant is just the constant itself and all the other limits are 0 (why?). Therefore $P_i(n) = \mathrm{Ord}(n^i)$, with c_i playing the role of c in Eq. (2). From this general result it follows that the sequence $\{6n^2-3n+4\}$ has order n^2 and that the constant c in Eq. (2) is 6. ∎

For an application of the result in Example 1, refer back to Table 1, Section 8.2, where we showed that the number of operations in Gaussian elimination for square systems is

$$(n^3+3n^2-n)/3.$$

Example 1 enables us to conclude that Gaussian elimination is an $\mathrm{Ord}(n^3)$ algorithm.

Often an execution sequence is not a single term, as in Example 1, or even a polynomial but rather a sum of different types of terms as in

$$\{n^3 + 2n^2 \log n\}. \tag{4}$$

How do we determine the order of such sequences? To answer this question, we need a definition and two theorems.

Definition 3. If $\{a_n\}$ and $\{b_n\}$ are two sequences such that

$$\lim_{n\to\infty}\frac{a_n}{b_n} = 0,$$

then we say that $\{b_n\}$ has a higher **order ranking** than $\{a_n\}$ and we write

$$\mathrm{Ord}(a_n) < \mathrm{Ord}(b_n).$$

Note: When two sequences have the relation described by Definition 3, it is sometimes denoted by

$$a_n = o(b_n),$$

or in words, "a_n is 'little oh' of b_n".

Theorem 1. For the sequences in (3) we have:

i) $\mathrm{Ord}(n^r) < \mathrm{Ord}(n^s)$, if $r < s$.
ii) $\mathrm{Ord}(n^r) < \mathrm{Ord}(n^r \log n) < \mathrm{Ord}(n^{r+1})$.
iii) $\mathrm{Ord}(a_1^n) < \mathrm{Ord}(a_2^n)$, if $a_1 < a_2$.
iv) $\mathrm{Ord}(n^r) < \mathrm{Ord}(a^n)$, for any r and any positive $a > 1$.

We won't prove this theorem although we'll ask you to prove it in [12, 13]. The second part of Theorem 1 enables us to write a string of inequalities [12c]:

$$\mathrm{Ord}(1) < \mathrm{Ord}(\log n) < \mathrm{Ord}(n) < \mathrm{Ord}(n \log n) < \mathrm{Ord}(n^2) < \mathrm{Ord}(n^2 \log n) < \cdots \tag{5}$$

The first term in (5), Ord(1), is the order of a sequence which does not grow with n but rather converges to some value [4]. Note that $\{1\}$ is, indeed, one of the sequences (3). Just set $r = 0$ in the second term. Note that it follows from (iv) of Theorem 1 that any term in string (5) is less than $\mathrm{Ord}(a^n)$ for any positive a [13b]. Figure 9.3 illustrates the growth rates of various functions in (3).

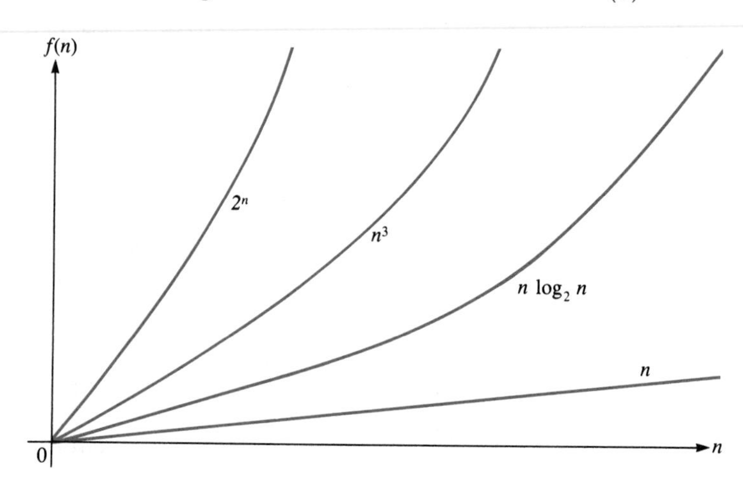

FIGURE 9.3

Rates of Growth of Various Standard Functions. The vertical and horizontal scales are not the same.

Theorem 2. If

$$\mathrm{Ord}(a_n) < \mathrm{Ord}(b_n), \tag{6}$$

then

$$a_n + b_n = \mathrm{Ord}(b_n).$$

PROOF We need to show that $\lim[(a_n+b_n)/b_n]$ is a nonzero constant (why?). Using Theorem 1, Section 9.2, we have

$$\begin{aligned}\lim\left(\frac{a_n+b_n}{b_n}\right) &= \lim\left(\frac{a_n}{b_n}+1\right)\\ &= \lim\left(\frac{a_n}{b_n}\right)+\lim 1 \qquad \text{[By (i) of Theorem 1, Section 9.2]}\\ &= 0+1 = 1, \qquad \text{[0 results from (6) and Definition 3]}\end{aligned}$$

which completes the proof. ∎

A related theorem, whose proof we leave to a problem [14], is:

Theorem 3. If

$$a_n = \mathrm{Ord}(b_n) \qquad \text{and} \qquad \mathrm{Ord}(b_n) < \mathrm{Ord}(c_n),$$

then

$$\mathrm{Ord}(a_n) < \mathrm{Ord}(c_n).$$

EXAMPLE 2 Express the order of $\{n^3+n^2\log n\}$ in terms of a sequence with a single term. Also compare the order of this sequence with n^4.

Solution From (5) it follows that

$$\mathrm{Ord}(n^2\log n) < \mathrm{Ord}(n^3).$$

Therefore from Theorem 2 we get

$$n^3+n^2\log n = \mathrm{Ord}(n^3\}.$$

Then, because $\mathrm{Ord}(n^3) < \mathrm{Ord}(n^4)$, it follows from Theorem 3 that

$$\mathrm{Ord}(n^3+n^2\log n) < \mathrm{Ord}(n^4). \quad ∎$$

The really important part of Theorem 1 is (ii) because it defines the relative rates of growth of powers and powers multiplied by logarithms. This information is crucial in the analysis of algorithms because most significant algorithms have execution sequences which are one of the latter two sequences in (3). If we could say that

$$\mathrm{Ord}(\log n) < \mathrm{Ord}(n), \tag{7}$$

we would be close to being able to prove (ii) of Theorem 1 (see [12b]). A proof of (7) would be quite lengthy using the tools we have developed, so we won't give it. But let's briefly try to convince you that (7) is correct. Indeed, its correctness is implied by Fig. 0.9 (Section 0.4), which shows that replacing $\log x$ by x in a function results in a more rapidly growing function. Figure 9.3 reemphasizes this fact. Actually, it is important for you to realize not just that $\{\log n\}$ grows more slowly than $\{n\}$ but that it grows *much* more slowly. The following tabulation which, for convenience, uses logarithms to the base 2 illustrates this point.

n	2	$2^{10} = 1024$	$2^{20} = 1048576$	$2^{30} = 1073741824$
$\log_2 n$	1	10	20	30

So, if you have two algorithms which have execution sequences of $\mathrm{Ord}(\log n)$ and $\mathrm{Ord}(n)$, respectively, the first isn't just faster than the second but much faster for all but perhaps quite small values of n. The same conclusion would be correct for two algorithms whose execution sequences have orders of $\mathrm{Ord}(n^r \log n)$ and $\mathrm{Ord}(n^{r+1})$ for any r.

Problems: Section 9.3

1. For each of the following sequences, whose nth term is given, state its order using one of the standard sequences (3).

a) $6n^3 + 4n + 1$
b) 3^{n+2}
c) $n^2 + \log n$
d) $n^2 + n \log n$
e) $n^2 + n^2/\log n$
f) $n^2 + n^3/\log n$
g) $(3n^2+1)/(2n-1)$
h) $n^3 \log n/(n^2+1)$
i) $n^2 + (1.1)^n$
j) $n^{50} + (1.01)^{n/100}$

2. a) Our definition of Ord is not, in fact, the same as the usual definition of $a_n = \Theta(b_n)$. That definition does not require a_n/b_n to have a limit but only that there be positive constants L and U such that, eventually,

$$L \leq \left|\frac{a_n}{b_n}\right| \leq U.$$

Show that if $a_n = \mathrm{Ord}(b_n)$, then, as defined here, $a_n = \Theta(b_n)$.

b) Consider $a_n = n(3+\cos n)$ and $b_n = n$. Does $a_n = \Theta(b_n)$, as defined in a)? Does $a_n = \mathrm{Ord}(b_n)$? Why?

3. In the long run are the following positive or negative? Why?

a) $n^2 - \log n$
b) $n^2 - n^3/\log n$
c) $n^2 - (1.1)^n$

4. Suppose that $\{a_n\}$ is such that $\lim a_n = L \neq 0$. Display a sequence $\{b_n\}$ such that $a_n = \mathrm{Ord}(b_n)$.

5. If $P(n)$ is a polynomial of degree i and $Q(n)$ is a polynomial of degree j with $i \geq j$, show that $P(n)/Q(n) = \mathrm{Ord}(n^{i-j})$. If $i < j$, is there still some order relation for $P(n)/Q(n)$? (See [4].)

6. We stated that the base of the logarithm in (3) was irrelevant. Show that we were correct, using Property 3 of logarithms in Section 0.4.

7. Is order also independent of the exponential base? That is, if $a_n = \mathrm{Ord}(a^n)$, is it also $\mathrm{Ord}(b^n)$ if $b \neq a$? Why?

8. Is the order of $2^{\log n}$ also independent of the base of the logarithm? Why?

9. For any k, what is $\lim(\log n)^k/n$? Why? *Hint*: Get rid of the log by making a change of variable.

10. (Knowledge of calculus required)

a) Argue that $\lim(\log n)/n = 0$.

b) Argue that $\lim(\log n)^j/n = 0$, for any positive integer j.

11. Let O_n be the sum of the reciprocals of all the odd integers less than or equal to n. Thus

$$O_{10} = \frac{1}{1} + \frac{1}{3} + \frac{1}{5} + \frac{1}{7} + \frac{1}{9}.$$

Show that $O_n = \mathrm{Ord}(H_n)$. (You may assume that $\lim O_n/H_n$ exists.)

12. a) Prove (i) and (iii) of Theorem 1.

b) Use (7) to prove (ii) of Theorem 1.

c) Justify the string of inequalities (5), using Theorem 1.

13. a) Use the result [10b] to prove (iv) of Theorem 1. *Hint*: Let $m = \log n$.

b) Thus argue that the order of any term in (5) is less than the order of a^n.

14. Prove Theorem 3.

15. By an order ranking of a set of sequences $\{a_{1n}\}$, $\{a_{2n}\}$, ..., $\{a_{mn}\}$ we mean a listing like (5) such that

$$\text{Ord}(a_{1n}) < \text{Ord}(a_{2n}) < \cdots < \text{Ord}(a_{mn})$$

where a $<$ may be replaced by $=$ if two sequences have the same order.

a) Find the order ranking for the set:

$$\{\{n^2\}, \{n^3\}, \{2^n\}, \{n^{\log_2 n}\}, \{2^{\log_2 n}\}\}.$$

b) Insert the following into the ranking of a).

$$\{\{(n+1)^2\}, \{n!\}, \{n^3/\log_2 n\},$$
$$\{1.1n^2 - \log_2 n\}, \{(\log_2 n)^{\log_2 n}\}\}$$

c) Are there any pairs of sequences in a) and b) combined which satisfy Eq. (2)?

d) Does every set of sequences have an order ranking? Why?

16. Prove or find a counterexample to the following claims.

a) If $a_n = \text{Ord}(b_n)$ and c is a positive constant, then $cb_n = \text{Ord}(a_n)$.

b) If $a_n = \text{Ord}(b_n)$ and $c_n = \text{Ord}(d_n)$, then $a_nc_n = \text{Ord}(b_nd_n)$.

c) If $a_n = \text{Ord}(b_n)$ and $c_n = \text{Ord}(d_n)$, then $a_n+c_n = \text{Ord}(b_n+d_n)$.

d) If $a_n = \text{Ord}(b_n)$ and $c_n = \text{Ord}(d_n)$, then $a_n/c_n = \text{Ord}(b_n/d_n)$.

e) If $a_n = \text{Ord}(b_n)$ and $b_n = \text{Ord}(c_n)$, then $a_n = \text{Ord}(c_n)$.

17. Suppose that you can rent time on a supercomputer that does 1 billion operations per second, and that the rental is \$1000/hr. Your company is willing to spend \$10,000 to solve the largest version of a certain problem it can.

a) If the best algorithm you know for solving this problem takes 2^n operations, where n is the size of the problem, what is the largest size you can solve?

b) Suppose that someone discovers an algorithm for the same problem that takes n^5 operations. What size problem can you solve now?

18. The original Towers of Hanoi problem was about a tower with 64 rings. Assume that you own the supercomputer in [17] and cost is of no concern. For how many years will you have to run the computer to complete the 64-ring problem? (Assume that each move is one computer operation.)

19. By the "relative effectiveness" of two algorithms for a specific instance of a problem, we simply mean the ratio of the number of operations the two algorithms need. (Divide the number of operations for the *second* algorithm by the number for the first, so that a ratio greater than 1 means the first algorithm is more effective.) Suppose that you have two algorithms for alphabetizing n names; the first takes $10n \log_2 n$ operations and the second takes $\frac{1}{2}n^2$.

a) What is the relative effectiveness of the first to the second if $n = 100$?

b) If $n = 10{,}000$?

20. Suppose that you have two algorithms for the same task. Let the execution time (in seconds) for the first algorithm be $n^2/10+n/5+1/3$ and let that for the second be $50n \log_2 n$.

a) Calculate the execution time for each algorithm for $n = 2^k$, $k = 0, 1, 2, \ldots, 15$.

b) Suppose that you knew that you had to do the task for which these two algorithms were written for 10 different but unknown (because they might depend on some other, not yet performed calculation) values of n which, with equal probability, might be any value from 1 to 50,000. Which algorithm would you choose?

9.4 Finite Differences and Factorial Functions

The natural topic to discuss after sequences is series since a **series** is just the sum of the terms in a sequence. And this topic is what we will discuss—in Section 9.5. First, however, in this section we are going to introduce some topics that are closely related to our discussion of sequences. These topics are also basic to a discussion of series and their summation and also form one of the essential bridges joining discrete and continuous mathematics. The main topic of this section—finite differences—is closely related to the subject of Chapter 5: difference equations.

At one time, finite differences played a major role in numerical analysis, the branch of mathematics whose chief concern is to solve problems in continuous mathematics by reformulating them in discrete terms. Numerical analysis, therefore, lies right at the interface between discrete and continuous mathematics. Although the role of differences in numerical analysis is much reduced from what it once was, differences are important to us for two reasons:

1. They provide an important tool for the evaluation of sums in closed form.
2. They provide a striking parallel to the calculus of polynomials and to the fundamental theorems linking differentiation and integration. If you have already studied calculus, you will see these parallels clearly; if you haven't, you will find the material of this section a useful entrée to calculus when you do study it.

Definition 1. Given a sequence $\{a_n\}$ defined on N, the **forward difference sequence** $\{\Delta a_n\}$ or, as we'll sometimes write it, $\Delta\{a_n\}$, is the sequence with domain N defined by

$$\Delta a_n = a_{n+1} - a_n.$$

The symbol Δ is called the forward difference operator, or just the **difference operator**, with "operator" implying that Δ "operates" on what follows it to transform it into something new.

EXAMPLE 1 If $a_n = c$, then

$$\Delta a_n = c - c = 0.$$

That is, the difference sequence of any constant sequence is the zero sequence. ∎

EXAMPLE 2 If $a_n = n$, then $\Delta a_n = (n+1) - n = 1$. More generally, if $\{a_n\}$ is the general linear sequence, $a_n = An+B$, where A and B are constants, then

$$\Delta a_n = [A(n+1) + B] - [An + B] = A(n+1-n) + (B-B) = A.$$

In short, the difference of any linear sequence is a constant sequence. ∎

We can generalize the result in Example 2 to illustrate an important principle: The difference operator "respects" sums and constant multiples. Here is a precise statement:

Theorem 1. If $\{a_n\}$, $\{b_n\}$, and $\{c_n\}$ are sequences such that

$$a_n = Bb_n + Cc_n$$

for all n, where B and C are constants, then

$$\Delta a_n = B\Delta b_n + C\Delta c_n. \tag{1}$$

In particular,

$$\Delta(b_n + c_n) = \Delta b_n + \Delta c_n \qquad \text{[The sum rule]}$$

and

$$\Delta(Bb_n) = B\Delta b_n. \qquad \text{[The constant multiple rule]}$$

PROOF Direct calculation gives

$$\begin{aligned}\Delta a_n &= a_{n+1} - a_n \\ &= [Bb_{n+1} + Cc_{n+1}] - [Bb_n + Cc_n] \\ &= B(b_{n+1} - b_n) + C(c_{n+1} - c_n) \\ &= B\Delta b_n + C\Delta c_n. \ \blacksquare\end{aligned}$$

Any operator which satisfies Eq. (1), when applied to the weighted sum (i.e. linear combination) of two sequences, is called a **linear operator**.

EXAMPLE 2 Continued

In light of Theorem 1, we may repeat Example 2 again with less algebra. With

$$a_n = An + B = An + B1,$$

we have, because $\Delta\{n\} = \{1\}$ and $\Delta\{1\} = \{0\}$,

$$\Delta a_n = A1 + B0 = A. \ \blacksquare$$

From Theorem 1, it also follows that:

Corollary 2. The difference of any polynomial of degree k is a polynomial of degree $k - 1$.

Can you prove this [3a]? The converse—if a difference of a sequence is a polynomial of degree $k-1$, then the sequence must be a polynomial of degree k—is also true [3b].

EXAMPLE 3 Find Δc^n, where c is an arbitrary constant.

Solution By direct calculation

$$\Delta c^n = c^{n+1} - c^n = (c-1)c^n.$$

From this, what do you conclude about the difference of an arbitrary exponential function (i.e., a function of the form Ac^n)? ∎

EXAMPLE 4 Calculate $\Delta n(n-1)(n-2)$.

Solution Usually one is told to rewrite every polynomial in standard form as a sum of powers: $\sum a_k n^k$. Mindlessly following this advice, you would get

$$n(n-1)(n-2) = n^3 - 3n^2 + 2n.$$

So

$$\begin{aligned}\Delta n(n-1)(n-2) &= [(n+1)^3 - n^3] - 3[(n+1)^2 - n^2] + 2[(n+1) - n] \\ &= (3n^2 + 3n + 1) - 3(2n+1) + 2 \\ &= 3n^2 - 3n \\ &= 3n(n-1).\end{aligned}$$

The answer, like the original form, is a simple factored form, so maybe the sum of powers form isn't such a good idea in this case. As we'll point out following Theorem 3, there's a much easier way to solve this problem. ∎

A sequence is simply a function whose domain is N. So, can we apply the difference operator to any function $f(x)$? Sure, as follows:

$$\Delta f(x) = f(x+1) - f(x).$$

This formulation is convenient if you want to emphasize functions defined not on N but on R. Essentially everything that follows in this section applies equally well to sequences and to general functions. For instance, when we talk about a polynomial P, we will, as convenient, mean either the function $P(x)$ or the sequence $P(0)$, $P(1)$, ...

Example 5 illustrates an important property of the difference operator.

EXAMPLE 5 Find Δn^k where n is the variable and k is a positive integer.

Solution We know from Corollary 2 that the answer is a polynomial of degree $k-1$. But what polynomial? By direct calculation using the binomial theorem, we obtain

$$\begin{aligned}(n+1)^k - n^k &= \left[\sum_{i=0}^{k} \binom{k}{i} n^i\right] - n^k \\ &= \sum_{i=0}^{k-1} \binom{k}{i} n^i. \quad \blacksquare\end{aligned}$$

The result in Example 5 is a bit unpleasant. The difference of a power function of degree k is a polynomial with *all* powers from 0 to $k-1$. It would be nice if there were an alternate set of building blocks which interact more nicely with differences. Fortunately, there is.

Definition 2. For any real number x and any positive integer k, define the kth **falling factorial function** $x_{(k)}$, as

$$\begin{aligned}x_{(k)} &= x(x-1)(x-2)\cdots(x-k+1) \qquad [k \text{ factors in all}] \\ &= \prod_{i=1}^{k}(x+1-i).\end{aligned}$$

By convention, $x_{(0)} = 1$.

Note that the last factor is $x-k+1$ not $x-k$. Also note that, when x is a positive integer $j \geq k$, then $j_{(k)} = j!/(j-k)! = P(j,k)$, or the number of permutations of j things taken k at a time (see Section 4.4). This form shows why the quantity defined in Definition 2 is called the kth falling factorial. In general, however, x need not be an integer, and $x_{(k)}$ is just a polynomial of degree k.

Caution. Our notation is not the only one used for these functions. Numerical analysis books often use $x^{(k)}$, but we will save this notation for **rising factorials** [15]. Another notation is $(x)_k$, but this is confusing whenever x is an expression which itself involves parentheses.

Theorem 3.

$$\Delta x_{(k)} = k x_{(k-1)}$$

with the difference taken with respect to x. In words, the difference of the kth falling factorial is k times the $(k-1)$st falling factorial.

PROOF From the definition of the difference operator for functions,

$$\Delta x_{(k)} = (x+1)_{(k)} - x_{(k)}.$$

The two products on the right-hand side have the same factors except for the first factor of $(x+1)_{(k)}$ and the last factor of $x_{(k)}$. Putting all the common factors to the right (after the second equals sign below), we find that

$$\begin{aligned}(x+1)_{(k)} - x_{(k)} &= [(x+1)x(x-1)\cdots(x+1-k+1)] \\ &\quad - [x(x-1)\cdots(x+1-k-1)(x-k+1)] \\ &= [(x+1)-(x-k+1)][x(x-1)\cdots(x+1-k+1)] \\ &= kx_{(k-1)}. \quad \blacksquare\end{aligned}$$

Note. The proof isn't right for $k=0$, because there are no factors to pull out. In fact the statement of the proof doesn't immediately make sense for $k=0$ because we haven't defined $x_{(-1)}$. But it doesn't matter: However we might define $x_{(-1)}$, the formula in Theorem 3 says that $\Delta x_{(0)} = 0$, which is correct from the definition of $x_{(0)}$. Thus we may use the identity of the theorem for any nonnegative k.

Look back at Example 4, which was actually an illustration of Theorem 3. The function $n(n-1)(n-2)$ is just $n_{(3)}$, and $3n(n-1)$ is $3n_{(2)}$.

Can we define differences of differences? Yes, by the inductive definition:

$$\Delta^k a_n = \begin{cases} a_{n+1} - a_n, & \text{if } k = 1; \\ \Delta[\Delta^{k-1} a_n], & \text{if } k > 1. \end{cases}$$

The operator Δ^1 is just another name for Δ. Naturally, we call Δ^k the **kth-order difference operator**. We'll explore some of the properties of higher order differences in the problems [4–5].

Antidifferencing

All right, what operation "undoes" differencing? That is, suppose we have a sequence $\{a_n\}$. Can we find a sequence $\{b_n\}$ such that $\Delta\{b_n\} = \{a_n\}$? Any such sequence $\{b_n\}$ is called an **antidifference sequence** of $\{a_n\}$, or just an **antidifference**.

Theorem 3 actually is a theorem about antidifferences. It says that

$$x_{(k-1)} = \frac{\Delta x_{(k)}}{k}.$$

This equation doesn't quite say yet that $\{x_{(k)}/k\}$ is an antidifference of $\{x_{(k-1)}\}$ because the difference operator doesn't operate on the $1/k$ factor. But Theorem 1 allows us to write

$$\frac{\Delta x_{(k)}}{k} = \Delta\left[\frac{x_{(k)}}{k}\right].$$

Therefore Theorem 3 says, in effect:

An antidifference of $\{x_{(k-1)}\}$ is $\left\{\dfrac{x_{(k)}}{k}\right\}$.

EXAMPLE 6 Find antidifferences for $\{a_n\}$ when $a_n = n$ and when $a_n = n^2$.

Solution So far we know only two large classes of difference–antidifference pairs: n^k and its polynomial difference (Example 5), and $\Delta n_{(k)}$ and its falling factorial difference (Theorem 3, as restated just before this example). The first class is quite complicated. Can we perhaps recognize n and n^2 as combinations of the factorial functions and use the second class? Yes, we can. Since $n = n_{(1)}$, an antidifference is

$$\tfrac{1}{2}n_{(2)},$$

which is our old friend, the sum of the first n integers. And as

$$n^2 = n(n-1) + n = n_{(2)} + n_{(1)},$$

using the preceding result and Theorem 1, an antidifference is

$$\tfrac{1}{3}n_{(3)} + \tfrac{1}{2}n_{(2)}. \blacksquare$$

We have carefully been saying *an* antidifference because we can add a constant to any antidifference without changing its difference (why?). Could two antidifferences of the same difference differ by other than a constant? No, as the following argument shows. Let $\{A_n\}$ be any antidifference of $\{a_n\}$ so that $\Delta A_n = a_n$. Therefore

$$A_1 = A_0 + \Delta A_0 = A_0 + a_0,$$

where A_0 can be chosen arbitrarily. Similarly,

$$\begin{aligned} A_2 &= A_1 + \Delta A_1 \\ &= A_0 + \Delta A_0 + \Delta A_1 && \text{[From equation for } A_1\text{]} \\ &= A_0 + a_0 + a_1, \end{aligned}$$

and

$$\begin{aligned} A_3 &= A_2 + \Delta A_2 \\ &= A_0 + \Delta A_0 + \Delta A_1 + \Delta A_2 && \text{[From equation for } A_2\text{]} \\ &= A_0 + a_0 + a_1 + a_2. \end{aligned}$$

Clearly, induction [13] should enable us to prove

Theorem 4. For any sequence $\{a_n\}$ and any number A, there is a unique sequence $\{A_n\}$ such that $\Delta A_n = a_n$ and $A_0 = A$. Specifically,

$$\begin{aligned} A_n &= A + \sum_{i=0}^{n-1} \Delta A_i && (2) \\ &= A + \sum_{i=0}^{n-1} a_i. \end{aligned}$$

Thus if we sum a sequence of differences of another sequence, we get back the original sequence except for an arbitrary constant. Or: *Summing undoes differencing.* More precisely, summing plus an arbitrary constant undoes differencing. The question we started with is answered. The only arbitrariness in an antidifference sequence is the constant A. This constant plays the same role in the difference calculus that the constant of integration serves in the differential calculus. Note, by the way, that the sum in Theorem 4 is to $n-1$, not n.

One interpretation of Theorem 4 is that it lets us find antidifferences by summing. Actually, we can seldom use Theorem 4 to find antidifferences that we don't otherwise know. The reason is that evaluating sums in closed form is, as you'll see in Section 9.5, generally quite difficult. Rather, it's quite the opposite interpretation of Theorem 4 which is useful: It lets us sum by using antidifferences. That is, if we know an antidifference of something we want to sum, then summing is easy. Theorem 5 is an almost immediate consequence of Theorem 4 and shows that more general sums than that in Eq. (2) can be calculated if we know an antidifference.

Theorem 5. Let $\{A_n\}$ be *any* antidifference sequence of $\{a_n\}$. Then, for $0 \le k \le l$,

$$\sum_{i=k}^{l} a_i = A_{l+1} - A_k. \tag{3}$$

PROOF Theorem 4 tells us that for some constant A (we don't care what it is), Eq. (2) is satisfied. Therefore

$$\begin{aligned} A_{l+1} - A_k &= \left[A + \sum_{i=0}^{(l+1)-1} a_i\right] - \left[A + \sum_{i=0}^{k-1} a_i\right] \\ &= \sum_{i=k}^{l} a_i. \quad \blacksquare \end{aligned}$$

Introducing the obvious notation for the antidifference, $\Delta^{-1}a_i = A_i$, we may rewrite Eq. (3) as

$$\sum_{i=k}^{l} a_i = \Delta^{-1}a_i \Big|_{k}^{l+1}, \tag{4}$$

where the long vertical line with superscript $l+1$ and subscript k means to evaluate the term to its left at these two values of the subscript and subtract.

Unlike most mathematical notation, which names a specific object, the Δ^{-1} notation is unspecific: It stands for any one of the (infinitely many) antidifference sequences of $\{a_n\}$. Therefore $\Delta^{-1}\{a_n\}$ is called the **indefinite antidifference sequence.** It is analogous to the indefinite integral in calculus.

EXAMPLE 7 Use Theorem 5 to find

$$\sum_{i=0}^{n} i \quad \text{and} \quad \sum_{i=0}^{n} i^2.$$

Solution We did all the necessary work in Example 6. An antidifference of $\{i\}$ is $\{i_{(2)}/2\}$. Therefore, using Eq. (3) or Eq. (4), we find that the first sum is

$$\frac{(n+1)_{(2)}}{2} - 0_{(2)} = \frac{(n+1)n}{2} - 0 = \frac{(n+1)n}{2},$$

which is an old friend. Similarly, $\{i^2\}$ has an antidifference $\{i_{(3)}/3 + i_{(2)}/2\}$. Therefore the second sum is

$$\frac{(n+1)_{(3)}}{3} + \frac{(n+1)_{(2)}}{2},$$

leaving out the zero terms. A little algebra simplifies this expression to

$$\frac{n(2n+1)(n+1)}{6},$$

which we've also derived earlier, but with much more work. ∎

We'll return to the use of antidifferences to calculate sums in Section 9.5.

Finite Differences and Difference Equations

It's easy to make up equations involving differences. Here are two:

$$\Delta a_n - a_n = 0, \quad n \geq 0; \qquad a_0 = 1; \tag{5}$$

and

$$b_n = 5n + 3(\Delta b_n)^2 + 2\Delta^2 b_n, \quad n \geq 0; \qquad b_0 = 4, \quad b_1 = 2. \tag{6}$$

Surely such equations should be called difference equations. So why did we use that name for the equations presented in Chapter 5? (Note that the index on the variable is always the same (n) in Eqs. (5) and (6) whereas in the recurrences in Chapter 5, there are no Δ's and a change of index is the essential aspect of these recurrences.) Before we try to answer this question, let's agree to call equations such as Eqs. (5) and (6) Δ-equations in contrast to the recurrences which we called difference equations in Chapter 5. (In practice, both kinds of equations are called difference equations.) The answer to our question is contained in the following theorem.

Theorem 6. Every Δ-equation is equivalent to a difference equation and vice versa.

That is, for every Δ-equation there is a difference equation with the same solution and vice versa. So, if we're given a Δ-equation, we simply turn it into a difference equation and solve it with the theory we already have.

We'll only sketch the outline of a proof of Theorem 6 and leave the rest to the problems [21, 22]. It's easy to show that every Δ-equation is equivalent to a difference equation. We know that

$$\Delta a_n = a_{n+1} - a_n, \tag{7}$$

so we can always eliminate Δ from the equation in favor of varying the index. Similarly, using the definition of higher differences, we have

$$\begin{aligned} \Delta^2 a_n &= \Delta(\Delta a_n) \\ &= \Delta(a_{n+1} - a_n) \\ &= (a_{n+2} - a_{n+1}) - (a_{n+1} - a_n) \\ &= a_{n+2} - 2a_{n+1} + a_n, \end{aligned} \tag{8}$$

so we can eliminate Δ^2. Proceeding inductively (can you spot the pattern?), we can eliminate every term $\Delta^k a_n$. As these eliminations are simply rewritings, they do not change the solutions.

EXAMPLE 8 Find the difference equations equivalent to Eq. (5) and Eq. (6).

Solution:

a) When we use Eq. (7), Eq. (5) becomes

$$\begin{aligned} \Delta a_n - a_n &= (a_{n+1} - a_n) - a_n \\ &= a_{n+1} - 2a_n, \end{aligned}$$

so the equivalent difference equation is

$$a_{n+1} - 2a_n = 0, \qquad n \geq 0.$$

The initial condition is unchanged.

b) When we use Eqs. (7) and (8), Eq. (6) becomes

$$b_n = 5n + 3(b_{n+1} - b_n)^2 + 2(b_{n+2} - 2b_{n+1} + b_n), \qquad n \geq 0. \tag{9}$$

Again, the initial conditions are unchanged. ∎

Both results in Example 8 may look a little strange because n is the lowest index instead of the highest as it was generally in Chapter 5, but there's nothing wrong with that. Note that the Δ-equation in Eq. (6) and the difference equation in Eq. (9) are both second-order and nonlinear.

What about going the other way? Note first that, using Eq. (7), we may write

$$a_{n+1} = a_n + \Delta a_n. \tag{10}$$

That shows how to eliminate a_{n+1}. To eliminate a_{n+2} from a difference equation, raise the index in Eq. (10) by 1 and substitute it into itself as follows:

$$\begin{aligned} a_{n+2} &= a_{n+1} + \Delta a_{n+1} \\ &= (a_n+\Delta a_n) + \Delta(a_n+\Delta a_n) \\ &= a_n + 2\Delta a_n + \Delta^2 a_n. \end{aligned}$$

Again, by induction, it follows that we can rewrite every term using Δ and a_n. Thus the translation method consists of rewriting the difference equation so that n is the lowest index and then using these eliminations.

EXAMPLE 9 Translate $r_n = 3r_{n-1}+n$ into a Δ-equation.

Solution Substituting Eq. (10) into

$$r_{n+1} = 3r_n + n + 1,$$

we obtain

$$r_n + \Delta r_n = 3r_n + (n+1), \qquad \text{or} \qquad \Delta r_n = 2r_n + n + 1. \blacksquare$$

Finite Differences and Calculus

"Finite" appeared in the title of this section but not again until now. Why? The answer is historical: to contrast these differences with "infinitesimal differences". We can "uglify" Δa_n without changing its value by writing it as

$$(a_{n+1}-a_n)/h, \qquad \text{where } h = 1.$$

This uglification, like many, has some use. You now see that Δa_n can be interpreted as not only the *amount* of change from n to $n+1$, but also as the *average rate* of change. More important, using functions instead of differences, h need not be 1, as in $[f(x+h) - f(x)]/h$. If we take h to be *infinitesimal*—infinitely tiny but not 0—we get the idea of the *derivative* in calculus. Indeed, differencing and antidifferencing play precisely analogous roles in discrete mathematics to the roles of differentiation and integration in calculus. Also, the falling factorial functions play precisely the same role with respect to differencing and antidifferencing that the power functions play with respect to differentiation and integration.

Problems: Section 9.4

1. Compute the following differences where n is the variable.

a) $\Delta(n-1)^2$ b) $\Delta n!$ c) $\Delta 2^n$
d) $\Delta n2^n$ e) $\Delta C(n,k)$ f) $\Delta 1/n$
g) $\Delta^2 n2^n$ h) $\Delta^2 C(n,k)$ i) $\Delta^j C(n,k)$
j) $\Delta^k n2^n$ k) $\Delta^k n^k$ l) $\Delta^k 1/n$

2. What is $\Delta c_{(n)}$? (c fixed, n variable)

3. a) Prove Corollary 2, which says that the difference of a polynomial of degree k is a polynomial of degree $k-1$.

b) Prove the converse that, if the difference of a function is a polynomial of degree $k-1$, then the function must be a polynomial of degree k.

4. Find and prove a formula for $\Delta^k a_n$ in terms of $a_{n+k}, a_{n+k-1}, \ldots$. *Hint*: Use the inductive paradigm.

5. a) Find and prove a formula for a_{n+2} in terms of

$\Delta^2 a_n$, Δa_n, and a_n.

b) Now find a formula for a_{n+3}.

c) Find a generalization and prove it.

6. If you succeeded with [4] and [5], you were probably struck by the similarities between them. The following idea makes both problems easier to do, and helps explain the similarities.

The parallel between differences and recursions has been somewhat obscured because we have represented differences by an operator Δ, but we have represented recurrences using varying subscripts. Let's define the (forward) **shift operator**, S, by

$$Sa_n = a_{n+1}.$$

That is, S acts on sequences by shifting every subscript by 1. Similarly,

$$S^2 a_n = S(Sa_n) = Sa_{n+1} = a_{n+2}.$$

For instance, using this operator, the recursion

$$r_n = r_{n-1} + r_{n-2}$$

becomes

$$S^2 r_{n-2} = Sr_{n-2} + r_{n-2}.$$

Furthermore, we have

$$\Delta a_n = Sa_n - a_n = (S-I)a_n, \qquad (11)$$

where $S-I$ refers to the operation of first applying S and then subtracting the effect of applying the **identity operator** (i.e., leaving a_n alone). Rearranging Eq. (11) we also obtain

$$Sa_n = (\Delta+I)a_n$$

What does all of this suggest about $\Delta^k a_n$? About relating a_{n+k} to a_n using Δs? Solve [4] and [5] again.

7. In some ways, an even better set of functions for forward differences than the factorial functions are the binomial coefficients viewed as polynomials: $P_k(x) = C(x,k)$. How are they better?

8. Evaluate the following sums using antidifferences and falling factorials. The limits in each case are 1 to n.

a) $k^2 - k$ b) $k^2 + k$
c) $k(k-2)(k-4)$

9. Evaluate the sum of the following in closed form by finding an antidifference of each. The limits in each case are 1 to n.

a) $1/[k(k+1)]$
b) $1/[m(m+1)(m+2)]$
c) $j(j!)$

10. a) Show that Δ^{-1} satisfies the same linearity condition shown for Δ in Theorem 1. The main difficulty here is to state appropriately the results you want. Remember, $\Delta^{-1}\{s_n\}$ is not uniquely defined. If I pick one sequence $\Delta^{-1}\{a_n\}$, you pick one sequence $\Delta^{-1}\{b_n\}$, and Jill picks one sequence $\Delta^{-1}\{a_n+b_n\}$, the sum of mine and yours probably *won't* equal hers.

b) We used the linearity of Δ^{-1} several times in the text. Find them.

11. In the solution of Example 6, we suggested that it's harder to find antidifferences using the formula for Δn^k than using the formula for $\Delta n_{(k)}$. Check this suggestion out as follows:

a) Find $\Delta^{-1} n$ again by piecing together $\Delta n^2 = 2n+1$ and $\Delta n = 1$.

b) Find $\Delta^{-1} n^2$ again by representing n^2 as a linear combination of Δn^3, Δn^2, and Δn.

12. Calculate $\sum_{i=1}^m i^2$, using antidifferencing but without using factorial functions.

13. Prove Theorem 4.

14. Calculate the following:

a) $5_{(3)}$ b) $4_{(6)}$ c) $(2/3)_{(4)}$
d) $(-1/4)_{(3)}$ e) $47_{(45)}$

15. The rising factorial $x^{(k)}$ is defined by

$$\prod_{i=1}^{k}(x-1+i)$$

for positive k with $x^{(0)} = 1$.

a) Express $x^{(k)}$ in terms of factorials.

b) State and prove the theorem analogous to Theorem 3 for rising factorials.

c) Use the result in b) to state the formula for the antidifference of a rising factorial.

16. Repeat the calculations of Example 6 using rising factorials. Verify that your results are the same as those in the text.

17. Calculate:

a) $4^{(5)}$ b) $(-3)^{(7)}$

c) $(-1/2)^{(4)}$ d) $(1.2)^{(3)}$

18. Find a relation for $x^{(k)}$ in terms of falling factorials. Verify that your relation is correct by repeating the calculations of [17] using the falling factorial form.

19. Solve the following Δ-equations.

a) $\Delta a_n = na_n,\ n \geq 0; \quad a_0 = 1.$

b) $\Delta^2 a_n - \Delta a_n - 6a_n = 0,\ n \geq 0;\ a_0 = 2,\ a_1 = 3.$

c) $r_n^2 + r_n \Delta r_n = 1,\ n \geq 0; \quad r_0 = 2.$

20. Translate the following recurrences into equivalent Δ-equations.

a) $f_{n+1} = f_n + f_{n-1}$

b) $r_n = r_{n-1} + 2r_{n-2}$

c) $r_n = r_{n-1} r_{n-2}$

21. Use the result of [4] to show that every Δ-equation is equivalent to a difference equation.

22. Use the result of [5] to show that every difference equation is equivalent to a Δ-equation.

9.5 Infinite Series and Their Summation

A series is the sum of (some or all) the terms of a sequence. Although an important part of discrete mathematics, series also play a crucial role in calculus. Therefore this topic, too, forms a natural bridge from discrete to continuous mathematics.

You have encountered series several times in this book, including the previous section. For example,

- arithmetic and geometric series were discussed in Sections 5.2 and 5.3, and
- at various times, such as in analyzing Algorithm SEQSEARCH (Section 1.6), we needed to sum the first n terms of $\{k\}$, that is the sum:

$$\sum_{k=1}^{n} k.$$

In this section we will introduce some elementary facts about series, discuss briefly the question of the convergence of a series whose upper limit is ∞, and introduce the most powerful method of summing both finite and infinite series in closed form. Then we'll show how we can use the factorial functions introduced in Section 9.4 to sum a certain class of functions and to accelerate the convergence of a related class.

A geometric series (see Section 5.2) is the sum of all the terms of the sequence $\{ar^k\}$. More generally, suppose that we have a sequence $\{a_k\}$ and want to find the sum of the infinite series

$$\sum_{k=1}^{\infty} a_k. \tag{1}$$

Right here, it's worth stopping for a moment to discuss the notation in expression (1), because the summation notation there really is doing double duty. On the one hand, it expresses the *process* of adding the terms of a sequence. Now, as the sum has an infinite number of terms, we can't actually carry out the process. But, perhaps, if we could carry it out there would be some easily expressible result. Therefore (1) also represents the *result* (if there is one) of carrying out the process. When we say the "result (if there is one)", we are asking implicitly: Can (1) can be represented in closed form by some symbolic or finite numerical quantity? In what follows in this section we will be most interested in the result represented by (1). However, you shouldn't forget the process which is implicit in (1).

In an important sense we have already discussed the summation of (1). Where? In Section 9.2 about sequences, as the following definition makes clear.

Definition 1. Let

$$S_n = \sum_{k=1}^{n} a_k, \tag{2}$$

where S_n is called the *partial sum* of the series (1) (see also Section 5.2). Then the series (1) *converges* or *diverges* according to whether the *sequence* $\{S_n\}$ of partial sums converges or diverges.

Definition 1 allows us to reduce any question about the convergence of a *series* to a question about the convergence of a related *sequence*, namely, the sequence of partial sums. Thus while series and sequences are two very different kinds of beasts, they are not different species but rather relatives. If the sum (1) does exist, that is, if the series converges, then Definition 1 says that

$$\lim_{n\to\infty} S_n = \lim_{n\to\infty} \sum_{i=1}^{n} a_i = \sum_{i=1}^{\infty} a_i. \tag{3}$$

How does Definition 1 apply to geometric series? The formula for the partial sum of a geometric series (with limits from 0 to n) is

$$\frac{a - ar^{n+1}}{1-r}.$$

If you don't remember this formula, refer back to Section 5.3. Now what happens as $n \to \infty$? In [11] of Section 9.2, we asked you to show that, if $|r| < 1$, the limit of r^n is 0. From this result, it follows that the sum of an infinite geometric series is $a/(1-r)$. But when $|r| \geq 1$, you should see that $\{r^n\}$ diverges and therefore so does the corresponding geometric series.

EXAMPLE 1 What is the sum of the series

$$\sum_{k=1}^{\infty} \frac{1}{k(k+1)}? \tag{4}$$

Solution Let's begin by using the inductive paradigm and compute some partial sums:

n	Sum of the first n terms
1	$1/2$
2	$1/2 + 1/6 = 2/3$
3	$2/3 + 1/12 = 3/4$
4	$3/4 + 1/20 = 4/5$
5	$4/5 + 1/30 = 5/6$

It looks like the sum of the first n terms is $n/(n+1)$. If so, in the limit as $n \to \infty$, the sum would be 1. Could the result be so simple without some simple way to find it directly? Yes, in general; but, no, in this case. All we need to see is that

$$\frac{1}{k(k+1)} = \frac{1}{k} - \frac{1}{(k+1)}. \tag{5}$$

(To determine whether Eq. (5) is correct, put the right-hand side over a common denominator. If you've ever seen the method of partial fractions (see [7–12, Supplementary]), you'll know how to derive the right-hand side of Eq. (5) from the left.) Now, using Eq. (3) and substituting Eq. (5) into Eq. (4), we obtain

$$\begin{aligned}\sum_{k=1}^{\infty}\left[\frac{1}{[k(k+1)]}\right] &= \sum_{k=1}^{\infty}\left[\frac{1}{k} - \frac{1}{(k+1)}\right] = \lim_{n\to\infty}\sum_{k=1}^{n}\left[\frac{1}{k} - \frac{1}{(k+1)}\right] \\ &= \lim_{n\to\infty}\left[\left(1-\frac{1}{2}\right) + \left(\frac{1}{2}-\frac{1}{3}\right) + \left(\frac{1}{3}-\frac{1}{4}\right) + \cdots + \left(\frac{1}{n}-\frac{1}{n+1}\right)\right],\end{aligned}$$

so that every term subtracted [except the very last one—$1/(n+1)$] is canceled by the next term that is added. Such a series is said to *telescope* (see [9, Section 2.2]). Therefore we may write the limit as

$$\begin{aligned}\lim_{n\to\infty}\left(1-\frac{1}{n+1}\right) &= \lim_{n\to\infty} 1 - \lim_{n\to\infty}\left(\frac{1}{n+1}\right) \quad \text{[By (i) of Theorem 1, Section 9.2]} \\ &= 1 - 0 = 1.\end{aligned}$$

Later in this section we'll show you another way to calculate sums like that in Eq. (4). ∎

EXAMPLE 2 Does the series

$$\frac{1}{2}+\frac{1}{4}+\frac{1}{4}+\frac{1}{8}+\frac{1}{8}+\frac{1}{8}+\frac{1}{8}+\frac{1}{16}+\frac{1}{16}+\frac{1}{16}+\frac{1}{16}+\frac{1}{16}+\frac{1}{16}+\frac{1}{16}+\frac{1}{16}+\cdots$$

converge or diverge?

Solution This series consists of groups of 2^{k-1} terms, $k = 1, 2, 3, \ldots$, each of which has the value $1/2^k$. We tabulate some partial sums:

$$\begin{aligned} S_1 &= \frac{1}{2}, \\ S_3 &= 1, \\ S_7 &= \frac{3}{2}, \\ S_{15} &= 2, \\ S_{31} &= \frac{5}{2}, \\ S_{63} &= 3, \end{aligned}$$

and, in general,

$$S_{2^j-1} = \frac{j}{2}.$$

Therefore the sequence of partial sums grows without bound. So, by definition, the series diverges. ∎

Even though asking a question about the convergence or divergence of a series is the same as asking a question about the related sequence of partial sums, several tests of convergence apply specifically to series. You may well have encountered some of them in a calculus course. Here, by means of two further examples, we will give you a feel for some of these tests.

EXAMPLE 3 **The Harmonic Series**

Does the **harmonic series**

$$\sum_{i=1}^{\infty} \frac{1}{i} = 1 + \frac{1}{2} + \frac{1}{3} + \frac{1}{4} + \frac{1}{5} + \frac{1}{6} + \cdots$$

converge?

Solution Let's compare this series with the series in Example 2 with a 1 added in front:

$$1 + \frac{1}{2} + \frac{1}{3} + \frac{1}{4} + \frac{1}{5} + \frac{1}{6} + \frac{1}{7} + \frac{1}{8} + \frac{1}{9} + \frac{1}{10} + \cdots$$
$$1 + \frac{1}{2} + \frac{1}{4} + \frac{1}{4} + \frac{1}{8} + \frac{1}{8} + \frac{1}{8} + \frac{1}{8} + \frac{1}{16} + \frac{1}{16} + \cdots$$

Each term in the harmonic series is greater than or equal to the corresponding term in the second series. The second series grows without bound, so you should be able to prove [6] that the harmonic series does also. This is a particular case of a **comparison test**, which sometimes enables us to determine the convergence or divergence of a series by comparing it with a series whose convergence or divergence is known [7]. ∎

Although the harmonic series diverges, the difference between two successive partial sums $[1/(n+1)]$ gets arbitrarily small. Thus Example 3 shows that a test of the difference of two successive terms of a sequence cannot tell you if the sequence converges (see [19], Section 9.2).

We would also like to know how *fast* the harmonic series diverges because this series shows up in the analysis of many algorithms. That is, we would like to know the order of H_n, the sum of the first n terms of the harmonic series. [In earlier sections we called this partial sum $H(n)$.] Deriving this result is beyond us here and, in fact, requires calculus. It is, however, at least clear that H_n grows more slowly than n. Indeed, it grows much more slowly because, before very long, the $1/k$ term is quite small compared to 1. Therefore you shouldn't be surprised that it can be proved that

$$H_n = \mathrm{Ord}(\log n).$$

EXAMPLE 4 Does the **alternating harmonic series**

$$\sum_{k=1}^{\infty}(-1)^{k+1}\left(\frac{1}{k}\right) = 1-\frac{1}{2}+\frac{1}{3}-\frac{1}{4}\cdots$$

converge?

Solution The alternating harmonic series derives its name from the fact that the signs of successive terms alternate from + to −. The alternation means that every other term cancels, to some degree, the contribution of its predecessor. If we let a_k represent successive terms without signs (so all a_k are positive), the successive terms of the alternating harmonic series also have the property that

$$\lim_{k\to\infty} a_k = 0. \tag{6}$$

It is a theorem that any **alternating series** (i.e., one whose terms alternate in sign) converges if its terms decrease in magnitude and satisfy Eq. (6). Therefore the alternating harmonic series converges [17]. ∎

Alternating series whose terms decrease in magnitude to 0 have a property which makes it easy to compute the sum of the series within any specified error. Any time you have just added a positive term, the sum thus far is *greater than* the infinite sum because all subsequent *pairs* of terms have a negative sum. Similarly, each time you have just added a negative term, the sum thus far is *less than* the infinite sum because all subsequent pairs of terms have a positive sum. So you approach the true sum from above and below. The resulting pincers let you calculate the true sum as accurately as you want, as in Algorithm 9.1, ALTSERIES.

Do you see why the result is correct to within $\epsilon/2$ rather than ϵ? If a_1 can be positive or negative, Algorithm ALTSERIES must be modified a bit; we leave the details to a problem [19b].

For the alternating harmonic series, Algorithm ALTSERIES works as follows with $\epsilon = 0.05$:

Algorithm 9.1

AltSeries

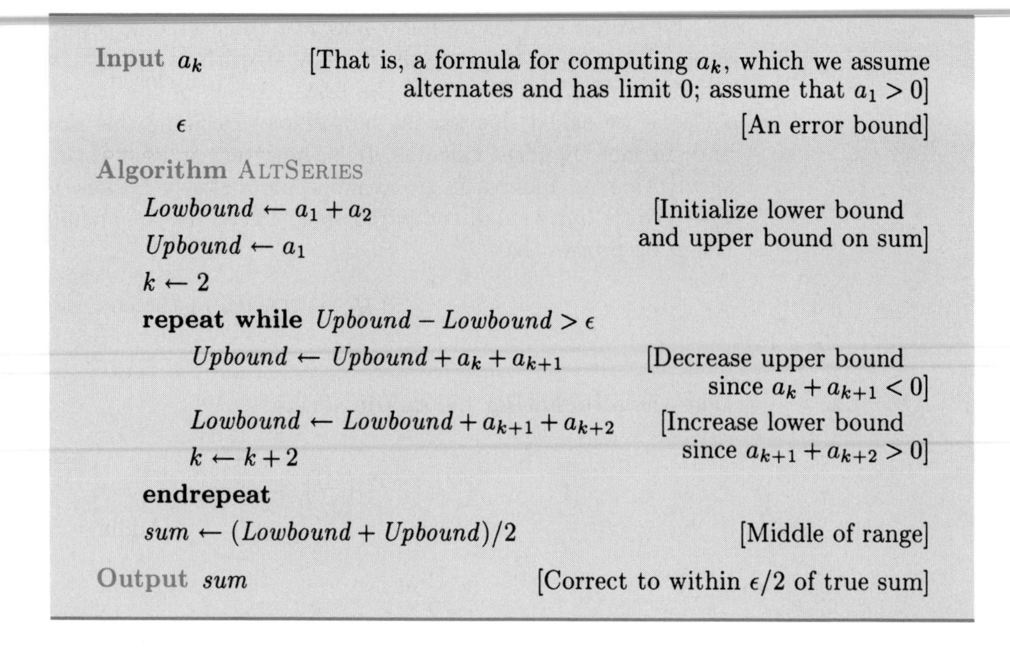

Input a_k [That is, a formula for computing a_k, which we assume alternates and has limit 0; assume that $a_1 > 0$]

ϵ [An error bound]

Algorithm AltSeries

```
Lowbound ← a_1 + a_2                         [Initialize lower bound
Upbound ← a_1                                 and upper bound on sum]
k ← 2
repeat while Upbound − Lowbound > ε
    Upbound ← Upbound + a_k + a_{k+1}        [Decrease upper bound
                                               since a_k + a_{k+1} < 0]
    Lowbound ← Lowbound + a_{k+1} + a_{k+2}  [Increase lower bound
    k ← k + 2                                  since a_{k+1} + a_{k+2} > 0]
endrepeat
sum ← (Lowbound + Upbound)/2                 [Middle of range]
```

Output *sum* [Correct to within $\epsilon/2$ of true sum]

k	Lowbound	Upbound	(Lowbound + Upbound)/2
(Initial values)	0.50000 (= 1−1/2)	1.00000	0.75000
2	0.58333	0.83333	0.70833
4	0.61666	0.78333	0.70000
6	0.63452	0.75952	0.69702
8	0.64563	0.74563	0.69563
10	0.65321	0.73654	0.69488
12	0.65870	0.73013	0.69442
14	0.66287	0.72537	0.69412
16	0.66613	0.72169	0.69391
18	0.66876	0.71876	0.69376

at which point the difference between Upbound and Lowbound is not greater than ϵ. The true value of this sum, by the way, is the logarithm of 2 to the base e ($\log_e 2$), which is 0.69315.... Note the excruciatingly slow convergence to the desired result. For much smaller values of ϵ, this slowness would become intolerable. Therefore in actual practice, methods are often needed to *accelerate* the convergence of alternating series.

The value of having a lower and upper bound on the sum of a series should not be underestimated. When you do not have two such bounds, as in the case of a series with all positive terms, you may be easily misled into thinking that a series converges when it doesn't, as in the case of the harmonic series (see Example 3).

Summation by Parts

Not very many infinite series can be summed in closed form. Geometric series are a notable exception. Thus to find the sum of most infinite series, we are forced to seek an approximate result by computation using algorithms such as Algorithm AltSeries. Still, there are enough series which can be summed in closed form to make it worthwhile to know the best method for doing so: **summation by parts.** Moreover, the more series we can sum in closed form, the more possibilities there are for finding approximate sums of other series by making use of series with known sums. We'll give an example of this approach at the end of this section.

To derive the formula for summation by parts we use the difference operator that we defined in Section 9.4. The essential observation is that, if $\{u_k\}$ and $\{v_k\}$ are two sequences, then

$$\begin{aligned}
\Delta(u_k v_k) &= u_{k+1}v_{k+1} - u_k v_k \qquad (7)\\
&= u_{k+1}v_{k+1} - u_k v_{k+1} + u_k v_{k+1} - u_k v_k \\
&\qquad\qquad [\text{Adding and subtracting } u_k v_{k+1}]\\
&= u_k(v_{k+1}-v_k) + v_{k+1}(u_{k+1}-u_k)\\
&= u_k\Delta v_k + v_{k+1}\Delta u_k.
\end{aligned}$$

Now let's rewrite Eq. (7) as

$$u_k\Delta v_k = \Delta(u_k v_k) - v_{k+1}\Delta u_k \qquad (8)$$

and sum both sides from 1 to m. We can simplify the sum of the first term on the right-hand side of Eq. (8) using Eq. (3) of Section 9.4:

$$\sum_{k=1}^{m} \Delta(u_k v_k) = u_{m+1}v_{m+1} - u_1 v_1.$$

When we use this equality, the sum of both sides of Eq. (8) becomes

$$\sum_{k=1}^{m} u_k\Delta v_k = u_{m+1}v_{m+1} - u_1 v_1 - \sum_{k=1}^{m} v_{k+1}\Delta u_k. \qquad (9)$$

which is the formula for summation by parts. Now suppose that the sequences $\{u_k\}$ and $\{v_k\}$ are such that

$$\lim_{k\to\infty} u_k v_k = 0. \qquad (10)$$

Then taking the limit of both sides of Eq. (9) as $m \to \infty$, we obtain

$$\sum_{k=1}^{\infty} u_k\Delta v_k = -u_1 v_1 - \sum_{k=1}^{\infty} v_{k+1}\Delta u_k. \qquad (11)$$

For those of you who know some calculus, you will recognize that Eq. (11) is the discrete analog of the integration by parts formula. It permits us to transform one series into another which may be easier to sum. We can illustrate this transformation most easily by examples.

EXAMPLE 5 Evaluate

$$\sum_{k=1}^{\infty} kr^k, \qquad 0 < r < 1.$$

Solution How do we choose u_k and v_k in order to use Eq. (11)? Making this choice is generally the hard part of using summation by parts. In this example we have essentially two choices. Either k or r^k is u_k and the other is v_k. Here, we'll try $u_k = k$ and we'll leave the other choice to a problem [23]. So we apply Eq. (11) with

$$u_k = k, \qquad \text{and} \qquad \Delta v_k = r^k.$$

Then

$$\Delta u_k = (k+1) - k = 1.$$

We want to find v_k which is the antidifference of r^k. The answer is given in Example 3, Section 9.4, as

$$\frac{r^k}{r-1}$$

With $u_k = k$ and $v_k = r^k/(r-1)$, is Eq. (10) satisfied? It is although we won't prove it. But your intuition should tell you that r^k gets small much more rapidly than k gets large. (Remember, $0 < r < 1$.)

Now, using Eq. (11), we obtain

$$\begin{aligned}
\sum_{k=1}^{\infty} kr^k &= \frac{-r}{r-1} - \sum_{k=1}^{\infty} \frac{r^{k+1}}{r-1} \\
&= \frac{r}{1-r} + \left(\frac{r}{1-r}\right) \sum_{k=1}^{\infty} r^k \quad \text{[Because } r/(1-r) \text{ doesn't depend on } k\text{]} \\
&= \frac{r}{1-r} + \left(\frac{r}{1-r}\right)\left(\frac{r}{1-r}\right) \quad \text{[Sum of geometric series]} \\
&= \frac{r(1-r) + r^2}{(1-r)^2} \\
&= \frac{r}{(1-r)^2}.
\end{aligned}$$

From this example you should be able to see how to sum $k^j r^k$ for $j > 1$ [26]. ∎

The series in Example 5 is one you've met before, although in a slightly different form. In Section 6.5 [see Eq. (8)], we derived the formula for the mean of the negative binomial distribution as

$$p \sum_{k=1}^{\infty} k(1-p)^{k-1} = \left(\frac{p}{1-p}\right) \sum_{k=1}^{\infty} k(1-p)^k.$$

Now, identifying $(1-p)$ with r, you can see that this sum is just $p/(1-p)$ times the sum in Example 5, or

$$\left(\frac{p}{1-p}\right)\left(\frac{1-p}{p^2}\right) = \frac{1}{p},$$

which is the result we gave in Section 6.5.

Here is a second example which illustrates the power of summation by parts.

EXAMPLE 6 Evaluate

$$\sum_{k=2}^{n} \frac{H_k}{k(k-1)},$$

where H_k is the kth harmonic number.

Solution We can solve this problem by summation by parts because the difference of two harmonic numbers has a simple form and because, suitably written, the factor $1/[k(k-1)]$ has the form of a difference. It is natural, therefore, to choose

$$u_k = H_k,$$

because then

$$\Delta u_k = H_{k+1} - H_k = \frac{1}{k+1}$$

(why?) and also to choose

$$\Delta v_k = \frac{1}{k(k-1)}$$

because, since

$$\frac{1}{k(k-1)} = \frac{1}{k-1} - \frac{1}{k}, \tag{12}$$

and it follows that we can choose

$$v_k = \frac{-1}{k-1}.$$

(We'll show you a way to derive this result directly just below.) Using Eq. (9) (replacing m with n and subscript 1's with subscript 2's because the lower limit in the sum is 2), the sum we wish to evaluate becomes

$$\sum_{k=2}^{n} \frac{H_k}{k(k-1)} = -\frac{H_{n+1}}{n} + H_2 + \sum_{k=2}^{n} \left(\frac{1}{k}\right)\left(\frac{1}{k+1}\right). \tag{13}$$

But, using Eq. (12), the series on the right telescopes and the right-hand side of Eq. (13) becomes

$$-\frac{H_{n+1}}{n} + \frac{3}{2} + \frac{1}{2} - \frac{1}{n+1} = 2 - \frac{1}{n+1} - \frac{H_{n+1}}{n},$$

where we have made use of the fact that $H_2 = 1+\frac{1}{2} = \frac{3}{2}$.

What happens as $n \to \infty$? Because, as we argued earlier, H_n grows as $\log n$, it's plausible that the limit of H_{n+1}/n is 0, as is the limit of $1/(n+1)$ also. Therefore, with the upper limit ∞, the sum is just 2. As you might expect (why?), the series converges rather slowly, with the first five terms summing to 1.425 and the first ten to 1.635. In a problem [37], we consider another approach to summing this series. ∎

It should be clear from these two examples that, to apply summation by parts, you must have a Δv_k which can be *inverted*, that is, for which you can readily find the corresponding v_k. In other words, we need difference–antidifference pairs. Table 9.2 contains a number of such pairs. In effect, in Section 9.4 we derived all but the last one, which we consider in a problem [24].

TABLE 9.2
$\Delta v_k - v_k$ Pairs for Summation by Parts

Δv_k	v_k	
a	ak	
a^k	$\dfrac{a^k}{a-1}$	$a \neq 1$
$\dfrac{1}{a^k}$	$\dfrac{1}{(1-a)a^{k-1}}$	$a \neq 1$
$k_{(j)}$	$\dfrac{k_{(j+1)}}{j+1}$	$j \neq -1$
$\binom{k}{j}$	$\binom{k}{j+1}$	

Note: a is a nonzero real constant, and j is an integer constant.

Summation of Rational Functions

Suppose that we have a series of the form

$$\sum_{n=j}^{1} \frac{1}{(n+r)_{(k)}}, \tag{14}$$

where $n_{(k)}$ is a falling factorial as defined in Section 9.4 and r is any constant. From Theorem 3, Section 9.4, we know that $(n+r)_{(k+1)}/(k+1)$ is an antidifference of $(n+r)_{(k)}$. However, this fact doesn't seem to do us any good here because the $(n+r)_{(k)}$ is in the denominator in (14). But all is not lost. Let's define falling factorials for negative values of the subscript as

$$n_{(-k)} = \frac{1}{(n+k)_{(k)}}, \qquad k > 0.$$

We get a useful alternative form of this definition by replacing n with $n-k$:

$$(n-k)_{(-k)} = \frac{1}{n_{(k)}}. \tag{15}$$

For example, the definition says that

$$4_{(-3)} = \frac{1}{7_{(3)}} = \frac{1}{7\cdot 6\cdot 5} = \frac{1}{210},$$

which we can also get from Eq. (15) with $n = 7$ and $k = 3$.

Why did we define negative falling factorials this way? (Remember, we *could* have defined them any way we wanted.) Because with this definition, Theorem 3, Section 9.4, is true for negative k also. That is, applying the difference operator to n and keeping k fixed,

$$\Delta n_{(k+1)} = (k+1)n_{(k)}$$

for *all* k (even $k = -1$). Or, alternatively, replacing k with $-k$ and n with $n-k$, we obtain

$$\Delta(n-k)_{(-k+1)} = (-k+1)(n-k)_{(-k)}, \tag{16}$$

a form we'll soon find convenient. You should be able to verify Eq. (16) for negative k using the definition in Eq. (15) [28]. Equation (16) says that $(n-k)_{(-k+1)}/(-k+1)$ is an antidifference of $(n-k)_{(-k)}$ (except when $k = 1$—why?). Using Eqs. (15) and (16) together, we may calculate the antidifference of $1/n_{(k)}$ as follows, when $k \neq 1$:

$$\begin{aligned}
\Delta^{-1}\left[\frac{1}{n_{(k)}}\right] &= \Delta^{-1}(n-k)_{(-k)} && \text{[From Eq. (15)]}\\
&= \frac{(n-k)_{(-k+1)}}{-k+1} && \text{[From Eq. (16)]}\\
&= \frac{(n-1-k+1)_{(-k+1)}}{-k+1}\\
&= \left(\frac{1}{-k+1}\right)\left(\frac{1}{(n-1)_{(k-1)}}\right). && \text{[From Eq. (15)]}
\end{aligned}$$

In particular, with $k = 2$, we have the result that $1/[n(n-1)] = 1/n_{(2)}$ has the antidifference $-1/(n-1)_{(1)} = -1/(n-1)$, a result we used in Example 6. This result also makes it easy to do the calculation in Example 1 [29b].

Another convenient form of Eq. (16) is found by replacing n with $n+r$ (with r a fixed integer) to obtain $(n+r-k)_{(-k+1)}/(-k+1)$ as an antidifference of $(n+r-k)_{(-k)}$. We'll use this form to sum series such as (14).

EXAMPLE 7 Calculate

$$\sum_{n=2}^{\infty} \frac{1}{(n-1)n(n+1)}$$

Solution From the definition of the factorial function for negative k, we have

$$\frac{1}{(n-1)n(n+1)} = \frac{1}{(n+1)_{(3)}} = (n-2)_{(-3)}$$

As $(n-2)_{(-3)} = (n+1-3)_{(-3)}$ has an antidifference $(n+1-3)_{(-2)}/(-2)$, and using Eq. (3) or (4) from Section 9.4, we get:

$$\sum_{n=2}^{l}(n-2)_{(-3)} = -\left(\frac{1}{2}\right)[(l+1+1-3)_{(-2)} - (2+1-3)_{(-2)}].$$

Now, you might think that the second term $(2+1-3)_{(-2)} = 0_{(-2)}$ is just 0. But it's not. Using our definition of the factorial function for negative arguments, we have

$$0_{(-2)} = \frac{1}{2_{(2)}} = \frac{1}{(2\times 1)} = \frac{1}{2}.$$

Now what happens to the first term as the upper limit l goes to ∞? Because

$$(l+2-3)_{(-2)} = \frac{1}{(l+1)_{(2)}} = \frac{1}{(l+1)l},$$

the first term converges to 0 as $l \to \infty$. Therefore the sum of the series is $(-1/2) \times (-1/2) = 1/4$. ∎

We can use the same approach to *accelerate* the convergence of series of the form

$$S = \sum_{k=1}^{\infty} \frac{P_b(k)}{Q_c(k)}, \tag{17}$$

where $P_b(k)$ and $Q_c(k)$ are polynomials of degrees b and c, respectively, and we assume that $c-b>1$ because otherwise the series in Eq. (17) doesn't converge [32].

EXAMPLE 8 Find a way to calculate an approximation to

$$\sum_{i=1}^{\infty} \frac{1}{k^3}, \tag{18}$$

which converges more rapidly than this series.

Solution Our object is to find a function which

a) when subtracted from $1/k^3$, has a greater difference between numerator and denominator degrees than 3, so that its sum will converge more rapidly than the sum in (18), and

b) can be summed in closed form.

If we can find such a function, call it $f(k)$ for now, then because

$$\frac{1}{k^3} = \left[\frac{1}{k^3} - f(k)\right] + f(k),$$

we will have changed the original problem to one of summing a more rapidly convergent series and summing a second series which we have assumed can be summed in closed form.

An obvious candidate for $f(k)$ is a falling factorial because we know how to sum this in closed form. Consider, for example,

$$f(k) = \frac{1}{(k-1)k(k+1)} = \frac{1}{(k+1)_{(3)}},$$

because it contains a k^3 term. We could also have chosen $1/k_{(3)}$ or some other form of the factorial. We'll explore in a problem what would have happened had we done so [33]. Using our choice for $f(k)$, we obtain

$$\frac{1}{k^3} - \frac{1}{(k-1)k(k+1)} = \frac{k^3 - k - k^3}{k^4(k^2-1)} = \frac{-k}{k^4(k^2-1)} = \frac{-1}{k^3(k^2-1)}.$$

This expression is meaningless when $k = 1$, so before trying to evaluate (18), we must separate out the $k = 1$ term. Doing this, we obtain

$$\begin{aligned}
\sum_{k=1}^{\infty} \frac{1}{k^3} - \sum_{k=2}^{\infty} \frac{1}{(k+1)_{(3)}} &= 1 + \sum_{k=2}^{\infty} \left[\frac{1}{k^3} - \frac{1}{(k+1)_{(3)}} \right] && \text{[Separating out the first term]} \\
&= 1 + \sum_{k=2}^{\infty} \frac{-1}{k^3(k^2-1)} && \text{[Common denominator]} \\
&= 1 - \sum_{k=2}^{\infty} \frac{1}{k^3(k^2-1)}.
\end{aligned}$$

From Example 7, we know that the sum of the factorial series is $\frac{1}{4}$. Thus,

$$\sum_{k=1}^{\infty} \frac{1}{k^3} = 1 + \frac{1}{4} - \sum_{k=2}^{\infty} \frac{1}{k^3(k^2-1)} \tag{19}$$

so that summing a series in which the numerator and denominator degree difference is 3 has been converted to summing one in which the difference is 5. The fact that we achieved a difference of 5 rather than 4 was just fortuitous and resulted from the fact that there was no k^2-term in the original sum or in $f(k)$. Comparing the original series and the result above, we have

$$\sum_{k=1}^{\infty} \frac{1}{k^3} = 1 + \frac{1}{8} + \frac{1}{27} + \frac{1}{64} + \frac{1}{125} + \frac{1}{216} + \cdots,$$

whereas from Eq. (19),

$$\sum_{k=1}^{\infty} \frac{1}{k^3} = \frac{5}{4} - \frac{1}{24} - \frac{1}{216} - \frac{1}{960} - \frac{1}{3000} - \cdots. \tag{20}$$

It should be clear that the new series converges much more rapidly than the original one. Note, by the way, that the original series converges from below (i.e., the value of the computed sum is always less than that of the true sum) while the converted one converges from above. How might this observation be useful [31]? ∎

The technique used in this problem can be used to accelerate the convergence of any series where the summand is a rational function [34].

Problems: Section 9.5

1. For each of the following series express the partial sum of the first n terms in closed form.

   a) $\sum_{k=0}^{\infty} \frac{1}{3^k}$

   b) $\sum_{k=0}^{\infty}(2k+1)$

   c) $\sum_{k=0}^{\infty} 4^k$

   d) $\sum_{k=0}^{\infty}(2^k+3k)$

   e) $1+1/3+1/3+1/6+1/6+1/6+1/6+1/9+1/9+1/9+1/9+1/9+1/9+1/9+1/9+\ldots$

2. By the definition of the convergence of a series, there is a sequence associated with every series. Use this definition to complete the following sentences:

   a) The series $\sum_{k=1}^{\infty} k$ diverges because the sequence ...

   b) The sequence $\{1-1/2^n\}$ has limit 1, which shows that the series ...

   c) The sequence $\{3-1/3^n\}$ has limit 3, which shows that ...

3. Suppose that

$$\sum_{k=0}^{\infty}(a_k+b_k)$$

converges. What, if anything, can you say about

$$\sum_{k=0}^{\infty} a_k \quad \text{and} \quad \sum_{k=0}^{\infty} b_k?$$

Illustrate your answer with examples. Does it matter whether the + in the first expression is changed to a $-$?

4. For $\sum 1/a_n$ to converge, is it necessary for the sequence $\{a_n\}$ to diverge? Is it sufficient?

5. In order for $\sum a_n$ to diverge, is it necessary for the sequence $\{a_n\}$ to diverge? Is it sufficient?

6. Complete the proof in Example 3 that the harmonic series diverges. *Hint*: Show that the partial sums grow without bound.

7. a) Suppose that you have two series of nonnegative terms such that each term of the first is less than a (fixed) multiple of each term of the second. Show that, if the second series converges, so does the first. *Hint*: Use Theorem 2, Section 9.2.

   b) Can you state and prove a similar *comparison* theorem which assures the divergence of one series if another diverges?

8. A ball is dropped on the floor and bounces up and down many times, each time less high than previously. According to the theory of elastic collisions, there is a constant $c < 1$ such that the height of each bounce is c times the previous height. It also follows that the time each bounce takes (from floor to maximum height or from maximum height back to the floor) is $\sqrt{c}$ times the previous time. Assume that the ball is dropped from a height of 4 feet. It follows from the laws of motion that the time the ball takes to hit the ground is 0.5 seconds.

   a) How high does the ball get on the nth bounce? How long does the nth bounce take (from floor to floor)?

   b) What is the total distance traveled by the ball before it finally rests on the ground? How much time elapses before this happens?

9. Johnny starts off from home to school. Halfway there he decides to play hooky and turns around to go home. Halfway back home from where he changed his mind, he changes his mind again and decides that he had better go to school after all. But halfway to school from there he changes his mind again and starts home. And so on.

   a) Describe the sequence of locations where he changes his mind. (Assume that home is at 0 on the number line, school is at 1, and the line segment between them is the path he walks on.)

   b) How far does he walk if he keeps on being indecisive?

10. A bug starts at $(0,0)$ on the coordinate plane. First it crawls one unit east to $(1,0)$. Then it makes a 90° left turn and crawls half a unit. Then it turns 90° left again and crawls $\frac{1}{4}$ unit. And so on. Where does it end up in the limit and how far did it crawl to get there?

11. Multiplier effect in economics. In general, people spend a certain fraction c of their income. Suppose that a government or a large corporation decides to spend an additional M dollars. This means an

additional M dollars that goes into people's pockets. Therefore they in turn spend Mc additional dollars. By spending it, they put money in other people's pockets. Those people then spend an additional $(Mc)c = Mc^2$ dollars. And so on. If the total money spent as a result of the original M is M', the ratio M'/M is called the *multiplier*.

a) What is the value of the multiplier?

b) A large multiplier means that a relatively small boost by the government can have a large effect on national income, which is generally thought to be a good thing. According to this theory, should people be encouraged to have a large c or a small c?

12. Present value of permanent resources. Suppose that you buy a resource, say a forest, that can go on producing salable products forever. Because it can provide an infinite income, you might think that the price would be infinite. But future income should be discounted by the interest rate. For instance, \$100 you receive a year from now is considered to be worth $100/(1+r)$ today, where r is the annual interest rate. [The reason? If you had $100/(1+r)$ today, you could put it in the bank for a year and then have \$100. Similarly, \$100 two years from now is worth $100/(1+r)^2$ today.]

a) Assuming that the interest rate stays the same from year to year, and assuming that the income from your resource stays the same from year to year, show that the present value of the resource (the sum of your discounted earnings over all time) is finite.

b) In fact interest rates vary. Assuming only that there is some positive constant c (say, 0.01) such that the interest rate is always at least c, show that the present value of the resource is still finite.

c) In fact, the income from the resource is likely to increase from year to year because of inflation. (The income in so-called constant dollars is likely to remain stable.) However, the bank interest rate is likely to be higher than the inflation rate. State some assumptions based on these remarks from which you can prove that the present value of the resource is finite, even if inflation will cause expected annual income to rise.

13. A pile driver drives a beam into the ground. The farther the beam goes in, the more the ground has been compacted and the harder it is to drive the beam farther. Suppose that the first "tap" by the driver drives the beam into the ground one foot. Assume also that the distance d_i that the ith tap drives in the beam is inversely proportional to the total distance driven so far. This can be modeled by

$$d_i = \frac{1}{1+\sum_{k=1}^{i-1} d_k}.$$

(The 1 in the denominator is there so that $d_1 = 1$.)

a) With a calculator or a computer, find out how far the beam has gone into the ground after 10 taps.

b) Will the ground ever compact so much that the beam cannot be driven in further? Show that (according to this model) the answer is No: The beam can be driven arbitrarily far.

c) Show that in order to drive the beam n feet, at most n^2 taps are necessary.

d) Show that to drive the beam n feet, at most $1+2+\cdots+(n+1) = (n+1)(n+2)/2$ taps are needed.

14. In a)–c) the general term and the limits of a series are given. Show in each case how the series can be summed (perhaps after some algebraic manipulation of its terms) because it telescopes.

a) $(2n-1)/[n^2(n-1)^2]$, n from 2 to ∞.

b) $\binom{n}{k}$, with n going from k to $2k$.
(*Hint*: Use a binomial coefficient identity and work from $2k$ back to k.)

c) $a(a-1)(a-2)\cdots(a-n)$, with a from $n+1$ to N. Here, you may find it helpful to refer back to the discussion about factorial functions in Section 9.4.

15. As with series, products can also telescope. Use this idea to calculate the following products for the stated general term and limits.

a) $(n+1)/n$, from 1 to N.

b) $(4-n)/(3-n)$, from 7 to 100.

16. As proved in [9, Section 2.2], the telescoping series $\sum_{k=1}^{n}(A_{k+1}-A_k)$ has value $A_{n+1}-A_1$. How is this fact related to Theorem 4, Section 9.4?

17. In Example 4, we gave a plausible argument that the alternating harmonic series (AHS) converges. But what does it converge to?

a) By considering pairs of consecutive terms in the series, show that the AHS is the same as the sum of $1/[(2i-1)2i]$ with the same limits.

b) From the result in a), show that the AHS has a sum less than the sum of the reciprocals of odd squares.

The sum of the reciprocals of all squares from 1 to ∞ is $\pi^2/6$. Actually, this is quite a crude upper bound for the sum of the AHS, as $\pi^2/6$ is about 1.65, whereas the sum of the AHS is about 0.69.

18. Apply Algorithm ALTSERIES to each of the following series whose general term is given. Start at $k=1$ and use $\epsilon = 0.01$ in each case.

a) $a_k = (-1)^k[1/(2k)!]$.

b) $a_k = (-1)^k/k^k$.

c) $a_k = (-1)^k 1/(2k+1)$.

d) $a_k = (-1)^k(k+2)/(k^2+1)$.

19. a) Modify Algorithm ALTSERIES so that the result it gives is only certain to be within ϵ of the true value.

b) Modify ALTSERIES so that it works whether a_1 is positive or negative.

20. Justify or give a counterexample:

a) If $\sum a_n$ converges, then so does $\sum a_n^2$.

b) If $\sum a_n$ converges and all $a_n \geq 0$, then $\sum a_n^2$ converges.

21. Some repeating decimals have the form

$$D = 0.R_kR_kR_k\ldots,$$

with the subscript k indicating that the repeating portion of the decimal has k digits. Thus in

$$0.264264264264\ldots,$$

$k=3$ and $R_3 = 264$.

a) Use the formula for the sum of a geometric series to express any repeating decimal of this form as a rational number (the ratio of two integers).

b) Instead, by calculating 10^kD and subtracting D from the result, show another way to find this rational number.

c) Apply this idea to

i) 0.345634563456...

ii) 0.147251472514725...

iii) 0.042042042042...

22. Other repeating decimals have the form

$$D = 0.S_jR_kR_kR_k\ldots,$$

where S_j is the portion which doesn't repeat and has j digits, and R_k is the repeating portion with k digits. Thus in

$$0.2637521752175217521\ldots,$$

$j=3$ with $S_j = 263$ and $k=4$ with $R_k = 7521$.

a) By calculating $10^{j+k}D$ and subtracting 10^jD from the result, show how the rational form of any repeating decimal can be found.

b) Apply this idea to

i) 0.45646464 ...

ii) 0.987878787 ...

iii) 0.654333333 ...

23. Try to solve Example 5 by letting $u_k = r^k$ and $\Delta v_k = k$. Explain what happens. What does this teach you about how to make good choices for u_k and Δv_k?

24. Verify the finite difference formulas in Table 9.2 or show where they are, in effect, verified elsewhere in this chapter.

25. Use summation by parts to sum the following series.

a) $\sum_{k=1}^{\infty} k/2^k$ b) $\sum_{n=1}^{\infty} n^2r^n$, $|r|<1$

c) $\sum_{k=1}^{n} kk_{(2)}$ d) $\sum_{n=k}^{N} \binom{n}{k}$

26. Use the approach of Example 5 and [25b] to express

$$\sum_{k=1}^{\infty} k^{j+1}r^k, \qquad |r| < 1,$$

in terms of the same sum, except replace $j+1$ with j for any $j > 1$.

27. Use Table 9.2 to show that

$$\sum_{i=k}^{j} \binom{i}{k} = \binom{j+1}{k+1}.$$

28. Show that the definition of the factorial function for negative subscript satisfies Eq. (16).

29. Use the technique of Example 7 to find closed form expressions for the sums from 1 to ∞ of the following expressions.

a) $1/[(k+1)(k+2)(k+3)]$
b) $1/(k+1)-1/k$
c) $1/[k(k+2)(k+4)]$

30. Apply the technique of Example 8 to each of the sums from 1 to ∞ whose general terms are given in a)–d). In each case:

i) Choose an appropriate falling factorial and calculate its sum.
ii) Compute the difference of the falling factorial and the term given.
iii) Calculate 10 terms of the original series and compare their sum with the sum of the factorial plus ten terms of the series obtained from (ii) to see the more rapid convergence of the converted series.

a) $1/k^2$
b) $1/(k^3+1)$
c) $k/(k^3+2k-1)$
d) $(k^2+1)/(k^4+k^2+1)$

31. In Example 8 we converted a series to a more rapidly convergent series. A feature of the result was that although the original series converged from below, the transformed series converged from above. How could you use this fact if you wanted to calculate the sum to within some given error tolerance?

32. Explain why the series (17) doesn't converge when $c-b\leq 1$.

33. Instead of using $(k+1)_{(3)}$ in Example 8, try using

a) $k_{(3)}$
b) $(k-1)_{(3)}$

In each case compare the result with Eq. (20).

34. Describe how you would generalize the technique of Example 8 to convert the sum of any rational function with numerator degree at least two less than denominator degree to a more rapidly convergent series. *Hint*: What falling factorial would you choose as in Example 8 to convert the problem to the sum of a rational function with greater difference of degrees between the numerator and denominator?

35. For a certain convergent series where n goes from 1 to ∞, you have determined that the tail from $n+1$ to ∞ has sum at most $1/(n^2+1)$. Based on this information, what is the least number of terms you need to add to approximate the infinite sum within 0.005?

36. Suppose that you wanted to evaluate a convergent infinite series and you knew an exact formula for the tail (not an inequality). What would you do next?

37. Find the same formula found in Example 6 by rewriting the sum as a double sum and changing the order of summation.

38. As with Fibonacci numbers, there are also lots of results about harmonic numbers. Show that

a) $\sum_{k=1}^{n} H_k = (n+1)H_n - n$. *Hint*: Express this equality in terms of a double summation and interchange the order of summation.
b) $H_{2^{m+1}} > H_{2^m} + 1/2$ and, therefore, $H_{2^m} > 1+m/2$. How could you use this result to show that the harmonic series diverges?

39. Find an explicit formula for

$$\sum_{k=1}^{n} kH_k,$$

a) using summation by parts;
b) by changing the order of summation in a double sum.

40. Evaluate

$$\sum_{n=1}^{1023} \log_2\left(1+\frac{1}{n}\right).$$

9.6 Generating Functions

This section is about another connection between sequences and series. We can use series as a mechanism to manipulate all the terms of a sequence at one time. This mechanism is called a **generating function** and is a powerful tool in many applications of discrete (and continuous) mathematics.

In Section 0.4, we mentioned that the symbol x in

$$P_n(x) = \sum_{k=0}^{n} a_k x^k \qquad (1)$$

need not be a variable which takes on (say) real values, but rather that it could be a **place holder** for the coefficients a_k. Since we didn't pursue this idea then, you may have ignored it. But you can't now. This idea forms the basis of our development of generating functions.

The essential idea is very simple: Let $\{a_k\}$ be any sequence. With this sequence we *associate* a polynomial

$$G(s) = \sum_{k=0}^{\infty} a_k s^k. \qquad (2)$$

The polynomial in Eq. (2), called the generating function of the sequence $\{a_k\}$, differs from that in Eq. (1) in the following respects.

- Its upper limit is ∞. (We could have used an infinite upper limit in Eq. (1), also, with the stipulation that $a_k = 0$ for $k > n$.)
- Instead of x in Eq. (1), we have used s in Eq. (2). Now, of course, the *name* of the variable in an expression like that in Eq. (2) is really irrelevant except as it may have suggestive value. In this case we've abandoned the usual notation x, which, subliminally anyhow, connotes a *variable*, in order to emphasize that s is not a variable but just a *symbol* which serves as a place holder for the elements of the sequence. And we have used $G(s)$ instead of $P(s)$ because this is the standard notation for a single generating function (although we'll use other letters later when dealing with more than one generating function at a time).

Actually, with the upper limit ∞, we shouldn't call $G(s)$ a polynomial. The usual mathematical term for the beast on the right-hand side of Eq. (2) is **power series** because it is an (infinite) series of powers (of s).

Why is the idea of a generating function a useful one? The answer is that a generating function, by *binding* a_k to the kth power of s, enables us to treat an infinite sequence as a single unit, namely, the summation which defines the generating function. Suppose that we are given a description of a_k (e.g., as some combinatorial function—see Example 7) and that we would like to find a closed form expression for a_k. If we can derive the generating function for the sequence $\{a_k\}$ in closed form and then, somehow, *expand* the closed form expression in powers of s, the coefficient of s^k will give us the desired expression for a_k.

Sometimes, using generating functions is also useful even when we know the form of each term of a sequence. For example, if an unknown sequence is defined by complicated operations on known, simpler sequences, it may be best to use the generating functions of the simpler sequences, perform some operations on them to get the generating function of the unknown sequence, and then expand this function to get the terms of the unknown sequence.

We'll illustrate the various uses of generating functions in Examples 4, 7, and 8, but first we give some simple examples to help you understand the concept of generating functions.

EXAMPLE 1 Given the closed form

$$(1+s)^n$$

of a generating function, what sequence does this correspond to?

Solution The answer here is immediate because you know how to expand this function using the Binomial Theorem. If you do so, the coefficient of s^k is

$$\begin{array}{ll} \binom{n}{k}, & 0 \le k \le n; \\ 0, & k > n. \end{array}$$ ∎

EXAMPLE 2 Given the closed form

$$(1+s+s^2)^n$$

of a generating function, what sequence does this correspond to?

Solution We could apply the binomial theorem again, expanding in terms of $s+s^2$. Doing so would just give us powers of $s+s^2$, which we could then expand to find the coefficients of the various powers of s. Not very easy or esthetic. Using the multinomial theorem (see Section 4.6) would be better. In Example 8 we'll show how this generating function arises in practice. ∎

EXAMPLE 3 What are the generating functions for the following sequences $\{a_k\}$?

i) $a_k = 1$, for all k.

ii) $a_k = 1$, when k is a multiple of $m > 0$ and 0 otherwise.

iii) $a_k = a^k$.

Solution The generating functions for all three of these sequences are geometric series. For (i), the series is

$$G(s) = \sum_{k=0}^{\infty} s^k,$$

whose sum is $1/(1-s)$.

For (ii), the series is

$$G(s) = \sum_{k=0}^{\infty} s^{mk},$$

whose sum is $1/(1-s^m)$.

For (iii), the series is

$$G(s) = \sum_{k=0}^{\infty} a^k s^k = \sum_{k=0}^{\infty} (as)^k,$$

whose sum is $1/(1-as)$. ∎

But wait a minute. Doesn't the convergence of a geometric series depend on the ratio r of successive terms of the series being less than 1 in absolute value? We seem to have ignored this condition in Example 3. Shouldn't we have said for (i) that $|s| < 1$, for (ii) that $|s^m| < 1$ and for (iii) that $|as| < 1$? Insofar as s is just a *symbol* and not a *variable* which takes on values, these inequalities don't make any sense. But at least it is true that, if we allow s to be a variable, then there are *some* values of s for which the series in Example 3 do converge.

A better way out of the dilemma in the previous paragraph—we only sketch the main idea—is this. In *formal* summation we ask only that the closed form sum be *algebraically* equivalent to the series. That is, if we declare infinite series of symbols to obey the same laws of algebra as finite sums (e.g., the commutative and distributive laws), we can show that the closed form is equal to the series. For example, to verify the solution of Example 3 (i), we need to show that

$$\frac{1}{1-s} = \sum_{k=0}^{\infty} s^k. \tag{3}$$

Multiplying both sides of Eq. (3) by $1-s$, we get 1 on the left and, on the right,

$$(1-s)(1+s+s^2+s^3+\cdots) = (1+s+s^2+\cdots) - (s+s^2+s^3+\cdots).$$

Except for the 1, all powers of s in the first series on the right are canceled by a corresponding power of s in the second series. Therefore the right-hand side is 1, as it should be for Eq. (3) to be an identity.

Now we'll illustrate the power of generating functions for solving two quite different types of problems. The first problem is the solution of difference equations. We can use generating functions to solve equations such as those discussed in Chapter 5 and, as well, certain classes of difference equations which we can't solve by the methods of Chapter 5. To keep things simple, we'll show how to use generating functions to solve a linear difference equation with constant coefficients, even though we know how to solve such equations by the methods of Chapter 5.

EXAMPLE 4 Use generating functions to solve

$$a_n - 3a_{n-1} + 2a_{n-2} = 0, \qquad a_0 = 0; \quad a_1 = 1. \tag{4}$$

Solution The generating function for the sequence $\{a_k\}$ defined by Eq. (4) is, by definition,

$$G(s) = \sum_{k=0}^{\infty} a_k s^k. \tag{5}$$

Using the coefficients 1, -3, and 2 in Eq. (4) and the powers of s corresponding to what we subtract from n in each subscript, let's write out $G(s)$, $-3sG(s)$, and $2s^2G(s)$, as follows:

$$\begin{aligned}
G(s) &= a_0 + a_1 s + a_2 s^2 + a_3 s^3 + a_4 s^4 + \cdots,\\
-3sG(s) &= - 3a_0 s - 3a_1 s^2 - 3a_2 s^3 - 3a_3 s^4 - \cdots,\\
2s^2G(s) &= 2a_0 s^2 + 2a_1 s^3 + 2a_2 s^4 + \cdots.
\end{aligned}$$

When we add these three equations, we get 0 for all columns on the right-hand side after the first two because of Eq. (4) (e.g., $a_3s^3 - 3a_2s^3 + 2a_1s^3 = (a_3 - 3a_2 + 2a_1)s^3 = 0$). Thus we have

$$(1-3s+2s^2)G(s) \;=\; a_0 + a_1 s - 3a_0 s \;=\; s$$

because of the initial conditions in (4). Solving for $G(s)$ we have

$$G(s) \;=\; \frac{s}{1-3s+2s^2} \tag{6}$$

Equation (6) gives in closed form the generating function for the sequence $\{a_k\}$ which is the solution of Eq. (4). Now we want to expand this closed form in powers of s so that we can see explicitly the form of a_k. The denominator is a quadratic, which factors into $(1-s)(1-2s)$. If you know the method of partial fractions (see [7, Supplementary]), you'll know how to expand the factors to get

$$\frac{s}{(1-s)(1-2s)} \;=\; \frac{-s}{1-s} + \frac{2s}{1-2s}.$$

At any rate you should be able to verify this identity by putting the terms on the right-hand side over a common denominator. Thus we have

$$G(s) \;=\; \frac{-s}{1-s} + \frac{2s}{1-2s}.$$

Then using identity (3) and its analog with s replaced by $2s$, we may write the preceding equality as

$$\begin{aligned} G(s) &= -s(1+s+s^2+s^3+\cdots) + 2s[1 + 2s + (2s)^2 + (2s)^3 + \cdots] \\ &= s + (2^2-1)s^2 + (2^3-1)s^3 + \cdots + (2^k-1)s^k + \cdots. \end{aligned} \tag{7}$$

The coefficient of s^n in $G(s)$ is just a_n, so from Eq. (7) it follows that

$$a_n \;=\; 2^n - 1. \quad \blacksquare$$

The result in Example 4 is the same result you would have gotten using the method of Section 5.5 to solve second-order linear difference equations with constant coefficients. We suspect that you found the method in Section 5.5 more congenial than the one in Example 4. Indeed, we would not recommend using generating functions to solve linear difference equations with constant coefficients in preference to the methods of Sections 5.5 and 5.6. But the idea which led to Eq. (6) did not depend on the equation having constant coefficients or even being linear. Thus the method illustrated by Example 4 is a powerful technique for solving a large variety of difference equations. We'll set you to work on some difficult ones in the problems [17, 18].

Our next two examples of the application of generating functions will take us back to a combinatorial problem we discussed in Chapters 4 and 5. Before getting to it, however, we need to introduce the idea of a **convolution.**

Definition 1. Let $\{a_k\}$ and $\{b_k\}$ be two sequences whose domain, as with all the sequences which follow, we assume to be N, the set of non-negative integers. Then the **convolution sequence** $\{c_k\}$ is defined by the relation

$$c_k = \sum_{j=0}^{k} a_j b_{k-j}. \tag{8}$$

Here are two examples to illustrate this definition.

EXAMPLE 5 Let $\{a_k\} = \{b_k\} = \{1\}$, the sequence of all 1's. Then $c_k = k+1$ because the sum in Eq. (8) has $k+1$ terms, each of which is $1 \cdot 1 = 1$. ∎

EXAMPLE 6 Let $\{a_k\} = \{\binom{k+2}{2}\}$ and $\{b_k\} = \{1\}$. Then from Eq. (8),

$$c_k = \sum_{j=0}^{k} \binom{j+2}{2} = \binom{k+3}{3},$$

a result which follows from [8], Section 4.5. ∎

Our first theorem about convolutions should help you to understand why this is a useful concept.

Theorem 1. Let $\{c_k\}$ be the convolution of $\{a_k\}$ and $\{b_k\}$. Let a_k and b_k represent the number of distinguishable ways to pick a subset of k objects from set A and from set B, respectively. Let A and B be disjoint and let anything in A be distinguishable from anything in B. Then c_k is the number of distinguishable ways to pick a subset of k objects from $A \cup B$.

PROOF Suppose that the set of k objects from $A \cup B$ contains j objects from A and $k-j$ objects from B. Then $a_j b_{k-j}$ is the number of ways to pick such a set (why?). But j can have any value from 0 to k. Summing $a_j b_{k-j}$ over j gives the number of distinguishable ways to form a subset of k objects from $A \cup B$. But this sum is precisely Eq. (8). ∎

Do you see how Theorem 1 relates to Example 5 and 6? In Example 5, let A be an infinite collection of red balls and B be an infinite collection of blue balls so that $\{1\}$ represents the number of ways to choose k balls from either A or B. The convolution sequence represents the fact that there are $k+1$ ways to choose k balls, some red ones from A and blue ones from B. For Example 6, let A contain an infinite number of red, green, and yellow balls and let B be as before. Then choosing j balls from A is equivalent to putting j (indistinguishable) balls in 3 distinguishable urns labeled red, green, and yellow. (Why are these two ways of handling j balls equivalent?) We solved the ball–urn problem in Section 4.7; the answer is $C(j+2, j) = C(j+2, 2)$. Again the convolution sequence represents the number of ways to choose k balls from A and B.

Our main result about convolution sequences is the following.

Theorem 2. Let $\{a_k\}$ and $\{b_k\}$ be two sequences with convolution $\{c_k\}$ and let $A(s)$, $B(s)$, and $C(s)$ be the generating functions of the three sequences. Then

$$C(s) = A(s)B(s).$$

In other words, generating functions turn the complicated process of convolution into the simple process of multiplication.

PROOF To calculate the product of two infinite series $A(s)$ and $B(s)$ we recall Eq. (19), Section 0.5, where we displayed a formula for the product of two polynomials. The idea there was to gather together all terms with the same power of the variable. Doing the same thing here, if we calculate the coefficient of s^k in

$$C(s) = A(s)B(s) = (a_0+a_1s+a_2s^2+\cdots)(b_0+b_1+b_2s^2+\cdots)$$

the result is

$$a_0b_k + a_1b_{k-1} + \cdots + a_kb_0. \tag{9}$$

But this is just c_k as given by Eq. (8). ∎

Now we're ready for some applications of convolutions. In Sections 4.4, 4.7, 5.2, and 5.3, we considered the problem of counting the number of combinations of n objects taken k at a time when unlimited repetitions of the objects are allowed. In Section 4.7 we showed that the answer is

$$a_{nk} = \binom{n+k-1}{k}. \tag{10}$$

Now we'll get the answer using convolutions.

EXAMPLE 7 **Combinations with Unlimited Repetition**

Find the number of combinations of n things, k at a time, with unlimited repetition allowed.

Solution We could obtain the answer by generalizing Theorem 2. Instead we'll leave that approach to a problem [12a] and use a different approach, which also uses a convolution.

Of the two parameters, n and k, which should we base the generating function on? With the advantage of hindsight we choose k and define

$$G_n(s) = \sum_{k=0}^{\infty} a_{nk}s^k, \tag{11}$$

with the n subscript on the generating function indicating that n is to be considered fixed. Thus Eq. (11) really defines a *family* of generating functions, one for each n.

The crucial part of our derivation is to use the recursive paradigm to write a_{nk} in terms of values of $a_{n-1,j}$ as follows:

$$a_{nk} = a_{n-1,k} + a_{n-1,k-1} + \cdots + a_{n-1,0}. \tag{12}$$

Why is Eq. (12) correct? Because it says that the number of combinations of n things, k at a time, is equal to the number of combinations of $n-1$ things, k at a time (with the nth thing not appearing at all) plus the number of combinations of $n-1$ things, $k-1$ at a time (with the nth thing appearing once), etc., with the last term representing the number of combinations of the $n-1$ things 0 at a time and (implicitly) the nth thing appearing k times. Now Eq. (12) doesn't look like a convolution because there's no product of terms from two different sequences. But suppose, as in Examples 5 and 6, that one of the sequences is just $\{1\}$, the sequence of all ones. Then the right side of Eq. (12) is a term from the convolution of this sequence and the sequence $\{a_{n-1,k}\}$ [make sure you see this by referring to Eq. (8)]. From Example 3 the generating function for the sequence $\{1\}$ is just $1/(1-s)$. Therefore, as Eq. (12) is a convolution, we get from Theorem 2

$$G_n(s) = \frac{G_{n-1}(s)}{1-s}.$$

But if this is true for n and $n-1$, the same relation must hold between the two generating functions in the $n-1$ and $n-2$ cases, etc. Thus we have

$$G_n(s) = \frac{G_{n-1}(s)}{1-s} = \frac{G_{n-2}(s)}{(1-s)^2} = \cdots = \frac{G_0(s)}{(1-s)^n}.$$

Setting $G_0(s) = 1s^0 = 1$, using the convention that there is one "empty" combination (i.e., 0 objects 0 at a time), we get

$$G_n(s) = \frac{1}{(1-s)^n} \tag{13}$$

To get a_{nk} we have to expand the right-hand side of Eq. (13) in powers of s. We already know that the answer is given by Eq. (10), so there's no reason to look here for a direct technique to expand Eq. (13) (but there are such ways—see [8]). Knowing what the answer is will be quite useful in Example 8. ∎

Now, let's consider the problem of limited repetitions in which we want to calculate the number of combinations of n objects k at a time but where only *up to m* repetitions of each object is allowed. For example, with $n = k = 3$, we have:

Combinations without repetition:

$$\binom{3}{3} = 1\text{—Combination is } abc.$$

Combinations with unlimited repetition:

$$\binom{5}{3} = 10\text{—}aaa, aab, aac, abb, abc, acc, bbb, bbc, bcc, ccc.$$

Combinations with restricted repetitions—let $m=2$:

$$7\text{—}aab, aac, abb, abc, acc, bbc, bcc.$$

(Is there a simple formula like a binomial coefficient for this case? No, but we'll derive a fairly convenient way to calculate the answer using generating functions.)

EXAMPLE 8 Calculate the number of combinations of n objects, k at a time, with up to m repetitions of each object.

Solution Let A_i, $i=1, \ldots, n$, be a set of m identical objects of the ith type. Our problem (cf. Theorem 1) is to find the number of distinguishable ways to choose k items from

$$A_1 \cup A_2 \cup \cdots \cup A_n.$$

Let's call this number a_{nmk}. Note that $a_{nmk}=0$ for $k>nm$ (why?).

Let $\{a_{ik}\}$ be the sequence representing the number of ways of choosing k objects from A_i. But there is only one way of doing this for $k=0, 1, \ldots, m$ and 0 ways for all $k>m$. Therefore for all i,

$$a_{ik} = \begin{cases} 1, & k = 0,1,\ldots,m; \\ 0, & k > m. \end{cases} \tag{14}$$

It follows from Eq. (14) that the generating function for $\{a_{ik}\}$ is

$$G_i(s) = 1+s+s^2+\cdots+s^m, \qquad i = 1,2,\ldots,n. \tag{15}$$

Equation (15) is just the partial sum of a geometric series, so we may rewrite it in the form:

$$G_i(s) = \frac{1-s^{m+1}}{1-s}. \tag{16}$$

If there were just two objects (i.e., $n=2$), there would be two generating functions. Then Theorems 1 and 2 together would say that the generating function of the sequence $\{a_{2mk}\}$, representing the number of ways of choosing k objects from m repetitions each of two objects would be

$$G_1(s)G_2(s) = (1+s+s^2+\cdots+s^m)^2 = \left(\frac{1-s^{m+1}}{1-s}\right)^2.$$

We'll leave it to a problem [12c] to prove that this result can be generalized for any n to give the generating function for the sequence $\{a_{nmk}\}$ (with m and n fixed) as

$$(1+s+s^2+\cdots+s^m)^n \;=\; \left[\frac{1-s^{m+1}}{1-s}\right]^n. \qquad (17)$$

How can we use Eq. (17) to obtain a_{nkm}, which is the coefficient of s^k? One way is just to expand the left-hand side. For example, with $n=3$ and $m=2$, we have

$$(1+s+s^2)^3 \;=\; 1+3s+6s^2+7s^3+6s^4+3s^5+s^6.$$

The coefficient of s^3 is $a_{323}=7$ and corresponds to the seven combinations shown just before this example.

Can we use the right-hand side of Eq. (17) to get the same result? Yes, as follows. The binomial theorem tells us that the first few terms of the expansion of $(1-s^3)^3$ are $1-3s^3+\cdots$. No more terms are needed because we are looking for the coefficient of s^3 in the expansion of the entire expression. Now how about the expansion of $1/(1-s)^3$. We need the constant term and the term in s^3. But these are given by Eq. (10) [recall the discussion after Eq. (13)], with $n=3$ and $k=0$ for the constant term and $k=3$ for the term in s^3:

$$\binom{2}{0} = 1, \qquad \text{and} \qquad \binom{5}{3} = 10.$$

Putting things together, we need the coefficient of s^3 in

$$(1-3s^3+\cdots)(1+\cdots+10s^3+\cdots),$$

which is $-3+10 = 7$, the same result as before. We'll ask you to apply both techniques used in this example in a problem [14]. ∎

Problems: Section 9.6

1. Express in closed form the generating functions for the following sequences, whose domain is N.
   a) $\{2^n\}$
   b) $\{1,0,1,0,1,0,\ldots\}$
   c) $\{0,1,0,1,0,1,\ldots\}$
   d) $\{1,0,0,-1,0,0,1,0,0,-1,\ldots\}$
   e) $\{k^2\}$ *Hint*: See [25b] and [26] in Section 9.5.
   f) $\{k(k+1)\}$

2. Let A be a set of 10 distinguishable balls. Let a_n be the number of distinguishable sets of n balls that can be chosen from A. Determine the closed form of the generating function for $\{a_n\}$.

3. Same as [2] except that the balls are indistinguishable.

4. Same as [2] except that A is infinite and consists of one black ball and otherwise indistinguishable red balls.

5. Give an expression for a_k when the generating function for the sequence $\{a_k\}$ is:

   a) $1/(1+s)$ b) $1/(1-2s)$ c) $s/(1-s^2)$
   d) $s^3/(1-s)$ e) $1/(1+s)^2$ f) $1/(2-s)$
   g) $2/(1-s)+3/(1-2s)$
   h) $1/[(1-s)(2-s)]$
   i) $1/(s^2-2s+1)$
   j) $1/(1-s-s^2)$

   Hint: Factor the denominator and then write the entire expression as the sum of two terms.)
   k) $s/(1-s)^3 - s/(1-s)^2$

6. Give a closed form expression for each of the following formal series. In each case, if the formal parameter is replaced by a real variable, for which range of values of the real variable does the closed form expression give the true value of the series?

a) $-1+2s-4s^2+8s^3-16s^4+\cdots$

b) $s^2+s^4+s^6+s^8+s^{10}+\cdots$

c) $3s+6s^2+12s^3+24s^4+\cdots$

7. Use generating functions to solve each of the following difference equations.

a) $r_n = 2r_{n-2}, \quad r_0 = 0; r_1 = 1.$

b) $2r_{n+1} = r_n + r_{n-1}, \quad r_0 = 0; r_1 = 1.$

c) $r_{n+1} = 2r_n - r_{n-1}, \quad r_0 = r_1 = 1.$

d) The TOH recursion:

$$h_{n+1} = 2h_n + 1, \qquad h_0 = 0.$$

8. (Requires knowledge of calculus) We know that

$$\frac{1}{1-s} = \sum_{k=0}^{\infty} s^k.$$

a) By differentiating, show that

$$\frac{1}{(1-s)^2} = \sum_{k=0}^{\infty}(k+1)s^k.$$

b) Continue differentiating and thereby derive Eq. (10) for all positive integers n.

9. True or false: The generating function for $(1+s)^n$ is

$$\sum_{k=0}^{\infty}\binom{n}{k}s^k$$

for all integers n (including negative integers).

10. For each pair of sequences, find a closed form expression for the general term of the convolution sequence.

a) $\{0,1,0,0,0,0,\ldots\}, \quad \{1,1,1,1,\ldots\}$

b) $\{1,1,0,0,0,0,\ldots\}, \quad \{0,1,2,3,4,\ldots\}$

c) $\{1,1,0,0,0,0,\ldots\}, \quad \{1,1,0,0,0,0,0,\ldots\}$

d) $\{0,0,1,0,0,1,0,0,1,\ldots\}$,
$\{1,0,0,1,0,0,1,0,0,\ldots\}$

11. Use the idea that the generating function of the convolution sequence is the product of the generating functions of the individual sequences to determine the sequences which correspond to the following generating functions.

a) $G(s) = (1+s+s^2)/(1-s)$.

b) $G(s) = (1+2s-3s^2)/(1+s)$.

12. Let $\{a_{1k}\}, \{a_{2k}\}, \ldots, \{a_{mk}\}$ be sequences and define

$$c_{ik} = \sum_{j=0}^{k} a_{ij}c_{i-1,k-j},$$

$$i = 2,\ldots,m; \quad c_{1k} = a_{1k}.$$

a) Generalize Theorem 2 to show that the generating function of $\{c_{mk}\}$ is

$$C_m(s) = \prod_{j=1}^{m} A_j(s),$$

where $A_i(s)$ is the generating function for the sequence $\{a_{ik}\}$.

b) Suppose that a_{ik} represents the number of ways of choosing k objects from an infinite set of identical objects (cf., Example 8). Use the result in a) to derive Eq. (13).

c) Show how a similar argument leads to Eq. (17).

13. Calculate the number of combinations of n things k at a time, with unlimited repetition when:

a) $n = 10, k = 3$.

b) $n = 6, k = 8$.

c) $n = 3, k = 10$.

14. Use the technique of Example 8 to calculate:

a) The number of combinations of 4 objects, 3 at a time, with up to 2 repetitions of each.

b) The number of combinations of 5 objects, 2 at a time, with up to 2 repetitions of each.

Do both parts two ways: By expanding the first part of Eq. (17) and by using the second part of Eq. (17), as in Example 8.

15. How many ways are there to pick 9 things from 3 types if there can be up to 4 things of types A and B but as many as we want of type C?

16. An experiment with success probability p is repeated until the first success. In Section 6.5 we showed that the expected number of trials is

$$\sum_{k=1}^{\infty} k(1-p)^{k-1}p,$$

and we gave various arguments to show that this simplifies to $1/p$. Obtain this result again, simply, using the methods of this section. *Hint*: Relate the expected value to the generating function $\sum ks^k$.

17. A **partial difference equation** is one which involves two or more subscript variables. Here is one which occurs in the analysis of Algorithm MAXNUMBER:

$$p_{nk} = \left(\frac{1}{n}\right) p_{n-1,k-1} + \left(\frac{n-1}{n}\right) p_{n-1,k}, \qquad (18)$$

with the initial conditions

$$p_{10} = 1; \qquad p_{1k} = 0, \quad k > 0;$$
$$p_{nk} = 0, \quad k < 0.$$

The generating function for $\{p_{nk}\}$ depends on which of the two subscripts we keep constant. Suppose that we let n be constant and define

$$G_n(s) = \sum_{k=0}^{\infty} p_{nk} s^k.$$

a) Use this generating function, Eq. (18), and the initial conditions to derive an equation relating $G_n(s)$ and $G_{n-1}(s)$.

b) Now rewrite this equation, replacing n with $n-1$ and $n-1$ with $n-2$ to get an equation for $G_{n-1}(s)$ in terms of $G_{n-2}(s)$.

c) Continue to apply the idea in b) to get $G_n(s)$ in terms of $G_1(s)$.

d) Give the explicit form of $G_1(s)$ using the initial conditions.

e) Why is $G_n(s)$ "essentially a binomial coefficient"?

18. We know the generating function for the binomial coefficients from the binomial theorem and Example 1. We now show how to obtain them from the recursive definition (really a partial difference equation) of the binomial coefficients, namely,

$$c_{nk} = c_{n-1,k} + c_{n-1,k-1}, \qquad (19)$$
$$n > 0,\ 0 < k < n;$$
$$c_{n0} = c_{nn} = 1;$$

Here, as in [17], we solve the problem by creating a family of generating functions.

a) Define

$$G_n(s) = \sum_{k=0}^{n} c_{nk} s^k.$$

Substitute Eq. (19) into this definition and show that

$$G_n(s) = (1+s)G_{n-1}(s), \quad n > 0;$$
$$G_0(s) = 1.$$

b) From the result in a), obtain an explicit formula for $G_n(s)$ that you already know.

9.7 Approximation Algorithms

Twice in Section 9.5—with Algorithm ALTSERIES and in Example 8—we discussed methods to compute approximate values of the sum of series which could not be summed in closed form. As we noted in Section 9.1, approximation methods lie at the heart of **numerical analysis**. Sometimes the problems to be solved in numerical analysis are expressed using the structures of algebra, as in the case of solving an equation of the form $f(x) = 0$. Sometimes the problems are from calculus, as with differential equations to be solved or as integrals to be evaluated. With few exceptions, such problems do not have solutions in closed form, so we must use approximate techniques in which, for example, a differential equation is expressed in the form of a difference equation, such as those we discussed in Chapter 5.

Sometimes the solution of a problem implies an infinite computation, as in summing a series, and we must approximate the solution by doing a finite amount of computation. Sometimes the problem does have a closed form solution, but the errors introduced in performing the arithmetic to compute the closed form solution can cause such difficulties that approximate techniques are worthy competitors to methods which find closed form solutions. We'll give an example of this latter behavior shortly.

A common thread running through all such problems is that the true solution would have the form of one or more real numbers. Except when the real numbers are integers or rational numbers (ratios of integers), they have infinite decimal expansions which do not repeat or have any discernible pattern. Therefore it stands to reason that, with a finite amount of computation, only a finite portion of the infinite decimal expansion can be computed. Both the examples in Section 9.5 had this character.

The algorithms used in numerical analysis to find approximations to the real number solutions of mathematical problems are normally expressed using the language and constructs of discrete mathematics. These algorithms therefore form a natural bridge from discrete mathematics to continuous mathematics. In this section we will consider two such algorithms.

First, however, we give a general structure for the approximation paradigm. The main portion of an approximation algorithm consists of a loop:

repeat
    improve approximation
    (stopping condition)
endrepeat

where the "stopping condition" is a test to see whether the approximation thus far obtained has a desired property. Usually this property is either that the current approximation is sufficiently near the true solution or that two successive approximations are sufficiently close to each other. (Does the latter criterion guarantee that the current approximation will be near the true solution? We'll return to this question later in this section.) The stopping condition may be in the interior of the loop or it may be attached to either the **repeat** or the **endrepeat**.

Our first illustration of an approximation algorithm concerns the problem of solving

$$f(x) = 0, \tag{1}$$

where $f(x)$ can be almost any mathematical function. Our only restriction on $f(x)$ is that it be continuous (roughly, that it can be drawn without lifting pencil from paper). If $f(x)$ is a linear function $ax+b$ or a quadratic function ax^2+bx+c you know how to solve Eq. (1) in closed form. But in the real world, there are very few examples of functions that have such closed form solutions, so usually we are forced to resort to an approximate method. Whole books have been written on approximate methods for solving Eq. (1), so you will realize that the method we're going to discuss is only one of many possibilities. It is, however, a powerful method, usable and reasonably efficient, for a wide variety of problems.

We assume that we know two points a and b, as shown in Fig. 9.4, where $f(a)$ and $f(b)$ have opposite signs. We may express this condition as

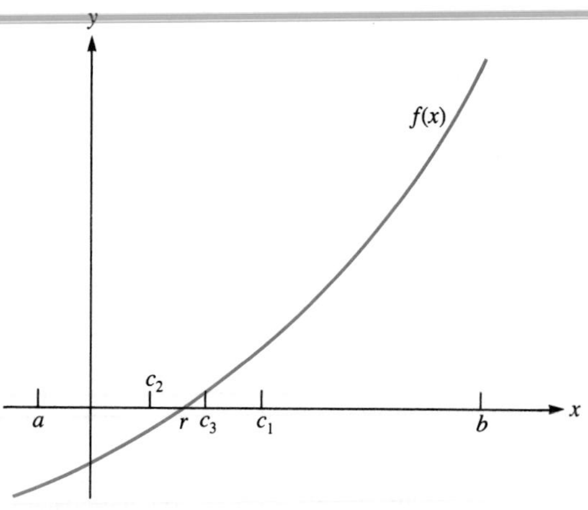

FIGURE 9.4

The Method of Bisection. The subscripts on c denote the successive midpoints of the bisected intervals. Thus c_1 is the midpoint of (a,b), c_2 is the midpoint of (a,c_1), and c_3 is the midpoint of (c_2,c_1). The root of $f(x)=0$ is denoted by r.

$$f(a)f(b) < 0. \tag{2}$$

This means that there is at least one solution in the interval (a,b) (why?). The assumption embodied in (2) is not a trivial one, because there are many examples of functions where two points with this property are not easy to find. We can try to find two such points by evaluating $f(x)$ at a sequence of points along the real line. However, this approach may fail if, for example, our sequence skips over all the points where $f(x)$ is, say, positive as illustrated in Fig. 9.5.

Still, suppose that we have managed to find two points such that (2) is satisfied. Our method, called the **method of bisection**, is based on the observation that, if we choose a point c halfway between a and b, at which $f(c)\neq 0$ (see Fig. 9.4), then $f(c)$ will have a sign different from $f(a)$ or different from $f(b)$. Therefore we can define a new interval containing a point such that $f(x)=0$ which is half the size of the original interval. That is, we have *bisected* the original interval. (Of course, $f(c)$ might be 0; that would be marvelous—we would have found the exact solution—but it is very unlikely.) Now we just iterate, applying this idea again and again, until the size of the interval is no more than some predefined error tolerance ϵ. We then know that the true solution is within ϵ of the endpoints of the final interval. Algorithm 9.2, BISECTION implements this idea.

Do you see why the output is actually within $\epsilon/2$ of the true solution [2a]? If you wanted to do only enough computation to ensure that the output would be within ϵ of the true solution, how might you modify the algorithm [2b]? It's worth remarking that Algorithm BISECTION is a continuous analog of Algorithm BINSEARCH in Section 1.6.

EXAMPLE 1 Apply Algorithm BISECTION to the function $f(x)=\sin x - x/2$, with $\epsilon = 10^{-4}$, to

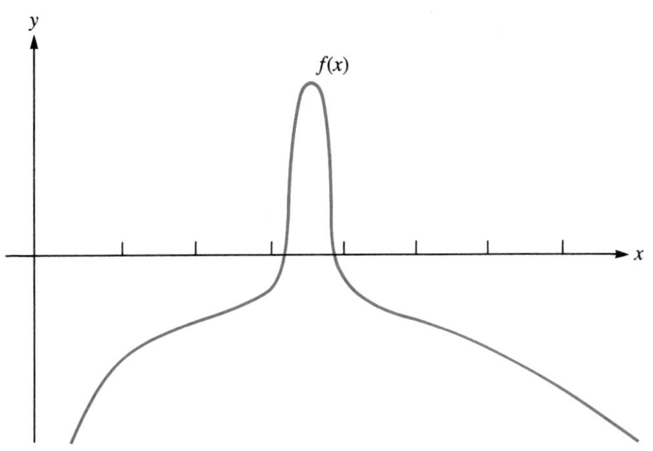

FIGURE 9.5

Missing a Solution in the Bisection Method. Points on the horizontal axis represent the values of the variable at which the values of the function are computed. All computed values will be negative, so that no two points a and b at which $f(a)f(b) < 0$ will be found.

Algorithm 9.2
Bisection

Input $f(x)$ [Really a procedure for computing $f(x)$]
a, b [Such that $a < b$ and $f(a)f(b) < 0$]
ϵ [Error tolerance]

Algorithm BISECTION
left ← *a*; *right* ← *b* [Initialize left and right endpoints]
repeat while (*right* − *left*) > ϵ
x ← (*left*+*right*)/2 [Midpoint]
if $f(x) = 0$ **then exit** [Solution found]
else
if $f(x)f(a) < 0$ **then**
right ← x [$f(x)$, $f(a)$ different signs]
else
left ← x [$f(x)$, $f(b)$ different signs]
endif
endif
endrepeat
answer ← (*left*+*right*)/2

Output *answer* [Within $\epsilon/2$ of true solution]

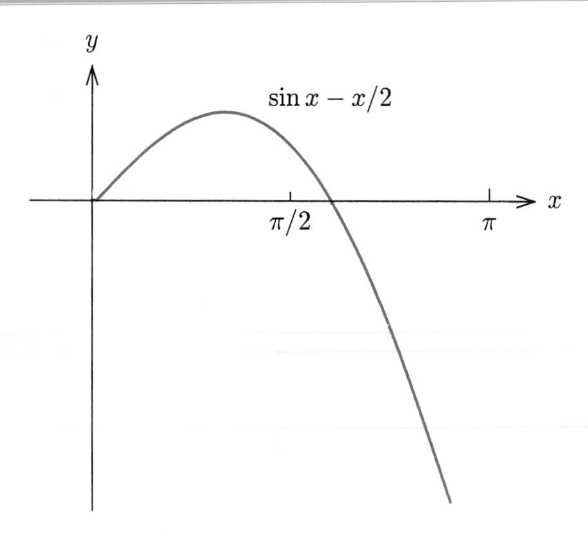

FIGURE 9.6

Graph of $\sin x - x/2$ for Example 1

TABLE 9.3
Algorithm BISECTION Applied to $f(x) = \sin x - x/2$

left	right	x	$f(x)$
1.570796	3.141593	2.356194	−0.470990
1.570796	2.356194	1.963495	−0.057868
1.570796	1.963495	1.767146	0.097213
1.767146	1.963495	1.865321	0.024280
1.865321	1.963495	1.914408	−0.015660
1.865321	1.914408	1.889865	0.004596
1.889865	1.914408	1.902137	−0.005461
1.889865	1.902137	1.896001	−0.000415
1.889865	1.896001	1.892933	0.002095
1.892933	1.896001	1.894467	0.000846
1.894467	1.896001	1.895234	0.000213
1.895234	1.896001	1.895618	−0.000101
1.895234	1.895618	1.895426	0.000056
1.895426	1.895618	1.895522	−0.000023
1.895426	1.895522		

find a root which is not 0.

Solution If you sketch $\sin x - x/2$, as in Fig. 9.6, you'll observe that there is a root between $\pi/2$ and π. When $x = \pi/2$, $f(x) = 1 - \pi/4 > 0$, and when $x = \pi$, $f(x) = 0 - \pi/2 < 0$. Now, using Algorithm BISECTION with $a = \pi/2$ and $b = \pi$, we compute the values shown in Table 9.3. The final *left* and *right* differ by less than 10^{-4}, so we may then calculate *answer* to be $(1.895426+1.895522)/2 = 1.895474$, which differs from the true solution by no more than $(1/2)10^{-4}$. Actually, to five decimal places, the true solution is 1.89549. While this convergence looks fast compared to, say, that in Algorithm ALTSERIES in Section 9.5, many other approximation algorithms for this problem would converge much more rapidly than Algorithm BISECTION. Note, by the way, that, while the lengths of successive intervals are monotonically decreasing, the magnitude of successive values of $f(x)$ does not decrease monotonically. ▮

But does Algorithm BISECTION always work? Yes, because 1) after n iterations of Algorithm BISECTION, the length of the interval is the original length $b - a$ times $(0.5)^n$, which is a geometric sequence with limit 0, and 2) it is a loop invariant that there is a root inside the interval (*left*, *right*). So, *theoretically* Algorithm BISECTION always terminates with a solution within $\epsilon/2$ of the true solution. In *practice*, however, if ϵ is chosen too small, **roundoff errors** (i.e., errors in arithmetic which result because computers operate on finite length numbers) may prevent *right* − *left* from ever being less than ϵ.

Our second illustration of an approximation algorithm is quite different. It concerns the most important special case of simultaneous linear equations in Chapter 8: square systems with unique solutions,

$$A\mathbf{x} = \mathbf{b}, \tag{3}$$

where A is an invertible $n \times n$ matrix. This problem may seem a strange candidate for an approximate algorithm. After all, it has an exact solution,

$$\mathbf{x} = A^{-1}\mathbf{b},$$

which we can compute in a finite number of steps by Gaussian elimination, as in Algorithm GAUSS-SQUARE in Section 8.2. And we've said that approximation algorithms are generally used for problems for which an exact solution would require an infinite number of steps. Still, we think we can convince you that such algorithms have a role to play in the solution of Eq. (3).

The method for solving Eq. (3) which we are going to discuss is called the Gauss–Seidel method. The idea behind it is quite simple. Suppose that we have an approximation to the solution, which we denote by

$$x_1, x_2, \ldots, x_n$$

and that we want to compute a new approximation, which we'll denote by

$$y_1, y_2, \ldots, y_n$$

Suppose also that none of the diagonal elements a_{ii} of the matrix A is 0. Now, remember that Eq. (3) is really a shorthand for the system of equations:

$$\begin{aligned} a_{11}x_1 + \cdots + a_{1n}x_n &= b_1, \\ a_{21}x_1 + \cdots + a_{2n}x_n &= b_2, \\ &\vdots \\ a_{n1}x_1 + \cdots + a_{nn}x_n &= b_n. \end{aligned} \tag{4}$$

We begin by using the first of Eqs. (4) to compute a new value of the first variable as follows:

$$y_1 = -\left(\frac{1}{a_{11}}\right)\left(\sum_{j=2}^{n} a_{1j}x_j - b_1\right) \tag{5}$$

All we have done in Eq. (5) is to solve the first of Eqs. (4) for x_1, compute the value given by the right-hand side, and call it y_1. Similarly, we can solve the second of Eqs. (4) for x_2, compute the right-hand side using the values of $x_3, x_4, \ldots, x_n$ and the value of y_1 from Eq. (5), and call this y_2:

$$y_2 = -\left(\frac{1}{a_{22}}\right)\left(a_{21}y_1 + \sum_{j=3}^{n} a_{2j}x_j - b_2\right).$$

Applying this idea to all other equations of (4), always using the y values already computed to get the next one, we obtain the general equation:

$$y_k = -\left(\frac{1}{a_{kk}}\right)\left(\sum_{j=1}^{k-1} a_{kj}y_j + \sum_{j=k+1}^{n} a_{kj}x_j - b_k\right), \qquad k = 1, \ldots, n. \tag{6}$$

This set of n equations computes a new set of y_i's from the given set of x_i's. Then, letting the y_i's now play the role of the x_i's we can repeat this iteration as many times as we like. This procedure probably seems rather strange, so let's look at an example before we proceed.

EXAMPLE 2 Use Eq. (6) to solve the system

$$\begin{aligned} 4x_1 - x_2 \qquad &= 2, \\ -x_1 + 4x_2 - x_3 &= 6, \\ -x_2 + 4x_3 &= 2. \end{aligned}$$

Solution These equations have a symmetry that is typical of the types of systems to which the Gauss–Seidel method is applied. Writing these equations in the form of Eq. (6) we get

$$\begin{aligned} y_1 &= (1/4)x_2 + 1/2. && \text{[Solving the first equation for } x_1\text{]} \\ y_2 &= (1/4)y_1 + (1/4)x_3 + 3/2. && \text{[Solving the second equation for } x_2\text{]} \\ y_3 &= (1/4)y_2 + 1/2. && \text{[Solving the third equation for } x_3\text{]} \end{aligned}$$

Let's start with $x_1 = x_2 = x_3 = 0$. (Note, by the way, that the starting value of x_1 doesn't matter. Why?) Then, applying these equations and remembering that, at the end of each iteration, the y_i's from one iteration become the x_i's in the next iteration, we get the following sequence of values for the x_i's.

	x_1	x_2	x_3
Starting values	0	0	0
Iteration 1	1/2	13/8	29/32
Iteration 2	29/32	125/64	253/256
Iteration 3	253/256	1021/512	2045/2048
Iteration 4	2045/2048	8189/4096	16381/16384

As you can see, the three columns are converging quite rapidly to $x_1 = 1$, $x_2 = 2$, and $x_3 = 1$, which by substituting into the original system, you can quickly verify is the true solution. ▮

So, at least, this method converges pretty rapidly for this example. We'll come back to the question of when it works and when it doesn't and how efficient it is. But first let's express this method in our algorithmic language. One definition before we do, however. If $\mathbf{r}$ is a vector with components $r_1, r_2, \ldots, r_n$, then the **norm** of $\mathbf{r}$, denoted by $\|\mathbf{r}\|$, is defined to be

$$\|\mathbf{r}\| = \left(\sum_{i=1}^{n} r_i^2\right)^{1/2}. \qquad (7)$$

Equation (7) is merely the definition of the length of the vector which generalizes the definition of length in R^2 and R^3.

Algorithm 9.3
GaussSeidel

Input A [Matrix of coefficients]
$\mathbf{b}$ [Right-hand side]
$x_i, \quad i = 1, \ldots, n$ [Initial guess at solution]
ϵ [Error tolerance]

Algorithm GaussSeidel

repeat
 $norm \leftarrow 0$
 for $k = 1$ **to** n
  $y_k \leftarrow -(1/a_{kk})\left(\sum_{j=1}^{k-1} a_{kj}y_j + \sum_{j=k+1}^{n} a_{kj}x_j - b_k\right)$
  $norm \leftarrow norm + (x_k - y_k)^2$ [Square of length of $x - y$]
 endfor
 for $i = 1$ **to** n $x_i \leftarrow y_i$ [y becomes x for next iteration]
endrepeat when $norm < \epsilon$

Output $x_i, \quad i = 1, \ldots, n$

Algorithm 9.3, GAUSSSEIDEL uses Eq. (6) repeatedly. During each iteration, the algorithm computes the square of the norm of the difference between **x** at the beginning of the iteration and **y** at the end of the iteration. The algorithm halts when this norm is less than the tolerance ϵ. After each iteration the newly computed **y** becomes the **x** for the next iteration. Going back to Example 2, you could compute the norm at the end of each of the four iterations—3.712, 0.775, 0.0085, and 0.00013, respectively—which also makes the rapid convergence clear.

But what can we say about the termination of Algorithm GAUSSSEIDEL? For what systems will the norm of the difference between **x** and **y** drop below ϵ at some point? Unfortunately, there is no simple answer to this question. However, for important classes of matrices, including many which occur in practical problems, it can be proved that Algorithm GAUSSSEIDEL terminates. But even when the algorithm does terminate, can we be sure that the closeness of two successive iterations means that each is close to the true solution? All we can say here is that *usually* closeness of successive iterations means that you are near the solution. However, for certain systems, called **ill-conditioned systems**, a small norm can nevertheless mean that you are a considerable distance from the true solution.

Is it useful to have a method like Gauss–Seidel when we have Gaussian elimination, which gives the exact value in a finite number of steps? It is, for the following reason. Although Gaussian elimination theoretically finds the exact solution in a finite number of steps, in practice the same roundoff errors discussed with respect to Algorithm BISECTION prevent finding an exact solution. So what? If you carry enough digits in the computation, you should still get very close to the true solution–closer than you would with a lot of iterations of Algorithm GAUSSSEIDEL. But each Gauss–Seidel iteration requires $\mathrm{Ord}(n^2)$ operations (do you see why?; if not, see [10]), whereas, as we noted in Section 9.3, Gaussian elimination requires $\mathrm{Ord}(n^3)$ iterations. So for large systems you can do many Gauss–Seidel iterations for the same effort required by Gaussian elimination. In fact, for very large systems (and systems with $n>100{,}000$ have been solved on computers) the Gauss–Seidel method or closely related methods are often more efficient and just as accurate as Gaussian elimination. This is particularly true for **sparse** systems, in which most of the elements are 0. Very large systems are often sparse. Taking advantage of sparseness using methods like Gauss–Seidel is easy. However, it usually isn't possible to do so using Gaussian elimination because the successive steps of the elimination generally turn zero elements into nonzero elements.

This completes our brief introduction to the very large topic of approximation algorithms. Anyone who wants to do scientific computing, that is, who wants to solve scientific problems on a computer, will need to know about approximation algorithms.

Problems: Section 9.7

1. Apply the method of bisection to the following functions with a and b as given and $\epsilon = 0.01$ in all cases.

a) $x^3 - 2x - 5$, with $a = 0$ and $b = 3.0$.

b) $\cos x - xe^x$, with $a = 0.2$ and $b = 1.0$.

c) $x^4 - 2x^2 + 3x - 7$, with $a = 0$ and $b = 2.0$.

d) $\tan x - \cos x - \frac{1}{2}$ with $a = 0.5$ and $b = 1.0$.

2. a) Why does Algorithm BISECTION actually find the root to within $\epsilon/2$ of the true value?

b) How would you modify BISECTION so that the result might only be within ϵ of the true solution?

3. Algorithm BISECTION actually provides a proof by algorithm that it does what it purports to do. Show why this is so.

4. Suppose that $f(x)$ is not continuous but that you know two points which satisfy (2). Assume also that you know (or can compute) the smaller and greater values of $f(x)$ at each point of discontinuity.

a) Under what conditions will the method of bisection still work?

b) How would your proof in [3] fail for those conditions under which Algorithm BISECTION would not work?

5. An alternative to the method of bisection is to take the two points

$$(a, f(a)) \qquad \text{and} \qquad (b, f(b)),$$

draw the line joining them, and use the point where this line crosses the x-axis (why must it?) as the next approximation. As with bisection, we use the new point x' and whichever of the two previous ones has sign opposite to $f(x')$ to define a new secant and then just iterate this procedure. This technique is called the **method of false position**.

a) Write the equation which gives x' as a function of a, b, $f(a)$, and $f(b)$.

b) Write an algorithm analogous to Algorithm BISECTION which implements the method of false position.

c) Apply this algorithm to each of the four functions of [1].

6. A fairly obvious generalization of the bisection idea is to trisect the interval in which the solution is known to lie. First, choose two points which are exactly one third and two thirds, respectively, of the way from a to b.

a) How should you choose the two points which are the endpoints of the new interval in which the solution must lie so that this interval will be as small as possible?

b) Embody the result of a) in an Algorithm TRISECTION to find a zero of $f(x)$.

c) Apply TRISECTION to each of the four functions in [1].

7. Which is the better algorithm, Algorithm BISECTION or Algorithm TRISECTION? To answer this you will need some criterion of "bestness". Let that criterion be the number of times that $f(x)$ must be evaluated in order to achieve a certain accuracy (i.e., to determine an interval of some predetermined smallness in which the solution must lie). Let's express the size of this interval as a fraction F of the size of the original interval (a, b). Thus $F = 0.01$ corresponds to an interval whose length is one hundredth of $b - a$.

a) How many evaluations of $f(x)$ are required in Algorithm BISECTION to ensure that the solution interval is no larger than F of the original interval?

b) How many evaluations of $f(x)$ are required in Algorithm TRISECTION to ensure that the solution interval is no larger than F of the original interval?

c) What do you conclude about which of the two algorithms is the more efficient?

8. Use Algorithm GAUSSSEIDEL to perform four iterations for each of the following systems. Use 0 as the starting value for all unknowns. In each case, what is the true solution?

a)
$$\begin{aligned} 3x_1 + x_2 &= 5; \\ x_1 + 2x_2 &= 5. \end{aligned}$$

b)
$$\begin{aligned} x_1 \qquad\qquad - x_3/4 - x_4/4 &= 1/2, \\ x_2 - x_3/4 - x_4/4 &= 1/2, \\ -x_1/4 - x_2/4 + x_3 \qquad\quad &= 1/2, \\ -x_1/4 - x_2/4 \qquad\quad + x_4 &= 1/2. \end{aligned}$$

9. A variant on the Gauss–Seidel idea is to use only the values from one iteration in order to determine the values in the next iteration. That is, in place of Eq. (6) we would have the equation

$$y_k = -\left(\frac{1}{a_{kk}}\right)\left(\sum_{\substack{j=1\\ j\neq k}}^{n} a_{kj}x_j - b_k\right), \qquad k = 1,\ldots,n. \tag{8}$$

This method is called the **Jacobi method**.

a) Write an Algorithm JACOBI analogous to Algorithm GAUSSSEIDEL to implement this idea.

b) Apply Algorithm JACOBI to the system of Example 2.

c) Apply Algorithm JACOBI to the two systems of [8].

10. Use Eq. (6) and Eq. (8) in [9] to count the number of multiplications required in each iteration of the Gauss–Seidel and Jacobi methods. Does the result favor one of these methods over the other?

11. Compare Algorithm GAUSSSEIDEL and Algorithm JACOBI based on the results of Example 2 and [8] and [9]. Do they both seem to work equally well? Does one appear to converge faster than the other? Why do you think this is? In fact, neither algorithm converges all the time, but we won't go into the conditions which determine when each converges.

12. Consider the two systems

$$\begin{aligned} 2x_1 + 6x_2 &= 8;\\ 2x_1 + 6.00001x_2 &= 8.00001; \end{aligned}$$

and

$$\begin{aligned} 2x_1 + 6x_2 &= 8;\\ 2x_1 + 5.99999x_2 &= 8.00002. \end{aligned}$$

a) Do 5 iterations of the Gauss–Seidel method on each of these systems. Start with $x_1 = x_2 = 0$.

b) Do the same for the Jacobi method.

c) Now find the exact solution of both using Gaussian elimination.

d) Can you explain the differences between the solutions of the two systems? *Hint*: Plot the straight lines represented by the two equations in either system. What would happen if you changed the coefficients just a bit?

Systems like these are called ill-conditioned because of the sensitivity of the solution to changes in the coefficients. Large ill-conditioned systems (i.e., many equations) are very difficult to solve accurately.

13. A **unimodal function** is one that rises to its maximum value and then decreases thereafter (or decreases to its minimum value and then increases thereafter). That is, a unimodal function changes direction only once. For instance, of the functions shown, only (a) is unimodal.

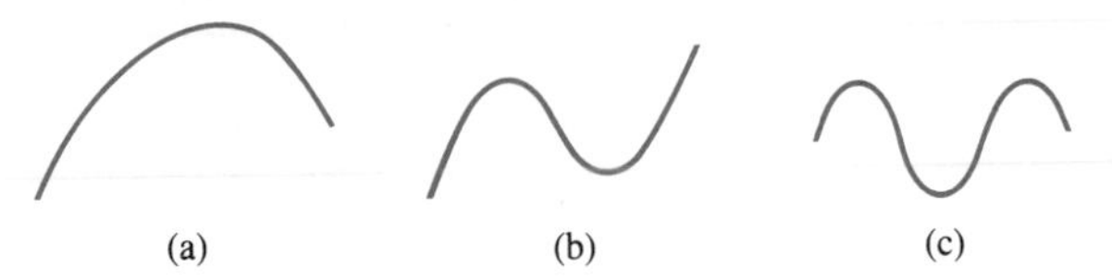

Suppose you know that $f(x)$ is unimodal and that its maximum or minimum lies between a and b.

a) Suppose that the values of $f(x)$ at a, b, and two points between them are as follows:

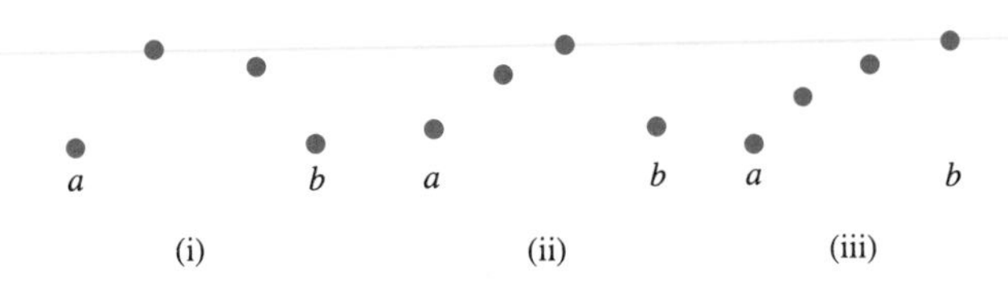

State where the maximum must be in each case. Explain.

b) Write an algorithm for finding the maximum (or minimum) of a unimodal function to within a given tolerance, assuming that you are given two points which bracket the extremum.

c) Apply your algorithm to find the maximum of $f(x) = x/2^x$. (Take our word for it that this function is unimodal. Those of you who know calculus might wish to verify this and also check the answer you get using the algorithm.)

Note 1: Your algorithm will require introducing one or two new points at each iteration. Where you put the points affects the speed at which the algorithm converges to the correct answer. Any ideas about the optimum placement of points? It turns out that the optimum placement involves a sequence that you have seen many times in this book.

Note 2: Those of you who know calculus may wonder why anyone should bother with algorithms like this when calculus finds maxima so easily. What's the answer?

Supplementary Problems: Chapter 9

1. Let $\{a_n\}$ and $\{b_n\}$ be sequences with positive values. Then using big Oh, little oh, and order, as defined in Section 9.3, consider the set of sequences:

$$S = \{\{n\log n\}, \{n^2\}, \{(n+10)^2\}, \{(3+\sin n)n^2\}\}.$$

Find all ordered pairs of functions from S such that the first function is:

a) Little oh of the second

b) Big Oh of the second

c) The order of the second

2. Repeat [1] with:

$$S = \left\{\left\{\sum_{k=1}^{n} k\right\}, \{n^2\}, \left\{\left(\frac{1}{2}\right)n^2\right\}, \{n^2 - \log n\}, \{n^3/\log^2 n\}\right\}.$$

3. Is $\log n = \text{Ord}(\lceil \log n \rceil)$? Justify.

4. Prove that, if $\lim(a_n/b_n)$ exists, then $\text{Ord}(a_n) < \text{Ord}(b_n)$ if and only if $a_n = O(b_n)$ and $a_n \neq \text{Ord}(b_n)$.

5. Prove: If S_n and T_n have the same kth difference for any n, then there exists a polynomial P of degree at most $k-1$ such that $T_n = S_n + P(n)$. *Hint*: Prove by induction, using Theorem 4, Section 9.4, and the result of [3b], Section 9.4.

6. Let a_{nk} be the coefficient of x^k when $x_{(n)}$ is expanded in powers of x. That is,

$$x_{(n)} = \sum_{k=0}^{n} a_{nk} x^k$$

Note that a_{nk} is certainly 0 except for $0 \le k \le n$. This observation will "smooth out the rough edges" in b).

a) Compute a_{nk} from its definition for $0 \le k \le n \le 3$.

b) Find a recursion for $a_{n+1,k}$ in terms of a_{nj} for various j (you figure out which j). *Hint*: $x_{(n+1)} = x_{(n)}(x-n)$.

c) Use b) to compute a_{nk} in a table for $0 \le k \le n \le 5$.

The number a_{nk} is called a **Stirling number of the first kind** and is usually written s_{nk}.

7. Partial fractions. Let $P(x)$ be a linear polynomial $ax+b$ and let $Q(x)$ be a quadratic polynomial cx^2+dx+e, which factors into $c(x-r)(x-s)$, with $r \neq s$. We want to find constants A and B such that

$$\frac{P(x)}{Q(x)} = \frac{A}{x-r} + \frac{B}{x-s},$$

with A and B expressed in terms of a, b, c, r, and s.

a) Do this first by putting the right-hand side over a common denominator and equating like powers of x in the two numerators. Show that what you have done *proves* that the expansion of $P(x)/Q(x)$ on the right-hand side, called the **partial fraction expansion**, actually exists.

b) Then find the partial fraction expansion by multiplying both sides of the equation by $(x-r)$ and then setting $x = r$ to find A. Similarly, multiply by $(x-s)$ to find B.

c) Apply the methods of both a) and b) to i) and ii).

i) $(2x-5)/(x^2-5x+6)$

ii) $(-3x+7)/(x^2+4x)$

8. Let $P_i(x)$ and $Q_j(x)$ be polynomials of degree i and j, respectively, with $i < j$ and suppose that $Q_j(x)$ factors into

$$c(x-r_1)(x-r_2)\cdots(x-r_j),$$

where all the r_k's are distinct.

a) By generalizing what you did in [7a], show that there exist constants A_k, $k = 1, \ldots, j$, such that

$$\frac{P_i(x)}{Q_j(x)} = \sum_{k=1}^{j} \frac{A_k}{x - r_k}.$$

This expansion is called the partial fraction expansion of $P_i(x)/Q_j(x)$.

b) Use the technique in [7b] to find the partial fraction expansions of:

i) $(2x^2-4x+7)/[(x-2)(x-3)(x-5)]$

ii) $\frac{-3x^3+4x^2-7}{(x-2)(x-4)(x-6)(x-8)(x-10)}$

9. Things get more complicated when $Q_j(x)$ has repeated factors or when it has quadratic factors which can't be decomposed into real linear factors. (Do you remember the fundamental theorem of algebra, which says that *any* polynomial with real coefficients can be factored into a product of real linear and real quadratic factors?)

a) If $Q(x)$ has a repeated linear factor such as $(x-r)^k$, the corresponding terms in the partial fraction expansion are

$$\frac{A_1}{x-r} + \frac{A_2}{(x-r)^2} + \cdots + \frac{A_k}{(x-r)^k}.$$

Show why the technique in [7b] doesn't work in this case.

b) But show that the technique in [8a] will still work.

c) Find the partial fraction expansions of:

i) $(x-5)/[(x-1)(x-2)^2]$

ii) $(2x^3-3x^2+4x-1)/[(x-3)(x-4)^3]$

10. If the denominator polynomial factors into quadratics which cannot be further factored into real linear factors and if these quadratic factors are not repeated, the corresponding term in the partial fraction expansion is

$$\frac{ax+b}{cx^2+dx+e}.$$

a) How can you determine whether cx^2+dx+e can be factored into real linear factors? (This question is really a test of whether you remember how to solve quadratic equations.)

b) Find the partial fraction expansion of:

i) $1/[(x-3)(2x^2-4x+5)]$

ii) $(2x+3)/[(x-1)^2(x^2+3x-7)]$

c) What do you suppose happens in the partial fraction expansion when the quadratic factor is repeated? Reason by analogy with what happens for real linear factors.

11. a) If the degree of the numerator polynomial is greater than or equal to the degree of the denominator polynomial, the method of partial fractions still works, but we can't apply it directly to $P(x)/Q(x)$. The argument you used in [7a] to show that the expansion exists won't work now. Why?

b) If we first perform a long division on $P(x)/Q(x)$, we can then do a partial fraction expansion of the remainder. Why?

c) Apply the idea in b) to:

i) $(2x^2-4x+5)/(x^2+3x+7)$

ii) $(4x^4-6x^2+3x-5)/(2x^2-3x+2)$

12. Use the results of [8–11] to state an algorithm to find the partial fraction expansion of *any* rational function (i.e., any ratio of two polynomials). Don't try to make your algorithm too detailed. Just state it in terms of the steps that must be carried out, taking into account the degrees of the numerator and denominator and all the possible factors which the denominator may have. Assume you have a procedure which can factor a polynomial.

13. Newton's interpolation formula. Let $P_m(x)$ be a polynomial of degree m, which we would like to express in the form:

$$P_m(x) = \sum_{k=0}^{m} a_k x_{(k)} \tag{1}$$

for suitable constants a_k.

a) Show that $a_0 = P_m(0) = \Delta^0 P_m(0)$.

b) By taking the difference of both sides of Eq. (1), show that $a_1 = \Delta P_m(0)$.

c) By continuing to take the difference of both sides of Eq. (1), show that $a_j = \Delta^j P_m(0)/j!$, for $j = 2, \ldots, m$. With the coefficients given this way, Eq. (1) is called **Newton's interpolation formula**.

d) a)–c) give the values of the coefficients in (1) if such a representation exists. But does it always? Show that any polynomial may be expressed as a linear combination of falling factorials. *Hint*: Use induction and the fact that the highest power x^m can be expressed as

$x_{(m)}$ – a polynomial of degree $m - 1$.)

e) Here is a short table of values of 2^x:

x	2^x
0	1
1	2
2	4
3	8

Pretend 2^x is a third degree polynomial $P_3(x)$ and calculate the a_k's in Eq. (1). For the second and third differences use the definition in Section 9.4, or the result of [4] of that section. Now evaluate the right hand side of Eq. (1) at $x = 2.5$. You might expect the result to be $2^{2.5}$ but it's not quite. How close is it? Why isn't it exact? Do the answers tell you why the formula developed in this problem is called an **interpolation formula**?

14. Counting binary trees. In [21, Section 3.8], you were asked to find a recursion for the number of binary trees with n vertices. The recursion we had in mind was

$$b_{n+1} = \sum_{k=0}^{n} b_k b_{n-k} \qquad (2)$$

because each such tree with $n + 1$ vertices has two binary trees dangling from its root (one of which may be empty) which, between them, have n vertices.

The right hand side of Eq. (2) is a convolution. Use this fact to translate Eq. (2) into a quadratic equation in the "unknown"

$$B(s) = \sum_{k=0}^{\infty} b_k s^k$$

Solve this quadratic using the quadratic formula and then expand the solution to find b_n. To do this expansion you will need the binomial series theorem of Section 4.6. To get the result in its most convenient form you will have to expand a binomial coefficient with upper index $\frac{1}{2}$ and use it to simplify the expression.

15. Exponential generating functions. Given a sequence $\{a_n\}$, its **exponential generating function** is defined as

$$E(s) = \sum_{n=0}^{\infty} \left(\frac{a_n}{n!}\right) s^n$$

a) Find the exponential generating function for $a_n = n!$.

b) (Requires knowledge of calculus) Find the exponential generating function for $a_n = 1$.

c) Suppose that $A(s)$ and $B(s)$ are the exponential generating functions for $\{a_n\}$ and $\{b_n\}$. Show that $A(s)B(s)$ is the exponential generating function for

$$c_n = \sum_{k=0}^{n} \binom{n}{k} a_k b_{n-k} \qquad (3)$$

d) Suppose that a_n is the number of distinguishable lists (ordered arrangements) of n things from some set A. Likewise, suppose that b_n is the number of distinguishable lists of n things from some set B. Show that if $A \cap B = \emptyset$ (with every item in A distinguishable from every item in B), then c_n as in Eq. (3) is the number of ways to pick a distinguishable list of n things from $A \cup B$.

e) Combine these results into a theorem about the interpretation of products of exponential generating functions. Also state the m-fold generalization.

f) Let A be an infinite set of identical blue balls. Let B be an infinite set of identical red balls. Use the earlier parts of this problem to determine the number of distinguishable n-sequences from $A \cup B$. (Of course, you already know the answer, but our goal here is merely to introduce exponential generating functions. As with ordinary generating functions, they can also be used to solve some problems that our previous methods couldn't.)

Epilogue

Sorting Things Out with Sorting

E.1 Comparison of Previous Methods

We appreciate the stamina you have shown to get this far. Your reward, you might think, is that we would have the grace to stop—finally. We almost do—but not quite. Our aim in this finale is to recapitulate the major themes of this book using a considerable amount—although certainly far from all—the discrete mathematics we have discussed.

Our vehicle for this recapitulation will be the problem of sorting a list into lexical (numerical or alphabetical) order. Building this Epilogue around a particular problem emphasizes yet again that an important motivation for studying discrete mathematics is its broad *applicability*. Sorting is one such application. Understanding it requires knowledge of much of what we have discussed in this book. At the end of this epilogue you will find some challenging Final Problems which give you the opportunity to recapitulate the themes of this book on problems other than sorting.

We have discussed sorting twice. In Section 5.9, we considered the merge sort method as an example of the use of difference equations to solve algorithmic problems. Then in Section 7.7, we introduced the insertion sort method to illustrate the use of quantifiers in verifying algorithms. In this Epilogue we will consider three questions:

1. How efficient is insertion sort compared to merge sort? We'll answer this question shortly, in this section.
2. What is the performance of the best possible algorithm for sorting by comparisons? This is a *complexity* question (see Section 1.6) which we'll answer for both the worst and average cases in Section E.2.
3. Are there methods which achieve the best possible performance, and if so, how do they compare with one another? We'll answer the first part of this

question in Sections E.2 and E.3, but we'll only be able to give a partial answer to the second part.

First, then, let's analyze insertion sort. For convenience, we reproduce Algorithm INSERTION from Section 7.7 as Algorithm E.1. As usual in such an analysis we focus on the key operation in the inner loop. At first glance this operation is the move

$$n_{j+1} \leftarrow n_j. \tag{1}$$

Focusing on (1) wouldn't really be wrong. But note that the test in the **until** portion of the **repeat** requires a *comparison* each time through the loop and one more comparison if the exit from the loop is not caused by $j = 0$. Counting comparisons will therefore give us a better measure of the efficiency of Algorithm INSERTION, and additionally, it will make possible a direct comparison (no pun intended) of this algorithm with the merge sort method described in Section 5.9.

Algorithm E.1

Insertion

```
Input m                                    [Integer > 0]
      n[1:m]                               [List to be sorted]
Algorithm INSERTION
  for i = 2 to m
      temp ← n_i
      j ← i − 1
      repeat until j = 0 or temp ≥ n_j
          n_{j+1} ← n_j
          j ← j − 1
      endrepeat
      n_{j+1} ← temp
  endfor
Output n[1:m]                              [In numerical order]
```

Let's begin with a worst case analysis. Algorithm INSERTION works worst when the exit from the inner loop is always caused by j getting to 0 because each n_i must then be compared with all the already sorted elements in the list. (What initial ordering of the elements of the list would give rise to this behavior [2b]?) Therefore the comparison at the beginning of the inner loop is done $i-1$ times for each value of i. (We assume that, when $j=0$, the expression "$j = 0$ **or** temp $\geq n_j$" is exited before an attempt is made to evaluate the nonexistent n_0.) Thus the total number of comparisons is

$$\sum_{i=2}^{m}(i-1), \tag{2}$$

which is our old friend, the sum of the first $m-1$ integers:

$$\frac{(m-1)m}{2}.$$

Therefore the number of comparisons in Algorithm INSERTION in the worst case is $\text{Ord}(m^2)$, using the order notation developed in Section 9.3.

What about the number of comparisons in the average case? This is one of those rarities when average case analysis isn't much more difficult than worst case analysis. We need to find the average value of the position at which to insert n_i because this will give us the average number of comparisons for each i. That is, we need to find the expectation of a random variable, whose value is the position to insert n_i, on the sample space consisting of i-tuples of values to be sorted (cf., Section 6.5). We'll assume that the probability measure on the sample space is the uniform measure (see Section 6.2) (i.e., any of the $i!$ orderings of the i values is equally likely). Then the position at which to insert n_i is equally likely to be any of the positions from the first (n_i less than any of the already sorted $i-1$ values) to the ith (n_i greater than n_{i-1}) [4].

If the correct place to insert n_i is the jth position, the number of comparisons is $i-j+1$ when $j=2,3,\ldots,i$. If n_i is to be inserted in position 1, then as with $j=2$, the number of comparisons is $i-1$. Therefore the *average* number of comparisons is

$$\begin{aligned}
\left(\frac{1}{i}\right)\left[i-1+\sum_{j=2}^{i}(i-j+1)\right] &= \left(\frac{1}{i}\right)\left[i-1+\sum_{j=1}^{i-1}j\right] \qquad \text{[New } j=i+1-\text{Old } j\text{]}\\
&= \left(\frac{1}{i}\right)\left[i-1+\frac{i(i-1)}{2}\right]\\
&= \frac{(i-1)(i+2)}{2i} = \frac{i^2+i-2}{2i}\\
&= \frac{i}{2}+\frac{1}{2}-\frac{1}{i}. \qquad (3)
\end{aligned}$$

Equation (3) gives the average or *expected* number of comparisons in the inner loop for any value of i. To find the overall expected number of comparisons, we need the expected value of the sum of the random variables for each i from 2 to m (see the outer loop in Algorithm E.1). But (see Theorem 1, Section 6.5) the expectation of the sum of random variables is the sum of the expectations, so we need only sum Eq. (3) from 2 to m. If we do so, the first term on the last right-hand side of Eq. (3) is the dominant one (why?) and its sum is [5]

$$\left(\frac{1}{2}\right)\left[\frac{m(m+1)}{2}-1\right].$$

Therefore the average case for sorting by insertion is also $\text{Ord}(m^2)$, except now the multiplicative constant c [cf., Eq. (2), Section 9.3] is $\frac{1}{4}$ rather than $\frac{1}{2}$. Thus, on average, Algorithm INSERTION is about twice as fast as the worst case. Why might you have expected that result intuitively without doing any formal analysis [6]?

We can make sorting by insertion more efficient, although its order won't change, if we search for the point to insert n_i more efficiently than in Algorithm INSERTION. We'll take this up in the problems [12].

What are the corresponding average and worst case results for merge sort? In Section 5.9 [Eq. (9)], we showed that merge sort is an $O(m \log m)$ algorithm in the worst case. Note that this was a Big Oh and not an Order result because we showed only that the number of comparisons was less than or equal to $m \log m$. However, if the worst case for merge sort is $O(m \log m)$, its order for both the worst and average cases can be no worse than $m \log m$ (why?). Therefore, although Algorithm INSERTION is effective for quite small values of m, merge sort will be much more effective for large values.

So we have one example of a sorting-by-comparisons method—merge sort—which has an order at least as small as $m \log m$. Are there algorithms for sorting by comparisons which have a smaller order (in the sense of an order ranking as defined in Section 9.3)? Answering that question is the subject of Section E.2.

Problems: Section E.1

1. Perform an insertion sort on each of the following lists, counting the number of comparisons you make. Is each list more nearly an average case or a worst case? Why?

a) 8 15 12 21 3 9 2

b) 4 9 11 17 13 21 3

c) 18 14 22 9 5 11 3

2. **a)** What is the best case for insertion sort for m items?

b) Describe the form of a list that gives the worst case behavior for insertion sort. Prove that the list you describe does, indeed, result in the predicted behavior.

3. What is the best case for merge sort with:

a) 8 items **b)** 2^k items

4. Recall our average case analysis of insertion sort. We asserted that assuming that each of the $i!$ orderings of i items is equally likely implies that the place to insert the ith item in insertion sort is equally likely to be any place from the first to the ith. Why is this assertion correct?

5. Sum the last right-hand side of Eq. (3) from $i = 2$ to m, thereby verifying the average case behavior for Algorithm INSERTION given in the text.

6. Our analysis of insertion sort showed that its average behavior is about twice as good as its worst behavior. Why should you have expected this result intuitively before doing any of the analysis of this section?

7. Inversion tables. An **inversion** in a list $S = s_1, s_2, \ldots, s_n$ is any pair (s_i, s_j) such that $i < j$ and $s_i > s_j$. An **inversion table** $T = t_1, t_2, \ldots, t_n$ corresponding to S has the property that t_j is the number of elements in S to the left of s_j which are greater than s_j.

a) Display the inversions in $S = 6, 3, 8, 10, 5, 12$.

b) Display the inversion table for this list.

c) Discover a relation between t_j in the table found in b) and the second elements in the inversions found in a).

d) Prove that the relation found in c) is valid for all inversion tables.

8. Recall the algorithm for bubble sort described in [5] of Section 7.7.

a) Calculate the inversion table for the list given in Fig. 7.12, Section 7.7.

b) Do one *pass* (i.e., proceeding from the beginning to the end of the list once) of bubble sort on this list and then calculate the inversion table for the resulting list.

c) Describe a relationship between the table calculated in a) and that calculated in b).

d) Prove that the relationship found in c) always holds after each pass of bubble sort when the lists before and after the pass are being compared. *Hint*: For each element e in the list, what happens when it is exchanged with a larger element to its left?

9. a) What initial form of a list leads to the worst possible behavior for bubble sort?

   b) Use the result in a) to do a worst case analysis of bubble sort.

10. Prove: The average number of passes in bubble sort for the pair consisting of a given list and its reversal (i.e., the list in reverse order) is at least $(m-1)/2$. *Hint*: The smallest number in the list will be to the right of each other number either in the list or its reversal. Consider the sum of the entries in the inversion tables for the list and its reversal corresponding to this element and use the result in [8d].

11. a) Suppose that you have a list of m elements and assume—because it makes the algebra a bit easier—that m is odd. Each pass of bubble sort requires one less comparison than its predecessor. Use this fact and the result in [10] to show that the average behavior of bubble sort is at least $\mathrm{Ord}(m^2)$.

    b) By comparing the result in a) with your result for the worst case behavior in [9], what can you say about the order of bubble sort in the average case?

12. Our description of insertion sort involved a *sequential search*, starting at the $(i-1)$st element of the list and proceeding down the list until we found the position at which to insert the next element. But we argued in Section 1.6 that sequential search was not nearly as efficient as binary search. Let's see if we can usefully apply binary search to insertion sort.

    a) Modify Algorithm INSERTION so that it uses binary search to find the insertion point. The resulting method is called binary insertion sort.

    b) How many comparisons are now required in the worst case? (Use the result we derived for binary search in Section 2.6.)

    c) The average case analysis of the number of comparisons in binary insertion sort is considerably more difficult than the worst case analysis. Why?

    d) Binary insertion sort is certainly an improvement over regular insertion sort. But for a list stored as an array as in Fig. 7.11, it's not as much of an improvement as the result in b) might imply. Why? *Hint*: Why is it wrong to focus on comparisons in the analysis of binary insertion sort?

E.2 The Complexity of Sorting by Comparisons

With merge sort we have an algorithm for sorting by comparisons which is $O(m \log m)$ in the worst case (and also therefore in the average case) where m is the length of the list. This gives us an *upper bound* on how well we can do with sorting by comparisons; that is, Algorithm MERGESORT requires no more than $cm \log m$ comparisons for some c and sufficiently large m. In this section, however, our concern is with *lower bounds*. The question we want to explore is this:

> Among all conceivable algorithms for sorting by comparisons, can we find some function of m, $f(m)$, such that no algorithm has a lower order ranking than $\mathrm{Ord}(f(m))$ (in the sense of Definition 3 in Section 9.3) in the worst case or the average case or—better still—both cases?

In this form the question is too easy. If $f(m) = 1$, we have found such a lower bound because no algorithm for sorting by comparisons can be such that the number of comparisons is no greater than some constant independent of m. (Why? See Theorem 1.) What we really want is the greatest possible lower bound so that we will really know how well we can possibly do.

In Section 1.6, we pointed out that complexity questions of this kind are inherently more difficult than analysis of algorithms questions because they require us to discover something about all possible algorithms for a task without knowing what all those algorithms may be. Only once before in this book—namely, for the Towers of Hanoi in Section 1.6—have we been able to answer the question of what the best possible algorithm for a task might be. As you'll see, sorting by comparisons is also a case where we can get a tight lower bound, with "tight" meaning that our lower bound will be the best possible one.

Note our emphasis on sorting by comparisons, which is the way both merge sort and insertion sort operate. Not all methods of sorting use comparisons of one list element with another.[†] So our result in this section will be incomplete as regards the overall problem of sorting. Still, it is fair to note that almost all sorting on computers is done using comparisons.

We also note that by "comparison" we mean "two-way comparison," such as in Algorithm INSERTION, in which one list element is compared with only one other. Any comparison of three or more elements may always be described in terms of a sequence of two-way comparisons [3] and, on current computers, must be so described.

Suppose that you have a list of two items. What's the smallest number of comparisons you need to sort them? Clearly, one; just compare the two items. How about three items? You can probably convince yourself rather easily that you need at least two comparisons to sort them. Theorem 1 generalizes this approach to give a lower bound for sorting m items (see also [31] in Section 5.9).

Theorem 1. Any valid algorithm that can sort all lists of m items using comparisons must use at least $m-1$ comparisons on each list.

Theorem 1 doesn't imply that there actually is an algorithm which *can* sort all lists of m items with $m-1$ comparisons. It means instead that no method of sorting by comparisons can possibly sort all lists with m items if it ever uses fewer than $m-1$ comparisons.

PROOF Let's represent any set of comparisons by a graph in which the items to be sorted are the vertices, and an edge between two vertices represents a comparison of those two items as shown in Fig. E.1(a). Suppose that the graph is not connected. Could an algorithm using the comparisons described by the graph possibly sort the

[†] One technique which does not is called *radix sorting*. Imagine each element to be sorted to be a decimal integer. Then, first separate the list into sublists according to the most significant digit in any number on the list. Then, within each sublist, separate into ten further groups based on the next most significant digit, etc. [1, 2].

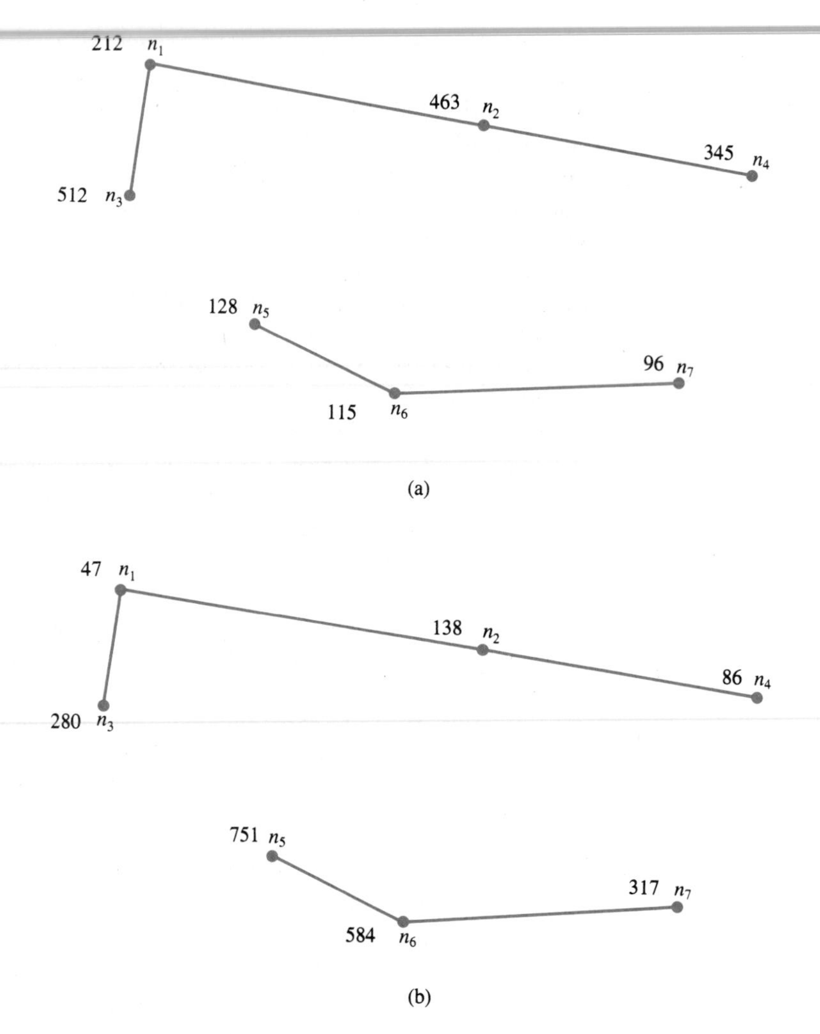

FIGURE E.1 A Graphic Model of Sorting by Comparisons. (a) All data in the upper component are greater than all data in the lower component; (b) all data in the upper component are less than all the data in the lower component. Each vertex represents an item to be sorted; each edge represents a comparison of the items at the two vertices it joins.

items at the m vertices? Well, if the graph isn't connected, there are at least two distinct components. Let A be the set of vertices in one component and B the set of vertices in another. Could comparisons just among the vertices of A and just among the vertices of B possibly tell you whether an item in a vertex in A goes before or after an item in a vertex in B? No, by the following argument. Suppose that the algorithm represented by the graph in Fig. E.1 is run twice, once with the data shown in Fig. E.1(a) and once with the data shown in Fig. E.1(b). Thus the first time all data in the upper component are greater than all data in the lower component—and vice versa the second time. But the claim made by any algorithm about how the data are ordered overall must be the same both times. The reason is that, in sorting by comparisons, all decisions are based on the relative sizes of the compared data and these relative sizes are the same in both cases. Because the true sorted order is different in the two cases, the algorithm must be incorrect for at least one of them. Therefore we conclude that the graph representation of a sorting-by-comparisons algorithm shown must be connected if the comparisons it represents can sort the m items.

Note: Since sorting-by-comparison algorithms are allowed to choose the next comparison on the basis of the results of previous comparisons, a given algorithm might produce a different graph for different data when modeled as in Fig. E.1. What we have shown is that *any* such graph must be connected.

Now, a connected graph on m vertices must have at least $m-1$ edges. Recall Algorithm SpanTree (Section 3.8), which finds a spanning tree on any connected graph. This algorithm starts with a single vertex and then repeatedly adds a new edge together with a new vertex until all vertices are included. Thus when the tree reaches all m vertices, it has $m-1$ edges. Hence the original connected graph had at least $m-1$ edges.

So

a) only a connected graph can represent the action of any sorting-by-comparisons algorithm, and
b) any connected graph with m vertices has at least $m-1$ edges.

Therefore it follows that any algorithm to sort all lists of m items must always use at least $m-1$ comparisons. ∎

Early in this section we argued that no algorithm for sorting by comparisons could be Ord(1). Theorem 1 does somewhat better: It says that any such algorithm must be at least Ord(m). Can we achieve a still better lower bound? Yes.

What we need is a better *model* of sorting by comparisons than the graph in Fig. E.1—one that captures the essence of sorting more effectively than that graph. Such a model is a **decision tree** of the kind introduced previously in Fig. 4.3, Section 4.3, in the discussion of the 13 coins problem. We show two such trees in Fig. E.2 that we can use to sort lists of two and three items. Each nonleaf node in a decision tree represents a binary decision (i.e., a decision based on a comparison of two items with a binary—yes or no, less-than or greater-than-or-equal—result). The two children of each nonleaf node contain the next decision to be made based on the result of the previous decision or, if either is a leaf, contain the sorted order determined by the previous decisions. As sorting by comparisons must involve a sequence of decisions, with the next one allowed to depend on the result of the prior one, any such method may be represented by a decision tree.

A decision tree as we have described it is just a *binary tree* (see Section 3.8), with the added restriction that each node has either two or zero children. In a

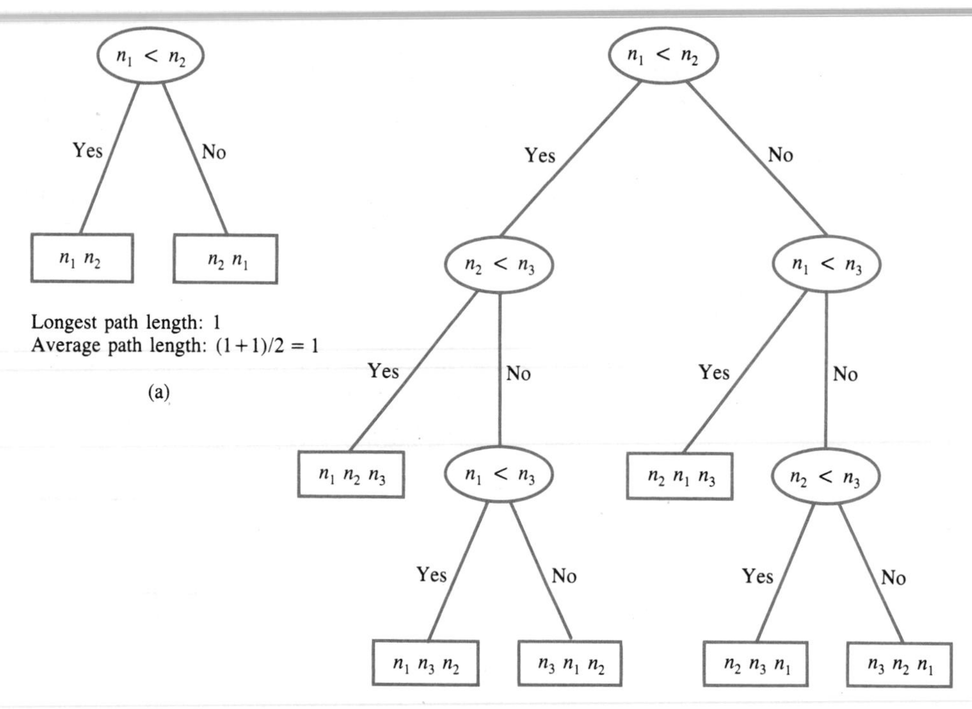

FIGURE E.2 A Decision Tree Model for Sorting by Comparisons. Nonleaf nodes represent comparisons; leaves show the correct sorted order determined by the comparisons.

decision tree describing sorting by comparisons, the leaves of the tree must describe all possible orderings, or *permutations* of the m elements. Thus the leaf you arrive at for a particular set of data gives you the sorted order for that data. Because a list of m items can have $m!$ possible permutations (see Section 4.4), the number of leaves in a decision tree for sorting by comparisons must be $m!$.

To use the decision tree model to get lower bounds on the average and worst cases of sorting by comparisons, it's useful to recall the notion of the *length* of a path in a tree (or in any graph): the number of edges on the path (see Section 3.6). In a decision tree which describes a method of sorting by comparisons, the length of a path from the root to a leaf, called the **height** of the leaf, represents the number of decisions (i.e., comparisons) made to sort the particular data represented by the leaf. The maximum of the heights of the leaves is called the height of the tree. Therefore the *longest* path from the root to a leaf gives the number of comparisons in the worst case, and the *average* of the lengths of the paths from the root to all the leaves gives the average number of comparisons. Therefore the two questions we want to answer are:

1. Among all possible decision trees for sorting m items by comparisons, which has a longest path which is shortest? This will give the best possible worst case and therefore an exact lower bound on the best you can hope to do in a worst case sense. (The longest path in Fig. E.2 is 3 when $m = 3$ and 1 when $m = 2$.)
2. Similarly, which decision tree has the smallest average path length? This will give a lower bound on how well you can do on the average. (The average path length in Fig. E.2 is $2\frac{2}{3}$ when $m = 3$ and 1 when $m = 2$.)

To answer the first question, the essential result we need is a bound on the number of leaves in any binary tree of height h (and hence in any decision tree of height h). This result will tell us how high a decision tree must be if it is to have $m!$ leaves. Before reading further, use the inductive paradigm to conjecture the answer. Then read on for a proof that you were (we hope) correct.

Theorem 2. A binary tree of height h has at most 2^h leaves.

PROOF By strong induction, if $h = 0$, the tree consists of only a root which is also a leaf. As $2^h = 2^0 = 1$, the basis case is correct. As the induction hypothesis, we assume that the theorem is true for trees of height h or less. Then suppose that we have a binary tree of height $h+1$. We may view this tree as a root with at most two subtrees, left and right. Since the entire tree has height $h+1$, each subtree has height at most h. Therefore by the induction hypothesis, each subtree has at most 2^h leaves. Because these leaves are also the leaves of the entire tree, the tree of height $h+1$ has at most

$$2^h + 2^h = 2^{h+1} \tag{1}$$

leaves, which proves the theorem. ∎

Now suppose that a decision tree has $m!$ leaves. Then it follows from Theorem 2 that its height must be at least

$$\lceil \log_2 m! \rceil. \tag{2}$$

Why is expression (2) correct? If you don't see why, work it out from Theorem 2. Turning around results like that in Theorem 2 is often not all that easy. Doing so in this case will test whether you really understand the ceiling function.

From (2), it follows that, to find the minimum possible height for a tree to sort m items—which will tell us how bad the worst case must be—we need to be able to calculate or, at least, estimate $\log_2 m!$. We do so as follows:

$$\begin{aligned} m! &= m(m-1)(m-2)\cdots 3\cdot 2\cdot 1 \\ &= m(m-1)\cdots(\lceil m/2\rceil)(\lceil m/2\rceil - 1)\cdots 3\cdot 2\cdot 1 \\ &\geq m(m-1)\cdots(\lceil m/2\rceil) \quad \text{[Because omitted terms are all } \geq 1] \\ &\geq (\lceil m/2\rceil)^{\lceil m/2\rceil} \quad \text{[Because there are at least } \lceil m/2\rceil \text{ terms, and each is at least } \lceil m/2\rceil] \end{aligned}$$

$$\geq \left(\frac{m}{2}\right)^{m/2}$$

Therefore

$$\lceil \log_2 m! \rceil \geq \left(\frac{m}{2}\right) \log_2 \left(\frac{m}{2}\right) = \left(\frac{m}{2}\right) (\log_2 m - 1). \quad (3)$$

It's convenient to get rid of the -1 in Eq. (3), which for $m \geq 4$, we can do by putting another factor of $\frac{1}{2}$ in front [16] to obtain

$$\lceil \log_2 m! \rceil \geq \left(\frac{m}{4}\right) \log_2 m, \qquad m \geq 4. \quad (4)$$

The bound in Eq. (4) is not a terribly good one, but it is quite good enough for our purposes. If you know calculus, you should be able to get a better one [17].

From Eq. (4), we can conclude that the number of comparisons for any algorithm for sorting by comparisons can, in the worst case, be no less than $\text{Ord}(m \log m)$ since the coefficient in Eq. (4) and the base of the logarithm don't affect the order of a sequence.

How good is this lower bound? Is it optimally tight? On the one hand, we have shown in Section 5.9 that merge sort is $O(m \log m)$ in the worst case. On the other hand, we have just shown that no algorithm for sorting by comparisons can be better than $\text{Ord}(m \log m)$ in the worst case. Together these two results show that, in fact, merge sort is $\text{Ord}(m \log m)$ in the worst case (why? [18]) and that no algorithm can have a lower order ranking than merge sort in the worst case. Therefore in the sense of the order of the worst case, merge sort is a best possible algorithm. Does that mean we should always use it, at least for large values of m? Not quite. We'll return to this question in Section E.3.

First, though, how about a lower bound for the average case? Just because we can't do better than $\text{Ord}(m \log m)$ in the worst case does not, in itself, mean that we can't do better than that for the average case. So, let's see if we can obtain a lower bound, valid for all decision trees, for the average length of paths from the root to all the leaves. We'll do so by showing that the average path length for all binary and thus all decision trees is at least as great as the average length for a particularly nice type of binary tree. Theorem 3 provides the essential result we need. First, however, a definition.

Definition 1. A binary tree is said to be **filled at level j** if it contains all possible nodes at that level.

What is the maximum number of nodes at level j for a binary tree? The answer is 2^j, which you should be able to prove by induction [19]. Also make sure you see that, if a binary tree is filled at level j, then it is filled at all higher levels also (see Fig. E.3). The tree in Fig. E.3 is filled at all levels except the bottom one. A tree is called **filled** if it is filled at each level except, perhaps, the bottom one.

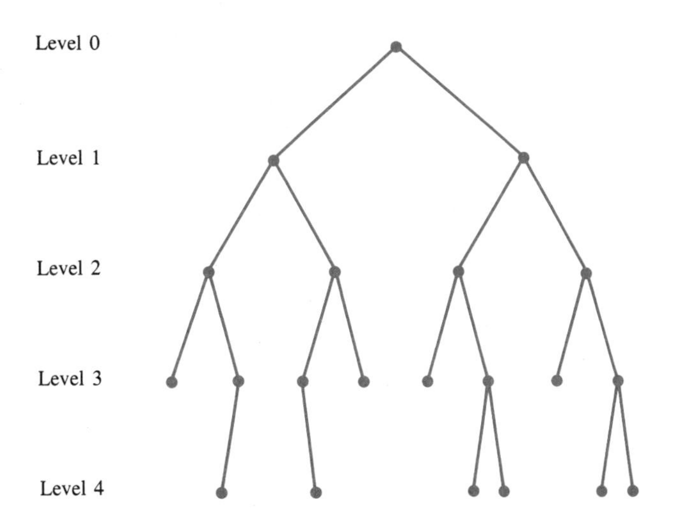

FIGURE E.3

A Binary Tree with All But the Last Level Filled

Theorem 3. Among all binary trees with k leaves, the sum of the path lengths to the leaves of some filled tree is at least as small as that to any nonfilled tree.

PROOF Note first that a filled tree of height L has all its leaves at heights L or $L-1$. For if there were a leaf two or more levels up from the bottom, the level below that leaf would not be filled (why?).

We'll prove the theorem by showing that for any hypothetical binary tree with k leaves, one of which is two or more levels up from the bottom, there is a corresponding filled binary tree with the same number of leaves which has a sum of path lengths at least as small as our hypothetical tree.

For a tree with one or more leaves two or more levels above the bottom level, we must consider two classes of leaves at the bottom level:

- The class of leaves which have a sibling; two nodes such as A and B in Fig. E.4(a) are siblings because they have the same parent, namely, C.
- The class of leaves which do not have a sibling.

For the first case refer to Fig. E.4(a), in which two leaves (e.g., A and D) have levels which differ by 2. Then, as shown, consider what happens if the children of C are moved to D (which is therefore no longer a leaf). This move doesn't change the number of leaves because C is now a leaf. What has happened to the sum of the heights of the leaves? We need only consider nodes A, B, C, and D because no other leaf heights have changed. We have

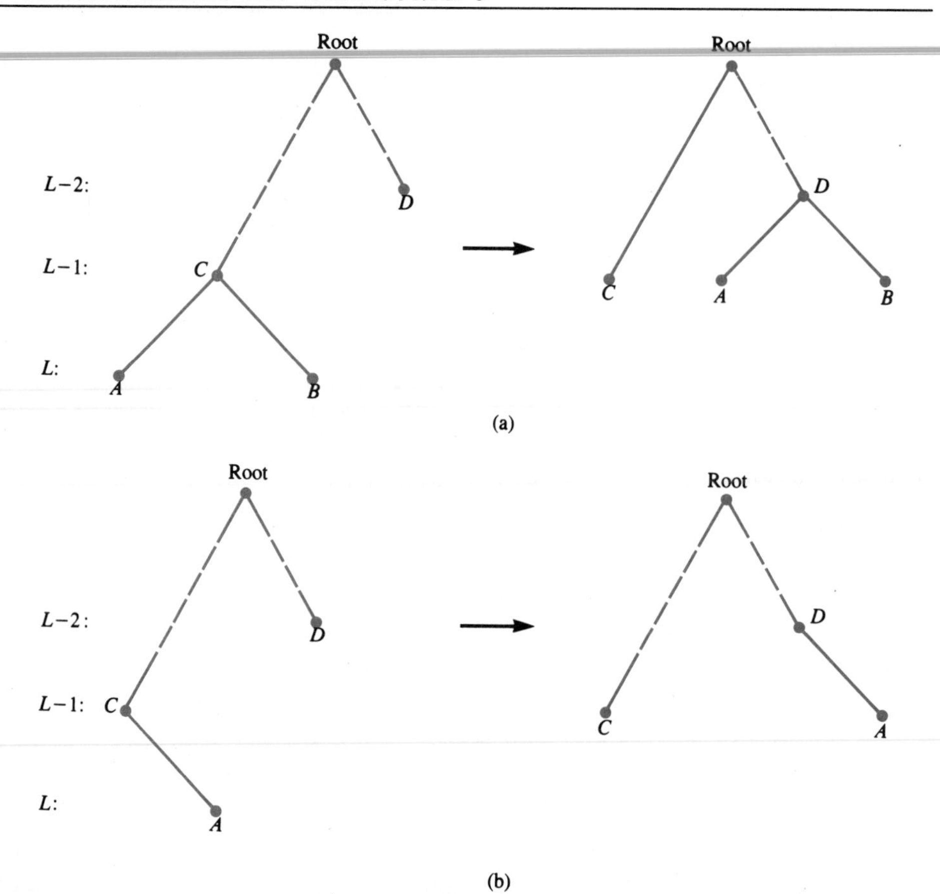

FIGURE E.4

Decision Trees with Minimal Average Path Length to the Leaves. (a) Leaves with levels which differ by 2, and the leaf at the lowest level has a sibling; (b) leaves with levels which differ by 2, but the leaf at the lowest level has no sibling. The lowest level is L in both (a) and (b).

$$\begin{aligned}
\text{Before move:} \quad & \text{Height}(A) + \text{Height}(B) + \text{Height}(D) &= L + L + L - 2 \\
& &= 3L - 2. \\
\text{After move:} \quad & \text{Height}(A) + \text{Height}(B) + \text{Height}(C) &= L - 1 + L - 1 + L - 1 \\
& &= 3L - 3.
\end{aligned}$$

Thus the sum of the heights has been reduced.

For the other case, consider Fig. E.4(b) in which A is the only child of C. This time we get

$$\begin{aligned}
\text{Before move:} \quad & \text{Height}(A) + \text{Height}(D) &= L + L - 2 = 2L - 2. \\
\text{After move:} \quad & \text{Height}(A) + \text{Height}(C) &= L - 1 + L - 1 \\
& &= 2L - 2.
\end{aligned}$$

So the sum of the heights is unchanged.

We can continue to make such moves until there are no more leaves whose levels differ by 2, that is, until we get some filled tree. Must this process stop? Sure, but we'll leave the proof to a problem [22]. As we never increase the height on any move, it follows that for every non-filled tree, there is a corresponding filled tree with the same or a smaller sum of heights. This proves the theorem. ∎

So a filled tree with k leaves has the lowest possible sum of path lengths and, therefore, the lowest possible average path length among all trees with k leaves. In fact, all filled decision trees with k leaves have the same average path length [21]. However, it suffices for our purposes to prove that every filled decision tree with k leaves—and hence any one with smallest average path length—has an average path length at least $h-1$, where h is the height of the tree. Let's assume that $k = m!$ because $m!$ will be the number of leaves in the decision tree corresponding to any algorithm to sort m items by comparisons (see also [23]). So, with h the height of a filled decision tree with $m!$ leaves, we have

$$2^{h-1} + 1 \;\leq\; m! \;\leq\; 2^h \tag{5}$$

because 2^h is the number of leaves if the bottom level is filled (see Theorem 2), and $2^{h-1}+1$ is the number of leaves if there is only one node at level $h-1$ which has children at level h. (Why does every other configuration have a number of leaves between these extremes [25]?)

Removing the 1 on the left-hand side of Eq. (5) changes the $\leq$ to a $<$. Then taking logarithms to the base 2, we obtain

$$h - 1 \;<\; \log_2 m! \;\leq\; h,$$

or

$$h - 1 \;<\; \lceil \log_2 m! \rceil \;=\; h. \tag{6}$$

Since all the leaves of the tree satisfying Theorem 3 are at height h or $h-1$, the maximum level or one less than this, it follows that the average path length to these leaves is at least $h-1$. And as h in Eq. (6) is exactly the same as in (2) (why might we have expected that? [26]), we can use Eq. (4) to state that the smallest average path length is greater than

$$\left(\frac{m}{4}\right) \log_2 m - 1, \qquad m > 4.$$

The -1 doesn't affect the order (why not?) so what we have found is that *the best possible average case behavior for sorting by comparisons is* $\mathrm{Ord}(m \log m)$.

This bound is tight because it is the same as the worst case bound, which is tight. Also, we have already found an algorithm—Merge Sort—which is at least as good as $\mathrm{Ord}(m \log m)$ in the average case. In Section E.3 we'll turn to the question of why we might wish to search for other algorithms even after we've found one which is best possible in both the worst and average cases.

Problems: Section E.2

1. Radix sorting. Design an algorithm for radix sorting as described in the footnote on p. 803. That is, suppose (in the decimal case) that you start with a list of length m in which each number has k or fewer digits. First, you put all the numbers whose kth (most significant) digit is 9 in one storage area, those whose kth most significant digit is 8 in another storage area, etc. Then you look at the $(k-1)$st digits in the 9 area and put all those with 9's in the $k-1$ position in one area, those with 8's in the $k-1$ position in another area, etc.

2. Now let's analyze radix sorting.

   a) If the length of the list is m, what is the order of radix sorting in both the average and worst cases, using as your measure the number of moves of a datum from one location to another?

   b) The result in a) is a time analysis. How about a space analysis? How many distinct storage locations are required by the algorithm you designed in [1]?

   Note: One way to overcome the space problems with radix sorting is to store the list in a linked array in which each storage element stores not only a value but also a *pointer* to the next element. But there are still lots of problems we haven't considered with a computer implementation of radix sorting. There we must deal with binary numbers and extract the bits of each number one by one (or group by group) to do the sorting.

3. a) Suppose that you wanted to do three-way comparisons to sort a list instead of the two-way comparisons used in this section. How could you do this, using a sequence of two-way comparisons?

   b) Generalize a) to derive an algorithm to do m-way comparisons using a sequence of two-way comparisons. What is the difference between the algorithm you derived and an algorithm to sort a list of m items?

4. Here is an algorithm that purports to sort any list of m items $n_1, n_2, \ldots, n_m$ in exactly $m-1$ comparisons. It reads the data in order. For each k from 2 to m, if n_k is larger than the largest value previously found, then n_k is declared to be larger than everything previously on the list; else n_k is declared to be less than everything previously on the list.

   a) Give an example to show that this algorithm is invalid.

   b) For how many possible orders of the m items will this algorithm give the correct sort? As $m \to \infty$, what is the probability that it will give a correct sort? (Assume that all orders of the data are equally likely.)

   c) For what values of m is there *any* algorithm which can sort all lists of m items using $m-1$ comparisons?

5. In [23, Section 3.7], you were asked to prove that an acyclic graph (and thus any free tree) with at least one edge has at least one vertex adjacent to exactly one other vertex. Now improve on this result: Show that any free tree with at least one edge has at least two leaves (vertices of degree 1). A proof by algorithm is preferred. Where does your proof fail for general graphs (i.e., for graphs which are not free trees)?

6. Suppose that a free tree contains one vertex of degree 6 and another of degree 9. What is the smallest number of leaves such a tree can have? Prove it.

7. Instead of (as in the text) using Algorithm SpanTree to show that a free tree with m vertices has $m-1$ edges, try to prove it by induction. Assume that a free tree with m vertices has $m-1$ edges and then try to prove it for free trees with $m+1$ vertices by adding an edge (and associated vertex) to a free tree with m vertices. Can you do it? If you can't, explain why not. *Hint*: Reread about buildup error in Section 2.5.

8. a) Compute the number of distinct filled binary trees of height h for $h=0$, 1, 2, and 3.

   b) Use the result in a) to conjecture the result for any h.

   c) Prove your conjecture in b) by deriving a difference equation which expresses the number of filled trees of height h in terms of the number with smaller heights. Then show that your conjecture in b) satisfies this difference equation. (If this seems difficult, see [9].)

d) In this book we have always proved conjectures derived from the inductive paradigm using mathematical induction. Was the proof you gave in c) an induction proof? Why or why not?

9. We don't know how you did the calculation in [8a], but if you had reasoned combinatorially rather than just counting trees, you would have obtained the conjecture in b) immediately and, effectively, a proof of that conjecture. If you didn't do so in [8], do it now. *Hint*: To build a filled binary tree of height h, start with a **complete** binary tree of height h—one which is filled at all levels including the bottom one—and for each leaf make a decision.

10. Determine the number of filled binary trees with precisely n nodes for any n.

11. Using Fig. E.2 as a model, display an algorithm to sort a four-item list in the form of a decision tree whose worst case is the best possible. What is the average number of comparisons for your algorithm?

12. In Sections 3.1 and 3.8, we discussed how to sort a list by building a binary tree and then traversing it. Such trees are often called **binary search trees**. They can be used to search for an item in a list (and for the data associated with it). At each node you first check to see whether the data at that node is what you're searching for. If it isn't, you make a binary decision about the direction in which to search. In this problem and [13], we assume that binary search trees are constructed as in Example 3, Section 3.1.

a) Suppose that in a binary search tree of m nodes, each node is at the end of a left branch of its parent. To find the node associated with a particular item, what is the maximum number of comparisons you might have to perform? What is the average number?

b) Suppose now that a binary search tree is filled (see Theorem 3). What are the maximum and average numbers of comparisons now?

13. The results in [12] imply the desirability of having binary search trees as nearly filled as possible. A binary search tree is said to be **balanced** if, for each node, the heights of the left and right subtrees differ by at most 1.

a) Draw all balanced binary search trees with four and five nodes.

b) Suppose that you want to add a new item to a balanced binary search tree and suppose further that it is equally likely that the new node should go at the end of any possible new branch to the tree. The new tree may therefore be unbalanced. For each of the trees in a), what is the probability that the new tree will still be balanced?

c) Consider the following unbalanced binary search tree, which we assume was built as in Example 3, Section 3.1.

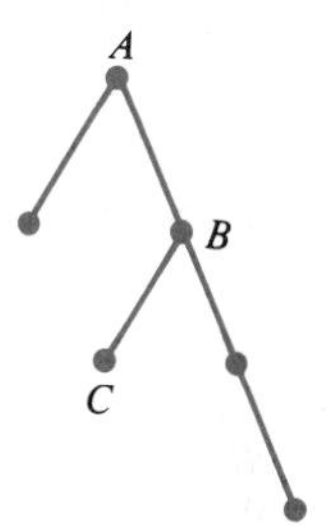

But if we (i) detach C and then rotate so that A becomes the left child of B and (ii) then make C the right child of A, with all other parent–child relationships unchanged, show that the new tree is balanced and that lexical relationships are preserved (i.e., left children are earlier lexically and right children are later lexically than their parents). This idea is part of a common computer algorithm for rebalancing trees.

14. a) Binary search itself can be represented as a binary search tree. Show how to do so. *Note*: Binary search seems to require a three-way decision: Is the search word less than, equal to, or greater than the datum? But it can be expressed as a test for equality at a node followed by a binary decision; cf., [12].

b) Use the result in a) to prove that, among all methods of searching a sorted list based on comparisons, binary search requires the minimum maximum number of comparisons (where one comparison consists of comparing the search word with the datum at a node and deciding which branch to take if they aren't equal). That is, show that in a worst case sense, binary search is the best of all search methods based on comparisons.

15. Internal and external path lengths. The internal path length (IPL) of a decision tree is the sum of the lengths of the paths from the root to all nodes which are *not* leaves. The external path length (EPL) is the sum of the lengths of the paths to all the leaves.

a) Use the inductive paradigm to try to find a relationship between the IPL and EPL which depends only on the number of nodes which are not leaves.

b) Prove the correctness of your conjecture in a).

16. Show why (4) follows from (3), with $m \geq 4$.

17. (Requires some knowledge of calculus)

a) Suppose that you want to compute

$$S = \sum_{k=r}^{s} f(k),$$

where the real-valued function $f(x)$ is increasing. Then show that

$$\int_{r-1}^{s} f(x)dx \leq S \leq \int_{r}^{s+1} f(x)dx.$$

(If you've seen the integral test for determining the convergence or divergence of a decreasing series, you'll recognize this inequality as a relative for determining sums of increasing functions.)

b) Use the result in a) to get a better bound on $\lceil \log_2 m! \rceil$ than that in Eq. (4).

18. Explain why our result about the minimum order for the worst case of sorting by comparisons implies that merge sort is $\mathrm{Ord}(m \log m)$ in the worst case.

19. Prove that the maximum number of nodes at level j of a binary tree is 2^j.

20. a) Compute the number of nodes in a complete binary tree (see [9]) of height h for $h = 0, 1, 2,$ and 3.

b) Conjecture the general case from the result in a) and prove your conjecture.

c) You could have derived the result in b) by summing a geometric series. How?

21. Prove that the sum of the path lengths to the leaves of any filled decision tree with k leaves is the same.

22. Prove that the process described in the proof of Theorem 3 must terminate.

23. In calculating the minimum possible average path length for a decision tree to sort m items, we assumed that the best possible such tree has $m!$ leaves. We know that no such search tree can have fewer than $m!$ leaves. But might it be possible for there to be an algorithm whose search tree has more than $m!$ leaves (so that some possible list orderings would correspond to more than one leaf) and which achieves a smaller average number of comparisons than the best tree with $m!$ leaves? Let's see.

a) Show that in any decision tree with $k > m!$ leaves, $k - m!$ of the leaves could not be reached during a sort of any possible list of m data items.

b) Conclude from a) that the right concept for determining the worst case for such an algorithm is not the average path length to all the leaves. What is the right concept?

c) Show how the $k - m!$ unused leaves in a decision tree with $k > m!$ leaves can be pruned off without changing the path length to any of the other $m!$ leaves.

d) Use the results in b) and c) to prove that no decision tree with more than $m!$ leaves can correspond to an algorithm which achieves the minimum average number of comparisons.

24. In proving Theorem 3, we did surgery on decision trees, moving parts of the trees around to "fill" them. We assumed that the decision trees we started with came from sorting algorithms, but we did not prove that the modified trees correspond to any algorithms. We claim that this omission is irrelevant. Why?

25. Prove more formally than we did in the text that any filled decision tree with $m!$ leaves has a height h such that Eq. (5) is satisfied.

26. Were you surprised that the lower bounds on the worst and average cases for sorting by comparisons were so nearly the same? Actually, you might have expected this. Use the decision tree model to argue why you would expect these results to be nearly the same without doing any calculations or analysis.

E.3 Quicksort

Why aren't we satisfied to have found a method for sorting by comparisons—merge sort—whose order is the best possible for both the worst and average cases? Two reasons:

1. All our analyses of algorithms in this book have been based, explicitly or implicitly, on analyzing the running time of an algorithm. But in Section 1.6 we mentioned the idea of space complexity as well as time complexity, although we said that we would focus, as we have, almost entirely on time complexity. Merge sort requires an amount of storage twice the length m of the list being sorted because it must be merged into a new space of length m. So what, you may think, as computers have plenty of memory, does it matter if you need $2m$ storage locations instead of m. Well, it could matter quite a lot if you want to sort very long lists on a microcomputer that has limited memory. And on mainframe computers, you're normally sharing the memory with lots of other users, so efficient use of memory is important. In any case, if the execution times of two algorithms for the same task are similar, choosing between them may depend on the efficiency with which they use memory.†
2. Just because a method for doing a task has the best possible order doesn't mean that it's the fastest possible method, even for very large values of its characteristic parameter. Don't forget the constant c in Eq. (2), Section 9.3. If two algorithms have the same order but one has a c less than the other, the one with the smaller c will probably be faster for large values of m, "probably" because the actual implementations of two algorithms on a computer may result in behavior which belies that predicted by their two c's if the constants are nearly equal. In any case, we should consider, if we can find them, more than just one algorithm which achieves the best possible order for a particular task.

In this section we'll consider a method of sorting by comparisons whose average case behavior has the same order as merge sort but which requires only m locations to sort a list of length m.

Quicksort or, as it is sometimes called, **partition exchange sort** is, like merge sort, a divide-and-conquer algorithm. In merge sort we just divided the list into two equal or nearly equal halves and applied this idea recursively. In Quicksort we relax the requirement that the list be partitioned into two nearly equal halves. Instead, we try to achieve two sublists in which all elements in one sublist are less than or equal to all elements in the other.

† It's possible to organize merge sort so that it uses only m storage locations, but the resulting complication adds considerably to the execution time.

As before, let the m items to be sorted be $n_1, n_2, \ldots, n_m$. Suppose then, in a classic application of the recursive paradigm, that we already know a method for replacing the original list by a list

$$S_1 \parallel n_j \parallel S_2, \tag{1}$$

in which

$$S_1 \leq n_j \leq S_2. \tag{2}$$

The notation in both Eqs. (1) and (2) needs some explanation. The $\parallel$ represents the **concatenation** of sublist S_1, element n_j, and sublist S_2. In Eq. (2), the notation $S_1 \leq n_j$ means that *every* element in S_1 is less than or equal to n_j and, similarly, $n_j \leq S_2$ means that n_j is less than or equal to every element in S_2. The transformation of the original data to the form in Eq. (1) is illustrated in Fig. E.5(a) and b).

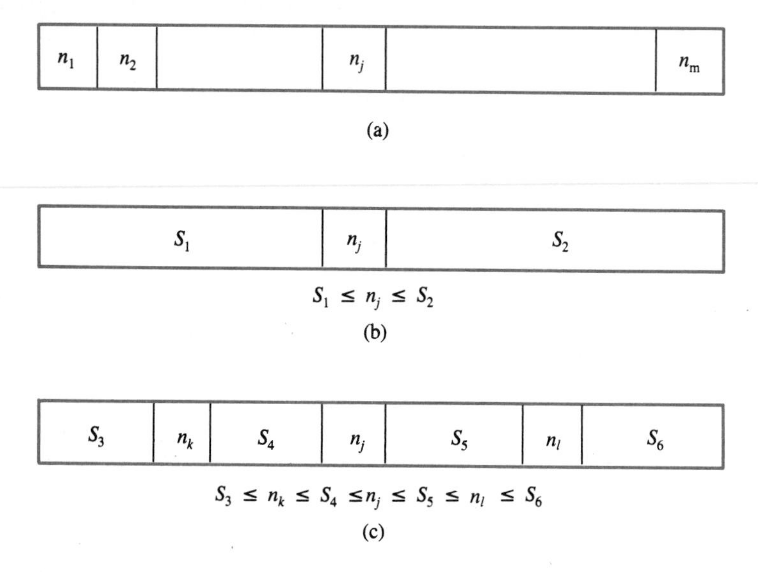

FIGURE E.5 The Essential Idea of Quicksort. (a) The original data; (b) The first partition; (c) The next partition.

Remember that our aim is to partition the list into two sublists, T_1 and T_2, such that

$$T_1 \leq T_2. \tag{3}$$

The list in Eq. (1) has this form if we define $T_1 = S_1$ and $T_2 = n_j \parallel S_2$. But why single out one of the elements from the original list as n_j? Because, as you'll see shortly, in order to divide the list into two sublists which satisfy Eq. (3), we find it convenient to choose an arbitrary element of the original list and then find all elements less than it, which go into S_1, and all elements greater than it, which go into S_2.

Now, suppose that we've gotten from our original list to Eq. (1). In usual recursive paradigm fashion, we apply whatever method we used to achieve Eq. (1) first to S_1 and then to S_2 to get

$$S_3 \parallel n_k \parallel S_4 \parallel n_j \parallel S_5 \parallel n_l \parallel S_6, \tag{4}$$

where n_k is an element from S_1, n_l is an element from S_2, and

$$S_3 \leq n_k \leq S_4 \leq n_j \leq S_5 \leq n_l \leq S_6. \tag{5}$$

This idea is illustrated in Fig. E.5(c).

A crucial point is that the number of elements in S_3 (and S_4) is less than the number in S_1, and the number in S_5 (and S_6) is less than the number in S_2. This is so because the elements n_k and n_l were elements of S_1 and S_2, respectively. It follows then that applying the method which got us (1) to successive S_i's must result, sooner or later, in sublists which have one element or are empty in which case the list will be in sorted order (why?). An example of the entire process is shown in Fig. E.7. Compared to merge sort, we have given up finding a division of the list into two roughly equal lists. But by getting relatively small items in one list and relatively large items in the other, we have dispensed with the need for merging two lists. Is this trade-off worth it? We'll see.

So, we have one hard problem to solve—finding a method of getting from the original list to Eq. (1)—and one easy one—how to choose the *pivot* element at each stage: n_j in Eq. (1), n_k and n_l in Eq. (4), etc. The second problem is easy because, although we would like for the two sublists at each stage to be as nearly equal as possible (why? [2]), to make them so requires, in effect, finding the *median* of the m values, which is an Ord(m) process with quite a large constant c. Although we won't prove it, such extra computation at each stage of the algorithm would not be worth the relatively small gain of getting two nearly equal sublists (and the Ord(m) algorithm to find medians is quite subtle). So, for convenience we'll just choose the *first* element of each sublist (n_1 at the first stage) and hope that this will work out OK.

Before solving the difficult problem, we should note that, in taking the original list and reforming it into a new list (1), we might just take the m elements and somehow reform them in a new array, thus using $2m$ total locations. This approach is essentially the one we use in merge sort. But one of our objectives here is to try to save memory by overwriting the original list with the new list in the form (1). Can we do this without a lot of extra manipulation? The answer is, yes, as Procedure Partition (see p. 818), which solves our hard problem, illustrates.

In order to get Eq. (1) in the first step of Quicksort we need to partition the full list which occupies positions 1 through m. But in subsequent steps we will want to partition sublists whose first and last elements may lie anywhere in the range from 1 to m. Therefore Procedure Partition takes a sublist whose first element is in position F and whose last element is in position L and partitions it into a new sublist:

(a)	8	12	3	9	1	14	6	12	10	15	11
	p	i									j
(b)	6	12	3	9	1	14	8	12	10	15	11
		i				j	p				
	T_1	n_p		T_2					T_3		
(c)	6	8	3	9	1	14	12	12	10	15	11
		p	i			j					
(d)	6	1	3	9	8	14	12	12	10	15	11
			i	j	p						
(e)	6	1	3	8	9	14	12	12	10	15	11
				p							
		S_1						S_2			

FIGURE E.6 An Example of Procedure Partition. (a) The original list; (b) the list after the pivot 8 has been compared with elements from the right and exchanged with 6; (c) the list after the pivot has been compared with elements from the left and exchanged with 12; (d) and (e) the list after the next two right and left passes. The values above p, i, and j denote the elements pointed to at each stage; the sublists S_1 and S_2 are shown after the partition is complete.

```
procedure Partition (F, L)
    p ← F                                               [p is the index of the pivot]
    i ← F + 1; j ← L                                    [Initialize pointers]
    repeat
        repeat until n_p > n_j or j = p
            j ← j − 1                                   [From the right, look for first n_j < n_p]
        endrepeat
        if p < j then n_p ↔ n_j; p ← j; j ← j − 1       [Interchange]
                else exit                               [p = j]
        repeat until n_p < n_i or i = p
            i ← i + 1                                   [From the left, look for first n_i > n_p]
        endrepeat
        if p > i then n_p ↔ n_i; p ← i; i ← i + 1       [Interchange]
                else exit                               [p = i]
    endrepeat
    return p                                            [Position of pivot returned to calling algorithm]
endpro
```

Algorithm E.2
Quicksort

```
Input n_1, n_2, ..., n_m

Algorithm QUICKSORT
    procedure Partition (F, L)
        (body of procedure)
    procedure Quick (s, t)
        if t - s > 0 then                [If t - s ≤ 0, nothing to sort]
            Partition (s, t)                               [Returns p]
            Quick (s, p-1)                                  [Sort S_1]
            Quick (p+1, t)                                  [Sort S_2]
        endif
        return
    endpro
    Quick (1, m)                                   [Main algorithm]
Output n[1, m]                                     [In sorted order]
```

$$S_1 \parallel n_F \parallel S_2, \tag{6}$$

where n_F is the element originally in position F, all elements in S_1 are less than or equal to n_F, and all elements in S_2 are greater than or equal to n_F. That is,

$$S_1 \;\leq\; n_F \;\leq\; S_2.$$

Procedure Partition is not easy to understand, although the job it does is simple enough. We'll explain what it does, using Fig. E.6, as well as the procedure itself. We begin by initializing pointers, i and j, to both ends of the list ($F+1$, not F for the beginning, because n_F is our pivot element, the one all others will be compared with). Because, as you'll see, the pivot element moves around, we also initialize (to F) the location p where the pivot resides.

Then in the first inner **repeat** ... **endrepeat** loop, we start at the right end and compare the pivot element with successive elements in the list. We continue moving left until either the pivot element n_p is greater than the element it is being compared with, or we reach the pivot element itself ($j = p$), in which case we're finished and we **exit** the outer loop. The situation at the end of this process is illustrated in Fig. E.6(b).

Upon exit from the first inner loop, if we haven't reached the pivot element (i.e., $p < j$), we exchange the pivot element with that element which caused the exit from the loop because it was less than or equal to the pivot [Fig. E.6(b)]. The pivot is now in position j, so we reset p to j and set the right end of the remaining list to $j-1$. Then in the second inner loop, we similarly start from the left and work right. So long as the pivot is greater than or equal to the element it is being compared with, we just go to the next element. But if it is less than this element or, as before, if we reach the pivot itself ($i = p$), we **exit** this loop. If $p > i$, we exchange the pivot with the element that caused the exit from the loop [Fig. E.6(c)].

At this point we may, in general, identify four portions of the list:

$$T_1 \parallel n_p \parallel T_2 \parallel T_3, \tag{7}$$

as shown in Fig. E.6(c). In (7) T_1 has elements all of which are less than or equal to the pivot, T_3 has elements all of which are greater than or equal to the pivot, and T_2 has elements which have not yet been examined. (At the beginning, T_2 is the entire list, except for the pivot, and T_1 and T_3 are empty.) When we reenter the outer loop, p still gives the location of the pivot and i and j point to the two ends of T_2, which is the sublist whose elements we now wish to compare with the pivot. Note that each pass through the outer loop reduces the size of T_2 by at least 1 because the first inner loop and the **if** which follows it always result in at least one element being moved from T_2 to T_3 (or to T_1 if n_p is greater than the last element of T_2). Therefore, sooner or later, T_2 must be empty. When T_2 is empty, Partition terminates. This occurs if $p = j$ at the end of the first inner loop or if $p = i$ at the end of the second inner loop. Or, if T_2 becomes empty after the exchange at the end of the first inner loop, then $i = p$ and the second inner loop and the outer loop both exit immediately. Finally, if T_2 becomes empty after the exchange at the end of the second inner loop, then $j = p$ and, upon reentry to the outer loop, the first inner loop and then the outer loop exit immediately. (Try it to make sure you understand what occurs.)

If you follow the path of the pivot you'll see that it moves back and forth along the list, always keeping known smaller (and, perhaps, equal) elements to its left and known larger (and, perhaps, equal) elements to its right.

Understanding Procedure Partition is the difficult part of understanding Algorithm QUICKSORT. This algorithm has the classic structure of a recursive algorithm.† Therefore if you understand how Procedure Partition works, you shouldn't have too much trouble understanding QUICKSORT. One point to note is how our method of passing values back and forth between procedures and their callers (see Section 1.5) makes the value of p found in each call of Partition available to the calls of Procedure Quick which need it. It is available because the value of p returned by Partition is known only *locally*; therefore it will never be erased by any other value of p from a recursive call at a different level.

Figure E.7 illustrates how QUICKSORT works. Although a large fraction of the calls to Procedure Quick result in no action because $t - s \leq 0$, such would not be the case for large values of m.

Our presentation and discussion of QUICKSORT may have convinced you that it really does sort a list. But because it is a recursive algorithm, we should be able

---

† Because the call of Procedure Partition precedes the two recursive calls, QUICKSORT has been called a "conquer-and-divide" algorithm.

8 12 3 9 1 14 6 12 10 15 11 [Original list]
Quick $(1, 11)$
    Partition $(1, 11)$
6 1 3 [8] 9 14 12 12 10 15 11 $[p = 4]$
    Quick $(1, 3)$
        Partition $(1, 3)$
3 1 [6] $[p = 3]$
        Quick $(1, 2)$
            Partition $(1, 2)$
1 [3] $[p = 2]$
            Quick $(1, 1)$
            Quick $(3, 2)$
        Quick $(4, 3)$
    Quick $(5, 11)$
        Partition $(5, 11)$
        [9] 14 12 12 10 15 11 $[p = 5]$
        Quick $(5, 4)$
        Quick $(6, 11)$
            Partition $(6, 11)$
        11 12 12 10 [14] 15 $[p = 10]$
            Quick $(6, 9)$
                Partition $(6, 9)$
        10 [11] 12 12 $[p = 7]$
                Quick $(6, 6)$
                Quick $(8, 9)$
                    Partition $(8, 9)$
                [12] 12 $[p = 8]$
                    Quick $(8, 7)$
                    Quick $(9, 9)$
            Quick $(11, 11)$

Final sorted list:

1 3 6 8 9 10 11 12 12 14 15

FIGURE E.7

An Example of Algorithm QUICKSORT. The successive lines show the original list, each call of procedures Partition and Quick, indented to show the level of recursion, the result after each call of Partition and the value of p returned by each call of Partition. The position of the pivot element after each call of Partition is indicated by a box. Note that, if the two arguments of Quick are equal or the first is one greater than the second, there is an immediate return.

to *prove* its correctness using, as in Chapter 2, mathematical induction. So we'll do it—almost—now. The "almost" is because we'll assume that Procedure Partition is correct, namely, that it partitions any list or sublist as shown in (1) and (2). (We'll discuss the verification of this procedure shortly.)

We want to prove that QUICKSORT sorts a list of length m. Recall from Chapter 2 (Sections 2.2 and 2.6) that to prove a recursive algorithm correct, it is sufficient to prove first that the recursive procedure within it is correct and then to prove that the main algorithm is correct because of the way it calls the recursive procedure. Therefore let's try to prove that Procedure Quick(s,t) is correct whenever $s \le t+1$. (Note from Fig. E.7 that we often call Quick with $s = t+1$.)

Our proof is, naturally, an induction proof. As the induction parameter we choose $t-s$ because each successive value of $t-s$ represents a list with one more element in it. What is the correct basis case? Since $t-s$ can be -1 (a vacuous list), we'd better include this value in the basis case. We'll also include the case $t-s=0$ (a list of length 1) in the basis because of the $t-s>0$ test in Quick. But when $t-s$ is -1 or 0, the assertion that Quick works is vacuously true. In both cases Quick does nothing, which is what it should do for vacuous lists or lists of length 1.

Our induction hypothesis is that Quick works whenever $t-s<k$. Using this hypothesis and strong induction we'll prove that the procedure works when $t-s=k$. When $t-s=k$, the call Quick(s,t) results in a call to Partition (s,t) (which we have assumed does its job correctly) and returns the pivot position p followed by recursive calls to Quick$(s,p{-}1)$ and Quick$(p{+}1,t)$, with $s \le p \le t$. Because

$$(p-1)-s \;\le\; t-1-s \;<\; t-s$$

and

$$t-(p+1) \;\le\; t-s-1 \;<\; t-s,$$

the induction hypothesis claims that both calls to Quick do their jobs correctly, sorting the two sublists into which Partition partitioned the original list. The result, as desired, is a completely sorted list, which completes the proof.

As our proof holds for any $t{-}s{>}{-1}$, it holds, in particular, when $s=1$ and $t=m$. This is the result that we're interested in because this case is the main algorithm call.

Incidentally, trying to prove that Quick$(1,m)$ works by induction on m would not have worked because some of the calls to Quick within Quick$(1,m)$ wouldn't have a first argument 1. Therefore we would have been forced to *strengthen the induction hypothesis* in order to prove that Quick(s,t) works.

Now, let's consider Procedure Partition. We won't verify it completely because the two inner loops inside the main loop make the verification slightly tedious. But let's verify the correctness of the outer loop by making appropriate assumptions that the inner loops do what we want them to. Each time an iteration of the outer loop of Partition is begun, we have the following situation:

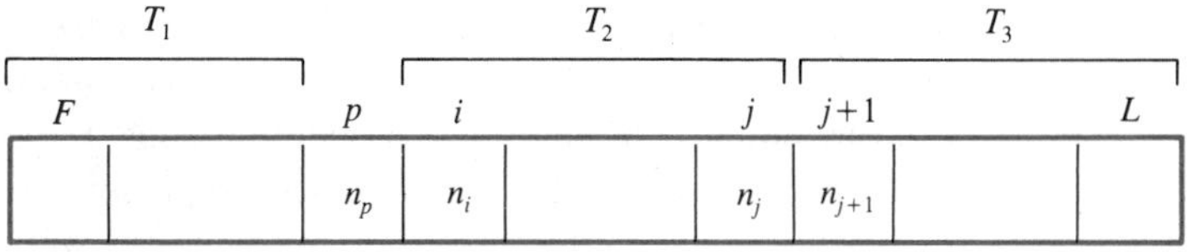

T_1 T_2 T_3

F p i j $j+1$ L

		n_p	n_i		n_j	n_{j+1}		

with F, L, p, i, and j as defined in the previous iteration (or initially when T_1 and T_3 are empty). The loop invariant is

$$\forall k : F \leq k \leq p-1 : n_k \leq n_p \wedge \forall k : j+1 \leq k \leq L : n_p \leq n_k.$$

The first part states that all elements of T_1 in (7) are less than or equal to the pivot. The second part states that all elements of T_3 are greater than or equal to the pivot. At the first entry into the outer loop $p = F$ and $j = L$, so both parts of the loop invariant are vacuous.

Is the loop invariant satisfied on each subsequent reentry into the outer loop? Well, assume that the first inner loop does what it is supposed to do—compares the pivot with elements of T_3 (i.e., elements from n_i to n_j) from the right and then places the pivot element just to the left of the last element greater than the pivot. Assume that the second inner loop does the corresponding task from the left of T_3 and then places the pivot element so that it is just to the right of the last element it was greater than. These two assumptions—and the assumption that the loop invariant was satisfied upon entry to the outer loop—mean that both portions of the invariant are again satisfied upon reentry. Therefore if the outer loop invariant is satisfied on one entry to the loop, it is satisfied on reentry. This conclusion completes our verification of the outer loop, except for the question of the loop termination condition, which we'll take up in a problem [3b].

To conclude this section—and this book—we'll analyze Algorithm QUICKSORT. First, what's the worst case of QUICKSORT? It occurs when, at every stage, the pivot is either smaller than all other items or larger than all of them [4, 5]. For then the pivot either remains where it is, or it moves from one end of the list to the other. The two lists S_1 and S_2 then have lengths 0 and $m-1$ at the first stage, 0 and $m-2$ at the next stage, etc. Can this phenomenon actually occur at every stage? Yes, if the original list is already sorted (!) or is in precisely reverse order. For example, if the list is already sorted, the first element will be less than or equal to all the others and will stay where it is. But all you *know* at the end of the pass from right to left is that the first element is the smallest. You don't know that the other $m-1$ elements are in order. So you have to look at the sublist of length $m-1$, do the partition, and continue in this way for $m-1$ total passes. A problem [6] considers the case when the original list is in reverse numerical order.

How many comparisons are there when the list is originally in numerical order? Answer: $m-1$ at the first stage, $m-2$ at the next, etc., until there is only one at the last stage. The total is

$$(m-1) + (m-2) + \cdots + 3 + 2 + 1 \;=\; \frac{m(m-1)}{2}.$$

Therefore in the worst case, QUICKSORT is $\mathrm{Ord}(m^2)$. Not a very pleasant result—no better, in fact, than insertion sort and worse than merge sort. Why have we been wasting your time with this method? If you persevere with us through the average case analysis, you'll see that we haven't been wasting your time.

Let q_k be the number of comparisons required on the average by a call Quick(s,t) when $t-s+1=k$. Thus k represents the length of the sublist on which Procedure Quick is operating so that what we want to calculate is q_m. Now look at Algorithm QUICKSORT and Procedure Partition. The call Quick$(1,m)$ results in:

- A call, Partition $(1,m)$, which returns the pivot position p.
- A recursive call, Quick$(1,p-1)$.
- A recursive call, Quick$(p+1,m)$.

How many comparisons does the call Partition$(1,m)$ require? The answer is $m-1$ because the pivot is compared to each of the other $m-1$ list elements once and only once. The two recursive calls to Quick require q_{p-1} and q_{m-p} comparisons, respectively, since $(p-1)-1+1=p-1$ and $m-(p+1)+1=m-p$. But what's p? It can be anything from 1 to m, depending on the result of Partition. So we should *average* the number of comparisons q_{p-1} and q_{m-p} over all possible values of p. Now, assuming as usual that any ordering of the m elements is equally likely implies that any value of p is equally likely (why?). The average we want is just

$$\left(\frac{1}{m}\right)\sum_{p=1}^{m}(q_{p-1}+q_{m-p}).$$

This average added to the $m-1$ comparisons for the call Partition $(1,m)$ is the average number of comparisons in the call Quick$(1,m)$. Therefore

$$q_m \;=\; m-1+\left(\frac{1}{m}\right)\sum_{p=1}^{m}(q_{p-1}+q_{m-p}). \tag{8}$$

Equation (8) is a difference equation but not one of the standard sorts you met in Chapter 5. Let's see whether we can turn it into something more familiar. Observe first that for both terms in the sum in Eq. (8), the range of values of the subscript is 0 to $m-1$. Thus if we rewrote this as the sum of two summations, each would represent the same sum. Therefore we may rewrite Eq. (8) as

$$q_m \;=\; m-1+\left(\frac{2}{m}\right)\sum_{i=0}^{m-1}q_i. \tag{9}$$

Equation (9), a linear, nonhomogeneous difference equation *without* constant coefficients, still doesn't look very tractable. It even has **variable order** because the difference between the smallest and largest subscripts is a function of m; (see [17–20, Supplementary, Chapter 5].) But let's persevere. One way to simplify Eq. (9) is to multiply both sides by m to obtain

$$mq_m \;=\; m(m-1)+2\sum_{i=0}^{m-1}q_i. \tag{10}$$

Equation (10) may not look much better—but it is. Let's write out the complete equation for the first few values of m:

$$\begin{aligned} m = 1: \quad & q_1 = 2q_0; \\ m = 2: \quad & 2q_2 = 2 + 2(q_1+q_0); \\ m = 3: \quad & 3q_3 = 6 + 2(q_2+q_1+q_0). \end{aligned}$$

Note that if we subtract each equation from its predecessor, a lot of cancellation occurs. Let's do this in Eq. (10) by first rewriting it, with m replaced by $m-1$, to get

$$(m-1)q_{m-1} = (m-1)(m-2) + 2\sum_{i=0}^{m-2} q_i.$$

Now, subtracting this equality from Eq. (10), all terms in the summation except that for $i = m-1$ cancel, and we get

$$mq_m - (m-1)q_{m-1} = m(m-1) - (m-1)(m-2) + 2q_{m-1},$$

which we can simplify to

$$mq_m = (m+1)q_{m-1} + 2(m-1),$$

or

$$q_m = \left(\frac{m+1}{m}\right) q_{m-1} + \frac{2(m-1)}{m}. \tag{11}$$

Lo and behold, Eq. (11) is a first order difference equation of just the type we solved in Section 5.8. What is an appropriate initial condition? Just $q_1 = 0$, as $m = 1$ represents a list with one element, which requires no comparisons to sort it.

To solve Eq. (11), we use Eq. (3), Section 5.8, replacing a with q, n with m, and 0 with 1, and with

$$c_i = \frac{i+1}{i} \quad \text{and} \quad d_j = \frac{2(j-1)}{j}.$$

This leads to one final algebraic marathon:

$$\begin{aligned} q_m &= \sum_{j=1}^{m} \left(\prod_{i=j+1}^{m} \frac{i+1}{i} \right) \left[\frac{2(j-1)}{j} \right] \\ &= \sum_{j=2}^{m} \left(\frac{m+1}{j+1} \right) \left[\frac{2(j-1)}{j} \right] && \text{[Because the term for } j = 1 \text{ is 0]} \\ &= 2(m+1) \left[\sum_{j=2}^{m} \frac{j-1}{j(j+1)} \right] \\ &= 2(m+1) \left[\sum_{j=2}^{m} \frac{1}{j+1} - \sum_{j=2}^{m} \frac{1}{j(j+1)} \right] && \text{[Separating into two terms]} \\ &= 2(m+1) \left[\left(\sum_{j=2}^{m} \frac{1}{j+1} \right) - \left(\frac{1}{2} - \frac{1}{m+1} \right) \right] && \text{[Because } 1/(j+1)j = 1/(j+1)_{(2)} \text{ has an antidifference } -1/j \text{ and then using Eq. (4), Section 9.4]} \end{aligned}$$

$$= 2(m+1)\left(H_{m+1}-1-\frac{1}{2}-\frac{1}{2}+\frac{1}{m+1}\right)$$ [From the definition of the harmonic numbers in Example 3, Section 9.5]

$$= 2(m+1)\left(H_m-2+\frac{2}{m+1}\right)$$ [Because $H_{m+1} = H_m + 1/(m+1)$]

$$= 2(m+1)H_m - 4m. \qquad (12)$$

Whew! Now recall that in Example 3, Section 9.5, we noted that $H_m = \text{Ord}(\log m)$. From that equality and Eq. (12), it follows that

$$q_m = \text{Ord}(m \log m).$$

Therefore, on the average, QUICKSORT has the same order as merge sort, an order as small as is possible for any method of sorting by comparison. Now, let's compare QUICKSORT and merge sort:

1. The worst case of QUICKSORT has an order greater than the worst case of merge sort. Does that matter? Generally, no, because in looking for general purpose methods on a computer, it is the average behavior which is important. But, if you are dealing with *time-critical* applications, such as on board a spacecraft, where incoming data may need to be sorted, then the worst case may be important.
2. QUICKSORT sorts *in place.* That is, except for the locations needed to store the indices p, i, and j, QUICKSORT uses only the m locations in which the original list was stored, compared with $2m$ for merge sort. This condition, as we noted, is important for large sorts on microcomputers with limited memory—or even on mainframes. It can also be important, for example, on a spacecraft computer where memory may be limited.
3. But which really is faster? Both QUICKSORT and merge sort are $\text{Ord}(m \log m)$ on average, which means there are constants c_1 and c_2 such that, for large m, QUICKSORT takes about $c_1 m \log m$ steps and merge sort takes about $c_2 m \log m$ steps. Which is smaller, c_1 or c_2? For merge sort, our analysis in Section 5.9 doesn't give the answer precisely. Rather, it suggests that the constant is somewhat less than 1 (with logarithms taken to the base 2). For QUICKSORT, we need the result (which requires calculus) that

$$\lim_{m\to\infty} \frac{H_m}{\log_2 m} = \log_e 2 \approx 0.69531.$$

Using this value in Eq. (12), we obtain a c of about 1.39. This analysis seems to favor merge sort. However, because the two values of c are not that different and because merge sort requires a lot of movement of the elements of the list in addition to the comparisons, it turns out that QUICKSORT generally executes faster than merge sort on computers.

Our conclusion is that both for reasons of speed and use of memory—that is, for both time complexity and space complexity reasons—Algorithm QUICKSORT is preferable to merge sort for main memory sorts on computers for applications which are not time-critical. (For sorting very long lists using both main and auxiliary memory, such as disk storage, variations on merge sort together with QUICKSORT for the main memory portion of the sort are often used.) QUICKSORT also is generally preferable to other $\mathrm{Ord}(m \log m)$ methods of sorting and is therefore the most widely used method for main memory sorting.

Concluding Note

This completes our Epilogue. We hope you agree that considerable portions of the mathematics introduced in this book, as well as the methodology we have emphasized, have been usefully employed in our analysis of sorting by comparisons. But sorting by comparisons is by no means unusual in this respect. Discrete mathematics—especially algorithms, mathematical induction, and the recursive and inductive paradigms—are generally a powerful combination with which to attack problems in modern applied mathematics.

Problems: Section E.3

1. a) Apply Procedure Partition to the three data sets of [1, Section E.1]. Show your work as in Fig. E.6.

b) Then, for each case in a), apply Algorithm QUICKSORT. Show your work as in Fig. E.7.

2. At each stage of Algorithm QUICKSORT a sublist (or, at the beginning, the original list) is partitioned into two sublists. In this problem suppose that the original data set is such that the two sublists at each stage are *always* of equal length.

a) What possible lengths could the original list have had for this to happen?

b) How many comparisons are required to sort the entire list for the lengths found in a)? That is, analyze QUICKSORT under the assumption that the conditions at the start of this problem are satisfied.

c) Give a heuristic argument that QUICKSORT will tend to be most efficient when the partitioned sublists at each stage are as nearly equal as possible.

3. a) State loop invariants for each of the two inner loops of Procedure Partition.

b) State a loop termination condition for the outer loop of Partition and prove that it is satisfied upon exit from this loop.

4. Let W_m be the worst case number of comparisons when QUICKSORT is applied to sort a list of m items.

a) Explain why

$$W_m = m-1+\max\{0 \le k < m : W_k + W_{m-1-k}\},\ W_0 = 0. \quad (13)$$

b) Show that $W_m = m(m-1)/2$ is the unique solution to Eq. (13). *Hint*: To show that it is a solution, you'll need to show that a particular value of k always gives the maximum value in the second term on the right-hand side of Eq. (13); to show uniqueness, you'll want to consider the possibility of another solution.

5. Argue that Algorithm QUICKSORT never compares the same two items from a list twice. Use this argument to show that the number of comparisons in the worst case of QUICKSORT cannot be greater than $m(m-1)/2$. This problem, together with [6], gives an alternative method to that in [4] for determining the worst case of QUICKSORT.

6. Analyze Algorithm QUICKSORT when the original list is in reverse numerical order.

The next seven problems consider another popular and efficient method of sorting called **heapsort**.

7. Consider a filled binary tree in which a data element to be sorted is associated with each node. If each node at the bottom level is as far left as possible (i.e., if v is the rightmost leaf on the bottom, then there is a leaf in every possible bottom position to the left of v; see the accompanying figure). And if each path from the root to a leaf passes through nodes whose data are in reverse lexical order (e.g., for numbers, each datum is less than or equal to its parent datum), then the filled tree is called a **heap**.

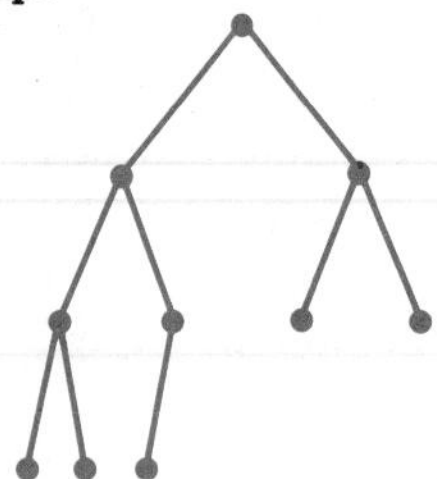

a) Suppose that the height of a node on the lowest level of a heap is h and that there are k nodes on the lowest level. Suppose also that the nodes are labeled 1, 2, 3, ..., starting at the root and proceeding top to bottom and left to right (so that the two nodes on level 1 are labeled 2 and 3). What are the labels of the nodes on the lowest level?

b) If j is the label of a node with two children, what are the labels of the children?

c) If j is the label of a node which is not the root, what is the label of its parent?

d) If the labels of the nodes in a heap are 1 to m, show that the data at the m nodes can be stored in an m-element list (with the item labeled 1 in the heap the first one in the list, the item labeled 2 the second one, etc.) without losing any of the parent–child information inherent in the tree. *Hint*: Use the results in b) and c) to show how the parent and child nodes of any element in the list can be determined.

8. In each of the following cases draw the heap as a tree which corresponds to the table shown.

a) 47 14 20 12 2 5 8 6

b) 65 33 19 25 31 4 7

9. a) Suppose that a table of $m-1$ items forms a heap and that you are presented with an mth item. Design a procedure which inserts this mth item appropriately (and moves other items, as necessary), so that all m items form a heap.

b) Apply this procedure to each of the following cases.

i) Heap: 47 14 12 2 5 8 6; 8th item, 20.

ii) Heap: 65 31 19 25 4 7; 7th item, 33.

10. Use the procedure in [9] to design a procedure whose input is a table of m data items and whose output is a table of the same items, which is a heap. *Hint*: Think inductively.

11. You can sort a heap of m elements as follows:

i) Exchange the root element with the mth element.

ii) Reform the first $m-1$ elements into a heap.

iii) Exchange the root element with the $(m-1)$st element, etc.

a) Give a proof that the final result is a sorted list.

b) Write a procedure to do step (ii). Be careful; sometimes a parent node will have two children and sometimes it will have only one child.

c) Use this procedure and the procedure in [10] to design an algorithm called HEAPSORT for sorting a list of m data items.

12. Apply Algorithm HEAPSORT to each of the data sets of [1, Section E.1].

13. a) Suppose that you want to add an ith element to a heap of $i-1$ elements, so that the result will be a heap as in the algorithm you designed in [9]. In the worst case, how many comparisons would your algorithm have to make? If your algorithm was well-designed, the answer is $\lfloor \log_2 i \rfloor$.

b) Suppose that the worst case for the algorithm in [9] is as given in a). Then the worst case for the algorithm designed in [10] should be $\sum_{i=2}^{m} \lfloor \log_2 i \rfloor$. To make things easier, suppose that $m = 2^{k+1}-1$. In this case show that the preceding sum may be written $\sum_{j=1}^{k} j2^j$.

c) Refer to Example 5, Section 9.5 to calculate the sum in b). What is the order of the result?

d) Argue heuristically that the entire Algorithm HEAPSORT has the order found in c).

e) Assuming that this order does not depend on our assumption that m is a power of 2—it doesn't—what is the order of HEAPSORT, on *average*?

f) Can you argue from the results in this problem that HEAPSORT is a better method of sorting than Algorithm QUICKSORT, or vice versa?

Final Problems

1. There are two versions of Straightline Towers of Hanoi (STOH), first discussed in [41–42, Section 2.2]. In one version you must move the tower between endpoles. In the other you must move it only from one end to the middle, or vice versa. Let us call the former LSTOH (L for long trip) and the latter S1STOH (S for short trip and 1 representing a move from the middle to the end) and S2STOH (2 for a move from the end to the middle).

   a) Prove again that all three versions can be won but use only previous cases of LSTOH to prove it and only previous cases of S1STOH and S2STOH to prove S1STOH, and similarly for S2STOH. For instance, if $P(n)$ is the statement that LSTOH can be won for n rings and $Q(n)$ is the statement that S1STOH can be won and $R(n)$ is the statement that S2STOH can be won, then the inductive step for $P(n)$ should use only earlier cases of P and not any earlier cases of Q or R. (In principle, it's perfectly legitimate, say, to first prove $P(n)$ for all n and then use $P(n)$ in the inductive step of $Q(n)$ or $R(n)$, but keeping the two versions completely separate leads directly to later parts of this problem.)

   b) Prove that the number of ring moves for S1STOH and S2STOH are the same.

   c) Turn your proofs in a) into recursions for the number of steps your method takes to win these games. Solve the recursions.

   d) From now on let's refer to both short-trip versions as SSTOH. No doubt your proof in a) for n-ring SSTOH at one point requires two calls of $(n-1)$-ring SSTOH to be performed one right after the other. Conceivably fewer ring moves are required if one call of LSTOH is made instead. This leads to a system of linked recursions, one for the minimum number of steps to solve n-ring LSTOH and one for minimal n-ring SSTOH. Set up and solve these linked recursions. (How can you solve linked recursions? One way is to do some algebra first to unlink them. Another is to use generating functions. After you turn the recursions into statements about two generating functions, the algebra to unlink is standard. In any event, this pair of linked difference equations is only "partially" linked and you can solve them without a general method.)

   e) Why does your answer in d) tell you that the minimum solution to LSTOH has the property that at one point the entire n-tower sits on the middle pole?

   f) Prove the statement in e) directly by induction (without using any knowledge of the number of moves in minimal solutions).

2. In Clockwise Towers of Hanoi (CTOH), the poles are on a circle and all the pieces are required to move clockwise but, as in STOH, always to an adjacent pole. There are two versions. In short-trip CTOH, you must move the tower to the next pole in clockwise order. In long-trip CTOH, you must move the tower to the next pole in *counterclockwise* order (but all the pieces must still move clockwise).

   a) Prove that both versions can be won. As in [1], do each proof by an induction that depends only on its own previous cases.

   b) Let L_n be the number of steps to win long-trip CTOH via the algorithm implicit in your proof in a). Similarly, let S_n be the number of steps in your short-trip algorithm. Write the recursions implicit in your proofs and solve them.

   c) The explicit formulas you got in b) do *not* give the minimum number of steps to win these games. For $n=2$, find this minimum by playing the game yourself. Then explain conceptually what's "wrong" with the recursions.

   d) Set up linked recursions for the minimum solutions for long- and short-trip CTOH. Justify why they are correct and solve them. (See the comments in [1d].

e) Write a recursive algorithm for solving both long-trip and short-trip CTOH in the minimum number of steps. Your algorithm will have two procedures that call both themselves and each other. This is sometimes called **corecursion.**

3. Again the poles of TOH are on a circle. Now, in addition to the basic rules, the smallest piece must move clockwise to the adjacent pole (others can move either way). Prove that the n-ring game can be won, from any pole to any other. Show that for each pair of start and finish poles, there is a unique minimum set of moves that wins this version of the game. Find a formula for the minimum number of moves.

4. Same as [3] except that the first move of the smallest ring must be clockwise, and thereafter each of its moves must be in the *opposite* direction from its previous move. (As in [3], there are no restrictions on the direction of movement of larger rings.)

5. Suppose that you have to solve the matrix problem $A\mathbf{x} = \mathbf{b}$, where A is a **tridiagonal matrix.** By this, we mean that all entries on the main diagonal are nonzero but that all other entries are 0, except on the diagonals just above and below the main diagonal. Can Algorithm GaussSquare be simplified and speeded up for tridiagonal matrices? Give an algorithm and explain why it works. What is the complexity of your simplified algorithm?

6. A society has n men and n women who want to get married. Each man has a preference ranking of all women. That is, he knows who is his first choice, his second choice, ..., his nth choice. Similarly, each woman has ranked all the men. Here is an algorithm for matching them up:

i) Simultaneously, each man proposes to his first choice woman.

ii) If each woman has exactly one proposal, all proposals are accepted and the matchmaking is over; else

iii) Simultaneously, each woman keeps the best of her current suitors "on a string" and tells any others to get lost forever.

iv) Simultaneously, each rejected man proposes to the next woman on his list; return to ii).

a) Show that this algorithm always terminates with a set of n marriages. Among other things, you must show that you never reach a state where some man has been rejected by all n women. For that case, the algorithm doesn't say what to do, so it would terminate abnormally.

b) A set of marriages is **stable** if there is no man–woman pair, not married to each other, who prefer each other to their spouses. Prove that the set of marriages constructed by the algorithm is stable.

c) This algorithm is inspired by the traditional convention that "man proposes, woman disposes." Let's turn things around and have the women propose and the men dispose. We must get stable marriages, but do we always get the same set of marriages as when the men propose?

d) Suppose that there are more men than women. Find and justify a modification of the algorithm that still obtains a stable set of marriages. Part of the job is to redefine stability as well; remember, there will be men who are unhappy because they are not married at all.

e) Can you modify the algorithm so that it works if there are more women than men?

f) The initial assumption was that every man was willing to marry any of the women (perhaps reluctantly) and vice versa. Can the algorithm be made to work if each person is allowed to exclude people from his or her preference list, that is, if people are allowed to regard remaining unmarried as preferable to marriage to some people?

g) Can you modify the algorithm to make it work for college admissions? College admissions are like marriage where bigamy is allowed: Both colleges and applicants have preference lists, but colleges are allowed to "marry" more than one student.

7. (Multiplying n-digit numbers) The traditional Arabic algorithm for doing this is brilliant: It breaks down a single multiplication of long numbers into many multiplications and additions of two single-digit numbers. Calculation (i) shows an example with two 3-digit numbers. Each line between the horizontal rules is a multiplication of 526 by one of the digits of 738. However, calculation (ii) shows that each such line is in fact the result of several single-digit multiplications merged into each other with the aid of carrying.

```
   526      526
   738        8
  ----     ----
  4208       48
 1578       16
3682       40
------    -----
388188    4208

  (i)      (ii)
```

a) In the worst case, how many multiplications of two single-digit numbers are involved in multiplying two n-digit numbers by the Arabic method? How many additions of two single-digit numbers? (Count handling each carry as one more addition.)

Let's try to do better by using the divide-and-conquer method. Suppose that $n = 2m$ and that we already know how to multiply two m-digit numbers. If A has n-digits, we can think of the leftmost m digits as a single "superdigit" P and the rightmost m digits as another superdigit Q. In standard notation, $A = P10^m + Q$. If B is another n-digit number, we can likewise think of it as written with two superdigits R and S, where $B = R10^m + S$. Then we can calculate the product AB by four superdigit multiplications analogously to the Arabic multiplication of two 2-digit numbers.

b) Justify the last claim. *Hint*: Use algebra to express AB in terms of two-way products of P, Q, R, and S and then interpret this algebra using a calculation like calculation (i).

c) Write down a recursion for the number of single-digit multiplications involved in multiplying two n-digit numbers by the divide-and-conquer method. Solve the recursion. (You may make the sort of simplifying assumptions explained in Section 5.9.) Has anything been saved in comparison with the Arabic multiplication method?

d) Similarly write down and solve a recursion for the maximum number of single-digit additions using the divide and conquer method. Assume that the number of additions to add together the results of the various n digit multiplications is some multiple of $n > 2$.

e) With A and B as given, verify that

$$AB = PR10^n + [(P+Q)(R+S) - (PR+QS)]10^m + QS. \tag{1}$$

Note that this formula shows that the n-digit numbers A and B can be calculated in terms of only *three* multiplications of m-digit superdigits. (Actually, in one case you might get $m+1$ digits, but we say you may ignore this. Why?) Devise a recursive procedure for multiplying A and B using this observation. Determine the number of single-digit multiplications involved.

f) Equation (1) appears to have more additions and subtractions than the standard method so maybe this undermines the reduction in multiplications. Devise and solve a recursion for the number of additions and subtractions. (Use the same assumption as in d).)

8. (Fast matrix multiplication) If A is a $2m \times 2m$ matrix, we may represent it as

$$A = \begin{bmatrix} A_{11} & A_{12} \\ A_{21} & A_{22} \end{bmatrix},$$

where A_{11}, A_{12}, A_{21}, and A_{22} are $m \times m$ matrices. Similarly, if B is $2m \times 2m$, we may write

$$B = \begin{bmatrix} B_{11} & B_{12} \\ B_{21} & B_{22} \end{bmatrix}.$$

a) Verify that $2m \times 2m$ matrices satisfy the same formula for multiplication as 2×2 matrices; that is, with the preceding notation,

$$AB = \begin{bmatrix} A_{11}B_{11} + A_{12}B_{21} & A_{11}B_{12} + A_{12}B_{22} \\ A_{21}B_{11} + A_{22}B_{21} & A_{21}B_{12} + A_{22}B_{22} \end{bmatrix}. \tag{2}$$

(Note, however, that the order of the factors in terms like $A_{11}B_{11}$ is now important, as matrices don't commute in general.)

b) Use this observation to describe a recursive procedure for multiplying $2^n \times 2^n$ matrices. Come up with a recursion for the total number of real-number multiplications and additions needed by your recursive method. How does the answer compare with the complexity of the standard method?

c) Equation (2) involved eight multiplications and four additions of $m \times m$ matrices. Suppose that someone came up with a different method to multiply $2m \times 2m$ matrices, call it procedure X, that involved seven $m \times m$ multiplications and 18 $m \times m$ additions. This doesn't sound promising, but determine the complexity of the recursive algorithm required to multiply $2^n \times 2^n$ matrices using procedure X. How does the answer compare with the standard method?

d) There is a procedure $X!$. Show this by verifying the following. Using the preceding notation, if

$$\begin{aligned} P &= (A_{12}-A_{22})(B_{21}+B_{22}),\\ Q &= (A_{11}+A_{22})(B_{11}+B_{22}),\\ R &= (A_{11}-A_{21})(B_{11}+B_{12}),\\ S &= (A_{11}+A_{12})B_{22},\\ T &= A_{11}(B_{21}-B_{22}),\\ U &= A_{22}(B_{21}-B_{11}),\\ V &= (A_{21}+A_{22})B_{11}, \end{aligned}$$

then

$$AB = \begin{bmatrix} P+Q-S+U & S+T \\ U+V & Q-R+T-V \end{bmatrix}.$$

How does this fact result in a procedure X?

e) How could you apply the method in d) to square matrices whose order is not a power of 2? Is the complexity the same for general n? What about using this method for multiplying an $m \times n$ matrix by a $n \times p$ matrix?

9. Way back in [7, Section 1.6], you gathered data about the complexity of Algorithm EUCLID for finding $\gcd(m, n)$. (What we say below applies to any of the versions of that algorithm we have discussed.) Actually, determining the worst case isn't too difficult, and the answer is a pleasant surprise. The key idea is to run the problem backward. The first line in the following display is the labeling we have used previously: m and n are the inputs, the r's are the remainders, and r_{k-1} is the gcd. The second line is the new labeling: That is, what used to be m is now s_{k+1}, ..., and what used to be r_k is now s_0.

$$\begin{array}{llllll} m, & n, & r_1, & r_2, \ldots, & r_{k-1}, & r_k = 0; \\ s_{k+1}, & s_k, & s_{k-1}, & s_{k-2}, \ldots, & s_1, & s_0 = 0. \end{array}$$

a) Assume that we always start with a pair (m, n), where $m \geq n$. Show that with the relabeling $\{s_0, s_1, \ldots, s_{k+1}\}$, there is a sequence of positive integers $\{Q_1, \ldots, Q_k\}$ so that

$$\begin{aligned} s_0 &= 0, \qquad s_1 \text{ a positive integer, and} \\ s_{j+1} &= Q_j s_j + s_{j-1}, \qquad j > 0. \end{aligned}$$

b) Argue from a) that if (m, n) is *any* pair that requires exactly k iterations of Euclid's algorithm to find the gcd (and $m \geq n$), then

$$m \geq f_k, \qquad n \geq f_{k-1},$$

where f_k is the kth Fibonacci number. Argue further that $(m, n) = (f_k, f_{k-1})$ takes exactly $k-1$ iterations. This means that consecutive Fibonacci numbers are the worst cases for the Euclidean algorithm!

c) What changes if we don't require that $m \geq n$?

You may be interested to know that, for large n, the *average* number of iterations to compute $\gcd(m, n)$ is $(12 \ln 2 \ln n)/\pi^2$. The proof of this expression is quite deep.

10. Suppose that $\{g_n\}$ satisfies the inequality

$$0 < g_n \leq cn^2 + g_{n-1} + g_{n-2}, \qquad n > 1,$$

where c is a constant. Prove that $g_n = O(f_n)$, where $\{f_n\}$ is the Fibonacci numbers. This result is useful in [11].

11. An independent set of vertices of a graph G is a set in which no two vertices are joined by an edge. We want an algorithm to find the **independence number** α for any graph G, that is, the maximum size of an independent set of vertices in G. When there is only 1 vertex, clearly $\alpha = 1$. More generally, suppose that we already know how to find α for graphs with $<n$ vertices and that we are now given G with n vertices. (When we say G is given, we mean that we have its adjacency matrix.) Pick any vertex v. Either a maximum independent set S contains v or it doesn't. If it doesn't, S is an independent set of $G - v$; if it does, $S - v$ is an independent set of $G - N(v)$, where by the **neighborhood** $N(v)$ of v we mean v and all adjacent vertices. (Of course, when you take vertices from a graph, you also remove all the edges incident to those vertices.)

a) State a recursive algorithm for finding α that corresponds to the preceding discussion.

b) Let g_n be the worst case amount of work for your algorithm when G has n vertices. (You have to stipulate what counts as work.) The preceding discussion did not ask you to take special care in picking v. You might pick an isolated vertex. In light of this, what is the best inequality recursion for g_n? Solve it. In the worst case, then, is your algorithm any better than checking each and every subset of the vertices of G for independence?

c) If G has no edges then, clearly, $\alpha = n$. (The only work in checking the adjacency matrix is to verify that there are no edges.) Otherwise, you may pick v with at least one incident edge. Modify your algorithm to incorporate these observations. Now devise and solve a new inequality recursion for g_n and thus verify that your modification improves the worst case behavior.

d) Apply the algorithms in a) and c) to G = a four-cycle. Show a tree of subcalls. Represent each node of the tree by a picture of the graph that the algorithm analyzes in that call.

The next two problems contain assertions and purported proofs. Decide whether the proofs are valid and give your reasons. If a proof is invalid but can be fixed without too many changes, fix it.

12. Assertion. Algorithm TELENET (Example 4, Section 1.2) is correct. (Recall that the purpose of the algorithm is to find the least-cost graph which connects all the vertices. The method involves always picking the least-cost edge between the vertices connected so far and all other vertices.)

"Proof". Induction on the number n of vertices. In the basis case, $n = 2$, there is only one edge, so the algorithm is clearly correct.

Inductive step. Assume that the algorithm is correct when there are n vertices. You are given a graph G with $n+1$ vertices. Let v be the vertex which TELENET connects last. Suppose that we had run TELENET on the subgraph G' connecting the other n vertices. Then the final tree from the run on G' would be the same tree as obtained after n steps on G. The reason is that the subset of edges from which TELENET picks a minimum at each stage in G' is a subset of the set from which it picks in G. However, the subset contains the element which is picked in G, so the same edge is picked when G' alone is considered. Therefore we conclude that after n vertices are joined to the tree when TELENET works on G, it has picked the minimum tree connecting those vertices. Now TELENET goes on to connect the final vertex v. It picks the minimum edge to v. Because it has picked the minimum set of edges apart from v, and the minimum edge to v, it has picked the minimum tree connecting all $n+1$ vertices. QED

13. Assertion: Every connected graph contains a spanning tree.

"Proof". Let $P(n)$ be: Every connected graph with n vertices contains a spanning tree.

Basis. $P(1)$ is obvious: The vertex alone, without any edges, reaches all vertices without using any cycles.

Inductive step. Imagine that you are given a connected graph G on $n+1$ vertices. Temporarily remove any one vertex. By $P(n)$, the remaining graph has a spanning tree. Now consider vertex $n+1$ again. As G is connected, there must at least one edge from $n+1$ to another vertex. Pick any such edge and add it to the n-tree. It cannot create a cycle because it is the only edge we use which touches vertex $n+1$. Thus we have found a spanning $(n+1)$-tree. QED

14. Do what you can to count the number of decision trees as a function of some parameter, such as height or the number of vertices. *Hint*: Try to relate the number of decision trees to the number of trees of some other kind which you know how to count.

15. Finally, coming full circle, consider again the minimum scheduling time problem from the Prologue in which a project consists of many tasks with a precedence relation among them. The project is represented by a vertex-weighted digraph with the weight of vertex k the time $d(k)$ required for task k. We want to know the minimum time to complete the whole project. Using the recursive paradigm we described in words an algorithm for this problem and argued informally that it would find the correct answer.

a) Write the algorithm in our algorithmic language. Then use mathematical induction to argue that it is correct.

b) Here's a different, classical approach to this problem. For each vertex in the precedence digraph, there is a longest directed path from the start vertex to vertex k. Let its length (the sum of the vertex weights on it) be $L(k)$. Let $T(k)$ be the minimum time to complete task k. The classical "minimax" theorem is: For each k, $L(k) = T(k)$. Prove this existentially as follows:

(i) Argue that surely $T(k) \geq L(k)$.

(ii) Show that the schedule "complete task k at time $L(k)$" is feasible, that is, when the last predecessor task to k is finished, there is enough time to finish task k before $L(k)$ time units are up.

Why do (i) and (ii) prove the theorem? Why is this proof existential?

c) It is valuable to know all longest paths from the start vertex to the finish vertex. One reason is because the length of any such path gives us, by b), the minimum completion time. Another reason is that the tasks on longest paths are the *critical* ones which, if they could be speeded up by any means, could result in speeding up the whole project. Use the recursive paradigm to devise an algorithm to find longest paths.

d) Use your algorithm from c) to give a second, constructive proof of the minimax theorem of b). *Hint*: Use induction to show that your algorithm simultaneously finds $L(k)$ and $T(k)$.

References

Discrete Mathematics Books Intended for a Mainly Freshman-Sophomore Audience

There have been more than 30 such books published in the past five years. The six books below are a sampling intended to suggest the different flavors of discrete math books.

Bogart, Kenneth P. 1988. *Discrete Mathematics*, D. C. Heath, Lexington, MA.

Similar in spirit to our book with a few differences in coverage and a somewhat different style; if our explanation of some topic doesn't work for you, try Bogart.

Doerr, Alan and Levasseur, Kenneth. 1985. *Applied Discrete Structures for Computer Science*, Science Research Associates, Chicago.

Aimed specifically at computer science students, this book emphasizes logic, automata and abstract algebra more than most discrete math books.

Dossey, John A., Otto, Albert D., Pence, Lawrence E. and Vanden Eynden, Charles. 1987. *Discrete Mathematics*, Scott, Foresman, Glenview, IL.

Covers many of the standard topics of discrete mathematics but from a more elementary perspective than many discrete math books; broad range of applications.

Epp, Susanna S. 1990. *Discrete Mathematics with Applications*, Wadsworth, Belmont, CA.

One of the most recent discrete mathematics books; well-written and with good coverage of classical discrete math topics.

Ross, Kenneth A. and Wright, Charles R. B. 1988. *Discrete Mathematics*, Second edition, Prentice Hall, Englewood Cliffs, NJ.

Covers a full range of discrete math topics from a mainly mathematical and fairly rigorous perspective.

Schagrin, Morton L., Rapaport, William J. and Dipert, Randall R. 1985. *Logic: A Computer Approach*, McGraw-Hill, New York.

One among a quite small number of books devoted to a single subject area in discrete mathematics at an elementary level.

Books for Your Next Course in Discrete Mathematics

A course from our book is preparation for a variety of later courses in discrete mathematics. The five books below are all more advanced and/or much more specialized in their subject matter than *Discrete Algorithmic Mathematics.*

Graham, Ronald L., Knuth, Donald E. and Patashnik, Oren. 1989. *Concrete Mathematics*, Addison-Wesley, Reading MA.

A spirited discussion of solution methods for hard problems in several areas of mathematics (mostly of particular interest in computer science) where a combination of CONtinuous and disCRETE mathematics is just what you need; nevertheless, this book has very much the flavor of discrete mathematics.

Purdom, Paul Walton, Jr. and Brown, Cynthia A. 1985. *The Analysis of Algorithms*, Holt, Rinehart and Winston, New York.

One among a number of books aimed specifically at algorithms and their analysis; covers many of the topics in our book at considerably more depth.

Roberts, Fred. 1976. *Discrete Mathematical Models*, Prentice Hall, Englewood Cliffs, NJ.

Contains a lot of graph theory (but largely disjoint from the topics in our book) and several topics rarely treated in introductory discrete math books (e.g. n-person games, group decision making); also contains many applications, primarily from the social and biological sciences.

Stanton, Dennis and White, Dennis. 1986. *Constructive Combinatorics*, Springer-Verlag, New York.

An advanced pure combinatorics book but with a distinctly modern, algorithmic flavor; as in our book, this book adheres firmly to the philosophy of using algorithms to generate both theorems and their proofs.

Tucker, Alan. 1984. *Applied Combinatorics*, John Wiley and Sons, New York.

A book devoted to only two of the topics of our book—Graph Theory and Combinatorics—which it explores at considerably more depth than we do.

Appendixes

1 Summary of Algorithmic Language

Name of Structure	Syntax or Description	Page Reference
Assignment	variable ← expression	89
Comment	Right justified in brackets	97
Compound Structure	Any sequence of statements or structures	96
Function	**function** Name(variable list) function body including **return** **endfunc**	116
Input	List in Input section of algorithm	89
Iteration	**repeat** compound structure **endrepeat**	92
	or	
	repeat until condition compound structure **endrepeat**	93
	or	
	repeat while condition compound structure **endrepeat**	94
	or	
	repeat compound structure **endrepeat when** condition	94
	or	
	for $C = Cstart$ **to** $Cstop$ compound structure **endfor**	94

<table>
<tr><td></td><td>or
for $C = \mathit{Cstart}$ **downto** Cstop
    compound structure
endfor</td><td>95</td></tr>
<tr><td></td><td>or
for $s \in$ Set
    compound structure
endfor</td><td>95</td></tr>
<tr><td>Loop termination</td><td>**exit**</td><td>92</td></tr>
<tr><td>Output</td><td>List in Output section of algorithm</td><td>89</td></tr>
<tr><td></td><td>or
print list of variables or literals</td><td>95</td></tr>
<tr><td>Procedure</td><td>**procedure** Name(variable list)
    procedure body including **return**
endpro</td><td>110</td></tr>
<tr><td>Selection</td><td>**if** condition-1 **then** compound structure-1
  condition-2 **then** compound structure-2
    ⋮
  condition-n **then** compound structure-n
        [**else** compound structure-n+1]
endif</td><td>90</td></tr>
<tr><td>Termination</td><td>**stop**</td><td>96</td></tr>
<tr><td>*Conventions*</td><td></td><td></td></tr>
<tr><td>Formatting</td><td></td><td>96</td></tr>
<tr><td>Local and global variables</td><td></td><td>111</td></tr>
<tr><td>Use of English</td><td></td><td>89</td></tr>
<tr><td>Use of mathematical notation</td><td></td><td>89</td></tr>
</table>

2 Abbreviations

Abbreviation	Meaning	Page First used
AGI	Arithmetic-Geometric Inequality	191
BFS	Breadth First Search	249
CNF	Conjunctive Normal Form	580 [29]
CTOH	Circular Towers of Hanoi	829 [2]
DFS	Depth First Search	249
DNF	Disjunctive Normal Form	570
IS	Input Specification	559
LHS	Left Hand Side	141
OS	Output Specification	559
PPN	Polish Prefix Notation	287
RHS	Right Hand Side	141
RPN	Reverse Polish Notation	287
STOH	Straightline Towers of Hanoi	829 [1]
TOH	Towers of Hanoi	109 [7]

Hints and Answers

Prologue

1. 50 **2.** 25 **3.** $T(j)=j+T(j-1)$ so $T(n)=n+(n-1)+\cdots+3+2+1=n(n+1)/2$. **4. a)** $S(j)=T(j)-d(j)$ **b)** $S(j)=\max[(i,j)\in E]\ S(i)+d(i)$. **5.** $T(j)=\max[h(e)=j]\{d(e)+T(t(e))\}$ where $d(e)$—duration of task, $h(e)$—vertex at head of e, $t(e)$—vertex at tail of e, $T(j)$—minimum time to get to vertex j and $h(e)=j$ means all tasks whose heads point to j. **6.** For n-vertex graph assume solution for $(n-1)$-vertex graph, assign 1 to some vertex with no arrow into it (why must there be such a vertex?), and then assign 2 to n to other vertices using $(n-1)$-vertex solution.

Chapter 0

Section 0.1

1. a) $\{\}, \{x\}, \{y\}, \{z\}, \{x,y\}, \{x,z\}, \{y,z\}, \{x,y,z\}$ **2.** $P(\{a\})$ is $\{\emptyset, \{a\}\}$; $P(P(\{a\}))$ is $\{\emptyset, \{\emptyset\}, \{\{a\}\}, \{\emptyset, \{a\}\}\}$ **3. c)** all non-prime positive integers > 2. **4. a)** $\{x \mid 100 < x < 200$ and $2 \not| x\}$ **5. a)** $F\cap N^+$ **c)** $Q-Z$ **e)** $F\cap N^+\cap E$ **g)** $Z-E-N$ **i)** $Z-N$ **6. c)** $R-S=\{2,4\}$, $R-T=\{1,2\}$, $S-T=\{1,9\}$ **e)** $\overline{\{1,3,4,5,6,7,8\}}$ **h)** $\{3\}$ **9.** No. (Why not?). **10. a)** $U-R$ is set of all things in U and not in R, which is $\overline{R}$. **c)** $R-(S\cap T)$ is the set of all things in R and not in both S and T and so is $(R-S)\cup(R-T)$ since only things in both S and T which are in R are not in $R-S$ or $R-T$. **11.** $|A\cup B|=|A|+|B|$ **13.** Nothing; give an example to show why. **16.** Least: m; Most: $m+n$ **18.** Build up a table with, for each value of p $(1\le p\le 10)$, the number of fractions p/q which have not already been counted. **20. b)** $(X\cap Y)\cap Z$ is the set of all things in both Z and in both X and Y, that is, the set of things in all three sets.

Section 0.2

1. a) $\{(x,y) \mid x,y\in R, x\le y\}$ **c)** $\{(x,y) \mid x,y\in Z, x\ne y\}$ **2. a)** reflexive, transitive **c)** symmetric **3. a)** $(1,5)$, $(2,6)$ are; $(1,6)$, $(2,5)$ are not. **b)** reflexive, symmetric and transitive **4. a)** $\in$ is not transitive; consider $a\in B$ and $B\in C$ where C is a set of sets. **b)** transitive **6. a)** $\{(x,y) \mid x<y\}$ **d)** R-reverse is the same as R. **e)** $\{(x,y) \mid x+1=y\}$ **7.** Yes for all three. **8. a)** $\{(s,t) \mid s<t\}$ **b)** itself **9. a)** child **b)** ancestor **10. a)** grandparent **c)** sibling (assuming you are your own sibling) **11.** 2: all even t, 36: all s which are factors of 36, namely 1, 2, 3, 4, 6, 9, 12, 18, 36 **13.** $\{(1,3),(2,4),(2,5),(4,6)\}$ **14.** Two pairs are (rhombus, quadrilateral), (square, parallelogram).

Section 0.3

1. b) 0, 1, 4, 9, 16, ... **2.** Injective **3.** If more than one student to a room, then not 1-to-1; if all rooms filled, then surjective. **5. a)** False **b)** True **6. a)** False **b)** False **7. a)** False **b)** False.

Section 0.4

3. a) 3 **6. a)** $0 \le x < 1$ **9. a)** Any integer **c)** All **e)** All **g)** Any integer or nonnegative real with fractional part $< 1/2$ or negative real with fractional part $> 1/2$ **i)** All multiples of n **10. a)** $\lfloor x+y \rfloor \ge \lfloor x \rfloor + \lfloor y \rfloor$ **c)** $\lfloor x-y \rfloor \ge \lfloor x \rfloor - \lceil y \rceil$ **e)** $n\lfloor x/n \rfloor \le \lfloor x \rfloor$, equal for all m, $nm \le x < nm+1$ **11.** Let $w = x+y$ and use [10b]. **15. a)** $R'(x) = \lceil x+.5 \rceil - 1$ **17.** $\lfloor 10x+.5 \rfloor/10$; rounds up if hundredths place is 5. **20. b)** 4 **c)** Need number of multiples of each power of 5; therefore $\lfloor n/5 \rfloor + \lfloor n/25 \rfloor + \lfloor n/125 \rfloor + \cdots$ **22.** Limit is 0. **23. a)** 15 **b)** 3; why are there no others? **25. b)** 7 **26. a)** 3 **28. a)** $\{x \mid x = 5n+2 \text{ and } n \in Z\}$ **c)** $\{x \mid x = 5n+4 \text{ and } n \in Z\}$ **30. a)** k even: $(2k)^2 = 4k^2$; k odd: $(2k+1)^2 = 4k(k+1)+1$. **31. a)** 3 **d)** 1 **g)** 10/3 **j)** 16 **32. a)** kL **c)** $-kL$ **33. c)** 10^{1000} **34. c)** 4 **36. a)** Domain: $x > 1$; Range: x real **38.** Fastest to slowest: $n^4/\log n$, n^3, n^2, $(\log n)^5$ **40.** Properties 1–3 hold but Property 4 does not; $\lim_{x\to\infty} \log_b x = -\infty$. **41.** Product: $x^6 - 4x^5$ **43.** $x^4 - 1$ **46. a)** $f(x+1)/f(x) = 1 + (2x+1)/x^2$; $f(2x)/f(x) = 4$. **48. c)** 3014

Section 0.5

1. d) $\prod_{i=0}^{49}(2i+1)$ **2. d)** $x + \sum_{i=1}^{7}(i+1)x^{2i}$ **3. b)** 7451/110592 **4. b)** 2742 **5. a)** $1 - 1/5$ for $n = 4$; telescopes **c)** 1/6 for $n = 5$; telescopes **e)** -8 for $n = 6$ **7.** $k = 0$ term is 0; index name is irrelevant. **9.** $i = 1$ to $n+1$ **11. a)** 498 **12. a)** -256 **13. a)** 3 for $n = 3$ **14.** 51; 200 **18. a)** $(0,10)$ **d)** $[0,1]$ **g)** $(0,1/10)$ **19. b)** $P_3 \wedge P_5 \wedge P_7$; true.

Section 0.6

1. $\begin{bmatrix} 6 & 8 \\ 10 & 12 \end{bmatrix}$ **3.** $\begin{bmatrix} 4 & 8 \\ 12 & 16 \end{bmatrix}$ **6. a)** $[5\,1\,9]$ **d)** 20 **7. a)** $A = \begin{bmatrix} 2 & 3 & 4 \\ 3 & 4 & 5 \\ 4 & 5 & 6 \end{bmatrix}$ $B = \begin{bmatrix} 1 & 2 & 3 \\ 2 & 4 & 6 \\ 3 & 6 & 9 \end{bmatrix}$ **d)** $\begin{bmatrix} 20 & 40 & 60 \\ 26 & 52 & 78 \\ 32 & 64 & 96 \end{bmatrix}$ **8. a)** 81 **9. a)** $\begin{bmatrix} a & b \\ c & d \end{bmatrix}$ **11.** $\begin{bmatrix} 285 \\ 335 \\ 560 \end{bmatrix}$ **12.** $n \times n$ matrix **13.** Only A and B **16.** $M^3 = \begin{bmatrix} 1 & 3 & 6 \\ 0 & 1 & 3 \\ 0 & 0 & 1 \end{bmatrix}$ M^n is like M^3 except that $(1,2)$ and $(2,3)$ elements are n and $(1,3)$ element is $n(n+1)/2$. **17.** $c = 22a_1 + 28a_2$. **19.** Columns of A = Rows of B, Columns of B = Rows of C; double sum over r and s of $a_{ir}b_{rs}c_{sj}$ to get (i,j) element of product. **20.** Generalize solution to [19]. **21.** Can compute all powers of any square matrix, no power of any non-square matrix.

Section 0.7

3. a) If n is even, then $2n$ is even. **c)** If there is a 29 Feb. this year, then this is a leap year. **4. a)** False **c)** True **e)** False **5. a)** If $ABCD$ is a parallelogram, then $AB \parallel CD$. **d)** If flowers wilt, then it's hot. **6. b)** If a computer program doesn't terminate, then it's not correct. **e)** If it's not cold, then it won't snow. **7. b)** If n is even, then $2n$ is odd. **9.** (a), (c) and (d) are equivalent; (b) and (e) are equivalent. **11. a)** True **d)** False **g)** True (but it would be false if Π were replaced by x). **13.** Pairs which are negations of each other are $a-d$, $b-c$, $c-e$, $d-f$. **14. a)** Same meaning (but not very good English!) **c,e)** Not good English **16. a)** Vacuously true since there is no k such that $1 \le k \le 0$. **e)** Not vacuously true and neither trivial nor obvious **h)** Not vacuously true but trivial

Supplementary Problems: Chapter 0

1. b) False since x could be -2 **e)** True. **2. b)** 44 **3. b)** 2145 **4. b)** The product of the two integers **c)** Show that if p^i is a factor of m and p^j is a factor of n (for p prime), then one of these is a factor of the gcd and the other is a factor of the lcm. **5. c)** Add $(-1)^k$ under summation. **6. c)** Substitute b/d into $f(x)$, then multiply by d^{n-1} to get result for d, multiply by d^n/b to get result for b. **d)** 1/2, 1, 3/2, 2, 3, 6 and their negations **8. a)** Number of variations in sign excluding zeros of $a_0, -a_1, a_2, \ldots, (-1)^n a_n$ **9. b)** 0 positive zeros, up to 10 negative zeros **10.** $n2^{n-1}$; correct but less helpful: $\sum_{k=1}^{n} k\,n!/[(n-k)!k!]$.

Chapter 1

Section 1.2

1. c) Be sure you understand why the computation doesn't terminate. **2. a)** What is *prod* at the end of each iteration and, therefore, at the termination of the algorithm? The algorithm terminates because u increases from zero to $x \geq 0$. **4. a)** Edges added: $(1,2)$, $(2,3)$, $(2,4)$; Total cost: 990 **5. b)** 44 **6.** Assume it works for $m \geq n$ when neither is zero (the case discussed in text); show how other cases are reduced to this or can be solved directly in some simple way. **7. a)** $P1$: Tasks: 7, 15, 14, 6, 11 (Time: 180); $P2$: 12, 5, 13, 10, 1 (175); $P3$: 3, 8, 2, 9, 4 (175).

Section 1.3

1.
```
if 2 | N then print N
       else if N > 0 then print N
endif
```

4.
```
if 2 | N and N > 0 then print N
                   else print '0'
endif
```

6. b)
```
if a > b then if a > c then L ← a
                       else L ← c
              endif
         else if b > c then L ← b
                       else L ← c
              endif
endif
```

8.
```
Q ← 10^6/N
repeat while Q = ⌊Q⌋
    print N
    Q ← Q/N
endrepeat
```

9. b)
```
i ← 1
repeat while i ≤ 26
    print L(i)
    i ← i+1
endrepeat
```

11. Use two nested **for** loops, one to choose the letter and one its repetition. **12.** Two nested loops, one to choose the last letter in group, other to print letters **13.** Quite similar to [11] **15.** You'll need **if** $L(i) \neq L$ somewhere. **17.** Remember $i \mid N$ is true only if i is a divisor of N. **19.** Did you remember to initialize the sum? **20.** If a number is even, what condition must be true? **22. a)** The value of the polynomial $a_0 + a_1x + \cdots + a_nx^n$ **c)** Counting done automatically in **for** must be done explicitly in **repeat** by initializing counter before **repeat**, testing for completion with **while** after **repeat**, and updating counter in loop. **23.** Fragment in solution to [6b] would work. **25.** Look from right for first nonzero digit. **27.** Let $S \leftarrow \text{PlusOne}(S)$ result in increasing S by 1; start with $S = n$ and use a loop which adds 1 to S m times. **28. b)** n vs. $2^n - 1$. **d)** Once you find the highest power of 2 equal to or less than m, call it $k = 2^i$, repeat the calculation for x^{m-k}; i.e. find the highest power of 2 in $m-k$ and so on until x^m has been computed.

29. The loop is:
```
repeat while d ≠ w
    if d < w then deposit more money
    if d > w then hit decrease button
endrepeat
```

31. To test if odd I is prime, we need not check any numbers greater than $\lfloor\sqrt{i}\rfloor$ to see if they divide N since, if $d > \lfloor\sqrt{i}\rfloor$ is such that $d \mid i$, then $i = dc$ where c is an integer less than $\sqrt{i}$; in fact, we need only check to see if i is divisible by all primes less than $\sqrt{i}$ since any composite d less than $\sqrt{i}$ is the product of smaller primes. **33. a)** Edges added: $\{2,4\}$, $\{2,3\}$, $\{3,6\}$, $\{4,5\}$, $\{1,5\}$; Total cost: 1050 **c)** Looking for maximum weight spanning tree is equivalent to negating all the weights and looking for a minimum weight tree; since TELENET finds a minimum weight tree, TELENET-MAX finds a maximum weight tree since it chooses the same edge with negated weights as TELENET with the original weights **34.** A poor method, in general, because the long tasks tend to be done by a few processors with the others idle.

Section 1.4

1. If $n > m$ the first argument increases at the first step and then decreases. **3. b)** After "gcd(m,n)" add "lcm $\leftarrow mn/$answer"; you should also add a check that neither m nor n is zero. **5. a)** $1 \to 2$; $1 \to 3$; $2 \to 3$ **6. a)** $1 \to 2$; $1 \to 3$; $1 \to 4$; $3 \to 4$; $2 \to 4$ **8.** Tail end: Move smallest disk to second pole, then move other $n-1$

disks.

Section 1.5

1. b) $S \leftarrow g_1$; $min \leftarrow g_1$
for $i = 2$ **to** n
  $S \leftarrow S + g_i$
  if $g_i < min$ **then** $min \leftarrow g_i$
endfor
$S \leftarrow S - min$
$A \leftarrow S/(n-1)$

3. a) function INT(x)
  if $x \geq 0$ **then** $int \leftarrow \lfloor x \rfloor$
    else $int \leftarrow -\lfloor -x \rfloor$
  return *int*
endfunc

5. A straightforward algorithm would just use the quadratic formula to calculate both roots; however, doing this x is undefined if $a = 0$, yet equations of the form $bx + c = 0$ do have roots. Moreover, the quadratic formula often leads to the subtraction of two nearly equal quantities for one of the roots which is bad practice. A better algorithm is:

if $a = 0$ **then if** $b = 0$ **then if** $c = 0$ **then print** 'all x are roots'
        else print 'no roots'
      endif
$b^2 < 4ac$ **then** $x_1 \leftarrow (-b + i\sqrt{4ac - b^2})/2a$ [Complex roots]
    $x_2 \leftarrow (-b - i\sqrt{4ac - b^2})/2a$
  else if $b \leq 0$ **then** $x_1 \leftarrow (-b + \sqrt{b^2 - 4ac})/2a$ [Real roots]
        $x_2 \leftarrow c/ax_1$
      else $x_1 \leftarrow (-b - \sqrt{b^2 - 4ac})/2a$
        $x_2 \leftarrow c/ax_1$
endif

7. Algorithm REVERSE
procedure Switch(x, y)
  $temp \leftarrow x$
  $x \leftarrow y$
  $y \leftarrow temp$
  return x, y
endpro
for $i = 1$ **to** $\lfloor n/2 \rfloor$
  Switch(a_i, a_{n+1-i})
endfor

8. a)for $j = S+1$ **to** F
  $T(j) \leftarrow d(j) + \max_{i \to j}\{T(i)\}$
endfor

9. b) function MR(p, q)
  if $q = 0$ **then** $mult \leftarrow 0$
    else $mult \leftarrow p + \text{MR}(p, q-1)$
  return *mult*
endfunc
$prod \leftarrow \text{MR}(x, y)$

10. Where is *prod* initialized?

11. a) Each level calls the next level once and not twice.

12. b) Iterative algorithm:
$prod \leftarrow a_1$
for $k = 2$ **to** n
  $prod \leftarrow prod \times a_k$
endfor
Recursive algorithm:
function Prod(k)
  if $k = 1$ **then** $prod \leftarrow a_1$
    else $prod \leftarrow a_k \times \text{Prod}(k-1)$
  return *prod*
endfunc
$P \leftarrow \text{Prod}(n)$

13. a) Iterative:
$F \leftarrow 1$
for $k = 2$ **to** n
  $F \leftarrow F \times k$
endfor
Recursive:
function Fact(k)
  if $k = 1$ **then** $fact \leftarrow 1$
    else $fact \leftarrow k \times \text{Fact}(k-1)$
  return *fact*
endfunc
$F \leftarrow \text{Fact}(n)$

c) Iterative:

```
S ← 1
for k = 2 to n
    S ← S + k
endfor
Recursive:
function T(k)
    if k = 1 then t ← 1
              else t ← k + T(k−1)
    return t
endfunc
S ← T(n)
```

15. Biggest problem is that Max(m) is a procedure, so that statement "**if** $a_m >$ Max($m-1$)" makes no sense; other problem is that many non-maximum values are printed. Better algorithm:

```
procedure Max(m)
    if m = 1 then big ← a_1
             else Max(m−1)
                  if a_m > big then big ← a_m
    endif
    return big
endpro
Max(n)
```

17. b)

```
procedure Factors(m)
    if m > 1 then if d|m then print d
                  repeat while d|m
                      m ← m/d
                  endrepeat
                  d ← d+1
                  Factors(m)
    endif
    return
endpro
d ← 2
Factors(N)
```

c)

```
d ← 2
repeat until N = 1
    repeat while d | N
        print d
        N ← N/d
    endrepeat
    d ← d+1
endrepeat
```

Note: If a table of primes is available, this can be made much more efficient.

19. a) Since every divisor of all members of a set of numbers divides its gcd, gcd of arguments on left divides both arguments on right and, therefore, their gcd; similarly gcd on right divides all arguments on left and, therefore, their gcd; since each gcd divides the other, they are the same. **21.** This "cheating" procedure stacks rings on an intermediate pole in reverse order and then restacks them on a final pole in correct order. **23. b)** Initialize *words* to 0 in main algorithm and then have single statement in processcharacter: **if** $s_{i+1} = \not{b}$ **then** $words \leftarrow words + 1$.

Section 1.6

1. a) $n+1$ additions; $n(n+1)/2$ multiplications **2. b)** Best case is 1 when first coin is bad. **c)** Worst case is 10, when last coin is bad. **4. a)** Best case is 1, when N is odd. **c)** Each odd number: 1 division; each even number: 1 division to determine that no more factors of 2 remain; each multiple of 2 which is not a multiple of 4: 1 more division; etc.

6. b)

```
for i = 1 to m
    c_i ← 0
    for j = 1 to n
        c_i ← c_i + a_ij b_j
    endfor
endfor
```

d) For b) algorithm: Best, worst and average cases same: mn

7. Here's part of the table: With $m=8$ and $n=1$ to 8: 1, 1, 3, 1, 4, 2, 2, 1 **10. a)** Don't do tests for $w < w_i$ or $w > w_i$. **c)** n **12. a)** Correct **13. a)** 2, 2, 3, 3,—not enough data to make a conjecture; see Section 2.6 **15. a)** Line 1: Replace x_1 by x_n; Line 2: $n-1$ **downto** 1 **b)** Initialize *msub* to 1 and update it whenever m is changed. **16.** By induction (see Chapter 2). **17.** 3: 5/6.

Supplementary Problems: Chapter 1

1. d)
```
function Dval(k)
    if k = 0 then dval ← a_n
             else dval ← Dval(k − 1) × b + a_{n−k}
    return dval
endfunc
D ← Dval(n)
```

3. b)
```
procedure Convert(m, i)
    if i < 0 then stop
             else v ← ⌊m/b^i⌋
                  m ← m − v × b^i
                  a_i ← v
                  Convert(m, i − 1)
    endif
    return a_i, a_{i−1}, ..., a_0
endpro
I ← ⌊log_b N⌋
Convert(N, I)
```

4. a) $K1$: 20, 4; $K2$: 15, 14; $K3$: 9, 8, 7, 3; item of size 2 won't fit **d)** $K1$: 14, 8, 3; $K2$: 20, 7, 2; $K3$: 15, 9; item of size 4 won't fit **5.** Pick the worker with the highest rating for any of the jobs and assign that worker to that job. Then of the remaining workers and jobs, find the worker who has the highest rating, and so on until everyone is assigned; for data given, worker 1 gets task 1 and worker 2 gets task 2, a very poor result. **7. a)** $(\sqrt{5}-1)/2$ **b)** .38 **8. b)** 25, the number of primes < 100 **9.** For average: 9×10^8 numbers require 1 operation; how many require 2, 3, 4 etc. operations?

Chapter 2

Note: For induction proofs, we usually just sketch the key step, typically the breakdown of the next case (usually $P(n+1)$) to obtain a previous case or cases (usually $P(n)$). Proofs submitted for grading should be complete and should follow the format in the text.

Section 2.2

1. Prove: $1+3+\cdots+(2n-1)=n^2$. Inductive step: $\sum_{k=1}^{n+1}(2k-1)=\sum_{k=1}^{n}(2k-1)+[2(n+1)-1]=n^2+[2(n+1)-1]=(n+1)^2$. **3.** $u_n=(2n-1)^2$. Key step: $(2n-1)^2+8n=[2(n+1)-1]^2$. **5. b)** $a=\frac{1}{2}$, $b=\frac{1}{2}$, $c=0$. **7.** $\sum_{k=1}^{n}[a+(k-1)d]=\sum_{k=1}^{n}a+d\sum_{k=1}^{n}(k-1)=na+d\sum_{j=0}^{n-1}j$. **9.** $(a_{n+2}-a_{n+1})+(a_{n+1}-a_1)=a_{(n+1)+1}-a_1$. **11.** $\prod_{n=2}^{m}=\frac{m+1}{2m}$. Key step of $P(m-1)$ to $P(m)$: $\frac{m}{2(m-1)}\cdot\frac{m^2-1}{m^2}=\frac{m+1}{2m}$. **13.** $\frac{a^{n+1}-b^{n+1}}{a-b}=a^n+\frac{b(a^n-b^n)}{a-b}$. **15.** The right triangle with base $\sqrt{n}$ and height 1 has hypotenuse $\sqrt{n+1}$. **17.** Basis is $n=2$. Also, $(1+x)^{n+1}=(1+x)^n(1+x)>(1+nx)(1+x)$. **19. a)** $c>0$ **c)** $ac<bc$ **21.** Let $P(n)$ be: $x_n^2>n$. For the inductive step, expand $x_{n+1}^2=\left(x_n+\frac{1}{x_n}\right)^2$. **23.** If $x_n=\sqrt{1+x_{n-1}}$ were rational, then $x_{n-1}=x_n^2-1$ would be rational. **25.** The difference between two consecutive sums is $(n+3)^3-n^3=9n^2+27n+27$. **27.** 2^n; first show that the number of sets doubles when a new element e is allowed—for each old set, include e or don't. **29. a)** 31254 (many others too) **b)** 25341 **31.** Every game is a tie. **32.** Put an arrow from i to $j \iff i$ wins the series with j. For each pair i, j there is still exactly one arrow between them, so the proof of Theorem 2 still holds. **34.** Given such a tournament with $2n-1$ teams, arbitrarily let A be any n of the teams and B the remaining $n-1$. Add new teams c and d. Have c beat all in A and lose to all in B; have d do the opposite. Between c and d who wins? **37.** For $n+1$ rings there are 2^n-1 moves from TOH$(n, r, 6-r-s)$, then 1 move, then 2^n-1 moves from TOH$(n, 6-r-s, s)$. **39.** Key idea: In a minimum sequence, the top $n-1$ rings must be together on pole $6-r-s$ exactly once, and the $(n-1)$-game sequence that gets them there (also the sequence by which they leave) must be minimal. Why? But these $(n-1)$-game sequences are unique by induction. **41. b)** For $n>1$ the call of STOH(n, r, s) is STOH$(n-1, r, s)$; robot$(r\to 2)$; STOH$(n-1, s, r)$; robot$(2\to s)$; STOH$(n-1, r, s)$. **d)** Key idea: The $(n-1)$-tower must be on the goal pole when ring n first moves, and must be on the start pole when ring n moves to the goal pole. **43.** For an n-clutter, assemble the top $n-1$ rings into a single pile by induction. If this pile is not already over ring

n, move it over using [41] or [42]. **45.** Let $a_n = 2+2(\frac{1}{2})^n$. Let $P(n)=$ "a=a_n after the nth trip through the loop". Key step: $\frac{[2+2(\frac{1}{2})^n]+2}{2} = 2+2(\frac{1}{2})^{n+1}$. **47.** $Q(n)$ asserts that the **repeat while** line is looked at n times and that $a \neq 2$ the nth time. Thus, at the end of the nth pass, new $a = \frac{a+2}{2} \neq 2$, so the loop is also entered an $(n+1)$st time. **49.** Termination iff $a=b$ originally.

Section 2.3

1. a) $\left(\sum^{n+1} x_i\right)/m = \left(\sum^n x_i + x_{n+1}\right)/m \overset{P(2)}{=} \left(\sum^n x_i\right)/m + (x_{n+1}/m) \overset{P(n)}{=} \sum^n (x_i/m) + (x_{n+1}/m) = \sum^{n+1}(x_i/m)$. **c)** Write $\sum^{n+1} a_i$ as $\sum^n a_i + a_{n+1}$, use $P(2)$ then $P(n)$. **3.** If s_i is the length of side i, then $s_i < \sum_{j\neq i} s_j$. For $n+1>3$, one approach is to slice off two adjacent sides (not including side i) and use $P(n)$, then $P(3)$. **5.** Write $\bigcup^{n+1} B_i$ as $\left(\bigcup^n B_i\right) \cup B_{n+1}$, use $P(2)$ then $P(n)$. **7.** Generalization: $\overline{\bigcup A_i} = \bigcap \overline{A_i}$. Use $\bigcup^{n+1} A_i = \left(\bigcup^n A_i\right) \cup A_{n+1}$. **9.** Start with the equality proved in [5] and complement both sides. **10.** To prove the given equality, first write $x = a+\epsilon$ where $a = \lfloor x \rfloor$ and $0 \leq \epsilon < 1$. Likewise, $y = b+\delta$. For $\lfloor x+y \rfloor \leq \lfloor x \rfloor + \lfloor y \rfloor + 1$, note that $\epsilon+\delta<2$. Generalization: $\sum^n \lfloor x_i \rfloor \leq \lfloor \sum^n x_i \rfloor \leq \sum^n \lfloor x_i \rfloor + (n-1)$. To prove it, write $\sum^{n+1} \lfloor x_i \rfloor$ as $\sum^n \lfloor x_i \rfloor + x_{n+1}$. **13.** $P(1)$ and $P(2)$ **15.** Prove $P(11)$ and $P(12)$ ($P(8)$ to $P(10)$ are in the text), then prove $P(n-5) \Longrightarrow P(n)$. **17. a)** If a_k, a_{k+1} are known to be integers, then the proof in Example 2 shows that $a_n \in N$ for $n \geq k$. For $n<k$ do downward induction: $[P(n+1), P(n)] \Longrightarrow P(n-1)$ because $a_{n-1} = a_{n+1} - 2a_n$ is an integer if a_{n+1} and a_n are. **18.** Let $\{b_n\}$ be the new sequence. Let $P(n)$ be: $b_n - a_n > b_{n-1} - a_{n-1} > 0$. **20.** By strong induction you can assume that $a_{\lfloor (n+1)/2 \rfloor} \geq a_{\lfloor n/2 \rfloor}$ and $a_{\lceil (n+1)/2 \rceil} \geq a_{\lceil n/2 \rceil}$. **21.** 2^{n-2} for $n \geq 2$. For the inductive step, note that each strictly increasing sequence from 1 to $n+1$ either includes n or it doesn't. **23.** To prove $Q(n) \Longrightarrow Q(n+1)$, if $n+1$ factors as rs, then $r, s \leq n$ and $Q(n)$ applies. **25.** Suppose t won (or tied for winning) the most games. If t' beat t, then t beat some t'' that beat t', for otherwise t' won more games than t. (Why and so what?) **27.** For $n>1$ do induction using formulas for $\sin(\alpha+\beta)$ and $\cos(\alpha+\beta)$ with $\alpha = n\theta$ and $\beta = \theta$. What about $n \leq 0$? **30.** Start with the Triangle Inequality, square, expand, subtract common terms, square again, and you get the Cauchy-Schwarz Inequality. Now verify that all steps reverse. **31.** Write $\prod_{i=1}^{n+1} f_i$ as $\left(\prod_{i=1}^n f_i\right) f_{n+1}$. The proof is a standard generalization by strong induction, as in [1], but the notation is harder to handle.

Section 2.4

1. Let $a_n = \sum_{k=1}^n k$. Guess $a_n/n^2 \to \frac{1}{2}$, then look at $a_n - n^2/2$. **2.** $(2n^3+3n^2+n)/6$. Prove by induction. **5. a)** 2^n+n **c)** $\frac{1}{2}[3^n+(-1)^n]$ **d)** $n^3-3n^2+2n = n(n-1)(n-2)$. **7.** $n/(3n+1)$ **9.** $n/(n+1)$ **11.** $a_n = n$. **12.** $L(n) = \lceil \log_2 n \rceil$. **14.** Two colors for any $n \geq 1$. For an inductive proof, ignore line n+1, 2-color the rest by assumption, then reverse the colors on one side of line n+1. **16.** In the assertion that line n+1 passes through *exactly* $n+1$ previous regions (because it intersects the other lines in n *distinct* points). **18.** $F(n,m) = nm$.

Section 2.5

1. Assertion false. By assumption, r and s are *either* squares or primes. **3.** Inductive step fine, but the basis wasn't shown and is false. **5.** Assertion false. To prove $P(n)$, case $P(n-2)$ was used, so two basis cases ($x_3 < x_2$ and $x_4 < x_3$) are needed; but $x_4 \not< x_3$. **7.** Assertion false (hardly obvious; smallest counterexample when $n=6$), breakdown error in proof. When t^* is removed, the remaining tournament may have a total winner or loser, so $P(n)$ need not apply. **9.** Assertion true but proof invalid: Side 1 may not be the longest in P', so $P(n)$ need not apply. Several easy fixes; e.g., strengthen $P(n)$ by replacing "the longest" by "any". **11.** Assertion true, proof passable, but buildup approach from n-long sequences to $(n+1)$-long sequences of dubious merit. Either argue that all allowed $(n+1)$-long sequences are obtained by this buildup, or start with $(n+1)$st case and break down. **13.** Assertion false. In proof, $P(n) \Longrightarrow P(n+1)$ valid only for $n>0$ (where $n \in M \Longrightarrow n = n+1$) and the case $n=0$ is needed. **16.** Assume $Q(n+1)$ and break it down to obtain $Q(n)$. Now $R(n)$ follows from assuming $P(n)$. Work back to $R(n+1)$. **17.** This $P(2k+1)$ does not say that all $(2k+1)$-tournaments have a certain property, but only that *one* $(2k+1)$-tournament does. Buildup validly produces one item. **19.** The "proof" in Example 1 needs little change. In the inductive step, put any one of the $n+1$ women in both n-sets. The common hair color is her hair color.

Section 2.6

1. BINSEARCH picks the middle item iff n is odd; otherwise it picks the lower-indexed item closest to the middle. **3.** In Eq. (2), $\lfloor x\rfloor - 1 = \lfloor x-1\rfloor$. In Eq. (3), $n - \lfloor x\rfloor = n + \lceil -x\rceil = \lceil n-x\rceil$. **5.** Treat three cases: $n = 2^k$; $n+1 = 2^k$; everything else. The equality if false for some real numbers. **6. b)** $C_n = 1 + C_{\lceil n/2\rceil} + C_{\lfloor n/2\rfloor}$. **7.** $L(n) = \lceil \log_2 n\rceil$. Prove with strong induction. **9.** Basis $P(1)$: add 1 once. $P(n) \Longrightarrow P(n+1)$: get to n by $P(n)$, then add 1 once more. **11.** The version of BINSEARCH in Section 2.6 correctly states that w is not found. The version in Section 1.6 crashes. See [12]. **14.** A good loop invariant is: The ith time the search loop is entered, either w is one of the words not yet considered or else w is not on the list at all." A complete proof requires care. **15.** The loop is: **for** $k = 1$ **to** n; $S \leftarrow S + a_k$; **endfor**. Loop invariant: at the end of the mth pass, $S = \sum_{k=1}^{m}$. **17. a)** Let $Q(n)$ be: For any $m \in N^+$, the value of "answer" returned by procedure gcd(m,n) is indeed the gcd of m and n. Use strong induction with basis $n = 0$. Algorithm EUCLID-REC is correct because it invokes gcd(m,n). **c)** Use strong induction with basis $j = 0$. The correctness of EUCLID follows by setting $j = n$. **19. a)** Loop: **for** $k = n-1$ **downto** 0; $d_k \leftarrow c'_{k+1}/a$; $c'_k \leftarrow c'_k - d_k b$; **endfor**, where initially $c'_k = c_k$. **b)** Let $P(x)$ be the polynomial. After i iterations of the loop, there is a partial quotient $Q_i(x) = d_{n-1}x^{n-1} + \cdots + d_{n-i}x^{n-i}$ and a remainder $R_i(x) = c'_{n-i}x^{n-i} + \cdots + c'_0$. A loop invariant is: $P(x) = (ax+b)Q_i(x) + R_i(x)$.

Section 2.7

1. $0! = 1$; $(n+1)! = (n+1)n!$ for $n > 0$. **3.** Key line: **if** $k = 0$ **then** *sum* $\leftarrow 0$ **else** *sum* $\leftarrow$ Sum$(k-1) + a_k$. **6.** Initially *prod* $\leftarrow 1$ (see solution for [7]). The loop body is *prod* $\leftarrow$ *prod* $* a_k$. **7.** $\prod_{k=1}^{0} a_k$ is the empty product that the algorithms in [5, 6] face initially. This product must be set to 1 for the algorithms to work, so generalize and define it to be 1 wherever. **9.** Because it's an empty product. **11.** The smallest set E such that i) $0 \in E$, and ii) $n \in E \Longrightarrow n+2 \in E$. **13.** $U = \{am+bm > 0 \mid a, b \in N\}$. When $m = 7$, $n = 11$, $U = \{7, 11, 14, 21, 22, \ldots, 58, 60, 61, \ldots\}$. Can you explain why every integer ≥ 60 is in U? **15.** Substitute $n = 0$ in ii), then use rules of addition and i) to obtain $1 \cdot m = m$. Now substitute $n = 1$ into ii) and then $n = 2$. **17.** Weak induction applies if $P(0)$ is true and $P(n) \Longrightarrow P(n+1)$. So by the definition of S, $0 \in S$ and $n \in S \Longrightarrow (n+1) \in S$. We conclude $N \subset S$, so $P(x)$ is true at least for $x \in N$.

Supplementary Problems: Chapter 2

1. Can be won. For $n+1$ rings, start by invoking case n twice. **3. a)** $f(x) = cx$ for any constant c. **d)** Strong induction on the number of summands. **5.** Tough! Maximum total overhang when the ith brick (counting from the top) hangs distance $1/2i$ over the brick below it. Just to prove this configuration (call it C) stable takes induction. To prove optimality, strength the hypothesis to $P(n)$: With n bricks, C is optimal and, for any n-brick configuration C' with d less total overhang, the center of gravity of C' is $< d$ farther to the right than in C. **7.** Don't try induction on "the n-game can be won" but rather on "Stage 1 of the n-game can be done", where Stage 1 gets the pieces to BXOXO$\cdots$XO (B = blank) or via mirror images to XO$\cdots$XOXOB. Stage 1 can be accomplished recursively—play the inner $(n-1)$-game to its Stage 1 first. After Stage 1 of the n-game is done, a brief Stage 2 gets you to OXOX$\cdots$OXB (or its mirror image). To finish and win, run the mirror image of Stage 1 backwards. **9. a)** One approach: $\frac{1}{2n}\sum_{i=1}^{2n} a_i = \frac{1}{2}(b_1+b_2)$, where $b_1 = \frac{1}{n}\sum_{i=1}^{n} a_i$ and $b_2 = \frac{1}{n}\sum_{i=n+1}^{2n} a_i$. Now apply $P(2)$ and then $P(n)$ twice. **11. a)** $\sum_{i=0}^{m} \frac{1}{2i+1} = \sum_{k=1}^{2m+1} \frac{1}{k} - \sum_{k \text{ even}}^{2m} \frac{1}{k}$. **c)** By induction using $H(2^{m+1}) = H(2^m) + (2^m$ terms all at least $1/2^{m+1})$. **13. a)** Key idea: Suppose $u_m + d_n < C$. Then $u_i + d_n < C$ for all i since $u_i \leq u_m$; so forget about d_n. Similarly, if $u_m + d_n > C$, forget u_m. **b)** Let $P(k,l)$ be: If the repeat loop is ever entered with $i = k$, $j = l$, then SUM-SEARCH correctly determines if $C = u_i + d_j$ for any $i \leq k$ and $j \leq l$ and otherwise returns (0,0). **16.** Invalid. Let $P(x)$ be the statement "$x \leq 1$". Then i) and ii) are true, but $P(x)$ is false for all $x > 1$. Remember, the implication in (ii) is true whenever the premise is false.

Chapter 3

Section 3.1

1. Can't even start; no trip with F gives legal group on near shore **3.** 12 legitimate states of H_1, H_2, W_1, W_2, B(oat) on near side of river; drawing graph and connecting states reachable from each other gives several solutions in 5 trips **5.** Only three possible states after first move **6.** Graph has nine states and is connected **7.** Traverse edges 1, 11–14, 10, 2–9, 15–16 **8.** 00010111 **10. a)** 6 trees; two have w_1 as root with one left and one right branch **b)** Two trees of height 1, four of height 2; average 5/3.

Section 3.2

1. b) $V=\{u,v,x,y\}$; $E=\{\{u,v\},\{u,x\},\{u,y\},\{v,x\},\{v,y\},\{v,v\}\}$. **2. b)** $V=\{u,v,x,y\}$; $E=\{(v,u),(v,x),(x,u)(x,y),(y,x)\}$. **3. a)** There is 1 with 0 edges; 1 with 1 edge, 2 with 2, 3 with 3, 2 with 4, 1 with 5, 1 with 6. **b)** 1 with 0 edges, 1 with 1, 4 with 2, 10 with 3 **4. a)** Three graphs **b)** Two graphs, one with edges in both directions between one pair of vertices **6. a)** Reflexive and transitive, not symmetric **b)** Divisibility **7. b)** Vertices—people; edges—people who shake hands; then apply a) **9.** Can't tell which of multiple edges is on path; definition must specify this **10.** What is the sum of the first $n-1$ integers? **11.** $\lfloor n/2 \rfloor$ **12.** $4!=24$ assuming it doesn't matter where you start. **14. b)** $\{v_1,v_2\}$, $\{v_1,v_4\}$, $\{v_3,v_2\}$, $\{v_3,v_6\}$, $\{v_5,v_2\}$, $\{v_5,v_4\}$, $\{v_5,v_6\}$ **16. a)** Two tournaments **c)** Yes; draw it **17. b)** Only trees which consist of a single path from the root to a leaf **18. b)** If connected and cyclic, each vertex on cycle would have indegree 1 but path from root to vertex on cycle would give that vertex indegree 2; if acyclic and not connected, a vertex in each component would have degree 0; otherwise could construct cycle **19. b)** Adjacent to $\rightarrow$ incident on; **d)** paths $\rightarrow$ edges; **21. a)** 8 graphs **b)** 4 graphs **23.** There are many groups of 5 integers (but none of 6 integers) between 1 and 25 such that the difference between any two in a group is greater than 4. **24. b)** mn.

Section 3.3

1. a) $\begin{bmatrix} 0&1&0&0&1\\ 1&0&1&0&1\\ 0&1&0&1&0\\ 0&0&1&0&1\\ 1&1&0&1&0 \end{bmatrix}$ **b)** $\begin{bmatrix} 0&1&0&0&0\\ 0&0&0&0&0\\ 1&0&0&1&0\\ 0&1&0&1&1\\ 0&0&0&0&0 \end{bmatrix}$ **3. a)** $\begin{bmatrix} 1&0&0&0&1&0\\ 1&1&0&0&0&1\\ 0&1&1&1&0&0\\ 0&0&1&1&1&1 \end{bmatrix}$ **3. c)** Sum of entries in row i is degree of vertex i; sum of columns is always two. **d)** $MM^T=A+D$ where A is adjacency matrix and D is a *diagonal* matrix all of whose entries are zero except on the diagonal where the ith element is the degree of vertex i. **e)** Compute MM^T and consider diagonal, non-diagonal terms separately; note that $m_{ik}m_{jk}=1$ iff e_k is incident on both v_i and v_j **4. a)** Try replacing 1's by -1's for vertices at the *tail* of an arrow. **6.** Build up adjacency matrix by processing edges one at a time **7. a)** $A^2=\begin{bmatrix} 3&1&1&0\\ 1&2&1&1\\ 1&1&2&1\\ 0&1&1&1 \end{bmatrix}$ **8. a)** $uvwuw$, $uwvuw$, $uvuvw$, $uwuvw$, $uxuvw$, $uvwvw$ **10. a)** Each path of length 3 from a vertex to itself is a triangle but this double counts since path can be in either direction; so compute 1/2 of diagonal element of A^3. **12. b)** uvw, vwx, wxu, xuv **13. a)** $uvwxu$, $vwxuv$, $wxuvw$, $xuvwx$ **15.** For any path between two distinct vertices, remove all cycles; remaining path can at most go through all vertices in which case length is $n-1$. **17. a)** When $m=0$, $A^0=I$, the identity matrix which gives the one null path for each vertex **b)** Change lower limit to 0. **18.** With modification WALSHALL says there is always a path from a vertex to itself; therefore output differs only when such path is null path for any vertex. **19.** Path matrix is all 1's. **21.** Matrix is all 1's (even on the diagonal). **23.** After "for $i=1$ to n", skip inner loop if row i is all zeros; after "for $j=1$ to n", if column j is all zeros, skip body of inner loop. **24.** Innermost loop: $j=i$ *to* n and after **then** set p_{ji} to p_{ij}. **25.** Suppose have shortest path from v_i to v_j using only first $k-1$ vertices; get next step from inner loop statement **if** $l_{ik}\neq\infty$ **and** $l_{kj}\neq\infty$ **then** $l_{ij}\leftarrow\min(l_{ij},l_{ik}+l_{kj})$

26. Final matrix from modified WARSHALL is: $\begin{bmatrix} 4&6&8&2&3&12\\ 6&4&2&4&3&6\\ 8&2&4&6&5&4\\ 2&4&6&2&1&10\\ 3&3&5&1&2&9\\ 12&6&4&10&9&8 \end{bmatrix}$ so answer is 12 (element in (1,6) position); diagonal elements are twice shortest edge from vertex; why? **28. b)** Need only start from i and follow from one vertex to the next until there is edge from current vertex to j **31.** Negative weights allow possibility of reducing weight by going around cycle; just not true that statement in solution to [25] gives the minimum length for the next vertex.

Section 3.4

1. Two more bridges between distinct pairs of vertices; three choices; one is $\{v_1, v_3\}$, $\{v_2, v_4\}$ **2. a)** Order of vertices: 1 2 5 6 2 3 6 7 8 3 4 1 **3.** Impossible; 2 vertices of odd degree; retrace two east-west blocks in center **5. a)** Graph has 3 vertices, 9 edges **d)** One is: 011022120 **7.** Always a de Bruijn cycle because graph always has Eulerian cycle; why? **8. a)** Yes; consider graph with 7 vertices 0, 1, ..., 6 and edges corresponding to dominoes (i.e., edge from 2 to 4 for (2,4) domino; each vertex has degree 8 counting loop **b)** No; now each vertex has degree 7. **9.** Consider multigraph with 3 vertices, A, B, C and edges as given (i.e., 2 from A to B); since degree of each vertex is even, there is an Eulerian cycle and circular hypothesis is consistent with data **10.** Same graph as [9] except edges from one test are distinguished from others, say, by being colored red and black; need to find Eulerian cycle with alternating red and black edges and there is one: $AAAACCABCBCB$ **12.** Use strong induction on number of vertices **14. a)** Need only consider 2 odd vertices case (why?); assume greater than 2 vertices; otherwise easy; Eulerian path must begin and end at odd-degree vertices; therefore choose v as in Proof 1 to be even-degree vertex and proceed as in Proof 1; both odd vertices must be in same component **c)** Get Eulerian cycle on G'; delete added edge; then have Eulerian path on G **16. a)** Same argument as [14a] **17.** G must have exactly 4 vertices of odd degree; add two edges to get all even vertices; get Eulerian cycle and then delete added edges **18.** Traverse Eulerian cycle and direct the edges in the direction on cycle. **20.** In each component add enough edges to G to give all vertices even degree; traverse Eulerian cycle to give directions; then delete added edges; all outdegrees same as indegrees or differ by 1. **22.** $b_1 \ldots b_n \rightarrow b_2 \ldots b_n c_1 \rightarrow b_3 \ldots b_n c_1 c_2$ etc. **24.** $(n-1)!$—Start at any v; $n-1$ choices for next vertex; $n-2$ for next etc. **25.** See Section 2.2, Theorem 2. **27. a,b)** Let each digit represent x, y, z coordinate; get 8 vertices of a cube; Hamiltonian path is 000, 001, 101, 100, 110, 111, 011, 010. **c)** Edge from 010 to 000 completes cycle. **d)** Think recursively; let graph in $(n+1)$-case consist of 2 copies of graph in n-case; join corresponding vertices of two graphs, add 0 to all labels of one, 1 to all labels of the other; path is sequence of pairs of edges, one from n-graph followed by one joining the two n-graphs. **e)** Yes; prove by induction on n; $P(n)$ is: Graph in (n,k)-case has Hamiltonian cycle for any k; take k copies of (n,k) graph, join corresponding labels, add 0 to all labels of one graph, 1 to all labels of next etc.; path consists of k edges joining successive (n,k) graphs followed by edge in an (n,k) graph etc.

Section 3.5

1. a) Shortest path: $v_0v_4v_3v_1v_2v_5$ **3. b)** Shortest path has length 6—along upper boundary of graph. **4.** Order in which vertices added: $v_1v_6v_7v_3v_2v_4v_8$; path length to v_8: 11; path is v_1, v_6, v_3, v_8 **6. a)** Apply SHORTESTPATH with all weights 1. **c)** Unique; working back from 15, prior vertex always uniquely determined **8.** For each edge delete it and find the length of the shortest path from the vertex at one end of the deleted edge to the vertex at the other end with all weights 1. **12.** Make bigraph into digraph by replacing each edge with two edges in opposite directions with appropriate weights; then apply SHORTESTPATH to digraph. **13. a)** No; in Prologue kth vertex could be reached only through vertices of smaller index but this is not true in general. **b)** No, next furthest vertex depends in no simple way on paths to further ones. **15.** With weights nonnegative any path with cycle at least as long as path without cycle **17.** No; algorithm may fail to find longest path; consider path from vertex 1 to vertex 4 in graph with edges (weights after semi-colon) $(1,2;5)$, $(1,3;7)$, $(2,4;3)$, $(2,3;4)$, $(3,4;3)$.

Section 3.6

1. a) By contradiction **3.** 6 vertices—length 1; 8 vertices—length 2 **4.** By contradiction **5. a)** Use DFS starting anywhere and test at end to see if all vertices visited. **6. a)** Edges connect squares mutually attacked by a queen. **7. c)** 1213123132.

Section 3.7

1. Ones with no edges **2. b)** 3 **3.** 3 **4.** 2 **6.** 2; no cycles so BIPARTITE succeeds **7.** 2 additional colors; use BIPARTITE ignoring red vertices; chromatic number is either 2 or 3 **9.** Vertices—courses; colors—exam periods; there is an edge $\{u, v\}$ when at least one student takes both u and v. **11.** An edge is needed between the Monday morning and the Physics 1 vertices and between the Tuesday evening and PoliSci 3 vertices; why? **13.** Claim is correct; if graph 3-partite, a 3-way partition of kind described must exist. **14. b)** O_3 is an octahedron. **d)** n; prove by induction **16. a)** 3 and 5 **b)** 3; color vertex i $1+(i-1) \bmod 3$ **19.** Color high degree vertices first so don't end up with high degree vertex to color with all its adjacent vertices having different colors. **20. b)** Both algorithms color v_{10} 4 and otherwise color v_i $1+(i-1) \bmod 3$. **21. b)** 3 **d)** n vertices: $n-1$ if n even, n if n odd in which case colors at each vertex are a different set of $n-1$ colors; proof by induction; odd $n \to$ even $n+1$: just color edge from new vertex different from all colors at each vertex of n-graph; even $n \to$ odd $n+1$: this is harder; change one edge at a time in n-vertex graph to color n or $n+1$ **22. a)** Planar; draw one diagonal outside square. **d)** Planar; draw one diagonal outside each square. **25.** Removal of any edge on cycle reduces number of regions by 1 but leaves G connected. **26. b)** Since $n=6$, $m=9$, from Euler $r=5$; show that structure of K_{33} requires that each region be surrounded by at least 4 edges; since each edge is boundary of two regions, would need 10 edges to get 5 regions. **27. a)** Q_{n+1} found from Q_n by making two copies of Q_n and putting an edge between each pair of corresponding vertices **b)** 4 **28. a)** Remember each edge on boundary of 2 regions **b)** use Euler's formula [25] and result in a) **29.** Suppose contrary so that $5n \leq 2m$ and use [28b] **30.** Same idea as [29]; at least one vertex of degree 3 **32.** 3; chemicals are vertices, colors are rooms.

Section 3.8

1. b) (i) Edges: $\{v_1, v_2\}$, $\{v_1, v_3\}$, $\{v_2, v_5\}$, $\{v_2, v_6\}$, $\{v_3, v_4\}$, $\{v_5, v_7\}$ **3. a)** Each vertex is on shortest path from root; if cross edge spanned two or more levels, there would be a shorter path to lower vertex using the cross edge **b)** Cross edges may span any number of levels. **5.** Algorithm will fail on line "Pick a ..." when tree spans initial component. **6.** Weights of edges on tree in order added: 1, 8, 2, 3, 5, 6 **10.** Lemma 3 not correct as stated because there may be no shortest edge **11.** Just replace "minimum" by "maximum" and "shortest" by "longest" everywhere in algorithm and proofs. **13.** Nothing in proof assures that MST on n nodes plus minimum edge to node $n+1$ gives an MST on $n+1$ nodes. **15.** Start with 1 node, 0 edges; at each step add 1 node, 1 edge. **16.** (i), (iii) $\Rightarrow$ (ii): Suppose not acyclic and delete edges until get acyclic graph but if $|E|<|V|-1$ resulting graph not connected; (i), (ii) $\to$ (iii); use [23] of Section 3.7 **18.** N_nH_{n+2}; by induction on number of N's; for $n+1$ case take N attached to only one other N (when will there not be one?) and replace it and H's attached to it by an H. **19. a)** Same parent node **c)** Both are descendants of node three levels higher. **e)** Have common ancestor two levels higher for one, four levels for the other **20. b)** There are 14. **21.** $B_n = B_{n-1}B_0 + B_{n-2}B_1 + \cdots + B_0B_{n-1}$. **23.** Sum of terms of form $T_iT_jT_{n-i-j-1}$ **24. b)** 1 2 4 5 8 9 3 6 7 10 **26. a)** Yes; for every node in Preorder, Inorder tells you whether it is on left or right subtree of parent **27. b)** W **29.** Tree has 7 nodes including root. **30.** Tree has 16 nodes including root. **32. a)** Tree has 12 nodes; first player wins **c)** After first move, game is reduced to 3- or 4-petal daisy.

Supplementary Problems: Chapter 3

1. b) If leaf, just delete; otherwise replace deleted node by maximum value on left subtree or minimum value on right subtree **3. a)** Need at least 5 since complete graph of 4 vertices not allowed; but with 5 vertices sum of degrees would be odd **5. b)** Leave out edge in $\langle\{1,2,3\}\rangle$ **7. a)** By induction; remember $B+B=B$ **9. a)** No **b)** 7 **c)** 2 **11. a)** Need to find if allowing $(k+1)$st edge gives any new paths; no simple way to determine this from $M^{(j)}$ matrices, $j \leq k$ **b)** No simple way to go from $N^{(k)}$ to $N^{(k+1)}$ **13. c)** If degree of vertex currently at is one, removal of edge will not increase number of components; otherwise degree must be 2 or more (3 or more if not at original vertex); removal of *any* edge cannot result in new component because, before removal, all vertices in new component must have had even degree (why?) and removal of edge would result in vertex of odd degree at end of removed edge and in new component which is impossible; therefore, never need to increase number of components and so find Eulerian cycle **15.** Argument in Section 2.2 assures that there is a Hamilton path; for $k=3$, use strong connectivity to get "back edge"; for inductive step use strong connectivity to insert one or two vertices into known cycle. **17.** Columns of all zeros represent vertices with indegree 0; number one of these 1; then delete its row and column and repeat each time numbering a vertex with indegree 0 in reduced graph.

20. Doesn't work because shortest edge from start vertex may not go to nearest vertex since negative weight could give smaller weight to multi-edge path **21. a)** No shortest path can have cycle since cycle weights are positive; must be a shortest path since number of simple paths is finite. **b)** SHORTESTPATH doesn't work for same reason as in [20]. **23. c)** Same values of distance function **d)** For Ford argue that, if distances to k nearest vertices already found, then algorithm next finds shortest distance to $(k+1)$st nearest vertex **26.** The complete graph of order 4 has 4 vertices of degree 3 and $\chi(G)=4$; for any graph with 3 or fewer vertices of degree 3 or more, color the vertices of degree 3 or more with distinct colors and apply K-COLORABLE to show $\chi(G)<4$ **29.** In the coloring with minimum possible number of colors, number vertices so that if $i<j$ color of $v_i \le$ color v_j; then apply ONE-COLOR-AT-A-TIME **31. a)** Those of odd length **b)** Suppose vertex of degree $n-2$ or less; delete it; if remaining graph is $(n-1)$-colorable, so was original graph **34. a)** There are 3. **b)** Same weights on each **c)** Suppose not; then if edges ordered by increasing size, Tree2 has at some point larger edge than Tree1; put all vertices on Tree2 up to this edge in one set, remaining vertices in another set and apply Lemma 3 to get contradiction **35. a)** When multiple edges of weight w, consider graph with new edge weights w, $w+\epsilon$, $w+2\epsilon$ etc. all less than next largest weight; apply Prim's algorithm; one such selection of weights must correspond to each minimum weight spanning tree **b)** Assume weights are integers; using idea in a) no two spanning trees could have weights which differ by as much as 1 if ϵ small enough; therefore, without ϵ trees must have same weight **37.** Stronger hypothesis: Free tree always has at least as many nodes of degree 1 as degree of vertex with maximum degree; for graph with $n+1$ vertices delete vertex of degree one (why does the induction hypothesis assure that there is one?) not adjacent to vertex of maximum degree (if there isn't such a vertex then $(n+1)$-graph has a very simple form) **39. a)** (i): If e is the internal edge of T, cut set is all internal edges. **41. a)** $+\times 42-3/\times 45+37$ **b)** Associate each operator with two most recently written down operands to form new operand. **43.** Player 1 wins if n is not a multiple of 3; Player 2 wins otherwise; if n is not multiple of 3, Player 1 can leave Player 2 with a multiple of 3; if n is a multiple of 3, Player 1 must leave Player 2 with a non-multiple of 3. **45. b)** Player 1 wins if n is not one more than a multiple of 3; Player 2 wins otherwise.

Chapter 4

Section 4.2

1. 21 **3.** 23 **5. a)** $26^3=17,576$ **c)** $21^2\cdot 5=2205$ **7. a)** $9!=362,880$ **b)** $20\cdot 19\cdot 18\cdots 12\approx 6.09\times 10^{10}$ **9.** Any digit may be 0, so $10^3-1=999$. (Why subtract 1?) **10. b)** $n(n-1)$ **11.** $26^3\cdot 10^3=17,576,000$. Counting commercial vehicles, this is probably not enough in CA or NY. Allowing digits to be first doubles the count and may be enough. **13.** Each route after the first takes $2n+2$ steps. Largest n is 13. **15.** 144 **17.** 8×10^6 7-digit phone numbers per area code $\times$ 144 area codes $=$ 1.152 billion phone numbers, which is enough. **19.** At most 50, fewer if some students play on both teams.

Section 4.3

1. 3^n (if one or both can be empty) **3.** Use Binary Search. Asker can win even with $2^{20}>10^6$ words. **5.** Two positions per ring, starting with the largest, so 2^n **7.** n for chairman, 2^{n-1} for members; therefore $n2^{n-1}$ ways. **9.** $9!/(2!4!3!)$ **11.** $5!$; $5!/5=4!$ **15.** Start by weighing coins 1,2,3,4,5 against 6,7,8,9,T, where T is the test coin. **17.** First weigh half the coins (all together). Lower bound: $\lceil \log_2 13\rceil=4$. **19.** $2^{\lceil n/2\rceil}$ symmetric sequences; therefore no. of sequences $=\frac{1}{2}(2^n-2^{\lceil n/2\rceil})+2^{\lceil n/2\rceil}=\frac{1}{2}(2^n+2^{\lceil n/2\rceil})$. **21. a)** 52^4 **b)** $52^4/2^4=26^4$ **23.** 0, 1, 6, 8, 9 make sense when flipped. The number of usable signs (first digit nonzero) that flip to a different usable sign is $4\cdot 5\cdot 5\cdot 4-4\cdot 5=380$, where $4\cdot 5$ is number of symmetric signs (same sign when flipped). So $9000-190=8810$ different signs. **25.** $(n+1)(m+1)-(m+n+1)=mn$ eliminations if no coefficients zero. **27.** 3 ways

Section 4.4

1. a) 5, 20, 60 **b)** 5, 10, 10 **3.** Same **5.** $\binom{n}{1}+\binom{n}{2}=n(n+1)/2$. **7.** $\binom{n-2}{k}+2\binom{n-2}{k-1}=\binom{n}{k}-\binom{n-2}{k-2}$. (Each side comes from a different approach.) **9.** $8!$, $C(8,5)\,P(8,5)$ **10.** $\binom{n}{3}$ **13. c)** LHS counts each intersection point on each line. So each intersection point is counted twice, once for each line that intersects there. **15.** Inefficient, also overflow possible **17.** $\binom{n}{2}$ **18.** Each of the $\binom{n}{2}$ possible edges is either in the graph of not; therefore $2^{\binom{n}{2}}$. **21.** $2^{\binom{n}{2}}$, $3^{\binom{n}{2}}$.

Section 4.5

1. Combinatorial: Choose a captain and k other team members various ways, one of which is to choose k members and then the captain. **3.** Combinatorial: Choose $k+j$ things from n various ways; e.g., for the LHS of a), choose k from n and then j from the rest. **5.** Line up k out of n things all at once, or choose the first thing and then worry about the rest. **8.** Induction on n: In the inductive step, rewrite $\binom{k}{m}$ on the LHS using Theorem 2. Combinatorial: On the LHS, choose $m+1$ numbers from $\{1,2,\ldots,n+1\}$ by first deciding the largest to be chosen. **9.** Second new row: 1, 6, 15, 20, 15, 6, 1. **11. a)** Define a function $C(n,k)$ with key line $C \leftarrow C(n-1,k)+C(n-1,k-1)$. **c)** The recursive algorithm (without flags) is much less efficient. To compute $\binom{n}{k}$ it calls itself $2\binom{n}{k}-1$ times (including main call), computing each value $\binom{n'}{k'}$ (for $n' \le n$ and $1 \ne k' \le k \ne n'$) $\binom{n-n'}{k-k'}$ times. **12.** For $0<k<p$, no factor in the denominator of $\frac{p!}{k!(p-k)!}$ divides the factor p in the numerator. **14.** For most cases where $k<0$ or $k>n$, all terms in the theorems are 0. The cases $k=0$ and $k=n$ in Theorem 2 need special attention because of the terms $\binom{n-1}{k}$ and $\binom{n-1}{k-1}$. **16.** In proving $P(n) \Longrightarrow P(n+1)$, rewrite $k\binom{n+1}{k}$ as $k\binom{n}{k}+k\binom{n}{k-1} = k\binom{n}{k}+(k-1)\binom{n}{k-1}+\binom{n}{k-1}$. **18.** There is a 1-to-1 correspondence of terms by $\binom{n}{k} \leftrightarrow \binom{n}{n-k}$. Both sums $=2^{n-1}$.

Section 4.6

1. b) $x^4-4x^3+6x^2-4x+1$ **2. b)** .9412 **3.** LHS $=(1+2)^N$. **4. b)** $(-3)^n$ **e)** $(-4)^n$ **5.** 12.68% **7. a)** For any set, the number of odd-sized subsets equals the number of even-sized subsets. **b)** It equals $\frac{1}{2}2^n$. **9.** Expand the LHS and look at the signs of the "additional" terms. **11.** Discovery: $\sqrt{a^2+\epsilon} = (a^2+\epsilon)^{1/2} = (a^2)^{1/2}+\frac{1}{2}(a^2)^{-1/2}\epsilon+\cdots$ **13.** When $j=0$ there is no y^{j-1} term in the first factor in the RHS of Eq. (5); when $j=n+1$ there is no y^j term. **15.** 30, 60 **17. a)** 3^n **19.** $(-1)^k$ **22.** No. The only case that both extensions treat is n,k integers with $0 \le n<k$. In that case, both extensions give $\binom{n}{k}=0$. **23. b)** $\sum_{k\ge0}(k+1)x^k$ **25.** No, picking out one i-set leaves a second $(n-i)$-set. **27.** Theorem 2 becomes $\binom{n}{i,j,k} = \binom{n-1}{i-1,j,k}+\binom{n-1}{i,j-1,k}+\binom{n-1}{i,j,k-1}$. **30.** The jth element goes into set $i \iff$ the letter in the jth position of the word is letter i. Counting permutations of words was discussed in Example 5, Section 4.3. **31.** Substitute $x=y=1$. **33. a)** Apply $\frac{d^2}{dy^2}$ to the Binomial Theorem, then substitute $x=y=1$. **b)** Apply $\frac{d}{dy}$ to the Binomial Theorem, multiply by y, then differentiate again.

Section 4.7

1. $\binom{10}{6}=210$ configurations **3.** 26^4 [dist. balls (positions) in dist. urns (letters)] **5.** Both dist., $3^4=81$; urns dist., $\binom{6}{2}=15$; balls dist., 14; both indist., 4. **7.** one way **8.** $\binom{u}{b}$ **10. a)** $\binom{13}{10}=286$. **c)** $\binom{7}{6}\binom{5}{4}=35$. **12.** Dist. balls (permutation positions) in dist. urns (repeatable symbols) **14.** For $n=4$: 4; 3,1; 2,2; 2,1,1; 1,1,1,1 **15. a)** 6^9 **17. a)** 12 **b)** 3 **19.** Each set of k elements from $\{1,2,\ldots,n+k-1\}$ can be put into a strictly increasing sequence exactly one way. **21.** Consider all sequences of b numbered balls and u numbered walls, starting with a wall. E.g., w_2, b_3, w_1, b_1, b_2 means urn 2 is first, containing ball 3, and urn 1 is next, containing ball 1 then ball 2. Answer: $u(b+u-1)!$. **22. e)** Dist. balls in dist. stacks, but order of stacks fixed. So divide answer to [21] by $u!$. **23.** $\binom{b-ku+u-1}{b-ku}$ **25. a)** $p^*(b,u)=p(b-u,u)$ **d)** $p(i,j)=\sum_{k=0}^{\min(i,j)} p(i-k,k)$. **26.** $p(6)=11$. In computing this from [25(d)], note that $p(b,u)=p(b)$ whenever $u>b$. **28. b)** $\lfloor b/2 \rfloor+1$ **30.** $P(b,u)u^{b-u}$ overcounts because it distinguishes the first ball into each urn. Try the case $b=2$, $u=1$. **32.** $S(0,0)=T(0,0)=1$

Section 4.8

1. 5 both; 35 field hockey only; 25 track only **2. a)** $|F|=40$, $|T|=30$, $|\overline{F}\cap\overline{T}|=1335$ **3. a)** 1291 **c)** 20 **5.** 48 **7.** 61; you need only consider the squares of primes **9.** If $r,s>1$ and $100\geq n=rs$, then one of $r,s\leq\sqrt{n}\leq\sqrt{100}=10$. 26 primes **10. a)** 0 **b)** $m!$ **12.** $\frac{10!}{3!4!}$ (treat aa as one block) **14.** $3\cdot 2^5$ **16.** Formula (9) always correct; (12) Correct $\iff b$ even **18. a)** 6400 **c)** 68,500 **19.** $|\cup_{i=1}^{n} A_i| = \sum_i N_i - \sum_{i,j} N_{ij} + \sum_{i,j,k} N_{ijk} - \cdots + (-1)^{n-1} N_{123\ldots n}$ **21.** One formula is $\sum_{k=0}^{10}(-1)^k\binom{10}{k}(10-k)^m - \sum_{k=0}^{9}(-1)^k\binom{9}{k}(10-k)^{m-1}$. **23. a)** Let A_i be the set of all elements satisfying property i. Then the theorem is just Theorem 1. **b)** Let A_i be the set of all elements that *don't* satisfy property i. Then the theorem is again Theorem 1. For instance, N_{12} is the set of elements satisfying at most properties $\{3,4,\ldots,n\}$. **25. b)** Decide which balls go with ball 1 and put the others recursively into $u-1$ urns. That is, $B(b,u)=\sum_{k=1}^{b}\binom{b-1}{k-1}B(b-k,u-1)$. $B(5,4)=51$.

Section 4.9

1. Each input does not have a unique output. **3.** It never stops if, say, RAND[1,n] always outputs 1. The probability that some number is never output by RAND goes to 0 as time goes on. See Ch. 6. **4. b)** 2, 5, 3, 1, 4 **5.** The second loop should be **for** $i=n$ **downto** 1 and should use $r\leftarrow$ RAND[1, i]. **6. a)** RAND[n,n] has to be n. **b)** Terminate the loop at $n-1$. **c)** Still linear **8.** The probability that Perm(n) $=n$ is too high $(\frac{n-1}{n})$. **10.** First *five*: 12345, 12354, 12435, 12453, 12534. Last five: 54132, 54213, 54231, 54312, 54321. **12. a)** 463 **b)** 43521 **15.** For any n-permutation P, $a_1,a_2,\ldots,a_n$ are the values in Eq. (3) for computing RAND(P) $\iff a_1{+}1, a_2{+}1,\ldots,a_n{+}1$ are the values of r in PERMUTE-2 that cause PERMUTE-2 to output P. **17.** Let $S\neq T$ be words. Let $F(S)$ be the first symbol of S and let $\overline{F}(S)$ be the remainder of the word. Assuming the order on individual symbols is known, then $S<T \iff \big[F(S)<F(T)$ or $(F(S)=F(T)$ and $\overline{F}(S)<\overline{F}(T))\big]$. If words can be different lengths, you also need to define the null word to come before all other words. **19. for** $s\in S$; Allperm($S{-}s$); insert s in front of the output of Allperm($S{-}s$); **endfor**. (Avoiding English is trickier; use more parameters in defining Allperm.) **21.** Cycle 34125, length 1200 **26.** $(i,j)\leftarrow\big(\text{RAND}[1,n],\text{RAND}[1,n]\big)$ **28.** All first entries are equally likely and they shouldn't be. As a result, the probability of getting $(1,n)=1/n^2$ (too low) and the probability of getting $(n,n)=1/n$ (too high). **29.** Many ways; here's a wasteful one. **repeat until** $y>x$; $x\leftarrow$ RAND[1, n], $y\leftarrow$ RAND[1, n]; **endrepeat** **31.** For PERMUTE-1 and PERMUTE-2, just end the for-loop at k. For PERMUTE-3, change n to k in the second for-loop and output only the first k entries of Perm. **33.** Any solution to [31] works. **36. a)** For k from 1 to $(n-1)!$, let P be the kth $(n-1)$-permutation. If k is odd, replace P by n permutations, each with the symbol n inserted in P, first at the right end, then next to the right, ..., finally on the left. If k is even, make the same replacements but insert symbol n starting on the left. **c)** Every number from 0 to $n!-1$ has a unique representation in the form $\sum_{k=1}^{n} a_k n!/k!$, $0\leq a_k\leq k-1$.

Section 4.10

1. $kn+1$ **3.** 26 socks **5.** All 10 socks **6. a)** There are 142 possible sums (6 to 147). **c)** $m=42$. **g)** Every sum from 3 to 47 must be obtained. 1+2 is the only way to get 3. 1+24 = 2+23. **7. a)** Compute and save the sums for all the 3-subsets, and set off an alarm if a sum is obtained twice. **b)** The algorithm forces you to set up a correspondence between 3-subsets and sums, forcing you to think about how many subsets and how many sums there can be. **9.** 5 trits **11.** To determine u_k (or d_k) you need to know u_j for each $j<k$. So procedure call k would have $k-1$ subcalls. Even just two subcalls per call (as in Algorithm FIBB) usually leads to exponential run time. In fact, recursive MONOTONIC on an n-sequence would have exactly 2^n calls. **13. b)** Define $U(k)$ to be the index of the term just before u_k in the longest up subsequence (found so far) ending at u_k. Include $U(k)$ and a similar $D(k)$ in the algorithm. For instance, in the line that begins "**if** $a_j<a_k$", when $u_j+1>u_k$ update $U(k)$. If at the end d_7 (say) is the largest d or u value (keep a running record of the biggest so far), then a longest sequence can be found by computing $D(7), D(D(7))$ and so on. **15.** If you still want a strictly monotonic subsequence, the conclusions are false. If any monotonic subsequence will do, the conclusions are true, but the inequalities in Algorithm MONOTONIC need to be changed. **17.** From a set of P pigeons, let subset P_i be in hole i. If there is at most one pigeon per hole, then $|P| \overset{\text{Sum Rule}}{=} \sum_{i=1}^{n} |P_i| \le \sum_{i=1}^{n} 1 = n$.

Supplementary Problems: Chapter 4

1. Let P_k be the 0–1 pattern in rows 0 through $2^k - 1$. Then P_{k+1} consists of three corner triangles and one center triangle. By induction show that all three corner triangles are P_k and the center triangle is all 0's. **3.** $\binom{n}{3} + 4\binom{n}{4} + 5\binom{n}{5} + \binom{n}{6}$. There are $\binom{n}{3}$ triangles with all three vertices on the polygon. There are $4\binom{4}{n}$ triangles with just two vertices on the polygon; the sides to the third triangle vertex extend to two other polygon vertices, and the chords from each set of 4 polygon vertices create four such triangles. $5\binom{n}{5}$ counts triangles with one polygon vertex; there are $\binom{n}{6}$ triangles with no polygon vertices. **5.** If a chord has i polygon vertices on one side (hence $n-2-i$ on the other), it contains $i(n-2-i)$ intersection points. The answer is $\frac{1}{4}\sum_{i=1}^{n-3} i(n-2-i)n$. (The sum counts each chord twice and thus each intersection point 4 times.) Find a closed form. **7. a)** 3 **c)** 10 **e)** 6 **9. b)** $A(x)B(x)=1 \Longleftrightarrow A(x)=B(x)=1 \Longleftrightarrow x \in A \cap B \Longleftrightarrow (A \cap B)(x)=1$. **d)** Direct from parts b) and c). **11.** $\binom{23}{3} - 4\binom{13}{3} + 6 = 633$

Chapter 5

Section 5.2

1. $a_n = a_{n-1} + n^2$, $a_1 = 1$. **3.** $a_n = ra_{n-1} + d$, $a_0 = a$. **5. a)** $Y_n = cY_{n-1} + I + G$. **7.** $N_n = \frac{3}{4}N_{n-1} + \frac{4}{5}L_{n-1}$; $L_n = \frac{1}{4}N_{n-1} + \frac{1}{5}L_{n-1}$. **9.** $r_n = r_{n-1} + 2r_{n-2}$; $r_0 = 3$; 3,3,9,15,33,... . **11.** The tempting analysis *is* wrong (why isn't r_{n-2} the number of rabbits born?), but $r_n = r_{n-1} + r_{n-2} - r_{n-12}$ is correct anyway! Start with $b_n = \sum_{i=2}^{11} b_{n-i}$, where b_n are the number born in month n. **13.** $c_n = c_{n-1}$[dot first] $+ c_{n-2}$[dash first]; $c_1 = 2$, $c_2 = 3$. **15.** $a_{n+1} = a_n + 2n$, $a_1 = 2$. **17.** $s_n = s_{n-1} + r_{n-1}$, where r_n is number of regions n *lines* divide a *plane* into. **18.** Formulas for $P_{k+1}, \hat{P}_{k+1}, V$ unchanged; $\hat{P}_0 = P/(1-t)$; $\hat{V} = \hat{P}_N - \hat{P}_N t$. **21.** $c_{u,b} = c_{u-1,b}$ [don't use first urn] $+ c_{u-1,b-1}$ [do use it]; $c_{u,0} = 1$; $c_{1,1} = 1$; $c_{1,b} = 0$ for $b \ge 1$. **23.** $P_k = 1$ for all k. **25.** $P_n = (1+r)P_{n-1} - p$; $P_0 = P$, $P_N = 0$. **27.** $P(n+1,1) = \sum_{i=10}^{50} c_i P(n,i)$, where c_i is the yearly birth rate for females age i. For $a > 1$, $P(n+1,a) = (1-d_a)P(n,a-1)$, where d_a is the yearly death rate for females age a. **29.** $N_n = \frac{7}{6}N_{n-1} - \frac{1}{6}N_{n-2}$; $N_0 = 1000$, $N_1 = 833\frac{1}{3}$.

Section 5.3

1. The pattern $1, 1, 0, -1, -1, 0$ cycles **3. a)** 1, 2, 4, 8, 16, 32 **c)** 1, 3, 9, 27, 81, 243 **e)** 0, 1, 5, 19, 65, 211 Note that a) and c) are geometric and e) is their difference. **5.** $E_{2n} = (n/2) + E_n$; $E_{2^k} = (2^k+1)/2$. **8.** Using Eq. (5), $\hat{V}/V$ is 1.02 in a), 1.62 in b) and 1.64 in c). **9.** Now $\hat{V} = \hat{P}_N - (\hat{P}_N - P)(t+.1)$. **a)** $\hat{V}/V = 1.11$. **b)** 12 years **11.** V unchanged, $\hat{V} = [P/(1-t)][(1+r)^N(1-t)] = P(1+r)^N$, so $\hat{V}/V = [(1+r)/(1+r(1-t))]^N > 1$. **14.** In the equations for [7, Section 5.2], substitute x for both N_n and N_{n-1} and $2000 - x$ for both L_n and L_{n-1}. The unique solution is $2000\left(\frac{16}{21}\right) \approx 1524$. **16.** $f(m,n) = (m-n+1)^n$. **18.** Same as the statement in [17] except, in the display, replace $P(m-1,n-1)$ by $P(m,n-1)$. **20. a)** $r_n = 1 + (1/r_{n-1})$. **b)** $r = 1 + (1/r) \Longrightarrow r^2 - r - 1 = 0$. The positive root is $(1+\sqrt{5})/2 \approx 1.618$. **21.** $\frac{L}{W} = \frac{L+W}{L}$, which with $r = \frac{L}{W}$ becomes $r = 1 + \frac{1}{r}$. Now see the solution to [20b]. **23.** The pattern is $P_n = P(1+r)^n - p\dfrac{1-(1+r)^n}{1-(1+r)}$. Since $P_N = 0$, $p = \dfrac{Pr(1+r)^N}{(1+r)^N - 1} = \dfrac{Pr}{1 - \left(\frac{1}{1+r}\right)^N}$.

25. If $a_0 > 0$, $a_n \to 2$; if $a_0 < 0$, $a_n \to -2$; if $a_0 = 0$, a_1 is undefined.

Section 5.4

1. first-order, nonlinear **3.** first-order, linear, nonhomogeneous, constant-coefficient **5.** second-order, linear, homogeneous, constant-coefficient **7.** second-order, linear, homogeneous **9.** linear, homogeneous, constant-coefficient (variable order) **11. a)** first-order, linear, homogeneous, constant-coefficient **b)** Same as in a) **13.** All sequences in which each term is either plus or minus 1. Theorem 1 says that a_1 uniquely defines a solution to a first-order recursion if a_n is given as a *function* of a_{n-1}, that is, if a_n is uniquely defined in terms of a_{n-1}. That's not true here: $a_n^2 = a_{n-1}^2$ leads to two values of a_n. **15.** Choose the kth term two different ways, and by Theorem 1 you already have two different sequences. **17.** In the second proof, the induction has been pushed back into the proof of Theorem 1.

Section 5.5

1. $a_n = \frac{5}{6}2^n - \frac{1}{3}(-1)^n$. **3.** $a_n = 3^n - 2^{n+1}$. **5.** $a_n = 2^{n+2} - 3^n$ **7. a)** 2^n **d)** $2^n + 3^n$ **8.** $a_n = Aw^n + \bar{A}\bar{w}^n$, where $A = \frac{1}{2} - \frac{i\sqrt{3}}{6}$, $w = \frac{1+i\sqrt{3}}{2}$ and bars denote complex conjugate. **11.** The trivial solution is obtained when $A = B = 0$. **13. a)** recursion **b)** closed form **c)** closed form **d)** probably recursion (pehaps even when fast computer algebra calculators are available) **14. a)** $\to\infty$ **b)** $\to 3$ **d)** Oscillates to ∞ like $(-1.2)^n$ **f)** Oscillates to 0 **g)** $\to 0$ **17.** $a_n = 2F_n$, where F_n is the nth Fibonacci number. **19.** $Ar_1^n + Br_2^n = Ar_1^n\left[1 + \frac{B}{A}\left(\frac{r_2}{r_1}\right)^n\right]$. **21.** In Eqs. (11) and (12), replace all exponents and subscripts 0 by k and all exponents and subscripts 1 by $k+1$; then solve for A and B. **23.** Show that the difference between the expression in the problem and Eq. (5) is $< 1/2$. **24.** The general solution is $a_n = Aw^n + B\bar{w}^n$, where $w, \bar{w}$, defined in the solution to [8], satisfy $w^6 = \bar{w}^6 = 1$. Thus $a_{n+6} = a_n$ for all n. **26.** From [24], $a_0 = a_6$ for any solution. **28.** $a_n = 2\cos\theta\, a_{n-1} - a_{n-2}$. **30.** Use formulas for $\sin(\alpha+\theta)$ and $\cos(\alpha+\theta)$ with $\alpha = n\theta$.

Section 5.6

1. $a_n = 1$. **3.** $e_n = \frac{2}{9}3^n$. **5.** $A(-1)^n + Bn(-1)^n + Cn^2(-1)^n$ **7.** $2^n - 3(-1)^n + 1$ **9. a)** $e_n = 1$. **b)** $e_n = n+1$. **10.** The degree of the characteristic polynomial is the difference between the indices of the highest and lowest indexed (nonzero) terms in the difference equation. By definition, this index difference is the order of the difference equation. **12.** $\sum_{i=1}^k m_i$ **14.** By induction. A linear combination of n things is a linear combination of the last thing with a linear combination of the $n-1$ other things. **15.** $a_n = -a_{n-1} + 12a_{n-2}$. **17.** $a_n = 2a_{n-1}$, $a_0 = 1$. **19.** $c_n = 5c_{n-1} - 6c_{n-2}$; $c_0 = 2$, $c_1 = 5$. **21.** $e_n = e_{n-1} + 4e_{n-2} - 4e_{n-3}$; $e_0 = 3$, $e_1 = 7$, $e_2 = 3$. **22.** $f_n = 5f_{n-1} - 3f_{n-2}$; $f_0 = 2$, $f_1 = r_1 + r_2 = 5$ since the sum of roots of a quadratic is negative the coefficient of x.

Section 5.7

1. a) $A3^n - 2^{n+1}$ **b)** $A3^n - 2^{n+1}(n+3)$ **c)** $A3^n + n3^n$ **d)** $A3^n - \frac{1}{2}$ **2. a)** $A + B2^n + n2^{n+1}$ **c)** $A + B2^n + 2\cdot 3^n$ **3. a)** $A2^n + Bn2^n + \frac{1}{4}n^2 2^n$ **c)** $A2^n + Bn2^n + 3^{n+1}$ **5.** $3^n - 1$ **8.** No, Yes **9.** $\{3a_n\}$ satisfies $s_n = s_{n-1} + s_{n-2} + 3n^2$. **10.** $f(n) = \sum_{i=1}^k P_i(n) r_i^n$, where k is any positive integer, each $P_i(n)$ is a polynomial, and each r_i is a constant. **11. b)** $(r-2)^2$ **d)** $(r-2)(r-1)$ **f)** $(r-1)^4(r+1)^3$ **13. b)** $\frac{1}{4}(n^4 + 2n^3 + n^2)$ **d)** $-\dfrac{a}{(a-1)^2}a^n + \dfrac{a}{a-1}na^n + \dfrac{a}{(a-1)^2} = \dfrac{na^{n+1}}{a-1} - \dfrac{a^{n+1} - a}{(a-1)^2}$ **f)** $\frac{1}{2}(-1)^n(n^2+n)$ **15.** The terms from $D(n)$ will cancel out in step 6); you get the same answer as before. **17.** When the poloynomials $p(r)$ and $q(r)$ in the general procedure have no common factors and also $q(r)$ has no multiple factors **19.** $b_n = A3^n + \sum_{k=2}^n 3^{n-k}\log k$.

Section 5.8

1. $a_n = n$. **3.** $c_n = 3^n$. **5.** $e_1 = 2$, $e_2 = 5$, and for $n \geq 2$, $e_n = \frac{5}{2}n!$ **7.** $g_n = n+1$. **9.** $m_n = 2^n(n+1)!$. **10. a)** $a_n = c^{\sum_{i=1}^k k} = c^{\binom{n+1}{2}}$. **b)** $a_n = \sum_{j=0}^n c^{\binom{n+1}{2} - \binom{j+1}{2}}$. **12.** Inductive paradigm! **16.** $S_n = 1/n^2$. **17.** $t_n = (n+1)/n!$. **19.** $k = d/(c-1)$, $b_n = b_0 c^n$. **20. a)** The IRA recursion is $\hat{P}_k = \hat{P}_{k-1}(1+r) + P$ for $k < N$ and $\hat{P}_N = \hat{P}_{N-1}(1+r)$ (because you don't deposit any more money just before you close the account). Also, $P_0 = P$. $\{\hat{P}_k\}$ is an arithmetic-geometric sequence (with $\hat{P}_N$ modified). The result: $\hat{P}_N = \frac{P}{r}\left[(1+r)^{N+1} - (1+r)\right]$. For the non-IRA, P_N is the same with $r^* = r(1-t)$ substituted for r. Finally, $\hat{V} = \hat{P}_N - [\hat{P}_N - PN]t = \hat{P}_N(1-t) + PNt$, and $V = P_N$. **b)** $\hat{V} = \$286,415$, $V = \$222,309$. **22. a)** $(s^{n+1} - r^{n+1})/(s-r)$ **b)** $(n+1)r^n$ **24.** $b_n = rb_{n-1} + a$ with $b_0 = a$. By Example 5 Continued, $b_n = r^n\left(a + \frac{a}{r-1}\right) - \frac{a}{r-1} = \frac{ar^{n+1} - a}{r-1}$.

Section 5.9

1. STRAIGHTFORWARD, $n+1$. HORNER, n. (Additions and subtractions in updating k have not been included.) See [22a, Section 1.3]. **3.** The savings are small for low-order polynomials. For people, HORNER requires more recording of partial results; e.g., many people can compute terms like $2\cdot 3^2$ (from STRAIGHTFORWARD) in their heads, but fewer can do $(2\cdot 3+4)\cdot 3$ mentally. Finally, STRAIGHTFORWARD emphasizes the standard algebraic form of polynomials, which is useful for other computations. **5.** Multiplications: $1+2(n-1) = 2n-1$. Additions: $1+n-1 = n$. Between the other two in efficiency. **7.** Induction on n **9.** Each call of Procedure TOH involves two comparisons of n to 1; each call of TOH* involves one. Both procedures make the same subcalls. **11.** FIBA uses $2(n-1)$ subtractions and $n-1$ additions ($n-1$ more if you count updating k). If FIBB uses a_n of these operations, then $a_n = a_{n-1}+a_{n-2}+3$; solution is $a_n = 3F_n - 3$. **13.** The initial nonhomogenous solution is $B_n = -1$. This implies $e_1 = e_2 = 2$, so C and D are twice what they were in Section 5.5. Thus $B_n = 2F_n - 1$. **15.** $3^k = (2^{\log 3})^k = (2^k)^{\log 3} = n^{\log 3}$. Similarly, $b^k = n^{\log b}$. **17.** $B_{2^k} = 3^k$, so $B_n = n^{\log 3}$ $(= 3^{\log n})$. **19.** $D_{2^k} = 3^{k+1} - 2^{k+1}$, so $D_n = 3n^{\log 3} - 2n$. **21.** $F_n = (1+\log n)n^2$. **23.** $H_n = \frac{1}{16}\left(21n^{\log 5} - 4\log n - 5\right)$. **25.** $J_n = 7n^{\log 3} - 2n(3+\log n)$. **27.** $L_n = 7n^{\log 7} - 6n^2$. **29.** The main line is **for** $k = m$ **downto** 0, **if** $a_k = 1$ **then** $p \leftarrow p^2 x$ **else** $p \leftarrow p^2$. **31.** See the proof of Theorem 1 in the Epilogue, Section 2. **33. a)** The recursion is $A_n = 1+A_{n/2}$, $A_1 = 1$. The solution is $A_n = 1+\log n$. **b)** Let $b_k = A_{2^k-1}$. Then the recursion is $b_k = 1+b_{k-1}$, $b_1 = 1$. The solution is $b_k = k$, or $A_n = \log(n+1)$ (for $n = 2^k - 1$). Except for the lack of ceilings (which are redundant for $n = 2^k - 1$), this is the answer obtained in [4, Section 2.6]. **35.** Least $= \min\{p, q\}$. Let the p-list be 1,2 and the q-list be 3,4,5,6. **37.** Solve $\overline{C}_n = \overline{C}_{n/2}+1$, $\overline{C}_1 = 0$, by Divide and Conquer to obtain $\overline{C}_n = \log n$. To prove $\overline{C}_n \le C_n$ for all $n \in N$, use strong induction and properties of logs in base 2. **39.** Sequentially search the first half, then if necessary the second half. If the search word is on the list in a random position, the simplified Divide and Conquer recursion is $E_n = \frac{1}{2}E_{n/2} + \frac{1}{2}(\frac{n}{2}+E_{n/2}) = E_{n/2} + \frac{n}{4}$, $E_1 = 1$. **41.** One can find the median by first sorting (which can be done by Divide and Conquer). But to use Divide and Conquer to find only the medians m_1, m_2 of two unsorted half lists tells you too little about the median of the whole list— only that it is m_1, or m_2, or between them. **43. a)** Example 6 shows how to get x^{2^k} in k multiplications. This is best possible since the largest power of x obtainable with $\le k-1$ multiplications is $x^{2^{k-1}}$ (induction!). **b)** Lower bound: Since k multiplications yields at most x^{2^k}, if $n \ge 2^k$ then $M(n) \ge k = \log n$. Upper bound: $C_n = 2\log n$ satisfies $C_n \ge C_{\lfloor n/2 \rfloor}+2$, $C_1 = 0$; see Eq. (5).

Supplementary Problems: Chapter 5

1. $F_{-n} = (-1)^n F_{n-2}$. **3.** At least one of the two roots of the characteristic equation will be periodic ($r^p = 1$); $a_n = Ar^n$ (or $Ar_1^n + Br_2^n$ if both roots have period p), so $\{a_n\}$ has period p. **5.** Your difference equation will be nonhomogeneous. First find a particular solution as with single-indexed difference equations. Solution: $2\binom{n}{k} - 1$. **7.** $p = P/\sum_{i=1}^{420} m_i(1+r)^{-i}$. To see why the difference equation is correct, it may help to multiply both sides by A, the number of original annuitants, including those who die by period n. AP_n represents the expected total amount of money in this "annuity pool" at the end of month n, that is, total taken in at the start, plus all accrued interest, minus all payouts through period n. **8.** The characteristic polynomial must have roots $(a \pm \sqrt{b})/2$ so the difference equation is $c_n = ac_{n-1} + \left(\frac{a^2-b}{4}\right)c_{n-2}$. By assumption the coefficients are integers. It's easy to check that c_0 and c_1 are integers for both sequences. Thus every c_n is an integer. **10.** $2F_n - 1$ **11.** $C_{n,k} = C_{n-1,k} + C_{n-2,k}$, and $C_{k,k} = C_{k+1,k} = 1$ (except $C_{1,0} = 0$). These are Fibonacci recursions (only the first index varies), so $C_{n,k} = F_{n-k}$ (except $C_{n,0} = F_{n-2}$). **14.** Very hard to prove directly, but its just a special case of the formula in [13]. **15.** General solution: $A(1+\sqrt{2})^n + B(1-\sqrt{2})^n + C$, so all possibilities mentioned possible, but explosive growth ($A \neq 0$) most likely. **18. a)** $r_n = 2^{n-1}$. **c)** For $n \ge 1$, $r_n = F_{n-1}$. **20. b)** $k = d/(-1+\sum c_j)$, so the denominator must be nonzero. **21. a)** $k2^{k-1}$, i.e., $\frac{1}{2}n\log n$ for $n = 2^k$. The recursion is $m_k = 2m_{k-1} + 2^{k-1}$. **b)** $2^{k+1} - k - 2 = 2n - 2 - \log n$ **23. c)** $F(n)$ takes $2n-1$ calls regardless of order: one call each for n, $n-1$ and 0, two calls for 1 to $n-2$. A difference equation paralleling the recursion in the algorithm ($c_n = c_{n-1} + c_{n-2}$) doesn't work; why? **25.** For $0 \le k \le n$, $a_{n,k} = \binom{n}{k}/2^n$. **27.** For almost all M, $\mathbf{v}_n$ converges. See the discussion of Algorithm MAX-EIGEN in Section 8.9. **29. a)** The roots r_1, r_2 are complex numbers with modulus $\sqrt{c}$ and arguments $\pm\theta$, where $\theta = \tan^{-1}\sqrt{(4c/b)-1}$. The basic solutions are $r_1^n = c^{n/2}e^{in\theta}$ and $r_2^n = c^{n/2}e^{-in\theta}$. Show that every real-valued linear combination of these sequences is a real linear combination of $\sin n\theta$ and $\cos n\theta$.

Chapter 6

Section 6.1

1. a) Out of 10,000 people, 92 of the 100 carriers will test positive, as will 396 (4%) of the 9900 noncarriers. If you test positive, you have only 92/(92+396) $\approx$ .189 probability of being a carrier. **3.** # dollars = # birds; make the coin biased.

Section 6.2

1. b) $\{1,4\}$ **2. a)** First toss a 2: $S = \{(2,1),(2,2),\ldots,(2,6)\}$; one toss a 2: $S \cup \{(1,2),(3,2),(4,2),(5,2),(6,2)\}$; sum is 5: $\{(1,4),(2,3),(3,2),(4,1)\}$ **3.** Done in Example 2 Continued. **4.** The atoms are the possible triplets of tosses. $E = \{\text{HHH, HHT, THH, THT}\}$. **5. a)** $A \cup B \cup C$ **c)** $\sim(A \cap B \cap C)$ **7.** Since $0 \le |A| \le |S|$, $0 \le |A|/|S| \le 1$; $|\emptyset|/|S| = 0/|S| = 0$, $|S|/|S| = 1$; if $A \cap B = \emptyset$, then $|A \cup B| = |A| + |B|$, so $\frac{|A \cup B|}{|S|} = \frac{|A|}{|S|} + \frac{|B|}{|S|}$. **9.** $\binom{2+6-1}{2} = 21$; ind. balls in dist. urns with repetition. **11.** 3) $\Longrightarrow \Pr\{a \cup b\} = \Pr(a) + \Pr(b) \Longrightarrow$ Eq. (2) by induction; (2) $\Longrightarrow \Pr(A \cup B) = \sum_{e \in A \cup B} \Pr(e)$ equals (if $A \cap B = \emptyset$) $\sum_{e \in A} \Pr(e) + \sum_{e \in B} \Pr(e) = \Pr(A) + \Pr(B)$. **13.** $\sim A \cap \sim B = \sim(A \cup B)$, so $\Pr(\sim A \cap \sim B) = 1 - \Pr(A \cup B)$. Now use Theorem 2. **15.** B and $A-B$ are disjoint, and $A \cup B = B \cup (A-B)$, so $\Pr(A \cup B) = \Pr(B) + \Pr(A-B) = \Pr(B) + \Pr(A) - \Pr(A \cap B)$. **17.** In Section 4.8 we could use any counting facts; now we can use only those probability facts in Def. 1. **19. a)** $\Pr(A) = \frac{1}{2}$, $\Pr(B) = \frac{1}{3}$, $\Pr(C) = 1 - \frac{1}{2} - \frac{1}{3} = \frac{1}{6}$, so odds against C are 5:1. **c)** $ps+qr+2qs : pr-qs$ **21. a)** Three cases: 0, 1 or both of A and B finite. **c)** $A_i = \{i\}$. **23.** If everyone in the population says "I have tuberculosis", then .001 of them are correct.

Section 6.3

1. a) $.6 \times .5 \ne .4$ so No. **b)** .8, 2/3 **4.** By Bayes $\Pr(A|B) = \frac{(.2)(.8)}{(.2)(.8)+(1-.2)(.3)} = .4$. **6. a)** Root has degree 2, neighbors of root have degree 3; each meal has Pr = 1/6. **c)** $\frac{1/6}{2/6} = 1/2$ (independent events). **7. c)** $\Pr(W \cap P)/\Pr(P) = 0/.25 = 0$.

9. a) With subscripts t, y for today and yesterday, $\Pr(G_y|G_t) = \frac{\Pr(G_t|G_y)\Pr(G_y)}{\Pr(G_t|G_y)\Pr(G_y)+\Pr(G_t|B_y)\Pr(B_y)} = .7$.
11. a) $\Pr(A|B)\Pr(B) = .2$ **c)** Not unique; In both diagrams, set $\Pr(A \cap B) = .2$, $\Pr(\sim A \cap B) = .3$, but in one set $A = A \cap B$ (so that $\Pr(A|\sim B) = 0$), in the other $A = (A \cap B) \cup \sim B$ (so that $\Pr(A|\sim B) = 1$). **13. a)** H (Pr=1/2), TH (1/4), TTH (1/8), TTT (1/8) **b)** 7/8, 1/4, 1/4 **14.** Let F = penny fair, H = get 10 heads; by Bayes $\Pr(\sim F|H) = \frac{10^{-6} \cdot 1}{10^{-6} \cdot 1 + (1-10^{-6})(1/2)^{10}} \approx .001$. **17.** Let C = cheated, R = last 10 right. Use Bayes; if $\Pr(C) = .001$, then $\Pr(C|R) = .001/(.001 + .999(\frac{1}{5})^{10}) \approx 99.99\%$. If $\Pr(C) = .0001$, then $\Pr(C|R) \approx 99.90\%$ **19. a)** Sample space $\{(B,B),(B,G),(G,B),(G,G)\}$, where first entry is sex of older child; equiprobable measure. Then B_o = older child a boy $= \{(B,B),(B,G)\}$; B_y = younger child a boy $= \{(B,B),(G,B)\}$. So $\Pr(B_y|B_o) = \Pr(B_y \cap B_o)/\Pr(B_o) = 1/2$. **c)** Same sample space and measure. Let B_i = first child introduced is a boy $= \{(B,B),(B,G),(G,B)\}$. So $\Pr(\text{other a boy}\,|\,B_i) = \Pr(\text{both boys})/\Pr(B_i) = 1/3$. **21.** $\frac{2}{3}$; $\frac{2}{3} - \frac{1}{4} = \frac{5}{12}$; Yes **22.** $\frac{5/12}{2/3} = \frac{5}{8} \le \Pr(B|A) \le 1$; $\frac{5}{9} \le \Pr(A|B) \le \frac{8}{9}$. **25. a)** To get B and C after A (LHS), first you have to get B after A and then C after B and A (RHS). **b)** LHS: $\frac{\Pr(A \cap (B \cap C))}{\Pr(A)}$; RHS: $\frac{\Pr(C \cap (A \cap B))}{\Pr(A \cap B)} \cdot \frac{\Pr(A \cap B)}{\Pr(A)}$. **26. a)** 1/6 (French) + 1/12

(German) + 1/12 (Russian) = 1/3. **c)** 1/3 **27. a)** $\prod_{n=30}^{49} p_n$ (note upper bound) **c)** $(\prod_{n=30}^{64} p_n)(1-p_{65})$ **e)** $(\prod_{n=60}^{64} p_n)(1-\prod_{n=65}^{69} p_n)$ **29. a)** l_{50}/l_{30} **c)** $(l_{65}-l_{66})/l_{30}$ **e)** $(l_{65}-l_{70})/l_{60}$ **31.** Proof for $\sim A$ and B: $\Pr(B) = \Pr(A\cap B)+\Pr(\sim A\cap B)$, so $\Pr(\sim A\cap B) = \Pr(B)-\Pr(A\cap B) = \Pr(B)-\Pr(A)\Pr(B) = (1-\Pr(A))\Pr(B) = \Pr(\sim A)\Pr(B)$. **33.** Condition 1): If $\emptyset \subset C \subset A$, then $0 \le \Pr(C) \le \Pr(A)$ [see 18, Section 6.2], so $0 \le \Pr_A(C) \le 1$. Condition 3): If $C\cap D = \emptyset$, then $\Pr(C\cup D) = \Pr(C)+\Pr(D)$, so divide by $\Pr(A)$ to get $\Pr_A(C\cup D) = \Pr_A(C)+\Pr_A(D)$ **35.** 1): Since $0 \le \Pr(A\cap B) \le \Pr(A)$, then $0 \le \Pr(A|B) = \Pr(A\cap B)/\Pr(A) \le 1$. 3): If B, C disjoint, so are $B\cap A$ and $C\cap A$. Thus $\Pr\big((B\cup C)\cap A\big) = \Pr(B\cap A)+\Pr(C\cap A)$. Now divide by $\Pr(A)$. **37.** $\Pr(\sim T|\sim P) = \frac{\Pr(\sim T)\Pr(\sim P|\sim T)}{\Pr(\sim T)\Pr(\sim P|\sim T)+\Pr(T)\Pr(\sim P|T)} = (.9925)(.96)/[(.9925)(.96)+(.0075)(.08)] \approx .99937$. **39.** $\Pr(T|P) \approx .802$; $\Pr(\sim T|\sim P) \approx .986$. **41.** $\Pr(\text{Dem}\,|\,\text{Wrote}) = .5$ (changed) **42.** Let $C =$ child blue-eyed; $N =$ neither parent blue-eyed. We want $\Pr(C\cap N)/\Pr(N)$. $\Pr(N) = (1-.2^2)^2$. Event $C\cap N$ requires both parents to be type bx or xb. Child gets both b's with Pr=1/4, so $\Pr(C\cap N) = (1/4)[2(.2)(.8)]^2$. Answer is $1/36 \approx .028$. **44. a)** Space $S =$ all 11-long HT-sequences, equiprobable. $E_1 =$ all atoms of S with exactly 8 H's in first 10 positions; $E_2 =$ all atoms with H in 11th position. $\Pr(E_2|E_1) = 1/2$. **b)** 10-long sequences; $\Pr = \binom{9}{7} \div \binom{10}{8} = 4/5$. **c)** Space is all 10-long HT-sequences. $\Pr = \binom{4}{1} \div \binom{5}{2} = 2/5$.

Section 6.4

1. a) 1/90 **b)** 11/90 **3. a)** $e^{-2}2^2/2! = 2e^{-2} \approx .271$. **b)** $5e^{-2} \approx .677$. **c)** $1-3e^{-2} \approx .594$. **4. a)** $B_{10,1/6}$ **c)** Poisson **e)** Negative Binomial (order 1), $p = 6/36$. **5.** Negative Binomial, $p = 1/2$. **7. a)** For $m > n$, $(\frac{3}{4})^{n-1}(\frac{1}{6})(\frac{11}{12})^{m-1-n}(\frac{1}{12})$; for $m = n$, 0; for $m < n$, $(\frac{3}{4})^{m-1}(\frac{1}{12})(\frac{5}{6})^{n-1-m}(\frac{1}{6})$. **c)** $\sum_{k\ge 1} f(k) < 1$; the "distribution" doesn't cover all possibilities. **9.** For $\lambda = 1$, $k = 0,\ldots,5$, the values of $e^{-\lambda}\lambda^k/k!$ are .368, .368, .184, .061, .015, .003. For $\lambda = 4$, $k = 2,\ldots,7$, the values are .147, .195, .195, .156, .104, .060. **11.** For $p = .3$, $k = 1,3,5,7,9$, the values of $(k-1)(1-p)^{k-2}p^2$ are 0, .126, .123, .091, .059. For $p = .5$, $k = 0,2,4,6,8,10$, the values are 0, .25, .187, .078, .027, .009. **13. a)** For $p = .5$, $n = 10, 20, 50$, the values of $B_{n,.5}(n/2) = \binom{n}{n/2}.5^n$ are .246, .176, .112 (getting smaller). **c)** For $p = .5$, $n = 10, 20, 50, 100, 200$ the values of $\sum_{k=-2}^{2} B_{n,.5}(\frac{n}{2}-k)$ are .891, .737, .520, .383, .276 (getting smaller). **16.** Substitute Stirling's Formula into $B_{n,p}(np) = \frac{n!}{(np)!(nq)!}p^{np}q^{nq}$ and simplify. Whatever constant p is, $[1/\sqrt{2\pi np(1-p)}\,] \to 0$ as $n \to \infty$. This shows that the probabilities in [12abc] go to 0. **17. a)** $1-F(x)$ **c)** $F(x_j)-F(x_i)$ **e)** $F(x_j)-F(x_{i-1})$ **19.** $B_{n,p}$ with $p = r/(r+b)$ **21. a)** $\frac{1}{10}\cdot 1$ [pick defective first] $+\frac{9}{10}\cdot\frac{1}{9} = \frac{1}{5}$. **b)** 7/15 **c)** Bayes says $(\frac{1}{4})(\frac{7}{15})/[(\frac{1}{4})(\frac{7}{15})+(\frac{3}{4})(\frac{1}{5})] = \frac{7}{16}$. **23.** $f_n(k,j) = \binom{n}{k,j,n-k-j}p^k q^j r^{n-k-j}$ [see trinomial coefficient, Section 4.6]. **25. a)** Down if $r < 2$; Pr = .376. Same if $r = 2$; Pr = .302. **b)** Up, .370; same, .218; down .411. Less likely to exactly break even; see [13]. **27.** .474 **30. a)** For $k = 0, 1, 2$ the approximations are .368, .368, .184; the actual values (to 3 decimals) are .366, .370, .185. **31.** $\Pr(X > k_0) = \Pr(\text{first } k_0 \text{ trials don't fail}) = q^{k+0}$. So $\Pr(X{=}k\,|\,X{>}k_0) = q^{k-1}p/q^{k_0} = q^{(k-k_0)-1}p = \Pr(X{=}k-k_0)$. **34.** For instance, $\Pr(X{+}Y{=}4) = \Pr(X{=}1)\Pr(Y{=}3)+\Pr(X{=}2)\Pr(Y{=}2)+\Pr(X{=}3)\Pr(Y{=}1) = 3(\frac{1}{6}\cdot\frac{1}{6}) = \frac{1}{12}$. So, if n is the number of ways to get sum of k on two dice, $\Pr(X{+}Y{=}k) = n/36$. **37.** $X+Y$ counts number of successes in one new period, with 2λ successes expected. So distribution should be $e^{-2\lambda}(2\lambda)^k/k!$. Indeed, $\sum_{i=0}^{k} f_\lambda(i)f_\lambda(k-i) = \sum_{i=0}^{k}\big(e^{-\lambda}\frac{\lambda^i}{i!}\big)\big(e^{-\lambda}\frac{\lambda^{k-i}}{(k-i)!}\big) = e^{-2\lambda}\frac{\lambda^k}{k!}\sum_{i=0}^{k}\frac{k!}{i!(k-i)!} = e^{-2\lambda}\frac{\lambda^k}{k!}2^k = e^{-2\lambda}\frac{(2\lambda)^k}{k!}$. **39.** Eq. (13) $\Longrightarrow$ Eq. (14): (14) is the special case of (13) with $A = \{x\}$, $B = \{y\}$. Converse: $\Pr(X{\in}A \cap Y{\in}B) = \sum_{x\in A, y\in B}\Pr(X{=}x \cap Y{=}y) \overset{(14)}{=} \sum_{x\in A, y\in B}\Pr(X{=}x)\Pr(Y{=}b) = \sum_{x\in A}\sum_{y\in B}\Pr(X{=}x)\Pr(Y{=}b) = \sum_{x\in A}\Pr(X{=}x)\sum_{y\in B}\Pr(Y{=}b)$ [factor] $= \Pr(X\in A)\Pr(Y\in B)$.

Section 6.5

1. $E(X_A) = \sum x\Pr(X_A{=}x) = 0\cdot\Pr(X_A{=}0)+1\cdot\Pr(X_A{=}1) = \Pr(A)$. **3.** $\overline{x} = (x_1+x_2+\cdots+x_n)/n$ where $x_i = a+(i-1)\frac{b-a}{n-1}$. Using summation formulas, $\overline{x}$ simplifies to $(a+b)/2$. **5. a)** -21.5¢ **b)** 10¢ **7.** $w = 1/p$.

9. a) A: -20¢; B: -10¢; C: -20¢; D: -5¢. **b)** $-13\frac{3}{4}$¢ **c)** If you split your dollar into bets of p,q,r,s on A,B,C,D, then your net return, respectively, if A, B, C, or D wins is $2p-1$, $3q-1$, $4r-1$, $9.5s-1$. Setting these equal, one finds $(p,q,r,s)=(114,76,57,24)/271$, and the guaranteed "winnings" are $-43/271$. **11. a)** Ann $2/3$, Bill $-2/3$. **b)** \$2/3 for each toss **c)** No payment; it's a fair game. **13.** Let $f(i)$ be the probability your ith guess is right. Then $f(1)=f(2)=f(3)=1/3$. For instance, $f(2)=\frac{2}{3}\cdot\frac{1}{2}$ (wrong on first guess $\times$ right on second). So $E(f)=2$. **14. c)** $\frac{1}{5}(1)+\frac{4}{5}(\frac{1}{4})=\frac{2}{5}$ point. **e)** For a), $\frac{1}{4}(1)+\frac{3}{4}(-\frac{1}{4})=\frac{1}{16}$. For b), $\frac{n}{16}$. **15.** $\sum_{k=0}^{n-1}\frac{n}{n-k}=n(1+\frac{1}{2}+\cdots+\frac{1}{n})$. **17.** \$1000 gain **19.** Certificate: \$1000 certain gain; Bond: \$1100 expected gain. **22.** $E(X)$ is a constant, and for any constant c, $E(c)=c$. So $E\big(X-E(X)\big)=E(X)-E(E(X))=0$. **23.** $\sum k\binom{n}{k}p^kq^{n-k}$ breaks into a sum of three terms: $q\sum k\binom{n-1}{k}p^kq^{(n-1)-k}$, $p\sum(k-1)\binom{n-1}{k-1}p^{k-1}q^{(n-1)-(k-1)}$, and $p\sum\binom{n-1}{k-1}p^{k-1}q^{(n-1)-(k-1)}$ **25.** Let $S_n=p\sum_{k=1}^n kq^{k-1}$. Then $S_n=S_{n-1}+\frac{p}{q}nq^n$. By Section 5.7 the solution is associated with the polynomial $(r-1)(r-q)^2$, so if $q\neq 1$ the solution is of the form $S_n=A+(B+Cn)q^n$. **28.** X_i satisfies the first-order negative binomial distribution, so $E(N_{n,p})=n\,E(N_{1,p})=n/p$. **31.** $\sum_{k\geq 0}k\lambda^k e^{-\lambda}/k!=\lambda e^{-\lambda}\sum_{k\geq 1}\lambda^{k-1}/(k-1)!=\lambda e^{-\lambda}e^{\lambda}=\lambda$. **33.** Both $=k(1+n)/2$. Let Z_i be the number on the ith ball picked. In both schemes, each Z_i is uniformly distributed, and $X_k=\sum Z_i$ (with Z's independent), $Y_k=\sum Z_i$ (with Z's not independent). But independence does not affect the mean of a sum. **35.** If $i<j$ then $\Pr((X_i{=}1)\cap(X_j{=}1))=\Pr(X_j{=}1)=q^{j-1}$, but $\Pr(X_i{=}1)\Pr(X_j{=}1)=q^{i+j-2}$. **37.** By definition: $0\cdot\frac{1}{8}+1\cdot\frac{3}{8}+2\cdot\frac{3}{8}+3\cdot\frac{1}{8}=\frac{3}{2}$. By Eq. (16): For each atom, $\Pr(e)=\frac{1}{8}$. If we list the atoms alphabetically, HHH, HHT,…, TTH, TTT, then $E(X)=(3+2+2+1+2+1+1+0)/8=3/2$. **39.** X takes on values 1 through 6 with equal probability; $(X-4)^2$ takes on values 0, 1, 4, 9 with probabilities $\frac{1}{6}$, $\frac{2}{6}$, $\frac{2}{6}$, $\frac{1}{6}$. By Eq. (17), $E[(X-4)^2]=0\cdot\frac{1}{6}+1\cdot\frac{2}{6}+4\cdot\frac{2}{6}+9\cdot\frac{1}{6}=\frac{19}{6}$. By Eq. (18), $E[(X-4)^2]=(9+4+1+0+1+4)/6=19/6$. **41.** $E(X|A)=\sum_{x\in\text{Range}(X)}x\Pr_A(X{=}x)=\sum_x x\Pr(\{X{=}x\}\cap A)/\Pr(A)=\sum_{e\in A}X(e)\Pr(e)/\Pr(A)$. **43. a)** $7/2$ **b)** $(1+4+9+16+25+36)/6=91/6\neq(7/2)^2$. **c)** $(0^2+0\cdot 1+\cdots+1\cdot 0+1\cdot 1+\cdots+6\cdot 6)/36=49/4$, which does equal $(7/2)^2$ (X,Y independent). **45.** $\text{Var}(X)=p(1-p)$, $\sigma_X=\sqrt{p(1-p)}$. **46.** $\text{Var}(Y)=E[(Y-\frac{7}{2})^2]=E(Y^2)-7E(Y)+\frac{49}{4}=\frac{91}{6}-\frac{49}{4}=\frac{35}{12}$. **49.** $\text{Var}(X{+}Y)=E[(X{+}Y-\bar{x}-\bar{y})^2]=E[(X-\bar{x})^2]+E[(Y-\bar{y})^2]+2\big(E(XY)-\bar{x}E(Y)-\bar{y}E(X)+\bar{x}\bar{y}\big)$. **51.** $\text{Var}\big(\frac{1}{n}\sum X_i\big)=(1/n^2)\,\text{Var}\big(\sum X_i\big)=\text{Var}(X)/n$.

Section 6.6

1. m is initialized to x_1. **2. a)** Both sides $=1/ij$. **4. a)** Best 2, worst $2n-2$, average $2+\sum_{k=3}^n(3/k)$ **b)** Best $n-1$, worst $2n-3$, average $2n-3-\sum_{k=3}^n(1/k)$ **6.** 1/3 cm; 365/3 cm **9.** With Y,Z as in the text discussion of sequential search, $X=Y$ when restricted to A (because $Z=0$) and $X=k+Z$ when restricted to $\sim\!A$ (because $Y=k$). So $E(X|{\sim}A)=k+E(Z|{\sim}A)=k+E_{n-k}$. **11.** In the last display in the proof of Theorem 1, expressions come in pairs, the first involving A, the second $\sim\!A$. Replace each pair with n expressions, the ith involving A_i. **13.** Let event A be "success on first trial". Use Theorem 1. **15. a)** With $a_k=A_{2^k}$ we get $a_k=1+a_{k-1}$; $a_0=1$. The solution is $a_k=k+1$, or $A_n=1+\log n$. This solution is high for all $n>1$, because actually $a_k<1+a_{k-1}$. **b)** Let $b_k=A_{2^k-1}$. Then $b_k=1+\frac{2^k-2}{2^k-1}b_{k-1}$, with $b_1=1$, is exact for $n=2^k-1$. **c)** $b_k=k-1+k/(2^k-1)$ **16.** It makes multisets with two distinct elements twice as likely as multisets with only one distinct element. **18. a)** Let $q=1-\sum_{i=1}^n\Pr(A_i)$ and $Q=\sum_{k\geq 1}q^{k-1}$. Then $p(A_i)=\sum_{k\geq 1}\Pr(A_i)q^{k-1}=\Pr(A_i)Q$. **b)** $Q=1/(1-q)$ so $p(A_i)=\Pr(A_i)/(1-q)=\Pr(A_i)/\sum_{j=1}^n\Pr(A_j)$. **19.** $\Pr(R{=}1\cap Y{=}2)=\frac{1}{3}P(2,2)/P(5,2)$; $\Pr(R{=}2\cap Y{=}1)=\frac{1}{3}(4\cdot 2)/P(5,2)$; $\Pr(R{=}3\cap Y{=}0)=\frac{1}{3}P(2,2)/P(5,2)$. So $\Pr(J=3)=(2+8+2)/60=1/5$. **21. a)** $k/2=(i-1)/2$ (each value of Y from 0 to $i-1$ equally likely). **b)** $E(R)=(n-k+1)/2$, so $E(J)=E(R)+E(Y)-(n+1)/2$. **23.** $\sum_{k\geq 1}\frac{(k-1)+1}{(k-1)!}=\sum_{k\geq 2}\frac{k-1}{(k-1)!}+\sum_{k\geq 1}\frac{k}{k!}=\sum_{k\geq 1}\frac{k}{k!}+\sum_{k\geq 1}\frac{k}{k!}$. **25.** For $k=n$ you cannot group the n-permutations by the rightmost $k+1$ entries. Also, there will only be $n-1$ comparisons, not $n=k$ (but the number of loop tests will be one more than the number of comparisons, compensating). **27.** $n!/(k+1)!$ too permutations, one for each choice of the rightmost $k+1$ digits ($k<n$). Average $=\frac{1}{n!}\sum_{k=1}^{n-1}\frac{n!}{(k+1)!}\sum_{j=1}^k j=$

$\sum_{k=1}^{n-1} \frac{1}{2(k-1)!} \approx e/2$. **29. a)** $12 \to 312, 132, 123$; $21 \to 321, 231, 213$. **b)** Putting n in the middle wipes out all temporary maxes to its right, so the number of maxes in the derived permutation does not have a constant relation to the number in the starting permutation. (A recursion *is* possible if we keep track of how many permutations have k temporary maxes in the leftmost j positions, but it's complicated: If $\text{Perm}(n, j)$ is the number of permutations of n things j at a time, one can obtain $P_{n,k} = \sum_{j=0}^{n-1} \text{Perm}(n-1, n-j-1) P_{j,k-1}$, and this can be reduced to the recursion in e) below.) **c)** $12 \to 231, 132, 123$; $21 \to 321, 312, 213$ **e)** The derived permutation has the same number of temporary maxes as the starting permutation if the appended digit is $< n$, and one more if n is appended. Thus $P_{n,k} = (n-1)P_{n-1,k} + P_{n-1,k-1}$. **31.** $p_{n,k} = \frac{n-1}{n} p_{n-1,k} + \frac{1}{n} p_{n-1,k-1}$.

Section 6.7

1. 2/3 **3.** $p_1 = \sum_{k=0}^{\infty} (5/6)^{2k}(1/6) = (1/6) \sum_{k=0}^{\infty} (25/36)^k$. **5.** $(p_1, p_2, p_3) = \frac{1}{91}(36, 30, 25)$. **7.** $E = (11/36)E + (25/36)(E+2)$, which implies $E = 61/11$. The probability that Player 1 gets 6 on the first toss *given* that she wins is $(1/6)/(6/11) = 11/36$. The probability that she does not get a 6 on the first toss given that she wins must be complementary: 25/36. **9.** Let p_i be the chance that Player i wins. Then $p_1 = q/(1+q^2)$, which is 2/5 when $p = q = 1/2$. Obtained from the recursion $p_1 = qp_2 + pqp_1$ and $p_2 = 1 - p_1$. **11.** Some values of (p, p_G) are: (.3, .099), (.4, .264), (.45, .377), (.49, .475), (.5, .5), (.51, .525), (.6, .736). **13.** $p^3 + C(3,1)p^3q + C(4,2)p^3q^2$ **15.** In all three cases, $E = 2/(p^2+q^2)$, even though the probabilities of winning are different. **17.** $\frac{1}{2}p_*^{21} + \frac{1}{2}pp_*^{21}$, where $p_* = p/(1-pq) = \sum_{k \geq 0} (qp)^k p$ is the probability that A is the first to win a point if A serves first. **19. a)** $\sum_{k \geq 0} (qp')^k p = p/(1-qp')$, where $q' = 1-p'$. **b)** $p'p/(1-qp')$. Or, solve both parts simultaneously by recursion. **21.** Let $g = (k-m)/2$, so that $g + (g+m) = k$. If $g \notin N$, then no ways; else $C(k, g)$ ways (choose which g games A wins). Probability is $C(k,g)p^g q^{g+m}$. **22.** $P_m = 1 - \frac{m}{m+n}$, so in Example 6, $P_3 = 5/8$. **24. b)** $P_m \approx 1 - \bar{r}^n$. **d)** $1 - \bar{r}^{m+n} < 1$, so $P_m > 1 - \bar{r}^n \geq 1 - \bar{r}^{n_0}$. For $p = .49$ and $n_0 = 100$, $P_m \geq 1 - (.49/.51)^{100} \approx .9817$. **25.** No. For P_m to be $\leq .01$, [22] shows that we need $m/(m+n) \geq .99$, or $m \geq 99n$. So the needed m increases with n. **27.** The recursion is $E_k = 1 + pE_{k+1} + qE_{k-1}$, $E_0 = E_{m+n} = 0$. Solution: E_m where $E_k = \frac{m+n}{(p-q)(1-r_*^{m+n})}(1 - r_*^k) + \frac{k}{q-p}$. **29. a)** Multiply top and bottom of Eq. (7) by p^{2n} and note that $p^{2n} - q^{2n}$ is the difference of two squares. **b)** Let p_i = probability Player i is ruined. There is a one-to-one correspondence between win-loss sequences where Player 1 is ruined and those where Player 2 is ruined: switch wins and losses. Therefore $p_1/p_2 = q^n/p^n$ (why?). Also, $p_1 + p_2 = 1$. **31. a)** $N_{k,m,n} = N_{k-1,m+1,n-1} + N_{k-1,m-1,n+1}$; $N_{k,0,n} = 0$ if $k > 0$, 1 if $k = 0$; $N_{k,m,0} = 0$ for $k \geq 0$ and $m > 0$. This implies $N_{k,m,n} = 0$ if $m > k$, which is convenient to use in calculations. **b)** $N_{5,3,5}$ expands to 3: A is ruined in five steps if she wins just once in the first three. **c)** 41 **d)** $\sum_{k \leq 10} N_{k,3,5}(.5)^k \approx .344$ (calculator). **33. a)** The diagonal $p_{4,0}$, $p_{3,1}$, $p_{2,2}$, $p_{1,3}$, $p_{0,4}$, for instance, is 0, 1/8, 4/8, 7/8, 1, which is 1/8 times the running sum of the row 1, 3, 3, 1 from Pascal's Triangle. **b)** We want $p_{N,N}$ where $p_{m,n} = \sum_{k \geq m} \binom{m+n-1}{k} p^k q^{m+n-1-k}$. **34. a)** The mapping truncates each scorecard in G' just after the point where A won for the mth time or B won for the nth. For each scorecard S in G, the set of scorecards in G' that map to S have together the same probability as S, because any sequence of results for the remaining points up to $m+n-1$ is possible. **35. a)** Make $p_{1,1}$ an initial case: $p_{1,1} = p^2/(p^2+q^2)$. **b)** To the sum in [34b] add $\binom{m+n-2}{m-1} p^{m-1} q^{n-1} [\frac{p^2}{p^2+q^2} - p]$. The case of interest is $m = n = N$. **37. a)** p_a and p_b in Example 4 are different. **c)** $p_{0,m,n} = pp_{0,m-1,n} + qp_{1,m,n}$; $p_{1,m,n} = pp_{0,m,n} + qp_{1,m,n-1}$. Use these to express $p_{0,m,n}$, $p_{1,m,n}$ solely in terms of smaller cases. **39. a)** Let p_k be the probability of being or returning home if he starts at point $k \in N$. So $p_0 = 1$ (he's already there), and $p_k = \frac{1}{2}(p_{k+1} + p_{k-1})$. All solutions for p_k are of the form $A + Bk$. The only solution which is a probability for every k and satisfies $p_0 = 1$ is $p_k = 1$. So the probability of *returning* home is $\frac{1}{2}(p_{-1} + p_1) = 1$. **b)** Infinity! Let E_k be the expected time until first being home if he starts at k. Then $E_k = 1 + \frac{1}{2}(E_{k+1} + E_{k-1})$, $E_0 = 0$. There are no finite solutions that are always nonnegative.

Supplementary Problems: Chapter 6

1. a) By Eq. (18) in [38, Section 6.5], $E(U(X))=\sum_x U(x)\Pr(X=x)=a\sum x\Pr(X=x)+b\sum \Pr(X=x)=aE(X)+b$. So if X represents, say, the possible monetary outcomes of investing your money in some stock, and Y represents an alternative, then $E(U(X)) > E(U(Y)) \iff E(X) > E(Y)$. **b)** The value of winning \$9999.75 would be relatively less and the value of losing 25¢ would be more negative. So n must be even smaller to make the contest worthwhile. **3.** $\sum_{k=51}^{N}\left(\prod_{n=51}^{k} p_n\right)5000$, where N is an upper bound on human age, usually set at 100. The most direct (but not efficient) translation of this formula into an algorithm is a double loop with $P \leftarrow P * p_n$ inside the inner loop and then $S \leftarrow S + 5000 * P$ in the outer loop. **5.** $E(f) = \sum_x xf(x) = cf(c) + \sum_{w>0}(c+w)f(c+w) + \sum_{w>0}(c-w)f(c-w) = \sum_x cf(x)$ because for each w, $wf(c+w)$ and $-wf(x-w)$ cancel. **7.** $E(X_i^2) = p$, $E(X_iX_j) = p^2$, so $E(B_{n,p}^2) = np + (n^2-n)p^2$. **9.** $E(B_{1,p}^2) = p$ and $E(B_{n+1,p}^2) = qE(B_{n,p}^2) + pE\big([1+B_{n,p}]^2\big) = E(B_{n,p}^2) + 2pE(B_{n,p}) + p = E(B_{n,p}^2) + 2np^2 + p$. By induction or sum formulas, $E(B_{n,p}^2) = n(n-1)p^2 + np$. **11. a)** 1/2 **b)** 1/4 **c)** 1/2 **d)** 1/3 **13. a)** 1/3 **c)** 1/6 **e)** 1/3 **g)** $\frac{1}{9}/(\frac{1}{9}+\frac{1}{3}) = 1/4$. **15.** From the text, $p_A = p/(1-pq)$, $p_B = p^2/(1-pq)$. Next, $p_A^* = p_A^2 + p_A(1-p_A)p_B^* + (1-p_A)p_Bp_A^*$. Similarly, $p_B^* = p_Bp_A + p_B(1-p_A)p_B^* + (1-p_B)p_Bp_A^*$. Now solve simultaneously. **17. a)** Let p_i = probability bug gets to C from vertex i without passing B. Then $p_A = \frac{1}{2}p_D$ and $p_D = \frac{1}{2}p_A + \frac{1}{2}$, so $p_A = 1/3$. Or $\sum_{k\ge 0}(1/4)^k(1/4) = 1/3$ [k trips ADA followed by DC]. **b)** 4 **c)** 3 **19. a)** Now LWW results in a win. **b)** $\sum_{k\ge 0} p^{\lfloor k/2\rfloor}q^{\lceil k/2\rceil}p^2 = \sum_{m\ge 0}(p^mq^m + p^mq^{m+1})p^2 = \left(\frac{1+q}{1-pq}\right)p^2$. **c)** $P_p = p(p+P_q)$, $P_q = qP_p$. **d)** Win-by-2 is better. Algebra shows that $\frac{1}{1-2pq} - \frac{1+q}{1-pq} = \frac{q^2(2p-1)}{(1-2pq)(1-pq)}$ and $2p-1>0$ for $p>.5$. Thus $\frac{1}{1-2pq} > \frac{1+q}{1-pq}$ for all $p>.5$. **21.** $\sum_{i=1}^{k} i2^{i-1} = (k-1)2^k + 1 = (k-1)(2^k-1) + k$. Thus $\sum_{i=1}^{k} in_i/(2^k-1) = k - 1 + \frac{k}{2^k-1}$.

Chapter 7

Section 7.2

1. I may have meat; otherwise I'll have fish. **2.** I deny that I like discrete math. **4.** $P\,|\,Q$ is true unless both P and Q are true. **5.** $P\,|\,Q = \neg(P\wedge Q)$; $P\downarrow Q = \neg(P\vee Q)$. **6. a)** Vacuously true because premise always false, **c)** Vacuously true because premise always false since quadratic can have only two solutions (assuming that x_1, x_2, x_3 must be distinct) **7. a)** $\neg J$ **d)** $J\wedge M$ **g)** $\neg J\wedge\neg M$ **j)** $\neg J\vee\neg M$ (if both are allowed not to be here); otherwise $\neg(J\Leftrightarrow M)$ **8. a,d)** $R\Rightarrow C$, **g)** $C\Rightarrow R$ (if interpretation is clouds are sufficient for rain but really ambiguous) **9. c)** $x=2\Rightarrow x^3=8$ **f)** $x^3=8\Rightarrow x=2$ **i)** $x^2\neq 4\Rightarrow x=2$ or $\neg(x^2=4)\Rightarrow x=2$ **10. b)** $(xy=0)\Leftrightarrow((x=0\vee(y=0))$ **d)** $x=3\Rightarrow x>0$ **11.** Here are two which require four applications: $\neg\neg P\wedge P$, $P\wedge\neg P\wedge P$. **13. b)** Is a wff **d)** Is not a wff **15. a)** $\Leftrightarrow$ is associative. **16. b)** Ambiguous **d)** Ambiguous **17. b)** For wff of [14b] only values which give T are $(P,Q,R) = (T,T,F)$, (F,T,T), (F,T,F), (F,F,T), (F,F,F) **c)** For wff of [16e] only values which give T are (T,T,T), (F,T,T), (F,T,F), (F,F,T), (F,F,F) **19.** Need to prove newrowatoms generates successive lines of table; with $T=1$, $F=0$ show each call of newrowatoms generates binary number for $P_nP_{n-1}\ldots P_1$ one less than previous value **21.** $P\leftarrow F$; **for** $i=1$ **to** 2; $P\leftarrow\neg P$; $Q\leftarrow F$; **for** $j=1$ **to** 2; $Q\leftarrow\neg Q$; $R\leftarrow F$; **for** $k=1$ **to** 2; $R\leftarrow\neg R$; **print** P, Q, R; **endfor**; **endfor**; **endfor**. **23.** Show $(A\Rightarrow B)\Leftrightarrow(\neg B\Rightarrow\neg A)$ is a tautology; $\neg B\Rightarrow\neg A$ is the contrapositive. **25. b)** Invalid because from what is given you don't know that Illinois is in USA. **26. b)** Invalid; if A, C true and B false, premise is true but conclusion false **27. a)** To go from n to $n+1$, apply induction hypothesis and then use basis case $(\neg\neg P\Leftrightarrow P)$ **28. b)** Not a tautology **d)** Tautology **29. a)** Not logically equivalent **c)** Logically equivalent. **30. a,c,g)** First doesn't logically imply second **e)** First does logically imply second **31. a)** Tautology—always true **c)** Tautology—premise always false **32. a)** Yes, just choose a tautology as the premise but this wouldn't be useful.

Section 7.3

1. a) Rule 1: $((A \Rightarrow B) \wedge A) \Rightarrow B$. **3. a)** A, B, $/A*A/$, $/A*B/$, $/B*A/$, $/B*B/$, $/A*/A*A//$ and 7 others of this form, $//A*A/*A/$ and 7 others of this form **b)** All of length $4k+1$; if put two such together with 2 slashes and a star, new length is also 1 more than multiple of 4 **c)** First is winning, next two are not **4. a)** Second rule: If A and B are true, then A is true. **5. a)** Rule 1 of [4] followed by Rule 2 of [4] **c)** Assume A; derive $B \wedge C$ (Premise 1 and modus ponens); derive B (Rule 2 of [4]); derive D (Premise 2 and modus ponens); derive $A \Rightarrow D$ (conditional proof) **7.** This line represents a statement "global" to anything in the subproof and, thus, is usable within the subproof.

Section 7.4

1. An assertion is normally a statement of something for which a proof is unnecessary while a hypothesis is something assumed true for the purposes of argument. **3.** IS: $x \neq 0$; OS: $y = 1/x$; Proof: Algorithm computes y as $1/x$ which can be done as long as $x \neq 0$ **5. b)** IS: $w_1 \leq w_2 \leq w_3 \leq \cdots \leq w_n$ (see Section 7.6 for a better way to express this) **6. c)** $1 \leq \mathrm{Perm}(i) \leq n$, $i = 1, \ldots, n \wedge \mathrm{Perm}(i) \neq \mathrm{Perm} j$ if $i \neq j$, **7. a)** IS: $\{n \geq 0\}$; $i \leftarrow n$; $even \leftarrow true$; $\{2 \mid (n-i) \wedge i \geq 0\}$; **repeat while** $i \geq 2$; $i \leftarrow i-2$; **endrepeat**; $\{2 \mid (n-i) \wedge 0 \leq i \leq 1\}$; **if** $i = 1$ **then** $even \leftarrow false$; OS: $\{(2|n \wedge even = true) \vee (2 \nmid n \wedge even = false)\}$; proof needs to show only that loop invariant is true on each entry and that loop termination condition is true.

Section 7.5

1. $1+a=1$; $0 \cdot a = 0$; use truth table or just note that results follow from definition of $+$ and $\cdot$. **3.** Just replace $\neg$ by complement, $\wedge$ by $\cdot$ and $\vee$ by $+$. **5.** For each just compute the rows of the truth table for the left and right sides and verify that they are the same. **6. c)** For [5b]: $\overline{(p+r)(q+\bar{r})} = (\bar{p}+r)(\bar{q}+\bar{r})$ **7. b)** $p(p+q) = pp+pq$ [(vii)]$=p+pq$ [Eq. (3)]$=p$ [7a] **d)** $p(\bar{p}q) = (p\bar{p})q$ [(vi)]$=0q$ [(x)]$=0$ [1] **9. a)** By induction **b)** $(p+q)(p+r) = (p+q)p+(p+q)r$ [(vii)]$=p+pr+qr${(iv), [7b]}, (vii)}$=p+qr$ [7a]; then generalize using induction; **11. b)** DNF: $pqr+p\bar{q}r+\bar{p}qr+\bar{p}\bar{q}r+pq\bar{r}+p\bar{q}\bar{r}$; first four terms give 2×2 region which simplifies to r, and last two terms give 2×1 region which simplies to $p\bar{r}$, so expression simplifies to $r+p\bar{r}$ **13. a)** For each 1 in right hand column, write product of variables (where there are 1's) or complements of variables (where there are 0's) in the other columns **14. c)** $\bar{p}\bar{q}\bar{r}+\bar{p}qr+p\bar{q}\bar{r}+pq\bar{r}+pqr$ **15.** $pq+p\bar{q}r$ or $pq\bar{r}+pr$; neither is simpler **17. a)** Expression already in DNF **c)** $pq\bar{r}+p\bar{q}\bar{r}+\bar{p}\bar{q}\bar{r}+\bar{p}qr+\bar{p}\bar{q}r$ **18. a)** No simplification possible; **c)** No simplification possible but can be rewritten $p\bar{r}+\bar{p}r+\bar{p}\bar{q}\bar{r}$ **19. a)** 4×8 array with left and right labels as on 4×4; on top duplicate q in 4×4 and similarly on bottom for s; add on bottom: $t\bar{t}\bar{t}t\bar{t}tt\bar{t}$ **b)** $2\times 2\times 2$ cube with $p, \bar{p}, q, \bar{q}, r, \bar{r}$ on the perpendicular sides; symmetric in each of the three variables whereas two-dimensional map is not; easier to determine simplification also **21. b)** $\bar{p}+pqr+pq\bar{r}\bar{s}$ **23.** Use [22] and (xii) of Theorem 1. **25.** Sum picks out just those rows for which value is 1; see [13a]. **27.** When a term in (13) is one, all terms in (14) are 0 and vice versa. **28. a)** $2n-1-i$ row is digit-by-digit complement of ith row; since, by (xii) of Theorem 1 (see [26]) complement of product is sum of complements of individual terms, identity follows **29.** Take complement twice, first using (14), then using [26] and [28]; need to pay attention to 0 lines in truth table since all other terms in product are 1 **31. a)** $(p+\bar{q}+r)(p+\bar{q}+\bar{r})(\bar{p}+q+r)$; **b)** Use $(a+b)(a+\bar{b}) = a$. **32.** [17a] : $(p+\bar{q})(\bar{p}+q)$ **35.** $c_1 = \bar{b}_1$; $c_2 = (b_1+b_2)(\bar{b}_1+\bar{b}_2)\bar{b}_8$; $c_4 = b_1 b_2 \bar{b}_4 + (\bar{b}_1+\bar{b}_2)b_4$; $c_8 = b_1 b_2 b_4 + \bar{b}_1 b_8$.

Section 7.6

1. a) $(n_1 \neq n_2) \wedge (n_1 \neq n_3) \wedge \cdots \wedge (n_1 \neq n_m) \wedge (n_2 \neq n_3) \wedge \cdots \wedge (n_2 \neq n_m) \wedge \cdots \wedge (n_{m-1} \neq n_m)$; **b)** $(\forall i : 1 \leq i \leq m)(\forall j : 1 \leq j \leq m) i \neq j \Rightarrow n_i \neq n_j$ **2. a)** $i = 0, 1, \ldots, 16$; **c)** Either $i = 0, 1, \ldots, 9$ or $j = 21, 22, \ldots$ or both **3. b)** $(\forall i : 1 \leq i \leq n)(1 \leq \mathrm{Perm}(i) \leq n \wedge (\forall j : i \leq j \leq n) i \neq j \Rightarrow \mathrm{Perm}(i) \neq \mathrm{Perm}(j)$ [cf. solution to [6c], Section 7.4] **4. a)** For each nonzero real number there is another such that the product of the two is a given real number a. **d)** There is a number [in fact, 0] such that x plus that number is x for any x; **g)** If two numbers are not equal, there is a number in between them. **5. a)** $(\forall x)(\exists y) x+y = 0$; **d)** $(\forall x)(x > 0 \Rightarrow (\exists y)(y > 0 \wedge y^2 = x))$; **g)** $(\forall x) x \neq 2x$ [of course, not true since false for $x = 0$] **6. a)** For each integer, there exists an integer one greater; **d)** For each integer, there exists an integer which is its square; **g)** If i divides j and j divides k, the floor of the floor of k/j divided by i is the same as the floor of k divided by the floor of j divided by i; false; for example, $i = 4$, $j = 8$, $k = 16$ **7. a)** $(\exists m : m \geq 0)(\forall n : n \geq 0)[n \geq m]$; **d)** $(\forall n : 0 \leq n \leq 100)(\exists i)\, 2^i < n < 2^{i+1}$;

g) $(\forall n)(\forall m)2\,|\,n \wedge 2\,|\,(m+1) \Rightarrow 2\,|\,mn$ **9. a)** x, y free; **d)** None free. **11. a)** $(\forall m : 2 < m < 10)(\forall n : 1 < n < m)P(m,n)$; **c)** $(\forall m : 1 \le m \le 5)(\forall n : m \le n \le 5)P(m,n) \wedge (\forall n : 1 \le n \le 5)P(0,n)$ **13.** Argue basis case (single quantifier) directly; $n \Rightarrow n+1$ quantifiers: Apply induction hypothesis to first n and then use basis case **14. a)** [4a]: $\neg(\forall x)(\exists y)(x \neq 0 \Rightarrow xy = a) \Leftrightarrow (\exists x)(\forall y)\neg(x \neq 0 \Rightarrow xy = a)$.

Section 7.7

1. When $j = 0$ value is smaller than all previous and is put in first position with other $i-1$ moved up one place which leaves outer loop invariant satisfied **3.** Outer loop is vacuous since $m = 1$ so algorithm does nothing **5. a)** Whatever element is largest initially, once it starts being compared with other elements, it always "wins"; **b)** Same argument; stop when comparing second to last and next to last elements; **c)** By induction. **7. a)** Should require $n-1$ passes; **b)** Stop whenever no interchanges are made during a pass. **9. a)** Outer loop: $0 < i \le m-1 \wedge (\forall j : i+2 \le j < m)n_j \le n_{j+1} \wedge (\forall k : 1 \le k \le i+1)n_k \le n_{i+2}$ (with $n_{m+1} = \infty$); Inner loop: $(\forall k : 1 \le k \le L)n_k \le n_{L+1}$ where L measures how far down the list the comparison has proceeded; **c)** Outer loop: $(\forall j : 2 \le j < m)n_j \le n_{j+1} \wedge n_1 \le n_2$; Inner loop: Replace L by i in loop invariant.

Section 7.8

1. Just show that every time a left parenthesis is added so is a right parenthesis. **3. a)** Calls Nextchar; returns after reading P; fails because next character not blank; **d)** Fails in Seekwffpair when character after Q is not ")" **5.** Each node with a left branch must also have a right branch. **7. a)** [6b]: T; [6d]: T; **b)** [6b]: T; [6d]: F; evaluate leaves and work up to root **10. a)** $\neg\neg PQRS \Leftrightarrow \neg\neg\wedge\vee$; **b)** Postorder traversal gives, for each operator, (possibly compound) left operand followed by (possibly compound) right operand followed by operator; thus any time operator is reached it is preceded by its left and right operands; procedure thus recovers original string.

Supplementary Problems: Chapter 7

1. a) $(P \Rightarrow Q) \Leftrightarrow \neg(P \wedge \neg Q)$; **d)** Since $\Leftrightarrow$ is associative, ignore parentheses; since $\neg P$ is false when P is true and any expression with just T's and $\Leftrightarrow$'s is true, can't explain $\neg$; for $\vee$, $\wedge$, $\Rightarrow$ show by induction that an expression with P, Q and $\Leftrightarrow$ has 0, 2 or 4 T's in truth table; but truth tables for $\vee$, $\wedge$, $\Rightarrow$ each have an odd number of T's. **3.** With P: The door I am pointing at is the red door and Q: Assistant tells the truth, ask: What is the truth value of $P \Leftrightarrow Q$? **4. c)** $(P\,|\,P)\,|\,(Q\,|\,Q)$ **5. b)** $(P \downarrow P) \downarrow (Q \downarrow Q)$ **7.** Since Peirce's arrow is Nor, just use formulas for $\wedge$, $\vee$, $\neg$, in terms of Peirce's arrow. **9.** Handle negation of $\wedge$ and $\vee$ using de Morgan's laws; handle $\Rightarrow$ by $\neg(A \Rightarrow B) \Leftrightarrow A \wedge \neg B$; handle $\Leftrightarrow$ by $\neg(A \Leftrightarrow B) \Leftrightarrow (A \wedge \neg B) \vee (\neg A \wedge B)$; $(\exists i)[((\exists j)P_j \wedge (\forall k)Q_{ik}) \wedge (R_i \wedge \neg S)]$.

Chapter 8

Section 8.2

1. a) $\left\{\begin{matrix} x+2y=3 \\ 2x+3y=4 \end{matrix}\right\} \to \left\{\begin{matrix} x+2y=\ \ 3 \\ -y=-2 \end{matrix}\right\} \to \left\{\begin{matrix} x+2y=3 \\ y=2 \end{matrix}\right\} \to \left\{\begin{matrix} x\ \ \ =-1 \\ y=\ \ 2 \end{matrix}\right\}$. **b)** $\left[\begin{array}{cc|c} 1 & 2 & 3 \\ 2 & 3 & 4 \end{array}\right] \to \left[\begin{array}{cc|c} 1 & 2 & 3 \\ 0 & -1 & -2 \end{array}\right] \to \left[\begin{array}{cc|c} 1 & 2 & 3 \\ 0 & 1 & 2 \end{array}\right] \to \left[\begin{array}{cc|c} 1 & 0 & -1 \\ 0 & 1 & 2 \end{array}\right]$.
3. $(x,y,z) = (-1,4,-2)$ **5.** Downsweep results in $\left[\begin{array}{ccc|c} 1 & 0 & -1 & 1 \\ 0 & 1 & 2 & -1 \\ 0 & 0 & 0 & a-1 \end{array}\right]$, so there are solutions $\Longleftrightarrow a=1$. **7.** 800 in NY, 1200 in LA. **9.** $n = 0$ gives $A + B = 1$; $n = 1$ gives $3A + B = 4$. Solution: $A = 3/2$, $B = -1/2$. So $A3^n + B = (3^{n+1}-1)/2$. **11.** $(A,B,C) = (2,-2,2)$ **13. a)** Following Example 1, for each row i, multiples of all lower rows would be subtracted from row i together. (Instead, for each k, multiples of row k were subtracted from all higher rows together.) **b)** Elementary row operations preserve solutions in any order, and the number of operations is unchanged. **c)** We think the Example 2 order is easier for people. In Upsweep for Example 1, the things done together all affect the same row, and there is too much arithmetic for people to keep in their heads. No difference for computers. **14. a-b)** Rational numbers can be obtained by division, but the rationals are closed under $+, -, \times, \div$, so no further types of numbers can be obtained. **c)** Real numbers **16.** Yes, whole row subtracted, but entries in columns 1 to $i-1$ don't change, so why go through the motions? The entry in column i is more quickly changed by assignment. **19.** Longer to run, because $a_{ki}/a_{ii} = m$ is recomputed for each l, that is, $n+1-i$ times instead of 1 time. Thus $\sum_{i=1}^{n}(n-i) = n(n-1)/2$ extra steps. **22. for** $i = n$ **downto** 1; $x_i \leftarrow \left(a_{i,n+1} - \sum_{k=i+1}^{n} a_{ik}x_k\right)/a_{ii}$;

endfor. (The $\sum$ expands to an inner loop.) **24.** Start matrix $\rightarrow \left[\begin{array}{ccc|c} 2 & -2 & 4 & -6 \\ -1 & 2 & 0 & 3 \\ 1/2 & 1/2 & 0 & 3/2 \end{array}\right] \rightarrow \left[\begin{array}{ccc|c} 2 & -2 & 4 & -6 \\ -1 & 2 & 0 & 3 \\ 3/4 & 0 & 0 & 3/4 \end{array}\right] \rightarrow$ $\left[\begin{array}{ccc|c} 0 & 0 & 1 & -1 \\ 0 & 1 & 0 & 2 \\ 1 & 0 & 0 & 1 \end{array}\right]$. **26.** "Downsweep" results in $\left[\begin{array}{ccc|c} 2/3 & 0 & 0 & 2/3 \\ 0 & 3 & 0 & 6 \\ -1 & 2 & -3 & 6 \end{array}\right]$. **27. a)** Scale($i$) must divide all of row i to the right of the pivot. No division needed to determine multipliers for Downsweep. Total steps: $(2n^3+6n^2-2n)/6$. **d)** Scale, Upsweep, and Downsweep must all be applied in all columns to the right of the pivot. No divisions needed to determine multipliers. Total steps: $(n^3+n^2)/2$. **28.** Equivalence is, for instance, transitive because if system 1 has the same solutions as system 2, and system 2 has the same solutions as system 3, then systems 1 and 3 have the same solutions. **29. a)** Yes **b)** No **c)** (iii) $\not\Leftrightarrow$ (ii).

Section 8.3

1. $(x,y,z)=(9/8,-5/4,1/8)$. **3.** $(x,y,z)=(-1,0,1)+y(1,1,0)$. **4. a)** Final matrix and solution remain the same (true also for parts b–d). **e)** No solution. **5. b)** $\begin{bmatrix} 1 & 0 & 0 & 0 \\ 0 & 1 & 3/2 & 0 \\ 0 & 0 & 0 & 1 \end{bmatrix}$ **c)** Same as part b). **e)** $\left[\begin{array}{cccc|c} 1 & 3 & 0 & 1 & -1 \\ 0 & 0 & 1 & 1/3 & 2/3 \\ 0 & 0 & 0 & 0 & -1 \end{array}\right]$ (no solution). **g)** no changes **7.** No solution, so why compute more? **9. a)** $\mathbf{x}=\mathbf{0}$ is a solution. **b)** The standard vector form for the solutions is $\mathbf{0}+x_i\mathbf{v}+\cdots$ with at least one free variable x_i. **11.** ScanSwitch(2,2) goes out of bounds for $\left[\begin{array}{cc|c} 1 & 0 & 0 \\ 0 & 0 & 0 \end{array}\right]$. **13.** False. If a pivot is found in the last row m , then the **repeat** loop in Main sets $r=m+1$, which causes exit from this loop. **15. for** $l=j$ to $n+1$ $a_{il} \leftrightarrow a_{rl}$. **17.** Loops over columns must be split. For instance, the loop in Procedure Scale(i,j) becomes: **for** $l=j+1$ to n; $a_{il} \leftarrow a_{il}/a_{ij}$; **endfor**; $b_i \leftarrow b_i/a_{ij}$. **18.** Incompressibility at u implies $-x_1+x_4=0$. Considering u,v,w,x in turn gives a homogeneous system with matrix $\begin{bmatrix} -1 & 0 & 0 & 1 & 0 \\ 1 & -1 & 0 & 0 & 1 \\ 0 & 1 & -1 & 0 & 0 \\ 0 & 0 & 1 & -1 & -1 \end{bmatrix}$. The solutions are $x_4\begin{pmatrix} 1\\1\\1\\1\\0 \end{pmatrix}+x_5\begin{pmatrix} 0\\1\\1\\0\\1 \end{pmatrix}$. **20.** $\begin{pmatrix} 1\\1\\1\\1\\0 \end{pmatrix}+x_5\begin{pmatrix} 0\\1\\1\\0\\1 \end{pmatrix}$. If x_1 was free when you solved [18], set $x_1=1$ in yourstandard vector form. If it wasn't free, solve again with negative the x_1 column on the right.(Why?) Or, exchange the x_1 column with a free-variable column and do Gaussian elimination again. **21.** $\mathbf{y}=y_3(0,-1,1,0,0)+y_4(-1,0,0,1,0)+y_5(1,-1,0,0,1)$ **23. b)** $(1,2,0,-3,-2)+y_3(0,-1,1,0,0)$ **25.** Trivial solution $\mathbf{0}$ only. **26. a)** $\mathbf{x}=(6,11,11,12,5)/37$.

Section 8.4

1. a) echelon **b)** echelon **c)** neither **d)** echelon (not reduced) **3.** All have rank 3. **5.** 15; $p(n-p+1)$ **8.** False if, after Downsweep, two or more rows have nonzero entries in the **b**-column only. **10.** Gaussian elimination produces at least one nonpivot column $\Longleftrightarrow$ rank$(A)<n$. **12.** All entries of A are 0. **14.** In Upsweep,a_{kj} is the wrong multiplier for row k if a_{ij} not set to 1 first by Scale. To switch the order, replace a_{kj} in Upsweep by $m=a_{kj}/a_{ij}$. **15.** If $a_{il}=0$ for one or more $l>j$ (e.g., if there are pivot columns to the right of column j, then Upsweep(i,j) and Scale(i,j) have no effect in any such column l. Scale-Upsweep(i,j) skips this vacuous work.

Section 8.5

1. a) $\left[\begin{array}{cc|cc} 1^* & 2 & 3 & -1 \\ 2 & 3 & 4 & 0 \end{array}\right] \rightarrow \left[\begin{array}{cc|cc} 1 & 2 & 3 & -1 \\ 0 & -1^* & -2 & 2 \end{array}\right] \rightarrow \left[\begin{array}{cc|cc} 1 & 0 & -1 & 3 \\ 0 & 1 & 2 & -2 \end{array}\right]$. **b)** For $\begin{bmatrix} 1 & 2 \\ 2 & 3 \end{bmatrix}$ L-and-U is $\begin{bmatrix} 1 & 2 \\ 2 & -1 \end{bmatrix}$. From $\begin{pmatrix} 3\\4 \end{pmatrix}$ we get $\begin{pmatrix} 3\\-2 \end{pmatrix} \rightarrow \begin{pmatrix} 3\\2 \end{pmatrix} \rightarrow \begin{pmatrix} -1\\2 \end{pmatrix}$. From $\begin{pmatrix} -1\\0 \end{pmatrix}$ we get $\begin{pmatrix} -1\\2 \end{pmatrix} \rightarrow \begin{pmatrix} -1\\-2 \end{pmatrix} \rightarrow \begin{pmatrix} 3\\-2 \end{pmatrix}$. **3.** L-and-U is $\begin{bmatrix} 1 & 2 & 3 \\ 2 & -1 & -2 \\ -1 & -3 & -1 \end{bmatrix}$; the solution is $(-1,4,-2)$. **5. a)** $\begin{bmatrix} 20 & 5 & 1 \\ 32 & -92 & -12 \\ 34 & -6518 & 229 \end{bmatrix}$; answer $(15,15,15)$. By hand easier to pivot starting in rightmost column, or make column switches first. **c)** Solution $(16,80,-280)$ so recruits must eat negative cabbage! **7.** $\begin{bmatrix} 1 & 1 & 1 \\ -1 & 2 & 4 \\ 2 & -32 & 8 \end{bmatrix}$; answer is $\begin{pmatrix} 9/8 \\ -10/8 \\ 1/8 \end{pmatrix}$. **8.** To get from $[u_{ij}]$ to $[v_{ij}]$ you do Scale and Upsweep on entries (k,k) with $k>i$.

Such operations do not affect the (i,i) entry since all entries below it in $[u_{ij}]$ (and subsequent matrices) are 0.

9. Scale factors still u_{ii}; Upsweep multiples are u_{ki}/u_{ii}. **12. a)** $\begin{bmatrix} 1^* & 2 \\ 3 & 4 \\ 5 & 6 \end{bmatrix} \to \begin{bmatrix} 1 & 2 \\ 3 & -2^* \\ 5 & -4 \end{bmatrix} \to \begin{bmatrix} 1 & 2 \\ 3 & -2 \\ 5 & 2 \end{bmatrix}$. Thus for first system $\begin{pmatrix} 1 \\ 2 \\ 3 \end{pmatrix} \to \begin{pmatrix} 1 \\ -1 \\ -2 \end{pmatrix} \to \begin{pmatrix} 1 \\ -1 \\ 0 \end{pmatrix} \to \begin{pmatrix} 0 \\ 1/2 \\ 0 \end{pmatrix}$. Bottom 0 says system consistent; solution: $(x,y)=(0,1/2)$. **b)** Change n to m in both lines **for** $k=1+i$ **to** n. **14. a)** L-and-U in the second problem is $\begin{bmatrix} 1 & 2 & 3 & 4 \\ 1 & 2 & 3 & 4 \\ 1 & -1 & 0 & 1 \end{bmatrix}$. From the nonpivot column, the general solution is $\begin{pmatrix} 0 \\ -3/2 \\ 1 \\ 0 \end{pmatrix} x_3$. The pivot columns would be used to reduce **b** if it weren't **0**.

15. Do the row switches, and obtain truly upper and lower triangular matrices. Keep a record of what switches were made and apply the same switches to **b** before applying L and U to them. **16.** LU-GAUSS makes $(n^3+3n^2-n)/3$ assignments, because each assignment is associated with a multiplication or division counted in Table 8.1. GAUSS-SQUARE has n^2 more assignments, since the final value of each a_{ij} in A is assigned (0 or 1). GAUSS has $(n^3+5n)/6$ more than GAUSS-SQUARE: $(n^3-n)/6$ more because of the additional entries computed in Scale and Upsweep(unnecessarily for matrices with no 0's in pivot positions)and n more because of the subscripted column index c_r. **18.** Save each column at the time it becomes a pivot column or a nonpivot column. For instance, Gauss-Jordan reduction on $\begin{bmatrix} 1^* & 3 \\ 2 & 4 \end{bmatrix}$ gives $\begin{bmatrix} 1 & 3 \\ 0 & -2^* \end{bmatrix} \to \begin{bmatrix} 1 & 0 \\ 0 & 1 \end{bmatrix}$, so save $\begin{bmatrix} 1 & 3 \\ 2 & -2 \end{bmatrix}$. Similarly, for $\begin{bmatrix} 1 & 2 & 3 \\ 2 & 4 & 4 \\ 3 & 6 & 5 \end{bmatrix}$ save $\begin{bmatrix} 1 & 2 & 3 \\ 2 & 0 & -2 \\ 3 & 0 & -4 \end{bmatrix}$. What should you do if there are row switches?

Section 8.6

1. A doesn't reduce to I (a 0 appears on the main diagonal in the final matrix). **2. a)** $\frac{1}{2}\begin{bmatrix} 1 & -1 & 1 \\ 1 & 1 & -1 \\ -1 & 1 & 1 \end{bmatrix}$ **b)** No inverse **3.** None have inverses (there's a theorem lurking here). **4. b)** $\frac{1}{ad-bc}\begin{bmatrix} d & -b \\ -c & a \end{bmatrix}$ **5.** $ad-bc \neq 0$. **7.** $\begin{bmatrix} 1 & 1 & 0 \\ 0 & 1 & -2 \\ 1 & -2 & 1 \end{bmatrix}$ **9. a)** $AB = \begin{bmatrix} 5 & 7 & 7 \\ 7 & 7 & 5 \\ -2 & 0 & 2 \end{bmatrix}$. Since A is simpler, think of AB as sums of rows of B. E.g., first row of AB is the sum of rows 1 and 2 of B. **b)** $\begin{bmatrix} 4 & 3 & -1 \\ 8 & 9 & 1 \\ 4 & 5 & 1 \end{bmatrix}$ **11.** To divide row i by d, $[a_{ij}]$ is I except $a_{ii}=1/d$; to subtract ith row m times from jth row, $[a_{ij}]=I$ except $a_{ji}=-m$; to switch rows i and j, $[a_{ij}]=I$ except $a_{ii}=a_{jj}=0$, $a_{ji}=a_{ij}=1$. **13.** Column k of AB is $(a_{11},a_{21},\ldots,a_{m1})^{\mathrm{T}}b_{1k}+(a_{12},\ldots,a_{m2})^{\mathrm{T}}b_{2k}+\cdots+(a_{1n},\ldots,a_{mn})^{\mathrm{T}}b_{nk} = \sum_{j=1}^{n} \mathbf{a}_j b_{jk}$. **15.** We are given $z_i = \sum_j a_{ij}y_j$ and $y_j = \sum_k b_{jk}x_k$. Set $\mathbf{x}=(x_1,\ldots,x_m)^{\mathrm{T}}$, $\mathbf{y}=(y_1,\ldots,y_n)^{\mathrm{T}}$, $\mathbf{z}=(z_1,\ldots,z_p)^{\mathrm{T}}$, $A=[a_{ij}]$, $B=[b_{jk}]$, $AB=[c_{ik}]$. Then $\mathbf{z}=A\mathbf{y}$ and $\mathbf{y}=B\mathbf{x}$, so $\mathbf{z}=A(B\mathbf{x})=(AB)\mathbf{x}$. Thus each z_i is a linear combination of the x's because $z_i=\sum_k c_{ik}x_k$. **17.** If you do the same row operations on $[A\,|\,I]$ you get $[PA\,|\,PI]=[B\,|\,P]$, revealing P. **18. a)** $\begin{bmatrix} 1 & 2 \\ 3 & 4 \end{bmatrix}^{-1} = \frac{1}{2}\begin{bmatrix} -4 & 2 \\ 3 & -1 \end{bmatrix}$, so $\begin{bmatrix} 1 & 2 \\ 3 & 4 \end{bmatrix}^{-1}\begin{bmatrix} 2 & 3 \\ 4 & 5 \end{bmatrix} = \begin{bmatrix} 0 & -1 \\ 1 & 2 \end{bmatrix}$, $\begin{bmatrix} 2 & 3 \\ 4 & 5 \end{bmatrix}\begin{bmatrix} 1 & 2 \\ 3 & 4 \end{bmatrix}^{-1} = \frac{1}{2}\begin{bmatrix} 1 & 1 \\ -1 & 3 \end{bmatrix}$. **b)** $\begin{bmatrix} 1 & 2 \\ 3 & 4 \end{bmatrix}^{-1}\begin{pmatrix} 2 \\ 4 \end{pmatrix} = \begin{pmatrix} 0 \\ 1 \end{pmatrix}$; $\begin{pmatrix} 2 \\ 4 \end{pmatrix}\begin{bmatrix} 1 & 2 \\ 3 & 4 \end{bmatrix}^{-1}$ undefined. **19.** Second paragraph of the proof. **21. a)** All $d_i \neq 0$. **b)** Inverse is $\begin{bmatrix} 1/d_1 & & \\ & \ddots & \\ & & 1/d_n \end{bmatrix}$. **23.** $\frac{4}{3}n^3$ [find A^{-1}] $+\, n^2$ [multiplications in $A^{-1}\mathbf{b}$], so n^3 more work (plus lower-order terms) to use $A^{-1}\mathbf{b}$. **25.** Ignore your godmother's gift. Both LU method and multiplying $A^{-1}\mathbf{b}$ take n^2 mult/div steps, but for $A^{-1}\mathbf{b}$ you would also have to input A^{-1}. **27. a)** $AB = \begin{bmatrix} 5 & 8 \\ 11 & 18 \end{bmatrix}$, $\mathbf{x} = \begin{pmatrix} -5 \\ 7/2 \end{pmatrix}$. **b)** $\mathbf{d} = \begin{pmatrix} 2 \\ 1/2 \end{pmatrix}$, **x** as before. **30.** When applying GAUSS to $[A\,|\,I]$, nothing changes in column j on the right until the jth downsweep. **31.** If A^{-1} exists, our version takes $(n^3-n)/3$ steps on A and $(4n^3+3n^2-n)/6$

on right, for $n^3+(n^2-n)/2$ total. **33. a)** $[A\,|\,I] \to \left[\begin{array}{ccc|ccc} 1 & -1 & 2 & 1/2 & 0 & 0 \\ 0 & 1 & 2 & -1 & 1 & 0 \\ 0 & 1 & -1 & 1/2 & 0 & 1 \end{array}\right] \to \left[\begin{array}{ccc|ccc} 1 & 0 & 4 & -1/2 & 1 & 0 \\ 0 & 1 & 2 & -1 & 1 & 0 \\ 0 & 0 & -3 & 3/2 & -1 & 1 \end{array}\right] \to$ $\left[\begin{array}{ccc|ccc} 1 & 0 & 0 & 3/2 & -1/3 & 4/3 \\ 0 & 1 & 0 & 0 & 1/3 & 2/3 \\ 0 & 0 & 1 & -1/2 & 1/3 & -1/3 \end{array}\right]$. **b)** jth column on right not changed until jth pivot **c)** Our version takes exactly n^3 steps, so slightly better than Gaussian elimination (by lower-order terms). **35. a)** $L = \begin{bmatrix} 1 & & \\ 2 & 1 & \\ -1 & 1 & 1 \end{bmatrix}$, $D = \begin{bmatrix} 2 & & \\ & 1 & \\ & & -3 \end{bmatrix}$, $U = \begin{bmatrix} 1 & -1 & 2 \\ & 1 & 2 \\ & & 1 \end{bmatrix}$. **b)** The new U is the old U with each row scaled. **c)** They are multipliers in Upsweep if all scaling done first. **37. a)** Reverse the order: Shoes off first! **b)** Think of matrices as acting on column vectors by premultiplication. Then $AB\mathbf{x}$ acts on $\mathbf{x}$ by first "putting on" B and then putting on A. To reverse the process, first reverse the order. **39.** Squareness still necessary for invertibility. With the new definition, the second paragraph of the proof of Theorem 1 still shows that $A\mathbf{x}=\mathbf{b}$ has exactly one solution for each $\mathbf{b}$. But this happens iff $A \to I$, which can only happen if A is square.

Section 8.7

1. a) Yes, vector space **c)** Yes **f)** No, closed under neither scalars nor sums **h)** No, not closed under sums **3. a)** Yes **c)** No, closed under neither property **e)** Yes **g)** Yes **5. a)** Yes **d)** Yes **f)** No, e.g., not closed under scalar multiplier 0. **6. a)** $\left[\begin{array}{ccc|c} 1 & 3 & 5 & 2 \\ 2 & 2 & 6 & 3 \\ 3 & 1 & 7 & 4 \end{array}\right] \to \left[\begin{array}{ccc|c} 1 & 3 & 5 & 2 \\ 0 & -4 & -4 & -1 \\ 0 & 0 & 0 & 0 \end{array}\right]$, so Yes. **b)** $\left[\begin{array}{ccc|c} 1 & 2 & 3 & 1 \\ 3 & 2 & 1 & 1 \\ 5 & 6 & 7 & 1 \end{array}\right] \to \left[\begin{array}{ccc|c} 1 & 2 & 3 & 1 \\ 0 & -4 & -8 & -2 \\ 0 & 0 & 0 & -2 \end{array}\right]$, so No. **7.** $\{\mathbf{0}\}$ closed under scalars and sums **9.** Yes **11.** Standard vector form for the solutions to $A\mathbf{x}=\mathbf{b}$ gives a fixed vector ($\mathbf{t}$) added to the span of several others vectors ($\mathbf{v} \in V$). **13. a)** There is a pivot in every row; rank $=m$. **b)** There is a pivot in every column; rank $=n$. **15.** [closed under linear combinations] $\Longrightarrow$ [closed under sums and scalars] because sums and scalars are simple types of linear combinations. For the converse, to get $\sum c_i\mathbf{v}_i \in V$ if each $\mathbf{v}_i \in V$, first use scalar closure to get each $c_i\mathbf{v}_i \in V$; then use repeated summing (induction) to get each of $c_1\mathbf{v}_1+c_2\mathbf{v}_2, \ldots, \sum c_i\mathbf{v}_i$ in V. **17.** First definition would make Span(S) closed under sums and scalars from S, but not under sums and scalars from itself. Similarly, $\{c\mathbf{u}+d\mathbf{v}\}$ not closed under sums. **18.** With $\mathbf{u},\mathbf{v},\mathbf{w},\mathbf{x}$ column vectors, $[\mathbf{w},\mathbf{x}] = [\mathbf{u},\mathbf{v}]\begin{bmatrix} 2 & 3 \\ 3 & -4 \end{bmatrix}$ and $\mathbf{y} = [\mathbf{w},\mathbf{x}]\begin{pmatrix} 5 \\ 1 \end{pmatrix} = [\mathbf{u},\mathbf{v}]\begin{bmatrix} 2 & 3 \\ 3 & -4 \end{bmatrix}\begin{pmatrix} 5 \\ 1 \end{pmatrix} = [\mathbf{u},\mathbf{v}]\begin{pmatrix} 13 \\ 11 \end{pmatrix}$. **20. b)** As the notation $\mathcal{N}_{A^{\mathrm{T}}}$ suggests, find a basis for the null space of A^{T}. **21. a)** Given matrix $[A\,|\,\mathbf{b}]$ reduces to $\left[\begin{array}{ccc|c} 1 & 4 & 7 & b_1 \\ 0 & -3 & -6 & b_2-2b_1 \\ 0 & 0 & 0 & b_1-2b_2+b_3 \end{array}\right]$, so $\mathbf{b} \in \mathcal{C}_A \iff A\mathbf{x}=\mathbf{b}$ solvable $\iff b_1-2b_2+b_3=0$.

Section 8.8

1. Yes, Yes, No, Yes. The null space of $[\mathbf{v}_1,\mathbf{v}_2,\mathbf{v}_3,\mathbf{v}_4]$ is all multiples of $(-1,0,-1,1)^{\mathrm{T}}$. Thus $\{\mathbf{v}_1,\mathbf{v}_2,\mathbf{v}_3, \mathbf{v}_4\}$ is dependent, indeed, each of $\mathbf{v}_1,\mathbf{v}_3,\mathbf{v}_4$ depends on the other two. But $\mathbf{v}_2$ does not depend on the others, for then there would be a solution to $\sum c_i\mathbf{v}_i=\mathbf{0}$ with $c_2 \neq 0$. **3.** Three vectors in R^2 are always dependent by Theorem 2. **4.** Parts a–c can be answered by reducing A. To solve all five problems at once, reduce the matrix $\left[\begin{array}{ccc|c} 1 & 3 & 5 & 2 \\ 2 & 2 & 6 & 3 \\ 3 & 1 & 7 & 4 \\ \hline 1 & 1 & 1 & * \end{array}\right]$, where it makes no difference what $*$ is. Two pivots yield $\left[\begin{array}{ccc|c} 1 & 3 & 5 & 2 \\ 0 & -4 & -4 & -1 \\ 0 & 0 & 0 & 0 \\ \hline 0 & 0 & -2 & * \end{array}\right]$, and you can stop. **a)** $(1,3,5)$, $(0,-4,-4)$ **b)** $(1,2,3)^{\mathrm{T}}$, $(3,2,1)^{\mathrm{T}}$ **c)** 1 **d)** Yes **e)** No, appending row (1,1,1) to A increased the rank. **6. a)** First, third, fourth columns of A **b)** $(-1,1,0,0)^{\mathrm{T}}$ **c)** First three rows of A. To show this from Theorem 5 you must reduce A^{T}, but can you justify it by reducing A? **d)** $(1,1,0,0)$, $(0,0,1,0)$, $(0,0,0,1)$ **8. a)** p **c)** $n-p$ **9. a)** Let the set be $\mathbf{v}_1=\mathbf{0}, \mathbf{v}_2, \ldots, \mathbf{v}_k$. Then $6\mathbf{v}_1+0\mathbf{v}_2+\cdots+0\mathbf{v}_k=\mathbf{0}$ and not all coefficients are 0. **11.** The $\mathbf{v}$'s in the solution to [9a]. **13.** Suppose $S=\{\mathbf{v}_1,\ldots,\mathbf{v}_k\}$ and $\mathbf{u}=\sum c_i\mathbf{v}_i$. Let $\mathbf{w}=\sum_{i=1}^k d_i\mathbf{v}_i+d_{k+1}\mathbf{u}$ be anything in Span($S \cup \{\mathbf{u}\}$). Then$\mathbf{w}=\sum_{i=1}^k (d_i+d_{k+1}c_i)\mathbf{v}_i \in$ Span(S). **15.** $\mathcal{C}_A=\mathcal{C}_B$ so $\dim(\mathcal{C}_A)=\dim(\mathcal{C}_B)=$ number of pivots. **17.** Let $W=\mathrm{Span}\{\mathbf{v}_1,\ldots,\mathbf{v}_m\}$. Then $\dim(W) \le m$. Apply Lemma 9. **19.** 6 **21.** They may not be inside the subspace of R^n that interests you, e.g., if it's $\mathcal{R}_A$ for $A=\begin{bmatrix}1&1&1\\1&2&3\end{bmatrix}$. **23.** Let W be the reduced echelon form of A. Part 1 (nonzero rows of W are a basis of $\mathcal{R}_A$): $\mathcal{R}_A=\mathcal{R}_W$ by Lemma 6. Independence of the nonzero rows easier than in text. Part 2: position of pivot columns doesn't change from U to W. Part 3: doesn't mention U. **25.** Similar to proof associated with Fig. 8.6. Suppose the matrix is upper triangular. If $\sum c_i\mathbf{r}_i=\mathbf{0}$. then $c_1=0$ because only $\mathbf{r}_1$ has nonzero first coordinate. Then ignoring $\mathbf{r}_1$, it follows $c_2=0$. And so on. **27.** Show that $\sum_{i=1}^{k+1} c_i\mathbf{v}_i=\mathbf{0} \Longrightarrow$ all $c_i=0$. First $c_{k+1}=0$; otherwise $\mathbf{v}_{k+1}$ depends on the others. Then $\sum_{i=1}^{k} c_i\mathbf{v}_i=\mathbf{0}$ so $c_1=\cdots=c_k=0$. **29. a)** At each iteration S is an independent set by [27]. GROW-BASIS must terminate since no independent set in R^n has $>n$ vectors. When it terminates, S also spans V. **31. a)** If not a basis, they wouldn't span, so GROW-BASIS would continue to a larger independent set; this contradicts Lemma 9. **b)** Write the vectors as columns of A. Since the columns are independent, there are n pivots, $A\rightarrow I$, and so $\mathcal{C}_A=R^n$. **32. a)** Columns span $R^m \Longleftrightarrow m$ pivots $\Longleftrightarrow \dim(\mathcal{R}_A)=m \Longleftrightarrow$ the rows of A independent. **b)** Columns independent $\Longleftrightarrow n$ pivots $\Longleftrightarrow \dim(\mathcal{R}_A)=n \Longleftrightarrow \mathcal{R}_A=R^n$. **34.** No. Let $A=\left[\frac{M}{N}\right]$. Then $\mathcal{R}_N\subset\mathcal{R}_M \Longleftrightarrow \mathcal{R}_A=\mathcal{R}_M$. **37.** $a\begin{pmatrix}1\\2\\3\end{pmatrix}+b\begin{pmatrix}2\\3\\1\end{pmatrix}$ is in $U\cap V \Longleftrightarrow$ there are c,d such that $\begin{bmatrix}1&2&3&1\\2&3&1&0\\3&1&2&-1\end{bmatrix}\begin{pmatrix}a\\b\\c\\d\end{pmatrix}=\mathbf{0}$. A basis of $U\cap V$ is $\begin{pmatrix}0\\1\\5\end{pmatrix}$. **39.** See solution for [4]. The extra row $\mathbf{v}$ receives a pivot iff $\mathbf{v}\notin\mathcal{R}_A$. **40.** Dependence preserved: Since $\sum c_i\mathbf{v}_i$ is computed coordinatewise, if each coordinate is 0 before truncation, the remaining coordinates are 0 after truncation. Independence not preserved: truncate $\begin{pmatrix}1\\0\end{pmatrix},\begin{pmatrix}1\\1\end{pmatrix}$ to one coordinate. **42.** Contrapositive: If an independent set S in V is not a basis, then $V-\text{Span}(S)\neq\emptyset$. By [27], for any $u\in V-\text{Span}(S)$, $S\cup\{\mathbf{u}\}$ is independent, so S not maximal.

Section 8.9

1. $A^3\begin{pmatrix}1000\\1000\end{pmatrix}=\frac{1000}{216}\begin{bmatrix}87&86\\129&130\end{bmatrix}\begin{pmatrix}1\\1\end{pmatrix}\approx\begin{pmatrix}801\\1199\end{pmatrix}$. **2. a)** $7\mathbf{v}$ **c)** $2^3\mathbf{u}+3\cdot7^3\mathbf{v}=8\mathbf{u}+1029\mathbf{v}$. **4.** No; No. **6. a)** $\lambda=2$, $\mathbf{v}=\begin{pmatrix}1\\1\end{pmatrix}$; $\lambda=0$, $\begin{pmatrix}-1\\1\end{pmatrix}$. **c)** Only one λ, with only one-dimensional eigenspace: $\lambda=1$, $\mathbf{v}=\begin{pmatrix}1\\0\end{pmatrix}$. **8.** For A square, A^{-1} exists $\Longleftrightarrow \mathcal{N}_A=\emptyset \Longleftrightarrow \mathcal{N}_{A-0I}=\emptyset \Longleftrightarrow$ 0 not an eigenvalue; $A-kI$ invertible $\Longleftrightarrow k$ not an eigenvalue. **9.** $\lambda=4$; if λ is an eigenvalue of A, then $\lambda-k$ is an eigenvalue of $A-kI$. **11.** $A\mathbf{0}=\mathbf{0}=\lambda\mathbf{0}$ for every λ, so if $A\mathbf{0}=\lambda\mathbf{0}$ made λ an eigenvalue, every scalar would be an eigenvalue. **13.** $\mathcal{N}_{A-\lambda I}\neq\emptyset \Longleftrightarrow \begin{pmatrix}a-\lambda\\c\end{pmatrix},\begin{pmatrix}b\\d-\lambda\end{pmatrix}$ dependent $\Longleftrightarrow \left[\text{either } k\begin{pmatrix}a-\lambda\\c\end{pmatrix}=\begin{pmatrix}b\\d-\lambda\end{pmatrix} \text{ or } \begin{pmatrix}a-\lambda\\c\end{pmatrix}=k\begin{pmatrix}b\\d-\lambda\end{pmatrix}\right]$. To show the last condition equivalent to $(a-\lambda)(d-\lambda)-bc=0$ is easy if all of $a-\lambda, b, c, d-\lambda$ nonzero: $(a-\lambda)(d-\lambda)-bc=0 \Longleftrightarrow b/(a-\lambda)=(d-\lambda)/c \Longleftrightarrow \left[k\begin{pmatrix}a-\lambda\\c\end{pmatrix}=\begin{pmatrix}b\\d-\lambda\end{pmatrix} \text{ with } k=b/(a-\lambda)\right]$. If one or more of $a-\lambda, b, c, d-\lambda$ is zero, you must consider cases. **15.** $A\mathbf{v}=\lambda\mathbf{v}\Longleftrightarrow\mathbf{v}=A^{-1}\lambda\mathbf{v}=\lambda A^{-1}\mathbf{v}\Longleftrightarrow\lambda^{-1}\mathbf{v}=A^{-1}\mathbf{v}$. **16.** Find the eigenvalue of A^{-1} with largest absolute value(MAX-EIGEN will find it because it will be unique) and reciprocate. **18. b)** Pivots are $-\lambda$, $(1-\lambda^2)/\lambda$ and $(\lambda^2-\lambda-2)/(1-\lambda)$ for a product of $-(\lambda+1)^2(\lambda-2)$. (For easier algebra, switch rows 1 and 3 before reduction; pivots are now 1, $-(\lambda+1)$, $-(\lambda^2-\lambda-2)$.) Eigenvalues are $-1, 2$. **20.** In Section 8.2, each multiplication or division of things like m_k and a_{ij} was considered a single real operation, but if m_k and a_{ij} are rationalfunctions of λ,then many real operations are involved in each such multiplication and division. **21. a)** $\lambda=1/6$, $\mathbf{v}=\begin{pmatrix}1\\-1\end{pmatrix}$; $\lambda=1$, $\mathbf{v}=\begin{pmatrix}2\\3\end{pmatrix}$. **c)** $\mathbf{u}_n=(200/6^n)\begin{pmatrix}1\\-1\end{pmatrix}+400\begin{pmatrix}2\\3\end{pmatrix}$. **22.** $\mathbf{u}_n=\begin{pmatrix}r_{n+1}\\r_n\end{pmatrix}$, $\mathbf{u}_n=\begin{bmatrix}5&-6\\1&0\end{bmatrix}\mathbf{u}_{n-1}$, from which $\mathbf{u}_n=5\begin{pmatrix}2\\1\end{pmatrix}2^n-\begin{pmatrix}3\\1\end{pmatrix}3^n$, so $r_n=5\cdot2^n-3^n$. **24.** Let $\mathbf{u}_n=(s_n,t_n,s_{n-1},t_{n-1},t_{n-2})^{\mathrm{T}}$. Then $\mathbf{u}_n=\begin{bmatrix}2&1&0&-3&0\\0&1&3&0&-1\\1&0&0&0&0\\0&1&0&0&0\\0&0&0&1&0\end{bmatrix}\mathbf{u}_{n-1}$. Five geometric solutions. **26.** They would tell us the (decimal) fraction of the total population in each age group. **28.** The relative sizes still stabilize—the limit vector is $\mathbf{x}=(1,.9034,.7255,.5462)$—and the animal still dies out, slightly more slowly: $\lambda=.9962$. **30.** In the first row (perhaps columns 10 to 50) and in the $(j+1,j)$ entries. **31.** Example 6: $\mathbf{u}_{25}=(1.0570,.9575,.7711,.5433)$, population 3329. Example 7: $\mathbf{u}_{25}=(.9782,.8861,.7136,.5028)$, population 3081. **33.** You were doing a variant of MAX-EIGEN (different scaling) and so the outputs converge to an eigenvector for the largest eigenvalue. **35.** $\mathbf{x}$ cycles: $(x_1,x_2)\to(x_1,-x_2)\to(x_1,x_2)$. **36. a)** $\lambda=i$, $\mathbf{v}=\binom{i}{1}$; $\lambda=-i$, $\mathbf{v}=\binom{-i}{1}$. **b)** $\mathbf{x}$ cycles $(x_1,x_2)\to(-x_2,x_1)\to(-x_1,-x_2)\to(x_2,-x_1)\to(x_1,x_2)$ because the eigenvalues have equal magnitude. **38. a)** Characteristic polynomial is $(1-\lambda)(-\lambda)-1=\lambda^2-\lambda-1$. **c)** MAX-EIGEN converges to $\mathbf{v}_1$ even when we use high precision. Unlike in [37], we cannot start with an exact numerical eigenvector for the smaller eigenvalue. **40.** $P(n-1)\Longrightarrow P(n)$: $\sum_{i=1}^n c_i\mathbf{v}_i=\mathbf{0}\Longrightarrow\sum_{i=2}^n c_i(\lambda_i-\lambda_1)\mathbf{v}_i=\mathbf{0}\overset{P(n-1)}{\Longrightarrow}$ for $i\ge2$, $c_i(\lambda_i-\lambda_1)=0\Longrightarrow c_2=c_3=\cdots=c_n=0$. What about c_1 and what about P(1)? **42.** $\lambda_1=$ the ratio of $A\mathbf{v}_1$ to $\mathbf{v}_1$. But $A\mathbf{v}_1$ has real entries since A and $\mathbf{v}_1$ do. So λ is real.

Section 8.10

1. For Example 1, $A-I=\begin{bmatrix}-1/2 & 1/3\\ 1/2 & -1/3\end{bmatrix}$ and clearly $(2,3)\in\mathcal{N}_{A-I}$. **2.** We want the second entry of $A^4(0,0,1,0)^{\mathrm{T}}$, which is $305/(6^4)\approx .2353$. **3.** $A=\dfrac{1}{3}\begin{bmatrix}0&1&1&1\\1&0&1&1\\1&1&0&1\\1&1&1&0\end{bmatrix}$; $\mathbf{w}=\dfrac{1}{4}\begin{pmatrix}1\\1\\1\\1\end{pmatrix}$. **5. a)** $A=\begin{bmatrix}.8&.2&.25\\.1&.6&.5\\.1&.2&.25\end{bmatrix}$. The states are the work classes; transition is from one generation to the next. **b)** $A^2(.1,.5,.4)^{\mathrm{T}}=(.3785,\ .4390,\ .1825)^{\mathrm{T}}$ **c)** $\mathbf{w}=\frac{1}{77}(40,25,12)^{\mathrm{T}}\approx(.5195,\ .3247,\ .1558)^{\mathrm{T}}$. **7.** The states are $0,1,\ldots,m+n$, representing how many dollars A has. All the matrix entries a_{ij} are 0 except $a_{00}=a_{m+n,m+n}=1$, and for $1\le i<m+n$, $a_{i-1,i}=1-p, a_{i+1,i}=p$. Theorem 4 relevant since two states are absorbing. **9.** Let gene G be dominant, g recessive.Each parent can be GG, Gg or gg, so the states are the $3^2=9$ male/female pairs GG/GG through gg/gg. Sample entries in the 9×9 matrix are: row GG/GG, column GG/GG: 1; row Gg/Gg, column Gg/Gg: 1/4; row Gg/gg, column Gg/Gg: 1/8. **10.** $\left[\begin{smallmatrix}0&1\\1&0\end{smallmatrix}\right]$. Unique matrix for those digraphs in which each vertex has a unique out-edge. **13.** regular, neither, neither **15. a)** Let $p_{ij}^{(n)}$ be the ij entry of A^n. Let $P(n)$ be the statement$\Pr(E_i^{(n)}\,|\,E_j^{(0)})=p_{ij}^{(n)}$. Inductive step: $\Pr(E_i^{(n+1)}\,|\,E_j^{(0)})=\sum_k\Pr(E_i^{(n+1)}\cap E_k^{(n)}\,|\,E_j^{(0)})=\sum_k\Pr(E_i^{(n+1)}\,|\,E_k^{(n)}\cap E_j^{(0)})\Pr(E_k^{(n)}\,|\,E_j^{(0)})$ [25, Section 6.3] $=\sum_k\Pr(E_i^{(n+1)}\,|\,E_k^{(n)})\Pr(E_k^{(n)}\,|\,E_j^{(0)})$ (Why?) $\overset{P(n)}{=}\sum_k p_{ik}p_{kj}^{(n)}=p_{ij}^{(n+1)}$. This is like the proof of Theorem 1, Section 3.3 (number of paths of length n from v_i to v_j). **b)** $p_i^{(n)}=\Pr(E_i^{(n)})=\sum_j\Pr(E_i^{(n)}\,|\,E_j^{(0)})\Pr(E_j^{(0)})=\sum_j p_{ij}^{(n)}p_j^{(0)}$. **16.** MAX-EIGEN; Gaussian elimination takes around $1000^3/3$ steps, whereas MAX-EIGEN takes essentially 1000^2i, where i is the number of times $\mathbf{x}=A\mathbf{x}$ must be iterated until convergence. i will be much less than 1000/3 (unless some $\lambda\ne1$ has $|\lambda|$ very close to 1) because entries of $\mathbf{x}$ are sums of termsfrom geometric series. **18.** Note that $(A-\lambda I)^{\mathrm{T}}=A^{\mathrm{T}}-\lambda I$. So if A is $n\times n$, then λ an eigenvalue of $A\iff\operatorname{rank}(A-\lambda I)<n\iff\operatorname{rank}(A^{\mathrm{T}}-\lambda I)<n\iff\lambda$ an eigenvalue of A^{T}. **20. a)** First: $S(A\mathbf{x})=\sum_i\sum_j a_{ij}x_j=\sum_j x_j\sum_i a_{ij}=\sum_j x_j$. Second: $S(A\mathbf{x})=\mathbf{y}(A\mathbf{x})=(\mathbf{y}A)\mathbf{x}\overset{[19a]}{=}\mathbf{y}\mathbf{x}=S(\mathbf{x})$. **22.** $N=\begin{bmatrix}36/11&30/11\\30/11&36/11\end{bmatrix}$. Expected number of tosses is sum of first column, 6.

Supplementary Problems: Chapter 8

1. a) $\left[\begin{smallmatrix}*&\times&\times\\0&*&\times\end{smallmatrix}\right],\left[\begin{smallmatrix}*&\times&\times\\0&0&*\end{smallmatrix}\right],\left[\begin{smallmatrix}*&\times&\times\\0&0&0\end{smallmatrix}\right],\left[\begin{smallmatrix}0&*&\times\\0&0&*\end{smallmatrix}\right],\left[\begin{smallmatrix}0&*&\times\\0&0&0\end{smallmatrix}\right],\left[\begin{smallmatrix}0&0&*\\0&0&0\end{smallmatrix}\right],\left[\begin{smallmatrix}0&0&0\\0&0&0\end{smallmatrix}\right]$, **c)** 2^n **2. a)** A has the same solutions as C but not the same as B or D. **c)** Suppose x_i is a free variable in system $\mathcal{S}$ and a pivot variable in$\mathcal{T}$. Set all free variables to the right of x_i to 0; hence for all $j>i$, $x_j=0$. Then in $\mathcal{T}$, x_i must be 0, but in $\mathcal{S}$ it may be anything. Now, what if the pivot variables in both systems are the same, as with A and B in part a)? **3.** Everything in L assigned; $n(n+1)/2$ steps on $\mathbf{b}$. **5.** $(n^3+3n^2+2n)/6$ steps from $[L\,|\,I]$ to $[I\,|\,L^{-1}]$. **7.** Yes

9. Suffices to check if $(1,-2,1)$ and $(1,1,2)$ are themselves in the null space; (1,1,2) is not. **11.** Suffices to test if a basis of $\mathcal{N}_N$ is in $\mathcal{N}_M$. Yes, $(1,1,1,1) \in \mathcal{N}_M$. **13. a)** Both claims equivalent to the appearance of a free variable during reduction of A. **b)** $A(\mathbf{x}-\mathbf{x}') = \mathbf{0}$, so $[\mathbf{x}, \mathbf{x}'$ solve $A\mathbf{x} = \mathbf{b}] \implies [(\mathbf{x}-\mathbf{x}'), \mathbf{0}$ solve $A\mathbf{x} = \mathbf{0}]$. This proves $\implies$. To prove $\impliedby$, define $\mathbf{b}$ to be $A\mathbf{x}$. Then $A\mathbf{x}' = A\mathbf{x} - A(\mathbf{x}-\mathbf{x}') = \mathbf{b}+\mathbf{0} = \mathbf{b}$. **15.** Let Perm($i$) be the index of the variable currently associated with column i. The final permutation is obtained from the identity by a number ofinterchanges, much like PERMUTE-3 in Section 4.9. **16. b)** $n \times m$ **c)** $\begin{bmatrix} -\frac{5}{3}+a & \frac{2}{3}+b \\ \frac{4}{3}-2a & -\frac{1}{3}-2b \\ a & b \end{bmatrix}$, where a, b arbitrary. **18.** Any all 0 matrix; also $\left[\begin{smallmatrix} 1 & 1 \\ 1 & 1 \end{smallmatrix}\right]$ and many others. **20.** For each right inverse A_R, $A_R\mathbf{b}$ is a solution, since $A(A_R\mathbf{b}) = (AA_R)\mathbf{b} = \mathbf{b}$. **23. a)** $d = 0$. **b)** Closed under scalars because if (x, y, z) satisfies $ax+by+cz = 0$, then $a(kx)+b(ky)+c(kz) = k(ax+by+cz) = k \cdot 0 = 0$. What about sums? **25.** $\mathbf{v}$ depends on S: $\mathbf{v} = \sum_{\mathbf{u} \in S} c_\mathbf{u}\mathbf{u}$ where only a finite number of the c's are nonzero. S dependent: $\sum_{\mathbf{u} \in S} c_\mathbf{u}\mathbf{u} = \mathbf{0}$ is true with a *finite, positive* number of nonzero c's. Proof of Theorem 1, Section 8.8 virtually unchanged. **27.** $n = 1$ for all m. **29.** Given $\mathbf{0} = c_1\mathbf{u}+c_2(\mathbf{u}+\mathbf{v})+c_3(\mathbf{u}+\mathbf{v}+\mathbf{w}) = (c_1+c_2+c_3)\mathbf{u}+(c_2+c_3)\mathbf{v}+c_3\mathbf{w}$, use the independence of $\mathbf{u}, \mathbf{v}, \mathbf{w}$ to get homogeneous equations in the c's and show that $c_1 = c_2 = c_3 = 0$. **31. b)** If $\sum c_i\mathbf{v}_i = \mathbf{0}$ with all $c_i \neq 0$, then the equation may be solved for each $\mathbf{v}_i$. **33.** One example: Theorem 5, Section 8.8, would be false if A is an all 0 matrix. Another example: $\{\mathbf{0}\}$ wouldn't have a basis (it's basis is $\emptyset$) unless $\text{Span}(\emptyset) = \{\mathbf{0}\}$.For $\{\mathbf{0}\}$ not to have a basis would contradict Corollary 7, Section 8.8. **35.** Let c be the entry in A with the largest absolute value. If A is $n \times n$, apply MAX-EIGEN to $A+n|c|I$.

Chapter 9

Section 9.1

2. a) $P = 1172.27$; 8.27%; **b)** $P = 670.05$; 10.25% **3.** $P = 1010.04$; effective interest rate: 1.004%; with 1% interest all terms after second in binomial expansion are too small to contribute anything significant **5.** Basic calculation in both parts: $P \leftarrow (1+.06/52)P$; **7. a)** Successive iterates are 1.5, 1.75, 1.875, 1.9375, ... and converge to 2 **8. a)** Get oscillation without convergence for values $< (1+\sqrt{5}/2$ or growth to ∞ for values $> (1+\sqrt{5}/2$; **b)** Converges rapidly to $(1+\sqrt{5}/2 \approx 1.618$ for any choice of starting value in [0,3]; to see why b) gives better results, consider $x_{n+1} - x_n$ in the two cases.

Section 9.2

1. b) 2; **d)** 1 **2. b)** Diverges; $(-1)^n$ term prevents convergence; **d)** Diverges; magnitude of successive terms grows without bound **5.** $t_n = (225/2)(.2)^n + (15/2)$; typos don't die out but converge to 15/2 (how come it's not an integer?) **6. b)** Counterexample: $a_n = (-1)^n$ **7.** Yes; it can oscillate as with $a_n = (-1)^n$. **9. a)** Just replace pluses by minuses on left hand side of (8); **c)** $||a_n| - |a|| = |a_n - a|$ for sufficiently large n since a_n and a will have the same sign (except when $a = 0$) **11.** Since monotonically decreasing and bounded below by 0, must converge; suppose converges to $L > 0$; then for any $\epsilon > 0$, $r^{n+1} - L = r^{n+1} - r^n + r^n - L = r^n(r-1) + r^n - L \leq (L+\epsilon)(r-1)+\epsilon = \epsilon r + L(r-1) < 0$ for sufficiently small ϵ but r^n converges to L from above; therefore $L = 0$, **13. a)** Behavior determined by $A_i r_i^n$ for first nonzero A_i; **b)** Behavior determined by $A_1 r_1^n + A_2(-r_1)^n$; **14. b)** If the sequence converges to L, then the second player can always win; even when the sequence does not converge to L, the second player may be able to win if the sequence gets sufficiently close to L. **15.** True; if *all* a_n after some N are *arbitrarily* close to L, then they must all equal L. **17. a)** Since $\{a_k\}$ is monotonically increasing, $\{A_k\}$ can never decrease; since $\{a_k\}$ is bounded above, so is $\{A_k\}$; therefore once A_k equals the integer part of L, it always equals this **19.** Difference of two successive terms by itself says *nothing* about how close you are to the true limit; in fact difference can get and stay arbitrarily small without sequence even converging; on the other hand, if sequence converges, difference will get arbitrarily small **21. a)** ≥ 1; **b)** Diverges if limit > 1; can't say anything (!) if limit is 1; [see Example 3, Section 9.5]; **c)** ≤ 1; **d)** converges whatever limit is since sequence bounded below by 0; **22. a)** Converges since monotonically decreasing after $n = 3$; limit 0; **b)** Diverges since $\lim a_{n+1}/a_n = \infty$ **23. a)** $a_n = 3 - 1/2^{n-1}$; proof by induction; **b)** Since

$1/n! \le 1/2^{n-1}$ for all $n \ge 1$, sequence is bounded above and since also monotonic, it converges; **c)** Limit is e, **24. a)** $a_n = n$; $b_n = n^2$; **c)** $a_n = -83.2n$; $b_n = n$, **25. b)** $a_n = 1/n$; $b_n = 1/(n+1)$; **d)** $a_n = 1/n$; $b_n = 1/n^2$, **27.** $|(1/\sqrt{5}[(1-\sqrt{5})/2]^{n+1}|$ is a decreasing sequence and first term is less than 1/2; therefore, since f_n is an integer, $(1\sqrt{5})[(1+\sqrt{5})/2]^{n+1}$ is always within 1/2 of actual f_n, **29. a)** 1; **b)** $a_n = [(1+1/n^2)^{n^2})]^{1/n}$ so $\log_e a_n = (1/n)\log_e(1+1/n^2)^{n^2}$ whose limit is $0 \cdot 1 = 0$ so limit of a_n is 1.

Section 9.3

1. a) n^3; **d)** n^2; **g)** n; **j)** $[(1.01)^{1/100}]^n$; **2. a)** Since for Ord lim $a_n/b_n = c$, for any $\epsilon > 0$ there is some N that $|c| - \epsilon \le |a_n/b_n| \le |c| + \epsilon$ for all $n > N$, **3. b)** Negative; use Theorem 1 to show that $\text{Ord}(n^2) < \text{Ord}(n^3/\log n)$, **5.** Divide numerator and denominator of $P(n)/Q(n)$ by n^j, then show $\lim[P(n)/Q(n)]/n^{i-j}$ is ratio of leading coefficients of $P(n)$ and $Q(n)$; if $i \le j$ there is an order relationship only if we allow negative exponents in second standard sequence in (3) **7.** No; consider $\lim a_n/b^n = \lim[a_n/a^n][a^n/b^n]$ and use (iii) of Theorem 1 of Section 9.2; **9.** Let $m = \log n$ and use (iv) of Theorem 1 **11.** $O_n = H_n - (1/2)H_{\lfloor n/2 \rfloor}$; therefore, $\lim O_n/H_n = \lim(1 - (1/2)H_{\lfloor n/2 \rfloor}/H_n)$; since $H_{\lfloor n/2 \rfloor} < H_n$ and we have assumed limit exists, it must be between 1/2 and 1 which shows $O_n = \text{Ord}(H_n)$ **12. b)** If (7) is true, $0 = \lim \log n/n = \lim n^r \log n/n^{r+1}$; also $\lim n^r/(n^r \log n) = \lim 1/\log n = 0$ **13. a)** With $m = \log_2 n$ limit in [10b] is m^j/a^m; since limit is 0, (iv) of Theorem 1 is proved; **b)** Follows from (ii) and (iv) of Theorem 1 **15. a,b)** $\text{Ord}(2^{\log_2 n}) < \text{Ord}(n^2) = \text{Ord}[(n+1)^2] = \text{Ord}(1.1n^2 - \log_2 n) < \text{Ord}(n^3/\log_2 n) < \text{Ord}(n^3) < \text{Ord}[(\log_2 n)^{\log_2 n}] < \text{Ord}(n^{\log_2 n}) < \text{Ord}(2^n) < \text{Ord}(n!)$; to see next to last $<$, take logarithms of both and use [10b]; last inequality follows by comparing product of n 2's with $n!$; **c)** Those connected by equal signs above; **d)** Yes, if $=$ allowed and if limit of ratio of any pair exists **16. b)** Prove using (iii) of Theorem 1, Section 9.2; **d)** Prove using (iv) of Theorem 1, Section 9.2. **17. a)** 45 since 2^{45} just less than 3.6×10^{13}, number of operations in 10 hours; **b)** 514 **19. a)** About .7; **b)** About 37.

Section 9.4

1. a) $2n-1$; **d)** $2^n(n+2)$; **g)** $2^n(n+4)$; **j)** $2^n(n+2k)$; prove by induction; **3. a)** Use Binomial Theorem to show that $\Delta n^j = (n+1)^j - n^j = n^{j-1} +$ terms with smaller exponents from which result follows **5. a)** $\Delta^2 a_n + 2\Delta a_n + a_n$; **b)** $\Delta^3 a_n + 3\Delta^2 a_n + 3\Delta a_n + a_n$; **c)** $a_{n+k} = \sum_{j=0}^{k} C(k,j)\Delta^j a_n$; prove by induction on k; express subscript $n+(k+1)$ as $(n+1)+k$ and express $j+1$ difference at a_{n+1} in terms of jth and $j+1$ differences at a_n, **7.** Because of the simple relationship $\Delta C(x,k) = C(x,k-1)$. **8. b)** $n(n+1)(n+2)/3$ **9. b)** $1/4 - 1/[2(n+1)(n+2)]$ **10. a)** Just show by taking differences that, if antidifference of $\{a_n\}$ is $\{A_n\}$ and of $\{b_n\}$ is $\{B_n\}$, then antidifference of $\{ca_n + db_n\}$ is $\{cA_n + dB_n\}$, **12.** Using [11b] get $(m+1)^3/3 - (m+1)^2/2 + (m+1)/6$, **14. b)** 0; **d)** $-45/64$, **15. b)** $\Delta x^{(k)} = k(x+1)^{(k-1)}$; prove by calculating difference **17. a)** 6720; **c)** $-15/16$, **18.** $x^{(k)} = (x+k-1)_{(k)}$ **19. b)** $4^n + (-1)^n$, **20. b)** $\Delta^2 r_n + \Delta r_n = 2r_n$ **21.** Just substitute for each Δ term its equivalent using a_i terms.

Section 9.5

1. b) n^2; **d)** $2^n - 1 + 3n(n-1)/2$, **2. b)** $\sum_{k=1}^{\infty} 1/2^k$ converges to 1. **3.** You can't say anything; example: $a_k = -b_k = (-1)^k$; doesn't matter if $+$ is changed to $-$, **5.** Not necessary; consider $a_n = 1/n$; but it is sufficient, **7. a)** Partial sums are monotone increasing and sum of first series is bounded above by multiple of sum of second series; **b)** If terms of a series of nonnegative terms are each greater than multiple of corresponding terms of a divergent series, the series diverges, **9. a)** $\{(2^{2i}-1)/3 \cdot 2^{2i}\}$, $\{(2^{2i}-1)/3 \cdot 2^{2i-1}\}$ **b)** Infinitely far **10.** How far should be easy; where requires the sum of two geometric series. **11. a)** $1/(1-c)$; **b)** Large c, **12. b)** Use [7a]. **13. b)** Assume contrary and show that each d_i always greater than some positive constant; **c)** By induction; assume amount added from n^2+1 tap to $(n+1)^2$ tap is not 1 or more, use induction hypothesis to show each d_i from n^2+1 on is greater than $1/(n+2)$ and use this to get a contradiction; **d)** By induction; use same idea as in c), **14. b)** Use the identity $C(n,k) = C(n+1,k+1) - C(n,k+1)$ and work from $2k$ term back to k term; sum is $C(2k+1,k+1)$ **15. b)** $N+1$, **17. a)** Since $1/(2i-1)-1/2i=1/[(2i-1)2i]$; **b)** Since $1/[(2i-1)2i] < 1/(2i-1)^2$, **19. a)** Just change ϵ to 2ϵ in **repeat** statement. **20. a)** False; $\sum(-1)i/\sqrt{i}$ converges since it is an alternating series whose terms decrease in magnitude to 0 but series with terms squared is harmonic series which diverges, **21. b)** $10^k D-D=R_k$; $D=R_k/(10^k-1)$; **c)** (ii): $D=147251/999999=11327/76923$, **22. b)** (ii): $D=978/990=163/165$, **23.** Sum on right hand side of (11) now involves k^2 which makes things worse; without care in choosing u_k and Δv_k, things are as likely to get worse as better using summation by parts, **25. a)** 2; just use $r=1/2$ in result of Example 5; **c)** $n(n^2-1)(3n+2)/12$. **27.** Using $C(i,k)=\Delta C(i,k+1)$ from Table 9.2, series telescopes to $C(j+1,k+1)-C(k,k+1)$ but latter term is 0, **29. a)** 1/12. **30. a)** Using $k_{(2)}$ faster converging sum is $2-\sum_{k=2}^{\infty} 1/[k^2(k-1)]$; **c)** Using $k_{(2)}$ faster converging sum is $3/2-\sum_{k=2}^{\infty}(k^2+2k-1)/(k^3+2k-1)k(k-1)$, **31.** True result must lie between results given by two series. **33. a)** Faster converging sum is $11/8-\sum_{k=3}^{\infty}(3k-2)/k^3(k-1)(k-2)$, **35.** Through $n=14$. **37.** Changing order of summation a little tricky because case $j=1$ must be treated separately, **39.** $n(n+1)H_n/2-n(n-1)/4$; **40.** Try to convert this into something which telescopes.

Section 9.6

1. a) $1/(1-2s)$; **d)** $1/(1+s^3)$, **3.** $(1-s^{11})/(1-s)$. **5. a)** $a_k=(-1)^k$; **d)** $a_k=0, k=0,1,2$; $a_k=1$, $k>2$; **g)** $a_k=2+3\cdot 2^k$; **j)** $1/(1-s-s^2)=(1+\sqrt{5})/\{2\sqrt{5}[1-(1+\sqrt{5})s/2]\}-(1-\sqrt{5})/\{2\sqrt{5}[1-(1-\sqrt{5})s/2]\}$ from which, by expanding each term in powers of s and collecting like powers of s, it follows that $a_k=f_k$, the kth Fibonacci number, **6. b)** $s^2/(1-s^2)$; correct for $|s|<1$, **7. a)** $r_n=0$ for even n and $2^{(n-1)/2}$ for odd n; **c)** $r_n=1$ for all n, **9.** True; for $n\geq 0$, $C(n,k)=0$ for $k>n$; for $n<0$ result is correct by Binomial Series Theorem, Eq. (6), Section 4.6, **10. a)** $c_0=0$, $c_k=1$, $k>0$; **c)** $c_0=c_2=1$; $c_1=2$; all other $c_k=0$, **11. a)** $a_k=1$, $k=0,1,2$; rest all 0; $b_k=1$ for all k; thus $c_0=1$, $c_1=2$, $c_k=3$, $k\geq 2$, **13. b)** Using (10), 1287, **14. a)** 16 from s^3 term in expansion of $(1+s+s^2)^4$ or from s^3 term in $(1-s^3)^4/(1-s)^4=(1-4s^3+\cdots)(1+\cdots+20s^3)$, **15.** 25; consider cases of 1 C through 9 C's and count combinations of A and B for each **17. a)** $G_n(s)=[(s+n-1)/n]G_{n-1}(s)$; **b)** $G_{n-1}(s)=[(s+n-2)/(n-1)]G_{n-2}(s)$; **c)** $G_n(s)=(s+n-1)(s+n-2)\ldots(s+1)G_1(s)/n!$; **d)** $G_1(s)=1$; **e)** Result of c) may be rewritten as $C(s+n,n)/(s+n)$.

Section 9.7

1. a) 2.0947 after 9 iterations; **c)** 1.6523 after 8 iterations **3.** The **if** ... **endif** assures that the sign of $f(x)$ is always different at *right* and *left*; since *answer* is half way between two values not more than ϵ apart such that the sign of $f(x)$ is different at each, the true solution is at most $\epsilon/2$ from *answer*, **5. a)** $x'=a\cdot f(b)/[f(b)-f(a)]+b\cdot f(a)/[f(a)-f(b)]$; **c)** For x^3-2x-5 with $a=0$ and $b=3.0$, after 4 iterations value of x' is 1.9589; converges more slowly than bisection because convexity of curve results in 3.0 always being one of the interval endpoints, **7. a)** Two evaluations of $f(x)$ for first iteration, then one more each subsequent iteration; need k iterations (i.e., $k+1$ evaluations) such that $1/2^k<F$ or $k>\log_2(1/F)$ so $1+\lceil\log_2(1/F)\rceil$ evaluations; **b)** Three at first iteration and two each subsequent iteration; $k>1+2\lceil(\log_2(1/F)/\log_2 3\rceil$; if take into account that 1/3 of time correct subinterval found after one evaluation, the multiplier 2 becomes 5/3; **c)** Bisection slightly more efficient and easier to program, **8. a)** After 4 iterations: $x_1=325/324$, $x_2=1295/648$; true solution: $x_1=1$, $x_2=2$, **9. c)** For [8a]: After 4 iterations: $x_1=35/36$, $x_2=35/18$, **11.** Gauss–Seidel generally (but not always) faster because it uses most recent (and, therefore, usually more accurate) values of the unknowns **13. a)** (ii) between the second and fourth points; **b)** Essential idea: Divide $[a,b]$ into n intervals; proceed from a to b evaluating $f(x)$ at each interval point until find three values such that $f(x)$ increases (or decreases) from the first to the second and does the opposite from the second to the third; then try n points in smaller interval in which extreme value must lie; etc.; if three points with desired property not found, try more points on interval; **c)** With $a=0$, $b=3$ and six points in interval find first that maximum between 1 and 2; then using six points on [1,2] find maximum between 4/3 and 5/3; etc.

Supplementary Problems: Chapter 9

1. a) One pair is: $n\log n=o(n^2)$; **b)** One pair is: $n^2=O((3+\sin n)n^2)$; **c)** One pair is: $n^2=\mathrm{Ord}((n+10)^2)$.

3. Since $\log n = \lceil \log n \rceil - a$ where $0 \le a < 1$ and since $\lim \log n = \infty$, $\lim \log n / \lceil \log n \rceil = 1$ and $\log n = \mathrm{Ord}(\lceil \log n \rceil)$ **5.** By Theorem 4, if $\Delta^k S_n = \Delta^k T_n$, then the antidifferences $\Delta^{k-1} S_n$ and $\Delta^{k-1} T_n$ differ by a constant; by [3b] of Section 9.4, $\Delta^{k-2}(S_n - T_n)$ is a polynomial of degree 1; working back to $S_n - T_n$ itself gives the desired result, **6. a)** $a_{20} = 0$, $a_{21} = -1$, $a_{22} = 1$; **b)** $a_{n+1,k} = a_{n,k-1} - n a_{nk}$ when $1 \le k \le n$; $a_{00} = 1$ and $a_{n0} = 0$, $a_{nn} = 1$ for $n > 0$; **c)** Last row of table is: 0 24 −50 35 −10 1 **7. a,b)** $A = (ar+b)/[c(r-s)]$, $B = (as+b)/[c(s-r)]$; since $r \ne s$, expansion is always possible; **c)** (i) $1/(x-3) + 1/(x-2)$ **8. b)** (i) $7/3(x-2) - 13/2(x-3) + 37/6(x-5)$ **9. a)** Multiplying by $(x-r)^j$ and letting $x = r$ doesn't get rid of all but one constant; **b)** Method of [8a] still creates algebraic identity; **c)** (i) $-4/(x-1) + 4/(x-2) - 3/(x-2)^2$, **10. b)** (i) $1/11(x-3) - 2(x+1)/11(2x^2 - 4x + 5)$, **11. a)** Putting right hand side over common denominator will not give numerator polynomial of degree of $P(x)$; **b)** Long division results in remainder with numerator degree less than denominator degree; **c)** (i) $2 - (10x + 9)/(x^2 + 3x + 7)$ **13. a)** For $k > 0$ $x_{(k)} = 0$ when $x = 0$; **b)** Just use $\Delta x_{(k)} = k x_{(k-1)}$; **c)** Just continue using formula in b); **e)** Result is 5.6875 while $2^{2.5} = 5.6569$; inexact because 2^x is *not* polynomial of degree 3; result would become exact only in limit as $m \to \infty$; $P_3(x)$ is cubic approximation to 2^x which enables approximate values of 2^x to be computed at values not in table, thereby *interpolating* in the table **14.** Result after much algebraic manipulation is $C(2n, n)/(n+1)$ **15. a)** $1/(1-s))$; **b)** e^s; **c)** Just calculate coefficient of $s^n/n!$ in $A(s)B(s)$; **d)** To choose from $A \vee B$ can choose ordered k items from A (a_k) and ordered $n-k$ from B (b_{n-k}) ; k elements from A can be placed in n positions in $C(n,k)$ ways; sum over k to get c_n; **e)** Interpretation of product of two terms is given in d); term in product of m exponential generating functions is given by $(m-1)$-fold summation involving $m-1$ binomial coefficients and represents number of distinguished lists of elements taken from each of m non-intersecting sets; **f)** Since $a_k = b_k = 1$, c_n is 2^n.

Epilogue

Section E.1

1. b) 12 comparisons; just about average (which is slightly less than 15) while worst case is 21, **2. a)** $m-1$ comparisons; **b)** Inverse order. **3. a)** In Eq. (6) of Section 5.9 replace $n-1$ by $n/2$ since this is the minimum number of comparisons for the merge; then compute $D_8 = 12$. **5.** Summing all terms in (3) get $m^2/4 + 3m/4 - H_m$, **7. a)** (6,3), (6,5), (8,5), (10, 5); ' **b)** 0 1 0 0 3 0; **c)** t_j represents the number of times s_j is the second element in an inversion; **d)** The j the element in the list, s_j, will be the second element in an inversion if and only if an element to its left is larger than s_j, **8. a)** 0 1 2 0 4 2 6 5 1; **b)** 0 1 0 3 1 5 4 0 0 **9. a)** Reverse order; **b)** Number of comparisons: $m(m-1)/2$ **11. a)** Minimum number of comparisons for list and its reversal occurs when one requires $m-2$ passes and the other 1 pass; total comparisons for two lists is $m(m+1)/2 - 2$; average is $m(m+1)/4 - 1 = O(m^2)$. **12. b)** $\sum_{i=2}^{m} \lceil \log_2 i \rceil$; **c)** Because average case for binary search much harder than worst case; **d)** Number of *moves* of data from one position to another same as for regular Insertion Sort and number of these is $\mathrm{Ord}(m^2)$.

Section E.2

2. a) mk is both worst and average cases; **b)** Should be about $20m$, **3. a)** See Figure E.2; **b)** An m-way comparison using 2-way comparisons involves the same operations as sorting m items using 2-way comparisons **5.** Start at a leaf (of which there must be one by [23, Section 3.7]) and grow a path which never uses any edge twice; since a free tree is acyclic, when you cannot proceed any further, you are at another leaf; in a general graph could end at vertex already visited because cycles are allowed. **7.** Yes, you can do it; you just need to show that any free tree with $m+1$ vertices can be found by adding an edge to a free tree with m vertices; again use [23, Section 3.7]. **9.** 2^h leaves at height h; each present or absent in filled tree except that all can't be absent; therefore, using the Product Rule, get $2^{2^h} - 1$; trees **11.** 24 leaves; best one has 16 at level 5, 8 at level 4 with average number of comparisons $(16 \times 5 + 8 \times 4)/24 = 14/3$, **13. c)** New tree has B at root, A at end of left branch from B and C at end of right branch from A with all others as before; as in original tree, $A < B$, $A < C$, $B > C$;

15. a) If k is the number of nodes which are not leaves, $EPL = IPL + 2k$; remember: each node in a decision tree has 0 or 2 children; **b)** By induction; for tree with $m+1$ nodes which are not leaves, delete two leaves which are children of same node to get tree with m non-leaf nodes; calculate change in EPL and IPL and use induction hypothesis to get desired result **17. a)** Use fact that integral from $i-1$ to i of $f(x) \le f_i \le$ integral from i to $i+1$; **b)** Integrating $\log_2 x$ get that $\lceil \log_2 m! \rceil \ge m \log_2 m - 1.44(m-1)$ **19.** By induction using fact that each node can have at most two children. **21.** Total number of leaves determines number of leaves at lowest and next to lowest levels which determines sum of path lengths. **23. a)** Each of $m!$ possible orderings must lead to a specific leaf; therefore, $k - m!$ leaves not reached; **b)** Average path length to leaves which can be reached; **c)** Deleting $k - m!$ leaves which cannot be reached does not affect any node reached on path to one of the $m!$ leaves which can be reached; **d)** Can always continue to prune as in c) until get tree with $m!$ leaves **25.** A filled tree of height h has at most 2^h leaves on level h; any leaf at level $h-1$ means 2 fewer leaves at level h; if only 2 leaves on level h, then $2^{h-1}-1$ at level $h-1$.

Section E.3

1. a) [1a]: First comparison from right : 2 15 12 21 3 9 8; first comparison from left: 2 8 12 21 3 9 15; second from right: 2 3 12 21 8 9 15; second from left: 2 3 8 21 12 9 15 **3. a)** $(\forall k : j < k \le L) n_p \le n_k$; $(\forall k : F \le k < i) n_p \ge n_k$; **b)** $(\forall k : F \le k \le p-1) n_p \ge n_k \wedge (\forall k : j+1 \le k \le L) n_p \le n_k$ **5.** Each application of Partition compares n_p with all other elements of the (sub)list, once this call is completed, n_p is never compared again with anything; thus maximum number of comparisons if $C(m,2) = m(m-1)/2$ **7. a)** $2^h, \ldots, 2^h + k - 1$; **b)** $2j, 2j+1$; **c)** $\lfloor j/2 \rfloor$; **d)** Contents of node i is stored in table location i; b,c) give locations of children, parents of each node **9. a)** Put new item in mth position; compare it with its parent in $\lfloor m/2 \rfloor$ position; if parent < child, interchange them; continue until parent ≥ child or new element arrives in position 1; **b)** (i) Parent(8) = 4, 20 > 2 so list becomes 47 14 12 20 5 8 6 2; parent(4) = 2, 20 > 14, so list becomes 47 20 12 14 5 8 6 2; parent(2) = 1, 20 < 47; therefore stop **11. a)** Since, as a result of (ii), root always contains largest element not yet in final position, each step brings next largest element into its proper position; **b)** Compare parent with maximum child (if two children); exchange if child larger than parent (it always will be at first step); then compare original root item in its new position with its maximum child; continue exchanging as long as parent less than maximum child or until original root element becomes a leaf; **c)** Build heap by [10], then sort it. **13. b)** The floor of the logarithm equals j for 2^j values of i; **c)** $(m+1)\log_2(m+1) - 2m$; order is $m \log m$; **d)** Forming heap and sorting it very similar operations; plausible, therefore, that order of both is the same; **e)** $m \log m$ since $m \log m$ in worst case and, from Section E.2, best possible average case is $m \log m$; **f)** No, depends on constant in Eq. (2) of Section 9.3 as well as on implementation details.

Final Problems

1. a) Induction hypothesis: LSTOH, S1STOH, S2STOH all work for $n-1$ disks; LSTOH: $n-1$ disks to end; largest to middle; $n-1$ disks to other end; largest to end; $n-1$ disks to end; S1STOH: $n-1$ disks to wrong end; largest to right end; $n-1$ disks to middle (by S2STOH); $n-1$ disks to right end; S2STOH: $n-1$ disks to middle; $n-1$ disks to other end (by S1STOH); largest to middle; $n-1$ disks to middle; **b)** By induction; **c)** LSTOH: $P(n) = 3P(n-1) + 2$; $P(1) = 2$; S1STOH, S2STOH: $Q(n) = 3Q(n-1) + 1$; $Q(1) = 1$; solutions: $P(n) = 3^n - 1$; $Q(n) = (3^n - 1)/2$; **d)** $Q(n) = Q(n-1) + P(n-1) + 1$, $Q(1) = 2$; same solution as c); **e)** Since short trip solution is unique (why?) and since using LSTOH gives exactly the same result as using short trip twice, LSTOH must in fact be made up of two short trips; **f)** Basis: LSTOH puts disk on middle pole first; with n disks, first $n-1$ disks go to far end; then after largest in middle, putting $n-1$ disks back on original end is same as LSTOH with $n-1$ disks; by induction hypothesis at some point $n-1$ disks are on top of largest disk on middle pole **3.** By induction; assume can win $n-1$ ring game to clockwise pole (CL) or counterclockwise pole (CCL); for n rings to CL: $n-1$ by CCL, largest to CL, $n-1$ by CCL; for n rings to CCL: $n-1$ by CL, largest to CCL, $n-1$ by CL; uniqueness by induction, too; never any choice in move of largest ring; CL moves: $(1/4)[5 \cdot 2^n + (-2)^n] - 1$; CCL moves: $(1/2)[5 \cdot 2^{n-1} + (-2)^{n-1}] - 1$, **5.** Downsweep: Only one element in ith column (in $i+1$ row) to eliminate; only two elements in $i+1$ row which change; therefore only three operations; Scale: Unchanged; Upsweep: Only one element in ith column to eliminate; therefore, only one operation; total operations: $3(n-1)$ from Downsweeps,

n from Scale, $n-1$ from Upsweeps; total: $5n-4$; therefore, order is n, **7. a)** Always n^2 multiplications; at most n^2+n-2 additions; **b)** $(P10^m+Q)(R10^m+S)=10^n PR+10^m(PS+QR)+QR)+QS$; **c)** $M_{2n}=4M_n$, $M_1=1$; for $n=2^k$ solution is $M_n=n^2$; no improvement; **d)** $A_{2n}=4A_n+cn$, $A_1=0$; for $n=2^k$ $A_n=(c/2)(n^2-n)$; since $c>2$ more additions than standard method needed for large n; **e)** Now $M_{2n}=3M_n$ so solution is $M_n=n^{\log_2 3}$; this is an improvement since $\log_2 3$ is about 1.56; extra digit in $P+Q$ or $R+S$ can be ignored if we assume m large; **f)** Now $A_{2n}=3A_n+cn$ and $A_n=c(n^{\log_2 3}-1)$ which is better than result in d) for large enough n **9. a)** With $s_j=r_{k-j}$, recursion for Euclidean algorithm becomes one given; **b)** By induction; $s_1\geq f_0$; since $Q_1\geq 1$, $s_2\geq f_1$; then since $Q_j\geq 1$, each $s_j\geq f_{j-1}$; with $(m,n)=(f_k,f_{k-1})$, each remainder is next smaller Fibonacci number; so $k-1$ iterations are required with last $2(=f_2)=2\cdot 1+0$; **c)** One more iteration needed; for $(m,n)=(f_{k-1},f_k)$, k iterations needed **11. a)** Pick a vertex v in n vertex graph G; on $G-v$ find α_1 and on $G-N(v)$ find α_2; size of maximum independent set is maximum of α_1 and $1+\alpha_2$; **b)** Let g_n be work to find α for an n-set; then $g_n>2g_{n-1}$; with $g_1=1$, this means $g_n>2^{n-1}$; since there are 2^n subsets of an n-set, this method is comparable to just checking each subset; **c)** If v connected to one other vertex, then $G-N(v)$ has at most $n-2$ vertices; therefore, $g_n>g_{n-1}+g_{n-2}$ so that $g_n>f_n$, the nth Fibonacci number; since Fibonacci numbers grow as about $(1.6)^n$, this is a better bound; **13.** Proof not OK because suppose removed vertex disconnects the graph; fix by always removing a vertex which does not disconnect the graph; this is always possible (why?); then proof is OK.

Index

Page references in boldface indicated the major reference to a term, usually to its definition; numbers in brackets refer to the problem where a reference can be found.

D

Meaning	Symbol or Example	Page or place first defined or used
5. Combinatorics		
Bell numbers	$B(b, u)$	331
Combinations (binomial coefficient)	$C(n, r)$ or $\binom{n}{r}$	303
Combinations with repetition	$C^*(n, r)$	308
Multinomial coefficient	$\binom{n}{k_1, k_2, \ldots, k_m}$	321
Permutations	$P(n, r)$	303
Permutations with repetition	$P^*(n, r)$	308
Trinomial coefficient	$\binom{n}{i, j, k}$	321
6. Probability		
Binomial distribution	$B_{n,p}(k)$	468
Conditional probability	$\Pr(B\|A)$	452
Expected value	$E(X)$ or $E(f)$	481
Event complement (see also Complement under 2. above)	$\sim E$	445
Poisson distribution	$f_\lambda(k)$	472
Probability measure	Pr	443
Random variable probability	$\Pr(X{=}c)$ or $\Pr_X(c)$	466
Standard deviation of random variable X	σ_X	488
Variance of X	$\text{Var}(X)$	488
7. Logic		
And	$\wedge$	532
	$\cdot$	565
Biconditional	$\Longleftrightarrow$	59
Existential quantifier	$\exists$	587
False	F	528
If and only if	iff	59
Implication	$\Longrightarrow$	57
Logical circuits:		
AND	[AND gate symbol]	576
NOT	[NOT gate symbol]	576
OR	[OR gate symbol]	576
Negation	$\neg$	58
	Overbar($^-$)	565
Nonequivalence	$\not\Longleftrightarrow$	534
Nonimplication	$\not\Longrightarrow$	534